Differential Equations

Solution Formula Techniques Involving First-Order ODEs

Explicit Techniques	
1. **Linear ODE** $$y' + p(t)y = q(t)$$	Multiply the ODE through by an integrating factor $e^{P(t)}$, where $P(t) = \int^t p$, and then use the Vanishing Derivative Theorem. (See Example 1.2.1.)
2. **Linear Cascade** $$x' = k_1 x + f(t)$$ $$y' = k_2 x + k_3 y + g(t)$$	The coefficients k_1, k_2, and k_3 may depend on t. Solve the first linear ODE for $x(t)$, insert $x(t)$ into second linear ODE, which then can be solved as a linear ODE for y(t). (See Example 1.6.2.)

Implicit Techniques	
1. **Variables Separate** $$N(y)y' + M(x) = 0$$	Find antiderivatives $F(x) = \int^x M$, $G(y) = \int^y N$. The level curves defined by $F(x) + G(y) = C$, for a constant C, are *integral curves* of the ODE in the xy-plane. An arc of an integral curve with no vertical tangents is a solution curve of the ODE. (See Examples 1.4.2, 1.4.3. For another way to plot integral curves see Example 1.4.5.)
2. **Exact ODE** $$N(x, y)y' + M(x, y) = 0$$ $$\frac{\partial M}{\partial y} = \frac{\partial N}{\partial x} \text{ in a simply-connected region}$$	Find a function $F(x, y)$ with $\partial F/\partial x = M$, $\partial F/\partial y = N$ in the region, then put $F(x, y) = C$, a constant, and solve for y. (See Examples 1.9.1, 1.9.3. To convert inexact ODEs into exact ODEs, see Problem 11 in Section 1.9.)
3. **Differential Form of ODE** $$M(x, y)\,dx + N(x, y)\,dy = 0$$	Can interchange independent and dependent variables to obtain an equivalent ODE (see Example 1.9.4): $$N\frac{dy}{dx} + M = 0, \text{ or } M\frac{dx}{dy} + N = 0$$

System Techniques	
1. **Planar Autonomous System** $$\frac{dx}{dt} = N(x, y), \quad \frac{dy}{dt} = -M(x, y)$$	Orbits of the system in the xy-plane with no vertical tangent lines are given by the solution curves of the first-order ODE $dy/dx = -M/N$. (See Examples 1.4.8, 1.7.7.)
2. **General ODE** $$N(x, y)y' + M(x, y) = 0$$	Convert the first-order ODE to a planar autonomous system $$\frac{dx}{dt} = N(x, y)$$ $$\frac{dy}{dt} = -M(x, y)$$ whose orbits are composed of solution curves of the ODE. (See Examples 1.9.6, 1.9.5.)

Change of Variables

1. Bernoulli Equation $y' + p(t)y = q(t)y^b$ $b \neq 0, 1$	Change the state variable to $z = y^{1-b}$, to obtain the new ODE $z' + (1-b)pz = (1-b)q$, which is linear in z. (See Example 1.7.1.) Problem 11 in Section 1.7 has more on Bernoulli's ODE.
2. Riccati Equation $y' = a(t)y + b(t)y^2 + F(t)$	If one solution, $g(t)$, of the Riccati equation is known, then every solution $y(t)$ has the form $y(t) = g(t) + 1/z(t)$ where $z(t)$ solves the linear ODE $z' + (a + 2bg)z = -b$. (See Problem 12 in Section 1.7.)
3. The ODE $y' = (a + by)(c(t) + d(t)y)$ where a, b are constants, $b \neq 0$	Change the state variable to $z = (a + by)^{-1}$, to obtain the new ODE $z' = [ad(t) - bc(t)]z - d(t)$, which is linear in z. (See Example 1.7.2.)
4. Homogeneous Rate Function of Order Zero $y' = f(x, y)$, where $f(kx, ky) = f(x, y)$, for all $k > 0$.	Change state variable to $z = y/x$, get new ODE, $xz' = f(1, z) - z$ whose variables separate. (See Examples 1.7.6, 1.7.8.) Also see Atlas Plate FIRST ORDER G.
5. ODEs in Polar Coordinates Write $y' = f(x, y)$ in polar form Look for solution curves of the new ODE $r = r(\theta)$	Change the variables x and y to polar form using the substitution $x = r\cos\theta$, $y = r\sin\theta$. If $r = r(\theta)$ describes a solution curve of the ODE, then $r(\theta)$ satisfies the ODE (see Example 1.7.4): $$\frac{dr}{d\theta}\sin\theta + r\cos\theta = f(r\cos\theta, r\sin\theta)\left(\frac{dr}{d\theta}\cos\theta - r\sin\theta\right)$$

Reduction to First-Order ODEs

1. Method of Varied Parameters $y'' + a(t)y' + b(t)y = f(t)$	Let $z(t)$ be a known solution of $z'' + a(t)z' + b(t)z = 0$. Then every solution $y(t)$ of $y'' + a(t)y' + b(t)y = f(t)$ is given by $y = uz$ where $u(t)$ solves the ODE $zu'' + (2z' + az)u' = f$, which is linear first-order for $v = u'$. (See Problems 3 and 4 in Section 1.8.)
2. $y'' = F(t, y')$ F is independent of y	The state variable $v = y'$ solves the first-order ODE $v' = F(t, v)$. The state variable $y = \int^t v$. (See Examples 1.3.2, 1.4.6, 1.8.1.)
3. $y'' = F(y, y')$ F is independent of t	Introduce y as a new independent variable and consider the state variable $v = y'$ as a function of y: $v = v(y)$. Since $y'' = (dv/dy)v$, it follows that v solves the first-order ODE $v(dv/dy) = F(y, v)$. Note that $y(t)$ solves the ODE $dy/dt = v(y)$. (See Example 1.8.2.)

Forced Oscillation

$y' + p_0 y = q(t)$ $0 \neq p_0 = $ constant, $q(t)$ is piecewise continuous and periodic with period T	Has a unique periodic solution with period T; generated by the initial condition $y(0) = \bar{y}_0$, where $\bar{y}_0$ is given by the equation $$\bar{y}_0 = (\exp(p_0 T) - 1)^{-1}\int_0^T \exp(p_0 s)q(s)\,ds$$ (See Problem 17 in Section 1.2. Also Atlas Plate FIRST ORDER K.)

ROBERT L. BORRELLI

COURTNEY S. COLEMAN

Harvey Mudd College

Differential Equations

A MODELING PERSPECTIVE

PRELIMINARY EDITION

JOHN WILEY & SONS, INC.
New York   Chichester   Brisbane
Toronto   Singapore

Preface

Why Study Differential Equations?

Differential equations are a powerful tool in constructing mathematical models for phenomena which evolve in time according to certain laws. So widespread is the use of differential equations in industry, research, and engineering, and so well do they perform their task that they are clearly one of the most successful of modeling tools. This is an exciting time to study differential equations because interactive computer solvers can quickly and easily generate striking graphical displays that can provide amazing insights to the system under study.

Course and Text Objectives

Differential Equations: A Modeling Perspective is an introductory textbook for students of science, mathematics, and engineering. Our approach is based on two principles: modeling and graphical visualization. We expect our readers to cover at least a few models in each chapter and to use a numerical solver to some extent to investigate the behavior of solutions of differential equations.

This text features many new ideas about teaching and learning DEs that have worked best for us and our students, while it retains crucial topics and objectives of a traditional course. One traditional objective is to explain how to obtain formulas for solutions, and much of this book is devoted to a presentation of these fundamental formulas and techniques. Other traditional objectives of this text are a brief presentation of the basic theory and an introduction to finding approximate numerical solutions of differential equations.

In a strange turn of events, computers spark heightened interest in the basic theory of DEs and details of modeling environments. This happens naturally because visual displays of solutions of differential equations can be interepreted with the insights gained from the theory and the modeling. We provide enough theory for a solid foundation for understanding the techniques and applications, and when proofs are given, they are as complete and self-contained as possible. The theoretical depth of this course is therefore flexible and can be tailored to student's needs and preferences.

Prerequisites

This text presumes a familiarity with calculus through single variable calculus. However, there are a few sections that require partial differentiation, multiple integration, or the divergence and Green's theorems. A course in linear algebra is not assumed; linear concepts are developed as needed. Some familiarity with elementary applications of Newton's Laws of Motion from a physics or a calculus course is presumed at the outset (but these Laws will be developed in later chapters as needed for other applications).

Highlights

- *Mathematical Modeling.* Every picture tells a story: building a model is like drawing a picture of the system, and interpreting the solution of the model equations is like telling a story. Scientists and engineers believe that a "good" mathematical model "describes" the system of interest. There are a great number of models in the text from which to choose, but they are all from three basic areas: rate processes, electrical circuits, and mechanics. Some sections are completely devoted to a single model, but for the most part models comprise only a portion of a section, and the text allows flexibility in the choice of how much the modeling environment affects the theory and solution techniques.

- *Graphical Visualization Emphasized.* The solutions of an ordinary differential equation are functions whose graphs are curves. These curves may be computer-generated and provide compelling visual evidence of theoretical deductions and a clear understanding of complicated solution formulas. Every graph in this text is accompanied by the data necessary to reproduce it. These graphs are the actual output of a solver package, not artist renderings. The text and the hundreds of graphs of solutions emphasize this visual connection with the theory. An Atlas of graphs of solutions appears in Appendix D. The Atlas graphs are annotated and are chosen for their insightful view of the properties of solutions of DEs.

- *Numerical Solutions Introduced Early.* With the ready availability of excellent and inexpensive computer solvers, it makes a great deal of sense to bring up numerical solution techniques very early so that students can begin early to examine the geometry of solutions and the way solutions change when the components of a differential equation are perturbed. An appendix on numerical methods provides more than enough material on numerical solutions for a first course.

- *Systems Introduced Early.* From the very start, simple systems of differential equations are treated matter-of-factly in the modeling process, because it is natural to do so. This does not present a problem since computer solvers can handle a system of first-order differential equations as easily as a single one. Early introduction of systems also makes it convenient to show the equivalence of an autonomous first-order planar system to a single first-order equation. Although we use systems early on, we postpone a serious study of them until Chapter 6.

- *Basic Theory.* The basic theory of DEs appears throughout the text, but the instructor can easily tailor the amount of theory to fit the course. Short proofs are given where appropriate, and longer proofs are given in Appendix A. The basic questions of *existence*, *uniqueness*, and *continuity with respect to the data* are treated as a recurring theme throughout the text. The appendix contains the background theory for treating each of these theoretical questions for first-order ODEs. To these basic theoretical questions we systemically add the more practical question of *solvability*. Ironically, the introduction of computers into a course often leads to heightened interest in the theory rather than less. Basic theory serves as a valuable tool for interpreting visual displays of solutions.

- *Problem Sets.* Problems are the heart of this book. Most sections contain some problems that require the use of a numerical ODE solver (they are marked with an icon). Many sections contain open-ended projects appropriate for a team of students (also marked with an icon). Answers to problems with underscored numbers are given at the end of the book.

Available Solvers

Computer programming is not required to complete the projects and problems in the text. There are many good quality differential equations solver packages available today which do not require the user to be computer literate. A listing and description of them is posted on the World Wide Web at the following URL address:

http://www.hmc.edu/codee/solvers.html

Solver packages can be found for nearly any computer platform, be it a hand-held calculator, personal computer, workstation, or large computer system. No specific solver package is presumed in this text.

A Course Based on this Text

Many courses are possible; here is a semester course that does not assume any prior exposure to differential equations:

Chapter 1:	Sections 1.1–1.4 plus one (or more) application sections selected from 1.5–1.9
Chapter 2:	Sections 2.1–2.3, 2.5, 2.6
Chapter 3:	Sections 3.1–3.4, plus 3.5 or 3.6
Chapter 4:	Section 4.1, 4.2, or 4.3 at the instructor's discretion
Chapter 6:	Sections 6.1, 6.2, plus one of 6.3, 6.4, 6.5
Chapter 7:	Sections 7.1–7.4, plus one of 7.5, 7.6, 7.7
	From this point on there are two tracks:
Track 1:	Chapter 8: Sections 8.1, 8.2, plus 8.3 or 8.4
	Chapter 9: Sections 9.1, 9.2, plus 9.3 or 9.4
Track 2:	Sections 10.2–10.6

This course covers about thirty sections of the book, leaving ample time for spending more than one lecture on some sections, and for exams, and so on.

Supplements

A Student Solution Manual gives complete solutions (along with graphs) to every other part of every odd-numbered problem (group problems not included).

The authors, together with William Boyce, have produced a collection of computer graphics experiments and modeling projects with the title *Differential Equations Laboratory Workbook*[1]. The Workbook supplements a course on ordinary differential equations. An appendix in the Workbook contains a telegraphic overview of three modeling environments: rate processes, electrical circuits and mechanics. The Workbook also contains many graphs of solutions of differential equations which serve as a handy visual reference. This text was written to be independent of the Workbook, although the two would obviously fit well together.

Supplements for the instructor are available from the publisher.

---

[1] John Wiley & Sons, Inc., New York, 1992. (ISBN 0-471-55142-2)

Acknowledgements

Using a layout by Kevin Murphy, this book was designed in LaTeX by David Richards, whose meticulous attention to detail deserves our warmest thanks.[2]

Very special thanks go to Tony Leneis, the principal author of ODETOOLKIT, a command script and user interface over the ODE solver MATHLIB DEQSOLVE,[3] which was used to produce all the figures in this text. Our very special thanks also go to Jenny Switkes, who put up with daily changes in the graphs and the text with good humor and remarkable efficiency as she set much of this textbook in LaTeX. We also extend our gratitude to our students who spent many hours of hard work on the project: Aaron Archer, Aaron Lamb, Xuemei Wu, and Kaiqi Xiong. We also thank our students Christie Lee, Susan McMains, Kal Wong, and Bob Zirpoli who in one way or another contributed to the production of this book.

The authors are also indebted to the reviewers of the early versions of this text whose comments and suggestions were extremely valuable. In particular we would like to thank Professors David Kraines, David Lerner, Roland di Franco, Ulrich Daepp, Karl E. Petersen, Zhongyuan Li, Mark Fuller, Bhagat Singh, Ben Fusaro, Michael Montano, Ed Spitznagel, Mike Pepe, David Voss, Steven R. Dunbar, Mark Farris, and Christina M. Yuengling.

The responsibility for any errors, of course, is ours. As this is a preliminary edition, we would particularly welcome comments about any topics that should be included and about the book in general. Please let us know about any mistakes or ambiguities so that they can be corrected before the regular edition is published.

And finally, we owe a debt of gratitude to Barbara Holland, our Wiley Mathematics Editor, for her generous support, advice, and encouragement during the writing of this text. Our warmest thanks, Barbara!

R. L. Borrelli   borrelli@hmc.edu
C. S. Coleman   coleman@hmc.edu

Claremont, October 1995

---

[2] The textbook was typeset in Times Roman using LaTeX 2_ε on a 486 PC, running Linux. Most figures were prepared in encapsulated postscript using Mathlib DEQSOLVE under VAX/VMS (Registered Trademark of Digital Equipment Corp., Maynard, MA) and imported using the epsf macros.

[3] DEQSOLVE, an ODE solver from the MATHLIB (Registered trademark of Innosoft International Inc., West Covina, CA.) software package is based on LSODA, a descendant of C. W. Gear's DIFFSUB, and is part of the package ODEPACK developed by Alan Hindmarsh at Lawrence Livermore National Laboratores. DEQSOLVE was written by Kevin Carosso, Ned Freed, and Dan Newman.

Students: Guarantee Your Success!

The tips below will help you to navigate your way successfully through this book.

- *Graphing.* Don't be afraid to use any means at your disposal to make sense of graphics output. Try rescaling, zooming, using different plot types, changing values of parameters and even using a ruler on graphics output to determine values when other means are not available.

- *Use the Computer for Visualization.* Almost all campuses today have computer laboratories that are accessible to students. It is likely that every laboratory has at least a few platforms running one differential equations solver package or another. The reader is urged to gain access to such a platform and learn how to use it. Every one of the graphs in this text is accompanied by the data necessary to reproduce it. These graphs are the actual output of a solver package, not artist renderings. Also, in the appendix there is an annotated collection of graphs, called the *Atlas*. The graphs in the Atlas were selected for their educational value, and much can be learned by using these graphs as a starting point. The Atlas entries give specific references to items in the text related to indicated Atlas figures.

- *CAS Users.* The grunge work of verifying computations can be relieved by using a Computer Algebra System (CAS) package. Some of these packages also contain solvers that will produce solution formulas for many first-order differential equations.

- *Skipping Around in the Book.* This text does not have to be covered in linear order, so don't worry if the instructor skips material and perhaps comes back to it later.

- *Basic Solution Techniques.* The basic techniques for finding solution formulas for first-order ODEs are listed, for reference purposes, in the tables at the front of the book. The tables refer to examples in the text that illustrate the use of the indicated technique.

- *Examples and Problems.* Answers to the problems with underlined numbers are at the back of the book. A Student Solution Manual that gives complete solutions (including graphs) of many of the problems is also available.

Contents

1 First-Order Equations: Solution and Modeling Techniques 1

1.1 Introduction 1

1.2 First-Order Linear Differential Equations 11

1.3 Introduction to Modeling 23

1.4 Separable Equations: Implicitly Defined Solutions 36

1.5 Growth Processes: The Logistic Model 49

1.6 Cold Pills and Cascades 58

1.7 Change of Variables: Pursuit Models 69

1.8 Reduction Methods: Escape Velocity 79

1.9 Exact Equations 86

2 Initial Value Problems and Their Approximate Solutions 97

2.1 Solution Curves and Direction Fields 97

2.2 Initial Value Problems: Basic Questions 104

2.3 Sensitivity 112

2.4 Population with Bounded Migration 120

2.5 Approximate Solutions 128

2.6 Computer Implementation 135

2.7 Euler's Method, The Logistic Equation, and Chaos 141

3 Second-Order Differential Equations 153

3.1 Springs: Linear and Nonlinear Models 153

3.2 Basic Theory: Second-Order ODEs 163

3.3 Basic Theory: Linear Second-Order ODEs 172

3.4 Constant-Coefficient Homogeneous Linear ODEs 185

3.5 Driven Linear ODEs: Annihilators 193

3.6 Driven Linear ODEs: Variation of Parameters 205

4 Applications of Second-Order Differential Equations 214

4.1 Newton's Laws: The Pendulum 214

4.2 Forced and Free Oscillations 225

4.3 Electrical Circuits: Kirchhoff's Laws 238

5 The Laplace Transform 251

5.1 Introduction to the Laplace Transform 251

5.2 The Calculus of the Transform 255

5.3 The Transform of a Periodic Function 267

5.4 A Model for Car-Following 273

5.5 Convolution 279

5.6 Convolution and the Delta Function 283

6 Introduction to Systems 290

6.1 First-Order Systems 290

6.2 Basic Theory of Systems 302

6.3 Interacting Species 313

6.4 Predator-Prey: Volterra's Principles 322

6.5 The New Zealand Possum Plague 331

7 Linear Systems of Differential Equations 336

7.1 Lead in the Human Body 336

7.2 Matrices, Eigenvalues, Eigenvectors 342

7.3 Homogeneous Linear Systems with Constant Coefficients 353

7.4 Orbital Portraits 362

7.5 Fundamental Matrices: The Matrix Exponential 372

7.6 Steady States of Linear Systems 384

7.7 Sensitivity to Parameter Changes 395

8 Nonlinear Systems: Stability 404

8.1 Stability 404

8.2 Lyapunov Functions 414

8.3 Conservative Systems 429

8.4 Stability and the Linear Approximation 437

9 Nonlinear Systems: Cycles, Bifurcations, and Chaos 452

9.1 Limit Cycles: van der Pol Systems 452

9.2 The Poincaré-Bendixson Alternatives 462

9.3 Bifurcations 473

9.4 The Lorenz System and Chaos 482

10 Fourier Series and Separation of Variables 497

10.1 Wave Motion: Vibrations of a String 497

10.2 Orthogonal Functions and Fourier Series 507

10.3 Fourier Trigonometric Series 520

10.4 Half-Range and Exponential Fourier Series 530

10.5 Sturm-Liouville Problems 535

10.6 Separation of Variables and Eigenfunction Expansions 538

11 Series Solutions: Bessel Functions, Legendre Polynomials 548

11.1 Introduction: Aging Springs and Steady Temperatures 548

11.2 Series Solutions Near an Ordinary Point 554

11.3 Legendre Polynomials 563

11.4 Regular Singular Points: Euler Equations 570

11.5 Series Solutions Near Regular Singular Points, I 578

11.6 Bessel Functions 585

11.7 Series Solutions Near Regular Singular Points, II 597

12   Partial Differential Equations   609

12.1   Introduction to Partial Differential Equations   609

12.2   Properties of Solutions of the Wave Equation   616

12.3   The Heat Equation: Optimal Depth for a Wine Cellar   624

12.4   Laplace's Equation and Harmonic Functions   635

12.5   Steady Temperatures in Spheres and Cylinders   640

A   Basic Theory of Initial Value Problems   A-1

A.1   Uniqueness   A-1

A.2   The Picard Process for Solving an Initial Value Problem   A-3

A.3   Extension of Solutions   A-11

A.4   Sensitivity of Solutions to the Data   A-12

B   Numerical Methods   B-1

B.1   One-Step Methods   B-1

B.2   Multistep Numerical Methods   B-4

B.3   Adams-Bashforth Methods: Interpolating Polynomials   B-8

C   Background Information   C-1

C.1   Engineering Functions   C-2

C.2   Power Series   C-4

C.3   Complex Numbers and Complex-Valued Functions   C-7

C.4   Useful Formulas   C-9

C.5   Theorems from Calculus   C-11

C.6   Review of Matrix Theory   C-15

C.7   Scaling and Units   C-21

D   Atlas   D-1

Answers to Selected Problems   S-1

Index   I-1

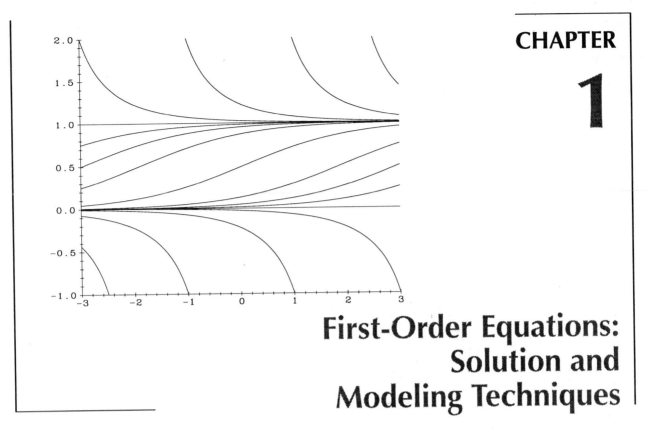

First-Order Equations: Solution and Modeling Techniques

Differential equations and the process of modeling have enjoyed a long history together—some 300 years now. When seventeenth and eigthteenth century scientists used the newly invented differential calculus to model and explain the motion of material bodies, a modeling tool of great power and scope came into being. In this introductory chapter we take a look at these partners. Along the way we shall develop a collection of solution formulas and techniques for treating those first-order differential equations that arise most often in applications. These solution formula techniques are given in the front cover sheets. The cover figure shows solution curves of a logisitc differential equation (see Example 1.5.1).

1.1 Introduction

Many phenomena in engineering and the sciences can be expressed by equations which involve the rates of change of quantities (such as position, temperature, and mass) that describe the state of the phenomena. The state of a system is characterized by *state variables* which give an instantaneous description of the system. The rates of change are derivatives with respect to time, and the equations involving the derivatives are called *differential equations*. The differential equations are said to "model" the phenomenon or process with which they are associated in the following sense: given an "initial state," unique time descriptions of the state variables can be found which "solve" (i.e., satisfy) the differential equations and take on the initial state at the initial time. An example from gravitational mechanics clarifies these concepts.

Vertical Motion

A ball is thrown in a vertical direction from a point near the earth's surface. Where will the ball be at any later time? Neglecting air resistance, and assuming the ball stays close to the earth's surface, numerous experiments have led to the formulation of a model for the ball's motion:

> Law of Vertical Motion.   An object thrown vertically moves in a vertical line with constant acceleration g, where g is the earth's gravitational constant.

The law suggests that suitable state variables are the distance $s(t)$ (measured in meters) of the ball's center above the ground at time t (measured in seconds), and the velocity $v(t) = s'(t)$ of the ball (measured in meters per second). The value of g near the ground is known to be about 9.8 meters/sec^2. Knowledge of the state variables at the time the ball is thrown, together with the stated law of motion, will determine the values of the state variables at later times (as we shall presently see).

EXAMPLE 1.1.1

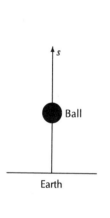

Differential Equation for a Moving Ball

The ball's acceleration is given by $a(t) = s''(t) = v'(t)$. Time is measured from the instant $t = 0$ that the ball is thrown. At that instant, suppose that $s(0) = s_0$ meters above the ground, and that $s'(0) = v(0) = v_0$ meters/sec. These are the initial conditions for the subsequent motion. The ball remains in motion until it hits the ground at some time $t = T > 0$. A model for the motion of the ball is given by the differential equation and initial conditions

$$s'' = -g, \qquad 0 \le t \le T$$
$$s(0) = s_0, \qquad s'(0) = v_0 \tag{1}$$

The minus sign before the constant g arises because s increases in the upward direction, while the gravitational force acting on the ball is directed downward. To say that $s''(t) = -g$ for $0 \le t \le T$ means that $s''(t) = -g$ for all t in the open interval $0 < t < T$ and that $s(t)$, $s'(t)$, and $s''(t)$ are continuous and have finite limits as $t \to 0^+$ and as $t \to T^-$. Also, as $t \to 0^+$, $s(t) \to s_0$, and $s'(t) \to v_0$. The differential equation $s'' = -g$ may not be valid for $t < 0$ or for $t > T$ because we do not know anything about the ball's motion before its initial state or after impact with the ground.

In this section we give examples of differential equations, define what is meant by solutions and solution curves, and give a simple technique for finding solutions. The modeling aspects are treated in the following sections.

ODEs and PDEs

Differential equations are equations expressed in terms of unknown functions and their derivatives. When all derivatives are taken with respect to the same independent variable, then the equations are *ordinary differential equations* (ODEs). When the unknown functions depend on several independent variables, then the equations are *partial differential equations* (PDEs). For simplicity we often use the "prime" notation for the derivatives in an ODE, occasionally dropping specific reference to the independent variable if the name of that variable is clear from the context; for example,

$$y' \quad \text{for} \quad \frac{dy(t)}{dt}, \qquad z'' \quad \text{for} \quad \frac{d^2 z(x)}{dx^2}, \qquad y^{(n)} \quad \text{for} \quad \frac{d^n y}{dt^n}$$

Here are some examples of ODEs and PDEs.

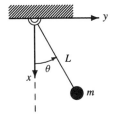

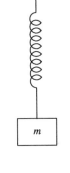

- The amount of a radioactive substance remaining in a sample at time t is denoted by $N(t)$, and, for a constant $k > 0$, satisfies the ODE

$$N'(t) = -kN(t) \qquad (2a)$$

- The position of a vertically moving body suspended by a spring is denoted by $y(t)$ and satisfies the ODE

$$my'' + cy' + ky = F(t) \qquad (2b)$$

where m is the mass of the body, the positive constants c and k measure the damping of the motion and the restoring force of the spring, respectively, and $F(t)$ is a given external force on the body.

- The angular displacement from the downward vertical of a pendulum bob moving in a plane is denoted by $\theta(t)$ and satisfies the ODE

$$mL\theta'' + cL\theta' + mg \sin\theta = 0 \qquad (2c)$$

where m is the mass of the bob, L is the length of the pendulum, c is a positive constant measuring damping of the motion, and g is the gravitational constant.

- The transverse displacement of a clamped, taut, and vibrating string at the point x and time t is denoted by $u(x, t)$ and, for a constant $c > 0$, satisfies the PDE

$$\frac{\partial^2 u(t, x)}{\partial t^2} = c^2 \frac{\partial^2 u(t, x)}{\partial x^2} \qquad (2d)$$

- There are ODEs, of course, which do not model anything; for example,

$$ty''' - y^2 = 0, \quad t > 0 \qquad (2e)$$

is an ODE in $y(t)$, but it does not model anything in particular (as far as we know).

The list above suggests the astonishing variety of differential equations. We cannot hope to find special techniques for solving all the "important" ODEs one by one. Instead, we shall group ODEs into natural categories and look for properties of solutions or solution techniques common to a given category.

Classification of ODEs: Order, Normal, Linear

Differential equations are classified in a number of ways. The *order* of an ODE is the order of the highest derivative of the unknown function that appears in the equation. For example, ODE (2a) is first order in the unknown function $N(t)$. ODEs (2b), (2c), and (2d) are second order and ODE (2e) is third order. ODEs of first or second order are particularly common in applications.

The general nth-order differential equation for the function $y(t)$ is

$$G(t, y(t), y'(t), y''(t), \ldots, y^{(n)}(t)) = 0 \qquad (3)$$

where $G(\cdot, \cdot, \ldots, \cdot)$ is a function of the variables $t, y, \ldots, y^{(n)}$. Because of the difficulty in constructing a theory for solutions of ODE (3), mathematicians often assume that it can be solved for $y^{(n)}(t)$ in terms of the other variables to obtain

$$y^{(n)}(t) = F(t, y(t), y'(t), \ldots, y^{(n-1)}(t)) \tag{4}$$

ODE (4) is in *normal form* over a region described by the variables $t, y, \ldots, y^{(n-1)}$ if F is continuous on that region. For example, the ODE $ty''' - y^2 = 0, t > 0$ has the equivalent normal form in the ty region indicated:

$$y''' = y^2/t, \quad t > 0, \quad -\infty < y < \infty$$

ODEs in normal form are much easier to handle theoretically. Also, computer solvers usually prefer to have ODEs entered in normal form.

If an ODE can be put in the form

$$a_n(t)y^{(n)} + a_{n-1}(t)y^{(n-1)} + \cdots + a_1(t)y' + a_0(t)y = f(t)$$

where $f(t)$ and the coefficients $a_j(t)$ are defined on a common t-interval, then it is called a *linear differential equation*. On a t-interval where the leading coefficient $a_n(t)$ is never zero, the *normal form of the linear ODE* is obtained by dividing by $a_n(t)$

$$y^{(n)} + \frac{a_{n-1}(t)}{a_n(t)}y^{(n-1)} + \cdots + \frac{a_0(t)}{a_n(t)}y = \frac{f(t)}{a_n(t)} \tag{5}$$

The ODE, $N' + kN = 0$, where k is a constant, is linear and in normal form. The ODE $my'' + cy' + ky = F(t)$ is also linear, but takes on normal form only after division by the leading coefficient m. On the other hand, the ODEs $mL\theta'' + cL\theta' + mg\sin\theta = 0$ and $ty''' - y^2 = 0$ are not linear because of the terms $\sin\theta$ and y^2.

Solutions and Solution Curves

We need a careful definition of a solution of a differential equation and its associated geometric characterization.

> ❖ **Solution of an ODE, Solution Curve**. A function $y(t)$ is a *solution* of the ODE, $G(t, y, y', \ldots, y^{(n)}) = 0$, if $y(t)$ and all its derivatives through order n exist and are continuous on an interval $a < t < b$ and, if upon substituting $y(t)$ for the unknown function, the ODE becomes an identity on that interval. The graph of a solution $y(t)$ in the ty-plane is called a *solution curve* for the ODE.

EXAMPLE 1.1.2

Exponential Growth: Solutions and Solution Curves
Direct substitution shows that $N(t) = e^{2t}$ is a solution of the first-order linear ODE

$$N'(t) = 2N(t)$$

on any interval $a < t < b$. In fact, for any constant C, $N = Ce^{2t}, a < t < b$, is a solution since $N' = 2Ce^{2t} = 2N$ on that interval. In Figure 1.1.1 solution curves are plotted for various values of C over the interval $-1 \le t \le 1$. Since C, a, and b are arbitrary, there is considerable flexibility in what we can accept as a solution. Even if the interval $a < t < b$ is fixed, there are infinitely many solutions on that interval—one for every value of the constant C.

See Atlas Plates FIRST-ORDER A, J for other examples of solution curves.

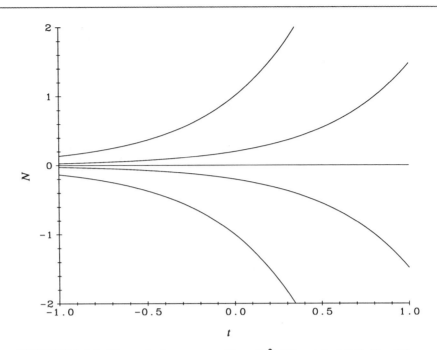

FIGURE 1.1.1 Five solution curves, $N = Ce^{2t}$ ($C = \pm 1, \pm 0.2, 0$), of the ODE, $N' = 2N$. See Example 1.1.2.

EXAMPLE 1.1.3

Exponential Solutions

Solutions of linear ODEs often involve exponential functions of the form e^{rt}, where r is a constant. For example, e^{rt} is a solution of the ODE

$$y' + 5y = 0$$

if and only if $r = -5$. To show this, insert $y = e^{rt}$ into the ODE to obtain

$$re^{rt} + 5e^{rt} = 0$$

which holds for all t if and only if $r = -5$ because e^{rt} is never 0.

Similarly, e^{rt} is a solution of the second-order linear ODE,

$$y'' + 3y' + 2y = 0$$

if and only if

$$r^2 e^{rt} + 3re^{rt} + 2e^{rt} = 0$$

for all t. Because e^{rt} is never 0, we must have that $r^2 + 3r + 2 = 0$, i.e., that $(r + 1)(r + 2) = 0$; hence, r must be -1 or -2.

The Geometry of Solution Curves

The geometry of solution curves gives insight into the properties of the solutions of an ODE. The function $f(t, y)$ in the first-order ODE

$$y' = f(t, y)$$

is a *rate function* because it gives the value of the rate of change of the solution $y(t)$ at any point $(t, y(t))$ on the solution curve. This has an important geometric interpreta-

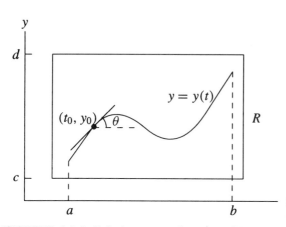

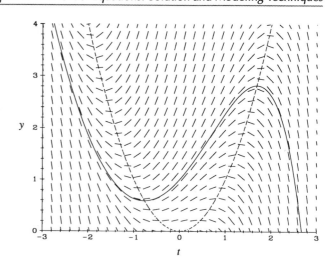

FIGURE 1.1.2 Solution curve for $y' = f(t, y)$. Note that $f(t_0, y_0) = \tan \theta$.

FIGURE 1.1.3 Direction field, a solution curve (solid), nullcline (dashed) for ODE $y' = y - t^2$ (Example 1.1.4).

tion as shown in Figure 1.1.2. Suppose that the rate function $f(t, y)$ is defined on the rectangle R in the ty-plane, and that $y = y(t)$ is a solution of the ODE whose graph lies in R. At the point (t_0, y_0) on the solution curve the slope of a line segment tangent to the curve is $f(t_0, y_0)$; that is, $f(t_0, y_0) = \tan \theta$, where θ is the angle the tangent line to the solution curve at (t_0, y_0) makes with the positive t-axis. So if the rate function $f(t, y)$ is positive at a point on a solution curve (i.e., if $\tan \theta$ is positive and the segment is tilted upwards), the curve is rising at the point and $y(t)$ is increasing. If $f(t, y)$ is negative, then the curve falls and $y(t)$ decreases. In Figure 1.1.1, the rate function for $N(t)$ is $2N(t)$, and solution curves rise if N is positive and fall if N is negative.

A line segment of slope $f(t, y)$ could be drawn centered at each point (t, y) in the rectangle R of Figure 1.1.2. To avoid overlapping segments, however, a grid of points is usually selected, and the correspondingly short line segments are drawn only at these points. The resulting field of line segments is called a *direction field*. A solution curve of the ODE, $y' = f(t, y)$, that passes through one of the grid points is tangent to the line segment centered at the point. Many ODE computer solvers also plot direction fields; all the direction fields in this book were constructed by a solver.

EXAMPLE 1.1.4

Solution Curves and a Direction Field
The linear ODE

$$y' = y - t^2$$

has solutions

$$y = t^2 + 2t + 2 + Ce^t$$

for every constant C. This is shown by the calculation

$$y' = 2t + 2 + Ce^t = (t^2 + 2t + 2 + Ce^t) - t^2 = y - t^2$$

Figure 1.1.3 shows a direction field of line segments for the ODE $y' = y - t^2$. The segment centered at the point (t, y) slants upward, and the solution curve rises if $y > t^2$ at the point. The segment slants downward and the curve falls if $y < t^2$. The dashed curve in Figure 1.1.3 is the *nullcline* (i.e., the curve of zero inclination) defined by $y = t^2$. The line segments centered on the nullcline are horizontal; the nullcline divides the ty-plane into a region where solution curves rise and another where they fall. The solid curve in the figure is the solution curve defined by $C = -1$:

$$y = t^2 + 2t + 2 - e^t$$

Observe how the solution curve "fits" the direction field; in particular, it crosses the nullcline with zero slope.

Atlas Plate FIRST ORDER B shows a direction field and a few solution curves for the ODE $y' = -y\cos t + \sin t$. Notice that the direction field appears to direct the "flow" of the solution curves as time advances.

A Simple Solution Technique: The Vanishing Derivative

We cannot continue to solve differential equations by guessing a solution and then verifying that our guess satisfies the equation, as we did in Examples 1.1.2 and 1.1.4. Surprisingly, once we find all the solutions of the simple differential equation

$$y'(t) = 0$$

we will have a technique for finding all solutions of many other first-order equations. Examples throughout the book demonstrate this fact. First, however, let us find all solutions of the ODEs $y'(t) = 0$ and $y'(t) = f(t)$.

THEOREM 1.1.1

> Vanishing Derivative Theorem. Every solution of the ODE $y'(t) = 0$ has the form $y(t) = C$, where C is any constant. If the function f is continuous, then all solutions of the ODE $y' = f(t)$ have the form $y(t) = F(t) + C$, where F is any antiderivative of f and C is an arbitrary constant.

Proof. If C is any constant, then the constant function $y(t) = C$ is a solution of $y' = 0$. We show that there are no other solutions. Suppose that $y(t)$ is a solution of the ODE, $y' = 0$, on a t-interval I. By the Mean Value Theorem (Theorem C.5.4 in Appendix C.5) if t_0 and t are both in I, then there is a number t^* between t_0 and t such that

$$y(t) - y(t_0) = y'(t^*)(t - t_0)$$

Since $y'(t^*) = 0$, we see that for all t in I

$$y(t) = y(t_0) = C$$

and the first part of the theorem is proved.

Now suppose that $y'(t) = f(t)$ where f is continuous on a t-interval I. Let F be an antiderivative of f, i.e., $F'(t) = f(t)$ for all t in I. Then the ODE, $y' = f$ can be rewritten as

$$y' - f = (y - F)' = 0$$

Applying the first part of the theorem, we see that for all t in I,

$$y(t) - F(t) = C$$

where C is an arbitrary constant; this proves the second part of the theorem.

The Vanishing Derivative Theorem, as simple as it is, is the source of many solution techniques for ODEs. Let us use it to show that we have already found all solutions of the ODE, $N' = 2N$, in Example 1.1.2

EXAMPLE 1.1.5

Exponential Growth: The Vanishing Derivative Theorem

Let $N(t)$, $a < t < b$, be any solution of ODE $N' = 2N$. Multiply each side of the ODE by the function e^{-2t} to obtain (after rearrangement)

$$e^{-2t}N' - 2e^{-2t}N = 0, \qquad a < t < b \tag{6}$$

Since e^{-2t} is never zero, every solution of ODE (6) is also a solution of $N' = 2N$, and conversely. Now since $N(t)$ is continuously differentiable, the product rule for differentiation may be used to rewrite (6) as

$$[e^{-2t}N(t)]' = 0, \qquad a < t < b$$

The Vanishing Derivative Theorem shows that $e^{-2t}N(t) = C$ for some constant C, and so $N(t) = Ce^{2t}$, $a < t < b$. Because a and b are arbitrary, we might as well set $a = -\infty$, $b = \infty$. Conversely, notice that $N(t) = Ce^{2t}$ solves ODE (6) for any constant C, and we have determined all of the solutions of the ODE $N' = 2N$.

The Vanishing Derivative Theorem can also be used to solve higher order ODEs, as the next example shows.

EXAMPLE 1.1.6

Solution of the Moving Ball Problem

To solve the problem of Example 1.1.1

$$\begin{aligned} s'' &= -g, & 0 \le t \le T \\ s(0) &= s_0, & s'(0) = v_0 \end{aligned} \tag{7}$$

write the ODE $s'' = -g$ as $(s' + gt)' = 0$. The Vanishing Derivative Theorem implies that

$$s' + gt = C$$

where C is a constant. The condition $s'(0) = v_0$ implies that $C = v_0$ and we have the first-order ODE

$$s' + gt = v_0$$

This ODE can be written in the form

$$(s + \frac{1}{2}gt^2 - v_0t)' = 0$$

and the Vanishing Derivative Theorem implies that

$$s + \frac{1}{2}gt^2 - v_0t = C$$

where C again is a constant. Since we require that $s = s_0$ when $t = 0$, we see that $C = s_0$. Thus, the solution $s(t)$ of the moving ball problem is

$$s(t) = s_0 + v_0t - \frac{1}{2}gt^2, \qquad 0 \le t \le T \tag{8}$$

Figure 1.1.4 shows solution curves with $s_0 = 10$ meters and $g = 9.8$ meters/sec^2 for various positive values of v_0. All of the curves are concave downward, because the second derivative $s''(t)$ is negative ($s'' = -g$). The curves show that the higher the initial velocity of the ball, the longer it takes the ball to reach the ground (recall that positive velocity corresponds to upward motion).

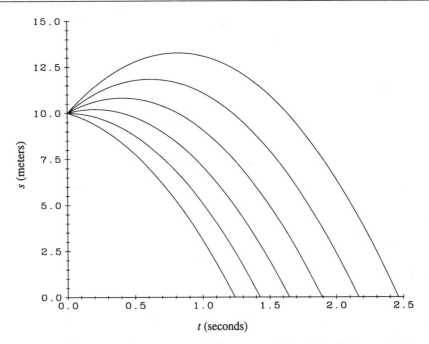

FIGURE 1.1.4 Solution curves of $s'' = -g$, $s(0) = 10$, $s'(0) = -2, 0,$ $2,\ldots, 8$, reading from the bottom curve up. See Example 1.1.6.

Systems of ODEs

Differential equations involving more than one unknown function occur quite often. A collection of differential equations that must be satisfied simultaneously by all the unknown functions involved is called a *differential system*. An example of a first-order system of two ODEs in two unknown functions $s(t)$ and $v(t)$ is given by

$$v' = -g$$
$$s' = v \tag{9}$$

where g is the gravitational constant. This is called a *first-order system* because no derivative of order higher than one is used. This system has already been normalized since the derivatives of the unknown functions have been isolated. For first-order systems that occur in the applications, the number of differential equations agrees with the number of unknown functions. Systems involving higher-order derivatives also occur, but we shall have little occasion to consider them.

System (9) is easily solved; indeed it has already been solved in Example 1.1.6, but the details are left to the reader (Problem 9).

PROBLEMS

1. Find the order of the ODEs below, and identify those which can be written as linear ODEs. Write each linear ODE in normal linear form.

    (a) $(t^2 + y^2)^{1/2} = y' + t$ **(b)** $dt/dy = 1/(t^2 - ty)$ [*Hint*: Find dy/dt.]

    (c) $y'' - t^2 y = \sin t + (y')^2$ **(d)** $y' = (1 + t^2)y'' - \cos t$

    (e) $e^t y'' + (\sin t)y' + 3y = 5e^t$ **(f)** $(y''')^2 = y^5$

In Problems 2–5, find all values of the constant so that the indicated functions are solutions of the accompanying ODEs. [*Hint*: Insert the proposed solutions into the ODEs.]

2. $y = e^{rt}$ (r is a constant)   [*Hint*: See Example 1.1.3.]

    (a) $y' + 3y = 0$ (b) $y'' + 5y' + 6y = 0$

    (c) $y^{(5)} - 3y^{(3)} + 2y' = 0$ (d) $y'' + 2y' + 2y = e^{-t}$

3. $y = ct^3$ (c is a constant)

    (a) $t^2 y'' + 6ty' + 5y = 0$ (b) $t^2 y'' + 6ty' + 5y = 2t^3$

    (c) $t^2 y'' + 6ty' + 5y = t^3$

4. $y = t^r$ (r is a constant)

    (a) $t^2 y'' + 4ty' + y = 0$ (b) $t^4 y^{(4)} + 7t^3 y''' + 3t^2 y'' - 6ty' + 6y = 0$

5. $x(t) = a \sin \omega t$, $y(t) = b \sin \omega t$ (a, b, and ω are constants).

$$x'' + 8x - 4y = 0, \quad y'' - 4y + 8x = 0$$

6.   Use the Vanishing Derivative Theorem to find all the solutions of the ODEs below. [*Hint*: In parts (a)–(c) multiply each side of the ODE by e^t.]

    (a) $y' + y = 1$ (b) $y' + y = t$

    (c) $y' + y = t + 1$ (d) $2yy' = 1$ [*Hint*: $(y^2)' = 2yy'$]

    (e) $2yy' = t$

7.   (*Solution Curves*). (a)–(e) Use a computer solver to plot solution curves of the ODEs in Problem 6 (a)–(e) in the rectangle R, $0 \le t \le 4$, $-4 \le y \le 4$, for the initial conditions $y(0) = -3, -1, 1, 3$. In each case, lightly shade by hand the part of R where solution curves rise as t advances.

8.   (*Vertical Motion of a Ball*). Review Examples 1.1.1 and 1.1.6 and answer the questions:

    (a) Graph some solution curves of the ODE in (1) in the ts-plane for various values of s_0, v_0. [*Hint*: Use the solution formula (8).]

    (b) From an initial height of 75 meters, what is the initial velocity of the ball so that it stays in the air 5 seconds before it hits the ground? Verify with a graph.

    (c) If the ball is thrown from the ground and reaches 30 meters, what is its initial velocity? Verify with a graph.

9.   (*System for Moving Ball*). Show that if the pair of functions $s = s(t)$, $v = v(t)$ is a solution of the system (9) satisfying the conditions $s(0) = s_0$, $v(0) = v_0$, then $s = s(t)$ is a solution of initial value problem (IVP) (7).

10.   (*Radioactive Decay*). A radioactive substance decays in such a way that the amount $N(t)$ remaining at time t (in years) satisfies the ODE, $N(t)' = -kN(t)$, where k is a positive constant.

    (a) Find $N(t)$. [*Hint*: Write the ODE as $N' + kN = 0$, multiply each side by e^{kt}, and use the Vanishing Derivative Theorem.]

    (b) After 100 years have elapsed, 90% of the substance remains. What is the value of k?

    (c) What percent of the substance remains after another 100 years have gone by?

11.   For each function $y(x)$ find an ODE, free of the constant C, for which $y = y(x)$ is a solution. [*Hint*: Eliminate C by using the equations for y and for y'.]

    (a) $y(x) = Ce^{x^2}$, where C is any constant.

    (b) $y(x) = 1/(x^2 + C)$, where C is any constant.

12. (*Direction Fields*). Use a computer ODE solver with a direction field plotting option to plot a direction field in the rectangle, $-2 \leq t \leq 2$, $-2 \leq y \leq 2$, for each of the following ODEs. Then use the solver to plot the solution curve through the point $t = 0$, $y = 0$, and then the solution curves through $t = 0$, $y = 2$ and through $t = -2$. Look at the line segments of the direction field that touch the plotted solution curves. Are they tangent? If not, explain why not. Plot the nullcline on the same graph. [*Hint*: See Figure 1.1.3.]

    (a) $y' = -y + 3t$ (b) $y' = y + \cos t$

    (c) $y' = \sin(ty)$ (d) $y' = \sin t \sin y$

    (e) $y' = \sin t + \sin y$ (f) $y' = \sin(t + y)$

13. (*Vanishing Derivative Theorem and Higher-Order Equations*). Use the Vanishing Derivative Theorem to find all solutions of each of the following problems. [*Hint*: See Example 1.1.6.]

    (a) $y'' = 0$ (b) $y'' = \sin t$, $y(0) = 0$, $y'(0) = 1$

    (c) $y''' = 2$ (d) $y'' + y' = e^t$ [*Hint*: Note that $(e^t y')' = e^t y'' + e^t y'$.]

1.2 First-Order Linear Differential Equations

We will find a formula for all solutions $y(t)$ of the *normal first-order linear ODE*

$$y' + p(t)y = q(t) \tag{1}$$

where the *coefficient* $p(t)$ and *driving term* (or *input*) $q(t)$ are continuous on a t-interval I. If $q(t)$ is not identically zero on I, then ODE (1) is said to be *nonhomogeneous* (or *driven*, or *forced*). We will see that ODE (1) has infinitely many solutions, all defined on the interval I. We shall show that a unique member of this infinitude of solutions can be selected by demanding that the solution $y(t)$ satisfy an *initial condition* $y = y_0$ at $t = t_0$, where t_0 is any (fixed) point in I, and y_0 is any given real number. Our main object of interest will be the *initial value problem* (abbreviated IVP)

$$y' + p(t)y = q(t), \qquad y(t_0) = y_0 \tag{2}$$

The solution $y(t)$ of IVP (2) is the *response* to the *input* $q(t)$ and to the *initial data* y_0. We shall illustrate the properties of ODE (1) and IVP (2) with examples and models.

The Integrating Factor Approach

In Section 1.1 we solved a particular case of the linear ODE,

$$y' = -ky \tag{3}$$

where k is a constant. A review of the way this ODE is solved provides the clue for solving ODE (1). Rearranging and multiplying by the "integrating factor" e^{kt}, we obtain the equivalent differential equation

$$e^{kt}[y' + ky] = [e^{kt}y]' = 0$$

The Vanishing Derivative Theorem (Theorem 1.1.1) implies that $e^{kt}y = C$, where C is an arbitrary constant, from which we obtain the solution formula,

$$y = Ce^{-kt}, \quad -\infty < t < \infty$$

This same process will work for ODE (1) as well, but we will need to be more skillful in the choice of an integrating factor.

Let $P(t)$ be any antiderivative of the coefficient $p(t)$ in ODE (1) on the interval I, i.e., $P'(t) = p(t)$ on I. The exponential function

$$e^{P(t)}$$

is called an *integrating factor* for the ODE (1). For any continuously differentiable function $y(t)$ on I, we have the identity

$$[e^{P(t)}y]' = e^{P(t)}y' + e^{P(t)}P'(t)y = e^{P(t)}[y' + p(t)y] \tag{4}$$

which is all we need to prove the following result.

THEOREM 1.2.1

General Solution Theorem. Consider the linear ODE, $y' + p(t)y = q(t)$, where $p(t)$ and $q(t)$ are continuous on an interval I. Let $P(t)$ be any antiderivative of $p(t)$, and let $R(t)$ be any antiderivative of $e^{P(t)}q(t)$ on I. Then every solution $y(t)$ of the ODE on the interval I has the form

$$y(t) = e^{-P(t)}C + e^{-P(t)}R(t) \tag{5}$$

for some constant C. Conversely, any function $y(t)$ of the form (5), where C is any constant, is a solution of the ODE on the interval I.

Proof. Let $y(t)$ be any solution of the ODE $y' + p(t)y = q(t)$ on the interval I. Multiplying the ODE by the integrating factor $e^{P(t)}$, we have

$$e^{P(t)}[y' + p(t)y] = e^{P(t)}q(t)$$

which, because of the identity (4), can be written as

$$[e^{P(t)}y]' = e^{P(t)}q(t) \tag{6}$$

Let $R(t)$ be an antiderivative of $e^{P(t)}q(t)$ on I (any one will do); applying the Vanishing Derivative Theorem to (6), we obtain

$$e^{P(t)}y = R(t) + C$$

where C is a constant. After multiplying each side of this equality by $e^{-P(t)}$, we obtain formula (5) for the solution $y(t)$.

The converse assertion that formula (5) for every value C defines a solution of the ODE follows by the product rule:

$$\begin{aligned} y'(t) &= \left[e^{-P(t)}C\right]' + \left[e^{-P(t)}R(t)\right]' \\ &= -p(t)e^{-P(t)}C - p(t)e^{-P(t)}R(t) + e^{-P(t)}e^{P(t)}q(t) \\ &= -p(t)\left[e^{-P(t)}C + e^{-P(t)}R(t)\right] + q(t) = -p(t)y(t) + q(t) \end{aligned}$$

where we used $P'(t) = p(t)$ and $R'(t) = e^{P(t)}q(t)$. This ends the proof.

An important consequence of Theorem 1.2.1 is that every solution of the ODE $y' + p(t)y = q(t)$ "lives" at least on any interval I where $p(t)$ and $q(t)$ are both continuous.

Formula (5) defines the family of all solutions of the linear ODE, $y' + p(t)y = q(t)$, over the interval I as the constant C takes on all possible values. For that reason (5) is called the *general solution* of that ODE. If $q(t) = 0$, then

$$y_h = Ce^{-P(t)}$$

is the general solution of the *homogeneous* (also called *free*, *reduced*, or *undriven*) *linear equation*

$$y' + p(t)y = 0$$

On the other hand, if $C = 0$ in formula (5), then

$$y_p = R(t)e^{-P(t)}$$

is a *particular solution* of the original nonhomogeneous ODE, $y' + p(t)y = q(t)$. Thus, the general solution (5) is displayed as

$$y = y_h(t) + y_p(t)$$

Actually, any particular solution can be used in this sum (see Problem 11).

The general solution formula (5) has a very important consequence: solution curves corresponding to different values of the constant C can never intersect! Why is that? Suppose that the solutions $y_1(t)$ and $y_2(t)$ of ODE (1) correspond to the different values C_1 and C_2 and intersect at $t = T$. Then

$$
\begin{aligned}
0 &= y_1(T) - y_2(T) \\
&= \left[e^{-P(T)}C_1 + e^{-P(T)}R(T)\right] - \left[e^{-P(T)}C_2 + e^{-P(T)}R(T)\right] \\
&= e^{-P(T)}[C_1 - C_2]
\end{aligned}
$$

and since $C_1 - C_2$ is not zero and exponentials are never zero, we have a contradiction. So our assumption that solution curves for distinct values of C can intersect was wrong.

EXAMPLE 1.2.1

Using an Integrating Factor, Solving an IVP

The ODE,

$$y' + 2y = 3e^t$$

is linear with $p(t) = 2$ and $q(t) = 3e^t$. Since $p(t)$ and $q(t)$ are continuous on the entire real line, each solution is also defined on the entire real line. Since $P(t) = 2t$ is an antiderivative of $p(t) = 2$, an integrating factor is $e^{P(t)} = e^{2t}$. Now $R(t) = e^{3t}$ is an antiderivative for $q(t)e^{P(t)} = 3e^{3t}$, and hence the general solution is

$$y = e^{-P(t)}C + e^{-P(t)}R(t) = e^{-2t}C + e^t, \qquad -\infty < t < \infty \qquad (7)$$

where C is an arbitrary constant.

Each value of C gives a distinct solution. Imposing an initial condition will determine C. Thus, to find the solution of the IVP

$$y' + 2y = 3e^t, \quad y(0) = -3 \qquad (8)$$

we see from (7) that $-3 = e^{-2 \cdot 0}C + e^0 = C + 1$. Hence $C = -4$ and

$$y = -4e^{-2t} + e^t, -\infty < t < \infty$$

The bottom curve in Figure 1.2.1 is the solution curve for IVP (8). The other solution curves correspond to initial conditions $y(0) = -2, -1 \dots, 5$ and are produced by solving forward and backward in time from the initial point. The solution curves in Figure 1.2.1 do not really touch; this is only a limitation of computer graphics. Zooming in on the upper right corner of Figure 1.2.1 separates the solution curves (Figure 1.2.2).

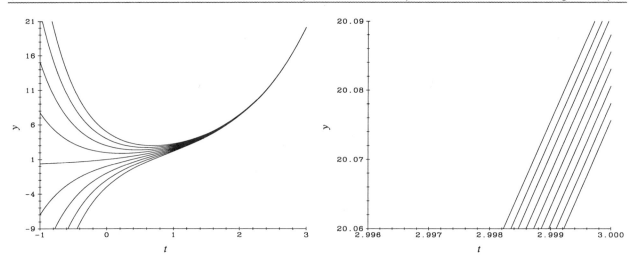

FIGURE 1.2.1 Solution curves of $y' + 2y = 3e^t$, $y(0) = -3, -2, \ldots, 5$. See Examples 1.2.1 and 1.2.2.

FIGURE 1.2.2 Zooming in on upper right of Figure 1.2.1 separates the solution curves (Example 1.2.1).

EXAMPLE 1.2.2

Long-Term Behavior of Solution Curves

The solution curves of $y' + 2y = 3e^t$ shown in Figure 1.2.1 seem to converge to a single curve as t increases. From formula (7) we see that as $t \to +\infty$ all solutions approach the particular solution

$$y_p(t) = e^t$$

whose graph is the middle of the nine curves in Figure 1.2.1. The reason for this is that the solutions

$$y_h(t) = Ce^{-2t}$$

of the homogeneous ODE, $y' + 2y = 0$, tend to 0 as $t \to +\infty$. On the other hand, as $t \to -\infty$, we see that $y_p(t) \to 0$, while for $C \neq 0$, $y_h(t) \to \pm\infty$. Thus, solutions $y = Ce^{-2t} + e^t$ of $y' + 2y = 3e^t$ for different values of C spread apart as t decreases.

The antiderivatives in formula (5) cannot always be evaluated in terms of elementary functions. Even so, these antiderivatives exist on any t-interval where p and q are continuous. In any case, the solution formula (5) provides a theoretically useful way to view solutions of first-order linear ODEs. An ODE solver can always be used to generate solution graphs.

EXAMPLE 1.2.3

What to do when the Solution Formula is Complicated

The linear ODE

$$y' + 2ty = (1 + t^2)^{-1}$$

has an integrating factor e^{t^2}. Multiplying by this factor, the ODE becomes

$$e^{t^2}[y' + 2ty] = [e^{t^2} y]' = e^{t^2}[1 + t^2]^{-1}$$

Antidifferentiation gives $e^{t^2} y = \int_{t_0}^t e^{s^2}(1 + s^2)^{-1}\, ds + C$, where t_0 is any (fixed) point on the real line. Thus the general solution of our equation is given by

$$y = e^{-t^2} C + e^{-t^2} \int_{t_0}^t e^{s^2}(1 + s^2)^{-1}\, ds, \qquad -\infty < t < \infty \tag{9}$$

$$= y_h(t) + y_p(t)$$

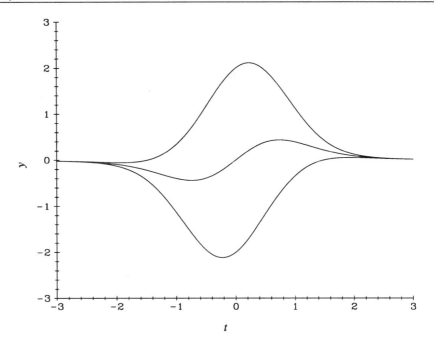

FIGURE 1.2.3 Solution curves of Example 1.2.3.

Since $p(t) = 2t$ and $q(t) = (1 + t^2)^{-1}$ are continuous for all t, the general solution is defined for all t. However, the integral in (9) is not expressible in terms of simple functions and we must leave the solutions in this form.

It is in a situation like this that a numerical ODE solver is particularly useful since it does not need solution formulas to do its work. The three solution curves of Figure 1.2.3 were generated by using a numerical solver to solve the ODE with $t_0 = 0$ and the three initial conditions $y(0) = 0$, ± 2, and moving backward and forward three units of time from $t = 0$. Formula (9) is useful, but the graphs of Figure 1.2.3 are even more helpful in understanding the behavior of solutions. From the graphs one might conjecture that as $t \to \pm\infty$, every solution tends to 0. Without the graphs, we could not easily make a conjecture about long-term behavior from formula (9) alone.

Initial Value Problems

The examples above show that one way to find a formula for the solution of the IVP

$$y' + p(t)y = q(t), \qquad y(t_0) = y_0$$

is to use solution formula (5) for the general solution of the ODE, and then use the initial data to determine the constant C. This method is straightforward and widely used, but there is another way that tells us a lot about how a linear IVP responds to changes in the input or the initial data.

THEOREM 1.2.2

Solution of Initial Value Problem. Let $p(t)$ and $q(t)$ be continuous on an interval I containing t_0, and let y_0 be any constant. Then the IVP

$$y' + p(t)y = q(t), \qquad y(t_0) = y_0 \tag{10}$$

has exactly one solution. If $P_0(t) = \int_{t_0}^{t} p(r)\, dr$, this solution is given by the formula

$$y(t) = e^{-P_0(t)} y_0 + e^{-P_0(t)} \int_{t_0}^{t} e^{P_0(s)} q(s)\, ds, \quad t \text{ in } I \tag{11}$$

Total Response = Response to initial data + Response to input

Proof. If the function $P(t)$ in formula (5) is replaced by $P_0(t)$, then the resulting function $y(t)$ is the general solution of the ODE in (10). Choosing $C = y_0$ yields the unique solution (11) of the given IVP (10) since $y(t_0) = y_0$ because $P_0(t_0) = 0$ and the integral in (11) is zero at $t = t_0$, so the assertion follows.

Rather than memorizing (11), the integrating factor approach outlined above should be followed to solve the IVP. The choice of the antiderivatives used in formula (11) is convenient for solving the IVP (10).

Balance Law

First-order linear ODEs often arise in a modeling environment as the result of an underlying basic principle: If $y(t)$ denotes the size of the population of a species (or the amount of a substance) at time t, then $y'(t)$ is the instantaneous rate of change of that population size (or substance amount) at time t. As a modeling aid, think of the population (or substance) as occupying a *compartment*, and then the rate $y'(t)$ can be calculated as the "rate in" minus the "rate out" of that compartment at the time t. We formalize this useful principle as the *Balance Law*.

Rate In

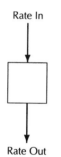

Rate Out

Balance Law.   Net rate of change of a substance = Rate in − Rate out.

EXAMPLE 1.2.4

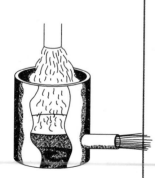

A Model IVP for Salt Accumulation in a Tank

A large tank contains 100 gal of brine in which, initially, S_0 pounds of salt are dissolved. More brine is allowed to run into the tank through an inlet pipe at the rate of 10 gal/min. The concentration $c(t)$ (pounds of salt per gallon) in this incoming brine varies with time. Suppose that the solution in the tank is thoroughly mixed and that brine flows out of the tank at the rate of 10 gal/min. Find the amount of salt $S(t)$ in the tank at any time t.

Thinking of the tank itself as the compartment that the salt occupies, we apply the Balance Law to find an expression for $S'(t)$, an expression that may involve $S(t)$ and t as well. Let the initial time $t_0 = 0$, and so S_0 is the amount of salt in the brine at $t = 0$. The rate at which salt is being added to the tank at time t is $10c(t)$ pounds per minute. The rate at which salt is leaving the tank is $10(S(t)/100) = S(t)/10$ because $S(t)/100$ pounds of salt per gallon is in the tank at time t. Thus, using the Balance Law we see that

$$S'(t) = \text{Rate in} - \text{Rate out} = 10c(t) - \frac{1}{10}S(t)$$

The corresponding initial value problem is

$$S'(t) + \frac{1}{10}S(t) = 10c(t), \qquad S(0) = S_0, \qquad t \geq 0 \tag{12}$$

Observe that the ODE in IVP (12) is linear, and so the integrating factor techniques are appropriate for constructing its solution.

EXAMPLE 1.2.5

How Much Salt Is in the Tank?

We now solve IVP (12). After multiplying through by the integrating factor

$$\exp\left[\int_0^t (1/10)\, ds\right] = e^{t/10}$$

the ODE in (12) becomes

$$[S(t)e^{t/10}]' = 10c(t)e^{t/10}$$

Integrating from 0 to t and using the initial condition $S(0) = S_0$, we have that

$$S(t) = e^{-t/10}S_0 + e^{-t/10}\int_0^t e^{s/10}[10c(s)]\, ds, \qquad t \geq 0 \tag{13}$$

The first term on the right-hand side of (13) represents the exponentially declining amount of salt in the tank if pure water runs into the tank and the brine runs out at the same rate. The second term measures the amount of salt in the tank at time t due solely to the incoming salt brine and assuming no salt in the tank initially. In the long run the second term will dominate, and the effect of the initial amount S_0 will be insignificant. It is the linearity of the ODE for $S(t)$ that allows us to decouple the effect of the initial amount of salt from the effect of the incoming stream; this generally cannot be done with nonlinear ODEs.

We can illustrate all this by graphing the terms of (13). To be specific, let us take $S_0 = 15$ lb and assume an input concentration $c(t)$ which varies periodically about a mean of 0.2 lb of salt per gallon. For example, suppose that

$$c(t) = 0.2 + 0.101 \sin t \quad \text{(lb of salt per gallon)}$$

Then using a table of integrals, (13) becomes

$$S(t) = [15e^{-t/10}] + [20 - 19e^{-t/10} + 0.1 \sin t - \cos t]$$

$$\text{Total Response} = \text{Response to initial data} + \text{Response to input}$$

The graphs of the responses are sketched in Figure 1.2.4. Observe that the response to the input eventually dominates.

From formula (13) we see precisely how $S(t)$ depends on the initial data $S_0 = 15$ and on the input $10c(t)$. From that formula it would be possible to calculate the effect on $S(t)$ of changes in S_0 and $c(t)$. See Atlas Plate FIRST ORDER D for the effect on $S(t)$ of changes in the amplitude of the sinusoid in the concentration $c(t)$ given above.

Discontinuous Driving Terms and Coefficients

Example 1.2.6 below suggests that the behavior of the solutions of a nonnormal ODE $a(t)y' + b(t)y = c(t)$ near values of t where $a(t)$ vanishes cannot easily be predicted.

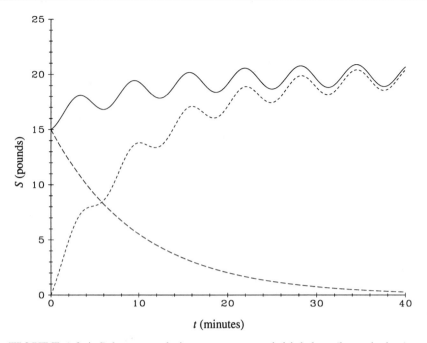

FIGURE 1.2.4 Salt accumulation: response to initial data (long dashes), response to input (short dashes), total response (solid) (Example 1.2.5).

To see what happens near a zero of $a(t)$ (or near a discontinuity of $p(t) = b(t)/a(t)$ or $q(t) = c(t)/a(t)$), one may first write the ODE in normal form, solve by the integrating factor technique on intervals where p and q are continuous, and then analyze the resulting solutions.

EXAMPLE 1.2.6

An IVP With Infinitely Many Solutions, Another with No Solutions
The ODE

$$ty' - y = t^2 \cos t \tag{14}$$

has the normal form

$$y' - y/t = t \cos t, \quad t < 0, \text{ or } t > 0 \tag{15}$$

The function $t \cos t$ is continuous for all t, but $1/t$ is discontinuous at $t = 0$. An integrating factor is given by $1/t$ on either interval, $t < 0$ or $t > 0$, and we have that

$$(y' - y/t)t = (y/t)' = \cos t$$

Antidifferentiation gives $y/t = \sin t + C$, or

$$y = t \sin t + Ct, \quad t < 0, \text{ or } t > 0 \tag{16}$$

Formula (16) defines the general solution of ODE (14) for all t, not just for $t \neq 0$, since

$$ty' - y = t(t \sin t - Ct)' - (t \sin t + Ct)$$

$$= t(t \cos t + \sin t + C) - (t \sin t + Ct) = t^2 \cos t$$

Thus it turns out that the "singularity" at $t = 0$, where the differential equation is not normal, does not cause trouble after all. All solutions (16) are defined for all t, and all satisfy $y = 0$ when $t = 0$. This means that all solution curves cut the y-axis at the

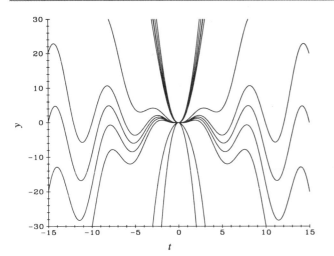

FIGURE 1.2.5 Solution curves of Example 1.2.6.

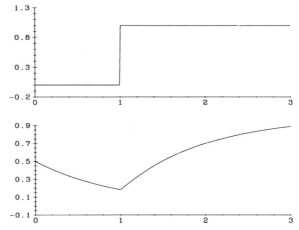

FIGURE 1.2.6 Response $y(t)$ (bottom) to step-function input at $t = 1$ (top). See Example 1.2.7.

origin, but nowhere else. Thus, the IVP,

$$ty' - y = t^2 \cos t, \qquad y(0) = 0$$

has infinitely many solutions, but the same ODE with initial condition $y(0) = y_0 \neq 0$ has no solution at all. Both results follow from the solution formula. The solution curves shown in Figure 1.2.5 were graphed by choosing initial points on the left, bottom, and right sides of the rectangle $|t| \leq 15$, $|y| \leq 30$ and solving forward or backward until the graph exited the rectangle.

Discontinuous coefficients and driving terms occur in many applications and must be taken into account. The following definition covers a common type of discontinuity.

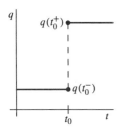

❖ **Piecewise Continuity.** A function $q(t)$ is *piecewise continuous* on the closed bounded interval I, $a \leq t \leq b$, if the one-sided limits $q(t_0^+)$ and $q(t_0^-)$ both exist for all t_0 in I and coincide for all but a finite number of points in I.[1] The function $q(t)$ is piecewise continuous on the real line if it is piecewise continuous on every interval $a \leq t \leq b$. The discontinuities of piecewise continuous functions are called *jump discontinuities*.

Step functions and *square waves* (see Appendix C.1) are notable examples of piecewise continuous functions which are not continuous. Continuous functions on closed intervals are also piecewise continuous, but the one-sided limits $q(t^+)$ and $q(t^-)$ have the same value.

If $p(t)$ is continuous on I, but $q(t)$ is only piecewise continuous on I, it is a remarkable fact that IVP (2) is still uniquely solvable if a solution is suitably defined. We will allow as solutions those continuous functions $y(t)$ on I whose derivatives $y'(t)$ are (perhaps) only piecewise continuous and satisfy ODE (1) at all points where $q(t)$ is continuous. Indeed, the solution formula (11) is still valid in this case and has the stated properties, but we omit the proof.

---

[1]$q(t_0^+)$ (respectively, $q(t_0^-)$) denotes the limit of $q(t)$ as t approaches t_0 from above (below).

EXAMPLE 1.2.7

Response to Piecewise-Continuous Input

The IVP

$$y' + y = \text{step}(t-1), \qquad y(0) = 1/2, \quad t \geq 0$$

has the piecewise continuous input

$$\text{step}(t-1) = \begin{cases} 0, & 0 \leq t < 1 \\ 1, & t \geq 1 \end{cases}$$

Since $p(t) = 1$, $P_0(t) = \int_0^t 1 \, ds = t$. Thus, we have that

$$y(t) = e^{-t}/2 + e^{-t} \int_0^t e^s \, \text{step}(s-1) \, ds$$

$$= \begin{cases} e^{-t}/2, & 0 \leq t \leq 1 \\ e^{-t}/2 + e^{-t} \int_1^t e^s \, ds = e^{-t}/2 + 1 - e^{1-t}, & t \geq 1 \end{cases}$$

where the definition of the step function has been used. Figure 1.2.6 shows how the solution suddenly changes both its slope and its concavity when the input function "turns on" at $t = 1$.

See Problem 15 for more problems involving step functions.

PROBLEMS

1. (*Integrating Factor*). Find the general solution of each ODE below by using an integrating factor. [*Hint*: First write the ODE in normal linear form.]

    (**a**) $y' - 2ty = t$ (**b**) $y' - y = e^{2t} - 1$

    (**c**) $y' = \sin t - y \sin t$ (**d**) $2y' + 3y = e^{-t}$

    (**e**) $t(2y - 1) + 2y' = 0$ (**f**) $y' + y = te^{-t} + 1$

2. (*Integrating Factor*). For each ODE below, find the general solution by using an integrating factor. [*Hint*: First write the ODE in normal linear form.]

    (**a**) $ty' + 2y = t^2$ (**b**) $(3t - y) + 2ty' = 0$

    (**c**) $y' = (\tan t)y + t \sin 2t$ (**d**) $y' = y^2(t + y)^{-1}$ [*Hint*: Find dt/dy.]

    (**e**) $(y - 2) + (3t - y)y' = 0$ [*Hint*: Find dt/dy.]

3. Find the general solution of the ODE $y' = y/t + t^n$, $t > 0$.

4. Find the solution of each IVP and give the largest t-interval in which the solution is valid. [*Hint*: First find the general solution and then use the initial condition to evaluate the constant C.]

    (**a**) $y' + y = e^{-t}$, $y(0) = 1$

    (**b**) $y' + 2y = 3$, $y(0) = -1$

    (**c**) $y' + 2ty = 2t$, $y(0) = 1$

    (**d**) $y' + (\cos t)y = \cos t$, $y(\pi) = 2$

    (**e**) $ty' + 2y = \sin t$, $y(\pi) = 1/\pi$

    (**f**) $(\sin t)y' + (\cos t)y = 0$, $y(3\pi/4) = 2$

    (**g**) $y' + y(\cot t) = 2\cos t$, $y(\pi/2) = 3$

    (**h**) $y' + (2/t)y = (\cos t)/t^2$, $y(\pi) = 0$

5. **(a)–(h).** Use a numerical solver to plot the solution of each IVP of Problem 4 **(a)–(h)** in a rectangle enclosing the initial point.

6. Consider the linear ODE, $y' + y = t\cos(t^2)$, $-5 \le t \le 10$.

   (a) Plot representative solutions $y_h(t)$ of the homogeneous ODE, $y' + y = 0$, where the initial data satisfies $-4 \le y(-5) \le 4$.

   (b) Using a numerical solver, plot the solution $y = y_p(t)$ of the IVP, $y' + y = t\cos(t^2)$, $y(-5) = 4$.

   (c) Use parts **(a)** and **(b)** to explain the behavior of the solution curves shown in Atlas Plate FIRST ORDER A. [*Hint*: The general solution has the form $y = y_h(t) + y_p(t) = Ce^{-t} + y_p(t)$. What happens as $t \to \infty$?]

7. Plot the solution curves of the given IVPs over the indicated time span.

   (a) $y' = (\sin t)(1 - y)$; $y(-\pi/2) = -1, 0, 1$; $-\pi/2 \le t \le 2\pi$

   (b) $y' + y = te^{-t} + 1$; $y(0) = -1, 0, 1$; $-1 \le t \le 3$

   (c) $ty' + 2y = t^2$; $y(2) = 0, 1, 2$; $0 < t < 4$. Repeat for $y(-2) = 0, 1, 2$; $-4 < t < 0$.

   (d) $y' = (\tan t)y + t\sin 2t$; $y(0) = -1, 0, 1$; $-\pi/2 < t < \pi/2$

8. (*Bounded Output*). The output $y(t)$ of a system, modeled by the IVP $y'(t) + p(t)y(t) = q(t)$, $y(0) = y_0$, is to be maintained at the constant level y_0 for $t \ge 0$. For a given $p(t)$, how can the term $q(t)$ be determined to accomplish the task?

9. (*Salt Solution*). A tank initially has 10 gallons of brine which contains 10 pounds of salt. A salt solution containing a constant c_0 pounds of salt per gallon flows into the tank at 1 gal/min and the well-stirred mixture flows out the bottom of the tank at the same rate. Denote the amount of salt in the tank at time t by $S(t)$.

   (a) Find a solution formula for $S(t)$. Find the value that $S(t)$ approaches as $t \to \infty$.

   (b) Plot $S(t)$ for values of c_0 in the range $0.1 \le c_0 \le 2$ over the interval $0 \le t \le 40$. Be sure to include $c_0 = 1.0$ and values of c_0 above and below 1.0. What happens to the salt solution as time goes on?

10. (*Salt Solution*). A tank initially contains 50 gallons of pure water. Starting at time $t = 0$, a salt solution containing 2 pounds of salt per gallon flows into the tank at a rate of 3 gal/min. The mixture is well-stirred. At time $t = 3$ min the mixture begins to flow out of the tank at a rate of 3 gal/min. [*Hint*: Divide the problem into two IVPs, one with $t_0 = 0$ and the other with $t_0 = 3$.]

    (a) How much salt is in the tank at $t = 2$ min?

    (b) How much salt is in the tank at $t = 25$ min?

    (c) How much salt remains in the tank as $t \to \infty$? Why could you use common sense to guess the amount without any calculation?

11. Show that if $y_p(t)$ is any particular solution of the ODE $y' + p(t)y = q(t)$, and if $P(t)$ is any antiderivative of the coefficient $p(t)$, then

$$y = Ce^{-P(t)} + y_p(t), \quad C \text{ an arbitrary constant}$$

    is the general solution of the ODE. [*Hint*: Let $y(t)$ be any solution of the ODE, and show that $z(t) = y(t) - y_p(t)$ solves the homogeneous ODE $z' + p(t)z = 0$.]

12. (*Industrial Waste*). Industrial waste is pumped into a tank containing 1000 gallons of water at the rate of 1 gallon of waste/min, and the well-stirred mixture leaves the tank at the same rate.

    (a) Find the concentration of waste in the tank at time t. What happens as $t \to \infty$?

    (b) How long does it take for the concentration to reach 20% of its maximum?

13. (*Compound Interest*). At the end of every month Wilbert makes a deposit in a savings account. He wants to buy a new car which will cost \$12,400. Currently, he has \$5800 saved. If he can earn 7% interest continuously compounded on his savings and has an income of \$2600 per month, to what amount must Wilbert limit his monthly expenses if he is to have enough saved to buy the car in 1 year? [*Hint*: Set up a sequence of initial value problems, $A_1' = 0.07A_1$, $A_1(0) = 5800$, $0 \le t \le 1$, $A_2' = 0.07A_2$, $A_2(0) = N$, $0 \le t \le 11/12$, ..., where N is Wilbert's monthly deposit. Then find N such that the total amount at the end of the year is \$12,400.]

14. (*Discontinuity in the Normalized ODE*). The linear ODE, $ty' + y = 2t$, has normal form $y' + y/t = 2$, and $p(t) = 1/t$ is discontinuous at $t = 0$.

(a) Find all the solutions of $ty' + y = 2t$.

(b) Show that the IVP, $ty' + y = 2t$, $y(0) = 0$ has exactly one solution, but if $y(0) = y_0 \ne 0$, there is no solution at all. Why does this not contradict the existence result in Theorem 1.2.2?

(c) Plot several solutions of the ODE over the interval $|t| \le 5$.

15. (*Discontinuity in the ODE*). Plot the solution of each IVP on the interval $0 \le t \le 2$. Explain any peculiar features of the graph in terms of a feature of the IVP itself. [*Hint*: When integrating a piecewise continuous function, recall that $\int_{t_0}^t = \int_{t_0}^T + \int_T^t$. This is especially helpful if the integrand has a discontinuity at T.]

(a) $y' + 2y = q(t)$, $y(0) = 0$, where $q(t) = \begin{cases} 1, & t \le 1 \\ 0, & t > 1 \end{cases}$

(b) $y' + p(t)y = 0$, $y(0) = 1$, where $p(t) = \begin{cases} 2, & t \le 1 \\ 1, & t > 1 \end{cases}$

16. (*Operator Formulation of an IVP*). Let $C^0(I)$ denote the set of all continuous functions on the interval I, and $C^1(I)$ the set of all differentiable functions on I with a continuous derivative. Let t_0 be an element of I, and $D(I)$ the set of all functions $y(t)$ in $C^1(I)$ with $y(t_0) = 0$. Finally, let L_0 be the mapping from $D(I)$ into $C^0(I)$ which carries $y(t)$ into $y'(t) + p(t)y(t)$; i.e., $L_0[y(t)] = y'(t) + p(t)y(t)$ for every function $y(t)$ in $D(I)$. Thus, the IVP $y'(t) + p(t)y(t) = q(t)$, $y(t_0) = 0$ can be formulated as $L_0[y(t)] = q(t)$ where $y(t)$ belongs to $D(I)$.

(a) (*Linearity*). Show that for any two functions $y(t)$ and $z(t)$ in $D(I)$ and any two constants a and b, we have that $L_0[ay(t) + bz(t)] = aL_0[y(t)] + bL_0[z(t)]$.

(b) (*Range, Solvability*). Show that the range of L_0 is $C^0(I)$ and that the equation $L_0y = q$ is uniquely solvable for each q in $C^0(I)$. [*Hint*: See Theorem 1.2.2.]

(c) (*Inverse of L_0*). Construct a mapping $G : C^0(I) \to D(I)$ with the property that

$$G[L_0[y]] = y \quad \text{for all } y \text{ in } D(I)$$

$$L_0[G[q]] = q \quad \text{for all } q \text{ in } C^0(I)$$

[*Hint*: The mapping G, called the *inverse* of L_0, involves an integral. G acts on the driving term q and returns the solution $y(t)$ of the IVP.]

17. (*Forced Oscillations*). The first order linear ODE, $y' + p_0 y = q(t)$, where p_0 is a nonzero constant and $q(t)$ is a periodic, piecewise continuous function of period T (see Appendix C.1 for examples) has some very interesting properties. Under such conditions, the ODE has a unique periodic solution of period T called the *forced oscillation* (i.e., the oscillation forced by the periodic driving term, $q(t)$). Show this by following the outline below:

- Show that there is exactly one value y_0 for which the IVP

$$y' + p_0 y = q(t), \qquad y(0) = y_0$$

  has a periodic solution of period T, and there is no other periodic solution of period T. [*Hint*: Use the solution formula (11), and the fact that a continuous function $y(t)$ is periodic with period T if and only if $y(t + T) = y(t)$, for all t.]

- Plot the solution of the IVP

$$y' + 0.5y = \mathrm{sww}(t, 50, 4), \qquad y(0) = 2$$

  over the interval $0 \le t \le 24$. See Appendix C.1 for the definition of the *sawtooth wave*, $\mathrm{sww}(t, 50, 4)$. Even though $y_0 = 2$ in this IVP is not the correct value to generate the forced oscillation, one can "see" the oscillation in the graph of the solution for "large" t. Why is this? Would this still be true for any nonzero value of the constant p_0? If not, Why not? (Examine Atlas Plate FIRST ORDER K for inspiration.)

- Replace p_0 in the IVP with a continuous periodic function $p(t)$ with the same period T as the periodic, piecewise continuous driving term, $q(t)$. If $\int_0^T p(s)\,ds \ne 0$, then show that there is a unique forced oscillation of period T. [*Hint*: You may proceed as in the first part, but using these facts: if $P_0(t) = \int_0^t p(v)\,dv$, then $P_0(t + T) = P_0(t) + P_0(T)$, for any t. Also, $\int_T^{t+T} e^{P_0(s)} q(s)\,ds = e^{P_0(T)} \int_0^t e^{P_0(s)} q(s)\,ds$ for any t value.]

- What can you say about the existence of forced oscillations for the driven ODE $y' + p(t)y = q(t)$, where p and q are as in the part above except that now $\int_0^T p(v)\,dv = 0$? Give some examples to illustrate your conclusions.

1.3 Introduction to Modeling

Modeling is a process loosely described as recasting a problem or concept from its natural environment into a form, called a model, that can be analyzed via techniques we understand and trust. A *model* is a device that aids the modeler in predicting or explaining the behavior of a phenomenon, experiment, or event. For example, say that a manned rocket is to be put into orbit about Mars. Physical intuition alone cannot give us more than a rough idea of where to aim the rocket and what guidance strategy to employ once the rocket is launched. Since accuracy will be critical for the success of this mission, a mathematical model of the problem is constructed utilizing applicable "laws" or "principles." The equations, constraints, and control elements in the model can then be treated by mathematical techniques and used to give a reasonably precise description of the orbital elements of the rocket in its course around Mars.

The use of models[2] in general to predict or explain outcomes of an observable situation is illustrated in Figure 1.3.1. Fairly broad portions of the natural environment

---

[2]Galileo Galilei (1564–1642) was the first modeler of modern times. Although he is better known for his work in astronomy, his use of inclined planes and pendulums led to a model for falling bodies relating the distance traveled to the time elapsed, eventually leading to the law of acceleration in 1604. Later, Galileo turned his telescope toward the heavens where he discovered four moons orbiting the planet Jupiter. His observations led him to support the Copernican model of the solar system. The natural connection between the physical and mathematical worlds are best said in his verse, "Philosophy is written in this grand book of the universe, which stands continually open to our gaze... It is written in the language of mathematics."

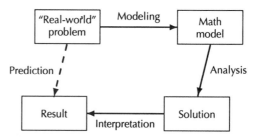

FIGURE 1.3.1 Problem solving via mathematical modeling.

are given a mathematical form by a general model in which all possible outcomes are described by a few basic principles. Figure 1.3.1 illustrates that a specific problem in the environment is translated into a specific mathematical problem in the model. That specific mathematical problem is solved (perhaps by a computer simulation) and the results are interpreted in the problem's natural environment.

Before discussing the modeling process in more detail, we present a few basic examples.

Radioactivity

Certain elements, or their isotopes, are known to be unstable, decaying into isotopes of other elements by emission of alpha particles (helium nuclei), beta particles (electrons), or photons. Such elements are said to be *radioactive*. For example, a radium atom might decay into a radon atom, giving up an alpha particle in the process, $^{226}\text{Ra} \rightarrow {}^{222}\text{Rn}$. The decay of a single radioactive nucleus is a random event, and the exact time of decay cannot be predicted with certainty. Nevertheless, something definite can be said about the decay process of a large number of radioactive nuclei.

Our system is the collection of radioactive nuclei in the sample, and the only measure of the system that concerns us is the number of radioactive nuclei present at any time. There is a great deal of experimental evidence to suggest that the following *decay law* is true.

> Radioactive Decay Law. In a sample containing a large number of radioactive nuclei, the decrease in the number of nuclei over a given interval of time is directly proportional to the length of the time interval and to the number of nuclei present at the start of the interval.

Denoting the number of radioactive nuclei in the sample at time t by $N(t)$, and a time interval by Δt, the law translates to the mathematical equation

$$N(t + \Delta t) - N(t) = -kN(t)\Delta t \tag{1}$$

where k is a positive constant of proportionality, which is independent of t and N. A reaction of this type is said to be of *first order* with *rate constant* k.

The mathematical model (1) of the decay law helps us to spot flaws in the law itself. $N(t)$ and $N(t + \Delta t)$ must be integers, but $k\Delta t$ need not be an integer, or even a fraction. If we want to keep the form of the law, we must transform the real phenomenon into an idealization in which a continuous rather than a discrete amount $N(t)$ undergoes decay. For example, measuring $N(t)$ in grams, say, will make it appear more "continuous" since the changes need not be multiples of unity.

The number of nuclei can be calculated from the mass of the sample at any given time. Thus there is no need to be specific about the units for N or t at this point. Even if $N(t)$ is continuous, (1) could not hold for arbitrarily large Δt since $N(t + \Delta t) \to 0$ as $\Delta t \to \infty$. Nor does (1) make sense if Δt is so small that no nucleus decays in the time span Δt. The failure of the law for large Δt can be ignored since we are interested only in local behavior (in time). The difficulty with small Δt is more troublesome. We can only hope that the mathematical procedures we now introduce will lead to a mathematical model from which reasonably accurate predictions can be made concerning $N(t)$ at any given time.[3]

If we divide both sides of formula (1) by Δt and let $\Delta t \to 0$ (ignoring the difficulty with small Δt mentioned above), we have the linear ODE

$$N'(t) = \lim_{\Delta t \to 0} \frac{N(t + \Delta t) - N(t)}{\Delta t} = -kN(t) \qquad (2)$$

Note that k is measured in units of reciprocal time.

EXAMPLE 1.3.1

Radioactive Decay

There are N_0 nuclei in a sample of a radioactive element at a given time t_0. How many radioactive nuclei are present at any later time? The state variable of the system is the number $N(t)$ of radioactive nuclei present. From ODE (2) we see that the law describing how the system evolves is given by the linear ODE, $N'(t) = -kN(t)$, where the rate constant k must be determined experimentally. In addition to the law of evolution, $N(t)$ must also satisfy the condition $N(t_0) = N_0$. Thus, the number of radioactive nuclei present in our particular sample must solve the initial value problem

$$N' = -kN, \qquad N(t_0) = N_0 \qquad (3)$$

Using the method of integrating factors (see Section 1.2), we see that IVP (3) has precisely one solution for any given value for N_0, namely,

$$N(t) = N_0 e^{-k(t-t_0)} \qquad (4)$$

Once k and N_0 are known, formula (4) can be used to predict the values of $N(t)$ at future times. See Figure 1.3.2 for an interpretation of Figure 1.3.1 in the context of the phenomenon of radioactive decay.

The *half-life* τ of the radioactive nuclei can be used to determine k. The number τ is the time required for half of the nuclei to decay. Curiously, τ is independent both of the time when the clock starts and of what the initial amount is. Indeed, formula (4) implies that

$$\frac{N(t + \tau)}{N(t)} = \frac{1}{2} = \frac{N_0 e^{-k(t+\tau-t_0)}}{N_0 e^{-k(t-t_0)}} = e^{-k\tau}$$

Taking logarithms of the second and the fourth terms, we have that

$$k = \frac{\ln 2}{\tau} \qquad (5)$$

---

[3]In this regard see the article by E. Wigner "The Unreasonable Effectiveness of Mathematics in the Physical Sciences," *Commun. Pure Appl. Math.* **13** (1960), pp 1–14.

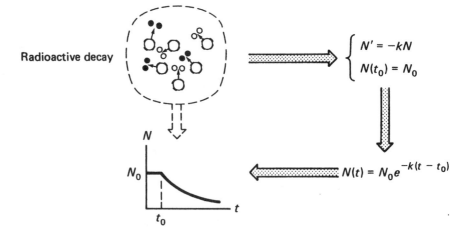

FIGURE 1.3.2 Modeling radioactive decay.

and τ is independent of N_0, t_0, and t, because k is.

In the case of ^{226}Ra, it is appropriate to measure time in years and thus the rate constant k has units of (years)$^{-1}$. Formula (4), with k given by (5) in terms of τ can be used to make predictions about the value of $N(t)$. These predictions can be checked against experimental determinations of $N(t)$ and the validity of the law and the model confirmed or questioned. Given the logical gaps mentioned above, it might seem surprising that (4) provides a remarkably accurate description of radioactive decay processes. But it is a fact of contemporary experimental physics that this is so, at least for time spans, $t - t_0$, which are neither too long nor too short. The leap of faith made in ignoring the flaws of the law and the model is justified by the results.

Motion Along the Local Vertical

The gravitational force on a body of mass m near the earth's surface has the magnitude mg, where g is a constant, and is directed downward along the local vertical (indeed, the direction of this force defines the local vertical). Recall that mg is the weight of the body. The motion of the body along the local vertical is governed by a law formulated by Isaac Newton[4] (a more general version of the law is given in Section 4.1):

---

[4]Isaac Newton (1642–1727) began his work in science and mathematics when he entered Trinity College in Cambridge, England in 1661 with a deficiency in geometry. He graduated without any special honors in 1665 and returned home to avoid the plague which was rapidly spreading through London. During the next two years, Newton discovered calculus, determined basic principles of gravity and planetary motion, and recognized that "white" light is composed of all colors, discoveries which he kept to himself at this time. In 1667, Newton returned to Trinity College, obtaining a Master's degree and then continuing there as a professor. Newton continued work on his earlier discoveries, formulating the law of gravitation, basic theories of light, thermodynamics, and hydrodynamics, and devising the first modern reflecting telescope. In 1687, he was finally persuaded to publish *Philosophae Naturalis Principia Mathematica*, which contains his basic laws of motion and is considered one of the most influential scientific books ever written. Despite his great accomplishments, Newton said of his work that, "I seem to have been only like a boy playing on the seashore and diverting myself in now and then finding a smoother pebble or prettier shell than ordinary, whilst the great ocean of truth lay all undiscovered before me."

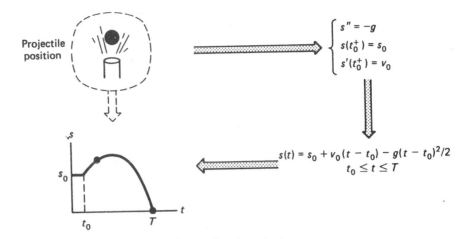

FIGURE 1.3.3 Modeling motion on local vertical.

Newton's Second Law. The time rate of change of the product of the mass m and the velocity v of a body (i.e., the body's momentum) is the sum of the external forces F_i acting vertically on the body:

$$(mv(t))' = \sum_i F_i(t)$$

If the gravitational force is the only force acting on the body, Newton's Second Law implies that the height $s(t)$ of the body above the earth's surface at time t satisfies the differential equation $(ms')' = -mg$, i.e., $s'' = -g$ if the mass m is constant. Thus if the earth had no atmosphere, then all bodies moving along the local vertical near the earth's surface would move in the same way regardless of their size, shape, or mass. In Section 1.1 we showed that if a vertically moving body is subjected only to the force of gravity, then the solution of the ODE, $s'' = -g$ is

$$s(t) = s_0 + v_0 t - gt^2/2$$

if the body is at height s_0 and moving with velocity v_0 at time $t_0 = 0$. This formula can be used to make predictions about the location and the velocity of the body at any time for which the model is valid. Figure 1.3.3 gives an interpretation of the modeling diagram in this case.

The validity of the laws of motion postulated for this problem can be checked against experimental observations. This has been done many times over the last three centuries. The model does have flaws. For example, it ignores air resistance and variations in the value of g with altitude and location on the earth. We next construct an improved model that takes the effect of air resistance into account.

Air Resistance: Viscous Damping

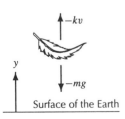

When a body falls toward the surface of the earth through a vacuum, the velocity steadily increases without limit until the time T of impact. The velocity of a body falling through a resistive medium such as our atmosphere, however, tends to a finite limit. Experiments show that for a body of low density and extended rough surface (e.g., a feather or a snowflake) the resistance of the air exerts a force on the body proportional to the magnitude of the velocity but, of course, acting opposite to the direction of motion. Let y be distance measured along the local vertical, with up as the positive direction, and let $y' = v$ be the velocity of the body. Thus if the body is falling, the force of resistance has magnitude $k|v|$, where k is a positive constant of proportionality, and the resistive force will act upward, opposing the fall. Similarly, if the body were rising through the air, the resistance would have magnitude $k|v|$ and would act in a downward direction. This kind of resistive force is called *viscous damping*, and k is the *viscous damping constant*.

EXAMPLE 1.3.2

Vertical Motion with Viscous Damping

Suppose that the object has constant mass m and is released from rest at a height h above the earth's surface. Then by Newton's Second Law the vertical location of the object solves the IVP

$$(my')' = my'' = -mg - kv, \qquad y(0) = h, \quad v(0) = 0 \tag{6}$$

where g is the gravitational constant, $y(t)$ denotes the height of the object above the ground, and $v = y'$. The minus sign in front of the kv term comes from the fact that the resistive force always acts opposite to the direction of motion. Replacing y'' by v', the differential equation in (6) becomes a first-order linear IVP in the velocity v:

$$v' + \left(\frac{k}{m}\right)v = -g, \qquad v(0) = 0$$

which can be solved via the integrating factor technique (see Section 1.2) to obtain

$$v(t) = \frac{mg}{k}(-1 + e^{-kt/m}) \tag{7}$$

From formula (7) we see that the velocity is negative for $t > 0$ (since $e^{-kt/m} < 1$), indicating that the body is falling. Also from (7) we see that as $t \to +\infty$, $v(t)$ approaches the limiting velocity $-mg/k$. Although the body never reaches the limiting velocity before it hits the ground, its velocity quickly approaches and remains very near this limiting value if k is large and m is small (e.g., if the object is a feather, or a snowflake). This is also visually evident from the slow fall of a feather or a snowflake at an apparently constant speed.

To determine $y(t)$, we replace $v(t)$ by $y'(t)$ in (7) and include initial data to obtain the first-order IVP

$$y' = \frac{mg}{k}(-1 + e^{-kt/m}), \qquad y(0) = h$$

whose solution is found by antidifferentiation to be

$$y(t) = h - \frac{mg}{k}t - \frac{m^2 g}{k^2}(-1 + e^{-kt/m}), \qquad 0 \le t \le T \tag{8}$$

The moment of impact T satisfies the equation $y(T) = 0$. However, the value of T can only be approximated because of the complicated form of the function $y(T)$. Naturally, the model IVP (6) ceases to have any validity after the time of impact.

Elements of a Model

The examples described above bring to light some essential components of a model.

System Variables: First, the natural process can be described by a collection of variables which all depend on a single independent variable. For example, in the radioactive decay problem (Example 1.3.1), the independent variable is time t, and the dependent variables are the number $N(t)$ of radioactive nuclei present at time t, and the instantaneous rate of change $N'(t)$ of that number at time t. For the falling body problem (Example 1.3.2), the independent variable is again time t, and the dependent variables are the position, velocity, and acceleration of the body at time t. Dependent and independent variables are called *system variables*.

Natural Laws: Next, there is an underlying mathematical structure in which *natural laws* or *principles* involving the dependent and independent variables can be expressed after appropriate notation is introduced. Sometimes these natural laws arise empirically as in the moving ball problem (Example 1.1.1) and the radioactive decay problem (Example 1.3.1), and sometimes the natural laws have an intrinsic significance, such as Newton's Second Law and the Gravitation Law in Example 1.3.2. Sometimes, a natural law is expressed in common sense terms, such as the Balance Law in the salt accumulation problem (Example 1.2.4).

System Parameters: The natural laws often contain *parameters* which must be experimentally determined; for example, the coefficient of air resistance k in Example 1.3.2, and the decay coefficient k in Example 1.3.1.

It is almost always the case that the laws or principles in a system describe the way in which the system evolves in time. Thus it does no great harm to think of the independent variable in most systems of interest to be time.

State Variables and Dynamical Systems

❖ **State Variables, Dynamical Systems.** A collection of dependent variables in a real-world system describes the state of the system at time t if the values of those variables at a given instant together with the natural laws of the system uniquely determine values of the variables for all times where the laws apply. The dependent variables which describe the state of a system are called *state variables*. The system itself is said to be a *dynamical system* in this case.

Some examples of state variables are position and velocity at time t of a body undergoing vertical motion (Example 1.1.6 and Example 1.3.2) and the number of radioactive nuclei present at time t in a radioactive decay system (Example 1.3.1).

The values of the state variables of a dynamical system at a given instant of time uniquely determine the values of the state variables at any time for which the system's evolution law is valid. The phenomena of radioactive decay and of the vertically moving body are examples of dynamical systems. The dynamical systems approach to modeling, involving as it does the evolution of the state of the system through time, is a concept that is modern in its scope but has been around in one form or another for centuries. It is the basic approach we shall take in this book.

Some conventional terminology has evolved in speaking about various components of a dynamical system. The effects of the external environment on a system are sometimes referred to as *input data* to the system. Input data are often called *driving terms* or *source terms*, and such systems are said to be *driven*. Values of the state variables at a given initial time t_0 are called *initial data*. For example, in a radioactive decay problem, assume that the radioactive material is being replenished at a known

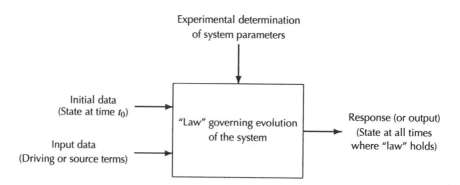

FIGURE 1.3.4 Schematic of a dynamical system.

rate, say $f(t)$. Then this problem is modeled by the differential equation

$$N' = -kN + f(t)$$

and $f(t)$ is the driving term in this case. The behavior of the state variables in time due to the input is called simply the *response* (or sometimes the *output*) of the system. In the radioactive decay model, $N(t)$ is the response to the driving term $f(t)$.

Figure 1.3.4 summarizes the basic components of a dynamical system; we will interpret this schematic in many specific cases throughout this text.

Modeling Process

We have so far only hinted at the process by which models are constructed. Obviously, we need tools, variables, and some natural laws to build a model, but one important ingredient in the modeling process has been overlooked. At the beginning of the modeling process the modeler needs to make some definitions and simplifying assumptions and to "discover" some "laws" or "principles" that "govern" or "explain" the behavior of the phenomenon at hand. Quotation marks are used around the terms above because the natural world neither knows nor cares about the modeler's efforts to understand its inner workings. The goal of the modeler is to generate a model that is general enough to explain the phenomenon at hand, but not so complicated as to preclude analysis. Thus there are many trade-offs along the way. To have any confidence at all in the model, the modeler will solve a variety of special problems and check the results against experimental evidence. In this way the modeler builds confidence in the validity of the model and also learns something about its limits of applicability. This last step in the modeling process is usually called *validation* of the model. For example, if the differential equation $s'' = -g$ were used to model the motion of a feather when dropped from a height of 100 feet, the inadequacy of the model would quickly be discovered. Thus modelers sometimes speak of *ranges* or *regimes of validity*. Figure 1.3.5 is a schematic describing the entire modeling process. The schematics shown in Figures 1.3.1, 1.3.4, and 1.3.5 are useful guides, but they are just guides, not rules for modeling. Modelers have these schematics in mind as they go through the modeling process, but they rarely follow them rigidly.

As experienced practitioners know well, models enjoy only a transitory existence. A model based on the empirical data of the day may become totally inadequate when a technological breakthrough produces better instruments, which, in turn, produce better data. In any field, moreover, models are constantly being examined for accuracy in predicting the phenomena modeled. This necessarily involves a careful reconsider-

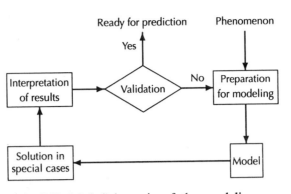

FIGURE 1.3.5 Schematic of the modeling process.

ation of the basic assumptions that produced the model, as well as an analysis of the mathematical approximations used in the course of computation. When models are found to be deficient they are modified or supplanted by other models. The literature of science is a chronicle of this process.

Initial Value Problems

It will come as hardly any surprise that differential equations form the foundation of many successful models for dynamical systems. Dynamical systems involve change over time, and instantaneous rates of change are interpreted conveniently by derivatives. As we have seen in the examples treated above, after selecting a frame of reference and appropriate coordinates and state variables for a dynamical system, a mathematical model arises which consists of a an ODE and conditions expressing values of the state variables at a given time t_0. These mathematical models are called *initial value problems* (IVPs), as we have previously noted, and the conditions involving the initial data are called *initial conditions*. For example,

$$s'' = -g, \qquad s(t_0) = s_0, \quad s'(t_0) = v_0$$

forms an initial value problem. Notice that the number of initial conditions in this IVP agrees with the order of the differential equation.

There is an important distinction between the IVPs for vertical motion of a moving ball and for radioactive decay. The thrown ball problem in its natural setting specifies initial data and requires the response of the system for times later than the initial time t_0 because the modeler has no confidence that the evolution law is valid for times prior to t_0. This problem is called a *forward initial value problem*. The initial conditions in such problems are expressed in terms of the limiting value of the state variables as $t \to t_0^+$. The radioactive decay problem in its natural setting, on the other hand, is a process that the modeler believes has persisted prior to an observation of the state variable N at time t_0. The mathematical model for this problem is a *"two-sided" initial value problem* in the sense that the initial time t_0 is in the interior of a time interval where the state variables are defined. To complete our characterization of initial value problems, we should also mention the *backward initial value problem*, the counterpart of the forward problem, where the initial conditions are imposed at a time t_0 and a time-description of the state variables is sought for $t < t_0$. There is an obvious relation among these three problems, but we will not make too much of these matters in a formal sense since the correct interpretation of initial value problems will be clear from the context.

An example will clarify these ideas.

Radiocarbon Dating

Living cells absorb carbon directly or indirectly from carbon dioxide (CO_2) in the air. The carbon atoms in some of this CO_2 are composed of a radioactive form of carbon, ^{14}C, rather than the common ^{12}C. ^{14}C is produced by the collisions of cosmic rays (neutrons) with nitrogen in the atmosphere. The ^{14}C nuclei decay back to nitrogen atoms by emitting β-particles. All living things, or things that were once alive, contain some radioactive carbon nuclei. In the late 1940s, Willard Libby[5] showed how a careful measurement of the ^{14}C decay rate in a fragment of dead tissue can be used to determine the number of years since its death. Let us consider a specific problem in order to focus our attention.

EXAMPLE 1.3.3

Age of the Lascaux Wall Paintings

A Geiger counter was used to measure the current decay rate of ^{14}C in charcoal fragments found in the cave of Lascaux, France, where there are prehistoric wall paintings. The counter recorded about 1.69 disintegrations per minute per gram of carbon, while for living tissue the number of disintegrations was measured in 1950 to be 13.5 per minute per gram of carbon. How long ago was the charcoal formed (and, presumably, the paintings painted)?

In any living organism the ratio of the amount of ^{14}C to the total amount of carbon in the cells is the same as that in the air. If the ratio in the air is constant in time and location, then so is the ratio in living tissue. After the organism is dead, ingestion of CO_2 ceases and only the radioactive decay continues. The half-life τ of ^{14}C is taken to be 5568 ± 30 years (the internationally agreed-upon value). Let $q(t)$ be the amount of ^{14}C per gram of carbon at time t in the charcoal sample. Time is measured in years, but $q(t)$ is dimensionless because it is the ratio of masses. Let $t = 0$ be the current time and suppose that $T < 0$ is the time that the wood tissue represented by the charcoal fragment died. Then $q(t) = q(T) = q_T$, a positive constant, for all $t \le T$. For $t > T$ the ^{14}C nuclei decay via the first-order rate equation $q' = -kq$, where, using formula (5), the rate constant k is computed via the half-life τ of ^{14}C by $k\tau = \ln 2$; in this case $k \approx 0.0001245$ (years)$^{-1}$. Thus we see that $q(t)$ satisfies the *backward* initial value problem

$$q' = -kq, \quad q(0) = q_0, \quad T \le t \le 0 \tag{9}$$

We solved IVP (9) earlier; its solution is $q = q_0 e^{-kt}$. Thus, at the time of death $t = T$ we have $q_T = q_0 e^{-kT}$, or solving for T, we obtain

$$T = -\frac{1}{k} \ln \frac{q_T}{q_0} \tag{10}$$

The problem is that we were not given the amounts q_T and q_0. We shall show, however, that the ratio q_T/q_0 can be calculated from the data in the problem statement.

---

[5]Willard Libby (1908–1980) was the American chemist who had the idea of dating once-living substances by measuring the amount of ^{14}C in samples. His book *Radiocarbon Dating*, 2nd ed. (Chicago: University of Chicago Press, 1955) discusses the techniques used to do the dating. The technique is most effective with material that is at least 200 years old, but not more than 70,000 years old. Libby received the 1960 Nobel Prize for Chemistry for his work.

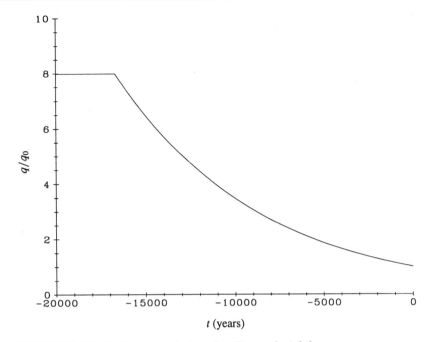

FIGURE 1.3.6 Radioactive dating. See Example 1.3.3.

According to the ODE in (9), the rate of disintegration of the radioactive nuclei at time t is proportional to the amount present. Thus

$$q'(T)/q'(0) = (kq(T))/(kq(0)) = q(T)/q(0)$$

The charcoal of the sample underwent 1.69 disintegrations per gram of carbon per minute, while in living tissue in 1950 (when the paintings were dated) there were 13.5 disintegrations per gram of carbon per minute. Since the number of disintegrations of ^{14}C per unit of time is proportional to the decay rate $q'(t)$, we have that $q_T/q_0 = q'(T)/q'(0) = 13.5/1.69$, and hence from formula (10)

$$T = -\frac{1}{k}\ln\frac{q_T}{q_0} = -\frac{\tau}{\ln 2}\ln\frac{q'(T)}{q'(0)} \tag{11}$$

$$= -\frac{5568 \pm 30}{\ln 2}\ln\frac{13.5}{1.69}$$

$$= -16,692 \pm 90 \text{ years}$$

Figure 1.3.6 shows the graph of $q(t)/q_0$ as a function of time t, $t \le 0$. The curve begins its exponential decay at $T = -16,692$ years.

The accuracy of the ^{14}C dating process depends on a knowledge of the exact ratio of radioactive ^{14}C to the carbon in the atmosphere. The ratio is now known to have changed over the years. First of all, there is a basic sinusoidal variation about the mean with a period of roughly 8000 years. On top of this variation, volcanic eruptions and industrial smoke, dump "dead" ^{14}C into the atmosphere and decrease the ratio. However, the most drastic change in recent times has occurred through the testing of nuclear weapons, resulting in an increase of 100% in the ratio, at least for some parts of the Northern Hemisphere. These events change the ratio in the atmosphere and, hence, in living tissue. These variations are now factored into the dating process.

The procedure described above can be used in conjunction with any radioactive process that satisfies a first-order rate law. Since the experimental error in determining the rate constants for decay processes with long half-lives is large, we would not expect to use this process to date events in the very recent past. Recent events can be dated if a radioactive substance with a short half-life is involved. White lead, a pigment used by painters, contains a small amount of a radioactive isotope of lead with a half-life of only 22 years. Using this fact, it has recently been shown that a number of paintings, purportedly from the seventeenth century, were actually painted in the 1940s. At the other extreme, radioactive substances such as uranium, with half-lives of billions of years, can be used to date the formation of the earth itself.

PROBLEMS

1. (*Radioactive Decay*). Radium decomposes at a rate k proportional to the quantity of radium present. Suppose it is found that in 25 years 1.1% of a certain quantity of radium has decomposed. What is the half-life of radium? [*Hint*: Use the data to find k; it is not necessary to know the amount N_0 of radium.]

2. (*Radioactive Decay*). If the half-life of a radioactive substance is 1000 years, what fraction of it is left after 100 years?

3. (*Radioactive Decay*). Radioactive phosphorus, ^{32}P, is used as a tracer in biochemical studies. It has a half-life of 14.2 days. Upon completion of a tracer experiment with 8 curies (Ci) of ^{32}P, the researchers wanted to dispose of the contents of the experiment after the level of radioactivity had decreased to an acceptable level (1.0×10^{-5} Ci). Calculate the length of time the contents must be stored until the radioactivity has reached an acceptable level. [*Note*: One curie is the quantity of radioactive isotope undergoing 3.7×10^{10} disintegrations per second.]

4. (*Population Growth*). A population grows exponentially for $0 \leq t \leq T$ with growth coefficient 0.03 per month. At the unknown time $T > 0$, circumstances cause the growth constant to change to 0.05 per month. After 20 months of growth, the population has doubled. At what time T did the growth constant change?

5. (*No Damping*). A 600-gram body is thrown vertically upward from the ground with an initial velocity of 2000 cm/sec. Ignore air resistance.

    (a) Find the highest point reached and the time required to reach that point. [*Hint*: See Example 1.1.6.]

    (b) Find the distance from the starting point and the velocity attained 3 sec after the motion begins.

    (c) Find the time of impact with the ground.

6. (*No Damping*). Suppose that a ball is thrown vertically upward, and at the time of release the ball is 2 m above the earth and traveling at 36 m/sec. Find the time it takes for the ball to strike the earth. Ignore air resistance. [*Hint*: See Example 1.1.6.]

7. (*No Damping*). A person drops a stone from the top of a building, waits 1.5 sec, then hurls a baseball downward with an initial speed of 20 m/sec. Ignore air resistance.

    (a) If the ball and stone reach the ground together, how high is the building?

    (b) For a given initial speed of the ball, show that there is a maximum waiting time after which the ball cannot reach the ground together with the stone, no matter how high the building.

8. (*No Damping: Longer to Rise or to Fall?*). Suppose that a ball is thrown vertically upward with initial velocity v_0 from ground level. Ignore air resistance. What is the time required for the ball to reach its maximum height? Does the ball spend as much time going up as it does coming down? What is the velocity of the ball on impact with the ground? Give reasons.

9. (*Vertical Motion: Longer to Rise or to Fall?*). Suppose that a whiffle ball of mass m at height h is given an upward velocity of v_0 by a sudden draft of air and moves subject to air resistance.

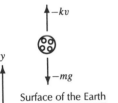

(a) Show that the whiffle ball's height y and velocity v at time t are governed, until the time of impact, by the IVP

$$v' = -g - \frac{k}{m}v, \quad v(0) = v_0$$
$$y' = v, \qquad\qquad y(0) = h$$

(b) Suppose $k/m = 1$ for the system in part **(a)**. Use a numerical solver to estimate the rise time and the fall time (back to the initial height) for various values of v_0. Are the fall times less than, equal to, or greater than the rise times? Answer this question when $k/m = 2$. Explain differences in the graphs for these two cases.

(c) (*Longer to Rise or to Fall?*). Suppose $k/m = 1$ in the IVP above. Find the time T required for the whiffle ball to reach its highest point. Let the time $\tau > 0$ be such that $y(\tau) = h$. Show that $\tau > 2T$. Thus, it takes longer for the whiffle ball to fall through a fixed distance than it does to rise. Discuss why you think this should be so. Compare with the results of Problem 8.

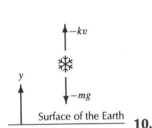

(d) Find the minimum speed of an updraft to keep a snowflake from descending.

10. (*Compound Interest*). Some financial institutions will cause the funds on deposit in a savings account to grow at a rate that is proportional to the instantaneous value of the account. If k is the constant of proportionality, this rate law is known as $100k\%$ *interest continuously compounded*.

(a) What rate of interest payable annually is equivalent to 9% interest continuously compounded? [*Hint*: $k = 0.09$.]

(b) If the funds in a savings account earn interest continuously compounded and the funds double in 8 years, what is the interest rate?

(c) How long will it take A dollars invested in a continuously compounded savings account to double if the interest rate is 5%; 9%; 12%?

11. (*Rule of 72*). Two common rules of thumb for business managers are the "Rule of 72" and the "Rule of 42." These rules say that the number of years it takes a sum of money invested at $r\%$ interest to double or to increase by 50%, respectively, is given by $72/r$ or $42/r$. Assuming continuous compounding [i.e., that $A'(t) = 0.01rA(t)$], show that these rules overestimate the time required.

12. (*Radiocarbon Dating: Scaling*). Follow the outline below to obtain a graphical solution for the radiocarbon dating problem.

(a) Define the new state variable $Q(t) = q(t)/q_0$ in Example 1.3.3. Show that this change of variables converts the IVP of that example into the form $Q' = -kQ$, $Q(0) = 1$. Show that the dating problem may be reformulated as follows: Find a value $T < 0$ such that $Q(T) = 13.5/1.69$.

(b) Plot the solution of the IVP of part **(a)** backward in time. Use this plot to find the value of T without using formula (11).

13. (*Dating Stonehenge*). In 1977, the rate of ^{14}C radioactivity of a piece of charcoal found at Stonehenge in southern England was 8.2 disintegrations per minute per gram. Given that in 1977 the current rate of ^{14}C radioactivity of a living tree was 13.5 disintegrations per minute per gram and assuming that the tree which was burned to produce the charcoal was cut during the construction of Stonehenge, estimate the date of construction. [*Hint*: Use (11) in Example 1.3.3.]

14. (*Dating a Sea Shell*). An archeologist finds a sea shell that contains 60% of the ^{14}C of a living shell. How old is the shell? [*Hint*: See Example 1.3.3.]

15. (*Newton's Law of Cooling*). When an object is placed in a relatively large surrounding medium of constant temperature, according to *Newton's Law of Cooling* (or *Warming*) the temperature of the object changes at a rate directly proportional to the difference between the object's temperature and the surrounding medium's temperature. In taking the temperature of a sick horse, a veterinarian notes that the thermometer reads 82°F at the time of insertion. After 3 min the veterinarian observes a reading of 90°F, and 3 min later a reading of 94°F; however, a sudden convulsion destroys the thermometer before a final reading can be obtained. What is the horse's temperature?

16. (*Shoveling Snow*). During a steady snowfall, a man starts clearing a sidewalk at noon, shoveling the snow at a constant rate and clearing a path of constant width. He shovels two blocks by 2 P.M., one block more by 4 P.M. When did the snow begin to fall? Explain your modeling process. [*Hint*: Additional assumptions are required to solve the problem. For example, assume that the man does not turn back to clear the newly fallen snow behind him.]

1.4 Separable Equations: Implicitly Defined Solutions

After linear equations, the next most commonly encountered first-order differential equations are the *separable equations*

$$N(y)y'(x) + M(x) = 0 \tag{1}$$

so named because the variables x and y are separated as indicated. We shall show how to find solutions of (1). The ODEs in this section use x as the independent variable and y as the dependent variable. The reason for the switch in notation is that later in this section we will introduce a system of ODEs (equivalent to ODE (1)) in which t is the independent variable.

Suppose that M is continuous on the x-interval I, and that N is continuous on the y-interval J. Then on the intervals I and J, M and N have antiderivatives denoted by

$$F(x) = \int^x M(s)\,ds, \qquad G(y) = \int^y N(s)\,ds$$

which is just another way of saying that $dF/dx = M(x)$, for all x in I, and $dG/dy = N(y)$, for all y in J. If $y(x)$ is any solution of ODE (1) whose solution curve remains in the xy-rectangle defined by I and J, the Chain Rule implies that ODE (1) can be rewritten as

$$N(y)y'(x) + M(x) = [G(y(x)) + F(x)]' = 0 \tag{2}$$

Thus the Vanishing Derivative Theorem of Section 1.1 implies via ODE (2) that there is a constant C such that

$$G(y(x)) + F(x) = C \tag{3}$$

for all x on the interval where $y(x)$ is defined.

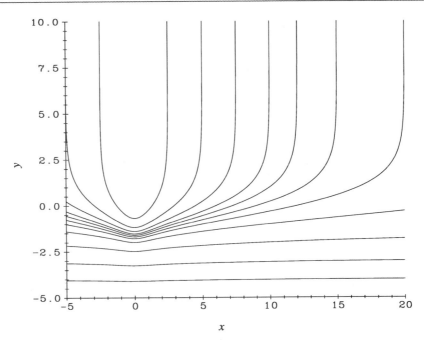

FIGURE 1.4.1 Solution curves for the separable ODE of Example 1.4.1.

Conversely, suppose that $y(x)$ is a continuously differentiable function whose graph lies in the xy-rectangle determined by I and J, and that $y(x)$ satisfies equation (3) for some constant C. Then differentiation of (3) via the Chain Rule easily shows that $y(x)$ is a solution of ODE (1). Hence (3), for C an arbitrary constant, can be regarded as defining (implicitly) the "general solution" $y = y(x)$ of ODE (1) in the xy-rectangle determined by I and J.

EXAMPLE 1.4.1

Separating Variables

The equation $y' = 2xe^y(1 + x^2)^{-1}$ in separated form is

$$-e^{-y}y' + 2x(1 + x^2)^{-1} = 0$$

Integrating, we find that any solution $y = y(x)$ of the ODE satisfies the equation

$$e^{-y} + \ln(x^2 + 1) = C$$

where C is some constant. Solving this equation for y in terms of x, we obtain

$$y = -\ln(C - \ln(x^2 + 1))$$

The solution $y(x)$ is defined on an x-interval for which $C > \ln(x^2 + 1)$. Observe that we must have $C > 0$. See Figure 1.4.1 for some solution curves.

It is not always possible to solve the relation (3) for y explicitly in terms of x, as the next example shows. In any case, the curves defined by (3) are called *integral curves* for ODE (1), and a solution curve is then an arc of an integral curve. The function $G(y) + F(x)$ is called an *integral* for ODE (1); thus, integral curves are the contours (or *level sets*) of the integral $G(y) + F(x)$. Sometimes a contour has several distinct *branches*.

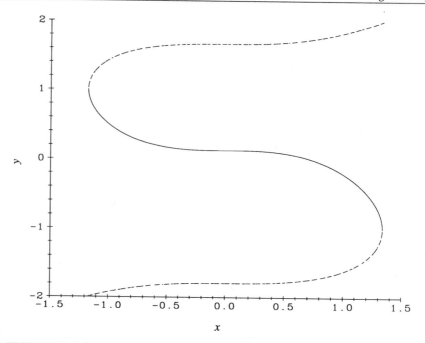

FIGURE 1.4.2 The integral curve of $(1 - y^2)y' + x^2 = 0$ passing through the point $(-1, 0.5)$. See Example 1.4.2.

EXAMPLE 1.4.2

Integral Curve, Solution Curve
The separable ODE

$$(1 - y^2)y' + x^2 = 0$$

has the integral curves

$$y - y^3/3 + x^3/3 = C$$

where C is a constant. The integral curve which passes through the point $(-1, 0.5)$ is the one for which $C = 1/8$. See Figure 1.4.2 for the *s*-shaped integral curve.

The solution curve for the IVP

$$(1 - y^2)y' + x^2 = 0, \qquad y(-1) = 0.5$$

(shown as solid line) is the arc of the integral curve in Figure 1.4.2 containing the point $(-1, 0.5)$ but no points with a vertical tangent (i.e., points where $|y'|$ is "infinite").

To find the largest x-interval on which this solution curve is defined use the ODE to find that y' is infinite when $y = \pm 1$. Putting $y = 1$ into the integral relation $y - y^3/3 + x^3/3 = 1/8$, we obtain $x^3 = -13/8$. Hence, $x_1 = -13^{1/3}/2$ is the x-coordinate of the left-hand endpoint of the largest interval where the solution curve is defined. Similarly putting $y = -1$ into the integral curve relation we obtain $x_2 = 19^{1/3}/2$. Thus, the largest x-interval on which this solution curve is defined is $x_1 < x < x_2$.

One small mystery still remains: How was the integral curve in Figure 1.4.2 plotted? We did not use a contour plotter! Read on and all will be revealed.

In general, we can use the procedure described in Example 1.4.2 to solve the initial value problem

$$N(y)y' + M(x) = 0, \qquad y(x_0) = y_0 \tag{4}$$

This may be done by finding any antiderivatives $G(y)$ and $F(x)$ of $N(y)$ and $M(x)$,

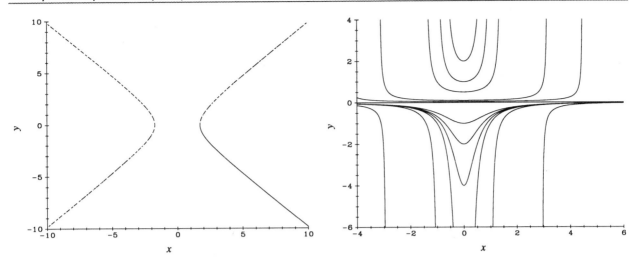

FIGURE 1.4.3 Four solution curves on one hyperbolic integral curve. See Example 1.4.3.   **FIGURE 1.4.4** Solution curves of Example 1.4.4.

respectively, and setting the costant C in formula (3) equal to $N(y_0) + M(x_0)$. The desired integral curve for solving IVP (4) has formula

$$G(y) + F(x) = G(y_0) + F(x_0) \qquad (5)$$

It is also instructive to derive the implicit solution formula (5) directly from (4) in the following way. Integrating both sides of the differential equation in (4) from x_0 to x, we have

$$G(y(x)) - G(y_0) + F(x) - F(x_0) = \int_{x_0}^{x} N(y(s))y'(s)\,ds + \int_{x_0}^{x} M(s)\,ds = 0$$

which is just (5) in a slightly different form. For this reason the ODE (1) is often written in the differential form

$$N(y)dy + M(x)dx = 0$$

just as if the derivative y' in ODE (1) is replaced by dy/dx and then the ODE "multiplied" through by the differential dx.

EXAMPLE 1.4.3

Four Solution Curves on One Integral Curve
Consider the initial value problem

$$yy' - x = 0, \qquad y(2) = -1$$

Any solution $y(x)$ must satisfy the equation

$$\int_{-1}^{y} s\,ds - \int_{2}^{x} s\,ds = 0$$

or $y^2/2 - 1/2 - x^2/2 + 2 = 0$, which defines a hyperbola through the initial point $(2, -1)$. The particular branch $y(x)$ of the hyperbola passing through $(2, -1)$ is the desired solution curve (solid line figure 1.4.3) and is defined by

$$y = -(x^2 - 3)^{1/2}, \qquad x \geq 3^{1/2}$$

The hyperbolic integral curve contains four solution curves, which are defined by $y = \pm(x^2 - 3)^{1/2}$, $x > 3^{1/2}$, or $x < -3^{1/2}$. These solution curves are the upper and the lower arcs of the two branches of the hyperbola (see Figure 1.4.3).

EXAMPLE 1.4.4

Loss of a Solution

The equation $y' = 2xy^2$ can be written in separable form as $y'y^{-2} = 2x$, but at the price of "losing" the solution $y(x) = 0$ of the original equation. For $y \neq 0$, we can solve the separated equation, obtaining $-y^{-1} = x^2 + C$ or $y = -(x^2 + C)^{-1}$ where C is an arbitrary constant. Thus the collection of all solution curves of $y' = 2xy^2$ is given via

$$y = \frac{-1}{x^2 + C}, \quad \text{and} \quad y = 0 \tag{6}$$

In the first case, if $C \leq 0$, x must be restricted to an interval on which $x^2 + C \neq 0$. For example, if $C = -1$, the first equation of (6) defines three solutions: one for $x < -1$, another for $|x| < 1$, and a third for $x > 1$. See Figure 1.4.4 for some solution curves.

With the above examples in mind, the method of solving separable equations should be used with some caution, being careful not to lose solutions or to choose the wrong arc of an integral curve when solving an initial value problem.

From a Separable ODE to a System: Planar Systems

Most of the dynamical systems considered so far have been modeled by single, first-order ODEs in one state variable. Some dynamical systems require two state variables, say x and y, and the laws of the system lead to rate equations in x, y, and time t:

$$\frac{dx}{dt} = x' = f(t, x, y), \quad \frac{dy}{dt} = y' = g(t, x, y) \tag{7}$$

The pair of differential equations is called a *first-order planar differential system*; the xy-plane is the *state* (or *phase*) *plane*, or *state space*. When the rate functions f and g do not depend on t, system (7) is said to be *autonomous*.

Systems first made their appearance at the end of Section 1.1, where we modeled the motion of a body of mass m moving along the local vertical, using position y and velocity v as state variables. Newton's laws imply that

$$y' = v, \quad v' = -g \tag{8}$$

which is a planar autonomous system in the state variables y and v. Example 1.3.2 gives a more realistic system model for the body's motion by including the force of friction:

$$y' = v, \quad v' = -g - \frac{k}{m}v \tag{9}$$

which is also a planar autonomous system since k and m are assumed to be constants independent of t.

There is some useful terminology for planar systems.

❖ **Solutions, Orbits, Component Graphs.** The functions $x = x(t)$, $y = y(t)$, t in an interval I, define a *solution* of system (7) if $x'(t) = f(t, x(t), y(t))$, $y'(t) = g(t, x(t), y(t))$, for all t in I. The parametric plot of $x(t)$ versus $y(t)$ in the xy-state plane is the *orbit* (or *trajectory*) of the solution. The graphs of $x(t)$ versus t in the tx-plane and $y(t)$ versus t in the ty-plane are the *component graphs* of the solution.

These terms can be applied to the examples mentioned above. For instance, the curves of Figure 1.1.4 in Section 1.1 would now be called y component graphs for system (8).

A planar differential system can be used to obtain graphical information about solutions of a first-order ODE. We illustrate the technique for a separable ODE. The solution curves of the separable ODE

$$N(y)y' + M(x) = 0$$

can be obtained by first plotting level curves of an integral. Here is another way to plot integral curves, even if an integral is not known: Consider the first-order system with two state variables, x and y,

$$\frac{dx}{dt} = N(y), \quad \frac{dy}{dt} = -M(x) \tag{10}$$

where t is the independent variable. Let $x(t)$, $y(t)$ be a solution of the above system for all t in an interval I; i.e., $dx(t)/dt = N(y(t))$ and $dy(t)/dt = -M(x(t))$ for all t in I. Then the orbit defined by $x = x(t)$, $y = y(t)$ in the xy-plane, t in I, is an integral curve of the ODE, $Ny' + M = 0$. For, if $G(y)$ and $F(x)$ are antiderivatives of $N(y)$ and $M(x)$, respectively, then the function $G(y(t)) + F(x(t))$ is constant on I because

$$\frac{d}{dt}[G(y(t)) + F(x(t))] = \frac{dG}{dy}\frac{dy}{dt} + \frac{dF}{dx}\frac{dx}{dt} \quad \text{(Chain Rule, Appendix C.5)}$$

$$= N(y)\frac{dy}{dt} + M(x)\frac{dx}{dt} \quad \begin{array}{l}\text{(Fundamental Theorem} \\ \text{of Calculus)}\end{array}$$

$$= N(-M) + MN \quad \text{(from system (10) above)}$$

$$= 0$$

The function $G(y(t)) + F(x(t))$ remains constant by the Vanishing Derivative Theorem of Section 1.1. Integral curves of $Ny' + M = 0$ can be obtained by using a computer ODE solver to plot orbits of the differential system in the xy-state plane. This is often the easiest way to plot the integral curves, because one need not worry about the denominator in $dy/dx = -M/N$ being zero, nor need one have explicit formulas for the antiderivatives of N and M.

EXAMPLE 1.4.5

Plotting Integral Curves of an ODE as Orbits of a System
Following the ideas above, we can plot the integral curve of the separable ODE

$$(1 - y^2)y' + x^2 = 0 \tag{11}$$

as it appears in Figure 1.4.2. We do this by plotting an orbit of the equivalent system of first-order ODEs

$$\frac{dx}{dt} = 1 - y^2$$
$$\frac{dy}{dt} = -x^2 \tag{12}$$

using the initial conditions

$$x(0) = -1, \quad y(0) = 0.5 \tag{13}$$

This is accomplished by using a numerical solver to solve system (12) subject to the initial data (13) and letting t run forward and then backward from $t = 0$ until the orbit exits the rectangle $-2 \le x \le 2$, $-3 \le y \le 3$. This is precisely the way we plotted Figure 1.4.2.

Raindrops and Baseballs: Newtonian Damping

y

kv^2

$-mg$

Earth

Separable ODEs occur frequently in applications. Here is an important example.

When a relatively dense body falls through air (e.g., a raindrop, baseball, or meteor), experimental evidence[6] seems to show that the resistive force of the air is not proportional to the velocity, but to the square of its magnitude. This is known as *Newtonian damping*; compare with viscous damping defined in Example 1.3.2. With "up" as the positive direction, we have the initial value problem

$$my'' = -mg - kv|v|, \qquad y(0) = h, \quad v(0) = 0 \qquad (14)$$

where $v = y'$ and the minus sign in the term $-kv|v|$ correctly models the fact that the drag acts in an opposite direction to the motion. The positive parameter k in (14) is called the *Newtonian damping constant*. In the margin sketch and in Example 1.4.6 below we use $+kv^2$ for the drag term because the body is falling.

EXAMPLE 1.4.6

Velocity of Ball Falling Against Air Resistance; Limiting Velocity

A ball is dropped from a height of h feet and is acted upon by air resistance. We will find the position and velocity of the ball at any time t before impact. Because the ball is falling we use the positive sign in the ODE in (14). Then replacing y'' by v' in (14), we obtain the first-order initial value problem

$$v' = -g + \left(\frac{k}{m}\right)v^2, \qquad v(0) = 0 \qquad (15)$$

Because the ball is falling, the graph of $v = v(t)$ is a falling curve with $v(0) = 0$ (hence, v' is negative), and we see from IVP (15) that $0 \geq v(t) \geq -(mg/k)^{1/2}$ for all $t \geq 0$. Now observe that the ODE in (15) is separable and can be solved using the solution formula (5). Indeed, if the variables are separated and integrals taken, then we would have

$$\int_0^v \frac{1}{-g + (k/m)s^2}\, ds = \int_0^t ds = t$$

Using partial fractions or a table of integrals to carry out the left-hand integration, we have that

$$\frac{1}{2}\left(\frac{m}{gk}\right)^{1/2} \ln\left|\frac{v - (mg/k)^{1/2}}{v + (mg/k)^{1/2}}\right| = t \qquad (16)$$

Setting $A = 2(gk/m)^{1/2}$ and taking exponentials of both sides of (16) we obtain

$$\left|\frac{v - (mg/k)^{1/2}}{v + (mg/k)^{1/2}}\right| = e^{At} \qquad (17)$$

Note that the quantity between the absolute value bars is negative for all $t \geq 0$, and so dropping the absolute value, we solve for v to obtain

$$v(t) = \left(\frac{mg}{k}\right)^{1/2} \frac{e^{-At} - 1}{e^{-At} + 1} \qquad (18)$$

Observe that as $t \to \infty$,

$$v(t) \to -(mg/k)^{1/2}$$

---

[6]See *Differential Equations and Applications* by J.B. Scarborough, Waverly Press, 1965.

This *limiting velocity* is not actually attained because the body hits the ground at some finite time and the model equations lose their validity.

EXAMPLE 1.4.7

Position of Ball Falling Against Air Resistance
To find y as a function of t, replace $v(t)$ by $y'(t)$ in equation (18) and solve the first-order IVP

$$y' = \left(\frac{mg}{k}\right)^{1/2}\frac{e^{-At}-1}{e^{-At}+1}, \qquad y(0) = h \tag{19}$$

by integration. Observe that

$$\frac{e^{-At}-1}{e^{-At}+1} = \frac{e^{-At/2}-e^{At/2}}{e^{At/2}+e^{-At/2}} = -\frac{\sinh At/2}{\cosh At/2} = -\tanh\frac{At}{2}$$

and thus (19) has the unique solution

$$y(t) = h - \left(\frac{mg}{k}\right)^{1/2}\int_0^t \tanh\left(\frac{1}{2}As\right) ds$$

or, since $A = 2(gk/m)^{1/2}$,

$$y(t) = h - \frac{m}{k}\ln\cosh\left[\left(\frac{gk}{m}\right)^{1/2}t\right] \tag{20}$$

where we have used a table of integrals and the fact that $\cosh 0 = 1$. We must, of course, restrict t to the interval $0 \le t \le T$, where T (the time of impact) can be calculated from (20) by setting $y = 0$. Although the height of the falling ball above the surface of the earth at any time t before impact is given by (20), the formula is not particularly simple and we shall carry our analysis no further.

Validation of Models for Falling Bodies

Why does a feather fall to earth more slowly than a large raindrop? For the feather (with viscous damping) we have shown in Section 1.3 that the limiting velocity v_f is mg/k_f, while for the raindrop (with Newtonian damping) the limiting velocity v_r is $(mg/k_r)^{1/2}$, where k_f and k_r are the respective k-values in the modes corresponding to these phenomena. Algebraic manipulation will show that

$$|v_f| < |v_r| \quad \text{if and only if} \quad k_r mg < k_f^2$$

One can argue about the validity of the last inequality in terms of the larger cross-sectional area of a feather and hence the larger k_f. However, the k-factor depends on too many other aspects of the falling body for us to be able to push the discussion much further.

The phenomenon of a free-falling human being (e.g., a parachutist before the chute opens) has been studied in great detail. For this case of a relatively dense body, Newtonian damping provides a good model. If m is approximately 120 kg (e.g., a parachutist with equipment), then k turns out to be approximately 0.1838 kg/m and the limiting velocity has magnitude approximately 80 m/sec. This value has been compared with the actual limiting velocity of free-falling parachutists, and remarkably good agreement has been observed. See Problem 7.

Destructive Competition: Models of Combat

In 1916, F.W. Lanchester[7] described some mathematical models for air warfare. These have been extended to a general combat situation, not just air warfare. We consider a particular case of the models here, others in the problem set.

An "x-force" and a "y-force" are engaged in combat. Let $x(t)$ and $y(t)$ denote the respective strengths of the forces at time t, where t is measured in days from the start of the combat. It is not easy to quantify "strength," including, as it does, the numbers of combatants, their battle readiness, the nature and number of the weapons, the quality of leadership, and a host of psychological and other intangible factors difficult even to describe, much less to turn into numbers. We shall take the easy way out and identify the strengths $x(t)$ and $y(t)$ with the numbers of combatants. The pair of values $x(t)$, $y(t)$, define the state of this "system" at time t.

We shall assume that $x(t)$ and $y(t)$ vary continuously and even differentiably as functions of time. This is, of course, an idealization of the true state of affairs since the strength must be an integer and change only by integer amounts as time goes on. However, as we did earlier with models of radioactive decay and growth, one might argue that when the numbers are large, an increase by one or two is infinitesimal compared to the total, and we might as well allow $x(t)$ and $y(t)$ to change continuously over small time spans. Once we have made the idealization to continuous functions, we "smooth off any corners" on the graphs of $x(t)$ and $y(t)$ versus t; thus we can reasonably take $x(t)$ and $y(t)$ to be continuous and differentiable.

Suppose that we can estimate the *noncombat loss rate* of the x-force (i.e., the loss rate due to the inevitable diseases, desertions, and other noncombat mishaps), the *combat loss rate* (due to encounters with the y-force), and the *reinforcement* (or *supply*) *rate*. Then the net rate of change in $x(t)$ is given by the Balance Law.

$$x'(t) = \text{reinforcement rate} - (\text{noncombat loss rate} + \text{combat loss rate})$$

A similar equation applies to the y-force. The problem is to analyze the solutions $x(t)$ and $y(t)$ of the respective differential equations to determine who "wins" the combat.

According to Lanchester, a model system of ODEs for a pair of conventional combat forces operating in the open with negligible noncombat losses is

$$x'(t) = -by(t) + R_1(t)$$
$$y'(t) = -ax(t) + R_2(t) \tag{21}$$

where a and b are positive constants and R_1 and R_2 are the rates of reinforcement (in numbers per unit time). In this model the reinforcement rates are assumed to depend only on time and not on the strength of either force (a dubious assumption). The combat loss rates, $-by(t)$ and $-ax(t)$, introduce actual combat into the model since without these terms neither force has any effect whatsoever on the other. Lanchester argues for the specific form of the combat loss rate terms in system (21) as follows. Every member of the conventional force is assumed to be within range of the enemy. It is also assumed that as soon as a conventional force suffers a loss, fire is concentrated on the remaining combatants. Thus the combat loss rate of the x-force is proportional

---

[7]Frederick William Lanchester (1868–1946) was an English engineer, mathematician, inventor, poet, and musical theorist who designed and produced some of the earliest automobiles and wrote the first theoretical treatise of any substance on flight, the book *Aerial Flight*, Constable, London, 1907–08. His book *Aircraft in Warfare*, Constable, London, 1916, contains the square law of conventional combat discussed in this section. Twenty-five years later (during the Second World War) Lanchester's scientific approach to military questions became the basis of a new science, Operations Research, which is now applied to problems of industrial management, production, and government procedures, as well as to military problems.

to the number of the enemy and is given by $-by(t)$. The coefficient b is a measure of the average effectiveness in combat of each member of the y-force. A similar argument applies to the term $-ax(t)$.

In the beginning of this section we showed how the solution curves of the ODE $N(y)y' + M(x) = 0$ can be characterized in terms of the orbits of the differential system

$$dx/dt = N(y), \quad dy/dt = -M(x)$$

The next example shows the usefulness of the reverse process.

EXAMPLE 1.4.8

From a System to a First-Order ODE: the Square Law

Consider the situation of a pair of isolated conventional forces with no reinforcements. In this setting system (21) reduces to

$$x' = -by, \qquad y' = -ax \tag{22}$$

a differential system in which time no longer explicitly appears; hence, the system is autonomous. Let $x = x(t)$ and $y = y(t)$ solve the system (22), and suppose that the orbit lies inside the first quadrant of the xy-plane (i.e., $x > 0$, $y > 0$). Since $x'(t) \neq 0$ in this quadrant, we can apply the Inverse Function Theorem of calculus and theoretically solve $x = x(t)$ for t in terms of x, obtaining $t = t(x)$. Moreover, $dt(x)/dx = [dx(t)/dt]^{-1}$. Thus, using the Chain Rule to find $d[y(t(x))]/dx$, we have

$$\frac{dy}{dx} = \frac{d}{dx}[y(t(x))] = \frac{dy}{dt} \cdot \frac{dt}{dx} = \frac{dy/dt}{dx/dt} = \frac{-ax}{-by}$$

or

$$by\frac{dy}{dx} = ax \tag{23}$$

which is a separable ODE. In effect, we have eliminated time as the independent variable and replaced it by x. Thus the integral curves of the "reduced" ODE (23) are just the orbits of system (22). The only assumptions made above are that $x > 0$ and $y > 0$ and, of course, that is no restriction at all for the phenomenon being modeled.

Let $x_0 > 0$ and $y_0 > 0$ be the strengths of the two forces at the start of combat. Integrating (23), we have that $\int_{y_0}^{y} by\,dy = \int_{x_0}^{x} ax\,dx$, or

$$by^2(t) - ax^2(t) = by_0^2 - ax_0^2 = k \tag{24}$$

where the constant k is determined by the initial data. Equation (24) is known as the *square law of conventional combat*. Although we cannot tell from (24) just how $x(t)$ and $y(t)$ change in time, (24) implies that the point $(x(t), y(t))$ moves along an arc of a hyperbola as time advances. These hyperbolas are orbits of the system (22) and are plotted in Figure 1.4.5 for various values of k (with $a = 0.064$ and $b = 0.1$). The arrowheads on the curves show the direction of changing strengths as time passes. Since $dx/dt < 0$ and $dy/dt < 0$ whenever $x(t) > 0$ and $y(t) > 0$, both $x(t)$ and $y(t)$ decrease as t increases (recall that this model system does not allow for reinforcements).

Who "wins" such a combat? Let us say that one force wins if the other force vanishes first. For example, y wins if $k > 0$, since according to equation (24), y never vanishes in this case, while the x-force has been annihilated by the time $y(t)$ has decreased to $(k/b)^{1/2}$. Thus the y-force seeks to establish a combat setting in which $k > 0$; that is, the y-force wants a large enough initial strength y_0 that the following inequality is valid:

$$\left(\frac{y_0}{x_0}\right)^2 > \frac{a}{b} \tag{25}$$

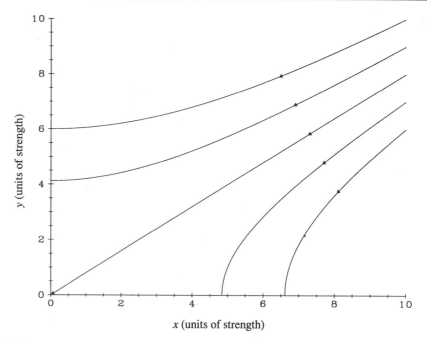

FIGURE 1.4.5 Arcs of hyperbolas of the square law of conventional combat. See Example 1.4.8.

Because of the quadratic magnification in (25), a small increase ϵ in y_0 could convert a predicted loss for the y-force $[(y_0/x_0)^2 < a/b]$ to a win $[(y_0 + \epsilon)^2/x_0^2 > a/b]$.

The simplified model solved above is unrealistic. If noncombat loss rates and reinforcement rates are included, however, a certain element of realism enters and one might actually compare the model with historical battles. Studies along these lines for the Battle of the Ardennes and the Battle of Iwo Jima of the Second World War have been carried out.[8] These studies give results reasonably close to the actual combat statistics once the coefficients a and b are determined. Whether the effectiveness coefficients could ever be estimated with any accuracy before an engagement is an open question. Terms such as "combat readiness" are used but are hard to quantify.

The Dark Side of Modeling

Combat models raise ethical questions about the uses of mathematics. G.H. Hardy, one of the best known English "pure" mathematicians of the first half of the twentieth century, has spoken on this matter:

> *So a real mathematician has his conscience clear; there is nothing to be set against any value his work may have; mathematics is... a harmless and innocent occupation.*[9]

---

[8] See, e.g., C.S. Coleman, "Combat Models," *Modules of Applied Mathematics*, vol. 1, Chap. 8, W. Lucas, M. Braun, C. Coleman, and D. Drew, eds. New York: Springer-Verlag, 1983.

[9] See G.H. Hardy, *A Mathematician's Apology*, 2nd ed., Cambridge: Cambridge University Press, 1967, pp. 140–141.

One wonders if it is that simple. Mathematics is not separate from the cultures and societies of the mathematicians who create or discover it. Since war and the preparation for war continue to be a preoccupation of humankind, it is not surprising that mathematics is applied to its study and analysis.

PROBLEMS

1. Find all solutions $y(x)$ of each separable ODE below. Plot representative integral curves in the xy-regions indicated by first replacing the ODE, $Ny' + M = 0$, by the system, $dx/dt = N$, $dy/dt = -M$. [*Hint*: Use Example 1.4.5 as a guide.]

   (a) $dy/dx = -4xy$, $|x| \le 5, -2 \le y \le 7$

   (b) $2y\,dx + 3x\,dy = 0$, $|x| \le 3, |y| \le 2$

   (c) $y' = -xe^{-x+y}$, $|x| \le 6, |y| \le 8$ [*Hint*: $e^{-x+y} = e^{-x}e^y$.]

   (d) $(1 - x)y' = y^2$, $|x| \le 5, |y| \le 5$

   (e) $y' = -y/(x^2 - 4)$, $|x| \le 5, |y| \le 5$

   (f) $y' = xe^{y-x^2}$, $|x| \le 5, -2 \le y \le 7$ [*Hint*: $e^{y-x^2} = e^y e^{-x^2}$]

   (g) $\cos(y^2)y' + x\sin(x^3) = 0$, $|x| \le 3, |y| \le 5$

2. (*The System Way to Plot Integral Curves*). Find an integral for the ODE in Atlas Plate FIRST ORDER C. Plot the integral curves displayed in that Atlas Plate by using the method of Example 1.4.5 to find an equivalent system and plotting some orbits. Repeat this process to reproduce the integral curve of Figure 1.4.2. [*Hint*: Choose initial points $(x(0), y(0))$ on the curves and then solve forward and backward in time.]

3. Solve each of the following initial value problems, and find the largest x-interval on which the solution $y(x)$ is defined.

   (a) $y' = (y+1)/(x+1)$, $y(1) = 1$ (b) $y' = y^2/x$, $y(1) = 1$

   (c) $y' = ye^{-x}$, $y(0) = e$ (d) $y' = 3x^2/(1+x^3)$, $y(0) = 1$

   (e) $y' = -x/y$, $y(1) = 2$ (f) $2xyy' = 1 + y^2$, $y(2) = 3$

4. For each ODE find a formula that defines solution curves implicitly. Do not solve for y as an explicit function of x. Use a computer to plot representative orbits of the equivalent system in the indicated rectangle R.

   (a) $y' = (x^2 + 2)(y + 1)/xy$. $R : |x| \le 5, |y| \le 5$

   (b) $(1 + \sin x)dx + (1 + \cos y)dy = 0$. $R : |x| \le 10, |y| \le 10$

   (c) $(\tan^2 y)dy = (\sin^3 x)dx$. $R : 0 \le x \le 8, 1 \le y \le 8$

5. (*Combat Model*).

   (a) Reproduce Figure 1.4.5 as closely as possible. Plot the orbit with initial condition $x_0 = 10$, $y_0 = 7$. Does y win or lose?

   (b) How much time does it take for the conflict to be resolved?

6. (*Projectile Motion: Newtonian Damping*). A spherical projectile weighing 100 lbs is observed to have a limiting speed of 400 ft/sec when dropped from a height.

   (a) Show that the velocity v of the projectile undergoing either upward or downward vertical motion and acted upon by air resistance is given by

   $$v' = -g - \frac{gk}{w}v|v|$$

where k is a positive constant and $w = mg$ is the weight of the projectile.

(b) Model the motion of the projectile as a differential system in state variables y and v, where $v = y'$.

(c) If the projectile is shot upward with an initial velocity of 500 ft/sec, what is its velocity when it passes its starting point on the way down? [*Hint*: Use a computer graphics/solver package to solve the system of part **(b)** with suitable initial conditions. Plot the orbit in the vy-plane.]

(d) (*Longer to Rise or to Fall?*). Does the projectile in **(c)** take the same amount of time going up as it does coming down? Give reasons. Repeat for other values of the initial velocity. Does the situation change? [*Hint*: Use a numerical solver and graph y vs. t.]

7. (*Parachutist: Newtonian Damping*). A parachutist and equipment weigh 240 lbs. (Note: weight $= mg$.) In free fall (i.e., before the parachute opens) the parachutist was observed to reach a limiting velocity of 250 ft/sec. Some time after the parachute opens the parachutist was observed to reach the limiting velocity of 17 ft/sec. Suppose that the same parachutist on another occasion jumps out of an airplane at 10,000 feet. Use $g = 32.2$ ft/sec^2. Answer the following questions:

(a) How much time must elapse before the parachutist falls at a speed of 100 ft/sec? [*Hint*: Use the data to determine the coefficient k in IVP (14); then use formula (16).]

(b) When falling at exactly 100 ft/sec the free-falling parachutist pulls the rip cord. How much longer does it take for the speed of the descent to drop to 25 ft/sec?

(c) If the parachutist follows the strategy described in **(b)**, then what is the total time duration of the jump? What is the parachutist's velocity on landing?

8. (*Newtonian versus Viscous Damping*). Design an experiment to determine whether Newtonian or viscous damping better describes the motion of the parachutist of Problem 7. Carry out a computer simulation and discuss the results. [*Hint*: Viscous damping is discussed in Example 1.3.2 and Newtonian damping in Example 1.4.6. In both cases, the damping constant can be determined from the terminal velocity.]

9. (*Conventional Combat with Reinforcements*). Consider the conventional combat model (21) where R_1 and R_2 are positive constants.

- Follow the procedure of Example 1.4.8, and obtain a separable first-order ODE in the variables x and y. Solve the newly obtained ODE to show that $x(t)$ and $y(t)$ satisfy the condition

$$b\left(y(t) - \frac{R_1}{b}\right)^2 - a\left(x(t) - \frac{R_2}{a}\right)^2 = C$$

where C is a constant determined by the initial data.

- Take $a = b = 1$, $R_1 = 2$, and $R_2 = 3$ and plot the orbits when (i) $x(t_0) = 3$, $y(t_0) = 1$; (ii) $x(t_0) = 2$, $y(t_0) = 2$; (iii) $x(t_0) = 4$, $y(t_0) = 2$.

- By considering the original differential system, assign arrowheads to the orbits indicating the direction of increasing time.

- Who wins in each of the cases described above?

- What happens when $x(t_0) = R_2/a$ and $y(t_0) = R_1/b$? Explain in terms of the development of the combat.

10. (*Conventional versus Guerrilla Combat*). Suppose the combat situation is modified so that one of the forces is a guerrilla force. Let $x(t)$ denote the strength of the conventional force and $y(t)$ the strength of the guerrilla force. Assume no reinforcements or noncombat losses. The loss rate inflicted on a conventional force by its guerrilla opponent is given by $x'(t) = -ay(t)$, but the situation is very different for a guerrilla force. A guerrilla force is usually invisible to its opponent and occupies a region of fixed area.

- Argue why the loss rate for the guerrilla force y may be modeled by cxy.

- Given the combat equations

$$x'(t) = -ay(t), \qquad y'(t) = -cx(t)y(t)$$

derive the *parabolic law*

$$cx^2 = 2ay + Q_0, \quad \text{where} \quad Q_0 = cx_0^2 - 2ay_0$$

- Let $a = 1$, $c > 0$. Plot some typical curves for various values of c. Indicate with arrows the direction of increasing time on each curve.

- Who wins when $Q_0 > 0$? $Q_0 = 0$? $Q_0 < 0$?

- In order for the conventional force to win, show that $cx_0^2 > ay_0$.

11. (*The System Way to Plot Solution Curves*). Consider the first-order ODE

$$N(x, y)y' + M(x, y) = 0$$

where M, M_y, N, N_y are continuous on a rectangle R in the xy-plane.

(a) Show that the solution curves of the ODE can be visualized by looking at the orbits of the planar system

$$\frac{dx}{dt} = N(x, y), \quad \frac{dy}{dt} = -M(x, y)$$

[*Hint*: Follow the proof (in this section) for the separable case.]

(b) Plot some solution curves for the ODE $(5y - x)y' + x - y = 0$.

1.5 Growth Processes: The Logistic Model

The U.S. Bureau of the Census often predicts population trends. To do this, demographers generally use known population data to formulate "laws" of population change. Each law is converted into a mathematical formula that can then be used to make the predictions. Since there is a good deal of uncertainty about which "law" best describes actual population shifts, several are derived, each proceeding from different assumptions about the conditions influencing the birth and death rates (and immigration and emigration). Similarly, ecologists make predictions about changes in the numbers of fish and animal populations, and biologists formulate "laws" for changing densities of bacteria growing in cultures. In this section we introduce some general laws of population change, solve the corresponding mathematical models, and interpret the results in terms of the growth, stabilization, or decline of a population. The effect of harvesting on population levels of a species is also discussed.

General Growth Models

Let $P(t)$ denote the population at time t of a species in a community. The values of $P(t)$ are integers and change by integer amounts as time goes on. However, for a large population an increase by one or two over a short time span is "infinitesimal" relative to the total, and we may think of the population as changing continuously instead of by discrete jumps. Once we assume that $P(t)$ is continuous, we might as well go all the way, smooth off any corners on the graph of $P(t)$, and assume that the function is differentiable. If we had let $P(t)$ denote the population *density* (i.e., the number per unit area or volume of habitat), the continuity and differentiability of $P(t)$ would have seemed more natural. However, we shall continue to interpret $P(t)$ as the size of the population, rather than density.

The underlying principle of the Balance Law of Section 1.2

$$\text{Net rate of change} = \text{Rate in} - \text{Rate out} \tag{1}$$

applies to the changing population $P(t)$, just as it does to the salt solution problem of Section 1.2. For a population, the "rate in" term is the sum of the birth and the immigration rates, while the "rate out" term is the sum of the death and the emigration rates. Let us regroup the rates into an "internal" rate I (birth minus death) and "external" rate M (immigration minus emigration). Averaged over all classes of age, sex, and fertility, a "typical" individual makes a net contribution R to the internal rate of change. The internal rate of change I at time t is, then, $RP(t)$, where

$$RP = (\text{Individual's contribution}) \times (\text{Number of individuals})$$

The *intrinsic rate coefficient R* will differ from species to species, but it always denotes the average individual contribution to the rate. Thus, given the rate coefficient R and the migration rate M, the Balance Law (1) becomes

$$P'(t) = I + M = RP(t) + M \tag{2}$$

and the study of population changes becomes the problem of solving ODE (2).

R and M may depend on P and t, but in the simplest cases that dependence will either disappear or be linear. If simple models suffice for observed growth processes, there is little need for more complex assumptions. A guiding principle of modeling in this, as in most situations, is that of Occam's Razor:

Occam's Razor. What can be accounted for by fewer assumptions is explained in vain by more.[10]

Exponential Growth

It may safely be pronounced, therefore, that population, when unchecked, goes on doubling itself every twenty-five years, or increases in a geometric ratio.

Malthus[11]

---

[10]William of Occam (1285–1349) was an English theologian and philosopher who applied the Razor to arguments of every kind. The principle is called the Razor because Occam used it so often and so sharply.

[11]Thomas Robert Malthus (1766–1834) was a professor of history and political economy in England. The quotation is from "An Essay on the Principle of Population As It Affects the Future Improvement of Society." Malthus's views have had a profound effect on nineteenth- and twentieth-century Western thought. Both Darwin and Wallace have said that it was reading Malthus which led them to the theory of evolution.

The Malthusian principle of explosive growth of human populations has become one of the classic "laws" of population change. The principle follows directly from (2) if we set $M = 0$ and $R = $ a positive constant, say r. In this case, (2) is linear and has the exponential solution

$$P(t) = P_0 e^{rt} \tag{3}$$

where P_0 is the population at the time $t = 0$. We encountered this ODE earlier in connection with radioactive decay, but now we have exponential growth instead of decay. The method of solution is exactly the same, however. We can see from (3) that the *doubling time* of a species is given by $T = (\ln 2)/r$ since if $P(t + T) = 2P(t)$, then $P_0 \exp[r(t + T)] = 2P_0 \exp[rt]$ implies that $\exp[rT] = 2$. Note the close connection with the half-life of a radioactive element (see Section 1.3). Malthus claimed a doubling time of 25 years for the human population, which implies that the rate coefficient $r = (1/T) \ln 2 = (1/25) \ln 2 \cong 0.02777$. The solution formula (3) implies that $P(t + 1)/P(t) = e^r$ for all t, and so Malthus's value of $r = 0.02777$ gives us that $P(t + 1)/P(t) \approx 1.0282$. Thus, Malthus's value for r would correspond to a 2.8% annual increase in population. Malthus's figure for r is too high for our late-twentieth-century world. However, individual countries, for example, Mexico and Sri Lanka, have intrinsic rate coefficients which exceed 0.02777 and may even be as high as 0.033. The corresponding doubling time shrinks from 25 years to 21 years when $r = 0.033$.

Logistic Growth

The positive checks to population are extremely various and include ... all unwholesome occupations, severe labor and exposure to the seasons, extreme poverty, bad nursing of children, great towns, excesses of all kinds, the whole train of common diseases and epidemics, wars, plague, and famine.

Malthus

The unbridled growth of a population as predicted by the simple Malthusian law of exponential increase cannot continue forever. Malthus claimed that resources grow at most arithmetically (i.e., the net increase in resources each year does not exceed a fixed constant). A geometric increase in the size of a population must soon outstrip the resources available to support that population. The resulting hardships would increase the death rate and put a damper on growth.

The simplest way to model restricted growth within the context of the rate equation (2) with no net migration is to assume that the rate coefficient R has the form $r_0 - r_1 P$, where r_0 and r_1 are positive constants. The term $-r_1 P$ represents a restraint on the growth rate. It is customary to write the rate coefficient as $r(1 - P/K)$, where r and K are positive constants, rather than as $r_0 - r_1 P$. We then have the *logistic equation*[12] with initial condition,

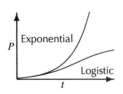

$$P' = r\left(1 - \frac{P}{K}\right)P, \qquad P(0) = P_0 \tag{4}$$

[12] The logistic equation and its solutions were introduced in the 1840s by the Belgian statistician Pierre-Francois Verhulst (1804–1849). He predicted that the population of Belgium would eventually level off at 9,5000,000. The estimated 1994 population of a little more than 10,000,000 is remarkably close to the predicted values.

Observe that $P(t) = 0$ and $P(t) = K$ are solutions of the logistic equation, the *stationary* or *equilibrium* solutions. K is the *carrying capacity* of the species, or the *saturation population*.

The logistic equation is not linear, but it is separable. Of course, in separating the variables we would have to avoid points where $P = 0$, but given the meaning of $P(t)$, it is no restriction to assume that $P > 0$. We must also avoid $P = K$. It will turn out that either $P(t) = K$ for all t, or else $P(t)$ never has the value of K.

Assuming that $P \neq 0$, K and separating the variables, we have

$$\frac{K}{(K - P)P} dP = r \, dt \tag{5}$$

Using partial fractions, notice that

$$\frac{K}{(K - P)P} = \frac{1}{K - P} + \frac{1}{P}$$

and hence integrating (5) we obtain

$$-\ln|K - P| + \ln|P| = rt + C$$

where C is an arbitrary constant. Exponentiating, we have

$$\left| \frac{P}{K - P} \right| = ce^{rt} \tag{6}$$

where $c = e^C$ is a positive constant. If the absolute value sign in (6) is dropped, then c can be either positive or negative, and solving for P we obtain (after putting $c = 1/d$)

$$P = \frac{K}{1 + de^{-rt}}$$

Evaluating the constant d from the initial condition, we have that

$$P(t) = \frac{K}{1 + de^{-rt}}, \quad \text{where} \quad d = \frac{K}{P_0} - 1 \tag{7}$$

Since $r > 0$, $P(t) \to K$ as $t \to \infty$. Note that $P(t)$ is strictly increasing if $0 < P_0 < K$ and strictly decreasing if $P_0 > K$. A population curve defined by (7) has an inflection point if $P = K/2$, since, upon differentiating each side of the ODE in (4), we have that

$$P'' = r(1 - P'/K)P + r(1 - P/K)P' = r\left(1 - \frac{2P}{K}\right)P' = r\left(1 - \frac{2P}{K}\right)r\left(1 - \frac{P}{K}\right)P$$

which changes sign at $P = 0$, $K/2$, and K.

The elongated S-shaped population curves of the logistic equation are called *logistic curves*. Figure 1.5.1 illustrates the appropriateness of the term "carrying capacity" or "saturation population" for $P = K$ ($K = 10$, $r = 1$ in Figure 1.5.1). The resources of the community can support a population of size K, and this is precisely the asymptotic limit of all the nonvanishing population curves. The figure on the cover page of Chapter 1 shows solution curves of a logistic equation in all four quadrants.

The example below shows how to scale the logistic equation to reduce the number of parameters, eliminate the dimensions of the variables, and make computing easier.

EXAMPLE 1.5.1

Scaling the Logistic Equation
Let us measure $P(t)$ in units of the carrying capacity K, $P = QK$. Substituting in IVP (4), we have the equivalent IVP in terms of Q,

$$Q' = r(1 - Q)Q, \quad Q(0) = Q_0 = P_0/K$$

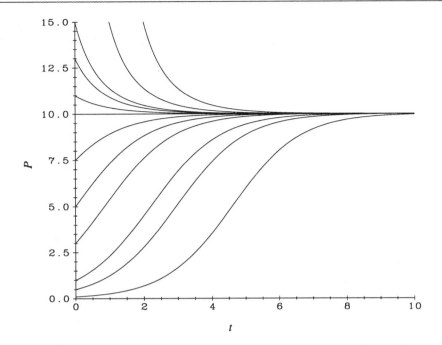

FIGURE 1.5.1 Logistic population curves with carrying capacity $K = 10$, $P' = (1 - P/10)P$.

Observe that Q is dimensionless, and that for computing purposes Q is much more convenient than P if K is a large number. We could also scale time in units of $1/r$ of the rate constant r, $t = s/r$. Then the IVP becomes

$$dQ/ds = (1 - Q)Q, \quad Q(0) = P_0/K$$

A marked advantage of this last IVP in dimensionless units of population and time is that there is only one parameter involved, the ratio P_0/K.

Although Example 1.5.1 shows how to rescale the logistic model into a dimensionless IVP with a single parameter, we continue to deal with unscaled problems so that the solution technique is emphasized.

Harvesting

What is the effect of harvesting on a logistically growing population? Obviously the numbers will decline, but by how much? Will a high rate of harvesting kill off the species? These questions can be answered for the simple model of a logistically growing population that is harvested at a constant rate.

EXAMPLE 1.5.2

Harvesting at a Constant Rate
Consider the logistic IVP of Figure 1.5.1, but with a constant harvest rate H of magnitude $9/10$ subtracted from the ODE:

$$P' = (1 - P/10)P - 9/10, \qquad P(0) = P_0 > 0 \qquad (8)$$

The rate function is the quadratic $-P^2/10 + P - 9/10$, which factors to yield

$$P' = -(P - 1)(P - 9)/10, \qquad P(0) = P_0 \qquad (9)$$

and we see that $P = 1$ and $P = 9$ are equilibrium populations. The population declines if $P > 9$ or if $0 < P < 1$ because P' is negative in those regions. The population increases if P lies between the two equilibria, $1 < P < 9$. A formula for $P(t)$ can be found by separating the variables in the ODE in (9) and integrating:

$$\frac{1}{P-1} \cdot \frac{1}{P-9} dP = -\frac{1}{10} dt, \quad P \neq 1, 9$$

$$\frac{1}{8}\left(\frac{1}{P-9} - \frac{1}{P-1}\right) dP = -\frac{1}{10} dt$$

$$\ln \left|\frac{P-9}{P-1}\right|^{1/8} = -\frac{1}{10} t + B, \quad B \text{ a constant}$$

$$\frac{P-9}{P-1} = be^{-4t/5}, \quad b = \pm e^{8B} = (P_0 - 9)/(P_0 - 1)$$

$$P = \frac{9 - be^{-4t/5}}{1 - be^{-4t/5}} \tag{10}$$

In addition to the equilibrium populations $P = 1, 9$, we have from (10) and the formula for the constant b that $P(t)$ decreases to $P = 9$ as $t \to \infty$ if $P_0 > 9$, and $P(t)$ increases to $P = 9$ as $t \to \infty$ if $1 < P_0 < 9$; $P(t)$ falls to the extinction level $P = 0$ at *extinction time* $t = (5/4)\ln(b/9)$ if $0 < P_0 < 1$. See Figure 1.5.2 for the graphs. The curves show the expected behavior; note the horizontal lines at the equilibrium population levels of $P = 1$ and $P = 9$.

The calculations require a fair amount of algebraic manipulation of the solution formulas if carried out by hand, but they could also be found by use of a CAS (Computer Algebra System) to solve IVP (8). The graphs in Figure 1.5.2 were found by applying a numerical solver directly to IVP (8) for various values of P_0.

Whatever solution method is used, we conclude that the addition of the harvesting rate term $H = 9/10$ lowers the carrying capacity from the level of 10 in the unharvested setting to 9. This is not much of a change, and one might say that at $H = 9/10$ harvesting is sustainable indefinitely. However, the unstable equilibrium $y(t) = 1$ might cause some concern, because if a disaster caused the population to drop below that level, then continued harvesting at the rate $H = 9/10$ would lead to extinction of the species in finite time.

See Problem 7 for a general treatment of the harvesting model and Problem 9 for a group project that explores the same model, but in dimensionless variables. We now address the inverse problem: knowing the solution, can we find the ODE?

Given the Solution Curves, Find the ODE

One of the challenges of science and engineering is to determine the specific mathematical model that best represents experimental results. This is far from easy, but a knowledge of a variety of mathematical models and their properties is of considerable help. The final example illustrates the process in the context of a harvested, logistically changing population.

EXAMPLE 1.5.3

A Harvested, Logistically Changing Population
Suppose that a researcher experimentally obtains the population growth curves shown in Figure 1.5.3. Let us try to construct a differential equation modeling the behavior shown. Since this plot looks qualitatively like Figure 1.5.2, we will assume that logistic

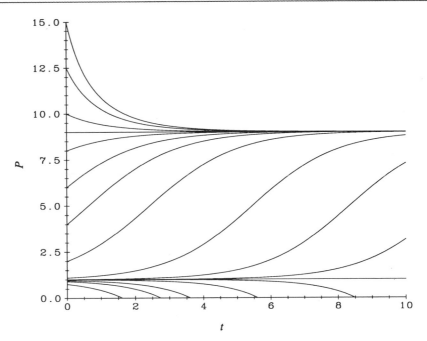

FIGURE 1.5.2 Logistic growth with harvesting. See Example 1.5.2.

growth with constant rate harvesting is occurring. We will look for an equation of the form

$$P' = f(P) = r(1 - P/K)P - H$$

for some positive constants r, K, H. Noting from Figure 1.5.3 that $f(P)$ must have roots $P = 8$ and $P = 4$ and that $f(P)$ is negative for $P > 8$, we obtain

$$P' = -c(P - 8)(P - 4)$$

where c is a positive constant. Writing this in logistic form with harvesting, we have

$$P' = 12c(1 - P/12)P - 32c$$

which provides us with new information about the population being investigated. Comparing this equation with equation (4), for example, we find that the carrying capacity of the population in the absence of harvesting would be 12. Whatever the value of $r = 12c$ may be, the rate of harvesting is $32c$.

Comments

Logistic laws may be best suited to laboratory or other isolated populations for which there is reason to believe that K and r are constants. Beginning with the experiments of the Soviet biologist G.F. Gauze in the early 1930s, there have been numerous experiments with colonies of protozoa growing under controlled laboratory conditions. The results of these experiments generally confirm the logistic model.[13]

---

[13] See J.H. Vandermeer, "The Competitive Structure of Communities: An Experimental Approach with Protozoa," *Ecology* **50** (1969), pp. 362–371.

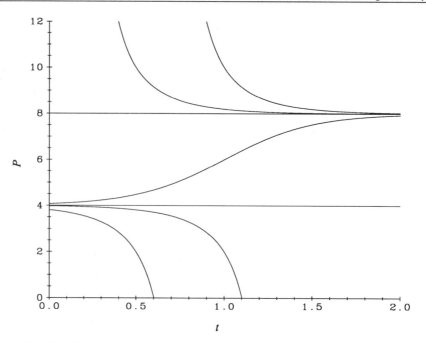

FIGURE 1.5.3 Population curves modeled in Example 1.5.3.

PROBLEMS

1. (*A Crowded Earth with Exponential Population Growth*). If the population of the earth in 1960 was 3 billion (3×10^9) and the population in 1970 was 4 billion, and if the earth can only support a maximum population of 10 billion, in what year will the limit be reached? In what year will the limit be reached if the maximum population that the earth can support is 20 billion? (Assume that the rate of growth is proportional to the population present.)

2. (*25 Year Doubling Time*). If the population of an island doubled in the past 25 years and the present population is 150,000, when will the island have a population of 560,000? (Assume exponential growth.)

3. (*Reproduction Stops*). Suppose that a population has a birth rate a and death rate b so that the growth of the population is $dN/dt = (a - b)N$. After 15 years of steady growth, the population stops reproducing; that is, a becomes zero. If initially $a = 0.06$ births/year and $b = 0.04$ deaths/year, how long after reproduction stops will it take for the population to reach the levels indicated below? [*Hint*: Solve for $0 \le t \le 15$; then solve for $t \ge 15$ in terms of $N(15)$.]

(a) The original level?

(b) 50% of its original level?

(c) 30% of the peak population level $N(15)$?

4. (*Constant Supply Rate*). Nutrients flow into a cell at a constant rate of R nutrient units per unit time, and leave it at a rate proportional to the concentration with constant of proportionality k. Let N be the number of units of nutrients in the cell at time t. Assume the cell volume V is fixed.
(a) Write the mathematical expression for the rate of change of the concentration $y = N/V$ of the nutrients in the cell.[*Hint*: First apply the Balance Law to the rate equation for N.]

(b) Will the concentration of nutrients approach an equilibrium? [*Hint*: Equilibrium occurs when the rate of change of the concentration is zero.]

(c) Plot graphs of $y(t)$ for $t \geq 0$ for various values of $y(0) > 0$, including $y(0) = R/k$ with $R/k = 10$.

5. (*Logistic Growth*). A colony of bacteria grows according to the logistic law, with a carrying capacity of 5×10^8 individuals and with natural growth coefficient $r = 0.01$ per day.

   (a) What will the population be after 2 days if it is initially 1×10^8 individuals? [*Hint*: You may want to rescale the problem (see Example 1.5.1).]

   (b) What will the population be after 3 days if it is initially 1×10^3 individuals?

6. (*Alternate Derivation of Solution Formula (7)*). The logistic equation can be solved by making the change of dependent variable $z = 1/P$, which transforms the logistic equation into a linear equation in z. Solve the transformed ODE, and show that the IVP (4) has the solution formula (7). [*Hint*: Set $P = 1/z$ and find P' in terms of z and z'. Substitute into the ODE for P to get a new ODE in z.]

7. (*Harvesting a Population*). The equation $dP/dt = r(1 - P/K)P - H$ models a population with natural rate coefficient r, carrying capacity K, and a constant harvest rate H. Assume that r, K, and H are positive constants, and that the population P_0 at time $t = 0$ is positive.

   (a) The value $rK/4$ is called the *critical harvesting rate*. Show that the population becomes extinct if H exceeds the critical harvesting rate. [*Hint*: First show that $dP/dt < 0$ for all $P > 0$ if $H > rK/4$.]

   (b) Plot portraits of solution curves in the quadrant $P > 0$, $t > 0$ for the three cases: no harvesting ($H = 0$); *subcritical harvesting* ($H = rK/8$); *supercritical harvesting* ($H = rK/2$). Let $r = 0.01$, $K = 1000$, and choose various values of P_0.

   (c) The rate equation for the harvesting model can be solved by separating the variables. Factor the quadratic rate function $r(1 - P/K)P - H$ as $-(r/K)(P - P_1)(P - P_2)$, where P_1 and P_2 denote the roots of the quadratic. Show that for $0 < H < rK/4$, P_1 and P_2 are positive numbers between 0 and K. What happens at $H = rK/4$? What happens if $H > rK/4$?

8. (*Constant Rate Harvesting*). Consider the logistic equation with constant-rate harvesting, $P' = r(1 - P/K)P - H$ where r, K, and H are positive constants.

   (a) Show that this ODE may be rewritten in the form $P' = -a(P - b)(P - c)$ where a, b, and c are positive constants.

   (b) Find the harvesting rate H and the carrying capacity K (in the absence of harvesting) as functions of a, b, and c.

   (c) (*Determination of System Parameters*). In each of the plots below, assume $a = 1$. Find the values of b and c. Find the harvesting rate, H, and the carrying capacity in the absence of harvesting, K. Write the differential equation in the form $P' = r(1 - P/K)P - H$. The ranges on the axes are $0 \leq t \leq 2$, $0 \leq P \leq 12$.

   (c1)   **(c2)**   **(c3)**

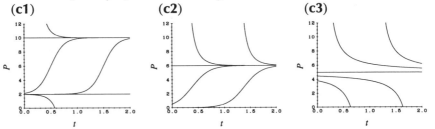

9. (*Scaling the Logistic ODE with Harvesting: Sensitivity*). After reading Example 1.5.1, show that the logistic IVP with harvesting,

$$P' = r(1 - P/K)P - H, \qquad P(0) = P_0,$$

can be rescaled to dimensionless form,

$$dQ/ds = Q(1 - Q) - H^*, \qquad Q(0) = Q_0,$$

where $H^* = H/(rK)$, $Q_0 = P_0/K$. Then for fixed Q_0 study the effects of changing H^*. Repeat with fixed H^* and varying Q_0. Your conclusions?

10. (*Seasonal Harvesting*). A high harvesting rate need not lead to extinction if the harvest season is restricted to a few months each year, a common practice for sports fishing and hunting. Consider the ODE $P' = (1 - P/12)P - H$ with saturation population 12 and harvest rate H. Show that if $P(0) = 20$ and $H = 6$, then $P(t) = 0$ for some $t < 4$. Now model seasonal harvesting by replacing $H = 6$ by $H = 6\,\text{sqw}(t, d, 1)$ for various values of d (see Appendix C.1 for the definition of the square wave train, $\text{sqw}(t, d, p)$). If $d = 8\frac{1}{3} = 100/12$ and $p = 1$, then the harvest season lasts for one month of the year (i.e., for $8\frac{1}{3}\%$ of the period). Atlas Plate FIRST ORDER E shows the population curves for $d = 0$ (no harvesting), $d = 100/3$ (a four-month season), and $d = 200/3$ (an eight-month season), and $d = 100$ (open season all year). To the best ability of your solver, find the shortest harvesting season that leads to population extinction if $P(0) = 20$. What is the longest season for which $P(t) \geq 10$ for all t if $P(0) = 20$? Estimate these values by using a computer solver to solve the corresponding IVP. Discuss the sensitivity of the population to changes in the length of the harvest season. Discuss this point.

1.6   Cold Pills and Cascades

At the first sign of a cold many of us start to take cold pills. These pills usually contain a decongestant to relieve stuffiness and an antihistamine to dry up a runny nose and watering eyes. The pill dissolves in the gastrointestinal tract, and the two medications diffuse into the bloodstream. The bloodstream takes the medications to the sites where they have therapeutic effect. Both medications are eventually removed from the blood by the kidneys and the liver. The dynamics of the rising and falling levels of the medications in the GI tract and in the bloodstream may be modeled by a system of first-order linear differential equations.

Pharmaceutical companies do extensive testing to determine the movement of medications through the body. This process is modeled by treating various components of the human body as *compartments* and following the medication as it enters and leaves these compartments. Examples of body compartments are the GI tract, the bloodstream, the tissues and the excretory system. Testing has shown that a typical cold medication leaves one compartment (e.g., the GI tract) and moves into another (e.g., the bloodstream) at a rate proportional to the amount present in the first compartment. The coefficient of proportionality depends upon the specific medication, the compartments involved, and the age and general health of the individual.

In this section we look at three models based on different dosage strategies.[14]

---

[14]The examples in this section are based on the work of the contemporary applied mathematician, Edward Spitznagel, Professor of Mathematics at Washington University, St. Louis, Missouri. His current research areas

Suppose that a single dose of a fast-dissolving cold pill is taken. The pill dissolves "instantaneously" in the GI tract and releases A milligrams each of decongestant and antihistamine. Each medication independently diffuses into the bloodstream.

EXAMPLE 1.6.1

Flow of medication

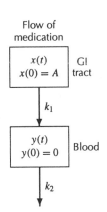

One Instantly Dissolving Dose: The Model IVP

Let $x(t)$ be the amount of medicine (either decongestant or antihistamine) in the GI tract at time t after the pill has dissolved. We shall use the Balance Law to compute the rate of change of the amount in each compartment. Because the medication moves out of the GI tract and into the bloodstream at a rate proportional to the amount in the GI tract, and because nothing is coming in, we have that

$$\frac{dx(t)}{dt} = -k_1 x(t), \qquad x(0) = A$$

where $k_1 > 0$ is the coefficient of proportionality, and A is the initial amount. The units for the rate constant k_1 are reciprocals of the units chosen to measure the time t. For example the units of k_1 are (hours)$^{-1}$ if time is given in hours.

The level of medication in the bloodstream will build up from zero (initially, there are no cold medications in the blood), but then will fall as the kidneys and liver do their job of removing "foreign" substances from the blood. If $y(t)$ denotes the amount of medication in the bloodstream at time t, then the Balance Law implies that

$$\frac{dy(t)}{dt} = k_1 x(t) - k_2 y(t), \qquad y(0) = 0$$

The first term on the right side of the rate equation models the observed fact that the medication leaving the GI tract goes directly into the bloodstream; the second rate term models the clearance of medication from the blood and into the excretory system. Consequently, the system of first-order ODEs and initial conditions that models the flow of medication is given by the IVP

$$\begin{aligned} \frac{dx}{dt} &= -k_1 x, & x(0) &= A \\ \frac{dy}{dt} &= k_1 x - k_2 y, & y(0) &= 0 \end{aligned} \tag{1}$$

where the (positive) coefficient k_2 is typically smaller than k_1.

System (1) models the flow of decongestant and also the flow of antihistamine. However, the flow rates differ because the experimentally determined values of the rate constants k_1 and k_2 are quite different for the two medications (see Table 1.6.1). These differences lead to very different levels of the two medications in the bloodstream.

EXAMPLE 1.6.2

Solving the Model IVP

Integrating factors may be used to solve the linear differential equations of IVP (1) one at a time, starting with the first. Rearrange the first rate equation, multiply by the integrating factor $e^{k_1 t}$, integrate, and then use the initial data to evaluate a constant of integration to obtain $x(t) = A e^{-k_1 t}$. This expression for $x(t)$ is inserted into the second

are pharmacokinetics (study of the movements of drugs through the body) and bioequivalence (study of the efficacy of medications), fields which use compartment models extensively. Professor Spitznagel started off his career in mathematics because he could not limit his interest to just one area of science. He does consulting work for a wide variety of clients, but principally for the pharmaceutical industry and medical schools. His advice to aspiring young applied mathematicians: learn as much mathematics as you can because it will make you versatile and able to pick up new ideas easily. Seek out opportunities to learn how to apply mathematics in the real world by "on the job" training.

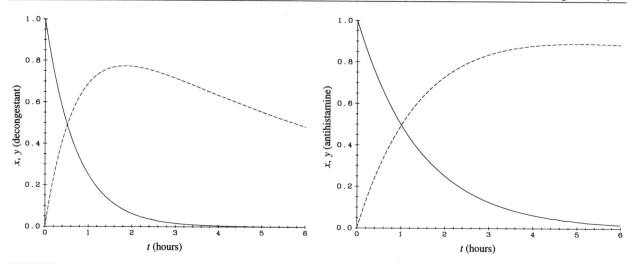

FIGURE 1.6.1 Decongestant in GI tract (solid) and bloodstream (dashed): single unit dose ($k_1 = 1.386$, $k_2 = 0.1386$). See Examples 1.6.1 and 1.6.2.

FIGURE 1.6.2 Antihistamine in GI tract (solid) and bloodstream (dashed): single unit dose ($k_1 = 0.6931$, $k_2 = 0.0231$). See Example 1.6.1 and 1.6.2.

rate equation, which may be solved using the same integrating factor approach, but with integrating factor $e^{k_2 t}$, to obtain

$$y(t) = \frac{k_1 A}{k_1 - k_2} \left(e^{-k_2 t} - e^{-k_1 t} \right)$$

Thus, the solution set for the linear system (1) is, for $k_1 \neq k_2$,

$$x(t) = A e^{-k_1 t}, \qquad y(t) = \frac{k_1 A}{k_1 - k_2} \left(e^{-k_2 t} - e^{-k_1 t} \right) \qquad (2)$$

where time t is measured forward from the instant the initial dose of A units of medication is released into the GI tract.

From the solution set (2) we see that, as expected, the levels of each medication in the GI tract and the bloodstream tend to zero as time increases. Since each medication is initially at the zero level in the bloodstream, these levels reach a maximum value at some later time (see the problem set).

Observe from Table 1.6.1 how much smaller k_2 is than k_1 for each medication, and how much smaller the coefficients for the antihistamine are than the corresponding coefficients for the decongestant. Consequently, each medication will stay in the bloodstream longer than in the GI tract, and it will take the antihistamine longer to reach its maximal level in the blood than the decongestant. Figures 1.6.1 and 1.6.2 show the levels of decongestant and antihistamine in the GI tract and the blood over a six-hour period if the values of k_1 and k_2 given in Table 1.6.1 are used, and $A = 1$.

TABLE 1.6.1 Rate Coefficients for Cold Medications

	Decongestant	Antihistamine
k_1	1.386 hour^{-1}	0.6931 hour^{-1}
k_2	0.1386 hour^{-1}	0.0231 hour^{-1}

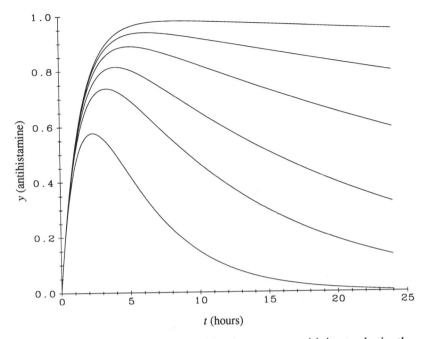

FIGURE 1.6.3 Antihistamine in bloodstream: sensitivity to k_2 in the model of Examples 1.6.1 and 1.6.2 ($k_1 = 0.6931$).

The *clearance coefficient* k_2 of medication from the bloodstream is often much lower for old and sick people than it is for the young and healthy. This means that for some people the medication levels in the blood may become dangerously high even with a "standard" dosage. Figure 1.6.3 displays the results of a parameter study in which the levels of antihistamine in the blood are plotted over a twenty-four hour period for six values of k_2 (top curve corresponds to the smallest value, bottom curve to largest):

$$k_2 = 0.00231, \ 0.01, \ 0.0231, \ 0.05, \ 0.09, \ 0.231$$

ranging from an order of magnitude smaller than the value $k_2 = 0.0231$ given in Table 1.6.1 to an order of magnitude larger.

Few of us stop with taking one cold pill. The models that follow treat the more realistic setting of taking medication over a period of time.

Continuous Doses

If a medication is embedded in resins that dissolve at varying rates, a fixed flow of medication can be assured. Tiny beads of the mixture are packed into a capsule, and the medications are released at a constant rate into the GI tract over a period of hours. One capsule of <u>con</u>tinuous-<u>ac</u>ting medication such as CONTAC (Smith-Kline Beecham) may need to be taken only twice a day. The mathematical model for the dynamics of this situation is slightly different from IVP (1). There is no medication in the GI tract at the start of the continuous dosage regimen, but instead there is a constant inflow rate I due to the gradually dissolving beads.

EXAMPLE 1.6.3

Continuous-Acting Capsules: Model IVP and its Solution
The model IVP for the amounts $x(t)$ and $y(t)$ of continuous-acting medication in the

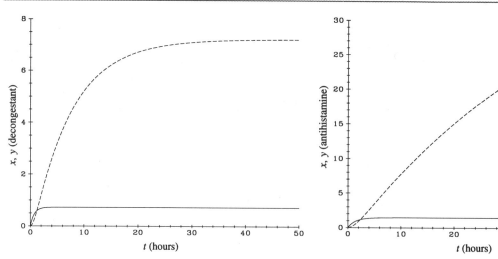

FIGURE 1.6.4 Decongestant in GI tract (solid) and bloodstream (dashed): continuous-acting capsules ($k_1 = 1.386$, $k_2 = 0.1386$). See Example 1.6.3.

FIGURE 1.6.5 Antihistamine in GI tract (solid) and bloodstream (dashed): continuous acting capsules ($k_1 = 0.6931$, $k_2 = 0.0231$). See Example 1.6.3.

GI tract and bloodstream, respectively, is

$$\frac{dx}{dt} = I - k_1 x, \qquad x(0) = 0$$
$$\frac{dy}{dt} = k_1 x - k_2 y, \quad y(0) = 0$$

(3)

where I is a positive constant modeling the rate of release of each medication into the GI tract.

The rate equations of IVP (3) may again be solved from the top down by using integrating factors. Using the integrating factor $e^{k_1 t}$ and the initial data to solve the first equation for $x(t)$, we see that $x(t) = I(1 - e^{-k_1 t})/k_1$. Inserting this expression for $x(t)$ into the second equation, using the integrating factor $e^{k_2 t}$ and the initial data, we obtain a formula for $y(t)$. The solution of IVP (3) is found to be

$$x(t) = \frac{I}{k_1}(1 - e^{-k_1 t})$$
$$y(t) = \frac{I}{k_2}\left[1 + \frac{1}{k_1 - k_2}\left(k_2 e^{-k_1 t} - k_1 e^{-k_2 t}\right)\right]$$

(4)

The derivations of these formulas are left to the reader.

Observe that as time increases, the level of medication in the GI tract tends to I/k_1, and that in the bloodstream to I/k_2. The solution formulas (4) are valid for any positive values of the coefficients k_1 and k_2, $k_1 \neq k_2$.

Suppose that the input rate I is taken to be one unit of each medication per hour. Thus, a total dose of twenty-four units is administered continuously over a period of one day. Figures 1.6.4 and 1.6.5 show that, if the values of k_1 and k_2 from Table 1.6.1 are used, the decongestant builds up quite rapidly to its equilibrium level in the blood. On the other hand, the values of k_1 and k_2 for the antihistamine are so much less than the decongestant values that the approach to equilibrium is much more gradual. Observe that even after fifty hours antihistamine levels in the blood are still far from equilibrium. Antihistamines make many people sleepy; that is why drowsiness is a problem long after the medications have knocked out the cold symptoms. Sniffles on Sunday, start to take cold pills on Monday, asleep in the math class on Friday!

Flow of medication

I

$x(t)$
$x(0) = 0$ | GI tract

k_1

$y(t)$
$y(0) = 0$ | Blood

k_2

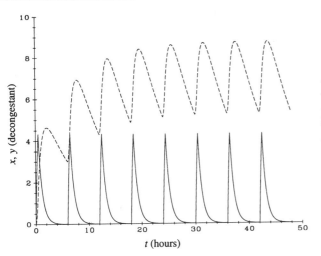

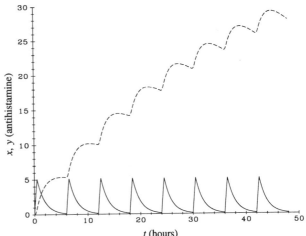
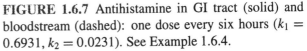

FIGURE 1.6.6 Decongestant in GI tract (solid) and bloodstream (dashed): one dose every six hours ($k_1 = 1.386$, $k_2 = 0.1386$). See Example 1.6.4.

FIGURE 1.6.7 Antihistamine in GI tract (solid) and bloodstream (dashed): one dose every six hours ($k_1 = 0.6931$, $k_2 = 0.0231$). See Example 1.6.4.

Repeated Doses of Fast-Dissolving Pills

Many cold pills dissolve fairly quickly in the GI tract, releasing their medications over a period of no more than half an hour. Repeated doses must then be taken every four to six hours in order to keep up the medication levels in the blood. In this case, the input rate I into the GI tract is not constant, but operates at a high level for a short time, then cuts off entirely until the next dose is taken four to six hours later.

EXAMPLE 1.6.4

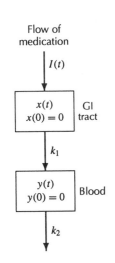

One Dose Every Six Hours: The Model IVP
In contrast to a continuous acting medication, other forms of medication dissolve very rapidly in the GI tract. This means that the medication is delivered at a high constant rate but only over a short time. Suppose that the rate is 12 units/hour, but delivered in half an hour, and then repeated at six-hour intervals. The model IVP is given below:

$$\frac{dx}{dt} = I(t) - k_1 x, \qquad x(0) = 0$$

$$\frac{dy}{dt} = k_1 x - k_2 y, \qquad y(0) = 0$$

$$I(t) = 12 \, \text{sqw}(t, 100/12, 6)$$

where $\text{sqw}(t, 100/12, 6)$ denotes the pulse function with period 6 hours, which is "on" for half an hour at the start of each period and "off" otherwise (1/2 hour $= 100/12\%$ of the 6-hour period.) See Appendix C.1 for a description of sqw and other piecewise continuous functions.

Although the integrating factor technique used before applies to this system, there is an awkward integration to carry out since $I(t)$ repeatedly turns on and off. The Laplace transform techniques of Chapter 5 may be a better method for producing a solution formula. A numerical solver that handles on-off functions is also appropriate if graphically defined solutions suffice, and that is what was used to produce Figures 1.6.6 and 1.6.7. The total dosage of 24 units in one day is the same as in the continuous-acting model. Figures 1.6.6 and 1.6.7 display the medication levels corresponding to this repeated on-off dosage pattern over a forty-eight-hour period if the

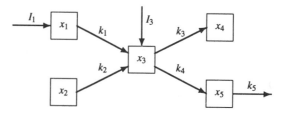

FIGURE 1.6.8 A joining-branching linear cascade. See Example 1.6.5.

values of k_1 and k_2 are taken from Table 1.6.1. Medication levels rise during the half hour the cold pill is dissolving in the GI tract, but then fall until the next dose. The amount of decongestant in the blood quickly reaches and then oscillates around an equilibrium level. However, the amount of antihistamine in the blood still has not reached equilibrium even after forty-eight hours.

These examples are typical of the many models of the passage of medication throughout the body. Mathematical models are an important part of the process by which dosage levels are set. A critical factor is the need to keep the levels of medication high enough to be effective, but not so high that they are dangerous.

Linear Cascades

A *compartment model* consists of a finite number of compartments, some of which are connected by arrows. The arrows indicate that the substance being tracked leaves the compartment at the arrow's foot and enters another compartment at the arrowhead.

> ❖ **Linear Cascade.** A compartment model is a *linear cascade* if: (a) there are no *loops*, i.e., there is no directed sequence of arrows that begins and ends at the same compartment, and (b) the substance exits one compartment at a rate proportional to the amount in that compartment and enters the next compartment at the same rate.

An arrow pointing toward one compartment, but not out of another compartment, indicates an external source of the substance into the compartmental system (i.e., an *input*). These arrows are labeled with an "*I*," where *I* is the input rate of the substance per unit time. An arrow that points out from a compartment, but not toward another compartment, indicates that the substance exits the system through that compartment. A "*k*" by an arrow leaving compartment *i*, say, and entering compartment *j*, means that the substance exits compartment *i* and enters compartment *j* at rate kx_i, where x_i is the amount of substance in compartment *i*. The models for the flow of cold medications are simple examples of linear cascades. However, the compartments of a cascade model can be joined in complex ways. Here is an example.

EXAMPLE 1.6.5

Joining-Branching Cascades

Figure 1.6.8 shows a joining-branching linear cascade with two inputs. The positive constants k_i are the rate coefficients for the linear rate terms. The corresponding set of differential equations can be constructed directly from the graphical model. Assuming

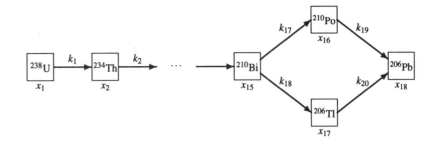

FIGURE 1.6.9 The uranium series of radioactive decay. See Example 1.6.6.

an initial amount c_i in compartment i, we have the following first-order system of IVPs based on the Balance Law for each compartment:

$$x_1' = I_1 - k_1 x_1, \qquad\qquad x_1(0) = c_1$$
$$x_2' = -k_2 x_2, \qquad\qquad x_2(0) = c_2$$
$$x_3' = I_3 + k_1 x_1 + k_2 x_2 - (k_3 + k_4) x_3, \quad x_3(0) = c_3$$
$$x_4' = k_3 x_3, \qquad\qquad x_4(0) = c_4$$
$$x_5' = k_4 x_3 - k_5 x_5, \qquad\qquad x_5(0) = c_5$$

Observe that this system of rate equations can be solved from the top down, one equation at a time. Conversely, given a set of linear rate equations like that above, a linear cascade model of boxes and arrows such as in Figure 1.6.8 may be constructed.

Other Applications of Compartment Models

Linear cascades may be used to model other physical phenomena besides the flow of a substance through a sequence of compartments. For example, in a radioactive decay process one element decays into another, which in turn decays into a third, and so on through a sequence of steps that terminates in a stable, nonradioactive substance. In this case each species can be modeled as a "compartment," although the species share a common physical space.

EXAMPLE 1.6.6

The Rate Equations of the Uranium Series
Uranium-238 (^{238}U) decays to stable lead-206 (^{206}Pb) in a decay cascade that involves sixteen intermediate elements and isotopes. Figure 1.6.9 shows the initial and the final steps of the uranium series. The corresponding rate equations are

$$x_1' = -k_1 x_1$$
$$x_2' = k_1 x_1 - k_2 x_2$$

$$\vdots$$

$$x_{16}' = k_{17} x_{15} - k_{19} x_{16}$$
$$x_{17}' = k_{18} x_{15} - k_{20} x_{17}$$
$$x_{18}' = k_{19} x_{16} + k_{20} x_{17}$$

Usually a radioactive decay process is expressed in terms of half-lives τ_i, instead of rate coefficients k_i. As noted in Section 1.3, however, the two are connected by the relation $\tau_i k_i = \ln 2$. The various half-lives of the radioactive elements in the uranium series differ by many orders of magnitude. For example, the half-life of uranium-238 is four and a half billion years, while that of the next element, thorium-234, is only twenty-four days. The corresponding ratio of k_1 to k_2 is calculated to be 1.5×10^{-11}. This extreme variation in the magnitudes of the rate coefficients suggests that a computer solution of the rate equations for the uranium series will be difficult to obtain unless some kind of scaling is done first.

Comments

We have seen that there are two ways to view compartment models. In the first interpretation, the system consists of distinct physical compartments with a substance making its way from one compartment to another. In the second, a single physical space contains several species, each of which is transformed into others. The mathematical models for these phenomena are often cascades of first-order linear ODEs which can be solved "from the top down" by using the integrating factor techniques of Section 1.2.

PROBLEMS

1. Write the system of first-order linear rate equations that models each of the following cascades. Then solve, using the given initial data. [*Hint*: See Example 1.6.5. Solve each system of ODEs from the top down as shown in Examples 1.6.1–1.6.3.]

    (a) $k_1 = 1$ to y, x, $k_2 = 2$ to z; $\quad x(0) = 1, \quad y(0) = z(0) = 0$

    (b) $I = 1 \to x$; $k_1 = 2 \to y \xrightarrow{k_3 = 1}$; $k_2 = 0.2 \to z$; $\quad x(0) = y(0) = z(0) = 0$

    (c) $I = 1 + \sin t \to x \xrightarrow{k_1 = 3} y \xrightarrow{k_2 = 1}$; $\quad x(0) = y(0) = 0$

    (d) $I_1 = 1 \to x$; $I_2 = 2 \downarrow$; $k_1 = 2 \to y \xrightarrow{k_3 = 1}$; $k_2 = 1 \to z$; $\quad x(0) = y(0) = z(0) = 0$

2. For each system sketch the corresponding linear cascade model. Label the compartments and the arrows appropriately (see Figures 1.6.8 and 1.6.9 for examples). Do not solve the rate equations.

(a) $x' = 5 - x$, $y' = 2 + x - 3y$ [*Hint*: This is a two-compartment cascade with inflow rates (from outside) of 5 into the first compartment, and 2 into the second.]

(b) $x' = -x/2$, $y' = 1 - y/3$, $z' = x/2 + y/3$

(c) $x' = -x$, $y' = x/2 - 3y$, $z' = x/2 + 3y - 2z$

3. (*Alternative Solution Techniques for a Single Dose Model*).

   (a) Solve system (1) with the condition $y(0) = 0$ replaced by $y(0) = B$, where B is a positive constant. Write out formulas for $x(t)$ and $y(t)$. What do these formulas tell you about the medication levels in the GI tract and bloodstream as time advances?

   (b) Use a computer ODE solver to solve the system (1) with $x(0) = 1$, $y(0) = 1$ and to graph the levels of decongestant and antihistamine in the GI tract and in the bloodstream over a six-hour period. Use the values of k_1 and k_2 given in Table 1.6.1. Estimate the highest level of each medication in the blood and the time it reaches that level.

   (c) Use a graphics package to plot the graphs of the solutions obtained in part (a) with $A = B = 1$, $0 \le t \le 6$ and k_1 and k_2 given in Table 1.6.1. Compare the graphs obtained this way with those obtained in part (b).

   (d) Use a Computer Algebra System to solve the system in part (a). Compare the results with the solution you derived by hand.

4. (*Cold Medication: Sensitivity to One Dose*).

   (a) Figure 1.6.3 shows how the antihistamine levels in the bloodstream depend upon the value of the rate coefficient k_2 for clearing the medication out of the blood and into the excretory system. Let $A = 1$ in system (1). Using a computer ODE solver, display the effects on the antihistamine levels in the bloodstream if k_2 is fixed at 0.0231/hour, but k_1 ranges through the six values $0.06931, 0.11, 0.3$, 0.6931, 1.0, and 1.5. Either use a computer numerical solver to solve IVP (1), or use a computer graphics package to plot the graphs of the antihistamine levels $y(t)$ given by the solution formulas (2). Plot the graphs over a twenty-four hour period. Explain and interpret your graphs.

   (b) It is desired to keep medication levels within a fixed range in order to be both effective and safe. The desired range for antihistamine levels in the blood is from 0.20 to 0.80 for a unit dose taken just once. With $k_2 = 0.0231$, find upper and lower bounds on k_1 so that the antihistamine levels in the blood reach 0.20 within two hours and remain below 0.80 for twenty-four hours after taking a single unit dose. Suggestion: Use your results from part (a).

5. (*Cold Medication: Continuous Doses*).

   (a) Using Table 1.6.1 for the values of k_1 and k_2, determine the equilibrium levels of decongestant and antihistamine in the GI tract and bloodstream with a continuous dosage of one unit/hour. [*Hint*: See Example 1.6.3.]

   (b) The coefficients k_1 and k_2 for the old and sick may be much less than those for the young and healthy. Use either a computer numerical solver to solve IVP (3) or a computer graphics package to plot $x(t)$ and $y(t)$ as given by (4) under the assumption that the coefficients k_1 and k_2 are one-third of those given in Table 1.6.1. Explain and interpret your graphs.

   (c) Assume that the clearance coefficients k_1 and k_2 have one-third of the values given in Table 1.6.1. Suppose that the levels of antihistamine in the blood should reach and remain between 25 and 50 units for a dose taken continuously at the rate of one unit per hour. Will this be possible during the second through the fifth days as an old and infirm patient takes a continuous-acting cold medication?

6. (*Cold Medication: One Dose Every Four Hours*).

(a) A dose of cold medication is taken every four hours, with the entire dose delivered in half an hour at the constant rate of 10 units/hour. Plot the medication levels in both compartments over a 48-hour period. [*Hint*: See Example 1.6.4.]

(b) (*Loading Dose*). Sometimes it is necessary to bring the medication level in the blood close to steady state more rapidly than by regular dosage. A *loading dosage* may be prescribed in which an extra amount of medication is given initially. This is followed by continuous or repeated dosage thereafter. Using the same dosage from part **(a)** and a numerical solver, graph the decongestant level in the blood, if a double, triple, or quadruple dose is taken initially. Which loading dosage(s) keep the level of decongestant in the blood below the safe limit of 10 units over a 48-hour period?

7. (*Tetracycline in the Body*). The antibiotic tetracycline is prescribed for ailments ranging from acne to acute infections. The drug is taken orally and absorbed through the intestinal wall into the bloodstream and, eventually, is removed from the blood by the kidneys and excreted in the urine.

(a) If the compartments are the intestinal tract, the bloodstream, and the urinary system, diagram the system and derive the compartment model, $x_1' = I - k_1 x_1$, $x_2' = k_1 x_1 - k_2 x_2$, $x_3' = k_2 x_2$, and explain the meaning of each term ($x_i(t)$ denotes the amount of tetracycline in milligrams at time t in compartment i). Why may the urinary system be considered an "absorbing" compartment?

(b) It has been shown experimentally that $k_1 = 0.72$ hour^{-1} and $k_2 = 0.15$ hour^{-1}. Plot the graph of $x_2(t)$ if $x_1(0) = 0.0001$, $x_2(0) = 0$. Show that the amount of tetracycline in the blood reaches a maximum about 2.75 hours after absorption from the GI tract begins (assuming that $x_1(0) > 0$ but that ingestion has ceased; i.e., $I = 0$).

(c) Show that the amount in the bloodstream declines to about 20% of the maximum after about 15 hours.

(d) Use a computer numerical solver to graph the amounts of tetracycline in the GI tract, bloodstream, and urinary system over a 24 hour period, if $I = 1$ unit/hour and the initial amounts in the three compartments are zero.

8. (*The Bones of Olduvai: Potassium-Argon Dating*). Olduvai Gorge, in Kenya, cuts through volcanic flows, tuff (volcanic ash), and sedimentary deposits. It is the site of bones and artifacts of early hominids, considered by some to be precursors of man. In 1959, Mary and Louis Leakey uncovered a fossil hominid skull and primitive stone tools of obviously great age, older by far than any hominid remains found up to that time. Carbon-14 dating methods being inappropriate for a specimen of that age and nature, dating had to be based on the ages of the underlying and overlying volcanic strata. The method used was that of potassium-argon decay. The potassium-argon clock is an accumulation clock, in contrast to the ^{14}C dating method. The potassium-argon method depends on measuring the accumulation of "daughter" argon atoms, which are decay products of radioactive potassium atoms. Specifically, potassium-40 (^{40}K) decays to argon-40 (^{40}Ar) and to calcium-40 (^{40}Ca) by the branching cascade illustrated in the margin. Potassium decays to calcium by emitting a β particle (i.e., an electron). Some of the potassium atoms, however, decay to argon by capturing an extranuclear electron and emitting a γ particle.

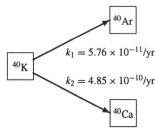

The rate equations for this decay process may be written in terms of the amounts $K(t)$, $A(t)$, and $C(t)$ of potassium, argon, and calcium in a sample of rock:

$$K' = -(k_1 + k_2)K$$

$$A' = k_1 K$$

$$C' = k_2 K$$

(a) Solve the system to find $K(t)$, $A(t)$ and $C(t)$ in terms of k_1, k_2, and $k = k_1 + k_2$. Set $K(0) = K_0$, $A(0) = C(0) = 0$. Why is $K(t) + A(t) + C(t) = K_0$ for all $t \geq 0$? Show that $K(t) \to 0$, $A(t) \to k_1 K_0 / k$, and $C(t) \to k_2 K_0 / k$ as $t \to \infty$.

(b) The age of the volcanic strata is the current value of the time variable t because the potassium-argon clock started when the volcanic material was laid down. This age is estimated by measuring the ratio of argon to potassium in a sample. Show that this ratio is $A/K = (k_1/k)(e^{kt} - 1)$. Then show that the age of the sample (in years) is $(1/k) \ln[(k/k_1)(A/K) + 1]$. When the actual measurements were made at the University of California at Berkeley a few years ago, the age of the volcanic material (and thus the age of the bones) was estimated to be 1.75 million years. What was the value of the measured ratio A/K?

1.7 Change of Variables: Pursuit Models

The ODE $y' = f(t, y)$ is written in terms of the state variable y and the independent variable t. A change of state variable may convert an apparently intractable ODE into another that can be solved by one of the techniques of this chapter. Suppose that we change from y to z by the formula $y = g(t, z)$. Then we must write the ODE in terms of z and t, solve it to obtain $z(t)$, and finally, write the solution in terms of the y variable by setting $y(t) = g(t, z(t))$. We shall illustrate this technique with some examples.

EXAMPLE 1.7.1

Bernoulli's Equation
The equation is

$$\frac{dy}{dt} + p(t)y = q(t)y^b \tag{1}$$

where b is a real number and the functions $p(t)$ and $q(t)$ are continuous on a common interval. The ODE is linear if $b = 0$ or 1, nonlinear otherwise. A change of variables reduces the ODE (1) to a first-order linear ODE which can be solved by an integrating factor technique. To see this, let $b \neq 0$, 1 and divide the ODE (1) through by y^b. Then multiply by $1 - b$ and observe that the ODE can be written as

$$\frac{d}{dt}(y^{1-b}) + (1-b)p(t)y^{1-b} = (1-b)q(t) \tag{2}$$

Now make the change of variables $z = y^{1-b}$ and observe that (2) takes the form

$$\frac{dz}{dt} + (1-b)p(t)z = (1-b)q(t) \tag{3}$$

which is a linear ODE. Thus, if $y(t)$ solves the nonlinear ODE (1), then $z(t) = (y(t))^{1-b}$ solves the linear ODE (3). Conversely, if z is any solution to ODE (3), then $y = z^{1/(1-b)}$ is a solution of ODE (1), or perhaps several solutions after taking the root $1/(1 - b)$. See Problem 11 for more on Bernoulli's equation.

EXAMPLE 1.7.2

Reduction of a Quadratic Rate Function to Linear Form
Let $y'(t) = (a + by)(c(t) + d(t)y)$, where a and b are constants, $b \neq 0$ and $c(t)$ and $d(t)$ are continuous on some t-interval I. The rate function is quadratic in y because

of the term $bd(t)y^2$. We shall introduce a new dependent variable by setting $y = (1/z - a)/b$, or, solving for z, $z(t) = [a + by(t)]^{-1}$. Using the Chain Rule, we have that

$$z' = \frac{-by'}{(a+by)^2} = \frac{-b(c+dy)}{a+by} = -bz\left[c + d\left(\frac{1}{bz} - \frac{a}{b}\right)\right]$$

since $z = [a + by]^{-1}$ and $y = 1/bz - a/b$. Note that $z(t)$ satisfies the linear ODE

$$\frac{dz(t)}{dt} = [ad(t) - bc(t)]z(t) - d(t) \tag{4}$$

The integration factor technique can be used to find a solution $z(t)$ of (4), and then $y(t) = 1/bz(t) - a/b$ is a solution of the original ODE. We must, of course, remain within a t-interval on which $z(t)$ does not vanish.

Sometimes we change variables, not to reduce an ODE to a form where there is a known solution formula, but to reduce the number of coefficients appearing in the ODE. This is usually done by scaling the variables, that is, replacing y and t in the ODE $y' = f(t, y)$ by $z = y/y_1$, $s = t/t_1$, for suitably chosen constants y_1 and t_1. The next example illustrates the process.

EXAMPLE 1.7.3

Scaling the State Variable and Time
The velocity $v(t)$ of a dense body moving against a resistive force along the local vertical is the solution of the IVP

$$v' = -g - \frac{k}{m}v|v|, \qquad v(0) = v_0 \tag{5}$$

(Problem 6**(a)** of Section 1.4). By a clever scaling of the state and time variables in the ODE of (5) we arrive at a transformed IVP in which the constants g, k, and m appear only in the ratio k/mg (see Problem 5). Let us put

$$t = \left(\frac{m}{kg}\right)^{1/2} s, \qquad v = \left(\frac{mg}{k}\right)^{1/2} w \tag{6}$$

and observe from the Chain Rule that

$$\frac{dw}{dt} = \frac{dw}{ds}\frac{ds}{dt} = \left(\frac{kg}{m}\right)^{1/2}\frac{dw}{ds} \tag{7}$$

Thus we see from (6) and (7) that

$$\frac{dv}{dt} = \left(\frac{mg}{k}\right)^{1/2}\frac{dw}{dt} = \left(\frac{mg}{k}\right)^{1/2}\left(\frac{kg}{m}\right)^{1/2}\frac{dw}{ds} = g\frac{dw}{ds} \tag{8}$$

Substitution into IVP (5) yields the new IVP

$$\frac{dw}{ds} = -1 - w|w|, \qquad w(0) = w_0 = \left(\frac{k}{mg}\right)^{1/2} v_0 \tag{9}$$

Some solution curves for the ODE (9) are plotted in Figure 1.7.1 for various values of w_0. Thus the behavior of the solution $v(t)$ of IVP (5) can be read from the graphs in Figure 1.7.1 no matter what the values of the parameters g, k, and m. All the solution curves in Figure 1.7.1 tend to the value $w = -1$ as $s \to \infty$. This means that for any values of v_0, g, k, and m, the solution of IVP (5) tends toward the limiting value $v = -(mg/k)^{1/2}$, a fact noted earlier in Section 1.4. See Problems 5 and 7 for more on scaling problems.

There are various reasons for scaling an ODE before solving it either by hand or by computer. One reason is that sometimes only combinations of parameters can

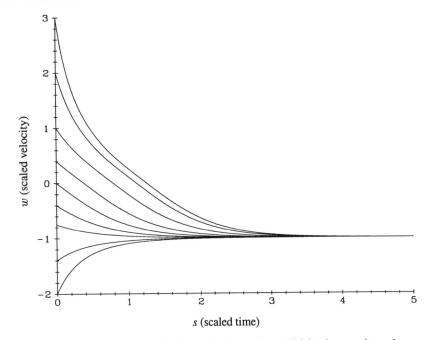

FIGURE 1.7.1 Scaled velocity and time of a solid body moving along local vertical. See Example 1.7.3.

be measured, not the individual parameters (see Example 1.3.3). In Example 1.7.3, rescaling has reduced the set of four original parameters, g, k, m, and v_0 to the single parameter $w(0) = w_0$.

ODEs in Polar Coordinates

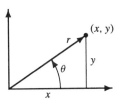

The ODE $dy/dx = f(x, y)$, expressed in terms of the rectangular coordinates x and y, can be transformed into an "equivalent" ODE in polar coordinates r and θ. To show this, first recall that the relation between polar and rectangular coordinates is

$$x = r\cos\theta, \qquad y = r\sin\theta$$

Now suppose that $y = g(x)$ describes a solution curve of the ODE $dy/dx = f(x, y)$. This solution curve can also be described (perhaps only in part) by specifying r as a function of the polar angle θ, i.e., $r = r(\theta)$. To find the ODE which $r(\theta)$ satisfies, we proceed as follows: In polar coordinates the relation $y = g(x)$ becomes

$$r\sin\theta = g(r\cos\theta) \tag{10}$$

which implicitly defines r as function of θ. Differentiating each side of (10) with respect to θ and using the Chain Rule, we have

$$\frac{dr}{d\theta}\sin\theta + r\cos\theta = g'(r\cos\theta)\left(\frac{dr}{d\theta}\cos\theta - r\sin\theta\right)$$

Now since $y = g(x)$ solves $y' = f(x, y)$, we see that $g'(r\cos\theta) = f(r\cos\theta, r\sin\theta)$, and so

$$\frac{dr}{d\theta}\sin\theta + r\cos\theta = f(r\cos\theta, r\sin\theta)\left(\frac{dr}{d\theta}\cos\theta - r\sin\theta\right) \tag{11}$$

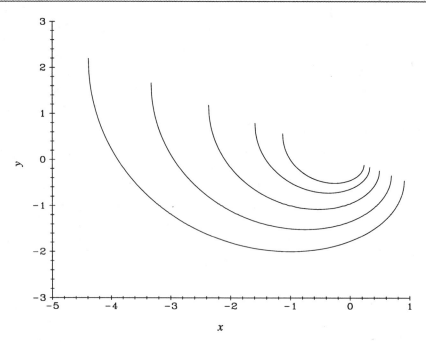

FIGURE 1.7.2 Solution curves for the ODE in Example 1.7.4

The ODE (11) is a first-order ODE in r and θ and is equivalent to $y' = f(x, y)$. ODE (11) seems more complicated than $y' = f(x, y)$, but it is actually simpler for certain functions $f(x, y)$.

EXAMPLE 1.7.4

Changing ODEs to Polar Form
Applying (11) to the ODE

$$\frac{dy}{dx} = \frac{y - 2x}{2y + x} \tag{12}$$

we have

$$\frac{dr}{d\theta} \sin\theta + r\cos\theta = \frac{r\sin\theta - 2r\cos\theta}{2r\sin\theta + r\cos\theta} \left(\frac{dr}{d\theta} \cos\theta - r\sin\theta \right) \tag{13}$$

After canceling the common factor r in the quotient, multiplying each side by $2\sin\theta + \cos\theta$, simplifying, and again canceling a common factor, ODE (13) reduces to

$$\frac{dr}{d\theta} = -\frac{r}{2} \tag{14}$$

The solutions of (14) are given by the decaying spirals $r = Ce^{-\theta/2}$, where C is any nonnegative constant. The solution curves of (12) are portions of those spirals that have no vertical tangent line because only such arcs of the spirals can be graphs of functions of x. Figure 1.7.2 shows a plot of some of these solution curves (the "top" arcs of the spiral are also solution curves but are not shown). Finally note that taking logarithms of each side of the equation $r = Ce^{-\theta/2}$ and returning to the xy-variables, we have that the solutions $y(x)$ of (12) satisfy the equation

$$\ln(x^2 + y^2) = \ln C - \frac{1}{2}\arctan(y/x) \tag{15}$$

Homogeneous Rate Functions

When the rate function $f(x, y)$ in the first-order ODE $dy/dx = f(x, y)$ is a homogeneous function of order zero (defined below), then there exists a simple change of the state variable which transforms the given ODE into a separable one.

❖ **Homogeneous Function of Order Zero.** A function $f(x, y)$ which is continuous on the xy-plane (except perhaps at the origin) is said to be homogeneous of order zero if for any $(x, y) \neq (0, 0)$ and any $k > 0$, we have $f(kx, ky) = f(x, y)$.

EXAMPLE 1.7.5

The function $f(x, y) = (x^2 + y^2)^{1/2}/(x + y)$ is homogeneous of order zero, because $f(kx, ky) = ((kx)^2 + (ky)^2)^{1/2}/(kx + ky) = f(x, y)$. On the other hand, $g(x, y) = (x + 1)/(x + y)$ does not have this property because $g(kx, ky) = (kx + 1)/(kx + ky) \neq g(x, y)$ if $k \neq 1$.

Now consider the ODE $y' = f(x, y)$, where $f(x, y)$ is homogeneous of order zero. Then the change of state variable $y = xz$ converts the given ODE to

$$xz' + z = y' = f(x, xz) = f(x \cdot 1, x \cdot z) = f(1, z)$$

since f is homogeneous of order zero (x is playing the role of k in this setting). Rearranging this differential equation, we have the separable equation

$$\frac{1}{f(1, z) - z} z' = \frac{1}{x} \tag{16}$$

Equation (16) can be solved by integrating each side:

$$\int^z \frac{1}{f(1, s) - s} ds = \ln|x| + C$$

where C is a constant of integration. Let $F(z)$ be an antiderivative of $[f(1, z) - z]^{-1}$. Then we have that $F(z) = \ln|x| + C$, and since $z = y/x$,

$$F\left(\frac{y}{x}\right) = \ln|x| + C \tag{17}$$

Thus a solution $y(x)$ of the first-order equation $y' = f(x, y)$ satisfies (17) on some x-interval for some choice of C.

This leaves open two questions: How can the antiderivative F be found, and how can (17) be solved for y in terms of x? For some equations F can be expressed in terms of elementary functions, but for others it cannot. When F is known, sometimes (17) can be easily solved for y in terms of x, but often this is not possible. The situation here resembles that encountered in Section 1.4 for separable equations: there are implicit formulas for solutions, but it may be hard to solve for y as an explicit function of x.

EXAMPLE 1.7.6

A Homogeneous Rate Function of Order Zero
The function $f(x, y) = (x^2 + y^2)/2xy$ is homogeneous of order zero. Hence, to solve

$$y' = \frac{x^2 + y^2}{2xy}, \qquad x \neq 0, \quad y \neq 0 \tag{18}$$

we introduce the new variable z by setting $y = xz$ to obtain the equation

$$xz' + z = \frac{x^2 + x^2 z^2}{2zx^2} = \frac{1 + z^2}{2z}$$

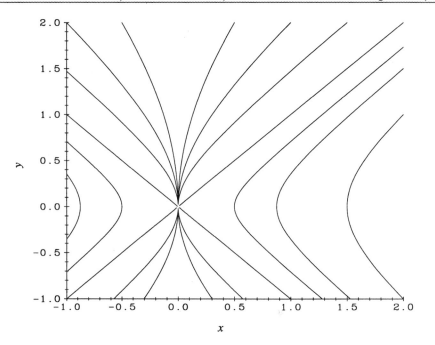

FIGURE 1.7.3 Integral curves for the ODE in Example 1.7.6.

Separating the variables, we have

$$\frac{2z}{1-z^2}z' = \frac{1}{x}, \qquad z \neq \pm 1$$

Antidifferentiating, we have

$$-\ln|1 - z^2| = \ln|x| + C, \qquad z \neq \pm 1$$

where C is the constant of integration. Rearranging and returning to the original variables y and x by setting $z = y/x$, we have

$$\ln|x| + \ln\left|1 - \frac{y^2}{x^2}\right| = \ln\left|x\left(1 - \frac{y^2}{x^2}\right)\right| = -C, \qquad y \neq \pm x$$

Exponentiating, we have

$$\left|x\left(1 - \frac{y^2}{x^2}\right)\right| = e^{-C}, \qquad y \neq \pm x \tag{19}$$

By direct substitution into (18) we see that $y = \pm x$ are also solutions of (18) on any interval not containing $x = 0$. Taking this fact into account, we can remove the absolute value signs from (19), replace e^{-C} by a constant K, which can have any real value, and obtain, after multiplying by x, the following equation defining solutions of (18):

$$x^2 - y^2 = Kx, \qquad x \neq 0, \quad y \neq 0 \tag{20}$$

Figure 1.7.3 shows some "integral curves" of the ODE (18) defined by (20); for each value of K, the curve is a hyperbola. Solution curves $y = y(x)$ of ODE (18) are the arcs of the hyperbolas above or below the x-axis (when $K \neq 0$), and the lines $y = x$, $y = -x$ ($K = 0$).

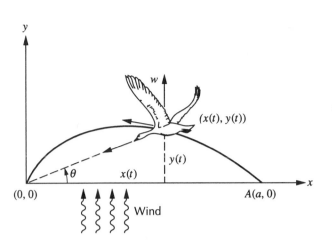

FIGURE 1.7.4 Path of the goose. See Example 1.7.7.

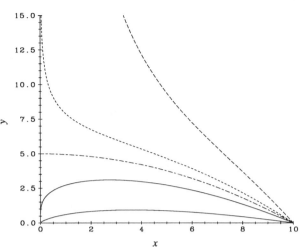

FIGURE 1.7.5 Flight path of a goose from the point $(10, 0)$ for selected values of $v = w/b$. From top to bottom, $v = 2$, 1.2, 1.0, 0.75, 0.25. See Example 1.7.8.

Flight Paths

In models of pursuit, one object pursues another (whose motion is known) by using a predetermined strategy, e.g., by deliberately aiming toward it at all times. For example, a ferryboat is set to sail directly across the river to its dock, but a strong current complicates the captain's decision making. Finally, the captain decides on the strategy of aiming the ferry toward the dock at all times. Will the ferry make it to the dock? The problem of finding the path of pursuit, where the pursued object is at rest, eventually comes down to solving a first-order ODE $dy/dx = f(x, y)$, where $f(x, y)$ is a homogeneous function of order zero.

EXAMPLE 1.7.7

A Goose Flies to its Roost: The Mathematical Model

A goose attempts to fly back to its roost, which is a units directly west of its position, but a constant crosswind is blowing. The goose keeps its heading to its roost, but the wind blows it off a direct course. What course is followed? Suppose that the wind is out of the south at constant speed w, while the goose's speed in still air is the constant b. Can the goose ever get home? If it does, what is its path? See Figure 1.7.4

After a little reflection, we realize that the goose does not have a chance of making it home unless its speed b exceeds the speed w of the wind. Hence we assume that $b > w$. Suppose that the path (as yet unknown) of the goose is given by the parametric equations $x = x(t)$, $y = y(t)$, where t is time, $(x(t), y(t))$ is the location of the goose at time t, $(x(0), y(0)) = A(a, 0)$, and $(x(T), y(T)) = (0, 0)$. Time T is the time of arrival at the roost, but, of course, it is not known. Let θ denote the angle of the heading of the goose as indicated in Figure 1.7.4; θ will change in time as the bird's position changes.

The rate of change of $x(t)$ is the component of the goose's velocity in the x direction,

$$\frac{dx(t)}{dt} = -b\cos\theta = \frac{-bx}{(x^2 + y^2)^{1/2}} \tag{21}$$

The rate of change of $y(t)$ is obtained similarly, except that the wind's effect must be included:

$$\frac{dy(t)}{dt} = -b\sin\theta + w = \frac{-by}{(x^2 + y^2)^{1/2}} + w \tag{22}$$

The differential system (21), (22) can be treated as we did the system in Example 1.4.8 because the rate functions do not contain time explicitly. Dividing ODE (22) by ODE (21) we have a first-order equation in x and y, where x is now the independent variable:

$$\frac{dy}{dx} = \frac{by - w(x^2 + y^2)^{1/2}}{bx} = f(x, y) \tag{23}$$

This first-order ODE has a rate function which is homogeneous of order zero, since

$$f(kx, ky) = \frac{bky - w(k^2x^2 + k^2y^2)^{1/2}}{bkx} = \frac{by - w(x^2 + y^2)^{1/2}}{bx} = f(x, y)$$

EXAMPLE 1.7.8

The Path of the Goose

Now we set $y = xz$ and obtain from ODE (23)

$$\frac{dy}{dx} = \frac{dz}{dx}x + z = \frac{zx - v(x^2 + x^2z^2)^{1/2}}{x} = z - v(1 + z^2)^{1/2}$$

where $v = w/b$. We have the separable ODE $(dz/dx)x = -v(1 + z^2)^{1/2}$ with solutions defined by $\ln[z + (1 + z^2)^{1/2}] = -v\ln x + C$, where C is a constant of integration. We do not need absolute value signs inside the logarithms since $x > 0$ and $z > 0$. Note that $C = v\ln a$ since $z = y = 0$ when $x = a$. Hence $\ln[z + (1 + z^2)^{1/2}] = -v\ln x + v\ln a = \ln(x/a)^{-v}$; exponentiating, we obtain

$$z + (1 + z^2)^{1/2} = \left(\frac{x}{a}\right)^{-v}$$

Solving for z involves writing the equation above as $(1 + z^2)^{1/2} = (x/a)^{-v} - z$, squaring, and then solving for z to obtain

$$z = \frac{1}{2}\left[\left(\frac{x}{a}\right)^{-v} - \left(\frac{x}{a}\right)^{v}\right]$$

Since $z = y/x$, we have that the equation of the path followed by the goose is

$$y = \frac{a}{2}\left[\left(\frac{x}{a}\right)^{1-v} - \left(\frac{x}{a}\right)^{1+v}\right]$$

In Figure 1.7.5 this path is plotted for each of several values of $v = w/b$. When $v > 1$, the goose is gone with the wind, but if $v < 1$, the bird will eventually reach its roost.

PROBLEMS

1. (*Homogeneous Rate Functions of Order Zero*). For each ODE find a formula that defines solution curves implicitly. [*Hint*: Change variables from x and y to x and z by $y = xz$, and solve as in Example 1.7.6.]

    (a) $(y + x)/x = y'$ (b) $(x - y)dx + (x - 4y)dy = 0$

    (c) $(x^2 - xy - y^2)dx - xy\,dy = 0$ (d) $(x^2 - 2y^2)dx + xy\,dy = 0$

    (e) $x^2y' = 4x^2 + 7xy + 2y^2$

2. **(a)–(e)** Use a computer to plot representative integral curves for parts **(a)–(e)** of Problem 1. Highlight arcs of the integral curves which are solution curves.

3. Nonseparable differential equations may become separable by changing the dependent variable. For each of the following cases, demonstrate this process and solve the resultant equation. Then find the solution of the original equation.

   (a) $dy/dx = \cos(x + y)$ [*Hint*: Let $z = x + y$.]

   (b) $(2x + y + 1)dx + (4x + 2y + 3)dy = 0$ [*Hint*: Let $z = 2x + y$.]

   (c) $(x + 2y - 1)dx + 3(x + 2y)dy = 0$

   (d) $e^{-y}(y' + 1) = xe^x$ [*Hint*: Let $z = x + y$.]

4. (*Logistic Equation*). Using the method of Example 1.7.2, show the solution of

   $$\frac{dP}{dt} = r\left(1 - \frac{P}{K}\right)P, \qquad P(0) = P_0 > 0$$

   where r and K are positive constants, takes the form (7) in Section 1.5.

5. (*Scaling Variables in Example 1.7.3*). The coefficients in equation (6) seem to have been picked out of thin air. In this problem, you are asked to show that these are natural choices.

   (a) Let $t = as$, $v = bw$, where a and b are positive scaling constants to be determined. Rewrite IVP (5) in terms of s, w, a, b, g, k, m.

   (b) Show that if we set the coefficients of w and of $w|w|$ equal to 1 in the ODE for w found in part **(a)**, then the values for a and b given in equation (6) are obtained.

6. Use the substitution $y = z^{1/2}$ to solve the initial value problem

   $$yy'' + (y')^2 = 1, \qquad y(0) = 1, \quad y'(0) = 0 \tag{24}$$

7. (*Falling Snowflake: Scaling*). As observed in Section 1.3, the velocity of a falling feather or snowflake is given by the ODE $v' = -g - (k/m)v$, where g, k, and m are positive constants. Rescale the state and time variables so that the scaled ODE is free of the constants g, k, and m. Solve the scaled ODE and draw a conclusion about the limiting velocity for the original ODE. [*Hint*: Let $t = aT$, $v = bV$; see Example 1.7.3.]

8. (*Polar Coordinates*). Let $f(x, y)$ be homogeneous of order zero. Use ODE (11) to introduce polar coordinates $x = r\cos\theta$ and $y = r\sin\theta$, and show that the differential equation $y' = f(x, y)$ in polar form is separable. Use this technique to solve $y' = (-x^2 + y^2)/xy$. [*Hint*: See Example 1.7.4.]

9. (*Polar Coordinates*). Convert each ODE below to polar form $dr/d\theta = h(r, \theta)$ and solve that ODE for $r = r(\theta)$. Use this information to plot some solution curves for the original ODE. Describe the behavior of these curves.

   (a) $\dfrac{dy}{dx} = \dfrac{x + 2y}{2x - y}$ 　　　　　 **(b)** $\dfrac{dy}{dx} = \dfrac{x + y - y(x^2 + y^2)}{x - y - x(x^2 + y^2)}$

   (c) $\dfrac{dy}{dx} = \dfrac{-x + 5y - y(x^2 + y^2)}{5x + y - x(x^2 + y^2)}$

10. (*Polar Coordinates*).

   (a) Let α, β be nonzero constants. Show that the ODE

   $$\frac{dy}{dx} = \frac{-\beta x + \alpha y}{\alpha x + \beta y}$$

   when converted to polar form is $dr/d\theta = -\alpha r/\beta$.

   (b) Solve the polar equation for r in terms of θ if $r(0) = 1$.

   (c) Choosing $\alpha = 1$, $\beta = 2$, plot the corresponding spiral in the xy-plane.

11. (*Bernoulli's Equation*). Bernoulli's equation is $dy/dt + p(t)y = q(t)y^b$, where b is a real number. See Example 1.7.1.

(**a**) Show that the logistic equation is a Bernoulli equation with $b = 2$.

(**b**) Find a formula for all solutions of $dy/dt + t^{-1}y = y^{-4}$, $t > 0$. Plot some solutions. [*Hint*: Set $z = y^{1-(-4)} = y^5$ and solve $dz/dt + 5t^{-1}z = 5$.]

(**c**) Find a formula for all solutions of $dy/dt - t^{-1}y = -y^{-1}/2$, $t > 0$. Plot some solutions.

(**d**) Find a formula (in terms of an integral) for all solutions of $t(dy/dt) + y = y^2 \ln t$, $t > 0$. Plot some solutions.

12. (*Riccati's Equation*). Riccati's equation is $dy/dt = a(t)y + b(t)y^2 + F(t)$. If $F(t) = 0$, Riccati's equation is a special case of Bernoulli's equation (Problem 11). If $F(t) \neq 0$, Riccati's equation may be reduced to a first-order linear ODE if one solution is known and a new variable is introduced as in (**a**) below. Parts (**b**)–(**g**) contain examples and additional properties for the Riccati equation.

(**a**) Let $y = g(t)$ be a known solution of Riccati's equation (i.e., $dg/dt = ag + bg^2 + F$). Let a new variable z be defined by $z = [y - g]^{-1}$. Show that $dz/dt + (a + 2bg)z = -b$ which is a first-order linear equation in z. Suppose that $z(t)$ has been found. How can $y(t)$ be determined?

(**b**) Show that the Riccati equation $dy/dt = (1 - 2t)y + ty^2 + t - 1$ has a particular solution $y = g(t) = 1$. Let $z = (y - 1)^{-1}$ and show that $dz/dt = -z - t$. Solve for all functions $z(t)$ and then find the general solution $y(t)$ of the Riccati equation.

(**c**) Find all solutions of $dy/dt = e^{-t}y^2 + y - e^t$. [*Hint*: First show that $y = e^t$ is a solution, and then use the techniques of (**a**).]

(**d**) Show that $y = t$ is one solution of $dy/dt = t^3(y - t)^2 + yt^{-1}$, $t > 0$, and then find all solutions of the ODE.

(**e**) Show that the ODE of Problem 7 of Section 1.5 is a Riccati equation. Solve the ODE if $K = 1600$, $r = 0.01$, $H = 3$. [*Hint*: Show that $P(t) = 1200$ is one solution of the ODE.]

(**f**) Suppose that $g_1(t)$ and $g_2(t)$ are two different solutions of the Riccati equation $y' = a(t)y + b(t)y^2 + F(t)$. Show that every other solution y, $y \neq g_1$, $y \neq g_2$ satisfies the equation

$$\frac{y - g_1}{y - g_2} = c \exp\left[\int^t b(g_1 - g_2)ds\right]$$

where c is a constant. [*Hint*: Calculate $(\ln[(y - g_1)/(y - g_2)])'$.]

(**g**) Use the technique of part (**f**) to solve $dy/dt = (t - 1)^{-1}y^2 - t(t - 1)^{-1}y + 1$, $t > 1$. [*Hint*: Show that $y = 1$ and $y = t$ are solutions.]

13. (*Flight Path of a Goose*).

(**a**) Set up the flight path problem for the goose, flying at speed b, if the wind is blowing from the southeast at a speed of $w = b/\sqrt{2}$ and the goose starts at $x = a > 0$, $y = 0$. Solve the IVP in implicit form (do not attempt to solve for y in terms of x). [*Hint*: See Example 1.7.7. Note that $x' = -b\cos\theta - b/2$, $y' = -b\sin\theta + b/2$. Use a table of integrals.]

(**b**) Use $b = 1$, $a = 1$, 2, 3, $\ldots$, 9 and plot the paths. [*Hint*: Use a numerical ODE solver on the planar system, $x' = \ldots$, $y' = \ldots$ that models the flight of the goose.] Does the goose reach the roost (the origin)? Does it overshoot?

14. (*Flight Path of a Goose*). In Example 1.7.7 it was shown that the path in the xy-plane followed by the goose in seeking its roost is the solution of the backwards

IVP

$$\frac{dy}{dx} = \frac{by - w(x^2 + y^2)^{1/2}}{bx}, \qquad y(a) = 0$$

where b is the goose's speed and w is the speed of the wind. Solve this IVP by first converting it to polar coordinates (see formula (11)). Setting $v = w/b$, verify the flight paths in Figure 1.7.5 for various values of v. This is a backwards IVP because it is solved over $0 \le x \le a$, where $y(a) = 0$ is the initial data.

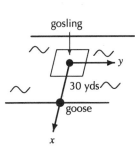

15. (*The Goose and a Moving Roost*). On a windless day a goose spies its gosling aboard a raft in the middle of a river which is moving at 8 yds/sec. When the raft is directly opposite the goose, the raft is 30 yards distant, and the goose instantly takes flight to save her gosling from going over a waterfall 60 yards further downstream. If the goose always flies directly toward the raft at the constant speed of 10 yds/sec, does she rescue her gosling before it tumbles over the falls? Follow the outline below to set up the model for this problem. At $t = 0$ place the raft and the goose in the xy-plane at the origin and at $(30, 0)$, respectively. Let the river flow in the positive y-direction. The parametric path $(x(t), y(t))$ followed by the goose in the xy-plane has the following properties: The goose's velocity vector at time t, $(x'(t), y'(t))$, always points toward the raft and $((x')^2 + (y')^2)^{1/2} = 10$ yds/sec at all times. Thus, if the goose is at the point (x, y) at time t, then the raft is at $(0, 8t)$ and there is a factor $k > 0$ (which may depend on x, y, and t) such that

$$x' = k(-x), \qquad y' = k(8t - y)$$

Now since $(x')^2 + (y')^2 = 100$ we find that $k = 10/(x^2 + (8t - y)^2)^{1/2}$, and hence we have the IVP

$$x' = \frac{-10x}{(x^2 + (8t - y)^2)^{1/2}}, \quad x(0) = 30$$

$$y' = \frac{10(8t - y)}{(x^2 + (8t - y)^2)^{1/2}}, \quad y(0) = 0$$

Observe that since the rate functions in this differential system explicitly depend on t we cannot directly apply the technique used in Example 1.7.7 to reduce the system to a first-order ODE of the form $dy/dx = f(x, y)$. Since the system is written in normal form, an ODE computer solver package can be used to plot an orbit of this system to determine if the goose rescues the gosling. If the goose does reach the gosling in time, how long does it take?

1.8   Reduction Methods: Escape Velocity

Some second-order differential equations can be solved by reducing them to first-order equations by a suitable choice of a new variable. The second-order ODE

$$y'' = F(t, y') \tag{1}$$

in which the dependent variable appears only in derivative form becomes a first-order ODE in $v = y'$

$$v' = F(t, v)$$

We may be able to solve this first-order ODE by using one of the solution techniques already introduced. If the initial conditions $y(t_0) = y_0$ and $v(t_0) = v_0$ are imposed, the corresponding IVP for (1) reduces to an initial value problem for the first-order system

$$v' = F(t, v), \quad v(t_0) = v_0$$
$$y' = v, \qquad y(t_0) = y_0$$

The special cases of falling bodies worked out in Sections 1.3 and 1.4 illustrate this reduction of order technique. Here is another example.

EXAMPLE 1.8.1

Solving $y'' = F(t, y')$
The ODE $y'' = y' - t$ can be solved by setting $v = y'$ and rewriting the ODE as $v' - v = -t$, a linear first-order ODE in v. Using the method of integrating factors, all solutions of this ODE are given by $v = C_1 e^t + 1 + t$, where C_1 is an arbitrary constant. Recalling that $y' = v$, an antidifferentiation yields the solutions $y = C_1 e^t + C_2 + t + t^2/2$ where C_1 and C_2 are arbitrary constants.

Another method applies when the forces acting on a body depend on its position and velocity, but not on time. The dependent and independent variables in the second-order initial value problem

$$y'' = F(y, y'), \qquad y(t_0) = y_0, \quad y'(t_0) = v_0 \tag{2}$$

are y and t. Since t does not appear explicitly in the differential equation, another independent variable might be more suitable. In fact, we shall introduce y itself as the new independent variable and $v = y'$ as a new dependent variable. Using the Chain Rule to rewrite y'' in the new variables, we have

$$y'' = \frac{d^2 y}{dt^2} = \frac{dv}{dt} = \frac{dv}{dy}\frac{dy}{dt} = \frac{dv}{dy} v$$

and IVP (2) is transformed to an initial value problem for the pair of first-order ODEs

$$v\frac{dv}{dy} = F(y, v), \qquad v = v_0 \quad \text{when} \quad y = y_0 \tag{3a}$$

$$\frac{dy}{dt} = v(y), \qquad y = y_0 \quad \text{when} \quad t = t_0 \tag{3b}$$

Notice that ODE (3a) decouples from ODE (3b) and hence can be solved for $v(y)$ separately as a first-order ODE. Once the solution $v(y)$ of ODE (3a) is known, ODE (3b) can be solved for $y(t)$ by separating variables and integrating.

EXAMPLE 1.8.2

Solving $y'' = F(y, y')$
The IVP

$$y'' = \frac{1}{y}(y')^2 - \frac{y'}{y}, \qquad y = 1 \quad \text{and} \quad y' = 2 \quad \text{when} \quad t = 0$$

can be solved for $y(t)$ by the method given above. We shall assume that $y > 0$. Introducing y and $v = y'$ as variables, we solve the first-order problem

$$v\frac{dv}{dy} = \frac{1}{y}v^2 - \frac{v}{y}, \qquad v = 2 \quad \text{when} \quad y = 1$$

The solution $v = 0$ is not of interest and we cancel the factor v to obtain the linear problem in v

$$\frac{dv}{dy} = \frac{1}{y}v - \frac{1}{y}, \qquad v = 2 \quad \text{when} \quad y = 1$$

The integrating factor technique of Section 1.2 leads to $v = 1 + y$ as the solution to this problem (y must be positive). The second IVP to be solved is

$$\frac{dy}{dt} = v(y) = 1 + y, \qquad y = 1 \quad \text{when} \quad t = 0$$

This linear initial value problem has the solution

$$y(t) = 2e^t - 1$$

where we must restrict t to be greater than $-\ln 2$ so that y remains positive. Curiously, $y(t) = 2e^t - 1$ solves the ODE $yy'' = (y')^2 - y'$ for all t on the real line, a reminder again of the difference between normalized and nonnormalized ODEs.

Example 1.8.2 illustrates the technique, but it is deceptive. Solution formulas for the two first-order ODEs (3a) and (3b) may not always be found quite so easily.

Gravitation

Using the extensive astronomical work and empirical laws of Tycho Brahe (1546–1601) and Johannes Kepler (1571-1603) concerning the orbits of the moon and of the planets, Newton focused attention on just one force, gravity. His law of universal gravitation deals with the gravitational effect of one body upon another:

Newton's Law of Universal Gravitation. The force **F** between any two particles having masses m_1 and m_2 and separated by a distance r is attractive, acts along the line joining the particles, and has magnitude

$$|\mathbf{F}| = G\frac{m_1 m_2}{r^2} \tag{4}$$

where G is a universal constant independent of the nature of the masses of the particles. This is known as the *Inverse Square Law of Gravitation*. In SI units, $G = 6.67 \times 10^{-11}$ Nm^2/kg^2.

Newton also showed that massive bodies affect one another as if the mass of each body were concentrated at its "center of mass," provided that the mass of each body is distributed in a spherically symmetric way. In this case r is the distance between the centers of mass.

A small object (e.g., a marble) is dropped and falls toward the surface of the earth. How do its velocity and location change with time? Suppose that the only significant force acting on the body is the gravitational attraction of the earth. We shall fix a z-coordinate line on the surface of the earth directly below the body and pointing upward. A positive value of z indicates a location above the surface of the earth. A negative value of $v = dz/dt$ means that the body is moving downward, that is, in the direction of the negative z-axis. A negative value of a force acting on the body means that the force acts in the direction of the negative z-axis, and so on. We shall suppose that at time $t = 0$ the body is at height $z(0) = h$ and has velocity v_0. If the body is falling, then $v_0 < 0$. In calculating the gravitational attraction of the earth on this body, we imagine that the mass of the earth is concentrated at its center. This means that the earth attracts the body with a force whose magnitude is

$$|\mathbf{F}| = \frac{GMm}{(R + z)^2}$$

where M is the mass of the earth, m the mass of the falling body, R the radius of the earth, and z the distance from the earth's surface to the center of mass of the falling body. During the course of the fall the factor $GM/(R + z)^2$ changes from $GM/(R + h)^2$ to GM/R^2. For motion near the surface of the earth with, say, h no more than $(0.001)R$, the relative change in this gravitational term is

$$\frac{GM/R^2 - GM/(R+h)^2}{GM/(R+h)^2} = \frac{(R+h)^2}{R^2} - 1 = \left(1 + \frac{h}{R}\right)^2 - 1$$

$$= \frac{2h}{R} + \left(\frac{h}{R}\right)^2 \leq 0.002 + 0.000001 = 0.002001$$

We shall ignore this small variation and set the gravitational coefficient equal to the constant $g = GM/R^2$. The gravitational force acting on the body is then $-mg$, where $g \approx 9.8$ m/sec^2 in SI units.

Escape Velocities

Hydrogen is one of the more abundant elements on the surface of the earth, yet there are today only minute traces of free hydrogen in the atmosphere. In itself that is not surprising since hydrogen gas is lighter than air and would tend to rise. What is surprising is that there is almost no hydrogen even in the topmost layers of the atmosphere. There must have been an abundance of atmospheric hydrogen in the early millennia of the earth's evolution. What happened?

The disappearance of the gas can be modeled in the following way. Let us suppose that a particle of mass m is in motion above the surface of the earth. Since we shall only be interested in the vertical components of the movements of the particle and the forces acting in a vertical direction, we shall employ a z-axis pointing upward from the earth's surface. According to Newton's Second Law and his Universal Law of Gravitation, ignoring all forces (frictional, magnetic, or other) except for gravity, and assuming no collisions, the vertical motion of a particle initially at the surface of the earth and moving upward is modeled by the second-order IVP

$$mz'' = \frac{-mMG}{(z + R)^2}, \qquad z(0) = 0, \quad z'(0) = v_0 \tag{5}$$

where M is the mass of the earth, G the universal gravitational constant, R the radius of the earth, and v_0 is positive.

Since the differential equation in IVP (5) does not involve t explicitly, we may use the second method of reduction of order. If we set $z'' = v\, dv/dz$, we have that

$$v\frac{dv}{dz} = -\frac{MG}{(z + R)^2}, \quad v = v_0, \quad \text{when } z = 0 \tag{6}$$

When the variables are separated and (6) is solved, we have that

$$v^2 = \left(v_0^2 - \frac{2MG}{R}\right) + \frac{2MG}{z + R} \tag{7}$$

From formula (7) we see that v^2 remains positive for all $z \geq 0$ as long as $v_0^2 \geq 2MG/R$. The velocity

$$v = (2MG/R)^{1/2}$$

is called the *escape velocity* of the particle; it is independent of the particle's mass m. The escape velocity from the earth's surface turns out to be roughly 11.179 km/sec.

TABLE 1.8.1 Escape Velocities from Bodies in the Solar System (Earth Units).

Body	Radius	Mass	Escape Velocity
Earth	1	1	1
Jupiter	11.21	317.84	5.321
Saturn	9.449	95.162	3.170
Neptune	3.882	17.205	2.0804
Uranus	4.007	14.500	1.9018
Venus	0.949	0.8150	0.9286
Mars	0.5326	0.1074	0.4482
Mercury	0.3825	0.0552	0.3795
Moon (of Earth)	0.2725	0.01230	0.2116
Pluto	0.180	0.0025	0.0882

If the particle has a positive initial velocity in excess of the escape velocity, its velocity never vanishes and the particle's upward motion never ceases. In addition, the particle moves infinitely far away from the earth as time increases. To see this, suppose that v_1 denotes the positive constant $(v_0^2 - 2MG/R)^{1/2}$. Then from (7) we have that

$$z'(t) = v(t) = \left[v_1^2 + \frac{2MG}{z(t) + R} \right]^{1/2} \geq v_1 \tag{8}$$

Hence upon integrating the differential inequality, $z'(t) \geq v_1$ from 0 to t, we have that $z(t) \geq v_1 t$ and $z(t) \to \infty$ as t increases. The particle escapes completely from the earth and moves into space.

A hydrogen molecule H_2 in the upper layers of the atmosphere at some point in the earth's history would have been subjected to thermal and other excitations. If an excitation transmits enough energy to the particle to raise its component of velocity in the upward direction above the escape velocity, the particle escapes. That is exactly what has happened over the millennia to free molecular hydrogen.

A lightweight molecule such as H_2 is particularly susceptible to this process. In contrast, the nitrogen molecule N_2, the major component of air and 14 times as heavy as H_2, is hardly affected. The kinetic energy of a particle of mass m and velocity of magnitude v are related by the equation $E = \frac{1}{2}mv^2$. A given amount of energy E_0 when entirely converted into kinetic energy produces the velocity $v_0 = (2E_0/m)^{1/2}$. If velocity v_0 corresponds to mass m and energy E_0, then the same amount of energy but a mass of $14m$ corresponds to a velocity of $0.27v_0$ since $(14)^{-1/2} \approx 0.27$. That is why the heavier elements such as nitrogen and oxygen move more slowly in our atmosphere and tend to stay, whereas the lighter elements with the same amount of energy have a higher velocity and escape.

The escape velocity $(2MG/R)^{1/2}$ from the surface of a body depends only on the body's mass and radius. Escape velocities for all the larger satellites in the solar system have been calculated. Some of these are listed in Table 1.8.1 together with the corresponding radii and masses, all expressed in "earth" units.[15]

---

[15]Based on data given in *Handbook of Chemistry and Physics*, 75th ed. (D.R. Lide, ed. CRC Press, Boca Raton, Fla., 1994).

Our moon has no atmosphere; the lunar escape velocity is low and whatever gases once were there have long since escaped. Titan, a moon of Saturn, does have a measurable atmosphere (probably methane gas, CH_4, eight times heavier than H_2) which is visible in the computer images produced by the *Voyager* spacecraft. Mercury has no atmosphere at all, even though it has a higher escape velocity than Titan. Mercury is so close to the sun that the solar wind (streams of protons from the sun) long ago imparted high energies to the particles in Mercury's atmosphere and all escaped. Pluto presumably has an atmosphere, but little is known about that distant planet. Mars has a thin atmosphere of carbon dioxide. All of the more massive planets have atmospheres (mostly of methane). The asteroids and other moons of the solar system have escape velocities much too low to support atmospheres.

Comments

We have interpreted escape velocities in terms of small particles, but we could also apply the results to projectiles or rockets. However, for the latter we should take into account the declining mass of the rocket as the fuel burns. For both projectiles and rockets friction in the form of air resistance plays a significant role. What this means is that the ODE in (5) should be modified to take these factors into account before escape velocities are calculated.

PROBLEMS

1. Solve the following differential equations by reduction of order. Plot solution curves $y = y(t)$ in the ty-plane using various sets of initial data. [*Hint*: If y does not appear explicitly, set $y' = v$, $y'' = v'$. If t does not appear explicitly, set $y' = v$ and $y'' = v\,dv/dy$.]

   (a) $ty'' - y' = 3t^2$ **(b)** $y'' - y = 0$

   (c) $yy'' + (y')^2 = 1$ **(d)** $y'' + 2ty' = 2t$

2. Solve the given initial value problems.

   (a) $2yy'' + (y')^2 = 0$, $y(0) = 1$, $y'(0) = -1$

   (b) $y'' = y'(1 + 4/y^2)$, $y(0) = 4$, $y'(0) = 3$

   (c) $y'' = -g - y'$, $y(0) = h$, $y'(0) = 0$, g, h are positive constants

3. (*Reduction of Order via Varied Parameters*). Let the coefficients $a(t)$, $b(t)$ be continuous on a common interval I, and suppose that $z(t)$ is a known solution of the linear second-order ODE

$$z'' + a(t)z' + b(t)z = 0$$

   Further, suppose that $z(t) \neq 0$ on I and that $f(t)$ is a given continuous function on I. Then show that $y(t) = u(t)z(t)$ is a solution on I of the ODE

$$y'' + a(t)y' + b(t)y = f(t)$$

   if $u(t)$ solves the first-order linear ODE in u'

$$zu'' + (2z' + az)u' = f$$

   which may be solved for $u'(t)$ by normalizing and using the integrating factor $z^2 e^{A(t)}$, where $A(t)$ is an antiderivative of $a(t)$.

4. Using the method described in Problem 3, find a solution of the ODE

$$t^2 y'' + 4ty' + 2y = \sin t, \quad t > 0$$

where it is known that $z(t) = t^{-2}$ solves the ODE $t^2 y'' + 4ty' + 2y = 0$. [*Hint*: First normalize the ODE.]

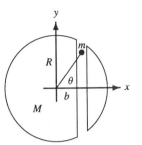

5. A straight tunnel is bored through the earth as pictured. An object of mass m is dropped into the tunnel from the earth's surface.

(a) Neglecting friction, write the equation of motion for the object if the force of gravity in the direction of motion is given by $f_y = GmM \sin\theta/(b^2 + y^2)$.

(b) Find the general solution of the equation of motion. [*Hint*: Use a reduction of order technique. Your answer should be left in implicit form with an integral.]

6. (*Escape Velocity in an Inverse Cube Law Universe*). Suppose that on a planet in another universe the magnitude of the force of gravity obeys the non-Newtonian Inverse Cube Law:

$$|\mathbf{F}| = \frac{\tilde{G}mM}{(y+R)^3}$$

where m is the mass of the object, y its location above the surface of the planet, M the mass of the planet, $\tilde{G}$ a new universal constant, and R the radius of the planet.

(a) What is the escape velocity from this planet? [*Hint*: Find a second order ODE for the motion using Newton's Second Law. Then use a reduction of order technique.]

(b) What is the ratio of this planet's escape velocity $\tilde{v}_0$ to the escape velocity on a planet where the Inverse Square Law holds?

7. (*Escape Velocity*). This problem examines the notion of an escape velocity from a geometrical viewpoint. Consider the IVP (5). Rescale it by defining $z = y - R$ and $t = s/(MG)^{1/2}$ and derive a new IVP using the new variables y and s. Using "earth units" for y, set $R = 1$. Let $y'(0) = a$, and use a as a scaled "velocity" parameter. Use a numerical solver to find as good a lower bound for the initial escape velocity a as you can. [*Hint*: You may want to refer to Atlas Plate FIRST ORDER F.]

8. (*Einstein's Field Equations of General Relativity*). These equations are a complicated system of nonlinear partial differential equations. In 1922, the Soviet mathematician Alexander Alexandrovich Friedmann succeeded in obtaining cosmological solutions governing the behavior of the universe as a whole. He assumed that on a broad scale the universe is homogeneous and isotropic and that pressure can be ignored (a dubious hypothesis if the universe is "small"). The PDEs reduce to a single nonlinear IVP

$$2RR'' + R'^2 + kc^2 = 0, \quad R(t_0) = R_0 > 0, \quad R'(t_0) = v_0 > 0 \quad (9)$$

where R is the "radius of the universe," $k = +1$ (spherical geometry), 0 (Euclidean geometry), or -1 (pseudospherical geometry), c is the speed of light, and the derivatives are with respect to time.

For each of the three cases, $k = 0, +1, -1$, decide whether the universe is born with a "big bang" (i.e., at some time t_1, as $t \to t_1^+$, $R(t) \to 0$, and $R'(t) \to +\infty$). In each case, what happens to the universe as t increases? Does the universe die in a "big crunch" (i.e., for some t_2, as $t \to t_2^-$, $R(t) \to 0$, and $R'(t) \to -\infty$). In carrying out this project, address the following points.

- Multiply the ODE in (9) by R' and rewrite as $[R(R')^2 + kc^2 R]' = 0$. Show that $R(R')^2 + kc^2 R = R_0 v_0^2 + kc^2 R_0 = C_0$.

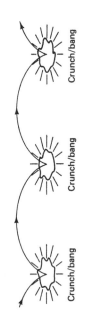

- Set $k = 0$ (a Euclidean universe) and show that $R^{3/2} = R_0^{3/2} \pm 1.5 C_0^{1/2}(t - t_0)$. Show that the minus sign leads to the eventual collapse of the universe (the "big crunch") at some time $T > t_0$, while the plus sign leads to perpetual expansion. What happens in both cases as $R_0 \to 0^+$?

- Set $k = 1$ (a spherical universe). Solve the ODE, $R(R')^2 + c^2 R = C_0$, in terms of a parameter u, by setting $R = C_0 c^{-2} \sin^2 u$, rewriting the ODE with u as the dependent variable, and solving to find t in terms of u. Explain why $t = t(u)$, $R = R(u)$ are the parametric equations of a cycloid in the tR-plane. Interpret each arch of the cycloid in terms of a big bang and big crunch cosmology.

- Set $k = -1$ (a pseudospherical universe). Follow the steps of the spherical case, but set $R = C_0 c^{-2} \sinh^2 u$. Interpret the solution in cosmological terms.

- Rewrite the original ODE as a system in dimensionless variables: $dx/ds = 2y$, $dy/ds = -x^{-1}(k + y^2)$, where $x(s_0) = 1$, $y(s_0) = v_0/c$, $x = R/R_0$, $y = R'/c$, $s = ct/2R_0$. Now use an ODE solver/graphics package to plot solutions $x = x(s)$ and $y = y(s)$ as functions of s for $k = 0, +1, -1$. Interpret the graphs in cosmological terms. Discuss the advantage of scaling the original variables as indicated before computing.

1.9   Exact Equations

The techniques introduced so far for solving first-order ODEs share a common feature: each method transforms a differential equation of some special form into a differential equation that can be solved by performing an integration. In this section we give a simple test that reveals whether a first-order ODE can be "solved" as it stands by performing an integration. If an ODE "passes" the test, we show how to find its solutions. If it fails the test, the ODE may perhaps be rewritten in another form that does "pass." Of course, a numerical solver package may always be used to plot approximate solution curves for the ODE.

Exact Equations and Integrals

Consider first-order ODEs of the form

$$N(x, y)y'(x) + M(x, y) = 0 \qquad (1)$$

where M and N are continuously differentiable functions in a rectangle R of the xy-plane formed by an interval I on the x-axis and an interval J on the y-axis (often written as $R = I \times J$). The form of (1) is not as special as it seems. Any normal first-order equation, $y' = f(x, y)$, can be written in the form (1), e.g., as $y' - f(x, y) = 0$. In fact, this can be done in many ways. For example, the equation

$$y' = \frac{1 + 2xy^2}{2y - 2x^2 y}$$

may be written as

$$(2x^2 y - 2y)y' + 1 + 2xy^2 = 0 \qquad (2)$$

with $N = 2x^2y - 2y$ and $M = 1 + 2xy^2$, or as

$$y' + \frac{1 + 2xy^2}{2x^2y - 2y} = 0 \tag{3}$$

with $N = 1$ and $M = (1 + 2xy^2)/(2x^2y - 2y)$, or in many other ways.

Differential equations of the form of (1) can be solved in implicit form if a certain "exactness" condition is met.

❖ **Exact Equation.** The ODE $N(x, y)y' + M(x, y) = 0$ is an *exact ODE* in a rectangle R of the xy-plane if M and N are continuously differentiable and if

$$\frac{\partial N}{\partial x} = \frac{\partial M}{\partial y}, \qquad \text{for all } (x, y) \text{ in } R \tag{4}$$

For example, ODE (2) is exact in the xy-plane because

$$\frac{\partial N}{\partial x} = \frac{\partial}{\partial x}(2x^2y - 2y) = 4xy = \frac{\partial}{\partial y}(1 + 2xy^2) = \frac{\partial M}{\partial y}$$

However, the same ODE written in the form (3) is not exact in any rectangle, as a straightforward calculation shows. Thus, whether or not a first-order differential equation is exact depends upon the form in which it is written. Note that if N is a function of y only and if M is a function of x only, then $N(y)y' + M(x) = 0$ is both separable and exact.

Formulas can be found which implicitly define solutions of an exact equation. The following theorem shows how this is done.

THEOREM 1.9.1

> *Implicit Solution Theorem.* Suppose that $N(x, y)y' + M(x, y) = 0$ is exact in a rectangle R. Then there is a twice continuously differentiable function $F(x, y)$ on R with the property that:
>
> $$\frac{\partial F}{\partial x} = M \quad \text{and} \quad \frac{\partial F}{\partial y} = N, \quad \text{all } (x, y) \text{ in } R \tag{5}$$
>
> If a differentiable function $y = y(x)$ with graph in R satisfies the equation
>
> $$F(x, y) = C \tag{6}$$
>
> for a constant C, then $y(x)$ is a solution of $Ny' + M = 0$. Conversely, any solution $y = y(x)$ of the ODE whose graph lies in R is also a solution of $F(x, y) = C$, for some constant C.

Proof. First suppose that $\partial M/\partial y = \partial N/\partial x$ throughout a rectangle R. We shall show that there is a function $F(x, y)$ with $\partial F/\partial x = M$, $\partial F/\partial y = N$ in R. Note that a function $F(x, y)$ will satisfy $\partial F/\partial x = M$ if and only if

$$F(x, y) = \int^x M(s, y)\, ds + g(y) \tag{7}$$

where $g(y)$ is an arbitrary differentiable function of y and the antidifferentiation is intended for fixed y. We now determine the function $g(y)$ so that $\partial F/\partial y = N$ as well. With $F(x, y)$ given by formula (7), observe that

$$\frac{\partial F}{\partial y} = \frac{\partial}{\partial y}\left[\int^x M(s, y)\, ds\right] + g'(y) = N(x, y)$$

and so

$$g'(y) = N(x, y) - \frac{\partial}{\partial y} \int^x M(s, y) \, ds \qquad (8)$$

The function $g(y)$ can be determined from equation (8) by an antidifferentiation, provided that the right-hand side is independent of x. This will be the case if the derivative of the right-hand side with respect to x is identically zero. Using the equality of the mixed partial derivatives and the Fundamental Theorem of Calculus and the exactness condition (4), we have that

$$\frac{\partial}{\partial x}\left[N - \frac{\partial}{\partial y}\int^x M(s, y)\, ds\right] = \frac{\partial N}{\partial x} - \frac{\partial}{\partial y}\left[\frac{\partial}{\partial x}\int^x M(s, y)\, ds\right] = \frac{\partial N}{\partial x} - \frac{\partial M}{\partial y} = 0$$

Hence, by the Vanishing Derivative Theorem (Theorem 1.1.1), the right-hand side of (8) really is independent of x. Hence integrating each side of (8) determines $g(y)$ up to an additive constant. Thus, a function $F(x, y)$ is determined (up to an additive constant) with $\partial F/\partial x = M$ and $\partial F/\partial y = N$ in R.

Now suppose that $y(x)$, $a < x < b$, is a solution of $Ny' + M = 0$ in R. Then using the identities in (5) and the Chain Rule [see Theorem C.5.14 of Appendix C.5], we have that

$$0 = M(x, y(x)) + N(x, y(x))y'(x) = \frac{d}{dx}[F(x, y(x))], \quad \text{for all} \quad a < x < b$$

Thus, from the Vanishing Derivative Theorem, there is a constant C such that $F(x, y(x)) = C$, for all $a < x < b$.

Conversely, suppose that $y(x)$, $a < x < b$, is a continuously differentiable function, that the curve $y = y(x)$ is in R, and that $F(x, y(x)) = C$, $a < x < b$, for some constant C. Reversing the above steps shows that $N(x, y(x))y'(x) + M(x, y(x)) = 0$, $a < x < b$, and so $y(x)$ is a solution of the ODE.

In other words, if $Ny' + M = 0$ passes the exactness test (4) in some rectangle R, it can be solved implicitly for solutions $y = y(x)$ that define solution curves in R. Each solution curve $y = y(x)$ in R lies on a *level curve* (or *contour*) of F defined by equation (6).

❖ **Integrals and Integral Curves.** A nonconstant function $F(x, y)$ on a rectangle R is said to be an *integral* of $Ny' + M = 0$ if for any solution $y(x)$ whose graph lies in R, $F(x, y(x)) = C$, for some constant C. If $F(x, y)$ is an integral, then a *level set* $F(x, y) = C$ is called an *integral curve* for the ODE.

This definition agrees with that given for integral curves in Section 1.4. Some observations: First, Theorem 1.9.1 states that an exact ODE has an integral, and the proof of the theorem shows how to find one. Second, the level set $F(x, y) = C$ may not be a single curve, but a collection of disjoint curves called *branches*. See Atlas Plate FIRST ORDER H.

Examples below show how to construct an integral F for an exact ODE.

EXAMPLE 1.9.1	**Constructing an Integral for an Exact ODE** As noted before, the ODE,

$$(2x^2 y - 2y)y' + 1 + 2xy^2 = 0 \qquad (9)$$

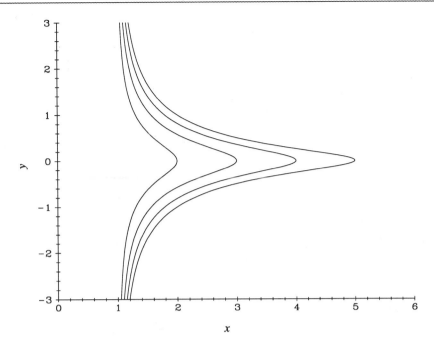

FIGURE 1.9.1 Integral curves for the exact ODE in Example 1.9.1.

is exact in the xy-plane. By Theorem 1.9.1, there is a function $F(x, y)$ such that

$$\frac{\partial F}{\partial x} = 1 + 2xy^2 \quad \text{and} \quad \frac{\partial F}{\partial y} = 2x^2y - 2y \tag{10}$$

Integration of the first of these equations with respect to x, holding y fixed, gives

$$F(x, y) = x + x^2y^2 + g(y) \tag{11}$$

where $g(y)$ is a differentiable function to be determined. Using (11) and the second equation of (10), we have that

$$\frac{\partial F}{\partial y} = 2x^2y + g'(y) = 2x^2y - 2y$$

Thus, $g'(y) = -2y$, and hence $g(y) = -y^2$; the constant of integration is omitted (it will come in later). Now that $g(y)$ is known, we have constructed an integral F,

$$F(x, y) = x + x^2y^2 - y^2$$

The solutions $y(x)$ of the exact ODE (9) are defined implicitly by

$$F(x, y) = x + x^2y^2 - y^2 = C \tag{12}$$

Solving (12) for y we obtain an explicit solution formula from which we could plot several solution curves if desired. Instead we use (12) to plot integral curves for $C = 2$, 3, 4, 5 and $x > 0$ (see Figure 1.9.1). In the next example we identify solution curves as certain arcs of these integral curves.

EXAMPLE 1.9.2

Solutions, Solution Curves, Integral Curves.
Equation (12) can be solved explicitly for y in terms of x:

$$y(x) = \left(\frac{C-x}{x^2-1}\right)^{1/2} \quad \text{or} \quad y(x) = -\left(\frac{C-x}{x^2-1}\right)^{1/2} \tag{13}$$

for values of x for which $(C-x)/(x^2-1)$ is positive, giving solutions of the exact ODE (9).

Suppose we want to find a solution that passes through a specific point, $x = 2$, $y = 1$. To find the appropriate value of C, insert $x = 2$ and $y = 1$ into the left side of (12) to find that $C = 5$. Using the first formula of (13), we see that

$$y(x) = \left(\frac{5-x}{x^2-1}\right)^{1/2}, \quad 1 < x < 5$$

is the desired solution. The graph of this solution lies along the upper half of the integral curve defined by (12) with $C = 5$. See the upper half of the rightmost curve in Figure 1.9.1. The lower half corresponds to the solution $y = -((5-x)/(x^2-1))^{1/2}$, $1 < x < 5$. The other curves in Figure 1.9.1 correspond to $C = 2, 3, 4$.

Often it is hard to imagine the shapes of the integral curves of an exact equation, and only computer graphics can reveal the unusual figures outlined by these curves. In principle, once an integral $F(x, y)$ for an exact ODE has been found, the integral curves $F(x, y) = C$ can be graphed with a contour-plotting software tool. Since software contour plotters sometimes have trouble rendering contours with high curvature, we have opted to use a different approach to plotting integral curves. See the later subsection on systems for details.

EXAMPLE 1.9.3

An Exact Equation with Exotic Integral Curves

Consider the first-order equation $Ny' + M = 0$ given by

$$(\sin y - 2\sin(x^2)\sin(2y))y' + \cos x + 2x\cos(x^2)\cos(2y) = 0 \tag{14}$$

Checking for exactness, we see that

$$\frac{\partial N}{\partial x} = \frac{\partial}{\partial x}(\sin y - 2\sin(x^2)\sin(2y)) = -4x\cos(x^2)\sin(2y)$$

$$\frac{\partial M}{\partial y} = \frac{\partial}{\partial y}(\cos x + 2x\cos(x^2)\cos(2y)) = -4x\cos(x^2)\sin(2y)$$

Thus (14) is exact in the xy-plane, and we set about finding an integral $F(x, y)$. Antidifferentiating $\partial F/\partial x = M$ with respect to x, holding y fixed, we see that

$$F(x, y) = \sin x + \sin(x^2)\cos(2y) + g(y)$$

where $g(y)$ is to be determined. Since $\partial F/\partial y = N$, we have

$$-2\sin(x^2)\sin(2y) + g'(y) = \sin y - 2\sin(x^2)\sin(2y)$$

Thus, $g' = \sin y$ and $g = -\cos y$. Hence, an integral F for the ODE (14) is

$$F(x, y) = \sin x + \sin(x^2)\cos(2y) - \cos y$$

Solutions of (14) are then defined implicitly by

$$\sin x + \sin(x^2)\cos(2y) - \cos y = C \tag{15}$$

where C is a constant. Solving (15) for y as an explicit function of x is possible, but not easy. In any case, the equation for $y(x)$ would be too intricate to be helpful. Suppose that we seek the integral curve that passes through the point $(0, \pi/2)$. The corresponding value of C is obtained by letting $x = 0$ and $y = \pi/2$ in the left side

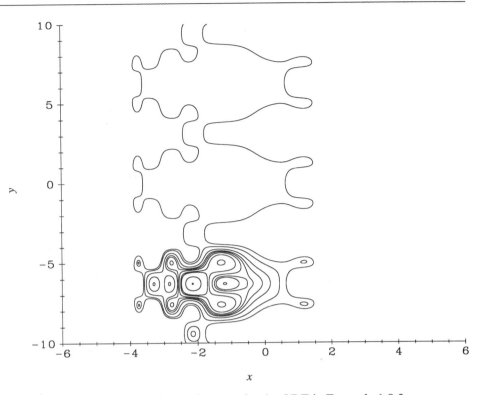

FIGURE 1.9.2 Teddy bear integral curves for the ODE in Example 1.9.3.

of (15); we see that $C = 0$. The rightmost curve in Figure 1.9.2 that outlines the torso and the legs of the line of "teddy bears" corresponds to that integral curve. More about these integral curves in Example 1.9.5.

Summarizing, the Implicit Solution Theorem states that if $Ny' + M = 0$ is exact, then every solution curve lies on some level curve of an integral $F(x, y)$. Conversely, every arc of a level curve that does not have vertical tangents is a solution curve. As noted in Example 1.9.3, we may have to be content with graphing the implicit equation $F(x, y) = C$ in order to identify solution curves $y(x)$ because of the difficulty of solving that equation for y in terms of x. Alternatively, a numerical solver may always be used to find solution curves.

Differential Form of the ODE $Ny' + M = 0$

Finally, we note that the ODE, $Ny' + M = 0$, is often written in the differential form of $M\,dx + N\,dy = 0$, as if the ODE were "multiplied" through by dx and the dx in $y' = dy/dx$ were "canceled." This form of the ODE is suggested by the fact that if $F(x, y)$ is an integral, then the so-called *total differential* of F is $dF = F_x\,dx + F_y\,dy = M\,dx + N\,dy$.

When we encounter the ODE, $Ny' + M = 0$, we seek solutions $y(x)$ that are functions of x. We could equally well rewrite the ODE as $M(x, y)x' + N(x, y) = 0$ with solutions $x(y)$ that are functions of y (recall that $dx/dy = (dy/dx)^{-1}$). Writing $Ny' + M = 0$ as $Mx' + N = 0$ (with x' denoting dx/dy) is appropriate if there are vertical tangents in the xy-plane. Example 1.9.4 illustrates that, although the integral curves remain the same when we replace $Ny' + M = 0$ by $Mx' + N = 0$, the solution curves may be different arcs of the integral curves. When $Ny' + M = 0$ and $Mx' + N = 0$ are written in the *differential form*

$$M(x, y)\,dx + N(x, y)\,dy = 0 \tag{16}$$

there is no preference given to x or to y as the independent variable. In fact, for more than two hundred years $M\,dx + N\,dy = 0$ has been the preferred way of writing $Ny' + M = 0$.

EXAMPLE 1.9.4

Interchanging the Roles of x and y
The ODE of Example 1.9.1 can be rewritten as

$$(1 + 2xy^2)x'(y) + 2x^2y - 2y = 0$$

An integral is $F(x, y) = x + y^2x^2 - y^2$, and solutions $x(y)$ are defined implicitly by

$$x + y^2x^2 - y^2 = C \tag{17}$$

which could be solved for x in terms of y, in general yielding two solutions $x(y)$ corresponding to the two roots of a quadratic. Using formula (12), some integral curves of the ODE in the region $x > 1$ are shown in Figure 1.9.1. But since the ODE is now written in the form $Mx' + N = 0$, each entire integral curve is also a solution curve. Observe from (17) that the lines $x = \pm 1$ are integral curves and hence are also solution curves of the ODE written in the form $Mx' + N = 0$. See Problem 9 and Atlas Plate FIRST ORDER H for other interesting properties of this ODE.

Inexact Equations with $N = 1$ and $M = p(x)y - q(x)$

ODEs such as $y' + x^2 + y^2 = 0$ are not exact, and the above process of constructing $F(x, y)$ breaks down because the expression defining $g'(y)$ is not a function of y alone. An inexact equation $Ny' + M = 0$ can always be made exact by multiplying by a nonvanishing and continuously differentiable function $\mu(x, y)$ to obtain an exact equation $\mu Ny' + \mu M = 0$. The solutions of $Ny' + M = 0$ and of the new equation are identical since the *integrating factor* μ never vanishes. For example, the linear equation, $y' + p(x)y - q(x) = 0$ is in general not exact since $\partial M/\partial y = p(x)$ while $\partial(N)/\partial x = 0$. However, the equation becomes exact if it is multiplied by the integrating factor $e^{P(x)}$, where $P(x) = \int^x p(s)\,ds$ (see Section 1.2 and Problem 11(c) in this section). In general, however, it is far from easy to find integrating factors for inexact equations. It is usually simpler to use a numerical solver to determine approximate solutions.

Planar Systems and First-Order ODEs

The ODE $Ny' + M = 0$ may be exact or inexact. In any case it can be written as $y' = -M/N$, but this ODE may be hard to solve numerically in regions where the denominator N vanishes. There is a way to get around this problem of the vanishing denominator, but at the cost of introducing a system of two first-order autonomous ODEs with t as the independent variable

$$\frac{dx(t)}{dt} = N(x(t), y(t))$$
$$\frac{dy(t)}{dt} = -M(x(t), y(t)) \tag{18}$$

Solutions of system (18) are pairs of functions $x = x(t)$, $y = y(t)$. We can think of these functions as giving parametric equations of a curve $(x(t), y(t))$ in the xy-plane where the parameter is t. The parametric plot in xy-space of a solution $x(t)$, $y(t)$ of (18) is called an *orbit* of system (18). At each point (x, y) of this curve the slope dy/dx is $-M(x, y)/N(x, y)$, as we see after dividing the second equation in (18) by the first. Thus, solution curves of $Ny' + M = 0$ are arcs of orbits of the system (18) which contain no vertical tangents, and conversely.

Integral curves of an exact ODE $Ny' + M = 0$ and orbits of the system $x' = N$, $y' = -M$ are essentially identical in any region where M and N do not vanish simultaneously. To see this, let $F(x, y)$ be an integral of the ODE and let $\{x(t), y(t)\}$ be an orbit of the system (18). Then $\partial F/\partial x = M$, $\partial F/\partial y = N$, and so $F(x(t), y(t)) = $ constant, for all t, because

$$\frac{d}{dt}(F(x(t), y(t))) = \frac{\partial F}{\partial x}x' + \frac{\partial F}{\partial y}y' = MN + N(-M) = 0, \text{ for all } t$$

EXAMPLE 1.9.5

Solution Curves for an Exact ODE

The first-order system that corresponds to the ODE (14) of Example 1.9.3 is

$$\frac{dx}{dt} = \sin y - 2\sin(x^2)\sin(2y)$$

$$\frac{dy}{dt} = -\cos x - 2x\cos(x^2)\cos(2y) \tag{19}$$

The integral curves of ODE (14) shown in Figure 1.9.2 were actually produced by using a numerical ODE solver to plot orbits of the system (19), given several initial points (x_0, y_0) located in the rectangle $|x| \le 6$, $|y| \le 10$. For example, the rightmost curve outlining the lower torso and legs of the teddy bears is the orbit of system (19) that passes through the point $(0, \pi/2)$ at $t = 0$. In each case, the orbit is run forward and backward in time from the initial point until either the orbit exits the rectangle or returns to its starting point (as happens with the "internal" orbits of the teddy bears).

In using the planar system (18) to plot solution curves of the first order ODE $N(x, y)y' + M(x, y) = 0$, we made no mention of whether that ODE was exact or inexact. So therefore the technique works in either case. Example 1.9.5 shows this for an exact ODE, and the next example shows the same for an inexact ODE.

EXAMPLE 1.9.6

Solution Curves for an Inexact ODE

Consider the inexact ODE $(x + y)y' + 2x = 0$. The planar system (18) takes the form

$$\frac{dx}{dt} = x + y, \quad \frac{dy}{dt} = -2x$$

and Figure 1.9.3 shows some orbits of this system in the rectangle $0.5 \le x \le 3$, $-7 \le y \le 2$. The portions of these orbits with no vertical tangents are solution curves of $(x + y)y' + 2x = 0$. Note that the vertical tangents to orbits occur on the line $x + y = 0$.

Comments

Observe that the replacement of $Ny' + M = 0$ by the system (18) can be made whether or not $Ny' + M = 0$ is exact. It is often desirable to study the behavior of solutions of $Ny' + M = 0$ by using a numerical ODE solver to plot orbits of system (18). This approach reveals a great deal of information about the extendability of solution curves. Thus, if the primary focus is on the geometry of solution curves, rather than solution

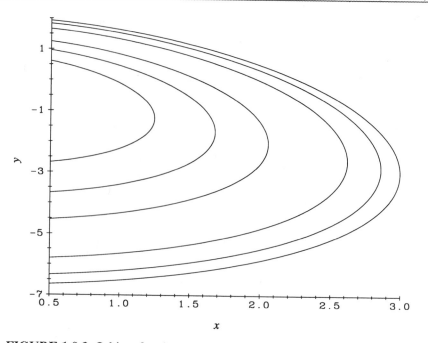

FIGURE 1.9.3 Orbits of an inexact ODE. See example 1.9.6.

formulas, one should use a numerical solver package to plot orbits of system (18). Atlas Plate FIRST ORDER I displays orbits of an inexact equation. These orbits are obtained by solving the particular system of form (18) that corresponds to an inexact ODE. Even if the ODE is exact and implicit formulas for the solutions are available, system (18) and a numerical solver may be the best choice for acquiring geometric insight into the solutions.

PROBLEMS

1. Show that each of the following ODEs $Ny' + M = 0$ is exact. Find an integral $F(x, y)$, and solve in the implicit form $F(x, y) = C$.

   (a) $(x^2 + 6xy + 3y^2)y' + 2xy + 3y^2 = 0$ (b) $xy' + y = 0$

   (c) $xe^y(y^2 + 2y)y' + y^2 e^y + 2x = 0$ (d) $y' \sin x \sin y - \cos x \cos y = 0$

2. (a)–(d) Plot several integral curves for the corresponding ODEs in Problem 1. Plot the curves in the rectangle $-5 \le x \le 5$, $-4 \le y \le 4$. [*Hint*: Replace the ODE by the corresponding system (18), and plot orbits of (18) for various initial points.]

3. Test each of the ODEs $M\,dx + N\,dy = 0$ for exactness and solve in the implicit form $F(x, y) = C$. (Solve the ODEs that are not exact by methods discussed in the preceding sections.)

   (a) $(x + 2y)\,dx + (2x + y)\,dy = 0$

   (b) $(2xy - 3x^2)\,dx + (x^2 + 2y)\,dy = 0$

   (c) $(\cos 2y - 3x^2 y^2)\,dx + (\cos 2y - 2x \sin 2y - 2x^3 y)\,dy = 0$

   (d) $(x^2 + y^2)\,dx - 2xy\,dy = 0$

   (e) $(2x - 3y)\,dx + (2y - 3x)\,dy = 0$

4. (a)–(e) Replace each ODE in Problem 3 by a planar system, and plot several orbits in the rectangle $|x| \leq 5$, $|y| \leq 4$. How are these orbits related to the solution curves of the corresponding first order ODE?

5. (*Teddy Bears*). Fill in more details in the picture of the teddy bears (Figure 1.9.2). Why do you expect to see the periodic repetition of the figures in the y-direction, but not in the x-direction? On your graph highlight several solution curves $y = y(x)$ for maximally extended solutions.

6. Solve (in implicit form) each of the following initial value problems.

   (a) $(2y - x)y' = y - 2x$, $y(1) = 2$

   (b) $2xydx + (x^2 + 1)dy = 0$, $y(1) = -3$

   (c) $2xy^3 + 3x^2y^2y' = 0$, $y(1) = 1$

   (d) $(ye^{xy} - 2y^3)\,dx + (xe^{xy} - 6xy^2 - 2y)\,dy = 0$, $y(0) = 2$

7. (a)–(d) Plot the solution curve of the corresponding IVP in Problem 6 in the rectangle $-5 \leq x \leq 5$, $-4 \leq y \leq 4$.

8. Consider the ODE $(y^2 + e^x \cos y + 2\cos x)y' + e^x \sin y - 2y \sin x = 0$.

   (a) Show that this ODE is exact in the xy-plane.

   (b) Find an integral $F(x, y)$ for the ODE.

   (c) Plot $F(x, y) = C$ for several values of the constant C. Alternatively, replace the differential equation by the corresponding system $dx/dt = N$, $dy/dt = -M$ and plot integral curves in the rectangle $-20 \leq x \leq 10$, $-4 \leq y \leq 6$. [*Hint*: See Atlas Plate FIRST ORDER L.]

9. (*Integral Curves, Orbits, Solution Curves*). This problem relates solution curves of an exact ODE, $Ny' + M = 0$, to its integral curves and to the orbits of the corresponding planar system, $x' = N$, $y' = -M$. Consider the ODE of Examples 1.9.1 and 1.9.2, $(2x^2y - 2y)y' + (1 + 2xy^2) = 0$, its integral $F(x, y) = x + x^2y^2 - y^2$, and the planar system, $x' = 2x^2y - 2y$, $y' = -1 - 2xy^2$.

   (a) Plot the integral curve $F(x, y) = C$, where $C = -3$, of the ODE in the rectangle $|x| \leq 5$, $|y| \leq 4$. How many smooth branches does the integral curve have? Repeat for $C = -1, 0, 1, 5$.

   (b) Each branch of the curve $F = -3$ plotted in part (a) is an orbit of the planar system. Choose an initial point on each branch of $F = -3$, and use a numerical solver to plot the orbit of the corresponding IVP. [*Hint*: Solve the IVP, $x' = 2x^2y - 2y$, $x(0) = -3$, $y' = -1 - 2xy^2$, $y(0) = 0$ forward and backward from $t = 0$ to get one of the three orbits.]

   (c) Find four distinct solution curves of the ODE on the integral curve $F = -3$. From parts (a), (b), (c), we see that each solution curve lies on an orbit and each orbit lies on an integral curve which may have several branches. [*Hint*: The solution $y = -[(-3 - x)/(x^2 - 1)]^{1/2}$, $x < -3$, defines one of the four curves.]

   (d) Show that the integral curve $F(x, y) = -1$ consists of the disjoint solution curves $y = \pm(1 - x)^{-1/2}$, $x < 1$, together with the line $x = -1$.

   (e) Show that the integral curve $F = -1$ contains all the "critical points" at which M and N both vanish.

   (f) Consider the IVP

$$\frac{dx}{dt} = 2x^2y - 2y, \quad x(0) = -3$$

$$\frac{dy}{dt} = -1 - 2xy^2, \quad y(0) = -\frac{1}{2}$$

Does the solution of this IVP generate the entire level curve $F = -1$ which passes through the point $(-3, -\frac{1}{2})$?

(g) Finally, reproduce Atlas Plate FIRST ORDER H. Label each branch with the appropriate value of C.

10. Consider the ODE $y' = (2xy + y^2)/(2x^2 + 2xy)$ whose right-hand side is a homogeneous function of order zero.

(a) Using the techniques of Section 1.7 find an implicit formula for the solutions.

 (b) Replace the ODE by the corresponding system of form (18) and use a computer solver to duplicate Atlas Plate FIRST ORDER G. Highlight some solution curves of the ODE on your graph.

11. (*Inexact ODEs*). Show that by multiplying through by the indicated integrating factor μ, the inexact ODE becomes exact. Solve the resulting exact ODE.

(a) $\cos x \, dy - (2y \sin x - 3) \, dx = 0, \qquad \mu(x) = \cos x, \quad -\pi/2 < x < \pi/2$

(b) $y \, dx + (4x - y^2) \, dy = 0, \qquad \mu(y) = y^3, \quad y > 0$

(c) Show that $\mu = \exp\left[\int^x p(s) \, ds\right]$ is an integrating factor for the general first-order linear equation $y' + p(x)y - q(x) = 0$.

12. (*The System Way to Plot a Level Curve*). Let $f(x, y)$ be continuously differentiable in a region R where f_x and f_y have no common zeros.

- For any point (x_0, y_0) in R show that the orbit of the autonomous planar IVP
$$x' = f_y(x, y), \qquad x(0) = x_0$$
$$y' = -f_x(x, y), \qquad y(0) = y_0$$
is a level curve $f(x, y) = f(x_0, y_0)$ of the function $f(x, y)$ through the point. [*Hint*: See the remark just before Example 1.9.5.]

- Consider the function $f(x, y) = -x + 2xy + x^2 + y^2$. Plot several level curves for f in the rectangle $|x| \le 6$, $|y| \le 6$.

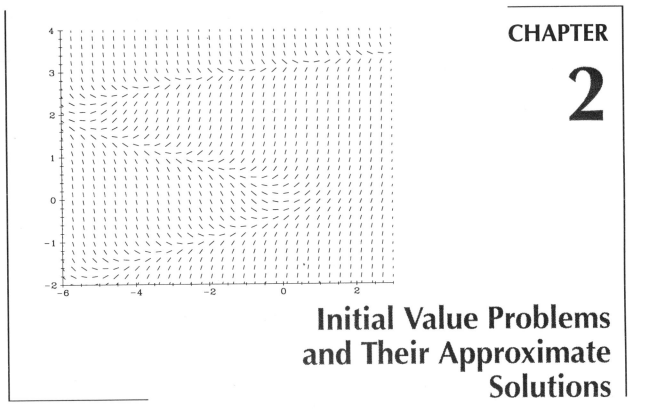

CHAPTER 2

Initial Value Problems and Their Approximate Solutions

The basic questions of existence, uniqueness, and sensitivity of solutions of the initial value problem $y' = f(t, y)$, $y(t_0) = y_0$, are outlined in this chapter. The geometry of solution curves, begun in Chapter 1, is extended and elaborated upon to provide ideas underlying the numerical methods for approximating solutions of initial value problems. Along the way, the art of using numerical solver packages to solve first-order initial value problems is discussed. Details and proofs of the basic questions and numerical methods are found in Appendices A and B. The direction field in the chapter cover figure appears in Problem 3, Section 2.1.

2.1 Solution Curves and Direction Fields

The general first-order differential equation in normal form is

$$y' = f(t, y) \tag{1}$$

where the function f is defined on some portion of the ty-plane. Chapter 1 is devoted primarily to devising techniques for constructing formulas for the solutions $y(t)$ of (1) when $f(t, y)$ has a variety of special forms. Some techniques are explicit, and all solutions $y(t)$ are given by a single formula involving t and an arbitrary constant. Other techniques are implicit and lead to equations involving y, t, and an arbitrary constant, equations that are often difficult to solve for y explicitly as a function of t and the constant.

How does one "find" a solution of the ODE (1) when no solution formula can be found, explicit or implicit? How do we even know that the IVP

$$y' = f(t, y), \qquad y(t_0) = y_0 \tag{2}$$

has a solution at all if we cannot construct a formula for it? How do we know that the IVP (2) has only one solution? How can we study behavior of solutions of ODE (1) if we can find no defining solution formula? In this chapter we will present some answers to these questions, answers that do not require solution formulas. This is fortunate because, except for a handful of rate functions $f(t, y)$ like those considered in Chapter 1, there are no known solution formulas for ODE (1) or IVP (2).

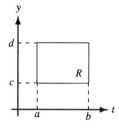

We first answer the basic questions of existence and uniqueness of the solution of the IVP (2) so that we can turn to the main topic of this section: the graphical characterization of solution curves. Let R be the closed rectangle in the ty-plane defined by $a \leq t \leq b$, $c \leq y \leq d$ (think of R as the user-defined screen on the user's solver platform). Assume that f and $\partial f/\partial y$ are both continuous on R. Then it is known that for any choice of (t_0, y_0) in R the IVP (2) has a unique solution whose graph extends to the boundary of R for both $t > t_0$ and $t < t_0$ (in other words, the solution curve does not "die" inside R). Moreover, the unique solvability of IVP (2) implies that no two solution curves of $y' = f(t, y)$ can intersect inside R. Thus, we can think of R as being completely covered by solution curves, each point of R on exactly one solution curve. We return to the basic questions in Sections 2.2 and 2.3.

EXAMPLE 2.1.1

Using a Numerical Computer Solver
Consider the ODE

$$y' = -\cos y + \frac{t}{(1 + y^2)} \tag{3}$$

and let R be the rectangle $-4 \leq t \leq 4$, $-3 \leq y \leq 3$. Observe that ODE (3) does not fall into any of the forms considered in Chapter 1, and hence we are left to our own devices to find solution curves. The rate function $f(t, y) = -\cos y + t/(1 + y^2)$ and its partial derivative with respect to the state variable, $\partial f/\partial y = \sin y - 2ty/(1 + y^2)^2$, are both continuous functions on R. Therefore, through each point of R there is exactly one solution curve of ODE (3). In Figure 2.1.1 we have chosen nine initial points on the left, lower, and right edges of R and then used our computer solver to graph the solution curves of ODE (3) through these points. Notice how each solution curve exits the rectangle R as t increases or decreases. Close inspection of Figure 2.1.1 appears to show that some solution curves touch inside R. We know that this cannot happen, and hence our comfortable world of theory collides with the practical problem of displaying data on a computer screen. Since the screen only contains a finite number of pixels, points that are less than a pixel apart are not distinguishable from one another, explaining the apparent contradiction with theory. We can separate the apparently touching curves by zooming in on them. See Figure 2.1.2 for a zoom on the upper left of Figure 2.1.1 where four solution curves seem to touch.

Nothing has been said so far about how computer solvers do their work. Later in this chapter we will see that numerical solvers use a step-by-step process to generate (often very accurate) approximate values for points on a solution curve, and then graph these approximate solutions. Before we say much about this process, it is helpful to know more about how the rate function and solution curves of ODE (1) are related.

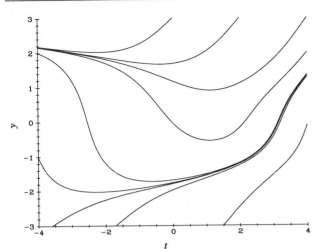

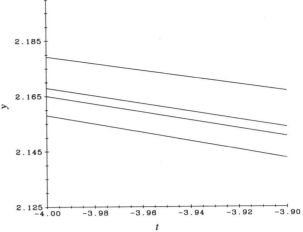

FIGURE 2.1.1 Solution curves for ODE in Example 2.1.1.

FIGURE 2.1.2 Zoom on upper left of Figure 2.1.1 separates solution curves. See Example 2.1.1.

Geometry of Solution Curves

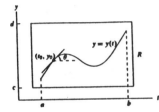

There is a way to view the solvability of the ODE $y' = f(t, y)$ which appeals to geometric intuition and hence lends itself to a graphical approach to finding solution curves. Let $f(t, y)$ be defined over a rectangle R in the ty-plane on which f and f_y are continuous. Then, as noted above, every point on R has a unique solution curve through it. Now the ODE expresses the fact that, at each point (t, y) in R, the value $f(t, y)$ gives the slope of the tangent line to the solution curve through that point. Conversely, suppose that the graph of the continuously differentiable function $y(t)$, $a < t < b$, lies in R. Then $y(t)$ defines a solution curve if at each point (t_0, y_0) on the graph, the slope of the tangent line has the value $f(t_0, y_0)$. Obviously, this is just the geometric equivalent of saying that $y'(t) = f(t, y(t))$, for each $a < t < b$, or that $y(t)$ is a solution for the ODE.

This change of viewpoint gives us a way to "construct" solution curves for the ODE. By drawing short line segments with slopes $f(t, y)$ and centered at a grid of points (t, y) distributed throughout the region R, we obtain a diagram, called a *direction field*. Most ODE solvers have the capacity to produce a direction field for the ODE over a given rectangle in the ty-plane. Some solvers allow the user to select horizontal and vertical grid densities as well as the length of the direction field lines. The direction field may "suggest" curves in R with the property that at each point of each curve the tangent line to the curve at that point lies along the direction field line at the point. This process reveals solution curves in much the same way as iron filings sprinkled on paper held over the poles of a magnet reveal magnetic field lines. If a particular direction field is not suggestive in this sense, then choosing a finer grid of points in R and repeating the process may help. In Figures 2.1.3 and 2.1.4 we search for solution curves by using this method.

EXAMPLE 2.1.2

Visualizing Solution Curves

Using the method of integrating factors, we see that all solutions of the linear differential equation $y' = (1 - t)y - t$ are given by the formula

$$y = e^{t - t^2/2} \left[C + \int_0^t (-s)e^{-s + s^2/2} \, ds \right] \tag{4}$$

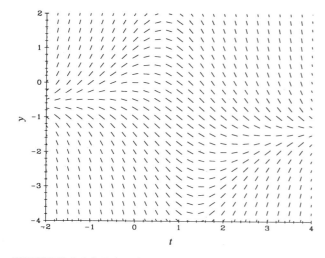

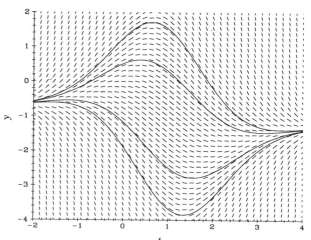

FIGURE 2.1.3 Direction field for the ODE in Example 2.1.2.

FIGURE 2.1.4 Solution curves for the ODE in Example 2.1.2.

where C is an arbitrary constant. Note that knowing the solution formula is not much help in understanding the behavior of the solution curves. Another approach to visualizing solution curves in a given rectangle R in the ty-plane might go something like this: First choose scales on the t-axis and the y-axis so as to construct R on your computer screen. Next, instruct your solver to construct a direction field in R. Your screen should look something like Figure 2.1.3. Since the rate function $f(t, y) = (1 - t)y - t$ is continuous on R, as is $\partial f/\partial y = 1 - t$, we know that the solution curves of the ODE are nonintersecting and completely cover R. Since the solution curves "follow" the field lines, Figure 2.1.3 gives a rough idea of what the curves look like. A finer grid in R and the direction field it generates are illustrated in Figure 2.1.4, along with a few solution curves generated by our solver.

Notice that as the direction field is refined by inserting shorter and shorter line segments and more and more grid points, solution curves become more clearly suggested. This technique does not, of course, produce exact solutions (and is not without its pitfalls), but we can often obtain a rough idea of how solution curves behave. The length of the direction field lines is, of course, arbitrary, but for aesthetic reasons no two field lines should touch. The field lines in Figure 2.1.3 are all the same length, even if it doesn't appear so at first glance. The reason for this is that the "screen"-length of a vertical unit is not the same as the "screen"-length of a horizontal unit. The ratio of the former "screen"-length to the latter is called the *aspect ratio* of the display. Our solver does not take aspect ratio into account, and field lines may appear to be of differing lengths.

Although a computer was used to plot the direction fields of Figure 2.1.3 and 2.1.4, basically the only tools needed are a straightedge and graph paper with a fine rectangular grid. The rate function is used to find the slopes of line segments at grid points, and then centered line segments with these slopes are drawn through the grid point. Some geometric shortcuts are described in the problems.

Isoclines, Equilibrium Solutions

For the ODE $y' = f(t, y)$ it is often useful to identify curves in the ty-plane where the direction field lines through any of its points have the same slope. These curves are

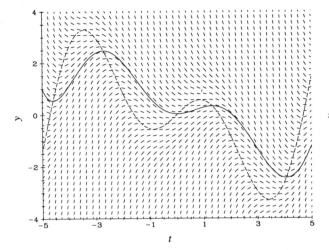
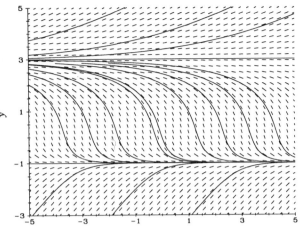
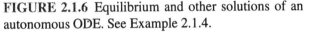

FIGURE 2.1.5 The nullcline (dashed) and a solution curve (solid). See Example 2.1.3.

FIGURE 2.1.6 Equilibrium and other solutions of an autonomous ODE. See Example 2.1.4.

called *isoclines* and are nothing but the level curves $f(t, y) = c$, for a given constant c, and may consist of disjoint branches. The isocline for $c = 0$, called the *nullcline*, has the property that if a solution curve intersects the nullcline, the solution curve has a horizontal tangent at the intersection point.

EXAMPLE 2.1.3

Nullclines and Solution Curves

The nullcline $y = t \cos t$ for the linear ODE $y' = -y + t \cos t$ in the rectangle $|t| \leq 5$, $|y| \leq 4$ is illustrated in Figure 2.1.5 by the dashed-line curve. Notice how the direction field "flattens out" at the nullcline. The solid line curve is a solution curve. Observe that the slope of the tangent line to the solution curve is zero at each point where the curve intersects the nullcline.

An ODE (or system of ODEs) is said to be *autonomous* if the rate functions do not depend on the independent variable. Consider the autonomous ODE $y' = f(y)$, where f is continuously differentiable on an interval I. Suppose that y_0 in I is such that $f(y_0) = 0$. Then the direction field line at any point of the horizontal line $y = y_0$, for all t, has zero slope. But the line $y = y_0$ also has zero slope everywhere and hence "fits" the direction field. Thus the line $y = y_0$, for all t, is a solution of the ODE $y' = f(y)$. Since this solution does not change with time, it is called an *equilibrium solution*.

The graphs of the equilibrium solutions of a first-order autonomous ODE separate the ty-plane into bandlike regions of solution curves. Each solution curve which is not an equilibrium solution is confined to a single band because otherwise some solution curve must cross the graph of an equilibrium solution, and that would violate uniqueness.

EXAMPLE 2.1.4

Equilibrium Solutions of Autonomous ODEs

The autonomous ODE $y' = (y - 3)(y + 1)/(1 + y^2)$ has two equilibrium solutions: $y = 3$, for all t, and $y = -1$, for all t. The bands are given by $y < -1$, $-1 < y < 3$, and $y > 3$. These and other solution curves are displayed in Figure 2.1.6.

PROBLEMS

1. For each linear ODE below, reproduce the displayed direction field and plot some solution curves suggested by that direction field. Then solve the ODE and compare the graphs of exact solution curves with your plot. [*Hint*: In case your solver does not plot direction fields, photocopy (and enlarge) those below.]

(a) $y' = -ty$ **(b)** $y' = 2|t|y$

(c) $y' = 3y/t$ **(d)** $y' = (2\,\mathrm{step}(t) - 1)y$

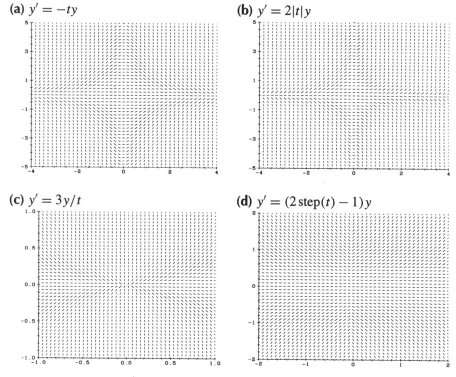

2. (*What to do When the Solution Formula Doesn't Help*). For some ODEs, one can find a general solution formula, but it still may not help you to find the limiting value of the solutions as $t \to \infty$.
   Consider the linear ODE

$$y' + 2ty = t\cos t$$

   (a) Using an integrating factor, find an expression for the general solution. Do not carry out the resulting integration which appears in the expression since the integral cannot be evaluated in terms of elementary functions.

   (b) Using an ODE solver, plot several solution curves for $|t| \leq 10$, $|y| \leq 4$.

   (c) What does the plot in part **(b)** suggest happens to solutions as $t \to \infty$?

   (d) Explain the symmetry in the graphs in part **(b)** obtained when t is replaced by $-t$. [*Hint*: Find the slope at $(-t, y)$ in terms of the slope at (t, y).]

3. (*What to do When There is No Solution Formula*). For the nonlinear ODE $y' = 3y\sin y + t$ of the chapter cover figure there is no known explicit solution formula, and we must turn to the line segments of the direction field or to a numerical ODE solver. Plot a direction field over the rectangle R: $-6 \leq t \leq 3$, $-2 \leq y \leq 4$. With a pencil, sketch some solution curves in R "suggested" by the direction field. Try refining your direction field in R to see if your sketched solution curves need any adjustment. Use a numerical solver to check the accuracy of your sketched solution curves. Record your observations about the approach of using direction fields to sketch these solution curves.

4. (*Solution Curves and Nullclines*). For each linear ODE below, use a solver to draw a direction field over the interval specified. Sketch the nullcline. Sketch a solution curve of the IVP. Verify your sketch by using a solver to graph the solution of the IVP. [*Hint*: In part (**d**), the IVP does *not* have a unique solution; draw several solution curves for this IVP over the indicated interval.]

(**a**) $y' = 2t$, $\quad\quad y(2) = 2$, $\quad |t| \leq 3, |y| \leq 3$

(**b**) $y' = t + y$, $\quad y(0) = 2$, $\quad -2 \leq t \leq 1, 0 \leq y \leq 5$

(**c**) $y' + ty = t^2$, $\quad y(-2) = 1$, $\quad -2 \leq t \leq 4, -2 \leq y \leq 18$

(**d**) $ty' = 3y$, $\quad\quad y(0) = 0$, $\quad |t| \leq 5, |y| \leq 3$

5. (*Method of Isoclines*). When finding direction fields for the first-order ODE $y' = f(t, y)$ other than by computer, laborsaving devices are a necessity. One such device is known as the *method of isoclines*. For any constant c, the implicitly defined level set $f(t, y) = c$ in the ty-plane is called the *c-isocline*. The isocline for $c = 0$ is the nullcline. The direction field line element at any point of the *c*-isocline has the slope c. After a few parallel field elements are drawn on a given *c*-isocline, the value of c is changed and the process is repeated. Isoclines can be useful in sketching solution curves.

(**a**) Reproduce Atlas Plate FIRST ORDER B as closely as possible. Determine the nullcline for the ODE and sketch it over your copy of the Atlas Plate. Describe what you observe to be the relation between the nullcline and the solution curves.

(**b**) For the autonomous ODE $y' = f(y)$ show that the *c*-isoclines are horizontal lines in the ty-plane. Use this fact to sketch solution curves for $y' = y(1 - y)$.

6. (*Rise, Fall*). The first derivative can be used to show that a function is increasing, decreasing, or constant over an interval.

(**a**) Let R be the rectangle $-6 \leq t \leq 6$, $-8 \leq y \leq 8$, in the ty-plane. Imagine R to be filled with the solution curves of the ODE $y' = -y\cos t$. For what regions of R are the solution curves: Rising ($y' > 0$)? Falling ($y' < 0$)?

(**b**) Make a rough sketch in R of the solution curves of the ODE. Sketch the nullclines. Verify your guess by using a computer solver.

7. Use direction fields and the method outlined in Problem 6 to sketch nullclines and some solution curves for the following ODEs. Your graphs should exhibit approximately the right slope. Verify your sketch by plotting solution curves.

(**a**) $y' = (y + 3)(y - 2)$ $\quad\quad$ (**b**) $y' = 2t - y$

(**c**) $y' = ty - 1$ $\quad\quad\quad\quad\quad$ (**d**) $y' = (1 - t)y$

(**e**) $y' = y - t^2$ $\quad\quad\quad\quad\quad$ (**f**) $y' + (\sin t)y = t\cos t$

8. (*Diverging and Converging Flows*). Consider the direction field for the equation $y' = f(t, y)$. The solution curves of the differential equation can be thought of as the *flow lines* of a fluid whose velocity at the point (t, y) is $f(t, y)$.

(**a**) If $\partial f/\partial y > 0$ at some $t = t^*$ and for all y, $c < y < d$, verify that the flow directions across the line $t = t^*$ diverge in the manner indicated in the diagram. If $\partial f/\partial y > 0$ on some region in the ty-plane, what can be said about the vertical distance between solution curves in that region as t increases? Another form of this result appears in Problem 4 of Appendix A.4.

(**b**) Formulate an analogous property of flows in regions where $\partial f/\partial y < 0$.

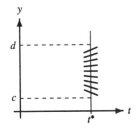

2.2 Initial Value Problems: Basic Questions

The intimate connection suggested in Chapter 1 between evolving physical systems and initial value problems indicates the reason for the central position played by such problems. Modelers exploit this connection by using their mathematical knowledge of initial value problems to predict the behavior of the systems, or to design special processes in the system's natural environment. Thus if our modeling efforts are to be successful, we need to learn as much as possible about the properties of IVPs.

The general normalized first-order initial value problem is

$$y' = f(t, y) \qquad y(t_0) = y_0 \tag{1}$$

where the function f is defined in some rectangular region R of the ty-plane, and the point (t_0, y_0) is in R. A function $y(t)$ is a solution to the IVP (1) if it is defined over an interval I containing t_0 and satisfies the conditions $y'(t) = f(t, y(t))$ for all t in the interior of I, and $y(t_0) = y_0$.

We shall sometimes refer to the rate function $f(t, y)$ and the value y_0 collectively as *data* for the IVP (1). The reason for using the term in this way is that both the function $f(t, y)$ and the initial value y_0 in IVP (1) may contain elements that are empirically determined or depend on approximations in the system's external environment.

Basic Questions for Initial Value Problems

The analysis of a mathematical model in this text amounts to an examination of the IVP associated with that model. There are several questions which always come up. Assuming that the model gives rise to a first-order ODE, the *Basic Questions* are:

> **Existence:** What conditions on the data guarantee that IVP (1) has at least one solution?
>
> **Uniqueness:** Under what conditions on the data will IVP (1) have at most one solution?
>
> **Description:** Assuming the data are such that IVP (1) has exactly one solution, how can it be described?
>
> **Sensitivity:** Assuming that IVP (1) has a unique solution for each data set within a class, how much does a solution change as the data change within that class?

Suppose that conditions on the data can be found which guarantee that IVP (1) always has a unique solution, and that this solution changes "continuously" when the data are changed continuously. Then IVP (1) is said to be *well-posed*. Our intuition tells us that IVPs which arise as models of natural phenomena are well-posed. Notice that well-posedness deals with three of the Basic Questions about IVPs: Existence, Uniqueness, and Sensitivity ("continuity in the data" is a mathematically precise expression of sensitivity).

Sometimes it is possible to describe the solution of IVP (1) by first finding an explicit expression for the "general" solution of $y' = f(t, y)$ and then using it to select a solution that meets the initial condition $y(t_0) = y_0$. This is the way initial value problems in Chapter 1 are solved. We shall use this technique whenever possible (and practical) to describe the solution of an IVP, but, with the exception of a few special classes of rate functions $f(t, y)$, it is not possible to find an explicit expression for the general solution of the differential equation $y' = f(t, y)$. Thus, other techniques for describing the solution of IVP (1) must be devised.

Simple conditions on the rate function f produce satisfactory answers to all four basic questions for IVP (1). The uniqueness and existence questions are treated in the theorem below. Solutions can be described by formulas as in Chapter 1 or by approximations using the numerical procedures of Section 2.5. An introduction to sensitivity appears in Section 2.3; a more general approach is in Appendix A.4.

THEOREM 2.2.1

> Existence and Uniqueness Theorem. Let $f(t, y)$ and $\partial f/\partial y$ be continuous on a closed rectangle R of the ty-plane, and suppose that (t_0, y_0) is an interior point in R. Then the IVP
>
> $$y' = f(t, y), \quad y(t_0) = y_0$$
>
> has a solution $y(t)$ on an interval I containing t_0 in its interior. Moreover, the IVP has at most one solution over any t-interval I containing t_0.

The proof of Theorem 2.2.1 is given in Appendices A.1 and A.2.

EXAMPLE 2.2.1

Computed Solution

Consider the IVP

$$y' = 3y \sin y + t, \qquad y(-4) = 1 \tag{2}$$

The nonlinear ODE in IVP (2) cannot be solved by any of the methods of Chapter 1. Notice that the rate function $f(t, y) = 3y \sin y + t$ and its derivative $f_y(t, y) = 3 \sin y + 3y \cos y$ are continuous on the entire ty-plane. Hence the conditions of the Existence and Uniqueness Theorem are verified for *any* rectangle R in the ty-plane. Therefore, IVP (2) has a solution $y = y(t)$ defined on some t-interval I containing $t_0 = -4$. See Figure 2.2.1 for an arc of the "forward" solution curve from the initial point $(-4, 1)$. Notice how well the solution curve "fits" the direction field. There seems to be nothing in the way of the solution curve for IVP (2) going on forever. A later example will show that this intuitive leap of faith may be deceptive.

The Existence and Uniqueness Theorem has a very important consequence which we shall often use.

THEOREM 2.2.2

> Non-Intersection of Solution Curves. Let the rate function $f(t, y)$ and its derivative $f_y(t, y)$ both be continuous on a rectangle R in the ty-plane. Then no two solution curves of the ODE $y' = f(t, y)$ can intersect in R.

Proof. Suppose that two distinct solution curves intersect at some point (t_0, y_0) in R. Then the IVP $y' = f(t, y)$, $y(t_0) = y_0$, would have two solutions on some interval I containing t_0, contradicting the Uniqueness part of Theorem 2.2.1.

As noted in Section 2.1, this means that if the conditions on the rate function f and f_y are met, then any apparent meeting of solution curves on a computer screen is

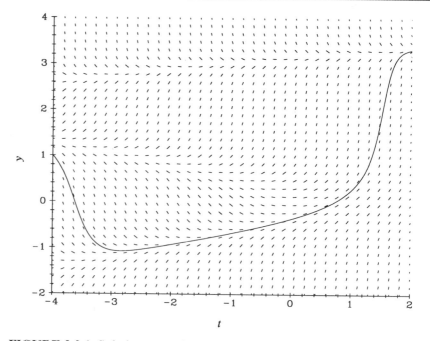

FIGURE 2.2.1 Solution curve for IVP (2).

a consequence of the finite resolution capability of the device; in theory the curves cannot touch. However, if the conditions are *not* met, then anything can happen—maybe there are several solutions (Example 2.2.2), maybe none at all (see Problem 5), or maybe exactly one solution.

EXAMPLE 2.2.2

Loss of Uniqueness
Consider the IVP

$$y' = 3y^{2/3}, \qquad y(t_0) = 0 \tag{3}$$

The rate function $f(t, y) = 3y^{2/3}$ is continuous on the whole ty-plane; however, $f_y(t, y) = 2y^{-1/3}$ is continuous everywhere but on the t-axis (that is, where $y = 0$). Thus, there is no rectangle R containing the initial point $(t_0, 0)$ over which f_y is continuous. Since the hypotheses of the Existence and Uniqueness Theorem are not satisfied, we should suspect trouble. In fact, we shall show that IVP (3) has more than one solution. Notice that $y = 0$, all t, is a solution of IVP (3). On the other hand, separating the variables in the ODE and integrating, we find that another solution of IVP (3) is $y = (t - t_0)^3$, $-\infty < t < \infty$. So in this case, the IVP has multiple solutions. If $y(t)$ is the state variable of a physical system modeled by IVP (3), how does the system "know" which solution curve to follow?

Extension of a Solution

There is a certain redundancy built into the definition of a solution of an ODE. Indeed, if $y(t)$, defined on an interval I, is a solution of an ODE, then $y(t)$ restricted to any subinterval of I is also a solution of the ODE. It is not useful to distinguish between these solutions, since they are all "pieces" of the original solution $y(t)$ on I. So far, so good. But what if there is a solution $z(t)$ on a larger interval that contains I with

$y(t) = z(t)$, for t in I? In this case the solution $z(t)$ is said to be an *extension* of the solution $y(t)$.

How can we be sure that a solution can be extended? Lacking a solution formula for a solution, here is how we might proceed: Let the rate function $f(t, y)$ and its derivative $f_y(t, y)$ both be continuous on a rectangle R in the ty-plane (think of R as that portion of the ty-plane displayed on your computer screen). Now suppose that $y(t)$, $a \leq t \leq b$, defines a solution curve of $y' = f(t, y)$ which lies entirely inside R. If the endpoint $(b, y(b))$ is not on the boundary of R, then there is a rectangle S inside R which contains $(b, y(b))$ in its interior. Then, the unique solution $z(t)$ to the IVP

$$z' = f(t, z), \quad z(b) = y(b)$$

extends the solution $y(t)$ to the right of $t = b$. A similar construction can be used to extend $y(t)$ to the left of $t = a$, if the point $(a, y(a))$ does not lie on the boundary of R. (This procedure is *not* recommended as a practical way for finding extensions when using numerical solvers—see Section 2.6.)

The observations above lead us to formulate the following theorem about the qualitative behavior of solution curves.

THEOREM 2.2.3

Extension Principle. Let the functions $f(t, y)$ and $f_y(t, y)$ be continuous on a closed rectangle R in the ty-plane. If (t_0, y_0) is any interior point in R, then any solution of the IVP

$$y' = f(t, y), \quad y(t_0) = y_0$$

can be extended forward and backward in t until its solution curve exits through the boundary of R.

Proof. See Appendix A.3.

Figure 2.2.1 and Atlas Plates FIRST ORDER A and B provide good illustrations of the Extension Principle.

Combining the Existence and Uniqueness Theorem and the Extension Principle, we have the following interesting observation: In a closed rectangle R where $f(t, y)$ and $f_y(t, y)$ are continuous, the ODE $y' = f(t, y)$ has a unique solution through each point (t_0, y_0) in R which can be extended forward and backward in t until the solution curve exits through the boundary of R. That is, solution curves of $y' = f(t, y)$ never "die" inside R.

An open question at this point is just how far a solution can be extended. To facilitate our discussion (and avoid making trivial distinctions between solutions) we give a name to solutions that cannot be extended.

❖ **Maximally Extended Solution.** A solution $y(t)$ of an ODE on an interval I is a *maximally extended solution* if it cannot be extended to an interval larger than I.

EXAMPLE 2.2.3

Extension of Computed Solutions
Consider the IVP

$$y' = 3t^2/(3y^2 - 4), \quad y(0) = 0 \tag{4}$$

The ODE in IVP (4) is nonlinear, but the rate function $f(t, y) = 3t^2/(3y^2 - 4)$ is

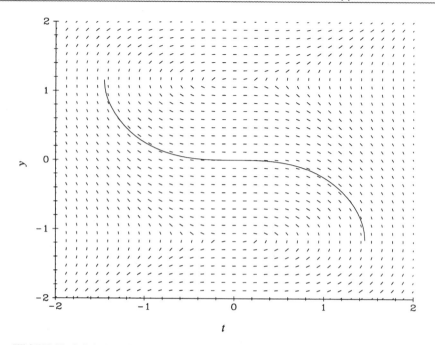

FIGURE 2.2.2 Maximally extended solution curve for IVP in Example 2.2.3

separable. So we have the option here of finding a formula which characterizes the solutions of IVP (4) implicitly, but instead we use a computer solver. We obtain the plot in Figure 2.2.2 by solving forward and backward from the initial time $t_0 = 0$, as far as we can go. Interestingly enough, the solution curve seems to stop at $t \approx \pm 1.45$ where $y \approx \mp 1.15$; the direction field shows that this observation makes sense. Is this consistent with the assertion of the Existence and Uniqueness Theorem? The answer is "yes," because the rate function $f(t, y)$ fails to be continuous on any rectangle R in the ty-plane which contains points on either of the horizontal lines $y = \pm\sqrt{4/3} \approx \pm 1.15$. Thus, when a solution curve reaches either line, then we have no idea what follows. See also Problem 9(**a**).

Based on the examples cited above it might appear that if the rate function $f(t, y)$ and its derivative $f_y(t, y)$ are continuous on the entire ty-plane, then every maximally extended solution would be defined for the entire t-axis. The next example shows that this is not the case.

EXAMPLE 2.2.4

Finite Escape Time
We will find all maximally extended solutions of the ODE

$$y' = y^2 \qquad (5)$$

First off, there is the *trivial solution*, $y = 0$, all t. If $y(t)$ is any other solution, then $y(t_0) \neq 0$ at some point t_0. Since $y(t)$ is continuous, there is an interval I containing t_0 on which $y(t) \neq 0$, and we divide each side of (5) by $y^2(t)$ to obtain

$$y^{-2}(t)y'(t) = 1, \quad \text{all } t \text{ in } I \qquad (6)$$

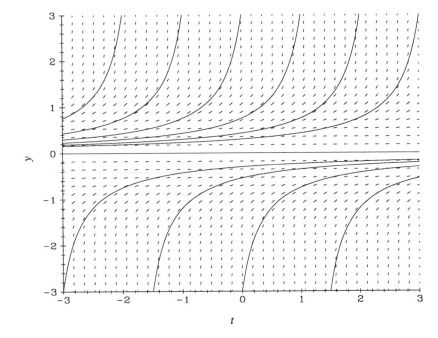

FIGURE 2.2.3 Solution curves for ODE (5).

Antidifferentiating each side of the separated ODE in (6), we have

$$-y^{-1} + C = t, \quad \text{all } t \text{ in } I \tag{7}$$

where C is some constant. Solving (7) for y, we have

$$y = \frac{1}{C - t}, \quad \text{all } t \text{ in } I \tag{8}$$

From (8) we see that the largest intervals I on which the solution formula (8) makes sense are $I = (-\infty, C)$ or $I = (C, \infty)$. In the former case $y \to +\infty$ as $t \to C^-$, and in the latter $y \to -\infty$ as $t \to C^+$. In either case the solution given by (8) is maximally extended and "escapes to infinity" as t tends to C; hence, the solution cannot pierce the "barrier" $t = C$. It would have been difficult to predict from ODE (5) that (except for $y = 0$) solutions are not defined for all time t. For all constants C, the solutions

$$y = 1/(C - t), \quad t < C, \quad \text{and} \quad y = 1/(C - t), \quad t > C$$

are maximally extended solutions. See Figure 2.2.3 for some solution curves.

Autonomous ODEs

Recall that if the rate function $f(t, y)$ does not depend on t, then it is usually written as $f(y)$, and the ODE $y' = f(y)$ is said to be *autonomous* (see Section 2.1). Solution curves of an autonomous ODE have a very interesting property. Suppose that $y(t)$ is a solution of the ODE $y' = f(y)$ on the interval $t_0 \le t \le t_1$. The curve defined by $y(t - T)$ is merely the solution curve defined by $y(t)$ translated to the right along the t-axis for T units if $T > 0$ and to the left T units if $T < 0$. Since the rate function is independent of t, the translated curve still fits the direction field, so $y(t - T)$ is also a solution of the ODE for any T.

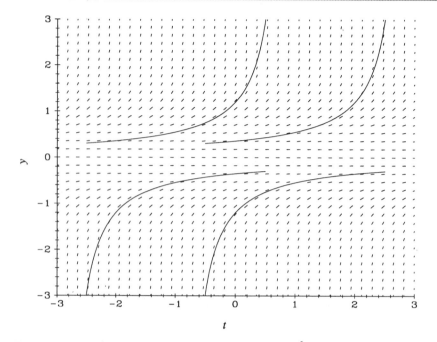

FIGURE 2.2.4 Translated solution curves for $y' = y^2$. See Example 2.2.5.

EXAMPLE 2.2.5

Translation Property of Solutions of Autonomous ODEs

In Figure 2.2.4 the direction field of the ODE $y' = y^2$ is constant along horizontal lines. The two solution curves in Figure 2.2.4 above the t-axis satisfy the initial conditions $y(-2.5) = 0.3$ and $y(-0.5) = 0.3$ and are graphed for 3 units of time. Note that the graph on the right is simply the graph on the left shifted to the right by 2 units. The solution curves below the t-axis are similarly related.

Another way of expressing this translation property is as follows: "The shape of a solution curve of an autonomous ODE does not depend on when the clock is started."

Sign Analysis

Consider the autonomous ODE $y' = f(y)$, where $f(y)$ and $f'(y)$ are continuous on the y-axis. It is a remarkable fact that by knowing no more than the (algebraic) sign of $f(y)$ on the y-axis, we can sketch qualitatively accurate solution curves in the ty-plane. This technique, often called *sign analysis*, is described below.

Observe from the translation property that if a solution curve $y(t)$ is known, then by translating it parallel to the t-axis, an entire horizontal strip of the ty-plane is covered by non-intersecting solution curves. Thus if we can find one solution curve in a horizontal strip, then we know what all solution curves in that strip look like.

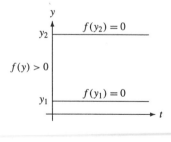

Suppose that $y_2 > y_1$ are two consecutive zeros of $f(y)$; i.e., $f(y_2) = f(y_1) = 0$, but $f(y)$ does not vanish in the interval $y_1 < y < y_2$. Then since $f(y)$ is continuous, $f(y)$ must be of one sign in the strip $y_1 < y < y_2$, say positive. Note that the lines $y = y_1$ and $y = y_2$ are solution curves for the ODE; they are called *equilibrium solutions* because the system is at rest in those states (see also Example 2.1.3). Now, because f is positive, the solution that passes through the point (t_0, y_0), $y_1 < y_0 < y_2$, must be increasing as t increases, if the solution curve remains in the strip $y_1 < y < y_2$. To leave this strip as t increases, the solution curve would have to cross the equilibrium solution curve $y = y_2$, but the Uniqueness Principle implies that this cannot happen.

The Extension Principle easily shows that our solution curve is defined for all $t \geq t_0$. Similarly, we see that for backward time our solution curve falls for decreasing t, but it never crosses the equilibrium solution $y = y_1$. Hence, the Extension Principle shows again that the solution is defined for all $t \leq t_0$. Thus, it follows that the limits

$$\lim_{t \to \infty} y(t) = b \leq y_2 \quad \text{and} \quad \lim_{t \to -\infty} y(t) = a \geq y_1$$

both exist. We now show that $a = y_1$ and $b = y_2$. By shifting this solution curve in t we can completely fill up the strip $a < y < b$ with the solution curves of the ODE. Now suppose that $b < y_2$, contrary to our assertion. Then, as noted above, the solution curve defined by the IVP $y' = f(y)$, $y(t_0) = b$, can be extended to the entire t-axis. Extending the solution backward, the solution must cross the line $y = b$ because $f(b) > 0$, and so must intersect a solution curve in the strip $a < y < b$, in contradiction to the Uniqueness Principle. Thus, it must be that $b = y_2$. Similarly, it can be shown that $a = y_1$.

EXAMPLE 2.2.6

The logistic ODE $y' = (1 - y/10)y$ is autonomous, with rate function $f(y) = (1 - y/10)y$. Note that $y_1 = 0$ and $y_2 = 10$ are consecutive zeros of $f(y)$, and that $f(y) < 0$ for $y > 10$ and $f(y) > 0$ for $0 < y < 10$. Thus, in the strip $0 < y < 10$ solution curves fall as $t \to -\infty$, approaching the equilibrium solution $y = 0$; the curves rise as $t \to +\infty$ and approach the equilibrium solution $y = 10$. In the strip $y > 10$ solution curves fall as $t \to +\infty$ approaching the equilibrium solution $y = 10$, and rise as t decreases. Figure 1.5.1 in Chapter 1 summarizes these observations.

PROBLEMS

1. Verify that there exists exactly one solution to each of the following IVPs by checking the hypotheses of the Existence and Uniqueness Theorem (do not solve).
   (a) $y' = e^t y - y^3$, $\quad y(0) = 0$ **(b)** $y' = |t|y^2 - 1/(3y + t)$, $\quad y(0) = 1$
   (c) $y' = |t||y|$, $\quad y(0) = 1$

2. **(a)** Why does the Existence Theorem not apply to the initial value problem $2tyy' = t^2 + y^2$, $y(0) = 1$?
   (b) Show that the IVP $ty' = 2y$, $y(0) = 0$ has one solution $y_1 = t^2$, $-\infty < t < \infty$, and another solution

   $$y_2 = \begin{cases} 0, & t < 0 \\ t^2, & t \geq 0 \end{cases}$$

   Why does this not contradict the Uniqueness Theorem?

3. In Example 2.2.2 two solutions of the IVP $y' = 3y^{2/3}$, $y(t_0) = 0$, are given. Find as many other solutions as you can. [*Hint*: Reread Example 2.2.5 on the translation property of solution curves of autonomous ODEs.]

4. Show that the IVP $y' = |y|$, $y(0) = 0$, has a unique solution even though the hypotheses of the Existence and Uniqueness Theorem cannot be satisfied.

5. (*Solvability of an IVP*). When the hypotheses of the Existence and Uniqueness Theorem are not satisfied, as in this problem, we must use other methods to determine the existence and the number of solutions.
   (a) Does the IVP $y' = y \operatorname{step}(t)$, $y(0) = 1$, have a solution? If so, what is it? If not, what goes wrong? Explain your answer.
   (b) Does the IVP $2yy' = -1$, $y(1) = 0$, have a solution? If it does, what is it? If not, why not? Explain your answer.

6. Does the initial value problem

$$y' = (1 - y^2)^{1/2}, \qquad y(0) = 1$$

have a unique solution? If there are multiple solutions, describe them.

7. (*Maximally Extended Solutions*). Find a solution formula for the maximally extended solution of each IVP below. Describe the t-interval on which the maximally extended solution is defined. Sketch the graph of this solution. Verify your sketch with a computer solver.

(a) $yy' = -t$, $\quad y(0) = 1$

(b) $2y' + y^3 = 0$, $\quad y(0) = -1$

8. (*Finite Escape Time*). Show that the solution of the IVP $y' = 1 - y^2$, $y(0) = y_0$, where $y_0 < 1$, escapes to infinity in finite time. Find the escape time as a function of y_0. [*Hint*: Separate variables and solve.]

9. (*Extension Principle*). Maximally extended solutions are discussed below.

(a) (*Solution Formula for Example 2.2.3*). By separating variables and solving, show that the solution $y = y(t)$ of the IVP, $y' = 3t^2/(3y^2 - 4)$, $y(0) = 0$, satisfies the relation $y^3 - 4y = t^3$. Show that $t \approx \pm 1.45$ if $y = \mp\sqrt{4/3}$. Explain why $y(t)$ cannot be extended beyond the interval $|t| < 1.45$.

(b) Use a computer solver to plot the maximally extended solution of the IVP

$$(y^2 - 1)y' = t^2 \qquad y(0) = 0.$$

Estimate the t-interval on which this maximally extended solution is defined. Find a precise description of this t-interval by first antidifferentiating each side of the ODE and using the initial data. Compare with your estimate.

10. (*Sign Analysis*). For each autonomous ODE below perform the following tasks: find all equilibrium solutions and sketch their graphs. Using sign analysis, sketch some representative solution curves above, below, and between the equilibrium curves. Use a computer solver to check the accuracy of your sketches.

(a) $y' = (1 - y)(y + 1)^2$ $\qquad$ (b) $y' = \sin(y/2)$

(c) $y' = y(y - 1)(y - 2)$ $\qquad$ (d) $y' = 3y - ye^{y^2}$

11. Examine the IVP, $ty' - 2y = t^3 \operatorname{step}(t)$, $y(1) = 0$. Explain why there is exactly one solution defined on the interval, $t \geq 0$, but infinitely many on the entire real line. Why doesn't this contradict the Existence and Uniqueness Theorem.

2.3 Sensitivity

In the previous section, the four Basic Questions for the IVP

$$y' = f(t, y), \qquad y(t_0) = y_0 \tag{1}$$

were introduced, and the issues of existence and uniqueness were discussed. In this section the question of sensitivity will be treated.

In many applications it is of great importance to know how the solution of IVP (1) changes with changes in f or y_0. In particular, if the rate function f is only changed by a small amount, can we be assured that the solution of the changed IVP will deviate no more than a prescribed amount from the solution of the original IVP?

EXAMPLE 2.3.1

Sensitivity to Changes in a Parameter
Look at the simple IVP

$$y' = c, \qquad y(0) = 0 \tag{2}$$

where c is a given constant. For $c = 1$, the solution is $y_1(t) = t$, and for an arbitrary value of c the solution is $y_c(t) = ct$. If $c \neq 1$, we see that the quantity $|y_c(t) - y_1(t)| = |c - 1||t|$ "blows up" as $|t| \to \infty$. So it would appear, in this case, that t must be restricted to a bounded interval if the solution $y_c(t)$ is to remain close to the solution $y_1(t)$. For example, if we require that for a fixed positive number T

$$|y_c(t) - y_1(t)| \leq 0.1, \quad 0 \leq t \leq T$$

then we must restrict c by the inequality

$$|c - 1| \leq 0.1/T$$

For IVP (2) there are simple inequalities that measure how much the solution changes on a fixed time interval if a rate parameter is changed. In the next example the rate function is fixed, but the initial data are changed.

EXAMPLE 2.3.2

Sensitivity to Changes in Initial Data
The IVP

$$y' = -0.05y, \quad y(0) = c, \quad c > 0$$

models radioactive decay with decay coefficient 0.05 and initial amount c. Suppose that the value of the decay coefficient is known to be accurate, but that the value of c is thought to be 1, but with a possible error of up to 10% either way; so $|c - 1| \leq 0.1$. How sensitive are future values of $y(t)$ to this uncertainty in the initial amount? Since the solutions are known to be the exponentials

$$y_c(t) = ce^{-0.05t}$$

we see that

$$|y_c(t) - y_1(t)| = |c - 1|e^{-0.05t} \leq 0.1e^{-0.05t} \tag{3}$$

From (3) we see that as t increases, the uncertainty in the amount decays exponentially. Thus, for $t \geq 0$ the solution is relatively insensitive to changes in the initial data.

Sensitivity: Computer Simulations

The above examples of sensitivity of the solution of an IVP to changes in a parameter or in the initial condition are straightforward to treat because simple solution formulas are available. What do we do if there are no solution formulas, or if the solution formulas are too complex to be of much help? The following two examples show what may be accomplished by using a numerical ODE solver.

EXAMPLE 2.3.3

Sensitivity to Changes in a Parameter: Using a Numerical ODE Solver
We examine the sensitivity of solutions of the nonlinear IVP

$$y' = -\cos y + \frac{t}{c + y^2}, \qquad y(-4) = 1 \tag{4}$$

to changes in the value of the constant c. The solution curves emerging from the initial point $(-4, 1)$ in Figure 2.3.1 correspond to $c = 1/2, 1, 3/2, 2$ (reading from

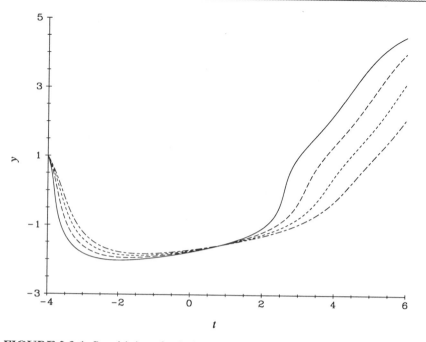

FIGURE 2.3.1 Sensitivity of solutions of IVP (4) to changes in c: $c = 1/2$ (solid), $c = 1$ (long dashes), $c = 3/2$ (short dashes), $c = 2$ (long/short dashes). See Example 2.3.3.

the bottom curve at the left). The solution curves diverge from one another at first, but after a while they converge together once more. Still later, the four curves diverge from one another again, perhaps more and more as time goes on. See Problem 3.

Denote the solution of IVP (4) for a given value of c by $y_c(t)$. Geometric intuition tells us that if t is restricted to a fixed interval, say $|t| \leq 4$, then for any value of $\epsilon > 0$ (say, $\epsilon = 0.1$) a value of $\delta > 0$ can be found such that if $|c - 1/2| < \delta$, then $|y_c(t) - y_{1/2}(t)| \leq 0.1$, for all $|t| \leq 4$. To show this rigorously is difficult, but if you "play" with the IVP on your solver you will be convinced that it is true. If a δ can be found for each ϵ, then we say that the solution is a continuous function of c at $c = 1/2$ on the indicated time interval.

Not all solvers have the capability of overlaying the graphs of a collection of IVPs on a common pair of axes. The example below shows how you can trick your solver into performing this task.

EXAMPLE 2.3.4

Using a Solver to Examine Sensitivity: Parameter as State Variable
If your solver is not set up to do sensitivity analysis directly, there is a way to coax your solver into being more cooperative. We illustrate this technique with an example: How does the solution of the IVP

$$y' = -y\sin t + ct\cos t, \qquad y(-4) = -6 \qquad (5)$$

change when the constant c changes? There is nothing to stop us from thinking of the constant c as a new state variable. Apparently the new state variable $c(t)$ satisfies the IVP $c' = 0$, $c(-4) = c_0$, where c_0 is the desired setting for the constant c. Your solver knows how to plot and overlay solutions $y = y(t)$, $c = c(t)$, of the planar system

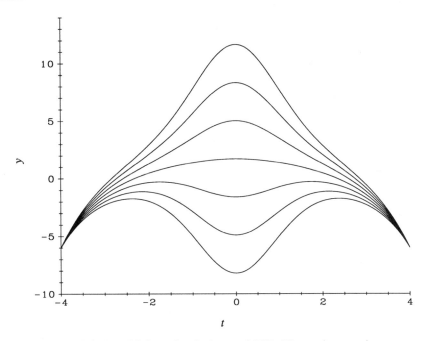

FIGURE 2.3.2 Sensitivity of solutions of IVP (5) to changes in a rate parameter.

$$y' = -y \sin t + ct \cos t, \quad y(-4) = -6$$
$$c' = 0, \quad\quad\quad\quad\quad c(-4) = c_0 \tag{6}$$

for various values of c_0. The component $y = y(t)$ of the solution pair $y(t)$, $c(t)$ of the IVP (6) must solve the IVP (5) for the value $c = c_0$. In Figure 2.3.2 we overlay graphs of $y(t)$ versus t for the values $c_0 = 0.7, 0.8, 0.9, 1.0, 1.1, 1.2$, and 1.3 (reading the graphs from the top down). Although the ODE here is linear, the solution formula involves an integral that cannot be expressed in elementary terms, and that is why we have turned to a numerical ODE solver. From Figure 2.3.2 we conjecture that as c is changed, the solution curves spread apart, but then come together again as t increases to 4. Thus, for values of t close to 4, $t < 4$, we might say that IVP (5) is not very sensitive to changes in the value of c in the range $0.7 \leq c \leq 1.3$. See also Problem 6.

Sensitivity for Linear ODEs: Inequality Estimates

Of the four Basic Questions about IVPs which were posed in the previous section, three have already been answered for the first-order linear IVP

$$y' + p(t)y = q(t), \quad\quad y(t_0) = y_0 \tag{7}$$

In Section 1.2 we saw that for any continuous $p(t)$ and $q(t)$ on an interval I and any initial data y_0, the IVP (7) has a unique solution given by the formula

$$y(t) = e^{-P_0(t)} y_0 + e^{-P_0(t)} \int_{t_0}^{t} e^{P_0(s)} q(s) \, ds, \quad\quad t \text{ in } I \tag{8}$$

where $P_0(t) = \int_{t_0}^t p(r)\,dr$. The data for IVP (7) are the initial data y_0 and the driving term (or input) $q(t)$. The solution $y(t)$ in (8) is the response (or output). Now we look at the fourth basic question about IVP (7); namely, how does the solution to IVP (7) change as the data change?

Before giving our answer, we analyze the two terms in the solution formula (8), much as we did in Section 1.2. If we put

$$z(t) = e^{P_0(t)} y_0 \quad \text{and} \quad Z(t) = e^{-P_0(t)} \int_{t_0}^t e^{P_0(s)} q(s)\,ds$$

then $y(t) = z(t) + Z(t)$, and the terms $z(t)$ and $Z(t)$ are the unique solutions of the respective IVPs

$$z' + p(t)z = 0, \quad z(t_0) = y_0; \qquad Z' + p(t)Z = q(t), \quad Z(t_0) = 0$$

Because of this, the solution formula (8) can be rephrased as follows:

$$\text{Total response} = \begin{array}{c} \text{Response due} \\ \text{to initial data} \\ \text{(with zero input)} \end{array} + \begin{array}{c} \text{Response due to} \\ \text{driving term (with} \\ \text{zero initial data)} \end{array}$$

The function $z(t)$ is the *initial data response* and measures the response of the system to a zero input and the initial state y_0. $Z(t)$ is the *driving term response* and denotes the system's response to zero initial state ($y(t_0) = 0$) and the driving term $q(t)$. We will often use this "response" terminology.

EXAMPLE 2.3.5

Adding the Responses

Using the initial data response and the driving term response notation described above, the solution of the IVP

$$y' + 0.01ty = 4 + \cos t, \qquad y(0) = 20 \tag{9}$$

can be written as $y(t) = z(t) + Z(t)$, where

$$z(t) = 20e^{-0.005t^2}, \quad Z(t) = e^{-0.005t^2} \int_0^t e^{0.005s^2} (4 + \cos s)\,ds$$

In Figure 2.3.3, the graphs of $z(t)$ (long dashes), $Z(t)$ (short dashes), and $y(t) = z(t) + Z(t)$ (solid) are plotted versus t.

The advantage of using this way of representing the solution of a linear IVP is that we can isolate the effects of changes in the initial data (changes in $z(t)$ only) from the effects of changes in the driving term (changes in $Z(t)$ only).

The result below is a satisfactory, if crude, answer to the sensitivity question for first-order linear IVPs.

THEOREM 2.3.1

Basic Sensitivity Estimate. Let $T > t_0$ in the interval I be given, and suppose that $w(t)$ is defined on I and solves the IVP

$$w' + p(t)w = q(t), \quad w(t_0) = w_0$$

where p and q are continuous functions on I. Then there are positive constants K_1 and K_2, independent of the data q and w_0, such that

$$|w(t)| \leq K_1|w_0| + K_2 \max_{t_0 \leq t \leq T} |q(t)| \quad \text{for all } t_0 \leq t \leq T \tag{10}$$

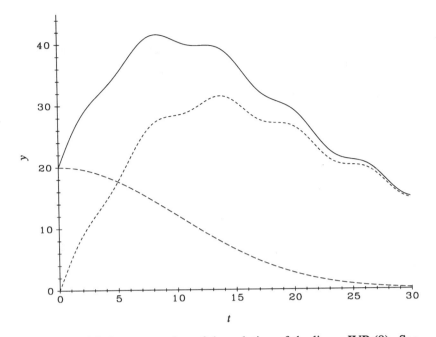

FIGURE 2.3.3 Representation of the solution of the linear IVP (9). See Example 2.3.5.

Proof. Let $P_0(t) = \int_{t_0}^{t} p(r)\, dr$, and put

$$m = \min_{t_0 \le t \le T} P_0(t), \qquad M = \max_{t_0 \le t \le T} P_0(t) \tag{11}$$

Using the solution formula (8), the Triangle Inequality (see Appendix C.6), and the Integral Estimate (see Appendix C.5), we derive the following inequalities valid on the interval $t_0 \le t \le T$:

$$|w(t)| \le |e^{-P_0(t)} w_0| + |e^{-P_0(t)} \int_{t_0}^{t} e^{P_0(s)} q(s)\, ds|$$

$$\le |e^{-P_0(t)}|\, |w_0| + |e^{-P_0(t)}| \int_{t_0}^{t} |e^{P_0(s)}|\, |q(s)|\, ds|$$

$$\le e^{-m} |w_0| + e^{M-m} \int_{t_0}^{t} |q(s)|\, ds$$

$$\le e^{-m} |w_0| + e^{M-m} |T - t_0| \max_{t_0 \le t \le T} |q(t)|$$

The estimate (10) now follows with

$$K_1 = e^{-m} \quad \text{and} \quad K_2 = e^{M-m} |T - t_0| \tag{12}$$

Using the Basic Estimate we can show that the solution of the IVP: $y' + p(t)y = q(t)$, $y(t_0) = y_0$ changes "continuously" as the data y_0 and $q(t)$ change "continuously."

THEOREM 2.3.2

Continuity with Respect to the Data. Let $T > t_0$ in I be given, and suppose that $y(t)$ and $\bar{y}(t)$ are the respective responses of the linear IVP $y' + p(t)y = q(t)$, $y(t_0) = y_0$ to the data y_0, $q(t)$ and $\bar{y}_0$, $\bar{q}(t)$. Then there are positive constants K_1 and K_2, independent of the data, such that

$$|y(t) - \bar{y}(t)| \leq K_1 |y_0 - \bar{y}_0| + K_2 \max_{t_0 \leq t \leq T} |q(t) - \bar{q}(t)| \qquad (13)$$

for all $t_0 \leq t \leq T$.

Proof. Put $w(t) = y(t) - \bar{y}(t)$ and note that $w(t)$ solves the IVP

$$w' + p(t)w = q(t) - \bar{q}(t), \qquad w(t_0) = y_0 - \bar{y}_0 = w_0$$

Applying Theorem 2.3.1,

$$w(t) \leq K_1 |w_0| + K_2 \max_{t_0 \leq t \leq T} |q(t) - \bar{q}(t)|$$

and the inequality (13) follows. ∎

Now since the constants K_1 and K_2 in (13) are independent of the data, we see that the more the data $\bar{y}_0$, $\bar{q}(t)$ look like y_0, $q(t)$, the more the response $\bar{y}(t)$ looks like $y(t)$. That is, "small" changes in y_0 and in $q(t)$ produce no more than "small" changes in $y(t)$ over a fixed time span. Such behavior is usually associated with the term "continuity," which explains the title of the theorem just proved. Thus a satisfactory answer to the fourth question posed in Section 2.2 has been found (at least for linear IVPs) by using the estimate (13) on $|y(t) - \bar{y}(t)|$.

Estimate (13) places a rough upper bound on the uncertainties in the solution of a linear IVP due to uncertainties in the initial data and driving term. It is possible to get tighter upper bounds for the uncertainties in specific cases.

EXAMPLE 2.3.6

Sensitivity of Salt Solution Model
In Section 1.3 we showed that the model IVP for the amount of salt $S(t)$ at time t in a tank holding 100 gallons of brine is

$$S'(t) = -S(t)/10 + 10c(t), \quad S(0) = S_0 \qquad (14)$$

where brine with a concentration of $c(t)$ pounds of salt per gallon runs into the tank at the rate of ten gallons per minute, the stirred mixture drains from the tank at the rate of ten gallons per minute, and the initial amount of salt in the brine in the tank is S_0. Suppose we want to estimate how much the uncertainties in the inflow concentration $c(t)$ and in the initial amount S_0 affect the value of $S(t)$. We could use Theorem 2.3.2 to do this, but in this case we can get better results by direct estimates. Suppose that $S(t)$ and $\bar{S}(t)$ are the respective solutions of IVP (14) corresponding to $c(t)$, S_0 and to $\bar{c}(t)$, $\bar{S}_0$. Using (8) for $S(t)$ and $\bar{S}(t)$, subtracting, and taking magnitudes we have (as in the proof of Theorem 2.3.1) that

$$|S(t) - \bar{S}(t)| \leq e^{-t/10}|S_0 - \bar{S}_0| + 10e^{-t/10} \int_0^t e^{s/10} |c(s) - \bar{c}(s)| \, ds \; .$$

Thus, if we estimate that $|S_0 - \bar{S}_0|$ is, say, no more than a positive constant k_1 and $|c(s) - \bar{c}(s)|$ is no more than a positive constant k_2, then

$$|S(t) - \bar{S}(t)| \leq e^{-t/10}k_1 + 100k_2 e^{-t/10}(e^{t/10} - 1)$$

$$= e^{-t/10}k_1 + 100k_2(1 - e^{-t/10})$$

and we see that we have a tight estimate on the amount of "error" caused by the uncertainties in S_0 and $c(s)$. Observe that as time goes on, the "k_1 error term" decays, but the "k_2 error term" increases toward the value $100k_2$.

Comments

The generalizations of Theorem 2.3.1 and 2.3.2 to nonlinear IVPs are presented in Appendix A.4. In any case, we can say that if the rate function $f(t, y, c)$ of the IVP

$$y' = f(t, y, c), \quad y(t_0) = y_0 \tag{15}$$

is jointly continuous in t, y, and c in a region R of tyc-space and if f_y and f_c are also continuous in R, then the solution of IVP (15) is a continuous function of t, y_0, c in R. Under these conditions, the solution remains within prescribed bounds in a given time interval when y_0 and c are changed within small enough bounds; in this case, the solution is "insensitive" to the changes, i.e., the solution is *robust* relative to the allowable changes. We say that the solution of IVP (15) is highly sensitive if even "small changes" in y_0 or in c over a given time interval $t_0 \leq t \leq t_0 + T$ lead to "large changes" in the solution at some time in the interval.

PROBLEMS

1. (*Sensitive or Insensitive?*). Solve each IVP below. Is the solution sensitive or insensitive on the given time interval to changes in the initial value a or parameter c? Explain your answers. [*Hint*: See Example 2.3.1 and 2.3.2.]

   (a) The solution of $y' = -y + e^{-t}$, $y(0) = a$ is denoted by $y(t, a)$. Find $|y(t, a) - y(t, y_1)|$ for $t \geq 0$, and answer the sensitivity questions. [*Hint*: First show that $y(t, a) = ae^{-t} + te^{-t}$.]

   (b) The solution of $y' = -y + c$, $y(0) = 1$ is denoted by $y_c(t)$. It is required that $|c - 5| \leq 0.1$. Find $|y_c(t) - y_5(t)|$ for $t \geq 0$, and answer the above questions about sensitivity.

   (c) The solution of $y' = -y + ct$, $y(0) = a$ is denoted by $y_c(t, a)$. Find $|y_c(t, a) - y_1(t, b)|$ for $0 \leq t \leq 10$ and answer the above questions about sensitivity.

2. (*Sensitivity Depends on Parameter Range, Time Span*). Graph the seven solution curves of the linear IVP, $y' = -cy/t$, $y(10) = 3$, $0 < t \leq 30$, where $c = -1.5$, $-1.0, -0.5, \ldots, 1.5$. Are the solutions sensitive to changes in c if $|c + 1| \leq 0.5$, $0 < t \leq 10$; $|c + 1| \leq 0.5$, but $10 \leq t \leq 30$; $|c - 1| \leq 0.5$, $0 < t \leq 10$; $|c - 1| \leq 0.5$, $10 \leq t \leq 30$?

3. (*Example 2.3.3 Continued*). Consider the IVP of Example 2.3.3: $y' = -\cos y + t/(c + y^2)$, $y(-4) = 1$.

   (a) What conjecture might you make about the long-term behavior of the solution curves if $1 \leq c \leq 4$ and all you have available is Figure 2.3.1?

   (b) Now extend the time interval to $-4 \leq t \leq 20$. A new conjecture?

   (c) Repeat (b) for $-4 \leq t \leq 100$. What do you conjecture now?

   (d) Extend the time interval to $-4 \leq t \leq 200$; what do you think now?

4. (*Linearity of Responses*). Let $z(t)$ and $\bar{z}(t)$ be the initial data responses of the IVP (7) with $q(t) = 0$ due to the initial states y_0, and $\bar{y}_0$, respectively. Show that for any constants a and b, $az(t) + b\bar{z}(t)$ is the initial data response to the initial

state $ay_0 + b\bar{y}_0$. Show that the driving term response $Z(t)$ of the IVP (7) with $y_0 = 0$ exhibits a similar property if the input is $aq(t) + b\bar{q}(t)$.

5. (*Sensitivity to Initial Data*). Graph the solutions of the IVP

$$y' = 3y \sin y - t, \quad y(0) = a, \quad 0 \le t \le 10$$

where $a = -3.13, -3.130119, -3.1301193, -3.13011939, -3.130119391,$ $-3.1301194, -3.1301196, -3.13014$. Describe what you see. What does this tell you about the accuracy of long-term solutions of IVPs? [*Hint*: See also Atlas Plate FIRST ORDER J.]

6. (*Example 2.3.4 Continued*). Consider the IVP

$$y' = -y \sin t + ct \cos t, \quad y(-4) = -6, \quad c = 0.7, 0.8, \dots, 1.3$$

Graph the seven solution curves for $-4 \le t \le 4$. Then for $-4 \le t \le 20$. Any conjectures about the long-term behavior of solutions if $|c - 1| \le 0.3$?

7. Recreate the graphs in Atlas Plate FIRST ORDER K. On this graph, overlay solution curves of the IVP for values of $c = 3, 1, 0.5$. What can you say about the long-term behavior of the graphs? What is the effect on the graphs of increasing the value of c?

8. On separate plots, graph some solution curves of the ODE $y' = y(2 - y) + c$ for $t \ge 0$ using several values of the constant c, with $0 \le c \le 1$, including $c = 0$. Use these graphs to estimate a bound for $|y(t) - 2|$, $t \ge 0$, where $y(t)$ is the solution of the IVP

$$y' = y(2 - y) + c, \quad y(0) = y_0$$

where $0 \le c \le 1$, and $0 \le y_0 \le 1 + \sqrt{1 + c}$. [*Hint*: Use sign analysis to characterize the behavior of the solution curves.]

2.4 Population with Bounded Migration

The character of the output of an IVP can change significantly when the data undergoes small changes. The effects of changing the data are of considerable importance since any system such as a chemical or nuclear reactor, an electrical circuit, a space probe, or an isolated population with bounded migration should be relatively impervious to the inevitable small disturbances and random shocks which will act upon it.

In this section we shall look at a different kind of sensitivity question, one that is important in practice. The question we will deal with can be put in this way:

> A persistent input disturbance, even though small in magnitude, may have a cumulative effect which in time will tend to destroy the system because the magnitude of the output could approach unacceptably high values. Can this response be predicted and then avoided?

This question can be neatly answered for first-order linear ODEs, and we present the result below as the *Bounded Input-Bounded Output Principle*. At the end of the section we show that the Bounded Input-Bounded Output Principle holds for some special nonlinear ODEs such as logistic growth with migration.

Linear ODEs with Bounded Inputs

Suppose that the "ideal" state $z(t)$ of a system is governed by the first-order linear IVP

$$z' + p(t)z = r(t), \qquad z(0) = z_0 \tag{1}$$

where $p(t)$ and $r(t)$ are continuous on the interval $I = [0, \infty)$. If the input $r(t)$ suffers an additive disturbance $q(t)$ and the initial data an additive disturbance y_0, then the response $w(t)$ of the system to these disturbances solves the IVP

$$w' + p(t)w = r(t) + q(t), \qquad w(0) = z_0 + y_0 \tag{2}$$

Our question can now be phrased as follows:

> What conditions on $p(t)$ and the disturbances y_0 and $q(t)$ will ensure that the deviation $y = w - z$ from the ideal solution $z(t)$ is bounded for all time $t \geq 0$?

The deviation $y(t)$ also solves a linear IVP because from (1) and (2) we have that

$$y' = w' - z' = (r + q - pw) - (r - pz)$$
$$= q - p(w - z) = q - py$$

Thus, y solves the IVP

$$y' + p(t)y = q, \qquad y(0) = w_0 - z_0 = y_0 \tag{3}$$

Suppose that the coefficient $p(t)$ is a design element under our control, but we have no control over the disturbances y_0 and $q(t)$. To answer the above question we need the following result.

THEOREM 2.4.1

> *Bounded Input-Bounded Output Principle.* Suppose that for $t \geq 0$, $p(t)$ and $q(t)$ are continuous functions satisfying (a) $p(t) \geq p_0 > 0$ and (b) $|q(t)| \leq M$ for some positive constants p_0 and M. Then for all $t \geq 0$ the solution $y(t)$ of IVP (3) satisfies
>
> $$|y(t)| \leq e^{-p_0 t}|y_0| + \frac{M}{p_0}(1 - e^{-p_0 t}) \leq \max\left\{|y_0|, \frac{M}{p_0}\right\} \tag{4}$$

Proof. First note that by the results of Section 1.2 the solution $y(t)$ of IVP (3) is given by

$$y(t) = e^{-P_0(t)}y_0 + e^{-P_0(t)} \int_0^t e^{P_0(s)}q(s)\,ds, \qquad t \geq 0 \tag{5}$$

where

$$P_0(t) = \int_0^t p(r)\,dr$$

Note that the hypothesis on $p(t)$ implies that $-p(t) \leq -p_0 < 0$, and hence that

$$-P_0(t) = \int_0^t -p(r)\,dr \leq \int_0^t -p_0\,dr = -p_0 t$$

Thus,

$$-P_0(t) + P_0(s) = \int_0^t -p(r)\,dr + \int_0^s p(r)\,dr = \int_s^t -p(r)\,dr$$

$$\leq \int_s^t -p_0\,dr = -p_0(t-s), \quad \text{if } 0 \leq s \leq t$$

We shall use these inequalities below. Since the solution $y(t)$ of IVP (3) is given by the solution formula (5), the following inequalities may be derived as in the proof of the Basic Sensitivity Estimate in Section 2.3:

$$|y(t)| = \left| e^{-P_0(t)} y_0 + e^{-P_0(t)} \int_0^t e^{P_0(s)} q(s)\,ds \right|$$

$$\leq e^{-P_0(t)} y_0 + \left| e^{-P_0(t)} \int_0^t e^{P_0(s)} q(s)\,ds \right| \quad \text{(by the Triangle Inequality)}$$

$$\leq e^{-P_0(t)} |y_0| + \int_0^t e^{-P_0(t)+P_0(s)} |q(s)|\,ds$$

$$\leq e^{-p_0 t} |y_0| + \int_0^t e^{-p_0(t-s)} M\,ds = e^{-p_0 t} |y_0| + \frac{M}{p_0} e^{-p_0 t} [e^{p_0 s}] \Big|_{s=0}^{s=t}$$

$$= e^{-p_0 t} |y_0| + \frac{M}{p_0} (1 - e^{-p_0 t})$$

This is the first inequality in (4).

The second inequality is obtained as follows: Put

$$B(t) = e^{-p_0 t} |y_0| + \frac{M}{p_0} (1 - e^{-p_0 t}), \quad t \geq 0$$

and so we have that

$$B'(t) = p_0 e^{-p_0 t} \left(\frac{M}{p_0} - |y_0| \right)$$

If $|y_0| \geq M/p_0$, $B(t)$ is nonincreasing and has maximal value $B(0) = |y_0|$, while if $|y_0| < M/p_0$, then $B(t)$ is increasing and tends to the upper bound M/p_0 as $t \to \infty$. This shows the second inequality in (4), finishing the proof.

The hypotheses of Theorem 2.4.1 require that the positive function $p(t)$ be "bounded away" from 0 by a positive constant p_0 and that the input $q(t)$ be bounded in magnitude by a constant M. The conclusion states that the output $y(t)$ is bounded in magnitude and, in addition, gives estimates for the bound.

EXAMPLE 2.4.1

Estimating the Size of the Output

We use the estimate (4) of the Bounded Input-Bounded Output Principle to estimate the maximum magnitude of the solutions of the IVP

$$y' + (1+t)y = 4\sin t, \qquad y(0) = y_0$$

over the interval $t \geq 0$, for any value of the initial data, y_0. Using the notation of Theorem 2.4.1 we see that $p(t) = 1+t \geq p_0 = 1$ and $|q(t)| = |4\sin t| \leq M = 4$, for $t \geq 0$. The estimate (4) becomes

$$|y(t)| \leq \max\{|y(0)|, 4\}, \quad t \geq 0$$

for any solution of the IVP. Figure 2.4.1 shows that $|y(t)| \leq 4$ for any solution with $|y(0)| \leq 4$, and $|y(t)| \leq |y(0)|$, for any solution with $|y(0)| \geq 4$.

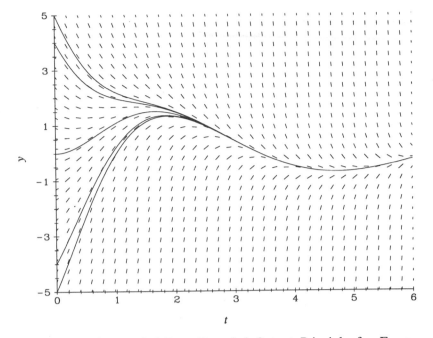

FIGURE 2.4.1 Bounded Input-Bounded Output Principle for Example 2.4.1.

The estimates of the Bounded Input-Bounded Output Principle are the best possible in the following senses:

- If $p(t) = p_0 > 0$, $q(t) = M > 0$, for all $t \geq 0$, the first inequality in conclusion (4) becomes an equality: $y(t) = e^{-p_0 t} y_0 + M(1 - e^{-p_0 t})/p_0$.

- If we have that $p(t) > 0$ but not $p(t) \geq p_0 > 0$, for all $t \geq 0$, IVP (3) may have an unbounded solution.

- If we have an unbounded input $q(t)$, the output $y(t)$ may also be unbounded.

See Problems 4–6 for examples that illustrate these points.

Recall that $y(t)$ measures the deviation from the ideal solution of IVP (1). From a design point of view we see that to obtain a tighter bound on the magnitude of $|y(t)|$, we should make $p(t)$ (hence p_0) as large as possible. Alternatively, if p_0 is large, the system can tolerate "shocks" (i.e., driving terms $q(t)$) of large magnitude. That is, the upper bound M on $|q(t)|$ may be large without running the risk of producing excessively large values of $|y(t)|$.

The Bounded Input-Bounded Output Principle has been interpreted in terms of the design of a dynamical system. As with any mathematical theorem, however, the principle can be applied to all phenomena that have the same mathematical model covered by the theorem. We have seen before just how effective this universality of mathematical theorems and models can be, and the following example illustrates the point once again.

---

[1]To a chemist, (6) is a *first-order* rate law, while rate laws such as $da/dt = -ka^2$ or $da/dt = -kab$ are *second-order*, involving products of two concentrations, a and b.

EXAMPLE 2.4.2

Chemical Reactor

A substance A is gradually converted into a different substance B in a chemical reactor, which is a tank of V liters filled to capacity with a solution of A, B, and perhaps, other chemicals and catalysts. How much of A remains in the reactor t units of time after the reaction begins?

Let $a(t)$ denote the *concentration* at time t of A in, say, moles per liter of solution, and suppose that a *first-order rate law* models the reaction $A \to B$,

$$\frac{da}{dt} = -ka \tag{6}$$

where k is the positive rate constant.[1] With the passage of time, $a(t)$ will decay exponentially from its initial value: $a(t) = a(0)e^{-kt}$.

In real life, however, we have to contend with leaky valves, open stopcocks, and so on, which permit small amounts of A to dribble into the reacting mixture even while the reaction is taking place. Let $r(t)$ denote the rate at which A drips into the reactor. For simplicity, we assume that the volume V of the substance in the reactor remains constant throughout the process. Then the basic Balance Law:

$$\text{Rate of accumulation} = \text{Rate in} - \text{Rate out}$$

implies that

$$\frac{da}{dt} = \frac{r(t)}{V} - ka$$

If we estimate that $0 \le r(t) \le r_0$, then the Bounded Input-Bounded Output Principle (4) implies that

$$a(t) \le e^{-kt}a(0) + \frac{r_0}{kV}(1 - e^{-kt}) \le \max\left\{a(0), \frac{r_0}{kV}\right\} \tag{7}$$

Suppose that we require that $a(t)$ never exceed some constant value K. Such a restriction might be necessary to ensure safe operation of the reactor. By (7), the restrictions $a(0) \le K$ and $r_0 \le kVK$ will guarantee that $a(t) \le K$. The first restriction is obvious, the second less so. In any event, the inequalities (7) give us practical limits on the concentration $a(t)$ over the course of the reaction.

The Bounded Input-Bounded Output Principle works for a linear ODE in large part because of the exponentially decaying term $e^{-p_0 t}$. What happens if the ODE is nonlinear?

Nonlinear Equations with Bounded Inputs

Every linear first-order ODE has the same form, but nonlinear ODEs have nothing in common but their nonlinearity. It is not surprising that there is no general extension of the Bounded Input-Bounded Output Principle to nonlinear ODEs. Restricted versions of the principle do apply, however, to certain nonlinear equations, as we shall see below. For other nonlinear equations, everything goes wrong and bounded inputs result in unbounded solutions. Both points may be illustrated with nonlinear ODEs of the form

$$y' = f(y) + q(t)$$

where we assume that $f(y)$ is not of the form $ay + b$, so that the ODE is not linear. We shall call $q(t)$ the *input* or *driving term* in accord with the terminology for the linear case. Example 2.4.3 shows that a Bounded Input-Output Principle applies to the logistic growth model.

EXAMPLE 2.4.3

Logistic Growth with Migration: A Cap on the Total Population

What happens to the population curves when a migration term $q(t)$ is added to the ODE of logistic growth of Section 1.5? Recall that $P(t)$ denotes the size of the population at time t and that the model is valid only for $P \geq 0$ and $t \geq 0$. The corresponding driven IVP is

$$P'(t) = r\left[1 - \frac{P(t)}{K}\right]P(t) + q(t), \qquad P(0) = P_0 \qquad (8)$$

where r, K, and P_0 are positive constants, and $q(t)$ denotes the net rate of migration, that is, the rate of immigration minus the rate of emigration. Observe from Section 1.5 that all solutions of the undriven logistic equation (i.e., IVP (8) with $q(t) = 0$) that enter the population quadrant $t \geq 0$, $P > 0$, tend to K as $t \to \infty$; these solutions are all bounded for $t \geq 0$. Are solutions of the driven IVP (8) also bounded in the quadrant for all $t \geq 0$? Let us assume that the migration rate $q(t)$ is bounded, $|q(t)| \leq Q$, where Q is a positive constant.

We shall show that for each $P_0 > 0$, either there is a positive constant P^* such that

$$0 < P(t) \leq P^*, \quad \text{for } t \geq 0 \qquad (9)$$

or, there are positive constants T^* and P^* such that

$$0 < P(t) \leq P^*, \quad \text{for } 0 \leq t < T^*, \quad \text{and } P(T^*) = 0 \qquad (10)$$

In other words, bounded migration implies that the population either remains forever bounded or else becomes extinct in finite time. This is a special nonlinear version of the Bounded Input-Bounded Output Principle.

To prove that the inequalities (9) and (10) hold, we first examine the direction field of the ODE in IVP (8). Since r, K, and Q are positive constants, the roots P_1 and P_2 of the quadratic polynomial

$$R(P) = r\left(1 - \frac{P}{K}\right)P + Q = -\frac{r}{K}P^2 + rP + Q \qquad (11)$$

are real and of opposite sign:

$$P_1 = \frac{K}{2} - \frac{1}{2}\left(K^2 + \frac{4KQ}{r}\right)^{1/2} < 0$$

$$P_2 = \frac{K}{2} + \frac{1}{2}\left(K^2 + \frac{4KQ}{r}\right)^{1/2} > 0$$

We ignore P_1 because it is negative and we are only interested in nonnegative values of P. Since the coefficient of P^2 in $R(P)$ is negative, $R(P)$ is negative for all values of P larger than P_2. For $P > P_2$, and $t \geq 0$ we see that

$$P'(t) = r\left[1 - \frac{P}{K}\right]P + q(t) \leq r\left[1 - \frac{P}{K}\right]P + Q = R(P) < 0$$

So, the slope of the direction field lines is negative in the entire sector $P > P_2$, $t \geq 0$, in the tP-plane. This means that for P_0 with $0 < P_0 \leq P_2$, the solution curve of IVP (8) must either remain in the strip $0 < P \leq P_2$ for all $t \geq 0$, or run into the t-axis at some $T^* > 0$ (there are no other possibilities). On the other hand, if $P_0 > P_2$, then the solution curve of IVP (8) must lie below the line $P = P_0$, for all $t \geq 0$. If the solution curve does not run into the t-axis, then $0 < P(t) \leq P_0$, for all $t \geq 0$. Hence the alternatives (9) and (10) hold in all cases.

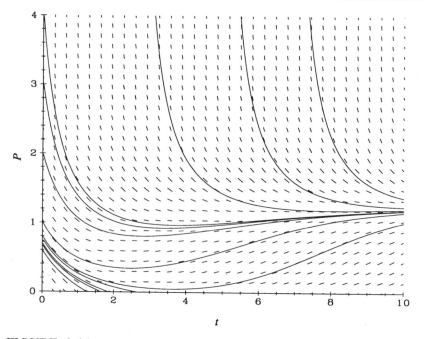

FIGURE 2.4.2 Logistic growth with limited migration.   See Example 2.4.4.

EXAMPLE 2.4.4

Limited Migration: Bounded Growth

Using the notation in Example 2.4.3 for the IVP

$$P' = \frac{1}{2}(1 - P)P + (-1 + t/4)/(1 + t), \quad t \geq 0, \quad P(0) = P_0 \qquad (12)$$

we see that $r = 1/2$, $K = 1$, and $q(t) = (-1 + t/4)/(1 + t)$. Note that $q(0) = -1$ and $q'(t) \geq 0$ for $t \geq 0$. Since $q(t) \to 1/4$ as $t \to \infty$, we see that the largest value of $|q(t)|$, $t \geq 0$ is $Q = |q(0)| = 1$. Then the polynomial $R(P)$ in (11) is $-P^2/2 + P/2 + 1$, and its largest root is $P_2 = 2$. See Figure 2.4.2 for a direction field and some solution curves for IVP (12) in the quadrant $P \geq 0$, $t \geq 0$. Notice that the direction field lines in the sector $P > P_2 = 2$, $t \geq 0$, all have negative slope, and that both alternatives (9) and (10) are present when $0 < P_0 \leq P_2 = 2$.

In spite of the positive results of Example 2.4.3, we cannot hope to have a general result. See Problem 7 for an example of just how badly things may go wrong in the presence of a different type of nonlinearity.

Comments

We have shown that the Bounded Input-Bounded Output Principle provides an effective estimate of the dependence of the solution of a linear initial value problem on the input. On the other hand, nonlinear ODEs can be problematic. Special versions of the principle are valid for special equations (see Examples 2.4.3, 2.4.4), but there is no general result.

PROBLEMS

1. (*Stocking a Population*). A decaying population is replenished from time to time with new stock. Suppose that the rate law is $P'(t) = -r(t)P(t) + R(t)$, where all we know about r and R is that for $t \geq 0$, $0 < r_0 \leq r(t)$ and $|R(t)| \leq R_0$, for some positive constants r_0 and R_0. If $P(0) = P_0 > 0$, find a reasonable upper bound for $P(t)$, $t \geq 0$. [*Hint*: Use the Bounded Input-Bounded Output Principle.]

2. (*Salt Solution*). A large vat contains 100 gallons of brine in which initially 5 pounds of salt is dissolved. More brine runs into the vat at a rate of r gallons per minute with a concentration of $c(t)$ pounds of salt per gallon. The solution is thoroughly mixed and runs out at a rate of r gallons per minute. For safety reasons, the concentration in the tank must never exceed 0.1 pound of salt per gallon. Suppose that for $t \geq 0$, $0 < r_0 \leq r \leq r_1$ and $0 \leq c(t) \leq c_0$, for some positive constants r_0, r_1, and c_0. What conditions must r_0, r_1, and c_0 satisfy to ensure safe operation?

3. (*Periodic Harvesting and Restocking a Logistic Population*). A logistically changing population is harvested/restocked sinusoidally:

$$P' = r(1 - P/K)P + H_0 \sin(2\pi t), \quad P(0) = P_0$$

where r, K, H_0, P_0 are positive constants. Assume that the model is valid only for $P(t) \geq 0$.

(a) Let $r = 0.4$, $K = 1000$, $H_0 = 121$. Plot solution curves for various values of $P_0 > 0$. Do you think there are values of P_0 for which the population eventually dies out? What do you conjecture the solution curves do as $t \to \infty$?

(b) Let P_2 be the larger of the two roots of the quadratic $-rP^2/K + rP + H_0$. Explain why $P(t)$, $t \geq 0$, never exceeds the larger of the two numbers P_0 and P_2. [*Hint*: Look at Example 2.4.3, particularly (11) and the discussion that follows.]

4. (*BIBO Principle Cannot be Improved*). Show that the inequality

$$y(t) \leq e^{-p_0 t}|y_0| + \frac{M}{p_0}(1 - e^{-p_0 t})$$

of the Bounded Input-Bounded Output Principle cannot be improved. [*Hint*: Solve the IVP $y' + p_0 y = M$, $y(0) = y_0$, where p_0, M, and y_0 are positive constants.]

5. (*Limited Validity of BIBO*). Show that the hypothesis $p(t) \geq p_0 > 0$ in the Bounded Input-Bounded Output Principle cannot be extended to the condition $p(t) > 0$. [*Hint*: Show that the problem $y' + (t + 1)^{-2}y = e^{1/(t+1)}$, $t \geq 0$, has unbounded solutions even though $p(t) > 0$ and $|q(t)| \leq e$ for $t \geq 0$.]

6. (*Limited Validity of BIBO*). Show that the equation $y' + y = t$ has unbounded solutions on the interval $0 \leq t$, thus showing that the hypothesis $|q(t)| \leq M$ cannot be dropped from the Bounded Input-Bounded Output Principle.

7. (*Failure of BIBO for a Nonlinear ODE*). The following shows how badly things may go wrong in the presence of a nonlinearity:

(a) Show that every solution of the undriven ODE $y' = -y/(1 + y^2)$ tends to zero as $t \to \infty$.

(b) Show that every solution $y(t)$ of $y' = -y/(1 + y^2) + 1$, the driven ODE, satisfies the inequality $y'(t) \geq 1/2$, for all $t \geq 0$. [*Hint*: Use calculus to find the minimum value of $-y/(1 + y^2) + 1$.]

(c) Upon integrating the inequality in part (b), we have that $y(t) \geq t/2 + C$, where C is a constant of integration. Show that even though all solutions of the undriven ODE in part (a) decay to 0 as $t \to \infty$, the addition of the bounded driving term

$q(t) = 1$ leads to unbounded solutions. [*Note*: The underlying cause of this behavior is that although all solutions of the undriven equation $y' = y/(1 + y^2)$ decay, they do so very slowly, and the addition of the bounded input $q(t) = 1$ is enough to destabilize the equation.]

(d) Plot some solution curves for the ODEs in parts (a) and (b).

2.5 Approximate Solutions

From the Existence and Uniqueness Theorem in Section 2.2 we know that the IVP

$$y' = f(t, y), \qquad y(t_0) = y_0 \tag{1}$$

has a unique solution on an interval containing t_0 if the rate function $f(t, y)$ is well enough behaved. How does one go about describing this solution? Unfortunately, the collection of rate functions for which a solution formula for IVP (1) can be found is remarkably small. As we have seen, even when a solution formula can be found, it is not always very informative. Lacking a solution formula, how does one then describe the solution? This is the third Basic Question for IVPs listed in Section 2.2.

In this section we present a few numerical procedures for finding approximate values for the solution of IVP (1) on a discrete set of points on the t-axis near t_0. Some things to watch for in implementing these procedures will be discussed in the next section. Appendix B discusses numerical approximation procedures in more detail.

Euler's Method

The direction field approach used in Section 2.1 to provide an intuitive notion of a solution of a first-order ODE also suggests techniques for finding approximate numerical solutions for IVP (1). Below, we describe one such method called *Euler's Method*, but leave the details to Appendix B. Say we wish to approximate $y(T)$, where $T > t_0$. First, partition the interval $t_0 \leq t \leq T$ with N equal *steps*

$$t_n = t_0 + nh, \quad n = 1, 2, \ldots, N$$

where the *step size* is

$$h = (T - t_0)/N$$

We know that (t_0, y_0) is on the solution curve. To find an approximation to $y(t_1)$ just follow the (known) tangent line to the solution curve through (t_0, y_0) out to t_1. Since the slope of the tangent line to the solution curve at (t_0, y_0) is $f(t_0, y_0)$, we see that $y_1 = y_0 + hf(t_0, y_0)$ is a reasonable approximation to $y(t_1)$, if h is small. Using (t_1, y_1) as a base point, and pretending that (t_1, y_1) is on the desired solution curve, we may construct an approximation y_2 to $y(t_2)$ in the same way: $y_2 = y_1 + hf(t_1, y_1)$. Of course, since (t_1, y_1) is most likely not on the desired solution curve, the calculated value y_2 also acquires an error from this source. This calculation can be repeated N times to produce an approximation y_N to the value $y(T)$ of the true solution of IVP (1) at $t = T$. Connecting the points (t_0, y_0), (t_1, y_1), ..., (t_N, y_N) by a broken line produces an approximation to the true solution curve of IVP (1). This approximation is called an *Euler Solution* or *Euler Polygon*. Summarizing, we have the following definition:

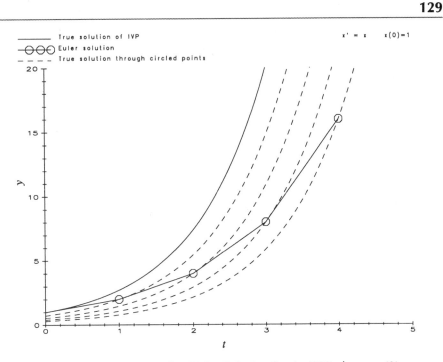

FIGURE 2.5.1 Geometry of an Euler Solution for the IVP $y' = y$, $y(0) = 1$: $h = 1$. See Example 2.5.1.

❖ **Euler's Method.** For the IVP, $y' = f(t, y)$, $y(t_0) = y_0$, the recursive scheme

$$y_n = y_{n-1} + hf(t_{n-1}, y_{n-1}), \quad t_n = t_{n-1} + h, \quad 1 \le n \le N \qquad (2)$$

is called *Euler's Method* with step size $h > 0$.

EXAMPLE 2.5.1

The Euler Polygon and the True Solution of an IVP
Figure 2.5.1 illustrates the geometry of Euler's Method for the IVP

$$y' = y, \quad y(0) = 1$$

with $h = 1$. Observe that the Euler approximation consistently underestimates the value of the true solution, $y = e^t$. The dashed curves in Figure 2.5.1 are the "true" solution curves of the corresponding IVPs,

$$y' = y, \quad y(t_n) = y_n$$

where y_n is the Euler estimate for e^{t_n}. The "badness" of the Euler approximation in this case is largely due to the fact that the step size $h = 1$ is much too large.

As shown in Appendix B, if $f(t, y)$ is smooth enough, then the error $|y(T) - y_N|$ can be made as small as desired by taking h small enough. Here is a second example of an Euler approximation.

EXAMPLE 2.5.2

Another Euler Polygon
We seek an approximate solution curve for the IVP

$$y' = y \sin 3t \qquad y(0) = 1, \qquad 0 \le t \le 4 \qquad (3)$$

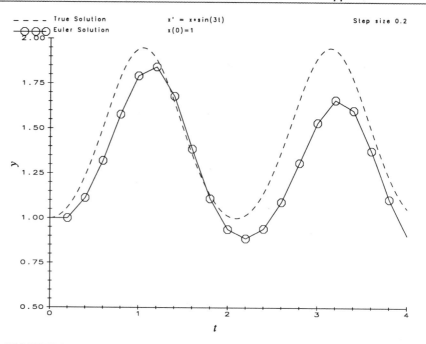

FIGURE 2.5.2 Comparison of the true solution to the Euler Solution. See Example 2.5.2.

Using Euler's Method, we choose $h = 0.2$ and $N = 20$, and put $t_n = (0.2)n$, $n = 0, 1, 2, \ldots, 20$. Then (2) becomes

$$y_n = y_{n-1} + (0.2)(y_{n-1} \sin 3t_{n-1}), \qquad n = 1, 2, \ldots, 20, \quad \text{with } y_0 = 1 \qquad (4)$$

The linear IVP (3) has the unique solution $y = \exp[(1 - \cos 3t)/3]$. The Euler Solution is generated by taking the points (t_0, y_0), (t_1, y_1), $\ldots$, (t_{20}, y_{20}) determined with the help of (4) and connecting them with linear segments. The results are graphed in Figure 2.5.2, where the exact solution of (3) is given by the dashed curve, and the solid line is the Euler Solution. Figure 2.5.2 shows that Euler's Method may provide only a rough approximation to the true solution of an IVP if h is not small enough.

One-Step Methods

Euler's Method for producing approximate solutions of IVP (1) is but one example of a class known as *one-step methods*. Such methods produce an approximate value for the solution of IVP (1) at a selected point T in the interval I in the following way. Suppose that $T > t_0$. Select an increasing sequence $t_1, t_2, \ldots, t_N$ with $t_N = T$ and $t_0 < t_1$ and define the *step size* $h_n = t_n - t_{n-1}$ at step n for $n = 1, 2, \ldots, N$. Then a one-step method computes an approximation y_n to $y(t_n)$ for each $n = 1, 2, \ldots, N$ using the scheme

$$y_n = y_{n-1} + h_n A(t_{n-1}, y_{n-1}, h_n), \quad n = 1, 2, \ldots, N \qquad (5)$$

with y_0 given, and $t_n = t_{n-1} + h_n$, where the function $A(t, y, h)$ is called the *approximate slope function* of the method. Notice that to compute y_n, only the value of y_{n-1} is required (and hence the reason for the name *one-step method*).

Method (5) uses the given value y_0 to generate y_1, and then y_1 to generate y_2, and so on, until the process terminates with the calculation of y_N, which is an approximation of $y(T)$. In Euler's Method for IVP(1) the increment function $A(t_{n-1}, y_{n-1}, h_n)$ is $f(t_{n-1}, y_{n-1})$. Since from (5) we have

$$(y_n - y_{n-1})/h_n = A(t_{n-1}, y_{n-1}, h_n)$$

the function $A(t_{n-1}, y_{n-1}, h_n)$ is also called an *approximate slope function* for $y(t)$ at t_{n-1}. In practice the points t_n are equally spaced with step size $h = t_{n+1} - t_n$ constant for all n. As we shall see, the slope function $f(t_{n-1}, y_{n-1})$ used in Euler's Method is not always the best choice for $A(t_{n-1}, y_{n-1}, h_n)$.

Errors

It is useful to estimate how much the approximations $y_1, y_2, \ldots, y_N$ generated by a one-step method deviate from the exact values $y(t_1), y(t_2), \ldots, y(t_N)$. If precise arithmetic (i.e., no rounding off or chopping of decimal strings) is used, the deviation

$$E_n = |y(t_n) - y_n|, \qquad n = 1, \ldots, N$$

is known as the *global* (or *accumulated*) *discretization error at the nth step.*

There is a local version of the error due to discretization. Assuming that y_{n-1} is exact [i.e., $y(t_{n-1}) = y_{n-1}$] and that $\tilde{y}_n$ is computed from formula (5) by

$$\tilde{y}_n = y(t_{n-1}) + h_n A(t_{n-1}, y(t_{n-1}), h_n)$$

then

$$e_n = |y(t_n) - \tilde{y}_n|, \qquad n = 1, \ldots, N$$

is known as the *formula* (or *local discretization* or *truncation*) *error*. Thus the global discretization error E_n at the nth step is due to two sources: the formula error e_n at that step coupled with the inexact value of y_{n-1} for $y(t_{n-1})$. In practice, the formula error e_n is easier than E_n to compute and is used in the analysis and control of errors.

In summary, the approximation to the solution $y(t)$ of IVP (1) by the discrete one-step scheme (5) has a simple interpretation. For each $j = 0, 1, \ldots, N - 1$, join the point (t_j, y_j) to (t_{j+1}, y_{j+1}) by a line segment. The broken-line path formed by these segments reaches from (t_0, y_0) to (t_N, y_N) and approximates the graph of the solution $y(t)$ (see Figure 2.5.3).

One-step methods may be classified according to the order of magnitude of the global discretization errors incurred when the method is applied.

❖ **Order of a One-Step Method.** Suppose that the one-step method (5) is used to approximate the solution of the IVP, $y' = f(t, y)$, $y(t_0) = y_0$, $t_0 \leq t \leq T$. For each positive integer N, fix the step size in (5) at $h^{(N)} = (T - t_0)/N$. Then method (5) is of *order p* if there is a positive constant M such that,

$$E_N \leq M[h^{(N)}]^p, \quad \text{for every positive integer } N$$

Observe that the upper bound on the global error E_N can be made as small as desired simply by choosing a small enough step size.

It is usually an easy matter to adjust the step size when implementing a one-step method on a computer. Thus, for a fourth-order method (i.e., $p = 4$), cutting the step size in half results in a 16-fold drop in the upper bound on E_N. A similar change from step size h to step size $h/2$ for a first-order method ($p = 1$) gives only a twofold decrease in the upper bound. This suggests that the higher the order of a method, the more accurately it will approximate the solution of IVP (1). In specific instances this may not hold true since the constant M may be larger for a higher-order method

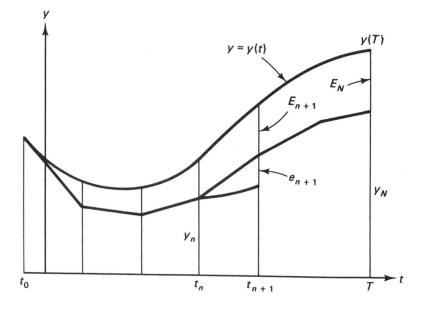

FIGURE 2.5.3 Broken-line approximation to the solution of IVP (1): discretization errors.

than for a lower-order algorithm. Also, higher-order methods usually involve more calculations and function evaluations, and the accompanying round-off errors may nullify the advantages of the higher order. Nevertheless, the order of a method is a significant indication of its accuracy. In Appendix B.1 it is shown that Euler's Method is of first- order.

Heun's Method

Euler's Method may be converted into a second-order method if we compute the approximate slope function by averaging the slopes at t_{n-1} and at t_n. Euler's Method is used to find a first approximation to y_n so that a slope at (t_n, y_n) can be calculated. We then have a new one-step method called

❖ **Heun's Method.** *Heun's Method* is the one-step method with constant step size h for the IVP, $y' = f(t, y)$, $y(t_0) = y_0$, given by

$$y_n = y_{n-1} + \frac{h}{2}[f(t_{n-1}, y_{n-1}) + f(t_n, y_{n-1} + hf(t_{n-1}, y_{n-1}))] \qquad (6)$$

Heun's Method can be shown to be of second-order. See Example 2.5.3 and Figure 2.5.4 for an illustration of the method.

Runge-Kutta Methods

One-step algorithms that use averages of the slope function $f(t, y)$ at two or more points over the interval $[t_{n-1}, t_n]$ in order to calculate y_n are said to be *Runge-Kutta*

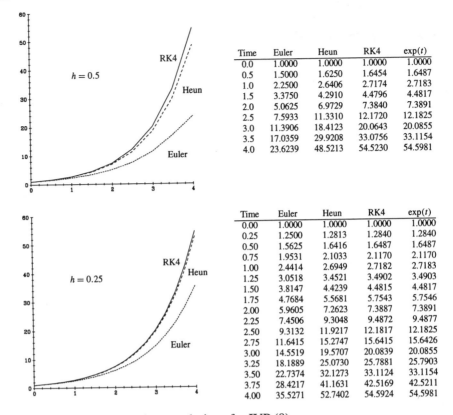

FIGURE 2.5.4 Approximate solutions for IVP (8).

Methods.[2] Heun's method is a second-order Runge-Kutta method. The fourth-order method given below is probably the most widely used of any of the one-step algorithms. It involves a weighted average of slopes at the midpoint, $t_{n-1} + h/2$, and at the endpoints t_{n-1} and $t_{n-1} + h$.

❖ **Fourth-Order Runge-Kutta Method.** For the IVP, $y' = f(t, y)$, $y(t_0) = y_0$, the *Fourth-Order Runge-Kutta Method* (RK4, for short) is the one-step method in the constant step size h given by

$$y_n = y_{n-1} + \frac{h}{6}(k_1 + 2k_2 + 2k_3 + k_4) \tag{7}$$

where

$$k_1 = f(t_{n-1}, y_{n-1}) \qquad\qquad k_2 = f\left(t_{n-1} + \frac{h}{2}, y_{n-1} + \frac{h}{2}k_1\right)$$

$$k_3 = f\left(t_{n-1} + \frac{h}{2}, y_{n-1} + \frac{h}{2}k_2\right) \qquad k_4 = f(t_n, y_{n-1} + hk_3)$$

It may be shown that RK4 generalizes Simpson's Rule for approximating an integral (see Problem 5). The next example compares the three methods discussed above.

---

[2]C.D.T. Runge (1856–1927) did notable work not only in numerical analysis, but also on the Zeeman effect (in physics) and in diophantine equations (in number theory). M.W. Kutta (1867–1944) was an applied mathematician who contributed to the early theory of airfoils.

EXAMPLE 2.5.3

Comparison of Numerical Methods
The initial value problem

$$y' = y, \qquad y(0) = 1 \tag{8}$$

has the unique solution $y = e^t$, $-\infty < t < \infty$. The approximations for the one-step methods of Euler, Heun, and RK4 are plotted in Figure 2.5.4 on the interval $0 \le t \le 4$. As expected, the higher-order methods give more accuracy than those of lower order. Accuracy is observed to improve as the step size is reduced from $h = 0.5$ to $h = 0.25$.

Comments

There are many methods for finding approximate numerical solutions of IVPs. Euler's Method has the virtue of being simple to visualize and implement, but it is not the most practical choice for IVP solvers. RK4 provides a combination of accuracy and efficiency which makes it a better choice than Euler's Method. Commercial solvers (like the one used to produce the plots in this book) generally use a variety of advanced methods in an adaptive environment. Some solvers will allow the user to select a particular method from a list that includes Euler's Method, RK4, and an adaptive method. We strongly urge the use of these already coded solvers.

The approximation methods given in this section may be extended to systems. See Appendix B for a more detailed treatment of approximation methods.[3]

PROBLEMS

1. Estimate $y(1)$ if $y' = -y$, $y(0) = 1$, $h = 0.1$ by using
   (a) Euler's Method   **(b)** Heun's Method   **(c)** RK4

   Plot the approximating polygons for Euler's Method and for RK4.

2. (*Comparison of Euler and RK4*). Estimate $y(1)$ if $y' = -y$, $y(0) = 1$, $h = 0.01$, $0.001, 0.0001$ and plot the approximating polygons.
   (a) Use Euler's Method.   **(b)** Use RK4.

3. (*Comparison of Euler and RK4*). Estimate $y(1)$ if $y' = -y^3 + t^2$, $y(0) = 0$, $h = 0.01$, $h = 0.001$, and $h = 0.0001$, and plot the approximating polygons.
   (a) Use Euler's Method.   **(b)** Use RK4.

4. Show that for each of the following IVPs, $y_N(T) \to y(T)$ as $N \to \infty$, where $\{y_N(T)\}_{N=1}^{N=\infty}$ is the sequence of Euler approximations to $y(T)$ and T is fixed.
   (a) $y' = 2y$, $y(0) = 1$ **(b)** $y' = -y$, $y(0) = 1$

<u>5.</u> Simpson's Formula for approximating the integral $\int_a^b f(t)\,dt$ is

$$\frac{b-a}{6}\left[f(a) + 4f\left(\frac{a+b}{2}\right) + f(b)\right]$$

Show that RK4 for solving the IVP $y' = f(t)$, $y(t_0) = y_0$, reduces to Simpson's Formula applied at each step.

---

[3]For a highly readable source on approximation methods for IVPs, see the book by L.F. Shampine, *Numerical Solution of Ordinary Differential Equations*, New York: Chapman & Hall, 1994.

2.6 Computer Implementation

Easy-to-use commercial software packages often contain very sophisticated differential equation solvers. The solvers have been designed by numerical analysts and have many automatic features of error control. These solvers are far more efficient and accurate than any casual user could hope to write on the spur of the moment.

ODE solver packages have reached such widespread acceptance that practically every computer services group subscribes to several of them. Thus, our best advice is: Before becoming bogged down in coding your own solver to meet an immediate need, check with your local computer services group for an already coded solver. Most graphs in this text could be reproduced in the reader's local environment using one of these solvers.

Using Solvers

In using a commercial solver for the initial value problem $y' = f(t, y)$, $y(t_0) = y_0$, we need to input the function $f(t, y)$, the initial data t_0 and y_0, and a rectangle in the ty-plane containing the point (t_0, y_0). Many solvers do not require a step size to be specified. Instead, the user specifies an error tolerance and the solver calculates a step size adaptively at each step which will achieve this tolerance. The control of errors lies at the heart of a good solver.

There are usually some optional inputs which are given default values (or ignored) by the solver if left unspecified. For example, to prevent the use of too much processing time, the user may be able to specify a minimum allowable step size or a maximum number of function evaluations. If accuracy requirements are too high, these constraints may be exceeded and an error message returned.

In addition to the actual solution, there may be output available from the solver which provides important information that could suggest changing some parameter. This information may include the actual step size used, the number of function evaluations, the number of interval steps used in one call to the solver, or an estimate of the local error in the solution accumulated during a call. Looking at these outputs may suggest that error tolerances be raised or lowered if the actual errors differ significantly from our worst-case estimates, or that a different technique be implemented which uses more or fewer function evaluations.

The inexperienced user is likely to find documentation for commercial software packages as confusing as it is helpful, simply because it was probably written by someone who understands perfectly all the intricacies of the program. Some solvers are "prepared for anything," and for normal, well-behaved ODEs many of their options can be ignored. Sorting out just what is what may require more than just reading the documentation.

Problems in Implementing Approximation Methods

Selecting one of the algorithms described in the preceding sections and coding it for a computer would seem to be an obvious way to find an approximate solution of an initial value problem. But this approach will not always produce satisfactory results unless one is prepared to spend considerable time validating the output. We give some reasons for these words of caution.

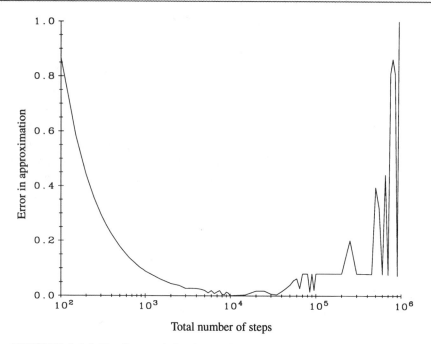

FIGURE 2.6.1 Decline and rise in total error in using Euler's Method to calculate $y(3)$ for the IVP $y' = y$, $y(0) = 1$. See Example 2.6.1.

Nonexact Arithmetic

A major source of errors in any step-by-step process arises because precise arithmetic or precise evaluation of functions is not possible in general. Every computation made by machine (or a human) chops long decimal strings, e.g., (1/3) might be carried as 0.33333, π as 3.14159, and e^x might be evaluated as 2.71828 when $x = 1$. These are examples of *round off errors*, individually small but often devastating in their cumulative effect. The *total error* of a computation includes the errors due to discretization and errors due to round off.

The values y_1, y_2, ..., generated in a one-step method

$$y_n = y_{n-1} + h_n A(t_{n-1}, y_{n-1}, h_n), \quad n = 1, 2, \ldots, N, \quad t_0 \text{ and } y_0 \text{ given}$$

cannot be computed with infinite precision, and hence the error $|y(t_n) - y_n|$ must be viewed as having a component due to round off error as well as to discretization. Round-off has the nasty feature that the more operations that are to be performed, the more likely it is for the error to grow as it propagates through the scheme. Thus if we wish to compute an approximate value y^* for the solution of the IVP

$$y' = f(t, y), \qquad y(t_0) = y_0$$

at $t = t^* > t_0$, we are faced with the following trade-off. By taking the step size $h = (t^* - t_0)/N$ very small, we can make the discretization error at t^* as small as desired. But since we have many more operations to perform in order to approximate $y(t^*)$, this will frequently be at the expense of a buildup of round-off errors. Thus it is not easy to say just what step size would minimize the total error.

EXAMPLE 2.6.1

Effect of Round-off Error

Suppose that Euler's Method is used to estimate $y(3)$, where $y(t)$ is the solution of the initial value problem $y' = y$, $y(0) = 1$. If the method is used for smaller and

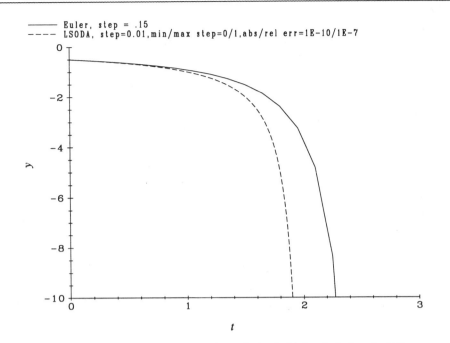

FIGURE 2.6.2 IVP (1): True solution (dashed), Euler Solution (solid).

smaller step sizes (hence, more and more steps), the total error decreases for a while. However, as we see in Figure 2.6.1, for the particular solver that we use the error begins to increase as the number of steps exceeds approximately 20,000, and round-off errors begin to overwhelm the drop in discretization error.

Problems in Choice of Step Size

As we saw in Example 2.6.1, too small a step size leads not only to inefficiency, but possibly also to large round-off errors. But as the next example shows, too large a step size may lead to sizeable global errors that obliterate solution features that are being sought. For an initial value problem whose solution is not well-behaved, the process of discretization may introduce some subtle errors that should make the modeler wary. For example, when the solution "escapes to infinity" in finite time (see Example 2.2.4), the straightforward methods considered so far may not be able to detect this fact. Some solvers have adaptive features which automatically adjust step size in order to better "track" misbehaving solution curves.

EXAMPLE 2.6.2

Tracking Solutions Which Escape to Infinity in Finite Time
Consider the initial value problem

$$y' = -y^2, \qquad y(0) = -\frac{1}{2} \tag{1}$$

which we solve numerically using Euler's Method with step size $h = 0.15$. The result is displayed in Figure 2.6.2, and appears to be a fairly smooth curve. Since the flow field of the differential equation is well-behaved, we are inclined to accept this as an approximate solution of IVP (1). But, in fact, IVP (1) can be solved exactly and we find the maximally extended solution, $y(t) = 1/(t-2)$, $t < 2$. Observe that the solution $y(t)$ escapes to $-\infty$ as $t \to 2^-$. Our numerical solution does not seem to be

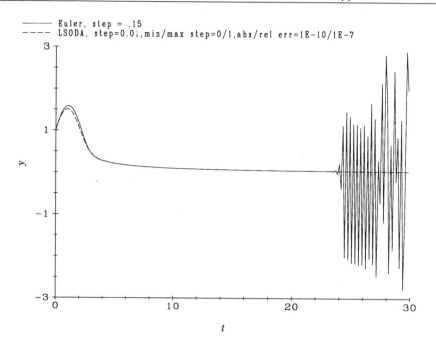

FIGURE 2.6.3 Euler solution accuracy in the long term. See Example 2.6.3.

aware of that fact. What happened, of course, is that Euler's Method "stepped" across $t = 2$ and connected up with a neighboring solution curve. Inspection of the ODE reveals nothing about "finite escape time," and if we did not know the true solution, we would have been inclined to accept the approximate "solution" of Figure 2.6.2. The implications of this simple observation for ODEs that cannot be explicitly solved are clear.

Long-Term Degradation

Step-by-step solvers may not track solution curves very well over large t-intervals. Sometimes the accumulated error grows so fast that it obscures the long-term features of the solution curve. The next example illustrates this fact.

EXAMPLE 2.6.3

If the IVP $y' = 1 - t \sin y$, $y(0) = 1$, is solved over the interval $0 \le t \le 30$ using Euler's Method with step size $h = 0.15$ and graphed in the ty-plane, a strange phenomenon is observed. See Figure 2.6.3. Comparing the Euler Solution to the "actual" solution (represented by the dashed curve in the figure), we see that the agreement is reasonably good, but not perfect, for $0 \le t \le 3$, but then it is dramatically good for a long time interval. Finally, at about $t = 24$ the accuracy of the Euler Solution seems to degenerate badly. The fact that Euler's Method is a first-order method tells us nothing about the *long-term* accuracy of Euler's Method. Choosing smaller step sizes seems only to delay the onset of degeneration of the Euler Solution.

Comments

To the observations above we can add that

- Computers can only recognize finitely many numbers. So a computer is confused by a number whose magnitude is smaller (or greater) than the numbers the computer recognizes—a condition called *underflow* (or *overflow*). Underflow or overflow may result if computations are not arranged appropriately.
- Inefficient code is unnecessarily expensive in terms of processing time.
- The methods described in Section 2.5 produce approximations to initial value problems which converge to the solution when exact arithmetic is used. But because exact arithmetic is impossible, practical techniques for finding such approximations require considerable ingenuity.

The reader should not be too discouraged in spite of the problematic behavior exhibited by the elementary solvers described in the examples above. Indeed, the plots of solutions of differential equations in this text were produced by a very powerful solver. It *is* possible to achieve quite satisfactory results using the approximation algorithms described in the foregoing sections, but designing a good general-purpose solver involves more than the coding of a one-step method.

PROBLEMS

1. (*High Rates*). Consider the IVP, $y' = y^3$, $y(0) = 1$.

    (a) Use separation of variables to find a formula for the maximally extended solution of the IVP. On what t-interval is the solution defined?

    (b) Plot an Euler Solution in the rectangle $0 \leq t \leq 1$, $0 \leq y \leq 20$. Use $h = 0.05$.

    (c) Plot the exact solution found in **(a)** and the Euler Solution found in **(b)** in the same rectangle. Explain what you see.

2. (*Reversibility of an Approximation Method*).

    (a) Assume that $f(t, y)$ satisfies the conditions of the Existence and Uniqueness Theorem and that $y = y(t)$, $t_0 \leq t \leq t_1$, solves the forward IVP

    $$y' = f(t, y), \qquad y(t_0) = y_0$$

    Show that this same function $y(t)$ defines a solution $z(t)$ of the backward IVP

    $$z' = f(t, z), \qquad z(t_1) = y(t_1), \qquad t_0 \leq t \leq t_1$$

    and that therefore $z(t_0) = y_0$

    (b) Use RK4 to find and plot an approximate solution of the IVP

    $$y' = 3y \sin y - t, \qquad y(0) = 0.4$$

    over the interval $0 \leq t \leq 8$, using the step size $h = 0.1$.

    (c) Using the value for $y(8)$ found in part **(b)**, solve the IVP

    $$z' = 3z \sin z - t, \qquad z(8) = y(8)$$

    backward from t-initial $= 8$ to t-final $= 0$ (use RK4). Plot the solution. How close is $z(0)$ to 0.4? Explain any significant difference. [*Hint*: See the merging/diverging solution curves in Atlas Plates FIRST ORDER J.]

3. (*More on Reversibility*). As noted in Problem 2, if $f(t, y)$ satisfies the conditions of the Existence and Uniqueness Theorem, then the IVP, $y' = f(t, y)$, $y(t_0) = y_0$, $t_0 \leq t \leq t_1$ is "reversible" in theory, but experience with numerical solvers shows that actual practice is different. In this problem, you will experiment with a linear IVP, and show that the practical difficulties of running an IVP backwards can be partly resolved by shortening the time step.

(a) Find all solutions of the linear ODE, $y' + 2y = \cos t$. Show that as $t \to \infty$, all solutions approach the particular solution $y_p = 0.4 \cos t + 0.2 \sin t$.

(b) Plot solution curves in the rectangle $0 \leq t \leq 20$, $|y| \leq 1.5$, using as initial points $(0, 0.4)$, $(0, \pm 1.5)$, and several other points on the top and bottom sides of the rectangle. Observe that solutions converge to the solution $y_p(t)$ as t increases.

(c) Use RK4 with step size $h = 0.1$ to solve the IVP with $y(0) = 0.4$ and plot an approximation to $y_p(t)$ for $0 \leq t \leq 20$.

(d) Using the value for $y_p(20)$ found in part (c), solve the IVP, $z' + 2z = \cos t$, $z(20) = y_p(20)$ backward from t-initial $= 20$ to t-final $= 0$, again using RK4 with $h = 0.1$. What happens? How would you explain the difficulty? [*Hint*: The solutions converge on $y_p(t)$ in forward time, but diverge from $y_p(t)$ in backward time. The approximate nature of computed solutions suggests that the computed point $(20, y_p(20))$ is not quite on the solution curve of $y = y_p(t)$.]

(e) Repeat part (d), but with $h = 0.01, 0.001, 0.0001$. Do you see any improvement in the backward solution?

4. (*Long-Term Behavior*). The problems below show the effects on the long-term behavior of approximate solutions using Euler's Method and RK4.

(a) Use Euler's Method with step size $h = 0.1$ to plot an approximate solution of the IVP $y' = 1 + ty \cos y$, $y(0) = 1$, $0 \leq t \leq 28$. Overlay on this plot an approximate solution with $h = 0.01$. Describe what you see.

(b) Repeat part (a) using RK4.

5. (*Qualitative Behavior of Numerical Approximations*). The problems show that approximate solutions do not always faithfully describe the qualitative properties of exact solutions.

(a) (*Logistic ODE*). Use separation of variables to solve the IVP $y' = y(1 - y)$, $y(0) = y_0$. Use a grapher to plot this solution over the interval $0 \leq t \leq 8$ when $y_0 = 2$. How does the solution behave as $t \to +\infty$, for any choice of $y_0 > 0$? Use your best computer solver to plot the solution of the IVP over the interval $0 \leq t \leq 8$, with $y_0 = 2$. How does this approximate solution compare to the true solution?

(b) Solve the IVP in part (a) using Euler's Method with $h = 0.75$ and $y_0 = 2$, and plot the Euler Solution on $0 \leq t \leq 8$. Does the Euler Solution behave qualitatively the same as the solutions in part (a)? Explain differences.

(c) Repeat part (b), but with $h = 1.5$ and $y_0 = 1.4$.

(d) Repeat part (b), but with $h = 2.5$ and $y_0 = 1.3$.

(e) (*Sign Analysis and Approximate Solutions*). Consider the IVP $y' = -y(1 - y)^2$, $y(0) = 1.5$. Show by sign analysis that solutions of the ODE above the solution $y = 1$ fall toward that solution, while solutions below fall away with increasing time. Graph the approximate solutions given by Euler's method with step size $h = 0.1, 0.5, 1, 1.5$, and 2, and comment on any noteworthy properties.

6. (*Sampling Rates and Aliasing*). All the solutions of the autonomous linear system of ODEs, $x' = y$, $y' = -4x$, are given by the solution formulas $x = A \sin(2t + \phi)$,

$y = 2A\cos(2t + \phi)$ where A and ϕ are arbitrary constants.

(a) In the tx-plane, the ty-plane, and the xy-plane, respectively, plot the tx and ty component graphs and the xy orbits of the solutions with initial points $x_0 = 0$, $y_0 = 1$, and then $x_0 = 0$, $y_0 = 2$. Describe these graphs in words.

(b) Using RK4 for systems (see (16) and (17) in Appendix B.1), solve the IVP

$$x' = y, \qquad x(0) = 0$$
$$y' = -4x, \quad y(0) = 1$$

over the time interval $0 \le t \le 100$ using step size $h = 0.1$, and plot the first component, $x(t)$, in the tx-plane.

(c) Using the step size $h = 0.1$ of part **(b)**, plot only every fifth computed point of $x(t)$. Repeat but plot only every tenth point, and then every twentieth point. Explain the differences in these plots from the one in part **(b)**.

7. (*High Rates*). Insert some integral curves in the blank spaces on the right side of Atlas Plate FIRST ORDER L. Do this by plotting orbits of the system corresponding to the first-order ODE $M dx + N dy = 0$: $x' = N$, $y' = -M$. In this case $N = y^2 + e^x \cos y + 2\cos x$ and $M = e^x \sin y + 2y \sin x$. Explain what you see.

2.7 Euler's Method, The Logistic Equation, and Chaos

The logistic ODE

$$P' = rP\left(1 - \frac{P}{K}\right) \tag{1}$$

where r and K are positive constants, governs the size of a single population $P(t)$, taking into account certain natural factors that affect population growth (see Section 1.5). Although the logistic ODE is nonlinear, there is a formula for all its solutions (see (7) in Section 1.5). Since the logistic ODE is autonomous, the sign analysis approach of Section 2.2 gives us a good idea of the long-term properties of solutions; every solution $P(t)$ of ODE (1) with $P(0)$ positive must tend monotonically to the equilibrium solution $P = K$ as $t \to +\infty$. See Figure 1.5.1 in Section 1.5.

As we will show in this section, some caution must be exercised when using an approximate numerical solution to describe qualitative features of the exact solution of an IVP. We will also show that the qualitative behavior of approximate solutions may be significantly affected by the choice of step size. We shall do this by using Euler's Method to generate approximate numerical solutions for the logistic ODE (1).

Euler Solutions for the Logistic ODE

For simplicity, we put $K = 1$ in the logistic equation (1) and consider the IVP

$$y' = ry(1 - y), \qquad y(0) = y_0 \tag{2}$$

where r is a positive constant. An application of Euler's Method to IVP (2) generates the sequence of approximate values $y_1, y_2, \ldots$, given by

$$y_{n+1} = y_n + hry_n(1 - y_n), \qquad n = 0, 1, 2, \ldots \tag{3}$$

where $h > 0$ is the step size and y_0 is the initial state for IVP (2). The iterates y_n generated by Euler's Method are plotted versus $t_n = nh$ in the ty-plane. When these "Euler

Points" are connected consecutively by straight line segments, the resulting broken-line graph is (as noted in Section 2.5) called an Euler Solution or Euler Polygon. We will examine the behavior of Euler Solutions of IVP (2) for initial values y_0 near 1, and for step sizes $h > 0$. These Euler Solutions exhibit some surprising properties as h changes.

We recall that Euler's Method is of first-order when exact arithmetic is used (i.e., no rounding off). For IVP (2) this means, in particular, that for any $T > 0$ and any tolerance $\epsilon > 0$, there is a step size h such that the values generated by (3) miss being the true values of the solution of IVP (2) by at most ϵ over the interval $0 \leq t \leq T$. It would seem reasonable to infer that if h is chosen small enough, the Euler Solution to IVP (2) "looks like" the exact solution for any initial value $y_0 > 0$.

That this is not quite the case comes as a shock to our intuition. To see this, look at the graphs in Figures 2.7.1–2.7.4. In each figure the true solution to IVP (2) for given values of r and y_0 is plotted (the dashed line) and then overlaid with the Euler Solution (solid) to that IVP for a given step size h,

$$y_{n+1} = y_n + ay_n(1 - y_n), \qquad y_0 \text{ given and } a = rh \qquad (4)$$

The first thing to notice is that all Euler Solutions defined by (4) would have the same shape in each figure, for any positive values of r and h, for which $rh = a$ has the indicated value. In other words, the Euler Solution depends on the value of the product rh, rather than on the individual values of r and h. Second, observe that if the only information we had about the solutions of IVP (2) were the Euler Solutions defined by (4) in Figures 2.7.1–2.7.4, then we would infer (wrongly) that the solution of IVP (2) has the following properties:

- The true solution with $y_0 = 2$ would fall initially and then rise monotonically toward the line $y = 1$, as $t \to +\infty$ (Figure 2.7.1, $a = 0.75, r = 1.5, h = 0.5$).

- The true solution with $y_0 = 1.4$ would oscillate about the line $y = 1$ with oscillations that die out as $t \to +\infty$ (Figure 2.7.2, $a = 1.5, r = 6.0, h = 0.25$).

- The true solution with $y_0 = 1.3$ would undergo regular oscillations about the line $y = 1$ which do not die out as $t \to +\infty$ (Figure 2.7.3, $a = 2.1, r = 1, h = 2.1$; Figure 2.7.4, $a = 2.5, r = 10, h = 0.25$).

As we see from Figures 2.7.1–2.7.4, none of these inferences about the true solution of IVP (2) is valid. This does not, however, contradict the fact that Euler's Method is of first-order for a *fixed* IVP (2), that is, for fixed values of r and y_0. Nevertheless, it does show that there is no "absolute" notion of a small step size—smallness for a step size is coupled with the size of the rate function, $f(t, y)$. In fact, we can make the following assertion (proof omitted):

> Given any step size $h > 0$, there exist values for the positive constant r and initial value y_0 with $\max\{1, 1/rh\} < y_0 < 1 + 1/rh$ such that the Euler Solution of IVP (2) with that step size exhibits any of the behaviors given in Figures 2.7.1–2.7.4.

So, it is hazardous to choose what appears to be a small step size and then claim that the resulting Euler Solution describes the true solution to IVP (2), even qualitatively.

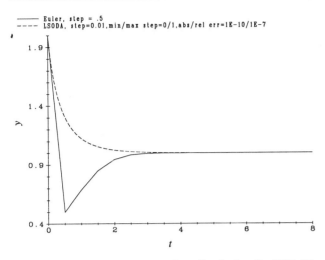

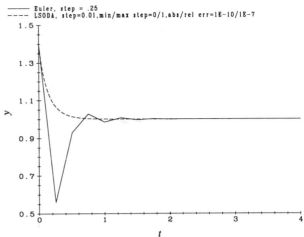

FIGURE 2.7.1 Euler versus "true" solution for IVP (2) with $a = rh = 0.75$, $y_0 = 2$.

FIGURE 2.7.2 Euler versus "true" solution for IVP (2) with $a = rh = 1.5$, $y_0 = 1.4$.

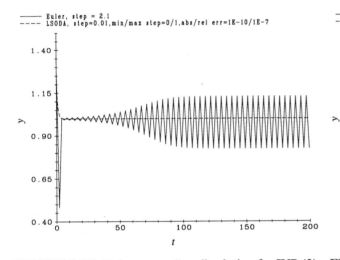

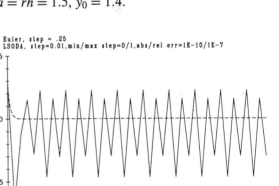

FIGURE 2.7.3 Euler versus "true" solution for IVP (2) with $a = rh = 2.1$, $y_0 = 1.3$.

FIGURE 2.7.4 Euler versus "true" solution for IVP (2) with $a = rh = 2.5$, $y_0 = 1.3$.

Cobweb Diagrams

To analyze the Euler Solutions of IVP (2) it is helpful to rewrite the right-hand side of Euler's Method (4) by completing the square in the quadratic expression for y_n, obtaining

$$y_{n+1} = -a \left(y_n - \frac{1+a}{2a} \right)^2 + \frac{(1+a)^2}{4a} \tag{5}$$

The graph of y_{n+1} versus y_n generated by (5) is the parabola that appears in Figure 2.7.5. Note that $y_n^{(1)}$ and $y_n^{(2)}$ in the figure are the values of y_n that make $y_{n+1} = 1$.

Observe that if $0 < a \le 3$, then $(1+a)^2/4a \le (1+a)/a$; hence, from Figure 2.7.5 it follows that the value y_n in the interval $[0, (1+a)/a]$ returns the value y_{n+1} in the same interval. Thus, if $0 \le y_0 \le (1+a)/a$, then Euler's Method (5) produces values $y_1, y_2, \ldots,$ all of which lie in the interval $[0, (1+a)/a]$ as well.

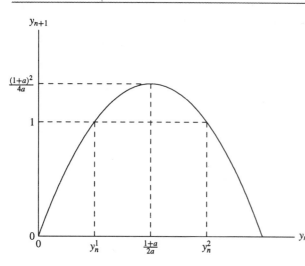

FIGURE 2.7.5 Graph of iteration scheme (5): $a = rh$.

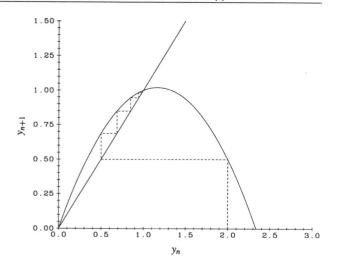

FIGURE 2.7.6 Generation of Euler Solution: $a = rh = 0.75$, $y_0 = 2$. Compare with Figure 2.7.1.

There is a simple geometric way of "seeing" what the Euler Solution will look like from a given starting point y_0 in $[0, (1 + a)/a]$ when $0 < a < 3$. First, superimpose the line $y_{n+1} = y_n$ over the graph of the parabola in Figure 2.7.5. Note that the point $(1, 1)$ on the parabola occurs to the *left* of the high point if $0 < a < 1$ and to the *right* of the high point if $a > 1$. Now, depending on the size of a, we describe a graphical procedure for constructing the Euler sequence $y_1, y_2, \ldots$, from the initial choice y_0 (refer to Figures 2.7.6, 2.7.7, 2.7.8):

- Follow the vertical dashed line upward from y_0 to the parabola to find y_1.
- Then move horizontally over to the line $y_{n+1} = y_n$.
- Move vertically to the parabola again to find y_2.
- Repeat these steps.

Note that this graphical process for generating the Euler Solution changes significantly when $a > 1$. The constructions and graphs in Figures 2.7.6, 2.7.7, and 2.7.8 are typical for the three cases $0 < a < 1$, $1 < a < 2$, and $2 < a < 3$, respectively. A great deal about the behavior of the Euler sequence $y_1, y_2, \ldots$ can be inferred from these graphs, often called *cobweb diagrams*.

Period Doubling

As we shall see, the longterm behavior of Euler Solutions of IVP (2) depends very strongly on the value of the parameter $a = rh$. For small values of a and for values of y_0 near 1, the Euler Solution tends to the constant solution $y = 1$ as the number N of Euler points increases. On the other hand, for large values of a and y_0 near 1, $y_0 \neq 1$, the Euler Solution apparently wanders chaotically about the equilibrium level $y = 1$ as N increases.

Problem 5 describes what happens to the Euler iterates defined by (4) if $0 < a \leq 2$, and $0 < y_0 < (1 + a)/a$. There are no big surprises here. In the long-term these sequences of Euler iterates tend to 1, as $n \to +\infty$. The manner in which this convergence to 1 occurs depends on the actual value of a in the interval $0 < a \leq 2$, and y_0 in the

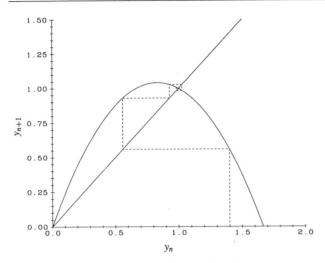

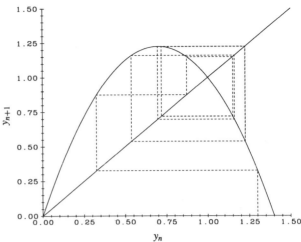

FIGURE 2.7.7 Generation of Euler Solution: $a = rh = 1.5$, $y_0 = 1.4$. Compare with Figure 2.7.2.

FIGURE 2.7.8 Generation of Euler Solution: $a = rh = 2.5$, $y_0 = 1.3$. Compare with Figure 2.7.4.

interval $0 < y_0 < (1+a)/a$. Figures 2.7.1 and 2.7.2 are about as weird as it gets for a-values in this regime.

Things really get interesting when $a > 2$ in the Euler sequence defined by (4). No longer do the Euler sequences converge to 1 as $n \to +\infty$. In fact, some strange things begin to happen.

EXAMPLE 2.7.1	**The Case** $a = 2.1$, $y_0 = 1.3$

Approximating the solution of IVP (2) using Euler's Method with step size h, where $a = rh = 2.1$, and $y_0 = 1.3$ results in the recursion relation

$$y_{n+1} = y_n + 2.1 y_n(1 - y_n), \qquad y_0 = 1.3 \tag{6}$$

Since our choice of $y_0 = 1.3$ places it in the interval $0 < y_0 < (1+a)/a = 3.1/2.1$, we would expect every Euler iterate y_n to lie in the same interval, and that is indeed the case. Plotting the Euler Solution generated by (6) over $0 \le t \le 200$, we obtain the graph in Figure 2.7.3. Remarkably, the Euler Solution in Figure 2.7.3 seems to settle down to a periodic function, after some initial confusion, eventually swinging back and forth between the values 1.13 and 0.83 (approximately). This periodic function is said to be a *2-cycle* since it repeats two values over and over again, forever.

A little calculation (see Problem 1) shows that the recursion relation (4) generates exactly two 2-cycles for any $a > 2$. The values of y_0 for the 2-cycles are

$$(y_0)_1 = \frac{a + 2 + \sqrt{a^2 - 4}}{2a}, \quad (y_0)_2 = \frac{a + 2 - \sqrt{a^2 - 4}}{2a} \tag{7}$$

The two 2-cycles are identical except that one is shifted a half-period from the other. Examining Figure 2.7.3 again, it appears that after sufficient time has elapsed (to let the "transients" die out) the Euler Solution looks like one of the 2-cycles described above. In fact, after some experimentation, it appears that for any y_0 near 1 (but not $= 1$) the resulting Euler Solution is "attracted" by one or the other of the two known 2-cycles. This attraction property is the reason that we can "see" these 2-cycles using computer graphics.

From this point on, each example begins with the appropriate Euler algorithm.

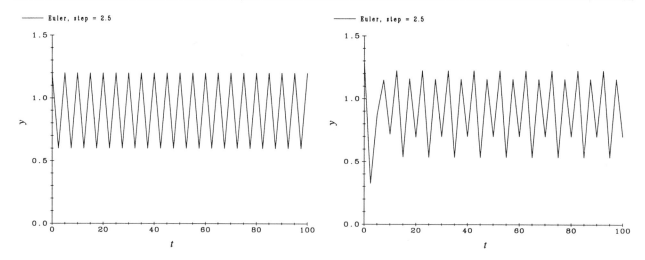

FIGURE 2.7.9 The Euler Solution generated by (8) with $y_0 = 1.2$.

FIGURE 2.7.10 The Euler Solution generated by (9) with $y_0 = 1.3$.

EXAMPLE 2.7.2

The Case $a = 2.5$, $y_0 = 1.2$

$$y_{n+1} = y_n + 2.5y_n(1 - y_n), \qquad y_0 = 1.2 \tag{8}$$

Plotting the Euler Solution generated by (8) over the interval $0 \le t \le 100$, we obtain the graph in Figure 2.7.9. This Euler Solution seems to be a 2-cycle, swinging back and forth between the values 1.2 and 0.6. To show that the Euler Solution in Figure 2.7.9 is indeed a 2-cycle we need only show that $y_2 = y_0 = 1.2$ because the recursion relation (8) is *autonomous*, i.e., does not contain the independent variable n explicitly. Since $y_0 = 1.2$, it follows from (8) that $y_1 = 0.6$ and that $y_2 = 1.2$. It is clear from (8) that all later y_n swing back and forth between 1.2 and 0.6, exactly as we conjectured from Figure 2.7.9.

EXAMPLE 2.7.3

The Cases $a = 2.5$, $y_0 = 1.3$, $y_0 = 1.201$

$$y_{n+1} = y_n + 2.5y_n(1 - y_n), \qquad y_0 = 1, 3, 1.201 \tag{9}$$

Putting first $y_0 = 1.3$ in (9) and plotting the resulting Euler Solution on $0 \le t \le 100$, we obtain the graph in Figure 2.7.10. After some initial confusion, the Euler solution seems to settle down to a 4-cycle, i.e., iterates which repeat in a cycle of four. These four values are 1.23, 0.54, 1.15, and 0.7 (approximately). The remarkable thing is that this same 4-cycle (or one of its translates) seems to come up in the long term when plotting an Euler Solution generated by (9) for *any* value of y_0 near, but not equal to, 1.2. Of course, to "see" the 4-cycle we need to wait long enough for the transients to "die out."

To verify this impression, let us plot the Euler Solution generated by (9), but this time with $y_0 = 1.201$, and plot it over $9900 \le t \le 10,000$. The result is the graph in Figure 2.7.11 which looks like the 4-cycle in Figure 2.7.10. Apparently, this 4-cycle attracts all nearby Euler Solutions generated by $y_0 \ne 1.2$; this is the reason we can "see" this 4-cycle at all. The 2-cycle in Figure 2.7.9 generated by $y_0 = 1.2$ is a fluke. Had y_0 not started exactly on the 2-cycle, we would never have seen it.

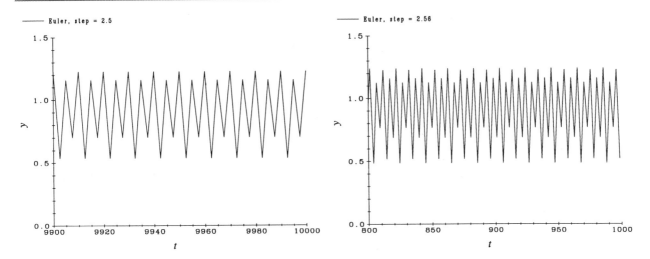

FIGURE 2.7.11 The Euler Solution generated by (9)   **FIGURE 2.7.12** The Euler Solution generated by (10). with $y_0 = 1.201$.

EXAMPLE 2.7.4

The Case $a = 2.56$, $y_0 = 1.3$

$$y_{n+1} = y_n + 2.56y_n(1 - y_n), \qquad y_0 = 1.3 \tag{10}$$

In Figure 2.7.12 the Euler Solution generated by (10) is plotted for $800 \le t \le 1000$. Looking at Figure 2.7.12 carefully, it is startling to discover that the graph appears to be an 8-cycle since the values of the iterates repeat after 8 steps. The broken-line graph in Figure 2.7.12 could be misleading, so we might try another approach: plot the Euler Solution, but leave out the connecting line segments. The graph in Figure 2.7.13 is more convincing that we are dealing with an 8-cycle. Changing y_0 slightly in (10) gives the same result in the long term, indicating that the 8-cycle is attracting.

EXAMPLE 2.7.5

The Case $a = 2.567$, $y_0 = 1.2$

$$y_{n+1} = y_n + 2.567y_n(1 - y_n), \qquad y_0 = 1.2 \tag{11}$$

The Euler iterates generated by (11) are plotted over $0 \le t \le 2000$ in Figure 2.7.14 without the connecting line segments. It may be difficult to see at first, but Figure 2.7.14 shows that the Euler Solution generated by (11) very quickly settles down to a 16-cycle. Close inspection shows that the iterates quickly align themselves in 16 lines, although this is difficult to see for the topmost pair. Zooming in on this region resolves the issue. Euler Solutions that originate at y_0-values close to 1.2 are eventually attracted to the same 16-cycle.

EXAMPLE 2.7.6

The Case $a = 2.569$, $y_0 = 1.2$

$$y_{n+1} = y_n + 2.569y_n(1 - y_n), \qquad y_0 = 1.2 \tag{12}$$

The Euler iterates generated by (12) are plotted over $0 \le t \le 2000$ in Figure 2.7.15 without the connecting line segments. On this scale, it is impossible to see that this Euler Solution quickly settles down to a 32-cycle. Zooming in on the groupings that

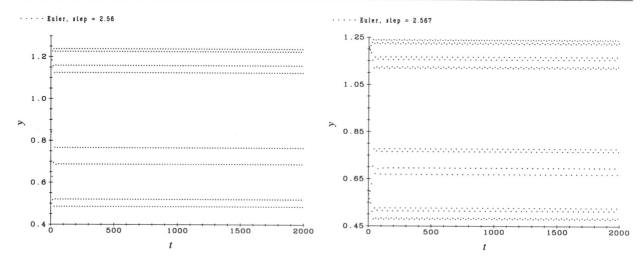

FIGURE 2.7.13 The Euler Solution generated by (10) but without connecting lines.   Compare with Figure 2.7.12.

FIGURE 2.7.14 The Euler Solution generated by (11) but without connecting lines.

make up the plot in Figure 2.7.15 produces the plots in Figures 2.7.16–2.7.19; the total of 32 horizontal lines in these plots suggests that we are dealing with a 32-cycle.

Chaotic Wandering

Figures 2.7.1–2.7.4 and 2.7.9–2.7.19 tell a story about how long-term values of the Euler iterates

$$y_{n+1} = y_n + ay_n(1 - y_n), \qquad y_0 \text{ given} \tag{13}$$

change as the parameter a changes. Recall that choosing y_0 in the interval $0 < y_0 < (1 + a)/a$ guarantees that all the y_n for that a-value also lie in this interval. Notice that for any $a > 0$, $y_0 = 1$ is a "fixed point" for (13) in that $y_n = 1$, $n = 1, 2, \ldots$. So let us agree to avoid choosing $y_0 = 1$ just so we can make the following observations:

- For $0 < a \le 2$, and y_0 near 1, the long-term values for y_n generate the 1-cycle: $1, 1, \ldots$.

- For $a > 2$, but somewhat less that 2.5, and y_0 near 1, the iterates approach a 2-cycle (see Figure 2.7.3 where $a = 2.1$).

- For a near 2.5, the iterates approach a 4-cycle (see Figures 2.7.10, 2.7.11 where $a = 2.5$).

- For a around 2.56, the iterates approach an 8-cycle (see Figures 2.7.12, 2.7.13 where $a = 2.56$).

- For a around 2.567, the iterates approach a 16-cycle (see Figure 2.7.14 where $a = 2.567$).

- For a around 2.569, the iterates approach a 32-cycle in the long term (see Figures 2.7.15–2.7.19 where $a = 2.569$).

Clearly, what is happening here is that, as a increases in size, the long-term behavior of the Euler iterates generated by (13) suddenly splits (bifurcates) from a 2^n-cycle

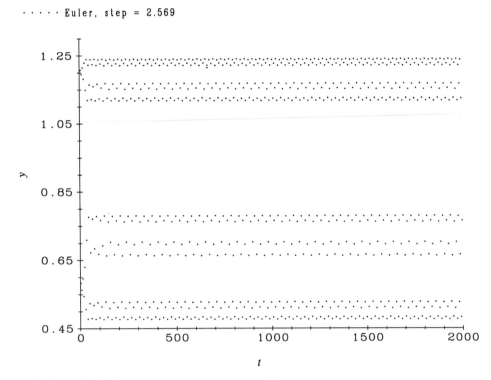

FIGURE 2.7.15 Euler iterates generated by (12).

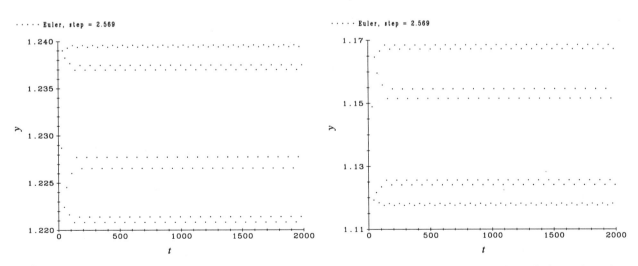

FIGURE 2.7.16 Zoomed in portion of Figure 2.7.15.   FIGURE 2.7.17 Zoomed in portion of Figure 2.7.15.

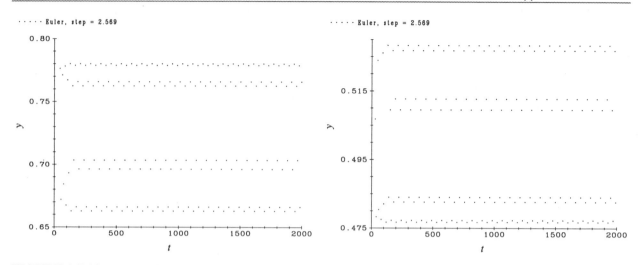

FIGURE 2.7.18 Zoomed in portion of Figure 2.7.15.   **FIGURE 2.7.19** Zoomed in portion of Figure 2.7.15.

to a 2^{n+1}-cycle. Notice that this period doubling takes place at an ever increasing rate, so that soon even tiny changes in a cause a bifurcation. It is interesting to see if the parameter a will get us beyond the period-doubling bifurcations.

EXAMPLE 2.7.7

The Case $a = 2.65$, $y_0 = 1.2$: Chaos

$$y_{n+1} = y_n + 2.65y_n(1 - y_n), \qquad y_0 = 1.2 \tag{14}$$

The Euler iterates generated by (14) are plotted in Figure 2.7.20 over the interval $0 \leq t \leq 2000$. Notice that the pattern of cycles no longer exists, and that instead the iterates wander around chaotically.

Comments

Approximate numerical solutions of IVPs are not always reliable for predicting the behavior of solutions as $t \to \infty$. Such techniques are designed to provide increasingly better numerical solutions with decreasing step size, but only over fixed, bounded t-intervals.

Additional Reading

The chaotic behavior illustrated in this section has been widely studied in recent years. Here are three references that provide more information:

- R.L. Devaney, *An Introduction to Chaotic Dynamical Systems*, 2nd ed., Addison-Wesley, New York, 1989.

- D. Gulick, *Encounters with Chaos*, McGraw-Hill, New York, 1992.

- M. Martelli, *Discrete Dynamical Systems and Chaos*, Longman/Wiley, New York, 1992.

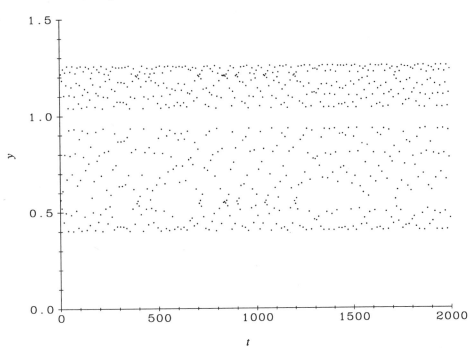

FIGURE 2.7.20 Euler iterates for $a = rh$ in the chaotic regime (see Example 2.7.7).

PROBLEMS

1. (*2-Cycles*). Show that for any $a > 2$ the recursion relation

$$y_{n+1} = y_n + ay_n(1 - y_n), \qquad y_0 \text{ given}$$

has precisely two values y_0 in the interval $0 < y_0 < (1 + a)/a$ which give rise to 2-cycles. Show that these two values are given by (7). [*Hint*: Let u be one such value and put $v = u + au(1 - u)$. Then for u to give rise to a 2-cycle we must have $u = v + av(1 - v)$.]

2. (*Convergent Euler Sequences*). For each of the parameter intervals $0 < a \leq 1$ and $1 < a \leq 2$, the Euler Solutions of IVP (2) exhibit qualitatively different properties in the long term. Figures 2.7.6–2.7.8 give a preview of this fact. Perform the following simulations. [*Hint*: In discussing the behavior of the Euler Solutions, use the fact that the rate function $ry(1 - y)$ is positive if $0 < y < 1$ and negative if $y > 1$ or $y < 0$.]

(**a**) Let $r = 10$, $h = 0.05$ (so $a = rh = 0.5$). Construct and graph Euler Solutions of IVP (2) for each of the initial values $y_0 = 0.3$, 2.0. Compare the Euler Solutions over $0 \leq t \leq 1$ (for $y_0 = 0.3$) and over $0 \leq t \leq 0.5$ (for $y_0 = 2.0$) to the exact solution curve, and describe and explain qualitative differences.

(**b**) Let $r = 100$, $h = 0.015$ (so $a = rh = 1.5$). Construct and graph Euler Solutions of IVP (2) for each of the initial values $y_0 = 0.5$, 1.5. Compare the Euler Solutions over $0 \leq t \leq 0.1$ (for $y_0 = 0.5$) and over $0 \leq t \leq 0.15$ (for $y_0 = 1.5$) to the exact solution curve. Describe and explain qualitative differences.

3. (*Period Doubling*). Let $r = 100$, $h = 0.023$ (so $a = rh = 2.3$). Construct the Euler Solution for $y' = ry(1 - y)$, $y(0) = y_0$ with $N = 50$ and the initial value

$y_0 = 1.3$. Graph this Euler Solution and compare it to the exact solution curve. Repeat this simulation with $r = 100$, $h = 0.025$ (hence $a = rh = 2.5$). Do the values of these Euler Solutions eventually settle down to any pattern? If so, how would you describe that pattern?

4. (*Chaotic Wandering*). Consider the Euler Solutions for $y' = ry(1 - y)$, $y(0) = 1.2$, with step size h, where $rh = a = 2.65$. In each case, graph the Euler Solution over the indicated intervals and interpret what you see. [See also Example 2.7.7.]

(a) Let $r = 1$ and $h = 2.65$. Graph the Euler polygon starting at $y(0) = 1.2$ over four intervals: $[0, 100]$, $[100, 200]$, $[200, 300]$, $[300, 400]$. Do you see any patterns?

(b) Let $r = 2.65$ and $h = 1$. Graph the Euler polygon starting at $y(0) = 1.2$ over the four intervals $[0, 100/2.65]$, $[100/2.65, 200/2.65]$, $[200/2.65, 300/2.65]$, $[300/2.65, 400/2.65]$. Why do these graphs look exactly like the corresponding graphs in part (a)? What would the graphs of the Euler polygons corresponding to $r = r_0 > 0$, $h = h_0 > 0$, $y(0) = 1.2$ look like if $r_0 h_0 = 2.65$ and the graphs are plotted over $[0, 100/r_0]$, $[100/r_0, 200/r_0]$, and so on? Explain.

5. (*Chaotic Wandering*). Construct the Euler iterates with $h = 2.75$ for the logistic IVP, $y' = y(1 - y)$, $y(0) = y_0$, over the interval $0 \le t \le 2000$, where $y_0 = 0.25$, $0.5, 0.75, 1.2$. Graph the iterates for each value of y_0 (use separate graphs). Do you see any pattern in the iterates? [Warning: Do this problem only if your solver will plot iterates without the connecting lines.]

6. Give a convincing justification for the following general description of the behavior of Euler Solutions for $y' = ry(1 - y)$, $y(0) = y_0$, in the indicated parameter regimes. [*Hint*: Use formula (4) and its geometric implementations in Figures 2.7.5–2.7.8.] What does this behavior imply about the choice of step size h?

- For $0 < a \le 1$ the Euler sequence $\{y_n\}$ generated by any choice of initial point y_0 in the interval $0 < y_0 < (1 + a)/a$ behaves as follows:

  (a) If $0 < y_0 < 1$, then $\{y_n\}$ rises steadily toward 1, as $n \to \infty$.
  (b) If $y_0 = 1$ or $1/a$, then $y_n = 1$, for $n \ge 1$.
  (c) If $1 < y_0 < 1/a$, then $\{y_n\}$ decreases steadily toward 1, as $n \to \infty$.
  (d) If $1/a < y_0 < (1 + a)/a$, then $0 < y_1 < 1$ and $\{y_n\}$ rises steadily toward 1, as $n \to \infty$.

- For $1 < a \le 2$ the Euler sequence $\{y_n\}$ generated by the initial choice of y_0 in the interval $0 < y_0 < (1 + a)/a$ behaves as follows:

  (a) If $0 < y_0 < 1/a$, the Euler sequence $\{y_n\}$ rises steadily until $y_N > 1$ for some N; for $n > N$, $\{y_n\}$ alternates about 1 while steadily approaching 1 as $n \to \infty$.
  (b) If $y_0 = 1/a$ or 1, then $y_n = 1$ for all $n \ge 1$.
  (c) If $1/a < y_0 < 1$, then $\{y_n\}$ alternates about 1 while steadily approaching 1 as $n \to \infty$.
  (d) If $1 < y_0 < (1 + a)/a$, then $0 < y_1 < 1$. Thereafter $\{y_n\}$ behaves as described in parts (a), (b), or (c), depending on the value of y_1.

- (*Chaotic Wandering and Period Doubling*). For $a > 2$ the Euler sequence $\{y_n\}$ generated by any initial value y_0 in the interval $0 < y_0 < (1 + a)/a$ does not approach 1 as $n \to \infty$. How would you describe the behavior of $\{y_n\}$?

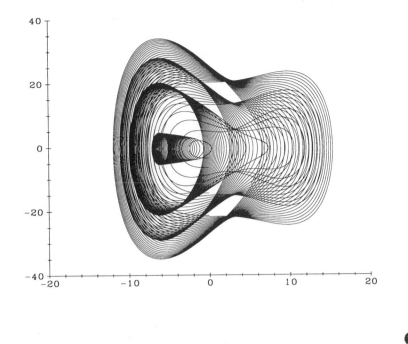

Second-Order Differential Equations

*Second-order differential equations model a great many important physical phenomena and therefore deserve a chapter of their own. The interplay between linear and nonlinear ODEs is examined in some detail in Section 3.1, but this chapter is devoted mostly to a treatment of linear second-order ODEs. We describe the solutions of both driven and undriven equations. The motion of a body suspended by a spring is modeled in Section 3.1, but other applications of second-order ODEs are postponed to Chapter 4. The chapter cover figure shows four orbits of a certain second-order ODE (see Problem 1(**h**), Section 3.2).*

3.1 Springs: Linear and Nonlinear Models

A body of mass m is attached to one end of a massless spring, the other end of which is fastened to the ceiling. If the spring is compressed or extended along its axis and then released, the body moves up and down below the spring's support as the spring vibrates. The nature of the motion depends on the strength of the spring force, on the frictional forces that tend to dampen motion, and on any other external forces (such as gravity) that act on the body. The system is at *equilibrium* when the spring is stretched but the body is at rest. Let y measure the displacement of a point on the body along the spring's axis from its position when the spring is unstretched; $y < 0$ corresponds to a stretched spring; $y > 0$ a compressed spring. The figure in the margin displays the spring-body system in a stretched position, subject to an external force $f(t)$ acting parallel to the y-axis, and with a frictional force represented by a dashpot or damper

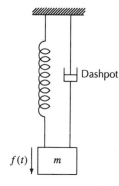

$f(t)$ m

(shown as a cylinder of fluid through which a piston moves). The gravitational force on the body acts in a vertical direction downward.

Newton's Second Law (Section 1.3) may be used to find the displacement $y(t)$ of the body under the action of the forces acting upon it. The following assumptions about the spring and damping forces are based on observations of real springs:

> Spring Force. The force exerted by a spring on a body acts parallel to the spring's axis (say, the y-axis) and has magnitude $S(y)$ where y is the displacement of the body from its position where the spring is neither stretched nor compressed.

> Viscous Damping Force. The viscous damping force on a body has magnitude proportional to the velocity of the body and acts in the direction opposite to the velocity.

Newton's Second Law states that the motion of a body must, at all times, satisfy the condition that the mass m of the body times the acceleration is equal to the (algebraic) sum of the forces acting on it. With the assumptions described above about the forces acting on the body, we see that the position $y(t)$ of the oscillating body at time t, must satisfy the second-order ordinary differential equation (ODE)

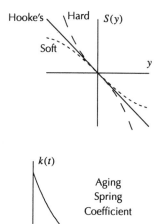

$$my'' = -cy' + S(y) + f(t) - mg \tag{1}$$

where c is the positive *damping constant* ($c = 0$ if there is no damping).

Careful measurements suggest that if the displacement from equilibrium is small, then the spring force $S(y)$ is an odd function (i.e., $S(-y) = -S(y)$), and points in the opposite direction to the displacement y. Four commonly used models with these properties are:

Hooke's Law spring	$S(y) = -ky$	(2)
Hard spring	$S(y) = -ky - jy^3$	(3)
Soft spring	$S(y) = -ky + jy^3$	(4)
Aging spring	$S(y, t) = -k(t)y$	(5)

where k and j are positive constants and $k(t)$ is a positive function that depend in no way on the environment. Observe that these forces act in the opposite direction to displacement (for small displacements from equilibrium). In comparison to the Hooke's Law force, the hard spring force strengthens with extension or compression, the soft spring force weakens with extension or compression, and the aging spring coefficient $k(t)$ weakens with time (see figures in margin). The constant k in (2)–(4) is often called the *spring constant*. Of course real springs can be compressed or stretched only so much, and so the region of validity of the model ODE (1) is limited.

Linear Models (Mostly)

Choosing Hooke's Law $S(y) = -ky$ for the force exerted by the spring on the body, we see from ODE (1) that the displacement $y(t)$ satisfies the second-order linear ODE

$$my'' + cy' + ky = f(t) - mg \tag{6}$$

The ODE (6) is a straightforward generalization of the first-order linear ODEs investigated in Section 1.2.

For any twice-differentiable function y, let us denote by the symbol $L[y]$ the function

$$L[y] = my'' + cy' + ky \tag{7}$$

Expressed in words, the operational formula (7) just says that: "When L acts on any twice-differentiable function $y(t)$, the result is $L[y] = my'' + cy' + ky$." The mathematical object L is called an *operator*. Observe that L "operates" on functions and generates other functions. Note that for any twice-differentiable functions $y_1(t)$ and $y_2(t)$ and any real numbers a, b, the function $ay_1 + by_2$ is twice-differentiable. Now a straightforward computation shows that for L as in (7)

$$L[ay_1 + by_2] = aL[y_1] + bL[y_2] \tag{8}$$

The operational formula (8) is the reason why the ODE (6), which can be written as

$$L[y] = f(t) - mg$$

is called a *linear* ODE, and L is called a *linear operator*. Observe that the operator L in (6) is linear even if the coefficients m, c, and k are functions of t. The reason for this terminology is the analogy with the linear function on the x-axis given by $M[x] = \alpha x$, where α is a given constant. For any two reals a and b, we have that

$$M[ax_1 + bx_2] = \alpha(ax_1 + bx_2) = aM[x_1] + bM[x_2]$$

which is the exact analog of the operational formula (8).

EXAMPLE 3.1.1

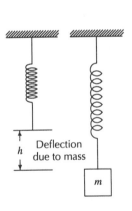

h Deflection due to mass

Hooke's Law Spring

Using Hooke's Law in ODE (1) gives rise to the linear ODE (6). We will show that a translation of the state variable y will eliminate the force of gravity term, mg. Let h be the magnitude of the *static deflection* due to hanging the body of mass m on the spring (see figure in the margin). It is easy to compute a value for h since, for the body to hang at rest at the end of the spring, we have that the force on the body exerted by the spring must be equal and opposite to the gravitational force on the body, i.e., $kh = mg$. Changing from y in ODE (6) to z, where $y = z - h$, we see that the differential equation modeling the motion of the body is given by

$$mz'' + cz' + kz = f(t) \tag{9}$$

or using the linear operator L, $L[z] = f$. It is interesting to observe that the gravitational constant has disappeared from our model. Thus, we are led to conclude that our spring-body system would behave the same way (relative to the equilibrium position) on the moon or on Mars as it would on the earth. However, since $h = mg/k$, the equilibrium position *does* depend upon the gravitational constant. See Problem 5 for the question of whether the motion of a hard or soft spring about an equilibrium position depends on the value of g.

EXAMPLE 3.1.2

A Lightly Damped, Driven Hooke's Law Spring

A massless Hooke's Law spring is suspended from a hook which is attached to a motor. When the motor is "off," a body weighing 1 lb is hung from the free end of the spring, and the spring stretches 15.36 in to its equilibrium position. A damping force is provided by a dashpot (see first margin sketch in this section) connected to the hook and to the body. The damping force is cv, where v is the velocity of the dashpot's plunger through the fluid (hence, the velocity of the body relative to the hook), and the

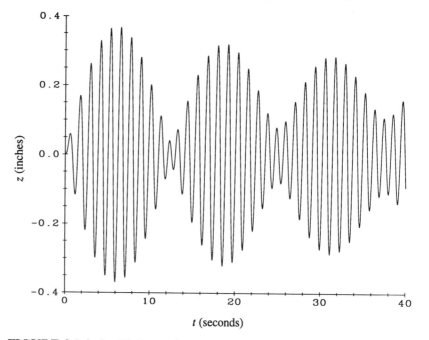

FIGURE 3.1.1 Oscillations of driven damped Hooke's Law spring. See Example 3.1.2.

value of c is 1.30×10^{-4} lb $\cdot$ sec /in. The motor is then turned on, providing a periodic driving force of $f(t) = 2.60 \times 10^{-3} \sin(5.6t)$ lb, where time t is measured in seconds. We will set up an initial value problem (IVP) that models the subsequent motion of the body as measured by $z(t)$, its displacement from equilibrium at time $t \geq 0$. The ODE is (by Example 3.1.1)

$$mz'' + cz' + kz = f(t)$$

but we must use the given data to quantify the various terms in the ODE. First since we are given the weight mg of the body rather than its mass m, we multiply the ODE by $g = 32$ ft/sec^2 = 384 in/ sec^2 to obtain

$$mgz'' + cgz' + kgz = z'' + cgz' + kgz = gf(t)$$

Using the given data we see that (since the weight mg is 1 lb)

$$cg = 1.30 \times 10^{-4} \times 384 = 0.05 \text{ lb/ sec}$$

$$kg = (mg)(g/h) = 384/15.36 = 25 \text{ lb/ sec}^2$$

$$gf(t) = \sin(5.6t) \text{ in lb/ sec}^2$$

and so the IVP is

$$z'' + 0.05z' + 25z = \sin(5.6t), \quad z(0) = 0, \quad z'(0) = 0$$

where z is measured in inches and time t in seconds. We show how to find an exact solution formula for this IVP in Sections 3.5 and 3.6. For now, we use a numerical solver to find the graph of the solution $z = z(t)$ (see Figure 3.1.1).

The calculations of Example 3.1.2 illustrate how to convert data given in British Engineering units (see also Appendix C.7) into usable data for the model ODE. A point to remember in doing this: pounds (lb) are a unit of force in the BE system, not of mass. Now we turn to some of the other spring models.

EXAMPLE 3.1.3

Aging Spring Equation (Linear)
We have seen that the ODE (6), the spring equation using Hooke's Law, is a linear second-order ODE. Using the aging spring restoring force, $-k(t)y$, in the general spring ODE (1), we have the linear *aging spring ODE*

$$my'' + cy' + k(t)y = f(t) - mg \tag{10}$$

Using the operational definition $L[y] = my'' + cy' + k(t)y$, we can rewrite ODE (10) as $L[y] = f(t) - mg$. Because the spring coefficient $k(t)$ is no longer a constant, there is no such thing as a static deflection, and we cannot introduce a new variable as we did earlier to eliminate the gravitational term $-mg$.

EXAMPLE 3.1.4

Hard or Soft Springs (Nonlinear)
Using the nonlinear spring restoring force $-ky \mp jy^3$ in the ODE (1), we obtain the *hard spring ODE* (if the upper sign is used) or the *soft spring ODE* (with the lower sign).

$$my'' + cy' + ky \pm jy^3 = f(t) - mg \tag{11}$$

Using the operational definition $K_\pm[y] = my'' + cy' + ky \pm jy^3$, we can write ODE (11) as $K_\pm[y] = g(t) - mg$. Since the operational formula (8) does not hold for K_+ or for K_- (because of the terms $\pm jy^3$), it follows that the ODEs (11) are not linear, and K_+ and K_- are not linear operators.

Second-Order ODEs and IVPs

Normalized second-order ODEs have the form

$$y'' = F(t, y, y') \tag{12}$$

Since the first-order ODE $y' = f(t, y)$ has the single state variable $y(t)$, it should not come as a surprise that the second-order ODE (12) has two state variables $y(t)$, $y'(t)$ (we shall discuss this in more detail in the next section). For the spring ODEs these state variables are just the position and velocity of the body as it undergoes its motion along the local vertical. In general, the IVP associated with ODE (12) is given by

$$y'' = F(t, y, y'), \qquad y(t_0) = y_0, \quad y'(t_0) = v_0 \tag{13}$$

A normalized linear second-order ODE is usually written in the form

$$y'' + a(t)y' + b(t)y = f(t) \tag{14}$$

where the functions $a(t)$, $b(t)$, and $f(t)$ are defined on a common t-interval, I (see (5) in Section 1.1). The linear ODE (14) is called *nonhomogeneous* (or *forced*, or *driven*) if the *forcing* (or *driving*) *term* $f(t)$ does not vanish identically, and *homogeneous* (or *free*, *undriven*, or *reduced*) if $f(t)$ does vanish identically on I. When the coefficients a and b are constants, ODE (14) is called a *linear second-order differential equation with constant coefficients*.

EXAMPLE 3.1.5

Second-Order IVPs from Chapter 1
Several second-order IVPs were derived in Chapter 1 as mathematical models of vertical motion. Some of these are listed below (with the notation slightly changed):

- $s'' = -g$, $s(0) = s_0$, $s'(0) = v_0$ (Example 1.1.1)
- $my'' + ky' = -mg$, $y(0) = h$, $y'(0) = 0$ (Example 1.3.2)
- $my'' = -ky'|y'| - mg$, $y(0) = h$, $y'(0) = 0$ (IVP (14), Section 1.4)
- $mz'' = -mMG(z+R)^{-2}$, $z(0) = 0$, $z'(0) = v_0$ (IVP (5), Section 1.8)

The first two ODEs are linear; the last two are not linear.

Autonomous ODEs; Equilibrium Solutions

The ODE in IVP (13) is said to be *autonomous* if F does not depend explicitly on t. For example, all of the ODEs cited in Example 3.1.5 are autonomous. For an autonomous ODE, $y'' = F(y, y')$, if y_0 is such that $F(y_0, 0) = 0$, then $y = y_0$, for all t, is a constant solution and is called an *equilibrium solution*. In other words, if the system starts out at rest (i.e., $y' = 0$) when $y(0) = y_0$, then the system stays at rest at this point forever.

EXAMPLE 3.1.6

Equilibrium Solutions

The Hooke's Law spring ODE (6) with no external driving force is an autonomous ODE. Writing this ODE as

$$y'' = -\frac{c}{m}y' - \frac{k}{m}y - g$$

we see that $F(y, y') = -(c/m)y' - (k/m)y - g$. There is only one equilibrium solution because $F(y_0, 0) = 0$ implies that $y_0 = -(mg)/k$, which agrees with the static deflection value h computed in Example 3.1.1.

The soft spring ODE (with $g = 9.8$ m/sec^2, $c/m = 0.2$ sec^{-1}, $k/m = 10$ sec^{-2}, $j/m = 0.2$ m^{-2} sec^{-2})

$$y'' = -0.2y' - 10y + 0.2y^3 - 9.8$$

has three equilibrium solutions. Since $F(y, y') = -0.2y' - 10y + 0.2y^3 - 9.8$ we see that the equilibrium solutions y_0 satisfy the condition $F(y_0, 0) = 0$, which gives the cubic equation $0.2y_0^3 - 10y_0 - 9.8 = 0$. After a bit of experimentation we discover that $y_0 = -1$ is a root and this cubic equation can be factored as $(y_0 + 1)(0.2y_0^2 - 0.2y_0 - 9.8) = 0$. So its three roots are $y_0 = -1$, and $y_0 = (1 \pm \sqrt{197})/2 \approx 7.52, -6.52$.

We know from experience that stretched or compressed springs always supply a force to unstretch or uncompress themselves. Thus, the soft spring law is valid only when the force on the spring, $y(-10 + 0.2y^2)$, has the opposite sign to the displacement y. This means, in particular, that the y-values must be restricted by $|y| \leq 50 \approx 7.07$. Note that the equilibrium position $y_0 = 7.52$ is outside the region of validity for the spring force. Although the position $y_0 = -6.52$ falls within the region of validity, it turns out that it makes little physical sense to study the spring motion near that equilibrium point.

Numerical Solution of an IVP: Systems

Some numerical solvers will not accept IVP (13) because the ODE is not first-order, but a simple trick will make it acceptable to any solver. Indeed, say that $y(t)$ solves IVP (13). If we let $v = y'$, then we see that $v(t)$ satisfies the ODE $v' = F(t, y, v)$ and the initial condition $v(t_0) = v_0$. Thus, the pair of functions $y(t)$, $v(t)$ solves the

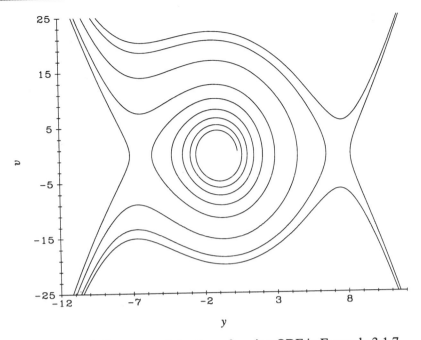

FIGURE 3.1.2 State portrait for the soft spring ODE in Example 3.1.7.

following IVP for a first-order differential system:

$$y' = v, \qquad\qquad y(t_0) = y_0$$
$$v' = F(t, y, v), \quad v(t_0) = v_0 \tag{15}$$

Conversely, if the pair of functions $y(t)$, $v(t)$ solves IVP (15), then we see that $y(t)$ solves IVP (13). Numerical solvers know how to solve IVPs for normalized first-order systems. The *state space* for the differential system in IVP (15) is the yv-plane. The solution, $y = y(t)$, $v = v(t)$, of IVP (15) for specific initial data y_0, v_0, generates a curve (parametrically) in the yv-plane which is called an *orbit* of the differential system in IVP (15). A collection of orbits is known as a *state portrait*. See Atlas Plate PLANAR PORTRAIT C for an example.

For autonomous ODEs the actual choice of the initial time t_0 in IVP (15) is immaterial when computing an orbit. Only the solve-time matters. Thus, given a fixed $T > 0$, the orbits generated by the IVP (15) over $t_0 \le t \le t_0 + T$ are the same for any value of t_0. A proof of this fact appears in Section 3.2. A similar property for autonomous first-order ODEs was noted in Section 2.2.

EXAMPLE 3.1.7	

State Portrait for a Soft Spring ODE

Consider the IVP for a soft spring

$$y'' + 0.2y' + 10y - 0.2y^3 = -9.8 \qquad y(0) = y_0, \quad y'(0) = v_0$$

Converting this IVP to an "equivalent" IVP for a first-order system as indicated above, we obtain

$$y' = v, \qquad\qquad\qquad y(0) = y_0$$
$$v' = -0.2v - 10y + 0.2y^3 - 9.8, \quad v(0) = v_0 \tag{16}$$

See Figure 3.1.2 for a state portrait of IVP (16), from which a number of interesting properties of the spring can be obtained. First observe that no two orbits intersect. Next observe that there are three equilibrium solutions, which show up as points in state space, and they are all located on the y-axis. The y-coordinates of these points are the roots of the cubic equation $-10y_0 + 0.2y_0^3 - 9.8 = 0$, and were calculated in Example 3.1.6 to be $y_0 = -1$, $y_0 = (1 + \sqrt{197})/2$, and $y_0 = (1 - \sqrt{197})/2$.

The state portrait shows that if the initial data are carefully chosen, then the solution $y = y(t)$, $v = v(t)$ tends to oscillate about the equilibrium point $(-1, 0)$ with amplitude that steadily decreases to zero. Finally, observe that orbits originating near the other two equilibrium points are very sensitive to the actual choice of initial data. The orbits that escape to infinity demonstrate the limits of the validity of the soft spring model. For large displacements from equilibrium the soft spring force changes sign and repels the body. The spring model treated here is valid only in the case of small displacements from the equilibrium point $(-1, 0)$.

Linear versus Nonlinear ODEs

Our experience with model building in earlier chapters indicates that no matter what natural system is considered, a model can almost always be found whose mathematical formulation is nonlinear. Practitioners sometimes describe this state of affairs by saying that "nature is inherently not linear." Interestingly enough, however, we will see that linear ODEs arise in the modeling of many physical systems and the results which they produce often agree rather well with physical reality. Thus linear ODEs occupy a very special place in the construction of mathematical models. But linear differential equations are also important theoretically because a knowledge of the behavior of their solutions is useful in the investigation of the behavior of solutions of more general differential equations (as we will see in Section 8.4).

For our next example it will be useful to use the definition of the linear approximation of a function $F(u, v)$ at a base point (u_0, v_0) where $F(u_0, v_0) = 0$.

> ❖ **Linear Approximation.** Suppose that $F(u, v)$ is continuously differentiable in a rectangle R in the uv-plane, and that (u_0, v_0) is a point inside R with $F(u_0, v_0) = 0$. Then
>
> $$G(u, v) = F_u(u_0, v_0)(u - u_0) + F_v(u_0, v_0)(v - v_0) \qquad (17)$$
>
> is called the *linear approximation* to $F(u, v)$ at the base point (u_0, v_0).

Observe that $G(u, v)$ is obtained by using Taylor's Theorem for $F(u, v)$ at the base point (u_0, v_0) and keeping only the linear terms (see Appendix C.5).

EXAMPLE 3.1.8

Comparison of Linear and Nonlinear ODEs

Consider again the IVP for a soft spring given in Example 3.1.7

$$y'' + 0.2y' + 10y - 0.2y^3 = -9.8, \qquad y(0) = y_0, \quad y'(0) = v_0 \qquad (18)$$

Writing the ODE in the form $y'' = F(y, y')$, we see that $F(y, y') = -0.2y' - 10y + 0.2y^3 - 9.8$, and that $F(-1, 0) = 0$. As the orbit of system (16), equivalent to IVP (18), spirals closer and closer to $(-1, 0)$, the difference between $F(y, y')$ and the linear approximation $G(y, y') = F_y(-1, 0)(y + 1) + F_{y'}(-1, 0)y'$ to $F(y, y')$ near $(-1, 0)$ becomes smaller and smaller. So we would expect the orbits of the systems equivalent to $y'' = G(y, y')$ and $y'' = F(y, y')$ to be close for initial points (y_0, v_0) close to the equilibrium point $(-1, 0)$.

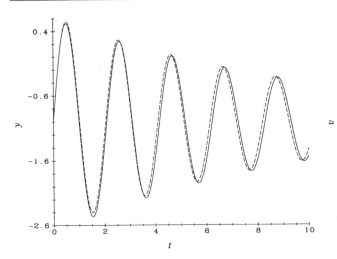

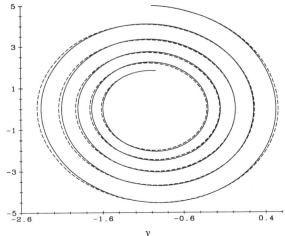

FIGURE 3.1.3 Plots of y vs. t for solutions of the nonlinear IVP (18) (solid) and the linear IVP (19) (dashed), with $y_0 = -1$, $v_0 = 5$. See Example 3.1.8.

FIGURE 3.1.4 Comparison of the nonlinear (solid) and linear (dashed) orbits in yv-plane. See Example 3.1.8.

To calculate the linear approximation $G(y, y')$ to $F(y, y') = -0.2y' - 10y + 0.2y^3 - 9.8$ at the base point $y_0 = -1$, $y_0' = 0$, observe that

$$\left.\frac{\partial F}{\partial y}\right|_{(-1,0)} = -9.4, \qquad \left.\frac{\partial F}{\partial y'}\right|_{(-1,0)} = -0.2$$

and so we see that the linear approximation to F at $(-1, 0)$ is

$$G(y, y') = -9.4(y + 1) - 0.2y'$$

The IVP

$$y'' = -9.4(y + 1) - 0.2y', \qquad y(0) = y_0, \quad y'(0) = v_0 \tag{19}$$

is called the *linearization* of IVP (18) at $(-1, 0)$.

In Figure 3.1.3 we plot y versus t for solutions of both IVP (18) and its linearization IVP (19) that use the same initial data $y_0 = -1$, $v_0 = 5$. Notice that as time increases the shapes of the two graphs look more and more similar but increasingly out of phase. Figure 3.1.4 shows orbits for both IVP (18) and IVP (19) that use the same initial data $y_0 = -1$, $v_0 = 5$. Note how well these orbits agree even though the initial data are not very close to the equilibrium point $(-1, 0)$, and in spite of the fact that the corresponding component graphs in Figure 3.1.3 are out of phase.

Comments

The second-order ODEs that model linear and nonlinear springs are widely used in science and engineering as guides to understanding the motions of real springs. The ODEs themselves may be used to model a host of other physical phenomena, and so there is good reason to study the ODEs independent of this physical interpretation. The remaining sections of the chapter focus on solving second-order linear ODEs; nonlinear ODEs are considered again in Section 4.1 and in Chapters 6, 8, and 9.

PROBLEMS

1. (*Hooke's Law Spring*). Consider a $1 - $ lb weight suspended from a Hooke's Law spring. Let z denote the deflection of the weight from its equilibrium position. Assume there is no damping or driving force acting on the weight.

    (a) If the static deflection of the weight is 24 inches, show that $z(t)$ satisfies the ODE, $z'' + 16z = 0$, where z is measured in inches. [*Hint*: Weight $= mg$ and $g = 384$ in/ sec^2]

    (b) Suppose that the weight is pushed up z_0 inches and released from rest, where z_0 takes in turn each of the values 15, 10, 5, and 2 inches. Plot the resulting orbits of the weight in state space, inserting arrowheads to indicate increasing time.

    (c) Plot the tz- and tz'-component graphs for each of the orbits in part (b). What can you say about the motion of this weight? If periodic, what is the period?

2. (*Damped Hooke's Law Spring*). Consider a $1 - $ lb weight suspended from a Hooke's Law spring. Let z denote the deflection of the weight from its equilibrium position, measured in inches. Assume damping but no driving force acting on the weight.

    (a) If the static deflection of the body is 24 inches and the damping coefficient $c = 2.60 \times 10^{-4}$ lb sec /in, show that $z(t)$ satisfies the ODE $z'' + 0.1z' + 16z = 0$. [*Hint*: Weight $= mg$, and $g = 384$ in/ sec^2.]

    (b) Suppose the weight is pushed up z_0 inches and released from rest, where y_0 takes, in turn, the values 15, 10, and 2 inches. Plot the resulting orbits in state space, inserting arrowheads to show the direction of increasing time.

    (c) Plot the component graph of y versus t for each of the orbits in part (b). What can you say about the motion of this weight and the period of the oscillations?

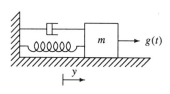

3. (*Nonlinear Springs*). A body of mass m is attached to a wall by means of a spring and moves back and forth along its axis on a horizontal frictionless surface. Let y measure the displacement of the body along the spring's axis from the equilibrium position; $y > 0$ corresponds to a stretched spring, $y < 0$ to a compressed spring. The margin figure shows an *external force* $g(t)$ acting on the body parallel to the y-axis, and damping forces represented by a damper. The downward gravitational force on the body is exactly cancelled by an upward force exerted by the supporting surface, so the mass moves only horizontally, unaffected by gravity. Under these assumptions, Newton's Second Law implies that

$$my''(t) = S(y(t)) - cy'(t) + g(t)$$

where $c > 0$ is the damping constant, and some possible formulas for the *spring restoring force* $S(y)$ are given by equations (2)—(5). In parts (a)—(c) below first write the equivelent system as in (15), and then plot the orbits in the yv-plane through the given initial points and the corresponding solution curves in the ty-plane. Interpret your graphs in terms of the motion of a spring, and discuss the "reality" of each orbit. Estimate the periods of periodic solutions; does the period increase, decrease, or stay the same as the amplitude increases?

    (a) (*Hard Spring*). $y'' = -0.2y - 0.02y^3$, $0 \le t \le 25$; $x_0 = 0$, $y_0 = 0, 1, 3, 9$.

    (b) (*Soft Spring*). $y'' = -0.2y + 0.02y^3$, $0 \le t \le 30$; $x_0 = 0$, $y_0 = 0, 0.4, 0.9$; $x_0 = -5$, $y_0 = 1.49, 1.51$; $x_0 = 5$, $y_0 = -1.49, -1.51$.

    (c) (*Damped Soft Spring*). $y'' = -y + 0.1y^3 - 0.1y'$, $0 \le t \le 40$; $x_0 = 0$, $y_0 = 0$, 2.44, 2.46.

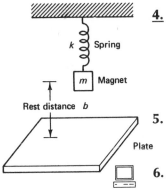

k ⛓ Spring

m Magnet

Rest distance b

Plate

4. (*Magnetically Driven Spring*). A magnet is suspended by a spring above a large fixed iron plate, as shown in the margin figure. Let z be the vertical displacement of the magnet from its rest position at distance b from the plate. Assume that the force of magnetic attraction is inversely proportional to the distance between the magnet and the plate, and that the spring force is directly proportional to the displacement z. Find the ODE describing the motion of the magnet.

5. (*Effect of Gravity on Spring Motion*). Does gravity affect the vertical motion of a weight on a hard or soft spring? Contrast your conclusion with the Hooke's Law spring, and give reasons. [*Hint*: Review Example 3.1.1.]

6. (*Driven Hooke's Law Spring*). In a Hooke's Law spring model system, let the spring constant be 1.01 N/m and the mass of the body be 1 kg. The body is acted on by damping (magnitude of damping force proportional to velocity) with coefficient equal to 0.2 Ns/m, and driven by the periodic force (measured in newtons) described by $F(t) = \text{sqw}(t, 50, 2\pi)$, where the function sqw is defined in Appendix C.1. Graph the response $z(t)$ of the system if at $t = 0$ it is at rest in its equilibrium state, and interpret what you see. [*Hint*: Use a numerical solver to solve an IVP that you construct to model the motion.]

7. (*Another Way to Generate Figures 3.1.3 and 3.1.4*). Some solvers may not be able to overlay plots from two different ODEs on the same set of axes. Follow the outline below to "trick" your solver into reproducing Figures 3.1.3 and 3.1.4.

- Consider the IVP

$$y'' = -0.2y' - 10y + 0.2y^3 - 9.8 + C(0.6y - 0.2y^3 + 0.4)$$
$$y(0) = -1, \quad y'(0) = 5$$

where C is a constant. Show that plotting solutions of this IVP for $C = 0$ and $C = 1$ over $0 \le t \le 10$ generates the graphs in Figure 3.1.3.

- Show that by denoting the state variable y' by v, the IVP above can be converted to the equivalent IVP for a first-order system:

$$y' = v, \qquad\qquad\qquad\qquad\qquad\qquad\qquad y(0) = -1$$
$$v' = -0.2v - 10y + 0.2y^3 - 9.8 + C(0.6y - 0.2y^3 + 0.4), \quad v(0) = 5$$

Show that the orbits generated in yv-state space over $0 \le t \le 10$ by this IVP for $C = 0$ and $C = 1$ give rise to the graph in Figure 3.1.4.

- Let $C = C(t)$ be another state variable and consider the following IVP for a first-order system with three state variables:

$$y' = v, \qquad\qquad\qquad\qquad\qquad\qquad\qquad y(0) = -1$$
$$v' = -0.2v - 10y + 0.2y^3 - 9.8 + C(0.6y - 0.2y^3 + 0.4), \quad v(0) = 5$$
$$C' = 0, \qquad\qquad\qquad\qquad\qquad\qquad\qquad C(0) = C_0$$

Show that if this IVP is solved over $0 \le t \le 10$ for $C_0 = 0$ and $C_0 = 1$ and y plotted against t in both cases, then Figure 3.1.3 results. How can one use this IVP to generate the graphs in Figure 3.1.4?

3.2 Basic Theory: Second-Order ODEs

We have already encountered several instances of the normalized second-order ODE

$$y'' = F(t, y, y') \tag{1}$$

in earlier sections. We saw in Section 1.8 that if the function F in the ODE (1) is free of either t or y, then certain substitutions reduce the search for a solution formula to solving a pair of first-order ODEs in succession (see Examples 1.8.1 and 1.8.2). In the previous section we saw how the vertical motion of a weight on a spring gives rise to a second-order IVP of the form

$$y'' = F(t, y, y'), \qquad y(t_0) = y_0, \quad y'(t_0) = v_0 \qquad (2)$$

and we showed how to solve this IVP numerically using a computer solver. So we have these two approaches: find a solution formula, use a numerical solver. Example 3.2.1 illustrates both.

EXAMPLE 3.2.1

Two Approaches to Solving an IVP

The second-order ODE in the IVP

$$t(y''y + y'^2) + y'y = 1, \qquad y(1) = 1, \quad y'(1) = -1 \qquad (3)$$

can be reduced by the substitution $u = y'y$ which converts the ODE into the first-order linear ODE $tu' + u = 1$. Solving this reduced ODE for $u(t)$ we obtain $u = 1 + c_1/t$, where c_1 is a constant. Recalling that $u = y'y$ yields the ODE for y: $y'y = 1 + c_1/t$. Writing $y'y = (y^2)'/2$ and using the Vanishing Derivative Theorem (Theorem 1.1.1) we obtain $y^2 = 2t + 2c_1 \ln |t| + c_2$ where c_1 and c_2 are constants for which $2t + 2c_1 \ln |t| + c_2 \geq 0$ on some interval I. Imposing the initial conditions now gives the values $c_1 = -2$ and $c_2 = -1$. The solution of IVP (3) is

$$y = (2t - 4 \ln t - 1)^{1/2}, \quad t \geq 1$$

since it can be shown (but not easily) that $2t > 4 \ln t + 1$ for all $t \geq 1$.

Rather than plot the solution of IVP (3) from this formula, we will solve the IVP numerically and plot the numerical solution. In preparation for this we first normalize the ODE by solving for y'', and then let $y' = v$ to obtain the first-order system

$$\begin{aligned} y' &= v & y(1) &= 1 \\ v' &= \frac{1 - yv - tv^2}{ty} & v(1) &= -1 \end{aligned} \qquad (4)$$

Using a computer solver to solve IVP (4) and plotting the component y versus t yields the graph of the solution of IVP (3). See Figure 3.2.1.

Existence and Uniqueness

Only very special IVPs of the form (2) can be solved as was done for IVP (3) in Example 3.2.1. When a solution formula for an IVP can be found, then questions of existence, uniqueness, and continuity of a solution are fairly easy to answer. Theorem 3.2.1 answers these basic questions even when no solution formula can be found, but first a few definitions are needed:

❖ **Continuity Sets.** Let I be an interval on the t-axis. The symbol $\mathbf{C}^0(I)$ denotes the set of all continuous functions on I. The symbol $\mathbf{C}^1(I)$ denotes the set of functions in $\mathbf{C}^0(I)$ with a continuous derivative on I. The symbol $\mathbf{C}^2(I)$ denotes the set of all functions in $\mathbf{C}^1(I)$ whose second derivative is continuous on I.

A subtle point in the above definition is easy to overlook. If I contains an endpoint, say the left-hand endpoint $t = a$, then to say that $y(t)$ is in the set $\mathbf{C}^1(I)$ means that

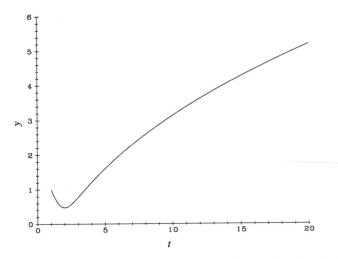

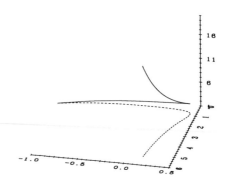

FIGURE 3.2.1 Graph of the solution of IVP (4). See Example 3.2.1

FIGURE 3.2.2 Time-state plot (solid line) for IVP (4), and associated orbit (dashed line) in yv-plane. See Example 3.2.2.

$y'(t)$ exists at all interior points of I and that $\lim_{t \to a^+} y'(t)$ exists. Similar statements hold for a right-hand endpoint and for functions in $C^2(I)$. For example, let I be the interval $-1 \leq t \leq 1$, and $y(t) = \sqrt{1 - t^2}$. Now $y' = -t/\sqrt{1 - t^2}$, for $-1 < t < 1$, and so $y'(t)$ does not have a limit as $t \to \pm 1$. Thus $y(t)$ does not belong to $C^1(I)$, but it does belong to $C^0(I)$ and to $C^1(J)$ where J is the open interval $(-1, 1)$.

Continuity sets allow us to be precise in defining a solution of an ODE without using awkward wording. The same goes for statements of the properties of solutions.

❖ **Solution and Time-State Curve.** A function $y(t)$ defined on an interval I containing t_0 is called a *solution* of the IVP

$$y'' = F(t, y, y'), \quad y(t_0) = y_0, \quad y'(t_0) = v_0$$

if y is in $C^2(I)$, $y(t_0) = y_0$ and $y'(t_0) = v_0$, and $y''(t) = F(t, y(t), y'(t))$, for all interior points t of I. For the solution $y(t)$ of the IVP, the curve in tyy'-space traced out by $(t, y(t), y'(t))$, t in I, is called a *time-state curve*.

The fundamental theorem can now be stated.

THEOREM 3.2.1

> Existence/Uniqueness/Continuity Theorem. Let $F(t, y, y')$ be defined on a box B in tyy'-space given by the inequalities: $t_1 \leq t \leq t_2$, $y_1 \leq y \leq y_2$, $v_1 \leq y' \leq v_2$, for some constants t_1, t_2, y_1, y_2, v_1, v_2. If F, F_y, and $F_{y'}$ are all continuous functions on B, then for any choice of (t_0, y_0, v_0) in B the IVP,
>
> $$y'' = F(t, y, y'), \quad y(t_0) = y_0, \quad y'(t_0) = v_0 \tag{5}$$
>
> has a unique solution defined on some interval I containing t_0, which is continuous in the parameters t_0, y_0, and v_0.

A proof of Theorem 3.2.1 can be fashioned along the lines of the proof for first-order IVPs given in Appendix A. A great deal more can be said about the solutions if ODE (1) is linear; i.e., has the form $y'' + a(t)y' + b(t)y = f(t)$. See Section 3.3.

As in Section 2.2, we usually assume that all solutions are extended forward and backward in time t as far as possible. We normally regard all solutions to be *maximally extended*. With this point in mind, the uniqueness conclusion of Theorem 3.2.1 implies that no two distinct time-state curves of $y'' = F(t, y, y')$ can touch in the box B. Moreover, no time-state curve can "die" inside the box; each curve reaches from one rectangular face of the box to another. Time-state curves of $y'' = F(t, y, y')$ "fill up" the box, with only one time-state curve passing through each point.

EXAMPLE 3.2.2

Time-State Curve for IVP (3)

The corresponding time-state curve for the solution of IVP (3) given in Example 3.2.1 appears in Figure 3.2.2 where time is plotted on the vertical axis. Writing the ODE in IVP (3) in the normalized form $y'' = F(t, y, y')$, we see that

$$F(t, y, y') = \frac{1 - yy' - t(y')^2}{ty}$$

Thus, to apply Theorem 3.2.1 to this ODE, the box B must not contain any points on the planes $t = 0$ or $y = 0$ in the 3-dimensional tyy'-space. That is why in IVP (3) we give the initial data at $t_0 = 1$ instead of at $t_0 = 0$.

Orbits, Solution Curves of Second-Order ODEs

Although the concept of a time-state curve for IVP (2) is convenient for stating existence and uniqueness properties, there are other graphs associated with solutions of ODE (1) that are more commonly encountered.

❖ **Solution Curve, Orbit.** Let $y(t)$ be a solution of ODE (1) defined on an interval I. The graph of $y(t)$ versus t in ty-space is called a *solution curve*. The parametric graph of $y = y(t)$, $y' = y'(t)$, t in I, in yy'-space is called an *orbit* (or *trajectory*).

Observe that a solution curve and a time-state curve of a first-order ODE are one and the same thing (see Section 1.1). This is not the case for higher-order ODEs because additional state variables are required.

The use of the term *orbit* here is consistent with the same term defined in Section 1.4 for a planar system of ODEs. To see this, suppose that $y(t)$ is a solution of ODE (1) and put $v = y'(t)$. Then the pair $y(t)$, $v(t)$ is a solution of the planar system

$$y' = v \qquad v' = F(t, y, v) \tag{6}$$

and so y and v can be regarded as state variables. That is why the curve traced out parametrically by $(t, y(t), v(t))$ is defined as a time-state curve. The graph of $(y(t), v(t))$ in the yv-state space is what we defined as an orbit of the planar system (6) in Section 1.4. Thus our definition of an orbit for ODE (1) is not out of line with the earlier use of the term. Note that orbits are just projections of time-state curves along the t-axis onto the yv-plane. Figure 3.2.2 shows a time-state curve for the ODE in Example 3.2.2 along with its associated orbit in the yy'-plane.

Properties of Time-State Curves, Orbits, and Solution Curves

Theorem 3.2.1 implies a number of important properties for time-state curves, orbits, and solution curves for the second-order ODE (1), which we summarize below. We shall assume that B is a box in tyy'-space in which $F(t, y, y')$, $F_y(t, y, y')$, and $F_{y'}(t, y, y')$ are continuous.

- *Time-state curves never intersect.* The uniqueness part of Theorem 3.2.1 says that through each point (t_0, y_0, y'_0) in B there is a unique time-state curve. Thus, two time-state curves cannot intersect at any point in B.

- *Solution curves may intersect.* The solution curve of a solution of ODE (1) is just the projection of a time-state curve onto the ty-plane "along the y'-axis." So solution curves might very well intersect, even though time-state curves never do. If two solution curves do intersect, they do so with different slopes.

- *Solution curves for autonomous ODEs.* If the function F in ODE (1) does not depend explicitly on t, it is usually written $F(y, y')$, and ODE (1) is then said to be *autonomous* (see Section 3.1). If $y(t)$ solves the autonomous ODE $y'' = F(y, y')$ over the interval $t_0 \leq t \leq t_1$, then, for any constant T, the function $y(t+T)$ solves the same ODE over the interval $t_0 - T \leq t \leq t_1 - T$. Clearly, the solution curve for $y(t+T)$ is just the solution curve $y(t)$ shifted T units backward along the t-axis if $T > 0$ and forward along the t-axis if $T < 0$. Thus the solution curve of an autonomous ODE over a t-interval of specified length would look the same no matter when the clock was started, except that it would be shifted to the right or the left along the t-axis to accommodate the new clock. The same property holds for the time-state curves in tyy'-space.

- *Orbits of autonomous ODEs.* Let R be the projection of the box B into the yy'-plane. When ODE (1) is autonomous it is an interesting fact that an orbit in R does not depend on when the clock is started (i.e., t_0), but only on the duration of time for which the orbit is defined. The reason for this is that time-state curves merely shift along the t-axis when the initial time is changed, and an orbit is just a projection of a time-state curve onto the yy'-plane "along the t-axis."

- *Orbits for autonomous ODEs do not intersect.* Two *distinct* orbits of an autonomous ODE (1) cannot intersect in R. To see this, suppose $(z(t), z'(t))$ and $(w(t), w'(t))$ are two orbits of ODE (1) which intersect at a point (y_0, y'_0) in R. Thus $z(t_1) = w(t_2) = y_0$, and $z'(t_1) = w'(t_2) = y'_0$, for times t_1 and t_2 (we cannot assume that $t_1 = t_2$). Shifting the w-orbit in time by $T = t_2 - t_1$, we find that $z(t)$ and $w(t+T)$ are both solutions of the same IVP

$$y'' = F(y, y'), \qquad y(t_1) = y_0, \quad y'(t_1) = y'_0$$

  Hence, from Theorem 3.2.1 the two orbits must be identical.

- *Closed orbit of autonomous system.* If an orbit of an autonomous system intersects itself, then by restarting the clock at that point we see that the closed orbit repeats itself forever. This is an example of a *periodic solution* of an autonomous ODE.

EXAMPLE 3.2.3

Body on an Undamped Hard Spring

Suppose that a body of mass m is attached to a wall by means of a hard spring and moves back and forth along its axis on a frictionless horizontal supporting surface. Denoting the displacement by y, then (see Problem 3, in Section 3.1) $y(t)$ satisfies the ODE $my'' = -ky - jy^3$, for some positive constants k and j. Choosing values for m,

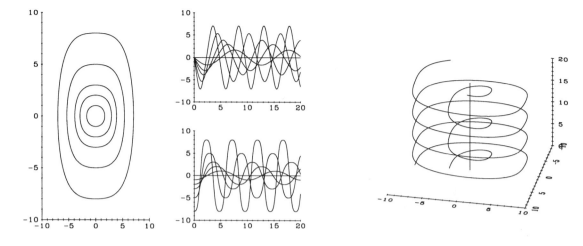

FIGURE 3.2.3 Orbits, solution curves (y vs. t) and velocity curves (y' vs. t) for ODE (7). See Example 3.2.3.

FIGURE 3.2.4 Three time-state curves for ODE (7). See Example 3.2.3.

k, and j such that $k/m = 0.3$, and $j/m = 0.04$, we see that $y(t)$ satisfies the ODE

$$y'' = -0.3y - 0.04y^3 \qquad (7)$$

The left-hand graph in Figure 3.2.3 shows closed orbits and an equilibrium point of ODE (7); these correspond to periodic solutions. Note that, as expected, they do not intersect since the function $F(y, y') = -0.3y - 0.04y^3$ is independent of t. The upper right-hand graph in Figure 3.2.3 shows the corresponding solution curves for ODE (7). Note that they all intersect at the initial point $y(0) = 0$ but have distinct initial slopes $y'(0)$. The lower right-hand graph shows the corresponding ty'-curves. Three time-state curves for ODE (7) (including the equilibrium solution) are plotted in Figure 3.2.4 (t-axis vertical). We see from the ty-graph in Figure 3.2.3 that the period seems to *decrease* as the initial velocity increases in magnitude.

EXAMPLE 3.2.4

Sliding Time-State Curves Up and Down the Time Axis

The autonomous ODE, $y'' + y = 0$, has among its solutions, $y = 4\cos(t - T)$, which defines a periodic solution for every value of T. Figure 3.2.5 shows the two time-state curves in tyy'-space (t-axis vertical) corresponding to

$$y = 4\cos(t - 40), \quad y' = -4\sin(t - 40), \quad 40 \le t \le 50$$

$$y = 4\cos(t - 90), \quad y' = -4\sin(t - 90), \quad 90 \le t \le 100$$

Each can be obtained form the other by a vertical translation parallel to the t-axis. Both project downward onto the circular orbit (dashed curve) in the yy'-plane; the equation of the orbit is $y^2 + (y')^2 = 16$.

EXAMPLE 3.2.5

Nonautonomous ODE: Graphs

In Figure 3.2.6 the time-state curve of the nonautonomous IVP

$$y'' + y = \cos(1.1t), \qquad y(50) = 5, \qquad y'(50) = 0$$

is plotted in tyy'-space (t-axis vertical) over the interval $50 \le t \le 125$. Notice that the time-state curve does not intersect itself. When this curve is projected "along

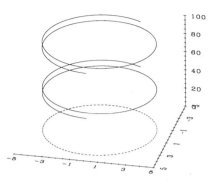

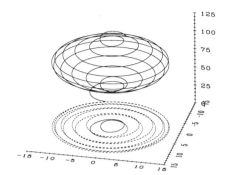

FIGURE 3.2.5 Two time-state curves (solid) and the corresponding orbit (dashed) for an autonomous ODE. See Example 3.2.4.

FIGURE 3.2.6 A time-state curve (solid) of a nonautonomous ODE and its orbit(dashed). See Example 3.2.5.

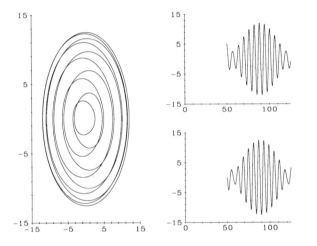

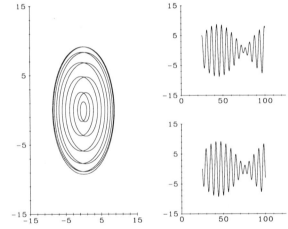

FIGURE 3.2.7 Orbit, solution curve (y vs. t), velocity curve (y' vs. t) of nonautonomous ODE. See Example 3.2.5.

FIGURE 3.2.8 The effects of changing the starting time when an ODE is nonautonomous. See Example 3.2.5.

the t-axis" into the yy'-plane the resulting curve (the dashed curve in Figure 3.2.6) is an orbit for the ODE, $y'' + y = \cos(1.1t)$, through the point $y_0 = 5$, $y_0' = 0$ with the clock starting at $t_0 = 50$. Notice that this orbit intersects itself (since the ODE is nonautonomous, this is no surprise). Figure 3.2.7 shows the orbit in the yy'-plane (the figure on the left), the solution curve (upper right), and the velocity curve (lower right).

If the clock were started at a different time, say $t_0 = 25$, then the orbit of $y'' + y = \cos(1.1t)$ through $y_0 = 5$, $y_0' = 0$, $25 \leq t \leq 100$, would be different from the one shown in Figure 3.2.7. Figure 3.2.8 shows the new orbit and the new solution and velocity curves.

Comments

The four questions posed in Section 2.2 for an IVP involving a first-order ODE could also be raised for an IVP involving a second-order ODE:

- Is there a solution?
- Is the solution unique?
- How can the solution be described?
- How sensitive is the solution to changes in the data?

In this section we have addressed the first three questions. The "sensitivity" question will arise in various contexts in later sections of this chapter and also in Chapter 4. Since solution formulas for general second-order ODEs are notoriously hard to come by, we shall devote the rest of this chapter to linear second-order ODEs, where solution formulas do indeed exist in a large number of cases.

PROBLEMS

1. Plot solution curves (y vs. t), velocity curves (y' vs. t), orbits (in the yy'-plane), and time-state curves (in tyy'-space) for each of the following IVPs. Use the given bounds for t, y, y'. If the ODE is autonomous, find all equilibrium solutions. Use your graphs to make conjectures about the long-term behavior of the solutions.

   (a) (*Undamped Hooke's Law Spring Model*). $y'' + 4y = 0$, $y(0) = 0.5, 1.0, 1.5$, $y'(0) = 0$; $0 \le t \le 20$, $|y| \le 2$, $|y'| \le 4$.

   (b) (*Damped Hooke's Law Spring Model*). $y'' + 0.1y' + 4y = 0$, $y(0) = 0.5, 1.0$, 1.5, $y'(0) = 0$; $0 \le t \le 40$, $|y| \le 2$, $|y'| \le 4$.

   (c) (*Going Backward in Time*). $y'' - 0.1y' + 4y = 0$, $y(0) = 0, 1, 2$, $y'(0) = 0$; $-100 \le t \le 0$, $|y| \le 2$, $|y'| \le 4$.

   (d) (*Damped Hard Spring*). $y'' + 0.2y' + 10y + 0.2y^3 = -9.8$, $y(0) = -4, 0$, $y'(0) = 0$; $0 \le t \le 40$, $-5 \le y \le 3$, $|y'| \le 9$.

   (e) (*Nonlinear ODE*). $y'' + (0.1y^2 - 0.08)y' + y^3 = 0$, $y(0) = 0$, $y'(0) = 0.5$; $0 \le t \le 100$, $-1.9 \le y \le 2.1$, $|y'| \le 1$.

   (f) (*Driven and Damped Hooke's Law Spring*). $y'' + 0.05y' + 25y = \sin(5.5t)$, $y(0) = 0$, $y'(0) = 0$; $0 \le t \le 40$, $|y| \le 0.4$, $|y'| \le 2.1$.

   (g) (*Painlevé Transcendent*). The ODE, $y'' - y^2 = -t$ is one of the *Painlevé transcendents*, a family of nonlinear ODEs extensively studied by many mathematicians, including Paul Painlevé (1863-1933). Use initial data $y(0) = 0$, $y'(0) = -5$, $-4.5, \ldots, 0, 0.5, 1$; $0 \le t \le 20$, $|y| \le 10$, $|y'| \le 10$.

   (h) (*Chapter 3 Cover Figure*). $y'' - y^2 + 0.1y^3 = -t$, $y(0) = 0$, $y'(0) = -5$; $0 \le t \le 50$, $|y| \le 15$, $|y'| \le 25$. The cover picture shows four orbits of this nonlinear ODE, which can be viewed as a modification of the Painlevé transcendent of part (g).

2. (*Concavity*). The sign of the second derivative of a function $y(t)$ determines whether the graph of $y = y(t)$ is concave up ($y'' > 0$) or down ($y'' < 0$). Thus, the concavity at (t, y) of a solution curve $y = y(t)$ of $y'' = F(t, y, y')$ is determined by the sign of $F(t, y, y')$. For each IVP below, find where solutions are concave up and where they are concave down. Then plot solutions and verify that the solution curves are concave up or down in the proper regions.

(a) $y'' = -1$; $y(0) = 1$; $y'(0) = 2, 0, -2$; $0 \le t \le 10$

(b) $y'' + y = 0$; $y(0) = 1, 0, -1$; $y'(0) = 0$; $0 \le t \le 20$

(c) (*Painlevé Transcendent*). $y'' - y^2 = -t$, $y(0) = 0$, $y'(0) = 0.75, 0.925$. [*Hint*: Solve over $0 \le t \le 10$. Then plot the parabola $t = y^2$, and note the change in concavity when a solution curve crosses the parabola.]

3. Plot solution curves of the IVP

$$t(y''y + (y')^2) + y'y = 1, \qquad y(1) = 1, \quad y'(1) = v_0$$

for several values of v_0 with $|v_0| \le 1.2$. Extend these solutions both forward and backward from the initial time $t = 1$. Use the ranges of Figure 3.2.1.

4. Reproduce the graphs in Figures 3.2.3 and 3.2.4. Estimate the periods of the solutions.

5. Find the solution formula for the IVP

$$y'' = 2y'y, \qquad y(0) = 1, \quad y'(0) = 1$$

Plot the orbit and solution graph of this IVP. What is the largest interval on which the solution of this IVP is defined? [*Hint*: Write $2y'y$ as $(y^2)'$.]

6. Let $F(y)$ be in the class $\mathbf{C}^1(\mathbf{R})$ ($\mathbf{R}$ is the set of all real numbers). Describe a method for finding all orbits of the ODE $y'' = F(y)$. [*Hint*: Multiply the ODE by y' and write $y'y''$ as $(y'^2)'/2$.] Repeat for the ODE $y'' = y'F(y)$.

7. Show that the IVP: $t^2y'' - 2ty' + 2y = 0$, $y(0) = 0$, $y'(0) = 0$, has infinitely many solutions. Why does this fact not contradict the assertion of uniqueness in Theorem 3.2.1? [*Hint*: Look for solutions in the form Ct^α, where C and α are constants.]

8. (*Alternate Way to Generate Figure 3.2.2*). Give a convincing explanation of why the procedure below generates the graphs in Figure 3.2.2: Consider the first-order differential system with four state variables

$$\begin{aligned} x_1' &= x_2 & x_3' &= x_4 \\ x_2' &= (1 - x_1x_2 - tx_2^2)/tx_1 & x_4' &= 0 \end{aligned}$$

- Solve the system over $1 \le t \le 20$ under the initial conditions $x_1(1) = 1$, $x_2(1) = -1$, $x_3(1) = 1$, $x_4(1) = 1$, and plot the solution parametrically in $x_1x_2x_3$-space. [*Hint*: Under the given conditions, $x_3(t) = t$, for all t.]

- Solve the system over $1 \le t \le 20$ under the initial conditions $x_1(1) = 1$, $x_2(1) = -1$, $x_3(1) = 1$, and $x_4(1) = 0$, and plot the solution parametrically in $x_1x_2x_3$-space. [*Hint*: Under the given conditions $x_3(t) = 1$, for all t.]

- Reproduce the graphs in Figure 3.2.2 by following the above procedure.

9. (*Alternate Way to Generate Figure 3.2.6*). Give a convincing explanation of why the procedure below generates the graphs in Figure 3.2.6: Consider the first-order differential system with three state variables:

$$x' = y$$

$$y' = -x + \cos(1.1t)$$

$$z' = f(z), \quad \text{where} \quad f(z) = \begin{cases} 0, & z = 0 \\ 1, & z \ne 0 \end{cases}$$

- Solve the system over $50 \le t \le 150$ under the initial conditions $x(50) = 5$, $y(50) = 0$, $z(50) = 0$, and plot the solution curve in xyz-space. [*Hint*: Under the conditions, $z(t) = 0$ for all t.]

- Solve the system over $50 \leq t \leq 150$ under the initial conditions $x(50) = 5$, $y(50) = 0$, $z(50) = 50$, and plot the solution curve in xyz-space. [*Hint*: Under the conditions $z(t) = t$, for all t.]

- Reproduce the graphs in Figure 3.2.6 by following the above procedure.

3.3   Basic Theory: Linear Second-Order ODEs

In this section we will look closely at the solution set of the normalized linear second-order ODE

$$y'' + a(t)y' + b(t)y = f(t) \tag{1}$$

where the *coefficients* $a(t)$ and $b(t)$, and the *driving term* $f(t)$ are all continuous on a common interval I on the t-axis. Such ODEs often arise in applying Newton's Second Law of Motion; see Example 1.3.2 in Section 1.3 (Vertical Motion with Viscous Damping), or Example 3.1.1 in Section 3.1 (Hooke's Law Spring). More examples are given in the next chapter.

Operator Notation

We will find it helpful to use the language and "calculus" of operators in searching for solutions of the linear ODE (1). An operator *acts* on an element in its *domain* to produce an element in another set called its *codomain*. The *range* of an operator is the (precise) set of all elements in the codomain which are actually produced by the operator acting on its domain. For example, let the domain and codomain of an operator L be $\mathbf{C}^2(I)$ and $\mathbf{C}^0(I)$, respectively, and let the action of L be given by

$$y \rightarrow L[y] = y'' + a(t)y' + b(t)y, \quad \text{for } y \text{ in } \mathbf{C}^2(I)$$

where $a(t)$, $b(t)$, and I are as given in ODE (1). Notice that square brackets enclose the element that L acts upon. $L[y]$ is an element of $\mathbf{C}^0(I)$ for any y in $\mathbf{C}^2(I)$; this is often indicated by the schematic $L: \mathbf{C}^2(I) \rightarrow \mathbf{C}^0(I)$. We will see later that $\mathbf{C}^0(I)$ is also the range of the operator L. An operator is fully defined when its domain and codomain are given and its action is known; thus, L is completely defined.

We shall reserve the symbol D for the operator which acts on an element $y(t)$ in $\mathbf{C}^1(I)$ and produces the element $y'(t)$ in $\mathbf{C}^0(I)$. Thus, $D: \mathbf{C}^1(t) \rightarrow \mathbf{C}^0(I)$. By the Fundamental Theorem of Calculus (see Theorem C.5.5) every continuous function $f(t)$ has an antiderivative $F(t)$ [i.e., $D[F](t) = f(t)$]. This means that the range of the operator D is all of $C^0(I)$. If $y(t)$ is in $\mathbf{C}^2(I)$, then D can be applied twice in succession to produce the action $y \rightarrow D[D[y]] = y''$. This operator is denoted by D^2 (for obvious reasons), and $D^2: \mathbf{C}^2(I) \rightarrow \mathbf{C}^0(I)$.

For any choice of functions $a_0(t)$, $a_1(t)$, and $a_2(t)$ in $\mathbf{C}^0(I)$ we shall use the notation

$$P(D) = a_2(t)D^2 + a_1(t)D + a_0(t) \tag{2}$$

to denote the *polynomial operator* $P(D): \mathbf{C}^2(I) \rightarrow \mathbf{C}^0(I)$ for which

$$P(D)[y] = a_2(t)D^2[y] + a_1(t)D[y] + a_0(t)y \tag{3}$$

A polynomial operator $P(D)$ could be defined for a polynomial of any degree. For simplicity, we restrict ourselves here to degree 2 or less.

EXAMPLE 3.3.1

Calculations with Polynomial Operators

Let r be a real number and observe that $D[e^{rt}] = re^{rt}$. Applying the operator D to both sides of this equation, we obtain $D^2[e^{rt}] = r^2 e^{rt}$. Thus for any polynomial operator $P(D)$ in (2) we have

$$P(D)[e^{rt}] = a_2 D^2[e^{rt}] + a_1 D[e^{rt}] + a_0 e^{rt} = a_2 r^2 + a_1 r + a_0 = P(r)e^{rt} \qquad (4)$$

where $P(r)$ is to be interpreted as the expression obtained by replacing the symbol D by r in $P(D)$. On the other hand, for any function $h(t)$ in $\mathbf{C}^2(\mathbf{R})$ we have that

$$(D - r)[h(t)e^{rt}] = [h(t)e^{rt}]' - rh(t)e^{rt} = h'(t)e^{rt}$$

Applying the operator $D - r$ to both sides of this equation we have that

$$(D - r)^2[he^{rt}] = h''e^{rt} \qquad (5)$$

for any constant r and any $\mathbf{C}^2$-function h.

Listed below are two basic properties of polynomial operators:

THEOREM 3.3.1

Linearity of $P(D)$. For any polynomial operator $P(D)$ as in (2), any constants c_1 and c_2, and any functions y_1 and y_2 in $\mathbf{C}^2(I)$, we have that

$$P(D)[c_1 y_1 + c_2 y_2] = c_1 P(D)[y_1] + c_2 P(D)[y_2]$$

Proof. Observe that the linearity property holds for D and D^2, so the general case for $P(D) = a_2 D^2 + a_1 D + a_0$ follows by direct calculation.

For a given function f in $\mathbf{C}^0(I)$, the second-order ODE $P(D)[y] = f$, with $P(D)$ as in (2), is precisely ODE (1), with $a_2 = 1$. ODE (1) is said to be linear because of the Linearity Property satisfied by $P(D)$.

❖ **Null Space of $P(D)$.** For $P(D)$ as in (2), the set of all functions in $\mathbf{C}^2(I)$ carried into the zero function by the action of $P(D)$ is said to be the *null space* of $P(D)$ and is denoted by $N(P(D))$.

Thus $y(t)$ is in $N(P(D))$ if and only if $P(D)[y] = 0$, and so $N(P(D))$ is the solution set of the equation $P(D)[y] = 0$. The linear second-order ODE $P(D)[y] = 0$ is said to be *homogeneous*. The Linearity Property of $P(D)$ has the following immediate consequence.

THEOREM 3.3.2

Null Space Closure Property for $P(D)$. For any $P(D)$ as in (2), for any functions $w(t)$ and $z(t)$ in $N(P(D))$, and for any constants c_1 and c_2, we have that $c_1 w + c_2 z$ is also in $N(P(D))$.

The expression $c_1 w + c_2 z$, where w and z are functions, c_1 and c_2 are constants, arises so often that it deserves a name: a *linear combination* of w and z. Theorem 3.3.2 can then be restated as follows: any linear combination of two elements in $N(P(D))$ lies in $N(P(D))$. Solutions of the homogeneous ODE $P(D)[y] = 0$ have another property which will help us to describe the solutions of the driven ODE $P(D)[y] = f$.

THEOREM 3.3.3

Abel's Theorem. Let $y_1(t)$, $y_2(t)$ be any two solutions of the ODE,

$$y'' + a(t)y' + b(t)y = 0$$

where $a(t)$, $b(t)$ are continuous on a common interval I. If

$$W = y_1 y_2' - y_2 y_1'$$

then W is a solution of the homogeneous first-order linear ODE,

$$W' + a(t)W = 0$$

Thus, either $W = 0$ for all t in I, or else $W(t)$ is never zero for t in I.

Proof. Differentiating W we obtain

$$W' = y_1' y_2' + y_1 y_2'' - y_2' y_1' - y_2 y_1'' = y_1 y_2'' - y_2 y_1''$$

$$= y_1(-ay_2' - by_2) - y_2(-ay_1' - by_1) = -aW$$

Now let $A(t)$ be an antiderivative of $a(t)$ on I. Then solving the ODE $W' + aW = 0$ by the method of integrating factors, we have $W(t) = C\exp(-A(t))$, where C is an arbitrary constant. Thus, the only way W can be zero is for $C = 0$, and in that case W is zero on all of I, finishing the proof.

EXAMPLE 3.3.2

The functions $y_1 = \sin 2t$ and $y_2 = \cos 2t$ are solutions of the linear ODE $y'' + 4y = 0$, for all t on the real line. Note that $W = y_1 y_2' - y_2 y_1' = -2$, for all t. Since $a(t) = 0$ in the ODE, we see from Abel's Theorem that $W' = 0$ for any pair of solutions y_1 and y_2 of the ODE. So the fact that W has a constant value in this case should be expected.

Initial Value Problems

We can sharpen the Existence/Uniqueness/Continuity Theorem of the previous section (Theorem 3.2.1) in the case of a linear ODE. Our next result is stated without proof.

THEOREM 3.3.4

Existence/Uniqueness/Continuity Theorem for Linear ODEs. For any functions $a(t)$, $b(t)$, and $f(t)$ that are continuous on a common interval I, for any choice of t_0 in I, and for any choice of the values y_0, v_0, the IVP

$$y'' + a(t)y' + b(t)y = f(t), \qquad y(t_0) = y_0, \quad y'(t_0) = v_0 \qquad (6)$$

has a unique solution $y(t)$ in $\mathbf{C}^2(I)$, and $y(t)$ is a continuous function of t, y_0, v_0.

There are several features of this result that are important to note:

- The unique solution of IVP (6), which is asserted to exist, is defined on the entire interval I. Solutions of the ODE in IVP (6) do not "escape to infinity in finite time" within the interval I where $a(t)$, $b(t)$, and $f(t)$ are continuous.
- The values y_0 in I and v_0 can be chosen arbitrarily.
- The assertions in Theorem 3.3.4 require the ODE in IVP (6) to be normalized.

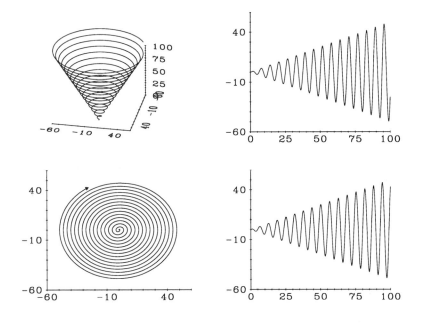

FIGURE 3.3.1 Graphs for IVP in Example 3.3.3.

- The polynomial operator $P(D)$: $\mathbf{C}^2(I) \rightarrow \mathbf{C}^0(I)$ whose action is $P(D)[y] = y'' + a(t)y' + b(t)y$ has as its range all of $\mathbf{C}^0(I)$.

EXAMPLE 3.3.3

Driven Hooke's Law Spring

Consider the IVP for an undamped, driven Hooke's Law spring initially at rest:

$$y'' + y = \cos t, \qquad y(0) = 0, \quad y'(0) = 0$$

From the Existence/Uniqueness/Continuity Theorem, this IVP has a unique solution defined for all t. Figure 3.3.1 shows the solution curve (upper right), the graph of v against t (lower right), the yv-orbit (lower left), and the time-state curve (upper left, t-axis vertical) of the corresponding system $y' = v$, $v' = -y + \cos(t)$, $y(0) = v(0) = 0$. The arrowhead on the orbit shows the direction of increasing t.

A consequence of Theorem 3.3.4 is often useful in simplifying calculations:

THEOREM 3.3.5

Vanishing Data Theorem. The IVP (6) with $f(t) = 0$ on I, $y_0 = v_0 = 0$, and t_0 in I, has only the trivial solution $y(t) = 0$, all t in I.

Proof. Since $y(t) = 0$ for all t in I is a solution of IVP (6) in this case, the Uniqueness Theorem asserts it is the only solution.

EXAMPLE 3.3.4

There is no nontrivial solution of IVP (6) with $f(t) = 0$, for all t, whose solution graph is tangent to the t-axis at some point. For if this were the case, then at the point t_0 of tangency, $y_0 = v_0 = 0$, and hence from Theorem 3.3.5 the only solution of the IVP is the trivial one.

EXAMPLE 3.3.5

Loss of Uniqueness

It is easy to check that the linear ODE, $t^2 y'' - 2ty' + 2y = 0$, has the solution $y = t^2$, and hence the IVP

$$t^2 y'' - 2ty' + 2y = 0, \qquad y(0) = 0, \quad y'(0) = 0 \tag{7}$$

has infinitely many solutions, $y = Ct^2$, where C is an arbitrary constant. This might appear to contradict the Uniqueness Theorem, but closer examination shows that is not the case. Putting the given ODE into normal form, we have $y'' - 2t^{-1}y' + 2t^{-2}y = 0$, and since the coefficients $a(t) = -2t^{-1}$ and $b(t) = 2t^{-2}$ are not continuous for any interval I containing the origin, the Uniqueness Theorem does not apply, and so there is no contradiction.

Structure of the Solution Set of a Linear ODE

There is a nice way to characterize the solution set of the linear ODE

$$y'' + a(t)y' + b(t)y = f(t)$$

where $a(t)$, $b(t)$, and $f(t)$ are continuous on some common interval I. To gain some insight on how to do this, let us first rewrite the ODE in operator form as $P(D)[y] = f$, with the polynomial operator $P(D) = D^2 + a(t)D + b(t)$. Suppose that by some means or another we have managed to find a solution $u(t)$ of $P(D)[y] = f$. The linearity of $P(D)$ implies that if $v(t)$ is any other solution of $P(D)[y] = f$, then $v - u$ is a solution of the homogeneous equation $P(D)[y] = 0$. In other words, $v - u$ is in $N(P(D))$, and hence v has the form $u(t)+$ an element in $N(P(D))$. On the other hand, linearity of $P(D)$ implies that $u(t)+$ any element in $N(P(D))$ is a solution of $P(D)[y] = f$. Thus all solutions of $P(D)[y] = f$ are known as soon as $N(P(D))$ and just one (any one) solution of $P(D)[y] = f$ are known.

We first turn to the task of characterizing $N(P(D))$.

❖ **Wronskian.** For any two functions f and g in $\mathbf{C}^1(I)$, the determinant

$$\begin{vmatrix} f(t) & g(t) \\ f'(t) & g'(t) \end{vmatrix} = f(t)g'(t) - g(t)f'(t)$$

is called the *Wronskian* [1] of the pair and is denoted by $W[f, g]$, or $W[f, g](t)$ if its dependence on the variable t is under consideration.

Now recall from Abel's Theorem (Theorem 3.3.3) that if $y_1(t)$ and $y_2(t)$ are any two solutions of the homogeneous ODE, $P(D)[y] = 0$, then either the Wronskian $W[y_1, y_2](t) = 0$ for all t in I, or $W[y_1, y_2](t) \neq 0$ for every t in I.

❖ **Fundamental Set.** The pair of solutions y_1 and y_2 of the same homogeneous ODE $y'' + a(t)y' + b(t)y = 0$, where a and b are in $C^0(I)$, is called a *fundamental set* if $W[y_1, y_2](t) \neq 0$ for t in I.

---

[1] Höene Wronski (1778–1853) started out as a soldier and then went on to become, successively, a mathematician, a philosopher, and insane.

Actually, since Abel's Theorem holds, we need only check the Wronskian at a single point in I to see if a pair of solutions of $P(D)[y] = 0$ is a fundamental set. Note that the trivial solution (i.e., $y = 0$, for all t in I) cannot belong to a fundamental set.

THEOREM 3.3.6

> **Fundamental Sets and Linear Combinations.** Consider the homogeneous ODE
>
> $$y'' + a(t)y' + b(t)y = 0$$
>
> where a and b are in $C^0(I)$. If $\{y_1, y_2\}$ is a fundamental set of solutions, then each solution y of the ODE may be written as a linear combination of y_1 and y_2,
>
> $$y = c_1 y_1 + c_2 y_2$$
>
> for some constants c_1 and c_2.

Proof. Suppose that we have somehow found a fundamental set, y_1 and y_2, for the ODE $P(D)[y] = 0$. If $y(t)$ is *any* solution of $P(D)[y] = 0$, we claim that there exist unique constants $\tilde{c}_1$ and $\tilde{c}_2$ such that $y = \tilde{c}_1 y_1 + \tilde{c}_2 y_2$. To show this, choose a point t_0 in I, and let $\tilde{c}_1$ and $\tilde{c}_2$ be the unique constants such that

$$\tilde{c}_1 y_1(t_0) + \tilde{c}_2 y_2(t_0) = y(t_0) = \alpha$$

$$\tilde{c}_1 y_1'(t_0) + \tilde{c}_2 y_2'(t_0) = y'(t_0) = \beta$$

That such constants $\tilde{c}_1$ and $\tilde{c}_2$ exist and are unique follows from the fact that the determinant of this system (see Appendix C.6) is just the Wronskian $W[y_1, y_2](t_0)$ which was assumed not to be zero. Thus the function $z(t) = \tilde{c}_1 y_1(t) + \tilde{c}_2 y_2(t)$ and the given function $y(t)$ both solve the IVP,

$$P(D)[w] = 0, \quad w(t_0) = \alpha, \quad w'(t_0) = \beta$$

The Uniqueness Theorem implies that

$$y(t) = z(t) = \tilde{c}_1 y_1(t) + \tilde{c}_2 y_2(t), \quad \text{for all } t \text{ in } I$$

establishing our claim. ∎

Does a fundamental set always exist? The answer to this question is "yes," but with a vengeance. For any choice of $a(t)$, $b(t)$ in $C^0(I)$ the ODE $P(D)[y] = 0$ has infinitely many fundamental sets. We show now how to find at least one fundamental set. Let t_0 be in I, and let y_1 be the solution of the IVP

$$P(D)[y] = 0, \qquad y(t_0) = 1, \quad y'(t_0) = 0$$

and y_2 the solution of the initial value problem

$$P(D)[y] = 0, \qquad y(t_0) = 0, \quad y'(t_0) = 1$$

guaranteed by the Existence/Uniqueness/Continuity Theorem (Theorem 3.3.4). Then because $W[y_1, y_2](t_0) = y_1(t_0)y_2'(t_0) - y_1'(t_0)y_2(t_0) = 1$, we see that the pair of solutions $\{y_1, y_2\}$ of $P(D)[y] = 0$ is a fundamental set.

Since fundamental sets are so important, it would be useful to have another way to recognize them that does not involve Wronskians. Problem 16 gives one such characterization which is often useful.

The following result brings together some of the above observations.

THEOREM 3.3.7

General Solution of $P(D)[y] = f(t)$. Let $\{y_1, y_2\}$ be a fundamental set of solutions for the homogeneous linear ODE,

$$y'' + a(t)y' + b(t)y = 0$$

where $a(t)$ and $b(t)$ are continuous on an interval I. For a given function $f(t)$ in $C^0(I)$, let $y_p(t)$ be a particular (fixed) solution of the corresponding driven ODE

$$y'' + a(t)y' + b(t)y = f(t) \tag{8}$$

Then all solutions of the driven ODE (8) are given by the general solution formula

$$y = y_p + c_1 y_1 + c_2 y_2, \quad c_1, c_2 \text{ arbitrary constants} \tag{9}$$

EXAMPLE 3.3.6

Finding the General Solution

One easily verifies that $y_p(t) = t$ is a particular solution of the ODE: $y'' + 4y = 4t$, and that the pair of functions $y_1(t) = \sin 2t$ and $y_2(t) = \cos 2t$ is a fundamental set for the corresponding homogeneous ODE $y'' + 4y = 0$ (because $W[y_1, y_2](t) = -2$). Thus by the General Solution Theorem 3.3.7, all solutions of the ODE,

$$y'' + 4y = 4t$$

are given by the formula

$$y = t + c_1 \sin 2t + c_2 \cos 2t, \quad c_1, c_2 \text{ arbitrary constants} \tag{10}$$

Notice that all solutions "live" on the entire t-axis, as predicted by Theorem 3.3.4.

EXAMPLE 3.3.7

Solving an IVP

To find a solution formula for the IVP

$$y'' + 4y = 4t, \qquad y(0) = 1, \quad y'(0) = 0 \tag{11}$$

we impose the initial conditions on the general solution formula (10) for the ODE to obtain $y(0) = 1 = 0 + (c_1)(0) + (c_2)(1)$, $y'(0) = 0 = 1 + (2c_1)(1) - (2c_2)(0)$. Solving this system we have $c_1 = -0.5$ and $c_2 = 1$. Thus the unique solution of IVP (11) is

$$y = t - 0.5 \sin 2t + \cos 2t$$

See Figure 3.3.2 for the solution curve (upper right), the graph of $v = y'$ against t (lower right), the directed yv-orbit (lower left), and the time-state curve (t-axis vertical) of $y' = v$, $v' = -4y + 4t$, $y(0) = 1$, $v(0) = 0$ (upper left). Observe that the orbit crosses itself, but this is no contradiction, because the ODE is nonautonomous and the crossing points correspond to different values of time.

EXAMPLE 3.3.8

An ODE with a Discontinuous Coefficient

The Existence/Uniqueness/Continuity Theorem 3.3.4 asserts that solutions of

$$t^2 y'' - 2y = 0$$

are defined at least over the interval $-\infty < t < 0$ or the interval $0 < t < +\infty$, because, after writing the ODE in normalized form $y'' - (2/t^2)y = 0$, the coefficient $b(t) = -2/t^2$ is continuous on any interval not containing the origin. We see by direct substitution that t^α, α a constant, is a solution of the ODE if $\alpha = -1$ or $\alpha = 2$. Put

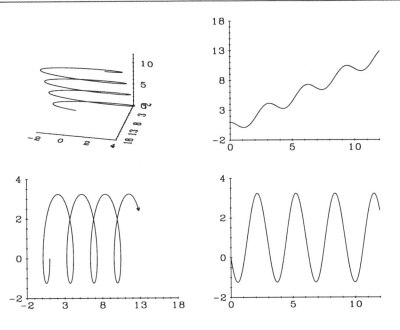

FIGURE 3.3.2 Graphs for IVP in Example 3.3.7.

$y_1(t) = t^2$ and $y_2(t) = 1/t$. Notice that $y_2(t)$ is a solution on either $-\infty < t < 0$ or on $0 < t < +\infty$, but that $y_1(t)$ is a solution of ODE on the entire t-axis. Since $W[y_1, y_2](t) = -3$, for $t \neq 0$, the pair of solutions is a fundamental set, and the general solution of ODE on the interval $0 < t < +\infty$ (to be specific) is

$$y = c_1 t^2 + c_2/t$$

where c_1 and c_2 are arbitrary constants.

The constants are determined from initial data at some point, say $t_0 = 1$. Thus if $y(1) = y_0$, and $y'(1) = v_0$, then $c_1 + c_2 = y_0$ and $2c_1 - c_2 = v_0$; or solving, we obtain that $c_1 = (y_0 + v_0)/3$ and $c_2 = (2y_0 - v_0)/3$. If initial data are chosen randomly at $t = 1$ and the resulting IVP is solved, then chances are that the solution will "blow up" as $t \to 0$ (because $c_2 \neq 0$ unless $v_0 = 2y_0$). Figure 3.3.3 contains some solution graphs of the ODE over $0 < t < +\infty$, including the elusive graph of the solution $2t^2$. Your solver may have difficulty in generating the graph of this solution because a small numerical error may cause your solver to track a solution that blows up as $t \to 0$.

Complex-Valued Solutions of a Linear ODE

If the differential equation $y'' + a(t)y' + b(t)y = f(t)$ arises in a mathematical model of a physical phenomenon, then the coefficients $a(t)$ and $b(t)$ and the driving term $f(t)$ are all real-valued functions, and the solutions we ultimately seek are real-valued. For that reason we have tacitly assumed (until now) that the coefficients, driving term, and data in IVP (6) are all real-valued. We can extend everything we have said so far about IVP (6) to cover the case where the coefficients, driving term, and data are complex-valued. In this case the solutions of IVP (6) are complex-valued, and the "constants" referred to in the definitions and theorems are complex numbers. As we shall presently see, the concept of complex-valued solutions of differential equations

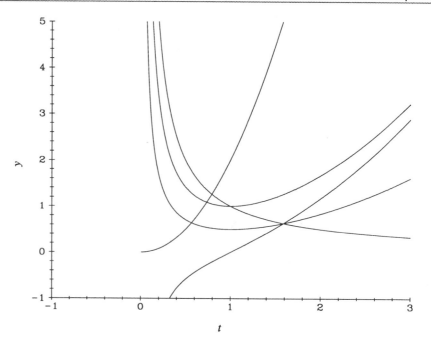

FIGURE 3.3.3 Solution graphs of the ODE in Example 3.3.8.

can be very useful as a computational device in much the same way that complex numbers make it easier to discuss the roots and factorability of polynomials.

But first, what does it mean to say that a complex-valued function $y(t)$ of a real variable t is a solution of the linear ODE $P(D)[y] = f$, where the coefficients in $P(D)$ and the driving term $f(t)$ are (possibly) complex-valued? To do this conveniently, we need to generalize slightly the operational notation introduced earlier in this section.

❖ **Derivative of a Complex-Valued Function.** If $y(t)$ is a complex-valued function with real and imaginary parts $f(t)$ and $g(t)$, then $y(t) = f(t) + ig(t)$, and

$$Dy = y' = D[f + ig] = Df + iDg$$

See Appendix C.3 for a review of complex numbers and the calculus of complex-valued functions of a real variable.

EXAMPLE 3.3.9

Differentiation Formula for e^{rt}

For any real number β the complex-valued function $e^{i\beta t}$ is defined as $e^{i\beta t} = \cos \beta t + i \sin \beta t$ (see Appendix C.3). Thus

$$D[e^{i\beta t}] = D[\cos \beta t] + iD[\sin \beta t] = -\beta \sin \beta t + i\beta \cos \beta t \tag{12}$$

$$= i\beta(\cos \beta t + i \sin \beta t) = i\beta e^{i\beta t} \tag{13}$$

Generally, let $r = \alpha + i\beta$ be any complex number and recall that $e^{rt} = e^{(\alpha + i\beta)t} = e^{\alpha t}e^{i\beta t}$. Then, using (13) and the product rule for differentiation, we have that

$$D[e^{rt}] = D[e^{\alpha t}e^{i\beta t}] = D[e^{\alpha t}]e^{i\beta t} + e^{\alpha t}D[e^{i\beta t}] = \alpha e^{\alpha t}e^{i\beta t} + e^{\alpha t}i\beta e^{i\beta t} = re^{rt} \tag{14}$$

The differentiation formula (14) has much to do with the reason e^{rt} is defined the way it is for complex r.

With this definition of D, it is not hard to show that all the usual differentiation rules are preserved for complex-valued functions of a real variable. In particular, the important operational formulas (4) and (5) hold even when r is a complex number and $z(t)$ is a complex-valued function. The concepts of Wronskian and fundamental set remain true exactly as stated, and so the general solution formula (9) is still valid, but the constants c_1 and c_2 are now complex numbers. The definitions of the continuity sets $\mathbf{C}^0(I)$, $\mathbf{C}^1(I)$, and $\mathbf{C}^2(I)$ for complex-valued functions remain exactly the same as before, and the polynomial operator $P(D)$: $\mathbf{C}^2(I) \to \mathbf{C}^0(I)$ is defined as before. Whether the domain $\mathbf{C}^2(I)$ of $P(D)$ consists of complex-valued or real-valued functions will be clear from the context, if not stated explicitly. Sometimes it is useful to find all the complex-valued solutions of the homogeneous ODE, $P(D)[y] = 0$, even when the coefficients $a(t)$ and $b(t)$ are real-valued. In this case, when the general solution formula (9) is used, one must be careful to distinguish the *general (complex-valued) solution* from the *general (real-valued) solution*.

EXAMPLE 3.3.10

Complex-Valued versus Real-Valued General Solutions
Consider $P(D) = D^2 + 4$. The differentiation formula (14) shows that $P(D)[e^{2it}] = 0$ and $P(D)[e^{-2it}] = 0$. Thus $\{e^{2it}, e^{-2it}\}$ is a pair of complex-valued solutions of the equation $y'' + 4y = 0$. Now since $W[e^{2it}, e^{-2it}] = -4i \neq 0$, we see that the pair is a fundamental set. Thus according to the General Solution Theorem 3.3.7, we see that *all* complex-valued solutions of the equation

$$y'' + 4y = 0$$

are given by

$$y = Ae^{2it} + Be^{-2it}$$

where A, B are arbitrary complex numbers. Observe that the real-valued solutions are all contained among the complex-valued ones [because the set of complex-valued functions $\mathbf{C}^2(I)$ contains the set of real-valued functions $\mathbf{C}^2(I)$]. Indeed, since $e^{2it} = \cos 2t + i \sin 2t$ and $e^{-2it} = \cos 2t - i \sin 2t$, we see that $\cos 2t = e^{2it}/2 + e^{-2it}/2$ and $\sin 2t = -(i/2)e^{2it} + (i/2)e^{-2it}$. Now since $W[\sin 2t, \cos 2t] = -2$, we see that $y_1 = \sin 2t$ and $y_2 = \cos 2t$ is a fundamental set and hence

$$y = c_1 \sin 2t + c_2 \cos 2t$$

for arbitrary reals, is the general (real-valued) solution of the ODE $y'' + 4y = 0$.

Another result that is useful in identifying real-valued solutions is as follows.

THEOREM 3.3.8

Real- and Complex-Valued Solutions. Let the polynomial operator

$$P(D) = D^2 + a(t)D + b(t)$$

have real-valued coefficients, and suppose that $f(t)$ and $g(t)$ are given real-valued continuous functions. Then

$$y(t) = u(t) + iv(t)$$

for real-valued functions $u(t)$ and $v(t)$ is a solution of

$$P(D)[y] = f(t) + ig(t)$$

if and only if

$$P(D)[u] = f \quad \text{and} \quad P(D)[v] = g$$

Proof. From the Linearity Property of $P(D)$ (Theorem 3.3.1) we have that $P(D)[u + iv] = P(D)[u] + iP(D)[v]$. Since the coefficients in $P(D)$ are real-valued, it follows that $P(D)[u]$ and $P(D)[v]$ are real-valued functions, and hence identification of the real and imaginary parts of $P(D)[u + iv]$ and $f + ig$ yields the result.

Thus if $P(D)$ has real-valued coefficients, any solution of the homogeneous ODE $P(D)[y] = 0$ has the property that its real and imaginary parts are real-valued solutions of that same ODE.

EXAMPLE 3.3.11

Getting Real-Valued Solutions from Complex-Valued Solutions
We saw in Example 3.3.10 that e^{2it} is a solution of $y'' + 4y = 0$. Since $e^{2it} = \cos 2t + i \sin 2t$, we see that $\cos 2t$ and $\sin 2t$ are real-valued solutions of $P(D)[y] = 0$. We saw in Example 3.3.10 that these real solutions also arise by choosing special values for the arbitrary constants in the family of complex-valued solutions.

PROBLEMS

1. (*Null Space, General Solution*). For the polynomial operator $P(D) = D^2$, find:

    (a) The null space of $P(D)$.

    (b) The general solution of $y'' = \sin t$.

2. (*Null Space, Uniqueness*). For the polynomial operator $P(D) = tD^2 + D$:

    (a) Find the null space of $P(D)$ over $0 < t < \infty$. [*Hint:* Write $P(D)[y]$ as $(ty')'$.]

    (b) Show that the IVP, $P(D)[y] = 2t$, $y(0) = a$, $y'(0) = b$, has a unique solution on any interval I containing the origin, but only provided that a and b are chosen suitably. [*Hint:* $t^2/2$ is a particular solution.]

    (c) Why does the result in part (b) not contradict Theorem 3.3.4?

3. Find the null space of the operator $P(D) = (1 - t^2)D^2 - 2tD$ and the general solution of $P(D)[y] = 0$ on the interval $|t| < 1$. [*Hint:* First let $v = Dy$, and solve $(1 - t^2)v' - 2tv = 0$.]

4. (*IVPs*). Solve the following IVPs, given the indicated information. [*Hint:* Refer to Theorem 3.3.7.]

    (a) $y'' - y' - 2y = 2e^t$, $y(0) = 0$, $y'(0) = 1$, given that $y_1 = e^{2t}$ and $y_2 = e^{-t}$ are solutions of the homogeneous ODE and that $y_p = -e^t$ is a solution of the nonhomogeneous ODE.

    (b) $y'' + 2y' + 2y = 2t$, $y(1) = 1$, $y'(1) = 0$, given that $y_1 = e^{-t} \cos t$ and $y_2 = e^{-t} \sin t$ are solutions of the homogeneous ODE and that $y_p = t - 1$ is a solution of the nonhomogeneous ODE.

5. (*IVPs*). First verify that the given functions y_1 and y_2 are solutions of the ODE. Then use y_1 and y_2 to find all solutions of the IVP. Plot the solution curves.

    (a) $t^2 y'' + ty' - y = 0$, $y_1 = t^{-1}$, $y(1) = 0$
                                       $y_2 = t$, $y'(1) = -1$

    (b) $t^2 y'' - ty' + y = 0$, $y_1 = t$, $y(1) = 1$
                                       $y_2 = t \ln t$, $y'(1) = 0$

    (c) $t^2 y'' + ty' - 4y = 0$, $y_1 = t^2$, $y(0) = 0$
                                       $y_2 = t^{-2}$, $y'(0) = 0$

6. (*Fundamental Sets*). Many pairs of solutions of the ODE $y'' - y' - 2y = 0$ are fundamental sets. One such pair is $\{e^{-t}, e^{2t}\}$. Show that the pair $\{ae^{-t} + be^{2t}, ce^{-t} + de^{2t}\}$ for any choice of constants a, b, c, d such that $ad - bc \neq 0$ is also a fundamental set.

7. (*Exponential Solutions*). Consider the polynomial operator $P(D) = D^2 + aD + b$, where a and b are real (or complex) constants. Show that for any real (or complex) constant r we have $P(D)[e^{rt}] = (r^2 + ar + b)e^{rt} = P(r)e^{rt}$. Therefore, $y = e^{rt}$ is a solution to $P(D)[y] = 0$ if and only if r is a root of $P(r)$.

8. (*Boundary Value Problems*). Find all solutions of the following *boundary value problems*. [*Hint*: See Problem 7 for part **(c)**.]

(a) $y'' = 2$, $\qquad$ $y(0) = 0$, $\qquad$ $y(1) + y'(1) = 0$

(b) $y'' = 2$, $\qquad$ $2y(0) + y'(0) = 0$, $\quad$ $2y(1) - y'(1) = 0$

(c) $y'' - 3y' + 2y = 0$, $\quad$ $y(0) - y'(0) = 1$, $\quad$ $y'(1) = -2$

9. (*Operator Identities*). The problems below give some useful identities.

(a) Show that if n is any positive integer, r_0 is any constant, and h is any n-times differentiable function, then $(D - r_0)^n[he^{r_0 t}] = e^{r_0 t}D^n[h]$. [*Hint*: Do for $n = 1$, and iterate.]

(b) Show for any polynomial $p(t)$ of degree $n - 1$ (or less) that $y = p(t)e^{r_0 t}$ is a solution for the nth-order ODE $(D - r_0)^n[y] = 0$ for any constant r_0.

10. (*Null Spaces*). Find as many elements as possible in the null spaces of the following operators. [*Hint*: See Problem 7; use De Moivre's Theorem (Appendix C.3) to find roots of a complex number.]

(a) $D^2 - 2D - 3$ $\qquad$ (b) $D^2 + 4D + 13$

(c) $D^2 - 4i$ $\qquad$ (d) $D^2 + (1 + i)D + i$

11. (*Real-Valued Solutions*). Find all real-valued solutions of the following ODEs. [*Hint*: Use Theorem 3.3.8, Euler's Formula, and Problem 7]. Plot several solution graphs and discuss solution behavior as $t \to \pm\infty$.

(a) $y'' - 2y' + 2y = 0$, $-1 \leq t \leq 4$; $t_0 = 0$

(b) $y'' + 2y' + 2y = \sin t$, $-5 \leq t \leq 20$; $t_0 = 0$ [*Hint*: Look for a particular solution of the form $A \sin t + B \cos t$.]

(c) $y'' + 4y = e^{-t} \cos 2t$, $|t| \leq 5$, $t_0 = 0$ [*Hint*: Look for a particular solution of the form $e^{-t}(A \sin 2t + B \cos 2t)$.]

12. (*Reduction of Order Technique*). Let $y = u(t)$ be a solution of $y'' + a(t)y' + b(t)y = 0$, where $a(t)$ and $b(t)$ are in $C^0(I)$.

(a) Show that $y = u(t)z(t)$ is also a solution of that equation if $z(t)$ satisfies the equation $uz'' + (2u' + a(t)u)z' = 0$.

(b) Show that

$$z(t) = \int_{t_0}^t \left\{ \frac{1}{(u(s))^2} \exp\left[-\int^s a(r)\, dr \right] \right\} ds$$

is a solution of the z-equation in part **(a)** if $u(t) \neq 0$ for any t in I.

(c) Show that the pair of solutions $\{u, uz\}$ constructed in parts **(a)** and **(b)** is a fundamental set.

(d) Find all solutions of $ty'' - (t + 2)y' + 2y = 0$ for $t > 0$, given that e^t is one solution.

13. (*Wronskian Reduction of Order*). Let $P(D) = D^2 + a(t)D + b(t)$, where $a(t)$, $b(t)$, are in $C^0(I)$. Let $W(t) = W[y_1, y_2](t)$ be the Wronskian of any pair of solutions $\{y_1, y_2\}$ of $P(D)[y] = 0$.

(a) Show that $W(t)$ satisfies the first-order linear equation $W' = -a(t)W$, and hence W is given by *Abel's Formula*

$$W(t) = W(t_0) \exp\left[-\int_{t_0}^{t} a(s)\, ds\right] \qquad \text{for } t_0, t \text{ in } I$$

[*Hint*: Differentiate W directly, and look at the proof of Theorem 3.3.3.]

(b) Use Abel's Formula to show that if $u(t) \neq 0$ is a solution of $P(D)[y] = 0$, then a second solution $v(t)$ of the ODE can be found by solving the following first-order linear ODE for v,

$$u(t)v' - u'(t)v = \exp\left[-\int^{t} a(s)\, ds\right]$$

Show that the pair $\{u(t), v(t)\}$ is a fundamental set for $P(D)[y] = 0$.

(c) Given that e^t is a solution of $ty'' - (t+2)y' + 2y = 0$, find a second solution v such that $\{e^t, v\}$ is a fundamental set. [*Hint*: Remember to normalize the ODE first.] Notice that by an appropriate choice of the limits of integration, the solution v can be made identical to the solution uz in Problem 12**(b)**.

14. (*Fundamental Sets*). Suppose that $P(D)$ is the second-order linear operator $D^2 + p(t)D + q(t)$, where $p(t)$ and $q(t)$ are real-valued and continuous for all t. Suppose further that $\{y_1(t), y_2(t)\}$ is a fundamental set for $P(D)[y] = 0$. Which of the following graphs of $y_1(t)$ and $y_2(t)$ are *not* logically possible? Explain. [*Hint*: Is the Wronskian zero for some value of t shown in the plot?]

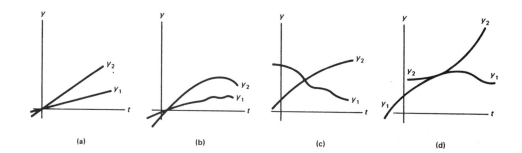

    (a)   (b)   (c)   (d)

15. (*Interlacing of Zeros of a Fundamental Pair*). Let $a(t)$ and $b(t)$ be continuous on $-\infty < t < \infty$, and suppose that $\{u(t), v(t)\}$ is a fundamental set of solutions for the ODE $y'' + a(t)y' + b(t)y = 0$. Show that between any two consecutive zeros of one solution there is precisely one zero of the other solution. Explain why the graphs in the margin figure are *not* graphs of a fundamental set of solutions.

16. (*Alternate Characterization of a Fundamental Set*). Let y_1, y_2 be two solutions of the homogeneous ODE, $P(D)[y] = 0$, where $P(D) = D^2 + a(t)D + b(t)$, with a and b in $C^0(I)$.

- Show that if the pair $\{y_1, y_2\}$ is not a fundamental set, then there exist constants c_1, c_2, not both zero, such that $c_1 y_1(t) + c_2 y_2(t) = 0$, for all t in I. [*Hint*: Choose a point t_0 in I and note that since $W[y_1, y_2](t_0) = 0$, the linear system $c_1 y_1(t_0) + c_2 y_2(t_0) = 0$, $c_1 y_1'(t_0) + c_2 y_2'(t_0) = 0$ has a solution $\tilde{c}_1$, $\tilde{c}_2$, not both zero.]

- Show that if c_1 and c_2 are constants, not both zero, such that $c_1 y_1(t) + c_2 y_2(t) = 0$ for all t in I, then the pair $\{y_1, y_2\}$ is not a fundamental set.

- Explain how the above properties characterize fundamental sets without using the Wronskian.

17. (*Linear Independence*). A collection of objects in which two operations are defined, addition and multiplication by a scalar, is called a *linear space* if these operations satisfy a certain set of axioms. (See Section 10.2 for the list of axioms and a formal definition of a linear space.) The continuity set of real-valued functions $C^2(I)$ is a linear space if we define addition of functions in the usual pointwise manner (that is, $(f + g)(t) = f(t) + g(t)$ for all t in I) and if scalar multiplication by reals is also defined pointwise. A fundamental property of subsets of a linear space is the notion of linear dependence and independence. A set of two functions y_1 and y_2 in $C^2(I)$ is *linearly dependent* if there exist scalars c_1 and c_2 such that $c_1 y_1(t) + c_2 y_2(t) = 0$ for all t.in I. A pair of functions which is not linearly dependent is *linearly independent*. One can show $\{y_1, y_2\}$ to be independent by showing that the assumption $c_1 y_1(t) + c_2 y_2(t) = 0$ for all t in I leads to the conclusion that $c_1 = c_2 = 0$.

- Show that if $r_1 \neq r_2$, then $\{e^{r_1 t}, e^{r_2 t}\}$ is independent. [*Hint*: Set $c_1 e^{r_1 t} + c_2 e^{r_2 t} = 0$, all t, and differentiate the identity to obtain a second identity. Choose an arbitrary $t = t_0$ in I and look at the system of homogeneous linear equations in c_1 and c_2 that you obtain.]

- Show that if y_1 and y_2 are different solutions to $y'' + a(t)y' + b(t)y = 0$, for some a and b in $C^0(I)$, then $\{y_1, y_2\}$ is linearly dependent on I if and only if $W[y_1, y_2](t)$ is identically zero for t in I. Compare with Problem 15 above. Combining this result with Abel's Theorem, we find that $\{y_1, y_2\}$ is linearly independent if and only if $W[y_1, y_2](t)$ is never zero for t in I.

- (*Independent Solutions, Vanishing Wronskians*). Show that $y_1 = t^3$, $y_2 = |t|^3$ is a pair of linearly independent solutions of the equation $t^2 y'' - 4ty' + 6y = 0$ on the interval $-\infty < t < \infty$. Now show that $W[t^3, |t|^3] = 0$ for all t. [*Hint*: Differentiate $|t|^3$ piecewise to show $(|t|^3)' = 3t|t|$, $(|t|^3)'' = 6|t|$.] Why does this not contradict the result above?

3.4 Constant-Coefficient Homogeneous Linear ODEs

We shall look at homogeneous second-order linear ODEs with constant coefficients. The Existence and Uniqueness Theorem (EUT) (Theorem 3.3.4) implies that every solution of the constant-coefficient linear ODE

$$y'' + ay' + by = 0$$

"lives" on the entire t-axis. It is also true that every solution of this ODE has derivatives of every order (notationally, we say that $y(t)$ belongs to the continuity class $C^\infty(\mathbf{R})$). To show this, recall that the EUT asserts that a solution is at least in $C^2(\mathbf{R})$, and write the ODE as $y'' = -ay' - by$. Thus, since $-ay' - by$ is at least of class $C^1(\mathbf{R})$ it follows that y'' is of class $C^1(\mathbf{R})$, or, equivalently, that y is at least of class $C^3(\mathbf{R})$. This argument can be repeated indefinitely.

Now consider the polynomial operator of the form $P(D) = D^2 + aD + b$, with a and b constants. If in $P(D)$ the symbol D is formally replaced by the algebraic variable r, the resulting polynomial

$$P(r) = r^2 + ar + b \tag{1}$$

is called the *characteristic polynomial* associated with the polynomial operator $P(D)$. Recall that if $P(D)$ is applied to e^{rt}, where r is any real (or complex) constant, then by direct calculation,

$$P(D)[e^{rt}] = (r^2 + ar + b)e^{rt} = P(r)e^{rt} \tag{2}$$

where $P(r)$ is precisely the polynomial (1) associated with the operator $P(D)$. The roots of the characteristic polynomial $P(r) = r^2 + ar + b$ are called the *characteristic roots* of the ODE: $y'' + ay' + by = 0$. Using (2), we will show that polynomial operators *with constant coefficients* can be factored in the same way as polynomials in r.

If r_1 is any constant and $y(t)$ is any function in $C^2(I)$, note that $(D - r_1)[y]$ is an element of $C^1(I)$. Thus if r_2 is constant, the operator $D - r_2$ may be applied to $(D - r_1)[y]$, and this particular succession of operators applied to a function y is denoted by $(D - r_2)(D - r_1)$. If $Q(D)$ is the "factorized" polynomial operator $(D - r_2)(D - r_1)$, then using successively the fact that

$$(D - s)[e^{rt}] = (r - s)e^{rt}, \qquad \text{for any constants } r, s \tag{3}$$

we see that

$$Q(D)[e^{rt}] = (D - r_2)[(r - r_1)e^{rt}] = (r - r_2)(r - r_1)e^{rt} = Q(r)e^{rt}$$

In fact, for every y in $C^2(\mathbf{R})$ a direct calculation shows that $P(D)[y] = Q(D)[y]$; hence, $Q(D) = (D - r_2)(D - r_1)$ is just the factored form of $P(D)$. In the same way we can show that $P(D) = (D - r_1)(D - r_2)$, and hence

$$(D - r_1)(D - r_2) = (D - r_2)(D - r_1), \qquad \text{for any constants } r_1 \text{ and } r_2$$

This "factored operator" approach is used to prove the following basic theorem.

THEOREM 3.4.1

Constant-Coefficient, Complex-Valued Solution Theorem.   Let a and b be any real (or complex) constants and suppose that r_1 and r_2 are the roots of the polynomial $r^2 + ar + b$. Then the general (complex-valued) solution of the homogeneous linear second-order ODE

$$y'' + ay' + b = 0 \tag{4}$$

is given by

$$y = K_1 e^{r_1 t} + K_2 e^{r_2 t}, \qquad \text{if } r_1 \neq r_2 \tag{5a}$$

$$y = K_1 e^{r_1 t} + K_2 t e^{r_1 t}, \qquad \text{if } r_1 = r_2 \tag{5b}$$

where K_1 and K_2 are arbitrary complex constants.

Proof. Writing (4) in the form $(D - r_1)(D - r_2)[y] = 0$ and using (3), we see that $e^{r_1 t}$ and $e^{r_2 t}$ are solutions. Since $W[e^{r_1 t}, e^{r_2 t}] = (r_2 - r_1)e^{(r_1 + r_2)t} \neq 0$ if $r_1 \neq r_2$, we see that $\{e^{r_1 t}, e^{r_2 t}\}$ is a fundamental set, and (5a) follows by the General Solution Theorem 3.3.7.

Now when $r_1 = r_2$ the procedure above fails to produce a second solution of ODE (4) that forms a fundamental set together with $e^{r_1 t}$. In this case we look for a second solution in the form $h(t)e^{r_1 t}$ for some $h(t)$ (see Problem 9 of Section 3.3). Using the identity $(D - r_1)[h(t)e^{r_1 t}] = e^{r_1 t}h'$ twice in succession, we see the $(D - r_1)^2[he^{r_1 t}] = e^{r_1 t}h''$, and hence $he^{r_1 t}$ solves ODE (4) if $h(t)$ is any solution of the ODE $h'' = 0$. We conclude that $(A + Bt)e^{r_1 t}$ must be a

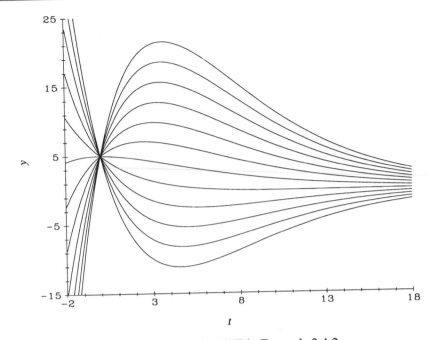

FIGURE 3.4.1 Solution curves for ODE in Example 3.4.2.

solution of (4) for any choice of constants A, B; in particular, te^{r_1t} is a solution. Since $W[e^{r_1t}, te^{r_1t}] \neq 0$, we see again from the General Solution Theorem 3.3.7 that this fundamental set of solutions generates all solutions as in (5b).

EXAMPLE 3.4.1	**Complex Characteristic Roots**

The characteristic polynomial of the linear ODE

$$y'' + y' + 100.25y = 0 \tag{6}$$

is $r^2 + r + 100.25$, which has the roots $r_1 = -1/2 + 10i$ and $r_2 = -1/2 - 10i$. Hence, by Theorem 3.4.1, the general (complex-valued) solution is

$$y = K_1 e^{(-1/2+10i)t} + K_2 e^{(-1/2-10i)t} \tag{7}$$

for arbitrary complex numbers K_1 and K_2.

EXAMPLE 3.4.2

Repeated Characteristic Roots

The equation

$$y'' + y'/2 + y/16 = 0$$

has the characteristic polynomial $r^2 + r/2 + 1/16$ with roots $r_1 = r_2 = -1/4$. Thus (5b) gives the general (complex-valued) solution as $y = (K_1 + K_2t)e^{-t/4}$, where K_1 and K_2 are arbitrary complex numbers. But since the solutions $e^{-t/4}$ and $te^{-t/4}$ are real-valued and form a fundamental set, the general (real-valued) solution is

$$y = (C_1 + C_2t)e^{-t/4}$$

where C_1 and C_2 are arbitrary *real* numbers. Figure 3.4.1 shows a "spray" of solution curves, where $y(0) = 5$, and $y'(0)$ lies between -10 and 12.

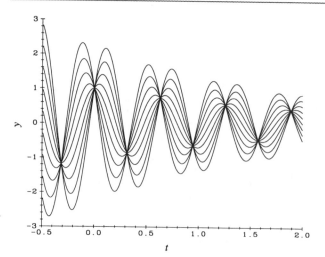

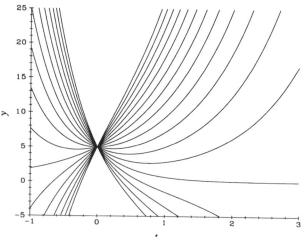

FIGURE 3.4.2 Solution curves of ODE (6), with $y(0) = 1$, $|y'(0)| \leq 20$. See Example 3.4.3.

FIGURE 3.4.3 Solution curves of $y'' - y' - 2y = 0$, $y(0) = 5$, $|y'(0)| \leq 20$. See Example 3.4.4.

Real-Valued Solutions

If ODE (4) has real coefficients, we want to find the *real-valued general solution* of that ODE, that is, all solutions which are *real-valued* functions. Of course, the complex-valued general solution must contain all real-valued solutions of ODE (4), and hence it is only a matter of recognizing them within this much larger class of solutions. Before a theorem is given, an example will help explain how this goes.

EXAMPLE 3.4.3

Real-Valued Solutions from Complex-Valued Solutions
We shall use the general complex-valued solution (7) for ODE (6) to find the general (real-valued) solution of that ODE. Recall that $e^{(\alpha + i\beta)t} = e^{\alpha t} \cos \beta t + i e^{\alpha t} \sin \beta t$, for any real numbers α and β, and hence the exponentials in (7) are complex conjugates of one another. Taking the constants $K_1 = K_2 = 1/2$ in formula (7), we have the real-valued solution $e^{-t/2} \cos 10t$, whereas if we put $K_1 = -i/2$ and $K_2 = i/2$ in (7), the real-valued solution $e^{-t/2} \sin 10t$ results. Now since $W[e^{-t/2} \cos 10t, e^{-t/2} \sin 10t] \neq 0$, we have a fundamental set of real-valued solutions of ODE (6) and hence by the General Solution Theorem 3.3.7 we have the general (real-valued) solution for ODE (6):

$$y = A e^{-t/2} \cos 10t + B e^{-t/2} \sin 10t, \qquad A, B \text{ arbitrary reals}$$

Figure 3.4.2 shows solution curves of ODE (6), where $y(0) = 1$, $|y'(0)| \leq 20$.

THEOREM 3.4.2

Constant-Coefficient, Real-Valued Solution Theorem. Suppose the coefficients of $P(D) = D^2 + aD + b$ are real numbers and that the characteristic polynomial $P(r)$ has the roots r_1 and r_2. Then the general real-valued solution of the homogeneous equation $P(D)[y] = 0$ is given by

$$\begin{array}{ll} k_1 e^{r_1 t} + k_2 e^{r_2 t}, & \text{if } r_1, r_2 \text{ real, } r_1 \neq r_2 \\ k_1 e^{\alpha t} \cos \beta t + k_2 e^{\alpha t} \sin \beta t, & \text{if } r_1 = \bar{r}_2 = \alpha + i\beta, \quad \beta \neq 0 \\ k_1 e^{r_1 t} + k_2 t e^{r_1 t}, & \text{if } r_1 = r_2 \end{array} \qquad (8)$$

where k_1 and k_2 are arbitrary real numbers.

Proof. Only the case $r_1 = \bar{r}_2 = \alpha + i\beta$, $\beta \neq 0$ needs any argument. Since $e^{r_1 t}$ is a complex-valued solution of $P(D)[y] = 0$, and since

$$e^{(\alpha + i\beta)t} = e^{\alpha t}\cos\beta t + ie^{\alpha t}\sin\beta t$$

we see from Theorem 3.3.8 that the functions $e^{\alpha t}\cos\beta t$ and $e^{\alpha t}\sin\beta t$ (the real and imaginary parts of the complex-valued solution) are real-valued solutions of that same ODE. Since $W[e^{\alpha t}\cos\beta t, e^{\alpha t}\sin\beta t] \neq 0$, the pair is a fundamental set and the assertion follows from the General Solution Theorem 3.3.7.

EXAMPLE 3.4.4

One Positive, One Negative Root of Characteristic Polynomial

The characteristic polynomial for the ODE $y'' + y' - 2y = 0$ is $r^2 + r - 2$ with roots $r_1 = 1$, $r_2 = -2$. The general real-valued solution is

$$y = k_1 e^t + k_2 e^{-2t}, \qquad k_1 \text{ and } k_2 \text{ arbitrary reals}$$

Figure 3.4.3 shows solutions with $y(0) = 5$ and values $|y'(0)| \leq 20$.

Periodic Solutions

A nonconstant function $f(t)$ defined on the real line is said to be a *periodic function* if there exists a value $T > 0$ such that $f(t + T) = f(t)$, for all t. Such a value T is called a *period* for the periodic function. Of course, any (positive) integer multiple of a period is also a period. The smallest period of a periodic function is said to be the *fundamental period*. The *amplitude* A of a periodic function f is given by half of the total back-and-forth swing. For example, $\sin 3t$ and $1 + 2\cos 2t$ are periodic functions; the first has the fundamental period $2\pi/3$ and amplitude 1, and the second has period π and amplitude 2. To see this recall that $\sin t$ and $\cos t$ each have the fundamental period 2π, and so the value T for which $3T = 2\pi$ must be the fundamental period of $\sin 3t$. Similarly, the value T for which $2T = 2\pi$ is the fundamental period for $\cos 2t$. In general, the trigonometric functions $\sin\omega t$ and $\cos\omega t$ have the fundamental period $T = 2\pi/\omega$.

The trigonometric functions are said to have the *circular frequency* ω (because ω measures the number of cycles that "fit" into 2π radians—a full circle). The *frequency* of the sinusoid $\sin\omega t$ is $\omega/(2\pi)$ (because it measures the number of cycles per radian).

If t is measured in seconds, then $\sin\omega t$ has the frequency of $\omega/2\pi$ cycles/sec. By convention, one cycle per second is a "unit" known as one *hertz* (Hz). Notice that the frequency and the fundamental period of a sinusoid are reciprocals of one another.

There are many functions other than sinusoids which are periodic. Some that occur in the applications are the square wave $sqw(t, d, p)$, the triangular wave $trw(t, d, p)$, and the sawtooth wave $sww(t, d, p)$, See Appendix C.1.

If $f(t)$ and $g(t)$ are periodic with the same period T, then $f(t)g(t)$ and $af(t) + bg(t)$, for arbitrary constants a and b, are both periodic with the same period T. Although it is less obvious, if $f(t)$ has period T and $g(t)$ has period S, then $f(t) + g(t)$ is periodic if and only if T/S is a rational number. From this point on, when we refer to the period of a periodic function, we mean the fundamental period.

Now let us examine an ODE all of whose solutions are periodic.

EXAMPLE 3.4.5

Simple Harmonic Motion

The undamped, undriven Hooke's Law spring ODE

$$y'' + \omega^2 y = 0$$

for any positive number ω, has the general solution $y = k_1 \sin \omega t + k_2 \cos \omega t$, where k_1 and k_2 are arbitrary reals. If k_1 and k_2 are not both zero, then $A = (k_1^2 + k_2^2)^{1/2} > 0$, and the general solution has the equivalent form

$$y = A \left(\frac{k_1}{A} \sin \omega t + \frac{k_2}{A} \cos \omega t \right)$$

Using the unit circle definition of the sine and cosine functions (see the margin figure), we define θ as the unique angle such that $\cos \theta = k_1/A$, and $\sin \theta = k_2/A$. Then we see that the general solution of $y'' + \omega^2 y = 0$ is

$$y = A \sin(\omega t + \theta) = A \cos(\omega t + \theta - \pi/2), \qquad A, \theta \text{ arbitrary reals}$$

Thus every solution of the ODE is a sinusoid of circular frequency ω and period $2\pi/\omega$. For that reason the ODE $y'' + \omega^2 y = 0$ is called the *simple harmonic oscillator* and its solution graph depicts *simple harmonic motion*.

EXAMPLE 3.4.6

Sampling Rates: Aliasing

From Example 3.4.5 we see that all of the solutions of the ODE

$$y'' + 4y = 0$$

are sinusoids

$$y = A \sin(2t + \theta)$$

with *amplitude* A and *phase angle* θ determined by the initial data. All solutions have period π. Figures 3.4.4—3.4.7 show four plots of the solution of the IVP

$$y'' + 4y = 0, \qquad y(0) = 0, \quad y'(0) = 1 \tag{9}$$

over the interval $0 \le t \le 100$. In Figure 3.4.4, fourth-order Runge-Kutta (RK4) (see Appendix B.1) is used to approximate the solution curve of IVP (8) at 1000 equally spaced points in the interval $0 \le t \le 100$. The result appears to be a sinusoid of period π (as it should be). The plots in Figures 3.4.5–3.4.7 are made using the exact same data points as in Figure 3.4.4, but plotting, respectively, every fifth point, every tenth point, and every twentieth point. Observe how the graphs degrade as fewer and fewer solution points are plotted. In Figures 3.4.5 and 3.4.6 the amplitudes of the sinusoids appear to be modulated, and in Figure 3.4.7, the period of the approximate solution curve appears to have doubled.

Since these graphs are all produced by sampling from the same list of 1000 fairly good approximate values, the accuracy of the solver is not to blame for this strange behavior. In fact, this phenomenon always occurs when an oscillatory curve is under sampled and the sample points are connected by line segments. The modulation appears to worsen as the number of sample points per unit time decreases. Finally, as Figure 3.4.7 shows, when there are fewer than two sample points per period the graphical representation appears to have a larger period (engineers call this phenomenon "aliasing"). Practically speaking, what can we do to correct the misrepresentation inherent in representing a continuous periodic function by a discrete point set? If the experimenter knows in advance the period of the oscillation, then the "sampling rate" can be set high enough to mitigate the problem. But in practice these periods are not known in advance, so some experimentation is called for before accepting the validity of a graphical result.

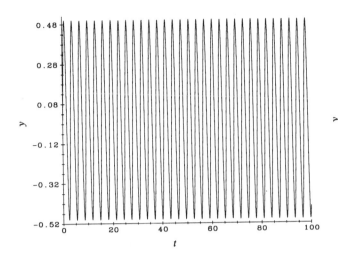

FIGURE 3.4.4 Sampling simple harmonic motion at 10 points/unit time. See Example 3.4.6.

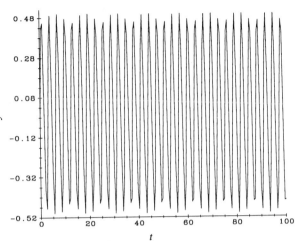

FIGURE 3.4.5 Sampling at 5 points/unit time. See Example 3.4.6.

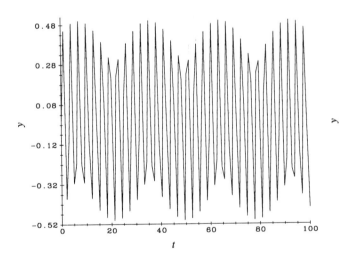

FIGURE 3.4.6 Sampling at 1 point/unit time. See Example 3.4.6.

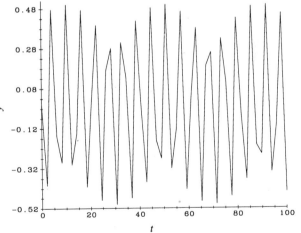

FIGURE 3.4.7 Sampling at 0.5 points/unit time: aliasing. See Example 3.4.6.

PROBLEMS

1. Find all real-valued solutions of $P(D)[y] = 0$ where
   (a) $P(D) = D^2 - D - 2$ (b) $P(D) = D^2 - 4D + 5$
   (c) $P(D) = D^2 - 4D + 4$

2. (*Using Solutions to Construct the ODE*). For each case below write the normalized linear ODE of lowest order with constant real coefficients for which the given mathematical expression is a solution.
   (a) $e^{-5t} - e^{-2t}$ [*Hint*: Consider $(D + 5)(D + 2)$.]
   (b) $3e^{4t} + 2te^{4t}$ [*Hint*: Consider $(D - 4)^2$.]
   (c) $8e^{-4t} \sin t$ (d) $e^{(3-5i)t}$
   (e) $\cos t + e^{(2-i)t}$ [*Hint*: First find $P(D)$ so that $P(D)[\cos t] = 0$. Then find $Q(D)$ so that $Q(D)[e^{(2-i)t}] = 0$. Look at $P(D)Q(D)$.]

3. Find the general real-valued solution of each ODE. Then plot solution curves in the ty-plane for $-1 \le t \le 5$, where $y(0) = 1$, $y'(0) = -6, -3, 0, 3, 6$. Plot the corresponding orbits in a rectangle in the yy'-plane that shows the main features of the orbits. Discuss your results.
   (a) $y'' + y' = 0$ (b) $y'' + 2y' + 65y = 0$
   (c) $y'' + 3y' + 2y = 0$ (d) $y'' + 10y = 0$
   (e) $y'' - y/9 = 0$ (f) $y'' - 3y'/4 + y/8 = 0$

4. (*IVPs*). Find the real-valued solution of each IVP below. Plot the solution curve in each case. What happens to solutions as $t \to +\infty$?
   (a) $y'' + y' + 2y = 0$, $\quad y(0) = 1$, $\quad y'(0) = 0$
   (b) $y'' + 4y' + 4y = 0$, $\quad y(1) = 2$, $\quad y'(1) = 0$
   (c) $y'' + y' - 6y = 0$, $\quad y(-1) = 1$, $\quad y'(-1) = -1$
   (d) $y'' + 2y' + 2y = e^t$, $\quad y(0) = 0$, $\quad y'(0) = 0$ [*Hint*: Use the identity $P(D)[e^{rt}] = P(r)e^{rt}$ to find a particular solution.]

5. (*Decaying Solutions*).
   (a) Show that if the real numbers a, b are such that $a > 0$, $b > 0$, then every solution of $y'' + ay' + by = 0$ tends to zero as $t \to \infty$.
   (b) Show that the converse of part (a) holds: If every solution tends to zero as $t \to \infty$, then $a > 0$ and $b > 0$.

6. (*Positively Bounded Solutions*).
   (a) A solution $y(t)$ of $y'' + ay' + by = 0$, a and b real numbers, is said to be *positively bounded* if there is a positive constant M such that $|y(t)| \le M$ for all $t \ge 0$. Show that all solutions are positively bounded if $a \ge 0$, $b \ge 0$, but not both a and b are zero.
   (b) Show the converse: If all solutions are positively bounded, then $a \ge 0$, $b \ge 0$, but not both a and b are zero.

7. (*ODEs with Complex Coefficients*).
   (a) Find the general solution of the ODE $y'' + iy' + 2y = 0$.
   (b) Does the ODE in part (a) have nontrivial real-valued solutions? (Recall that Theorem 3.4.2 does not apply because the coefficients are complex.)
   (c) Find the general solution of $y'' + iy = 0$. [*Hint*: Use De Moivre's Formula to find the roots of the characteristic polynomial.]

8. (*More ODES with Complex Coefficients*). Give all the real-valued solutions (if any) of the following ODEs. [*Hint*: Factor the characteristic polynomial by inspection.]

   (a) $y'' + (1 + i)y' + iy = 0$ **(b)** $y'' + (-1 + 2i)y' - (1 + i)y = 0$

9. (*Amplitude Modulation*). Explain the modulation phenomenon in Figures 3.4.5 and 3.4.6. Formulate a prediction about how the modulation varies with the sampling rate and verify your prediction experimentally.

3.5 Driven Linear ODEs: Annihilators

In Section 3.3 we investigated the polynomial operator

$$P(D) = D^2 + a(t)D + b(t) \tag{1}$$

where the coefficients are in $\mathbf{C}^0(I)$ for some interval I on the t-axis. In Section 3.4 we saw that it is easy to find a fundamental set for $P(D)[y] = 0$ when the coefficients $a(t)$ and $b(t)$ are constants. For nonconstant coefficients it is usually difficult to find a fundamental set. We will come back to this question in Chapter 11.

In this section we present a method for finding a particular solution for $P(D)[y] = f$ when $P(D)$ has constant coefficients and f is a polynomial-exponential function. It is called the *Method of Annihilators*, and it is always successful (even if the calculations are laborious).[2] First we outline how the *Method of Annihilators* works and then later a recipe for how to find annihilators. The method can only be used when $P(D)$ has constant coefficients and for driving terms $f(t)$ that can be written as a finite sum $f(t) = g_1(t) + g_2(t) + \cdots + g_m(t)$, where each g_j has the form $p(t)e^{rt}$ for a polynomial $p(t)$ and a complex number r. Here is how the method works:

1. For each $g_j(t)$ find a polynomial operator $Q_j(D)$ with constant coefficients, called an *annihilator*, such that $Q_j(D)[g_j] = 0$. Annihilators are easier to find than you would think.

2. Apply the operator $Q_j(D)$ to both sides of the ODE $P(D)[y] = g_j$ to obtain the homogeneous ODE $Q_j(D)P(D)[y] = 0$.

3. Find the general solution y of the homogeneous ODE $Q_j(D)P(D)[y] = 0$ (easy to do since $Q_j(D)P(D)$ has constant coefficients). Substitute it into the driven ODE $P(D)[y] = q_j(t)$ to find a particular solution $y_j(t)$ of this ODE.

4. The sum $y = y_1 + y_2 + \cdots + y_m$ will be a particular solution of $P(D)[y] = f(t)$.

Annihilators

For any complex number r, any positive integer n, and any function $h(t)$ in $\mathbf{C}^n(\mathbf{R})$, it is straightforward to show that

$$(D - r)^n[he^{rt}] = e^{rt}D^n[h], \qquad n = 1, 2, \ldots \tag{2}$$

---

[2] Another method, called the *Method of Undetermined Coefficients*, is equivalent to the Annihilator Method but is often awkward to use (in practice it requires some judicious guessing but see Problem 12).

(see Problem 9 in Section 3.3). Thus for any polynomial $p(t)$ with $\deg(p) \le n - 1$ we see that $(D - r)^n[p(t)e^{rt}] = 0$, for all t.

❖ **Annihilator.** The function $z(t)$ is said to be *annihilated* by the polynomial operator $P(D)$ if $P(D)[z] = 0$. $P(D)$ is called an *annihilator* for $z(t)$.

According to (2), $(D - r)^n$ is an annihilator for $p(t)e^{rt}$ for any polynomial $p(t)$ with $\deg(p) \le n - 1$. This also means that $p(t)e^{rt}$ is a solution of the ODE, $(D - r)^n[y] = 0$, for any polynomial $p(t)$ of degree $n - 1$ or less.

Let r_1 and r_2 be any complex numbers, and n_1, n_2 be any positive integers. Since $(D - r_1)(D - r_2) = (D - r_2)(D - r_1)$ it follows that

$$(D - r_1)^{n_1}(D - r_2)^{n_2} = (D - r_2)^{n_2}(D - r_1)^{n_1} \tag{3}$$

Identities (2) and (3), and the fact that $P(D) = (D - r_1)^{n_1}(D - r_2)^{n_2}$ is a linear operator, imply that $P(D)$ annihilates $p_1(t)e^{r_1 t} + p_2(t)e^{r_2 t}$, where $p_1(t)$ and $p_2(t)$ are any two polynomials with $\deg(p_1) \le n_1 - 1$ and $\deg(p_2) \le n_2 - 1$. Thus, $y = p_1(t)e^{r_1 t} + p_2(t)e^{r_2 t}$ is a solution of $P(D)[y] = 0$.

❖ **Polynomial-Exponential Functions.** The function $f(t) = \sum_{j=1}^{n} p_j(t)e^{r_j t}$ where the r_j are any real or complex numbers and the $p_j(t)$ are polynomials with real or complex coefficients is called a *polynomial-exponential function*.

EXAMPLE 3.5.1

Polynomial-Exponential Functions and Complex Exponentials
The function $t^2 e^{-t} \sin 2t$ is a polynomial-exponential function even though it may not appear so at first. To show this, recall (see Appendix C.4) that

$$\cos \theta = \frac{1}{2}(e^{i\theta} + e^{-i\theta}), \qquad \sin \theta = \frac{i}{2}(e^{-i\theta} - e^{i\theta})$$

Replacing θ by $2t$ in the second indentity we have $\sin 2t = i(e^{-2it} - e^{2it})/2$ and so

$$t^2 e^{-t} \sin 2t = \frac{i}{2}t^2 e^{(-1-2i)t} - \frac{i}{2}t^2 e^{(-1+2i)t}$$

which is a polynomial-exponential function.

The collection of polynomial-exponential functions has several properties.

THEOREM 3.5.1

Properties of Polynomial-Exponential Functions. Let $f(t) = \sum_{j=1}^{n} p_j(t)e^{r_j t}$. Then for every $k \ge 0$ $D^k f(t)$ is also a polynomial-exponential function,

$$D^k f(t) = \sum_{j=1}^{n} q_{jk}(t)e^{r_j t}$$

where q_{jk} is a polynomial of degree $\le$ degree p_j. In addition, if $f(t)$ is not identically zero, then $\lim_{t \to \infty} f(t) = 0$ if and only if $\text{Re}(\lambda_j) < 0$ for all j.

Proof. The first property is evident since $D(p_j(t)e^{r_j t}) = p_j' e^{r_j t} + r_j p_j e^{r_j t}$. Next note that if $\text{Re}(\lambda_j) = \alpha_j < 0$, then

$$\lim_{t \to \infty} |p_j(t)e^{r_j t}| = \lim_{t \to \infty} c_m t^m e^{\alpha_j t} = 0$$

where $c_m t^m$ is the term of p_j of highest degree and we have applied L'Hopital's Rule [Appendix C.5] m times to $t^m/e^{-\alpha_j t}$. Conversely, if $\text{Re}\,\alpha_j = \alpha_j \ge 0$ and

$p_j(t)$ is not identically zero, then $\lim_{t \to \infty} |p_j(t)e^{r_j t}| = \lim_{t \to \infty} c_m t^m e^{\alpha_j t} \neq 0$.

The opening remarks show how to find an annihilator for a polynomial-exponential function. Here is another useful technique for finding annihilators.

EXAMPLE 3.5.2

Finding an Annihilator

Let us find an annihilator for the polynomial-exponential function $t^2 e^{-t} \sin 2t$ given in Example 3.5.1. Rather than converting this function into a recognizable polynomial-exponential form (as was done in Example 3.5.1), we proceed as follows: Note that

$$t^2 e^{-t} \sin 2t = \text{Im}[t^2 e^{(-1+2i)t}]$$

because $e^{(-1+2i)t} = e^{-t} e^{2it} = e^{-t}(\cos 2t + i \sin 2t)$. Thus, by Theorem 3.3.8, if $P(D)$ is a polynomial operator with *real* coefficients such that $P(D)[t^2 e^{(-1+2i)t}] = 0$, then also $P(D)[t^2 e^{-t} \sin 2t] = 0$. From the identity (2) we have

$$(D - (-1 + 2i))^3 [t^2 e^{(-1+2i)t}] = 0, \qquad \text{for all } t \qquad (4)$$

But we want an annihilator with real coefficients. For any complex number r, note that $(D - r)(D - \bar{r}) = D^2 - (r + \bar{r})D + r\bar{r}$, where $\bar{r}$ is the complex conjugate of r. Thus, with $r = -1 + 2i$, $\bar{r} = -1 - 2i$, we have that $(D - r)(D - \bar{r}) = D^2 + 2D + 5$. Operating on both sides of (4) with $(D - (-1 - 2i))^3$ we obtain

$$(D^2 + 2D + 5)^3 [t^2 e^{(-1+2i)t}] = 0, \qquad \text{for all } t$$

Thus, $(D^2 + 2D + 5)^3$ is an annihilator with real coefficients for $t^2 e^{-t} \sin 2t$.

Higher-Order Homogeneous Linear ODEs

Consider the *monic* operator (i.e., the coefficient of D^n is one)

$$P(D) = D^n + a_{n-1}(t)D^{n-1} + \cdots + a_1(t)D + a_0(t) \qquad (5)$$

In the theorem below, we note that if the coefficients $a_j(t)$ are all of class $\mathbf{C}^0(I)$, on some common interval I, and the function $f(t)$ is also of class $\mathbf{C}^0(I)$, then every solution of $P(D)[y] = f$ "lives" in I, and the general solution depends upon n arbitrary constants. It is also noted that a unique member of the general solution can be selected by specifying initial data at a point t_0 in I.

THEOREM 3.5.2

> **Existence and Uniqueness Theorem.** Let the functions $f(t)$ and $a_j(t)$, $j = 0, 1, \ldots, n-1$, all be continuous on a common interval, I. Then for any values of the constants $b_0, b_1, \ldots, b_{n-1}$, and any t_0 in I, the initial value problem
> $$P(D)[y] = y^{(n)} + a_{n-1}(t)y^{(n-1)} + \cdots + a_1(t)y' + a_0(t)y = f(t)$$
> $$y^{(k)}(t_0) = b_k, \qquad k = 0, 1, \ldots, n-1 \qquad (6)$$
> has a unique solution $y(t)$ in $\mathbf{C}^n(I)$. If the coefficients $a_j(t)$, the driving term $f(t)$, and the initial data b_k are all real-valued, the solution of IVP (6) is real-valued.

A proof of the existence of a solution $y(t)$ for IVP (6) can be patterned after the existence proof for first-order IVPs in Appendix A, but we omit it.

As was done for the case $n = 2$ in Section 3.3, we will find it convenient to use the Wronskian determinant to identify a fundamental set of solutions for $P(D)[y] = 0$.

❖ **Wronskian Determinant.** If $y_1(t), \ldots, y_n(t)$ are any n functions in the continuity class $\mathbf{C}^{n-1}(I)$ for some interval I, then their *Wronskian Determinant* (or simply *Wronskian*) is defined by

$$W[y_1, \ldots, y_n](t) = \det \begin{bmatrix} y_1 & \cdots & y_n \\ \vdots & & \vdots \\ y_1^{(n-1)} & \cdots & y_n^{(n-1)} \end{bmatrix}$$

A basic property of Wronskians is given below without proof.

THEOREM 3.5.3

> **Abel's Formula.** For $P(D) = D^n + a_{n-1}(t)D^{n-1} + \cdots + a_1(t)D + a_0(t)$, let $y_1, \ldots, y_n$, be any n solutions of $P(D)[y] = 0$. Then the Wronskian $W(t) = W[y_1, \ldots, y_n](t)$ solves the ODE, $W' = -a_{n-1}W$, and so
>
> $$W(t) = Ce^{-\int^t a_{n-1}(s)\,ds} \quad \text{where } C \text{ is a constant.}$$

Thus, for n solutions $y_1, \ldots, y_n$ of $P(D)[y] = 0$, either $W[y_1, \ldots, y_n](t) = 0$, for all t, or $W[y_1, \ldots, y_n](t) \neq 0$, for all t.

❖ **Fundamental Set.** For $P(D)$ as in (5), a set of n solutions, $y_1, \ldots, y_n$, of $P(D)[y] = 0$ is called a *fundamental set* if $W[y_1, \ldots, y_n](t) \neq 0$ for all t.

As before, the importance of a fundamental set is that it can be used to write down the general solution of $P(D)[y] = 0$.

THEOREM 3.5.4

> **Fundamental Set Property.** For $P(D)$ as in (5), if $y_1, \ldots, y_n$ is a fundamental set for the ODE $P(D)[y] = 0$, then every complex-valued solution $y(t)$ of the ODE has the form
>
> $$y = c_1 y_1 + \cdots + c_n y_n, \qquad c_1, \ldots, c_n \text{ arbitrary complex numbers} \quad (7)$$
>
> If all the solutions in the fundamental set, $y_1, \ldots, y_n$, are real-valued, then (7) describes the general (real-valued) solution of the ODE, where $c_1, \ldots, c_n$ are arbitrary reals.

Proof. The proof uses the Existence and Uniqueness Theorem just as in the case $n = 2$ derived in Section 3.3.

Constant Coefficient Higher-Order Homogeneous Linear ODEs

If the coefficients in $P(D)$ are constant, we know much more. From Theorem 3.5.2, every solution of $P(D)[y] = 0$ "lives" on the entire t-axis, and it can be shown as well that every solution has derivatives of every order.

Since $P(D)$ has constant coefficients, when $P(D)$ is applied to the function e^{rt}, where r is any real (or complex) constant, then

$$P(D)[e^{rt}] = (r^n + a_{n-1}r^{n-1} + \cdots + a_1 r + a_0)e^{rt} = P(r)e^{rt} \quad (8)$$

The polynomial $P(r)$ in (8) is called the *characteristic polynomial* of $P(D)$.

A fundamental result of algebra (see Appendix C.4) is that any polynomial $P(r)$ can be factored completely into linear factors if complex numbers are used, and in only one way if $P(r)$ is a monic polynomial (i.e., the coefficient of the top degree term in P is unity). For example,

$$r^3 - 6r^2 + 11r - 6 = (r-1)(r-2)(r-3)$$

$$r^5 + 8r^3 + 16r = r(r^2+4)^2 = r(r-2i)^2(r+2i)^2$$

Note that the order of the factors is immaterial.[3] Since everything we have to say hinges on it, we will write the monic polynomial $P(r)$ in (8) in the factored form

$$P(r) = (r-r_1)^{m_1}(r-r_2)^{m_2}\cdots(r-r_k)^{m_k} \qquad (9)$$

where the distinct roots r_j may be real or complex numbers, and the m_j are nonnegative integers with $\sum m_j = n$. By convention, m_j is called the *multiplicity of the root r_j*.

If $P(r)$ has the factored form (9), it follows, much as in the second-order case, that the corresponding polynomial operator $P(D)$ has the similar factorization

$$P(D) = (D-r_1)^{m_1}(D-r_2)^{m_2}\cdots(D-r_k)^{m_k} \qquad (10)$$

Remember that the order of the factors is immaterial.

We now use our knowledge of annihilators to solve the ODE $P(D)[y] = 0$.

THEOREM 3.5.5

> Polynomial-Exponential Solutions of $P(D)[y] = 0$. For $P(D)$ as in (10), we have that for each $j = 1, 2, \ldots, k$,
>
> $$P(D)[p_j(t)e^{r_j t}] = 0, \qquad \text{for any polynomial } p_j \text{ of degree } \le m_j - 1$$

Proof. Since the order of factors in $P(D)$ is arbitrary, then for each j, $P(D)$ can be written so that the factor $(D-r_j)^{m_j}$ appears as the rightmost factor. Since $p_j(t)e^{r_j t}$ is annihilated by $(D-r_j)^{m_j}$ if the polynomial $p_j(t)$ has degree $\le m_j - 1$, the result immediately follows.

Now we have seen that every constant-coefficient linear homogeneous differential equation has polynomial-exponential solutions. Our next result shows that there are no other kinds of solutions.

THEOREM 3.5.6

> Constant-Coefficient General Complex-Valued Solution. For $P(D)$ as in (10) the general complex-valued solution of $P(D)[y] = 0$ is given by
>
> $$y = p_1(t)e^{r_1 t} + p_2(t)e^{r_2 t} + \cdots + p_k(t)e^{r_k t} \qquad (11)$$
>
> where each p_j, $j = 1, 2, \ldots, k$, is an arbitrary polynomial of degree $\le m_k - 1$ with complex coefficients.

[3]To factor polynomials one must have a way of precisely computing roots. If $n = 1$ or 2, there is no difficulty in finding these roots. If $n = 3$ or 4, it can still be done explicitly, but not easily, in terms of radicals. The formulas were first discovered in the sixteenth century and are given in most mathematical handbooks. However, if $n \ge 5$, it has been proven that no general formulas for expressing the roots in terms of radicals can exist.

Proof. For $P(D)$ as in (10) we see from Theorem 3.5.5 that $P(D)[y] = 0$ has the set of n solutions (with $n = \sum m_j$)

$$B = \{e^{r_1 t}, te^{r_1 t}, \ldots, t^{(n_1-1)}e^{r_1 t}, \ldots, e^{r_k t}, te^{r_k t}, \ldots, t^{(m_k-1)}e^{r_k t}\}$$

It can be shown that the solution set B is a fundamental set and so by Theorem 3.5.4 the asserted representation (11) follows.

EXAMPLE 3.5.3

Finding a Fundamental Set for a Fifth-Order ODE
We shall determine all solutions of the homogeneous ODE,

$$y^{(5)} + 8y^{(3)} + 16y' = 0$$

The characteristic polynomial of this equation is $P(r) = r^5 + 8r^3 + 16r$, which has the factorization

$$P(r) = r(r - 2i)^2(r + 2i)^2$$

Thus the ODE can be written in factored operator form $D(D - 2i)^2(D + 2i)^2[y] = 0$. From Theorem 3.5.6 above, the solutions of this ODE are determined by the fundamental set $B = \{1, e^{2it}, te^{2it}, e^{-2it}, te^{-2it}\}$, that is, the general solution of the ODE is

$$y = c_1 + c_2 e^{2it} + c_3 te^{2it} + c_4 e^{-2it} + c_5 te^{-2it}$$

where c_j, $j = 1, \ldots, 5$, are arbitrary complex numbers.

EXAMPLE 3.5.4

Finding a Fundamental Set for a Higher Order ODE
Let $P(D) = (D^2 + 2D + 5)^3(D - 1)^2 D^5$. To put $P(D)$ in the form (10) we need the roots of the quadratic factor, which are $-1 \pm 2i$. Thus

$$P(D) = (D - (-1 + 2i))^3(D - (-1 - 2i))^3(D - 1)^2 D^5$$

So, $r_1 = -1 + 2i$ and $r_2 = \bar{r}_1 = -1 - 2i$ are both roots of multiplicity $m_1 = m_2 = 3$. The root $r_3 = 1$ has multiplicity $m_3 = 2$. Finally, the root $r_4 = 0$ has multiplicity $m_4 = 5$. So the ODE $P(D)[y] = 0$ has the following complex-valued solutions

$$(a + bt + ct^2)e^{(-1+2i)t} \qquad a, b, c \text{ arbitrary complex numbers}$$
$$(a + bt + ct^2)e^{(-1-2i)t} \qquad a, b, c \text{ arbitrary complex numbers}$$
$$(a + bt)e^t \qquad a, b \text{ arbitrary complex numbers}$$
$$a + bt + ct^2 + dt^3 + et^4 \qquad a, \ldots, e \text{ arbitrary complex numbers}$$

By choosing the constants in the above list appropriately we obtain the special solutions

$$\begin{array}{llll}
y_1 = e^{(-1+2i)t}, & y_4 = e^{(-1-2i)t}, & y_7 = e^t, & y_{10} = t \\
y_2 = te^{(-1+2i)t}, & y_5 = te^{(-1-2i)t}, & y_8 = te^t, & y_{11} = t^2 \\
y_3 = t^2 e^{(-1+2i)t}, & y_6 = t^2 e^{(-1-2i)t}, & y_9 = 1, & y_{12} = t^3 \\
& & & y_{13} = t^4
\end{array}$$

The collection of solutions $y_1, \ldots, y_{13}$ is a fundamental set for the ODE

$$(D^2 + 2D + 5)^3(D - 1)^2 D^5[y] = 0$$

Thus, from Theorem 3.5.4 above, the general (complex-valued) solution is given by $y = c_1 y_1 + \cdots + c_{13} y_{13}$, where $c_1, \ldots, c_{13}$ are arbitrary complex numbers. Or, using

Theorem 3.5.6, the general solution can also be written as

$$y = p_1(t)e^{(-1+2i)t} + p_2(t)e^{(-1-2i)t} + p_3(t)e^t + p_4(t)$$

where the polynomials are such that $\deg p_1 \leq 2$, $\deg p_2 \leq 2$, $\deg p_3 \leq 1$, $\deg p_4 \leq 4$, and have arbitrary complex coefficients.

Real-Valued Solutions of $P(D)[y] = 0$

When $P(D)$ has real coefficients we would prefer to have the general real-valued solution of the ODE $P(D)[y] = 0$, and not the general complex-valued solution described in Theorem 3.5.6. We turn now to this task.

The nonreal roots of a polynomial with real coefficients always occur in conjugate pairs and with the same multiplicities. Let P have the nonreal conjugate pairs r_j, $\bar{r}_j$, each with multiplicity n_j, $j = 1, 2, \ldots, k$, and the real roots s_j with multiplicity m_j, $j = 1, 2, \ldots l$. Then the differential equation $P(D)[y] = 0$ takes the form

$$(D - r_1)^{n_1}(D - \bar{r}_1)^{n_1} \cdots (D - r_k)^{n_k}(D - \bar{r}_k)^{n_k}(D - s_1)^{m_1} \cdots (D - s_l)^{m_l}[y] = 0 \quad (12)$$

where $2(n_1 + \cdots + n_k) + (m_1 + \cdots + m_l) = n$. Thus, from Theorem 3.5.4 we need only find a fundamental set of n real-valued solutions of the ODE $P(D)[y] = 0$. As the next result shows, the real and imaginary parts of complex-valued solutions provide enough real-valued solutions to do the job.

THEOREM 3.5.7

Constant-Coefficient General Real-Valued Solution. Let $P(D)$ have the factorization (12) and put $r_j = \alpha_j + i\beta_j$, $j = 1, 2, \ldots, k$. Then the real-valued general solution of $P(D)[y] = 0$ is given by

$$\begin{aligned} y = h_1(t)e^{s_1 t} + &\cdots + h_l(t)e^{s_l t} \\ &+ p_1(t)e^{\alpha_1 t}\cos\beta_1 t + q_1(t)e^{\alpha_1 t}\sin\beta_1 t + \cdots \qquad (13) \\ &+ p_k(t)e^{\alpha_k t}\cos\beta_k t + q_k(t)e^{\alpha_k t}\sin\beta_k t \end{aligned}$$

for all polynomials h_j, p_j, q_j with real coefficients such that $\deg h_j \leq m_j - 1$, for $j = 1, 2, \ldots, l$, and $\deg p_j \leq n_j - 1$, $\deg q_j \leq n_j - 1$, for $j = 1, 2, \ldots, k$. There are precisely n arbitrary constants implicit in the polynomials in the representation (13).

Proof. That the n real-valued functions

$$e^{s_1 t}, \ldots, t^{m_l - 1}e^{s_l t}$$

$$e^{\alpha_1 t}\cos\beta_1 t, e^{\alpha_1 t}\sin\beta_1 t, \ldots, t^{n_1 - 1}e^{\alpha_1 t}\cos\beta_1 t, t^{n_1 - 1}e^{\alpha_1 t}\sin\beta_1 t$$

$$\cdots \qquad (14)$$

$$\cdots$$

$$e^{\alpha_k t}\cos\beta_k t, e^{\alpha_k t}\sin\beta_k t, \ldots, t^{n_k - 1}e^{\alpha_k t}\cos\beta_k t, t^{n_k - 1}e^{\alpha_k t}\sin\beta_k t$$

are each real-valued solutions of $P(D)[y] = 0$ is immediate from Theorem 3.3.8. It can be shown that the solution set (14) is a fundamental set; the description (13) of the general real-valued solution follows by Theorem 3.5.4.

EXAMPLE 3.5.5

$P(D)$ in Example 3.5.3 has real coefficients, and using the notation of the theorem above: $s_1 = 0$, with multiplicity $m_1 = 1$, and $r_1 = 2i$ with multiplicity $n_1 = 2$. Thus the real-valued general solution of $P(D)[y] = 0$ is

$$y = c_1 + c_2 \cos 2t + c_2 \sin 2t + c_4 t \cos 2t + c_5 t \sin 2t$$

where c_1, c_2, c_3, c_4, and c_5 are arbitrary real numbers.

EXAMPLE 3.5.6

Finding an Annihilator with Real Coefficients
Turning the problem around, we find a polynomial $P(r)$ with real coefficients such that the function $t^3 e^{-t} \cos 2t$ is a solution of $P(D)[y] = 0$. Observe that $t^3 e^{-t} \cos 2t = \text{Re}[t^3 e^{(-1+2i)t}]$. Thus $r_1 = -1 + 2i$ needs to be a root of $P(r)$ with multiplicity at least 4. To ensure that $P(r)$ has real coefficients $\bar{r}_1 = -1 - 2i$ must also be a root of $P(r)$ with the same multiplicity. Thus

$$P(r) = [r - (-1 + 2i)]^4 [r - (-1 - 2i)]^4 = (r^2 + 2r + 5)^4$$

will suffice for the stated purpose.

EXAMPLE 3.5.7

Solving an IVP
To solve the initial value problem

$$P(D)[y] = y^{(5)} + 8y^{(3)} + 16y' = 0$$
$$y(0) = 1, \quad y'(0) = 1/2, \quad y^{(2)}(0) = y^{(3)}(0) = y^{(4)}(0) = 0 \tag{15}$$

first observe that all real solutions of $P(D)[y] = 0$ are given by the fundamental set $\{1, \cos 2t, \sin 2t, t \cos 2t, t \sin 2t\}$ (see Examples 3.5.3 and 3.5.5). Thus we need to find coefficients $c_1, \dots, c_5$ such that $y = c_1 + c_2 \cos 2t + c_3 \sin 2t + c_4 t \cos 2t + c_5 t \sin 2t$ satisfies the initial conditions. That is, we must solve the linear system $y(0) = c_1 + c_2 = 1$, $y'(0) = 2c_3 + c_4 = 1/2$, $y^{(2)}(0) = -4c_2 + 4c_5 = 0$, $y^{(3)} = -8c_3 - 12c_4 = 0$, $y^{(4)} = 16c_2 - 32c_5 = 0$. Solving, we have $c_1 = 1$, $c_2 = c_5 = 0$, $c_3 = 3/8$, $c_4 = -1/4$, and hence the IVP (15) has the unique solution

$$y = 1 + (3 \sin 2t)/8 - (t \cos 2t)/4$$

Method of Annihilators

Consider the ODE

$$P(D)[y] = f(t) = p_1(t)e^{r_1 t} + \cdots + p_m(t)e^{r_m t} \tag{16}$$

where $P(D)$ is a polynomial operator, the $p_j(t)$ are polynomials, and the r_j are distinct complex numbers. Because $P(D)$ is a linear operator, we see that if $P(D)[y_j] = p_j e^{r_j t}$ for each $j = 1, 2, \dots, m$, then $y = y_1 + \cdots + y_m$ is a solution of $P(D)[y] = f$. Thus it suffices to describe the method when f has the special form $p(t)e^{rt}$. If $Q(D)$ is any polynomial operator such that $Q(D)[p(t)e^{rt}] = 0$ (i.e., $Q(D)$ is an annihilator for $p(t)e^{rt}$), then for any particular solution y_p of $P(D)[y] = p(t)e^{rt}$, it follows that $Q(D)P(D)[y_p] = 0$. Thus y_p is a function in the solution space of the ODE $Q(D)P(D)[y] = 0$, and a fact that we can use to identify a solution y_p of $P(D)[y] = p(t)e^{rt}$. Some examples will clarify the procedure.

EXAMPLE 3.5.8

Finding a Particular Solution and the General Solution

Suppose that $y_p(t)$ is a particular solution of the ODE

$$(D+1)(D-2)[y] = (1+t)e^t \tag{17}$$

Then, since $(D-1)^2$ annihilates $(1+t)e^t$, we see that $(D-1)^2(D+1)(D-2)[y_p] = 0$. It follows from Theorem 3.5.7 that the collection of real-valued solutions of the ODE $(D-1)^2(D+1)(D-2)[y] = 0$, and thus y_p as well, has the form

$$(A+Bt)e^t + Ce^{-t} + Ee^{2t} \tag{18}$$

for any choice of the real constants A, B, C, E. So we plug (18) for y into the ODE (17) and solve for the constants. Notice that $(D+1)(D-2)$ annihilates e^{-t} and e^{2t}, so the values of C and E do not matter. Therefore, we might as well take $C = E = 0$. We are left with

$$(D+1)(D-2)[(A+Bt)e^t] = (D^2 - D - 2)[(A+Bt)e^t]$$
$$= e^t(B - 2A - 2Bt)$$
$$= (B-2A)e^t - 2Bte^t = e^t(1+t)$$

Matching coefficients of like terms across the last equality, we have $B - 2A = 1$ and $-2B = 1$, so $A = -3/4$ and $B = -1/2$. Thus

$$y_p = (-3/4 - t/2)e^t$$

is a particular solution. Therefore, the general solution of this ODE is

$$y = c_1 e^{-t} + c_2 e^{2t} + (-3/4 - t/2)e^t$$

because $\{e^{-t}, e^{2t}\}$ is a fundamental set for the ODE $(D+1)(D-2)[y] = 0$.

EXAMPLE 3.5.9

Solving an IVP

Let us find the unique solution of the IVP

$$y'' - y' - 2y = (1+t)e^t, \qquad y(0) = 0, \quad y'(0) = 0 \tag{19}$$

From Example 3.5.8 we see that the general solution of the ODE in (19) is

$$y = c_1 e^{-t} + c_2 e^{2t} + (-3/4 - t/2)e^t$$

Applying the initial conditions in IVP (19) we have that $y(0) = c_1 + c_2 - 3/4 = 0$ and $y'(0) = -c_1 + 2c_2 - 5/4 = 0$. Solving this system we obtain $c_1 = 1/12$ and $c_2 = 2/3$. So the unique solution of IVP (19) is

$$y = e^{-t}/12 + 2e^{2t}/3 + (-3/4 - t/2)e^t$$

The Method of Annihilators also works for higher-order linear ODEs with constant coefficients.

EXAMPLE 3.5.10

Particular Solution for Higher-Order ODE

Let us find a particular solution of the ODE

$$y''' - 3y' + 2y = 6(3t - 1)e^t \tag{20}$$

Putting $P(D) = D^3 - 3D + 2$, the ODE becomes $P(D)[y] = 6(3t - 1)e^t$. Note that $P(D) = (D-1)^2(D+2)$, and that $Q(D) = (D-1)^2$ annihilates $6(3t -$

$1)e^t$. Thus any particular solution of ODE (20) satisfies the homogeneous ODE, $Q(D)P(D)[y] = 0$. Hence, any particular solution y_p of ODE (20) has the form

$$y_p = h(t)e^t + Ae^{-2t}$$

for some constant A and a polynomial h of degree ≤ 3. Using the operator identity $(D - r_0)^n[he^{r_0t}] = e^{r_0t}h^{(n)}$ and the fact that $P(D)$ is a linear operator we have that

$$P(D)[y_p] = (D-1)^2(D+2)[he^t] + (D-1)^2(D+2)[Ae^{-2t}] = (D+2)[h''e^t]$$

If we put $h(t) = k_0 + k_1 t + k_2 t^2 + k_3 t^3$, then we have

$$P(D)[y_p] = (D+2)[(2k_2 + 6k_3 t)e^t] = e^t(6k_2 + 6k_3 + 18k_3 t)$$

and since k_0, k_1, and A are of no use at all in generating $P(D)[y_p] = 6(3t-1)e^t$ we may as well take $k_0 = k_1 = A = 0$. Comparing coefficients we must have that $6k_2 + 6k_3 = -6$, and $18k_3 = 18$. Solving, we obtain $k_2 = -2$ and $k_3 = 1$, and hence

$$y_p = (-2t^2 + t^3)e^t$$

Exponential Driving Terms

For any constant-coefficient polynomial operator $P(D)$ there is a direct way to find a particular solution of $P(D)[y] = f$ when the driving term has the form $f(t) = Ae^{\omega t}$ where A and ω are any real or complex constants and $P(\omega) \neq 0$. From the operational formula $P(D)[Ce^{\omega t}] = CP(\omega)e^{\omega t}$ for any constant C (see (8)), we could easily find a particular solution of the ODE,

$$P(D)[y] = Ae^{\omega t} \tag{21}$$

by taking

$$y_p = (A/P(\omega))e^{\omega t} \tag{22}$$

provided that $P(\omega) \neq 0$.

EXAMPLE 3.5.11

Consider $P(D) = D^2 - D + 1$ and the equation $P(D)[y] = -2e^{-it}$. Now since $P(-i) = (-i)^2 - (-i) + 1 = i$, we see that

$$y_p = \frac{-2}{i}e^{-it} = 2ie^{-it}$$

is a particular solution for the ODE, $y'' - y' + y = -2e^{-it}$.

To find a real-valued particular solution for the equation $P(D)[y] = Ae^{\alpha t}\cos\beta t$ when A, α, and β are real constants and $P(D)$ has real coefficients, we proceed as follows. First note that $Ae^{\alpha t}\cos\beta t$ is the real part of $Ae^{\omega t}$, where $\omega = \alpha + i\beta$. Then from the relationship between real and complex solutions we see that if z_p is any solution of the ODE, $P(D)[z] = Ae^{\omega t}$, then $y_p = \text{Re}[z_p]$ is a real-valued solution to the ODE $P(D)[y] = Ae^{\alpha t}\cos\beta t$.

EXAMPLE 3.5.12

Response of Hooke's Law Spring to Oscillatory Driving Force
Consider the ODE that models the response of a damped Hooke's Law spring to an oscillatory driving force:

$$y'' + 4y' + 3y = 3e^{-t}\cos 3t \tag{23}$$

Note that $3e^{-t}\cos 3t = \text{Re}[3e^{(-1+3i)t}]$, and hence we look at the ODE

$$(D^2 + 4D + 3)[z] = 3e^{(-1+3i)t} \tag{24}$$

Putting $P(D) = D^2 + 4D + 3$, we see that $P(-1 + 3i) = -9 + 6i \neq 0$. From (22) we see that ODE (24) has the particular solution

$$\begin{aligned}
z_p &= \frac{3}{P(-1 + 3i)} e^{(-1+3i)t} = \frac{3}{-9 + 6i} e^{(-1+3i)t} \\
&= -\left(\frac{3}{13} + \frac{2}{13}i\right) e^{-t}(\cos 3t + i\sin 3t)
\end{aligned}$$

Thus

$$y_p = \text{Re}[z_p] = \left(-\frac{3}{13}\cos 3t + \frac{2}{13}\sin 3t\right) e^{-t}$$

is a particular solution of ODE (23).

Initial Value Problems

Now that we have some practical methods for finding particular solutions of nonhomogeneous linear differential equations, we have a convenient way of finding the solution of the initial value problem

$$P(D)[y] = f, \qquad y(t_0) = y_0, \quad y'(t_0) = v_0 \tag{25}$$

where $P(D)$ has constant coefficients, f is a polynomial-exponential function, t_0 is in I, and y_0 and v_0 are arbitrary constants. Our approach is as follows:

- First find a fundamental set $\{y_1, y_2\}$ for $P(D)[y] = 0$.

- Then find a solution y_p of $P(D)[y] = f$ using the Method of Annihilators.

- Then the general solution of $P(D)[y] = f$ is given by $y = k_1 y_1 + k_2 y_2 + y_p$, where k_1 and k_2 are arbitrary constants.

- The constants k_1 and k_2 are then determined so that the initial conditions are satisfied.

EXAMPLE 3.5.13

Solving An IVP

To solve the initial value problem

$$y'' + y = t^3, \qquad y(0) = 1, \quad y'(0) = -1$$

we first use annihilators to find the particular solution $y_p = t^3 - 6t$ of the ODE. Now $\{\cos t, \sin t\}$ is a fundamental set of real-valued solutions of $(D^2 + 1)[y] = 0$. Thus $y = k_1 \cos t + k_2 \sin t + (t^3 - 6t)$, with k_1 and k_2 arbitrary reals, is the real-valued general solution of $y'' + y = t^3$. Imposing the initial conditions we find that $k_1 = 1$, $k_2 = 5$, and hence the solution is

$$y = \cos t + 5\sin t + (t^3 - 6t)$$

Comments

A tacit assumption in our approach has been that the roots of the characteristic polynomial $P(r)$ are known precisely, together with their multiplicities. As observed earlier in a footnote, such precise information is, in general, not possible if $\deg P \geq 5$. So what is the value of our solution formulas and procedures if we cannot explicitly find the precise information about the characteristic polynomial required by our approach? Under appropriate circumstances, if we have suitably close bounds on the characteristic roots, the solution formulas are useful in obtaining approximate solutions or in obtaining information on the qualitative properties of solutions of $P(D)[y] = f$. Many commercial software packages contain efficient solvers for finding close approximations of all roots of a polynomial. Hence we consider the case where $P(D)$ has constant coefficients and f is a polynomial-exponential function as "solved."

PROBLEMS

1. (*Polynomial-Exponential Functions*). Express the functions below as polynomial-exponential functions. [*Hint*: See Example 3.5.1.]

    (a) $\cos 2t - \sin t$ (b) $t \sin^2 t$

    (c) $t^2 \sin 2t - (1+t)\cos^2 t$ (d) $\sin^3 t$

    (e) $(1-t)e^{it}\cos 3t$ (f) $(i+t-t^2)e^{(3+i)t}\sin^2 3t$

2. Express each of the functions in Problem 1(a)–(c) as the real part of some polynomial exponential function. [*Hint*: See Example 3.5.2.]

3. (*Annihilators*). Find an annihilator with real coefficients for each polynomial-exponential function below. [*Hint*: See 3.5.6.]

    (a) $t^2 e^{-it}$ (b) $te^{-t}\cos 2t$

    (c) $t + \sin t$ (d) $\sin t + \cos 2t + t^2$

4. Find a constant-coefficient homogeneous differential equation with real coefficients which includes $t^3 e^{-3t}$ and $te^t \sin 2t$ among its solutions. [*Hint*: Figure out what the roots (and their multiplicities) of the characteristic polynomial must be.]

5. Find the solution of each of the following initial value problems

    (a) $(D^3 + 1)y = 0$, $y(0) = 0$, $y'(0) = 0$, $y''(0) = 1$

    (b) $y''' - y'' + 3y' + 5y = 0$, $y(0) = 1$, $y'(0) = y''(0) = 0$

    (c) $(D^2 + 1)(D+2)^2 y = 0$, $y(0) = y'(0) = 0$, $y''(0) = y'''(0) = 1$

6. Find the general real-valued solution for each ODE below.

    (a) $y^{(5)} + 2y^{(3)} + y' = 0$ [*Hint*: See Example 3.5.3 and 3.5.5.]

    (b) $(D^2 + 2D + 5)^3 (D-1)^2 D^4[y] = 0$ [*Hint*: See Example 3.5.4 and 3.5.5.]

7. Find the real-valued general solution of each of the following ODEs.

    (a) $y^{(4)} + 81y = 0$ (b) $(D^3 + 1)[y] = 0$.

    (c) $y''' + 3y'' - y' - 3y = 0$ (d) $y^{(4)} + 18y'' + 81y = 0$

    (e) $(D-1)^3 (D+1)^3[y] = t$ (f) $D^2(D-1)^3[y] = 1$

8. Solve each of the following initial value problems.

    (a) $y'' - 4y = 2 - 8t$, $y(0) = 0$, $y'(0) = 5$

    (b) $y'' + 9y = 81t^2 + 14\cos 4t$, $y(0) = 0$, $y'(0) = 3$

    (c) $y'' + y = 10e^{2t}$, $y(0) = 0$, $y'(0) = 0$

    (d) $y'' - y = e^{-t}(2\sin t + 4\cos t)$, $y(0) = 1$, $y'(0) = 1$

(e) $y'' - 3y' + 2y = 8t^2 + 12e^{-t}$, $\quad y(0) = 0$, $\quad y'(0) = 2$

9. (*Using Annihilators to Solve ODEs*). Use the Method of Annihilators to find all solutions of the following differential equations. Plot solution graphs of the corresponding IVPs with $y(0) = 0$, $y'(0) = -1, 0, 1$ for (a)–(g).

(a) $y'' - y' - 2y = 2\sin 2t$ (b) $y'' - y' - 2y = t^2 + 4t$

(c) $y'' - 2y' + y = -te^t$ (d) $y'' - 2y' + y = 2e^t$

(e) $y'' + 2y' + y = e^t \cos t$ (f) $y'' + y' + y = \sin^2 t$

(g) $y'' + 4y' + 5y = e^{-t} + 15t$ (h) $y'' + 4y = e^{2it}$

(i) $y'' + y' + 2y = 2e^{it}$

10. (*Boundary Value Problem*). Find all solutions of the boundary value problem

$$y'' - 3y' + 2y = \sin t, \qquad y(0) - y'(0) = 1, \quad y'(1) = -2$$

11. (*An Operator Identity*). For any polynomial operator $P(D)$ and any real or complex number r show the operator identity

$$P(D)[e^{rt}y] = e^{rt}P(D+r)[y]$$

for all y in the domain of $P(D)$. [*Hint*: First show the identity for $P(D) = D^k$, and then use linearity.]

12. Prove the following result:

THEOREM 3.5.8

Undetermined Coefficients Theorem. Let the constant-coefficient operator $P(D) = D^2 + aD + b$ have the factorization $(D - r_1)(D - r_2)$. Then the ODE $P(D)[y] = p(t)e^{\lambda t}$, where p is a polynomial and λ a constant, has a particular solution $y_p = q(t)e^{\lambda t}$, with polynomial q as below.

If $\deg(p) = n$, and:	Then take
$\lambda \neq r_1, \lambda \neq r_2$	$q(t) = \displaystyle\sum_{k=0}^{n} a_k t^k$, and $q'' + (a + 2\lambda)q' + P(\lambda)q = p$
$\lambda = r_1 \neq r_2$	$q(t) = t\displaystyle\sum_{k=0}^{n} a_k t^k$, and $q'' + (r_1 - r_2)q' = p$
$\lambda = r_1 = r_2$	$q(t) = t^2 \displaystyle\sum_{k=0}^{n} a_k t^k$, and $q'' = p$

3.6 Driven Linear ODEs: Variation of Parameters

Now we turn to the task of finding a particular solution of the linear ODE $P(D)[y] = f$ where $P(D) = D^2 + a(t)D + b(t)$, and $a(t)$, $b(t)$, $f(t)$ belong to the continuity class $C^0(I)$, for some t-interval I. The Method of Annihilators is successful only when $P(D)$ has constant coefficients and $f(t)$ is a polynomial-exponential function. Thus the method we are about to describe is more general than that of annihilators. Our approach assumes that we are given a fundamental set for the homogeneous ODE $P(D)[y] = 0$, and then uses this pair to construct a particular solution of $P(D)[y] = f$. The construction of the general solution then follows, as before, from the General Solution Theorem 3.3.7.

Variation of Parameters

For any fundamental set $\{y_1, y_2\}$ of solutions to the homogeneous ODE $P(D)[y] = 0$, we know that $c_1 y_1 + c_2 y_2$ is also a solution. We shall show that it is possible to "vary the parameters" c_1 and c_2 (i.e., make them functions of t) in such a way that $c_1(t)y_1 + c_2(t)y_2$ becomes a solution of the nonhomogeneous equation $P(D)[y] = f$. This is the essence of the *Method of Variation of Parameters*.

THEOREM 3.6.1

Variation of Parameters. Let $P(D) = D^2 + a(t)D + b(t)$ with $a(t)$, $b(t)$ in $\mathbf{C}^0(I)$, and suppose that $\{y_1, y_2\}$ is a given fundamental set for $P(D)[y] = 0$ with Wronskian $W[y_1, y_2](t) = W(t)$. Let t_0 be any point in I, and f any function in $\mathbf{C}^0(I)$. Then the function

$$y_p = c_1(t)y_1(t) + c_2(t)y_2(t) \tag{1}$$

where

$$c_1(t) = \int_{t_0}^{t} -\frac{y_2(s)f(s)}{W(s)}\, ds, \qquad c_2(t) = \int_{t_0}^{t} \frac{y_1(s)f(s)}{W(s)}\, ds \tag{2}$$

is the unique solution of the initial value problem

$$P(D)[y] = f, \qquad y(t_0) = 0, \quad y'(t_0) = 0 \tag{3}$$

Proof. Our approach is straightforward. We substitute y_p as given in (1) into the ODE $P(D)[y] = f$ and determine the "varied parameters" c_1 and c_2 so that the ODE is satisfied. As we shall see, it is helpful to require that $c_1(t)$ and $c_2(t)$ satisfy the further condition that $c_1' y_1 + c_2' y_2 = 0$ on I, because this simplifies y_p' greatly. Substituting y_p into the equation $P(D)[y] = f$, we obtain (after some calculation) that $c_1' y_1' + c_2' y_2' = f$. Thus the coefficients c_1 and c_2 must be chosen to satisfy the conditions

$$c_1' y_1 + c_2' y_2 = 0, \qquad c_1' y_1' + c_2' y_2' = f, \qquad \text{for all } t \text{ in } I \tag{4}$$

Since the coefficient matrix of this system of equations in the unknown functions c_1' and c_2' is $W(t) = W[y_1, y_2](t)$, which we know is never zero on I, we can solve (4) algebraically for c_1' and c_2' to obtain the unique solution

$$c_1' = \frac{-y_2 f}{W}, \qquad c_2' = \frac{y_1 f}{W}$$

Thus, any antiderivatives of $-y_2 f/W$ and $y_1 f/W$ will yield candidates for c_1 and c_2 which, when substituted into (1), will provide a particular solution for $P(D)[y] = f$. The expressions in (2) are the special antiderivatives that vanish at $t = t_0$. Using (2) to construct y_p, a direct calculation shows that the initial conditions $y_p(t_0) = 0$, $y_p'(t_0) = 0$ are satisfied, and hence y_p is the unique solution of the IVP (3).

Notice that if the expressions for c_1 and c_2 in (2) are substituted into (1), we have the alternative expression for y_p,

$$
\begin{aligned}
y_p(t) &= y_1(t) \int_{t_0}^{t} \frac{-y_2(s)f(s)}{W(s)}\, ds + y_2(t) \int_{t_0}^{t} \frac{y_1(s)f(s)}{W(s)}\, ds \\
&= \int_{t_0}^{t} \frac{y_1(s)y_2(t) - y_2(s)y_1(t)}{W(s)}\, f(s)\, ds
\end{aligned}
$$

$$= \int_{t_0}^{t} K(t, s) f(s) \, ds \tag{5}$$

where the *kernel function* $K(t, s)$ [also called *Green's Kernel*] is the ratio of two determinants

$$K(t, s) = \frac{\det \begin{bmatrix} y_1(s) & y_2(s) \\ y_1(t) & y_2(t) \end{bmatrix}}{\det \begin{bmatrix} y_1(s) & y_2(s) \\ y_1'(s) & y_2'(s) \end{bmatrix}} \tag{6}$$

The denominator in $K(t, s)$ is the Wronskian $W[y_1, y_2]$. Observe that the kernel $K(t, s)$ in (6) does *not* depend on the driving term $f(t)$. Thus if $g(t)$ is another driving term, a particular solution for $P(D)[y] = g$ can be calculated via (5) *without* recalculating $K(t, s)$, a significant advantage for some applications. Also note that $K(t, s)$ is the same no matter what fundamental set $\{y_1, y_2\}$ for $P(D)[y] = 0$ is used in the definition. This fact is far from obvious (see Problem 7), but allows for considerable flexibility in the calculation of K.

EXAMPLE 3.6.1

Finding a Particular Solution: Using Green's Kernel

Let us find a particular solution of the equation

$$t^2 u'' - t u' + u = F(t) \tag{7}$$

on $t > 0$ for F in $C^0(I)$, where I is the interval $0 < t < \infty$. As we shall see below, the function $y_p(t)$ given by (5) would be such a solution if $t_0 = 1$ (for example), $f(t) = F(t)/t^2$, and $\{y_1, y_2\}$ is any fundamental set of solutions of $t^2 u'' - t u' + u = 0$. To apply the formulas (5) and (6) we must first normalize ODE (7) to obtain

$$u'' - t^{-1} u' + t^{-2} u = \frac{F(t)}{t^2}$$

A direct calculation shows that we can take the fundamental set $\{y_1, y_2\}$ to be the pair $\{t, t \ln t\}$. Thus (5) and (6) provide a particular solution of (7) with $K(t, s) = t \ln(t/s)$:

$$u_p(t) = \int_1^t \frac{\det \begin{bmatrix} s & s \ln s \\ t & t \ln t \end{bmatrix}}{\det \begin{bmatrix} s & s \ln s \\ 1 & 1 + \ln s \end{bmatrix}} \frac{F(s)}{s^2} \, ds = \int_1^t t \ln \left(\frac{t}{s} \right) \frac{F(s)}{s^2} \, ds \tag{8}$$

EXAMPLE 3.6.2

Solving an IVP

Let us solve the IVP

$$y'' + y = \tan t. \qquad y(0) = 0, \quad y'(0) = 0$$

The Existence and Uniqueness Theorem guarantees that the solution of this IVP is defined on $|t| < \pi/2$. Note that the driving term $f(t) = \tan t$ is not a polynomial-exponential function and so we cannot use the Method of Annihilators to find a particular solution. The particular solution $y_p(t)$ defined by (5), (6), is precisely the solution of the IVP that we seek. A fundamental set of solutions of the homogeneous ODE $y'' + y = 0$ is given by the pair $y_1 = \cos t$, $y_2 = \sin t$. From (5) and (6) we have that $K(t, s) = \cos s \sin t - \sin s \cos t$, and so

$$y_p = \int_0^t (\cos s \sin t - \sin s \cos t) \tan s \, ds$$

$$= \sin t \int_0^t \sin s \, ds - \cos t \int_0^t \sin s \tan s \, ds$$

$$= \sin t (-\cos s)|_0^t + \cos t \int_0^t . (\cos s)' \tan s \, ds$$

$$= -\sin t \cos t + \sin t + \cos t \left[(\cos s \tan s)|_0^t - \int_0^t \cos s \sec^2 s \, ds \right]$$

where we have used integration-by-parts on the last integral. Continuing, we have

$$y_p = -\sin t \cos t + \sin t + \cos t \sin t - \cos t \int_0^t \sec s \, ds$$

$$= \sin t - \cos t \ln|\sec t + \tan t|$$

Note that $y_p(0) = 0$, $y_p'(0) = 0$.

EXAMPLE 3.6.3

Solving an IVP for ODE (7)
Using the particular solution (8) we see that the general solution of (7) is

$$u = k_1 t + k_2 t \ln t + \int_1^t \ln\left(\frac{t}{s}\right) \frac{F(s)}{s^2} \, ds$$

where k_1, k_2 are arbitrary constants. If we impose the initial conditions $u(1) = -1$, $u'(1) = 1$, then we must choose k_1, k_2 such that $-1 = k_1$, and $1 = k_1 + k_2$. Hence the solution of the initial value problem

$$t^2 u'' - t u' + u = F(t), \qquad u(1) = -1, \quad u'(1) = 1$$

is given by

$$u = -t + 2t \ln t + \int_1^t t \ln\left(\frac{t}{s}\right) \frac{F(s)}{s^2} \, ds$$

$$= \frac{\text{Response to}}{\text{initial data}} \quad + \quad \text{Response to input}$$

This decomposition of the solution into the solution of the homogeneous ODE that satisfies the initial data and a particular solution that satisfies zero initial conditions is one of the advantages of Variation of Parameters. It allows us to identify just how the system responds to the initial data whatever the input may be and how the system reponds to the input independently of the values of the initial data.

Although Variation of Parameters has many advantages, other methods of finding a particular solution are often easier to use for particular input functions.

EXAMPLE 3.6.4

Comparison of Methods
Consider the initial value problem

$$y'' + 2y' + 10y = 3\cos 3t, \qquad y(0) = y_0, \quad y'(0) = y_0' \tag{9}$$

The roots of the characteristic polynomial $r^2 + 2r + 10$ are $r_1 = -1 + 3i$ and $r_2 = -1 - 3i$. Thus the real-valued solutions of $y'' + 2y' + 10y = 0$ are described by the fundamental set $\{e^{-t}\cos 3t, e^{-t}\sin 3t\}$. Hence using (5), (6) to find a particular solution of the driven equation, we see that the general solution of the driven equation is

$$y(t) = k_1 e^{-t}\cos 3t + k_2 e^{-t}\sin 3t + \int_0^t e^{-(t-s)}\sin 3(t-s)\cos 3s \, ds \tag{10}$$

where Green's Kernel is given by

$$K = e^{-(t-s)}[\sin 3t \cos 3s - \cos 3t \sin 3s]/3$$

$$= e^{-(t-s)}[\sin 3(t-s)]/3$$

Using the initial conditions, we have $k_1 = y_0$, $k_2 = (y_0' + y_0)/3$, and when inserted into (10), we have the solution to IVP (9). It is interesting to carry out the integration to calculate explicitly the integral in (10). Using the identity $\sin\alpha\cos\beta = [\sin(\alpha + \beta) + \sin(\alpha - \beta)]/2$, we have

$$y_p = \frac{1}{2}e^{-t}\int_0^t e^s\{\sin 3t + \sin(3t - 6s)\}\,ds$$

$$= \frac{1}{2}e^{-t}(e^t - 1)\sin 3t + \frac{1}{2}e^{-t}\int_0^t e^s \sin(3t - 6s)\,ds$$

Using integral tables to evaluate the last integral, we obtain

$$y_p = \frac{1}{37}(18\sin 3t + 3\cos 3t - 19e^{-t}\sin 3t - 3e^{-t}\cos 3t) \tag{11}$$

Notice that the last two terms of (11) are actually solutions of the corresponding homogeneous equation and that the sum of the first two terms is another particular solution of the driven equation (and is the particular solution the Method of Annihilators would have produced).

The form of (11) is typical if Variation of Parameters is used to construct a solution of a driven ODE. The particular solution found this way will be the sum of another particular solution and an element of the null space needed so that the sum will both satisfy the differential equation and have trivial initial data.

Piecewise-Continuous Driving Terms

For a polynomial operator $P(D)$ we sometimes encounter the ODE $P(D)[y] = f$, where the driving term is only piecewise continuous on the interval I and not continuous. In this case $y(t)$ is said to be a solution of the ODE if y is in $\mathbf{C}^1(I)$, y'' is piecewise continuous on I, and the equation $P(D)[y] = f$ is satisfied wherever f is continuous. (Observe that the value of f at a discontinuity is immaterial.)

An advantage of the particular solution y_p to IVP (3) given by (5), (6) is that it still provides a solution to IVP (3) even when $f(t)$ is only piecewise continuous, albeit in the generalized sense described above. The proof is a straightforward but long calculation and we omit it.

EXAMPLE 3.6.5

An IVP with Discontinuous Data
The IVP

$$u'' = \begin{cases} 2, & t < 1 \\ -2, & t > 1 \end{cases}, \qquad u(0) = u'(0) = 0$$

has the solution

$$u(t) = \begin{cases} t^2, & t < 1 \\ -t^2 + 4t - 2, & t \geq 1 \end{cases}$$

Note that $u(t)$ and $u'(t)$ are continuous for all t, while $u''(t)$ is discontinuous at $t = 1$ but continuous otherwise [note that $u''(1^-) = 2$, $u''(1^+) = -2$].

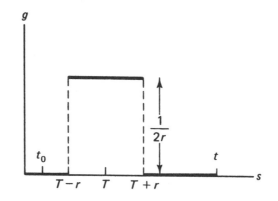

FIGURE 3.6.1 Unit function.

Properties of the Kernel Function

The kernel function $K(t, s)$ can be interpreted as the response at t of the system determined by the operator $P(D)$ to a "unit impulse" input located at s, where s is between t_0 and t, $t_0 \leq s \leq t$. We can make this more precise as follows. Fix T, $t_0 < T < t$, and let r be such that $t_0 < T - r < T < T + r < t$. Let $g(r, T, s)$ be the "unit function" [see Figure 6.3.1]

$$g = \begin{cases} \dfrac{1}{2r} & \text{if } T - r \leq s \leq T + r \\ 0 & \text{all other values of } s \end{cases} \tag{12}$$

If $K(t, s)$ is the Green's Kernel, then the response to the input g is

$$y_P(t) = \int_{t_0}^{t} K(t, s) g(r, T, s) \, ds = \int_{T-r}^{T+r} K(t, s) \frac{1}{2r} \, ds$$

where (12) has been used. Now the *Mean Value Theorem for Integrals* (see Appendix C.5) implies that for some value s^* between $T - r$ and $T + r$, we have that

$$y_P(t) = \int_{T-r}^{T+r} K(t, s) \frac{1}{2r} \, ds = K(t, s^*) \frac{1}{2r} [(T + r) - (T - r)] = K(t, s^*)$$

Now let $r \to 0$, which forces s^* to approach T. Thus since $K(t, s)$ is continuous,

$$\lim_{r \to 0} y_P(t) = \lim_{r \to 0} K(t, s^*) = K(t, T)$$

Note that as $r \to 0$, the graph of the unit function $g(r, T, s)$ narrows and elongates so that the area under the graph remains 1. In the limit as $r \to 0$, it has "infinite height," "infinitesimal width," but retains the area 1 property. Obviously no longer a function in the conventional sense, this limit is called a *unit impulse* located at T. Thus $K(t, T)$ is the response at t to a unit impulse located at T.

This result has practical implications. Often the underlying action operator $P(D)$ of a physical system is not known. However, the Green's Kernel $K(t, s)$ can be numerically approximated by driving the system with unit impulses at selected values of s and measuring the responses at selected values of $t > s$.

Note that the solution of $P(D)[y] = f$, $y(t_0) = 0$, $y'(t_0) = 0$ is obtained by multiplying $K(t, s)$ by $f(s)$ and then "summing" (i.e., integrating) as s ranges from t_0 to t. This can be interpreted as asserting that the response of the system at t to the input $f(s)$, $t_0 \leq s \leq t$, is the "superposition" (i.e., integral) of the responses at t to a continuum of impulsive inputs at s scaled by $f(s)$ as s ranges from t_0 to t.

PROBLEMS

1. Find the general real-valued solution of the following ODEs using Variation of Parameters. [*Hint*: Leave y_p as an integral if integration is difficult.]

    (a) $y'' + y = 1/\cos t$, $|t| < \pi/2$ (b) $4y'' + 4y' + y = te^{-t/2} \sin t$

    (c) $y'' + y' - 12y = (e^{2t} + 1)^2/e^{2t}$ (d) $y'' + 2y' + y = \sin 3t + te^{-t}$

    (e) $y'' + 4y = t^2 + 3t \cos 2t$

2. For which of the ODEs in Problem 1 does the Method of Annihilators fail? Give reasons.

3. Solve the IVP $y'' + y = f(t)$, $y(0) = y'(0) = 0$, using Variation of Parameters. [*Hint*: Your answer will have an integral in it.]

4. (*Euler Equation*). The ODE $t^2 y'' + pty' + qy = 0$, for $t > 0$ or $t < 0$, is called an *Euler equation* if p and q are constants.

    (a) Show that the Euler equation has the solution $y = t^r$, where r is a constant, if r solves the quadratic equation $r^2 + (p-1)r + q = 0$. If a root of this quadratic equation is the complex number $r = \alpha + i\beta$, then show that $y_1 = t^\alpha \cos(\beta \ln t)$ and $y_2 = t^\alpha \sin(\beta \ln t)$ are a fundamental set of solutions of the Euler equation. [*Hint*: Write $t^{\alpha + i\beta}$ as $e^{(\alpha + i\beta)\ln t}$.]

    (b) Searching for solutions on $t > 0$, show that the change of independent variable $t = e^s$ converts the Euler equation into

    $$\frac{d^2 y}{ds^2} + (p-1)\frac{dy}{ds} + qy = 0$$

    (c) Use the techniques presented in parts (a) or (b) to solve the IVP $t^2 y'' - 2ty' + 2y = 0$, $y(1) = 0$, $y'(1) = -1$. What is the largest interval on which this solution is defined?

    (d) Use the technique presented in part (b) to solve the IVP $t^2 y'' - 2ty' + 2y = t^2$, $y(1) = 0$, $y'(1) = 0$.

    (e) Use the method of Variation of Parameters to solve the IVP in part (d).

5. (*Green's Kernel for Constant Coefficient ODE*). Show that if a, b are real numbers and f is in $C^0(I)$, the real-valued general solution of $(D^2 + aD + b)[y] = f$ is

    $$y(t) = k_1 y_1(t) + k_2 y_2(t) + \int_{t_0}^t K(t, s) f(s) \, ds, \qquad t \text{ in } I$$

    where t_0 is any point in I, and the functions $y_1(t)$, $y_2(t)$ and the kernel $K(t, s)$ are determined as follows. Let r_1 and r_2 be the roots of the characteristic polynomial $r^2 + ar + b$.

    Case 1: $r_1 \neq r_2$, r_1 and r_2 real numbers

    $$y_1(t) = e^{r_1 t}, \quad y_2(t) = e^{r_2 t}, \quad K(t, s) = \frac{e^{r_2(t-s)} - e^{r_1(t-s)}}{r_2 - r_1}$$

    Case 2: $r_1 = r_2$ (and hence real)

    $$y_1(t) = e^{r_1 t}, \quad y_2(t) = te^{r_1 t}, \quad K(t, s) = (t - s)e^{r_1(t-s)}$$

    Case 3: $r_1 = \alpha + i\beta$, $r_2 = \alpha - i\beta$, $\beta \neq 0$

    $$y_1(t) = e^{\alpha t} \cos \beta t, \quad y_2(t) = e^{\alpha t} \sin \beta t, \quad K(t, s) = \frac{e^{\alpha(t-s)} \sin \beta(t - s)}{\beta}$$

6. How does the solution curve in Figure 3.6.1 from Example 3.6.2 behave as $t \to \frac{\pi}{2}^-$ and $t \to -\frac{\pi}{2}^+$?

7. (*Properties of Green's Kernel*). Let the function $K(t, s)$ be the kernel defined in (6) for the polynomial operator $P(D) = D^2 + a(t)D + b(t)$.

(**a**) What is the domain of $K(t, s)$ in the ts-plane?

(**b**) Prove that $K(t, t) = 0$ and $K_t(t, t) = 1$, for all t in I. [K_t denotes the partial derivative of $K(t, s)$ with respect to t.]

(**c**) Show that for each fixed s in I the function $y(t) = K(t, s)$ is a solution on I of the initial value problem $P(D)[y] = 0$, $y(s) = 0$, $y'(s) = 1$.

(**d**) Use part (**c**) and the uniqueness part of the EUT (Theorem 3.3.4) to show that $K(t, s)$ is the same for every fundamental set $\{y_1(t), y_2(t)\}$.

8. (*Forced Oscillations*). Consider the ODE $y'' + ay' + by = f(t)$, where a and b are constants and $f(t)$ is a periodic continuous function on $\mathbf{R}$ with period T. Show that if the ODE $y'' + ay' + by = 0$ does not have a nontrivial periodic solution with period T, then the driven ODE $y'' + ay' + by = f(t)$ has a unique periodic solution of period T, called the *forced oscillation*. [*Hint*: Follow the outline below.]

- Show that a solution $y(t)$ of the ODE is a periodic solution with period T if and only if $y(0) = y(T)$ and $y'(0) = y'(T)$. [*Hint*: To show that $y(t)$ repeats in $T \le t \le 2T$ when the conditions hold, replace t by $s + T$ in the ODE and start the clock at $s = 0$.]

- Let y_1, y_2 be solutions of the homogeneous ODE $y'' + ay' + by = 0$ with $y_1(0) = 1$, $y_1'(0) = 0$; $y_2(0) = 0$, $y_2'(0) = 1$. Suppose that $y_p(t)$ is the particular solution of the driven ODE $y'' + ay' + by = f(t)$ given by formulas (1) and (2) with $t_0 = 0$. Show that $y = y_0 y_1 + y_0' y_2 + y_p$ is the solution of the IVP

$$y'' + ay' + by = f(t) \qquad y(0) = y_0, \quad y'(0) = y_0'$$

- From before, we know that the conditions $y(0) = y(T)$, and $y'(0) = y'(T)$ applied to the solution $y(t)$ above guarantee $y(t)$ is periodic, and vice versa. Show that these conditions on y are equivalent to the matrix equation

$$(I - M(T)) \begin{bmatrix} y_0 \\ y_0' \end{bmatrix} = \begin{bmatrix} y_p(T) \\ y_p'(T) \end{bmatrix} \qquad (*)$$

where I is the matrix $\begin{bmatrix} 1 & 0 \\ 0 & 1 \end{bmatrix}$ and the matrix $M(T) = \begin{bmatrix} y_1(T) & y_2(T) \\ y_1'(T) & y_2'(T) \end{bmatrix}$

- Suppose that $I - M(T)$ is not invertible. Then there exists a nonzero two-vector V with components α and β such that $(I - M(t))V = 0$ (see Appendix C.6 on matrix theory), so

$$\begin{bmatrix} \alpha \\ \beta \end{bmatrix} = M(T) \begin{bmatrix} \alpha \\ \beta \end{bmatrix}$$

Show that $z(t) = \alpha y_1(t) + \beta y_2(t)$ is a nontrivial periodic solution of the homogeneous ODE, $y'' + ay' + by = 0$, contradicting the hypothesis of the theorem.

- The contradiction above to our hypothesis establishes that $I - M(T)$ is invertible. Show that the ODE $y'' + ay' + by = f(t)$ has a unique periodic solution with period T.

- Show that the above result holds even when $f(t)$ is a piecewise continuous periodic function, provided we allow the second derivative of a solution to be piecewise continuous.

- Show that the ODE, $y'' + 4y' + 3y = \text{sqw}(t, 50, 4)$ has a unique periodic solution of $T = 4$. Without actually calculating the y_0 and y_0' necessary to generate the periodic solution, try to produce a graph approximating this periodic solution. How do you know your graph stays close to the periodic solution for large t?

Applications of Second-Order Differential Equations

The models treated in this chapter lead to second-order ODEs, both linear and nonlinear, and involve motion in a force field or the flow of electrical energy in a circuit. Along the way we give a brief introduction to the vector concept in mechanics and to Kirchhoff's laws for circuits. Some important properties of driven linear ODEs are presented: beats, resonance, and frequency modeling. The principal application in mechanics is to the motion of a pendulum. The cover figure shows orbits of the ODE that models the motion of a simple damped pendulum.

4.1 Newton's Laws: The Pendulum

What principles will help us to model the motion of a falling body, the up and down vibrations of a body suspended by a spring, the oscillations of a pendulum, the course of the earth in its orbit around the sun? Principles are needed that relate these changes in the motion of a body to the action of its surroundings. Building on the work of Galileo, Isaac Newton saw to the heart of the matter and formulated three fundamental laws of motion which relate the acceleration of a material body to the "mass" of the body and the "resultant force" acting on that body. To do this, Newton in effect introduced the vector concept as a modeling device in order to express his Laws of Dynamics with an elegant simplicity that transcends any particular frame of reference or coordinates in these frames.

The Vector Concept in Mechanics

In introductory physics courses, students are introduced to the geometric vector approach in mechanics where basic physical concepts such as "velocity" and "acceleration" are thought of as directed line segments, or arrows (each with a "head" and a "tail"), in the familiar 3-space of Euclid. It is common to reinforce the idea that a symbol stands for a geometric vector by placing a small arrow over it, but for simplicity we shall drop the arrow and only use boldface letters in this section. Two vectors **v** and **w** are considered equivalent if and only if they can be made to coincide by translations (which preserve length and direction of vectors). Thus two parallel vectors of equal length, but directed oppositely, are not equivalent. Two equivalent vectors with different initial points can share the same name. The *length* (or *magnitude* or *norm*) of a vector **v** is denoted by $\|\mathbf{v}\|$. To be complete, we define the *zero vector*, denoted by **0**, to be the vector of zero length (note **0** has no direction).

The velocity and acceleration vectors of a material body each have the following property: If a body is known to have the velocities (or accelerations) **v** and **w** in two directions at the same time, then it has the resultant velocity (or acceleration), denoted by **v** + **w**. The resultant is defined by the "parallelogram law" as follows: Find a vector equivalent to **v** whose tail coincides with **w**'s head, then **v** + **w** is the vector whose tail is **w**'s and whose head is **v**'s. The vector **v** + **w** is a diagonal of the parallelogram formed by **v** and **w** (see the margin figure). This inspires the following definition:

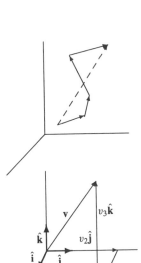

> ❖ **Basic Operations for Geometric Vectors.** Let **v** and **w** be any two geometric vectors and r any real number. The *sum* **v** + **w** is the vector produced by the parallelogram law. The *product* $r\mathbf{v}$ is the vector of length $|r|\,\|\mathbf{v}\|$ which points in the direction of **v** if $r > 0$, and in the opposite direction to **v** if $r < 0$.

Note that $r\mathbf{v} = \mathbf{0}$ if either $r = 0$, or $\mathbf{v} = \mathbf{0}$. Also note that $(+1)\mathbf{v} = \mathbf{v}$, and that $(-1)\mathbf{v}$ has the property that $\mathbf{v} + (-1)\mathbf{v} = \mathbf{0}$, for any **v**. The vector $(-1)\mathbf{v}$ is commonly denoted by $-\mathbf{v}$. Note that $-\mathbf{v}$ has the same length as **v** but points in the opposite direction. When a number of vectors are to be added together, the parallelogram law is used in succession—the order in which this is done can be shown to be immaterial (see the margin figure).

When geometric notions have served their purpose in constructing a model, the idea of coordinates for vectors is introduced to facilitate computation. A *coordinate frame* is a triple of vectors, denoted by $\{\hat{\mathbf{i}}, \hat{\mathbf{j}}, \hat{\mathbf{k}}\}$, which are mutually orthogonal and all of unit length. Now a little thought (and some trigonometry) reveals that every vector can be uniquely written as the sum of vectors parallel to $\hat{\mathbf{i}}, \hat{\mathbf{j}}$, and $\hat{\mathbf{k}}$. Thus for each vector **v** there is a unique set of real numbers v_1, v_2, and v_3 such that $\mathbf{v} = v_1\hat{\mathbf{i}} + v_2\hat{\mathbf{j}} + v_3\hat{\mathbf{k}}$. The elements of the ordered triple (v_1, v_2, v_3) are called the *coordinates* (or *components*) of **v** in the frame $\{\hat{\mathbf{i}}, \hat{\mathbf{j}}, \hat{\mathbf{k}}\}$ (see the margin figures). To put together the vector **v** from its coordinates in a given frame, one must have a way of associating each coordinate with the appropriate frame vector. This is normally done by writing coordinates in the same order of occurrence as vectors in the frame $\{\hat{\mathbf{i}}, \hat{\mathbf{j}}, \hat{\mathbf{k}}\}$.

Other computations with vectors are defined geometrically as follows:

- Two nonzero, nonparallel vectors **u** and **v** define a plane. In this plane two vectors equivalent to **u** and **v** can be found which have coincident "tails." The angle less than or equal to 180° between these vectors is said to be the *angle* between **u** and **v**. The angle between nonzero parallel vectors is either 0° or 180°.

- The *dot product* (or *scalar product*) of two vectors **u** and **v**, denoted by $\mathbf{u} \cdot \mathbf{v}$, is the real number $\|\mathbf{u}\| \, \|\mathbf{v}\| \cos\theta$ if neither $\mathbf{u}$ nor $\mathbf{v}$ is the zero vector and θ is the angle between them. When either $\mathbf{u}$ or $\mathbf{v}$ is the zero vector, the dot product is zero. Two nonzero vectors are *perpendicular* (or *orthogonal*) if and only if their dot product is zero. See Problems 1 and 2 for additional properties.

• If the vector $\mathbf{u} = \mathbf{u}(t)$ depends on t in an interval of the real axis, then the *derivative* $d\mathbf{u}/dt$ (or $\mathbf{u}'(t)$)[1] is defined in the usual way as the limit of a difference quotient:

$$\mathbf{u}'(t) = \frac{d\mathbf{u}}{dt} = \lim_{h \to 0} \frac{\mathbf{u}(t+h) - \mathbf{u}(t)}{h} \tag{1}$$

where the limit exists if the magnitude of the difference between the vectors $d\mathbf{u}/dt$ and $[\mathbf{u}(t+h) - \mathbf{u}(t)]/h$ tends to zero as $h \to 0$. If $\mathbf{u}$ is a constant vector, note that $\mathbf{u}' = \mathbf{0}$. If $\mathbf{u}(t)$ and $\mathbf{v}(t)$ are two differentiable vector functions and $r(t)$ is a real-valued differentiable function, then we have the identities

$$(\mathbf{u} \cdot \mathbf{v})' = \mathbf{u}' \cdot \mathbf{v} + \mathbf{u} \cdot \mathbf{v}', \qquad [r(t)\mathbf{u}(t)]' = r'\mathbf{u} + r\mathbf{u}'$$

which resembles the usual product differentiation rule.

There are many frames of reference in the three-dimensional space of our experience. One can imagine frames that move in space, or frames that are fixed. Suppose it is known that $\{\hat{\mathbf{i}}, \hat{\mathbf{j}}, \hat{\mathbf{k}}\}$ is a fixed frame and suppose that a particle moves in a manner described by the *position vector*

$$\mathbf{R} = \mathbf{R}(t) = x(t)\hat{\mathbf{i}} + y(t)\hat{\mathbf{j}} + z(t)\hat{\mathbf{k}}$$

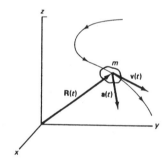

where the *xyz*-coordinates are defined relative to the fixed frame (see the margin figures). If $\mathbf{R}$ is differentiable,

$$\mathbf{R}'(t) = x'(t)\hat{\mathbf{i}} + y'(t)\hat{\mathbf{j}} + z'(t)\hat{\mathbf{k}} \tag{2}$$

where the primes indicate time derivatives. $\mathbf{R}'(t) = \mathbf{v}(t)$ is the *velocity vector* of the particle at time t, and $\mathbf{v}(t)$ is tangential to the path of the particle's motion at the point $\mathbf{R}(t)$. Furthermore, if $\mathbf{R}'(t)$ is differentiable,

$$\mathbf{R}''(t) = \mathbf{v}'(t) = \mathbf{a}(t) = x''(t)\hat{\mathbf{i}} + y''(t)\hat{\mathbf{j}} + z''(t)\hat{\mathbf{k}} \tag{3}$$

is the *acceleration vector* for the particle (see the margin figure). The study of the velocity and acceleration vectors in given coordinate frames when the path of the motion is known is called *kinematics*.

Forces, Newton's Laws

Newton developed Galileo's central idea that the environment creates "forces" which act on bodies causing them to accelerate. Galileo's principle (formulated also by Newton as his *First Law*) states that

---

Newton's First Law. A body remains in a state of rest or of uniform motion in a straight line if there is no net external force acting on it.

---

[1] In the physics and engineering literature it is conventional to use the "dot" notation, $\dot{\mathbf{u}}$, instead of the "prime" notation, $\mathbf{u}'$, to denote the derivative of the vector function of time, $\mathbf{u}(t)$. For simplicity, however, we will always use the prime notation for derivative of a function of one variable.

Forces can be measured (or computed, as we shall see) without taking into account frames of reference. Scientists of Newton's time were able to show experimentally that forces behave like geometric vectors (i.e., that they satisfy the parallelogram law).

The effect of Newton's First Law is to identify frames of reference which are either fixed in space or undergoing a translation at constant velocity with respect to a fixed frame. Such frames are called *inertial*. Practically speaking, how do we know when we are dealing with an inertial frame? According to Newton's First Law, a frame is inertial if and only if a body is unaccelerated with respect to that frame whenever the vector sum of all the forces acting on the body vanishes.

Next, we have the following basic principle in dynamics:

Newton's Second Law. For a body in any inertial frame we have that

$$m\mathbf{a} = \mathbf{F} \qquad (4)$$

where m denotes the mass (assumed to be constant), $\mathbf{a}$ is the acceleration of the body, and $\mathbf{F}$ is the resultant (i.e., the vector sum) of all external forces acting on the body.[2]

The terms in Newton's Second Law must be quantified in some system of units. In the *MKS system* (meter, kilogram, second), also called the *SI system*, the unit of mass is the kilogram; (4) is then used to define one unit of force as that amount of force which will accelerate a 1-kg mass at 1 m/sec^2. This unit of force is called a *newton*. In the *CGS system* (centimeter, gram, second) the unit of force is a *dyne* (1 dyne = 10^{-5} newton) and is that force which accelerates a 1-g mass at 1 cm/sec^2. In these metric units, mass, length, and time are considered to be fundamental quantities, while force is secondary. In this text we usually avoid the use of any particular system of units (see also Appendix C.7). When units must be used, they will be explicitly labeled.

Newton's Third Law gives some insight on how to treat multiple body systems.

Newton's Third Law. If body A exerts a force $\mathbf{F}$ on body B, then body B exerts a force $-\mathbf{F}$ on body A.

Since acceleration is defined to be the rate of change of velocity, Newton's Second Law can be written in the form

$$m\mathbf{a} = m\frac{d\mathbf{v}}{dt} = \frac{d(m\mathbf{v})}{dt} = \mathbf{F} \qquad (5)$$

where v is the velocity of the body and, once again, the mass is assumed to be constant. The product $m\mathbf{v}$ is called the *momentum* of the body. Newton himself expressed the Second Law in terms of the rate of change of the "quantity of motion," his term for momentum. Indeed, $(m\mathbf{v})' = \mathbf{F}$ holds whether or not the mass is constant. Thus Newton's Second Law states that "the rate of change of the momentum of a body equals the resultant external force acting on the body." To give this simply stated principle some operational content, we would need to specify ways of calculating the "momentum"

---

[2]This form (4) of Newton's Second Law is for so-called "point" masses. For some "distributed" masses it is permissible to apply (4) as if all the mass of the body were concentrated at its center of mass.

of a body and the "external force" acting on a body in an inertial frame of reference. These techniques form the central core of the field known as *dynamics*, and we shall give a brief introduction to this, the oldest branch of physics. As we shall see, a major consideration in the application of Newton's Second Law is the choice of a frame of reference and a coordinate system within that frame.

Nothing can be done to solve the first-order differential equation (5) and find the velocity and, eventually, the position of a body as functions of time until some functional form is given to **F**. This, of course, depends on the nature of the forces acting on the body—gravitational or electromagnetic forces, for example. Many other forces, such as spring forces, tensions in cables, and the normal force exerted by surfaces in contact with a body, are actually electromagnetic forces at a fundamental level. Nuclear forces, on the other hand, are so short-ranged that they cannot be used in Newton's Laws (quantum mechanics must be used for these forces).

Newton's Second Law is applied below to the motion of a pendulum.

The Simple Pendulum

The *simple pendulum* consists of a bob of mass m hanging on a (presumed massless) rigid rod of fixed length L firmly attached to a horizontal support. The pendulum is in equilibrium when the bob and rod are aligned with the local vertical and are at rest. If the pivot is positioned at the origin of a Cartesian frame (the standard xyz-coordinate system) and the center of mass of the bob is in the xy-plane (see the margin figure), the bob will oscillate back and forth in the xy-plane when it is released with an initial velocity in that plane (see Problem 10). Thus the polar angle θ is sufficient to track the motion of the bob in this case. Making this assumption now, we have converted a three-dimensional problem into a two-dimensional one.

Let us assume that the moving pendulum experiences a viscous damping force **f** proportional to its linear velocity and opposing the motion, and that the other forces acting on the bob are gravity and the tension **T** in the rod (see the margin figure). Now to apply Newton's Second Law we shall derive an expression for the acceleration of the bob relative to the inertial $\hat{\mathbf{x}}\hat{\mathbf{y}}$-frame, compute the sum of the external forces acting on the bob, and then substitute these quantities into Newton's Second Law. This procedure is not as easy to carry out as it sounds if one proceeds in a mindless way. First, it may not be so easy to compute the acceleration and external force vectors in terms of Cartesian coordinates in the $\hat{\mathbf{x}}\hat{\mathbf{y}}$-frame, [3] and before the problem can be solved it must be converted to a statement about the components of these vectors in some frame of reference. Any frame will do for this purpose (the virtue of vectors!), and hence, the original $\hat{\mathbf{x}}\hat{\mathbf{y}}$-frame need not be used. The real skill of the modeler comes through in the choice of a frame of reference in which to express Newton's Second Law—a poor choice could present some formidable technical difficulties.

To derive the equation of motion consider the position vector **R** from the support of the pendulum to the center of the bob. Note that $\mathbf{R}(t)$ tracks the motion of the center of the bob and that the vectors $\mathbf{R}'(t)$ and $\mathbf{R}''(t)$ are the velocity **v** and acceleration **a**, respectively, for this motion. Introducing polar coordinates in the plane of oscillation with the origin at the support and θ measured positively in the counterclockwise direction from the downward position, we can write

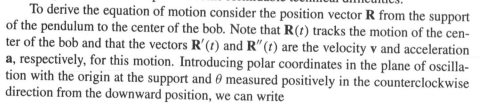

$$\mathbf{R}(t) = r(t)\hat{\mathbf{r}}(t)$$

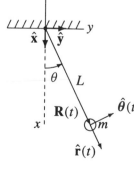

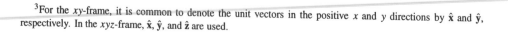

---

[3]For the xy-frame, it is common to denote the unit vectors in the positive x and y directions by $\hat{\mathbf{x}}$ and $\hat{\mathbf{y}}$, respectively. In the xyz-frame, $\hat{\mathbf{x}}$, $\hat{\mathbf{y}}$, and $\hat{\mathbf{z}}$ are used.

where $r(t)$ is the length of $\mathbf{R}(t)$ and $\hat{\mathbf{r}}(t)$ is a unit vector directed along $\mathbf{R}(t)$. Note that $r(t) = L$ since the rod has fixed length. Denoting by $\hat{\boldsymbol{\theta}}(t)$ the unit vector orthogonal to $\mathbf{R}(t)$ and pointing in the direction of increasing θ, we can calculate $\mathbf{R}'(t)$ and $\mathbf{R}''(t)$ in terms of $\hat{\mathbf{r}}(t)$ and $\hat{\boldsymbol{\theta}}(t)$ as follows. First note that the unit vectors $\hat{\mathbf{r}}$, $\hat{\boldsymbol{\theta}}$ are related to the Cartesian unit vectors $\hat{\mathbf{x}}$, $\hat{\mathbf{y}}$ [and $\hat{\mathbf{x}}$, $\hat{\mathbf{y}}$ to $\hat{\mathbf{r}}$, $\hat{\boldsymbol{\theta}}$] by the equations

$$\hat{\mathbf{r}} = \cos\theta\hat{\mathbf{x}} + \sin\theta\hat{\mathbf{y}}, \quad \hat{\boldsymbol{\theta}} = -\sin\theta\hat{\mathbf{x}} + \cos\theta\hat{\mathbf{y}}$$
$$\hat{\mathbf{x}} = \cos\theta\hat{\mathbf{r}} - \sin\theta\hat{\boldsymbol{\theta}}, \quad \hat{\mathbf{y}} = \sin\theta\hat{\mathbf{r}} + \cos\theta\hat{\boldsymbol{\theta}} \tag{6}$$

Using (6) and the differentiation rule (2) for vectors expressed in an inertial frame, we see that

$$\frac{d(\hat{\mathbf{r}})}{dt} = (-\sin\theta\hat{\mathbf{x}} + \cos\theta\hat{\mathbf{y}})\theta' = \theta'\hat{\boldsymbol{\theta}}$$

$$\frac{d(\hat{\boldsymbol{\theta}})}{dt} = (-\cos\theta\hat{\mathbf{x}} - \sin\theta\hat{\mathbf{y}})\theta' = -\theta'\hat{\mathbf{r}}$$

Hence using the fact that $r(t) = L$ we see that

$$\mathbf{R}' = (L\hat{\mathbf{r}})' = L\theta'\hat{\boldsymbol{\theta}}, \qquad \mathbf{R}'' = (L\theta'\hat{\boldsymbol{\theta}})' = L\theta''\hat{\boldsymbol{\theta}} - L(\theta')^2\hat{\mathbf{r}} \tag{7}$$

Now for external forces note that

$$\mathbf{T} = -T\hat{\mathbf{r}} \quad \text{for some scalar } T > 0$$
$$\mathbf{f} = -c\mathbf{v} = -c\mathbf{R}' = -cL\theta'\hat{\boldsymbol{\theta}} \quad \text{for some constant } c > 0 \tag{8}$$
$$mg\hat{\mathbf{x}} = mg(\cos\theta\hat{\mathbf{r}} - \sin\theta\hat{\boldsymbol{\theta}}) = \text{force of gravity on bob}$$

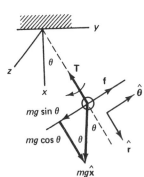

where we have used (6) and (7) to obtain expressions for $\mathbf{v}$ and $\hat{\mathbf{x}}$ in terms of θ and r. See the margin figure for a geometric visualization of these forces. Now substituting (7) and (8) into Newton's Second Law

$$m\mathbf{R}'' = \mathbf{T} + \mathbf{f} + m\hat{\mathbf{x}}$$

and equating coefficients of $\hat{\mathbf{r}}$ and $\hat{\boldsymbol{\theta}}$ on both sides of the equation we obtain

$$-mL(\theta')^2 = -T + mg\cos\theta \qquad \text{(from } \hat{\mathbf{r}}\text{)}$$
$$mL\theta'' = -cL\theta' - mg\sin\theta \qquad \text{(from } \hat{\boldsymbol{\theta}}\text{)} \tag{9}$$

If the pendulum were driven by a given external force $\mathbf{F}$ of the form $F(t)\hat{\boldsymbol{\theta}}$, the second equation in (9) would be

$$mL\theta'' + cL\theta' + mg\sin\theta = F(t) \tag{10}$$

called the *equation of the driven damped simple pendulum*. Although the polar angle θ is usually restricted to the range $0 \leq \theta \leq 2\pi$, or to $|\theta| \leq \pi$, we make no such restriction here. Indeed, we want to consider situations in which the pendulum is given so much energy that it rotates end-over-end about its pivot (counterclockwise, say), and the only way this motion can be recorded is to allow the angular variable θ to increase or decrease without bound.

EXAMPLE 4.1.1

Orbits of an Undamped Simple Pendulum

The ordinary differential equation (ODE) (10) with $c = 0$, $g/L = 4$, and $F = 0$ describes the motion of an undamped, undriven simple pendulum. It is interesting to examine the orbits in the state space (i.e., $\theta\theta'$-space) of the corresponding ODE,

$$\theta'' + 4\sin\theta = 0$$

Figure 4.1.1 shows some typical orbits. The equilibrium solutions $\theta = 2n\pi$, $\theta' = 0$, all t, where n is any integer, correspond to the pendulum hanging at rest straight down, and give us points in $\theta\theta'$-state space (called appropriately *equilibrium points*). Surrounding

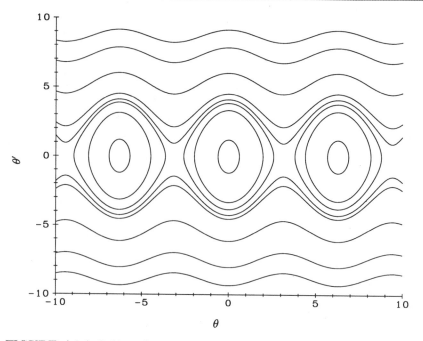

FIGURE 4.1.1 Orbits of an undamped simple pendulum. See Example 4.1.1.

each of these equilibrium points is a region of closed orbits which, since the ODE is autonomous, correspond to periodic solutions of the ODE. The periodic solutions are nothing but the familiar back-and-forth motions of the pendulum about a downward equilibrium position.

If the pendulum starts out with high enough θ_0', the pendulum will forever tumble end-over-end about its pivot. The orbits are the wavy lines above and below the regions of periodic solutions. More difficult to see are the equilibrium points $\theta = (2n + 1)\pi$, $\theta' = 0$, all t, where n is an integer; these points correspond to the pendulum at rest and perfectly balanced over its pivot. Finally, the orbits most difficult of all to see are those that separate the periodic solutions from the tumbling solutions. These orbits tend to the equilibrium points $\theta = (2n + 1)\pi$, $\theta' = 0$ as $t \to +\infty$, and $\theta = (2n - 1)\pi$, $\theta' = 0$ as $t \to -\infty$, from above the θ-axis (or the other way around from below the θ-axis). The corresponding orbits are called *separatrices* and represent the pendulum moving in one direction toward the upended position, but never arriving. The ODE is autonomous, and its orbits never intersect.

The above treatment of the orbits of $\theta'' + 4 \sin \theta = 0$ is based on an examination of the graphs in Figure 4.1.1, graphs which were generated by a numerical solver. Computer graphics suggest, but cannot prove. So how do we know that the properties ascribed above to the orbits of the ODE are real? Fortunately, these properties can be proved mathematically; those proofs are outlined in Problems 14–16.

The Linearized Pendulum

Notice that ODE (10) is a nonlinear second-order ODE in the variable θ. For "small" motions, ODE (10) is closely approximated by a linear ODE. Indeed, Taylor's Theorem implies that $\sin \theta$ is approximately θ for small $|\theta|$. Hence for motions where $|\theta|$ is small, the term $\sin\theta$ in ODE (10) may be replaced by θ without appreciably affecting its solutions over limited time periods. When this approximation is made we have the equation of the *driven linearized pendulum with damping*,

$$mL\theta'' + cL\theta' + mg\theta = F(t) \tag{11}$$

Note the similarity of ODE (11) with ODE (7) in Section 3.1 (for the driven damped Hooke's Law spring). Thus we see again the power of the modeling approach: Even though the linearized pendulum is different physical system from the Hooke's Law spring-body system considered earlier, both have ODE models of the same form. Thus, the methods of Chapter 3 give a satisfactory explanation of the motions of both systems.

Comments

Just as in Section 3.1 where the second-order ODE modeling the motion of a spring was also written as an equivalent system of two first-order ODEs, we can do the same thing for the ODE (10) of the driven and damped simple pendulum. In terms of state variables θ and $\theta' = v$, we have the system

$$\begin{aligned} \theta' &= v \\ v' &= -\frac{g}{L}\sin\theta - \frac{c}{m}v + \frac{1}{mL}F(t) \end{aligned} \tag{12}$$

It is often convenient to use system (12) rather than ODE (10) to study the motion of the pendulum. Indeed, both the chapter cover figure and Figure 4.1.1 are most easily interpreted in terms of systems having the form of (12). See also Problem 13.

PROBLEMS

1. (*Coordinate Version of the Dot Product*). Let $\{\hat{\mathbf{i}}, \hat{\mathbf{j}}, \hat{\mathbf{k}}\}$ be a fixed frame in Euclidean 3-space. Then every vector $\mathbf{u}$ can be written uniquely as a sum $u_1\hat{\mathbf{i}} + u_2\hat{\mathbf{j}} + u_3\hat{\mathbf{k}}$, where the scalars are known as the *coordinates* of $\mathbf{u}$ in the given frame. Care should be taken to use the subscript 1 always to denote $\hat{\mathbf{i}}$-coordinates, the subscript 2 for $\hat{\mathbf{j}}$-coordinates, and so on. Thus, when coordinates are listed as an ordered triple (u_1, u_2, u_3), the vector $\mathbf{u}$ can be recovered from its coordinates.

    (a) Show that $\|\mathbf{u}\|^2 = u_1^2 + u_2^2 + u_3^2$, for any vector $\mathbf{u}$.

    (b) Show that $\mathbf{u} \cdot \mathbf{v} = u_1v_1 + u_2v_2 + u_3v_3$, for any vectors $\mathbf{u}$ and $\mathbf{v}$. [*Hint*: Imagine $\mathbf{u}$ and $\mathbf{v}$ to have their tails at the origin, then use the Cosine Law (see Appendix C.4), the definition of $\mathbf{u} \cdot \mathbf{v}$, and part (a).]
    [*Remark*: Use parts (a) and (b) to calculate $\|\mathbf{u}\|$ and $\mathbf{u} \cdot \mathbf{v}$ in *any* reference frame.]

2. (*Properties of the Dot Product*). Verify that the dot product $\mathbf{u} \cdot \mathbf{v}$ satisfies the following properties. [*Hint*: Use the geometric definition of $\mathbf{u} \cdot \mathbf{v}$ or the coordinate version in Problem 1.]

    (a) (*Symmetry*). $\mathbf{u} \cdot \mathbf{v} = \mathbf{v} \cdot \mathbf{u}$ for all $\mathbf{u}, \mathbf{v}$.

    (b) (*Bilinearity*). $(\alpha\mathbf{u} + \beta\mathbf{w}) \cdot \mathbf{v} = \alpha\mathbf{u} \cdot \mathbf{v} + \beta\mathbf{w} \cdot \mathbf{v}$ for all scalars α, β, and vectors $\mathbf{u}, \mathbf{v}$, and $\mathbf{w}$.

    (c) (*Positive Definiteness*). $\mathbf{u} \cdot \mathbf{u} \geq 0$ for all $\mathbf{u}$; $\mathbf{u} \cdot \mathbf{u} = 0$ if and only if $\mathbf{u} = 0$.

    (d) (*Length*). $\mathbf{u} \cdot \mathbf{u} = \|\mathbf{u}\|^2$ for all $\mathbf{u}$.

    (e) (*Parallelogram Property*). $\|\mathbf{u} + \mathbf{v}\|^2 + \|\mathbf{u} - \mathbf{v}\|^2 = 2\|\mathbf{u}\|^2 + 2\|\mathbf{v}\|^2$ for all $\mathbf{u}, \mathbf{v}$. Why is this called the Parallelogram Property?

(f) *(Cauchy-Schwarz Inequality)* . $|\mathbf{u} \cdot \mathbf{v}| \leq \|\mathbf{u}\| \, \|\mathbf{v}\|$ for all vectors $\mathbf{u}$ and $\mathbf{v}$.

(g) *(Difference of Squares of Diagonals)*. $\mathbf{u} \cdot \mathbf{v} = \frac{1}{4}(\|\mathbf{u} + \mathbf{v}\|^2 - \|\mathbf{u} - \mathbf{v}\|^2)$ for all $\mathbf{u}$, $\mathbf{v}$. Note that $\|u + v\|$ and $\|u - v\|$ are the lengths of the two diagonals of the parallelogram defined by $\mathbf{u}$ and $\mathbf{v}$.

(h) Show that if $\mathbf{u} + \mathbf{v} + \mathbf{w} = 0$, and if each vector is orthogonal to the other two, then $\mathbf{u} = \mathbf{v} = \mathbf{w} = 0$.

3.   *(Orthogonality)*. Let $\{\hat{\mathbf{i}}, \hat{\mathbf{j}}, \hat{\mathbf{k}}\}$ be a reference frame, and consider the two vectors

$$\mathbf{u} = 3\hat{\mathbf{i}} - \hat{\mathbf{j}} + 2\hat{\mathbf{k}}, \quad \mathbf{v} = 2\hat{\mathbf{i}} + \hat{\mathbf{j}} - \hat{\mathbf{k}}$$

(a) Determine the conditions on the coordinates of $\mathbf{w} = w_1\hat{\mathbf{i}} + w_2\hat{\mathbf{j}} + w_3\hat{\mathbf{k}}$ which guarantee that $\mathbf{w}$ is orthogonal to $\mathbf{u}$. Find conditions on w_1, w_2, w_3 that imply that $\mathbf{w}$ lies in the plane determined by $\mathbf{u}$ and $\mathbf{v}$. [*Hint*: $\mathbf{w}$ lies in the plane if and only if $\mathbf{w}$ is a linear combination of $\mathbf{u}$ and $\mathbf{v}$.]

(b) Find a scalar α such that the vector $\mathbf{w} = \mathbf{u} - \alpha\mathbf{v}$ is orthogonal to $\mathbf{v}$.

4.   *(Vector Addition)*. A small airplane, during a part of its flight that begins at a point P in space, flies 20 mi due south, then turns left 90° and goes into a climb 8 mi long at an angle of 10° with the horizontal, turns left again, and flies horizontally 42 mi due north. Let $\hat{\mathbf{i}}, \hat{\mathbf{j}}, \hat{\mathbf{k}}$ point north, west, and upward from P. What is the final position of the airplane relative to P?

5.   *(Vector Components)*. An aircraft is observed by radar to be 80 mi away at 60°10′ east of north and 83°40′ down from the local vertical. Find the northward, eastward, and vertical components of its positions.

6.   *(Gravitational Attraction)*. Two particles of respective masses $m_1 = 10$ grams and $m_2 = 40$ grams are located in a rectangular coordinate system with length measured in centimeters. The location of the first particle is $(4, 0, 0)$, and the location of the second particle is $(-4, 0, 0)$.

(a) Find the rectangular components of the local gravitational force exerted by the two particles on a 1.0-gram particle placed at the point $(-5, 0, 0)$; at the point $(0, 0, 6)$. [*Hint*: See Section 1.8 for the gravitational law and the constant G.]

(b) At what point on the x-axis is the force on a 1.0 gram particle zero?

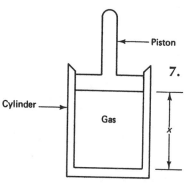

7.   *(Ideal Gas Law)*. According to the Ideal Gas Law, the pressure P, volume V, temperature T, and the number of moles (1 mol= 6.02×10^{23} molecules) n of a gas in a closed container satisfy the equation $PV = nRT$, where R is a universal constant. Now suppose that we have a cylinder containing an ideal gas with a piston of mass m on top (see the diagram). Assume that the only forces acting on the piston are gravity and gas pressure. Find the differential equation for the position of the piston measured from the bottom of the cylinder. Assume that T is constant. Note that P is defined to be the magnitude of the gas pressure per unit area of the piston.

8.   *(Creeping Bugs)*. Four bugs are at the corners of a square table whose sides are of length a. The bugs begin to move at the same instant, each crawling at the same constant speed directly toward the bug on its right. Find the path of each bug. Do the bugs ever meet? If so, when? [*Hint*: Let $\mathbf{R}(t) = r(t)\hat{\mathbf{r}}$ point from the center of the table to one of the bugs, where $r(0) = a/\sqrt{2}$ and $\theta(0) = 0$. Explain why the velocity vector $\mathbf{R}'(t)$ makes a 45° angle with $\mathbf{R}(t)$ for $t \geq 0$.]

9.   *(The Smugglers and the Coast Guard)*. A Coast Guard boat is hunting a smuggler's launch in a dense fog. The fog lifts momentarily and the the smugglers are spotted 4 mi away. The fog returns and the smugglers flee along a straight path in

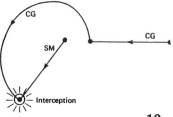

an unknown direction. If the Coast Guard boat has a top speed of three times that of the smuggler's launch, what path should the boat follow for the best chance to intercept the smugglers? Explain your model and sketch the path. [*Hint*: Initially the Coast Guard boat travels directly toward the sighting for 3 mi. Use polar coordinates to describe the path of the Coast Guard boat after the first 3 mi (see the sketch). Assume that both boats operate at top speed.]

10. (*Planar Pendulum Motion*). Referring to the margin figure, let $\mathbf{R}(t)$ denote the radius vector from the pendulum support to the bob and suppose that at $t = 0$, $\hat{\mathbf{z}} \cdot \mathbf{R}(0) = 0$ and $\hat{\mathbf{z}} \cdot \mathbf{R}'(0) = 0$. Show that for $t \geq 0$ the bob will oscillate back and forth in the xy-plane [i.e., $\hat{\mathbf{z}} \cdot \mathbf{R}(t) = 0$ for all t]. [*Hint*: Show that $w(t) = \hat{\mathbf{z}} \cdot \mathbf{R}(t)$ satisfies a homogeneous linear differential equation and has vanishing initial data.]

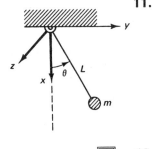

11. (*Linearized Pendulum*). Consider a simple, undamped, linearized and undriven pendulum described by ODE (11) (with $c = 0$ and $F = 0$).

(a) Find the period T of the pendulum in terms of its length L and the earth's gravitational constant g.

(b) Find the length of a pendulum whose period is exactly 1 sec.

(c) If the pendulum is 1 meter long and swings with an amplitude of 1 radian, compute the angular velocity of the pendulum at its lowest point. Then find the accelerations at the ends of each swing.

12. (*Chapter Cover Picture; Damped Pendulum*). The ODE for the simple undriven pendulum with viscous damping is $mL\theta'' + cL\theta' + mg\sin\theta = 0$.

(a) Set $m = 1$, $c = 0$, $g/L = 10$ and plot a portrait of the orbits of the equivalent system in θ, θ' state variables, $\theta' = v$, $v' = -10\sin\theta$. Use the screen size, $|\theta| \leq 15$, $|\theta'| \leq 15$. [*Hint*: See Figure 4.1.1.]

(b) Now set $c = 1$ so that the ODE becomes: $\theta'' + \theta' + 10\sin\theta = 0$. Plot a portrait of the orbits of the equivalent system $\theta' = v$, $v' = -10\sin\theta - v$, on a screen defined by $|\theta| \leq 15$, $|\theta'| \leq 15$. [*Hint*: See the chapter cover figure.]

(c) Compare the portrait in part (b) with that in part (a). Explain what you see.

13. (*Variable-Length Pendulum*). Show that if the length of the pendulum is a function of time $L(t)$, then the equation of motion of the pendulum is

$$mL\theta'' + (2mL' + cL)\theta' + mg\sin\theta = F$$

[*Hint*: Use $\mathbf{R}' = (L\hat{\mathbf{r}})'$, but do not assume L constant.]

14. (*Orbits of Undriven, Undamped Simple Pendulum*). The ODE of the orbits of the undriven, undamped simple pendulum may be derived directly from the ODE, $mL\theta'' + mg\sin\theta = 0$.

(a) Multiply each side of the ODE by θ' and observe that

$$mL\theta'\theta'' + mg\theta'\sin\theta = \left[\frac{1}{2}mL(\theta')^2 - mg\cos\theta\right]'$$

Show that $(1/2)mL(\theta')^2 - mg\cos\theta = c$, where c is a constant; these are the equations of the orbits in the $\theta\theta'$-plane. Let $g/L = 4$ and try to reproduce Figure 4.1.1 by plotting orbits for several values of c.

(b) (*Conservation of Energy*). The *kinetic energy* of the moving pendulum is $(1/2)m(L\theta')^2$ and the *potential energy of position* is $-mgL(1 - \cos\theta)$. Show that, given θ and θ' at time 0, the total energy $E(t)$ at time t is the same as $E(0)$. Explain why $E(t) = E(0)$ is the equation of an orbit.

15. (*Periods of the Pendulum*). Consider the undriven, undamped simple pendulum ODE, $\theta'' = -4\sin\theta$, some of whose orbits are plotted in Figure 4.1.1

- Reproduce Figure 4.1.1. Plot solution curves of θ vs. t.

- Make a table of the periods of the periodic solutions surrounding the equilibrium point $\theta_0 = 0$, $\theta_0' = 0$. Describe the behavior of these periods as the orbits approach the equilibrium point, and as orbits approach the separatrices.

- Linearize the ODE at the equilibrium point, $\theta_0 = 0$, $\theta_0' = 0$, and produce a state-space portrait of orbits near that equilibrium point. Is there a separatrix? Find the period of any periodic orbit, and describe how these periods change.

- Numerically solve the nonlinear and linearized ODEs for a variety of initial conditions, and compare the solution graphs and orbits.

16. (*Closed Orbits of an Undriven, Undamped Simple Pendulum*).The ODEs (10) and (11) with $c = 0$, $F(t) = 0$, model the motion of an undriven, undamped simple pendulum, and of a linearized pendulum, respectively. State portraits of each ODE show a region of closed orbits encircling the origin (see Figure 4.1.1, for example). These closed orbits (or *cycles*) correspond to periodic solutions because the ODEs are autonomous. Find and compare the periods of the cycles for the simple and for the linearized pendulum. Follow the outline below in proving the existence of cycles and in studying the periods. [*Hint*: See also Problems 14 and 15.]

- (*Linearized Pendulum*). The equation of the linearized pendulum is $mL\theta'' + mg\theta = 0$. Show that all nonconstant periods are periodic of period $2\pi\sqrt{L/g}$. Show that the corresponding orbits in the $\theta\theta'$-state space are elliptical cycles of the same period. Choose a value for L, plot a portrait of cycles in the state space, plot component graphs and verify graphically the formula for the period.

- (*Closed Orbits of Simple (Nonlinear) Pendulum*). Suppose the simple pendulum (modeled by $mL\theta'' + mg\sin\theta = 0$) is released from rest when $\theta = \theta_0$, where $0 < \theta_0 < \pi$. Show that the subsequent motion is periodic. [*Hint*: Use Problem 14 to show that orbits are described implicitly by the relation $(\theta')^2 = (2g/L)(\cos\theta - \cos\theta_0)$, and use symmetries in this relation to show that orbits are closed, and hence represent periodic solutions.]

- (*Periods of Simple Pendulum*). Let T be the period of the orbit with $\theta(0) = \theta_0$, $\theta'(0) = 0$. Show that T is given by

$$T = 4\sqrt{\frac{L}{2g}} \int_0^{\theta_0} \frac{d\theta}{\sqrt{\cos\theta - \cos\theta_0}}$$

  The integral is improper since the integrand becomes infinite at $\theta = \theta_0$, but it is known that T is finite if $0 < \theta_0 < \pi$. [*Hint*: First show that $\theta(t)$ initially decreases as t increases, and so $\theta' = -(2g/L)^{1/2}(\cos\theta - \cos\theta_0)$. Note that θ continues to decrease until the time $t = S$ for which $\theta(S) = -\theta_0$.]

- (*Elliptic Integrals and the Periods of the Simple Pendulum*). Show that the change of variables $k = \sin(\theta_0/2)$, $\sin\phi = (1/k)\sin(\theta/2)$, give

$$T = 4\sqrt{\frac{L}{g}} \int_0^{\pi/2} \frac{d\phi}{\sqrt{1 - k^2\sin^2\phi}}$$

The integral is called an *elliptic integral of the first kind*. Its approximate values have been tabulated for various values of k.[4] For example, if $\theta_0 = 2\pi/3$, then $k = \sqrt{3}/2$, and the value of the integral is ≈ 2.157. The corresponding period is $\approx 8.626\sqrt{L/g}$, quite different from the period of the linearized pendulum, which is $2\pi/\sqrt{L/g} \approx 6.282\sqrt{L/g}$. Why would you expect the period of the nonlinear pendulum to be greater than the period of the linearized pendulum? Choose various values for L and use an ODE solver to verify the above estimate for the periods.

- (*Asymptotic Values of the Periods of the Simple Pendulum*). Argue that $T \to 2\pi\sqrt{L/g}$ as $\theta_0 \to 0$. Why would one expect this to be true? It is known that $T \to \infty$ as $\theta_0 \to \pi$, although a complete mathematical proof of this fact is not given here. Why is this result expected on physical grounds? [*Hint*: Think of the motion of a pendulum whose initial angle is almost π.]

4.2 Forced and Free Oscillations

In Section 3.1 we found that the vertical motion of a body attached to a Hooke's Law spring satisfies the ODE

$$my'' + cy' + ky = f(t) \tag{1}$$

where y measures displacement from static equilibrium, m is the mass of the body, c the damping coefficient, k the coefficient in the linear spring force, and $f(t)$ the external force acting on the weight. In Section 4.1 we found that the motion of the linearized pendulum also satisfies a differential equation that has the same form as ODE (1), with the constants m, c, k, and the external force appropriately identified. Thus, the techniques of Sections 3.4, 3.5, and 3.6 can be applied to show that all possible motions of the body at the end of a spring (or oscillations of a linearized pendulum) are determined when all solutions of the homogeneous ODE $my'' + cy' + ky = 0$ (the *free solutions*) and a single particular solution (a *forced solution*) of ODE (1) are known.

Free Oscillations: Periodic and Damped

Free solutions behave differently when damping is present than when it is not. In the undamped case free solutions satisfy the *harmonic oscillator* equation

$$my'' + ky = 0 \tag{2}$$

In Example 3.4.5 of Section 3.4, all solutions of (2) are found to be the sinusoids

$$y(t) = A\cos(\omega_0 t + \delta)$$

where A and δ are arbitrary real numbers, and $\omega_0 = \sqrt{k/m}$; ω_0 is called the *natural circular frequency* of the harmonic oscillator. Thus motions described by ODE (2) are periodic with circular frequency ω_0, and persist forever. Such motion is called *simple harmonic motion* or a *free periodic oscillation*.

---

[4]M. Abramowitz and I.A. Stegun, eds, *Handbook of Mathematical Functions*, Washington D.C.: National Bureau of Standards, 1964; Dover (reprint), New York, 1965.

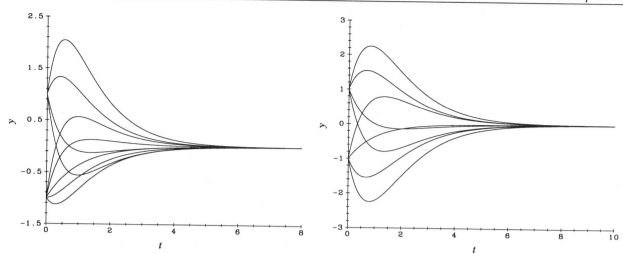

FIGURE 4.2.1 Solution curves for $y'' + 3y' + 2y = 0$ (overdamping).

FIGURE 4.2.2 Solution curves for $y'' + 2y' + y = 0$ (critical damping).

Of course, no physical spring or linearized pendulum can sustain simple harmonic motion since some damping is always present because of frictional forces. Free solutions in the damped case

$$my'' + cy' + ky = 0 \tag{3}$$

can be found by computing the roots r_1 and r_2 of the characteristic polynomial $P(r) = mr^2 + cr + k$. Using the quadratic formula we have

$$r_1, r_2 = \left(-c \pm \sqrt{c^2 - 4mk}\right)/2m \tag{4}$$

Since m, c, and k are all positive real numbers, we see that the real parts of r_1 and r_2 are always negative, whatever the actual values of m, c, and k. Thus all damped free solutions are *transients* (i.e., die out with increasing time), but their exact nature depends on the relative sizes of the constants m, c, and k.

There are three cases:

- *Overdamped.* If $c^2 > 4mk$, then r_1 and r_2 are real, negative, and distinct. Every solution of ODE (3) is a linear combination of the decaying real exponentials $\exp[r_1 t]$ and $\exp[r_2 t]$. See Figure 4.2.1.

- *Critically damped.* If $c^2 = 4mk$, then $r_1 = r_2 = -c/(2m) < 0$ and every solution of ODE (3) is the product of the decaying exponential $\exp[r_1 t]$ and the polynomial $k_1 + k_2 t$, where k_1 and k_2 are constants. See Figure 4.2.2.

- *Underdamped.* If $c^2 < 4mk$, then $r_1 = \alpha + i\beta = \bar{r}_2$, where $\alpha = -c/(2m) < 0$ and $\beta = \sqrt{4mk - c^2}/(2m)$. Every solution is the product of a decaying exponential $A \exp[\alpha t]$ and a sinusoid $\cos(\beta t + \gamma)$, where A and γ are constants. In this case the solutions are *free damped oscillations* with circular frequency β. See Figure 4.2.3.

Graphically, there is not much difference between the overdamped and critically damped solution curves, but the oscillating behavior of the underdamped solution curves clearly sets them apart.

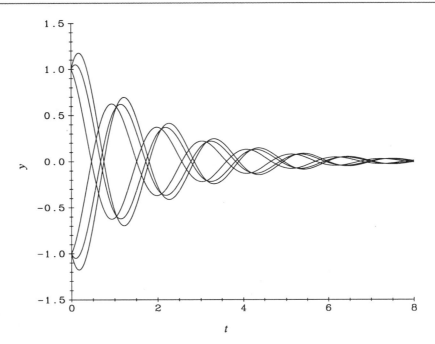

FIGURE 4.2.3 Free damped oscillations: solution curves for $y'' + y' + 9.25y = 0$ (underdamping).

Forced Oscillations: Beats and Resonance in Undamped Systems

In Section 3.5 we developed techniques for finding the response of a constant-coefficient linear second-order ODE to a sinusoidal driving force. When the free solutions of the linear ODE are oscillatory in character, sinusoidal driving forces may produce peculiar responses which scientists call *resonance*. The excitation of a tuning fork at rest in response to the vibrations of another tuning fork is an example of resonance. Tuning the circuits in radio receivers is also a familiar example. The phenomenon of resonance is central to the operation of a great many devices and instruments. Resonance can be explained mathematically in terms of the analysis of Section 3.4. For simplicity, we take up undamped systems first.

Let us consider the differential equation

$$y'' + \omega_0^2 y = A \cos \omega t \tag{5}$$

where ω_0, A, and ω are positive constants. As we have seen in Section 3.1, ODE (5) arises in modeling an undamped spring-body system driven by a sinusoidal force of circular frequency ω. The ODE can arise in many other contexts. Since the analysis will use only the ODE and no other feature of the physical system that underlies the model, the conclusions apply to any system modeled by the ODE, a very important attribute of the modeling approach to problem solving.

For convenience we distinguish two cases in treating ODE (5). Observe that every free solution of the ODE is periodic with circular frequency ω_0. The driving term has circular frequency ω, and the character of the solutions of the ODE depends on whether or not $\omega = \omega_0$.

Case I: Pure Resonance: $\omega = \omega_0$
Using the Method of Annihilators (see Section 3.5), we first note that $\text{Re}[Ae^{i\omega t}] = A \cos \omega t$, and so we look for a real-valued particular solution of ODE (5) by first find-

ing a (complex-valued) particular solution of the ODE

$$z'' + \omega^2 z = A e^{i\omega t} \qquad (6)$$

Putting $P(D) = D^2 + \omega^2$, note that $P(D) = (D - i\omega)(D + i\omega)$. The operator $D - i\omega$ annihilates $e^{i\omega t}$ and so any particular solution z_p of ODE (6) must solve the homogeneous linear ODE $(D - i\omega)^2 (D + i\omega)[z] = 0$. Thus z_p has the form $z_p = a e^{i\omega t} + b t e^{i\omega t} + c e^{-i\omega t}$ for some constants a, b, c. Now

$$P(D)[z_p] = P(D)[b t e^{i\omega t}] = 2 b i \omega e^{i\omega t}$$

and so we must take $2 b i \omega = A$. We may as well take $a = c = 0$ since their values have no effect on the outcome. Thus

$$z_p = \frac{A}{2i\omega} t e^{i\omega t}$$

Now since we seek a real-valued solution of ODE (5), and since $P(D)$ has real coefficients we see that

$$y_p = \text{Re}[z_p] = \frac{A}{2\omega} t \sin \omega t$$

is the particular solution we seek. So the general solution of ODE (5) is

$$y = C_1 \cos \omega t + C_2 \sin \omega t + \frac{A}{2\omega} t \sin \omega t$$

where C_1 and C_2 are arbitrary real constants.

Every solution of ODE (6) is the superposition of a free periodic oscillation, $C_1 \cos \omega t + C_2 \sin \omega t$, with a natural circular frequency ω, and a *forced oscillation*, $(A/2\omega) t \sin \omega t$, which has the form of a sinusoid of circular frequency ω with amplitude which grows over time. Observe that the forced oscillation is the response of the system to a periodic external force whose frequency exactly matches the natural frequency of the system. When an undamped system is driven with a sinusoidal external force having the system's natural frequency, the unbounded oscillation that results is said to be due to *pure resonance*. In the case of pure resonance, the system responds to a bounded input with an unbounded output. Figures 4.2.4 and 3.3.1 give graphic illustrations of pure resonance.

Case II: Beats: $\omega \neq \omega_0$

Using the Method of Annihilators, we note that $\text{Re}[A e^{i\omega t}] = A \cos \omega t$, and so we look for a real-valued particular solution of $y'' + \omega_0^2 y = A \cos \omega t$ by first finding a complex-valued particular solution of the ODE

$$z'' + \omega_0^2 z = A e^{i\omega t} \qquad (7)$$

Again putting $P(D) = D^2 + \omega_0^2$, we note that $P(i\omega) = -\omega^2 + \omega_0^2 \neq 0$ and so we may look for a particular solution of ODE (7) by using the operational formula $P(D)[e^{rt}] = e^{rt} P(r)$, for any complex number r. Taking $z_p = c e^{i\omega t}$, for some constant c, we see that $P(D)[c e^{i\omega t}] = c e^{i\omega t} P(i\omega)$, and we must take $c P(i\omega) = A$, or $c = A/(\omega_0^2 - \omega^2)$ and

$$z_p = \frac{A}{\omega_0^2 - \omega^2} e^{i\omega t} \qquad (8)$$

So the particular solution y_p of ODE (7) we seek is

$$y_p = \text{Re}[z_p] = \left(A/(\omega_0^2 - \omega^2) \right) \cos \omega t$$

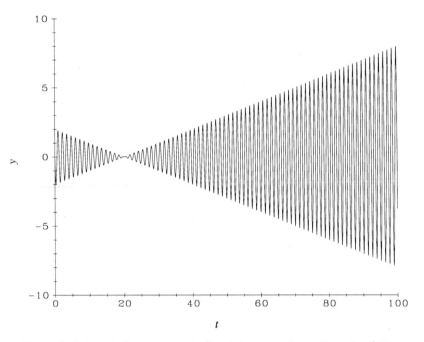

FIGURE 4.2.4 Solution curve of $y'' + 25y = \cos 5t$, $y(0) = 0$, $y'(0) = -10$: pure resonance.

The general solution of ODE (7) is given by

$$y(t) = C_1 \cos \omega_0 t + C_2 \sin \omega_0 t + \frac{A}{\omega_0^2 - \omega^2} \cos \omega t \tag{9}$$

where C_1 and C_2 are arbitrary real constants.

Thus, the solution $y(t)$ in (9) is the superposition of the *free periodic oscillation* $C_1 \cos \omega_0 t + C_2 \sin \omega_0 t$ with *natural circular frequency* ω_0, and a *forced periodic oscillation* $[A/(\omega_0^2 - \omega^2)] \cos \omega t$ representing the system's response to the externally applied force, $A \cos \omega t$. Observe that the frequency of the forced oscillation exactly matches that of the external applied force. Also note that if the ratio ω/ω_0 is a rational number, then every solution of ODE (5) is periodic. Indeed if m and n are positive integers such that $\omega/\omega_0 = m/n$, then $m/\omega = n/\omega_0$, and so each term in the general solution (9) has the period $(2\pi/\omega)m = (2\pi/\omega_0)n$.

Let us find that unique solution of $y'' + \omega_0^2 y = A \cos \omega t$ which starts out at $y = 0$ from rest, that is, a solution $y(t)$ such that $y(0) = y'(0) = 0$. Using the general solution (9), we see that these initial conditions imply that $C_1 = A(\omega^2 - \omega_0^2)^{-1}$, $C_2 = 0$ and we obtain the solution

$$y(t) = \frac{A}{\omega_0^2 - \omega^2} (\cos \omega t - \cos \omega_0 t)$$

Using the identity

$$\cos \alpha - \cos \beta = 2 \sin \frac{\beta - \alpha}{2} \sin \frac{\beta + \alpha}{2}$$

to rewrite this solution, we have

$$y(t) = \left(\frac{2A}{\omega_0^2 - \omega^2} \sin \frac{\omega_0 - \omega}{2} t \right) \sin \frac{\omega_0 + \omega}{2} t \tag{10}$$

which has the form of a sinusoid of circular frequency $(\omega_0 + \omega)/2$ and amplitude that

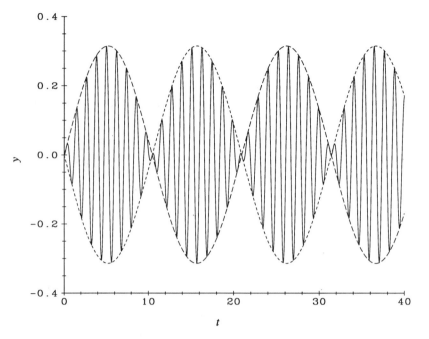

FIGURE 4.2.5 Beats for the IVP: $y'' + 25y = \cos(5.6t)$, $y(0) = y'(0) = 0$. See Example 4.2.1.

varies periodically in time with circular frequency $|\omega_0 - \omega|/2$. This observation leads to a very interesting interpretation when $|\omega_0 - \omega|$ is small with respect to $\omega_0 + \omega$, the phenomenon of *beats*. For in this case, the system's response to being driven by a sinusoid of frequency ω is a sinusoid of frequency $(\omega_0 + \omega)/2$ but with a relatively slowly varying amplitude, called a *beat*, which is a sinusoid of circular *beat frequency* $|\omega_0 - \omega|/2$.

EXAMPLE 4.2.1

Beats from Initial Rest State
Following the approach described above for producing beats, consider the IVP

$$y'' + 25y = \cos(5.6t), \qquad y(0) = 0, \quad y'(0) = 0 \tag{11}$$

where $\omega_0 = 5.0$, $\omega = 5.6$, and $A = 1$. This sinusoidally driven model IVP has a response which exhibits the beat phenomenon. According to formula (10), the solution of IVP (11) is

$$y(t) = \left(\frac{1}{3.18} \sin 0.3t \right) \sin 5.3t$$

The solution curve is plotted in Figure 4.2.5 (solid line). Notice that the beat is given by $y = (1/3.18) \sin 0.3t$ and has circular frequency 0.3 and period $2\pi/0.3 \approx 20.94$. The graphs of $y = \pm(1/3.18) \sin 0.3t$ (dashed lines) envelop the solution curve of IVP (11). Each "hump" in the solution curve is $\pi/0.3 \approx 10.47$ units of time wide.

EXAMPLE 4.2.2

Beats from an Initial State not at Rest
When a harmonic oscillator, not initially at equilibrium, is driven by a sinusoid, then beats have a different appearance than those in Figure 4.2.5. Consider the IVP

$$y'' + 25y = \cos 5.6t, \qquad y(0) = y_0, \quad y'(0) = y'_0 \tag{12}$$

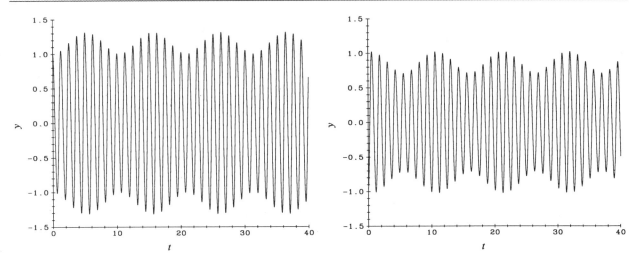

FIGURE 4.2.6 Beat pattern for $y'' + 25y = \cos(5.6t)$, $y(0) = 1$, $y'(0) = 0$. See Example 4.2.2.

FIGURE 4.2.7 Beat pattern for $y'' + 25y = \cos(5.6t)$, $y(0) = -1$, $y'(0) = 1$. See Example 4.2.2.

When $y_0 = 1$, $y_0' = 0$, the solution graph of IVP (12) over $0 \leq t \leq 40$ is given in Figure 4.2.6. For $y_0 = -1$, $y_0' = 1$, the solution graph of IVP (12) is given in Figure 4.2.7. Notice that in each case the basic structure of the beats is the same as in Figure 4.2.5. The "humps" are approximately 10.47 units wide, but the oscillations are not "pinched" off between the humps. Also notice that the response appears to translate along the t-axis and the height of the "humps" changes as the initial data in IVP (12) change. The modulation of the response, however, always retains its most important feature: its frequency.

Beats can be experienced, for example, as the alternative fading and swelling in amplitude heard when two tuning forks vibrate at not quite the same frequency; this simple device permits the human ear to detect frequency differences as low as 0.06%. The beat phenomenon is used in radio signal detection, where an incoming radio signal is mixed with the signal of an oscillator in the receiver to give a new signal with a frequency down in the audio range. That signal is then applied to a detector whose output will drive a loudspeaker. Finally observe that the maximum amplitude, $|2A/(\omega_0^2 - \omega^2)|$, of the sinusoid in (10) tends to ∞ as $\omega \to \omega_0$; this is another manifestation of the phenomenon of resonance.

Forced Oscillations in Damped Systems

Now let us add some damping to the system, but not enough to destroy the oscillatory character of the free solutions. Anticipating the application which follows, the ODE we now consider has the form

$$y'' + 2cy' + k^2 y = F_0 \sin \omega t \tag{13}$$

where the positive constants c and k are such that $c < k$. Setting $P(D) = D^2 + 2cD + k^2$, we see that the roots of $P(r) = r^2 + 2cr + k^2$ are

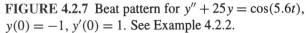

$$r_1 = -c + i\sqrt{k^2 - c^2}, \qquad r_2 = -c - i\sqrt{k^2 - c^2}$$

and so all the free solutions of ODE (13) have the form

$$y = Ae^{-ct}\cos(k^2 - c^2)^{1/2}t + Be^{-ct}\sin(k^2 - c^2)^{1/2}t \qquad (14)$$

where A, B are arbitrary constants. These free solutions are transients because they decay to zero exponentially as $t \to \infty$.

As we saw in Section 3.5, ODE (13) has a particular solution $y_p(t)$ which is a periodic sinusoid with the same circular frequency ω as the driving term; y_p is called a *forced periodic oscillation*. Since the general solution of ODE (13) is this particular solution $y_p(t)$ plus the transients in (14), we see that ODE (13) has the unique periodic solution $y_p(t)$. When engineers study how the amplitude of this unique periodic response to the sinusoidal input changes as ω changes, it is called *frequency response modeling*. Before we turn to the task of finding a periodic particular solution of the general ODE (13), we give two examples.

EXAMPLE 4.2.3

A Forced Periodic Oscillation
The ODE with damping,

$$y'' + y' + y = 13\sin 2t$$

has the unique periodic solution (after using the identity $\sin\alpha + \beta = \sin\alpha\cos\beta + \cos\alpha\sin\beta$)

$$y_p = -3\sin 2t - 2\cos 2t = \sqrt{13}\sin(2t + \varphi)$$

where $\cos\varphi = -3/13$, $\sin\varphi = -2/13$. This solution is the periodic forced oscillation; it has the circular frequency 2 of the driving function. The free solutions of the ODE have the form $e^{-t/2}[C_1\cos(\sqrt{3}t/2) + C_2\sin(\sqrt{3}t/2)]$ and are transients. The general solution

$$y = e^{-t/2}\left[C_1\cos(\sqrt{3}t/2) + C_2\sin(\sqrt{3}t/2)\right] + \sqrt{13}\sin(2t + \varphi)$$

is the sum of the transients and the forced periodic oscillation. The forced oscillation attracts all solutions as $t \to \infty$. Observe that the transient solutions oscillate with circular frequency $\sqrt{3}/2$; they are *not* periodic because their amplitude is not constant, but decays. On the other hand, the forced oscillation *is* periodic. See Figure 4.2.8 for the graphs of the forced oscillation [initial data, $y(0) = -2$, $y'(0) = -6$] and three other solution curves [initial points $(-8, -6)$, $(0, 0)$, $(2, 6)$], all of which decay to the forced oscillation as t increases.

Now let us see how the general response to a sinusoidal input is the sum of a transient and a forced periodic oscillation.

EXAMPLE 4.2.4

Response of a Damped System to a Sinusoidal Input
It can be directly verified that $y_p = (60\cos t + 2\sin t)/9.01$ is the unique periodic solution of the ODE $y'' + y'/2 + 257y/16 = 100\cos t$, and that the transients are given by $y_{tr} = C_1e^{-t/4}\cos 4t + C_2e^{-t/4}\sin 4t$. In Figure 4.2.9 we plot $y_p(t)$ (long-dashed curve) and the unique transient y_{tr} (short-dashed curve) such that $y = y_p + y_{tr}$ solves the IVP

$$y'' + y'/2 + 257y/16 = 100\cos t, \quad y(0) = y'(0) = 0$$

along with the sum $y = y_p + y_{tr}$ itself (solid curve). Notice that y_p is a sinusoid with period 2π, and that y_{tr} is indeed an oscillatory transient.

How do we find a forced periodic oscillation in the general case of ODE (13)? We shall use the Method of Annihilators (see Section 3.5). The first step is to replace the

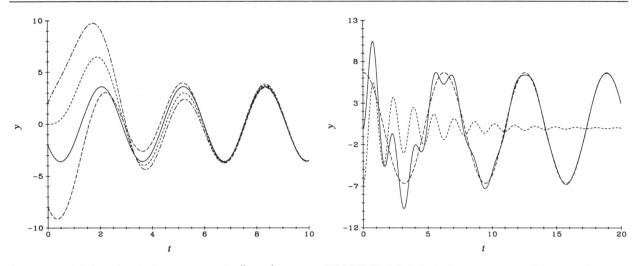

FIGURE 4.2.8 All solution curves of $y'' + y' + y = 13 \sin t$ approach the periodic forced oscillation (solid curve). See Example 4.2.3.

FIGURE 4.2.9 Solution curves (solid), steady-state (long dashes), transient (short dashes). See Example 4.2.4.

driving term in ODE (13) by a complex exponential function whose imaginary part is $F_0 \sin \omega t$, and then find a response to this new input function. The imaginary part of the response will then solve the ODE because the ODE has real coefficients. Note that $\text{Im}[F_0 e^{i\omega t}] = F_0 \sin \omega t$ and that ODE (13) assumes the complex form

$$P(D)[z] = F_0 e^{i\omega t} \tag{15}$$

where $P(D) = D^2 + 2cD + k^2$. Now since $P(D)[e^{rt}] = P(r)e^{rt}$ for any real (or complex) constant r, we are led to look for a particular solution of ODE (15) in the form $z_p(t) = Ae^{i\omega t}$, where A is a constant to be determined. Observe that $P(D)[Ae^{i\omega t}] = AP(i\omega)e^{i\omega t}$ and that $P(i\omega) = (i\omega)^2 + 2c(i\omega) + k^2 = k^2 - \omega^2 + i2c\omega$, which does not vanish unless $k = \omega$ and $c = 0$. Hence, except for that special case, we see that if $H(r) = 1/P(r)$, then

$$z_p(t) = F_0 e^{i\omega t}/P(i\omega) = H(i\omega)F_0 e^{i\omega t} \tag{16}$$

solves the ODE (15). $H(r)$ is called the *transfer function* for ODE (13).

To extract the imaginary part of the solution in (16) we first write $H(i\omega)$ in Cartesian form

$$H(i\omega) = \frac{1}{k^2 - \omega^2 + 2ic\omega} = \frac{k^2 - \omega^2}{(k^2 - \omega^2)^2 + 4c^2\omega^2} + i\frac{-2c\omega}{(k^2 - \omega^2)^2 + 4c^2\omega^2} \tag{17}$$

and then in polar form as follows:

$$H(i\omega) = M(\omega)e^{i\varphi(\omega)} \tag{18}$$

$$M(\omega) = \frac{1}{\sqrt{(k^2 - \omega^2)^2 + 4c^2\omega^2}} \qquad \varphi(\omega) = \cot^{-1}\left(\frac{\omega^2 - k^2}{2c\omega}\right) \tag{19}$$

with $-\pi \le \varphi(\omega) \le 0$. [*Note*: From (17) the complex number $H(i\omega)$ always "points" downward as a vector in the complex plane because the coefficient of i is negative, and so the polar angle $\varphi(\omega)$ is in the indicated range.] Now using (18) in formula (16) we see that

$$H(i\omega)F_0 e^{i\omega t} = F_0 M(\omega)e^{i(\omega t + \varphi(\omega))}$$

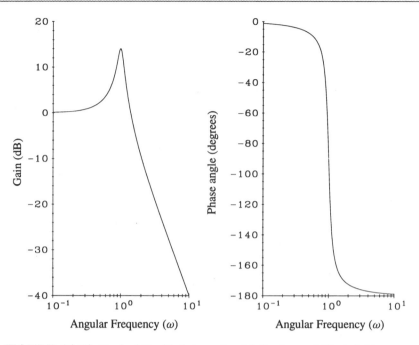

FIGURE 4.2.10 Typical Bodé plots: gain at left, phase shift at right.

$$= F_0 M(\omega)\{\cos[\omega t + \varphi(\omega)] + i\sin[\omega t + \varphi(\omega)]\}$$

and extracting the imaginary part of this function we find that the sinusoid

$$y_p = F_0 M(\omega)\sin[\omega t + \varphi(\omega)] \tag{20}$$

is a real-valued solution of ODE (13). Note that this particular solution is a periodic forced oscillation with circular frequency ω, and all other solutions of ODE (13) decay to this solution as $t \to +\infty$. The solution y_p in (20) is known as the *steady-state solution*.

The functions $M(\omega)$ and $\varphi(\omega)$ do not depend on the initial data. Comparing (20) with the input $F_0 \sin \omega t$, we see that the steady-state solution has the same sinusoidal form as the input, but with a different amplitude and phase. The ratio of the steady-state amplitude $F_0 M(\omega)$ to the input amplitude is $M(\omega)$ and is called the *gain*. The steady-state solution is shifted in time by $|\varphi(\omega)|/\omega$ radians to the right, so it is common to refer to $\varphi(\omega)$ as the *phase shift*. The graphs of gain and phase shift against ω (using a $\log_{10}$ scale on the ω-axis) are called *Bodé plots* and give valuable information about the system. Engineers call the quantity $20 \log_{10} M(\omega)$ the gain in *decibels* (dB), and use decibel units on the gain axis. Degrees are used on the φ-axis instead of radians. These two graphs constitute the *frequency response curves* for the system. See Figure 4.2.10 for typical Bodé plots.

EXAMPLE 4.2.5

Parameter Identification

The graph of the phase-shifted sinusoid $\sin(\omega t + \varphi)$ against t can be obtained from the graph of $\sin \omega t$ against t as follows: If φ is positive (resp., negative), then shifting the graph of $\sin \omega t$ to the left (resp., right) by φ/ω radians results in the graph of $\sin(\omega t + \varphi)$. This clearly follows from writing $\sin(\omega t + \varphi) = \sin(\omega(t + \varphi/\omega))$. Sometimes it is known that a system can be modeled by an ODE of the type given in (13), but the system parameters cannot be accurately measured (i.e., the constants c and k are known only approximately). If a sinusoid $\sin \omega t$ is used as input in ODE (13) and if the

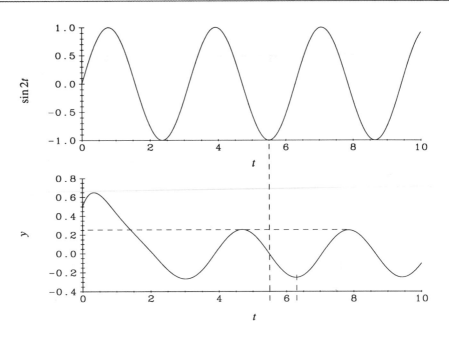

FIGURE 4.2.11 Graphical determination of steady-state response. See Example 4.2.5.

gain $M(\omega)$ and phase shift $\varphi(\omega)$ in the steady-state output (20) can be experimentally determined, then formulas (19) form a basis for determining the system constants c and k.

Consider the ODE $y'' + 2y' + 4y = \sin 2t$ and initial data $y(0) = 0.5$, $y'(0) = 1$. We use a solver to produce the input-output graphs in Figure 4.2.11 and use these graphs to compute $M(2)$ and $\varphi(2)$ and compare these values to the theoretical values given by (19). Observe that the output quickly settles down to a periodic steady-state. Using the vertical dashed lines, we estimate that the output is shifted to the right by about 0.75 unit, hence $\varphi(2)$ is about 1.5 radians. From (19) we have that $\cot(\varphi(2)) = 0$, and since $-\pi \le \varphi(\omega) \le 0$, it follows that $\varphi(2) = -\pi/2$. Since $\pi/2 \approx 1.57$, our graphical procedure has done well. Using the horizontal dashed lines, we estimate that the gain $M(2) \approx 0.26$. From (19) we see that $M(2) = 0.25$, exactly.

Resonance in Damped Systems

The shapes of the Bodé plots in Figure 4.2.10 change as the parameters c and k of the system change. Notice from (19) in all cases (under the general assumption that $c < k$) that $M(\omega)$ is a bounded continuous function on $\omega \ge 0$. It can be shown (see Problem 11) that if $k^2 > 2c^2$, then the plot of $M(\omega)$ versus ω looks like the one on the left in Figure 4.2.10 with a unique value ω_r where $M(\omega)$ achieves a maximum value:

$$M(\omega_r) = [2c\sqrt{k^2 - c^2}]^{-1}, \quad \omega_r = \sqrt{k^2 - 2c^2} \tag{21}$$

The unique value ω_r is called the *resonant frequency* of the damped system. Observe that if there is no damping (i.e., $c = 0$), then $M \to +\infty$ as $\omega \to k$, where k is the natural frequency of the undamped system. If damping is present in the system, then $M(\omega_r)$ can be made arbitrarily large if c is made small enough.

The applications of pure resonance carry over in an obvious way to the damped case: It has been observed that certain structures such as bridges or buildings have resonant frequencies. An external oscillating force could conceivably induce an oscillation of destructive amplitude. One theory of the collapse of the Tacoma Narrows bridge in 1940 while a gentle wind was blowing is that the support work along the roadway created sinusoidal eddies whose frequency matched the natural frequency of the bridge. Tachometers, seismometers, and vibrometers all depend on the resonance phenomenon for their operation. For example, the *Frahm tachometer* is a box containing rows of cantilever-mounted steel rods of varying natural frequencies. When the box is placed on a vibrating machine, only the reeds with natural frequencies close to the machine's vibrational frequency will be seen to move.

PROBLEMS

1. (*Beats*). For $\omega \neq \omega_0$ show that the solution of the IVP

    $$y'' + \omega_0^2 y = A \sin \omega t, \qquad y(0) = 0, \quad y'(0) = -\frac{A}{\omega_0 + \omega}$$

    exhibits the phenomenon of beats similiar to the graph in Figure 4.2.5. [*Hint*: Note that $\sin \alpha - \sin \beta = 2 \sin[(\alpha - \beta)/2] \cos[(\alpha + \beta)/2]$, for all α, β.]

2. Solve the IVP $y'' + 25y = \sin 5.6t$ for four sets of initial points $(y(0), y'(0))$. Plot the results on separate graphs. Compare these graphs and record your observations. Choose initial points which give quite different graphs.

3. (*Archimedes' Buoyancy Principle*). According to Archimedes' Principle, the buoyant force acting on a body wholly or partially immersed in a fluid equals the weight of the fluid displaced.

    (a) Ignoring friction, show that a flat wooden block of face L^2 square feet and thickness h feet floating half-submerged in water will act as a harmonic oscillator of period $2\pi\sqrt{h/2g}$ if it is depressed slightly. [*Hint*: Show that if x is a vertical coordinate from the midpoint of the block, then $L^2 h(\rho/2)x'' + L^2 \rho g x = 0$, where ρ is the density of water.]

    (b) Assuming no friction, replace the block in part (a) by a buoyant sphere of radius R floating half-submerged, and show that if the initial displacement is slight, the sphere will act as a harmonic oscillator of period $2\pi\sqrt{2R/3g}$. Show that if the displacement is large, the nonlinearities in the ODE modeling the motion cannot be ignored, and the ODE of soft-spring is obtained (see Section 3.1).

4. (*Periodic Forced Oscillation*). Find the unique periodic solution of the ODE $y'' + 2cy' + k^2 y = F_0 \cos \omega t$, where $c, k, F_0,$ and ω are positive constants. [*Hint*: See the text for the technique of solving ODE (13).]

5. Consider the IVP: $y'' + 0.5y' + 16y = 100 \sin t$, $y(0) = 0$, $y'(0) = 0$. Identify the periodic response and the transient components of the unique solution of the IVP. On a single pair of axes plot the solution graph of the IVP along with the periodic and transient components of the solution of the IVP. [*Hint*: See Example 4.2.4.]

6. (*Gain and Phase Shift*). Use the solution of the IVP

    $$y'' + 2y' + 4y = 3 \sin 2t, \qquad y(0) = 0.5, \quad y'(0) = 0.5$$

    to estimate the gain $M(2)$ and phase shift $\varphi(2)$ for the steady-state forced response. [*Hint*: See Figure 4.2.11 and Example 4.2.5] Compare these estimates with the exact values calculated from (19). Find the system transfer function.

7. (*Gain and Phase Shift*). Consider the ODE $y'' + y' + 2y = F_0 \sin \omega t$.

(a) Use an ODE solver to find enough values of the gain $M(\omega)$ and the phase shift $\varphi(\omega)$ to make rough sketches of M and φ vs. ω.

(b) Use the formulas in (19) to plot the graphs of $M(\omega)$ and $\varphi(\omega)$. Compare these plots with the rough sketches drawn in part (a).

8. (*Parameter Identification*). Find the parameters c and k of the system whose Bodé plots are shown in Figure 4.2.10. [*Hint*: Pick a frequency ω and find M and φ from the Bodé plots. The equations in (19) are then to be solved for $2c$ and k^2.]

9. (*Parameter Identification*). In a damped Hooke's Law spring system modeled by $y'' + 2cy' + ky = 0$, the static deflection is 0.5 in, while damped vibrations decay from amplitude 0.4 to amplitude 0.1 in. in 20 cycles. Assume a mass of 1 kg.

(a) Find the damping constant $2c$. [*Hint*: First find k, then find c such that amplitudes decay by a factor of 4 over a $20T$ time span, where $T = 2\pi/(k - c^2)^{1/2}$.]

(b) Find the resonant frequency ω_r and calculate the maximal amplitude of the steady-state response of the system to a driving force $A \cos \omega_r t$.

10. (*Detecting a Periodic Forced Oscillation*). Solutions of the ODE $y'' + \omega_0^2 y = A \cos \omega t$, where A, ω_0, ω ($\omega_0 \neq \omega$) are positive constants, are superpositions of sinusoids of period $2\pi/\omega_0$ and of period $2\pi/\omega$. Given the graph of a solution, is it possible to detect the forced oscillation (i.e., the unique solution of period $2\pi/\omega$)? Does the forced oscillation have a phase shift? Are the nonconstant solutions periodic, and if they are what is their period? In addressing these questions, start out by solving graphically the two IVPs $y'' + 4y = \sin \omega t$, $y(0) = 0$, $y'(0) = 0$, $\omega = 3$, π. You may also want to find solution formulas.

11. (*Gain and Phase Shift*). The steady-state response of the ODE $y'' + 2cy' + k^2 y = F_0 \sin \omega t$ with $0 < c \leq k$ can be characterized in terms of an amplitude magnification and a phase shift applied to the sinusoidal driving term $f(t)$. Explore this connection by performing the tasks below:

- Verify [using (19)] that the general features of the graphs of $M(\omega)$ and $\varphi(\omega)$, for fixed positive constants k and c with $k^2 > 2c^2$, are similiar to the plots in Figure 4.2.10. What is the value ω_r (the resonant frequency) where $M(\omega)$ achieves a maximum? What are $M(0)$, $M(\omega_r)$? What is the value ω_0 where $\varphi(\omega)$ changes inflection?

- What happens to the frequency ω_r and the maximum value $M(\omega_r)$ determined in part (a), for fixed $k > 0$ as $c \to 0$? Interpret this observation for a vibrating mechanical system generated by ODE (13) with very small damping constant c and a sinusoidal driving term with frequency close to a natural (i.e., undamped) frequency of the system.

12. (*Gain and Phase Shift*). Consider the IVP $y'' + 0.2y' + y = \cos \omega t$, $y(0) = 0$, $y'(0) = 0$. Do some frequency response modeling for this system by performing the tasks below:

- Let $\omega = 0.5$. Plot the solution of the IVP over a time sufficiently long that the transient part of the solution has become negligible and the stead-state response is clearly visible. From the graph, determine the amplitude of the stead-state response, compare it to the amplitude of the driving function, and then determine the gain M.

- Now determine the phase shift φ between the driving function and the steady-state response. [*Hint*: See Figure 4.2.11.]

- The amplitude M and the phase shift φ do not depend on initial conditions.

Explain why this is so. Confirm this fact by repeating the above steps for several sets of initial conditions.

- Determine the amplitude M and the phase shift φ for various other values of ω for the IVP. For instance, choose $\omega = 0.25, 0.75, 1, 1.5, 2$, and 3. Try values of ω very close to 1. Sketch the graphs of M vs. ω, and φ vs. ω.

- Solve the ODE analytically and determine the steady-state response. Find M and φ as functions of ω. Compare these results with those found above.

4.3 Electrical Circuits: Kirchhoff's Laws

Among the many physical processes that can be modeled by second-order linear ordinary differential equations, one of the most important is the flow of electrical energy in a circuit.

Electrical Units

A *circuit* consists of basic elements called *resistors*, *capacitors*, and *inductors* (described below) connected by conducting wires. See the margin for a schematic diagram of a simple circuit.

The *current* I in a circuit is proportional to the number of free electrons (each with a constant negative charge) moving per second through (any) given point in the conductor. According to common practice, however, current flow is described in terms of positive charge carriers whose movement is opposite to that of electrons; that is, if I is positive when the circuit orientation is clockwise, the actual movement of the electrons is counterclockwise. Current is measured in *amperes*,[5] the basic electrical unit in the MKS system, in terms of which all other units are defined. (Actually, one ampere corresponds to 6.2420×10^{18} positive carriers moving past a given point in one second.)

The unit of charge is the *coulomb*, defined to be the amount of charge which flows through a cross section of wire in one second when a current of one ampere is flowing. Thus one ampere is one coulomb per second. If the current I is not constant in time, an instantaneous current $I(t)$ flowing past a point in a circuit is defined in the same way as other physical quantities (e.g., velocity). Thus if $I(t)$ is continuous, the amount of charge $q(t)$ flowing past a point in the time interval $[a, b]$ is given by $\int_a^b I(t)\, dt$.

As current moves through a circuit, the charge carriers either impart energy to a circuit element or receive energy from a circuit element. The way in which physicists usually describe this phenomenon is by defining a potential function V throughout a circuit. The energy per coulomb of charge that has been imparted (or received) by the charge carriers as they flow from point a to point b is computed as

$$V_{ab} = V_a - V_b$$

---

[5]André Marie Ampère (1775–1836), a French mathematician and natural philosopher, is known for his contributions to electrodynamics. The names of many early researchers in electricity are used for electrical units. Thus we have ohms [Georg Simon Ohm (1787–1854), German physicist], henries [Joseph Henry (1797–1878), American physicist], farads [Michael Faraday (1791–1867), English scientist], volts [Alessandro Volta (1745–1827), Italian physicist], and coulombs [Charles Auguste de Coulomb (1736–1806), French physicist].

where V_a and V_b are the values of the potential function V at points a and b of the circuit. The difference V_{ab} is called the *voltage drop* or *potential difference* from a to b and is measured in joules per coulomb, or *volts*.

Some circuit elements are devices which have the peculiar property that they can maintain a prescribed potential difference (or voltage), denoted by E, between two terminals. Such devices behave like an external force when connected to a circuit, and in fact are known as sources of *electromotive force* (EMF). Electromotive force E is also measured in volts. Examples are batteries and electric generators. For batteries, the higher potential terminal is labeled with a plus sign and the lower potential terminal with a minus sign. The internal chemical energy supplied by the battery imparts a constant amount of energy per coulomb as positive charge carriers move through the battery, thus raising the potential function V by the voltage rating of the device.

Circuit Elements

Three basic elements can be used to build a simple circuit: the resistor, capacitor, and inductor, which we describe briefly below.

Resistor

As a current flows through a substance the charge carriers lose energy (which then appears as other energy forms such as heat and light), and hence the potential where the current emerges is lower than the potential where the current entered. Those substances which dissipate large amounts of energy while still allowing a current to flow are called *resistors* and are symbolized by the schematic shown in the margin. Common examples of resistors are heating elements and lightbulb filaments. Materials such as wires which dissipate a negligible amount of energy when current flows are called *conductors*.

The voltage drop across a resistor and the current flowing through it are usually modeled in a particularly simple way.

> Ohm's Law.   The instantaneous voltage drop V_{ab} between the two terminals a and b of a resistor is found experimentally to be proportional to the current flowing from a to b through the resistor at that instant. If the current is measured in amperes, then the constant of proportionality, denoted by R, is said to be measured in *ohms* (the ohm is denoted by the capital Greek letter Ω). Thus we have
>
> $$V_{ab} = RI$$
>
> R is called the *resistance* of the resistor.

If the current is directed from a to b, then the voltage at b is lower than the voltage at a, and hence $V_{ab} = RI$ is positive.

Inductor

A changing electrical current $I(t)$ through a conductor creates a changing magnetic field near the conductor. This changing magnetic field induces a voltage between the ends of the conductor which opposes the change in current. This effect can be quite large in conductors arranged in certain geometries (such as coils). These devices are called *inductors* and are denoted by the schematic shown here.

> Faraday's Law.   It has been found experimentally that the voltage drop V_{ab} across an inductor is proportional to the rate of change of the current from a to b through the inductor. If V_{ab} is measured in volts and dI/dt in amperes per second, then this constant of proportionality, denoted by L, is said to be measured in *henries* (denoted by H). Thus we have
>
> $$V_{ab} = L\frac{dI}{dt}$$
>
> L is called the *inductance* of the inductor.

One consequence of Faraday's Law is that the potential drop does *not* directly depend on the current I, but on its rate of change, dI/dt.

Capacitor

A *capacitor* consists of two conductors separated by an insulator (such as air). Capacitors are often visualized as a pair of conducting plates separated by a gap and represented schematically as shown. If terminals a and b of a capacitor are connected to a voltage source, a negative charge will build up on the plate connected to the negative terminal, and a positive charge will build up on the plate connected to the positive terminal. We speak of the total *charge* $q(t)$ on the capacitor and observe that if $q(t_0)$ is the initial charge, then [since $dq/dt = I(t)$]

$$q(t) = q(t_0) + \int_{t_0}^{t} I(s)\,ds \quad \text{for } t \geq t_0 \tag{1}$$

> Coulomb's Law.   It has been observed experimentally that the instantaneous voltage drop V_{ab} as the current flows from a to b across a capacitor is proportional to the charge on the capacitor. When the charge is measured in coulombs and V_{ab} in volts, the constant of proportionality is $1/C$, where C is measured in *farads* (denoted by F). Thus we have
>
> $$V_{ab}(t) = \frac{1}{C}q(t) = \frac{1}{C}\left\{q(t_0) + \int_{t_0}^{t} I(s)\,ds\right\}$$
>
> C is called the *capacitance* of the capacitor.

Note again that the voltage drops in the direction of the current flow. The capacitance C depends on the area of the plates, their separation, and the nature of the insulator between. Because the coulomb is a rather large amount of charge, a typical capacitor will store only a small charge at typical voltages. Hence C is usually very small (on the order of 10^{-5} or 10^{-6} farads).

Strictly speaking, the coefficients of resistance, inductance, and capacitance are not constant but depend on the magnitudes of the current and the voltage, and hence their values can vary in time. We shall ignore this fact, however, since over the range of standard engineering applications these coefficients are essentially constant.

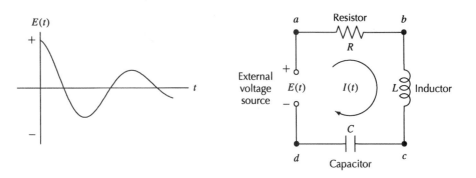

FIGURE 4.3.1 Simple *RLC* series loop prepared for modeling.

Kirchhoff's Voltage Law; Single-Loop Circuits

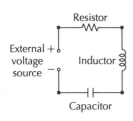

One of the simplest circuits involving resistors, capacitors, and inductors is the series circuit shown in the margin, called the *simple RLC series circuit*, which is driven by a given external voltage source. More complicated circuits are easy to imagine (see Example 4.3.4 and the problem set), but we will not treat them very extensively here. The modeler should keep in mind that resistance, inductance, and capacitance are circuit characteristics which are not always accounted for by devices called resistors, inductors, and capacitors. Sometimes these characteristics arise through the physical properties of a circuit as a whole (e.g., all conducting wire offers some resistance to current flow). Therefore elementary circuit elements sometimes appear in a circuit description in an effort to "lump" together the effects of these distributed characteristics. *Lumping* in this sense is an important and frequently used modeling tool.

What is needed now for finding the current flowing in a circuit is a relation between the voltage drops across the various components of a circuit. The following "conservation" law has been observed to hold:

> Kirchhoff's Voltage Law. First choose a direction for the current I in the circuit. Select the points $a_1, a_2, \ldots, a_n$ in the circuit such that the current flows from a_i to a_{i+1}, $i = 1, 2, \ldots, n$ ($a_{n+1} = a_1$). Then
>
> $$V_{a_1 a_2} + V_{a_2 a_3} + \cdots + V_{a_n a_1} = 0$$
>
> where $V_{a_i a_{i+1}} = V_{a_i} - V_{a_{i+1}}$ is the voltage drop between points a_i and a_{i+1}.

For example, in Figure 4.3.1 the sequence of points *abcda* forms an oriented closed loop for the simple *RLC* series circuit.

Because of the way voltage is defined, Kirchhoff's Voltage Law[6] is a conservation-of-energy law. The potential function V around the circuit and the current flowing in the circuit may vary with time, but at any given instant t_0, $V(t_0)$ is a piecewise constant function of position in the circuit. Ohm's Law, Faraday's Law, and Coulomb's Law

---

[6]Gustav Robert Kirchhoff (1824–1887) was a German physicist who studied the properties of the first electrical circuits.

give the magnitude of the jump discontinuity of the potential function at any instant for each circuit element (idealized as a point). On the other hand, $I(t_0)$ is a constant function at all points in the circuit.

We now apply Kirchhoff's Voltage Law to the simple RLC series circuit to derive an equation that governs how the current in the simple loop evolves with time. For this purpose, it will be helpful to introduce some notation into the simple loop. In what follows refer to Figure 4.3.1. First we label the polarity of the external source with "plus" and "minus" signs in order to be precise about the way the external source is connected to the circuit (the labeling shown indicates how the voltage source $E(t)$ graphed in Figure 4.3.1 is connected to the circuit). Next observe that the current flows in the same direction and has the same value through every circuit element, so we may as well *assume* that the positively directed current $I(t)$ moves clockwise around the circuit (as indicated in Figure 4.3.1). Should it turn out that the current actually moves counterclockwise at a certain t_0, then the value of $I(t_0)$ will turn out to be a negative number. Now as the current flows through the external voltage source the voltage will increase, but on each of the other circuit elements the voltage drops in the (assumed) direction of the current. Finally, we introduce reference points a, b, c, and d as shown. Now observe that (refer to Figure 4.3.1) Ohm's Law, Faraday's Law, and Coulomb's Law yield the voltage drops

$$V_{ab} = RI(t), \quad V_{bc} = L\frac{dI}{dt}, \quad V_{cd} = \frac{1}{C}\left[q(t_0) + \int_{t_0}^{t} I(s)\,ds\right]$$

Now since the voltage increases as the current flows through an external source, it follows that $V_{da} = -E(t)$, and hence Kirchhoff's Voltage Law yields the integrodifferential equation

$$RI(t) + L\frac{dI}{dt} + \frac{1}{C}\left[q(t_0) + \int_{t_0}^{t} I(s)\,ds\right] = E(t) \tag{2}$$

for the unknown current $I(t)$.

We can convert (2) to a differential equation in two ways. In terms of the charge $q(t)$ on the capacitor, we have that

$$Lq'' + Rq' + \frac{1}{C}q = E(t) \tag{3}$$

since $q = q(t_0) + \int_{t_0}^{t} I(s)\,ds$, and so $q' = I$, and $q'' = I'$.

It is often more convenient to use $I(t)$ rather than $q(t)$ as the unknown. If $E(t)$ is differentiable, we can differentiate (2) with respect to t and obtain (since $q' = I$)

$$LI'' + RI' + \frac{1}{C}I = E' \tag{4}$$

The relevant initial value problem (IVP) for the circuit equation in the form (4) is derived as follows: If q_0, the initial charge, and I_0, the initial current, are known, then from (2) we have that $RI_0 + LI'(t_0) + (1/C)q_0 = E(t_0)$ which immediately determines $I'(t_0)$. Denoting $I'(t_0)$ by I_0' we are thus led to solve the IVP

$$LI'' + RI' + \frac{1}{C}I = E'(t)$$
$$I(t_0) = I_0, \quad I'(t_0) = I_0' = \frac{E(t_0) - RI_0 - q_0/C}{L} \tag{5}$$

In practice, t_0 is usually taken to be 0.

$E = $ Constant: Solving the Homogeneous Current Equation

If $E(t)$ is constant in time, then we can use the techniques of Section 3.4 to find all solutions of ODE (4). Indeed, let us assume that $E = E_0$, a constant. Then ODE (4) becomes the homogeneous equation

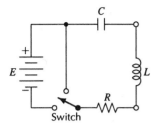

$$LI'' + RI' + \frac{1}{C}I = 0 \qquad (6)$$

where we assume that L, R, and C are positive constants. We can think of this as modeling a circuit such as the one at left in which a battery is allowed to charge up a capacitor and then at some time (say $t_0 = 0$) the battery is removed from the circuit. Assume that the battery had been "in line" long enough that the circuit has reached equilibrium with $q_0 = E_0/C$ and $I_0 = 0$ when the switch is thrown to the vertical position. Thus IVP (5) becomes

$$LI'' + RI' + \frac{1}{C}I = 0, \qquad I_0 = 0, \quad I'(0) = I_0' = -q_0 C/L \qquad (7)$$

The roots of the characteristic polynomial $P(r) = Lr^2 + Rr + 1/C$ are

$$r_1, r_2 = [-R \pm (R^2 - 4L/C)^{1/2}]/2L$$

Thus r_1 and r_2 have negative real parts for any positive constants R, L, and C. Hence, all solutions of ODE (6) are transients (i.e., die out with increasing time), but their exact nature depends on the relative sizes of the constants R, L, and C. We use the General Solution Theorem 3.4.2 to distinguish three qualitatively distinct cases:

- If $R^2 > 4L/C$, the circuit is overdamped and every solution I is a linear combination of the decaying real exponentials $\exp[r_1 t]$ and $\exp[r_2 t]$:

$$I = k_1 e^{r_1 t} + k_2 e^{r_2 t}$$

- If $R^2 = 4L/C$, the circuit is critically damped and every solution is the product of the decaying exponential $\exp[-Rt/2L]$ and a linear polynomial $k_1 + k_2 t$:

$$I = (k_1 + k_2 t)e^{-Rt/2L}$$

- If $R^2 < 4L/C$, the circuit is underdamped, and since r_1 and r_2 have the form $\alpha \pm i\beta$ where $\alpha = -R/2L$, $\beta = (4LC^{-1} - R^2)^{1/2}/2L$, every solution of ODE (6) is the product of sinusoids with a frequency of β cycles per second and a decaying exponential:

$$I = (k_1 \cos \beta t + k_2 \sin \beta t)e^{\alpha t}$$

Contrasting these three cases, we observe that in the overdamped and critically damped cases, solutions essentially decay steadily to zero, whereas in the underdamped case, solutions oscillate back and forth about the zero solution as they decay. For all values of I_0 and I_0' the solution of IVP (7) is a transient that behaves in one of the three ways described above.

Finally, observe that if the resistance in our constant-EMF-driven *RLC* circuit is negligible, we might expect that the current could be modeled by the equation

$$LI'' + (1/C)I = 0$$

Since the characteristic roots of this equation are $\pm i\beta$, with $\beta = 1/(LC)^{1/2}$, every solution has the sinusoidal form

$$I = A\cos(\beta t + \varphi)$$

where A and φ are arbitrary constants. Note that all these solutions are periodic and hence do not decay as $t \to \infty$. Thus such a circuit would be able to sustain undiminished oscillations forever, and for this reason the circuit is known as an *LC oscillator*.

In practice, however, the resistance in a circuit may be very small, but it can never be zero. Hence from what we have just seen, it would be a serious mistake to infer too much about an *RLC* circuit with very small R by setting $R = 0$ in ODE (6). Thus an *LC* oscillator is an idealization of a physically unrealizable phenomenon.

Driven Circuits

We have discussed the free solutions of the simple $R\overset{'}{L}C$ series circuit modeled by ODE (6) for the current I (where $E' = 0$). Now we turn to the forced solutions, that is, solutions of ODE (3) for the charge q or the ODE (4) for the current I when the external voltage source $E(t)$ is nonconstant, and the circuit is driven (or forced). We know from the General Solution Theorem 3.3.7 that all forced solutions are known if just one forced solution and all free solutions are known. The following example uses the charge ODE (3).

EXAMPLE 4.3.1

Finding the Charge on the Capacitor of a Driven Simple *RLC* Circuit

We shall determine all solutions of the simple *RLC* series circuit ODE (3) when $L = 20$ H, $R = 80\ \Omega$, $C = 10^{-2}$ F, and the external voltage is $E(t) = 100\cos 2t$. Thus ODE (3) becomes

$$q'' + 4q' + 5q = 5\cos 2t \tag{8}$$

with characteristic polynomial $P(r) = r^2 + 4r + 5$, which has the roots $r_1 = -2 + i$ and $r_2 = -2 - i$. By the General Solution Theorem 3.4.2, all of the free solutions of ODE (8) (i. e., solutions of $q'' + 4q' + 5q = 0$) have the form

$$k_1 e^{-2t}\cos t + k_2 e^{-2t}\sin t$$

where k_1 and k_2 are arbitrary real constants. Now we need to find a forced solution for ODE (8). We use the Method of Annihilators (see Section 3.5). Observe that $5\cos 2t = \text{Re}[5e^{2it}]$, so if z_p is any (complex-valued) solution of the equation

$$P(D)[z] = z'' + 4z' + 5z = 5e^{2it}$$

then $\text{Re}[z_p] = q_p$ is a (real-valued) solution of ODE (8).

We try a solution of the form $z_p = ce^{2it}$ for some complex number c. Now $P(D)[ce^{2it}] = cP(D)[e^{2it}] = cP(2i)e^{2it}$ by property (2) of Section 3.4. Thus since we require that $P(D)[ce^{2it}] = 5e^{2it}$, we must choose $c = 5/P(2i) = 5/(1 + 8i)$. We see that

$$c = \frac{5}{1 + 8i}\frac{1 - 8i}{1 - 8i} = \frac{1}{13}(1 - 8i)$$

since $(1 + 8i)(1 - 8i) = 65$. Hence

$$q_p(t) = \text{Re}[ce^{2it}] = \text{Re}\left[\frac{1}{13}(1 - 8i)e^{2it}\right] = \frac{1}{13}(\cos 2t + 8\sin 2t) \tag{9}$$

is a particular solution of ODE (8) which is a periodic forced oscillation. Thus, at any time t the charge on the capacitor has the form

$$q(t) = k_1 e^{-2t}\cos t + k_2 e^{-2t}\sin t + \frac{1}{13}(\cos 2t + 8\sin 2t)$$

where k_1 and k_2 are constants.

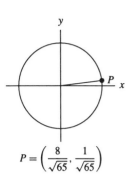

$$P = \left(\frac{8}{\sqrt{65}}, \frac{1}{\sqrt{65}}\right)$$

We present an often-used trick to rewrite $q_p(t)$ as a single sinusoid. Picking out the coefficients of $\cos 2t$ and $\sin 2t$ in the parentheses, we form the point $(8, 1)$ in the Cartesian plane. Multiplying the coordinates by the scalar $1/(1^2 + 8^2)^{1/2} = 1/\sqrt{65}$, we obtain $(8/\sqrt{65}, 1/\sqrt{65})$, a point P on the unit circle. From trigonometry we know there is a unique angle φ with $-\pi < \varphi \le \pi$ such that

$$\cos \varphi = \frac{8}{\sqrt{65}}, \quad \sin \varphi = \frac{1}{\sqrt{65}} \tag{10}$$

Inserting (10) into (9) and using the addition formula for the sine function, we have

$$q_p = \frac{5}{\sqrt{65}}(\sin \varphi \cos 2t + \cos \varphi \sin 2t) = \frac{5}{\sqrt{65}} \sin(2t + \varphi) \tag{11}$$

Thus q_p is revealed to be sinusoidal with the same frequency as the driving term but out of phase slightly (use a calculator to find that $\varphi \approx 7.1°$). Now since every solution of ODE (3) can be written as a free solution (which decays to zero and thus is a transient) plus q_p, we infer that after a while the only charge we "see" in the circuit is the steady-state periodic forced oscillation $q_p(t)$. From (11) we see that

$$I_p = \frac{dq_p}{dt} = \frac{10}{\sqrt{65}} \cos(2t + \varphi)$$

is a particular solution of the current equation ODE (4).

Using Example 4.3.1 as a guide, we find all solutions $q(t)$ of ODE (3), where $R \ne 0$ and the driving term $E(t) = E_0 \cos \omega t$ for given real numbers E_0 and ω. The free solutions are transients and have been treated earlier, so we need only a single particular solution to complete our analysis. Using the approach given in Example 4.3.1 we first find a particular (complex-valued) solution z_p of

$$Lq'' + Rq' + \frac{1}{C}q = E_0 e^{i\omega t} \tag{12}$$

and then observe that $q_p = \text{Re}[z_p]$ is a particular real-valued solution of ODE (3) with $E(t) = E_0 \cos \omega t$. Since the roots of the characteristic polynomial have strictly negative real parts, $P(i\omega) \ne 0$, and so $z_p = E_0 e^{i\omega t}/P(i\omega)$ is a particular solution of ODE (12). Since $P(i\omega) = -L\omega^2 + 1/C + iR\omega$, we have that

$$1/P(i\omega) = [(-L\omega^2 + 1/C) - iR\omega]/[(-L\omega^2 + 1/C)^2 + R^2\omega^2]$$

and hence

$$q_p = \text{Re}[z_p] = E_0 \frac{(1/C - L\omega^2)\cos \omega t + \omega R \sin \omega t}{(1/C - L\omega^2)^2 + \omega^2 R^2} \tag{13}$$

As was done to derive (10), let φ be an angle with $-\pi < \varphi \le \pi$ such that

$$\sin \varphi = \frac{1/C - L\omega^2}{[(1/C - L\omega^2)^2 + \omega^2 R^2]^{1/2}}, \quad \cos \varphi = \frac{\omega R}{[(1/C - L\omega^2)^2 + \omega^2 R^2]^{1/2}}$$

Then solution formula (13) for the steady-state charge on the capacitor can be written as the sinusoid

$$q_p = \frac{E_0 \sin(\omega t + \varphi)}{[(1/C - L\omega^2)^2 + \omega^2 R^2]^{1/2}} \tag{14}$$

Differentiating (14) to find the steady-state current in the circuit, we find that

$$I_p = \frac{E_0 \omega \cos(\omega t + \varphi)}{[(1/C - L\omega^2)^2 + \omega^2 R^2]^{1/2}} \tag{15}$$

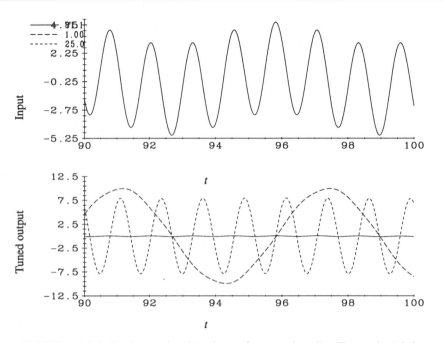

FIGURE 4.3.2 Tuning a circuit to input frequencies. See Example 4.3.2.

Observe that the steady state response has the same frequency as the driving term, but is shifted in phase. The denominator $[(1/C - L\omega^2)^2 + \omega^2 R^2]^{1/2}$ is called the *impedance* of the circuit; it plays a role similar to the one played by the resistance R in Ohm's Law. Note that (14) could have been written with cosine instead of sine since $\sin(\omega t + \varphi) = \cos[\omega t + \varphi - (\pi/2)]$. Both are called sinusoids.

For a fixed resistance R, the amplitude

$$A(\omega) = \frac{E_0 \omega}{[(1/C - L\omega^2)^2 + \omega^2 R^2]^{1/2}}$$

of the steady-state response current (15) to a given driving frequency ω depends very much on the values of L and C. In particular, for a fixed positive R, $I_p(t)$ (and $q_p(t)$) has its maximum amplitude if the term $1/C - L\omega^2$ in the denominator of $A(\omega)$ is zero, i.e., when

$$\frac{1}{LC} = \omega^2 \tag{16}$$

This observation makes it possible to "tune" the RLC circuit to receive a signal of a given frequency. The next example illustrates this property.

EXAMPLE 4.3.2

Tuning a Circuit
In the circuit shown in Figure 4.3.1 take $L = 1$ H, $R = 0.1$ Ω, and suppose that the input voltage is $E(t) = \cos t - (4/5)\cos 5t$. The ODE (4) for the current becomes

$$I'' + 0.1 I' + \frac{1}{C} I = E'(t) = \sin t + 4 \sin 5t$$

The input $E'(t)$ is the sum of the two circular frequencies 1 and 5. The component of frequency 5 is evident in the upper graph of Figure 4.3.2; the frequency 1 component is there as well, but not so obvious. Note that the ODE is solved over $0 \le t \le 100$, but plotting is done for $90 \le t \le 100$. If we "tune" the circuit badly by setting $1/LC$ equal to, say, 81 (which is far away from $\omega^2 = 1^2$ or 5^2), we would not expect to hear either input frequency, and indeed we do not (see the nearly flat line (solid) in the lower

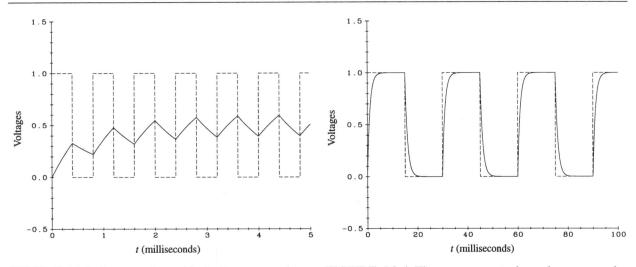

FIGURE 4.3.3 Source voltage (dashed), output voltage (solid); $f = 1250$ Hertz ($T = 0.8$ milliseconds). See Example 4.3.3.

FIGURE 4.3.4 The message gets through: source voltage (dashed), output voltage (solid); $f = 33$ Hertz ($T = 30$ milliseconds). See Example 4.3.3.

graph of Figure 4.3.2). On the other hand, if we tune the circuit correctly by choosing C so that $1/LC = \omega^2$ we can easily pick up a component of frequency ω in the input. The long-dashed curve of frequency 1 in the lower graph of Figure 4.3.2 corresponds to setting $C = 1$ (so that $1/LC = 1$, since $L = 1$) and represents the circuit's response to the input $\sin t$. The amplitude of 8 is what one expects when the circuit is tuned to maximize the amplitude of a particular frequency (see formula (21) in Section 4.2 for maximal gain). In the same way, if C is set at $1/25$ farad so that $1/LC = 25$, then the circuit picks up the frequency 5 component of the input and amplifies it (see the short dashed curve in Figure 4.3.2). See also Problem 7.

The following example is adapted by permission from an article by A. Felzer, in *CODEE*, Winter 1993, pp. 7–9.

EXAMPLE 4.3.3

A Simple Low-Pass Circuit: Will the Message Get Through?

A bread-and-butter problem in electrical engineering is to find the band of frequencies that can be transmitted by a simple communications channel and still be recognizable at the receiving end. The channel can be modeled by the electrical circuit shown in the margin. The source voltage $V_S(t)$ is the "message sent," the output voltage $V_O(t)$ is the "message received"; the constants R and C are the (lumped) resistance and capacitance. Assuming that the current $I(t)$ traverses the circuit clockwise, Kirchhoff's Voltage Law says that $V_S = V_R + V_O$, where V_R is the voltage drop across the resistor, and V_O the voltage drop across the capacitor. Ohm's Law says that $V_R = RI$, and Coulomb's Law says that $V_O = (q_0 + \int_0^t I(s)ds)/C$, where q_O is the initial charge on the capacitor plate. Thus, $V_O' = I/C$, and so $V_R = RCV_O'$. Putting these facts together we obtain the model ODE

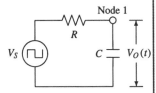

$$RCV_O' + V_O = V_S$$

For an example, measure time in milliseconds and set $RC = 1$, $V_O(0) = 0$ to obtain the IVP:

$$V_O'(t) + V_O(t) = V_S(t), \quad V_O(0) = 0$$

Suppose that the source voltage is an on-off pulse of period T milliseconds (i.e., of frequency $f = 1000/T$ cycles per second, or Hertz) with amplitude 1, and is "on" for the

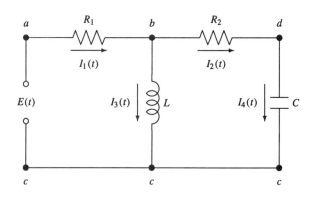

FIGURE 4.3.5 Two-loop circuit.

first half of each cycle and "off" for the second half (i.e., $V_S(t) = \text{sqw}(t, 50, T)$). Using a numerical solver to solve the IVP for frequencies $f = 1250$ Hertz (Figure 4.3.3) and 33 Hertz (Figure 4.3.4), we conclude that the communication channel does best with low frequency signals. It is called a *low-pass channel*, because it lets low frequency signals pass through with little distortion. See Problem 9.

Kirchhoff's Current Law: Multiloop Circuits

Much more complicated circuits than the simple *RLC* series circuit can be modeled via differential equations for the currents flowing in loops that compose the circuit. In addition to Kirchhoff's Voltage Law, there is another law that governs currents in circuit elements.

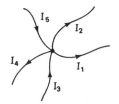

> Kirchhoff's Current Law.   At each point of a circuit the sum of the currents entering is equal to the sum of the currents leaving.

To apply Kirchhoff's laws to multiloop circuits it is important to note that the currents through the various circuit elements of a loop may have different directions. Thus, the first thing we must do is *assume* a direction for the current through each circuit element; the current through the element will be positive if it flows in the direction specified, and negative otherwise. If our assumption about (positive) current flow is incorrect for a given current element, then our circuit equations will ultimately inform us that the current has a negative value for the element, indicating current flow in the opposite direction. Next we introduce reference points a, b, c, ..., and finally we decide on the polarity of the given external source by indicating the positive and negative connectors by "+" and "−" signs. An example best illustrates the use of Kirchhoff's laws in a multiloop circuit.

EXAMPLE 4.3.4

A Two-Loop Circuit

Consider the two-loop circuit pictured in Figure 4.3.5 where I_1, I_2, I_3, and I_4 are the positively directed currents through the elements indicated. From Kirchhoff's Current Law at node points b and d, the current I_3 through the inductor with inductance L

must equal $I_1 - I_2$, while the currents I_2 and I_4 must be equal. Applying Kirchhoff's Voltage Law to the two inner loops, we have that (assuming $q_0 = 0$ on the capacitor)

$$R_1 I_1 + L(I_1' - I_2') = E(t)$$

$$R_2 I_2 + \frac{1}{C} \int_0^t I_2(s)\, ds + L(I_2' - I_1') = 0$$

Rearranging both equations and differentiating the second to get rid of the integration, we have that

$$L(I_1' - I_2') + R_1 I_1 = E(t)$$

$$L(I_2'' - I_1'') + R_2 I_2' + \frac{1}{C} I_2 = 0$$

We leave the mathematical treatment of such coupled linear systems to later chapters.

Comments

If $E(t)$ is not a continuous and piecewise-differentiable function, then ODE (4) is not an alternative way to solve ODE (3). To see this, consider the case where E is piecewise constant. If the initial data $I(0) = 0$, $I'(0) = 0$, then the only solution of ODE (4) that satisfies these inital data would be $I = 0$, for all $t \geq 0$, which we know cannot be the case. We shall pursue this situation further in the problems.

Although second-order linear constant coefficient ODEs model circuits quite well, there are some difficulties. For example, what if the circuit elements are not passive but change in time? What if the elements are distributed through the circuit and cannot be lumped as we have assumed? The latter case is exemplified by the current I and voltage drop V in a long transmission line; a partial differential equation model is required in this case because I and V depend on location as well as time.

PROBLEMS

1. **(a)** Solve the IVP (5) if $R = 20$ Ω, $L = 10$ H, $C = 0.05$ F, $E(t) = 12$ volts, $q_0 = 0.6$ coulomb, and $I_0 = 1$ amp.

    (b) Show that a damped oscillation occurs. Plot the graph of $I = I(t)$.

    (c) Suppose the circuit sketched in the margin opposite IVP (6) is switched to the battery. What is the limiting value of $q(t)$? Plot $q = q(t)$.

    (d) To what value must the resistance be increased in order to reach the over-damped case?

2. Suppose that a simple *RLC* circuit is "charged up" by a battery so that the charge on the capacitor is $q_0 = 10^{-3}$ coulomb, while $I_0 = 0$. The battery is then removed. In each case find the charge $q(t)$ on the capacitor and the current $I(t)$ in the circuit for $t \geq 0$. [*Hint*: Use IVP (5) with $I_0 = 0$, $E(t) = 0$, $t_0 = 0$, $q_0 = 10^{-3}$, and find the current. Then find the charge by using formula (1).]

    (a) $L = 0.3$ H, $R = 15$ Ω, $C = 3 \times 10^{-2}$ F ·

    (b) $L = 1$ H, $R = 1000$ Ω, $C = 4 \times 10^{-4}$ F

    (c) $L = 2.5$ H, $R = 500$ Ω, $C = 10^{-6}$ F

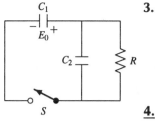

3. Two capacitors ($C_1 = 10^{-6}$ F and $C_2 = 2 \times 10^{-6}$ F) and a resistor ($R = 3 \times 10^6$ Ω) are arranged in a circuit as shown. The capacitor C_1 is initially "charged" to E_0 volts with polarity as shown. The switch S is closed at time $t = 0$. Determine the current that flows through the capacitor C_1 as a function of time. [*Hint*: Apply Kirchhoff's Voltage Law to the left loop, then to the right loop. Note that $q_1(0) = C_1 E_0$, $q_2(0) = -C_2 E_0$.]

4. Find the charge $q(t)$ on the capacitor of the *RLC* circuit in Figure 4.3.1. Assume zero initial charge and current, $R = 20$ Ω, $L = 10$ H, $C = 0.01$ F, and $E(t) = 30 \cos 2t$ volts.

5. A simple *RLC* circuit has a capacitor with $C = 0.25$ F, a resistor with $R = 7 \times 10^4$ Ω , and an inductor with $L = 0.25$ H. The initial charge on the capacitor is zero. If an impressed voltage of 60 V is connected to the circuit, and the circuit is closed at $t = 0$, determine the charge on the capacitor for $t > 0$. Estimate the charge when $t = 0.1$ sec. [*Hint*: First solve the current ODE for $I(t)$, then integrate to get $q(t)$.]

6. **(a)** Show that if $R = 0$ in the simple *RLC* circuit and the impressed voltage is of the form $E_0 \cos \omega t$, the charge on the capacitor will become unbounded as $t \to \infty$ if $\omega = 1/\sqrt{LC}$. This is the phenomenon of resonance. [*Hint*: Look at the form of particular solutions.]

   (b) Show that the charge will always be bounded, no matter what the choice of ω, provided that there is some resistance in the circuit.

7. (*Tuning a Circuit: Sensitivity to Resistance*). Using the ODE for $I(t)$ from Example 4.3.2 with $L = 1$ H, $C = 1/25$ F, $R = 1.0, 0.1, 0.01$ Ω, graphically study the effect of the resistance on the output current when tuning the circuit to the input frequency 5. What do you conclude? Explain. [*Hint*: Look at the formulas (21) in Section 4.2.]

8. Set up, but do not solve, three ODEs for I_1, I_2, I_3 for the circuit below.

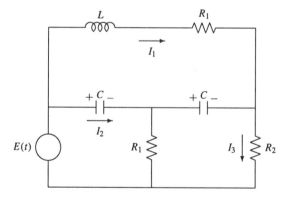

9. (*Low-Pass Circuit; Sensitivity to Resistance*). Carry out a sensitivity analysis for the low-pass circuit of Example 4.3.3, where the sensitivity parameter is the resistor R [set $C = 1$ as in the example]. How do changes in R widen or narrow the pass-band of the circuit? By pass-band we mean the broadest range of input frequencies $f = 1000/T$ which pass through the circuit with "minimal" distortion. Explain your results.

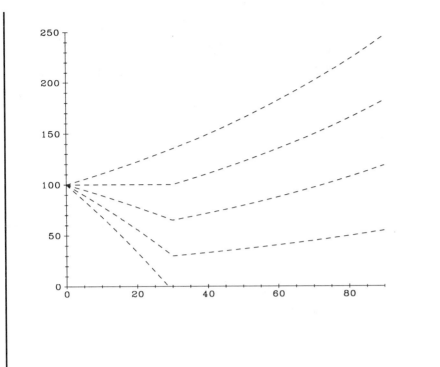

CHAPTER 5

The Laplace Transform

In this chapter we study a special linear operator that is widely used to construct solution formulas for differential, difference, and integral equations. The operator involves an improper integral and "transforms" problems involving differential equations into problems with a simpler structure—hence the name "integral transform." The method of Laplace transforms is applied to linear ODEs with constant coefficients and with a wide variety of driving terms. The cover figure shows the effects on a population of seasonal harvesting at various rates (see Section 5.2).

5.1 Introduction to the Laplace Transform

An operator is defined by describing its domain (the collection of elements on which it acts), its action, and its range (the collection of elements resulting from the action). In this section we describe the *Laplace transform* operator and its inverse, and show how the transform may be used to solve initial value problems.

Action of the Transform

The action of the Laplace transform operator is defined by an improper integral and acts on a class of functions defined over the half-line $0 \leq t < \infty$. This class includes polynomials, simple exponentials, and bounded piecewise continuous functions.

❖ **Action of the Laplace Transform.** The Laplace transform of the function $f(t), 0 \leq t < \infty$, is the function $\mathcal{L}[f]$ given by

$$\mathcal{L}[f](s) = \int_0^\infty e^{-st} f(t)\, dt \qquad (1)$$

defined for all real s where the integral converges.

Note that $f(t)$ may also be defined for $t < 0$, but these values are not used in calculating the transform. The Laplace transforms of some simple functions are calculated in the following examples. In each case the reader should keep in mind that the transform $\mathcal{L}[f]$ of a function $f(t)$ is itself a function, but in the transform variable s.

EXAMPLE 5.1.1

Transform of a Constant
Suppose that $f(t)$ is the constant function c for all $t \geq 0$ and that the transform variable s is positive. Then

$$\mathcal{L}[f](s) = \int_0^\infty e^{-st} c\, dt = -\frac{c}{s} e^{-st}\Big|_{t=0}^{t=\infty} = \frac{c}{s}, \qquad s > 0$$

since $\lim_{t \to \infty} (e^{-st})/s = 0$, whenever $s > 0$.

EXAMPLE 5.1.2

Transform of t
Let $f(t) = t$ for all $t \geq 0$, and suppose that $s > 0$. Then integration by parts (or a table of integrals) gives us

$$\mathcal{L}[f](s) = \int_0^\infty e^{-st} t\, dt = -\frac{1}{s^2} e^{-st}(st + 1)\Big|_{t=0}^{t=\infty}$$

Since s is assumed to be positive, L'Hôpital's Rule (see Appendix C.5) implies that

$$\lim_{t \to \infty} e^{-st}(st + 1) = \lim_{t \to \infty} \frac{st + 1}{e^{st}} = \lim_{t \to \infty} \frac{s}{s e^{st}} = 0$$

and so

$$\mathcal{L}[f](s) = -\frac{1}{s^2} e^{-st}(st + 1)\Big|_{t=0}^{t=\infty} = 0 + \frac{1}{s^2} = \frac{1}{s^2}, \qquad s > 0$$

EXAMPLE 5.1.3

Transform of e^{at}
Let $f(t) = e^{at}$ for all $t \geq 0$, where $a \neq 0$. Then for all $s > a$, we have

$$\mathcal{L}[f](s) = \int_0^\infty e^{-st} e^{at}\, dt = \frac{1}{a - s} e^{(a-s)t}\Big|_{t=0}^{t=\infty} = \frac{1}{s - a}, \qquad s > a$$

EXAMPLE 5.1.4

Transform of a Square Pulse
Let $f(t)$ be the piecewise-continuous function defined by

$$f(t) = \begin{cases} 1, & 0 \leq t \leq 1 \\ 0, & t > 1 \end{cases}$$

The function f is also call a *square pulse*, and is denoted by $\text{sqp}(t, 1)$ as noted in Appendix C.1. Then for all $s > 0$,

$$\mathcal{L}[f](s) = \int_0^\infty e^{-st} f(t)\, ds = \int_0^1 e^{-st}\, dt = -\frac{1}{s} e^{-st}\Big|_{t=0}^{t=1} = \frac{1}{s}[1 - e^{-s}], \qquad s > 0$$

which like the previous three transforms, decays to zero as $s \to \infty$. Note that $\mathcal{L}[f](s)$ is an infinitely differentiable function of s for $s > 0$, although the original function $f(t)$ is not even continuous on the interval $0 \le t < \infty$.

Recovering a Function From Its Transform

If $\mathcal{L}[f]$ and $\mathcal{L}[g]$ are defined on an interval, $s > s_0$, so is $\mathcal{L}[\alpha f + \beta g]$ for any real α, β. Moreover, $\mathcal{L}$ is a linear operator since

$$\mathcal{L}[\alpha f + \beta g] = \alpha \mathcal{L}[f] + \beta \mathcal{L}[g]$$

Of course, the linear operator $\mathcal{L}$ would not be very useful unless a function can be recovered from its transform. The operator that does this is linear; it is denoted by $\mathcal{L}^{-1}$ and is called the *inverse Laplace transform*. It is not easy to construct an explicit formula describing the action of $\mathcal{L}^{-1}$, but as we shall show, we can get along without a formula by using tables of transforms such as those in the endpapers. Suppose that $f(t)$ and $g(t)$ are both continuous for $t \ge 0$, and suppose that both transforms exist and are identical on a common interval:

$$\mathcal{L}[f](s) = \mathcal{L}[g](s), \qquad s \ge s_0$$

Then it can be shown that $f(t) = g(t)$, $t \ge 0$. Thus if $\mathcal{L}[f](s)$ is, say, known to be $1/s^2$ for $s > 0$, the only continuous function f with this transform is $f(t) = t$, $t \ge 0$ (Example 5.1.2).

From the definition of the integral, if two functions f and g differ only at a finite number of points on any finite interval, then $\mathcal{L}[f] = \mathcal{L}[g]$. For example, both the function $f(t)$ of Example 5.1.4 and the function g defined by

$$g(t) = \begin{cases} 1, & 0 \le t < 1 \\ 2, & t = 1 \\ 0, & t > 1 \end{cases}$$

have as their transforms

$$\mathcal{L}[f](s) = \mathcal{L}[g](s) = \frac{1}{s}[1 - e^{-s}]$$

Thus, we recapture the original function from a transform even if the original function is only piecewise continuous, as long as we are willing to tolerate ambiguities at the points of discontinuity.

An Initial Value Problem

The usefulness of the Laplace transforms in solving linear initial value problems is suggested by the following example. This example shows that a differential equation and initial data can be transformed into an algebraic equation which involves no derivatives at all.

EXAMPLE 5.1.5

Solving an Initial Value Problem

Consider the initial value problem

$$y'(t) + ay(t) = f(t), \qquad y(0) = y_0 \tag{2}$$

where a is a real constant and f is, say, piecewise continuous on $[0, \infty)$. We can, of course, solve (2) by the integrating factor approach of Section 1.2. The method of Laplace transforms may also be applied to solve (2) in the following way. Assuming that $L[y']$, $L[y]$, and $L[f]$ are defined on a common interval (s_0, ∞) for some s_0, apply L to each side of (2) and use the linearity of L to obtain

$$L[y'](s) + aL[y](s) = L[f](s), \qquad s > s_0 \tag{3}$$

Now, assuming that integration by parts is valid in this setting and that $\lim_{t \to \infty} e^{-st} y'(t) = 0$ for each $s > s_0$, we have that

$$L[y'](s) = \int_0^\infty e^{-st} y'(t)\, dt = e^{-st} y(t) \Big|_{t=0}^{t=\infty} + s \int_0^\infty e^{-st} y(t)\, dt = -y(0) + sL[y](s)$$

Thus (3) becomes an algebraic equation for the transform $L[y]$,

$$sL[y](s) - y(0) + aL[y](s) = L[f](s) \tag{4}$$

Solving (4) algebraically for $L[y](s)$, we have that

$$L[y](s) = \frac{y(0)}{s+a} + \frac{L[f](s)}{s+a} \tag{5}$$

At this point, we know the Laplace transform of the solution of (2) even if we do not yet have the solution itself. Assuming that the operator L^{-1} exists, then it is linear, and we can apply L^{-1} to each side of (5) to obtain

$$y(t) = y(0) L^{-1} \left[\frac{1}{s+a} \right](t) + L^{-1} \left[\frac{L[f](s)}{s+a} \right](t) \tag{6}$$

The inverse terms in (6) may be determined by inspection of the Laplace transform tables in the endpapers,

$$L^{-1} \left[\frac{1}{s+a} \right](t) = e^{-at} \qquad \text{[Use number 3 of Table II with } n = 1 \text{ and } a \text{ replaced by } -a]$$

$$L^{-1} \left[\frac{L[f]}{s+a} \right](t) = \int_0^t e^{-a(t-u)} f(u)\, du \qquad \text{[Use number 8 of Table I with } g_1(s) = 1/(s+a),\ g_2(s) = L[f](s)]$$

Thus we have that (6) becomes

$$y(t) = y_0 e^{-at} + \int_0^t e^{-a(t-u)} f(u)\, du \tag{7}$$

The integrating factor technique of Section 1.2 gives the same result.

The advantage of the transform approach is that it reduces the problem of solving an initial value problem such as (2) to the inspection of a table of transforms. In later sections of this chapter we give a number of applications of this approach.

Comments

Example 5.1.5 showed that the transform of a derivative does not involve a derivative. It is this single fact that makes the Laplace transform a useful alternative method for solving linear differential equations. But the method is practical only if a sufficiently large table of transforms is available. The examples of this and later sections show how some of the entries in the Laplace transform tables in the endpapers are determined.

PROBLEMS

1. Find the Laplace transform of each of the following functions; a and b are real constants and n is a positive integer. Specify the largest interval $s > s_0$ on which the transform is defined.

    (a) $3t - 5$ **(b)** t^2 **(c)** t^n **(d)** $\cos at$ **(e)** te^{at} **(f)** $t \sin at$

2. Compute the Laplace transform for each of the following functions and determine the largest interval (s_0, ∞) on which the transform is defined.

    (a) $f(t) = \begin{cases} \sin t, & 0 \le t \le \pi \\ 0, & t > \pi \end{cases}$ **(b)** $f(t) = \begin{cases} 0, & 0 \le t \le 1 \\ t, & t > 1 \end{cases}$

    (c) $f(t) = \begin{cases} t, & 0 \le t \le 1 \\ 0, & t > 1 \end{cases}$ **(d)** $f(t) = \begin{cases} t, & 0 \le t \le 1 \\ 2 - t, & t > 1 \end{cases}$

3. As in Example 5.1.5, solve the following IVPs using the Laplace transform.

    (a) $y' + 2y = 0$, $y(0) = 1$

    (b) $y' + 2y = e^{-3t}$, $y(0) = 5$ [*Hint*: When working with $\mathcal{L}[y]$, write $1/(s + 2)(s + 3)$ as $1/(s + 2) - 1/(s + 3)$.]

    (c) $y' + 2y = e^{it}$, $y(0) = 0$ [*Hint*: Treat i as a constant.]

5.2   The Calculus of the Transform

It is not always efficient to use the definition of the Laplace transform in order to transform a function. To speed up the process we shall develop a "calculus" with which we may express the transform of a combination of functions in terms of the known transforms of the component functions. Thus we need to develop a list of the transforms of elementary functions and some general rules for computing Laplace transforms and for passing back from the s-domain to the t-domain (i.e., for calculating the inverse transform). In this respect the transform calculus resembles the construction of the differential and integral calculus—first the underlying ideas and definitions are presented, and then the formulas and rules are developed.

By the end of this section a number of transforms will have been calculated and several rules of the transform and the inverse transform formulated and proved. For ease of reference, we have placed these transforms and rules, as well as many others, in the tables in the endpapers.

Smoothness and Decay of the Transform

From this point on we need to be more specific about the functions that the Laplace transform acts upon.

❖ **Piecewise Continuous Functions of Exponential Order: $\mathcal{E}$.** A function f defined on $[0, \infty)$ is of *exponential order* if there are positive constants M and a such that $|f(t)| \le Me^{at}$ for all $0 \le t < \infty$. We denote by $\mathcal{E}$ the collection of all piecewise continuous functions on $[0, \infty)$ which are of exponential order.

EXAMPLE 5.2.1

Exponential Order

The function $f(t) = -10t \cos t e^{70t}$ is in $\mathcal{E}$ because it is continuous on $[0, \infty)$ and

$$|f(t)| \le 10te^{70t} \le 10e^t e^{70t} = 10e^{71t}$$

Hence, we can set $M = 10$ and $a = 71$. We have used the fact that $t < e^t$ if $t \ge 0$, which follows from the Taylor series for e^t about $t = 0$,

$$e^t = 1 + t + t^2/2 + \cdots + t^n/n! + \cdots$$

On the other hand, $g(t) = e^{t^2}$ is not in $\mathcal{E}$. For if it were, then there would be positive constants M and a such that $e^{t^2} \le Me^{at}$ for $t \ge 0$. Thus, $e^{t^2-at} \le M$ for $t \ge 0$, but this is false since $e^{t(t-a)} \to \infty$ as $t \to \infty$. Thus, functions in $\mathcal{E}$ can become unbounded as $t \to \infty$, but they cannot "blow up" too fast.

It can be shown that sums, products, and antiderivatives of functions in $\mathcal{E}$, are also in $\mathcal{E}$. Unless stated otherwise, every function f considered in this chapter will be assumed to be in $\mathcal{E}$. This assumption has a very important consequence.

THEOREM 5.2.1

> Smoothness and Decay Property for $\mathcal{L}[f]$. For f in $\mathcal{E}$, the Laplace transform $\phi(s) = \mathcal{L}[f](s)$ exists and is infinitely differentiable on some half-line (s_0, ∞). Moreover, $\phi(s) \to 0$ as $s \to \infty$.

Proof. We give only the basic ideas of the proof. First of all, let f belong to $\mathcal{E}$. To show that $\mathcal{L}[f](s)$ exists for all $s > s_0$ for some s_0, observe that for some positive constants a, M

$$\left| e^{-st} f(t) \right| \le Me^{-(s-a)t}$$

If $A > 0$, then the Integral Estimate of Appendix C.5 implies that

$$\left| \int_0^A e^{-st} f(t)\, dt \right| \le \int_0^A e^{-st} |f(t)|\, dt \le \int_0^A e^{-st} Me^{at}\, dt$$

$$\le \frac{M}{a-s} e^{(a-s)t} \Big|_{t=0}^{t=A} = \frac{M}{s-a} \left[1 - e^{(a-s)A} \right]$$

For $s > a$, we can let $A \to \infty$, and the improper integral defining $\phi(s) = \mathcal{L}[f](s)$ converges. To show that $\phi'(s)$ exists, use Leibniz's Rule (Appendix C.5)

$$\frac{d}{ds} \int_0^\infty e^{-st} f(t)\, dt = \int_0^\infty \frac{\partial e^{-st}}{\partial s} f(t)\, dt = -\int_0^\infty e^{-st} t f(t)\, dt$$

where we have assumed that $(d/ds) \int_0^\infty = \int_0^\infty (\partial/\partial s)$. The functions t and $f(t)$ are in $\mathcal{E}$ and, hence, so is $tf(t)$. Thus, $\phi'(s)$ exists for all $s >$ some s_0. In fact, we can take $s_0 = a$. By the same kind of proof it can be shown that $\phi^{(n)}(s)$ exists for all n.

Thus the Laplace transforms of functions in $\mathcal{E}$ are very well-behaved, even though the functions themselves may have many discontinuities.

Transforms of Derivatives

We shall begin by finding conditions on a function f in $\mathcal{E}$ which guarantee that $\mathcal{L}[f'](s) = s\mathcal{L}[f](s) - f(0)$ [see Example 5.1.5].

Denote by $\mathbf{C}^0([0, \infty))$ the class of all continuous functions on the interval $0 \leq t < \infty$. First we show that if $f(t)$ is in $\mathbf{C}^0([0, \infty))$ and f' is in $\mathcal{E}$, then f is in $\mathcal{E}$. Since f' is piecewise continuous, then the Fundamental Theorem of Calculus (Appendix C.5) states that

$$f(t) = f(0) + \int_0^t f'(r)\, dr, \quad \text{for all } t \geq 0 \tag{1}$$

If $f'(t)$ is also of exponential order, then for some positive constants M, a, we have that $|f'(t)| \leq Me^{at}$, for all $t \geq 0$. Using (1) and the Integral Estimate (Appendix C.6) we obtain the estimate

$$|f(t)| \leq |f(0)| + \int_0^t |f'(r)|\, dr \leq |f(0)| + M \int_0^t e^{ar}\, dr$$

$$= |f(0)| + \frac{M}{a}(e^{at} - 1) \leq \left(|f(0)| + \frac{M}{a}\right)e^{at}, \quad \text{for all } t \geq 0$$

and so $f(t)$ is also of exponential order. Thus $\mathcal{L}[f](s)$ and $\mathcal{L}[f'](s)$ are both defined on a half-line $s \geq s_0$, where $s_0 > a$. Using integration by parts, note that for $s \geq s_0$,

$$\mathcal{L}[f'](s) = \int_0^\infty e^{-st} f'(t)\, dt = e^{-st} f(t)\big|_{t=0}^{t=\infty} + s \int_0^\infty e^{-st} f(t)\, dt = -f(0) + s\mathcal{L}[f](s)$$

where we have used the fact that $e^{-st} f(t) \to 0$, as $t \to \infty$, for any $s > s_0$. Thus, when f is continuous on $0 \leq t < \infty$ and f' is a piecewise continuous function of exponential order, then $\mathcal{L}[f']$ can be computed from $\mathcal{L}[f]$ via the formula (2)

$$\mathcal{L}[f'] = s\mathcal{L}[f] - f(0) \tag{2}$$

We now extend formula (2) so that we can find the transform of higher-order derivatives in terms of the transform of the function itself.

THEOREM 5.2.2

> **Transform of the nth Derivative.** For some integer $n \geq 1$, suppose that $f^{(n-1)}$ and $f^{(n)}$ are in $\mathcal{E}$. Then $f^{(k)}$ is in $C^0([0, \infty))$ and in $\mathcal{E}$, for $k = 0, 1, \ldots, n - 1$, and
>
> $$\mathcal{L}[f^{(n)}] = s^n \mathcal{L}[f] - s^{n-1} f(0) - \cdots - sf^{(n-2)}(0) - f^{(n-1)}(0) \tag{3}$$

Proof. Since $f^{(n)}$ is in $\mathcal{E}$, we see from the Fundamental Theorem of Calculus (Appendix C.5) that

$$f^{(n-1)}(t) = f^{(n-1)}(0) + \int_0^t f^{(n)}(r)\, dr$$

and hence $f^{(n-1)}$ is continuous on $0 \leq t < \infty$. Since $f^{(n)}$ is in $\mathcal{E}$, we can use the argument applied to (1) to establish that $f^{(n-1)}$ is in $\mathcal{E}$ as well. Repeating this argument n times, we see that $f^{(k)}$ is in $\mathcal{E}$ for $k = 0, 1, \cdots, n$, and that $f^{(k)}$ is in $C^0([0, \infty))$ for $k = 0, 1, \cdots, n - 1$. Formula (3) is proved by induction, after integration by parts is used to establish the induction step:

$$\mathcal{L}[f^{(N)}] = \int_0^\infty e^{-st} f^{(N)}(t)\, dt = e^{-st} f^{(N-1)}(t)\Big|_{t=0}^{t=\infty} + s \int_0^\infty e^{-st} f^{(N-1)}(t)\, dt$$

$$= -f^{(N-1)}(0) + s\mathcal{L}[f^{(N-1)}]$$

If (3) holds for all $n \leq N - 1$, the above calculation shows that (3) also holds when $n = N$. Formula (2) establishes the anchor step $n = 1$. By induction, the theorem is proven.

Many applications involve second-order ODEs, so it is useful to write out formula (3) for $n = 2$

$$\mathcal{L}[f''](s) = s^2 \mathcal{L}[f] - sf(0) - f'(0) \tag{4}$$

Inverse Transforms of Derivatives

There is a dual formula in the transform calculus, a formula which tells us what transforms into $\phi^{(n)}(s)$, where $\phi = \mathcal{L}[f]$, rather than what $f^{(n)}(t)$ transforms into:

THEOREM 5.2.3

Inverse Transform of the nth Derivative. Let f be in $\mathcal{E}$ and let $\phi = \mathcal{L}[f]$. Then

$$\phi^{(n)}(s) = \mathcal{L}[(-1)^n t^n f(t)] \tag{5}$$

Theorem 5.2.3 is proved by differentiating $\phi(s) = \int_0^\infty e^{-st} f(t)dt$, n times. Before we see how we can apply these results, let us work out some examples.

EXAMPLE 5.2.2

Using formula (4) for $f(t) = \cos at$, we see that

$$\mathcal{L}[-a^2 \cos at] = s^2 \mathcal{L}[\cos at] - s \cdot \cos 0 - 0$$
$$-a^2 \mathcal{L}[\cos at] = s^2 \mathcal{L}[\cos at] - s$$

or, solving for $\mathcal{L}[\cos at]$,

$$\mathcal{L}[\cos at] = \frac{s}{s^2 + a^2}$$

EXAMPLE 5.2.3

Let $f(t) = -(1/a)\cos at$. Now since $f'(t) = \sin at$, we have by formula (3) and Example 5.2.2 that for all sufficiently large s,

$$\mathcal{L}[\sin at] = s\mathcal{L}\left[-\frac{1}{a}\cos at\right] - \left(-\frac{1}{a}\cos 0\right)$$
$$= -\frac{s}{a}\frac{s}{s^2 + a^2} + \frac{1}{a} = \frac{a}{s^2 + a^2}$$

EXAMPLE 5.2.4

Let $f(t) = t^n$, $n = 0, 1, 2, \ldots$; then $f^{(n)}(t) = n!$. Solving for $\mathcal{L}[f]$ in formula (3), we see that

$$\mathcal{L}[f] = \frac{1}{s^n}\{\mathcal{L}[f^{(n)}] + s^{n-1}f(0) + \cdots + f^{(n-1)}(0)\}$$

and, because $\mathcal{L}[n!] = n!/s$, we have that

$$\mathcal{L}[t^n] = \frac{1}{s^n}\left\{\frac{n!}{s} + 0 + \cdots + 0\right\} = \frac{n!}{s^{n+1}}$$

EXAMPLE 5.2.5

As shown in Example 5.1.3, $\mathcal{L}[e^{at}] = (s-a)^{-1}$, $s > a$. This result can also be shown using formula (3) with $f(t) = e^{at}$ because $\mathcal{L}[ae^{at}] = a\mathcal{L}[e^{at}] = s\mathcal{L}[e^{at}] - 1$.

EXAMPLE 5.2.6

Initial Value Problem

We shall solve the IVP

$$y'' - y = 1, \quad y(0) = 0, \quad y'(0) = 1$$

We know that the problem is uniquely solvable, and that the solution and its derivatives of all orders are in $\mathcal{E}$. We may transform both sides of the ODE using formula (4) and the initial data to obtain that

$$\mathcal{L}[y'' - y] = \mathcal{L}[1]$$

$$s^2 \mathcal{L}[y] - sy(0) - y'(0) - \mathcal{L}[y] = 1/s$$

$$\mathcal{L}[y] = \frac{sy(0) + y'(0) + 1/s}{s^2 - 1} = \frac{1 + 1/s}{s^2 - 1} = \frac{1}{s(s-1)} = \frac{1}{s-1} - \frac{1}{s}, \quad s > 1$$

Thus from Example 5.2.4 with $n = 0$ and Example 5.2.5 with $a = 1$, we have that

$$y(t) = e^t - 1$$

The methods of Chapter 3 could have been used to solve the initial value problem of Example 5.2.6 just as easily. However, we shall soon have examples where the transform technique is much more efficient than the characteristic polynomial, variation of parameters, or the Green's Kernel approach of earlier chapters. Whatever method is used, the same solution will result as long as the Existence and Uniqueness Theorem applies. Note that the transform method automatically involves the initial data, unlike the situation with earlier methods, where the data were used almost as an afterthought to evaluate constants of integration. Of course, the transform approach will be successful only if we can carry the transform of the solution of the initial value problem back into the time domain.

Transforms of Integrals

We may transform an integral of a function as easily as a derivative.

THEOREM 5.2.4

Transform of an Integral. Let f be in $\mathcal{E}$, and $a \geq 0$. Then $\int_a^t f(x)\, dx$ is in $\mathcal{E}$, and

$$\mathcal{L}\left[\int_a^t f(x)\, dx \right] = \frac{1}{s} \mathcal{L}[f] - \frac{1}{s} \int_0^a f(x)\, dx \tag{6}$$

Proof. Since antiderivatives of piecewise continuous functions are continuous, $\int_a^t f(x)\, dx$ is a continuous function of t on $[0, \infty)$. Since f is in $\mathcal{E}$ there exist positive constants, M and b, such that $|f(t)| \leq M e^{bt}$, for all $t \geq 0$. Thus, we have the estimate

$$\left| \int_a^t f(x)\, dx \right| \leq \int_a^t |f(x)|\, dx \leq M \int_a^t e^{bx}\, dx \leq \frac{M}{b} e^{bt}, \quad \text{all } t \geq 0$$

and hence $\int_a^t f$ is of exponential order. Integration by parts yields formula (6) as follows:

$$\mathcal{L}\left[\int_a^t f(x)\, dx \right] = \int_0^\infty e^{-st} \left(\int_a^t f(x)\, dx \right) dt$$

$$= \left[-\frac{1}{s} e^{-st} \int_a^t f(x)\, dx \right]_{t=0}^{t=\infty} + \frac{1}{s} \int_0^\infty e^{-st} f(t)\, dt$$

$$= -\frac{1}{s} \int_0^a f(x)\, dx + \frac{1}{s} \mathcal{L}[f]$$

EXAMPLE 5.2.7

Observe that $\int_0^t x e^x dx = t e^t - e^t + 1$. Thus, from formula (6), with $a = 0$

$$\mathcal{L}\left[\int_0^t x e^x\, dx \right] = \frac{1}{s} \mathcal{L}[t e^t] - \frac{1}{s} \int_0^0 x e^x\, dx = \frac{1}{s} \mathcal{L}[t e^t]$$

on the one hand, and $\mathcal{L}[\int_0^t x e^x\, dx] = \mathcal{L}[t e^t] - \mathcal{L}[e^t] + \mathcal{L}[1]$, on the other hand. Equating these two expressions, we can solve for $\mathcal{L}[t e^t]$ to find that

$$\mathcal{L}[t e^t] = \frac{1}{(s-1)^2}, \qquad s > 1$$

This result also follows immediately from formula (5) and Example 5.2.5 by letting $\phi = \mathcal{L}[-e^t] = -1/(s-1)$. Then set $n = 1$ in (5) and obtain $\mathcal{L}[t e^t] = \phi'(s) = 1/(s-1)^2$.

Shifting Theorems: The Heaviside Function

One of the more useful functions in the applications and in the transform calculus is the function defined below.

❖ **Heaviside Function.** The *Heaviside* or *unit step function* is given by

$$H(t) = \begin{cases} 1, & t \geq 0 \\ 0, & t < 0 \end{cases} \tag{7}$$

Observe that $H(t) = \text{step}(t)$ as defined in Appendix C.1.[1]

For any real number a,

$$H(t-a) = \begin{cases} 1, & t \geq a \\ 0, & t < a \end{cases}, \qquad H(a-t) = \begin{cases} 0, & t > a \\ 1, & t \leq a \end{cases}$$

Many on-off functions can be written in terms of Heaviside functions. For example, the *pulse function*

$$p(t) = \begin{cases} 1, & a \leq t \leq b \\ 0, & \text{all other } t \end{cases} = H(t-a) - H(t-b) \tag{8}$$

If we agree that any function f in $\mathcal{E}$ is defined for all real numbers t by simply defining $f(t)$ arbitrarily for $t < 0$, then $H(t-a) f(t-a)$ is just f translated a units along the t-axis (to the right if $a > 0$, to the left if $a < 0$), and reduced to zero for $t < a$.

---

[1] Oliver Heaviside (1850–1925) was a self-taught Englishman who invented his own operational calculus to solve differential equations many years before Laplace transforms were introduced into electrical engineering for the same purpose. Heaviside coined the terms "inductance" and "capacitance" and introduced the telegrapher's equation, a partial differential equation for the voltage in a cable as a function of time, location, resistance, inductance, and capacitance. He also predicted the existence of a reflecting ionized region around the earth (the ionosphere). Poverty stricken and scorned in his early years because of his lack of education, his genius was eventually recognized, and he received many honors and awards.

The next theorem is often used in computing transforms.

THEOREM 5.2.5

The Shifting Theorem. Suppose that f is in $\mathcal{E}$. Then $e^{at} f(t)$ is in $\mathcal{E}$ and

$$\mathcal{L}[e^{at} f(t)](s) = \mathcal{L}[f](s - a) \tag{9}$$

$$\mathcal{L}[H(t-a)f(t-a)] = e^{-as}\mathcal{L}[f(t)], \qquad a \geq 0 \tag{10}$$

$$\mathcal{L}[H(t-a)f(t)] = e^{-as}\mathcal{L}[f(t+a)], \qquad a \geq 0 \tag{11}$$

Proof. Formula (9) follows directly from the definition of the Laplace transform operator. Note from (9) that multiplying a function in the time domain by e^{at} shifts the transform variable s by the amount a. To prove (10) note that

$$\mathcal{L}[H(t-a)f(t-a)] = \int_0^\infty e^{-st} H(t-a)f(t-a)\, dt = \int_a^\infty e^{-st} f(t-a)\, dt$$

$$= \int_0^\infty e^{-s(x+a)} f(x)\, dx = e^{-as}\mathcal{L}[f]$$

where the variable change $x = t - a$ was used in the last integral. Formula (10) can be viewed this way: Define f to be 0 for $t < 0$, and shift the graph of f to the left by a; the effect of transforming this translated function is to multiply the transform of the original function by e^{-as}. To derive formula (11), set $g(t) = f(t-a)$, so $g(t+a) = f(t)$ in formula (10); then rename g as f.

EXAMPLE 5.2.8

With the help of Example 5.2.2 and formula (9) with $a = -2$ we see that

$$\mathcal{L}[e^{-2t} \cos 3t] = \frac{s+2}{(s+2)^2 + 9}$$

EXAMPLE 5.2.9

Inverting a Transform
Let us try to find a solution $f(t)$ to the equation

$$\mathcal{L}[f] = \frac{(2s+3)}{(s^2 - 4s + 20)}$$

We write

$$\frac{2s+3}{s^2 - 4s + 20} = \frac{2(s-2) + 7}{(s-2)^2 + 16} = 2\left[\frac{s-2}{(s-2)^2 + 16}\right] + \frac{7}{4}\left[\frac{4}{(s-2)^2 + 16}\right]$$

and hence from (9) of the Shifting Theorem with $a = 2$ and Examples 5.2.2 and 5.2.3 we see that $f(t) = 2e^{2t} \cos 4t + (7/4)e^{2t} \sin 4t$.

EXAMPLE 5.2.10

Consider the function

$$f(t) = \begin{cases} 1, & 0 < t < 1 \\ 2 - t, & 1 < t < 2 \\ 0, & t > 2 \end{cases}$$

Because by (8) $H(t-a) - H(t-b)$ is the pulse function whose value is 1 for $a \leq t < b$ and 0 otherwise, we see that

$$f(t) = \{H(t) - H(t-1)\} + (2-t)\{H(t-1) - H(t-2)\}$$
$$= H(t) - H(t-1)(t-1) + H(t-2)(t-2)$$

Applying the Laplace transform we have that

$$\mathcal{L}[f] = \mathcal{L}[H(t)] - \mathcal{L}[H(t-1)(t-1)] + \mathcal{L}[H(t-2)(t-2)]$$

and using formula (10) in the Shifting Theorem, we see that,

$$\mathcal{L}[f] = \frac{1}{s} - e^{-s}\frac{1}{s^2} + e^{-2s}\frac{1}{s^2} = \frac{s - e^{-s} + e^{-2s}}{s^2}, \qquad s > 0$$

EXAMPLE 5.2.11

According to formula (11) in the Shifting Theorem

$$\mathcal{L}[H(t-a)] = \mathcal{L}[H(t-a) \cdot 1] = e^{-as}\frac{1}{s} \tag{12}$$

which could also be obtained by direct integration.

Incidentally, it is useful to have an expression for the transform of $H(a-t)$. By integration we have, for $a > 0$,

$$\mathcal{L}[H(a-t)] = \int_0^a e^{-st}\, dt = \frac{1 - e^{-as}}{s}, \qquad s > 0 \tag{13}$$

EXAMPLE 5.2.12

According to (11) (and a trigonometric identity in Appendix C.4) we have that

$$\mathcal{L}[H(t-a)\sin t] = e^{-as}\mathcal{L}[\sin(t+a)] = e^{-as}\mathcal{L}[\sin t \cos a + \cos t \sin a]$$
$$= e^{-as}\left\{\frac{\cos a}{s^2+1} + \frac{s\sin a}{s^2+1}\right\}$$

where we also used the results of Examples 5.2.2 and 5.2.3.

Partial Fractions

As we saw in Example 5.2.6, before we could conveniently calculate the inverse Laplace transform of $(s(s-1))^{-1}$ we had to first write it as a sum of simpler terms, $(s-1)^{-1} - s^{-1}$. The procedure which produced this result is called the *method of partial fractions*. The method is often used in the process of solving initial value problems (IVPs) with the Laplace transform technique. The few special cases needed to solve the IVPs of this section are given below.

For any constants A, B, a, b, with $a \neq b$, we can find constants C and D so that

$$\frac{As+B}{(s-a)(s-b)} = \frac{C}{s-a} + \frac{D}{s-b}$$

Multiplying this equation through by $(s-a)(s-b)$, we obtain the identity

$$As+B = C(s-b) + D(s-a), \quad \text{for all } s$$

Evaluating this identity at $s = a$ and $s = b$, we obtain that $Aa + B = C(a-b)$, $Ab +$

$B = D(b - a)$, from which C and D are easily determined. Hence,

$$\frac{As + B}{(s - a)(s - b)} = \frac{(Aa + B)/(a - b)}{s - a} + \frac{(Ab + B)/(b - a)}{s - b} \tag{14}$$

The method used to derive (14) extends to the general case where the polynomial in the denominator has only simple roots. Suppose

$$Q(s) = (s - \alpha_1) \cdots (s - \alpha_n), \qquad \alpha_1, \ldots, \alpha_n \text{ distinct}$$

and $P(s)$ is a polynomial of degree less than n. Then we have the expansion

$$\frac{P(s)}{Q(s)} = \sum_{k=1}^{n} \frac{P(\alpha_k)}{Q_k(\alpha_k)} \cdot \frac{1}{s - \alpha_k}, \quad \text{where } Q_k(\alpha_k) = \prod_{\substack{i=1 \\ i \neq k}}^{n} (\alpha_k - \alpha_i) \tag{15}$$

Observe that $Q_k(\alpha_k)$ is just $Q(s)$ with the factor $s - \alpha_k$ omitted, and then the result evaluated at α_k. For example, if A, B, C are given (not all zero) and a, b, c are distinct, then

$$\frac{As^2 + Bs + C}{(s - a)(s - b)(s - c)} = \frac{Aa^2 + Ba + C}{(a - b)(a - c)} \cdot \frac{1}{s - a} + \frac{Ab^2 + Bb + C}{(b - a)(b - c)} \cdot \frac{1}{s - b}$$
$$+ \frac{Ac^2 + Bc + C}{(c - a)(c - b)} \cdot \frac{1}{s - c} \tag{16}$$

There are similar (but more complicated) expansions for the cases where $Q(s)$ has multiple roots. Rather than write them out in detail, we look at two examples. For constants A, B, and a, we can find unique constants C, D such that

$$\frac{As + b}{(s - a)^2} = \frac{C}{s - a} + \frac{D}{(s - a)^2}$$

Multiplying this equation through by $(s - a)^2$, we obtain the identity

$$As + B = C(s - a) + D, \quad \text{for all } s$$

Evaluating this identity and its derivative at $s = a$, we obtain $Aa + B = D$, $A = C$. Hence

$$\frac{As + B}{(s - a)^2} = \frac{A}{s - a} + \frac{Aa + B}{(s - a)^2} \tag{17}$$

Suppose that A, B, C are constants (not all zero), and that a and b are distinct constants. Then

$$\frac{As^2 + Bs + C}{(s - a)(s - b)^2} = \frac{D}{s - a} + \frac{E}{s - b} + \frac{F}{(s - b)^2} \tag{18}$$

where D, E, and F are found by multiplying through by $(s - a)(s - b)^2$, and then setting $s = a, b$ to obtain D, F, and finally differentiating twice to find E

$$D = \frac{Aa^2 + Ba + C}{(a - b)^2}, \quad F = \frac{Ab^2 + Bb + C}{b - a}, \quad E = A - D \tag{19}$$

The values of D, F, and E given in (19) can be inserted into the partial fraction decomposition (18).

As noted earlier, these partial fraction formulas are enough to handle all the expansions needed in this section. See Section 5.3 for additional partial fraction expansions and Appendix C.4 for a general partial fractions expansion theorem.

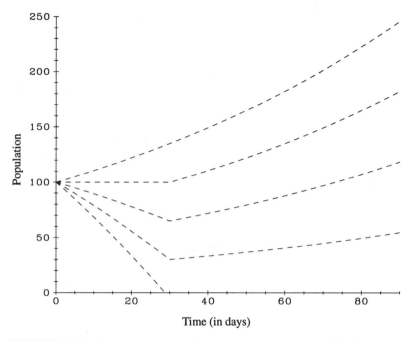

FIGURE 5.2.1 Harvested population as h changes. See IVP (20) where $k = 0.01$, $A = 100$, $h = 0, 1, 2, 3, 4$ (from top curve down).

The Effect of a Thirty-Day Harvest on a Population

Suppose that the growth of a population may be modeled by a law of exponential change. What will be the effect of "harvesting" the population at a constant rate for a fixed span of time? (Refer to the problem set of Section 1.5 for a similar problem, but with perpetual harvesting and logistic growth.)

Let $y(t)$ denote the size of the population at time t, where time is measured in days, and suppose that $y(0) = A$, a positive constant. Let us suppose that the rate coefficient is another positive constant k measured in $(\text{days})^{-1}$, and that harvesting is carried out at the rate of h "individuals" per day over the 30 days of the harvest season. A model of this is given by the initial value problem

$$y'(t) = ky(t) - hH(30 - t), \qquad y(0) = A \tag{20}$$

Our goal is to find an expression for $y(t)$ for all $t \geq 0$.

Applying the Laplace transform to (20), we have from formula (13) that for $s > k$

$$s\mathcal{L}[y] - A = k\mathcal{L}[y] - \frac{h}{s}(1 - e^{-30s})$$

$$\mathcal{L}[y] = \frac{A}{s-k} - \frac{h}{s(s-k)}(1 - e^{-30s})$$

$$= \frac{A}{s-k} + \frac{h}{k}\left(\frac{1}{s} - \frac{1}{s-k}\right)(1 - e^{-30s}) \tag{21}$$

where we have used partial fractions to write

$$\frac{1}{s(s-k)} = -\frac{1}{k}\left(\frac{1}{s} - \frac{1}{s-k}\right)$$

Rearranging the terms on the right of (21) so that we may find the inverse transforms more easily, we have that

$$\mathcal{L}[y] = \frac{A - h/k}{s - k} + \frac{h}{k}\frac{1}{s}(1 - e^{-30s}) + \frac{h}{k}e^{-30s}\frac{1}{s - k}$$

Using Example 5.2.5 and formulas (13) and (10), we may move back into the t-domain:

$$y(t) = \left(A - \frac{h}{k}\right)e^{kt} + \frac{h}{k}H(30 - t) + \frac{h}{k}H(t - 30)e^{k(t-30)} \qquad (22)$$

Equation (22) represents the changing population only for $y(t) > 0$. Thus, the solution makes sense only for those values of the parameters A, h, and k for which $y(t)$ remains positive for all $t \geq 0$. If $h \leq Ak$, there clearly is no difficulty. If $h > Ak$, it may happen that at some time while harvesting is going on, the population becomes extinct. It may be shown that extinction within 30 days occurs only if the harvesting rate and the initial population satisfy the inequality

$$h \geq \frac{kA}{1 - e^{-30k}} \qquad (23)$$

In this case, the model is not valid after the time of extinction. In Figure 5.2.1 we have sketched the population curves for the harvesting rates $h = 0, 1, 2, 3, 4$ (reading from the top down). Observe that the highest harvesting rate, $h = 4$, leads to extinction.

PROBLEMS

1. Find the Laplace transform of each of the following functions.

    (a) $\sinh at$ (b) $\cosh at$

    (c) $t^2 e^{at}$ (d) $(1 + 6t)e^{at}$

    (e) $te^{2t}f'(t)$ (f) $(D^2 + 1)f(t)$

    (g) $(t + 1)H(t - 1)$ (h) $(t - 2)[H(t - 1) - H(t - 3)]$

    (i) $e^{at}[H(t - 1) - H(t - 2)]$ (j) $te^{at}H(t - 1)$

2. (*Using Transform Tables*). Use the transform tables in the end covers to find $\mathcal{L}[f]$ if f is given, and f if $\mathcal{L}[f]$ is given. Which formula did you use?

    (a) $f(t) = \begin{cases} 0, & 0 \leq t < 2, t > 5 \\ 3, & 2 \leq t \leq 5 \end{cases}$ (b) $f(t) = \begin{cases} 2\sin t, & 0 \leq t \leq \pi \\ 0, & t > \pi \end{cases}$

    (c) $f(t) = t^{12}e^{5t}$ (d) $f(t) = 6t\sin 3t$

    (e) $f(t) = J_0(t)$ (f) $\mathcal{L}[f] = \sqrt{s + 2} - \sqrt{s}$

    (g) $\mathcal{L}[f] = \ln(s + 5) - \ln(s - 2)$ (h) $\mathcal{L}[f] = s^{-1}(e^s - 1)^{-1}$

3. Use the Laplace transform to solve each IVP. [*Hint*: See end cover tables, the partial fraction formulas, and the Shifting Theorem (Theorem 5.2.5).]

    (a) $y'' - y' - 6y = 0$, $y(0) = 1$, $y'(0) = -1$

    (b) $y'' + y = \sin t$, $y(0) = 0$, $y'(0) = 1$

    (c) $y'' - 2y' + 2y = 0$, $y(0) = 0$, $y'(0) = 1$

    (d) $y'' + 4y' + 4y = e^t$, $y(0) = 1$, $y'(0) = 1$

    (e) $y'' - 2y' + y = H(t - 1)$, $y(0) = 1$, $y'(0) = 0$

    (f) $y'' + 2y' - 3y = H(1 - t)$, $y(0) = 1$, $y'(0) = 0$

4. Find the Laplace transform of each of the following.

(a) $\int_0^t (t-1)e^t\, dt$ (b) $\int_0^t (t^2 - 2t)\, dt$

(c) $\int_0^t \sin(t - \pi/4)e^t\, dt$ (d) $\int_0^t e^{(t-a)}\cos t\, dt$

5. (*Square Pulse*). Show that for $0 < a < b$, $H(t-a) - H(t-b) = \text{sqp}(t - a, b - a)$, and find $\mathcal{L}[\text{sqp}(t - 1, 3)]$. See Appendix C.1 for the function sqp.

6. (*Radioactive Decay*). A sample of a radioactive element decays at a rate proportional to the amount y present, where k_1 is the proportionality constant. Suppose that at some time, $t = a$, more of the element is allowed to enter the sample from outside at a constant rate k_2, and at a later time, $t = b$, this flow is stopped.

(a) Show that this system can be modeled by

$$y' = -k_1 y + k_2[H(t-a) - H(t-b)], \qquad 0 < a < b$$

(b) Use the Laplace transform to find $y(t)$ when $y(0) = y_0 > 0$. Plot the graph of $y(t)$ with $k_1 = 1$, $k_2 = 0.5$, $a = 1$, $b = 10$, $y_0 = 1$, $0 \le t \le 20$.

7. (*Circuits*). Consider the basic LC or RLC circuit whose source of electromotive force can be turned on and off by a switch. In each problem below denote the roots of the appropriate characteristic polynomial by r_1 and r_2 (assumed distinct, but not necessarily real). Express your results as fuctions of t, a, r_1, and r_2.

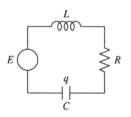

(a) Use the Laplace transform to find the charge $q(t)$ if the switch is turned on at time $t = a > 0$:

$$L\frac{d^2q}{dt^2} + \frac{1}{C}q = E_0 H(t-a), \qquad q(0) = 0, \quad q'(0) = 0$$

(b) Find $q(t)$ if the switch is turned on at $t = a$, and off at $t = b$, $0 < a < b$:

$$L\frac{d^2q}{dt^2} + \frac{1}{C}q = E_0[H(t-a) - H(t-b)], \qquad q(0) = 0, \quad q'(0) = 0$$

(c) Repeat part (a) for the RLC circuit modeled by the IVP:

$$L\frac{d^2q}{st^2} + R\frac{dq}{dt} + \frac{1}{C}q = E_0 H(t-a), \qquad q(0) = 0, \quad q'(0) = 0$$

(d) Repeat part (b) for the RLC circuit modeled by the IVP:

$$L\frac{d^2q}{dt^2} + R\frac{dq}{dt} + \frac{1}{C}q = E_0[H(t-a) - H(t-b)], \qquad q(0) = 0, \quad q'(0) = 0$$

8. (*Maintenance of a Game Species*). In the model of the 30-day harvest, find a relation among A, h, and k which will ensure that exactly 330 days after the end of the harvest, the population will once more be at the initial level A.

9. (*Difference Equations*). Solve the difference equation $3x(t) - 4x(t-1) = 1$, where $x(t) = 0$ if $t \le 0$. [*Hint:* First show that $\mathcal{L}[x(t-1)](s) = e^{-s}\mathcal{L}[x]$. Next show that

$$\mathcal{L}[x] = 1/[s(3 - 4e^{-s})] = (1/3s)(1 + 4/3e^{-s} + \cdots + (4/3e^{-s})^n + \cdots)$$

by using a geometric series expansion. Use the transform tables to find $x(t)$.]

5.3 The Transform of a Periodic Function

The driving force of a mechanical or electrical system often has the form of a periodic function. Any practical calculus of transforms must have a transform formula adequate to handle periodic functions ranging from smoothly contoured sinusoids to the angular pulses of a square or triangular wave train. Fortunately, there is such a formula, and we shall derive it and then apply it to determine the responses of an electrical circuit to a wave train of rectangular pulses. Before looking at the circuit problem, we will show how to transform a periodic function. At the end of the section we give some more partial fraction expansions that are useful in preparing transforms for inversion.

Transform of a Periodic Function

If $|x| < 1$, then the series $1 + x + \cdots + x^n + \cdots = \sum_{n=0}^{\infty} x^n$ sums to $1/(1 - x)$. The series is said to be the *geometric series expansion* of the function $(1 - x)^{-1}$. We shall use the expansion primarily with $x = \pm e^{-as}$, where a and s are both positive, and hence, $|x| < 1$. Under these circumstances, we have that

$$\frac{1}{1 - e^{-as}} = \sum_{n=0}^{\infty} e^{-ans} \tag{1}$$

$$\frac{1}{1 + e^{-as}} = \sum_{n=0}^{\infty} (-1)^n e^{-ans} \tag{2}$$

These formulas will be useful later on. Next, we derive the transform formula for a periodic function.

THEOREM 5.3.1

> Transforming Periodic Functions. **Let** f be piecewise continuous and periodic with period p. Then
>
> $$\mathcal{L}[f] = \frac{1}{1 - e^{-ps}} \int_0^p e^{-sv} f(v) \, dv \tag{3}$$

Proof. We have that

$$\mathcal{L}[f] = \int_0^p e^{-sv} f(v) \, dv + \int_p^{2p} e^{-sv} f(v) \, dv + \cdots$$

Making the variable change $v = u + np$ in the nth integral, $n = 0, 1, 2, \ldots$, and using the fact that $f(u + np) = f(u)$, we have

$$\mathcal{L}[f] = \sum_{n=0}^{\infty} \left\{ \int_0^p e^{-su} f(u) \, du \right\} e^{-psn} = \frac{\int_0^p e^{-su} f(u) \, du}{1 - e^{-ps}}$$

since $(1 - e^{-ps})^{-1} = \sum_{n=0}^{\infty} e^{-psn}$ by formula (1).

Let us see how the theorem applies to a specific periodic function.

EXAMPLE 5.3.1

Transform of a Sinusoid
Although we have already shown that $\mathcal{L}[\sin at] = a(s^2 + a^2)^{-1}$, we shall now derive

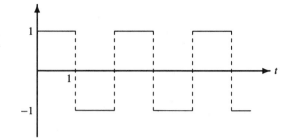

FIGURE 5.3.1 Square wave train (Example 5.3.2).

the formula from Theorem 5.3.1. The period here is $p = 2\pi/a$. Hence, by formula (3) and using a table of integrals

$$\mathcal{L}[\sin at] = \frac{1}{1 - e^{2\pi s/a}} \int_0^{2\pi/a} e^{-sv} \sin av \, dv = \frac{a}{s^2 + a^2}$$

The more interesting situation is with the on-off or angled wave trains of science and engineering.

EXAMPLE 5.3.2

Transform of a Square Wave

Let us find the transform of the periodic function $f = 2 \, \text{sqw}(t, 50, 2) - 1$ shown in Figure 5.3.1. In this case the period is 2 and we have that

$$\int_0^2 e^{-sv} f(v) \, dv = \int_0^1 e^{-sv} \, dv - \int_1^2 e^{-sv} \, dv = \frac{1}{s}(1 - 2e^{-s} + e^{-2s}) = \frac{1}{s}(1 - e^{-s})^2$$

Hence by Theorem 5.3.1,

$$\mathcal{L}[f] = \frac{1}{1 - e^{-2s}} \frac{1}{s}(1 - e^{-s})^2 = \frac{(1 - e^{-s})^2}{s(1 - e^{-s})(1 + e^{-s})}$$

$$= \frac{1 - e^{-s}}{s(1 + e^{-s})} = \frac{1}{s} \tanh\left(\frac{s}{2}\right), \quad s > 0 \tag{4}$$

Expanding $(1 + e^{-s})^{-1}$ in a geometric series (see formula (2)), we can write $\mathcal{L}[f]$ in another way as

$$\mathcal{L}[f] = \frac{1 - e^{-s}}{s} \cdot \frac{1}{1 + e^{-s}} = \frac{1 - e^{-s}}{s} \sum_{n=0}^{\infty} (-1)^n e^{-ns}$$

$$= \frac{1}{s}(1 - e^{-s})(1 - e^{-s} + e^{-2s} - e^{-3s} + \cdots) = \frac{1}{s}\left[1 + 2\sum_{n=1}^{\infty} (-1)^n e^{-ns}\right] \tag{5}$$

a form which is often easier to use in the applications.

A Driven *LC* Circuit

The methods given above may be used to determine the current in an *LC* circuit where the input voltage has the form of the triangular wave train:

$$E(t) = 2 \, \text{trw}(t, 100, 2) - 1$$

The model for the response current $I(t)$ of the circuit to the input voltage $E(t)$ is given by the initial value problem

$$I''(t) + \omega^2 I(t) = \frac{E'(t)}{L} = \frac{4}{L} \operatorname{sqw}(t, 50, 2) - 2, \qquad I(0) = 0, \quad I'(0) = 0 \quad (6)$$

where we have set $\omega^2 = 1/LC$ and have used the definition of the function sqw given in Appendix C.1. We assume that I is measured in amperes, the inductance L in henries, C in farads, time in seconds, and E in volts. Sketches of the circuit and of the functions E/L and E'/L are given in the margin.

Using Example 5.3.2 and other transform formulas of this and the preceding section, we have that

$$s^2 \mathcal{L}[I] + \omega^2 \mathcal{L}[I] = \frac{2}{L} \frac{1 - e^s}{s(1 + e^{-s})}, \quad s > 0$$

or, solving for the transform and using formula (5),

$$\mathcal{L}[I] = \frac{2/L}{s(s^2 + \omega^2)} \left\{ 1 + 2 \sum_{n=1}^{\infty} (-1)^n e^{-ns} \right\}$$

$$= \frac{2/L}{s(s^2 + \omega^2)} + \frac{4}{L} \sum_{n=1}^{\infty} (-1)^n \frac{e^{-ns}}{s(s^2 + \omega^2)} \quad (7)$$

Thus if we determine $\mathcal{L}^{-1}[1/s(s^2 + \omega^2)]$, we may use formula I.3 in the tables to find $I(t)$.

Let us decompose $1/s(s^2 + \omega^2)$ into its partial fractions (see Section 5.2):

$$\frac{1}{s(s^2 + \omega^2)} = \frac{1}{s(s + \omega i)(s - \omega i)} = \frac{1/\omega^2}{s} - \frac{1/(2\omega^2)}{s + \omega i} - \frac{1/(2\omega^2)}{s - \omega i} = \frac{1/w^2}{s} - \frac{s/\omega^2}{s^2 + \omega^2}$$

Hence, using II.6 in the tables,

$$\mathcal{L}^{-1} \left[\frac{1}{s(s^2 + \omega^2)} \right] = \frac{1}{\omega^2} \mathcal{L}^{-1} \left[\frac{1}{s} \right] - \frac{1}{\omega^2} \mathcal{L}^{-1} \left[\frac{s}{s^2 + \omega^2} \right]$$

$$= \frac{1}{\omega^2} - \frac{1}{\omega^2} \cos \omega t \quad (8)$$

From (7) and (8) above, and (10) of the Shifting Theorem (Theorem 5.2.5) and assuming that $\mathcal{L}^{-1} \sum = \sum \mathcal{L}^{-1}$, we have that

$$I(t) = \frac{2}{L} \mathcal{L}^{-1} \left[\frac{1}{s(s^2 + \omega^2)} \right] + \frac{4}{L} \sum_{n=1}^{\infty} (-1) \mathcal{L}^{-1} \left[\frac{e^{-ns}}{s(s^2 + \omega^2)} \right]$$

$$= \frac{2}{L\omega^2} (1 - \cos \omega t) + \frac{4}{L\omega^2} \sum_{n=1}^{\infty} (-1)^n H(t - n)[1 - \cos \omega(t - n)] \quad (9)$$

The form of the solution is somewhat daunting, but for any given positive value of t, there are only a finite number of terms in the sum, the remaining terms being annihilated by the Heaviside functions. For example, we have that

$$I(t) = \begin{cases} \dfrac{2}{L\omega^2}(1 - \cos \omega t), & 0 \le t < 1 \\[2mm] \dfrac{2}{L\omega^2}(1 - \cos \omega t) - \dfrac{4}{L\omega^2}(1 - \cos \omega(t - 1)), & 1 \le t < 2 \\[2mm] \dfrac{2}{L\omega^2}(1 - \cos \omega t) - \dfrac{4}{L\omega^2}(1 - \cos \omega(t - 1)) + \dfrac{4}{L\omega^2}(1 - \cos \omega(t - 2)), & 2 \le t < 3 \end{cases}$$

See Figure 5.3.2 for the plot of the square wave input E'/L, the solution $I(t)$ of IVP (6), and $I'(t)$, where $\omega^2 = 400$ and $L = 1$.

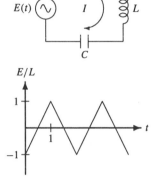

E/L

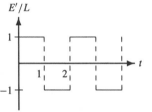

E'/L

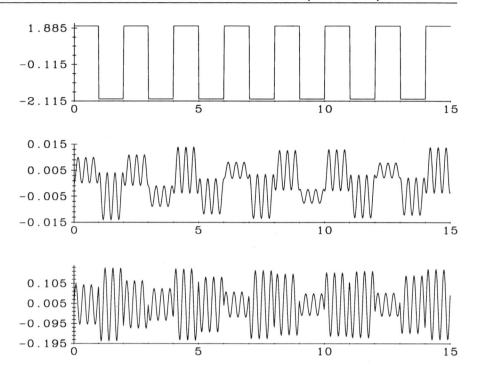

FIGURE 5.3.2 Input-response curves for driven *LC* circuit: input (upper), $I(t)$ (middle), $I'(t)$ (lower). See IVP (6).

More Partial Fractions

Several partial fraction expansions were given in Section 5.2. Here are a few more that are of some use when preparing a Laplace transform for inversion.

Suppose that A, B, C, D, a, b, and c are real constants. Then there exist unique real numbers α, β, γ, δ such that for all s

$$\frac{As^3 + Bs^2 + Cs + D}{(s^2 + a^2)((s-b)^2 + c^2)} = \frac{\alpha s}{s^2 + a^2} + \frac{\beta}{s^2 + a^2} + \frac{\gamma(s-b) + \delta}{(s-b)^2 + c^2} \tag{10}$$

To find α, β, γ, and δ, put the terms on the right over the common denominator $(s^2 + a^2)((s-b)^2 + c^2)$, and obtain (after canceling the denominator)

$$As^3 + Bs^2 + Cs + D = \alpha s((s-b)^2 + c^2) + \beta((s-b)^2 + c^2)$$
$$+ \gamma(s-b)(s^2 + a^2) + \delta(s^2 + a^2)$$

There are several ways to find α, β, γ, δ in terms of A, B, C, D, a, b, and c but the simplest may be to multiply out the terms on the right and then match coefficients of like powers of s on the left and right to obtain four linear equations for α, β, γ, δ, which then must be solved for the unknowns. For example, matching coefficients of s^3 and of s^2, we have that

$$A = \alpha + \gamma$$
$$B = -2b\alpha + \beta - b\gamma + \delta$$

We shall go no farther here in actually calculating α, β, γ, δ.

The right side of (10) is written in a form that is particularly suitable for inversion. For example, by formula (9) in the Shifting Theorem of Section 5.2

$$\mathcal{L}^{-1}\left[\frac{\gamma(s-b)}{(s-b)^2+c^2}\right]=\gamma e^{bt}\cos ct$$

Here is another useful expansion. Given A, B, C, D, a, b, c ($a \neq b$), there exist α, β, γ, δ so that

$$\frac{As^3+Bs^2+Cs+D}{(s-a)(s-b)(s^2+c^2)}=\frac{\alpha}{s-a}+\frac{\beta}{s-b}+\frac{\gamma s}{s^2+c^2}+\frac{\delta}{s^2+c^2} \qquad (11)$$

We omit the details, except to note that after putting the terms on the right over the common denominator $(s-a)(s-b)(s^2+c^2)$ and equating the numerators on the left and right we may set $s=a$ to find α, $s=b$ to find β, and then either match coefficients of like powers of s or else differentiate with respect to s in order to determine γ and δ.

Comments

With the additional partial fractions decompositions given above and with the formula for transforming a periodic function, we have now presented most of the elementary techniques of the Laplace transform. These techniques may be applied to a wide class of constant-coefficient linear ordinary differential equations, and several of the problems below ask the reader to do just that.

PROBLEMS

1. Using the material in Sections 5.1–5.3 and integration tables, find the Laplace transform of each of the following functions. You may use the tables in the end covers only to check your answers.

   (a) $e^t \sin t$ **(b)** $\sin^2 t$

   (c) $\cos^2 t$ **(d)** $t \sin at$

   (e) $t^2 \sin at$ **(f)** $\cos^3 t$

   (g) $e^{-3t}\cos(2t+\pi/4)$ **(h)** $t^2 e^t \cos t$

2. Find the inverse Laplace transform of each of the following without using the tables in the end covers.

   (a) $\dfrac{1+e^{-s}}{s}$ **(b)** $\dfrac{3e^{-2s}}{3s^2+1}$

   (c) $\dfrac{1}{(s-a)^n}$ **(d)** $\dfrac{1-e^{-s}}{s^2}$

   (e) $\dfrac{(s-1)e^{-s}+1}{s^2}$ **(f)** $\ln\dfrac{s+3}{s+2}$ [*Hint:* $\dfrac{d}{ds}\left(\ln\dfrac{s+3}{s+2}\right)=\dfrac{-1}{(s+2)(s+3)}$.]

3. Use partial fractions to find the inverse Laplace transform of each of the following:

   (a) $\dfrac{1}{s(s+1)}$ **(b)** $\dfrac{1}{s(s+2)^2}$

   (c) $\dfrac{1}{(s-a)(s-b)}$, $a \neq b$ **(d)** $\dfrac{s^2+3}{(s-1)^2(s+1)}$

   (e) $\dfrac{3s+1}{(s^2+2s+2)(s-1)}$ **(f)** $\dfrac{s+1}{(s-2)(s^2+9)}$

4. Use Theorem 5.2.2, partial franctions, and the end cover tables to solve the following initial value problems. Plot the solution curves.

   (a) $y'' + 6y' + 5y = t$, $y(0) = y'(0) = 0$

   (b) $y'' + 2y' + y = e^t$, $y(0) = y'(0) = 0$

   (c) $y'' - 2y' + 2y = \sin t$, $y(0) = y'(0) = 0$

   (d) $y'' - 4y' + 4y = 2e^t + \cos t$, $y(0) = 3/25,\ y'(0) = -4/25$

   (e) $y'' - 2y' + y = e^t \sin t$, $y(0) = y'(0) = 0$

   (f) $y'' + 2y' + y = te^{-t}$, $y(0) = 1,\ y'(0) = -2$

   (g) $y''' + y'' + 4y' + 4y = -2$, $y(0) = 0,\ y'(0) = 1,\ y''(0) = -1$

5. (*LC Circuit*). The response $I(t)$ of the IVP $I'' + \omega^2 I = 4\,\mathrm{sqw}(t, 50, p) - 2,\ I(0) = 0,\ I'(0) = 0$, is sensitive to changes in ω and in p. Plot the input and the response for $0 \le t \le 50$ [$0 \le t \le 100$ in part(**b**)] if ω and p are as given below. Does the response appear to be periodic? If it is, what is the period? If not periodic, describe and explain what the response is doing.

   (a) $\omega^2 = 1,\ p = 5\pi$ (b) $\omega^2 = 20,\ p = 5\pi$ (c) $\omega^2 = 1,\ p = 2\pi$

6. (*Transform of Square Wave*). Let $f(t) = A\,\mathrm{sqw}(t, 50, 2a) - B$, A and B constants. Show that

$$\mathcal{L}[f] = \frac{A}{s}\left[1 + \sum_1^\infty (-1)^n e^{-nas}\right] - B/s$$

[*Hint*: See Example 5.3.2 and III.2 in the tables; then use the linearity of $\mathcal{L}$.]

7. (*Square Wave Inputs*). Solve each IVP below. Plot the input and the output. [*Hint*: Use the expansion given in Problem 6.]

   (a) (*Low Pass Filter*). $y' + y = \mathrm{sqw}(t, 50, p)$, $y(0) = 0$. First let $p = 30,\ 0 \le t \le 100$; then $p = 1,\ 0 \le t \le 10$. [Note that the IVPs model a low-pass communication channel that transmits low-frequency signals with much less distortion than high-frequency signals. See Example 4.3.3.]

   (b) (*Beats*). $y'' + 4y = \mathrm{sqw}(t, 50, p)$, $y(0) = 0,\ y'(0) = 0,\ 0 \le t \le 60$. First let $p = 7$ and then $p = 3$. Plot the input, $y(t)$, and $y'(t)$. Explain why pronounced "beats" appear when $p = 3$. [*Hint*: The natural period is $\pi = 3.14159\ldots$, while the period of the driving term is 3. See Section 4.2 for a discussion of beats.]

8. (a) (*Transform of Triangular Wave*). Let $f(t) = A\,\mathrm{trw}(t, 100, a) - B$, A and B constants. Show that

$$\mathcal{L}[f] = \frac{2A}{as^2}\left[1 + 2\sum_{n=1}^\infty (-1)^n e^{-nas/2}\right] - B/s$$

   (b) (*Transform of Sawtooth Wave*). Let $f(t) = A\,\mathrm{sww}(t, 100, a) - B$, A and B constants. Show that

$$\mathcal{L}[f] = \frac{A}{as^2} - A\sum_{n=1}^\infty e^{-nas}/s - B/s$$

9. (*Triangular and Sawtooth Wave Inputs*). Solve each IVP below. Plot the input and the output. [*Hint*: See the expansions in Problem 8.]

   (a) (*Low Pass Filter*). $y' + y = \mathrm{trw}(t, 100, p)$, $y(0) = 0$. First let $p = 30,\ 0 \le t \le 10$; then set $p = 1,\ 0 \le t \le 10$. If the IVP models a communication channel, which input ($p = 30$ or $p = 1$) is transmitted with less distortion? Explanation? [*Hint*: See Example 4.3.3.]

(b) (*Beats*). $y'' + 4y = \text{sww}(t, 100, p)$, $y(0) = 0$, $y'(0) = 0$. First let $p = 30$, $0 \leq t \leq 200$; then set $p = 3$, $0 \leq t \leq 150$. Plot input and the two outputs y and y'. Any explanation of the "beats"? [*Hint*: The natural period is π and $p = 3$ is pretty close to π. See the explanation of beats in Section 4.2.]

5.4 A Model for Car-Following

In heavy traffic or in constricted passages such as bridges or tunnels, cars follow one another with little or no lane changing or passing. Rear-end collisions are a common occurrence in this setting. Bad driving is the cause of many of these accidents, but some seem to occur for no apparent reason. In recent years mathematical models of car-following have been created and solved in an attempt to understand the phenomenon and to design traffic controls and driving codes that promote safe and efficient traffic flow. We construct and analyze a simple car-following model in this section, and then specialize it to the situation of a line of cars stopped at a red light. What happens when the light changes and the lead car accelerates? Under what circumstances will there be a rear-end collision somewhere down the line as each car accelerates in turn? We give some tentative answers to these questions.

Stimulus and Response: Velocity Control Models

Many stimuli may cause the driver in a line of cars to respond with an acceleration or deceleration of his or her own car: a speeding car to the rear, a changing traffic light ahead, a variation in the velocity of the car directly in front. Of course, the driver's response does not occur instantaneously. The driver must first detect the stimulus and then decide whether to ease up on, or to depress the accelerator or the brake pedal. The car itself does not respond instantaneously and contributes to the lag in the response of the car-driver system. We shall use the word "car" to denote the car-driver system, and refer to the delay in the response of the "car" as the *response time* of the car.

We shall assume that the dominant stimulus to the car is the difference between the velocity of the car ahead and its own velocity. The car's response, we shall assume, is an acceleration if the car ahead has a positive relative velocity, and a deceleration if the relative velocity is negative, hence the term *velocity control*. The magnitude of the car's acceleration or deceleration (after the appropriate response time) will be assumed to be proportional to the magnitude of the relative velocity. The constant of proportionality (assumed positive) is called the *sensitivity* of the car to the stimulus.

Our first goal is to quantify these somewhat vague notions of sensitivity and response time and then include them with the more sharply defined concepts of velocity and acceleration in a mathematical model. Hereafter, we shall use only the word "acceleration"; deceleration is simply a negative acceleration.

A Mathematical Model

Let us suppose that there are N cars in a line, with no passing or lane changing possible. Let $v_j(t)$ and $a_j(t)$, $j = 1, \ldots, N$, denote the velocity and the acceleration at time t of the jth car in line. Then the simplest mathematical model of the stimulus/response, velocity control system above is given by the system of *differential-delay equations*

$$v_1(t) = f(t)$$
$$a_j(t + T_j) = v'_j(t + T_j) = \lambda_j[v_{j-1}(t) - v_j(t)], \quad j = 2, \dots, N \tag{1}$$

where $f(t)$ is the velocity of the lead car, λ_j is the *sensitivity* coefficient of the jth car, and T_j is the *response time* of the jth car. We assume that λ_j and T_j are positive. Distances are measured in feet, time in seconds, and sensitivity in (second)$^{-1}$. Note that if the relative velocity $v_{j-1} - v_j$ is positive, then the car ahead is moving faster than the jth car, which responds after a time delay of T_j seconds by accelerating. The distinctive features of model (1) are the time delay, sensitivity, and velocity control factors.

System (1) does not contain enough information to be very useful in treating a traffic-flow problem. For instance, nothing has been said about the initial data for each car. Moreover, we are more interested in the locations of the cars than in their velocities. We shall add this information to (1) while setting up a model of a specific situation.

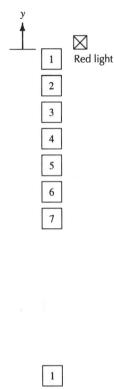

Cars at a Stoplight: A Mathematical Model

Let us suppose that there are N identical cars in a line stopped at a red light. The light turns green and after two seconds the first car accelerates. Three cars back and seven seconds later there is a rear-end collision. Why? We shall show that events such as this may be predicted by the model.

First, let us measure time forward from the moment the lead car accelerates, which is two seconds after the light turns green. The acceleration will be taken to be a positive constant a_0 ft/sec^2. We shall assume that each car is 15 ft long and that there is a gap of 5 ft between successive cars in the line. The location of the jth car will be measured from the stoplight to its front bumper.

Assuming identical cars (and drivers) and thus a common delay time T and sensitivity λ, we obtain the specific mathematical model involving differential-delay equations:

$$
\begin{aligned}
v_1(t) &= a_0 t, & t &\geq 0 \\
v'_j(t + T) &= \lambda[v_{j-1}(t) - v_j(t)], \quad j = 2, \dots, N & t &\geq 0 \\
v_j(t) &= 0, \quad j = 1, \dots, N & t &\leq 0
\end{aligned}
\tag{2}
$$

The equations of position are

$$
\begin{aligned}
y_1(t) &= \frac{1}{2}a_0 t^2 & t &\geq 0 \\
y_j(t) &= -20(j - 1) + \int_0^t v_j(\tau)\,d\tau, \quad j = 2, \dots, N & t &\geq 0
\end{aligned}
\tag{3}
$$

Our goal is to find the functions $y_j(t)$ and, in particular, to determine if a collision occurs, that is, if for some t and j, $y_j(t) = y_{j-1}(t) - 15$, which corresponds to the jth and $(j-1)$st car colliding.

Solving the Equations of the Model

The Laplace transform is ideally suited for solving a linear system with time delays. Applying the transform to (2), we have

$$\mathcal{L}[v_1] = \frac{a_0}{s^2}$$
$$\mathcal{L}[v'_j(t + T)] = \lambda\{\mathcal{L}[v_{j-1}] - \mathcal{L}[v_j]\} = e^{Ts}s\mathcal{L}[v_j], \quad j = 2, \dots, N \tag{4}$$

where we have used, first, the fact that $v_j'(t+T) = H(t+T)v_j'(t+T)$ since $t+T \geq 0$ and, second, the formula

$$\mathcal{L}[v_j'(t+T)] = \mathcal{L}[H(t+T)v_j'(t+T)] = e^{Ts}\mathcal{L}[v_j'(t)] = e^{Ts}s\mathcal{L}[v_j] \tag{5}$$

Formula (5) follows from (10) in the Shifting Theorem (Theorem 5.2.5).

We have that for $j = 2, \ldots, N$,

$$\mathcal{L}[v_j] = \frac{\lambda}{\lambda + se^{Ts}}\mathcal{L}[v_{j-1}] = \left\{\frac{\lambda}{\lambda + se^{Ts}}\right\}^2 \mathcal{L}[v_{j-2}] = \cdots = \left\{\frac{\lambda}{\lambda + se^{Ts}}\right\}^{j-1}\mathcal{L}[v_1]$$

$$= \left\{\frac{\lambda}{\lambda + se^{Ts}}\right\}^{j-1}\frac{a_0}{s^2} \tag{6}$$

where the first line of equalities is obtained from the second line of (4) by solving for $\mathcal{L}[v_j]$ in terms of $\mathcal{L}[v_{j-1}]$ and then using iteration to express $\mathcal{L}[v_j]$ in terms of $\mathcal{L}[v_1]$.

The velocity of the jth car in line may be found by applying $\mathcal{L}^{-1}$ to (6). First, however, we shall rewrite $\mathcal{L}[v_j]$ in (6) by using a binomial expansion (see Appendix C.2):

$$\mathcal{L}[v_j] = \frac{a_0}{s^2}\left(\frac{\lambda}{se^{Ts}}\right)^{j-1}\left(1 + \frac{\lambda}{s}e^{-Ts}\right)^{-(j-1)}$$

$$= a_0\lambda^{j-1}\left\{\frac{e^{-(j-1)Ts}}{s^{j+1}} - \frac{\lambda(j-1)je^{-jTs}}{s^{j+2}} + \frac{\lambda^2(j-1)je^{-(j+1)Ts}}{2!s^{j+3}} - \cdots\right\} \tag{7}$$

Assuming that $\mathcal{L}^{-1}\{\cdot + \cdot + \cdots\} = \mathcal{L}^{-1}[\cdot] + \mathcal{L}^{-1}[\cdot] + \cdots$ and making use of the formula (16) of the Shifting Theorem (Theorem 5.2.5), we have from (7) that

$$v_j(t) = a_0\lambda^{j-1}\left\{\frac{1}{j!}H[t-(j-1)T][t-(j-1)T]^j\right.$$

$$-\lambda\frac{j-1}{(j+1)!}H[t-jT][t-jT]^{j+1} \tag{8}$$

$$\left. +\lambda^2\frac{(j-1)j}{2!(j+2)!}H[t-(j+1)T][t-(j+1)T]^{j+2} - \cdots\right\}$$

Integrating (8) and using the formula in (3), we have that

$$y_j(t) = -20(j-1) + \frac{a_0\lambda^{j-1}}{(j-2)!}\sum_{n=0}^{\infty}\frac{(-\lambda)^nH(t-T_{nj})(t-T_{nj})^{n+j+1}}{n!(n+j-1)(n+j)(n+j+1)} \tag{9}$$

where $t \geq 0$ and $T_{nj} = (n+j-1)T$, $j = 2, \ldots, N$. We also have that the location of the lead car is given by

$$y_1(t) = \frac{1}{2}a_0t^2, \quad t \geq 0 \tag{10}$$

Formula (9) looks formidable, but appearances are deceiving, and it is not hard to use. For a fixed time $t > 0$, there are only a finite number of nonvanishing terms in each formula, the remaining terms being annihilated by the Heaviside functions. In fact, the summation in (9) has effective range from $n = 0$ only to $n = [t]$, the greatest integer not larger than t. For example, we have for the second car in line that its location at time t is given by

$$y_2(t) = \begin{cases} -20, & 0 \leq t < T \\ -20 + \dfrac{a_0\lambda}{6}(t-T)^3, & T \leq t < 2T \\ -20 + \dfrac{a_0\lambda}{6}(t-T)^3 - \dfrac{a_0\lambda^2}{24}(t-2T)^4, & 2T \leq t < 3T \end{cases} \tag{11}$$

See Figure 5.4.1 for a sketch of the positions of the first two cars in line.

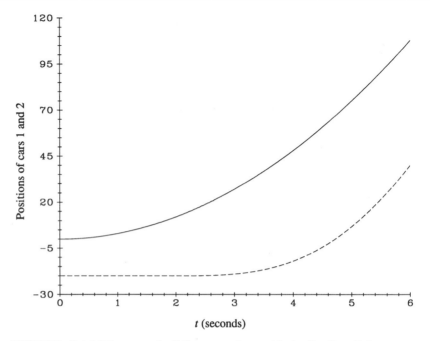

FIGURE 5.4.1 First car (solid), second car (dashed) after light turns green. See (10), (11) with $a_0 = 6$, $\lambda = 1$, $T = 2$.

We have reached the stage of model building and testing at which we need to select specific values for the parameters and interpret the results. Do the functions $y_j(t)$, $j = 1, 2, \ldots, N$, give reasonable estimates for the locations of the car as they move away from the light? Will there be collisions?

Bad Driving and a Rear-End Collision

Suppose that the second car in line is a high performance model, but driven by a drunk. It will take the driver a long time to respond to the acceleration of the lead car, but when he does, he overreacts. Thus T and λ may both be "large" in a sense that will become apparent.

Let us suppose that the lead car accelerates at $a_0 = 6$ ft/sec^2. Then the distance between the first two cars in line can be obtained by using $y_1(t) = 3t^2$, $t \geq 0$, and $y_2(t)$ from (11):

$$y_1(t) - y_2(t) = \begin{cases} 3t^2 + 20, & 0 \leq t \leq T \\ 3t^2 + 20 - \lambda(t-T)^3, & T \leq t \leq 2T \\ 3t^2 + 20 - \lambda(t-T)^3 + \lambda^2(t-2T)^4/4, & 2T \leq t \leq 3T \end{cases} \tag{12}$$

If $y_1(t^*) - y_2(t^*) = 15$, the first and second cars collide at time t^*. This cannot happen in the first time span of T seconds, but there will be a crash during the second span of T seconds if

$$y_1(2T) - y_2(2T) = 12T^2 + 20 - \lambda T^3 \leq 15$$

that is, if

$$5 \leq T^2(\lambda T - 12), \qquad \text{or} \qquad \lambda T \geq 12 + 5/T^2 \tag{13}$$

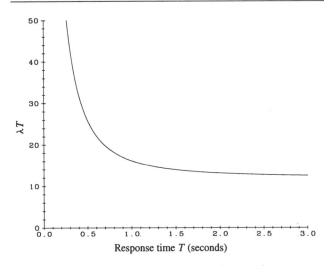

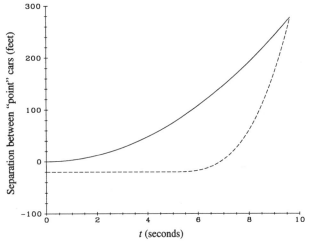

FIGURE 5.4.2 If $(T, \lambda T)$ lies above the curve $\lambda T = 12 + 5/T^2$, crash occurs in time interval $[T, 2T]$. See equation (13).

FIGURE 5.4.3 Collision between first (solid) and second car (dashed) at $t = 9.6$ sec: $\lambda = 3$, $T = 5$.

Inequality (13) suggests that T and λT are the important parameters, rather than T and λ. In Figure 5.4.2 we have sketched the curve defined by $\lambda T = 12 + 5/T^2$ using T and λT axes. If the point $(T, \lambda T)$ lies anywhere above the curve, the second car hits the first in the second span of T seconds.

Suppose that $T = 5$ seconds and that $\lambda = 3$ (seconds)$^{-1}$. Then $\lambda T = 15$ and $12 + 5/T^2 = 12.2$, and thus by condition (13) the second car "rear-ends" the first at some time t, $0 \le t \le 10$. In Figure 5.4.3 we have reduced the cars to "points" by reducing their length from 15 ft to 0 ft. The expected collision occurs approximately 9.6 seconds after the light turns green.

A response time of $T = 5$ sec and sensitivity $\lambda = 3$ sec^{-1} (as above) leads to a crash, while the same response time but the lower sensitivity of $\lambda = 1$ sec^{-1} does not lead to an accident (according to (13)), or, more accurately, there is no crash in the first 15 seconds. One might imagine a lower sensitivity to be inherently more dangerous. In fact, just the opposite is true since a time lag is also involved. The response does not occur at the time t an observation is made, but T seconds later, at which time the situation may have changed. High sensitivity may lead to a large response to a "deviation" that no longer exists.

Comments

Mathematical models of traffic flow are of relatively recent vintage and have not acquired the acceptance of, say, Newton's laws or the laws of electrical circuits. There is too much of the unknown, unknowable, and random in traffic flow, as in most human systems, for there to be a single comprehensive model. There are a good many other models of car-following, models which include other stimuli, such as the separation distance between cars or the relative velocity or separation of the car behind as well as the car ahead. Some of these models are given in the Problem Set.

PROBLEMS

1. (*Velocity Control Model*). Using the velocity control model of the text, which of the following sets of values of T and λT will result in a collision between the first two cars within the second span of T seconds? [*Hint*: See inequality (13).]

(a) $T = 1$, $\lambda T = 20$ (b) $T = 5$, $\lambda T = 12.3$ (c) $T = 2$, $\lambda T = 14$

2. (*Collision Between Second and Third Car*). The velocity control model of the text is used here to give conditions that the second and third cars collide.

(a) Use (9) to show that

$$
y_3(t) =
\begin{cases}
-40, & 0 \le t \le 2T \\
-40 + \dfrac{a_0 \lambda^2}{24}(t - 2T)^4, & 2T \le t < 3T
\end{cases}
$$

(b) Use (11) and the formula in part (a) above to show that

$$
y_2(t) - y_3(t) =
\begin{cases}
20, & 0 \le t < T \\
20 + \dfrac{1}{6} a_0 \lambda (t - T)^3, & T \le t < 2T \\
20 + \dfrac{1}{6} a_0 \lambda (t - T)^3 - \dfrac{1}{12} a_0 \lambda^2 (t - 2T)^4, & 2T \le t < 3T
\end{cases}
$$

(c) Let $a_0 = 6\,\text{ft/sec}^2$. Show that the third car rear-ends the second car for some value of t between $2T$ and $3T$ if the first and second cars have not collided and if $10 \le T^2(\lambda T - 16)\lambda T$.

(d) Show that if the first two cars don't collide during the time span $0 \le t \le 3T$, then $8 + 8(1 + 10/(64T^2))^{1/2} < \lambda T < 12 + 5/T^2$. [*Hint*: Use (12).] Plot the following three curves in the T, λT-plane: $\lambda T = 8 + 8(1 + 10/(64T^2))^{1/2}$, $\lambda T = 12 + 5/T^2$, and $T^2(\lambda T - 16)\lambda T = 10$ over the time span $0.4 \le T \le 0.7$. Shade the region of points $(T, \lambda T)$ where the second and third cars crash during the time span $0 \le t \le 3T$, if the first two cars don't. Find three points in the region and discuss the reality of the corresponding physical situation.

3. (*Separation Control Model*). The distance between cars (i.e., their separation) rather than the relative velocity may be taken as the control mechanism for cars at a stoplight. Assume for simplicity that the cars are "points" rather than the 15-foot cars used in the text.

(a) Write out a car-following model in which the stimulus to the $(j + 1)$st driver is the separation from the car ahead. Assume a response time of T seconds and a sensitivity of λ (seconds)$^{-1}$. [*Hint*: Compare with the model of (1), and make the needed changes.]

(b) Repeat part (a) but in a situation where the stimulus is a linear combination of the velocity control model treated in the text and the separation control model of part (a). Use sensitivity coefficients λ_1 and λ_2 for the respective models.

(c) For the model in part (a), assume that the lead car accelerates from a stop at the constant rate $a_0 = 6\,\text{ft/sec}^2$, and that the initial separation between cars is 5 feet. Find the motion of the following car. Are there values of λ, given the delay time $T = 2$, that will result in a rear-end collision for some t, $2 \le t \le 4$? [*Hint*: Use (5) and (6) to find $\mathcal{L}[v_2'(t + T)]$. As in (7), you will have to use a geometric series expansion before taking the inverse transform to get $v_2(t)$.]

 4. (*Reciprocal Separation Model*). Suppose that the acceleration of the $(j + 1)$st car in a line, $j \ge 1$, is directly proportional to the relative velocity of the car

with respect to that of the car ahead and inversely proportional to some power of the separation distance. Assume a response time of T seconds and sensitivity of λ (seconds)$^{-1}$. Discuss what happens when the light turns green in this model. Address the following points.

- Justify the model equation

$$v'_{j+1}(t + T) = \lambda \frac{v_j(t) - v_{j+1}(t)}{y_j(t) - y_{j+1}(t)}$$

- Let l be the *effective length* of a car (i.e., the reciprocal of the *jam concentration*, which is the number of cars per unit length of road when the traffic is completely jammed). Experimental evidence suggests that l is approximately 23 ft. Show that

$$v_{j+1}(t + T) = \lambda \ln\left[\frac{y_j(t) - y_{j+1}(t)}{l}\right]$$

- Let $m > 1$ and consider the model equation

$$v'_{j+1}(t + T) = \lambda \frac{v_j(t) - v_{j+1}(t)}{\left[y_j(t) - y_{j+1}(t)\right]^m}$$

Show that

$$v_{j+1}(t + T) = \frac{\lambda}{m - 1}\left\{l^{1-m} - \left[y_j(t) - y_{j+1}(t)\right]^{1-m}\right\}$$

- The *steady-state velocity* corresponds to a steady flow at constant velocity. Hence $v_j(t) = v_0$, $j = 1, 2, \ldots$, for all t in this case. Find the steady-state velocity for the models above.

- Use a computer to find $y_j(t)$, $j = 1, 2, 3$, for several values of λ and T for $m = 1, 2$.

5.5 Convolution

When the Laplace transform is used to solve an initial value problem, it often happens that the transform of the solution involves a product of functions. Although the Shifting Theorem (Theorem 5.2.5) allows us to find the inverse transform of certain special products, we do not yet have a general procedure for inverting a product. The far-reaching notion of the convolution product enables us to fill this gap. At the same time the convolution casts a new light on the meaning of the Green's Kernel of Section 3.6.

The Convolution Product

Suppose that $F(s)$ and $G(s)$ are the Lapace transforms of the known functions $f(t)$ and $g(t)$, respectively. We shall derive an elegant formula for $\mathcal{L}^{-1}[F(s)G(s)]$.

First we define a new kind of product.

❖ Convolution. The *convolution product* $f * g$ of f and g is given by the formula

$$(f * g)(t) = \int_0^t f(t - u)g(u)\, du \tag{1}$$

where f and g are assumed to be in $\mathcal{E}$.

The basic properties of the convolution product are simple enough.

THEOREM 5.5.1

> Properties of Convolution. Let f, g, and h be in $\mathcal{E}$. Then
>
> 1. (*Closure*) $f * g$ belongs to $\mathcal{E}$.
> 2. (*Commutativity*) $f * g = g * f$.
> 3. (*Associativity*) $(f * g) * h = f * (g * h)$.
> 4. (*Distributivity*) $(f + g) * h = f * h + g * h$.

Proof. We shall show Property 2; see Problem 6 for Properties 1, 3, and 4. To show commutativity, we set $v = t - u$ in the integral appearing in the definition of the convolution product and obtain, for each fixed $t \geq 0$,

$$(f * g)(t) = \int_0^t f(t - u)g(u)\, du = \int_t^0 f(v)g(t - v)(-dv)$$

$$= \int_0^t f(v)g(t - v)\, dv = \int_0^t g(t - v)f(v)\, dv = (g * f)(t)$$

which completes the proof of commutativity.

As the reader may have guessed, the convolution product is the sought-for inverse transform of a product.

THEOREM 5.5.2

> Convolution Theorem. Let f and g be in $\mathcal{E}$, and let F and G be their respective transforms. Then
>
> $$\mathcal{L}[f * g] = FG \tag{2}$$
>
> or, equivalently,
>
> $$\mathcal{L}^{-1}[FG] = f * g \tag{3}$$

Proof. We shall verify (2). Let R be the region in the tu-plane defined by $\{(t, u) : 0 \leq u \leq t, 0 \leq t\}$; see figure in margin. We have that

$$\mathcal{L}[f * g] = \int_0^\infty e^{-st} \left\{ \int_0^t f(t - u)g(u)\, du \right\} dt$$

$$= \int_0^\infty \left\{ \int_0^t e^{-st} f(t - u)g(u)\, du \right\} dt \tag{4}$$

Because f and g are in $\mathcal{E}$, they are piecewise continuous. The double integral in (4) may be evaluated by an iterated integral in either order, i.e., in the order of the iterated integral (4), or in the reverse order. In the reverse order we have

$$\mathcal{L}[f * g] = \int_0^\infty g(u) \left\{ \int_u^\infty e^{-st} f(t - u)\, dt \right\} du \tag{5}$$

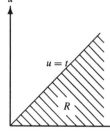

Making the change of variable $v = t - u$ in (5), we have that

$$\mathcal{L}[f * g] = \int_0^\infty g(u) \left\{ \int_0^\infty e^{-s(v+u)} f(v)\, dv \right\} du$$

$$= \int_0^\infty g(u)e^{-su}\left\{\int_0^\infty e^{-sv}f(v)\,dv\right\}du$$

$$= \left\{\int_0^\infty e^{-su}g(u)\,du\right\}\left\{\int_0^\infty e^{-sv}f(v)\,dv\right\} = GF$$

and the theorem is proved since $GF = FG$.

Note that the inverse transform of the product is *not* the product of the inverse transforms. With the Convolution Theorem in hand, we may find inverse transforms of a wide variety of functions without having to use partial fraction expansions.

EXAMPLE 5.5.1

Since $\mathcal{L}[1] = 1/s$ and $\mathcal{L}[\sin t] = (s^2+1)^{-1}$,

$$\mathcal{L}^{-1}\left[\frac{1}{s(s^2+1)}\right] = \mathcal{L}^{-1}\left[\frac{1}{s}\right] * \mathcal{L}^{-1}\left[\frac{1}{s^2+1}\right]$$

$$= \int_0^t 1 \cdot \sin u\,du = 1 - \cos t$$

a formula found earlier by the lengthier method of partial fractions (see (8) in Section 5.3).

Convolution and Green's Kernel

An advantage of the convolution product is that it allows us to construct the Green's Kernel of a constant-coefficient differential operator very quickly. Before stating and proving a general result to that effect, we shall illustrate the process by an example.

EXAMPLE 5.5.2

Solving an IVP by Convolution

Let $h(t)$ belong to $\mathcal{E}$ and consider the initial value problem

$$y'' + y' - 6y = f(t), \qquad y(0) = y'(0) = 0 \tag{6}$$

Applying the Laplace transform, we have that $(s^2+s-6)\mathcal{L}[y] = \mathcal{L}[f]$, and hence $\mathcal{L}[y] = \mathcal{L}[f]/(s^2+s-6)$. Taking inverse transforms and using the Convolution Theorem, we have that

$$y = f * \mathcal{L}^{-1}\left[\frac{1}{s^2+s-6}\right] \tag{7}$$

Now, since by the method of partial fractions

$$\frac{1}{s^2+s-6} = \frac{1}{5}\left[\frac{1}{s-2} - \frac{1}{s+3}\right]$$

it follows that

$$\mathcal{L}^{-1}\left[\frac{1}{s^2+s-6}\right] = \frac{1}{5}e^{2t} - \frac{1}{5}e^{-3t}$$

Hence, (7) becomes

$$y(t) = \int_0^t \left\{\frac{1}{5}e^{2(t-u)} - \frac{1}{5}e^{-3(t-u)}\right\}f(u)\,du \tag{8}$$

Since the function $y(t)$ in (8) is continuous, we conclude that it must be the unique solution of the initial value problem (6).

Writing the ODE in (6) in operator form, we have $P(D)y = (D^2 + D - 6)y = f$. Denoting the (continuous) inverse transform of $1/P(s)$ by $g(t)$, we see that (8) may be written as

$$y(t) = \int_0^t g(t - u)f(u)\,du = g * f \qquad (9)$$

Comparing (9) with formula (5) of Section 3.6 we observe that the function $g(t - u)$ is the Green's Kernel of the IVP (6).

This reasoning still holds in solving the initial value problem

$$P(D)[y] = y^{(n)} + a_{n-1}y^{(n-1)} + \cdots + a_0 y = f$$
$$y(0) = y'(0) = \cdots = y^{(n-1)}(0) = 0 \qquad (10)$$

where the a_i are real constants. Indeed, transforming (10), we can easily obtain

$$\mathcal{L}[y] = \frac{\mathcal{L}[f]}{P(s)} \qquad (11)$$

where $P(s)$ is the characteristic polynomial for the differential operator $P(D)$ in (10). Thus if we set $g(t) = \mathcal{L}^{-1}[1/P(s)]$, then (9) gives the solution of initial value problem (10), where $g(t - u)$ is the Green's function for the problem. We have proved the following theorem.

THEOREM 5.5.3

Convolution and the Green's Kernel. The Green's Kernel for the operator $P(D)$ of IVP (10) is

$$g(t) = \mathcal{L}^{-1}[1/P(s)] \qquad (12)$$

The solution of (10) is

$$y(t) = (g * f)(t) = \int_0^t g(t - u)f(u)\,du \qquad (13)$$

The advantages of the Green's Kernel approach for solving IVPs have been argued in Section 3.6. The new result of this section is that the Green's Kernel of the operator may be calculated by means of the transform calculus, and then convoluted with the driving force of the initial value problem to obtain the solution.

PROBLEMS

1. Use convolution to find the Laplace transform of each function. [*Hint*: Each integral is the convolution of two functions, so the transform of the integral is the product of two transforms.]

   (a) $\int_0^t (t - u)u\,du$ (b) $\int_0^t \sin u\,du$ [*Hint*: $\sin u = 1 \cdot \sin u$.]

   (c) $\int_0^t (t^2 - 2tu + u^2)\,du$ (d) $\int_0^t (\sin t \sin u \cos u - \cos t \sin^2 u)\,du$

2. Find the inverse Laplace transform of each function by using the Convolution Theorem. Write your answers as convolution products, but do not evaluate the integrals. [*Hint*: Write each expansion as the product of two functions whose inverse transforms you can find.]

(a) $\dfrac{s}{(s^2+1)^2}$ (b) $\dfrac{s}{(s^2+10)^2}$

(c) $\dfrac{1}{s^2(s+1)}$ (d) $\dfrac{s}{(s^2+9)^3}$

(e) $\dfrac{s}{(s+1)(s+2)^3}$ (f) $\dfrac{s^2+4s+4}{(s^2+4s+13)^2}$

(g) $\dfrac{\mathcal{L}[f]}{s^2+1}$ (h) $\dfrac{e^{-3s}\mathcal{L}[f]}{s^3}$ [*Hint*: Use a formula in Theorem 5.2.5.]

3. Write the solution of each IVP as a convolution product (do not evaluate the integral). [*Hint*: Theorem 5.2.5 and partial fraction expansions may be useful.]

   (a) $y'' + y = tH(t-1)$, $y(0) = y'(0) = 0$

   (b) $y'' + y = f(t)$, $y(0) = y'(0) = 0$

   (c) $2y'' + y' - y = f(t)$, $y(0) = y'(0) = 0$

   (d) $y'' + 2y' + y = f(t)$, $y(0) = y'(0) = 0$

   (e) $y'' + 4y = \begin{cases} 0, & 0 \le t \le 1 \\ t-1, & 1 \le t \le 2, \\ 1, & t > 2 \end{cases}$ $y(0) = y'(0) = 0$

4. Use formula (12) to construct the Green's Kernel $g(t)$ for each of the following differential operators, where $D = d/dt$.

   (a) $D^2 + 6D + 13$ (b) $D^2 + (1/3)D + 1/36$

   (c) $D^3 + 1$

5. Use Theorem 5.5.3 to solve the following IVPs.

   (a) $y'' + 6y' + 13y = f(t)$, $y(0) = y'(0) = 0$

   (b) $y'' + y'/3 + y/36 = f(t)$, $y(0) = y'(0) = 0$

   (c) $y''' + y = t$, $y(0) = y'(0) = y''(0) = 0$

6. Suppose that f, g, and h are in $\mathcal{E}$. Show that $f * g$ is in $\mathcal{E}$, that $f * (g+h) = f*g + f*h$, and that $f * (g*h) = (f*g)*h$.

7. Show that $e^{at}(f*g) = (e^{at}f)*(e^{at}g)$.

5.6 Convolution and the Delta Function

The convolution product may be used to find the response of a dynamical system to a sudden force of enormous amplitude but short duration (i.e., to an impulsive force). Impulsive forces may be modeled by the Dirac delta "function." Green's Kernel appears once more, this time as the response of the dynamical system to the delta function.

We shall begin on a hypothetical level, "defining" the delta function by a property we want it to possess. We then see how far we can develop the theory and the applications of the delta function from this somewhat shaky beginning. Next we take up the question of the meaning of the delta function. Finally, we conclude with an interpretation of delta functions in terms of "window" functions and impulsive forces.

The Dirac Delta Function and Its Properties

We have been working with elements in the set $\mathcal{E}$ of functions of exponential order on the interval $[0, \infty)$. We extend the domains of all the functions f in $\mathcal{E}$ to the entire real line by taking the value $f(t)$ to be zero for all negative t. The new collection of functions is denoted by $\mathcal{E}_0$. We may now give the definition of a very strange object, which Dirac[2] called a "function."

> ❖ **Dirac Delta Function.** Suppose there is an element δ in $\mathcal{E}_0$ such that for every function f in $\mathcal{E}_0$ which is continuous at t_0
>
> $$\int_{-\infty}^{\infty} \delta(t_0 - u) f(u) \, du = f(t_0) \tag{1}$$
>
> Any such element δ is called a *Dirac delta function.*

We shall assume below that $t_0 > 0$. For the next several paragraphs we shall assume that a Dirac delta function exists in $\mathcal{E}_0$. The reader may have doubts about the development of the calculus of a function that may not exist, but we shall plunge right ahead anyway.

THEOREM 5.6.1

Properties of the Delta Function. We have that

$$\int_{-\infty}^{\infty} \delta(t) \, dt = 1 \tag{2}$$

$$\mathcal{L}[\delta](s) = 1 \tag{3}$$

$$\mathcal{L}[\delta(t - u)] = e^{-us} \tag{4}$$

Proof. To prove (2), let $f(t) = H(t)$, the Heaviside function. From (1) we have that for every t

$$H(t) = \int_{-\infty}^{\infty} \delta(t - u) H(u) \, du = \int_{-\infty}^{\infty} \delta(v) H(t - v) \, dv$$

$$= \int_{-\infty}^{t} \delta(v) \, dv \qquad (\text{since } H(t - v) = 0, \text{ if } v > t)$$

where we have used the variable change $u = t - v$. Thus $\int_{-\infty}^{t} \delta(v) \, dv = 1$ for all $t > 0$, and the desired result follows.

We prove (3) as follows. By the Convolution Theorem (Theorem 5.5.2), we have that

$$\mathcal{L}[\delta * H] = \mathcal{L}[\delta]\mathcal{L}[H] \tag{5}$$

On the other hand,

$$\mathcal{L}[H] = \mathcal{L}[1] \tag{6}$$

---

[2]Paul Adrien Maurice Dirac (1902-1984), the British theoretical physicist, was awarded (jointly with Erwin Schrödinger) the 1933 Nobel Prize in physics for his work in quantum mechanics. Among other achievements, Dirac described the motion of an electron by four simultaneous differential equations. Finding negative energy states predicted by this mathematical model, Dirac hypothesized the existence of positrons, or electron anti-particles, whose existence was later experimentally confirmed.

We also have that

$$\mathcal{L}[\delta * H] = \mathcal{L}\left[\int_0^t \delta(t-u)H(u)\,du\right] = \mathcal{L}\left[\int_0^t \delta(v)H(t-v)\,dv\right]$$
$$= \mathcal{L}\left[\int_{-\infty}^\infty \delta(v)H(t-v)\,dv\right] = \mathcal{L}[1] \tag{7}$$

where we have used identities from the proof of (2) and the facts that $\delta(v) = 0$, $v < 0$, and $H(t-v) = 0$, $v > t$. From (5), (6), and (7) we conclude that $\mathcal{L}[\delta] = 1$. We leave the proof of (4) to Problem 4.

One of the many applications of δ is in solving initial value problems. Suppose that $P(D) = D^n + a_{n-1}D^{n-1} + \cdots + a_1 D + a_0$, where $a_{n-1}, \ldots, a_0$ are real.

THEOREM 5.6.2

> Solving Inital Value Problems. The solution of the IVP
>
> $$P(D)[y] = f, \qquad y^{(k)}(0) = 0, \quad k = 0, 1, \ldots, n-1 \tag{8}$$
>
> where f lies in $\mathcal{E}_0$ and is given by
>
> $$y = \int_{-\infty}^\infty G(t, u)f(u)\,du \tag{9}$$
>
> where the function $G(t, u)$ is the solution of the problem
>
> $$P(D)z = \delta(t-u), \qquad z^{(k)}(0) = 0, \quad k = 0, 1, \ldots, n-1 \tag{10}$$
>
> for each value of u.

Proof. Apply the Laplace transform to (10) and use formula (4) and formula (11) of Section 5.5 to obtain $\mathcal{L}[G(t, u)] = e^{-us}/P(s)$. From (10) of the Shifting Theorem (Theorem 5.2.5) and the discussion concerning the Green's Kernel g given in the Section 5.5, we have that $G(t, u) = H(t-u)g(t-u)$, and hence

$$y = \int_0^t g(t-u)f(u)\,du = \int_{-\infty}^t g(t-u)f(u)\,du \tag{11}$$
$$= \int_{-\infty}^\infty H(t-u)g(t-u)f(u)\,du = \int_{-\infty}^\infty G(t, u)f(u)\,du$$

finishing the proof of formula (9)..

Thus we have come up with yet another way to find the unique solution of an IVP. However, this is nothing really new since $G(t, u) = H(t-u)g(t-u)$, and we have already seen in Section 5.5 that the Green's Kernel g may be used to solve IVP (8).

Does the Dirac Delta Function Exist?

So far we have not shown that the delta function actually exists as an element of $\mathcal{E}_0$. Indeed, we shall now show that there is no function in $\mathcal{E}_0$ which satisfies (1). Let $t_0 > 0$ be a point of continuity for δ and assume that $\delta(t_0) > 0$. Therefore, there is an interval $t_0 - T \le t \le t_0$, for some positive T, on which $\delta(t)$ is positive and continuous.

Define the function f_0 in $\mathcal{E}_0$ as follows:

$$f_0(t) = \begin{cases} 1, & 0 \leq t \leq T \\ 0, & \text{all other real } t \end{cases}$$

Now from (1) we must have that

$$0 = f_0(t_0) = \int_{-\infty}^{\infty} \delta(t_0 - u) f_0(u)\, du = \int_0^T \delta(t_0 - u)\, du$$

Hence from the conditions placed on δ we conclude that

$$\delta(t_0 - u) = 0 \quad \text{for all} \quad 0 \leq u \leq T$$

and, in particular, $\delta(t_0) = 0$. Thus, δ vanishes at *every* point of continuity. The left-hand side of (1) vanishes for all t, and so the delta "function" can be nothing more than a convenient device for constructing the function $G(t, u)$ in (9). We already had an inkling from (3) that δ cannot be a function, since any function in $\mathcal{E}_0$ must have a transformation that decays to 0 as $s \to \infty$.

Since Heaviside's time, "functions" such as δ have been very important in the applications. In advanced treatments of modern applied mathematics, a logically rigorous theory is constructed which contains objects, called *distributions* or *generalized functions*, that behave just like the "delta function" should. We shall do no more here than give a tiny glimpse of how the "delta function" idea can be embedded in a larger framework.

δ as a Limit

We might try to salvage the "delta function" identity (1) in the following way. Consider the function $\phi(t, \tau)$ defined for positive τ and all real t by

$$\phi(t, \tau) = \begin{cases} 1/\tau, & 0 < t < \tau \\ 0, & \text{otherwise} \end{cases} \tag{12}$$

Observe that $\int_{-\infty}^{\infty} \phi(t, \tau)\, dt = 1$ for all $\tau > 0$ and that $\phi \geq 0$, and so the set $\{\phi(t, \tau) : \tau > 0\}$ is a family of density functions for computing average values of other functions. For example, let f belong to $\mathcal{E}_0$ and let f be continuous at $t_0 > 0$. From (12) we see that

$$\int_{-\infty}^{\infty} \phi(t - t_0, \tau) f(t)\, dt = \frac{1}{\tau} \int_{t_0}^{t_0 + \tau} f(t)\, dt \tag{13}$$

Hence ϕ does indeed act as a weighted average of f with the weight entirely concentrated in the interval $J = [t_0, t_0 + \tau]$.

Now the Mean Value Theorem for Integrals (see Appendix C.5) asserts that if f is continuous on J, there is a number ξ between t_0 and $t_0 + \tau$ such that

$$\int_{t_0}^{t_0 + \tau} f(t)\, dt = f(\xi)\tau \tag{14}$$

From (13) and (14) we see that

$$\lim_{\tau \to 0} \int_{-\infty}^{\infty} \phi(t - t_0, \tau) f(t)\, dt = \lim_{\tau \to 0} f(\xi) = f(t_0) \tag{15}$$

which shows that the family $\{\phi(t, \tau) : \tau > 0\}$ "behaves like the delta function" in the limit. See Figures 5.6.1 and 5.6.2 for sketches of ϕ and of ϕf. Observe that the graph

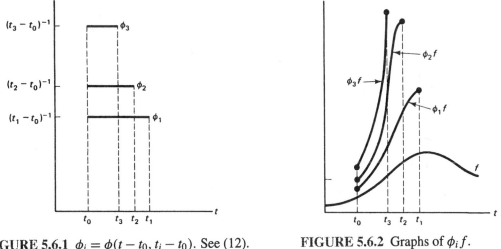

FIGURE 5.6.1 $\phi_i = \phi(t - t_0, t_i - t_0)$. See (12).   FIGURE 5.6.2 Graphs of $\phi_i f$.

of each function ϕ is a "window" through which a segment of the graph of f may be seen, but in distorted form.

Green's Function as a Limit

In view of the preceeding remarks, we might try to define the solution of IVP (10) in the following way. Denote by $z(t, u, \tau)$ the unique solution of the problem

$$P(D)z = \phi(t - u, \tau), \qquad z^{(k)}(0) = 0, \quad k = 0, 1, \ldots, n - 1$$

Then we shall define the solution $z(t, u)$ of (10) to be the limit of the family $z(t, u, \tau)$ as $\tau \to 0$; that is,

$$z(t, u) = \lim_{\tau \to 0^+} z(t, u, \tau)$$

We shall show that this procedure leads to the same result as we had earlier for the solution of (10): namely, that $z(t, u) = H(t - u)g(t - u)$, where $g = \mathcal{L}^{-1}[1/P(s)]$. First observe from (12) and Figure 5.6.1 that

$$z(t, u, \tau) = \int_0^t g(t - v)\phi(v - u, \tau)\, dv$$

$$= \begin{cases} 0, & t \le u \\ 1/\tau \int_u^t g(t - v)\, dv, & u \le t \le u + \tau \\ 1/\tau \int_u^{u+\tau} g(t - v)\, dv, & t \ge u + \tau \end{cases}$$

Now let $t_0 > u$ be a fixed value of t; then for all sufficiently small positive τ, we have by the Mean Value Theorem for Integrals that for some ξ, $u < \xi < u + \tau$,

$$z(t_0, u, \tau) = \frac{1}{\tau}\int_u^{u+\tau} g(t - v)\, dv = g(t - \xi)$$

and hence, $z(t_0, u, \tau) \to g(t - u)$, as $\tau \to 0^+$, proving the desired result.

Thus the main use of the "δ-function" in this context is operational. Its use allows us to replace rather lengthy limit arguments by very brief symbolic "proofs."

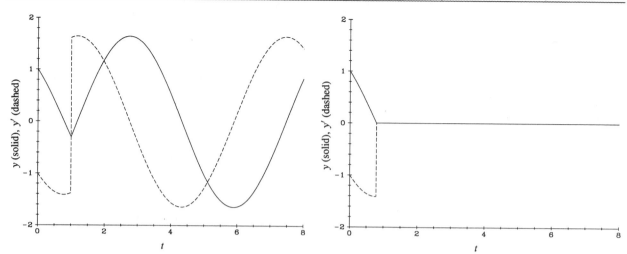

FIGURE 5.6.3 Response of body on spring to blow at time $T = 1$. See Example 5.6.1.

FIGURE 5.6.4 Stopping the oscillations with a well-timed blow at $T = \pi/4$. See Example 5.6.2.

Impulsive Forces

Another way to view the δ-function is in connection with a sudden sharp force acting on a dynamical system.

EXAMPLE 5.6.1

What Happens When You Hit an Oscillating Spring?

Suppose that the weighted end of an oscillating Hooke's Law spring is struck a hard blow. How does the system respond? If y measures the displacement of the weight from equilibrium, then we may model the system by

$$y''(t) + \omega^2 y(t) = A\delta(t - T), \quad y(0) = \alpha, \quad y'(0) = \beta \tag{16}$$

where $\omega^2 = k/m$, A is a positive constant, $T > 0$ is the time when the spring is struck, and α and β are the initial data. The "force" $A\delta(t - T)$ is called an *impulsive force*, while its integral over time is the resulting *impulse*. Taking the transform of (16),

$$(s^2 + \omega^2)\mathcal{L}[y] - s\alpha - \beta = Ae^{-Ts}$$

$$\mathcal{L}[y] = \frac{s\alpha + \beta}{s^2 + \omega^2} + \frac{Ae^{-Ts}}{s^2 + \omega^2} \tag{17}$$

Inverting (17), we have that

$$y(t) = \alpha \cos \omega t + \frac{\beta}{\omega} \sin \omega t + \frac{A}{\omega} H(t - T) \sin \omega(t - T) \tag{18}$$

Note that $y(t)$ is continuous for all t, but $y'(t)$ has a jump discontinuity at $t = T$. See Figure 5.6.3, where $\alpha = 1$, $\beta = -1$, $\omega = 1$, $A = 3$, $T = 1$.

The spring in Example 5.6.1 continues to oscillate after the impulsive blow. Can we choose the amplitude and timing so that the blow stops the oscillations altogether?

EXAMPLE 5.6.2

Stopping the Oscillations

Now if we want to stop the vibrations of the spring in Example 5.6.1, then we need to choose A and T so that $y(t) = 0$ for $t > T$. Since $\sin \omega(t - T) = \sin \omega t \cos \omega T - $

$\cos \omega t \sin \omega T$, we see from (18) that $y(t) = 0$ for $t > T$ if

$$\frac{A}{\omega} \sin \omega T = \alpha, \quad \frac{A}{\omega} \cos \omega T = -\frac{\beta}{\omega}$$

From these two equations we see that if $\beta \neq 0$, we must time the blow and choose its amplitude so that

$$\tan \omega T = -\frac{\alpha \omega}{\beta}, \quad A = \frac{\alpha \omega}{\sin \omega T}$$

where we assume that $\omega T \neq k\pi$. With this choice of T and A, the vibrations of the spring suddenly stop for $t \geq T$. See Figure 5.6.4 where $\alpha = 1, \beta = -1, \omega = 1, T = \pi/4$, $A = \sqrt{2}$.

PROBLEMS

1. Use the Laplace transform to find the solution of each initial value problem, where in each case, $y(0) = y'(0) = 0$.

   (a) $y'' + 2y' + 2y = \delta(t - \pi)$ (b) $y'' + 4y = \delta(t - \pi) - \delta(t - 2\pi)$

   (c) $y'' + 3y' + 2y = \sin t + \delta(t - \pi)$ (d) $y'' + y = \delta(t - \pi) \cos t$

   (e) $y'' + y = e^t + \delta(t - 1)$

2. (*Driven Spring*). A Hooke's Law spring with spring constant k supports an object of mass 1 and is subject to a force $f(t) = A \sin \omega t, \omega^2 \neq k$, for $t \geq 0$. The mass is given a sharp blow at $t = 2$ that gives an impulse of 2 units, and so the force of the blow $= 2\delta(t - 2)$. Use the Laplace transform to determine the motion of the system if the mass is at rest at its natural equilibrium at $t = 0$. Plot the solution when $k = 1, A = 1, \omega = 2$.

3. (*Driven LC Circuit*). Suppose that an LC circuit is subject to a constant electromotive force (EMF) F_0. At time $t = 1$ the circuit is dealt a sharp burst of EMF of size $2F_0$ (i.e., impulsive EMF $= 2F_0 \delta(t - 1)$). Find the charge as a function of time if $q(0) = q'(0) = 0$. Plot the charge as a function of time if $F_0 = 10$.

4. Assume that $\delta(t)$ exists. Show that $\mathcal{L}[\delta(t - u)](s) = e^{-us}$.

5. Show that $\delta(at) = (1/|a|)\delta(t), a \neq 0$. [*Hint*: Replace t by at in equation (2).]

6. Let $f(t)$ be in $\mathcal{E}_0$. Prove that $\delta(t) * f(t) = f(t)$.

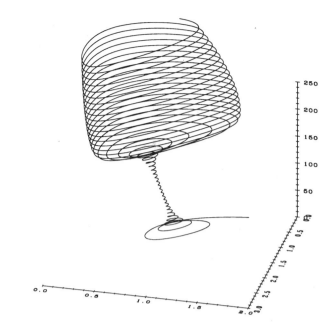

Introduction to Systems

The processes of nature are complex and interrelated. Few dynamical systems are isolated and complete in themselves, although we may imagine them to be so when we attempt to create mathematical models of their behavior. In the earlier chapters that is just what we have done. We have modeled the oscillations of a simple pendulum, the transient current in a simple electrical circuit, the growth or decline in the population of a solitary species. We now broaden our scope to include interacting systems: the strange oscillations of a double pendulum, voltages and currents in a multiloop circuit, the changing populations of a predator species and its prey. In this chapter several physical processes are described, the model ODE systems are constructed, a vocabulary for talking about systems and their solutions is introduced, and mathematical theory and computer-graphics simulations are used to give information about the behavior of these solutions. Chapter 7 goes more deeply into linear systems, while Chapters 8 and 9 address nonlinear systems and questions of stability and asymptotic behavior. The chapter cover figure shows a time-state curve for the two-dimensional autocatalator of Example 6.1.6.

6.1   First-Order Systems

A physical system evolves in time according to certain "laws." A mathematical model of the system represents the laws by equations in appropriate variables. Time is a natural independent variable to be used in the ordinary differential equations (ODEs) of the model. The other variables in the equations, the state variables, are functions that depend on time; their values portray the essential characteristics of the system. Systems of ODEs that involve the first derivatives of the state variables, but no higher derivatives, are called *first-order systems*. We have already seen a few first-order systems of ODEs, e.g., the position-velocity system in Section 1.1 that models the motion of a falling ball, and the cascade systems of Section 1.6. Now we shall focus entirely on first-order systems and construct a vocabulary for discussing their solutions.

From a Scalar Equation to a System

A system of first-order ODEs can be found which is "equivalent" to a scalar nth-order ODE in normal form

$$y^{(n)} = g(t, y, y', \dots, y^{(n-1)}) \tag{1}$$

where the derivatives are with respect to t. State variables $x_1, x_2, \dots, x_n$ are introduced in the following way,

$$x_1 = y, \quad x_2 = y', \dots, x_{n-1} = y^{(n-2)}, \quad x_n = y^{(n-1)} \tag{2}$$

Differentiate each side of each equation in (2) and use (1) to obtain the first-order system in the state variables $x_1, \dots, x_n$:

$$
\begin{aligned}
x_1' &= x_2 \\
x_2' &= x_3 \\
&\ \vdots \\
x_{n-1}' &= x_n \\
x_n' &= g(t, x_1, x_2, \dots, x_n)
\end{aligned}
\tag{3}
$$

System (3) is equivalent to the scalar ODE (1) in the following sense. Suppose that $y = y(t)$ is a solution of ODE (1). Then $x_1 = y(t), x_2 = y'(t), \dots, x_n = y^{(n-1)}(t)$ defines a solution of system (3) because of the way the state variables $x_1, \dots, x_n$ are defined in (2). Conversely, if $x_1 = x_1(t), \dots, x_n = x_n(t)$ solves system (3), then $y = x_1(t)$ solves ODE (1), again because of (2). Similarly, initial data for ODE (1) may be transformed to initial data for system (3), and conversely.

EXAMPLE 6.1.1

Motion of a Damped Hooke's Law Spring

The motion of a body suspended at the end of a Hooke's law spring was shown in Section 3.1 to be modeled by the scalar, second-order initial value problem (IVP),

$$mz'' + cz' + kz = 0, \qquad z(t_0) = z_0, \quad z'(t_0) = v_0 \tag{4}$$

where m is the mass of the body, c is a positive damping coefficient, k is the positive Hooke's law spring constant, and z_0 and v_0 are the position and velocity of the body at the initial time t_0. Introducing state variables, $x_1 = z$, $x_2 = z'$, we have an IVP for a system of two first-order ODEs,

$$
\begin{aligned}
x_1' &= x_2, & x_1(t_0) &= z_0 \\
x_2' &= -\frac{k}{m}x_1 - \frac{c}{m}x_2, & x_2(t_0) &= v_0
\end{aligned}
\tag{5}
$$

If there are only two or three state variables, then we shall usually employ x and y or x, y, and z as the state variables instead of subscripted x's. In xy-state variables system (5) has the form

$$
\begin{aligned}
x' &= y, & x(t_0) &= x_0 \\
y' &= -\frac{k}{m}x - \frac{c}{m}y, & y(t_0) &= y_0
\end{aligned}
\tag{6}
$$

Physical Systems

Why do we replace the single ODE of (4) by the set of two ODEs of (5)? One wonders what the purpose is, and, so far as finding solution formulas is concerned, replacing a scalar ODE by a system of ODEs is not particularly helpful. However, we are after much more than solution formulas, which in any case are available only for a small (but important) handful of ODEs. We want to use numerical solvers to help understand how solutions change as the coefficients of the ODEs and the initial data change—and most solvers "prefer" first-order ODEs even if it means dealing with several ODEs simultaneously. Secondly, many physical phenomena are modeled quite naturally by systems of first-order ODEs. For the two-loop electrical circuit of the next example the Circuit Laws of Section 4.3 lead us directly to a system of three first-order ODEs.

EXAMPLE 6.1.2

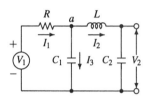

A Low-Pass Filter

The electrical circuit sketched in the margin is a common component in communication equipment. The input is a known voltage $V_1(t)$, and the circuit responds with the output voltage V_2 across the second capacitor (see Section 4.3 for background on circuits and circuit elements). Candidates for state variables are the currents into and out of the junction a and the respective voltages across the resistor, the inductor, and the capacitors. However, several of these quantities are interrelated, and so we don't need them all. For example, by Kirchhoff's Current Law applied at node a, we have that $I_3 = I_1 - I_2$, so we don't need I_3, as long as we keep the currents I_1 and I_2 as state variables. As it happens V_2, I_1, I_2 will serve as state variables for the circuit.

It follows from the derivative form of Coulomb's Law that the rates of change of the voltages at the two capacitors are, respectively,

$$I_3/C_1 = (I_1 - I_2)/C_1, \quad V_2' = I_2/C_2$$

respectively. Kirchhoff's Voltage Law applied to the outer loop through the input, resistor, inductor, and second capacitor asserts that

$$V_1 = RI_1 + LI_2' + V_2$$

where RI_1 is the voltage across the resistor (Ohm's Law) and LI_2' is the voltage across the inductor (Faraday's Law). Kirchhoff's Voltage Law (in derivative form) for the loop through the input, resistor, and first capacitor implies that

$$V_1' = RI_1' + (I_1 - I_2)/C_1$$

The relationships above can be arranged into a first-order system of ODEs in the state variables V_2, I_1, and I_2:

$$
\begin{aligned}
V_2' &= \frac{1}{C_2}I_2 \\
I_1' &= -\frac{1}{RC_1}I_1 + \frac{1}{RC_1}I_2 + \frac{1}{R}V_1' \\
I_2' &= -\frac{1}{L}V_2 - \frac{R}{L}I_1 + \frac{1}{L}V_1
\end{aligned}
\tag{7}
$$

Given the input voltage $V_1(t)$, given the circuit constants R, L, C_1, and C_2, and given the initial voltages and currents, $V_2(0)$, $I_1(0)$, and $I_2(0)$, we would like to determine the state variables $V_2(t)$, $I_1(t)$, and $I_2(t)$ from system (7). In Section 7.7 we show that if the input voltage $V_1(t)$ is sinusoidal, then there is a steady-state response voltage V_2 with the same period as $V_1(t)$ (i.e., a forced oscillation). We also show there that the circuit filters out high-frequency "noise" ($V_1 = a_0 e^{i\omega t}$, $|\omega|$ large) in the sense that the amplitude of the response voltage V_2 to such an input is a small fraction of $|a_0|$. In contrast, we shall show in Section 7.7 that low-frequency voltages pass through the circuit without much change.

States, Systems of ODEs, Solutions

The mathematical model of the low-pass filter contains three fairly simple ODEs in the three state variables of the circuit. More complex phenomena may lead to more equations, more state variables, and more intricate relations between the rates and the variables. An astounding number and variety of phenomena may be modeled by *first-order systems of ODEs* of the form

$$
\begin{aligned}
x_1' &= f_1(t, x_1, x_2, \ldots, x_n) \\
x_2' &= f_2(t, x_1, x_2, \ldots, x_n) \\
&\;\;\vdots \\
x_n' &= f_n(t, x_1, x_2, \ldots, x_n)
\end{aligned}
\tag{8}
$$

in the *state variables* $x_1, \ldots, x_n$ and the independent variable t, which represents time. All derivatives are with respect to time, and the *rate functions* $f_1, \ldots, f_n$ are assumed to be known functions of time and state. If the rate functions do not depend explicitly on the time variable t, the system is said to be *autonomous*. *Initial conditions* of the form $x_1(t_0) = b_1$, $x_2(t_0) = b_2, \ldots, x_n(t_0) = b_n$ may also be prescribed. A *solution* is a set of functions $x_1 = x_1(t)$, $x_2 = x_2(t), \ldots, x_n = x_n(t)$, which are defined on a common interval I of time containing t_0, and which satisfy the equations of (8) for all t in I, and which satisfy given initial conditions.

The definitions above may be considerably abbreviated by the use of vector notation. Let the *state vector* x, the *initial state* x^0, and the *vector rate function* f be defined as column vectors,

$$
x = \begin{bmatrix} x_1 \\ \vdots \\ x_n \end{bmatrix}, \qquad x^0 = \begin{bmatrix} b_1 \\ \vdots \\ b_n \end{bmatrix}, \qquad f = \begin{bmatrix} f_1 \\ \vdots \\ f_n \end{bmatrix}
$$

Then the problem stated above for a first-order system with initial conditions may be written in compact vector form. Note that in Section 4.1 state vectors are written in bold-faced type (e.g., **x**). For simplicity, we now use ordinary type fonts for vectors (e.g., x rather then **x**).

❖ **Initial Value Problem for a First-Order System.** Find a vector function $x(t)$, defined on a nontrivial t-interval I containing t_0, that satisfies the conditions

$$
x'(t) = f(t, x(t)), \qquad x(t_0) = x^0 \tag{9}
$$

for all t in I, where t_0 is the initial time and x^0 is the initial state.

To avoid ambiguity it is usually assumed that I is the largest interval on which the system and the solution $x(t)$ are both defined (i.e., the solution is *maximally extended*). The vector function $x(t)$ is a *forward solution* for IVP (9) if it is defined for $t \geq t_0$; a forward solution is associated with the future evolution of the state of the system and is said to result from a *forward initial value problem*. A *backward solution* for IVP (9) is defined for $t \leq t_0$, depicts the history of the system, and is said to result from a *backward IVP*. It is usually clear in the applications whether a forward, or a backward, or a *two-sided solution* (defined on an interval containing t_0 in its interior) is required.

The systems of Examples 6.1.1 and 6.1.2 have two and three state variables, respectively; each can be written in the vector form above. These two systems are also linear in the state variables. Linear systems are defined as follows.

❖ **Linear Systems.** The general form of a *linear system* is

$$x'_1 = a_{11}x_1 + \cdots + a_{1n}x_n + F_1$$

$$\vdots$$

$$x'_n = a_{n1}x_1 + \cdots + a_{nn}x_n + F_n$$

where the *coefficients* a_{ij} and the *input* (or *driving*) *terms* $F_1, \ldots, F_n$ are constants or functions of t (but not of $x_1, \ldots, x_n$).

Special methods are available for constructing solution formulas for linear systems. These are taken up in Chapter 7.

State Space, Orbits, Time-State Curves, Component Graphs

A solution of a system of first-order ODEs determines several curves whose behavior reflects properties of the solution. The definition below defines these curves and the spaces they "live" in.

❖ **State Space, Time-State Curves, Orbits, Component Graphs.** Let $x = x(t)$ be a solution of the system $x' = f(t, x)$, t in an interval I. As t traverses I, the point $x(t)$ traces the *orbit* (or *trajectory*) of the solution in the n-dimensional *state* (or *phase*) *space* of the state vectors. Arrowheads on orbits are sometimes used to show the direction of increasing time. A collection of several orbits is a *portrait* of orbits. The point $(t, x(t))$ traces a *time-state curve* in the $n + 1$-dimensional space of the time and state variables $(t, x_1, x_2, \ldots, x_n)$. The projection of a time-state curve onto the tx_j-plane is the x_j-*component graph*.

Many of the systems we consider are autonomous . Autonomous systems have the property that the clock can be started at any time without having any effect on the orbits in state space. Indeed, if $x = x(t)$, $a < t < b$, is a solution of the autonomous system $x' = f(x)$, then $x = x(t + c)$, $a - c < t < b - c$, is another solution for any constant c, and the orbits of these solutions coincide. Thus, for an orbit of an autonomous system it is the initial position and the elapsed time that matter, not the initial time. This property of autonomous ODEs was first noted in the context of scalar ODEs in Section 2.2.

We will illustrate the above concepts by using a numerical ODE solver to plot orbits, time-state curves, and component graphs of the systems of Examples 6.1.1 and 6.1.2 for particular values of the parameters.

EXAMPLE 6.1.3

Damped Hooke's Law Spring

In the autonomous linear IVP (6) let us set $k/m = 64$, $c/m = 0.2$, $x_0 = 4.5$, $y_0 = -0.45$, and use a numerical ODE solver to plot curves associated with the system. Figure 6.1.1 shows the decaying oscillations of the tx and ty component graphs (upper right and lower right, respectively) and the inward spiraling orbit in the xy-state plane (left). The time-state curve in the three-dimensional txy-space resembles a peaked hat (Figure 6.1.2). Since the system is autonomous, we can take t_0 to be any convenient value. In Figure 6.1.2, $t_0 = 40$ and the time-state curve is plotted for $40 \leq t \leq 90$;

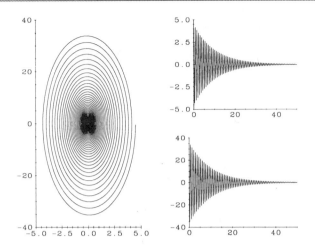

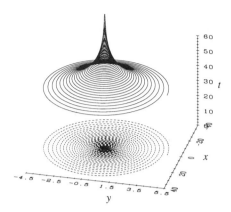

FIGURE 6.1.1 Orbit and component graphs for a damped Hooke's law spring. See Example 6.1.3.

FIGURE 6.1.2 Time-state curve (solid) and orbit (dashed) for a damped Hooke's law spring. See Example 6.1.3.

the t-axis is vertical. The corresponding orbit is found by projecting the solution curve downward parallel to the vertical t-axis onto the xy-plane. The equivalent scalar IVP (4) may be solved explicitly using the data given above and setting $t_0 = 40$ to obtain $x(t) = 4.5 \exp(4 - t/10) \cos(\sqrt{63.99}(t - 40))$. The function $y(t)$ may be obtained by differentiating $x(t)$. In this case, a graphics plotter could have been used to plot orbits, component graphs, and time-state curves directly from the formulas for $x(t)$ and $y(t)$.

EXAMPLE 6.1.4

Another Look at the Low-Pass Filter

Suppose that the circuit parameters and input voltage of the low-pass filter circuit modeled by the driven linear system (7) are given by

$$R = 1, \quad C_1 = 1/2, \quad C_2 = 3/2, \quad L = 4/3, \quad V_1 = \sin(3t/2)$$

where all quantities are measured in standard units, i.e., ohms, farads, ... The initial data at $t_0 = 0$ are taken to be $V_1(0) = 0$, $I_1(0) = 0$, $I_2(0) = 0$. Figure 6.1.3 shows the component graphs for the output voltage $V_2(t)$ (the solid curve), the current $I_1(t)$ (long dashes), and the current $I_2(t)$ (short dashes). It certainly appears that the state variables tend to periodic functions with the same period $4\pi/3$ as the driving voltage $V_1(t)$. Figure 6.1.4 displays the corresponding orbit in the three-dimensional space of the state variables V_2, I_1 (the vertical axis), and I_2. As time increases, the orbit appears to settle down to a closèd curve representing a periodic solution. We cannot show the time-state curve because it lies in the four-dimensional space of t, V_2, I_1, and I_2 variables. The graphs in Figures 6.1.3 and 6.1.4 are obtained from a numerical ODE solver.

Modeling the Dynamics in a Chemical Reactor

In a chemical reaction molecules of various species (*reactants*) interact and generate other species (*intermediates*) which interact in turn to create additional species (*prod-*

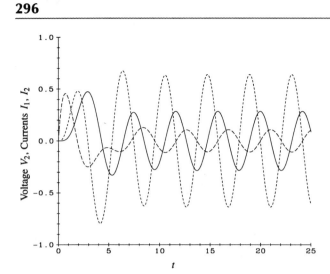

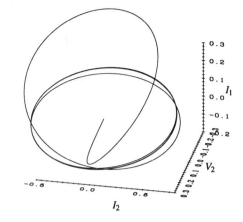

FIGURE 6.1.3 Output voltage (solid) and currents (dashed) in the low-pass filter. See Example 6.1.4.

FIGURE 6.1.4 Orbit of low-pass filter. See Example 6.1.4.

ucts). A diagram such as

$$X + Y \xrightarrow{k_1} Z + 2Y \xrightarrow{k_2} W \tag{10}$$

is chemical shorthand for a pair of reactions that begins with reactants X and Y. One molecule of X interacts with one molecule of Y to produce one molecule of the intermediate Z and two molecules of Y (a so-called *autocatalytic* reaction in which more of Y is produced than is used up). In the second reaction of (10) one molecule of Z and two of Y react to produce one molecule of the product W. We shall suppose that the volume of the chemical reactor is constant and that the amount of each species is specified by its concentrations; we shall use lowercase letters for concentrations, e.g., $x(t)$ for the concentration of species X at time t.

The velocity of a reaction is the rate at which a species is produced (or the negative of the rate at which it is destroyed). To calculate the velocity of a reaction we need the following empirical "law":

❖ **Chemical Law of Mass Action.** Suppose that n *reactant* molecules X_1, X_2, ..., X_n react to produce the m *product* molecules U_1, U_2, ..., U_m (the X_i's and U_j's need not all be distinct). Let x_1, x_2, ..., x_n and u_1, u_2, ..., u_m be the concentrations at time t of the corresponding molecules. The rates at which the x_i decrease at time t are identical for all $i = 1, 2, ..., n$; each rate is proportional to the product $x_1 x_2 \cdots x_n$. Denoting the constant of proportionality by $k > 0$, the reaction is written as

$$X_1 + X_2 + \cdots + X_n \xrightarrow{k} U_1 + U_2 + \cdots + U_m$$

(sometimes rX_1, where r is a positive integer, is written instead of $X_1 + X_1 + \cdots + X_1$, r-times). The rate at which U_j is being produced at time t, for $j = 1, 2, ..., m$, is equal to $kx_1 x_2 \cdots x_n$ (the exact rate of decrease of each X_i). This rate law is known as the *Chemical Law of Mass Action*.

With these ideas in mind, we can write a rate equation (i.e., a first-order ODE) for each chemical species that is involved in reaction (10). The best way to do this is to think of the species X, Y, Z, and W as occupying separate compartments and to apply

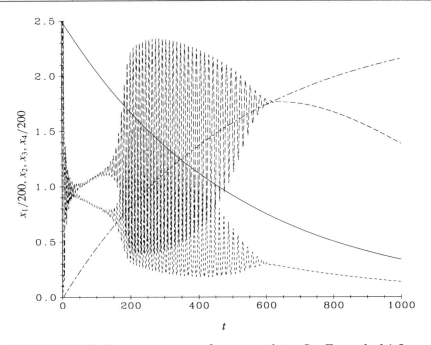

FIGURE 6.1.5 Component curves for autocatalator. See Example 6.1.5.

the Balance Law to each species, calculating rates from the Chemical Law of Mass Action. Doing this, we have for reaction (10):

$$
\begin{aligned}
x' &= -k_1 xy \\
y' &= -k_1 xy + 2k_1 xy - k_2 zy^2 \\
z' &= k_1 xy - k_2 zy^2 \\
w' &= k_2 zy^2
\end{aligned}
\tag{11}
$$

The first two terms on the right in the rate equation for y represent the destruction of one molecule of Y in the reaction, but the creation of two more. The term zy^2 models the mass action effect of one molecule of Z reacting with two molecules of Y to create a molecule of W. (Think of this as $Z + Y + Y \rightarrow W$.)

Real chemical reactions are more complex than this. The rate coefficients k_i are hard to measure, and they change as the temperature in the reactor changes. Some reactions go both forward and backward and at different rates. In addition, the intermediate steps and chemical species are often unknown and can only be guessed at. Nevertheless, the simplifications of the above approach have been remarkably successful in helping scientists to think (and to make predictions) about the behavior of chemical reactions.

EXAMPLE 6.1.5

The Autocatalator

A chemical X_1 decays at a rate proportional to its concentration, generating an intermediate species X_2 in the process. The intermediate X_2 decays to an intermediate species X_3 at a rate proportional to the concentration of X_2. In addition, one molecule of X_2 reacts with two molecules of X_3 to produce three molecules of X_3 in an autocatalytic reaction. Finally, X_3 decays to the product X_4 at a rate proportional to its concentration. The schematic for all this is

$$
X_1 \xrightarrow{k_1} X_2, \quad X_2 \xrightarrow{k_2} X_3, \quad X_2 + 2X_3 \xrightarrow{k_3} 3X_3, \quad X_3 \xrightarrow{k_4} X_4
\tag{12}
$$

The corresponding first-order system of ODEs in the concentrations x_1, x_2, x_3, x_4 is

$$\begin{aligned}
x_1' &= -k_1 x_1 \\
x_2' &= k_1 x_1 - k_2 x_2 - k_3 x_2 x_3^2 \\
x_3' &= k_2 x_2 - k_4 x_3 + k_3 x_2 x_3^2 \\
x_4' &= k_4 x_3
\end{aligned} \tag{13}$$

Observe that it is $k_3 x_2 x_3^2$, not $3k_3 x_2 x_3^2$ in the second and third rate equations because it takes two molecules of x_3 to produce three molecules of x_3, and the net increase is one, not three.

We shall suppose that the concentrations and time have been scaled to dimensionless quantities, and that we have the following data set

$$\begin{aligned}
k_1 &= 0.002, \quad k_2 = 0.08, \quad k_3 = 1, \quad k_4 = 1 \\
x_1(0) &= 500, \quad x_2(0) = x_3(0) = x_4(0) = 0
\end{aligned} \tag{14}$$

Figure 6.1.5 displays the component graphs of a solution of system (13) with data given by (14); these graphs are obtained by a numerical ODE solver. The dimensionless concentrations $x_1(t)$ and $x_4(t)$ are divided by 200 before plotting in order to accommodate the scales of x_2 and x_3. The concentration of the reactant X_1 decays exponentially in time as expected (solid line), and the concentration of the product X_4 increases (long-short dashes).

The startling feature of Figure 6.1.5 is the onset and disappearance of oscillations in the concentrations of the intermediates X_2 (long dashes) and X_3 (short dashes). The oscillations so surprised the chemists who first detected them in real chemical reactions that they thought their measurement instruments were defective. It had long been almost an article of chemical faith that the concentrations of intermediates would steadily rise to peak values and then steadily decline. But the oscillatory behavior seen here is now well-established experimentally, and the chemical theory is (at least partly) understood.[1]

EXAMPLE 6.1.6

The Two-Dimensional Autocatalator

The orbit and the time-state curve of ODE (13), with data set given by (14), "live," respectively, in four-dimensional state space and five-dimensional time-state space. If we want to "see" either curve, we must somehow reduce the number of dimensions. Fortunately, this can be done because the product concentration x_4 appears in none of the rate functions of (13), and $x_1(t)$ is easily seen from the first ODE of (13) to be $x_1(0)e^{-k_1 t}$. This leaves a two-dimensional nonautonomous IVP in the two intermediates:

$$\begin{aligned}
x_2' &= k_1 x_1(0)e^{-k_1 t} - k_2 x_2 - k_3 x_2 x_3^2, \quad x_2(0) = 0 \\
x_3' &= k_2 x_2 - k_4 x_3 + k_3 x_2 x_3^2, \quad\quad\quad\; x_3(0) = 0
\end{aligned} \tag{15}$$

With the data given in (14), a numerical solver is applied to (15) yielding the orbit shown in Figure 6.1.6. The orbit starts out at the origin in $x_2 x_3$-state space at time zero, runs off the screen of the figure, eventually reenters, and begins its oscillatory phase. Finally as dimensionless time approaches 1000, the orbit stops oscillating and heads back toward the origin.

---

[1] For more on the chemistry and mathematics of these reactions, see P. Gray and S.K. Scott, *Chemical Oscillations and Instabilities* (Oxford: Clarendon Press, 1990), and S.K. Scott, *Chemical Chaos*, (Oxford: Clarendon Press, 1991).

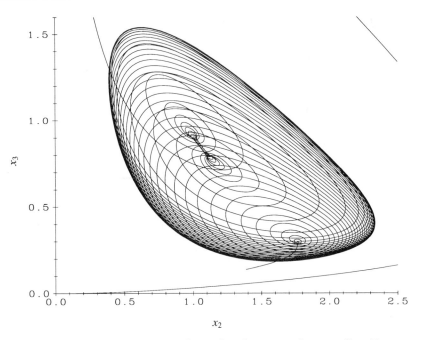

FIGURE 6.1.6 Orbit of two-dimensional autocatalator. See Example 6.1.6.

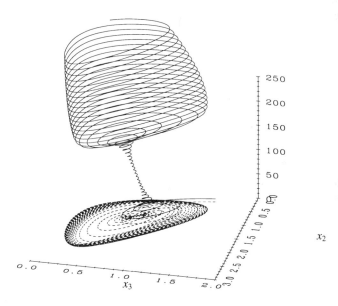

FIGURE 6.1.7 Time-state curve and orbit for two-dimensional autocatalator. See Example 6.1.6.

The corresponding time-state curve lies in the three-dimensional tx_2x_3-space. The "wineglass" of Figure 6.1.7 and the cover page of this chapter is a segment of the solution curve corresponding to the time span $6.5 \leq t \leq 377$. If the wineglass is projected downward along the vertical t-axis onto the x_1x_2-plane, then the orbit appears (the dashed curve below the wineglass).

Comments

Our aim has been to introduce first-order systems of ODEs, the graphs they generate, and some of the host of physical processes that can be modeled by these systems. The next section has more of the theory of systems of ODEs.

PROBLEMS

1. *(From Scalar ODEs to Systems).* For each scalar linear ODE below construct an equivalent first-order system. For each first-order system of linear ODEs below find an equivalent scalar ODE; solve the scalar ODE, then use these solutions to construct all solutions of the system. [*Hint:* If $y = Ay_1 + By_2$ is the general solution of $y'' + ay' + by = 0$, then $x_1 = Ay_1 + By_2$, $x_2 = Ay_1' + By_2'$ is the general solution of the equivalent system $x_1' = x_2$, $x_2' = -bx_1 - ax_2$.]

   (a) $y'' - 4y = 0$ (b) $y'' + 9y = 0$

   (c) $y'' + 5y' + 4y = 0$ (d) $x_1' = x_2$, $x_2' = -2x_1 - 2x_2$

   (e) $x_1' = x_2$, $x_2' = -16x_1$ (f) $y''' + 6y'' + 11y' + 6y = 0$

   (g) $x_1' = x_2$, $x_2' = x_3$, $x_3' = 4x_1 - 4x_2 + x_3$

2. *(Hooke's Law Springs).* The system $x' = y$, $y' = -bx - ay + A\cos\omega t$ is equivalent to the scalar ODE, $x'' + ax' + bx = A\cos\omega t$, that models the motion of a body on a driven Hooke's law spring. In each case, find $x(t)$ and $y(t)$ explicitly. Then use a graphics package or a numerical ODE solver to plot x and y component graphs, the orbit, and the time-state curve. Use the time interval $0 \le t \le T$, where T is chosen so that the graphs display the main features of the solution. Interpret what you see in terms of the behavior of the spring.

   (a) $x' = y$, $y' = -x - y$, $x(0) = 1$, $y(0) = 0$

   (b) $x' = y$, $y' = -x - 2y$, $x(0) = 1$, $y(0) = 0$

   (c) $x' = y$, $y' = -x - 3y$, $x(0) = 1$, $y(0) = 0$

   (d) $x' = y$, $y = -25x + \cos 5t$, $x(0) = 0$, $y(0) = 0$

   (e) $x' = y$, $y' = -25x + \cos(5.5t)$, $x(0) = 0$, $y(0) = 0$

3. *(Cascades).* The linear cascade systems of Section 1.6 can be solved one equation at a time. Solve the cascade IVPs below for each state variable $x_i(t)$ as a function of time. Plot component graphs. If the number of state variables is no more than three, plot the corresponding orbit. If there are no more than two state variables, plot the time-state curves as well. In each case find the limiting value of each state variable as $t \to \infty$.

   (a) $x_1' = -3x_1$, $x_2' = x_1$, $x_3' = 2x_1$, $x_1(0) = 10$, $x_2(0) = x_3(0) = 0$

   (b) $x_1' = -2x_1$, $x_2' = 2x_1 - x_2$, $x_3' = x_2$, $x_1(0) = 10$, $x_2(0) = x_3(0) = 0$

   (c) $x_1' = -x_1$, $x_2' = x_1 - 3x_2$, $x_1(0) = 10$, $x_2(0) = 20$

   (d) $x_1' = -2x_1$, $x_2' = -3x_2$, $x_3' = 2x_1 + 3x_2$, $x_1(0) = 1$, $x_2(0) = 2$, $x_3(0) = 0$

   (e) $x_1' = -2x_1$, $x_2' = -x_2$, $x_3' = 2x_1 + x_2 - 3x_3$, $x_4' = x_3$, $x_5' = 2x_3$,
   $x_1(0) = 10$, $x_2(0) = 5$, $x_3(0) = x_4(0) = x_5(0) = 0$

4. *(Low-Pass Filter).* In system (7) let $R = 1\ \Omega$, $C_1 = 1/2$ coulomb, $C_2 = 3/2$ coulomb, $L = 4/3\ F$, and $V_1 = A\sin\omega t$ volts where A and ω are positive parameters. Use a numerical ODE solver to solve (7) with $V_2(0) = 0$, $I_1(0) = I_2(0) = 0$, $A = 1$, and $\omega = 0.5, 1.0, 10.0, 100.0$. Plot the V_2 component graph, and ex-

plain why the circuit is called a low-pass filter. Now set $\omega = 1.0$ and repeat with $A = 1, 10, 100$. What do you conclude is the effect of increasing the amplitude of the input?

5. (*Orbits of Autonomous Systems*). Sometimes formulas for orbits can be found without first finding formulas for solutions. In each of the following problems find formulas for the orbits by solving $dx_2/dx_1 = f_2/f_1$. Plot some orbits in the x_1x_2-state plane: $|x_1| \le 10$, $|x_2| \le 10$.

(a) $x_1' = x_1$, $x_2' = -3x_2$

(b) $x_1' = x_1x_2$, $x_2' = x_1^2 + x_2^2$ [*Hint*: Apply the method of homogeneous functions (Section 1.7) to $dx_2/dx_1 = (x_1^2 + x_2^2)/x_1x_2$.]

(c) $x_1' = x_2$, $x_2' = -e^{-x_1}$

6. (*Solving a Linear System*). The solutions of the scalar ODE $x' = ax$ have the form $x = x_0e^{at}$. This suggests that solutions of the system, $x_1' = ax_1 + bx_2$, $x_2' = cx_1 + dx_2$, where a, b, c, and d are constants, might have the form $x_1 = \alpha e^{rt} + \beta e^{st}$, $x_2 = \gamma e^{rt} + \delta e^{st}$ where r, s, α, β, γ, and δ must be determined by inserting x_1 and x_2 (as given) into the differential system and matching coefficients of like exponentials. Using this method, solve the following systems.

(a) $x_1' = x_1 + 3x_2$, $x_2' = x_1 - x_2$ (b) $x_1' = 2x_1 - x_2$, $x_2' = 3x_1 - 2x_2$

7. (*Laplace Transform Technique for Linear Systems*). Laplace transforms may be used to solve some systems. For example, if $x_1' = ax_1 + bx_2 + f_1(t)$, $x_2' = cx_1 + dx_2 + f_2(t)$, with a, b, c, and d real constants, then applying the transform to each side of the equation, we have that $s\mathcal{L}[x_1](s) - x_1(0) = a\mathcal{L}[x_1](s) + b\mathcal{L}[x_2](s) + \mathcal{L}[f_1](s)$. There is a similar equation for $s\mathcal{L}[x_2](s) - x_2(0)$. These two linear equations may be solved for $\mathcal{L}[x_1]$ and $\mathcal{L}[x_2]$. Use transform tables and formulas from Chapter 5 to find $x_1(t)$ and $x_2(t)$. Solve the following systems this way.

(a) $x_1' = 3x_1 - 2x_2 + t$, $x_2' = 5x_1 - 3x_2 + 5$, $x_1(0) = x_2(0) = 0$

(b) $x_1' = x_1 + 3x_2 + \sin t$, $x_2' = x_1 - x_2$, $x_1(0) = 0$, $x_2(0) = 1$

8. (*A Chemical Reaction*). Describe the behavior of the solutions of system (11) for various values of the rate constants and various values of $x(0)$ and $y(0)$. Assume that $z(0) = w(0) = 0$. Why are only the first three rate equations of (11) needed?

9. (*Two-Dimensional Autocatalator, Turning Oscillations On and Off*). The model system (15) for the two-dimensional autocatalator has five parameters, $x_1(0)$ and k_1, k_2, k_3, k_4. It may be possible to turn the oscillations on or off by changing any one of the five. That is the aim of the project: change a parameter until the oscillations of the intermediates x_2 and x_3 disappear. Then explain why you think that happens. Keep the following points in mind.

- Plot component curves of system (15) with the data given by (14).

- If your solver supports 3-dimensional graphs, duplicate the wineglass on the first page of the chapter and in Figure 6.1.7. What happens if you plot the solution curve for $0 \le t \le 1000$, rather than the truncated curve of the wineglass, $6.5 \le t \le 377$?

- Look at the Atlas Plates AUTOCATALATOR A–C and explain what you see.

- What happens when $x_1(0) = 500$ is replaced by larger and then smaller values, but keeping the values of the rate constants k_i fixed at the values given in (14). Now let $k_2 = 0.08, 0.10, 0.14, 0.14$. What happens? Any explanation?

- Try varying k_1 or k_3, keeping the other parameters fixed at the values given in (14). Explain what you see.

6.2   Basic Theory of Systems

The normal form of an IVP for a first-order differential system is

$$x' = f(t, x), \qquad x = x^0 \quad \text{when} \quad t = t_0 \tag{1}$$

where x, x^0, and f are n-vectors. Sometimes solutions can be found explicitly, but for any system which models a physical phenomenon of any complexity, that is rarely the case. Before proceeding, we define the useful notion of a general region

> ❖ **Region, Interior Point.** A *region R* in R^k is a connected set (i.e., not in disjoint pieces) consisting only of *interior points*. A point P is an interior point if there is a ball of positive radius centered at P and contained in R.

The following fundamental result provides assurance that under certain conditions on the rate function f, the IVP (1) has exactly one solution for each initial point (t_0, x^0) in some region of tx-space. Moreover, the solution may be extended to the edge of the domain of f, and the solution is a continuous function of time and initial data.

THEOREM 6.2.1

> Existence, Uniqueness, Extension, Continuity. **Consider the vector IVP**
>
> $$x' = f(t, x), \quad x(t_0) = x^0$$
>
> where the functions f_i and $\partial f_i/\partial x_j$ $(i, j = 1, \dots, n)$ are continuous throughout a region R of $(n+1)$-dimensional tx-space, and (t_0, x^0) is a point inside R.
>
> **Existence:** The IVP has a solution $x = x(t, t_0, x^0)$ in a t-interval I containing t_0.
>
> **Uniqueness:** The IVP has exactly one solution defined on I.
>
> **Extension:** The solution of the IVP can be extended to a t-interval for which the time-state curve tends to the boundary of R (which may be "at infinity") as t tends to either endpoint of the interval.
>
> **Continuity (Sensitivity):** The solution of the IVP is a continuous function of t, t_0, and x^0.

The proof is omitted since the basic ideas are those of the proof given in Appendix A for an IVP in one state variable.

From now on, we consider only systems for which the Existence and Uniqueness Theorem applies. Usually, the solution of IVP (1) will be written $x = x(t)$, unless there is some reason to emphasize its dependence on the initial data, in which case we shall write out $x = x(t, t_0, x^0)$. To avoid ambiguity, we always assume that the solution of (1) is maximally extended. Observe that a maximally extended time-state curve cannot "disappear" inside a region R as t approaches either endpoint of the t-interval on which the maximally extended solution is defined. The following simple example illustrates the conclusions of the Existence and Uniqueness Theorem.

EXAMPLE 6.2.1

Existence, Uniqueness, Extension, Continuity
The system

$$\begin{aligned} x' &= x^2, & x(t_0) &= b_1 \\ y' &= 2y, & y(t_0) &= b_2 \end{aligned} \tag{2}$$

meets the conditions of the Existence and Uniqueness Theorem for two state variables. The unique solution of IVP (2) may be obtained by separating variables in each ODE and solving. For example, the solution formula for y is found in this way to be

$$y(t) = b_2 e^{2(t-t_0)} \quad -\infty < t < \infty \tag{3}$$

Solving the ODE for x is slightly more complicated. First note that $x(t) = 0$ is the solution corresponding to $b_1 = 0$. If $x \neq 0$, the variables may be separated. After integrating, using the initial data, and solving for x in terms of t, we have that

$$x(t) = \frac{b_1}{1 - b_1(t - t_0)}, \quad \text{where} \begin{cases} t < t_0 + 1/b_1 & \text{if } b_1 > 0 \\ t > t_0 + 1/b_1 & \text{if } b_1 < 0 \end{cases} \tag{4}$$

Observe that formula (4) also gives the solution $x(t) = 0$, $-\infty < t < \infty$, if $b_1 = 0$. The change in the t-interval depending on the sign of b_1 is dictated by the need to keep t_0 inside the interval. In either case, the solution has a finite escape time since $x(t)$ becomes unbounded as t tends to the finite value $t_0 + 1/b_1$. This restriction of the t-interval in the definition of $x(t)$ forces the same restriction on the interval in the definition of $y(t)$ since the full solution of (2) requires both $x(t)$ and $y(t)$ to be defined on a common interval. IVP (2) has exactly one maximally extended solution, and this solution is seen through formulas (3) and (4) to be a continuous function of t, t_0, b_1, and b_2. Small changes in any or in all of these parameters lead to only small changes in the function $x(t)$ and $y(t)$.

Parameters as State Variables

Physical systems depend upon parameters; for example, the motion of a body suspended by a damped Hooke's Law spring (see Example 6.1.1) depends upon the ratios of the spring constant and the damping coefficient to the mass, k/m and c/m, respectively. How do the solutions of the model IVP depend upon those parameters? The question can be partially answered by a "trick": rename each parameter as a new state variable whose rate of change is zero. Then apply Theorem 6.2.1 to the augmented system and conclude that the solution depends continuously on the parameters as long as the rate functions are continuous and continuously differentiable functions of the parameters. An example shows how this is done.

EXAMPLE 6.2.2

Changing Parameters into State Variables

The IVP for the Hooke's Law spring in Example 6.1.1 is

$$\begin{aligned} x_1' &= x_2, & x_1(t_0) &= z_0 \\ x_2' &= -(k/m)x_1 - (c/m)x_2, & x_2(t_0) &= v_0 \end{aligned} \tag{5}$$

Rename k/m as x_3 and c/m as x_4 and create the IVP

$$\begin{aligned} x_1' &= x_2, & x_1(t_0) &= z_0 \\ x_2' &= -x_3 x_1 - x_4 x_2, & x_2(t_0) &= v_0 \\ x_3' &= 0, & x_3(t_0) &= k/m \\ x_4' &= 0, & x_4(t_0) &= c/m \end{aligned} \tag{6}$$

Then, we see from Theorem 6.2.1, that the components x_1 and x_2 of the solution of (6) are continuous functions of t, t_0, and the initial data z_0, v_0, k/m, and c/m.

As noted in Chapter 2, this trick is useful in the computer study of how solutions change as parameters change, i.e., in "sensitivity studies."

Autonomous Systems

The rate functions of a first-order system may change with time, or they may be independent of time (i.e. autonomous). Thus the system $x' = f(x)$ is autonomous. The corresponding IVP is

$$x' = f(x), \qquad x(t_0) = x^0 \tag{7}$$

where the conditions of the Existence and Uniqueness Theorem (Theorem 6.2.1) are assumed to hold in a region R of tx-space. Since f does not depend on t, we may replace R by a region S in x-space. Let x^0 be in S and let t_0 be any real number. The four conclusions of Theorem 6.2.1 may now be interpreted in terms of S instead of R. For example, the Maximal Extension Property may be recast in the following form.

THEOREM 6.2.2

> Maximal Extension (Autonomous Systems). Let the functions $f_i(x)$ and $\partial f_i / \partial x_j$, i and $j = 1, \ldots, n$, be continuous in the region S of $\mathbf{R}^n$, and suppose x^0 is in S. The solution $x = x(t)$ of IVP (7) may be extended to a unique maximal open interval (a, b) with the following property: Either $b = \infty$ ($a = -\infty$), or else b is finite (a is finite) and $x(t)$ tends to the boundary of S as $t \to b^-$ ($t \to a^+$).

EXAMPLE 6.2.3

Maximal Extension
Set $t_0 = 1, b_1 = 1, b_2 = 1$ in system (2). The solution is $x = (2 - t)^{-1}$, $y = e^{2(t-1)}$, $t < 2$ [use formulas (3) and (4)]. Observe that the maximal interval is $(-\infty, 2)$. The point (x, y) tends to the "infinite boundary" of $S = \mathbf{R}^2$ as $t \to 2^-$.

Time is only an incidental variable in an autonomous system. As noted in Section 6.1, if $x = x(t)$, $a < t < b$, is a solution of the autonomous system, $x' = f(x)$, then $x = x(t + c)$, $a - c < t < b - c$, is also a solution. The two solutions have exactly the same orbit in state space.

As we proved in Chapters 2 and 3 for first- and second-order scalar autonomous ODEs, distinct orbits of an autonomous system never meet.

THEOREM 6.2.3

> Separation of Orbits. Suppose that f_i and $\partial f_i / \partial x_j$ $(i, j = 1, \ldots, n)$ are continuous on a region S in state space. If $x^1(t)$ and $x^2(t)$ are solutions of $x' = f(x)$, whose orbits lie in S, then the orbits either never touch or else they coincide.

Computer-generated orbits of autonomous systems often appear to touch, but, as noted before, that is only because of the limitations of the computer and the display screen. See Atlas Plates BIFURCATION B, PLANAR PORTRAIT B, and STRANGE ATTRACTOR A for three of many examples of portraits of the orbits of an autonomous system.

Equilibrium Points and Cycles of Autonomous Systems

Constant orbits (*equilibrium points*) and periodic orbits (*cycles*) are special types of orbits of autonomous systems which often seem to play a fundamental organizational role in the portrait of the orbits. Equilibrium points are simply the zeros of $f(x)$.

There are several computer techniques for solving $f(x) = 0$; often the zeros can be determined by inspection. If x_0 is an equilibrium point for the autonomous system $x' = f(x)$, then the vector function $x(t) = x^0$, all t, defines a constant solution of the system, a so-called *equilibrium solution*. Note that a non-constant orbit of the system cannot pass through an equilibrium point since that would violate the Separation of Orbits Theorem 6.2.3. Finally, observe that if Γ is a cycle of an autonomous system, then no other orbit can touch it at a finite value of t (that would violate uniqueness), but that other orbits may approach Γ as $t \to +\infty$ or as $t \to -\infty$. See, for example, Atlas Plates BIFURCATION J, L, and PENDULUM A.

Periodic solutions and their corresponding orbits (*cycles*) cannot be found quite so easily. Sometimes they can be found by inspection of the system, sometimes by actually constructing solution formulas, sometimes by more intricate theoretical means (see, for example, the techniques of Sections 6.4, 9.1–9.3). Frequently, however, we can only "see" cycles on the computer screen after using a numerical solver to compute and graph orbits. See Figure 6.2.1 and Atlas Plates BIFURCATION I and L, CONSERVATIVE SYSTEM E, FIRST ORDER C, and STRANGE ATTRACTOR A for examples of cycles.

Planar Autonomous Systems

There are several specialized techniques that help us understand the behavior of orbits of the planar autonomous system

$$x' = f(x, y)$$
$$y' = g(x, y) \tag{8}$$

To fix ideas, let us make the blanket assumption that the rate functions f and g in (8) and their derivatives f_x, f_y, g_x, g_y are continuous on a region S in xy-space. Then we know that every point (x_0, y_0) has precisely one orbit passing through it.

The state space of this system is the xy-plane, and an orbit is a curve described by the endpoint of the *position vector*

$$\mathbf{R}(t) = x(t)\hat{\mathbf{x}} + y(t)\hat{\mathbf{y}}$$

where $x(t)$, $y(t)$ is a solution of the system (8) and $\hat{\mathbf{x}}$, $\hat{\mathbf{y}}$ are unit vectors in the positive x and y directions, respectively. If you imagine now that the endpoint of the position vector describes the motion of a particle, then the velocity $\mathbf{v}(t)$ of the particle is

$$\mathbf{v}(t) = \mathbf{R}'(t) = x'(t)\hat{\mathbf{x}} + y'(t)\hat{\mathbf{y}} = f(x, y)\hat{\mathbf{x}} + g(x, y)\hat{\mathbf{y}} \tag{9}$$

Since the velocity vector is tangent to the path of the particle, we see from (9) that the path itself need not be known to find $\mathbf{v}(t)$ at the point (x_0, y_0) on the "flight" path. From (9) we see that $\mathbf{v}$ at this point is $f(x_0, y_0)\hat{\mathbf{x}} + g(x_0, y_0)\hat{\mathbf{y}}$, and since the time t that the particle arrives at (x_0, y_0) is irrelevant (remember that the system is autonomous) we may as well think of $\mathbf{v}$ as a function of x and y and not t. Note that the length $||\mathbf{v}||$ of $\mathbf{v}$ gives the *speed* of the point mass as it traverses an orbit.

Now suppose that the region S is a rectangle $a \le x \le b$, $c \le y \le d$, and that a grid of equally spaced (both horizontally and vertically) points has been selected inside S. At each grid point (x_0, y_0) where $\mathbf{v}(x_0, y_0) \ne 0$ imagine the unit vector $\mathbf{u}(x_0, y_0) = \mathbf{v}(x_0, y_0)/||\mathbf{v}(x_0, y_0)||$ centered at the point (x_0, y_0). A constant $\varepsilon > 0$ can be determined from the mesh size of the grid such that for any two adjacent grid points (x_1, y_1), (x_2, y_2), the vectors through them, $\varepsilon\mathbf{u}(x_1, y_1)$ and $\varepsilon\mathbf{u}(x_2, y_2)$, do not intersect. Such a field of vectors is called a *direction field*. Direction fields for the system (8) are obviously not unique, but their usefulness is that the field vectors seem

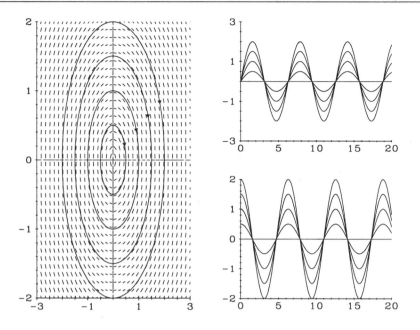

FIGURE 6.2.1 Direction field, nullclines, orbits, component graphs. See Example 6.2.4.

to guide the "flow" of the orbits. Examples follow. [Note: A direction field for a scalar first-order ODE, $y' = f(t, y)$ was defined in Sections 1.1 and 2.1, but that field was plotted in a *ty*-rectangle. Thus the direction fields of an autonomous system, $x' = f(x, y)$, $y' = g(x, y)$, and the scalar ODE, $y' = f(t, y)$ share the same name, but are conceptually different.]

For system (8), the curves defined by $f(x, y) = 0$ in the *xy*-plane are called *x*-*nullclines*. A non-constant orbit that intersects an *x*-nullcline must do so vertically. Similarly, the *y*-*nullclines* are defined by $g(x, y) = 0$, and non-constant orbits intersect these nullclines horizontally. The nullclines divide the *xy*-state plane into regions where orbits rise or fall, move to the right or to the left. Knowing where these regions are can be of considerable help in visualizing the portrait of the orbits even before the actual orbits are constructed. The equilibrium points of system (8) are located at the crossing points of the *x*-nullclines with the *y*-nullclines.

The following examples show how direction fields can be applied in particular cases to gain a good sense of what the orbits of a planar autonomous system look like.

EXAMPLE 6.2.4

The Cycles of the Harmonic Oscillator

The harmonic oscillator system

$$x' = y, \quad y' = -x \tag{10}$$

has solutions $x = A \sin(t + \phi)$, $y = A \cos(t + \phi)$ where A and ϕ are any constants. There is a single equilibrium point at the origin. The *x*-nullcline is the *x*-axis ($y = 0$), and the *y*-nullcline is the *y*-axis ($x = 0$). Figure 6.2.1 shows a direction field, the nullclines (the *x* and *y* axes), orbits, and the *x* and *y* component graphs of the orbits. Observe that the orbits in the first quadrant move to the right (because $x' > 0$) and downward (because $y' < 0$) as t increases. Orbits and line elements cross the nullclines horizontally or vertically, as expected. Although we have solution formulas for the system (10), they were not actually used in the above discussion or in constructing the graphs in Figure 6.2.1.

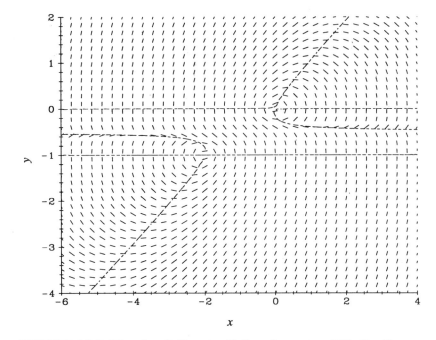

FIGURE 6.2.2 Direction field and nullclines for system (11). See Example 6.2.5.

System (10) is linear, and can be solved explicitly. The next system is neither linear nor simple, but the direction field and nullcline approach can still be used.

EXAMPLE 6.2.5

A System with Quadratic Rate Functions

The system

$$x' = -y - y^2$$
$$y' = 0.5x - 0.2y + xy - 1.2y^2 \qquad (11)$$

cannot be solved in terms of known functions. The equilibrium points are found by setting the rate fuctions equal to zero. From the first equation we see that the y-coordinate of an equilibrium point must be 0 or -1. Inserting first one of these values and then the other into the rate function for y' and then equating to zero, we determine that the two equilibrium points are $(0, 0)$ and $(-2, -1)$. The x-nullclines are the lines $y = 0$ and $y = -1$, while the y-nullclines are defined by

$$0.5x - 0.2y + xy - 1.2y^2 = 0$$

and are the two branches of a tilted hyperbola. As expected, the equilibrium points $(0, 0)$ and $(-2, -1)$ are precisely the points where the x-nullclines and the y-nullclines intersect. Figure 6.2.2 shows these nullclines and a direction field defined by the system (11). One can see from Figure 6.2.2 that some orbits must rotate around each equilibrium point, but it is not very clear just what the orbital portrait looks like. See Atlas Plate FIRST ORDER I for the surprising picture.

EXAMPLE 6.2.6

A Symmetric Example

The quadratic system

$$x' = y^2 - x^2$$
$$y' = x^2 - a \qquad (12)$$

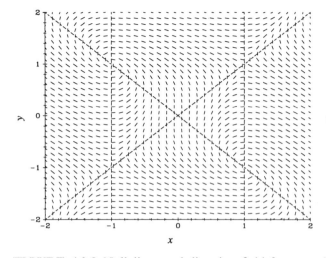

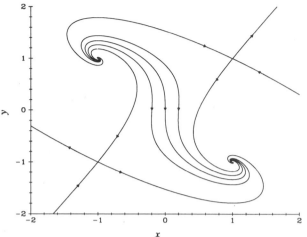

FIGURE 6.2.3 Nullclines and direction field for system (12). See Example 6.2.6.

FIGURE 6.2.4 Orbits of system (12). See Example 6.2.6.

where a is a positive constant, has x-nullclines $y = \pm x$, and y-nullclines $x = \pm\sqrt{a}$. The equilibrium points are the four points $(\pm\sqrt{a}, \pm\sqrt{a})$. Figure 6.2.3 shows the nullclines and the direction field when $a = 1$. The figure suggests rotational motion around the equilibrium points $(\sqrt{a}, -\sqrt{a})$ and $(-\sqrt{a}, \sqrt{a})$, and some kind of "hyperbolic" motion near the other two equilibrium points. Since the rate functions of system (12) are invariant when x and y are replaced by $-x$ and $-y$, we expect the orbits to be symmetric through the origin. Figure 6.2.4 shows this symmetry. Arrowheads show time's increase.

Sensitivity

Finally, we show how orbits change as some parameter in the system changes.

EXAMPLE 6.2.7

Changing the Parameter in System (12)

Consider the orbit of system (12) that passes through the origin. As suggested by Figure 6.2.4, that orbit approaches the equilibrium point $(1, -1)$ if $a = 1$. Now suppose that a takes on the values 1, 0.9, 0.8, 0.4, 0.1, and 0.025. Figure 6.2.5 shows how the orbit with x_0, $y_0 = 0$ (and $0 \le t \le 50$) changes in response to these changes in the parameter a. Apparently, the orbit approaches the equilibrium point $(\sqrt{a}, -\sqrt{a})$ in more or less the same fashion for every value of $a > 0$, but the length of the orbit shortens and the equilibrium point nears the origin as a approaches zero.

Figure 6.2.6 shows the x, y, and z component curves (reading from the top) of the augmented system when the parameter a is converted to a third state variable z:

$$\begin{aligned}
x' &= y^2 - x^2, & x(0) &= 0 \\
y' &= x^2 - z, & y(0) &= 0 \\
z' &= 0, & z(0) &= 1,\ 0.9,\ 0.8,\ 0.4,\ 0.1,\ 0.025
\end{aligned} \tag{13}$$

The component curves correspond to the orbits of Figure 6.2.5, but plotted over the time interval $0 \le t \le 10$ instead of $0 \le t \le 50$.

The mathematical model of the damped simple pendulum was introduced in Section 4.1 and some of its orbits were portrayed in the cover figure for Chapter 4. Con-

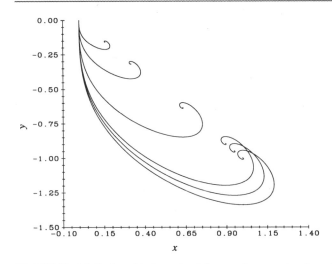

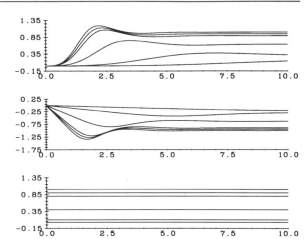

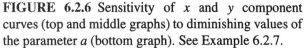

FIGURE 6.2.5 Sensitivity of orbits to changes in the parameter *a*. See Example 6.2.7.

FIGURE 6.2.6 Sensitivity of *x* and *y* component curves (top and middle graphs) to diminishing values of the parameter *a* (bottom graph). See Example 6.2.7.

sider the second-order ODE that models the motion of an undriven pendulum of length *L* supporting a metal bob of mass *m* and subject to a damping force:

$$mLx'' + cLx' + mg \sin x = 0 \tag{14}$$

where *c* is the damping coefficient, *x* is the angle of the pendulum measured counterclockwise from the vertical, and x' is the angular velocity. Divide by mL, set $c/m = \alpha$ and $g/L = 10$. Using *x* and $x' = y$ as state variables and imposing specific initial conditions, we have the IVP

$$\begin{aligned} x' &= y, & x(0) &= 0 \\ y' &= -10 \sin x - \alpha y, & y(0) &= 10 \end{aligned} \tag{15}$$

The initial conditions model the pendulum hanging downward at angle 0 radians and with an initial counterclockwise angular velocity of 10 radians/sec.

EXAMPLE 6.2.8

How Changing the Mass Affects the Motion of the Pendulum

Because $\alpha = c/m$, and *c* is assumed to be a fixed positive number, we see that changing the mass in the pendulum model (15) has an inverse effect on the damping force. So the pendulum with a large mass (hence, light "friction") might whirl all the way up and over the pivot several times before frictional forces gradually compel the pendulum to settle down to back-and-forth decaying oscillations about a downward vertical position. Figure 6.2.7 shows four orbits: from top to bottom, $\alpha = 0$ (no damping), $\alpha = 0.3$, $\alpha = 0.5$, and $\alpha = 1.75$. The orbits are graphed over the time interval $0 \le t \le 30$. The top orbit (solid line) with $\alpha = 0$ corresponds to the pendulum whirling counterclockwise forever, i.e., the angle *x* goes to ∞ as $t \to +\infty$. This is because there is no friction and so energy is never dissipated.

The big change occurs when α switches from value 0 to a positive number, i.e., when the damping is "turned on." The graph, with $\alpha = 0.3$ (long dashes), shows how light friction (or a fairly large mass) results in a couple of over-the-top turns followed by decaying oscillations that approach the equilibrium point $(4\pi, 0)$. If $\alpha = 0.5$ (short dashes), the somewhat heavier friction (or smaller mass) produces an orbit that settles down toward the equilibrium point $(2\pi, 0)$. If $\alpha = 1.75$ (long dashes), the heavy friction (or very small mass) prevents the pendulum from going over-the-top at all; in this case the angle *x* increases from its initial value of 0 to a maximal value

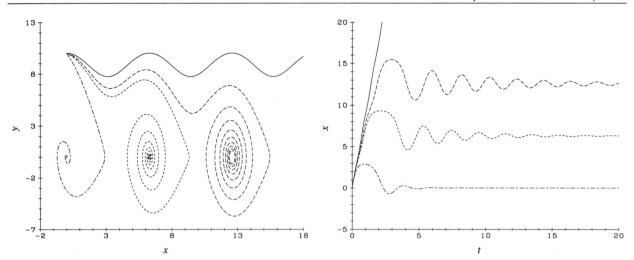

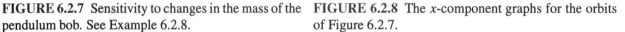

FIGURE 6.2.7 Sensitivity to changes in the mass of the pendulum bob. See Example 6.2.8.

FIGURE 6.2.8 The x-component graphs for the orbits of Figure 6.2.7.

of about 3 radians. See the rightmost point on the leftmost orbit (long/short dashes) in Figure 6.2.7. The pendulum then stops and falls back into a mode of decaying oscillations about $x = 0$, $y = 0$. Figure 6.2.8 shows the graphs of the x component of these orbits, but over the time span, $0 \leq t \leq 20$.

It would be interesting to determine those crucial values of α which divide orbits that approach the equilibrium point $(2n\pi, 0)$ from those approaching the equilibrium point $((2n - 2)\pi, 0)$. Problem 9 considers a related problem. The system's behavior is quite sensitive to small changes in the value of α near these critical values.

Planar Autonomous Systems in Polar Coordinates

The orbits of the planar autonomous system

$$\begin{aligned} x' &= f(x, y) \\ y' &= g(x, y) \end{aligned} \tag{16}$$

can be described in polar coordinates r, θ, where

$$x = r \cos \theta, \quad y = r \sin \theta \tag{17}$$

To transform the ODEs in (16) into polar coordinates, we proceed as follows. Let $x = x(t)$, $y = y(t)$ be a solution of (16). Then differentiating each side of (17) with respect to t, we have

$$\begin{aligned} x' &= r' \cos \theta - r\theta' \sin \theta = f(r \cos \theta, r \sin \theta) \\ y' &= r' \sin \theta + r\theta' \cos \theta = g(r \cos \theta, r \sin \theta) \end{aligned} \tag{18}$$

Solving for r' and θ', we have

$$\begin{aligned} r' &= \cos \theta f(r \cos \theta, r \sin \theta) + \sin \theta g(r \cos \theta, r \sin \theta) = F(r, \theta) \\ \theta' &= (1/r)[\cos \theta g(r \cos \theta, r \sin \theta) - \sin \theta f(r \cos \theta, r \sin \theta)] = G(r, \theta) \end{aligned} \tag{19}$$

Observe that $x_0 = r_0 \cos \theta_0$, $y_0 = r_0 \sin \theta_0$ with $r_0 \neq 0$ is an equilibrium point of (16) if and only if $F(r_0, \theta_0) = 0$, $G(r_0, \theta_0) = 0$ in (19).

How is the velocity field $\mathbf{v} = f(x, y)\hat{\mathbf{x}} + g(x, y)\hat{\mathbf{y}}$ of system (16) related to the rate functions $F(r, \theta)$ and $G(r, \theta)$ defined in (19)? Recall that, at each point (r, θ) of the plane, $\hat{\mathbf{r}}$, $\hat{\boldsymbol{\theta}}$ are unit vectors with $\hat{\mathbf{r}}$ pointing outward along r and $\hat{\boldsymbol{\theta}}$ orthogonal to $\hat{\mathbf{r}}$ pointing in the direction of increasing θ (i.e., counterclockwise). See the margin figure. A little trigonometry shows that $\hat{\mathbf{r}} = \cos\theta\hat{\mathbf{x}} + \sin\theta\hat{\mathbf{y}}$, and $\hat{\boldsymbol{\theta}} = -\sin\theta\hat{\mathbf{x}} + \cos\theta\hat{\mathbf{y}}$, which can be solved for $\hat{\mathbf{x}}$ and $\hat{\mathbf{y}}$ to obtain

$$\hat{\mathbf{x}} = \cos\theta\hat{\mathbf{r}} - \sin\theta\hat{\boldsymbol{\theta}}, \quad \hat{\mathbf{y}} = \sin\theta\hat{\mathbf{r}} + \cos\theta\hat{\boldsymbol{\theta}}$$

Thus we see that the velocity field $\mathbf{v}$ for the system (16) takes the form

$$\begin{aligned}
\mathbf{v} = f\hat{\mathbf{x}} + g\hat{\mathbf{y}} &= (\cos\theta\hat{\mathbf{r}} - \sin\theta\hat{\boldsymbol{\theta}})f + (\sin\theta\hat{\mathbf{r}} + \cos\theta\hat{\boldsymbol{\theta}})g \\
&= (f\cos\theta + g\sin\theta)\hat{\mathbf{r}} + (g\cos\theta - f\sin\theta)\hat{\boldsymbol{\theta}} \\
&= F(r,\theta)\hat{\mathbf{r}} + rG(r,\theta)\hat{\boldsymbol{\theta}}
\end{aligned}$$

which shows how to express $\mathbf{v}$ in terms of $\hat{\mathbf{r}}$ and $\hat{\boldsymbol{\theta}}$.

EXAMPLE 6.2.9

From Rectangular to Polar Coordinates
The system

$$\begin{aligned}
x' &= x - 10y \\
y' &= 10x + y
\end{aligned} \tag{20}$$

can be rewritten in terms of polar coordinates as

$$\begin{aligned}
r' &= \cos\theta(r\cos\theta - 10r\sin\theta) + \sin\theta(10r\cos\theta + r\sin\theta) = r \\
\theta' &= (1/r)[\cos\theta(10r\cos\theta + r\sin\theta) - \sin\theta(r\cos\theta - 10r\sin\theta)] = 10
\end{aligned} \tag{21}$$

Thus,

$$r = r_0e^t, r_0 \geq 0, \quad \theta = 10t + \theta_0 \tag{22}$$

define the general solution of system (21). We see that the velocity field $\mathbf{v}$ for system (20) can be written as $\mathbf{v} = r\hat{\mathbf{r}} + 10r\hat{\boldsymbol{\theta}}$. Hence we see that all orbits spiral counterclockwise away from the origin with ever-increasing speed since from (22) $\mathbf{v} = r_0e^t(\hat{\mathbf{r}} + 10\hat{\boldsymbol{\theta}})$, and so the speed is $||\mathbf{v}|| = r_0e^t||\hat{\mathbf{r}} + 10\hat{\boldsymbol{\theta}}|| = \sqrt{101}r_0e^t$. The general solution of system (20) in Cartesian coordinates is

$$x = r_0e^t\cos(10t + \theta_0)$$
$$y = r_0e^t\sin(10t + \theta_0)$$

where r_0 and θ_0 are arbitrary constants.

Polar coordinates are particularly suitable for planar autonomous systems where rotational behavior is suspected.

Comments

Planar autonomous systems may be analyzed using direction fields and nullclines, but these techniques are not available for nonautonomous planar systems or for systems with more than two state variables. Chapters 7, 8, and 9 introduce techniques that may enable us to determine the behavior of orbits, time-state curves, and component graphs in some cases. The basic properties given in Theorem 6.2.1 are usually taken for granted and not specifically mentioned when dealing with a system. That does not mean that they are unimportant, but just the opposite. Without the assurances given in Theorem 6.2.1, there could be no real theory of systems of ODEs because we would never be sure that an IVP even had a solution, or if it did, that the solution would be unique, or that the solution could be extended forward and backward, or that the solution would change continuously as the data are changed. So, even though we do not always mention Theorem 6.2.1 in the sections and chapters that follow, it is the foundation on which all of our work is based.

PROBLEMS

1. Verify that each IVP satisfies the hypotheses of Theorem 6.2.1 for all values of t and the state variables. Find formulas for the solutions. For **(a)–(e)** plot the nullclines, and plot the nine orbits corresponding to all possible combinations of $a, b = -1, 0, 2$. Identify any orbits that are equilibrium points or cycles. For **(a)–(e)** plot the orbits on the ranges $|x| \leq 3$, $|y| \leq 3$.

    (a) $x' = y$, $y' = -x - 2y$; $x(0) = a$, $y(0) = b$ [*Hint*: Note that $x'' + 2x' + x = 0$.]

    (b) $x' = 2x$, $y' = -4y$; $x(0) = a$, $y(0) = b$

    (c) $x' = y$, $y' = -9x$; $x(0) = a$, $y(0) = b$ [*Hint*: Note that $x'' + 9x = 0$.]

    (d) $x' = y^3$, $y' = -x^3$; $x(0) = a$, $y(0) = b$ [*Hint*: Write as $dy/dx = -x^3/y^3$ and separate variables.]

    (e) $x' = -x^3$, $y' = -y$; $x(0) = a$, $y(0) = b$

    (f) $x' = y$, $y' = -26x - 2y$, $z' = z/2$; $x(0) = y(0) = z(0) = 1$. Sketch the projection of the orbit in each coordinate plane. Use the ranges $|x| \leq 1$, $-4 \leq y \leq 3$, $1 \leq z \leq 15$, and the time interval $0 \leq t \leq 5$.

2. Find the equilibrium points of each system. Plot the nullclines, direction fields, and several orbits in a rectangle large enough to contain several equilibrium points.

    (a) $x' = x - y^2$, $y' = x - y$ **(b)** $x' = y \sin x$, $y' = xy$

    (c) $x' = 2 + \sin(x + y)$, $y' = x - y^3 + 27$ **(d)** $x' = y + 1$, $y' = \sin^2 3x$

    (e) $x' = 3(x - y)$, $y' = y - x$

3. Find all solutions of each system. Verify that if $x = x(t)$, $y = y(t)$ is a solution, then so is $x = x(t - T)$, $y = y(t - T)$, where T is any constant. Specify the interval on which each solution is defined.

    (a) $x' = 3x$, $y' = -y$ **(b)** $x' = 1/x$, $y' = -y$

    (c) $x' = -x^3$, $y' = 1$ **(d)** $x' = x^2(1 + y)$, $y' = -y$

4. Consider the planar autonomous system

    $$x' = 1 - y^2, \quad y' = 1 - x^2$$

    (a) Locate all the equilibrium points of the system.

    (b) Create a state portrait of orbits for the system over the region $|x| \leq 3$, $|y| \leq 3$. Use arrowheads on orbits to indicate the direction of increasing time. [*Hint*: See Atlas Plate FIRST ORDER C.]

    (c) Use geometric reasoning with the direction field to show that the line $y = x$ is composed of five orbits. Identify these orbits and use arrowheads to show the direction of increasing time. [*Hint*: Plotting a direction field first might help.]

    (d) For each nonconstant orbit in part **(c)** above, find a function $f(t)$ such that $x = f(t)$, $y = f(t)$ describes the orbit. Show that the orbit that originates at $x(0) = a$, $y(0) = a$, for any value of $a < -1$ escapes to infinity in finite time.

5. (*Attraction and Repulsion*). The system below illustrates the attraction and repelling properties of cylces and equilibrium points.

    (a) Explain why the unit circle, $x^2 + y^2 = 1$, is a cycle of the system

    $$x' = x - y - x(x^2 + y^2), \quad y' = x + y - y(x^2 + y^2) \qquad (*)$$

and the origin is the only equilibrium point. [*Hint*: Use (19) to write ODEs for r' and for θ'.]

(b) Solve the ODEs in r and θ. Then explain why all nonconstant orbits approach the cycle as time $t \to +\infty$.

(c) Draw a direction field for system $(*)$, plot x- and y-nullclines, and plot orbits inside, on, and outside the unit circle for $|x| \le 2$, $|y| \le 2$. Why could you call the origin a "repeller" and the unit circle an "attractor."

(d) Plot x- and y-component graphs for the orbits plotted in part **(c)**. What is the period of the cycle?

6. (*Sensitivity*). Refer to Example 6.2.6 in the parts below.

(a) Plot direction fields and x- and y-nullclines and some orbits like those shown in Figures 6.2.3 and 6.2.4 for system (12) with $a = 1, 0.4$, and 0.1.

(b) Consider the related system $x' = by^2 - x^2$, $y' = x^2 - 1$ for positive values of the parameter b. Plot direction fields, x- and y-nullclines, and some orbits for $b = 0.1, 1, 10$.

7. (*Orbits of Nonautonomous Systems*). For a nonautonomous system, an orbit may intersect itself and still not be a cycle. Illustrate this fact by solving the IVP $x' = \pi \cos \pi t$, $y' = y \cos t$, $x(0) = 0$, $y(0) = 1$, and plotting the orbit.

8. (*Cycles and Equilibrium Points*). Suppose the conditions of Theorem 6.2.2 are met for the autonomous system $x' = f(x)$. Properties of equilibrium points and cycles are given below. The converses are also true, but not proved here.

(a) Suppose that $x(t)$ is a nonconstant solution for which $x(t) \to P$ as $t \to T$. Show that if P is an equilibrium point, then $|T| = \infty$. [*Hint*: Since $x(t)$ is a continuous function of t, $x(T)$ must be P if T is finite. This violates the Uniqueness Principle. (Why?) Use the Extension Principle (Theorem 6.2.2) to show that P lies on the boundary of S, or else $|T| = \infty$.]

(b) Show that if $x = x(t)$ is a nonconstant periodic solution, then the corresponding orbit is a simple closed curve. [*Hint*: Suppose that the least period of $x(t)$ is T and that $x(t_1) = x(0)$ for some t_1, $0 < t_1 \le T$. Then $y(t) = x(t + t_1)$ is also a solution (why?) and $y(t) = x(t)$ since $y(0) = x(0)$. Hence $x(t)$ has period t_1. Show that $t_1 = T$.]

9. (*Sensitivity of Pendulum Motion*). Consider the pendulum system $x' = y$, $y' = -A \sin x - y$. Use a solver, a variety of initial points, and forward and backward solutions (in time) to give a complete analysis of what happens to the motion of the pendulum as the parameter A varies from close to zero to very large. In every case use the rectangle $|x| \le 20$, $|y| \le 20$ and justify all your assertions. Now in a different direction altogether, return to the IVP in Example 6.2.8 and try to determine a value α_0 for α such that the solution tends to the equilibrium point $(\pi, 0)$ as $t \to \infty$. What happens if the value of α is just slightly different from α_0? Interpret everything in terms of the motion of a pendulum.

6.3   Interacting Species

In the dynamics of life, no species is alone. The survival or the flourishing of a species rests on strategies of cooperation, competition, and predation. We may gain an understanding of a small part of the biological universe by considering the interactions of just two species. We shall begin with a general discussion of the model building process for these interactions. Then we shall construct prototype models for specific kinds of interactions and display computer simulations of the orbits and solution graphs of the model systems.[2]

Rate Equations

Let $x(t)$ and $y(t)$ denote the population sizes of two species at time t. The Balance Law of Section 1.2 applies to each:

$$\text{Net rate of change of a population} = \text{Rate in} - \text{Rate out}$$

We suppose that the net migration into the community of the two species is negligible, and so the net rate of change is just the difference between the birth and death rates. One model for the changing rates is the IVP,

$$x' = R_1(x, y)x, \quad x(0) = x_0$$
$$y' = R_2(x, y)y, \quad y(0) = y_0 \tag{1}$$

where the functions R_1 and R_2 are the *intrinsic rate coefficients*. Each coefficient measures the contribution of the average individual of the species to the overall growth rate of the population of that species. This contribution generally is not constant, but depends on time, population sizes, individual ages, and a host of other factors. We shall average out the time variation in the coefficients. We also suppose that the two species form a community isolated from all other influences—not very realistic, but simplification (even oversimplification) is always an important first step in modeling.

Different choices for the coefficients R_1 and R_2 model different kinds of interactions. We shall assume that R_1 and R_2 are continuously differentiable autonomous functions of x and y in a region S that includes the *population quadrant*, $x \geq 0$, $y \geq 0$. This means that the Existence and Uniqueness Theorem of Section 6.2 applies, and IVP (1) has a unique solution for each choice of initial point $x_0 \geq 0$, $y_0 \geq 0$. This has an important consequence.

THEOREM 6.3.1

> **Isolation of the Population Quadrant.** The solution $x = x(t)$, $y = y(t)$ of IVP (1), lies entirely in the population quadrant for all values of t for which it is defined if and only if $x_0 \geq 0$, $y_0 \geq 0$. The orbit lies entirely in the interior of the quadrant if and only if $x_0 > 0$, $y_0 > 0$.

Proof. The x and the y axes are unions of orbits of system (1) because $x(t) = 0$, for all t, solves the first ODE of (1), and $y(t) = 0$, for all t, solves the second. The Separation of Orbits Theorem 6.2.3 implies that no orbit of IVP (1) can enter or leave the population quadrant across the axes; to do so would imply that two distinct orbits have a point on one of the axes in common.

The observations of real populations are consistent with much of what Theorem 6.3.1 says. "Negative" populations do not exist, and species cannot be created from nothing.

---

[2]An excellent source book and introduction to the general area of modeling biological systems is *Mathematical Models in Biology* (New York: Random House, 1988). The book was written by the contemporary applied mathematician and researcher, Leah Edelstein-Keshet, University of British Columbia. Her areas of research include pattern formation, population growth, and medical modeling. Another basic reference in mathematical biology is the encyclopedic volume (and advanced text) by the renowned mathematical biologist J.D. Murray, *Mathematical Biology* (New York: Springer-Verlag, 1989).

However, there is much to criticize on biological grounds. For example, the model ODEs (1) surely cannot be correct for very small populations. At low population levels, species often become extinct in finite time, although Theorem 6.3.1 implies that extinction, if it occurs at all, takes an infinitely long time. Nevertheless we shall see what can be concluded from various forms of the intrinsic rate coefficients, even if the biological interpretation may be not all that accurate for real species.

Population Models

Different choices for R_1 and R_2 in system (1) model different kinds of interactions. We shall usually assume that R_1 and R_2 are first degree polynomials in the populations x and y, and that any time dependence has been averaged out so that system (1) is autonomous. The coefficients $a, b, \alpha, \ldots$ in the models below are assumed nonnegative. The model ODEs are usually defined for all x and y, but make biological sense only in the population quadrant, $x \geq 0$, $y \geq 0$.

- *No Interspecies Interaction.* The system

$$\begin{aligned} x' &= R_1(x)x \\ y' &= R_2(y)y \end{aligned} \tag{2}$$

  models a two-species community in which each species has no effect at all on the growth rate of the other. The exponential and logistic models of Section 1.5 are special examples of single species models like those in (2).

- *Predator-Prey Interaction.* Alfred Lotka and Vito Volterra independently introduced the system

$$\begin{aligned} x' &= (-\alpha + by)x = -\alpha x + bxy \\ y' &= (\beta - cx)y = \beta y - cxy \end{aligned} \tag{3}$$

  as a model for a community of predators and their prey. The predator population is $x(t)$, the prey is $y(t)$; α and β are the *natural decay* and *growth coefficients* of each species if the other species is extinct. Lotka and Volterra presumed that the number of predator-prey encounters is jointly proportional to the population of each species (*Law of Mass Action*). Mass action coefficients b and c measure, respectively, predator efficiency in converting food into fertility, and the probability that a predator-prey encounter removes one of the prey. System (3) is one of the rare systems of nonlinear ODEs for which there is a formula for the orbits (see Section 6.4). See also Atlas Plates PREDATOR-PREY A–C.

- *Overcrowding.* Excess population may lower the intrinsic rate coefficients just as in the logistic model of Section 1.5. For example, too many predators may lead to overcrowding effects that lower the growth rate of the predator species. If the effects of overcrowding on each of the predator and prey species of system (3) are taken into account, then we have the model system

$$\begin{aligned} x' &= (-\alpha - ax + by)x \\ y' &= (\beta - cx - dy)y \end{aligned} \tag{4}$$

  Overcrowding is modeled by intraspecies mass action terms, $-ax^2$ for the predator and $-dy^2$ for the prey.

- *Harvesting.* One or both of the species may be harvested. For example, suppose that the x-species in (1) is harvested at a *constant rate*. The corresponding rate equation is

$$x' = R_1 x - H_1 \tag{5}$$

where H_1 is the harvest rate. In *constant effort* harvesting the harvest rate is proportional to the size of the population:

$$x' = R_1 x - H_1 x \tag{6}$$

where H_1 is a positive constant. In *seasonal harvesting*, the harvest rate is modeled by an on-off function that is "on" for the season, then "off" for the rest of the year;

$$x' = R_1 x - H_1 \,\mathrm{sqw}(t, d, 1)$$

where $\mathrm{sqw}(t, d, 1)$ is the square wave of period one year whose formula is

$$\mathrm{sqw}(t, d, 1) = \begin{cases} 1, & 0 \le t \le d/100 \\ 0, & d/100 < t < 1 \\ \text{extended periodically from } [0,1) \text{ to } [1,2), [2,3), \dots \end{cases}$$

The *duty cycle d* is the percentage of the period that harvesting is "on." For example, if the harvest season lasts for four months out of each year, then $d = 25$ (i.e., 25% of the period). See also Appendix C.1. It is assumed here that the harvesting season always begins at the start of the year.

- *Cooperation.* A large number of biological interactions are of mutual benefit. An example is an ant-aphids community. Watch ants on a rose bush tending their aphids, or "ant-cows." The aphids feed off of the internal sap of the plant, and the ants dine on a fluid secreted by the aphids. As one part of the plant withers, the ants carry their cows to fresh "pasture." This kind of interaction is termed *cooperation* or *mutualism*. A model for cooperation is

$$\begin{aligned} x' &= (\alpha + by)x \\ y' &= (\beta + cx)y \end{aligned} \tag{7}$$

where each species alone has a natural, exponentially growing population (modeled by the terms αx and βy.) The mass action terms bxy and cxy show that each species promotes the growth of the other. Systems such as (7) cannot model real populations for any length of time because of the explosive growth in x and y. A better model would include overcrowding terms.

- *Competition.* Two species may compete for a resource in short supply. One model for competitive interaction with overcrowding is

$$\begin{aligned} x' &= (\alpha - ax - by)x \\ y' &= (\beta - cx - dy)y \end{aligned} \tag{8}$$

The mass-action terms $-ax^2 - bxy$ and $-cxy - dy^2$ model the negative effects of large populations on growth rates.

- *Predator Satiation.* The predator's appetite may be satiated by a glut of prey. This phenomenon may be modeled by the following modification of system (3):

$$\begin{aligned} x' &= -\alpha x + \frac{b}{m + ky} \cdot xy \\ y' &= \beta y - \frac{c}{m + ky} \cdot xy \end{aligned} \tag{9}$$

The mass action coefficients b and c of system (3) are modified by division by the positive term $m + ky$; the new coefficients tend to zero as the prey population tends to infinity. This model is analyzed in Section 9.3.

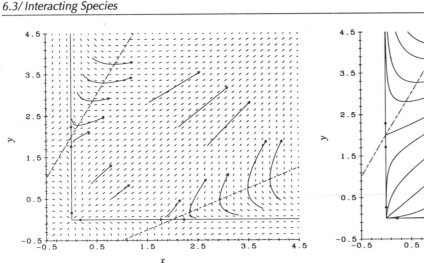

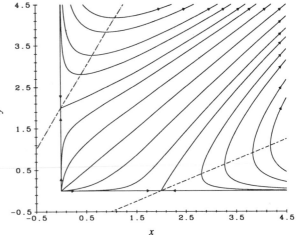

FIGURE 6.3.1 Direction field for unstable coopera-tion. See system (10).

FIGURE 6.3.2 Explosively growing orbits. See sys-tem (10).

Computer Simulations

Concrete examples (given below) and the computer-generated pictures of their orbits put some meat on the bones of the abstract models of the first part of this section. Detailed mathematical explanations and model interpretations in terms of real species are given in Section 6.4 and Chapters 8 and 9.

EXAMPLE 6.3.1

Cooperation Leading to Explosive Growth
The model ODEs

$$x' = (2 - x + 2y)x = 2x - x^2 + 2xy$$
$$y' = (2 + 2x - y)y = 2y - y^2 + 2xy \tag{10}$$

represent both mutually beneficial interactions (the terms $2xy$) and self-limitation (the terms $-x^2$ and $-y^2$). The symmetries in the rate equations suggest that we have a single species divided into two cooperating groups, each having the same dynamics as the other. The nullclines for the x-species (i.e., the curves on which $x'(t)$ is zero) are defined by $(2 - x + 2y)x = 0$. Thus, the x-nullclines are the y-axis and the line of slope $1/2$ through the point $(2, 0)$. Similarly, the y-nullclines are the x-axis and the line through the point $(0, 2)$ of slope 2. See Figure 6.3.1 for the nullclines (the dashed lines and the coordinate axes). The equilibrium points are the intersection points of the x-nullclines and the y-nullclines. In this example, one of the equilibrium points is down in the third quadrant and of no importance for the population model. The other three are on the edge of the population quadrant: $(0, 0)$, $(2, 0)$, $(0, 2)$. Figure 6.3.1 shows a direction field for the system (10); arrows indicate the direction of motion for increasing time. Sign Analysis (see Section 2.2) may be used to orient the arrows. For example, if the point (x_0, y_0) in the population quadrant is below the x-nullcline, $2 - x + 2y = 0$, and above the y-nullcline, $2 + 2x - y = 0$, then x' and y' are both positive, and the arrow through (x_0, y_0) is directed to the right and upward.

The line elements and the arrows of Figure 6.3.1 suggest that orbits inside the population quadrant move into the region between the two dashed line nullclines with increasing time, and then continue to move upward and to the right in that region. Figure 6.3.2 shows some of the orbits in the population quadrant $x \geq 0$, $y \geq 0$. It certainly appears that the populations multiply explosively.

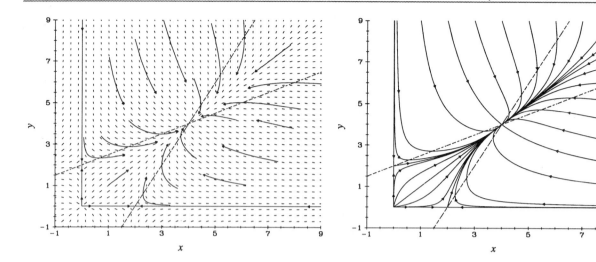

FIGURE 6.3.3 Direction field for stable cooperation. See system (11).

FIGURE 6.3.4 Orbits for stable cooperation. See system (11).

All of the figures in this section have a border around the population quadrant to emphasize that quadrant's isolation from orbits in other quadrants.

EXAMPLE 6.3.2

A Stable Cooperation Model

System (10) cannot be valid for a real species because the populations get too large too fast. The model and the population can be stabilized by increasing the "cost" of overcrowding within each species or by decreasing the "benefits" of cooperation. Here is a model that does both:

$$\begin{aligned} x' &= (4 - 2x + y)x \\ y' &= (4 + x - 2y)y \end{aligned} \tag{11}$$

The increase in the overcrowding coefficient from 1 to 2 and the decrease in the mutualism coefficient from 2 to 1 (compared to the model of Example 6.3.1) gives us a model that is stable and a lot more realistic than the unstable growth model of system (10). The increase in the natural growth coefficient from the value 2 in system (10) to 4 in system (11) just shifts the nullclines somewhat, and does not affect stability. Figure 6.3.3 shows the line element field, the nullclines, and the arrows corresponding to the system (11). The internal population orbits appear to be pulled into a pair of wedge-shaped regions whose apex is at the internal equilibrium point (4,4). Figure 6.3.4 shows the orbits themselves. The graphs give compelling visual evidence that the cooperating species tend to a stable state of coexistence.

Can populations stabilize and survive if both species do not cooperate, but compete for resources in limited supply? The next two examples give contrasting answers to this question.

EXAMPLE 6.3.3

A Model for Stable Competition

Suppose that the population of each of two species would change logistically if left alone, but that the two species compete with one another for scarce resources when they share the same habitat. The following system models one kind of competition dynamics:

$$\begin{aligned} x' &= (2 - 2x - y)x \\ y' &= (2 - x - 2y)y \end{aligned} \tag{12}$$

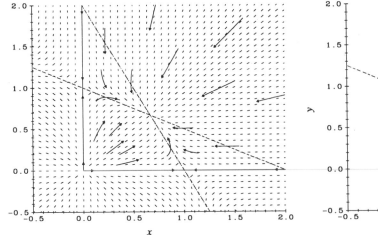

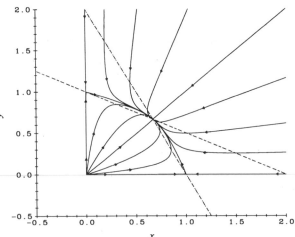

FIGURE 6.3.5 Direction field for stable competition. See system (12).

FIGURE 6.3.6 Orbits of stable competition. See system (12).

The x-nullclines are the y-axis and the line $2 - 2x - y = 0$; the y-nullclines are the x-axis and the line $2 - x - 2y = 0$ (see Figure 6.3.5) The x-nullclines intersect the y-nullclines in the equilibrium points $(0,0)$, $(1,0)$, $(0,1)$, and $(2/3,2/3)$. Sign Analysis verifies the validity of the line element field and the direction of the arrows shown in Figure 6.3.5. As in Example 6.3.2, orbits seem to be drawn into the two wedge-shaped regions bounded by the nullclines (dashed lines), and then are attracted toward the equilibrium point $(2/3,2/3)$. Figure 6.3.6 displays orbits. It appears from the graphs that the species of this model of competition tend to a stable, attracting equilibrium: $x(t) \rightarrow 2/3$ and $y(t) \rightarrow 2/3$ as time advances.

EXAMPLE 6.3.4

Competitive Exclusion

Life is not always favorable for species in the competition for resources. Suppose that the slanting nullclines (dashed lines) in Figure 6.3.5 are exchanged, the x-nullcline becoming a y-nullcline; and vice versa. The new model of competition is

$$x' = (2 - x - 2y)x$$
$$y' = (2 - 2x - y)y \tag{13}$$

Sign Analysis and the new position of the dashed nullclines suggest a surprising change in the ultimate fate of each species. Referring to Figure 6.3.7, we see that some population orbits seem to be attracted into the lower right wedge between the null clines; these orbits then approach the point $(2,0)$ of x-species equilibrium and y-species extinction. Other orbits are attracted into the upper left wedge-shaped region and approach the point $(0,2)$ of x-extinction and y-equilibrium. It can be shown (but not here) that exactly two nonconstant orbits (called *separatrices*) tend to the internal equilibrium point $(2/3,2/3)$; other orbits may move near the point, but they always turn away and tend to $(2,0)$ or to $(0,2)$ as time advances. Figure 6.3.8 shows orbits of the system. The two separatrices divide the population quadrant into a region above the separating line $y = x$ that is favorable to the y-species but devastating to the x-species, and a complementary region that favors the x-species. This feature is called the *Principle of Competitive Exclusion*, because the population quadrant is split into two regions in each of which one of the species thrives, but the other dies out. Thus, the two competing species in this model are unlikely to coexist.

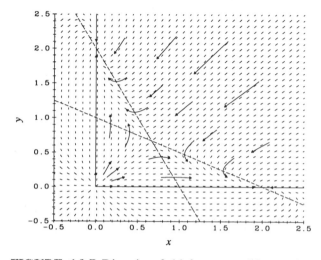

FIGURE 6.3.7 Direction field for competitive exclusion. See system (13).

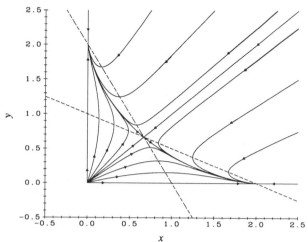

FIGURE 6.3.8 Orbits of competitive exclusion. See system (13).

Comments

The models of interacting species introduced in this section should be critically studied. Are they reasonably accurate models of the changing populations of real interacting species? What conclusions are possible through Sign Analysis and computer simulations? Can those conclusions be verified by mathematical arguments? We have answered the second question for a few specific model systems. The first and third questions are addressed for predator-prey interactions in Section 6.4. Several examples in Chapters 8 and 9 take up some of the mathematical issues raised here, questions of stability, for example.

In general, none of the models have the explicit predictive character of, say, the models for the motion of a pendulum. However, ever since the models were introduced in the 1920s and 1930s, they have directed the way scientists think about species interactions. Studies of population biology now begin with these models, if only to refute and to replace them by something better.

PROBLEMS

1. Explain the biological dynamics modeled by the systems below. Identify predators, prey, competitors, cooperators; explain population behavior in terms of the rate functions; identify harvesting and restocking terms and their nature. For example, if a two-species interaction is modeled by the system

$$x' = (-4 - x + 2y)x + 2\,\mathrm{sqw}(t, 50, 1)$$
$$y' = (2 - x - y)y$$

then x is a predator on the prey species y (the rate terms $+2xy$ and $-xy$), and the predator population's natural decline in the absence of prey (the term $-4x$) is accelerated by overcrowding (the term $-x^2$). The predator's growth is promoted by restocking during the first six months of each year (the term $+2\,\mathrm{sqw}(t, 50, 1)$). The coefficients in the rate equations are assumed to be positive.

(a) $x' = (\alpha - by)x, \quad y' = (\beta - cx)y + H$

(b) $x' = (\alpha - ax - by)x, \quad y' = (\beta - cx - dy)y$

(c) $x' = (\alpha + by)x, \quad y' = (-\beta + cx - dy)y$

(d) $x' = (\alpha - ax - by)x, \quad y' = (-\beta + cx - dy)y + (2 + \cos t)$

(e) $x' = (\alpha + by)x - H\,\mathrm{sqw}(t, 25, 1), \quad y' = (\beta + cx - dy)y$

(f) $x' = (\alpha - ax)x, \quad y' = (-\beta + \dfrac{cx}{m + kx})y + H$

(g) $x' = (d - bz)x, \quad y' = (\beta - dy - kz)y, \quad z' = (-\gamma + ax + cy)z - Hz$

2. For each system below, find the equations of the x-nullclines and the y-nullclines and sketch their graphs in the population quadrant. Find the equilibrium points in the population quadrant. Use Sign Analysis to determine the direction of orbital motion on or across the nullclines as time advances. This can be done with a straightedge since the nullclines are straight lines. Identify the nature of the x and y species, e.g., predator, prey, cooperator, competitor. Now suppose that at time $t_0 = 0$ the populations are at a nonequilibrium point (x_0, y_0) inside the population quadrant. What do you think happens to $x(t)$ and to $y(t)$ as time increases? Use your sketch to justify your answers.

   (a) $x' = (5 - x + y)x, \quad y' = (10 + x - 5y)y$ [Hint: See Example 6.3.3.]

   (b) $x' = (10 - x + 5y)x, \quad y' = (5 + x - y)y$ [Hint: See Example 6.3.2.]

   (c) $x' = (5 - x - y)x, \quad y' = (10 - x - 2y)y$

   (d) $x' = (10 - x - 5y)x, \quad y' = (5 - x - y)y$ [Hint: See Example 6.3.4.]

3. (a)–(d). Use a computer solver to sketch the orbits of the systems of 2(a)–(d). Describe what you see in terms of the long-term behavior of the two species.

4. (*Models of Cooperation*). Consider the cooperation model with self-limiting terms

$$x' = (2 - x + 2y)x, \quad y' = (2 + 2ax - y)y$$

   where a is a positive parameter that can be adjusted to "steer" the system.

   (a) What is the meaning of the coefficient $2a$?

   (b) Find the critical value a_0 of a which divides the explosive growth model of Example 6.3.1 from a stable model in which both populations tend to an equilibrium inside the population quadrant as time increases. Plot orbits in the population quadrant if a is less than the critical value. Repeat with a at the critical value, and then with a above the critical value. [*Hint*: Find the maximal value a_0 of a such that if $0 < a < a_0$, then there is an equilibrium point inside the population quadrant.]

5. (*Knocking Out the Competition*). The system

$$x' = (2 - x - 2y)x + Hx$$
$$y' = (2 - 2x - y)y$$

   models a competition model where the x-species can be restocked ($H > 0$) at a rate proportional to the population. Using a numerical solver to plot solution curves, find the restocking coefficient H of minimal magnitude so that the y-species becomes extinct regardless of its initial population. Plot orbits for values of H below, at, and above the critical value and explain what you see. [*Hint*: Find the value H_0 of the restocking coefficient which has the property that as H increases through H_0, the equilibrium point inside the population quadrant exits across an axis.]

6. (*Competitive Exclusion*). In the competition model

$$x' = (\alpha - ax - by)x$$
$$y' = (\beta - cx)y$$

only the x species has a self-limitation term. Regardless of the values of the positive coefficients, α, β, a, b, c, the Principle of Competitive Exclusion applies (see Example 6.3.4). Explain why. [*Hint*: Consider the two cases, $\alpha/a > \beta/c$ and $\alpha/a \le \beta/c$.] Plot orbits for your choices of values for the coefficients. Explain what you see.

7. (*A Competition Model*). Suppose that you belong to the x-species and are able to change the rate constants in the x' equation of system (8), but can have no effect on the rate constants in the rate equation of your competitor y. Analyze the effects on the competition of changing each of your rate constants. First, however, nondimensionalize the system by letting $x = ku$, $y = mv$, and $t = nT$ and then choosing the positive scaling constants k, m, and n so that the new rate equation for your competitor is $dv/dT = (1 - u - v)v$. Your rescaled rate equation is $du/dT = (\alpha^* - a^*u - b^*v)u$. Explain what happens as you "tune" the system by changing α^*, a^*, and b^* in turn, and what values of these parameters might be both "realistic" and "optimal" for your species. Plot (and interpret) orbits.

6.4 Predator-Prey: Volterra's Principles

Charles Darwin had this to say about predator and prey dynamics:

> *A struggle for existence inevitably follows from the high rate at which all organic beings tend to increase. Every being, which during its natural lifetime produces several eggs or seeds, must suffer destruction during some period of its life, and during some season or occasional year; otherwise, on the principle of geometrical increase, its numbers would quickly become so inordinately great that no country could support the product. Hence, as more individuals are produced than can possibly survive, there must in every case be a struggle for existence, either one individual with another of the same species, or with the individuals of distinct species, or with the physical conditions of life. It is the doctrine of Malthus applied with manifold force to the whole animal and vegetable kingdoms; for in this case there can be no artificial increase of food, and no prudential restraint from marriage. Although some species may be now increasing, more or less rapidly, in numbers, all cannot do so, for the world would not hold them....*
>
> *The amount of food for each species of course gives the extreme limit to which each can increase; but very frequently it is not the obtaining food, but the serving as prey to other animals, which determines the average numbers of a species.*[3]

---

[3]Charles Darwin, "Struggle for Existence," *The Origin of Species*, new ed., Chap. 3 (from 6th English ed.) (New York: Appleton, 1882).

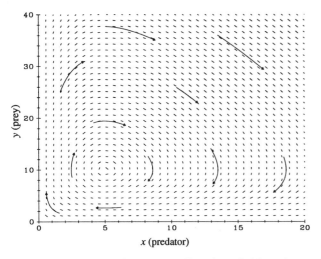

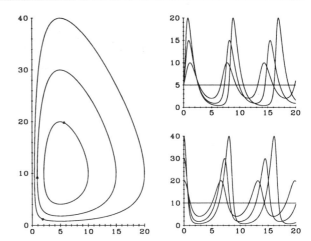

FIGURE 6.4.1 Predator-prey direction field and arrows: $x' = (-1 + 0.1y)x$, $y' = (1 - 0.2x)y$.

FIGURE 6.4.2 Predator-prey component graphs (tx at top right, ty at bottom right), and orbits (left): $x' = (-1 + 0.1y)x$, $y' = (1 - 0.2x)y$.

In the 1920s and 1930s, Vito Volterra and Alfred Lotka independently reduced Darwin's predator-prey interactions to mathematical models, and it is these models that are presented in this section.

Predator and Prey

The simplest model of predator and prey association (introduced in Section 6.3) includes only natural growth or decay and the predator-prey interaction itself. All other relationships are assumed to be negligible. We shall assume that the prey population grows exponentially in the absence of predation, while the predator population declines exponentially if there are no prey to consume. The predator-prey interation is modeled by mass-action terms proportional to the product of the two populations. The model system of ODEs for the predator and prey populations is

$$x' = -ax + bxy$$
$$y' = cy - dxy \tag{1}$$

where $x(t)$ is the predator population and $y(t)$ the prey population at time t, and the rate constants a, b, c, d are positive. The linear terms $-ax$ and cy model the natural decay and growth, respectively, of the predator and the prey, while the quadratic terms bxy and $-dxy$ model the effects of interaction on the rates of change of the two species. System (1) is called the *predator-prey* or *Lotka-Volterra* system.

The coordinates of the equilibrium points are found by simultaneously solving the two equations

$$-ax + bxy = 0, \quad cy - dxy = 0$$

The equilibrium points are the origin and the point $P(c/d, a/b)$ inside the population quadrant, $x \geq 0$, $y \geq 0$.

Figure 6.4.1 shows a direction field for system (1). The field lines suggest that population orbits turn around the point P, but it is impossible to tell whether orbits spiral toward or away from P, or whether they are closed curves. The computed orbits

TABLE 6.4.1 Percentages of predator species (sharks, skates, rays, etc.) out of the total fish catch

Port	1914	1915	1916	1917	1918	1919	1920	1921	1922	1923
Fiume	12%	21%	22%	21%	36%	27%	16%	16%	15%	11%
Trieste	14%	7%	16%	15%	-	18%	15%	13%	11%	10%

of Figure 6.4.2 imply that the orbits (on the left) are indeed closed and the solutions periodic. The component curves (x-curves upper right, y-curves lower right) of Figure 6.4.2 also suggest that the solutions are periodic, that predator peaks lag behind prey peaks (as expected), and that the period of an orbit increases with the amplitude (measured from the equilibrium point P). Because $x' > 0$ if $y > a/b$ and $x' < 0$ if $y < a/b$, while $y' > 0$ if $x < c/d$ and $y' < 0$ if $x > c/d$, the orbits turn clockwise about P as time increases.

The computational results are simple, interesting, and curious, but do they have anything to do with the populations of real species?

Adriatic Fisheries and the First World War

In 1926, Humberto D'Ancona, an Italian biologist, completed a statistical study of the changing populations of various species of fish in the upper reaches of the Adriatic Sea. His estimates of the populations during the years 1910 to 1923 were based on the numbers of each species sold on the fish markets of the three ports, Trieste, Fiume, and Venice. D'Ancona assumed, as we shall, that the relative numbers of the various species in the market reflected the relative abundance of those species in the marine community. A part of the data is given in Table 6.4.1

As often happens, the data do not provide overwhelming support for any particular theory of changing fish populations. D'Ancona observed, however, that the percentages of predator species were generally higher during and immediately after World War I (1914–1918). Fishing was drastically curtailed during the war years as the fishermen abandoned their nets to fight in the war, and D'Ancona concluded that the decline in fishing caused the change in the proportions of predator to prey. He formulated the hypothesis that during the war the predator-prey community was close to its natural state of a relatively high porportion of predator fish, while the more intensive fishing of the pre- and postwar years disturbed that equilibrium to the advantage of the prey species. Unable to give a biological or ecological reason for the phenomenon, D'Ancona asked his father-in-law, the noted Italian mathematician Vito Volterra (1860–1940), if there was a mathematical model that might cast some light on the matter. Within a few months, Volterra had outlined a series of models for the interactions of two or more species. System (1) is the simplest of Volterra's models.[4]

---

[4]Volterra eventually wrote a book on his theories, *Leçons sur la théorie mathématique de la lutte pour la vie* (Paris: Gauthier-Villars, 1931; reproduced by University Microtexts, Ann Arbor, Mich., 1976). The Soviet biologist G.F. Gause (or Gauze) tested Volterra's theories by carrying out numerous laboratory experiments with various competing and predatory microorganisms. He published his results in *The Struggle for Existence* (Baltimore: Williams & Wilkins, 1934; reissued by Hafner, New York, 1964). D'Ancona defended Volterra's work in a book which again used the memorable phrase of Malthus and Darwin, *The Struggle for Existence* (Leiden: E. J. Brill, 1954). A. J. Lotka, an American biologist and, later in life, an actuary, arrived at many of the same conclusions independently of Volterra; see his book, *Elements of Physical Biology* (Baltimore: Williams & Wilkins,

Volterra's Laws

Volterra summarized his conclusions about the solutions and orbits of system (1) in the form of three laws, which we formulate as theorems.

THEOREM 6.4.1

> The Law of the Periodic Cycle. The fluctuations of the populations of the predator and its prey are periodic. The period depends on the rate coefficients of system (1) and on the initial data; this period increases with the amplitude of the corresponding cycle.

We shall find a formula for the orbits of system (1). Since we are interested only in the orbits inside the population quadrant, we may divide the second rate equation of system (1) by the first and obtain

$$\frac{dy}{dx} = \frac{cy - dxy}{-ax + bxy} = \frac{y}{-a + by}\left(\frac{c}{x} - d\right)$$

Separating the variables, we have that

$$\left(\frac{c}{x} - d\right) dx = \left(b - \frac{a}{y}\right) dy$$

and integrating,

$$c \ln x - dx + a \ln y - by = C \tag{2}$$

where C is a constant. Exponentiate each side of (2) and apply the laws of exponents and logarithms to find a formula for the orbit through (x_0, y_0):

$$(x^c e^{-dx})(y^a e^{-by}) = K_0 \tag{3}$$

where $K_0 = (x_0^c e^{-dx_0})(y_0^a e^{-by_0})$ and (x_0, y_0) is the initial point of the orbit. It is not at all clear that equation (3) defines a closed orbit about the point P. That (3) defines a simple closed curve for each $(x_0, y_0) \neq (c/d, a/b)$ inside the population quadrant is not proved here (but see Problem 6). In any event, the computed orbits of Figure 6.4.2 offer strong evidence that the orbits are indeed cycles and that the period increases with the amplitude of the cycle. See Problems 2 and 3.

The Law of the Periodic Cycle shows that the visual evidence of periodic orbits shown in Figure 6.4.2 is accurate. Whether actual predator and prey populations change periodically is another matter, but the solutions of system (1) in the first quadrant are indeed periodic. Remarkably, the average population of each species over any of the cycles is a fixed constant. This is Volterra's second principle.

THEOREM 6.4.2

> The Law of the Averages. In system (1), the average predator population over the period of a cycle is c/d; the average prey population over the period of a cycle is a/b.

1925); reprinted as *Elements of Mathematical Biology* (New York: Dover, 1956).

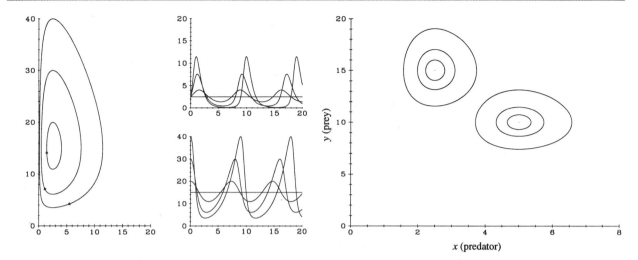

FIGURE 6.4.3 Harvesting harms predator, helps prey: $H_1 = H_2 = 0.5$ in system (7). Compare with unharvested model shown in Figure 6.4.2.

FIGURE 6.4.4 Predator-prey orbits for system (7): no harvesting, $H_1 = H_2 = 0$ (orbits on right); light harvesting, $H_1 = H_2 = 0.5$ (orbits on left).

Proof. Suppose that $x = x(t)$, $y = y(t)$ is a nonconstant solution of system (1) that defines a cycle of period T inside the population quadrant. The average predator population over one period is defined to be

$$\bar{x} = \frac{1}{T} \int_0^T x(t)\, dt \tag{4}$$

Here is how we can show that $\bar{x} = c/d$. Rearranging the terms of the second rate equation of (1), we have that

$$x(t) = \frac{c}{d} - \frac{1}{d} \frac{y'(t)}{y(t)} \tag{5}$$

Integrating each side of the equation from 0 to T, and then dividing by T:

$$\bar{x} = \frac{1}{T} \int_0^T x(t)\, dt = \frac{1}{T} \int_0^T \frac{c}{d}\, dt - \frac{1}{T} \int_0^T \frac{1}{d} \frac{y'(t)}{y(t)}\, dt$$

$$= \frac{c}{d} - \frac{1}{T} \cdot \frac{\ln y(T) - \ln y(0)}{d} = \frac{c}{d}$$

because $y(T) = y(0)$. A similar argument shows that $\bar{y} = a/b$.

Volterra's Law of the Averages carries us closer to real populations since averages are easy to calculate from numerical counts over time. In fact, D'Ancona's data from the fish markets are averages, although not necessarily taken over the same period as that of the presumed population cycles.

Why Harvesting Hurts the Predator and Helps the Prey

Volterra's third principle explains what happens when the two species are harvested. The simplest model is that of *constant-effort harvesting*, in which the amount caught per week, say, is proportional to the number of fish present. Consequently, we have Volterra's model of a predator-prey community of fish subjected to constant-effort harvesting:

$$x' = -ax + bxy - H_1x$$
$$y' = cy - dxy - H_2y \tag{6}$$

where the nonnegative numbers H_1 and H_2 are the harvesting coefficients. The equilibrium point inside the population quadrant shifts from $(c/d, a/b)$ to the point $((c - H_2)/d, (a + H_1)/b)$. It is assumed that $c > H_2$; otherwise the equilibrium point is not inside the population quadrant. By the Law of the Averages (Theorem 6.4.2) the averages of the populations around any cycle are given by the coodinates of the equilibrium point. Since harvesting causes the equilibrium point to move toward the upper left of the population quadrant, harvesting helps the prey and harms the predator. Consider the specific example

$$x' = (-1 + 0.1y)x - H_1x$$
$$y' = (1 - 0.2x)y - H_2y \tag{7}$$

A comparison of Figures 6.4.2 and 6.4.3 illustrates the motion of the equilibrium point for system (7). Figure 6.4.4 shows how the population averages shift for system (7) when $H_1 = H_2 = 0$ (no harvesting), and $H_1 = H_2 = 0.5$ (light harvesting). This leads to Volterra's third principle.

THEOREM 6.4.3

> The Law of Harvesting. Constant-effort harvesting raises the average number of prey per cycle and lowers the average number of predators per cycle.

Before the data of Table 6.4.1 can be compared with the predictions of Volterra's Law of Harvesting, that law must be reformulated in terms of percentages.

THEOREM 6.4.4

> Corollary to Theorem 6.4.3. Constant-effort harvesting raises the average percentage of prey per cycle in the total fish population and lowers the average percentage of predators per cycle.

The proof is left to the reader (Problem 4).

If the harvesting coefficients in system (6) are too large, the internal equilibrium point $((c - H_2)/d, (a + H_1)/b)$ passes through the positive y-axis, and one (or both) species becomes extinct. Figure 6.4.5 shows four orbits for system (7) through the initial point $x_0 = 10$, $y_0 = 20$. The solid orbits in Figure 6.4.5 show that with no or light harvesting both species survive. The dashed orbits show that if the harvesting is too heavy then one or both species may not survive.

Volterra's Laws of Harvesting answer D'Ancona's question. If Volterra's model represents conditions in the upper Adriatic, the predator population and percentage should indeed increase with a drop in fishing and then decline when fishing resumes. The situation is the other way around for the prey population.

Validity of Volterra's Law of Harvesting

Volterra's models have been both challenged and supported many times in the years since their formulation, but the model continues to be the starting point for most seri-

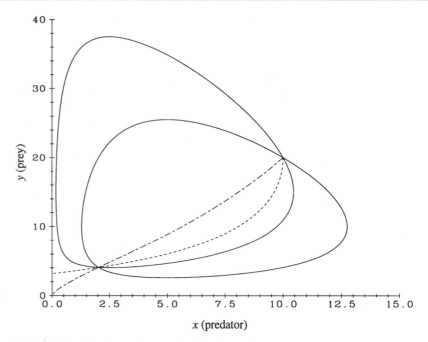

FIGURE 6.4.5 Heavy harvesting kills species in system (7): $H_1 = H_2 = 0$ (solid, lower right), 0.5 (solid, upper left), 1.0 (short dashes), 5.0 (long/short dashes).

ous attempts to understand just how predator-prey communities evolve with or without harvesting.

One dramatic confirmation of the general validity of Volterra's Law of Harvesting occurred when the insecticide DDT was applied to control the cottony cushion scale insect which infested American citrus orchards. The scale insect had been accidentally introduced from Australia in 1868. Its numbers were controlled (but not eliminated) by subsequent importation from Australia of the insect's natural predator, a particular kind of ladybird beetle. When DDT was first introduced as the "harvesting" agent, it was hoped that the scale insect could be completely wiped out. But DDT acts indiscriminately, killing all insects it touches. The consequence of the widespread application of DDT was that the numbers of the ladybirds drastically dropped, while the population of the scale insects, freed from the extensive ladybird predation, increased.

It has since been observed that the effect of insecticides such as DDT is even more drastic than Volterra's model suggests. DDT does more damage to the inhabitants at the upper trophic levels, and generally speaking, predators are higher on the trophic scale than their prey. Thus DDT has more effect on the predators than on their prey. In addition, the regeneration time is usually more rapid for the prey than for the predator, and the prey develop resistance to pesticides much faster than do their predators. For all these reasons, there is now a more cautious approach to the use of pesticides than there was in earlier years.

Comments

Darwin's observation that it is "the serving as prey to other animals which determines the average numbers of a species" is supported by Volterra's model and, apparently, by the data of the fish catches in the Adriatic. The model has its flaws: No account is taken of the delay in time between an action and its effect on population numbers, the averaging over all categories of age, fertility, and sex is dubious, the parameters of the model probably depend on time. Nonetheless, Occam's Razor applies here as it does to most models: "What can be accounted for by fewer assumptions is explained in vain by more."

PROBLEMS

1. (*Lotka-Volterra Systems*). For each system identify which is the predator population and which is the prey. Find the average predator and prey populations, and determine whether the population cycles turn clockwise around the equilibrium point inside the population quadrant. [*Hint*: Use Theorem 6.4.2. For the direction of the "flow," determine the sign of x' when $x > 0$, $y = 0$ and the sign of y' when $y > 0$, $x = 0$.]

   (a) $x' = -x + xy$, $\quad y' = y - xy$

   (b) $x' = 0.2x - 0.02xy$, $\quad y' = -0.01y + 0.001xy$

   (c) $x' = (-1 + 0.09y)x$, $\quad y' = (5 - x)y$

 2. (*Estimating the Periods of the Cycles*). (a)–(c) Plot the equilibrium point and the cycles that pass through the points (5,10) and (10,5) for the systems of Problem 1. Plot component graphs for the orbits and use these graphs to estimate the period of each cycle.

3. (*Estimating the Periods of Small-Amplitude Cycles*). In Problem 2 the periods of the predator-prey cycles are estimated "visually." Here is another way to estimate periods of small-amplitude cycles. The periods of the cycles near the equilibrium point $(c/d, a/b)$ of system (1) may be approximated by first approximating system (1) by a linear system near the equilibrium point.

   (a) Rewrite system (1) in X, Y variables, where $x = X + c/d$, $y = Y + a/b$.

   (b) Show that if the quadratic terms in the XY-system are dropped, then $X' = bcY/d$, $Y' = -adX/b$, a linear system that approximates (1) near $(c/d, a/b)$.

   (c) Show that the orbits of the linearized XY-system are ellipses centered at $X = Y = 0$ (i.e., $x = c/d$, $y = a/b$), and that the period of the corresponding linearized solutions is $2\pi/\sqrt{ac}$, which is a good approximation to the true period of small amplitude cycles near the equilibrium point. [*Hint*: Show that solutions of the linearized XY-system have the form $X = A\cos\omega t$, $Y = B\sin\omega t$, by inserting these functions into the XY-system and matching coefficients. Then determine ω.]

   (d) Set $a = b = c = d = 1$ and plot orbits of system (1) and orbits of the linearized system, $x' = y - 1$, $y' = -x + 1$, near the equilibrium point (1,1). Use initial data $x_0 = 1$, $y_0 = 1$, 1.1, 1.3, 1.5, 1.9; plot over the interval $0 \le t \le 20$. Then plot the corresponding component graphs of both systems. Compare the graphs, and explain what you see.

   (e) (*Return to Problem 2*). For each system of Problem 2, find the period of the linearized cycles. If you have already done the corresponding part of Problem 2, compare this period with your previous estimates.

4. (*Proof of the Law of the Percentages*). Prove Theorem 6.4.4 for the predator percentages. [*Hint*: The harvested system is $x' = -ax + bxy - H_1 x$, $y' = cy - dxy -$

$H_2 y$, with $H_2 < c$. Explain why the predator fraction F of the total average catch is

$$F = \frac{1}{1 + \dfrac{d(a + H_1)}{b(c - H_2)}}$$

Explain why F decreases as H_1 and H_2 increase.]

5. (*Harvesting to Extinction*). High harvesting coefficients H_1 and H_2 may lead to extinction of the species. This problem explores what happens.

   (a) Explain the graphs in Figure 6.4.5, which arise from system (7).

   (b) Set H_1 at a fixed positive value in system (6). Show that as $H_2 \to c^-$, the equilibrium point approaches the point $(0, (a + H_1)/b)$ on the y-axis, and that, if $H_2 = c$, all points on the y-axis are equilibrium points of system (6).

   (c) Let $H_2 = c$ in system (6). Show that $dy/dx = (dxy)((a + H_1)x - bxy)^{-1}$. Separate the variables, solve, and find x as a function of y and the initial data (x_0, y_0), where $x_0 > 0$, $y_0 > 0$.

    (d) What do you think happens to the harvested predator and prey species as time increases for the case $H_2 = c$? Give an informal explanation, but with reasons for your conclusions. As an aid, set $a = b = c = d = 1$, $H_1 = 0.1$, and plot orbits.

6. (*The Orbits of the Lotka-Volterra System are Cycles*). Follow the steps below to show that for $0 < K \leq K_0$ the graph of (3) is a simple closed curve (i.e., a cycle) in the interior of the population quadrant, unless $K = K_0 = a^a(be)^{-a}c^c(de)^{-c}$ when the graph is the equilibrium point $(c/d, a/b)$.

   (a) Show that $f(x) = x^c e^{-dx}$ is defined for $x \geq 0$, rises from the value of 0 at $x = 0$ to its maximum value of $M_1 = c^c(de)^{-c}$ at $x = c/d$, falls as x increases beyond c/d, and $f(x) \to 0$ as $x \to \infty$. The function $g(y) = y^a e^{-by}$ has the same properties, but its maximum value is $M_2 = a^a(be)^{-a}$ attained at $y = a/b$.

   (b) Show that equation (3) has no positive solutions if $K > M_1 M_2 = K_0$ but that it has the unique solution $(c/d, a/b)$ if $K = K_0$.

   (c) Let Δ be a positive number, $\Delta < M_1$. Show that $f(x) = \Delta$ has two solutions, x_1 and x_2, $x_1 < c/d < x_2$. Show that $g(y) = \Delta M_2/f(x)$ has no solution y if $x > x_2$ or $x < x_1$; exactly one solution, $y = a/b$, if $x = x_2$ or if $x = x_1$; and two solutions, $y_1(x) < a/b$ and $y_2(x) > a/b$, if $x_1 < x < x_2$. Show that $y_1(x) \to a/b$ and $y_2(x) \to a/b$ if $x \to x_1$ or $x \to x_2$.

   (d) Show that (3) defines a simple closed curve inside the population quadrant if $0 < K < K_0$.

7. (*Harvesting Strategies*).

   (a) Suppose that system (6) models the interactions of a predator-prey system and that you are the harvestor. Describe how you would choose the positive coefficients H_1 and H_2 to maintain each species within prescribed bounds. That is, suppose that it is required that $0 < x_m \leq x(t) \leq x_M$ and $0 \leq y_m \leq y(t) \leq y_M$ for all t, where the bounds x_m, x_M, y_m, y_M are prescribed positive numbers and $x(0) = c/d$ and $y(0) = a/b$. Justify your arguments.

   (b) (*Seasonal Harvesting*). High harvesting rates can be maintained if the harvesting season is sufficiently short. After analyzing Atlas Plates PREDATOR-PREY A–C, construct and justify your own strategy for maximizing the yield, while maintaining the predator and the prey populations within reasonable bounds.

6.5 The New Zealand Possum Plague

The ecological balance in New Zealand has been disturbed by the introduction of the Australian possum, a marsupial the size of a domestic cat with the proper name *Trichosurus vulpecula*. The animal was introduced in the 1830s for a planned fur trade. However, in the lush forests of New Zealand there were no natural predators, and the possum population rapidly increased. Today only a few areas are possum-free, and the estimated possum population is 70 million. The possums have become a reservoir of bovine tuberculosis with about half of the possum population infected. This poses a real threat of the disease being transferred to the livestock that form an important part of the New Zealand economy. An intense effort is now under way to understand possum ecology and the disease. This effort includes the construction of a series of mathematical models for the possum/disease dynamics. The simplest of these is treated here.[5]

Our approach will be somewhat telegraphic, and we leave the detailed analysis to the reader.

Modeling the Population/Disease Dynamics

Suppose that $P(t)$ is the population of possums at time t and $I(t)$ is the subpopulation of infected possums, where time is, say, measured in years, and $0 \leq I(t) \leq P(t)$. Note that $P(t) - I(t)$ is the population of healthy possums at time t. The simplest dynamical model is

$$
\begin{aligned}
P' &= (a - b)P - \alpha I \\
I' &= \beta I(P - I) - (\alpha + b)I
\end{aligned}
\tag{1}
$$

where a and b are the natural birth and death rate coefficients, respectively, α is the disease-induced death rate coefficient, and β is the mass-action coefficient that measures the effectiveness of interactions between the population I of the diseased and the population $P - I$ of healthy animals in transmitting the disease. Once a possum contracts bovine tuberculosis, it never recovers and so there is no subpopulation of "recovered." For simplicity, all of these rate coefficients are assumed to be positive constants.

This model is only a crude approximation to reality and neglects such factors as patchiness (i.e., spatial variations in the population levels) and differences due to age-structure in the possum population. But the model is surprisingly effective in giving a macroscopic picture of the situation and in giving some indication of how much effort is going to be needed to change the long-term outcome.

---

[5]This section is adapted from the note "Percy Possum Plunders" by Graeme Wake, Professor of Applied Mathematics at the University of Aukland, New Zealand. The note appeared in the newsletter CODEE (Winter 1995). A detailed paper, "Thresholds and Stability Analysis of Models for the Spatial Spread of a Fatal Disease," that includes other mathematical models of the possum population dynamics, was written by Wake and his colleagues K. Louie and M.G. Roberts, and appeared in *IMA Journal for Mathematics Applied to Medicine and Biology*, **10**, (1993), pp. 207–226. Wake is a distinguished applied mathematician who enjoys applying mathematics to "real" problems like the possum problem discussed here. In the past he has applied mathematical techniques to help solve the mysteries of why piles of wool being shipped overseas began to smolder, and why a number of fish-and-chips shops in Australia suddenly caught fire. An excellent general reference on mathematical models for the spread of disease is an article by another distinguished applied mathematician, H.W. Hethcote, "Qualitative Analysis of Communicable Disease Models," *Math. Biosci.* **28** (1976), pp. 335–356.

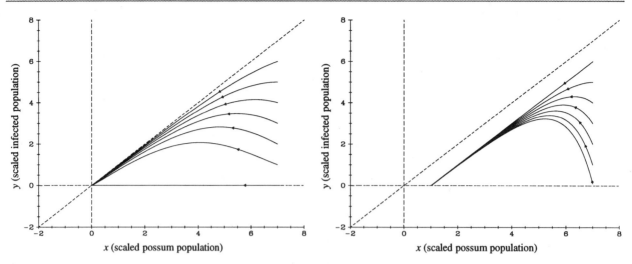

FIGURE 6.5.1 Possum extinction: $c = -1$, $r = 2$. See system (5).

FIGURE 6.5.2 The disease disappears: $c = 0$, $r = 2$. See system (5).

There is no analytical formula for the solution of system (1), and we must rely on mathematical theory and computer simulation. Our approach here is through the latter; see Sections 8.4 and 9.3 for applications of stability and bifurcation theory to this model.

Scaling the Variables

Suppose that we introduce scaled state variables x and y and scaled time s by

$$P = hx, \quad I = jy, \quad t = ks \tag{2}$$

where h, j, and k are positive constants to be determined. Replacing P, I, and t, respectively, by hx, jy, and ks in system (1), we have

$$\frac{h}{k}\frac{dx}{ds} = (a - b)hx - \alpha jy$$
$$\frac{j}{k}\frac{dy}{ds} = \beta jy(hx - jy) - (\alpha + b)jy \tag{3}$$

The scale constants will be chosen to focus attention on the coefficients of the x-term in the first rate equation of (3) and the y-term in the second. If we set

$$h = j = \alpha/\beta, \quad k = 1/\alpha$$

and introduce c and r as follows

$$c = (a - b)/\alpha, \quad r = (\alpha + b)/\alpha \tag{4}$$

then we have the simplified system in dimensionless variables and parameters x, y, s, c, r:

$$\frac{dx}{ds} = cx - y$$
$$\frac{dy}{ds} = y(x - y) - ry \tag{5}$$

We restrict attention to the region R, $0 \leq y \leq x$; it is assumed that $r > 1$ and that $|c|$ is not too large (c may be positive or negative).

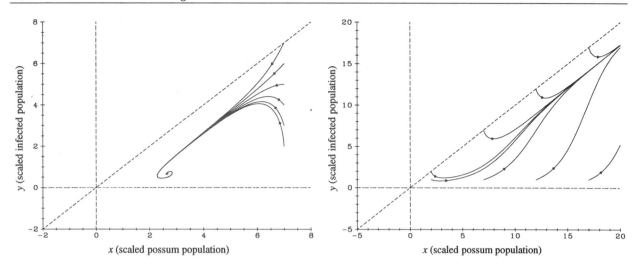

FIGURE 6.5.3 Disease becomes endemic in possum population: $c = 0.25$, $r = 2$. See system (5).

FIGURE 6.5.4 Explosive growth of healthy and diseased populations: $c = 1.5$, $r = 2$. See system (5).

The equilibrium points of system (5) are the origin ($x = 0$, $y = 0$) and the point E whose coordinates are $x = r/(1 - c)$, $y = cr/(1 - c)$. Observe that E is nonexistent if $c = 1$. Observe also that if $c = 0$, then $(x, 0)$ is an equilibrium point in the region R for every $x \geq 0$. One goal of the computer simulation will be to determine whether the population orbits $x = x(t)$, $y = y(t)$ of system (5) tend to an equilibrium point or become unbounded as time increases. This can be accomplished by a parameter study of the orbits of (5) for various values of c and r.

Computer Simulation

Without being concerned about the "reality" of the values of the parameters at this point, it is useful to examine orbital behavior for what we may imagine to be typical values.

EXAMPLE 6.5.1

A Model in the Making

We shall set $r = 2$ in system (5) and plot orbits in the region R for $c = -1$, 0, 0.25, and 1.5. Figures 6.5.1–6.5.4 show the results. In Figure 6.5.1 all orbits in R tend to the origin; that is, as time advances, the possums die out (and the disease, presumably, with them). Despite appearances, the orbits in Figure 6.5.2 are not all asymptotic to a single equilibrium state $(x, 0)$. Zooming in on the region near $(1, 0)$ shows that the seven orbits tend to seven different equilibrium points. Each equilibrium point models a disease-free steady-state possum population. We leave it to the reader to interpret the disastrous situation shown in Figure 6.5.4.

Finally, we shall explore the intriguing situation exemplified by Figure 6.5.3, where the possum population and its diseased cohort approach the equilibrium state E. This may be interpreted in terms of the disease being endemic in an equilibrium population of possums. Figures 6.5.5 and 6.5.6 show an oscillating orbit and the corresponding component graphs in the case $c = 0.05$ and $r = 1.1$ (somewhat more realistic than the values $c = 0.25$, $r = 2$ of Figure 6.5.3). However, the period of the decaying oscillations is so large that it would take years of observation before one could decide whether the data support the validity of a model like this.

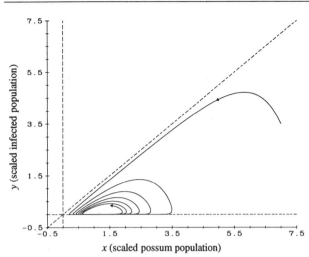

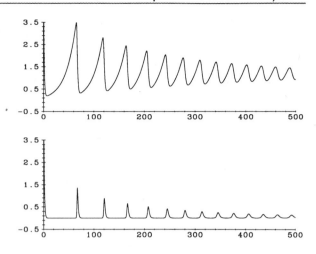

FIGURE 6.5.5 Spiral approach to endemic equilibrium: $c = 0.05$, $r = 1.1$. See system (5).

FIGURE 6.5.6 Slow approach to endemic equilibrium, x-component above and y-component below: $c = 0.05$, $r = 1.1$. See system (5).

Comments

The possum/bovine tuberculosis model is one of a large number of disease models that have been formulated over the last seventy-five years as aids in understanding the dynamics of the spread of disease in a population. For many diseases, the population may be divided into three "cohorts," the susceptible, the infected, and the recovered. These models are widely used, and one of them is outlined in the problem set. In the possum/tuberculosis model, there is no subpopulation of the recovereds.

PROBLEMS

1. (*Outbreak of a Measles Epidemic: Kermack-McKendrick SIR Model*). We may model the evolution of an epidemic in a fixed population by dividing the population into three distinct classes:

   $S =$ *Susceptibles*, those who have never had the illness and so can catch it

   $I =$ *Infectives*, those who are infected and so are contagious

   $R =$ *Recovered*, those who already had the illness and so are immune

   We will assume that the disease is mild enough that everyone eventually recovers, that the disease confers permanent immunity on those who have recovered, and that those with the disease are infective until they recover.

   (a) Argue that the evolution of the epidemic may be modeled by the following nonlinear system of first-order ODEs:

   $$S' = -aSI$$
   $$I' = aSI - bI$$
   $$R' = bI$$

where the parameter b represents the reciprocal of the period of infection and a represents the reciprocal of the level of exposure of a typical person.   [*Hint*: Compare with models for predator-prey interaction.]

(b) Show that the onset of an epidemic can only occur with the presence of a large susceptible population. Specifically, find the threshold value for S above which more people are infected each day than recover.

(c) Suppose that a German measleslike illness lasts for four days. Suppose also that the typical susceptible person meets 0.3% of the infected population each day and that the disease in transmitted in 1 out of every 6 contacts with an infected person. Find the values of the parameters a and b in the SIR model. How small must the susceptible population be for this illness to fade away without becoming an epidemic? Verify this by plotting the component graph for $I(t)$ for your choice of $I(0)$ (keep this fixed) and values of $S(0)$ at, 50% above, and 50% below the threshold value found in part **(b)**. Plot over $0 \leq t \leq 30$, and discuss what you see.

(d) Suppose that for another measleslike illness $a = 0.001$ and $b = 0.08$ and suppose that we introduce 100 infected individuals into a population. Investigate how the spread of the infection depends on the size of the population by plotting S, I, and R component graphs for $0 \leq t \leq 50$, $I(0) = 100$, $R(0) = 0$, and $S(0)$ ranging from 0 to 2,000 in increments of 500. Comment on how the initial number of susceptible persons affects the speed with which the epidemic runs its course.

(e) Find I as a function of S, I_0, and S_0, and plot for various values of I_0 and S_0, using $a = 0.001$ and $b = 0.08$. Interpret your graphs in terms of the model. [*Hint*: $dI/dS = b/(aS) - 1$.]

2. (*Scaling the Possum Plague*). Suppose that the possum populations S and I are measured in units of a million and that time is measured in years. Show that if the scale constants h and j, k of (2) and (3) are measured, respectively, in units of $(\text{millions})^{-1}$ and $(\text{years})^{-1}$, then x, y, and s are dimensionless variables.

3. Write the possum model in terms of the suceptibles, $S = P - I$, and the infected, I. Then introduce scaled variables $z = x - y$, y, and s (where x, y, and s are as in the text), write the z, y rate equations, find the equilibria, and plot orbits for $z \geq 0$, $y \geq 0$ with the values of c and r as given in Figures 6.5.1–6.5.5. Interpret your graphs in terms of the model.

4. (*Parameter Study of the Possum Plague*). Make a parameter study of the possum plague model, using any parameters you want to. Justify your choices, explain your results, and relate your conclusions to a "realistic" assessment of the situation in New Zealand. Can you see any plausible suggestions for controlling the problem?

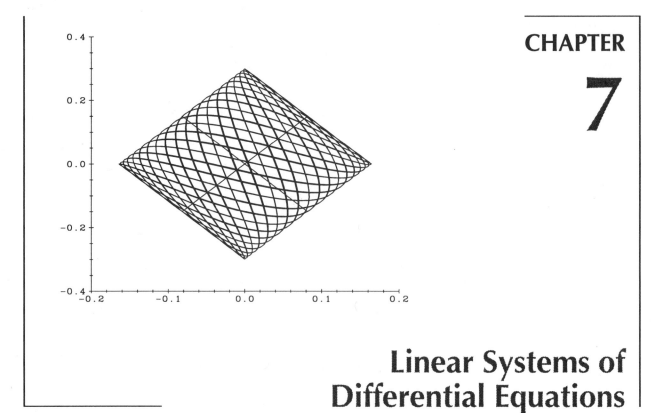

CHAPTER

7

Linear Systems of Differential Equations

Coupled systems of linear ODEs arise in the modeling of multiloop electrical circuits, interconnected systems of shock absorbers, diffusion of a substance back and forth through the compartments of the body, and in many other settings. Systems of linear ODEs are good models of the corresponding physical phenomena if the state variables remain near equilibrium levels. Matrices and linear algebraic concepts are basic to the characterization of the solutions of a system of linear ODEs, and so we develop what is needed along these lines in Sections 7.2 and 7.3 and in Appendix C.6. In the other sections of the chapter we construct and develop models and apply the theory of linear systems of ODEs, particularly in the case that the coefficients in the system are constants. Orbits and time-state curves of a linear system often have patterns of regularity lacking in a nonlinear system, and many of the figures of this chapter show these regular patterns. The chapter cover figure shows a Lissajous orbit (curved lines) and two normal mode orbits (straight lines) traced out by the two angular variables of the system modeling a linearized double pendulum (see Section 7.4).

7.1 Lead in the Human Body

Lead enters the bloodstream via food, air, or water contaminated by industrial emissions, leaded products of everyday use, and leaded gasoline. It accumulates in the blood, in tissues, and especially in the bones. Some lead is excreted into the external environment through the urinary system and by hair, nails, and sweat, but enough

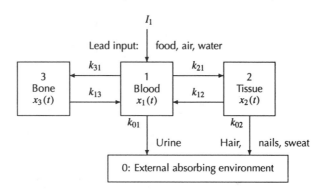

FIGURE 7.1.1 The ingestion, distribution, and excretion of lead.

may remain in the body to impair mental and motor capacity. A mathematical model for this system is based on tracking the concentrations of lead in each of three body compartments: blood, tissue, and bones. Numbering these compartments as 1, 2, 3, respectively, the schematic diagram of Figure 7.1.1 represents the flow of lead through the compartments.

The amount of lead in compartment i at time t is denoted by $x_i(t)$. The rate of transfer of lead into compartment i from compartment j is proportional to $x_j(t)$ (a *first-order rate law*), and the constant of proportionality is denoted by the constant k_{ij}. It is always assumed that $k_{ij} \geq 0$; $k_{ij} = 0$ if there is no transfer of lead into compartment i from compartment j. A reverse transfer from compartment i into compartment j may also occur, but the rate constant k_{ji} need not be equal to k_{ij}.

A mathematical model for the flow of lead is based on the Balance Law:

$$\begin{array}{c} \text{Net rate of change of amount} \\ \text{of lead in a compartment} \end{array} = \text{Rate in} - \text{Rate out} \qquad (1)$$

Applying the Balance Law to the blood, tissue, and bone compartments, we have a system of three rate equations

$$\begin{aligned} \text{(blood)} \quad & x_1' = -(k_{01} + k_{21} + k_{31})x_1 + k_{12}x_2 + k_{13}x_3 + I_1 \\ \text{(tissue)} \quad & x_2' = k_{21}x_1 - (k_{02} + k_{12})x_2 \\ \text{(bone)} \quad & x_3' = k_{31}x_1 - k_{13}x_3 \end{aligned} \qquad (2)$$

The intake rate I_1 of lead into the bloodstream is assumed to be a piecewise continuous function of time. The rate equation for the external absorbing compartment

$$x_0' = k_{01}x_1 + k_{02}x_2 \qquad (3)$$

is omitted from system (2) because the external compartment does not send lead back into the other body compartments.

System (2) is an interesting model, but what does it have to do with the specifics of real people ingesting contaminated food and drink, and breathing air laced with industrial pollution and exhaust fumes?

EXAMPLE 7.1.1

Lead in the Human Body: A Case Study
Rabinowitz, Wetherill, and Kopple made a carefully controlled study of the lead intake and excretion of a healthy volunteer in an industrial urban setting.[1] The data from this study were used to estimate the rate constants for the compartment model (2). Lead is measured in micrograms and time in days. For example, the rate term 49.3 in system (4) below is the ingestion rate I_1 of lead in micrograms per day, while the

coefficient 0.0361 (day)$^{-1}$ in the first rate equation of (4) is the sum of the three compartment transfer coefficients k_{01}, k_{21}, and k_{31}, of lead from the blood into, respectively, the excretory system, tissue, and bones. The full IVP is given by

$$x_1' = -0.0361x_1 + 0.0124x_2 + 0.000035x_3 + 49.3, \quad x_1(0) = 0$$
$$x_2' = 0.0111x_1 - 0.0286x_2, \quad x_2(0) = 0 \tag{4}$$
$$x_3' = 0.0039x_1 - 0.000035x_3, \quad x_3(0) = 0$$

where we have assumed that initially there is no lead in the compartments. With system (4) we can study the effect of changing the input rate $I_1 = 49.3$ micrograms/day of lead into the bloodstream, or the effect of a medication that increases the diffusion coefficient $k_{13} = 0.000035$ (day)$^{-1}$ of lead out of the bones. Figure 7.1.2 displays the buildup of lead in the body compartments over a period of 800 days. Lead levels in the bloodstream and the tissue appear to have nearly reached steady state after the first 200 days, but the lead level in the bones is far from a steady state. The transfer coefficients $k_{13} = 0.000035$ (day)$^{-1}$ of lead from the bones back into the blood stream is so small that the skeleton acts like a storage reservoir for lead.

Now, suppose that after 400 days the subject is placed in a completely lead-free environment (i.e., the term 49.3 in system (4) becomes 0 for $t \geq 400$). Figure 7.1.3 shows that the lead levels in the blood and tissue plunge dramatically, but the amount of lead in the bones does not seem to drop very much, at least not in the next 400 days.

Another way to remove lead from the bones is to administer an antilead medication that increases the rate at which lead leaves the skeletal system. In particular, suppose that the rate coefficient $k_{13} = 0.000035$ increases by an order of magnitude to 0.00035. Figure 7.1.4 shows the very small effect this change has if the medication is administered from the 400th day onward. However, if the subject moves to a lead-free environment and takes the medication from the 400th day onward, then there is a noticeable drop in the skeletal lead levels (Figure 7.1.5). Let us see what happens when a massive dose is given, so large that k_{13} increases from 0.000035 to 0.035. Suppose from the 400th day on that the heavy dose is given and the subject is in a lead-free environment. The lead exits the bones very quickly. Following a slight rise after the 400th day, lead exits the blood and tissue as well (Figure 7.1.6). Of course that much medication is probably more harmful than the lead!

The component graphs in Figures 7.1.2–7.1.6 for the amounts of lead in the bones, blood, and tissue are generated by a numerical solver because we have not yet presented a mathematical theory for handling systems such as (4). After the theory has been developed in Sections 7.2–7.6, we shall return to this model in Section 7.7.

Linear Systems

Systems (2) and (4) are examples of linear systems. The *general linear system* in the n state variables $x_1, \ldots, x_n$ has the form

$$x_1' = a_{11}x_1 + \cdots + a_{1n}x_n + F_1(t)$$
$$\vdots \tag{5}$$
$$x_n' = a_{n1}x_1 + \cdots + a_{nn}x_n + F_n(t)$$

---

[1]Their work was reported in *Science* **182** (1973),pp. 725–727, and later extended by Batschelet, Brand, and Steiner in *J. Math. Biol.* **8** (1979),pp. 15–23.

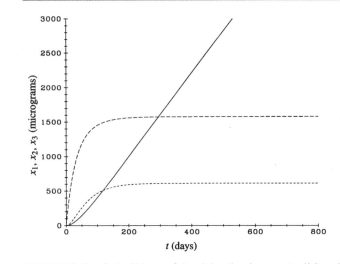

FIGURE 7.1.2 Buildup of lead in the bones (solid), blood (long dashes), tissue (short dashes). See Example 7.1.1.

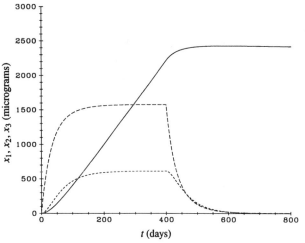

FIGURE 7.1.3 Lead intake stops on the 400th day. See Example 7.1.1.

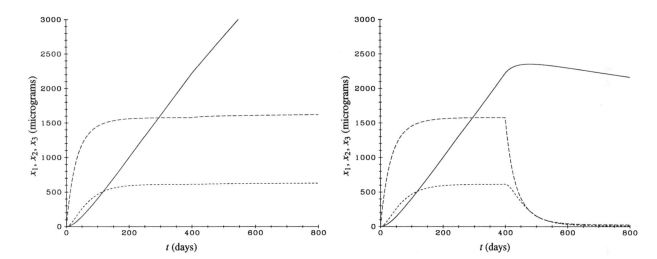

FIGURE 7.1.4 Antilead medication taken from the 400th day on does not help much. See Example 7.1.1.

FIGURE 7.1.5 Lead-free environment and medication from the 400th day on help a lot. See Example 7.1.1.

where the *coefficients* a_{ij} and the *driving functions* F_i are constants or functions of t, but they do not depend on the state variables. If the coefficients a_{ij} are not constants, then we assume they are all defined and continuous (or, perhaps, piecewise continuous) on a common t-interval I. System (5) can be written in matrix form as

$$x' = Ax + F \tag{6}$$

where

$$A = \begin{bmatrix} a_{11} & \cdots & a_{1n} \\ \vdots & & \vdots \\ a_{n1} & \cdots & a_{nn} \end{bmatrix}, \quad x = \begin{bmatrix} x_1 \\ \vdots \\ x_n \end{bmatrix}, \quad F = \begin{bmatrix} F_1 \\ \vdots \\ F_n \end{bmatrix} \tag{7}$$

The matrix A (also denoted by $[a_{ij}]$) is the *system matrix*, x is the *state vector*, and F is the *input* or *driving force*. The entries a_{ii} are the *diagonal* terms of A, and a_{ij}, $i \neq j$,

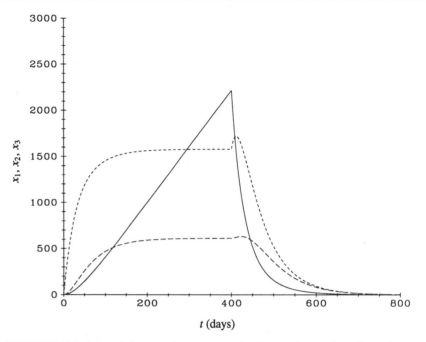

FIGURE 7.1.6 Lead-free environment and a heavy dose of antilead drug from the 400th day on. Is the subject still alive? See Example 7.1.1.

are the *off-diagonal* terms. See Appendix C.6 for a review of matrix-vector algebra. If an initial condition is imposed, then we have the initial value problem (IVP)

$$x' = Ax + F, \quad x(t_0) = x^0 = \begin{bmatrix} x_1^0 \\ \vdots \\ x_n^0 \end{bmatrix} \tag{8}$$

where $x_1^0, \ldots, x_n^0$ are any given constants and t_0 any point in I. In most cases A will be assumed to be a constant matrix and I to be the real line. If $F(t)$ is the zero vector for all t, the system is said to be *homogeneous, free, undriven,* or *unforced.*

EXAMPLE 7.1.2

The Lead System is a Linear System

System (2) is a linear system with

$$
\begin{array}{llll}
a_{11} = -(k_{01} + k_{21} + k_{31}) & a_{12} = k_{12} & a_{13} = k_{13} & F_1 = I_1 \\
a_{21} = k_{21} & a_{22} = -(k_{02} + k_{12}) & a_{23} = 0 & F_2 = 0 \\
a_{31} = k_{31} & a_{32} = 0 & a_{33} = -k_{13} & F_3 = 0
\end{array} \tag{9}
$$

For the particular values of these coefficients given in system (4)

$$A = \begin{bmatrix} -0.0361 & 0.0124 & 0.000035 \\ 0.0111 & -0.0286 & 0 \\ 0.0039 & 0 & -0.000035 \end{bmatrix}, \quad F = \begin{bmatrix} 49.3 \\ 0 \\ 0 \end{bmatrix} \tag{10}$$

See Figure 7.1.2. If the subject is placed in a lead-free environment after 400 days (see Figure 7.1.3), then

$$F_1(t) = 49.3\,\text{step}(400 - t) = \begin{cases} 49.3, & 0 \le t < 400 \\ 0, & t \ge 400 \end{cases} \tag{11}$$

Comments

Linear compartment models were introduced in Section 1.6, but the discussion was restricted to cascades where the flow through the compartments is one-way. The rate equations for a cascade model can be solved one ordinary differential equation (ODE) at a time. However, the lead system models back-and-forth flow between the compartments, and the corresponding rate equations are coupled in such a way that it is not possible to solve them one at a time. Matrix methods are needed to explain the behavior of solutions in this more general setting. We shall return to the lead system and general compartment models in Section 7.7 after matrix techniques for linear systems of ODEs have been introduced.

PROBLEMS

1. (*Lead in the Body: Equilibrium Levels*). Figure 7.1.2 shows how the amounts of lead in the body compartments behave as time advances. It appears that the amounts of lead in the blood and tissue approach equilibrium levels (i.e. steady states). The same is true for lead in the bones, but it takes much longer to reach equilibrium.

   (a) Find the equilibrium levels by setting the rate functions of IVP (4) equal to zero and solving for x_1, x_2, and x_3.

   (b) Using IVP (4) and a numerical solver, estimate how long it takes for the lead in the blood to reach 85% of its equilibrium level. Repeat for the lead in the tissue. You may try it for the lead in the bones, but the time span is incredibly long.

   (c) Replace the lead ingestion rate of 49.3 micrograms per day used in IVP (4) by the positive constant I_1 (unspecified). Now find the equilibrium levels in terms of I_1. Explain why the equilibrium levels are cut in half if I_1 is cut in half.

   (d) For $I_1 = 10, 20, 30, 40, 50$, plot the $x_3(t)$ component curve and estimate the time it takes the amount of lead in the bones to reach 1000 micrograms.

2. (*Lead in the Body: Lead-Free Environment*). Suppose that the initially lead-free subject of Example 7.1.1 is exposed to lead for 400 days, then is removed to a lead-free environment (Figure 7.1.3). Use a computer to estimate how long it takes for the amount of lead in the bones to decline to 50% of its maximum; repeat for 25% and 10%.

3. (*Antilead Medication*). The graphs in Figure 7.1.4 seem to indicate that administering an antilead medication does not by itself do much good in reducing the amount of lead in the bones. What happens if an extra heavy amount is given and the coefficient k_{13} is changed from 0.000035 to 0.35, or 3.5? Use your computer to solve the system and draw the component graphs for $x_3(t)$. Assume that $x_1(0) = x_2(0) = x_3(0) = 0$ and that the medication is taken from the 400th day on; plot for $0 \le t \le 800$.

4. (*A Compartment System*). The sketch below shows a compartment system where a substance Q is injected at a rate of $I_3 = 1$ unit of substance per unit of time into compartment 3. The substance Q moves out of compartment j into compartment i at a rate proportional to the amount $x_j(t)$ of Q in compartment j; the constant of proportionality is k_{ij}.

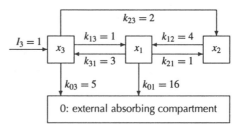

(a) Write the rate equations for $x_1(t)$, $x_2(t)$, $x_3(t)$.

(b) Find the equilibrium levels of Q in each of the three compartments.

(c) Plot the three component curves over a long enough time span that each is within 90% of its equilibrium level. Assume that $t \geq 0$ and $x_1(0) = x_2(0) = x_3(0) = 0$.

5. (*The Tracer Inulin*). Water molecules move back and forth between blood plasma (compartment 1) and intercellular area (compartment 2) and from the plasma into the urinary system (compartment 0). This motion may be tracked by injecting the tracer inulin into the blood at a constant rate of $I_1 > 0$ gram/hour. Molecules of inulin attach to molecules of water and may be followed by x-ray.

(a) Draw a "boxes and arrows" diagram like that for the lead system, then write the rate equations that govern the system. Assume that inulin leaves compartment j and enters compartment i at a rate $k_{ij}x_j(t)$; i.e., the rate is proportional to the amount $x_j(t)$ in compartment j at time t, where k_{ij} is a nonnegative constant.

(b) Find the equilibrium levels of inulin in the blood and in the intercellular areas.

(c) Let $I_1 = 1$, $k_{21} = 0.01$, $k_{12} = 0.02$, $k_{01} = 0.005$, $x_1(0) = x_2(0) = 0$. Plot the tx_1 and tx_2 component curves over a long enough time span that the levels of inulin in the blood and in the intercellular area have reached at least 90% of their equilibrium values.

7.2   Matrices, Eigenvalues, Eigenvectors

As we saw in the previous section, the general linear system in n state variables has the form

$$x' = Ax + F$$

where x is a column vector with n entries, A is an $n \times n$ matrix whose elements a_{ij} do not depend on the state variables x_i, and F is a column vector whose n entries F_i also do not depend on the state variables. When the entries of A are constants and the entries of F are all zero, it will turn out that all solutions of the differential system $x' = Ax$ can be constructed from certain algebraic elements defined by the matrix A. In this section we shall introduce these elements and discuss some of their underlying concepts (see also Appendix C.6 for matrix operations). The next section will show how to use these algebraic elements to find a general solution formula for constant-coefficient systems. If you are already familiar with such things as linear spaces, bases, eigenvectors and eigenvalues, then you may safely skip forward to the next section.

Linear Spaces

A *linear space* is a collection of objects, called *vectors*, on which two operations are defined satisfying certain properties: the operations are called "addition" and "multiplication by a scalar." Denoting vectors by symbols such as x, y, u, or v, and scalars by symbols such as a, b, or c, the sum of x and y is denoted by $x + y$, and multiplication by a scalar by ax. The properties look very much like the ones encountered in geometric vectors (see Section 4.1), but are given more precisely in Section 10.2. The scalars are either the real numbers $\mathbf{R}$, or the complex numbers $\mathbf{C}$. A linear space where the scalars are the real numbers is called a *real linear space*. Examples are the set of geometric vectors from Section 4.1, and $\mathbf{R}^n$, the set of n-tuples of reals. The continuity classes of real-valued functions $\mathbf{C}^0(I)$, $\mathbf{C}^1(I)$, and $\mathbf{C}^2(I)$ are also real linear spaces.[2] A *complex linear space* has the complex numbers $\mathbf{C}$ as the set of scalars. Examples are $\mathbf{C}^n$, the set of n-tuples of complex numbers, as well as the continuity classes $\mathbf{C}^0(I)$, $\mathbf{C}^1(I)$, and $\mathbf{C}^2(I)$ when they consist of complex-valued functions. We shall use the same symbol $\mathbf{C}^k(I)$ to denote the linear space whose elements $x(t)$ are n-tuples of complex-valued functions written as a column with components $x_1(t), \ldots, x_n(t)$ in $\mathbf{C}^k(I)$.

Let A be a subset of a linear space V. Then for any finite collection[3] of vectors $v^1, v^2, \ldots, v^m$ in A and any scalars $a_1, a_2, \ldots, a_m$, the sum $a_1 v^1 + a_2 v^2 + \cdots + a_m v^m$ is said to be a *finite linear combination* over A. The *span* of A, denoted by Span(A), is the collection of all finite linear combinations over A. Span(A) is contained in V, and if we use the operations of addition and multiplication by scalars from V, we see that Span(A) is itself a linear space. Span(A) is called the *subspace of V spanned by A*. The set A is said to *span* V if Span(A) = V. In general, the subset A of a linear space V is a *subspace* of V if Span(A) = A. Thus, a subset A of a linear space V is a subspace of V if and only if any linear combination of any two elements of A is a vector in A. Every linear space has a trivial subspace, the set consisting of only the zero vector.

EXAMPLE 7.2.1

Visualizing Vectors, Subspaces

There is a simple device for "visualizing" vectors in $\mathbf{R}^3$. Associate with each vector (x_1, x_2, x_3) in $\mathbf{R}^3$ the geometric vector $x_1 \mathbf{i} + x_2 \mathbf{j} + x_3 \mathbf{k}$ where $\{\mathbf{i}, \mathbf{j}, \mathbf{k}\}$ is an orthogonal frame in Euclidean 3-space. Then in the Cartesian coordinates generated by that frame, x_1, x_2, and x_3 are the coordinates of the head of the geometric vector, and the origin is the tail. The sum of vectors in $\mathbf{R}^3$ is associated with the sum of the associated geometric vectors via the parallelogram law. Thus, we can use our geometric intuition in Euclidean 3-space to "see" the following: (a) The span of a single nonzero vector x is the line through the origin determined by x; (b) the span of two nonparallel vectors x and y is the plane through the origin determined by those two vectors; (c) let three vectors have the property that one vector does not lie in the plane determined by the other two. Then the span of those three vectors must be all of $\mathbf{R}^3$. Thus it follows that every nontrivial subspace of $\mathbf{R}^3$ is either a line or a plane through the origin, or $\mathbf{R}^3$ itself. Lines or planes *not* containing the origin are *not* subspaces of $\mathbf{R}^3$.

---

[2] Note that the same symbol $\mathbf{C}^k(I)$ is used to denote a class of real-valued functions *and* a class of complex-valued functions—context reveals which one is intended.

[3] For clarity, vectors in this chapter are sometimes indexed with superscripts.

EXAMPLE 7.2.2

Subspaces in $\mathbf{R}^3$

Observe that the vectors in $\mathbf{R}^3$ which solve the homogeneous linear system

$$3x_1 - x_2 + 2x_3 = 0$$

$$x_1 + 4x_2 - x_3 = 0$$

form a subspace of $\mathbf{R}^3$ because any linear combination of solutions of these equations is again a solution of these same equations. Thus the solution space of these equations must either be the trivial subspace consisting only of the origin, or a line or plane through the origin, or $\mathbf{R}^3$ itself—there are no other possibilities (see Example 7.2.1).

Linear Independence

If B is a collection of vectors in the linear space V, then we saw that $W = \text{Span}(B)$ is a subspace of V. Now there are many sets of vectors in W which, like B, span W. Thus, if we demand that each vector in W be written uniquely as a finite linear combination over B, then we must choose the spanning set B in a special way. In particular, we see that if there are two different finite linear combinations over B which add up to the same vector in W, then by subtracting them we obtain a finite linear combination over B, not all of whose scalar coefficients are zero, which adds up to the zero vector. We refer to a finite linear combination of vectors with at least one nonzero scalar coefficient as a *nontrivial* finite linear combination, and *trivial* otherwise.

     ❖ **Linear Independence.** The set B in a linear space is said to be *linearly independent* if and only if there exists no nontrivial finite linear combination over B whose sum is the zero vector.

Thus we see that a set B in a linear space is linearly independent if and only if each vector in $\text{Span}(B)$ can be written as a finite linear combination over B in exactly one way. Any subset of an independent set is also independent.

     A frequently used result that we state without proof is that the set of n vectors $\{v^1, \ldots, v^n\}$ in $\mathbf{R}^n$ (or $\mathbf{C}^n$) is linearly independent if and only if $\det[v^1 \cdots v^n] \neq 0$. See Appendix C.6 for properties of determinants.

     ❖ **Basis.** A subset B of a linear space V is a *basis* for V if B is linearly independent and $V = \text{Span}(B)$.

Every vector in V can be written as a finite linear combination over the basis B in exactly one way.

EXAMPLE 7.2.3

Bases and Linear Independence

The infinite set of polynomials $B = \{1, x, x^2, \ldots, x^n, \ldots\}$ in the real linear space $C^0(I)$ does not span $C^0(I)$ because there are continuous functions which are not polynomials. $\text{Span}(B)$ is the set of all polynomials with real coefficients. We define the zero polynomial as the polynomial which is zero for all x. The set B is linearly independent because no finite linear combination of the vectors in B can be the zero polynomial, for then we would have a nontrivial polynomial with infinitely many roots—an impossibility. Thus B is a basis for $\text{Span}(B)$.

EXAMPLE 7.2.4

Linear Independence in $\mathbf{C}^1(\mathbf{R})$
The vectors v^1, v^2, and v^3 in $\mathbf{C}^1(\mathbf{R})$ given by

$$v^1 = \begin{bmatrix} e^t \\ -e^t \\ 0 \end{bmatrix} \quad v^2 = \begin{bmatrix} e^{-t} \\ 0 \\ e^{-t} \end{bmatrix} \quad v^3 = \begin{bmatrix} e^{-2t} \\ 2e^{-2t} \\ -e^{-2t} \end{bmatrix}$$

are linearly independent. To prove this, let c_1, c_2, and c_3 be real numbers such that $c_1 v^1 + c_2 v^2 + c_3 v^3 = [0\ 0\ 0]^T$, for all t in $\mathbf{R}$.[4] Thus, for $t = 0$ we must have that $c_1 v^1(0) + c_2 v^2(0) + c_3 v^3(0) = [0\ 0\ 0]^T$, which can be written in the matrix form as

$$\begin{bmatrix} 1 & 1 & 1 \\ -1 & 0 & 2 \\ 0 & 1 & -1 \end{bmatrix} \begin{bmatrix} c_1 \\ c_2 \\ c_3 \end{bmatrix} = \begin{bmatrix} 0 \\ 0 \\ 0 \end{bmatrix}$$

Since the determinant of this system is nonzero, it follows that $c_1 = c_2 = c_3 = 0$ is the only solution, and so $\{v^1, v^2, v^3\}$ is an independent set.

Dimension

A linear space V is said to be *finite-dimensional* if there is a finite set B in V which spans V. If B is also linearly independent, then B is a basis for V. It is a fact that every basis for a finite-dimensional linear space V contains the same number of elements. This number, denoted by dim V is called the *dimension* of V. The dimension of the trivial linear space (which consists only of the zero vector) is zero. A linear space which is not finite-dimensional is said to be *infinite dimensional*. Notice that a linear space is infinite dimensional if it contains a linearly independent set with infinitely many vectors.

EXAMPLE 7.2.5

Dimensions of Some Linear Spaces
$\mathbf{R}^n$ and $\mathbf{C}^n$, $n = 1, 2, \ldots$ have dimension n because the set

$$B = \{(1, 0, \ldots, 0), (0, 1, 0, \ldots, 0), \ldots, (0, \ldots, 0, 1)\}$$

is a basis for each of them. The linear space $C^1(I)$ of real-valued continuously differentiable functions on an interval I is infinite dimensional because it contains the independent set $B = \{1, x, x^2, \ldots, x^n, \ldots\}$ which has infinitely many elements. The solution set of the ODE $y'' - 3y' + 2y = 0$ is a two-dimensional subspace of $C^2(\mathbf{R})$ because it is spanned by the independent set $\{e^t, e^{2t}\}$.

Eigenvalues, Eigenvectors

For an $n \times n$ constant matrix A, there is a great deal of interest in finding a nonzero column vector v for which $Av = \lambda v$, for some constant λ. Evidently, the action of multiplying such a vector v by the matrix A is to return v again, but multiplied by the constant λ. It is a remarkable fact that any matrix A has at least one such v, λ pair.

---

[4] $[a\ b]^T$ denotes the transpose of the row vector $[a\ b]$; i.e., the column vector $\begin{bmatrix} a \\ b \end{bmatrix}$.

Vectors and associated scalars with this property play a very important role in the solution of a linear differential system $x' = Ax$ where A is a constant matrix. It's time for some terminology:

❖ **Eigenvalues and Eigenvectors.** A scalar λ is an *eigenvalue* of the $n \times n$ constant matrix A if there is a nonzero vector v such that

$$Av = \lambda v \tag{1}$$

The vector v is called an *eigenvector* of A corresponding to the eigenvalue λ. For any eigenvalue λ of A, the set V_λ of all solutions v of the equation $Av = \lambda v$ is called the *eigenspace* of A corresponding to λ.

It is easy to verify that the eigenspace V_λ is actually a subspace of $\mathbf{R}^n$ (or $\mathbf{C}^n$, if appropriate), and that any nonzero element in V_λ is an eigenvector of A corresponding to λ. Indeed, if v^1 and v^2 are any two elements in the eigenspace V_λ of a matrix A, then $Av^1 = \lambda v^1$ and $Av^2 = \lambda v^2$; and so for any constants c_1 and c_2,

$$A(c_1 v^1 + c_2 v^2) = c_1 Av^1 + c_2 Av^2 = \lambda(c_1 v^1 + c_2 v^2)$$

Hence, $c_1 v^1 + c_2 v^2$ is in V_λ, proving that V_λ is a subspace. Note that two eigenspaces V_λ and V_μ of a matrix A can have only the zero vector in common if $\lambda \neq \mu$.

EXAMPLE 7.2.6

Finding Eigenvalues, Eigenspaces
Denoting the 2-vector v by $[a\ b]^T$, let us determine all scalars λ for which the matrix equation

$$\begin{bmatrix} 5 & 3 \\ -6 & -4 \end{bmatrix} \begin{bmatrix} a \\ b \end{bmatrix} = \lambda \begin{bmatrix} a \\ b \end{bmatrix} \tag{2}$$

has a solution where a and b are *not* both zero.

With the 2×2 identity matrix I, this linear system can be written as

$$(A - \lambda I) \begin{bmatrix} a \\ b \end{bmatrix} = \begin{bmatrix} 0 \\ 0 \end{bmatrix} \tag{3}$$

Now (3) has the solution $a = b = 0$, but this corresponds to the trivial solution. Equation (3) has nontrivial solutions if and only if the determinant of the coefficient matrix vanishes (see Appendix C.6). This means that λ must satisfy the condition

$$\det(A - \lambda I) = \det \begin{bmatrix} 5 - \lambda & 3 \\ -6 & -4 - \lambda \end{bmatrix} = \lambda^2 - \lambda - 2 = 0$$

This happens for just two values of λ: $\lambda_1 = -1$ and $\lambda_2 = 2$. First setting $\lambda = -1$, we see that (2) reduces to the single equation $6a + 3b = 0$, one of whose solutions can be constructed if b is assigned the value -2 and then the value $a = 1$ is computed.

Next setting $\lambda = 2$, equation (2) reduces to the single equation $3a + 3b = 0$, one of whose solutions can be constructed if b is assigned the value 1 and then the value -1 is computed for a. Thus, we have that

$$A \begin{bmatrix} 1 \\ -2 \end{bmatrix} = (-1) \begin{bmatrix} 1 \\ -2 \end{bmatrix} \quad \text{and} \quad A \begin{bmatrix} 1 \\ -1 \end{bmatrix} = 2 \begin{bmatrix} 1 \\ -1 \end{bmatrix} \tag{4}$$

In other words, the matrix A applied to the vectors $[1\ -2]^T$ and $[1\ -1]^T$ produces those same vectors again but multiplied by the scalars -1 and 2, respectively.

Note that in Example 7.2.6 the eigenvalue equation (3) can be solved if $\lambda = -1$ or 2, and yields, in particular, the formulas (4).

EXAMPLE 7.2.7

Eigenspaces

The equations (4) show that $\lambda_1 = -1$ and $\lambda_2 = 2$ are eigenvalues of the matrix A in Example 7.2.6 and that $v = [1 \ -2]^T$, and $w = [1 \ -1]^T$ are eigenvectors corresponding to λ_1 and λ_2, respectively. From the calculations in Example 7.2.6 it is also clear that A has no other eigenvalues and that V_{λ_1} is spanned by $[1 \ -2]^T$ and that V_{λ_2} is spanned by $[1 \ -1]^T$. Note that $\{[1 \ -2]^T, \ [1 \ -1]^T\}$ is a basis for $\mathbf{R}^2$.

Even though the entries of A are real, the eigenvalues and hence the eigenvectors may be complex. Recall that when dealing with vectors in $\mathbf{C}^n$, the scalars are complex numbers; when dealing with vectors in $\mathbf{R}^n$ the scalars are real numbers. The next example deals with vectors in $\mathbf{C}^2$.

EXAMPLE 7.2.8

Complex Eigenvalues

To find the eigenvalues and eigenvectors of $A = \begin{bmatrix} 1 & 2 \\ -2 & 1 \end{bmatrix}$ we need to find scalars λ for which there are nonzero vectors v such that

$$(A - \lambda I)v = \begin{bmatrix} 1-\lambda & 2 \\ -2 & 1-\lambda \end{bmatrix} \begin{bmatrix} v_1 \\ v_2 \end{bmatrix} = \begin{bmatrix} 0 \\ 0 \end{bmatrix} \tag{5}$$

System (5) has nonzero solutions v when λ is chosen so that

$$\det \begin{bmatrix} 1-\lambda & 2 \\ -2 & 1-\lambda \end{bmatrix} = \lambda^2 - 2\lambda + 5 = 0$$

The values of λ which satisfy this condition are $\lambda_1 = 1 + 2i$ and $\lambda_2 = \bar{\lambda}_1 = 1 - 2i$. The eigenspace V_{1+2i} corresponding to λ_1 is the solution set of the equation $(A - \lambda_1 I)v = 0$. Observe that $v = [a \ \ b]^T$ is a solution of $(A - \lambda_1 I)v = 0$ if and only if

$$(1 - \lambda_1)a + 2b = -2ia + 2b = 0$$

or simply, $b = ia$. Thus, $v = [a \ ia]^T$, where a is an arbitrary complex number, is the solution set we seek, and hence $V_{1+2i} = \text{Span}([1 \ \ i]^T)$. Similarly, we obtain that $V_{1-2i} = \text{Span}([1 \ -i]^T)$.

A collection of eigenvectors, each corresponding to a different eigenvalue, forms an independent set. In fact, we have the following useful result.

THEOREM 7.2.1

Eigenspace Property. Let V_i, $i = 1, \ldots, p$, be eigenspaces of an $n \times n$ matrix A corresponding to the distinct eigenvalues λ_i, $i = 1, \ldots, p$. For $i = 1, \ldots, p$ let B_i be an independent subset of V_i. Then the set B consisting of all the elements of $B_1, \ldots, B_p$ is an independent set.

The proof is by mathematical induction on p and is outlined in Problem 8.

An important consequence of the Eigenspace Property is that if an $n \times n$ matrix A has n distinct eigenvalues, $\lambda_1, \ldots, \lambda_n$, then there is a set of n independent eigenvectors. Indeed, let B be a set of eigenvectors, one from each V_{λ_i}, $i = 1, 2, \ldots, n$, and apply the Eigenspace Property to show that B is an independent set. The set is actually a basis for $\mathbf{R}^n$ and is often called an *eigenbasis*.

EXAMPLE 7.2.9

Eigenbases

The matrix A of Example 7.2.6 has eigenvectors $[1 \ -2]^T$ and $[1 \ -1]^T$ corresponding, respectively, to the eigenvalues -1 and 2. The set $\{[1 \ -2]^T, [1 \ -1]^T\}$ is an independent subset of $\mathbf{R}^2$. Similarly, the set $\{[1 \; i]^T, [1 \; -i]^T\}$ of eigenvectors corresponding to the eigenvalues $1 + 2i$ and $1 - 2i$, respectively, of the matrix of Example 7.2.8 is an independent subset of $\mathbf{C}^2$.

The Characteristic Polynomial

Examples 7.2.6 and 7.2.8 suggest a method for finding all the eigenvalues of a matrix. Let A be an $n \times n$ matrix with real or complex entries. If λ is an eigenvalue of A, there must be a nontrivial column vector v such that $Av = \lambda v$. Thus, the matrix equation $(A - \lambda I)v = 0$ has a nontrivial solution v, and furthermore, the set of all solutions of this equation forms the eigenspace V_λ. Recall that the linear system $(A - \lambda I)v = 0$ has a nonzero solution v if and only if the determinant of the system $\det(A - \lambda I) = 0$. Note that $p(\lambda) = \det(A - \lambda I)$ is a polynomial of degree n in λ:

$$p(\lambda) = \det(A - \lambda I) = \det \begin{bmatrix} a_{11} - \lambda & a_{12} & \cdots & a_{1n} \\ a_{21} & a_{22} - \lambda & \cdots & a_{2n} \\ \vdots & \vdots & \ddots & \vdots \\ a_{n1} & a_{n2} & \cdots & a_{nn} - \lambda \end{bmatrix} \tag{6}$$

$$= (-1)^n \lambda^n + (-1)^{n-1}(a_{11} + \cdots + a_{nn})\lambda^{n-1} + \cdots + \det A$$

The form of the coefficients of λ^n and λ^{n-1} may be established by expanding the determinant in (6) by cofactors (see Appendix C.6) and using induction on n. That the constant term of $p(\lambda)$ is $\det A$ follows from the definition if λ is equated to 0. The polynomial p has a name:

❖ **Characteristic Polynomial.** The polynomial $p(\lambda) = \det(A - \lambda I)$ given in (6) is called the *characteristic polynomial* of the square matrix A.

From this definition and the discussion above, we have the following result.

THEOREM 7.2.2

> Characteristic Polynomial Theorem. The scalar λ is an eigenvalue of an $n \times n$ matrix A if and only if λ is a root of the characteristic polynomial $p(\lambda) = \det(A - \lambda I)$.

Thus the calculation of the eigenvalues of an $n \times n$ matrix reduces to the problem of finding the roots of a polynomial of degree n. If $n = 2$, the quadratic formula may be used, but for $n > 2$ it may be a difficult problem to find the roots. Computer software exists for finding approximations to the eigenvalues of a matrix, and for $n > 2$ this may be the only practical method available. Sometimes, factoring techniques of algebra allow the determination of all the eigenvalues even if $n > 2$.

EXAMPLE 7.2.10

Finding the Eigenvalues
The characteristic polynomial of

$$A = \begin{bmatrix} -3 & 1 & -2 \\ 0 & -1 & -1 \\ 2 & 0 & 0 \end{bmatrix}$$

is given by

$$p(\lambda) = \det \begin{bmatrix} -3-\lambda & 1 & -2 \\ 0 & -1-\lambda & -1 \\ 2 & 0 & -\lambda \end{bmatrix}$$

$$= -\lambda^3 - 4\lambda^2 - 7\lambda - 6$$

Substitution of small integers for λ quickly shows that -2 is a root of $p(\lambda)$. Thus $p(\lambda) = -(\lambda+2)q(\lambda)$ where the quadratic $q(\lambda)$ is found by division to be $\lambda^2 + 2\lambda + 3$. The roots of $q(\lambda)$ are found by the quadratic formula to be the complex conjugates $-1 + \sqrt{2}i$ and $-1 - \sqrt{2}i$. Thus the eigenvalues of A are $-2, -1 + \sqrt{2}i$, and $-1 - \sqrt{2}i$.

A polynomial may have a multiple root, and so we have the following definition. If m_0 is the largest positive integer for which $(\lambda - \lambda_0)^{m_0}$ is a factor of the polynomial $p(\lambda)$, then m_0 is said to be the *multiplicity* of the root λ_0.

❖ **Multiplicity of an Eigenvalue.** A scalar λ_0 is an eigenvalue of A of *multiplicity k* if λ_0 is a root of multiplicity k of the characteristic polynomial $p(\lambda) = \det(A - \lambda I)$. Eigenvalues of multiplicity 1 are called *simple* eigenvalues.

EXAMPLE 7.2.11

An Eigenvalue of Multiplicity Two
Let

$$A = \begin{bmatrix} 2 & 2 & 1 \\ 1 & 3 & 1 \\ 1 & 2 & 2 \end{bmatrix}$$

Then we have that

$$p(\lambda) = \det[A - \lambda I] = \det \begin{bmatrix} 2-\lambda & 2 & 1 \\ 1 & 3-\lambda & 1 \\ 1 & 2 & 2-\lambda \end{bmatrix}$$

$$= -\lambda^3 + 7\lambda^2 - 11\lambda + 5 = -(\lambda-1)^2(\lambda-5)$$

The roots are $\lambda_1 = 1, \lambda_2 = 1, \lambda_3 = 5$; $\lambda = 1$ is an eigenvalue of multiplicity 2, and $\lambda = 5$ is a simple eigenvalue. To find the eigenspace V_1, we must find all solutions of

$$(A - I)v = \begin{bmatrix} 1 & 2 & 1 \\ 1 & 2 & 1 \\ 1 & 2 & 1 \end{bmatrix} \begin{bmatrix} v_1 \\ v_2 \\ v_3 \end{bmatrix} = \begin{bmatrix} 0 \\ 0 \\ 0 \end{bmatrix}$$

Thus $v_1 + 2v_2 + v_3 = 0$, which is the equation of a plane through the origin and defines the eigenspace V_1. To find a basis of V_1, let $v_1 = r, v_2 = s$, so $v_3 = -r - 2s$, yielding $v = [r \quad s \quad -r-2s]^T = r[1 \; 0 \; -1]^T + s[0 \; 1 \; -2]^T$ as the general vector in V_1. So $\{[1 \; 0 \; -1]^T, [0 \; 1 \; -2]^T\}$ spans V_1; it is clearly an independent set, so it is a basis for V_1, which is therefore a two-dimensional subspace of $\mathbf{R}^3$.

It is *not* always the case that the multiplicity of an eigenvalue equals the dimension of its eigenspace. The matrix $A = \begin{bmatrix} 3 & 1 \\ 0 & 3 \end{bmatrix}$ is a counterexample. This matrix has a double eigenvalue 3, while the corresponding eigenspace has dimension 1. We do have the following result, which we shall not prove, concerning the dimension of an eigenspace and the multiplicity of the corresponding eigenvalue.

THEOREM 7.2.3

> Eigenspace Dimension Theorem. Let A be an $n \times n$ matrix and λ an eigenvalue of A of multiplicity m. Then the dimension of the eigenspace V_λ is no larger than m, but at least 1.

Thus, if an eigenvalue is simple, its eigenspace is one-dimensional.

Invertibility of $n \times n$ Matrices, Similarity

An $n \times n$ matrix A is said to be *invertible* if there is an $n \times n$ matrix B such that $BA = I$, where I is the $n \times n$ identity matrix. A matrix which is not invertible is said to be *singular*. It turns out that a matrix B with the property that $BA = I$ is unique (if it exists), and B is denoted by A^{-1}. It is also a fact that $AA^{-1} = A^{-1}A$. A general property of determinants is that if A and B are any $n \times n$ matrices, then $\det(AB) = \det A \cdot \det B$. Thus, in particular, $\det A^{-1} \cdot \det A = 1$ and so A is invertible if and only if $\det A \neq 0$, which from (6) is the case if and only if $\lambda = 0$ is *not* an eigenvalue for A. Extending the equivalence further, we see that A is invertible if and only if the equation $Av = w$ is solvable for all vectors w in $\mathbf{R}^n$.

If P is any invertible $n \times n$ matrix, then the matrix $C = P^{-1}AP$ is said to arise from A by a *similarity transformation*, and A is said to be *similar* to C. Now since

$$\det(C - \lambda I) = \det(P^{-1}AP - \lambda I) = \det(P^{-1}(A - \lambda I)P)$$
$$= (\det P^{-1})(\det(A - \lambda I))(\det P) = \det(A - \lambda I)$$

it follows that A and $P^{-1}AP$ have the same characteristic polynomial, and hence, the same eigenvalues with the same multiplicities.

Deficient Eigenspaces, Generalized Eigenvectors

When an eigenspace of an $n \times n$ matrix has the maximum possible dimension (equal to the multiplicity of the corresponding eigenvalue) the eigenspace is said to be *nondeficient*; otherwise it is said to be *deficient*. Symmetric matrices[5] have the property that all of their eigenspaces are nondeficient. Unfortunately, not all matrices have nondeficient eigenspaces as noted above and in the following example.

EXAMPLE 7.2.12

A Deficient Eigenspace
Let

$$\begin{bmatrix} 1 & -2 & -2 \\ -2 & 2 & 3 \\ 2 & -3 & -4 \end{bmatrix}$$

The characteristic polynomial is $p(\lambda) = (1 + \lambda)^2(1 - \lambda)$, and the eigenvalues are

---

[5]A matrix A is symmetric if $A = A^T$.

$\lambda_1 = -1$ (double) and $\lambda_2 = 1$ (simple). However, the eigenspace V_{-1} is only one-dimensional. For we have that the only solutions of the system

$$[A - \lambda_1 I]\begin{bmatrix} a \\ b \\ c \end{bmatrix} = \begin{bmatrix} 2 & -2 & -2 \\ -2 & 3 & 3 \\ 2 & -3 & -3 \end{bmatrix}\begin{bmatrix} a \\ b \\ c \end{bmatrix} = \begin{bmatrix} 0 \\ 0 \\ 0 \end{bmatrix}$$

are $a = 0$, b arbitrary, $c = -b$. Hence V_{-1} is spanned by $\{[0\ 1\ -1]^T\}$ and is one-dimensional. V_1 is spanned by $\{[1\ -1\ 1]^T\}$. Hence there does *not* exist an independent set of three eigenvectors of A.

In Example 7.2.12 we see that the eigenspace V_{-1} of A corresponding to the eigenvalue $\lambda = -1$ is deficient because the multiplicity of that eigenvalue is 2 but $\dim V_{-1} = 1$. We now define a concept which is useful if an eigenspace is deficient.

❖ **Generalized Eigenvectors.** Let u^1 be an eigenvector for the matrix A corresponding to the eigenvalue λ. The vectors $u^1, u^2, u^3, \ldots$ are called *generalized eigenvectors* based on u^1 if $(A - \lambda I)u^{j+1} = u^j$, $j = 1, 2, \ldots$.

Notice that if $(A - \lambda I)u^2 = u^1$ has *no* solution, then the chain of generalized eigenvectors based on u^1 has length 1. Although it is not evident, a chain of generalized eigenvectors is linearly independent. If we construct a chain of generalized eigenvectors from each of the elements of a basis for V_λ, then the totality of the chains is also an independent set. Call this set B and let $|B|$ denote the number of elements in B. It is always possible to choose a basis for V_λ such that $|B| = $ multiplicity of λ, but it is never possible for $|B|$ to exceed the multiplicity of λ.

EXAMPLE 7.2.13

Generalized Eigenvectors
In Example 7.2.12 we saw that the eigenspace V_{-1} corresponding to the eigenvalue $\lambda = -1$ is deficient because $\dim V_{-1} = 1$, whereas the multiplicity of $\lambda = -1$ is 2. The eigenvector $u^1 = [0\ 1\ -1]^T$ is a basis for the eigenspace V_{-1} and thus may support a chain of generalized eigenvectors. We find that $u^2 = [1\ 1\ 0]^T$ is a solution of $(A + I)u^2 = [0\ 1\ -1]^T$ (the solution u^2 is not unique since $A + I$ is singular). Notice that the equation $(A + I)u^3 = u^2$ has no solution, and the chain ends with two vectors, u^1 and u^2, as expected.

PROBLEMS

1. Find the eigenvalues, their multiplicities, and a basis for each eigenspace of the following matrices.

(a) $\begin{bmatrix} 1 & 0 \\ 2 & 1 \end{bmatrix}$ (b) $\begin{bmatrix} 1 & 1 \\ 1 & 1 \end{bmatrix}$

(c) $\begin{bmatrix} 0 & 3 \\ 3 & 0 \end{bmatrix}$ (d) $\begin{bmatrix} 6 & -7 \\ 1 & -2 \end{bmatrix}$

(e) $\begin{bmatrix} 3 & -5 \\ 5 & 3 \end{bmatrix}$ (f) $\begin{bmatrix} 1 & 0 & 1 \\ 0 & 1 & 0 \\ 1 & 0 & 1 \end{bmatrix}$

$$\text{(g)} \begin{bmatrix} 5 & -6 & -6 \\ -1 & 4 & 2 \\ 3 & -6 & -4 \end{bmatrix} \qquad \text{(h)} \begin{bmatrix} \cos\theta & -\sin\theta \\ \sin\theta & \cos\theta \end{bmatrix}, \text{ real } \theta$$

$$\text{(i)} \begin{bmatrix} -1 & 36 & 100 \\ 0 & -1 & 27 \\ 0 & 0 & 5 \end{bmatrix} \qquad \text{(j)} \begin{bmatrix} a & b \\ -b & a \end{bmatrix}, \text{ real } a, b$$

2. (*Triangular Matrix*). A matrix is *triangular* if all its entries below (or above) the diagonal are zero. Show that the eigenvalues of a triangular matrix are the diagonal entries, the multiplicity of each eigenvalue equaling the number of times the eigenvalue appears on the diagonal.

3. (*Eigenvalues of A^k*). Show that if λ is an eigenvalue of A, then λ^k is an eigenvalue of A^k. If μ is an eigenvalue of A^k, is $\mu^{1/k}$ always an eigenvalue of A?

4. (*Eigenvalues of A^T*). Show that a square matrix A and its transpose A^T have the same characteristic polynomial, hence the same eigenvalues with the same multiplicities.

5. (*Using Eigenvalues to Find* det A). Let $\lambda_1, \ldots, \lambda_n$ be the eigenvalues of a matrix A, each repeated according to its multiplicity. Show that $\det A = \lambda_1 \cdots \lambda_n$. [*Hint*: Write $p(\lambda)$ as $(-1)^n(\lambda - \lambda_1) \cdots (\lambda - \lambda_n)$, expand, and compare with the form of $p(\lambda)$ in (6).]

6. (*Trace*). The *trace* of an $n \times n$ matrix A is the sum of its diagonal entries: tr $A = a_{11} + \cdots + a_{nn}$.

   (a) Show that tr A is the sum of the eigenvalues of A. [*Hint*: See the hint in Problem 5.]

   (b) Use part (a) to show that the matrix $\begin{bmatrix} 1 & 5 & 3 \\ 6 & -7 & 10 \\ 37 & 56 & 2 \end{bmatrix}$ must have an eigenvalue with a negative real part.

7. (*Eigenvalues and Nonsingularity*). Show that the square matrix A is nonsingular if and only if all its eigenvalues are nonzero. [*Hint*: Use Problem 5.]

8. (*Proof of the Eigenspace Property*). Use induction on p to show that the collection B of all vectors in $B_1, \ldots, B_p$ is an independent set if each B_i is an independent subset of V_i, where the V_i are eigenspaces of an $n \times n$ matrix A and corresponding to distinct eigenvalues.

   (a) First assume that each B_i contains a single vector v^i. Verify that the Eigenspace Property holds for all p by showing (1) that it holds for $p = 1$ (the anchor step), and (2) that if it holds whenever $p \leq k - 1$, then it also holds for $p = k$ (the induction step).

   (b) Show the Eigenspace Property for the general case where not all the set B_i are singleton sets. [*Hint*: Reduce to the case considered in part (b).]

 9. (*Generalized Eigenspace Property*). Let A be an $n \times n$ matrix with eigenvalues $\lambda_1, \lambda_2, \ldots, \lambda_k$ with respective multiplicities $m_1, m_2, \ldots, m_k$. Furthermore, let $V_{\lambda_1}, V_{\lambda_2}, \ldots, V_{\lambda_k}$ be the eigenspaces of A corresponding to $\lambda_1, \ldots, \lambda_k$. For each j, if V_{λ_j} is nondeficient, let B_j be a basis for V_{λ_j}. If V_{λ_j} is deficient, there is a basis for V_{λ_j} such that the total number of independent generalized eigenvectors using each element of this basis as a starting element, is the multiplicity m_j. Denote the totality of these generalized eigenvectors by B_j. Show the set B consisting of all the elements in all the B_j is a basis for $\mathbf{R}^n$ (or $\mathbf{C}^n$, as the case may be). [*Hint*: Use induction on k. Apply matrices of the form $(A - \lambda_1 I)$ to show first that B_1 is linearly independent no matter how many chains of generalized eigenvectors it contains. Then show by induction that B is independent. Use $(A - \lambda_i I)$ repeatedly.]

7.3 Homogeneous Linear Systems with Constant Coefficients

The general form of a *linear differential system* in the n state variables $x_1, x_2, \ldots, x_n$ is

$$
\begin{aligned}
x_1' &= a_{11}x_1 + \cdots + a_{1n}x_n + F_1(t) \\
&\ \ \vdots \\
x_n' &= a_{n1}x_1 + \cdots + a_{nn}x_n + F_n(t)
\end{aligned}
\tag{1}
$$

where the *coefficients* a_{ij} are real constants or real-valued continuous functions on some common t-interval I, and the *input functions* F_i are continuous on I as well. Using the standard notation for matrices and column vectors, and their algebraic properties (see Appendix C.6), system (1) can be written in the compact form

$$
x' = Ax + F
\tag{2}
$$

where

$$
A = A(t) = \begin{bmatrix} a_{11} \cdots a_{1n} \\ \vdots \quad \vdots \\ a_{n1} \cdots a_{nn} \end{bmatrix} = [a_{ij}], \quad x = x(t) = \begin{bmatrix} x_1 \\ \vdots \\ x_n \end{bmatrix}, \quad F = F(t) = \begin{bmatrix} F_1 \\ \vdots \\ F_n \end{bmatrix}
$$

and the column vector x' is produced by differentiating the components x_k of the column vector x one-by-one. The column vector x is called the *state vector*, and A is the *system* (or *coefficient*) *matrix* of the differential system. The column vector F is called the *input vector* or *driving term*. The cascade models in Section 1.6 and the lead model in Section 7.1 are examples of linear differential systems.

Associated with system (2) is the initial value problem

$$
x' = Ax + F, \qquad x(t_0) = x^0
\tag{3}
$$

where t_0 belongs to the interval I on which $A(t)$ and $F(t)$ are continuous, and x^0 is a point in $\mathbf{R}^n$. The properties of the solution of IVP (3) are essentially given by Theorem 6.2.1 which is reformulated and stated here for convenience.

THEOREM 7.3.1

> Existence, Uniqueness, Extension. IVP (3) has exactly one maximally extended solution $x(t)$. This solution is defined at least for all t in the interval I, and is a continuous function of t_0 and x^0.

Existence and uniqueness follow from Theorem 6.2.1 since the vector function $f = A(t)x + F(t)$ and all the first partial derivatives $\partial f_i/\partial x_j = a_{ij}(t), i, j = 1, \ldots, n$, are continuous for all t in I and x in $\mathbf{R}^n$ by assumption. We omit the proof that the solution $x(t)$ can be extended to I and varies continuously with t_0 and x^0.

Theorem 7.3.1 above does not give a formula or process for finding the unique solution of IVP (3) or for finding all solutions of the system (2). Our goal in this section is to show how this may be done, if A is a matrix of constants and $F = 0$. We will see that the eigenvalues, eigenvectors, and generalized eigenvectors of A will greatly simplify our task.

Linear Operators

It will be helpful to recast the ODE (2) in the language of operators. You may first want to review our initial encounter with operators in Section 3.3. If you are a bit shaky on linear spaces, you may want to read about them in the previous section.

Let V and W be two linear spaces (both with real, or both with complex scalars). A mapping $L: V \to W$ is said to be a *linear operator* if for any two elements v^1, v^2 in V and any two scalars a_1, a_2, we have that $L[a_1 v^1 + a_2 v^2] = a_1 L[v^1] + a_2 L[v^2]$.

As an example, let us reformulate system (2) by introducing an operator which acts on the continuity class $\mathbf{C}^1(I)$, which is the collection of column vectors $x(t)$ with continuously differentiable components. Let the operator L have the action $L[x] = x' - Ax$. Then $L[x]$ is in the continuity class $\mathbf{C}^0(I)$ and system (2) assumes the form $L[x] = F$. Notice that for any two constants, a and b, and any two continuously differentiable column vectors, x and y, we have that $L[ax + by] = aL[x] + bL[y]$. Thus, L is a linear operator. As was done for linear first-order and second-order ODEs in Sections 1.2 and 3.3, we can characterize all solutions of system (2) as the sum of one particular solution and the null space of the operator L. This is a direct consequence of the linearity of the operator L. Indeed, if x^p is a particular solution of $L[x] = F$, then for any solution x of $L[x] = F$ we have from the linearity of L that $L[x - x^p] = L[x] - L[x^p] = F - F = 0$, and so $x - x^p$ belongs to the null space of L. Conversely, if x^h belongs to the null space of L, then

$$L[x^h + x^p] = L[x^h] + L[x^p] = 0 + F = F$$

so $x^h + x^p$ is a solution of $L[x] = F$. Hence we can break up our search for all solutions for system (2) into two parts: Finding a particular solution of $x' = Ax + F$, and finding the general solution of the homogeneous linear system $x' = Ax$. We turn now to the latter task, leaving the former to Section 7.5.

Constant-Coefficient Matrix, Zero Input Vector

The simplest case of system (2) is when F is identically zero (no input) and A is a matrix of constants. An example illustrates an approach we shall use to construct solutions of linear differential systems in this case.

EXAMPLE 7.3.1

Finding Solutions of $x' = Ax$

By analogy with the one state variable case, the linear system

$$x' = Ax, \quad \text{with} \quad A = \begin{bmatrix} 5 & 3 \\ -6 & -4 \end{bmatrix} \tag{4}$$

might be expected to have solutions x of the form $x = ve^{\lambda t}$, for some constant λ and constant vector $v = [a \, b]^T$. Substituting this into system (4) yields

$$\lambda v e^{\lambda t} = A v e^{\lambda t}$$

$$\lambda v = A v \tag{5}$$

Thus, $x = ve^{\lambda t}$ solves system (4) if λ is an eigenvalue of A and v is a corresponding eigenvector. The eigenvalue problem (5) was solved in Example 7.2.6 (see (4) of that example), and so we have the two solutions

$$x^1(t) = \begin{bmatrix} 1 \\ -2 \end{bmatrix} e^{-t}, \quad x^2(t) = \begin{bmatrix} 1 \\ -1 \end{bmatrix} e^{2t} \tag{6}$$

Using the linearity of the operator L defined by $L[x] = x' - Ax$, we see that

$$x = c_1 \begin{bmatrix} 1 \\ -2 \end{bmatrix} e^{-t} + c_2 \begin{bmatrix} 1 \\ -1 \end{bmatrix} e^{2t}, \quad c_1, c_2 \text{ arbitrary constants} \qquad (7)$$

is also a solution of system (4).

Actually, formula (7) gives all the solutions of the homogeneous system (4). To see this, let $x(t)$ be any solution of system (4), and note from Theorem 7.3.1 that $x(t)$ must be defined on the whole real line **R**. Now set $x(0) = x^0$. There are unique values c_1^0 and c_2^0 such that $c_1^0 x^1(0) + c_2^0 x^2(0) = x(0) = x^0$ because the determinant of this algebraic system is

$$\det[x^1(0) \quad x^2(0)] = \det \begin{bmatrix} 1 & 1 \\ -2 & -1 \end{bmatrix} = 1 \neq 0 \qquad (8)$$

Now $z(t) = c_1^0 x^1(t) + c_2^0 x^2(t)$ and $x(t)$ both solve system (4) and also have the same initial data $z(0) = x^0 = x(0)$. So by the uniqueness part of Theorem 7.3.1 we see that $z(t)$ is identical to $x(t)$. Thus every solution of system (4) can be written as a linear combination of the two solutions x^1 and x^2 given in (6), establishing our claim. It is also a fact that each solution $x(t)$ of system (4) can be written as a linear combination of x^1 and x^2 in only one way. Indeed, say that $x(t) = ax^1 + bx^2$, and also $x(t) = cx^1 + dx^2$. Thus, the numbers $a - c$ and $b - d$ satisfy the identity

$$(a - c)x^1(t) + (b - d)x^2(t) = 0, \quad \text{for all } t$$

and because of the determinant condition (8), it follows that $a - c = 0$ and $b - d = 0$, establishing our claim.

EXAMPLE 7.3.2

Solving an IVP, $x' = Ax$, $x(0) = x^0$
Suppose that A is the matrix of Example 7.3.1 and that $x^0 = [2 \quad 3]^T$. Then since formula (7) is the general solution of $x' = Ax$, we know that there are constants c_1 and c_2 so that $x(0) = x^0$. Thus c_1 and c_2 must be found so that

$$x(0) = \begin{bmatrix} 2 \\ 3 \end{bmatrix} = c_1 \begin{bmatrix} 1 \\ -2 \end{bmatrix} + c_2 \begin{bmatrix} 1 \\ -1 \end{bmatrix} = \begin{bmatrix} c_1 + c_2 \\ -2c_1 - c_2 \end{bmatrix}$$

That is,

$$c_1 + c_2 = 2$$
$$-2c_1 - c_2 = 3$$

Hence, $c_1 = -5$ and $c_2 = 7$, and the desired solution is

$$x = -5 \begin{bmatrix} 1 \\ -2 \end{bmatrix} e^{-t} + 7 \begin{bmatrix} 1 \\ -1 \end{bmatrix} e^{2t}$$

Example 7.3.1 and the above discussion suggest that in general all solutions of a constant coefficient linear system of first-order ODEs with no input can be found by pure matrix-algebraic means, as we now show. We will freely use the terminology of eigenvalues and eigenspaces of matrices throughout what follows, including the notion of generalized eigenvectors. If you are a bit rusty on these topics (or perhaps never studied them at all), then you should look over the previous section before proceeding any further. All the relevant matrix-algebraic concepts we will need are developed in that section.

Nondeficient Eigenspaces

Recall that an eigenspace is nondeficient if its dimension is equal to the multiplicity of its corresponding eigenvalue. If all eigenspaces of a matrix A are nondeficient, it is a straightforward matter to construct the general solution of $x' = Ax$.

THEOREM 7.3.2

> **All Eigenspaces Nondeficient.** Let A be a $n \times n$ constant matrix all of whose eigenspaces are nondeficient. Then the general solution of $x' = Ax$ is given by
>
> $$x = c_1 v^1 e^{\lambda_1 t} + \cdots + c_n v^n e^{\lambda_n t} \tag{9}$$
>
> where $c_1, \ldots, c_n$ are arbitrary constants, $\{v^1, \ldots, v^n\}$ is any independent set of eigenvectors of A, and $\lambda_1, \ldots, \lambda_n$ are the corresponding eigenvalues (repeated as multiplicities require).

Proof. Let $\lambda_1, \ldots, \lambda_k$, the distinct eigenvalues of A, have respective multiplicities $m_1, m_2, \ldots, m_k$, and let $V_{\lambda_1}, \ldots, V_{\lambda_k}$ be their corresponding eigenspaces. Then nondeficiency implies that $\dim V_{\lambda_i} = m_i$, $i = 1, 2, \ldots, k$. Now if for each i, B_i is a basis for V_{λ_i}, then by the Eigenspace Property (Theorem 7.2.1) the collection B of all the vectors in the B_i is an independent set in $\mathbf{R}^n$ (or $\mathbf{C}^n$, if appropriate). Labeling the eigenvectors in B as $v^1, v^2, \ldots, v^n$, with corresponding eigenvalues repeated as multiplicities require, we see that formula (9) defines a solution of the system $x' = Ax$ for any constants $c_1, \ldots, c_n$.

Now suppose that $x(t)$ is *any* solution of $x' = Ax$. Theorem 7.3.1 implies that $x(t)$ is defined on the entire real line $\mathbf{R}$. Since $B = \{v^1, \ldots, v^n\}$ is a basis for $\mathbf{R}^n$ (or $\mathbf{C}^n$, as the case may be), there exist unique constants $c_1^0, \ldots, c_n^0$ such that $x(0) = c_1^0 v^1 + \cdots + c_n^0 v^n$. Putting $z(t) = c_1^0 v^1 e^{\lambda_1 t} + \cdots + c_n^0 v^n e^{\lambda_n t}$, we see that $z(t)$ and $x(t)$ solve the same IVP, so by uniqueness they must be the same function. Thus formula (9) includes all solutions of $x' = Ax$. ∎

Wronskians and Independence

Analyzing the proof of Theorem 7.3.2 more closely, we see that the key to showing that the solutions $x^1(t), x^2(t), \ldots, x^n(t)$ generate all solutions of the system $x' = Ax$ on the interval I by the formula $x = c_1 x^1 + \cdots + c_n x^n$, for arbitrary scalars c_j, is that for some t_0 in I, the matrix-vector system

$$[x^1(t_0) \quad x^2(t_0) \quad \cdots \quad x^n(t_0)] \begin{bmatrix} a_1 \\ \vdots \\ a_n \end{bmatrix} = \begin{bmatrix} b_1 \\ \vdots \\ b_n \end{bmatrix}$$

be solvable for all vectors b on the right-hand side. From Appendix C.6 we see that this is the case if and only if the determinant $\det[x^1(t_0) \cdots x^n(t_0)] \neq 0$. This motivates the following definition.

❖ **Wronskian.** Let $x^1, x^2, \ldots, x^n$ be n solutions of the system $x' = A(t)x$ where $A(t)$ is an $n \times n$ matrix. The determinant

$$W[x^1, x^2, \ldots, x^n] = \det[\, x^1(t) \quad x^2(t) \quad \cdots \quad x^n(t) \,]$$

is called the *Wronskian* of the n solutions $x^1(t), \ldots, x^n(t)$.

THEOREM 7.3.3

> Wronskian Property for $x' = Ax$. If the entries $a_{ij}(t)$ in the $n \times n$ matrix $A(t)$ are all continuous on some interval I, and if $x^1, x^2, \ldots, x^n$ are n solutions of the system $x' = A(t)x$, then the Wronskian $W = W[x^1, \ldots, x^n]$ satisfies the first-order ODE
>
> $$W' = [a_{11}(t) + a_{22}(t) + \cdots + a_{nn}(t)] W \qquad (10)$$
>
> on the interval I.

Proof. We shall take $n = 2$. Let $y(t)$ and $z(t)$ be two solutions of $x' = Ax$. Then, by differentiation of the determinant $W = W[y, z]$, and the properties of determinants,

$$
W' = \det \begin{bmatrix} y_1' & z_1' \\ y_2 & z_2 \end{bmatrix} + \det \begin{bmatrix} y_1 & z_1 \\ y_2' & z_2' \end{bmatrix}
$$

$$
= \det \begin{bmatrix} a_{11}y_1 + a_{12}y_2 & a_{11}z_1 + a_{12}z_2 \\ y_2 & z_2 \end{bmatrix} + \det \begin{bmatrix} y_1 & z_1 \\ a_{21}y_1 + a_{22}y_2 & a_{21}z_1 + a_{22}z_2 \end{bmatrix}
$$

$$
= \det \begin{bmatrix} a_{11}y_1 & a_{11}z_1 \\ y_2 & z_2 \end{bmatrix} + \det \begin{bmatrix} a_{12}y_2 & a_{12}z_2 \\ y_2 & z_2 \end{bmatrix}
$$

$$
+ \det \begin{bmatrix} y_1 & z_1 \\ a_{21}y_1 & a_{21}z_1 \end{bmatrix} + \det \begin{bmatrix} y_1 & z_1 \\ a_{22}y_2 & a_{22}z_2 \end{bmatrix}
$$

$$
= (a_{11} + a_{22})W
$$

as was to be shown. The proof for general n proceeds in a similar fashion.

Note that the ODE (10) implies that the Wronskian of n solutions of $x' = Ax$ is either always zero in I or never zero in I. Again drawing on the proof of Theorem 7.3.2 for inspiration we define the concept of a fundamental set as follows:

❖ **Fundamental Set.** Let the entries in the $n \times n$ matrix $A(t)$ be continuous on some t-interval I. A collection of n solutions $x^1, x^2, \ldots, x^n$ of $x' = A(t)x$ is a *fundamental set* if the Wronskian $W[x^1, x^2, \ldots, x^n]$ is nonzero on I.

It can be shown (as in the proof of Theorem 7.3.2) that n solutions of $x' = Ax$ form a fundamental set if and only if the set of solutions is linearly independent. Another form of this statement is this: the solution set $\{x^1(t), x^2(t), \ldots, x^n(t)\}$ of the system $x' = A(t)x$ is fundamental if and only if $\det[x^1(t_0) \cdots x^n(t_0)] \neq 0$ for any t_0 in the interval I.

THEOREM 7.3.4

> Fundamental Sets and the General Solution. Let $x^1, x^2, \ldots, x^n$ form a fundamental set of solutions of $x' = Ax$, where A is an $n \times n$ matrix of constants or functions of t. Then the general solution of $x' = Ax$ is $x = c_1 x^1(t) + \cdots + c_n x^n(t)$ where $c_1, \ldots, c_n$ are arbitrary constants.

Proof. If $\{x^1, x^2, \ldots, x^n\}$ is a fundamental set of solutions for $x' = Ax$, then the $x^i(t)$ are all defined on an interval I where $A(t)$ is continuous. Now if $x(t)$ is any solution defined on I, pick a t_0 in I. The linear system $x(t_0) = c_1 x^1(t_0) + \cdots + c_n x^n(t_0)$ has a unique solution $c_1^0, c_2^0, \ldots, c_n^0$, because $\det[x^1(t_0) \cdots x^n(t_0)] \neq$

0. Since $z(t) = c_1^0 x^1(t) + \cdots + c_n^0 x^n(t)$ and $x(t)$ both solve $x' = Ax$ and have the same data at t_0, i.e., $x(t_0) = z(t_0)$, it follows that $x = c_1 x^1 + \cdots + c_n x^n$, where $c_1, c_2, \ldots, c_n$ are arbitrary scalars, is the general solution of $x' = Ax$.

The solution formula (9) verifies this assertion with the solutions $x^j = v^j e^{\lambda_j t}$, $j = 1, 2, \ldots, n$.

EXAMPLE 7.3.3

Finding a General Solution

Consider the system $x' = Ax$ where

$$\begin{bmatrix} 1 & -2 & -2 \\ -2 & 1 & 2 \\ 2 & -2 & -3 \end{bmatrix}$$

The characteristic polynomial of A is $p(\lambda) = (1 + \lambda)^2 (1 - \lambda)$, and the eigenvalues are -1 (a double eigenvalue) and 1 (a simple eigenvalue). Direct calculation shows that V_{-1} is two dimensional and is spanned by $\{[1\ 1\ 0]^T, [1\ 0\ 1]^T\}$, while V_1 is spanned by $\{[1\ -1\ 1]^T\}$. Thus the eigenspaces of A are both nondeficient and so $\{[1\ 1\ 0]^T, [1\ 0\ 1]^T, [1\ -1\ 1]^T\}$ is a set of three independent eigenvectors. It follows that $W[x^1, x^2, x^3]|_{t=0} \neq 0$ where x^1, x^2, x^3 are the solutions

$$x^1 = \begin{bmatrix} 1 \\ 1 \\ 0 \end{bmatrix} e^{-t}, \qquad x^2 = \begin{bmatrix} 1 \\ 0 \\ 1 \end{bmatrix} e^{-t}, \qquad x^3 = \begin{bmatrix} 1 \\ -1 \\ 1 \end{bmatrix} e^{t}$$

So, $\{x^1, x^2, x^3\}$ is a fundamental set. Hence the general solution of $x' = Ax$ is

$$x = c_1 \begin{bmatrix} 1 \\ 1 \\ 0 \end{bmatrix} e^{-t} + c_2 \begin{bmatrix} 1 \\ 0 \\ 1 \end{bmatrix} e^{-t} + c_3 \begin{bmatrix} 1 \\ -1 \\ 1 \end{bmatrix} e^{t}$$

Note that if the eigenvalues of A are all distinct, then the eigenvalues of A are simple and so the eigenspaces are nondeficient since they must be one dimensional. Thus, there is an independent set of n eigenvectors of A. The only problematic matrices are those with some multiple eigenvalues, and as Example 7.3.3 shows, even then it may still be possible to find a basis of eigenvectors.

Deficient Eigenspaces

How can we use generalized eigenvectors to produce solutions to the system $x' = Ax$ when the $n \times n$ constant system matrix A has deficient eigenspaces? The following identity will be useful to us along the way. Let L be the linear operator $L[x] = x' - Ax$ which was introduced earlier in this section. Then for any constant vector v, any differentiable real-valued function $h(t)$, and any constant λ, we have the identity

$$L[vh(t)e^{\lambda t}] = vh'e^{\lambda t} - he^{\lambda t}(A - \lambda I)v \tag{11}$$

or, after rearranging the terms,

$$vh'e^{\lambda t} = L[vhe^{\lambda t}] + he^{\lambda t}(A - \lambda I)v \tag{12}$$

Now, for example, suppose that u^1, u^2, u^3 is a chain of generalized eigenvectors for the matrix A based on the eigenvector u^1 and the eigenvalue λ. Then we have that

$$(A - \lambda I)u^1 = 0, \quad (A - \lambda I)u^2 = u^1, \quad (A - \lambda I)u^3 = u^2 \tag{13}$$

Now, using (11) with $v = u^2$ and $h = 1$ and then (13), we have $L[u^3 e^{\lambda t}] = -e^{\lambda t} u^2$. Using (12) with $v = u^2$ and $h(t) = t$, we see that $u^2 e^{\lambda t} = L[u^2 t e^{\lambda t}] + t e^{\lambda t} u^1$. Using (11) and (13) again with $v = u^1$ and $h(t) = t^2/2$, we have that $u^1 t e^{\lambda t} = L[u^1 (t^2/2) e^{\lambda t}]$. Putting these facts together we have

$$L[u^3 e^{\lambda t}] = -L[u^2 t e^{\lambda t}] - L[u^1 \frac{t^2}{2} e^{\lambda t}]$$

Using the fact that L is a linear operator,

$$L[u^3 e^{\lambda t} + u^2 t e^{\lambda t} + u^1 \frac{t^2}{2} e^{\lambda t}] = 0$$

Similarly, it can be shown that

$$L[u^2 e^{\lambda t} + u^1 t e^{\lambda t}] = 0$$

Thus, for this example we have three solutions of the system $x' = Ax$

$$x^1 = u^1 e^{\lambda t}, \quad x^2 = (u^2 + t u^1) e^{\lambda t}, \quad x^3 = (u^3 + t u^2 + \frac{t^2}{2} u^1) e^{\lambda t} \tag{14}$$

These solutions are linearly independent because the set of vectors $\{x^1(0), x^2(0), x^3(0)\}$ is independent.

EXAMPLE 7.3.4

Matrix with a Deficient Eigenspace
Consider the system $x' = Ax$ where A is the 3×3 matrix in Example 7.2.12,

$$A = \begin{bmatrix} 1 & -2 & -2 \\ -2 & 2 & 3 \\ 2 & -3 & -4 \end{bmatrix}$$

We found that A has double eigenvalue $\lambda_1 = -1$, and a simple eigenvalue $\lambda_2 = 1$. The deficient eigenspace V_{-1} is spanned by the single eigenvector $[0 \ 1 \ -1]^T$ and the nondeficient eigenspace V_1 by the eigenvector $[1 \ -1 \ 1]^T$. In Example 7.2.13 we found a chain of generalized eigenvectors based on the eigenvector $v^1 = [0 \ 1 \ -1]^T$ in V_{-1}:

$$u^1 = [0 \quad 1 \quad -1]^T \quad \text{and} \quad u^2 = [1 \quad 1 \quad 0]^T$$

Using the technique described immediately above we find the three linearly independent solutions of $x' = Ax$

$$x^1 = \begin{bmatrix} 0 \\ 1 \\ -1 \end{bmatrix} e^{-t}, \quad x^2 = \begin{bmatrix} 1 \\ 1 \\ 0 \end{bmatrix} e^{-t} + \begin{bmatrix} 0 \\ 1 \\ -1 \end{bmatrix} t e^{-t}, \quad x^3 = \begin{bmatrix} 1 \\ -1 \\ 1 \end{bmatrix} e^t$$

Hence, the general solution of $x' = Ax$ is $x = c_1 x^1 + c_2 x^2 + c_3 x^3$, where c_1, c_2, and c_3 are arbitrary reals.

We have seen in Examples 7.2.13 and 7.3.4 that if there are deficient eigenspaces, then some solutions have components containing polynomial exponential functions such as $t e^{\lambda t}$ or $t^2 e^{\lambda t}$. If λ had been a complex number $\alpha + i\beta$ with $\beta \neq 0$, this would mean terms such as $t e^{\alpha t} \cos \beta t$ or $t e^{\alpha t} \sin \beta t$, and so on. In fact we have the following general result (whose proof is omitted) concerning the form of the components of solutions of $x' = Ax$.

THEOREM 7.3.5

> Components of Solutions of $x' = Ax$. Every component of every real-valued solution of $x' = Ax$, where A is an $n \times n$ matrix of real constants, is a linear combination of terms of the form $t^k e^{\alpha t} \cos \beta t$ and $t^k e^{\alpha t} \sin \beta t$, where $\lambda = \alpha + i\beta$ is an eigenvalue of A and k is a nonnegative integer less than the multiplicity of λ.

Using the terminology of Section 3.5, every component of every solution of $x' = Ax$ is a polynomial-exponential.

Comments

The eigenvalues and eigenvectors of a square constant matrix A may be used to construct solutions of the linear differential system $x' = Ax$. The eigenvalues and eigenvectors themselves are determined purely algebraically, without considering the differential system at all. This process of finding solutions of a constant-coefficient linear differential system by algebraic methods somewhat resembles the corresponding process outlined in Chapter 3 for solving a second-order constant-coefficient homogeneous ODE. In fact, the term "characteristic polynomial" is used in both settings.

PROBLEMS

1. (*General Solutions*). Find the general solution for each of the systems below.

    (a) $x_1' = 5x_1 + 3x_2; \quad x_2' = -x_1 + x_2$ (b) $x_1' = 2x_1 - x_2; \quad x_2' = x_1 + 2x_2$

    (c) $x_1' = 5x_1 - 2x_2; \quad x_2' = -2x_1 + 8x_2$ (d) $x_1' = x_1 - x_2; \quad x_2' = x_1 + 3x_2$

2. (*Solving IVPs*). (a)–(d). Solve the IVP consisting of the rate equations given in 1(a)–(d) and the initial condition $x_1(0) = 1$, $x_2(0) = 1$.

3. (*Nondeficient Eigenspaces*). Find all solutions of $x' = Ax$, where the matrix A is as given. Write your answers in the form given in (9) in the text.

    (a) $\begin{bmatrix} 0 & 1 \\ 8 & -2 \end{bmatrix}$ (b) $\begin{bmatrix} 0 & 1 \\ -64 & 0 \end{bmatrix}$ (c) $\begin{bmatrix} 3 & -2 \\ 2 & -2 \end{bmatrix}$

    (d) $\begin{bmatrix} 2 & -1 \\ 8 & -2 \end{bmatrix}$ (e) $\begin{bmatrix} 2 & 1 \\ -3 & 6 \end{bmatrix}$ (f) $\begin{bmatrix} -1 & 4 \\ 2 & 1 \end{bmatrix}$

    (g) $\begin{bmatrix} 0 & 1 & 0 \\ 0 & 0 & 1 \\ 1 & 0 & 0 \end{bmatrix}$ (h) $\begin{bmatrix} 3 & 2 & 2 \\ 1 & 4 & 1 \\ -2 & -4 & -1 \end{bmatrix}$ (i) $\begin{bmatrix} 2 & 2 & 1 \\ 1 & 3 & 1 \\ 1 & 2 & 2 \end{bmatrix}$

4. (*Solving IVPs*). (a)–(i). Solve the IVP, $x' = Ax$, $x(0) = [-1 \quad 1]^T$ where A is the matrix given in 3(a)–(f). For (g)–(i), $x(0) = [1 \quad 1 \quad 1]^T$.

5. (*Complex Eigenvalues of a Real Matrix*). Let A be a real matrix with a nonreal eigenvalue λ and a corresponding eigenvector v.

    (a) Show that v must have at least one nonreal component and that $\bar{v}$ is an eigenvector of A with $\bar{\lambda}$ its corresponding eigenvalue.

    (b) Show that $\{ve^{\lambda t}, \bar{v}e^{\bar{\lambda}t}\}$ is an independent set of complex-valued solutions of $x' = Ax$.

    (c) Show that $\{\text{Re}[ve^{\lambda t}], \text{Im}[ve^{\lambda t}]\}$ is an independent set of real-valued solutions. [*Hint*: $\text{Re}[z] = \frac{1}{2}(z + \bar{z})$, $\text{Im}[z] = \frac{-i}{2}(z - \bar{z})$.]

(d) Use the results of part **(c)** to find the general real-valued solutions of each system $x' = Ax$, where

$$A = \begin{bmatrix} 1 & 1 \\ -5 & 3 \end{bmatrix}, \qquad A = \begin{bmatrix} 1 & 2 \\ -2 & 1 \end{bmatrix}, \qquad A = \begin{bmatrix} -3 & 1 & -2 \\ 0 & -1 & -1 \\ 2 & 0 & 0 \end{bmatrix}$$

[*Hint*: Review Examples 7.2.8, 7.2.9.]

6. (*Scalar ODE and its Equivalent System*). The "characteristic polynomial" of the nth-order constant-coefficient ODE $y^{(n)} + a_{n-1}y^{(n-1)} + \cdots + a_1 y' + a_0 y = 0$ is $r^n + a_{n-1}r^{n-1} + \cdots + a_1 r + a_0$.

(a) Show that if $x_1 = y$, $x_2 = y'$, ..., $x_n = y^{(n-1)}$, the ODE above becomes the system $x' = Ax$ where

$$A = \begin{bmatrix} 0 & 1 & 0 & \cdots & \cdots & 0 \\ 0 & 0 & 1 & & & \\ \vdots & & \ddots & & & \vdots \\ \vdots & & & \ddots & & \vdots \\ & & & & 0 & 1 \\ -a_0 & -a_1 & -a_2 & \cdots & -a_{n-2} & -a_{n-1} \end{bmatrix}$$

(b) Show that the characteristic polynomial $p(\lambda)$ of matrix A is $(-1)^n(\lambda^n + a_{n-1}\lambda^{n-1} + \cdots + a_1\lambda + a_0)$. [*Hint*: Using column operations, first transform $A - \lambda I$ into a matrix whose top $(n-1)$ rows look like a column of zeros followed by an $(n-1) \times (n-1)$ triangular matrix with ones on the diagonal. Then calculate the determinant by cofactors.]

7. (*Deficient Eigenspace*). Find the general solution of each system below.

(a)

$$x' = \begin{bmatrix} 1 & -2 & -2 \\ -2 & 2 & 3 \\ 2 & -3 & -4 \end{bmatrix} x$$

[*Hint*: See Example 7.2.12.]

(b)

$$x' = \begin{bmatrix} 1 & 0 & -2 \\ 0 & -1 & 1 \\ 0 & 0 & -1 \end{bmatrix} x$$

8. (*Deficient Eigenspaces*). Find the general solution of the system

$$x' = Ax = \begin{bmatrix} -1 & 0 & 2 \\ 0 & -1 & 1 \\ 0 & 0 & -1 \end{bmatrix} x$$

by following the outline below:

- Show that the system matrix has only one eigenvalue and that the corresponding eigenspace has dimension 2.
- To find an eigenvector that yields a chain of generalized eigenvectors of length 2, proceed as follows: Let λ be the eigenvalue of A and let v^1 and v^2 be a basis of the eigenspace of A corresponding to λ. Find conditions on the scalars c_1 and c_2 such that the system $(A - \lambda I)u^2 = c_1 v^1 + c_2 v^2$ has a nonzero solution.

- Using a set of values c_1^0 and c_2^0 as determined above, show that $u^1 = c_1^0 v^1 + c_2^0 v^2$ and a nonzero solution u^2 of $(A - \lambda I)u^2 = u^1$ is a chain of generalized eigenvectors.

- Using v^1, u^1, and u^2, construct the general solution of $x' = Ax$.

7.4 Orbital Portraits

We showed in Section 7.3 how to use the eigenvalues, eigenvectors, and generalized eigenvectors of the $n \times n$ matrix A of real constants to construct the general solution of the homogeneous linear system with constant coefficients.

$$x' = Ax, \quad \text{where} \quad A = \begin{bmatrix} a_{11} \cdots a_{1n} \\ \vdots \qquad \vdots \\ a_{n1} \cdots a_{nn} \end{bmatrix}, \quad x = \begin{bmatrix} x_1 \\ \vdots \\ x_n \end{bmatrix} \tag{1}$$

The examples in Section 7.3 illustrate the solution process. In this section we shall construct orbital portraits for planar autonomous systems ($n = 2$), give an example that shows how an orbital portrait changes as an entry in the system matrix A changes (i.e., discuss the sensitivity of a system to changes in the data), and conclude with the general solution and orbits of a system that models the motions of a double pendulum.

Gallery of Planar Orbital Portraits

We shall build a gallery of portraits near the origin for the system $x' = Ax$, where A is a 2×2 matrix of real constants. We assume that $\lambda = 0$ is not an eigenvalue of A, and so the origin is the only equilibrium point of the system. There are six distinct types of orbital portraits, and these are given in Table 7.4.1 along with their commonly used names, the nature of the eigenvalues and a fundamental set of solutions. In the table, eigenvectors corresponding to the respective eigenvalues λ_1 and λ_2 are denoted by v^1 and v^2, and w denotes a generalized eigenvector based on v^1. The names "improper node," "saddle," and so on, are explained in the examples below. These names apply to both the orbital portrait and the equilibrium point at the origin.

The six cases cover all of the possibilities for a real 2×2 matrix A with nonzero eigenvalues. The examples and orbital portraits below are representative of the six cases. There are straight line orbits in some of the portraits. These correspond to "eigenspace orbits" because if λ is a real eigenvalue of A with v a corresponding eigenvector, and c any real constant, then $x(t) = cve^{\lambda t}$ is a solution that generates an orbit that lies on the straight line $x = sv$, $-\infty < s < \infty$, through the eigenvector v. If $c > 0$, or $c < 0$, the orbit lies on the ray generated by, respectively, $0 < s < \infty$ or $-\infty < s < 0$; if $c = 0$, the solution is the trivial solution $x = 0$ whose orbit is the equilibrium point at the origin. Recall that the uniqueness property implies that no two distinct maximally extended orbits may cross. Thus, two of these "eigenspace lines" in the xy-plane divide the plane into sectors, and no orbit can cross from one sector into another. Eigenspace lines are present in Figures 7.4.1–7.4.4. We use x and y as the names of the state variables in all of the examples, rather than x_1 and x_2.

EXAMPLE 7.4.1

Improper Node

The matrix $\begin{bmatrix} 3 & -1 \\ 0 & 1 \end{bmatrix}$ of the system $x' = 3x - y$, $y' = y$ has eigenvalues $\lambda_1 = 1$ and

$\lambda_2 = 3$ with respective eigenvectors $[1 \quad 2]^T$ and $[1 \quad 0]^T$. The general solution is

$$\begin{bmatrix} x \\ y \end{bmatrix} = c_1 \begin{bmatrix} 1 \\ 2 \end{bmatrix} e^t + c_2 \begin{bmatrix} 1 \\ 0 \end{bmatrix} e^{3t} = e^t \left(c_1 \begin{bmatrix} 1 \\ 2 \end{bmatrix} + c_2 \begin{bmatrix} 1 \\ 0 \end{bmatrix} e^{2t} \right) \qquad (2)$$

The corresponding eigenspace lines are $y = x$ and the x-axis. Observe from (2) that all orbits emerge from the origin as t increases from $-\infty$, that only the eigenspace solutions $c_2 e^{3t} [1 \; 0]^T$ emerge tangent to the x-axis, and that all other solutions leave the origin tangent to the line $y = 2x$ as t increases from $-\infty$ because the term in parentheses in (2) tends to $c_1 [1 \quad 2]^T$ as $t \to -\infty$. Figure 7.4.1 shows the two eigenspace lines as well as a selection of other orbits. If the eigenvalues are negative and distinct, the orbital portrait has the same form, but the orbits approach the origin as $t \to \infty$.

The term "node" is used in geometry for a point where several curves intersect with different tangents. In Example 7.4.1 there are just two tangents at the nodal point at the origin; the node is said to be "improper" because not every direction is a direction of tangency.

EXAMPLE 7.4.2

Deficient Improper Node

The system matrix $\begin{bmatrix} 1 & -2 \\ 0 & 1 \end{bmatrix}$ of the system $x' = x - 2y$, $y' = y$ has the double eigenvalue $\lambda = 1$ and an eigenvector $v = [1 \quad 0]^T$, but there is no second independent eigenvector. Figure 7.4.2 shows the "eigenspace line," which is the x-axis in this case. After finding a generalized eigenvector $w = [0 \;\; -1/2]^T$ by solving $(A - I)w = v$ for w, we see that the general solution is

$$\begin{bmatrix} x \\ y \end{bmatrix} = c_1 \begin{bmatrix} 1 \\ 0 \end{bmatrix} e^t + c_2 \left(\begin{bmatrix} 0 \\ -1/2 \end{bmatrix} + t \begin{bmatrix} 1 \\ 0 \end{bmatrix} \right) e^t \qquad (3)$$

Observe from Figure 7.4.2 that the generalized eigenvector w does not define a straight line of orbits. All orbits emerge from the origin tangent to the x-axis as t increases from $-\infty$ because for large negative values of t the dominant terms in (3) are $c_1 [1 \quad 0]^T e^t$

TABLE 7.4.1 Classification of Orbital Portraits for the Planar System $x' = Ax$. The vectors v^1, v^2 are eigenvectors of A corresponding to eigenvalues λ_1, λ_2; w is a generalized eigenvector based on v^1; $\lambda = 0$ is not an eigenvalue of A.

Classification	Fundamental Set
Improper node λ_1 and λ_2 real, distinct, and same sign	$\{v^1 e^{\lambda_1 t}, \; v^2 e^{\lambda_2 t}\}$
Deficient improper node $\lambda_1 = \lambda_2$; deficient eigenspace	$\{v^1 e^{\lambda_1 t}, \; (w + t v^1) e^{\lambda_1 t}\}$
Proper (or star) node $\lambda_1 = \lambda_2$; nondeficient eigenspace	$\{v^1 e^{\lambda_1 t}, \; v^2 e^{\lambda_1 t}\}$
Saddle $\lambda_1 < 0 < \lambda_2$	$\{v^1 e^{\lambda_1 t}, \; v^2 e^{\lambda_2 t}\}$
Center (or vortex) $\lambda_1 = i\beta = \bar{\lambda}_2$, $\beta \neq 0$, β real	$\{\mathrm{Re}[v^1 e^{i\beta t}], \; \mathrm{Im}[v^1 e^{i\beta t}]\}$
Focus (or spiral) $\lambda_1 = \alpha + i\beta = \bar{\lambda}_2$, α, β real, nonzero	$\{\mathrm{Re}[v^1 e^{(\alpha + i\beta)t}], \; \mathrm{Im}[v^1 e^{(\alpha + i\beta)t}]\}$

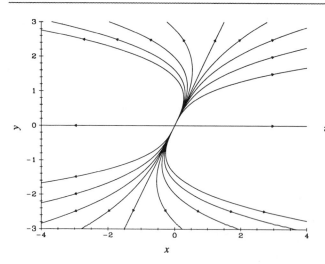

FIGURE 7.4.1 An improper node. Two eigenspace lines: $y = 0$ and $y = 2x$. See Example 7.4.1.

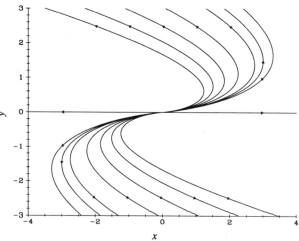

FIGURE 7.4.2 A deficient improper node. A single eigenspace line: $y = 0$. See Example 7.4.2.

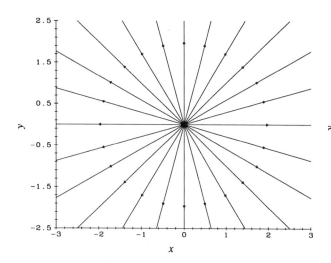

FIGURE 7.4.3 Proper (or star) node. Double eigenvalue, all lines are eigenspace lines. See Example 7.4.3.

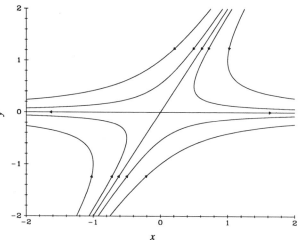

FIGURE 7.4.4 A saddle. Eigenvalues have opposite sign, two eigenspace lines: $y = 0$, $y = 2x$ See Example 7.4.4.

if $c_2 = 0$, and $c_2 t [1 \quad 0]^T e^t$ if $c_2 \neq 0$. If the eigenvalue is negative, the orbits look the same except that they approach the origin as $t \to \infty$, instead of moving away.

In this case the improper node is "deficient" because (unlike the system of Example 7.4.1) there is only one orbital tangent line at the origin.

EXAMPLE 7.4.3

Proper or Star Node
The system matrix $A = I$ of the system $x' = x$, $y' = y$, has the double eigenvalue $\lambda = 1$, and any independent pair of vectors are eigenvectors. Solutions have the form $v e^t$ for any vector v. If the eigenvalue is negative, then all solutions tend to the origin as $t \to \infty$.

Figure 7.4.3 makes the term "star node" obvious, while the node is said to be "proper" because every direction is a direction of tangency at the origin.

EXAMPLE 7.4.4

Saddle

The eigenvalues of the system matrix $\begin{bmatrix} 1 & -1 \\ 0 & -1 \end{bmatrix}$ of $x' = x - y$, $y' = -y$ are $\lambda_1 = 1$ and $\lambda_2 = -1$ with corresponding eigenvectors $[1\ 0]^T$ and $[1\ 2]^T$. The general solution is

$$\begin{bmatrix} x \\ y \end{bmatrix} = c_1 \begin{bmatrix} 1 \\ 0 \end{bmatrix} e^t + c_2 \begin{bmatrix} 1 \\ 2 \end{bmatrix} e^{-t}$$

where c_1, c_2 are arbitrary reals. The "eigenspace" orbits $c_1[1\ 0]^T e^t$ emerge from the origin on the x-axis as t increases from $-\infty$, while the other eigenspace orbits $c_2[1\ 2]^T e^{-t}$ approach the origin on the line $y = 2x$ as $t \to +\infty$. Figure 7.4.4 shows the eigenspace orbits as well as other orbits that resemble branches of hyperbolas which emerge from "infinity" asymptotically tangent to the eigenspace line $y = 2x$ as t increases from $-\infty$ and then become asymptotically tangent to the other eigenspace line, $y = 0$, as $t \to +\infty$.

Curves in Figure 7.4.4 may be interpreted as contour curves on a map of a mountainous region where there is a peak in the lower left and another in the upper right. The "saddle point" is the low point of the ridge joining two higher peaks.

EXAMPLE 7.4.5

Center, or Vortex

The eigenvalues of the matrix $A = \begin{bmatrix} 3 & -2 \\ 5 & -3 \end{bmatrix}$ of the system $x' = 3x - 2y$, $y' = 5x - 3y$ are $\lambda_1 = i$, $\lambda_2 = -i$ and corresponding eigenvectors are $[2\quad 3 - i]^T$ and $[2\quad 3 + i]^T$. As noted in Section 7.3, the general (real-valued) solution is given by linear combinations of the special real-valued solutions

$$\text{Re}\left[\begin{bmatrix} 2 \\ 3 - i \end{bmatrix} e^{it}\right] = \text{Re}\begin{bmatrix} 2\cos t + 2i\sin t \\ 3\cos t + \sin t + i(3\sin t - \cos t) \end{bmatrix} = \begin{bmatrix} 2\cos t \\ 3\cos t + \sin t \end{bmatrix}$$

and

$$\text{Im}\left(\begin{bmatrix} 2 \\ 3 - i \end{bmatrix} e^{it}\right) = \begin{bmatrix} 2\sin t \\ 3\sin t - \cos t \end{bmatrix}$$

The orbits are tilted ellipses (see Figure 7.4.5). Note that all solutions (other than $x = 0$) are periodic with period 2π. In this example the orientation on the orbits induced by increasing time is counterclockwise. This is not evident from the solution formula. One way to "see" this is to plot a direction field vector on a coordinate axis and look at the direction in which it points. If we replaced the two rate functions of the system by their negatives, the orbits would look exactly like those of Figure 7.4.5, except the orientation would be clockwise.

A "center" or a "vortex" is a point that is surrounded by whirling masses (e.g., the eye of a hurricane or a tornado), hence the suitability of the name in this context.

EXAMPLE 7.4.6

Focus, or Spiral

The matrix of the system $x' = x - 4y$, $y' = 5x - 3y$ has the eigenvalues $\lambda_1 = -1 + 4i$, and $\lambda_2 = -1 - 4i$. The eigenspace corresponding to λ_1 consists of all vectors $v = [v_1\ v_2]^T$ with $(2 - 4i)v_1 - 4v_2 = 0$. So if we put $v_1 = 2s$, for any complex number s, then $v_2 = (1 - 2i)s$, and we see that $[2\ 1 - 2i]^T$ is a basis for this eigenspace. Put

$$x^1 = \text{Re}\left[\begin{bmatrix} 2 \\ 1 - 2i \end{bmatrix} e^{(-1+4i)t}\right] = e^{-t}\begin{bmatrix} 2\cos 4t \\ \cos 4t + 2\sin 4t \end{bmatrix}$$

$$x^2 = \text{Im}\left[\begin{bmatrix} 2 \\ 1 - 2i \end{bmatrix} e^{(-1+4i)t}\right] = e^{-t}\begin{bmatrix} 2\sin 4t \\ \sin 4t - 2\cos 4t \end{bmatrix}$$

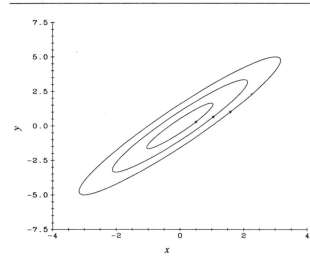

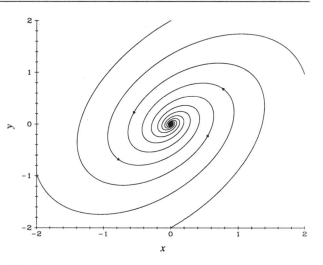

FIGURE 7.4.5 A center, or vortex: pure imaginary eigenvalues. See Example 7.4.5.

FIGURE 7.4.6 Focus or spiral: complex eigenvalues with nonzero real parts. See Example 7.4.6.

The Wronskian $W[x^1, x^2]$ evaluated at $t = 0$ is -4 and so x^1, x^2 is a fundamental set and the general real-valued solution is $x = c_1 x^1 + c_2 x^2$, for arbitrary reals c_1, c_2. It is hard to see from the solution formula, but the orbits spiral counterclockwise and inward as they curl around the equilibrium point at the origin (see Figure 7.4.6).

The word "spiral" is appropriate here, but "focus" is ambiguous. In fact, a focal point is a point where rays meet (as in the focal point of a lens), and so the term would be more appropriate for a proper node. No matter how unsuitable the term, systems with eigenvalues $\lambda = \alpha \pm i\beta$, $\alpha \neq 0$, and their equilibrium points at the origin, are commonly called foci. We use "focus" and "spiral" interchangeably.

The orbit of any planar linear system $x' = Ax$ where the eigenvalues of A are $\lambda = \alpha \pm i\beta, \alpha \neq 0$, $\beta \neq 0$, behave in essentially the same way as the orbits in Figure 7.4.6, but the spiraling may be clockwise or counterclockwise and either inward or outward. For example, if the rate functions in the system of Figure 7.4.6 are replaced by their negatives, then the orbits of the new system will be exactly the same, but spiraling clockwise outward. In case there is any doubt, one can check on the direction of spiraling by plotting the direction field at any point of the x-axis or y-axis (except the origin) and looking at the direction in which a field vector points. The sign of $\text{Re}[\lambda] = \alpha$ determines whether the spiral is inward or outward as t increases.

Each planar linear autonomous system whose system matrix has nonzero eigenvalues has an orbital portrait that resembles one of Figures 7.4.1–7.4.6. The arrowheads may be reversed, spirals may be tighter or looser, the angles between the lines of eigenspaces may change, but the general features will be the same.

Sensitivity

If entries in the system matrix of a linear system are changed, one expects to see changes in the solutions and the orbits. Changes in the orbital structure may be modest even though the changes in the matrix entries are large (thus showing insensitivity to parameter changes). Sometimes the orbits change in drastic ways, and in this case the structure is highly sensitive to a parameter change. The next example shows both cases.

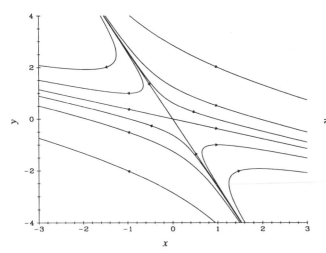

FIGURE 7.4.7 Saddle: $\alpha = -2$. See Example 7.4.7.

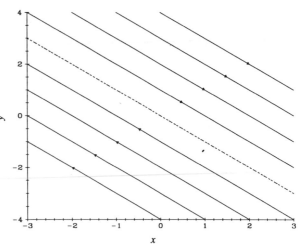

FIGURE 7.4.8 Parallel lines, line of equilibrium points: $\alpha = -1$. See Example 7.4.7.

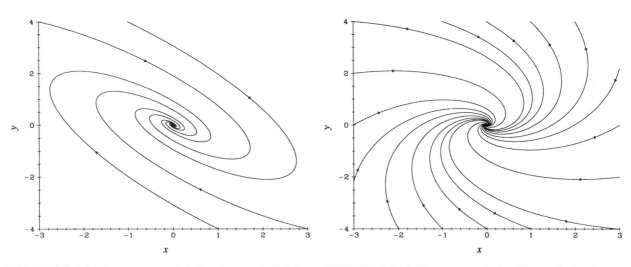

FIGURE 7.4.9 Focus: $\alpha = -0.5$. See Example 7.4.7.   FIGURE 7.4.10 Focus: $\alpha = 1$. See Example 7.4.7.

EXAMPLE 7.4.7

Sensitivity to a Parameter Change
The planar autonomous system

$$\begin{aligned} x' &= -\alpha x + y \\ y' &= -x - y \end{aligned} \tag{4}$$

has an adjustable real parameter α in the system matrix. The characteristic polynomial is $p(\lambda) = \lambda^2 + (1 + \alpha)\lambda + 1 + \alpha$ and the eigenvalues are

$$\lambda_{1,2} = -\frac{1}{2}(1 + \alpha) \pm \frac{1}{2}\sqrt{(1 + \alpha)^2 - 4 - 4\alpha} \tag{5}$$

Figures 7.4.7– 7.4.12 show orbital portraits of system (4) for six values of α: $-2, -1, -0.5, 1, 3, 5$. If $\alpha = -2$, the system has a saddle point at the origin (Figure 7.4.7). If $\alpha = -1$, $y = -x$ is a line of equilibrium points, and all other orbits are parallel to this line (Figure 7.4.8). If $\alpha = -0.5$ or 1, then there is a focus at the origin (Figures 7.4.9, 7.4.10). There is a deficient improper node with a single eigenspace line if $\alpha = 3$ (Figure 7.4.11) and an ordinary improper node with two eigenspace lines when

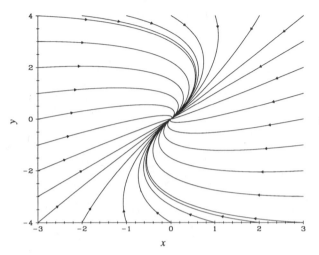

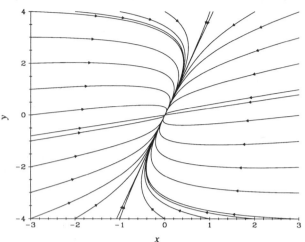

FIGURE 7.4.11 Deficient improper node with single eigenspace line: $\alpha = 3$. See Example 7.4.7.

FIGURE 7.4.12 Improper node: $\alpha = 5$. See Example 7.4.7.

$\alpha = 5$ (Figure 7.4.12). Thus, as α changes from -2 to $+5$ the orbital structure changes quite drastically. However, there is always a saddle point if $\alpha < -1$, and always an ordinary improper node if $\alpha > 3$. (See also Problem 1.)

Orbits in Higher Dimensions

If the number of state variables in a linear autonomous system is n, then the orbits "live" in an n-dimensional state space. If $n \geq 3$, the problem of "seeing" the orbits is far from simple. The following example illustrates one way to see how solutions, component curves, and orbits behave in higher dimensions.

EXAMPLE 7.4.8

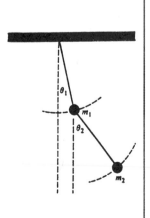

The Equations of a Double Pendulum

Suppose that one pendulum is suspended from another (see the margin sketch). It can be shown (after appropriate rescaling of time and the state variables to make them dimensionless) that the following linear system models the small-amplitude oscillations of the two pendulums:

$$
\begin{aligned}
x_1' &= x_2 \\
x_2' &= -x_1 + \alpha x_3 \\
x_3' &= x_4 \\
x_4' &= x_1 - x_3
\end{aligned}
\tag{6}
$$

where $\alpha = (m_2/m_1)(1 + m_2/m_1)^{-1}$, x_1 and x_3 are the scaled angles θ_1 and θ_2, and x_2 and x_4 are the scaled angular velocities, θ_1' and θ_2'. Observe that $0 < \alpha < 1$. (See also Example 8.3.4.) The system matrix A and characteristic polynomial $p(\lambda)$ are

$$
A = \begin{bmatrix} 0 & 1 & 0 & 0 \\ -1 & 0 & \alpha & 0 \\ 0 & 0 & 0 & 1 \\ 1 & 0 & -1 & 0 \end{bmatrix}, \quad p(\lambda) = \lambda^4 + 2\lambda^2 + 1 - \alpha
$$

The squares of the eigenvalues are given by $\lambda^2 = -1 \pm \sqrt{\alpha}$. Since $0 < \alpha < 1$, the four eigenvalues are each pure imaginary:

$$\lambda_1 = i\sqrt{1 + \sqrt{\alpha}} = \bar{\lambda}_2 = \beta i, \quad \text{where } \beta = \sqrt{1 + \sqrt{\alpha}}$$

$$\lambda_3 = i\sqrt{1 - \sqrt{\alpha}} = \bar{\lambda}_4 = \gamma i, \quad \text{where } \gamma = \sqrt{1 - \sqrt{\alpha}}$$

Corresponding eigenvectors, v^1, $v^2 = \bar{v}^1$, v^3, and $v^4 = \bar{v}^3$, have complex components, but we can follow the technique of Section 7.3 to generate the real solution space by taking the real and the imaginary parts of $e^{\lambda_1 t} v^1$ and $e^{\lambda_3 t} v^3$. After carrying out the lengthy calculations, we find that the general real solution of system (6) is given by

$$\begin{bmatrix} x_1 \\ x_2 \\ x_3 \\ x_4 \end{bmatrix} = c_1 \begin{bmatrix} (1 + \lambda_1^2) \cos \beta t \\ -\beta(1 + \lambda_1^2) \sin \beta t \\ \cos \beta t \\ -\beta \sin \beta t \end{bmatrix} + c_2 \begin{bmatrix} (1 + \lambda_1^2) \sin \beta t \\ \beta(1 + \lambda_1^2) \cos \beta t \\ \sin \beta t \\ \beta \cos \beta t \end{bmatrix}$$

$$+ c_3 \begin{bmatrix} (1 + \lambda_3^2) \cos \gamma t \\ -\gamma(1 + \lambda_3^2) \sin \gamma t \\ \cos \gamma t \\ -\gamma \sin \gamma t \end{bmatrix} + c_4 \begin{bmatrix} (1 + \lambda_3^2) \sin \gamma t \\ \gamma(1 + \lambda_3^2) \cos \gamma t \\ \sin \gamma t \\ \gamma \cos \gamma t \end{bmatrix} \tag{7}$$

where c_1, c_2, c_3, and c_4 are arbitrary real constants. Of special importance are the *normal mode* vectors in the $x_1 x_3$-plane. These are the vectors $N^1 = [1 + \lambda_1^2 \quad 1]^T = [-\sqrt{\alpha} \quad 1]^T$ and $N^2 = [1 + \lambda_3^2 \quad 1] = [\sqrt{\alpha} \quad 1]^T$. If the initial angular velocities are zero and if the initial angle vector $[x_1(0) \quad x_3(0)]^T$ lies along the line of either N^1 or N^2, then the double pendulum will oscillate periodically along that direction. For example, if $[x_1(0) \quad x_3(0)]^T = N^1$, and if the initial angular velocities are zero, then the solution vector is given by formula (7) with $c_1 = 1$, $c_2 = c_3 = c_4 = 0$.

In Figures 7.4.13 and 7.4.14, $\alpha = 0.3$, and normal mode oscillations are given by (7) with $c_1 = -0.15$ and $c_2 = c_3 = c_4 = 0$ (long-dashed curves), $c_3 = 0.15$ and $c_1 = c_2 = c_4 = 0$ (short-dashed curves). The solid curves correspond to the sum of the two normal mode oscillations: $c_1 = -0.15 = -c_3$, $c_2 = c_4 = 0$. Since the dashed x_1 and x_3 component curves oscillate with respective periods $2\pi/\beta$ and $2\pi/\gamma$, and since β/γ is not rational, the solid curves are not periodic. Figure 7.4.15 shows the projection of the three corresponding orbits from the $x_1 x_2 x_3 x_4$-state space onto the three-dimensional $x_1 x_3 x_2$-space (x_2-axis vertical). The two "normal mode" orbital projections appear to be elliptical but the third orbit seems to lie in a torus. Figure 7.4.16 shows three orbits, but projected onto the $x_1 x_3$-plane of the angular variables x_1 and x_3. The "normal mode" oscillations project onto straight line orbital segments, while the third orbit projects onto the oscillating curve lying in the rectangular region generated by the normal mode segments. The curved orbit in Figure 7.4.16 is called a Lissajous curve; these curves often appear in orbital analysis of coupled oscillators.

PROBLEMS

1. The system for the following problems is $x' = -\alpha x + y$, $y' = -x - y$ (see Example 7.4.7).

(a) Explain why the equilibrium point at the origin is a saddle if $\alpha < -1$.

(b) Explain why the origin is a focus if $-1 < \alpha < 3$.

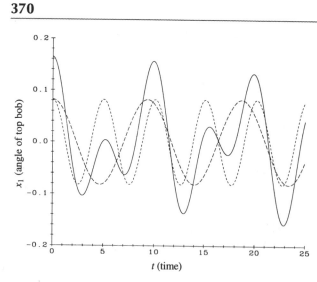

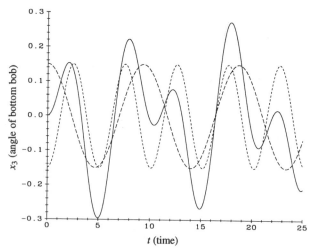

FIGURE 7.4.13 Angle of top pendulum varies periodically in normal modes (dashed), aperiodically otherwise (solid curve). See Example 7.4.8.

FIGURE 7.4.14 Angle of lower pendulum varies periodically in normal modes (dashed), aperiodically otherwise (solid curve). See Example 7.4.8.

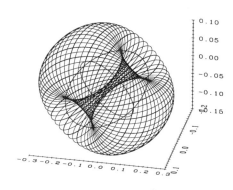

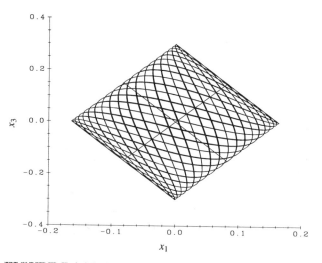

FIGURE 7.4.15 Three orbital projections into $x_1 x_3 x_2$-space. See Example 7.4.8.

FIGURE 7.4.16 Oscillations along normal modes (the crossed lines) and a Lissajous figure. See Example 7.4.8.

(c) Why is the origin an improper node if $\alpha > 3$? Explain why the system matrix has a pair of independent real eigenvectors in this case.

(d) Plot orbital portraits in the frame $|x| \leq 3, |y| \leq 4$ for the three cases $\alpha = -4, 0, 4$.

2. Find the characteristic polynomial of the system matrix, the eigenvalues, a set of two independent eigenvectors (or generalized eigenvectors, if necessary), and the general real-valued solution of each system. Then classify each as a saddle, center, spiral, or one of the three types of nodes.

(a) $x' = x + y \quad y' = 4x - 2y$ **(b)** $x' = 6x - 8y, \quad y' = 2x - 2y$

(c) $x' = 7x + 6y, \quad y' = 2x + 6y$ **(d)** $x' = 4y, \quad y' = -x$

(e) $x' = -x - 4y, \quad y' = x - y$ **(f)** $x' = -2x, \quad y' = -2y$

3. **(a)–(f)** Plot orbital portraits in the frame $|x| \le 3, |y| \le 4$ for each system in Problem 2.

4. *(Sensitivity to Changes in System Parameters)*. Describe how the orbital portrait of the system, $x' = -x + \alpha y, \quad y' = -x - y$ changes as the system parameter α ranges from $-\infty$ to ∞. Find the "critical values" of α where the portrait suddenly changes its nature. Plot the orbital portraits in the frame $|x| \le 3, |y| \le 4$ for $\alpha = -2, -1, 0, 1$ and identify the type of each portrait (e.g., a focus if $\alpha = 1$). [*Hint*: Read Example 7.4.7 for the general approach to a problem like this.]

5. *(Planar Affine Systems)*. The system $x' = ax + by + r, \quad y' = cx + dy + s$, where a, b, r, c, d, s are real constants is said to be an *affine system*. Suppose that

$$\det \begin{bmatrix} a & b \\ c & d \end{bmatrix} \ne 0$$

(a) Show that the affine system has a unique equilibrium point.

(b) If the equilibrium point is denoted by (x_0, y_0), show that the change of variables from x and y to u and v given by $x = u + x_0, y = v + y_0$ yields the homogeneous system, $u' = au + bv, v' = cu + dv$.

(c) Show that if $u = u(t), v = v(t)$ is the general solution of the u', v' system in part **(b)**, then the general solution of the affine system is $x(t) = u(t) + x_0$, $y(t) = v(t) + y_0$.

(d) Find the general solution of the system $x' = x + y + 1, y' = 4x - 2y - 1$. What is the nature of the system (i.e., is it nodal, focal, ...)? Plot an orbital portrait in the frame $|x - x_0| \le 3, |y - y_0| \le 4$, where (x_0, y_0) is the equilibrium point; be sure to plot any "eigenspace lines" through the equilibrium point.

6. *(Second-Order ODE)*. The scalar ODE $y'' + 2ay' + by = 0$, where a and b are real constants, is equivalent to the system $x_1' = x_2, x_2' = -bx_1 - 2ax_2$.

(a) Explain why the system has a saddle point if $a = 1, b < 0$, while if $a = 1$, $0 < b < 1$, the system has an improper node.

(b) Explain why if $a = 1, b = 1$, the system has a deficient improper node, while if $a = 1, b > 1$, the system has a spiral point (focus).

(c) Find all values for a and b for which the system has a center.

(d) Show that no values of a and b lead to a system which has a proper node.

7. *(An Inverse Problem)*. A homogeneous 2×2 linear system with constant coefficients has the form $x' = Ax$, where A is a 2×2 constant matrix and $x(t)$ is a column vector with two components $x_1(t)$ and $x_2(t)$. It is known that $x_1 = 2 \sin(2t - \pi), x_2 = \cos(2t - \pi)$ is a solution of such a system.

(a) Find the general, real-valued solution of the system.

(b) Plot some representative orbits of the system near the origin.

8. *(Double Pendulum Model)*. System (6) has a parameter α where $0 < \alpha < 1$. For each of the following values of α find the normal mode vectors, N^1 and N^2, in the $x_1 x_3$-plane. Plot x_1 and x_3 component graphs for a normal mode oscillation in the direction of N^1, and also the x_1 and x_3 component graphs in the direction of N^2. Then plot $x_1 x_3 x_2$ and $x_1 x_3$ orbital projections like those shown in Figures 7.4.15 and 7.4.16. For the normal mode oscillations use initial data $(\pm\sqrt{\alpha}, 0, 1, 0)$. For a third orbit use initial data $(2\sqrt{\alpha}, 0, 0, 0)$.

(a) $\alpha = 0.05$ **(b)** $\alpha = 0.5$ **(c)** $\alpha = 0.95$

9. (*Zero Eigenvalues*). Except for system (4) with $\alpha = -1$, we have avoided considering the case of a planar autonomous system where the real system matrix A has at least one zero eigenvalue. What happens if A has a zero eigenvalue? Explain why either there is one straight line consisting solely of equilibrium points or else all points are equilibrium points if at least one eigenvalue is zero. Plot orbital portraits of the various distinct cases where at least one of the eigenvalues is zero, and explain why your "gallery" of portraits represents all the possible cases.

10. (*Orbital Portraits in Three Dimensions*). Consider the system $x' = Ax$, where $x = [x_1 \ x_2 \ x_3]^T$ and A is a 3×3 matrix of real constants with nonzero eigenvalues. Construct a gallery of representative orbital portraits in $x_1 x_2 x_3$ state space that covers all of the cases. For example, if λ_1, λ_2, and λ_3 denote the eigenvalues of A, you will need to consider the case where the three eigenvalues are real, distinct, and have a common sign, and plot orbits in 3D of a system such as $x_1' = -x_1$, $x_2' = -2x_2$, $x_3' = -3x_3$. Another one of the several cases you must consider is $\lambda_1 = \alpha + i\beta$, $\lambda_2 = \alpha - i\beta$, α, β, and λ_3 real and nonzero; the representative system in this case might be $x_1' = -x_1 + 5x_2$, $x_2' = -5x_1 - x_2$, $x_3' = -x_3$. Since any system with eigenvalues λ_1, λ_2, and λ_3 behaves as $t \to \infty$ exactly the same as the system with eigenvalues $-\lambda_1, -\lambda_2$, and $-\lambda_3$ as $t \to -\infty$, their orbital portraits are identical, so do not treat them as distinct cases. For example, consider the case with three distinct positive real eigenvalues and the case with three distinct negative real eigenvalues as the same. Your gallery should have more than ten distinct orbital portraits. [*Hint*: To construct your example system, start with the six planar cases of Examples 7.4.1–7.4.6 and add a third ODE such as $x_3' = \alpha x_3$.]

7.5   Fundamental Matrices: The Matrix Exponential

The general real linear system has the form

$$x' = A(t)x + F(t) \tag{1}$$

where the real-valued entries $a_{ij}(t)$ of the system matrix $A(t)$ and the components $F_i(t)$ of the input vector $F(t)$ are all continuous on a common interval I. Although many of the calculations of the earlier sections are no longer valid if the system matrix is not constant, the main theoretical ideas can be extended to system (1).

Define the linear operator $L = D - A(t)$, whose action on a vector function $x(t)$ in $C^1(I)$ produces $L[x] = x' - A(t)x$, a vector function in $C^0(I)$. Then (1) becomes $Lx = F$. The solution set of (1) consists of all vector functions

$$x(t) = v(t) + x^p(t) \tag{2}$$

where $v(t)$ is any solution of the homogeneous ODE $L[x] = 0$, and $x^p(t)$ is any one particular solution of system (1). Thus, all solutions of system (1) are obtained by finding the general solution $v(t)$ of $x' = A(t)x$ (the undriven system) and just one solution $x^p(t)$ of the driven system (1).

Solving $x' = A(t)x$: The Transition Matrix

Again, let the entries $a_{ij}(t)$ of the system matrix A be continuous on an interval I. We focus first on the undriven system with general initial data,

$$x' = A(t)x, \qquad x(t_0) = x^0, \qquad t_0 \text{ in } I \text{ and } x^0 \text{ in } \mathbf{R}^n \qquad (3)$$

The conditions of the Existence and Uniqueness Theorem (Theorem 7.3.1) are satisfied, and IVP (3) has a unique solution defined for all t in the interval I for any choice of t_0 in I and x^0 in $\mathbf{R}^n$. In particular, the IVP

$$x' = A(t)x, \qquad x(t_0) = \begin{bmatrix} 0 \\ \vdots \\ 1 \\ \vdots \\ 0 \end{bmatrix}, \qquad 1 \text{ in the } j\text{th component of } x(t_0), \qquad t_0 \text{ in } I \quad (4)$$

has a unique solution $x = u^j(t)$ for each $j = 1, \ldots, n$.

THEOREM 7.5.1

> **Spanning the Solution Set.** The set of functions $\{u^1(t), \ldots, u^n(t)\}$, where $u^j(t)$ solves IVP (4), $j = 1, \ldots, n$, is a fundamental set for $x' = A(t)x$. For arbitrary data x^0 in $\mathbf{R}^n$, the solution of the IVP
>
> $$x' = A(t)x, \quad x(t_0) = x^0,$$
>
> is given by the solution formula
>
> $$x = v(t) = x_1^0 u^1(t) + \cdots + x_n^0 u^n(t) \qquad (5)$$

Proof. The determinant of the matrix $\{u^1(t_0) \cdots u^n(t_0)\}$ is $\det I = 1$ because $u^j(t_0) = [0 \cdots 1 \cdots 0]^T$, $j = 1, \ldots, n$. Hence, as noted in Section 7.3, the Wronskian of the set of solutions $\{u^1(t), \ldots, u^n(t)\}$ is nonzero, so that the set is fundamental. Thus $v(t)$ in (5) is a solution of $x' = A(t)x$. Since $v(t_0) = x_1^0 u^1(t_0) + \cdots + x_n^0 u^n(t_0) = x^0$, $v(t)$ also satisfies the given initial data. So by the Uniqueness Property, $x = v(t)$ is the unique solution to the IVP.

There is nothing special about the functions $u^1(t), \ldots, u^n(t)$ used to form a fundamental solution set for $x' = A(t)x$. Any set of n solutions of $x' = A(t)x$ will do, as long as the Wronskian of the set is nonzero at some point t_0 of I (i.e., the set is fundamental). The examples of Section 7.3 give techniques for constructing fundamental sets when the system matrix A is constant.

Solutions of $x' = A(t)x$ may be entered as columns in a matrix. In particular, a *solution matrix* is an $n \times n$ matrix $X(t)$ whose columns $x^j(t)$, $j = 1, \ldots, n$, are solutions of $x' = A(t)x$. Thus a solution matrix $X(t)$ satisfies the linear *matrix differential equation*

$$X' = A(t)X \qquad (6)$$

since $(x^j)' = A(t)x^j$ for each column x^j of X. There is a corresponding matrix IVP

$$X' = A(t)X, \qquad X(t_0) = B \qquad (7)$$

where B is an $n \times n$ matrix of constants and t_0 is in I. By the Existence and Uniqueness Theorem (Theorem 7.3.1), IVP (7) has a unique solution matrix $X = X(t)$.

EXAMPLE 7.5.1

A Solution Matrix

The constant system matrix for the system

$$x' = 5x + 3y, \quad y' = -6x - 4y$$

has eigenvalues -1 and 2 and respective eigenvectors $[1 \ -2]^T$ and $[1 \ -1]^T$. The pair of solutions $e^{-t}[1 \ -2]^T$ and $e^{2t}[1 \ -1]^T$ form a fundamental set for the system. The following are all solution matrices:

$$\begin{bmatrix} e^{-t} & e^{2t} \\ -2e^{-t} & -e^{2t} \end{bmatrix}, \quad \begin{bmatrix} e^{-t} & 2e^{-t} \\ -2e^{-t} & -4e^{-t} \end{bmatrix}, \quad \begin{bmatrix} e^{2t} & e^{-t}+e^{2t} \\ -e^{2t} & -2e^{-t}-e^{2t} \end{bmatrix} \quad (8)$$

If the columns of a solution matrix $X(t)$ form a fundamental set of solutions of $x' = A(t)x$, then $X(t)$ is a *fundamental (solution) matrix*. Each fundamental solution matrix $X(t)$ is nonsingular for all t in I since $\det X(t)$ is the Wronskian, which cannot vanish because the columns of $X(t)$ form a fundamental set of solutions of $x' = A(t)x$. Observe that the first and third matrices in (8) are fundamental because their determinants (which are the Wronskians of their columns) are nonzero at $t = 0$.

If C is any $n \times n$ nonsingular matrix of constants and $X(t)$ is a fundamental solution matrix, then $X(t)C$ is also a fundamental solution matrix since $X(t)C$ is nonsingular and its columns are linear combinations of the columns of $X(t)$. One type of fundamental matrix is basic to our needs.

❖ **Transition Matrix.** The *transition matrix* $\Phi(t, t_0)$, t, t_0 in I, for the system $x' = A(t)x$ is the unique fundamental matrix which solves the matrix IVP

$$X' = A(t)X, \quad X(t_0) = I \quad (9)$$

Note that the columns u^j of Φ are the solutions of the j distinct IVPs specified by (4). We can derive Φ from any fundamental solution matrix $X(t)$ as follows: Since the matrix

$$\Phi = X(t)X^{-1}(t_0), \quad t, t_0 \text{ in } I \quad (10)$$

solves the matrix IVP (9), it must be the transition matrix $\Phi(t, t_0)$ by uniqueness.

EXAMPLE 7.5.2

A Transition Matrix

The transition matrix for the system of Example 7.5.1 can be constructed by using formula (10) and the first matrix in (8). We have that the transition matrix is

$$\begin{aligned} \Phi(t, t_0) &= \begin{bmatrix} e^{-t} & e^{2t} \\ -2e^{-t} & -e^{2t} \end{bmatrix} \begin{bmatrix} e^{-t_0} & e^{2t_0} \\ -2e^{-t_0} & -e^{2t_0} \end{bmatrix}^{-1} \\ &= \begin{bmatrix} e^{-t} & e^{2t} \\ -2e^{-t} & -e^{2t} \end{bmatrix} \cdot \frac{1}{e^{t_0}} \cdot \begin{bmatrix} -e^{2t_0} & -e^{2t_0} \\ 2e^{-t_0} & e^{-t_0} \end{bmatrix} \\ &= \begin{bmatrix} -e^{-t+t_0}+2e^{2t-2t_0} & -e^{-t+t_0}+e^{2t-2t_0} \\ 2e^{-t+t_0}-2e^{2t-2t_0} & 2e^{-t+t_0}-e^{2t-2t_0} \end{bmatrix} \end{aligned} \quad (11)$$

where t and t_0 are any real numbers. Observe that $\Phi(t_0, t_0) = I$.

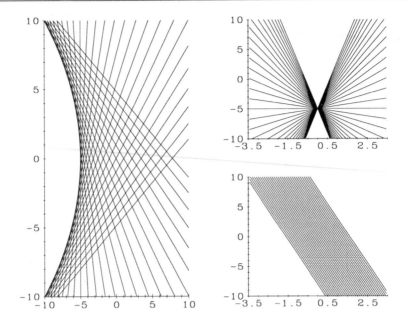

FIGURE 7.5.1 Orbits (on the left) and x and y component curves (upper and lower right, respectively) of linear system (12), with $x(0) = -5$ and $y(0) = -8, -7.5, \ldots, 7.5, 8$.

EXAMPLE 7.5.3

Transition Matrix If $A(t)$ Is Nonconstant

The nonautonomous, undriven system (defined for $|t| < 1$)

$$
\begin{aligned}
x' &= \frac{1}{1-t^2}(-tx + y) \\
y' &= \frac{1}{1-t^2}(x - ty)
\end{aligned}
\tag{12}
$$

has a solution, $x = 1$, $y = t$, and a second, independent solution, $x = t$, $y = 1$, as a direct calculation shows. The two solutions have a nonvanishing Wronskian for $|t| < 1$ and thus form a fundamental set. According to formula (10) the transition matrix for system (12) is

$$
\Phi(t, t_0) = \begin{bmatrix} 1 & t \\ t & 1 \end{bmatrix} \begin{bmatrix} 1 & t_0 \\ t_0 & 1 \end{bmatrix}^{-1} = \frac{1}{1 - t_0^2} \begin{bmatrix} 1 - tt_0 & t - t_0 \\ t - t_0 & 1 - tt_0 \end{bmatrix}
\tag{13}
$$

where t and t_0 lie in the interval $(-1, 1)$. Observe that $\Phi(t_0, t_0) = I$. See Figure 7.5.1 for graphs of component curves and orbits of (12) where $x(0) = -5$ and $y(0) = -8, -7.5, -7, \ldots, 7.5, 8$. Observe that the solver (and the solutions) do not "see" the singularities at $t = \pm 1$.

The transition matrix has a number of important properties.

THEOREM 7.5.2

Properties of the Transition Matrix. Let $\Phi(t, t_0)$ be the transition matrix for $x' = A(t)x$ where t, t_0 are any points in the interval I. Then,

- (*Identity*): $\Phi(t_0, t_0) = I$, the identity matrix.
- (*Derivative*): $\Phi'(t, t_0) = A\Phi(t, t_0)$.
- (*Fundamental Matrices*): $\Phi(t, t_0)B$ is a fundamental matrix for every $n \times n$ nonsingular matrix B.
- (*Transition from x^0 to $x(t)$*): The unique solution of the IVP

$$x' = A(t)x, \qquad x(t_0) = x^0 \tag{14}$$

is given by

$$x(t) = \Phi(t, t_0)x^0 \tag{15}$$

- (*Product*): $\Phi(t, t_0) = \Phi(t, t_1)\Phi(t_1, t_0)$, for all t, t_0, t_1 in I.
- (*Inverse*): $\Phi^{-1}(t, s) = \Phi(s, t)$, for all s, t in I.

Proof. The first four properties follow immediately from the definition of Φ. The remaining properties are proven below:

- (*Product*): $X_1 = \Phi(t, t_0)$ and $X_2 = \Phi(t, t_1)\Phi(t_1, t_0)$ solve the same matrix IVP,

$$X' = A(t)X, \quad X(t_1) = \Phi(t_1, t_0)$$

By uniqueness, $X_1 = X_2$.

- (*Inverse*): Use the Product and Identity properties to show that $\Phi(t, s)\Phi(s, t)$ is the identity matrix for all s, t in the interval I.

We have now characterized the solution set of $x' = Ax$ in terms of the transition matrix by formula (15). Because of its importance, we rephrase the result as a theorem.

THEOREM 7.5.3

General Solution of $x' = A(t)x$. The general solution of $x' = A(t)x$ is

$$x = \Phi(t, t_0)c \tag{16}$$

where $\Phi(t, t_0)$ is the transition matrix, t belongs to I, t_0 is any fixed number in I, and c is an arbitrary constant vector.

EXAMPLE 7.5.4

Solving $x' = Ax$

Using formula (11) for the transition matrix of the planar system

$$x' = 5x + 3y, \quad y' = -6x - 4y$$

we see that the general (real-valued) solution of that system is

$$\begin{bmatrix} x(t) \\ y(t) \end{bmatrix} = \Phi(t, t_0)c = \begin{bmatrix} -e^{-(t-t_0)} + 2e^{2(t-t_0)} & -e^{-(t-t_0)} + e^{2(t-t_0)} \\ 2e^{-(t-t_0)} - 2e^{2(t-t_0)} & 2e^{-(t-t_0)} - e^{2(t-t_0)} \end{bmatrix} \begin{bmatrix} c_1 \\ c_2 \end{bmatrix} \tag{17}$$

where c_1 and c_2 are arbitrary real constants, and t_0 is any fixed number. Since the system is autonomous, it is customary to set $t_0 = 0$. If $x(0) = a$ and $y(0) = b$, then the above formula solves the corresponding IVP with $c_1 = a$, $c_2 = b$.

Note that the unique solution of the IVP,

$$x' = A(t)x, \qquad x(t_0) = x^0$$

is $x(t) = \Phi(t, t_0)x^0$. Finding the transition matrix $\Phi(t, t_0)$ explicitly is another story. In the important case where the system matrix A has constant entries, the methods of Section 7.3 suffice for finding $\Phi(t, t_0)$.

Variation of Parameters

According to our plan for solving the driven IVP $x' = Ax + F$, $x(t_0) = x^0$, the next step is to find a particular solution $x^P(t)$ of $x' = A(t)x + F(t)$. Any solution will do, and the transition matrix constructed above can be used to carry out the construction.

THEOREM 7.5.4

> Variation of Parameters for Driven Systems. Let $\Phi(t, t_0)$ be the transition matrix for $x' = A(t)x$, for t, t_0 in the interval I, where the entries of $A(t)$ are continuous on I. Then for any vector function $F(t)$ whose components are continuous on I, the vector function
>
> $$x^P(t) = \Phi(t, t_0) \int_{t_0}^{t} \Phi(t_0, s) F(s)\, ds = \int_{t_0}^{t} \Phi(t, s) F(s)\, ds \qquad (18)$$
>
> is the unique solution of the IVP
>
> $$x' = A(t)x + F(t), \qquad x(t_0) = 0 \qquad (19)$$
>
> [The integrals are performed on vector functions component by component.]

Proof. Note that the second equality in formula (18) is a consequence of the product property of transition matrices: $\Phi(t, t_0)\Phi(t_0, s) = \Phi(t, s)$. To verify that formula (18) defines a solution, we proceed as follows

$$(x^P)' = \frac{d}{dt} \left\{ \Phi(t, t_0) \int_{t_0}^{t} \Phi(t_0, s) F(s)\, ds \right\}$$

$$= \left\{ \Phi'(t, t_0) \int_{t_0}^{t} \Phi(t_0, s) F(s)\, ds \right\} + \Phi(t, t_0)\Phi(t_0, t) F(t)$$

(by the product rule and the Fundamental Theorem of Calculus) and so

$$(x^P)' = \left\{ A(t)\Phi(t, t_0) \int_{t_0}^{t} \Phi(t_0, s) F(s)\, ds \right\} + \Phi(t, t) F(t)$$

$$= A(t)x^P(t) + F(t)$$

where we have used several properties of the transition matrix.

When using the formula (18), note that the system must be written in the normalized form (19). The origin of the title "Variation of Parameters" is outlined in Problem 9. We can now solve the central problem for systems of linear ODEs.

THEOREM 7.5.5

Solution of a Linear System. Let $\Phi(t, t_0)$, t, t_0 in I, be the transition matrix for $x' = A(t)x$. Then the general solution of $x' = A(t)x + F(t)$, t in I, is

$$x(t) = \Phi(t, t_0)c + \Phi(t, t_0) \int_{t_0}^{t} \Phi(t_0, s) F(s) \, ds \tag{20}$$

where c is any constant vector and t_0 is any point in I. The solution of the initial value problem $x' = A(t)x + F(t)$, $x(t_0) = x^0$, t_0 in I, is

$$
\begin{array}{ccccc}
x(t) & = & \Phi(t, t_0)x^0 & + & \Phi(t, t_0) \displaystyle\int_{t_0}^{t} \Phi(t_0, s) F(s) \, ds \\[2mm]
\text{Total} & = & \text{Response to} & + & \text{Response to driving} \\
\text{response} & & \text{initial data} & & \text{force}
\end{array}
\tag{21}
$$

Observe that in formulas (20) and (21) the matrix factor $\Phi(t, t_0)$ can be brought inside the integral and combined with $\Phi(t_0, s)$ to obtain $\Phi(t, s)$. Formulas (20) and (21) do everything we want them to do, although they require the prior construction of the transition matrix (not always an easy task) and then integration of the components of the vector ΦF (also not always easy). Nevertheless, solution formulas like (20) and (21) for whole classes of systems of ODEs are rare, and whenever they exist they provide the foundation for much theoretical and practical work. Finally, note that one can replace the term $\Phi(t, t_0)c$ in formula (20) by $X(t)c$, where $X(t)$ is any fundamental solution matrix for $x' = A(t)x$.

EXAMPLE 7.5.5

Total Response

Formula (21) represents the total response of a driven linear system with given initial data as the sum of the responses to the initial data and to the driving force (see also Sections 1.2, 2.3). Figure 7.5.2 illustrates this principle for the IVP

$$
\begin{aligned}
x' &= -x + 10y + 5\cos t, & x(0) &= 0 \\
y' &= -10x - y + 5\cos \pi t, & y(0) &= 0.5
\end{aligned}
\tag{22}
$$

The short-dashed curve in Figure 7.5.2 is the y-component of the response of the undriven system to the initial data, the long-dashed curve is the y-component of the response to the driving cosine terms (the initial data here are $x(0) = y(0) = 0$), and the solid curve is the y-component of the sum of the two responses.

EXAMPLE 7.5.6

Solving $x' = Ax + F(t)$

Using formula (21) for the IVP, $x' = 5x + 3y + F_1(t)$, $y' = -6x - 4y + F_2(t)$, $x(0) = a$, $y(0) = b$, where the transition matrix $\Phi(t, t_0)$ for this system is given by (17), we have that [using the identity $\Phi(t, 0)\Phi(0, s) = \Phi(t, s)$]

$$
\begin{bmatrix} x(t) \\ y(t) \end{bmatrix} = \Phi(t, 0) \begin{bmatrix} a \\ b \end{bmatrix} + \int_{0}^{t} \Phi(t, s) \begin{bmatrix} F_1(s) \\ F_2(s) \end{bmatrix} ds
\tag{23}
$$

The matrices $\Phi(t, 0)$ and $\Phi(t, s)$ are obtained from the matrix $\Phi(t, t_0)$ in formula (17) by replacing t_0 by 0 and by s, respectively.

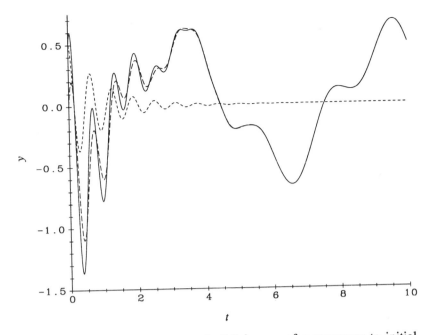

FIGURE 7.5.2 Total y-response (solid) is sum of y-responses to initial data (short dashes) and to driving term (long dashes). See Example 7.5.5.

EXAMPLE 7.5.7

Solving $x' = A(t)x + F(t)$

Referring to Example 7.5.3, we see from (21) that the solution of the IVP

$$x' = \frac{1}{1 - t^2}(-tx + y) + F_1(t), \quad x(0) = a$$

$$y' = \frac{1}{1 - t^2}(x - ty) + F_2(t) \qquad y(0) = b$$

for $|t| < 1$ is

$$\begin{bmatrix} x(t) \\ y(t) \end{bmatrix} = \Phi(t, 0) \begin{bmatrix} a \\ b \end{bmatrix} + \int_0^t \Phi(t, s) \begin{bmatrix} F_1(s) \\ F_2(s) \end{bmatrix} ds \tag{24}$$

where by formula (13) (with t_0 replaced first by 0 and then by s)

$$\Phi(t, 0) = \begin{bmatrix} 1 & t \\ t & 1 \end{bmatrix}, \quad \Phi(t, s) = \frac{1}{1 - s^2} \begin{bmatrix} 1 - ts & t - s \\ t - s & 1 - ts \end{bmatrix}$$

The Matrix Exponential, e^{tA}

The transition matrix has particularly simple properties if the system matrix A is a matrix of constants. These properties are so much like those of a scalar exponential function that we use the same notation.

❖ **Matrix Exponential** e^{tA}. Let A be an $n \times n$ matrix of constants. The transition matrix for $x' = Ax$ with $t_0 = 0$ is called the *matrix exponential* and is denoted by e^{tA}, and so $e^{tA} = \Phi(t, 0)$.

The following result justifies this suggestive notation.

THEOREM 7.5.6

Properties of e^{tA}. Let A be an $n \times n$ matrix of real constants. Let $e^{tA} = \Phi(t, 0)$. Then

- *(Derivative)*: $(e^{tA})' = Ae^{tA} = e^{tA}A$, all t.
- *(Transition)*: $e^{(t-t_0)A}$ is the transition matrix $\Phi(t, t_0)$ for $x' = Ax$.
- *(Product)*: $e^{tA}e^{sA} = e^{(t+s)A}$, all t and s.
- *(Inverse)*: $e^{tA}e^{-tA} = I$, so $(e^{tA})^{-1} = e^{-tA}$, all t.
- *(Series)*: $e^{tA} = I + tA + \frac{t^2}{2!}A^2 + \cdots + \frac{t^k}{k!}A^k + \cdots$, all t.
- *(Real)*: e^{tA} is a real matrix if A is real.
- *(Solving an IVP)*: The solution of the IVP $x' = Ax + F(t)$, $x(0) = x^0$, is

$$x = e^{tA}x^0 + e^{tA}\int_0^t e^{-sA}F(s)\,ds \qquad (25)$$

Proof.

- *(Derivative)*: The first equality is immediate since $e^{tA} = \Phi(t, 0)$ is a solution of the matrix system $X' = AX$. To show the second equality, note that $Y = e^{tA}A$ and $Z = Ae^{tA}$ are both solutions of the matrix IVP, $X' = AX$, $X(0) = I$. For, $Y' = (Ae^{tA})A = A(e^{tA}A) = AY$, $Z' = A(Ae^{tA}) = AZ$, and $Y(0) = Z(0) = I$. Thus, by uniqueness, $Y = Z$.

- *(Transition)*: Since the system $x' = Ax$ is autonomous, if $x(t)$ is a solution, then so is $x(t - t_0)$. So $\Phi(t, t_0)$ and $\Phi(t - t_0, 0) = e^{(t-t_0)A}$ are both solution matrices which evaluate to I at $t = t_0$, and so they are equal by uniqueness.

- *(Product)*: Since the system is autonomous, the solution $x(t, x^0)$ to the IVP $x' = Ax$, $x(0) = x^0$ has the property $x(t + s, x^0) = x(t, x(s, x^0))$. Expressing x in terms of Φ we have $\Phi(t + s, 0)x^0 = \Phi(t, 0)(\Phi(s, 0)x^0)$ for all x^0, and the result follows since $e^{(t+s)A} = \Phi(t + s, 0)$.

- *(Inverse)*: Using the Product formula, $e^{tA}e^{-tA} = e^{0A} = \Phi(0, 0) = I$.

- *(Series)*: The series $S(t)$ of matrices on the right side of the equality can be shown to converge absolutely for all t and all matrices A (meaning that each element of the matrix sum converges absolutely; see Problem 11). Under these circumstances, we note that $S(0) = I$ and that $S'(t)$ may be obtained by differentiating the series term by term,

$$\begin{aligned} S'(t) &= A + tA^2 + \cdots + \frac{t^{(n-1)}}{(n-1)!}A^n + \cdots \\ &= A\left[I + tA + \cdots + \frac{t^{n-1}}{(n-1)!}A^{n-1} + \cdots\right] \qquad (26) \\ &= AS(t) \end{aligned}$$

Thus, the solution matrices e^{tA} and $S(t)$ both solve the matrix IVP, $X' = AX$, $X(0) = I$. By the Uniqueness Theorem, $e^{tA} = S(t)$.

- *(Real)*: This follows from the Series Property since t and A are real.

- *(Solving an IVP)*: Formula (25) is a restatement of formula (21) with $t_0 = 0$, $\Phi(t, 0) = e^{tA}$, $\Phi(t, s) = e^{(t-s)A} = e^{tA}e^{-sA}$.

The properties of the matrix exponential e^{tA} are analogous to familiar properties for the scalar exponential e^t, but see the next examples for some unexpected results.

EXAMPLE 7.5.8

The Matrix Exponential May Not Be the Matrix of the Exponentials

From Example 7.5.4, we see that if $A = \begin{bmatrix} 5 & 3 \\ -6 & -4 \end{bmatrix}$ then

$$e^{tA} = \Phi(t, 0) = \begin{bmatrix} -e^{-t} + 2e^{2t} & -e^{-t} + e^{2t} \\ 2e^{-t} - 2e^{2t} & 2e^{-t} - e^{2t} \end{bmatrix}$$

Note that e^{tA} is *not* obtained by exponentiating each entry in the matrix tA:

$$e^{tA} \neq \begin{bmatrix} e^{5t} & e^{3t} \\ e^{-6t} & e^{-4t} \end{bmatrix}$$

EXAMPLE 7.5.9

Diagonal Matrices

Suppose that $D = \begin{bmatrix} \lambda_1 & 0 \\ 0 & \lambda_2 \end{bmatrix}$ is a diagonal matrix. Then by the Series Property in Theorem 7.5.6,

$$
\begin{aligned}
e^{tD} &= \begin{bmatrix} 1 & 0 \\ 0 & 1 \end{bmatrix} + t\begin{bmatrix} \lambda_1 & 0 \\ 0 & \lambda_2 \end{bmatrix} + \frac{t^2}{2!}\begin{bmatrix} \lambda_1^2 & 0 \\ 0 & \lambda_2^2 \end{bmatrix} + \cdots + \frac{t^n}{n!}\begin{bmatrix} \lambda_1^n & 0 \\ 0 & \lambda_2^n \end{bmatrix} + \cdots \\
&= \begin{bmatrix} 1 + t\lambda_1 + \cdots + \dfrac{(t\lambda_1)^n}{n!} + \cdots & 0 \\ 0 & 1 + t\lambda_2 + \cdots + \dfrac{(t\lambda_2)^n}{n!} + \cdots \end{bmatrix} \qquad (27) \\
&= \begin{bmatrix} e^{\lambda_1 t} & 0 \\ 0 & e^{\lambda_2 t} \end{bmatrix}
\end{aligned}
$$

In general, if D is an $n \times n$ diagonal matrix with $a_{jj} = \lambda_j$, $a_{ij} = 0$, $i \neq j$, then e^{tD} is the diagonal matrix with $a_{jj} = e^{t\lambda_j}$, $a_{ij} = 0$ for $i \neq j$.

EXAMPLE 7.5.10

Nilpotent Matrices

A matrix A is *nilpotent* if some power A^k is the zero matrix. Then e^{tA} may be calculated from the expression in the Series Property. For we have that $A^k = A^{k+1} = \cdots = 0$, so

$$e^{tA} = I + tA + \frac{t^2}{2!}A^2 + \cdots + \frac{t^{k-1}}{k!}A^{k-1}$$

For example, with the matrix A given below we have

$$A = \begin{bmatrix} 0 & 1 & 0 \\ 0 & 0 & 1 \\ 0 & 0 & 0 \end{bmatrix}, \quad A^2 = \begin{bmatrix} 0 & 0 & 1 \\ 0 & 0 & 0 \\ 0 & 0 & 0 \end{bmatrix}, \quad A^3 = \begin{bmatrix} 0 & 0 & 0 \\ 0 & 0 & 0 \\ 0 & 0 & 0 \end{bmatrix}$$

and so

$$e^{tA} = I + tA + \frac{t^2}{2!}A^2 = \begin{bmatrix} 1 & t & t^2/2! \\ 0 & 1 & t \\ 0 & 0 & 1 \end{bmatrix}$$

Comments

Observe that Examples 7.5.8, 7.5.9, and 7.5.10 show that e^{tA} *cannot* generally be calculated by exponentiating each term of the matrix tA. If A is a constant matrix, then the matrix exponential e^{tA} *can* always be calculated from any fundamental matrix $X(t)$ for $x' = Ax$ by using formula (10):

$$\Phi(t, 0) = e^{tA} = X(t)X^{-1}(0)$$

Finally, the results of this section apply even if some of the components of $F(t)$ are only piecewise continuous, although solution curves, component curves, and orbits may have "corners" at the discontinuities.

PROBLEMS

1. Find e^{tA}, where A is given. Then solve $x' = Ax$, $x(0) = [1\ 2]^T$. [*Hint*: Find the eigenvalues of A and a spanning set for each eigenspace. Use them to construct a fundamental solution matrix $X(t)$ as in Examples 7.5.1, 7.5.4, 7.5.8. Then $e^{tA} = X(t)X^{-1}(0)$. In part (c), solve the system directly first.]

(a) $\begin{bmatrix} 1 & 3 \\ 1 & -1 \end{bmatrix}$ (b) $\begin{bmatrix} 0 & 1 \\ -9 & 0 \end{bmatrix}$ (c) $\begin{bmatrix} -1 & 1 \\ 0 & -1 \end{bmatrix}$ (d) $\begin{bmatrix} 1 & 1 \\ -1 & 1 \end{bmatrix}$

2. Find e^{tA}, A given below. Then solve $x' = Ax$, $x(0) = [1\ 2\ 3]^T$. [*Hint*: See Examples 7.5.9, 7.5.10.]

(a) $\begin{bmatrix} 0 & 1 & 1 \\ 0 & 0 & 1 \\ 0 & 0 & 0 \end{bmatrix}$ (b) $\begin{bmatrix} 2 & 0 & 0 \\ 0 & -3 & 0 \\ 0 & 0 & 7 \end{bmatrix}$ (c) $\begin{bmatrix} 0 & 0 & 0 \\ 2 & 0 & 0 \\ 3 & 4 & 0 \end{bmatrix}$

3. (a) Show that if A is the block matrix $\begin{bmatrix} B & 0 \\ 0 & C \end{bmatrix}$, then e^{tA} is the block matrix $\begin{bmatrix} e^{tB} & 0 \\ 0 & e^{tC} \end{bmatrix}$. [*Hint*: The system $x' = Ax$ "decouples" into two subsystems.]

(b) Use part (a) to find e^{tA} if

$$A = \begin{bmatrix} 1 & 3 & 0 \\ 1 & -1 & 0 \\ 0 & 0 & 2 \end{bmatrix}, \quad \text{and} \quad A = \begin{bmatrix} 1 & 3 & 0 & 0 \\ 1 & -1 & 0 & 0 \\ 0 & 0 & 0 & 1 \\ 0 & 0 & 0 & 0 \end{bmatrix}$$

[*Hint*: The upper-left block in both matrices is the same matrix as in 1(a). The lower-right block in the 4×4 matrix is nilpotent.]

4. Solve the following driven systems. Leave your answers in the form of equation (25); calculate e^{tA}, $e^{tA}x^0$, $e^{-sA}F(s)$, but don't evaluate the integrals.

(a) $x' = \begin{bmatrix} 0 & 2 \\ -2 & 0 \end{bmatrix} x + \begin{bmatrix} 1 \\ 0 \end{bmatrix}$, $x(0) = \begin{bmatrix} a \\ b \end{bmatrix}$

(b) $x' = \begin{bmatrix} 2 & -1 \\ 3 & -2 \end{bmatrix} x + \begin{bmatrix} 3e^t \\ t \end{bmatrix}$, $x(0) = \begin{bmatrix} 1 \\ 2 \end{bmatrix}$ Plot the component graphs.

(c) $x' = \begin{bmatrix} 2 & -5 \\ 1 & -2 \end{bmatrix} x + \begin{bmatrix} \cos t \\ 0 \end{bmatrix}$, $x(0) = \begin{bmatrix} a \\ b \end{bmatrix}$

(d) $x' = \begin{bmatrix} -1 & -4 \\ 1 & -1 \end{bmatrix} x + \begin{bmatrix} e^{-3t} \\ 1 \end{bmatrix}$, $x(0) = \begin{bmatrix} 0 \\ 0 \end{bmatrix}$ Plot the component graphs.

(e) $x' = \begin{bmatrix} 3 & -1 & 1 \\ 2 & 0 & 1 \\ 1 & -1 & 2 \end{bmatrix} x + \begin{bmatrix} f_1(t) \\ f_2(t) \\ f_3(t) \end{bmatrix}, \quad x(0) = \begin{bmatrix} a \\ b \\ c \end{bmatrix}$

5. (*Undetermined Coefficients*). Sometimes a particular solution $x^p(t)$ of $x' = Ax + F$ can be found by a method of undetermined coefficients. Find the solution set for each driven system below.

   (a) $x' = \begin{bmatrix} 2 & -1 \\ 5 & -2 \end{bmatrix} x + \begin{bmatrix} e^t \\ 1 \end{bmatrix}$ [*Hint*: Assume a particular solution of the form $x_1 = ae^t + b$, $x_2 = ce^t + d$ and find the coefficients by inserting x_1 and x_2 into the ODEs and matching coefficients of corresponding terms. Then add this particular solution to the general solution of the undriven system.]

   (b) $x' = \begin{bmatrix} 1 & 3 \\ 1 & -1 \end{bmatrix} x + \begin{bmatrix} \cos t \\ 2t \end{bmatrix}$ [*Hint*: Assume a particular solution of the form $x_1 = a_1 \cos t + b_1 \sin t + c_1 + d_1 t$, $x_2 = a_2 \cos t + b_2 \sin t + c_2 + d_2 t$.]

6. (*Exponential Matrices May Not Commute*). Here are some unexpected properties of e^{tA}.

   (a) Show that $e^{tA}e^{tB}$ need not be either $e^{t(A+B)}$ or $e^{tB}e^{tA}$ by calculating all three where $A = \begin{bmatrix} 0 & 1 \\ 0 & 0 \end{bmatrix}$ and $B = \begin{bmatrix} 1 & 0 \\ 0 & 0 \end{bmatrix}$.

   (b) Suppose that $AB = BA$. Show that $e^{t(A+B)} = e^{tA}e^{tB} = e^{tB}e^{tA}$ for all t. [*Hint*: Show that if $P(t) = e^{t(A+B)}e^{-tA}e^{-tB}$, then $P'(t) = 0$ for all t. Since $P(0) = I$, we must have $P(t) = I$.]

7. (*Nonconstant System Matrix*). Consider the system $x' = y$, $y' = -2x/t^2 + 2y/t$, $t > 0$.

   (a) Show that $X(t) = \begin{bmatrix} t^2 & t \\ 2t & 1 \end{bmatrix}$ is a fundamental solution matrix.

   (b) Find $\Phi(t, t_0)$, and verify that $\Phi(t_0, t_0) = I$, $\Phi^{-1}(t, t_0) = \Phi(t_0, t)$, $t_0 > 0$.

   (c) Plot the component curves and the orbits for the system in part (a), given the initial data $x(1) = 1$, $y(1) = -5, -4, \ldots, 4, 5$.

8. Show that Theorem 3.6.1 in Section 3.6 is a special case of the Variation of Parameters Theorem (Theorem 7.5.4). [*Hint*: Transform the second-order scalar ODE, $y'' + a(t)y' + b(t)y = f$ to the system $u' = v$, $v' = -bu - av + f$, where $u = y$, $v = y'$.]

9. (*Another Approach to Variation of Parameters*). Prove Theorem 7.5.4 by assuming that $x' = A(t)x + F$, $x(t_0) = 0$, has a solution of the form $x = \Phi(t, t_0)c(t)$, where Φ is the transition matrix for $x' = A(t)x$ and $c(t)$ is the unknown "varied parameter" vector. [*Hint*: Insert Φc for x in the ODE, use properties of Φ, and show that $c' = \Phi(t_0, t)F(t)$, from which $c(t) = \int_{t_0}^t \Phi(t_0, s)F(s)\,ds$ if $c(t_0) = 0$.]

10. (*Laplace Transforms for Systems*). The following steps outline the method of Laplace transforms for solving the IVP

$$x' = Ax + F, \qquad x(0) = x^0$$

where A is a matrix of constants. The Laplace Transform $\mathcal{L}[x(t)]$ of a vector function is the vector of the Laplace Transform of the individual components. The definitions and results of Chapter 5 are assumed.

   (a) Apply the operator $\mathcal{L}$ to the above system and obtain $[sI - A]\mathcal{L}[x] = x^0 + \mathcal{L}[F]$.

   (b) Show that the matrix $sI - A$ is invertible for all real s which are large enough. [*Hint*: $sI - A$ is singular precisely when s is an eigenvalue of A. Recall that A has a finite number of eigenvalues.]

(c) Show that the solution of the IVP above is $x = \mathcal{L}^{-1}[(sI - A)^{-1}x^0)] + \mathcal{L}^{-1}[(sI - A)^{-1}\mathcal{L}[F]]$. Compare with (21) and show that $\mathcal{L}^{-1}[(sI - A)^{-1}] = e^{tA}$, $\mathcal{L}^{-1}[(sI - A)^{-1}\mathcal{L}[F]] = e^{tA} * F(t)$, the convolution of e^{tA} and $F(t)$.

(d) Use the Laplace Transform to solve the IVP of Problem 1**(b)**.

11. (*Defining the e^{tA} by Series*). Prove that the matrix power series

$$I + tA + \frac{t^2}{2!}A^2 + \cdots + \frac{t^k}{k!}A^k + \cdots$$

converges absolutely for all t, where A is any $n \times n$ matrix of constants (possibly complex). That is, if $A^k = [a_{ij}^{(k)}]$, prove that

$$\sum_{k=0}^{\infty} \left| \frac{a_{ij}^{(k)} t^k}{k!} \right|$$

converges for all t. [*Hint*: In Appendix C.6, the matrix norm $\|A\|$ is defined to be $\sum_{ij} |a_{ij}|$. It follows that $\|A^k\| \le \|A\|^k$, $|a_{ij}| \le \|A\|$, and $|a_{ij}^{(k)}| \le \|A\|^k$. Hence, $|a_{ij}^{(k)} t^k / k!| \le \|A\|^k |t|^k / k!$. Use the comparison test for convergence.]

7.6 Steady States of Linear Systems

A well-designed physical system should be relatively impervious to disturbances. A disturbance may drive the system away from its steady state, but not far away if the disturbance is small. Once the disturbance ends, one would expect the system to return to the steady state. In this section we formulate the mathematical equivalent of these ideas for a linear, constant-coefficient system of ODEs

$$x' = Ax + F(t) \tag{1}$$

where A is an $n \times n$ matrix of real constants, x is an n-vector, and F is an n-vector of constants, or else F is periodic.

What is a steady state? Following common engineering paractice, we distinguish two kinds.

❖ **Constant Steady State.** A constant solution x^s of system (1) is a *constant steady state* if all solutions of (1) tend to x^s as $t \to \infty$.

❖ **Periodic Steady State.** A periodic solution $x^s(t)$ of system (1) is a *periodic steady state* if all solutions of (1) tend to $x^s(t)$ as $t \to \infty$.

These are commonsense definitions that correspond in the first case to a single constant equilibrium solution, and in the second case to a single periodic solution. In both cases the steady state "attracts" all solutions.

Constant Steady State

Suppose that A is a nonsingular matrix of real constants and that F_0 is a constant vector. Direct substitution shows that

$$x(t) = -A^{-1}F_0 \tag{2}$$

is a constant solution of the system

$$x' = Ax + F_0 \tag{3}$$

Since A is nonsingular, system (3) has no other constant solutions.

THEOREM 7.6.1

> Constant Steady State. Suppose that A is nonsingular. Then the constant solution $x^s = -A^{-1}F_0$ is the (unique) steady state of system (3) if and only if all eigenvalues of A have negative real parts.

Proof. The general solution of system (3) is

$$x(t) = e^{tA}c - A^{-1}F_0$$

where c is any constant column vector. Since all entries in the matrix e^{tA} are polynomial-exponential functions containing exponentials of the form $e^{\alpha t}$, where α is any of the real parts of the eigenvalues of A, we see that if all eigenvalues have negative real parts, then as $t \to \infty$ all entries in e^{tA} tend to 0 as $t \to \infty$. Thus, every solution $x(t)$ of (3) tends to $x^s = -A^{-1}F_0$ as $t \to \infty$; hence, x^s is the unique steady state. The proof of the converse is omitted.

EXAMPLE 7.6.1

A Constant Steady State
The system matrix of the system

$$\begin{aligned} x' &= -x + 4y + 14 \\ y' &= -3x - 2y + 28 \end{aligned} \tag{4}$$

has eigenvalues $-3/2 \pm i\sqrt{47}/2$, while $-A^{-1}F_0$ is calculated to be $x^s = [10 \;\; -1]^T$. According to Theorem 7.6.1, all solutions of system (3) tend to the unique equilibrium solution and steady state x^s. See Figure 7.6.1 for orbits (at left) and the corresponding x and y component curves at top right and bottom right, respectively.

Periodic Steady State, Forced Oscillations

Now suppose that the driving vector $F(t)$ is periodic of least period $T > 0$. A *forced oscillation* of the system

$$x' = Ax + F(t) \tag{5}$$

is a periodic solution of period T. We have the following result on the existence of a forced oscillation (see Problem 14 for the proof).

THEOREM 7.6.2

> Forced Oscillation. Suppose that the driving term $F(t)$ is periodic with least period $T > 0$. If $2\pi i/T$ is *not* an eigenvalue of A, then system (5) has a unique forced oscillation.

The reason that we require that $2\pi i/T$ not be an eigenvalue is that if it were, then the undriven system $x' = Ax$ would have solutions containing terms such as $\cos(2\pi t/T)$ or $\sin(2\pi t/T)$ of period T. Since $F(t)$ has period T, this could lead

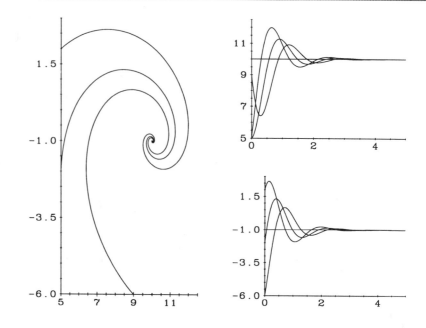

FIGURE 7.6.1 Solutions of system (4) approach the constant steady state $x = 10$, $y = -1$. See Example 7.6.1.

to resonance and the presence of unbounded oscillations (see Section 4.1). The next theorem gives conditions where a forced oscillation is actually a steady state, i.e., a forced oscillation that "attracts" all solutions of system (5) as $t \to \infty$.

THEOREM 7.6.3

A Steady State Forced Oscillation. Suppose that $F(t)$ is periodic of least period $T > 0$. Then system (5) has a unique periodic steady state if and only if all eigenvalues of A have negative real parts.

Proof. Suppose that the eigenvalues of A have negative real parts. Then $2\pi i / T$ is not an eigenvalue of A. The conditions of Theorem 7.6.2 are met, and system (5) has a forced oscillation (of period T); call it $x^s(t)$. Since all solutions of system (5) have the form

$$x(t) = e^{tA}c + x^s(t)$$

all solutions tend to $x^s(t)$ as $t \to \infty$ because $e^{tA}c$ tends to the zero vector as $t \to \infty$. Hence, $x^s(t)$ is the unique periodic steady state. The proof of the converse is omitted.

EXAMPLE 7.6.2

A Periodic Steady State
The eigenvalues of the system matrix of the system

$$x' = -x + 4y + 5z + 20\,\text{sqw}\,(t, 50, \pi)$$

$$y' = -10x - y + 20\,\text{trw}\,(t, 50, \pi) \tag{6}$$

$$z' = -z + 20\cos 2t$$

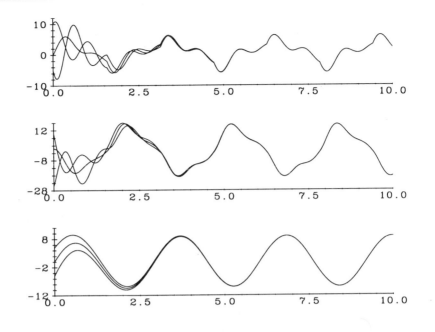

FIGURE 7.6.2 The x, y, z, components (top to bottom) of three solutions of system (6) approach a periodic steady state of period π. See Example 7.6.2.

are -1, $-1 + 2i\sqrt{10}$, $-1 - 2i\sqrt{10}$, and the driving vector has period π. The conditions of Theorem 7.6.3 are met, and system (6) has a unique periodic steady state. Figure 7.6.2 displays the x, y, and z component curves (reading from the top and then down) of three solutions of system (6). Each of these solutions settles down quite quickly into the periodic steady state, or (more accurately) gets so close to that state that the differences are too small to be seen on the computer screen.

Decay, BIBO

We might guess that all solutions of system (3) or (5) decay exponentially to the steady state if all eigenvalues of A have negative real parts. That is just what happens as the following theorem implies (proof omitted):

THEOREM 7.6.4

> Decay Estimate. Let all eigenvalues of A have real parts less than $-\alpha$, where α is a positive constant. Then there is a positive constant M such that the solution $x = x(t, x^0)$ of $x' = Ax$, $x(0) = x^0$, satisfies
>
> $$|x(t, x^0)|_{\max} \leq Me^{-\alpha t}|x^0|_{\max}$$
>
> for all $t \geq 0$ and each x^0 in $\mathbf{R}^n$, where $|x|_{\max} = \max_{1 \leq j \leq n} |x_j|$.

Suppose, finally, that we know little about the input vector $F(t)$ except that it is bounded for $t \geq 0$, i.e., that for some positive constant M, $\|F(t)\| \leq M$ for all $t > 0$. The following result summarizes a bounded-input, bounded-output (BIBO) property first observed for a first-order scalar linear ODE in Section 2.4.

THEOREM 7.6.5

> **Bounded Input, Bounded Output.** Suppose that the eigenvalues of the matrix A have negative real parts and that the input vector $F(t)$ is bounded for $t \geq 0$. Then every solution of $x' = Ax + F(t)$ is bounded for $t \geq 0$.

The proof follows from Theorem 7.6.4, but we omit the details. This result is of considerable importance in science and engineering. It implies that as long as the eigenvalues of A have negative real parts, then persistent disturbances, i.e., the components of the bounded input vector $F(t)$, cannot "destabilize" the system, and unbounded solutions cannot arise.

EXAMPLE 7.6.3

BIBO

The system below is defined for $t \geq 0$

$$x' = -0.2x + 5y + 10\sin(0.05t^2) \tag{7}$$
$$y' = -5x - 0.2y + 10\cos\sqrt{5}t$$

The input vector is $[10\sin(0.05t^2) \quad 10\cos\sqrt{5}t]^T$, which is bounded in magnitude by $10\sqrt{2}$ for all t. The system matrix has eigenvalues $-0.2 \pm 5i$. All conditions of the BIBO theorem are met; hence, all solutions are bounded for $t \geq 0$.

We have restricted the term "steady state" to an attracting constant solution or an attracting forced oscillation and to linear systems with a constant or periodic input. If we try to define a steady state for the system $x' = Ax + F(t)$ for an input F that is neither constant nor periodic, then we run into some trouble. For, even if all solutions of the system are bounded for $t \geq 0$ and the eigenvalues of A have negative real parts, it is not clear just how to define the steady state. The reason is that if $x^1(t)$ and $x^2(t)$ are any two solutions of the system, then $x^1 - x^2$ solves $x' = Ax$, and hence

$$\lim_{t \to \infty} \|x^1(t) - x^2(t)\| = 0$$

Thus, all solutions of the driven system approach a "steady state." Since all solutions tend to one another, any solution could be called the steady state. And that is why we have restricted the term "steady state" to the specific situations where F is either constant or periodic.

The last part of this section outlines some useful ways to determine whether all the roots of a given characteristic polynomial have negative real parts, and whether all the eigenvalues of a given matrix have negative real parts.

The Coefficient Test and the Routh Criterion

The following tests give information about the signs of the real parts of the roots of a polynomial (e.g., the eigenvalues of the matrix A as roots of its characteristic polynomial) without the necessity of actually finding the roots.

THEOREM 7.6.6

> Coefficient Test. Let $P(r) = r^n + a_{n-1}r^{n-1} + \cdots + a_0$, where the coefficients are real. If any coefficient of $P(r)$ is either zero or negative, then at least one root has a nonnegative real part.

Proof. We shall prove the contrapositive. Suppose that all roots are negative or have negative real parts. Thus $P(r)$ is a product of factors of the form $(r + r_j)$ and $(r + \alpha_k + i\beta_k)(r + \alpha_k - i\beta_k) = (r^2 + 2\alpha_k r + \alpha_k^2 + \beta_k^2)$, where r_j and α_k are positive (i.e., $-r_j$ and $-\alpha_k \pm i\beta_k$ are roots of the type assumed). Multiplying out all of these factors, we see that all of the coefficients of $P(r)$ are positive. This proves the theorem.

EXAMPLE 7.6.4

Applying the Coefficient Test
The Coefficient Test implies that the polynomials $r^3 + 3r + 10$ and $r^3 - r^2 + 5r + 7$ each have at least one root with nonnegative real part. The test does not apply to the polynomial $r^3 + r^2 + r + 1$ because all coefficients are positive.

As the example $r^3 + r^2 + r + 1 = (r + 1)(r^2 + 1) = (r + 1)(r + i)(r - i)$ shows, the positivity of all the coefficients of a polynomial does not necessarily guarantee that all the roots have negative real parts. However, the Routh Criterion will handle this case, and, in fact, it can be used with any polynomial whose leading coefficient is $+1$.

THEOREM 7.6.7

> Routh Criterion. Suppose that the coefficients of $P(r) = r^n + a_{n-1}r^{n-1} + \cdots + a_0$ are real. Then all roots of $P(r)$ have negative real parts if and only if all entries in the first column of the *Routh Array* are positive:
>
> $$\begin{bmatrix} 1 & a_{n-2} & a_{n-4} & \cdots & a_{n-2p+2} \\ a_{n-1} & a_{n-3} & a_{n-5} & \cdots & a_{n-2p+1} \\ b_{11} & b_{12} & b_{13} & \cdots & b_{1p} \\ b_{21} & b_{22} & b_{23} & \cdots & b_{2p} \\ \vdots & \vdots & \vdots & & \vdots \\ b_{n-1,1} & b_{n-1,2} & b_{n-1,3} & \cdots & b_{n-1,p} \end{bmatrix}$$
>
> where $n = 2p - 2$ or $2p - 1$ (that is, a_0 is in the last column), $a_k = 0$ for $k < 0$, $a_n = 1$, $b_{ip} = 0$ for all i, and
>
> $$b_{1j} = \frac{a_{n-1}a_{n-2j} - a_n a_{n-2j-1}}{a_{n-1}}, \quad b_{2j} = \frac{b_{11}a_{n-2j-1} - a_{n-1}b_{1,j+1}}{b_{11}}, \quad j < p$$
>
> $$b_{ij} = \frac{b_{i-1,1}b_{i-2,j+1} - b_{i-2,1}b_{i-1,j+1}}{b_{i-1,1}}, \quad i > 2, \ j < p$$

The proof of the Routh Criterion is omitted. The algorithm for constructing the Routh Array looks formidable, but for any particular polynomial it is easy to use. The rows are constructed from the top down and from left to right. If a zero or a negative number ever appears in the left-hand column of the Routh Array during its construction, there is no point in proceeding further since the criterion has already been violated, and the polynomial must have at least one root with nonnegative real part. The Routh Criterion need only be used for a polynomial with positive coefficients since the Coefficient Test handles the other cases.

EXAMPLE 7.6.5

Applying the Routh Array
The Routh Array for the polynomial $r^3 + 2r^2 + 3r + 5$ is a 4×2 array:

$$\begin{bmatrix} 1 & 3 \\ 2 & 5 \\ 1/2 & 0 \\ 5 & 0 \end{bmatrix} \begin{array}{l} \} \text{ (coefficient rows)} \\ \\ \} \ (b_{ij} \text{ rows}) \end{array}$$

Since the entries in the left-hand column are positive, all roots of the polynomial have negative real parts.

Traces and Gerschgorin Disks

The Coefficient Test and the Routh Criterion are effective tests for deciding whether all the roots of the characteristic polynomial of a real matrix A have negative real parts. But what do we do if the matrix is so large that it would be hard to write out the characteristic polynomial? The next two results are helpful in this regard. The first uses the trace of an $n \times n$ matrix A, where trace $A = a_{11} + a_{22} + \cdots + a_{nn}$.

THEOREM 7.6.8

> **The Trace Test.** Suppose that A is an $n \times n$ matrix of real constants and that trace A is negative (positive). Then at least one eigenvalue of A has a negative (positive) real part.

Proof. The result follows from the fact that if $\lambda_1, \ldots, \lambda_n$ are the eigenvalues of A (not necessarily distinct), then (according to Problem 6, Section 7.2)

$$\lambda_1 + \cdots + \lambda_n = \text{trace } A \tag{8}$$

So, if trace $A < 0 \, (> 0)$ then $\text{Re}[\lambda_i] < 0 \, (> 0)$ for at least one eigenvalue λ_i.

If tr $A = 0$, the proof above can be modified to show that either all eigenvalues have zero real parts, or else there are eigenvalues whose real parts have opposite signs.

EXAMPLE 7.6.6

Using the Trace Test
The matrix A_1 below must have an eigenvalue with non-negative real part because the trace is $+1$. The trace of the matrix A_2 below is negative (tr $= -2$), but A_2 has $+9$ as an eigenvalue. This does not contradict the Trace Test because that theorem asserts that if the trace is negative, then there is at least one eigenvalue whose real part is negative, not that all eigenvalues have negative real parts.

$$A_1 = \begin{bmatrix} -165 & 13 & 360 \\ -200 & 15 & 267 \\ -150 & -8 & 151 \end{bmatrix}, \qquad A_2 = \begin{bmatrix} -10 & 0 & 0 \\ 0 & -1 & 0 \\ 0 & 0 & 9 \end{bmatrix}$$

Finally we give a method first discovered by the Russian mathematician S. Gerschgorin in the 1930s. Let A be an $n \times n$ matrix of constants (which may be complex).

The *Gerschgorin row disk* R_i in the complex plane is defined by

$$R_i = \text{The set of all complex numbers } z \text{ such that } |z - a_{ii}| \leq r_i, \text{ where } r_i = \sum_{\substack{k=1 \\ k \neq i}}^{n} |a_{ik}|$$

Thus R_i is the circular disk of radius r_i centered at a_{ii} in the complex plane. Note that r_i is the sum of the magnitudes of all entries in the ith row of A, except for $|a_{ii}|$. Similarly, the *Gerschgorin column disk* C_i is defined by

$$C_i = \text{The set of all } z \text{ such that } |z - a_{ii}| \leq c_i = \sum_{\substack{k=1 \\ k \neq i}}^{n} |a_{ki}|$$

These disks contain the eigenvalues.

THEOREM 7.6.9

> Eigenvalues and the Gerschgorin Disks. Let A be an $n \times n$ matrix of constants, and let R_i and C_i, $i = 1, \ldots, n$, be the respective Gerschgorin row and column disks of A. Then all eigenvalues of A lie in the union $R_1 \cup R_2 \cup \cdots \cup R_n$ of the row disks, and also in the union $C_1 \cup C_2 \cup \cdots \cup C_n$ of the column disks. Moreover, a union of k of the row disks (column disks) that is disjoint from the other $n - k$ row disks (column disks) contains exactly k eigenvalues, counting multiplicities.

Proof. Let λ be an eigenvalue of $A = [a_{ij}]$, $v = [v_1 \cdots v_n]^T$ a corresponding eigenvector. Since $\lambda v = Av$, we have for each i, $i = 1, \ldots, n$, that

$$\lambda v_i = \sum_{j=1}^{n} a_{ij} v_j, \quad \text{and hence,} \quad (\lambda - a_{ii}) v_i = \sum_{j=1, j \neq i}^{n} a_{ij} v_j \qquad (9)$$

Suppose that v_k is the component of v of maximal magnitude; i.e.,

$$|v_k| = \max_{1 \leq r \leq n} |v_r| \neq 0$$

Let $i = k$ in (9), divide $(\lambda - a_{kk}) v_k = \sum_{j=1, j \neq k}^{n} a_{kj} v_j$ by v_k, and take magnitudes:

$$|\lambda - a_{kk}| = \left| \sum_{\substack{j=1 \\ j \neq k}}^{n} a_{kj} \frac{v_j}{v_k} \right| \leq \sum_{\substack{j=1 \\ j \neq k}}^{n} |a_{kj}| \left| \frac{v_j}{v_k} \right| \leq \sum_{\substack{j=1 \\ j \neq k}}^{n} |a_{kj}| = r_k$$

where we have used the Triangle Inequality of Appendix C.6 and the fact that $|v_j/v_k| \leq 1$. Thus, λ lies in the kth Gerschgorin row disk.

Because A and A^T have the same eigenvalues, λ also lies in at least one Gerschgorin column disk. The collection of all eigenvalues lies in the union of all the row disks and also in the union of all the column disks. Note that this does *not* imply that each disk has one or more eigenvalues in it. The proof of the final conclusion about disjoint unions of disks is omitted.

EXAMPLE 7.6.7

Gerschgorin Disks

Let $A = \begin{bmatrix} -1+i & 1.5 & 1 \\ 0.5 & -2 & -1 \\ 0.25 & 0 & -7 \end{bmatrix}$. We shall use the Gerschgorin disks to get some idea

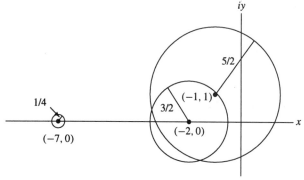

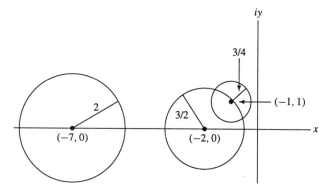

FIGURE 7.6.3 Gerschgorin row disks. See Example 7.6.7.

FIGURE 7.6.4 Gerschgorin column disks. See Example 7.6.7.

of the location of the eigenvalues of A without finding the characteristic polynomial. The row disks of A are given by

$$|z + 1 - i| \le 2.5, \quad |z + 2| \le 1.5, \quad |z + 7| \le 0.25$$

while the column disks are defined by

$$|z + 1 - i| \le 0.75, \quad |z + 2| \le 1.5, \quad |z + 7| \le 2$$

These are sketched in Figures 7.6.3 and 7.6.4.

Using both unions of disks, we see that there is a simple eigenvalue in the row disk, $|z + 7| \le 0.25$, and two eigenvalues in the union of the two column disks, $|z + 2| \le 1.5$, $|z + 1 - i| \le 0.75$. Each eigenvalue of A has negative real part because all three of the column disks lie entirely to the left of the imaginary axis in the complex plane.

If P is a nonsingular $n \times n$ matrix, then A and the similar $P^{-1}AP$ have the same eigenvalues (see Section 7.2). Sometimes the Gerschgorin disks for $P^{-1}AP$ give more useful information about the eigenvalues of A than do the disks for A itself. In particular, let P be the $n \times n$ diagonal matrix with 1s on the diagonal except that $p_{jj} = b \ne 0$ for some one j. Then the rows and columns of $B = P^{-1}AP$ are the same as those of A except that the jth row has been divided by b and the jth column has been multiplied by b. The Gerschgorin row [column] disks of B have the same centers as those of A, but the radii are different if $b \ne 1$. For example, the second row disk for $A = \begin{bmatrix} -7 & 1 \\ 2 & -1 \end{bmatrix}$ overlaps the right half of the complex plane, while if $P = \begin{bmatrix} 1 & 0 \\ 0 & 3 \end{bmatrix}$, we have that $B = P^{-1}AP = \begin{bmatrix} -7 & 3 \\ 2/3 & -1 \end{bmatrix}$, whose row disks lie inside the left half-plane. Hence, the eigenvalues of B (and so of A) have negative real parts.

EXAMPLE 7.6.8

More Gerschgorin Disks: Testing for Decay

A certain dynamical system can be modeled by an autonomous system $x' = Ax$, where A is a constant matrix. It is known that all solutions tend to die out in time. Can either A_1 or A_2 be the system matrix?

$$A_1 = \begin{bmatrix} -5 & 1 & 2 & -1 \\ 0 & 8 & 9 & -1 \\ 2 & 7 & -1 & 0 \\ 0 & -1 & 5 & -1 \end{bmatrix} \quad A_2 = \begin{bmatrix} -3 & 1 & 2 & 0 \\ 1 & -10 & 3 & 3 \\ 2 & 3 & -11 & 1 \\ 1/10 & 1/10 & 0 & -1 \end{bmatrix}$$

A_1 cannot be the system matrix because tr $A_1 = +1$; hence, the sum of the eigenvalues is $+1$, and at least one eigenvalue has positive real part. The trace of A_2 is negative, but that does not guarantee that all eigenvalues have negative real parts. The Gerschgorin row disks of A_2 do not overlap the right half of the complex plane, but the first row disk is centered at -3 and has radius 3. Thus, the origin is on its boundary and 0 could conceivably be an eigenvalue. If det $A_2 \neq 0$, then 0 is not an eigenvalue, but the calculation of the determinant would be tedious. Instead, we will look at the matrix $B = P^{-1}A_2P$, where P and B are

$$P = \begin{bmatrix} 3 & 0 & 0 & 0 \\ 0 & 1 & 0 & 0 \\ 0 & 0 & 1 & 0 \\ 0 & 0 & 0 & 1 \end{bmatrix}, \qquad B = \begin{bmatrix} -3 & 1/3 & 2/3 & 0 \\ 3 & -10 & 3 & 3 \\ 6 & 3 & -11 & 1 \\ 3/10 & 1/10 & 0 & -1 \end{bmatrix}$$

The Gerschgorin row disks of B lie entirely to the left of the imaginary axis. Hence, all eigenvalues of B (and of A_2) have negative real parts. Thus A_2 is a system matrix such that all solutions of $x' = A_2x$ tend to the origin as $t \to \infty$.

Comments

There are several good numerical root finders and eigenvalue solvers that can be used to do the algebra mentioned in this section. But the techniques outlined here are simple, pencil-and-paper techniques that are so easy to use that they should be what one does first, before turning to a computer program.

PROBLEMS

1. (*Constant Steady States*). Is the constant solution $x = 0$ a steady state for $x' = Ax$ if A is as given? Give reasons.

   (a) $\begin{bmatrix} -10 & 1 & 2 \\ -3 & 4 & 5 \\ -7 & -6 & 7 \end{bmatrix}$ (b) $\begin{bmatrix} -10 & 1 & 2 \\ -3 & 4 & -2 \\ -7 & -6 & -40 \end{bmatrix}$ (c) $\begin{bmatrix} -10 & 1 & 2 \\ 0 & -1 & 100 \\ 0 & 0 & -1 \end{bmatrix}$

2. (*Tuning a System to Create a Steady State*). Find the values of the real constant α for which $x = 0$ is the steady state of $x' = Ax$. [*Hint*: In (b) and (c), examine the equivalent third-order scalar equation.]

   (a) $\begin{bmatrix} \alpha & 2 \\ 3 & -4 \end{bmatrix}$ (b) $\begin{bmatrix} 0 & 1 & 0 \\ 0 & 0 & 1 \\ \alpha & -1 & -1 \end{bmatrix}$ (c) $\begin{bmatrix} 0 & 1 & 0 \\ 0 & 0 & 1 \\ -2 & -3 & \alpha \end{bmatrix}$

3. (*BIBO*). The system $x'_1 = x_2$, $x'_2 = -bx_1 - ax_2 + f(t)$ is equivalent to the scalar equation $x'' + ax' + bx = f(t)$. Show that every solution of the system is bounded for $t \geq 0$ if $a > 0$, $b > 0$, and $f(t)$ is continuous and bounded for $t \geq 0$.

4. (*Unbounded Solutions*). Show that the system $x'_1 = x_2$, $x'_2 = -x_1 + \cos t$ has unbounded solutions, but that this does not contradict the Bounded Input, Bounded Output Theorem, Theorem 7.6.5.

5. (*Unbounded Solutions*). Show that the system $x'_1 = -x_1 + e^{3t}x_2$, $x'_2 = -x_2$ has unbounded solutions for $t \geq 0$ even though the eigenvalues of the system matrix are negative constants. Why is this not a contradiction of the Bounded Input, Bounded Output Theorem, Theorem 7.6.5?

6. (*Unbounded Solutions*). If the matrix A has an eigenvalue with nonnegative real part, then there always exists a bounded driving function $F(t)$ such that $x' = Ax + F$ has unbounded solutions. Parts (a) and (b) show why this is so.

    (a) Show that $x' = Ax$ has unbounded solutions if A has an eigenvalue with positive real part.

    (b) Let the matrix A have an eigenvalue λ with zero real part. Show that if v is an eigenvector corresponding to λ, the system $x' = Ax + e^{\lambda t}v$ has the unbounded solution $x = te^{\lambda t}v$, $t \geq 0$, even though the input $e^{\lambda t}v$ is bounded.

7. (*Applying the Routh Criterion*). Let $P(r) = r^n + a_{n-1}r^{n-1} + \cdots + a_0$ have positive coefficients. Use the Routh Criterion to show that the roots have negative real parts if:

    (a) $n = 2$ (b) $n = 3$, and $a_2 a_1 > a_0$

    (c) $n = 4$, and $a_3 a_2 a_1 > a_3^2 a_0 + a_1^2$, and $a_3 a_2 > a_1$

8. (*Applying the Routh Criterion*). For which of the polynomials below do all the roots have negative real parts?

    (a) $z^3 + z^2 + 2z + 1$

    (b) $z^4 + z^3 + 2z^2 + 2z + 3$

    (c) $z^6 + 10z^4 + z + 1$

    (d) $z^6 + z^5 + 6z^4 + 6z^3 + 11z^2 + 6z + 6$

    (e) $z^5 + 6z^4 + 12z^3 + 12z^2 + 11z + 6$

9. (*Applying the Routh Criterion*). Show that the roots of $z^4 + 2z^3 + 3z^2 + z + a$ have negative real parts if and only if $0 < a < 1.25$.

10. (*Steady State*). Explain why each system has a steady state. Then find it, and plot component graphs of the steady state and of several other solutions.

    (a) $x' = y + 5$, $y' = -2x - 3y + 10$

    (b) $x' = y$, $y' = -x - 2y + \cos t$ [*Hint*: Write as an equivalent second order ODE.]

    (c) $x' = -x + y + \cos t$, $y' = -x - y$ [*Hint*: Assume a solution of the form $x = A\cos t + B\sin t$, $y = C\cos t + D\sin t$ and match coefficients.]

11. (*Periodic Steady State*). Explain why each system has a periodic steady state. Then plot the components of several solutions over a long enough time period that the periodic steady state can be seen. Highlight as much of the periodic steady state as you can see.

    (a) $x' = -x + y/2 + \cos^2 t$, $y' = -x - 2y + \sin t$

    (b) $x' = -2x + 5y + \text{sqw}(t, 50, 2)$, $y' = -5x - 2y + \cos \pi t$

    (c) $x' = y + \text{trw}(t, 50, 1)$, $y' = -x - y + \text{sqw}(t, 50, 2)$

    (d) $x' = y + (\sin t)\,\text{sqw}(t, 50, 2\pi)$, $y' = -4x - y + 1$

12. (*Unbounded Solutions*). Show that if the eigenvalues of A have negative real parts and if one solution of $x' = Ax + F(t)$ is unbounded as $t \to \infty$, then all solutions are unbounded as $t \to \infty$.

13. (*Gerschgorin Disks*). Use Gerschgorin disks to explain why the eigenvalues of the matrices of parts (a) and (b) have negative real parts, but the matrix of part (c) has an eigenvalue with positive real part. [*Hint*: For the matrix A of part (b) use the technique illustrated in Example 7.6.8.]

$$\text{(a) } A = \begin{bmatrix} -10 & 1 & 8 \\ 1 & -3 & 2 \\ 8 & 1 & -11 \end{bmatrix} \qquad \text{(b) } A = \begin{bmatrix} -12 & 4 & 10 \\ -1 & -5 & 1 \\ 1 & 1 & -6 \end{bmatrix}$$

$$\text{(c) } A = \begin{bmatrix} 10 & 1 & 10 \\ 1 & -5 & 8 \\ 1 & 1 & -20 \end{bmatrix}$$

14. (*Forced Oscillations*). A forced oscillation of the system $x' = Ax + F(t)$ is a nonconstant solution $x(t)$ of period T, where A is an $n \times n$ matrix of real constants and $F(t)$ has period T. Show that the system has a unique forced oscillation if $2\pi i / T$ is *not* an eigenvalue of A. Note that F may be a constant vector in this setting (in which case F is periodic of every period). [*Hint*: $x(t) = e^{tA}x^0 + e^{tA}\int_0^t e^{-sA}F(s)\,ds$ is periodic of period T if and only if $x(0) = x(T) = e^{TA}x^0 + e^{TA}\int_0^T e^{-sA}F(s)ds$, where $T > 0$.]

7.7 Sensitivity to Parameter Changes

The data of a dynamical physical system often include parameters that can be adjusted in order to "tune" the system. What happens to the solutions of the equations of the mathematical model of the physical system as these parameters are changed? This "sensitivity to the parameters" has been a recurrent theme of the earlier sections, and we return to the subject now in the context of linear systems. In particular, we shall study the sensitivity of the model of the low-pass filter introduced in Example 6.1.2 and the sensitivity of the model of lead in the human body of Example 7.1.1.

Sensitivity of a Low-Pass Filter to Input Frequencies

An electrical filter is a network of coils, capacitors, and resistors designed to allow signals in certain bands of frequencies to pass through relatively unchanged (*passbands*), while attenuating or suppressing signals in other frequency bands (*stopbands*). A *low-pass* filter passes low-frequency signals but stops those with high frequency. This is particularly important in audio equipment where high-frequency "noise" can be annoying.

The circuit sketched in the margin was introduced and modeled in Example 6.1.2. The model equation of the circuit is:

$$x' = Ax + F(t) \tag{1}$$

where

$$x = \begin{bmatrix} V_2 \\ I_1 \\ I_2 \end{bmatrix}, \qquad A = \begin{bmatrix} 0 & 0 & 1/C_2 \\ 0 & -1/RC_1 & 1/RC_1 \\ -1/L & -R/L & 0 \end{bmatrix}, \qquad F = \begin{bmatrix} 0 \\ V_1'/R \\ V_1/L \end{bmatrix} \tag{2}$$

The characteristic polynomial of A, $p(\lambda)$, is

$$p(\lambda) = -\left[\lambda^3 + \frac{1}{RC_1}\lambda^2 + \frac{1}{L}\left(\frac{1}{C_1} + \frac{1}{C_2}\right)\lambda + \frac{1}{LRC_1C_2}\right] \tag{3}$$

Since the circuit parameters R, L, C_1, C_2 are assumed to be positive, we may proceed directly to the Routh Array to verify that the roots of $-p(\lambda)$ [hence of $p(\lambda)$] have negative real parts. The Routh Array for $-p(\lambda)$ is

$$
\begin{bmatrix}
1 & (C_1 + C_2)/(LC_1C_2) \\
1/(RC_1) & 1/(LRC_1C_2) \\
1/(LC_2) & 0 \\
1/(LRC_1C_2) & 0
\end{bmatrix}
$$

Since all entries in the first column are positive, the Routh Criterion implies that the eigenvalues of A have negative real parts.

Suppose that the input voltage has the form $V_1 = a_0 e^{i\omega t}$. Then

$$
F = \begin{bmatrix} 0 \\ V_1'/R \\ V_1/L \end{bmatrix} = \alpha e^{i\omega t}, \quad \text{where } \alpha = \begin{bmatrix} 0 \\ a_0 i\omega/R \\ a_0/L \end{bmatrix} \tag{4}
$$

Since the eigenvalues of A have negative real parts, Theorem 7.6.3 implies that there is a unique periodic steady state $x^s(t)$ of period $2\pi/\omega$. We expect the steady state response to have the form $x^s = \beta e^{i\omega t}$ for some constant vector β. To find the vector β, insert x^s into system (1):

$$
i\omega e^{i\omega t} \beta = e^{i\omega t} A\beta + e^{i\omega t}\alpha \tag{5}
$$

and, solving (5) for β,

$$
\beta = [i\omega I - A]^{-1}\alpha, \quad \text{and} \quad x^s(t) = [i\omega I - A]^{-1}\alpha e^{i\omega t} \tag{6}
$$

The matrix inverse in (6) exists because every eigenvalue of A has negative real part, and, hence, the pure imaginary number $i\omega$ cannot be an eigenvalue of A.

The voltage $V_2(t)$ (i.e., the first component of $x^s(t)$) is considered to be the steady-state output voltage. A straightforward (but long) calculation using the formulas (4) and (6) shows that the response V_2 to the input $V_1 = a_0 e^{i\omega t}$ is

$$
V_2(t) = \frac{a_0}{p(i\omega)LRC_1C_2} e^{i\omega t} \tag{7}
$$

where p is the characteristic polynomial (3). Thus, the ratio of the amplitude of the steady state output voltage V_2 to amplitude of the input voltage V_1 is

$$
\left|\frac{V_2}{V_1}\right| = \frac{1}{|p(i\omega)LRC_1C_2|}
$$

$$
= \left|1 - LC_2\omega^2 + iR\omega(C_1 + C_2 - \omega^2 LC_1C_2)\right|^{-1} \tag{8}
$$

If ω is small, the ratio $|V_2/V_1| \approx 1$ and the circuit "passes" the input voltage with little change in its amplitude. However, if ω is large, then $|V_2/V_1|$ is small and the circuit "stops" the input voltage. For these reasons, the circuit is said to be a low-pass filter. Note that we considered another kind of circuit filter in Section 4.3; that circuit also acted as a low-pass filter.

EXAMPLE 7.7.1

A Particular Low-Pass Filter

Let $R = 1$, $L = 1$, $C_1 = C_2 = 0.001$, $V_1(t) = \cos \omega t$, $V_2(0) = 0$, $I_1(0) = I_2(0) = 0$. Figures 7.7.1 and 7.7.2 show that a voltage of circular frequency $\omega = 1$ passes through the circuit with almost no distortion (after an initial period when the transient volt-

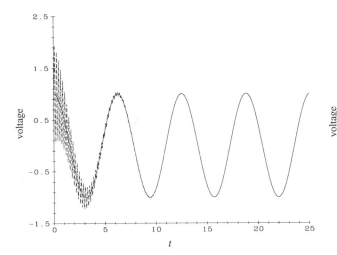

FIGURE 7.7.1 Input voltage $\cos t$ (solid), output voltage (dashed). See Example 7.7.1.

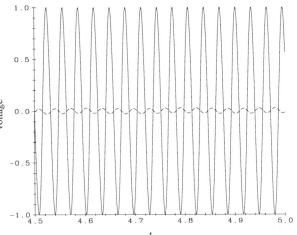
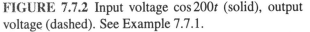

FIGURE 7.7.2 Input voltage $\cos 200t$ (solid), output voltage (dashed). See Example 7.7.1.

ages are still significant), while an input voltage of circular frequency 200 is severely attenuated after 4.5 units of time have elapsed.

The filtering property of the circuit applies not only to a periodic input of the form $F = \alpha e^{i\omega t}$, but to any periodic input. See Problem 4 for the case of a triangular wave input voltage.

Suppose that a low-frequency input voltage is "contaminated" by high-frequency noise. The low-pass circuit acts as a filter, effectively stops the noise, and returns a voltage virtually identical to the "pure" input voltage.

EXAMPLE 7.7.2

Filtering Out Noise

Suppose that the input voltage consists of the terms $10 \cos t - 5 \cos 2t$, plus the "noise" voltage $0.5 \cos 200t$. In Figure 7.7.3, the solid curve is the graph of $10 \cos t - 5 \cos 2t$, while the dashed curve is the output voltage $V_2(t)$, given the contaminated input, $V_1 = 10 \cos t - 5 \cos 2t + 0.5 \cos 200t$. The "noise" in the input is clearly filtered out in a very short time.

The circuit acts as a low-pass filter for all positive values of the circuit parameters, R, L, C_1, and C_2, but the amount of attenuation of the output voltage will vary with the values of the parameters.

Sensitivity of Lead Levels to Antilead Medication

A system of ODEs that models the flow of lead through the blood, tissue, and bones is introduced in Section 7.1 (see system (2) of Section 7.1). The rate coefficients and the intake rate of lead into the blood are specified in system (4) of Section 7.1. Figure 7.1.2 of Section 7.1 shows how the lead levels build up over an 800-day period if the initial levels are zero.

What happens if an antilead medication is administered? The medication speeds up the removal rate of lead from the bones by increasing a rate constant. The model

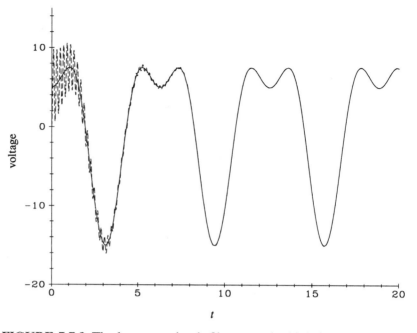

FIGURE 7.7.3 The low-pass circuit filters out the high-frequency noise in the input voltage. See Example 7.7.2.

system of linear ODEs is

$$x'_1 = -0.0361x_1 + 0.0124x_2 + k_{13}x_3 + 49.3$$
$$x'_2 = 0.0111x_1 - 0.0286x_2 \tag{9}$$
$$x'_3 = 0.0039x_1 - k_{13}x_3$$

where k_{13} is the rate constant that is increased by dosages of antilead medication, and $x_1(t)$, $x_2(t)$, and $x_3(t)$ are the respective amounts of lead (in micrograms) in the blood, tissue, and bones at time t (in days).

First, let us use Gerschgorin disks to show that the eigenvalues of the system matrix

$$A = \begin{bmatrix} -0.0361 & 0.0124 & k_{13} \\ 0.0111 & -0.0286 & 0 \\ 0.0039 & 0 & -k_{13} \end{bmatrix}$$

all have negative real parts. The three Gerschgorin row disks in the complex plane are defined by

$$|z + 0.0361| \leq 0.0124 + k_{13}, \qquad |z + 0.0286| \leq 0.0111, \qquad |z + k_{13}| \leq 0.0039$$

and the three Gerschgorin column disks are defined by

$$|z + 0.0361| \leq 0.0111 + 0.0039, \qquad |z + 0.0286| \leq 0.0124, \qquad |z + k_{13}| \leq k_{13}$$

For values of k_{13} in the range between 0.035 and 0.000035, the row disks overlap the right half of the complex plane, and so we cannot conclude from examination of the row disks that the eigenvalues lie inside the left half-plane.

What about the column disks? The first two column disks lie inside the left half-plane, but the third disk is centered at $(-k_{13}, 0)$ and just touches the origin, thus raising the possibility that 0 is an eigenvalue. We can try the same technique used in Example 7.6.7 in Section 7.6, that is pull the third column disk into the left half-plane by

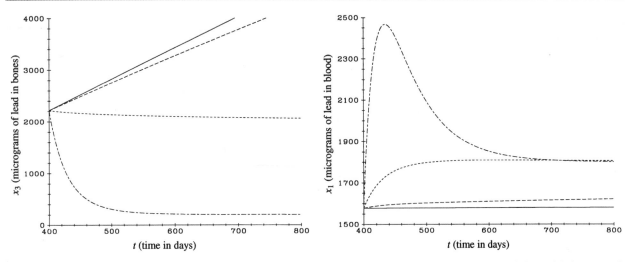

FIGURE 7.7.4 Lead levels in bones drop with increased levels of antilead medication after 400 days. See Example 7.7.3.

FIGURE 7.7.5 Lead levels in blood rise with increased levels of antilead medication after 400 days. See Example 7.7.3.

shrinking its radius a little, at the same time as the radii of the other two disks are slightly increased. For example, if the matrix P is as given below, then the matrix $B = P^{-1}AP$ has all of its Gerschgorin column disks inside the left half-plane:

$$P = \begin{bmatrix} 1 & 0 & 0 \\ 0 & 1 & 0 \\ 0 & 0 & 0.5 \end{bmatrix}, \qquad B = P^{-1}AP = \begin{bmatrix} -0.0361 & 0.0124 & 0.5k_{13} \\ 0.0111 & -0.0286 & 0 \\ 0.0078 & 0 & -k_{13} \end{bmatrix}$$

The three Gerschgorin column disks of B are defined by

$$|z + 0.0361| \le 0.0111 + 0.0078, \qquad |z + 0.0286| \le 0.0124, \qquad |z + k_{13}| \le 0.5k_{13}$$

The Gerschgorin column disks of B are inside the left half of the complex plane for every positive value of k_{13}. Because B and A have the same eigenvalues, all eigenvalues of A have negative real parts. Thus, as time $t \to \infty$, all solutions of the system tend to the constant steady state,

$$x^s = -A^{-1} \begin{bmatrix} 49.3 \\ 0 \\ 0 \end{bmatrix} \qquad (10)$$

EXAMPLE 7.7.3

Getting the Lead Out
What happens to the lead levels if the amount of antilead medication increases? Suppose that antilead medication is taken from the 400th day on. Lead levels in the bones and in the blood are plotted (Figures 7.7.4 and 7.7.5, respectively). The solid curves in the two figures show the response if $k_{13} = 0.000035$, while the long-dashed, short-dashed, and long/short-dashed curves show the responses if heavier doses of medication increase the transfer coefficient k_{13} to 0.00035, 0.0035, 0.035, respectively. These increased levels of medication help get the lead out of the bones, but at the expense of more lead in the blood (also in the tissue).

Comments

We have shown how the solutions of the linear mathematical models of a low-pass filter and of lead in the human body are sensitive to changes in parameters of the model. We have not constructed the exact solution formulas for the model equations. Instead, we have used indirect techniques such as the Routh Criterion and Gerschgorin disks to check that the eigenvalues of the system matrices have negative real parts, and we have used a numerical ODE solver to solve and plot solution component graphs.

PROBLEMS

1. (*Low-Pass Filter*). Explain why the circuit sketched in the margin of this section acts as a low-pass filter for all positive values of L, R, C_1, and C_2 if the input F is given by (4). [*Hint*: Show that for $|\omega|$ small, $|V_2(t)|$ is approximately $|a_0|$, while if $|\omega|$ is large, $|V_2(t)|$ is approximately $|a_0|/\omega^3 LRC_1C_2$.]

2. (*Low-Pass Filter: Computer Simulations*). The model equations for the circuit discussed in the text are

$$x'_1 = x_3/C_2$$
$$x'_2 = -x_2/(RC_1) + x_3/(RC_1) + V'_1/R$$
$$x'_3 = -x_1/L - Rx_2/L + V_1/L$$

Let $V_1 = \cos\omega t$. We shall say that the circuit *passes* (amplifies) the input voltage V_1 if the output voltage $x_1 = V_2(t)$ eventually has amplitude of at least 90% (110%) of the amplitude of V_1. We say that the circuit *stops* the input voltage if the amplitude of the output voltage is eventually less than 10% of the amplitude of V_1. Otherwise, we say that the circuit is a *partial filter* for the input. In each case, use a numerical solver to plot a component graph of $x_1(t) = V_2(t)$ and to decide whether the circuit amplifies, passes, stops, or acts as a partial filter for $V_1 = \cos\omega t$. Use $x_1(0) = x_2(0) = x_3(0) = 0$ as initial data; solve over a sufficiently long time span that the solution is essentially in the steady state at the end of the time interval. See also Problem 3.

   (a) $L = 1$, $R = 1$, $C_1 = C_2 = 1$; $\omega = 1, 2, 50$

   (b) $L = 1$, $R = 0.1$, $C_1 = C_2 = 1$; $\omega = 1, 2, 10$. The circuit amplifies one of these input frequencies. Which one, and by how much?

   (c) $L = 1$, $R = 100$, $C_1 = C_2 = 0.01$; $\omega = 1, 2, 50$

3. (*Low-Pass Filter: Amplitude Analysis*). In the text it is shown that if the input voltage $V_1(t)$ for the system of Problem 2 is $a_0 e^{i\omega t}$, then the ratio of the output voltage $V_2(t)$ to $V_1(t)$ is given by (8). Use this formula to justify the following claims (see the text of Problem 2 for the terminology).

   (a) If $L = 1$, $R = 1$, $C_1 = C_2 = 1$, then the frequencies $\omega = 1, 2, 50$ are, respectively, passed, partially filtered, and stopped.

   (b) If $L = 1$, $R = 0.1$, $C_1 = C_2 = 1$, then the frequencies $\omega = 1, 1.5, 10$ are, respectively, passed and amplified, passed, and stopped.

   (c) If $L = 1$, $R = 100$, $C_1 = C_2 = 0.01$, then the frequencies $\omega = 0.1, 1, 50$ are, respectively, passed, partially filtered, and stopped.

4. (*Triangular Wave Input*). In the circuit system of Problem 2, set $L = 1$, $R = 1$, $C_1 = C_2 = 1000$, $V_1(t) = 2\,\mathrm{trw}(t, 100, p) - 1$.

   (a) Explain why $V'(t) = (8/p)\,\mathrm{sqw}(t, 50, p) - 4/p$.

(b) Plot $V_1(t)$ and $x_1 = V_2(t)$ in the two cases $p = 2\pi, \pi/100$. From the graphics decide whether each frequency is passed and amplified, passed but not amplified, partially filtered, or stopped.

5. *(Sensitivity of Lead System to Lead Intake Levels).* The intake rate of lead into the blood is 49.3 micrograms per day in system (9). If the lead levels in food, air, and water are lowered (thus reducing the rate below 49.3), the steady-state levels in the blood, tissues, and bones should diminish.

   (a) Explain why if the intake rate is lowered from I to αI, where α is a constant between 0 and 1, then the equilibrium levels in each of the compartments are lowered by the same proportion α. [*Hint:* According to Section 7.6, the steady-state vector of $x' = Ax + F_0$, where F_0 is constant and all eigenvalues of A have negative real parts, is $-A^{-1}F_0$.]

   (b) Plot graphs of the amount of lead in the blood, tissues, and bones over a 1200-day period if $x_1(0) = x_2(0) = x_3(0) = 0$, if the system matrix A is that of system (9), and if the intake rate is reduced from 49.3 micrograms per day to (49.3)/2 per day from the 400th day on. Then repeat if the intake rate is cut in half again from the 800th day on to (49.3)/4 micrograms per day. Compare the computed results with predictions based on part **(a)**.

   (c) Suppose that the intake rate I of lead into the blood diminishes exponentially from the 400th day onward:

   $$I = \begin{cases} 49.3, & 0 \le t \le 400 \\ 49.3\exp(a(400 - t)), & t > 400 \end{cases}$$

   where a is a positive parameter. Use a solver/plotter package to solve system (9) with this intake function for various positive values of a, $0 < a < 0.1$. Estimate the smallest value of a that will ensure that the lead level in the bones will never exceed 3000 micrograms.

6. *(Seasonally Varying Lead Levels).* Replace $I = 49.3$ in system (9) by $I = 49.3[1 + 9\sin^2(2\pi t/365)]/10$. Explain why this models seasonal variations in lead intake rates. Then plot $x_3(t)$ over a long enough time span that you can approximate the maximal lead level in the bones. What is this level and when is it first reached?

7. *(Open and Closed Compartment Systems).* The lead system is an example of an *open compartment system*, i.e., a system in which every component opens directly or indirectly through other compartments to the "environmental compartment," Other linear compartment systems are closed. In this project, you are asked to explain and give examples of the various kinds of compartment models. Here are the definitions you are to use. In every case the system considered is $x' = Ax + F$ where $x = [x_1 \cdots x_n]^T$, $F = [F_1 \cdots F_n]^T$, and $A = [a_{ij}]$ is an $n \times n$ matrix of real constants.

   ❖ **Linear Compartment System.** The linear system $x' = Ax + F$ where the constant matrix A and the vectors x and F are as given above, is a *linear compartment system* if

   $$a_{ii} \le -\sum_{k=1, k \neq i}^{n} a_{ki}, \quad i = 1, \ldots, n$$

   $$a_{ij} \ge 0, \quad i, j = 1, \ldots, n; \quad i \neq j$$

   $$F_i(t) \ge 0, \quad i = 1, \ldots, n$$

   ❖ **Open Compartment System.** A linear compartment system is *open to the external absorbing compartment* if for every $j = 1, \ldots, n$ there is a sequence $i_1, i_2, \ldots, i_r$ such that

$$a_{i_1 j} > 0, \quad a_{i_2 i_1} > 0, \ldots, a_{i_r i_{r-1}} > 0, \quad \sum_{i=1}^{n} a_{ii_r} < 0$$

❖ **Closed Compartment System.** A linear compartment system is *closed* if each column of A sums to zero:

$$\sum_{i=1}^{n} a_{ij} = 0, \quad j = 1, \ldots, n$$

Use the general lead system (2) in Section 7.1 and the particular system (9) of this section as examples in your discussion of the meaning of the definitions of open and of closed linear compartment systems. Address the following points:

- Why are the diagonal elements of A nonpositive?

- Why is the magnitude of $|a_{ii}|$ of the ith diagonal entry at least as large as the sum of all other entries in column i?

- What does it mean if $a_{ii} < -\sum_{k=1, k\neq i}^{n} a_{ki}$? [*Hint:* Think of flow from compartment i directly to an external absorbing compartment 0.]

- Why is $a_{ij} \geq 0$ if $i \neq j$?

- For an open system, what does the condition $\sum_{i=1}^{n} a_{ii_r} < 0$ mean? What about the other conditions for openness?

- Explain why the augmented lead system which consists of the rate equations of (2), (3) in Section 7.1 is closed.

- Can you close any open system by including an external absorbing compartment?

- After reading Section 1.6, draw several "boxes and arrows" diagrams with specific numbers for the rate constants k_{ij}. Write out the corresponding linear compartment systems. Are they open? Closed?

- Now write down some new open and closed compartment systems with specific numbers for the rate constants k_{ij}. Draw appropriate "boxes and arrows" diagrams. Begin by explaining how you get from the system

$$x_1' = -20x_1 + 4x_2 + x_3$$
$$x_2' = x_1 - 4x_2 + 2x_3$$
$$x_3' = 3x_1 - 8x_3 + 1$$

to the diagram

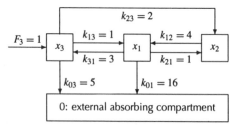

- Explain why the inulin system of Section 7.1, Problem 5, is open.

- Use your computer solver to graph component curves for several driven open and closed systems. The "Washout Theorem" asserts that if the driving function F of an open system is "turned off," then the substance levels in all compartments tend to zero as $t \to \infty$. Is this observed for your open systems? The "Equilibrium Theorem" asserts that if the driving function F of a closed system is turned off, then the amount of the substance in each compartment tends to a constant level. Is this observed to be true for your closed systems? Can you explain why the system matrix A of a closed system must be singular?

- Explain why the Aleutian carbon ecosystem for carbon flow (see below) is closed, but first write out the full system of rate equations using $x_i(t)$ for the amount of carbon in compartment i at time t and, as usual, k_{ij} for the rate coefficient for flow from compartment j into compartment i. Is there an external absorbing compartment?

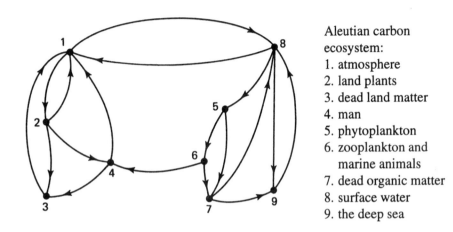

Aleutian carbon ecosystem:
1. atmosphere
2. land plants
3. dead land matter
4. man
5. phytoplankton
6. zooplankton and marine animals
7. dead organic matter
8. surface water
9. the deep sea

You may want to consult the text by M.R. Cullen, *Linear Models in Biology*, Ellis Horwood Ltd., Chichester, England (1985).

Nonlinear Systems: Stability

Questions about differential systems and their solutions fall into two categories:

1. *Is there any solution at all? If there are solutions, how many are there?*

2. *Is the system stable? What happens to solutions when system parameters change? What is the asymptotic behavior of solutions? Does the system have any periodic solutions? Is there chaotic behavior?*

Questions of type 2 dominate this chapter and the next. These questions arise in the design and analysis of real systems. Qualitative properties of systems can be extracted without having solution formulas—a fortunate circumstance since most real systems are not linear and solution formulas are hard to come by. For simplicity, we deal only with autonomous systems. The cover figure shows an integral surface and some orbits associated with a rotating rigid body (see Problem 8, Section 8.3).

8.1 Stability

The behavior of the orbits of a dynamical system is strongly affected by the location and the nature of the constant and the periodic solutions, called *equilibrium points* and *cycles*, respectively. Equilibrium points and their stability characteristics are studied in this chapter; cycles are taken up in the next chapter. It is assumed throughout that the systems considered are autonomous and meet the conditions of the Existence and Uniqueness Theorem (Theorem 6.2.1) in a region S of state space. By Theorem 6.2.3

distinct (maximally extended) solutions of an autonomous system in S define distinct orbits in state space that never touch or cross one another. Because an individual orbit can thus be tracked in state space, portraits of the orbits provide a good understanding of the general behavior of the solutions. These portraits, usually constructed by numerical solvers, are a basic tool in the study of autonomous systems. Alternatively, time-state curves may be plotted in time-state space, which is state space augmented by a time line, and thus is of dimension $n + 1$ if state space is n-dimensional. Of course, if the dimension of state space or time-state space is more than three, then we have to turn to other graphical alternatives. First of all, if $x(t)$ denotes a (vector) solution of a system, then each of its scalar components $x_1(t), \ldots, x_n(t)$ can be plotted against t. These component graphs tell us a lot about the behavior of the solution. Second, one can plot any two or three of the components $x_i(t)$ of a solution in a projection of the full state space onto a subspace. For example, if $x(t) = (x_1(t), x_2(t), x_3(t), x_4(t))$, then $(x_1(t), x_2(t))$ and $(x_1(t), x_2(t), x_3(t))$ trace out curves in $\mathbf{R}^2$ and $\mathbf{R}^3$, respectively, while the full orbit $x(t)$ would require $\mathbf{R}^4$. A deeper understanding of the behavior of a solution may require inspection of several of these component graphs and of several projections into subspaces of state space.

Although computer graphics allow us a marvelous glimpse into the mysteries of solution behavior, we need theoretical guides to tell us what is possible and what to look for, and the graphs of one or two solutions may not do that. Even more, we need a collection of useful definitions that encode properties of systems and their solutions, and that represent important physical ideas.

We begin with the central idea of the stability of a system at an equilibrium point, an idea based on what we already know of the behavior of the orbits of a constant-coefficient homogeneous linear system. Suppose that the n-vector function $f(x)$ is continuously differentiable on a region S in $\mathbf{R}^n$. Under these conditions, the initial value problem (IVP)

$$x' = F(x), \quad x(0) = x^0, \quad \text{for } x^0 \text{ in } S \tag{1}$$

has a unique, maximally extended solution which we denote by $x(t, x^0)$. The simplest of these solutions are the equilibrium points.

❖ **Equilibrium Point.** Suppose that the point p in S is such that $F(p) = 0$. Then p is an *equilibrium point* for the autonomous system

$$x' = F(x) \tag{2}$$

Note that if p is an equilibrium point, then $x = p$, for all t, is a constant solution.

The term "stability of the system at the equilibrium point p" has virtually the same technical meaning as it has in common usage: the system (2) is stable at the equilibrium point p if the solution $x(t, x^0)$ of IVP (1) originating at any point x^0 near enough to p does not stray more than a prescribed distance from p as t increases. The Euclidean distance function $\|q - p\|$ is used for the distance between points q and p. We have the following definition of stability:

❖ **Stability.** The system $x' = F(x)$ is *stable* at an equilibrium point p if for each positive number ε there is a positive number δ such that for all $t \geq 0$, $\|x(t, x^0) - p\| < \varepsilon$ whenever $\|x^0 - p\| < \delta$. A system that is not stable at p is said to be *unstable* at p.

Stability means that as time increases each orbit remains in a prescribed neighborhood of p [i.e., $\|x(t, x^0) - p\| < \varepsilon$, for all $t \geq 0$] provided that the initial displacement x^0

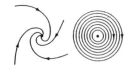

is sufficiently close to p (i.e., $\|x^0 - p\| < \delta$). The central idea of stability is that for *every* positive number ε, there is *some* δ with the desired property.

Stability may be *local* in the sense that only orbits for which $\|x^0 - p\|$ is small remain near p with advancing time. The orbit defined by $x(t, x^0)$ may wander away from the neighborhood of p with increasing t if $\|x^0 - p\|$ is too large, or even become unbounded. See the margin figures for sketches of the orbits of some stable and some unstable systems at an equilibrium point.

The following examples illustrate these ideas and show how the definition of stability is used. The variables x, y, and z are used as names of state variables.

EXAMPLE 8.1.1

A Stable System

We show that the conditions for the stability of a system at an equilibrium point (the origin in this case) are satisfied by the linear system

$$x' = -x, \qquad y' = -2y \tag{3}$$

The general solution is

$$x(t) = c_1 e^{-t}, \quad y(t) = c_2 e^{-2t}, \quad c_1, c_2 \text{ any constants} \tag{4}$$

The square of the distance V from the point $(x(t), y(t))$ to the origin is

$$V = x^2(t) + y^2(t) = c_1^2 e^{-2t} + c_2 e^{-4t}$$

Since the rate of change of V, $V' = -2c_1^2 e^{-2t} - 4c_2^2 e^{-4t}$, is negative if c_1, c_2 are not both zero, the distance from point $(x(t), y(t))$ to the origin decreases as time increases. Thus, for each positive ε, let $\delta = \varepsilon$, and we have that if $(x^2(0) + y^2(0))^{1/2} < \delta$, then $(x^2(t) + y^2(t))^{1/2} < \varepsilon$, $t \geq 0$, and so system (3) is stable at the origin. Figure 8.1.1 shows the unit circle (dashed line) and several orbits with initial points inside the circle. As time increases the orbits remain inside the circle.

Example 8.1.1 is simple to analyze, but the next example is harder.

EXAMPLE 8.1.2

The Stability of a Hooke's Law Spring

In Section 3.1 it is shown that the ODE that models the motion of a body of mass m suspended by a Hooke's Law vertical spring with no damping is

$$mx'' + kx = 0$$

where x is the displacement from equilibrium, and k is a positive Hooke's Law constant. The equivalent first-order system is

$$x' = y, \qquad y' = -(k/m)x \tag{5}$$

whose general solution is given by

$$x = A\cos(\sqrt{k/m}\, t + \phi), \qquad y = -A\sqrt{k/m}\sin(\sqrt{k/m}\, t + \phi) \tag{6}$$

where A and ϕ are any constants. Suppose that $k/m = 4$, then the orbits defined by (6) form the family of ellipses

$$x^2 + y^2/4 = A^2 \tag{7}$$

whose minimal distance from the origin is $|A|$ and maximal distance is $2|A|$. Note that if $0 < A_1 < A_2$, the ellipse defined by A_2 encloses the ellipse defined by A_1. Thus, if we interpret the ε in the Stability Definition as $2|A|$, and if we set $\delta = \varepsilon/2$, then we see that, if $(x^2(t) + y^2(t))^{1/2} < \delta$, then $(x^2(t) + y^2(t))^{1/2} < \varepsilon$ for all $t > 0$; system (5) for the Hooke's Law spring is stable at the origin. Figure 8.1.2 shows $\varepsilon = 1$ and $\delta = 1/2$ circles (both dashed). Shown are three nonconstant orbits whose initial points on the

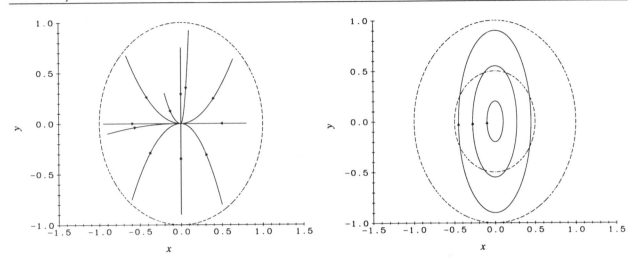

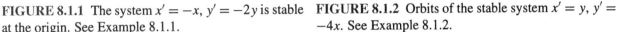

FIGURE 8.1.1 The system $x' = -x$, $y' = -2y$ is stable at the origin. See Example 8.1.1.

FIGURE 8.1.2 Orbits of the stable system $x' = y$, $y' = -4x$. See Example 8.1.2.

x-axis, $x_0 = 0.1$, 0.275, 0.45, all lie inside the δ-circle; these orbits remain within the ε circle for all $t \geq 0$.

EXAMPLE 8.1.3

Instability of a Planar System at the Origin
The general solution of the linear system $x' = x$, $y' = -2y$, is $x = ae^t$, $y = be^{-2t}$, where a and b are any real constants. The system is unstable at the origin $(0, 0)$ since as $t \to \infty$ the solution $x = ae^t$, $y = 0$ becomes unbounded for every nonzero a, no matter how small (see Figure 8.1.3 for some orbits).

The systems of the first three examples are linear and planar, but the concepts of stability and instability also apply to nonlinear systems, and in any number of dimensions.

EXAMPLE 8.1.4

Stable Nonlinear System in $\mathbf{R}^3$
The origin is an equilibrium point for the nonlinear system
$$\begin{aligned} x' &= -ax - by - cx(x^2 + y^2 + z^2) \\ y' &= bx - ay - cy(x^2 + y^2 + z^2) \\ z' &= -az - cz(x^2 + y^2 + z^2) \end{aligned} \qquad (8)$$
and the system is stable at the origin if a, b, and c are positive constants. One way to see this is to calculate the rate of change with respect to time of the distance of an orbital point $(x(t), y(t), z(t)) \neq (0, 0, 0)$ of system (8) from the origin. Actually, it is easier to find the rate of change of the square of the distance, $V = x^2 + y^2 + z^2$:
$$\frac{dV}{dt} = 2xx' + 2yy' + 2zz' = -2(x^2 + y^2 + z^2)(a + c(x^2 + y^2 + z^2)) \qquad (9)$$
where some algebraic manipulation is needed to get the last equality. From (9), we see that the rate of change of V is negative except at the origin where $V' = 0$. That means that the orbit moves closer and closer to the origin as time advances. In terms of the definition of stability, given any positive number ε, if $[x^2(0) + y^2(0) + z^2(0)]^{1/2} < \varepsilon$, then $[x^2(t) + y^2(t) + z^2(t)]^{1/2} < \varepsilon$ for all $t \geq 0$, i.e., δ may be chosen to be ε in the definition of stability. Thus, system (8) is stable at the origin. Figure 8.1.4 displays

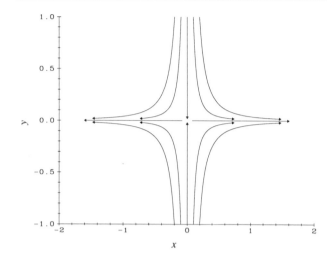

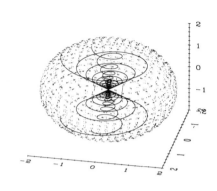

FIGURE 8.1.3 Orbits of the unstable system $x' = x$, $y' = -2y$. See Example 8.1.3.

FIGURE 8.1.4 Orbits of the stable system (8) that start inside the sphere of radius ε stay inside for $t \geq 0$ ($\varepsilon = 2$ in the graph). See Example 8.1.4.

four orbits of (8) with initial points just inside the sphere of radius 2 (shaded surface), where $a = 1.0$, $b = 20$, $c = 0.1$. The four orbits lie inside this sphere for all $t \geq 0$.

A system may have several equilibrium points, and so the system may be stable at some of these points and unstable at others. Here is a one-dimensional example.

EXAMPLE 8.1.5

An ODE with a Stable and an Unstable Equilibrium Point
The scalar ODE, $x' = x(1 - x)$, is a logistic equation (see Section 1.5) with the x-axis as the state space. The points $x = 0$ and $x = 1$ are the only equilibrium points. The equation is separable and solutions $x(t, x^0)$, where $x^0 \geq 0$, are given by

$$x(t, x^0) = \frac{x^0}{x^0 + (1 - x^0)e^{-t}}, \qquad t \geq 0$$

Note that the above formula also defines a solution for $x^0 < 0$, but in that case the solution escapes to $-\infty$ in finite (positive) time. Observe that for every $x^0 > 0$, $x(t, x^0)$ rises (falls) toward 1 if $0 < x_0 < 1$ ($x_0 > 1$) as t increases from 0 to $+\infty$. Thus, we see that the ODE is stable at the equilibrium point 1, but unstable at the equilibrium point 0.

The stability properties of the systems of Examples 8.1.1, 8.1.2, 8.1.3, and 8.1.5 at equilibrium points are determined by examining formulas for the solutions. That approach is not possible for the many systems for which there are no solution formulas. The system of Example 8.1.4 illustrates another approach to stability questions, an approach which does not require solution formulas and turns out to be of vital importance (see Sections 8.2–8.4). But first more terminology is needed.

Asymptotic Stability

System (3) in Example 8.1.1 is stable at the origin, but the stability is considerably stronger than that for system (5) of Example 8.1.2. The orbits of system (3) that start out near the origin not only stay nearby as time advances, but also approach the equilibrium point as $t \to \infty$. This is also true for the ODEs of Examples 8.1.4 (at the origin) and 8.1.5 (at the point $x = 1$). This stronger kind of stability is called asymptotic stability.

❖ **Asymptotic Stability.** The system $x' = F(x)$ is *asymptotically stable* at an equilibrium point p if the system is stable at p, and if for all x^0 in some region S containing p, we have $\lim_{t \to \infty} \| x(t, x^0) - p \| = 0$. The asymptotic stability at p is *global* if $\lim_{t \to \infty} \| x(t, x^0) - p \| = 0$ for all x^0 in $\mathbf{R}^n$, and *local* otherwise. The *region of asymptotic stability* for the system at p is the set of all points x^0 such that $\lim_{t \to \infty} \| x(t, x^0) - p \| = 0$.

EXAMPLE 8.1.6

Global Asymptotic Stability
Look back at the system of Example 8.1.1, $x' = -x$, $y' = -2y$, which has already been shown to be stable at the origin. In fact it is globally asymptotically stable since we showed in Example 8.1.1 that for every solution $x = x(t)$, $y = y(t)$, we have $x^2(t) + y^2(t) \to 0$ as $t \to \infty$.

EXAMPLE 8.1.7

Global Asymptotic Stability in $\mathbf{R}^3$
Consider system (8) of Example 8.1.4. For $V = x^2 + y^2 + z^2$, we have from (9) that

$$\frac{d(V)}{dt} = -2V(a + cV) \le -2aV$$

since a and c are positive. Dividing by V, we have that

$$\frac{1}{V}\frac{d(V)}{dt} \le -2a$$

Integrating each side of the last inequality from 0 to t and then exponentiating:

$$\ln V(t) - \ln V(0) \le -2at$$

$$V(t) \le V(0)e^{-2at}$$

It follows that $V(t) \to 0$ as $t \to +\infty$, and so $x(t), y(t), z(t) \to (0, 0, 0)$. System (8) is asymptotically stable at the origin, and this asymptotic stability is global.

Note that the region of asymptotic stability for the scalar ODE of Example 8.1.5 at the point $x = 1$ is the interval $x > 0$, so that the ODE is only locally asymptotically stable at $x = 1$. Here is another example.

EXAMPLE 8.1.8

Local Asymptotic Stability
Following the techniques given in Section 6.2, the system

$$\begin{aligned} x' &= (-x - 10y)(1 - x^2 - y^2) \\ y' &= (10x - y)(1 - x^2 - y^2) \end{aligned} \tag{10}$$

may be written in polar coordinates r and θ as

$$r' = -r(1 - r^2)$$

$$\theta' = 10(1 - r^2)$$

The reader may use sign analysis (see Section 2.2) on the rate equation for r to show that if $r(t)$ is a solution of the rate equation for r and if $0 < r(0) < 1$, then $r(t)$ is a decreasing function of t. Certainly, $r' < 0$ if $0 < r < 1$, and orbits inside the unit circle move closer to the origin with increasing time. Alternatively, the rate equation for r

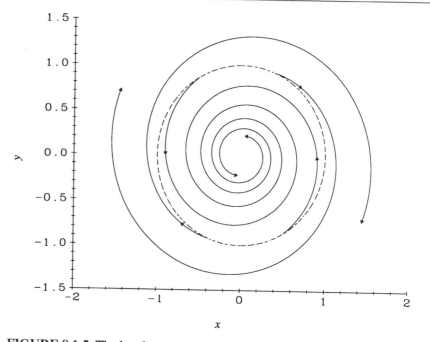

FIGURE 8.1.5 The local asymptotic stability of system (10) at the origin; the dashed curve is the unit circle of equilibrium points. See Example 8.1.8.

can be solved by separating variables, integrating, and then solving for r as a function of t and the initial value r_0. From that formula it can be seen that $r \to 0$ as $t \to \infty$ if $0 < r_0 < 1$. However, the asymptotic stability of the system at the origin cannot be global because all points on the unit circle are equilibrium points. In fact, the region of asymptotic stability for the origin is the interior of the unit circle. See Figure 8.1.5 for four nonconstant orbits (solid curves) of this system; the dashed curve is the unit circle of equilibrium points.

Attractors

An asymptotically stable equilibrium point p attracts all nearby points in the sense of the following definition:

> ❖ **Attractors.** An equilibrium point p of $x' = F(x)$ is an *attractor* if for all x^0 in a region S containing p, $\lim_{t \to \infty} x(t, x^0) = p$, i.e., $\lim_{t \to \infty} \|x(t, x^0) - p\| = 0$.

Thus, the system $x' = F(x)$ is asymptotically stable at p if (i) it is stable at p, and if (ii) p is an attractor. The terms *local attractor*, *global attractor*, and *region of attraction* (or *basin of attraction*) are used with the obvious meanings. The origin is a global attractor for the system of Examples 8.1.1 and 8.1.4, but only a local attractor for the system of Example 8.1.8. Warning! An attractor need not be stable, and the next example illustrates this curious phenomenon.

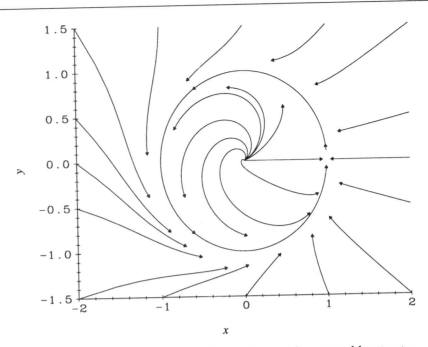

FIGURE 8.1.6 The equilibrium point $x = 1$, $y = 0$ is an unstable attractor for system (11). See Example 8.1.9.

EXAMPLE 8.1.9

An Unstable Attractor

The system

$$x' = x - y - x^3 - xy^2 + xy(x^2 + y^2)^{-1/2}$$
$$y' = x + y - x^2y - y^3 - x^2(x^2 + y^2)^{-1/2} \qquad (11)$$

has the following form in polar coordinates:

$$r' = r(1 - r^2) \qquad (12)$$
$$\theta' = 1 - \cos\theta$$

There are two equilibrium points, $x = y = 0$ (i.e., $r = 0$), and $x = 1$, $y = 0$ (i.e., $r = 1$, $\theta = 2k\pi$). The equilibrium point $(1, 0)$ attracts all orbits (except the equilibrium point at the origin), but it is not stable because if $x_0 = 1$, $y_0 = \theta_0 > 0$, then the corresponding orbit exits $(1,0)$ as t increases from $-\infty$, goes counterclockwise once around the unit circle and approaches $(1,0)$ as $t \to \infty$. See Figure 8.1.6 for an orbit portrait, and Problem 6 for the details and another example of an unstable attractor.

We use the expression "the equilibrium point repels the orbit" to mean that for a specific equilibrium point the orbit exits the equilibrium point as t increases from $-\infty$. The equilibrium point is a *repeller* if *all* nearby orbits exit the point as t increases from $-\infty$.

Neutral Stability

The stability properties of the system of Example 8.1.2 at the equilibrium point at the origin are weaker than those of the systems of Examples 8.1.1 and 8.1.4. There is a term for that weak kind of stability.

❖ **Neutral Stability.** The system $x' = F(x)$ is *neutrally stable* at an equilibrium point p if the system is stable at p, but not asymptotically stable.

The system of Example 8.1.2 is only neutrally stable because its orbits do not approach the origin. The system is stable, but the origin is not an attractor.

From now on, the terminology will frequently be simplified or inverted if there is no danger of ambiguity. For example, the statement "system (2) is stable at an equilibrium point" may be truncated to "system (2) is stable" or inverted to "the equilibrium point is stable." Warning: This simplification should be used very carefully if a system has several equilibrium points.

Linear Systems and Stability

It is no accident that Examples 8.1.1, 8.1.2, and 8.1.3 are constant-coefficient homogeneous linear systems since such systems provide the simplest examples of the varieties of stability and instability. The following result is a consequence of the analysis in Sections 7.3 and 7.6.

THEOREM 8.1.1

> Stability Properties of a Linear System. Let A be an $n \times n$ matrix of real constants. If λ is an eigenvalue of A, let m_λ and d_λ denote, respectively, the multiplicity of λ as a zero of the characteristic polynomial of A and the dimension of λ's eigenspace. The system $x' = Ax$ is:
>
> 1. Globally asymptotically stable at the origin if and only if every eigenvalue of A has negative real part.
>
> 2. Neutrally stable at the origin if and only if every eigenvalue of A has nonpositive real part, there is at least one zero or pure imaginary eigenvalue, and $m_\lambda = d_\lambda$ for every zero or pure imaginary eigenvalue λ.
>
> 3. Unstable at the origin if and only if some eigenvalue of A has positive real part, or else there is an eigenvalue λ with zero real part for which $m_\lambda > d_\lambda$.

Proof. We shall consider only Cases 1 and 3 here. In addition, we shall prove the "if" part but not the "only if." If all eigenvalues of A have negative real parts, then by the Decay Estimates (Theorem 7.6.4) there are positive constants α and M such that for all $t \geq 0$, $\|x(t, x^0)\| \leq M\|x^0\|e^{-\alpha t}$ for every solution $x(t, x^0)$ of $x' = Ax$. In the definition of stability, take $\delta = \varepsilon/M$ for each $\varepsilon > 0$ since $\|x(t, x^0)\| \leq M\|x^0\|e^{-\alpha t} < M(\varepsilon/M)e^{-\alpha t} \leq \varepsilon$ for $t \geq 0$ if $\|x^0\| < \varepsilon/M$. Thus, the origin is stable. The origin is also a global attractor since $e^{-\alpha t} \to 0$ as $t \to \infty$; the system is globally asymptotically stable.

Turning to Case 3 of the theorem, $x' = Ax$ is unstable at the origin if A has an eigenvalue $\lambda = \alpha + i\beta$ with positive real part α. For, if v is a corresponding eigenvector, then $x = e^{\alpha t}\text{Re}\left[e^{i\beta t}v\right]$ is a real-valued solution and $\|x(t)\| = e^{\alpha t}\|v\| \to \infty$ as $t \to \infty$. If $\alpha = 0$ and $m_\lambda > d_\lambda$, there are solutions with components of the form $t^k \cos(\beta t)v$, where $1 \leq k \leq m_\lambda - d_\lambda$. Thus, the system has solutions that are unbounded as $t \to +\infty$, and the system is again unstable.

Observe that Theorem 8.1.1 applies to the linear systems of Examples 8.1.1, 8.1.2, and 8.1.3. By Cases 1, 2, and 3 of the theorem the respective systems are globally asymptotically stable, neutrally stable, and unstable at the origin since the eigenvalues are, respectively, negative, simple and pure imaginary, and one positive.

Comments

Tests are needed to determine the stability, asymptotic stability, or instability of a nonlinear system at an equilibrium point without having to use solution formulas. The Lyapunov Method of Section 8.2 provides these tests.

PROBLEMS

1. (*Visual Stability Properties*). Decide whether the systems in the Altas Plates in Appendix D are asymptotically stable, neutrally stable, or unstable at the equilibrium points in the figures. Give *brief* explanations (no ε's or δ's needed!). If necessary, use a numerical solver near the equilibrium points to help you decide about the stability properties. For part (e) write the ODE as a system.

    (a) BIFURCATION A, B   (b) BIFURCATION D, E, F

    (c) BIFURCATION G, H, I   (d) DUFFING A

    (e) FIRST ORDER C   (f) PENDULUM A

    (g) PLANAR PORTRAIT C   (h) POINCARÉ-BENDIXSON B

2. (*Asymptotic Stability*). Why is each of the following systems asymptotically stable at the origin? Plot a portrait of the orbits (time-state curves for (a)–(c)) near the origin. Insert arrowheads on orbits to show increasing time. [*Hint*: Use Theorem 8.1.1 for the linear ODEs and systems. Solve the nonlinear ODEs explicitly, and use the solution formulas to determine the stability properties.]

    (a) $x' = -x$ (b) $x' = -x^3$

    (c) $x' = -\sin x$ (d) $x' = -4x, \quad y' = -3y$

    (e) $x' = x - 3y, \quad y' = 4x - 6y$ (f) $x' = -x + 4y, \quad y' = -3x - 2y$

    (g) $x' = -2x^3, \quad y' = -5y$ (h) $x' = -x + y + z, \; y' = -2y, \; z' = -3z$

    (i) $x' = -x^3, \quad y' = -y^3$ (j) $x' = -x - y, \quad y' = x - y, \quad z' = -z^3$

3. (*Neutral Stability*). Show that each of the following systems is neutrally stable at the origin. Plot a portrait of orbits near the origin. Insert arrowheads on orbits to show the direction of motion as time increases. [*Hint*: Use Theorem 8.1.1 for the linear ODEs and systems. Solve the nonlinear ODEs explicitly, and use the solution formulas to determine the stability properties.]

    (a) $x' = 2y, \quad y' = -8x$ (b) $x' = 0, \quad y' = -y^3$

    (c) $x' = y, \quad y' = -x, \quad z' = -z^3$

4. (*Instability*). The following systems are unstable at the origin. Explain why. Plot orbits (time-state curves for parts (a)–(c)) near the origin. Insert arrowheads on orbits to show the direction of motion as time increases. [*Hint*: Use Theorem 8.1.1 for the linear ODEs and systems. Solve the nonlinear ODEs explicitly, and use the solution formulas to determine the stability properties.]

    (a) $x' = x^2$ (b) $x' = \sin x$

    (c) $x' = |x|$ (d) $x' = 3x - 2y, \quad y' = 4x - y$

    (e) $x' = 3x - 2y, \quad y' = 2x - 2y$ (f) $x' = x + y, \quad y' = -x - y$

    (g) $x' = x^3, \quad y' = -3y$

5. (*A Stable Attractor*). Show that the system $x' = -x - 10y - x(x^2 + y^2), \; y' = 10x - y - y(x^2 + y^2)$ is globally asymptotically stable at the origin. [*Hint*: In polar coordinates the system becomes $r' = -r - r^3, \theta' = 10$.]

6.   (*Unstable Attractors*). Parts **(a)**–**(c)** outline the steps needed to show that system (11) has an unstable attractor at $x = 1$, $y = 0$. Part **(d)** gives another (but closely related) system with an unstable attractor.

(a) Separate variables and solve the decoupled system in polar coordinates: $r' = r(1 - r^2)$, $\theta' = 1 - \cos\theta$. Let $x = r\cos\theta$, $y = r\sin\theta$ ($r > 0$), and show that the equivalent Cartesian system is given by system (11)

(b) Show that the point $x = 1$, $y = 0$ is an unstable attractor for the system. [*Hint*: The unit circle consists of the equilibrium point $x = 1$, $y = 0$ and an orbit that leaves the point as t increases from $-\infty$, but returns to the point as $t \to +\infty$; that shows the instability of the equilibrium point.]

(c) Plot a portrait of orbits for the system in **(a)** using the xy-coordinates.

(d) Plot a portrait of the orbits of the system

$$x' = x - y - x^3 - x^2y + xy(x^2 + y^2)^{-1/2}$$
$$y' = x + y - x^2y - y^3 - x^2(x^2 + y^2)^{-1/2}$$

in the rectangle $|x| \le 3$, $|y| \le 3$, and describe what you see (but if you don't see an unstable attractor somewhere you need to plot more orbits!).

7.   Using "if . . . then . . . else" statements for defining functions, or other programming tricks, construct a single planar autonomous system that has all of the properties below, and then plot orbits:

- A curve of equilibrium points.

- A locally asymptotically stable equilibrium point.

- An unstable equilibrium point.

- A neutrally stable equilibrium point.

- An unstable attractor (see Example 8.1.9).

8.2   Lyapunov Functions

How can a system be tested for stability? If the system is linear, $x' = Ax$, the signs of the real parts of the eigenvalues of A generally determine the stability properties (see Theorem 8.1.1). It is not quite so simple if the system is nonlinear. In this section, the method of Lyapunov functions is introduced for this purpose. Lyapunov[1] defined the scalar functions (now named in his honor) as test functions for the stability characteristics of a system, linear or nonlinear, autonomous or nonautonomous. For simplicity, only autonomous systems are considered here.

Lyapunov observed that engineers and scientists determined the stability characteristics of complex physical systems by one of two kinds of measurements:

---

[1] Aleksandr Mikhailovich Lyapunov (1857–1918) introduced the functions in his doctoral dissertation for the University of Moscow. The dissertation was published in 1892 in the research journal *Comm. Soc. Math Khar'kov* and has been reprinted many times in the last 50 years; see *Stability of Motion*, Academic Press, New York, 1966, for an English translation. At first little attention was paid to Lyapunov's work, but in the 1930s mathematicians, physicists, and engineers of the former Soviet Union suddenly took up the question of stability of motion. The work of these researchers, based as it was on Lyapunov's ideas, dominated the field for decades. It is no exaggeration to say that Lyapunov's dissertation and the memoirs of Henri Poincaré are the foundation of most of the contemporary developments in the theory and applications of ordinary differential equations.

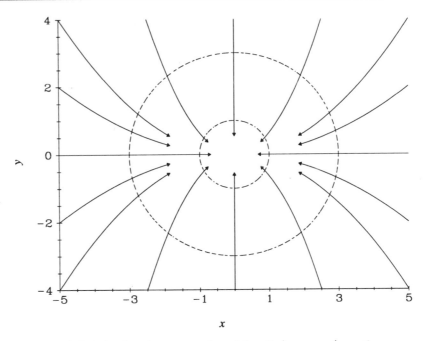

FIGURE 8.2.1 As time increases, the orbits of $x' = -x$, $y' = -2y$ move closer to the origin. See Example 8.2.1.

1. Measuring the "distances" of several evolving states of the system from a known equilibrium state and, over time, determining whether these distances diminish (or at least do not increase).

2. Measuring the energies of evolving states and determining whether these energies decline to a minimum level (or at least do not increase).

Here are three examples that illustrate Lyapunov's ideas.

EXAMPLE 8.2.1

Decreasing Distances
The square of the distance V of an orbital point $(x(t), y(t))$ from the equilibrium point $(0,0)$ of the system

$$x' = -x, \qquad y' = -2y \tag{1}$$

is

$$V = x^2(t) + y^2(t)$$

The rate of change of V is

$$V' = \frac{d}{dt} V(x(t), y(t)) = 2x\frac{dx}{dt} + 2y\frac{dy}{dt} = 2x(-x) + 2y(-2y) = -2x^2 - 4y^2$$

which is negative (except at the origin). Thus, without even using a solution formula, we expect the system to be stable (even asymptotically stable) at the origin. See Figure 8.2.1 for the graphs of the level sets $V = 9, 1$ (dashed) and of several orbits crossing these level sets and moving inward closer to the origin as time increases.

We used the Euclidean distance in this example, but there are other measures of distance and the next example shows one of these.

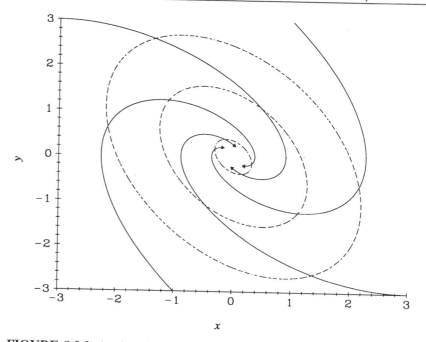

FIGURE 8.2.2 As time increases, orbits of $x' = y$, $y' = -x - y$ move inward across the level sets of $V = 3x^2 + 2xy + 2y^2$. See Example 8.2.2.

EXAMPLE 8.2.2

Another Measure of Distance: Damped Hooke's Law Spring

Suppose that $(x(t), y(t))$ is an orbital point of the damped Hooke's Law spring system

$$x' = y$$
$$y' = -x - y \tag{2}$$

Note that this is the system equivalent of the ODE, $mx'' + cx' + kx = 0$, where x is the displacement from equilibrium of a body of mass m suspended by a vertical spring whose motion is governed by Hooke's Law with damping (Section 3.1). In this case, we have set $m = 1$, $c = 1$, $k = 1$. Now measure the "square distance" from the orbit point $(x(t), y(t))$ to the equilibrium point $(0, 0)$ by

$$V = 3x^2 + 2xy + 2y^2 = 5(x + y)^2/3 + (2x - y)^2/3$$

which has positive values everywhere, except at $x = 0$, $y = 0$ where its value is zero. How does $V(x(t), y(t))$ change as $(x(t), y(t))$ moves along an orbit with advancing time? We have by the Chain Rule that

$$V' = \frac{d}{dt}V(x(t), y(t)) = 6x\frac{dx}{dt} + 2y\frac{dx}{dt} + 2x\frac{dy}{dt} + 4y\frac{dy}{dt}$$
$$= (6x + 2y)y + (2x + 4y)(-x - y) = -2x^2 - 2y^2$$

which is negative (except at $x = 0$, $y = 0$). So it appears that the positive-valued scalar function $V(x, y)$ diminishes along nonconstant orbits of system (2). The level sets of V, i.e., the curves

$$3x^2 + 2xy + 2y^2 = C$$

form a family of nested ellipses, where if $0 < C_1 < C_2$, then the C_2-ellipse encloses the C_1-ellipse. See Figure 8.2.2 for the graphs of the level sets $V = 12, 4, 1/4$, and of four orbits. Observe, in particular, how the orbits cut across each ellipse, always moving toward an ellipse with a lower C-value. So one can think of orbits moving "closer" to

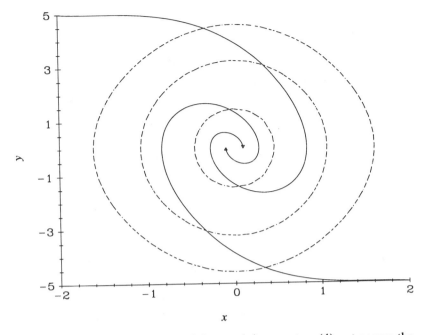

FIGURE 8.2.3 Orbits (solid) of the pendulum system (4) cut across the level curves (dashed) of the energy V, moving inward from higher to lower V-values as t increases. See Examples 8.2.3, 8.2.4.

the origin in terms of the function $V^{1/2} = (3x^2 + 2xy + 2y^2)^{1/2}$ used to measure the distance between the points (x, y) and $(0, 0)$.

Energy is a scalar quantity that can often be measured, and the next example illustrates how the changing energy along the orbit of a system can be used to determine stability properties.

EXAMPLE 8.2.3

The Energy of a Damped Simple Pendulum

The ODE model for the motion of the damped simple pendulum is

$$mLx'' + cLx' + mg \sin x = 0 \tag{3}$$

or, in system form

$$x' = y, \qquad y' = -\frac{g}{L} \sin x - \frac{c}{m} y \tag{4}$$

where x is the angle of the pendulum measured counterclockwise from the downward vertical, y the angular velocity, g the gravitational constant, L the length of the pendulum, c a positive frictional constant, and m the mass of the pendulum bob. We shall consider the motion of the pendulum "near" the downward equilibrium point $x = 0$, $y = 0$. The *gravitational potential energy* (measured from the downward position at $x = 0$) and the *kinetic energy* of the pendulum system in the state (x, y) are, respectively, $mgL(1 - \cos x)$ and $m(Ly)^2/2$. The *total energy* of the system in state (x, y) is

$$V = mgL(1 - \cos x) + m(Ly)^2/2 \tag{5}$$

The scalar function V reaches its minimum value of 0 at the equilibrium point $(0, 0)$. V is positive for all other values of x and y if $|x| < 2\pi$.

How do the values of the total energy of the pendulum change as time goes on? We expect the frictional forces modeled by the term $-cy/m$ in system (4) to dissipate energy, and thus lead to ever lower levels of energy in the pendulum system itself.

EXAMPLE 8.2.4

Decaying Energy of the Damped Simple Pendulum

On an orbit $x = x(t)$, $y = y(t)$ of system (4), V may be regarded as a function of time since x and y are. The time derivative of V is calculated from (5) using (4) and the Chain Rule:

$$\frac{d}{dt}V(x(t), y(t)) = \frac{\partial V}{\partial x}\frac{dx}{dt} + \frac{\partial V}{\partial y}\frac{dy}{dt}$$

$$= (mgL\sin x)y + (mL^2 y)\left(-\frac{g}{L}\sin x - \frac{c}{m}y\right)$$

$$= -cL^2 y^2(t) \leq 0 \tag{6}$$

Near the origin (where $V = 0$), the level sets of V defined by $V = C$, where C is a positive constant, are oval-shaped curves with the property that $V = C_2$ encloses the curve $V = C_1$ if $C_2 > C_1$, while as $C \to 0$ the curve $V = C$ shrinks onto the origin.

Suppose that $g = 9.8$ m/sec^2, $L = 0.98$ m, while the mass of the bob is 10 kg, and $c = 20$ kg· m/sec. Then system (4) for the motion of the pendulum is

$$x' = y, \quad y' = -10\sin x - 2y$$

while from (5) the total energy function V is

$$V(x, y) = 96.04(1 - \cos x) + 4.802y^2$$

Figure 8.2.3 shows the curves $V = 100, 50, 10$ (dashed lines). Since $dV/dt \leq 0$ by (6), orbits of system (4) move across these level curves so that V is a nonincreasing function of time (see the two orbits in Figure 8.2.3). This suggests that system (4) is stable at the origin. See Problem 6 for a proof that, not only is the pendulum stable at the origin, but the stability is asymptotic.

The striking aspect of the calculations of Examples 8.2.1–8.2.4 is that, aside from the equilibrium points, none of the solutions of the respective systems of ODEs need to be known. It is precisely this fact that makes the method of Lyapunov so important.

Definiteness, Derivative Following the Motion

The ideas introduced above may be extended to any autonomous system with an equilibrium point at the origin,

$$x' = F(x), \qquad F(0) = 0 \tag{7}$$

where x is an n-vector, and $F(x)$ is continuously differentiable in a region S of $\mathbf{R}^n$ that contains the origin 0. All of the analysis that follows applies equally well at any equilibrium point p if p is first shifted to the origin by the change of variables $x \to x + p$.

First we define a general type of "distance" or "energy" that has the properties of the V-functions in the previous examples.

❖ *V-functions and Definiteness.* Suppose that $V(x)$ is a real-valued function that is continuously differentiable on a region S of $\mathbf{R}^n$ containing the origin, and suppose that $V(0) = 0$. V is *positive definite* [*negative definite*] on S if $V(x) > 0$

$[V(x) < 0]$ for $x \neq 0$, x in S. V is *positive semidefinite* [*negative semidefinite*] on S if $V(x) \geq 0$ [$V(x) \leq 0$], x in S. Otherwise, V is said to be *indefinite* on S.

Note that a definite function is also semidefinite of the same type, but not conversely. Also note that an indefinite function takes on both positive and negative values in every subset of S containing the origin.

EXAMPLE 8.2.5

Positive Definite V-functions
The V-functions of Examples 8.2.1 and 8.2.2, $V = x^2 + y^2$ and $V = 3x^2 + 2xy + 2y^2$, are both positive definite on all of $\mathbf{R}^2$.

EXAMPLE 8.2.6

Total Energy of the Pendulum: Positive Definiteness
The function $V(x, y)$ defined by (5) is positive definite in the vertical strip S defined by $|x| < 2\pi$, $-\infty < y < \infty$, because $V(0, 0) = 0$ and $mgL(1 - \cos x)$ is positive in S except at $x = 0$, while $m(Ly)^2/2 \geq 0$ for all y.

Quadratic forms in x and y have definiteness properties that are simple to characterize, and the following theorem does just that.

THEOREM 8.2.1

> **Planar Quadratic Forms.** Let $V(x, y) = ax^2 + 2bxy + cy^2$, where a, b, and c are real numbers. Then $V(x, y)$ is positive definite (positive semidefinite) on the xy-plane if and only if $a > 0$ and $b^2 < ac$ ($a \geq 0$ and $b^2 \leq ac$). $V(x, y)$ is negative definite (negative semidefinite) if and only if $a < 0$ and $b^2 < ac$ [$a \leq 0$ and $b^2 \leq ac$]. Otherwise, $V(x, y)$ is indefinite.

The quadratic formula may be used to prove the above criteria for definiteness (see Problem 8). Observe that the conditions on the coefficient a may be replaced by the same conditions placed on c.

EXAMPLE 8.2.7

Testing for Definiteness
The tests of Theorem 8.2.1 show that the functions $x^2 + y^2$ and $3x^2 + 2xy + 2y^2$ are positive definite, $x^2 + 2xy + y^2$ is positive semidefinite, $x^2 - y^2$ is indefinite, while $-x^2 + xy - y^2$ is negative definite.

The usefulness of the time derivative of the V functions of Examples 8.2.1– 8.2.4 may be extended to a more general setting. First, we introduce the gradient vector ∇V of a scalar function $V(x_1, \ldots, x_n)$. The gradient vector ∇V is the row vector

$$\nabla V = [\frac{\partial V}{\partial x_1} \quad \frac{\partial V}{\partial x_2} \quad \cdots \quad \frac{\partial V}{\partial x_n}]$$

❖ **Derivative Following the Motion.** Suppose that $V(x)$ is a real-valued function that is continuously differentiable on a region S of $\mathbf{R}^n$. Then the *derivative of V following the motion* of the system $x' = F(x)$ is defined to be $\nabla V(x) \cdot F(x)$, where F and the gradient vector ∇V are written as row vectors and the standard dot product is used.

The next theorem indicates why the derivative of V following the motion of the system is so important.

THEOREM 8.2.2

Time Rate of Change of $V(x(t))$. Suppose that $V(x)$ is a real-valued and continuously differentiable function of x on a region S of $\mathbf{R}^n$, and suppose that the vector function $F(x)$ satisfies the conditions of the Existence and Uniqueness Theorem (Theorem 6.2.1) in S. If $x(t)$ is any solution of the system $x' = F(x)$ whose orbit lies in S, then

$$V' = \frac{d}{dt} V(x(t)) = \nabla V(x(t)) \cdot F(x(t)) \tag{8}$$

is the derivative of $V(x(t))$ following the motion of the system at the point $x(t)$.

Proof. Using the Chain Rule, we have that

$$V' = \frac{d}{dt} V(x(t)) = \sum_{i=1}^{n} \frac{\partial V}{\partial x_i} x_i' = \sum_{i=1}^{n} \frac{\partial V}{\partial x_i} F_i(x) = \nabla V(x) \cdot F(x)$$

and the theorem is proved.

Sometimes V' is called the *total time derivative* of V with respect to an orbit $x(t)$ of $x' = F(x)$. The total time derivative $V'(x(t))$ may be calculated from (8) without having a formula for $x(t)$, and that fact turns out to be critically important. Note that the time derivative of $V(x(t))$ depends solely on the state x, and *not* on the time t when the system is in the state x.

Testing for Stability

Lyapunov took these concepts of definiteness and built a theory of a stability. We shall begin with stability itself, without distinguishing (at first) between neutral and asymptotic stability.

❖ **Lyapunov Stability Functions.** A real-valued scalar function $V(x)$ that is continuously differentiable on a region S containing the origin is a *Lyapunov stability function* for the system $x' = F(x)$, where $F(0) = 0$, at the origin if V is positive definite on S, while the derivative V' following the motion of the system is negative semidefinite on S.

The functions V in Examples 8.2.1–4 are all Lyapunov stability functions for their respective systems. Observe that if a Lyapunov stability function V exists at all for a system, it cannot be unique since cV, c a positive constant, is also a Lyapunov stability function.

The existence of a Lyapunov stability function for a system at the origin is intimately connected with the stability of the system.

THEOREM 8.2.3

Test for Stability. The system, $x' = F(x)$, where $F(0) = 0$, is stable at the origin if and only if there is a Lyapunov stability function for the system at the origin.

Proof. We shall prove the sufficiency of the test. Let V, defined in a region S of state space, be a Lyapunov stability function for $x' = F(x)$, where $F(0) = 0$, at the origin. Suppose that $x(t, x^0)$ solves the IVP, $x' = F(x)$, $x(0) = x^0$.

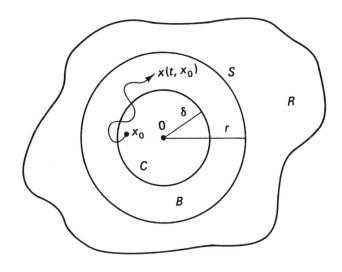

FIGURE 8.2.4 Sets used in the proof of the Stability Test. (Theorem 8.2.3).

Recall from Section 8.1 that we must show that for each positive number ε, there is a positive number δ such that $\|x(t, x^0)\| < \varepsilon$ for all $t \geq 0$ whenever $\|x^0\| < \delta$. Let $\varepsilon > 0$ be given and suppose that B is a closed ball of radius $r < \varepsilon$ centered at 0, and contained in S. The Lyapunov stability function $V(x)$ is continuous and positive on the closed and bounded "spherical" boundary S of B. By the Minimum Value Theorem for continuous functions, $V(x)$ has a positive minimum value V_0 on S:

$$0 < V_0 \leq V(x), \qquad \text{all } x \text{ in } S \tag{9}$$

Since $V(x)$ is continuous and has value 0 at the origin, there is an open ball C of radius δ, say, which is centered at 0, lies entirely inside B, and for which

$$0 \leq V(x) < \frac{1}{2}V_0, \qquad \text{for all } x \text{ in } C$$

See Figure 8.2.4 for a sketch of the sets B, S, and C in two dimensions. Let x^0 be in C (i.e., $\|x^0\| < \delta$). Since $V(x^0) \leq V_0/2$ and since V does not increase along an orbit as time increases (recall that V' is negative semidefinite because V is a Lyapunov stability function), we have that

$$V(x(t, x^0)) \leq V_0/2 \tag{10}$$

for all $t \geq 0$ for which $x(t, x^0)$ is defined.

If we can show that $x(t, x^0)$ is defined for all $t \geq 0$, then $x' = F(x)$ is stable at the origin. First we show that if the maximally extended solution $x(t, x^0)$ is defined on $0 \leq t < T$ for some positive $T \leq +\infty$, then $x(t, x^0)$ lies in B, $0 \leq t < T$ (i.e., $\|x(t, x^0)\| < r < \varepsilon$). Then we shall show that we can take $T = +\infty$.

Suppose, to the contrary, that for some t_1, $0 < t_1 < T$, the point $x(t_1, x^0)$ lies outside the region B. Then we have for $t = 0$ and $t = t_1$ that

$$0 < \|x(0, x^0)\| = \|x^0\| < \delta \leq r < \|x(t_1, x^0)\|$$

Since $\|x(t, x^0)\|$ is a continuous function of t, the Intermediate Value Theorem implies that there is a number $\bar{t}$ between 0 and t_1 such that

$$\|x(\bar{t}, x^0)\| = r$$

Thus, at time $\bar{t}$ the orbit defined by $x(t, x^0)$ lies on the surface S of the region B. From (9) and (10), we are led to the absurd conclusion that

$$0 < V_0 \le V(x(\bar{t}, x^0)) < V_0/2$$

Hence, our earlier assumption that the point $x(t_1, x^0)$ is outside B is wrong and $\|x(t, x^0)\| \le r < \varepsilon$ for $0 \le t < T$. Now T must be $+\infty$, for if that were not so, $x(t, x^0)$ must approach the boundary of S as $t \to T^-$, a consequence of the Extension Property (Theorem 6.2.2). Thus, $\|x(t, x^0)\|$ exceeds r for some values of t, $0 < t < T$. This cannot be so; hence, $T = +\infty$, and the proof is finished.

The key to applying the Stability Test is finding a Lyapunov stability function. If $x' = F(x)$ models a mechanical or other physical system for which total energy can be defined, the energy function is a reasonable candidate for a Lyapunov stability function. However, any function V will do if meets the criteria for a Lyapunov stability function (e.g., the quadratic forms of Examples 8.2.1 and 8.2.2).

EXAMPLE 8.2.8

Stability of the Damped Pendulum at the Origin
As shown in Examples 8.2.3 and 8.2.4, the function

$$V = mgL(1 - \cos x) + mL^2 y^2/2$$

is a Lyapunov stability function for system (4) at the origin because V is positive definite and V' is negative semidefinite on the region $|x| < 2\pi$, $|y| < \infty$ containing $(0, 0)$. Lyapunov's Stability Test then implies that system (4) is stable at $(0, 0)$. In fact, it is asymptotically stable at $(0, 0)$, but a different kind of V function is needed to prove it (see Problem 6).

Lyapunov's Stability Test does not distinguish between asymptotic and neutral stability, as illustrated in the next example.

EXAMPLE 8.2.9

Neutral Stability of an Oscillating Spring
The system $x' = y$, $y' = -kx$, where $k > 0$, models a body of unit mass oscillating without friction at the end of a spring. The quadratic function $V = kx^2/2 + y^2/2$ is positive definite. We have $V' = kxx' + yy' = kxy + y(-kx) = 0$, which is (trivially) negative semidefinite. Thus, the system is stable at the equilibrium point. In this case, the elliptical level sets of V defined by $V(x, y) = $ constant are the orbits themselves (see Figure 8.2.5). The system is stable, but not asymptotically stable (i.e., it is neutrally stable). Observe that $V(x, y) = kx^2/2 + y^2/2$ has the property that if $x(t)$, $y(t)$ is any solution of the system $x' = y$, $y' = -kx$, then $V(x(t), y(t)) = $ constant, for all t. Such functions are said to be integrals of the system. More on this in Section 8.3.

Testing for Asymptotic Stability

If a system is stable at an equilibrium point, then nearby orbits remain nearby with increasing time. But these orbits do not necessarily approach the equilibrium point. If the system is asymptotically stable, however, all nearby orbits decay with increasing time to the equilibrium point. This more stringent kind of stability is associated with a

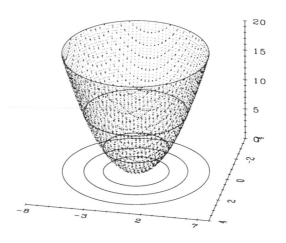

FIGURE 8.2.5 Orbits, constant energy curves, and the graph of the energy function for Example 8.2.9.

decrease in "energy" along the nonconstant orbits as they are traced out with increasing time.

❖ **Lyapunov Asymptotic Stability Functions.** Let V be a Lyapunov stability function for the system $x' = F(x)$, where $F(0) = 0$, at the origin. V is a *Lyapunov asymptotic stability function* if V' is negative definite on some region containing the origin.

Obviously, every Lyapunov asymptotic stability function of a system at the origin must also be a Lyapunov stability function, but the converse need not be true (e.g., the function $V(x, y)$ of Example 8.2.9). The V-functions of Examples 8.2.1– 8.2.3 are all Lyapunov asymptotic stability functions.

THEOREM 8.2.4

> Test for Asymptotic Stability. The system $x' = F(x)$, where $F(0) = 0$, is asymptotically stable at the origin if and only if there is a Lyapunov asymptotic stability function for the system at the origin.

The proof of Theorem 8.2.4 is omitted. Thinking of a Lyapunov asymptotic stability function V as a measure of "energy," one can imagine the negative definiteness of V' as implying that the system is losing or dissipating energy along every nonconstant orbit in some region containing the origin. It is then reasonable to expect orbits in such a region gradually to "slow down" and to approach the equilibrium position at the origin where the energy V has its smallest value. The full proof of the asymptotic stability test is based on this notion of decaying "energy." For physical systems, asymptotic stability is often more significant than stability.

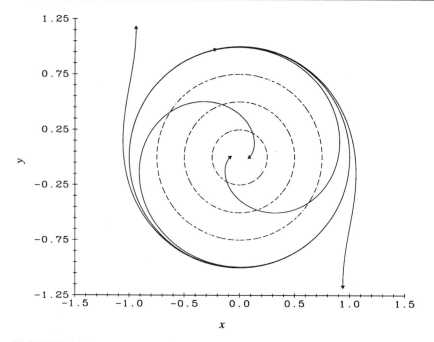

FIGURE 8.2.6 Local asymptotic stability of system of Example 8.2.10 at origin.

Geometrically, the gradient vector ∇V of a Lyapunov asymptotic stability function V and the field vector F of the system $x' = F(x)$ form an obtuse angle θ at each point near (but not at) the origin. The reason for this is that

$$V' = \sum_i \frac{\partial V}{\partial x_i} F_i = \nabla V \cdot \mathbf{F} = \|\nabla V\| \, \|F\| \cos \theta < 0$$

Thus, $\cos \theta$ must be negative and θ obtuse. The gradient vector ∇V at a point P on a level set $V = C$ points "outward" in the direction of increasing values of V. The obtuseness of the angle between F and ∇V then implies that the field vector F must point "inward" toward lower values of V. Thus, the orbit must also move "inward" toward the origin, where the value of V in minimal.

EXAMPLE 8.2.10

Another Lyapunov Asymptotic Stability Function
The system

$$x' = -x + y + x(x^2 + y^2)$$
$$y' = -x - y + y(x^2 + y^2)$$

has an equilibrium point at the origin. The positive definite function $V = x^2 + y^2$ is a Lyapunov asymptotic stability function at the origin if we restrict V to the open disk defined by $x^2 + y^2 < 1$. For, V is positive definite and

$$V' = 2xx' + 2yy' = -2(x^2 + y^2) + 2(x^2 + y^2)^2 = -2r^2(1 - r^2)$$

where $r = (x^2 + y^2)^{1/2}$. Hence, V' is negative definite for $r < 1$. The system is locally asymptotically stable at the origin, and the domain of attraction is the open disk, $r < 1$. Figure 8.2.6 shows two orbits inside the disk, orbits that cut across the (dashed) circles $V = 9/16, \, 1/4, \, 1/16$ moving inward. Note that $x^2 + y^2 = 1$ is itself an orbit and separates the domains of stability and instability of the system.

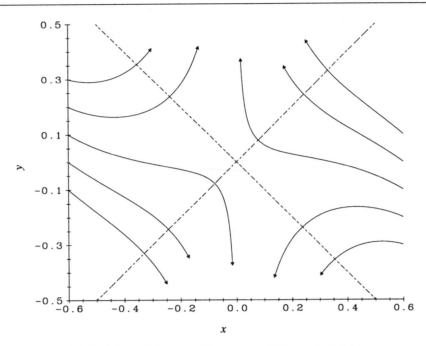

FIGURE 8.2.7 Orbits of the unstable system of Example 8.2.11.

Testing for Instability

Lyapunov not only constructed tests for stability, but also a test for instability. The test involves a scalar V-function, but now the values of V increase along orbits, or at least along some of them.

❖ **Lyapunov Instability Functions.** A real-valued scalar function $V(x)$ that is continuously differentiable on a region S containing the origin is a *Lyapunov instability function* for the system $x' = F(x)$, where $F(0) = 0$, at the origin if $V(0) = 0$, $V(x)$ is positive for at least one point in every region containing the origin, and V' is positive definite on some region containing the origin.

As we might expect, a test for instability can now be formulated.

THEOREM 8.2.5

> Test for Instability. The system $x' = F(x)$, where $F(0) = 0$, is unstable at the origin if there is a Lyapunov instability function for the system at the origin.

The idea of the Instability Test is this: An orbit starting at a point where V is positive is carried away from the origin with increasing time since V' is positive definite. The proof of Theorem 8.2.5 is omitted, but the next example illustrates the main point.

EXAMPLE 8.2.11

A Lyapunov Instability Function
The system to be considered is

$$x' = -x - x^2 y$$
$$y' = y + 2x^3$$

The function $V = y^2 - x^2$ is positive on the region $|y| > |x|$. We have that

$$V' = 2y(y + 2x^3) - 2x(-x - x^2 y) = 2x^2 + 2y^2 + 6x^3 y$$
$$= 2r^2(1 + 3r^2 \cos^3 \theta \sin \theta) \geq 2r^2(1 - 3r^2)$$

where $x = r \cos \theta$, $y = r \sin \theta$, and the inequality $\cos^3 \theta \sin \theta \geq -1$ has been used. Thus, V' is positive definite on the region $S: r^2 < 1/3$. The system is unstable at the origin. See Figure 8.2.7 for a graph of $V = 0$ (dashed lines) and some orbits of the system. The values of V are positive in the upper and lower triangular regions.

The Lyapunov Method

Lyapunov's tests for the stability characteristics of a system involve the use of scalar functions V with the properties listed in one or another of the three definitions given earlier. There is a generic name for these V functions.

> ❖ **Lyapunov Functions.** A function $V(x)$ with the properties given by any one of the definitions preceding Theorems 8.2.3–8.2.5 is called a *Lyapunov function* for the system $x' = F(x)$, where $F(0) = 0$, at the origin.

The *Lyapunov Method* for determining whether a system is stable, asymptotically stable, or unstable at an equilibrium point consists of finding a suitable Lyapunov function that "works" for the problem at hand. This is not always easy, but the boxed statements below may help in the search for a suitable Lyapunov function.

> Implementing the Lyapunov Method. Suppose that the system $x' = F(x)$ has an equilibrium point at the origin. Then:
>
> 1. Guess whether the system is stable, asymptotically stable, or unstable at the origin. You may want to use your computer to plot a few orbits near the origin before you make your guess, or else compare the system with others whose stability properties you know.
>
> 2. Construct a Lyapunov function V to verify your guess. You may want to use an energy function, an integral, or a quadratic form for this purpose.
>
> 3. Check that your Lyapunov function meets the conditions of one of Theorems 8.2.3–8.2.5. If it does, then the system has the stability characteristics given in the statement of that particular theorem.

Of course, if you make a bad guess about the stability or instability of the system, then no matter how hard you try to construct a Lyapunov function to confirm your guess, nothing will work. There is a way to "hedge your bets." The next example illustrates how this might be done.

EXAMPLE 8.2.12

Quadratic Forms as Lyapunov Functions
To test for the stability properties of the system

$$x' = y - x^3, \qquad y' = -x - y^3$$

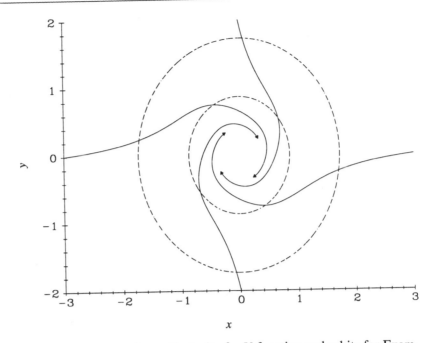

FIGURE 8.2.8 Level sets (dashed) of a V function and orbits for Example 8.2.12.

at the equilibrium point at the origin, introduce the generic quadratic form $V = ax^2 + 2bxy + cy^2$, where the constants a, b, and c will be determined later. We have

$$V' = 2axx' + 2byx' + 2bxy' + 2cyy'$$
$$= -2bx^2 + 2(a-c)xy + 2by^2 - 2(ax^4 + bx^3y + bxy^3 + cy^4)$$

The minus sign before the last term of V' suggests that we may have asymptotic stability. To prove that fact, we will try to choose values of a, b, and c for which V is positive definite and V' is the negative definite. We see that if we choose $a = c = 1$, $b = 0$, then V is the positive definite quadratic form $V = x^2 + y^2$ and V' is the negative definite biquadratic form, $-(x^4 + y^4)$. Hence, the system is asymptotically stable at the origin. Figure 8.2.8 shows orbits of the system cutting across the circular level sets $V = 3$, $3/4$ and spiraling toward the origin.

Quadratic forms are probably the simplest Lyapunov functions. Theorem 8.2.1 lists the criteria for definiteness if only two variables are involved. It is also possible to characterize the definiteness properties of a quadratic form in any number of variables:

$$V = \sum_{i,j=1}^{n} b_{ij}x_i x_j$$

where the coefficients b_{ij} are real constants. There is no loss of generality if it is assumed that $b_{ji} = b_{ij}$, since $b_{ij}x_i x_j + b_{ji}x_j x_i = (b_{ij} + b_{ji})x_i x_j$, and we may replace each of b_{ij} and b_{ji} by $(b_{ij} + b_{ji})/2$ without changing the values of V. Observe that if B is the symmetric matrix $[b_{ij}]$ of coefficients, V may be written as $V = x^T Bx$.

THEOREM 8.2.6

> Quadratic Forms and Definiteness. Let $V = x^T B x$, where B is an $n \times n$ real symmetric matrix. Then V is positive definite, positive semidefinite, negative semidefinite, or negative definite if and only if all eigenvalues of B are, respectively, positive, nonnegative, nonpositive, or negative. Otherwise, V is indefinite.

The proof is omitted.

Comments

There is a certain amount of arbitrariness in constructing Lyapunov functions. This can be an asset because it means that you don't have to find the "unique" V-function that will work for your system. You are free to try energy functions, integrals, quadratic forms, or whatever is at hand.

PROBLEMS

1. (*Using Lyapunov Functions to Test Stability*). Test the stability properties of each system at the origin by using a function of the form $V = ax^2 + cy^2$, adding the mixed term $2bxy$ only if necessary. Distinguish between neutral and asymptotic stability, if possible.

   (a) $x' = -4y - x^3$, $y' = 3x - y^3$

   (b) $x' = y$, $y' = -9x$

   (c) $x' = -x + 3y$, $y' = -3x - y$

   (d) $x' = 2x + 2y$, $y' = 5x + y$

   (e) $x' = (-x + y)(x^2 + y^2)$, $y' = -(x + y)(x^2 + y^2)$

   (f) $x' = -2x - xe^{xy}$, $y' = -y - ye^{xy}$

   (g) $x' = -x^3 + x^3 y - x^5$, $y' = y + y^3 + x^4$

   (h) $x' = y$, $y' = -2x - 3y$

2. (*Plotting Level Sets of Lyapunov Asymptotic Stability Functions*). Plot several level sets of the Lyapunov functions near the origin of the xy-plane. Then plot several orbits of the indicated system from Problem 1; note how the orbits cut across the level sets and tend to the origin as t increases.

   (a) $V = 3x^2 + 4y^2$; system of Problem 1(a)

   (b) $V = x^2 + y^2$; system of Problem 1(c)

   (c) $V = x^2 + y^2$; system of Problem 1(e)

   (d) $V = x^2 + y^2$; system of Problem 1(f)

   (e) $V = 5x^2 + 2xy + y^2$; system of Problem 1(h)

3. (*Testing for Stability*). Each system of the indicated Atlas Plates has an equilibrium point at the origin. Using a Lyapunov function of the form $V = ax^2 + 2bxy + cy^2$, determine whether the system is asymptotically stable or unstable at the origin.

    (a) BIFURCATION A   **(b)** BIFURCATION B

    (c) BIFURCATION D   **(d)** BIFURCATION F

    (e) BIFURCATION G

4. (*More Lyapunov Functions*). Use $V = ax^{2m} + by^{2n}$, m and n positive integers, to determine the stability properties of the following systems at the origin. Plot some level sets of V and some orbits of the systems. [*Hint*: Calculate V', and then pick strategic values for a, b, m, n.]

    (a) $x' = -x^3 + y^3$, $y' = -x^3 - y^3$

    (b) $x' = -2y^3$, $y' = 2x - y^3$

5. (*Neutral Stability*). Show that $x' = (x - y)^2(-x + y)$, $y' = (x - y)^2(-x - y)$ is neutrally stable at the origin. [*Hint*: Use $V = x^2 + y^2$. Note the line of equilibrium points.]

6. (*Stability Properties of the Damped Pendulum*). The assertions below fully describe the stability properties of the system $x' = y$, $y' = -(g/L)\sin x - cy/m$ for a damped pendulum. (See Atlas Plates PENDULUM A–C.)

    (a) Show that this system is (locally) asymptotically stable at every equilibrium point $(2k\pi, 0)$. [*Hint*: Put

$$V(x, y) = \frac{4g}{L}[1 - \cos(x - 2k\pi)] + 2y^2 + \frac{2c}{m}(x - 2k\pi)y + \frac{c^2}{m^2}(x - 2k\pi)^2$$

and determine the type of definiteness of V and V' at $(2k\pi, 0)$.]

    (b) Show that system (4) is unstable at every point $(x_0, 0)$, where $x_0 = (2k + 1)\pi$. [*Hint*: Use $V = -y\sin(x - x_0) + (c/m)(1 - \cos(x - x_0))$.]

7. (*Lotka-Volterra Model*). The system $x' = (a - by)x$, $y' = (-c + dx)y$, where the coefficients are positive, models the interactions of a predator-prey community (see Section 6.4). Show that the system is neutrally stable at the equilibrium point $(c/d, a/b)$ by using the function $Q = (y^a e^{-by})(x^c e^{-dx})$. [*Hint*: Let $V = Q(c/d, a/b) - Q(x, y)$. See Problem 6(**a**) of Section 6.4.]

8. Derive the definiteness properties given in Theorem 8.2.1 for the planar quadratic form $V = ax^2 + 2bxy + cy^2$. [*Hint*: The value of V is zero if and only if either $a \neq 0$ and $x = a^{-1}y(-b \pm \sqrt{b^2 - ac})$, or $a = 0$ and either $y = 0$ or $2bx + cy = 0$. Show that V is definite if and only if $b^2 < ac$.]

9. Consider the system $x' = y - xf(x, y)$, $y' = -x - yf(x, y)$.

    (a) Show that the origin is stable if $f(x, y)$ is positive semidefinite in some neighborhood of $(0, 0)$.

    (b) Show that the origin is asymptotically stable if $f(x, y)$ is positive definite in a neighborhood of $(0, 0)$.

    (c) Show that the origin is unstable if $f(x, y)$ is negative definite in a neighborhood of $(0, 0)$.

    (d) Illustrate parts **(b)** and **(c)** above by the respective functions $f = |x| + |y|$ and $f = -\cos(x^2 + y^2)$. Plot orbits of each system in a region containing the origin.

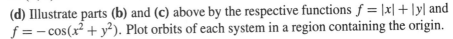

8.3 Conservative Systems

Testing the orbital properties of a general system is not easy if the process depends on having exact formulas for the solutions of the system. There are indirect ways of determining the behavior of the orbits, ways which do not require solution formulas. For example, the Lyapunov functions introduced in Section 8.2 can be used to determine the stability of a system near an equilibrium point by examining the values of a scalar function and its derivative following the motion of the sysem. In this section, a somewhat different, but related, approach is taken for investigating properties of conservative systems.

Integrals and Exactness

First, consider the planar system,

$$\frac{dx}{dt} = N(x, y), \qquad \frac{dy}{dt} = -M(x, y) \tag{1}$$

and the corresponding scalar equation $dy/dx = -M/N$, or in another form

$$N(x, y)\frac{dy}{dx} + M(x, y) = 0 \tag{2}$$

According to Theorem 1.9.1, if ODE (2) is exact in a region S of the xy-plane, then there is a nonconstant function $K(x, y)$, called an *integral*, with the property that $\partial K/\partial x = M$ and $\partial K/\partial y = N$ in S. Orbits in S of system (1) lie on level sets, $K(x, y) = C$, where C a real constant. Thus, the value of $K(x(t), y(t))$ remains constant on the orbit $x = x(t)$, $y = y(t)$ of system (1) as t varies; that is, K is "conserved" along each orbit. Observe that the derivative K' following the motion of system (1) is

$$K' = \frac{\partial K}{\partial x}x' + \frac{\partial K}{\partial y}y' = M \cdot N + N \cdot (-M) = 0$$

Thus, as expected, the value of K does not change as we move along a specific orbit of system (1). This means that knowledge of the scalar function K will convey considerable information about the behavior of the orbits of system (1).

Integrals and Conservative Systems

The above ideas may be extended to the system

$$x' = F(x) \tag{3}$$

where x is a real n-vector and F is continuously differentiable on a region S of $\mathbf{R}^n$. The following definition is based on properties of the function K above.

> ❖ **Integrals and Conservative Systems.** Let $K(x)$ be a continuously differentiable, real-valued, scalar function on a region S of $\mathbf{R}^n$. The function K is said to be an *integral* on S of the system $x' = F(x)$ if K is nonconstant on every region in S and if for each solution $x(t)$ of the system $K(x(t))$ is constant for all t for which the solution $x(t)$ remains in S. If an integral exists on S, then the system is said to be *conservative* on S.

The condition that K be nonconstant on regions avoids the trivial case where K is constant throughout a region. The condition that $K(x(t))$ remains constant along each orbit of the system $x' = F(x)$ is equivalent to requiring that the derivative of K following the motion, $K'(x(t)) = \nabla K \cdot F$, must be 0 for all t for which $x(t)$ lies in S. Thus, ∇K (which is orthogonal to the level set $K = $ constant) is perpendicular to the tangent vector F to the orbit at each point. The term "integral" is used because K is often

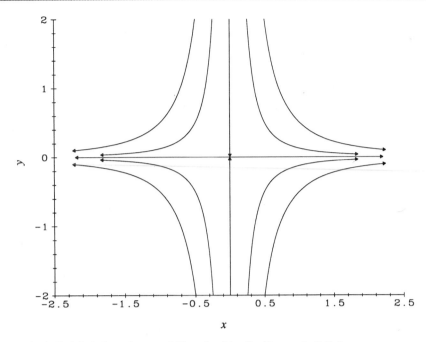

FIGURE 8.3.1 Level sets of K and orbits for Example 8.3.1.

obtained by an integration, while the word "conservation" alludes to the constancy of K on an orbit (i.e., K is "conserved"). If an integral K of system (3) on S exists at all, it is not unique since cK, for any constant $c \neq 0$, is also an integral.

   The condition that the derivative of K following the motion be zero on orbits of the system (3) is equivalent to the condition that K be constant on each orbit. Thus, the following result is not surprising.

THEOREM 8.3.1

Integrals and Orbits. Suppose that the system $x' = F(x)$ has an integral $K(x)$ on S. Then each orbit of the system in S lies on a level set of K, and every level set of K in S is a union of orbits.

   Proof. For x^0 in S, denote the value of $K(x^0)$ by K_0, where $K(x)$ is an integral of $x' = F(x)$ in S. By the definition of an integral, $K(x(t, x^0)) = K_0$ for all t for which the solution $x(t, x^0)$ of the IVP $x' = F(x)$, $x(0) = x^0$ is defined. Hence the orbit of $x = x(t, x^0)$ lies on the level set $K(x) = K_0$. Next, suppose that x^1 is any point in the level set $K(x) = K_0$. Then by the same argument $K(x(t, x^1)) = K_0$ for all t for which $x(t, x^1)$ is in S. But since x^1 is arbitrary it follows that $K(x) = K_0$ is a union of orbits.

EXAMPLE 8.3.1

An Unstable Conservative System
The planar linear system

$$x' = x, \quad y' = -2y \tag{4}$$

has an integral

$$K = x^2 y$$

which may be found by dividing y' by x' so that $dy/dx = y'/x' = -2y/x$. This ODE can be solved by separating the variables to obtain $x^2 y = $ constant. The derivative of K following the motion of (4) is zero since

$$\frac{dK}{dt} = \frac{d}{dt}(x^2 y) = \frac{\partial}{\partial x}(x^2 y)\frac{dx}{dt} + \frac{\partial}{\partial y}(x^2 y)\frac{dy}{dt}$$

$$= 2xy(x) + x^2(-2y) = 0$$

It is clear that K is not constant on regions. Thus, K meets all the conditions for being an integral, and so the system (4) is conservative. The system is unstable at the origin, which is a saddle point (see Section 7.4). Figure 8.3.1 shows arcs of several level sets of K, each of which contains orbits of the system.

EXAMPLE 8.3.2

A Neutrally Stable Conservative System
The equation of motion of a body of mass m suspended by an undamped Hooke's Law spring is $mx'' + kx = 0$. It is equivalent to the linear system

$$dx/dt = y, \qquad dy/dt = -kx/m \tag{5}$$

If we eliminate dt, we obtain the differential equation of the orbits of system (5), $dy/dx = -kx/(my)$ and this can be written in the form $mydy + kxdx = 0$, which can be integrated to give the positive definite integral

$$E = my^2/2 + kx^2/2$$

This is precisely the total energy of the body of mass m, where the first term in the formula for E is the kinetic energy and the second is the potential energy due to the spring force. The derivative of E following the motion of system (5) is zero (hence, energy is conserved) since

$$\frac{dE}{dt} = myy' + kxx' = my(-kx/m) + kxy = 0$$

Hence, E is a Lyapunov stability function, and from Theorem 8.2.3 we know that system (5) is stable at the origin. Actually, the stability is neutral and the orbits are ellipses enclosing the origin and defined by $E = $ constant.

Here is an example in $\mathbf{R}^3$. The level sets of an integral are now surfaces.

EXAMPLE 8.3.3

A Neutrally Stable and Conservative System
The system

$$\begin{aligned} x' &= xz - 10yz \\ y' &= 10xz + yz \\ z' &= -x^2 - y^2 \end{aligned} \tag{6}$$

is conservative on $\mathbf{R}^3$. An integral is

$$K(x, y, z) = x^2 + y^2 + z^2$$

and the level sets of K are spheres. The derivative of K following the motion of system (6) is zero because

$$\frac{dK}{dt} = 2xx' + 2yy' + 2zz'$$

$$= 2x(xz - 10yz) + 2y(10xz + yz) + 2z(-x^2 - y^2) = 0$$

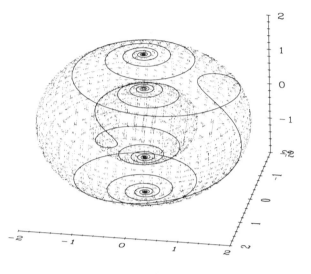

FIGURE 8.3.2 Orbits and integral surfaces for system (6) in Example 8.3.3.

Observe that K is nonconstant on regions of $\mathbf{R}^3$. System (6) is thus conservative on $\mathbf{R}^3$, and K is an integral. Because the values of K cannot change along an orbit, each orbit of (6) remains, for all t, on a spherical level set of K. See Figure 8.3.2 for a view of the level sets $K = 4, 1$ and of the two orbits of (6) with initial points $(2/\sqrt{3}, 2/\sqrt{3}, 2/\sqrt{3})$ and $(-1/\sqrt{3}, -1/\sqrt{3}, -1/\sqrt{3})$; the z-axis is vertical. See Atlas Plate CONSERVATIVE A for component plots of one of the orbits. Because of the scales on the axes, the spherical surfaces in Figure 8.3.2 appear as ellipsoids. With increasing t, the nonconstant orbits spiral away from the equilibrium point of (6) at a sphere's north pole $(0, 0, z)$, $z > 0$, and spiral toward the equilibrium point at the south pole. In this example, K is positive definite and K' is negative semidefinite (because $K' = 0$), i.e., K is a Lyapunov stability function. By Theorem 8.2.3, system (6) is stable at the origin, but the stability is neutral because the origin is not an attractor.

Examples 8.3.1, 8.3.2, and 8.3.3 illustrate conservative systems that are unstable or neutrally stable. One might expect an example of a conservative system that is asymptotically stable, but no such system exists. First, recall that an equilibrium point is said to be an *attractor* if all orbits in some region containing the point tend to the point as $t \to \infty$, and that asymptotically stable equilibrium points are attractors.

THEOREM 8.3.2

No Attractors. Suppose that the system $x' = F(x)$ is conservative on a region S in $\mathbf{R}^n$. Then S contains no attracting equilibrium points of the system.

Proof. Let K be an integral of the system $x' = F(x)$ on S and suppose, to the contrary, that S contains an attracting equilibrium point, which may be taken to be the origin. For some $r_0 > 0$, the solution $x(t, x^0)$ of the IVP, $x' = F(x)$, $x(t_0) = x^0$, has the property that $x(t, x^0) \to 0$ as $t \to \infty$ if $||x^0|| < r_0$. Since $K(x)$ is continuous, $K(x(t, x^0)) \to K(0)$ as $t \to 0$, which implies that $K(x(t, x^0)) = K(0)$ for all t for which $x(t, x^0)$ is in S since K must remain constant on the orbit. But this reasoning applies to every orbit whose initial point x^0 satisfies $||x^0|| < r_0$. Hence $K(x)$ has the constant value $K(0)$ on the open ball $||x|| < r_0$, and this violates the condition of the nonconstancy of an integral on open sets. So, S cannot contain an attracting equilibrium point.

The existence of an integral is a counterindicator for asymptotic stability, but is consistent with neutral stability or instability. The problem set gives several examples of conservative systems. The next examples give further instances of systems where the integral is the total energy of the system.

EXAMPLE 8.3.4

Simple Undamped Pendulum

The equation of motion of the simple undamped pendulum is $mLx'' + mg \sin x = 0$. Its total energy is $K = \text{kinetic} + \text{potential} = m(Lx')^2/2 + mgL(1 - \cos x)$. It is easy to show that the derivative of $K(x, y)$ following the motion of the equivalent system, $x' = y$, $y' = -\omega^2 \sin x$, where $\omega^2 = g/L$, is zero. Hence K is an integral since K is clearly not constant on any planar region. Figure 4.1.1 in Section 4.1 shows several level sets of this integral. Observe that in the setting of conserved total energy there are both neutrally stable and unstable equilibrium points.

EXAMPLE 8.3.5

Double Pendulum

The equations of motion of an undamped, linearized double pendulum, where each (massless) support rod has length L and the respective masses are m_1 and m_2, may be shown to be (see Example 7.4.8 for more on the double pendulum):

$$L^2(m_1 + m_2)\theta_1'' + L^2 m_2 \theta_2'' + Lg(m_1 + m_2)\theta_1 = 0$$
$$L^2 m_2 \theta_1'' + L^2 m_2 \theta_2'' + Lgm_2\theta_2 = 0 \tag{7}$$

For simplicity, let the magnitudes of L and g coincide. Rescale time by

$$\tau = t(1 + m_2/m_1)^{1/2}$$

and let $\alpha = (m_2/m_1)(1 + m_2/m_1)^{-1}$ to obtain, after some algebraic manipulation,

$$\frac{d^2\theta_1}{d\tau^2} + \theta_1 - \alpha\theta_2 = 0$$

$$\frac{d^2\theta_2}{d\tau^2} - \theta_1 + \theta_2 = 0$$

The definition of α implies that $0 < \alpha < 1$ (α is called the *reduced mass* of the system). Let $x_1 = \theta_1$, $x_2 = d\theta_1/d\tau$, $x_3 = \theta_2$, $x_4 = d\theta_2/d\tau$ and obtain the linear system in $\mathbf{R}^4$

$$\frac{dx_1}{d\tau} = x_2 \qquad \frac{dx_2}{d\tau} = -x_1 + \alpha x_3$$
$$\frac{dx_3}{d\tau} = x_4 \qquad \frac{dx_4}{d\tau} = x_1 - x_3 \tag{8}$$

which is the system previously considered in Section 7.4. After considerable effort we find the quadratic function of the state variable

$$K = \left(\frac{1}{\alpha} - 1\right)x_1^2 + (1-\alpha)x_3^2 + \frac{1}{\alpha}x_2^2 + 2x_2x_4 + x_4^2 \tag{9}$$

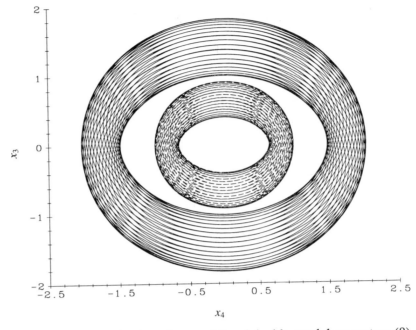

FIGURE 8.3.3 Projection of two orbits of double pendulum system (8) (with $\alpha = 3/10$) onto $x_4 x_3$-plane. See Example 8.3.5.

which is an integral of the system (calculate the derivative K' following the motion of system (8) and show that it is zero). In fact, since $0 < \alpha < 1$, and since

$$x_2^2/\alpha + 2x_2 x_4 + x_4^2 = (x_2 + x_4)^2 + (1/\alpha - 1)x_2^2$$

K is a sum of squares and is positive definite. The level sets of K are "ellipsoidal hypersurfaces" in $\mathbf{R}^4$. Each level set is a union of the orbits of system (8) that share a common value of K. Because these level sets cannot be plotted in our three-dimensional world, we must be satisfied with projections into lower dimensions. Figure 8.3.3 displays the projection of two orbits onto the $x_4 x_3$-plane. The orbit with $\alpha = 3/10$ and initial point $(1, 0, -1, 0)$ in $\mathbf{R}^4$ lies on the hypersurface $K = K(1, 0, -1, 0) = 91/30$, while the orbit with initial point $(0.5, 0, 0.4, 0)$ and the same value of α lies on the hypersurface $K = K(0.5, 0, -0.4, 0) = 1043/1500$. The positive definiteness of K implies that system (8) is neutrally stable at the unique equilibrium point at the origin, since dK/dt is zero. See Atlas Plates PENDULUM D–F for component graphs for three values of α.

PROBLEMS

1. (*Simple Integrals*). For each system below, find an integral and plot some orbits. What are the stability properties of the equilibrium points? [*Hint*: Write each as $dy/dx = F_2/F_1$, and solve to find an integral.]

   (a) $x' = 3x, \quad y' = -y$ **(b)** $x' = -y, \quad y' = 25x$

   (c) $x' = y^3, \quad y' = -x^3$ **(d)** $x' = x, \quad y' = y + x^2 \cos x$

2. (*A Conservative System with an Attractor?*). Show that the linear system $x' = -x$, $y' = -2y$, is asymptotically stable at the origin and that the function $K = yx^{-2}$

is constant on each orbit not touching the y-axis. Why doesn't this contradict Theorem 8.3.2. [*Hint*: The integral K is not continuous on any region containing the attractor at the origin. Observe that the system *is* conservative on any region disjoint from the y-axis. One must specify a region before determining whether a system is conservative.]

3. Prove that the system $x' = f(y)$, $y' = g(x)$, where f, g are continuously differentiable, is conservative in the xy-plane. Does the system have any attractors?

4. **(a)** (*Hard Spring*). Find an integral and sketch the orbits of $x' = y$, $y' = -10x - x^3$. What are the stability properties of the system at the origin? [*Hint*: See Section 3.1.]

   (b) (*Soft Spring*). Repeat part **(a)** for $x' = y$, $y' = -10x + x^3$ and discuss the stability properties of each of the three equilibrium points. [*Hint*: See Section 3.1.]

5. (*Skew Symmetry*). A real $n \times n$ matrix A is *skew symmetric* if $A^T = -A$.

   (a) Show that the system $x' = Ax$ is conservative on $\mathbf{R}^n$ and neutrally stable at the origin if A is skew symmetric. [*Hint*: Let $K = x^T x = ||x||^2$.]

   (b) Prove that if all of the nonzero eigenvalues of the skew symmetric matrix A are pure imaginary numbers, then every integral surface, $K(x) = $ constant of the system $x' = Ax$ is a union of equilibrium points and nonconstant orbits defined by solutions whose components are linear combinations of periodic functions. Show that each orbit lies on the surface of a "sphere," $K(x) = $ constant, in $\mathbf{R}^n$.

6. (*Hamiltonian Systems*). Parts **(a)–(d)** below refer to the *Hamiltonian system*

$$x_i' = \frac{\partial H}{\partial y_i}, \qquad y_i' = -\frac{\partial H}{\partial x_i}, \qquad i = 1, \ldots, k$$

   where the *Hamiltonian function* $H(x_1, \ldots, x_k, y_1, \ldots, y_k)$ is a twice continuously differentiable, real-valued scalar function on a region S that is a subset of $\mathbf{R}^{2k}$, and H is nonconstant on every region of S.

   (a) Show that H is an integral of the Hamiltonian system.

   (b) Show that a Hamiltonian system cannot be asymptotically stable at an equilibrium point.

   (c) Show that the system $x' = y$, $y' = -g(x)$ [equivalent to $x'' + g(x) = 0$] has no asymptotically stable equilibrium points. [*Hint*: The system has Hamiltonian $H = y^2/2 + \int_0^x g(s)\, ds$, the "total energy" of the system.]

   (d) What does the result in part **(c)** imply about the possibility of constructing a "restoring force" $-g(x)$ such that all solutions of the system $x' = y$, $y' = -g(x)$ tend toward a single equilibrium state?

7. (*Double Pendulum*). Make a detailed study of the motion of the double pendulum of Example 8.3.5. See also the double pendulum model in Section 7.4. In your study, be sure to address the following points:

   - Carry out the derivation of (8) from (7).

   - Do a parameter study of the behavior of the orbits as a function of α for values of α throughout the range $0 < \alpha < 1$. Use a computer for this study. Refer to Atlas Plates PENDULUM D–F and explain what you see there. What are the two pendulums doing in each case?

   - Be creative in projecting orbits into two or three dimensions chosen from the set $\{x_1, x_2, x_3, x_4, t\}$. Explain what you see in Figure 8.3.3 and in your projections. What are the corresponding pendulum motions?

8. (*Rotational Stability of a Tennis Racket*). A tennis racket has a triple of orthogonal body axes, called the *principal axes*, that are convenient for examining the rotational motion of the body. Denote the components of the body's angular velocity vector about the respective body axes by ω_1, ω_2, and ω_3. There are positive constants I_1, I_2, and I_3 (called *principal inertias*) such that the system of ODEs

$$\omega_1' = I_1^{32}\omega_2\omega_3$$
$$\omega_2' = I_2^{13}\omega_1\omega_3$$
$$\omega_3' = I_3^{21}\omega_1\omega_2$$

with $I_\alpha^{\beta\gamma} = (I_\beta - I_\gamma)/I_\alpha$, describes the evolution of the angular velocity vector ω of the body when no external torques are acting. The constant angular velocity vector $(1, 0, 0)$ represents a pure steady rotation of the body about its first principal axis. The constant vectors $(0, 1, 0)$ and $(0, 0, 1)$ are similarly interpreted. The body axes for a tennis racket are just the axes of symmetry. Assuming that $I_3 < I_1 < I_2$, it can be shown that the principal axis parallel to the face of the racket has the principal inertia I_1. Toss the racket into the air, trying to produce a steady and pure rotation about each principal axis in turn. What happens? Carry out a computer graphics study of the orbits of angular velocity. Address the following points:

- Find all the equilibrium points.
- Show that $K = a\omega_1^2 + b\omega_2^2 + c\omega_3^2$ is an integral of the system if a, b, and c satisfy a certain equation involving I_1, I_2, and I_3.
- Look at the cover figure for Chapter 8 and the Atlas Plates CONSERVATIVE SYSTEM D–F and explain what you see in light of your analysis.
- Determine the stability properties of the orbits.
- Explain why there are periodic orbits in $\omega_1\omega_2\omega_3$-space. Are the periods all the same?

9. Construct four nonlinear conservative systems in three state variables, which have $K = x^2 \pm 2y^2 \pm 3z^2$ (four cases) as integrals. Plot orbits on level sets of K, and plot component curves. Describe what you see. [*Hint*: Study Example 8.3.3, Figure 8.3.2, and Atlas Plates CONSERVATIVE SYSTEM A, B.]

8.4 Stability and the Linear Approximation

The Lyapunov tests provide ways to settle stability questions for a differential system in the neighborhood of an equilibrium point without having to find the solutions. This is a distinct advantage both in the analysis of a given physical system modeled by differential equations and in the design of a dynamical system to operate in or near the equilibrium state. In this section, the Lyapunov Method is applied to a system whose rate function can be written as a linear part plus a "higher-order" perturbation.

The Linear Approximation to a System

Suppose that a physical system with an equilibrium state at the origin can be modeled by the differential system in n state variables,

$$x' = F(x) \tag{1}$$

where F is twice continuously differentiable in a region R of $\mathbf{R}^n$ containing the origin 0, and $F(0) = 0$. It is no restriction to place the equilibrium point at the origin because this can always be achieved by a simple translation in x-space. There is no way to determine the stability of system (1) at 0 unless we have more information about F. A rule frequently followed in engineering and applied analysis may be used: *linearize F about the origin*. This suggests expanding $F(x)$ about the origin by using the multidimensional version of Taylor's Theorem (see Appendix C.5) and looking at the linear terms. Since $F(0) = 0$, we have

$$x' = F(0) + \left.\frac{\partial F}{\partial x}\right|_{x=0} x + P(x) = Ax + P(x) \tag{2}$$

where $\partial F/\partial x|_{x=0}$ is the real, $n \times n$ matrix A of the first-order partial derivatives $\partial F_i/\partial x_j$ evaluated at $x = 0$, and $P(x)$ is the remainder term in the Taylor expansion of $F(x)$ about $x = 0$. The matrix $[\partial F_i/\partial x_j]$ is called the *Jacobian matrix* of F. Since F is twice continuously differentiable, $P(x)$ is at least second-order in x, where "second-order" is one case of the following general definition of order:

> ❖ **Order of a Function.** Suppose that $P(x)$ is a function for which $P(0) = 0$. $P(x)$ is said to be of *order* $k \geq 0$ in x at $x = 0$ if there is a positive constant c such that
>
> $$\|P(x)\| \leq c\|x\|^k \tag{3}$$
>
> for all x in some region containing the origin.

EXAMPLE 8.4.1

Order of a Scalar Function
The scalar function of the scalar variable x, $P(x) = x^2 - x^3$, is of order 2 at the origin because (applying the Triangle Inequality to $|1 - x|$)

$$|x^2 - x^3| = |x|^2|1 - x| \leq |x|^2[1 + |x|] \leq 2|x|^2$$

for all x, $|x| \leq 1$.

EXAMPLE 8.4.2

Order of a Vector Function
The vector function $P(x, y) = (x^2 + xy, 3y^2 - x^2)$ is of order 2 at the origin because (using polar coordinates):

$$\begin{aligned}
\|P(x, y)\| &= [(x^2 + xy)^2 + (3y^2 - x^2)^2]^{1/2} \\
&= [r^4(\cos^2\theta + \cos\theta\sin\theta)^2 + r^4(3\sin^2\theta - \cos^2\theta)^2]^{1/2} \\
&\leq r^2[(1 + 1)^2 + 3^2]^{1/2} \\
&= \sqrt{13}\, r^2 = \sqrt{13}\, \|(x, y)\|^2
\end{aligned}$$

where we have used the fact that $|\cos\theta| \leq 1$, $|\sin\theta| \leq 1$, and the Triangle Inequality.

The vector $P(x)$ in system (2) may be viewed as a higher-order disturbance, or *perturbation*, of the *linear system of first approximation*

$$x' = Ax \tag{4}$$

System (4) is asymptotically stable at the origin if every eigenvalue of A has a negative real part, and is unstable if there is an eigenvalue with a positive real part. The Lyapunov Method will be used to show that, if either one of these eigenvalue conditions holds, then the perturbed system (2) has exactly the same stability properties at the origin as the corresponding system (4) of linear approximation.

Asymptotic Stability and the Linear Approximation

If the eigenvalues of the matrix A have negative real parts, then the asymptotic stability of the "linearized" system (4) at the origin carries over to the original system (2), although, as later examples show, the region of attraction may be restricted. The following theorem shows how to construct a Lyapunov asymptotic stability function for the system (4) which may be used for the system (2) as well. See Section 7.5 to review the definition of the matrix exponential e^{tA}.

THEOREM 8.4.1

> Asymptotic Stability and the Linear Approximation. Suppose that A is a real $n \times n$ matrix whose eigenvalues have negative real parts. Let $P(x)$ be a vector function of order 2 in some open set S containing the origin. Then the system $x' = Ax + P(x)$ is asymptotically stable at the origin, and the function
>
> $$V = x^T Bx, \quad \text{where} \quad B = \int_0^\infty e^{tA^T} e^{tA} \, dt$$
>
> is a Lyapunov asymptotic stability function for $x' = Ax + P(x)$ and for $x' = Ax$ at the origin. [The integration is carried out componentwise.]

Proof. First it must be shown that the improper integral defining B converges. This is a consequence of the fact that every entry in the matrix e^{tA} (hence also in the product matrix $e^{tA^T} e^{tA}$) is a linear combination of polynomial-exponential terms $p(t)e^{-at}k(t)$, where $p(t)$ is a polynomial, a is a negative real number, and $k(t)$ is $\sin bt$ or $\cos bt$ for some real constant b. Antiderivatives of these terms are again polynomial-exponentials of the same form. Thus, the antiderivatives tend to 0 as $t \to +\infty$. Hence, the matrix B is obtained by evaluating the antiderivatives of the entries of the matrix $e^{tA^T} e^{tA}$ at $t = 0$.

The matrix B has two other properties to be used in the proof below. First, $V(x) = x^T Bx$ is positive definite on $\mathbf{R}^n$, because we have that for $y(t) = e^{tA}x$,

$$x^T Bx = \int_0^\infty x^T e^{tA^T} e^{tA} x \, dt = \int_0^\infty y^T(t)y(t) \, dt = \int_0^\infty \|y(t)\|^2 \, dt \qquad (5)$$

which is positive for $x \neq 0$. Second, B is a solution of the matrix equation

$$A^T B + BA = -I \qquad (6)$$

where I is the $n \times n$ identity matrix. Equation (6), which is called the *Lyapunov equation*, may be derived as follows:

$$A^T B + BA = \int_0^\infty (A^T e^{tA^T} e^{tA} + e^{tA^T} e^{tA} A) \, dt = \int_0^\infty (e^{tA^T} e^{tA})' \, dt$$

$$= [e^{tA^T} e^{tA}]_{t=0}^{t=\infty} = -I$$

because the entries in the matrix $e^{tA^T} e^{tA}$ all vanish at $t = \infty$, while $e^Z = I$, where Z is the $n \times n$ matrix of zeros. B is also symmetric, but the proof of that fact is left to Problem 14.

The properties of B may now be used to prove the theorem. Calculate the derivative of $V(x) = x^T B x$ following the motion of $x' = Ax + P$:

$$
\begin{aligned}
V' &= (x')^T B x + x^T B x' = (x^T A^T + P^T) B x + x^T B (Ax + P) \\
&= x^T (A^T B + BA) x + P^T B x + x^T B P = x^T (-I) x + g(x) \qquad (7) \\
&= -\|x\|^2 + g(x)
\end{aligned}
$$

where $g(x)$ is the scalar function $P^T B x + x^T B P$. Use the Triangle Inequality, the Cauchy-Schwarz Inequality, properties of matrix norms (Appendix C.6), and the estimate (3) with $k = 2$ to estimate $|g(x)|$:

$$
\begin{aligned}
|g(x)| = |P^T B x + x^T B P| &\leq |P^T B x| + |x^T B P| = 2|P^T B x| \\
&\leq 2\|P\| \cdot \|Bx\| \leq 2c\|x\|^2 \|Bx\| \qquad (8) \\
&\leq 2c\|x\|^2 \|B\| \cdot \|x\| \leq C\|x\|^3
\end{aligned}
$$

where C is a positive constant. An upper bound for V' is found from (7) and (8):

$$
V'(x) \leq -\|x\|^2 [1 - C\|x\|] \qquad (9)
$$

If x is restricted to the region S inside the "sphere" defined by $\|x\| = 1/C$, (9) implies that V' is negative definite on S. Thus, $V(x)$ is a (local) Lyapunov asymptotic stability function at the origin for the system $x' = Ax + P$ and that system is (locally) asymptotically stable. $V(x)$ is a (global) Lyapunov asymptotic stability function for the linear system $x' = Ax$. The theorem is proved.

Theorem 8.4.1 gives a way to construct a quadratic Lyapunov asymptotic stability function $V = x^T B x$ for any perturbed system $x' = Ax + P(x)$ whose system of linear approximation $x' = Ax$ is asymptotically stable. Both methods are based on the matrix A and are independent of the perturbation $P(x)$. Actually, Theorem 8.4.1 gives not one, but two ways to construct V. The two examples below illustrate these methods. First, according to (5),

$$
V = x^T B x = \int_0^\infty \|y(t)\|^2 \, dt
$$

where $y(t) = e^{tA} x$. In some cases, the construction of e^{tA} is straightforward, particularly if A is a diagonal matrix, as in the following example.

EXAMPLE 8.4.3

Constructing a Lyapunov Function, I
The asymptotically stable, linear system $x_1' = -x_1$, $x_2' = -2x_2$, has the diagonal system matrix

$$
A = \begin{bmatrix} -1 & 0 \\ 0 & -2 \end{bmatrix}
$$

In this case

$$
e^{tA} = \begin{bmatrix} e^{-t} & 0 \\ 0 & e^{-2t} \end{bmatrix}, \qquad y(t) = e^{tA} \begin{bmatrix} x_1 \\ x_2 \end{bmatrix} = \begin{bmatrix} e^{-t} x_1 \\ e^{-2t} x_2 \end{bmatrix}
$$

and so a Lyapunov asymptotic stability function for the system is

$$
V = x^T B x = \int_0^\infty \|y(t)\|^2 \, dt = \int_0^\infty (e^{-2t} x_1^2 + e^{-4t} x_2^2) \, dt = \frac{1}{2} x_1^2 + \frac{1}{4} x_2^2
$$

It is not usually this easy to calculate e^{tA}. The Lyapunov equation $A^T B + BA = -I$ offers another way to calculate B by purely algebraic methods. We have already shown that the matrix $B = \int_0^\infty e^{tA^T} e^{tA} \, dt$ solves the Lyapunov equation. But Lyapunov

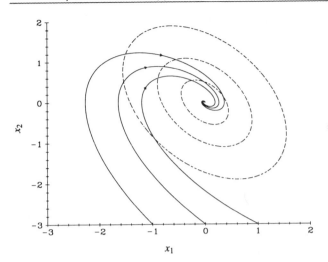

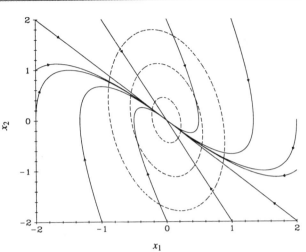

FIGURE 8.4.1 Level sets of V and orbits of (10): $a = 1, b = 1$. See Example 8.4.4.

FIGURE 8.4.2 Level sets of V and orbits of (10): $a = 2, b = 3$. See Example 8.4.4.

showed that, given a matrix A whose eigenvalues have negative real parts, equation (6) has a unique matrix solution B, and B is symmetric. So this solution can be none other than $B = \int_0^\infty e^{tA^T} e^{tA} \, dt$, giving us another way to calculate B via the Lyapunov equation (6).

EXAMPLE 8.4.4

Constructing a Lyapunov Function, II

Suppose that a and b are positive constants. The scalar ODE $x'' + ax' + bx = 0$ generates the globally asymptotically stable system with system matrix A:

$$x_1' = x_2 \qquad\qquad A = \begin{bmatrix} 0 & 1 \\ -b & -a \end{bmatrix} \tag{10}$$
$$x_2' = -bx_1 - ax_2$$

To construct the Lyapunov asymptotic stability function $V = x^T Bx$ for system (10) using Lyapunov's equation (6), we proceed as follows:

$$A^T B + BA = \begin{bmatrix} 0 & -b \\ 1 & -a \end{bmatrix} \begin{bmatrix} b_{11} & b_{12} \\ b_{12} & b_{22} \end{bmatrix} + \begin{bmatrix} b_{11} & b_{12} \\ b_{12} & b_{22} \end{bmatrix} \begin{bmatrix} 0 & 1 \\ -b & -a \end{bmatrix} = \begin{bmatrix} -1 & 0 \\ 0 & -1 \end{bmatrix}$$

where $B = [b_{ij}]$ is to be determined. There are three linear equations for the unknowns b_{11}, b_{12}, b_{22}:

$$-bb_{12} - bb_{12} = -1, \quad -bb_{22} + b_{11} - ab_{12} = 0, \quad 2b_{12} - 2ab_{22} = -1$$

The unique solutions are

$$b_{11} = \frac{1}{2ab}(a^2 + b^2 + b), \quad b_{12} = \frac{1}{2b}, \quad b_{22} = \frac{1}{2ab}(b + 1)$$

A Lyapunov asymptotic stability function for (10) is then

$$V = x^T Bx = \frac{1}{2ab}(a^2 + b^2 + b)x_1^2 + \frac{1}{b}x_1 x_2 + \frac{1}{2ab}(b+1)x_2^2 \tag{11}$$

See Figure 8.4.1 for the elliptical level sets (dashed) $V = 3, 1, 1/4$ and several orbits (solid) of system (10) when $a = 1$ and $b = 1$. Observe that the mixed product term $x_1 x_2$ cannot be eliminated from equation (11). Figure 8.4.2 shows the level sets (dashed) $V = 1, 25/64, 1/16$, and some orbits (solid) when $a = 2$ and $b = 3$. Note that in one case we have a focus and in the other case an improper node.

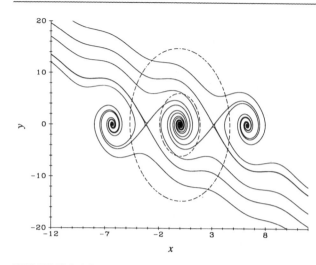

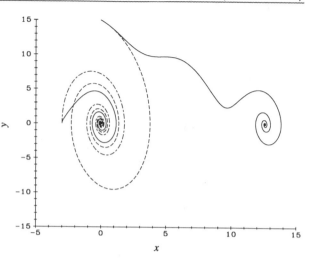

FIGURE 8.4.3 Orbits (solid) of damped pendulum system: $x' = y$, $y' = -10 \sin x - y$; level sets (dashed) $V = 20, 120$. See Example 8.4.5.

FIGURE 8.4.4 Orbits of pendulum system (13) (solid), orbits of linearized system (14) (dashed): initial points $(0, 15)$ and $(-3, 0)$. See Example 8.4.5.

Theorem 8.4.1 asserts that the Lyapunov asymptotic stability function given in (11) for the linear system (10) also "works" for the system when higher-order perturbations are added to the rate functions, although the domain of attraction of the origin may be drastically shrunk. The next example shows what may be done along these lines.

EXAMPLE 8.4.5

Simple Damped Pendulum System and Its Linear Approximation

The ODE which models the simple damped pendulum has the form (see Section 4.1),

$$x'' + ax' + b \sin x = 0 \tag{12}$$

where a and b are positive constants.

An equivalent system for (12) is

$$\begin{aligned} x' &= y \\ y' &= -b \sin x - ay \end{aligned} \tag{13}$$

Because the Maclaurin expansion for $\sin x$ is $x - x^3/3! + x^5/5! - \cdots$, the linearized version of system (13) about the equilibrium point at the origin is

$$\begin{aligned} x' &= y \\ y' &= -bx - ay \end{aligned} \tag{14}$$

According to Example 8.4.4, a Lyapunov asymptotic stability function for (13) and (14) is

$$V = \frac{1}{2ab}(a^2 + b^2 + b)x^2 + \frac{1}{b}xy + \frac{1}{2ab}(b+1)y^2 \tag{15}$$

The level sets of V are slightly tilted ellipses. For all C small enough, orbits of (13) that enter the ellipse $V = C$ are attracted to the equilibrium point at the origin. But if C is large, then the nonlinear terms in the Taylor expansion of $\sin x$ strongly influence the behavior of the orbits passing through points inside the ellipse, points that are distant from the origin. These points may not be in the region of attraction of the equilibrium point of (13) at the origin $(0, 0)$, although they may be in the region of attraction of some other asymptotically stable equilibrium point $(2k\pi, 0)$. See Figure 8.4.3 where $a = 1$, $b = 10$, and the level sets are given by $V = 20$ and $V = 120$. All orbits crossing the ellipse $V = 20$ move inward toward $(0, 0)$ as t increases. Some of the orbits entering the larger ellipse later exit the ellipse and go elsewhere.

This is in marked contrast to the situation with the system of linear approximation for which the region of attraction of $(0, 0)$ is the entire xy-plane. In addition, system (13) has other equilibrium points besides $(0, 0)$, but (14) does not. Figure 8.4.4 shows that, if the initial positions are far from the origin (for example, $x_0 = 0$, $y_0 = 15$ as in the figure), then orbits of (13) and (14) may have very different asymptotic properties. On the other hand, if the initial point is close to the origin (for example, $x_0 = -3$, $y_0 = 0$ as in Figure 8.4.4), the two orbits show similar behavior as $t \to \infty$. See also Atlas Plates PENDULUM A–C.

Instability and the Linear Approximation

Theorem 8.4.1 settles the question of how the higher-order terms in the system $x' = Ax + P(x)$ affect stability properties if the eigenvalues of A all have negative real parts. Suppose now that at least one eigenvalue of A has positive real part. Then the linearized system $x' = Ax$ is unstable at the origin (Theorem 8.1.1). The practical question in this situation is whether a physical system modeled by $x' = Ax + P(x)$ can be stabilized at the origin by a suitable choice of the higher-order term $P(x)$.

THEOREM 8.4.2

> Instability and the Linear Approximation. Suppose that A is a real $n \times n$ matrix with at least one eigenvalue whose real part is positive. Then the system, $x' = Ax + P(x)$, is unstable at the origin for all perturbations $P(x)$ which are at least of order 2 at the origin.

The proof of the theorem is omitted. We see from Theorem 8.4.2 that higher-order perturbations cannot stabilize the system. The "intuitive" reason is that in every region containing the origin there is at least one point x^0 such that the solution $x(t, x^0)$ of the unstable linear system (4) moves out of the region exponentially fast. The addition of the higher order terms of $P(x)$ cannot appreciably diminish these exponential exit rates.

EXAMPLE 8.4.6

Unstable Systems

The system

$$x' = y, \qquad y' = x - 2ay$$

where a is a positive number, is unstable at the origin because the eigenvalues of the system matrix are $-a \pm (a^2 + 1)^{1/2}$, one of which is positive. The system

$$x' = y + P_1, \qquad y' = x - 2ay + P_2$$

remains unstable at the origin for every choice of perturbations P_1 and P_2 which are of order 2 in a region containing the origin.

Stability and Jacobian Matrices

Theorems 8.4.1 and 8.4.2 may be combined into a result about the stability properties of the system $x' = F(x)$ at any equilibrium point p.

THEOREM 8.4.3

Stability Properties and the Jacobian Matrix. Suppose that $F(x)$ is twice continuously differentiable in a neighborhood of an equilibrium point p of the system, $x' = F(x)$. Suppose that A is the Jacobian matrix of the function F at the point p. Then:

1. If all eigenvalues of A have negative real parts, the system $x' = F(x)$ is asymptotically stable at p.

2. If A has at least one eigenvalue with a positive real part, the system $x' = F(x)$ is unstable at p.

Proof. Translate the equilibrium point p to the origin and apply Theorems 8.4.1 and 8.4.2.

The next example shows how Theorem 8.4.3 may be used to determine the stability properties of the damped pendulum system at each of its equilibrium points. The stability properties were determined in Problem 6 of Section 8.2 by constructing complicated Lyapunov functions, but now all we need to do is calculate eigenvalues of Jacobian matrices.

EXAMPLE 8.4.7

Stability Properties of Damped Simple Pendulum: Region of Attractor

System (13) has equilibrium points at $(n\pi, 0)$. The Jacobian matrices of the system at a general point (x, y) and at the specific equilibrium points $p = (2k\pi, 0)$, and $((2k + 1)\pi, 0)$ are, respectively,

$$\frac{\partial F}{\partial x} = \begin{bmatrix} 0 & 1 \\ -b\cos x & -a \end{bmatrix}, \quad \frac{\partial F}{\partial x}\bigg|_{(2k\pi,0)} = \begin{bmatrix} 0 & 1 \\ -b & -a \end{bmatrix}, \quad \frac{\partial F}{\partial x}\bigg|_{(2k+1)\pi,0)} = \begin{bmatrix} 0 & 1 \\ b & -a \end{bmatrix}$$

The eigenvalues of the second matrix above are the roots of the characteristic polynomial, $\lambda^2 + a\lambda + b$, and, hence, have negative real parts because a and b are positive. Thus, by Theorem 8.4.3 the system is asymptotically stable at every equilibrium point $(2k\pi, 0)$. However, the eigenvalues of the Jacobian matrix at the equilibrium points $((2k + 1)\pi, 0)$ are the roots of $\lambda^2 + a\lambda - b$, and, hence, one of them is positive and the other negative. By Theorem 8.4.3 the damped pendulum system is unstable at each of these equilibrium points, a fact which is hardly surprising because they correspond to the pendulum balanced in a vertically upward position. Figure 8.4.5 shows three asymptotically stable, and two unstable equilibrium points. The region of attraction of the origin is a wavy band running downward from the upper left of the xy-plane. Nonconstant orbits on the boundary of the region are called *separating orbits*.

EXAMPLE 8.4.8

A Stable Cooperation Model

The system of ODEs introduced in Example 6.3.2

$$\begin{aligned} x' &= (4 - 2x + y)x \\ y' &= (4 + x - 2y)y \end{aligned} \tag{16}$$

models cooperative interaction between two species (the term xy in each rate equation) subject to self-limitation (the terms $-2x^2$ and $-2y^2$). Computer simulations of this system (see Figures 6.3.3 and 6.3.4) suggest that the system is unstable at the equilibrium points (0,0), (0,2), and (2,0), but asymptotically stable at (4,4). To verify these stability properties we shall find the eigenvalues of the Jacobian matrices of system (16) at each of the equilibrium points and use Theorem 8.4.3. The Jacobian matrix

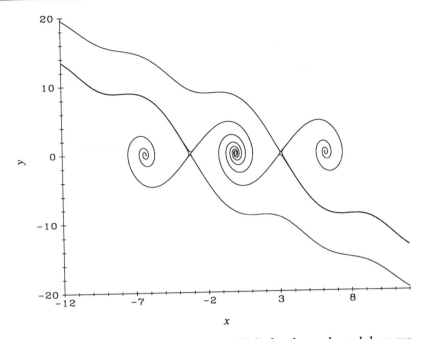

FIGURE 8.4.5 Region of attraction of (0,0) for damped pendulum system; the "stalks" of the "tendril" orbits are near the separating orbits. See Example 8.4.7.

at a general point is

$$\begin{bmatrix} 4 - 4x + y & x \\ y & 4 - 4y + x \end{bmatrix}$$

The Jacobian matrices at the respective equilibrium points (0,0), (2,0), (0,2), and (4,4) are

$$\begin{bmatrix} 4 & 0 \\ 0 & 4 \end{bmatrix}, \quad \begin{bmatrix} -4 & 2 \\ 0 & 6 \end{bmatrix}, \quad \begin{bmatrix} 6 & 0 \\ 2 & -4 \end{bmatrix}, \quad \begin{bmatrix} -8 & 4 \\ 4 & -8 \end{bmatrix}$$

and the respective sets of eigenvalues are

$$\{4, 4\}, \quad \{-4, 6\}, \quad \{-4, 6\}, \quad \{-12, -4\}$$

Theorem 8.4.3 then implies that the system is indeed unstable at the first three equilibrium points and asymptotically stable at the fourth. See also Problem 15.

None of the theorems have taken up the case when the Jacobian matrix A has some eigenvalues with zero real parts. Indeed, there are no general results in this case because the stability of the perturbed system depends very much on the specific perturbation. Examples 8.4.9 and 8.4.10 below show what can happen.

EXAMPLE 8.4.9

The system

$$x' = y^3, \qquad y' = x^3$$

is unstable at the origin because the indefinite function $V(x, y) = xy$ has a positive definite total derivative $dV/dt = x^4 + y^4$. See Figure 8.4.6. However, the linearized system

$$x' = 0, \qquad y' = 0$$

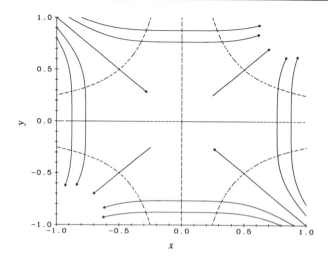

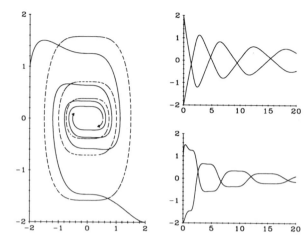

FIGURE 8.4.6 Orbits of $x' = y^3$, $y' = x^3$; level sets of $V = xy = -0.25, 0, 0.25$ (dashed). See Example 8.4.9.

FIGURE 8.4.7 Orbits and component graphs of $x' = y$, $y' = -x^3 - x^2y$; level sets (dashed) of $V = x^4/4 + y^2/2 = 0.3, 1, 5$. See Example 8.4.10.

is neutrally stable at the origin because there is no motion at all: all points are at rest. Thus, the nonlinear system and its linear approximation at the origin have completely different stability properties.

EXAMPLE 8.4.10

The system

$$x' = y, \qquad y' = -x^3 - x^2y$$

is stable at the origin because $V = x^4/4 + y^2/2$ is a positive definite function and $dV/dt = -x^2y^2$ is negative semidefinite. See Figure 8.4.7. The linear approximation

$$x' = y, \qquad y' = 0$$

has solutions $x = y_0t + x_0$, $y = y_0$, and is unstable at the origin. The stability properties of the system and of its linear approximation are very different.

These two examples show the difficulty in reaching conclusions about the stability properties of a nonlinear system if the linear approximation has an eigenvalue whose real part is zero. See also Problems 5 and 11.

In Section 7.4 planar linear autonomous systems with nonsingular system matrices are classified according to the nature of the eigenvalues and eigenspaces of the system matrix. With Theorem 8.4.3 in hand, we can add higher-order perturbations to these planar systems, and in many cases the basic structure of the orbits near the equilibrium point at the origin does not change much. Table 8.1 at the end of the chapter lists the possibilities.

PROBLEMS

1. (*Jacobian Matrices and Stability*). Determine the stability properties of each system at the origin. [*Hint*: Find the eigenvalues of the Jacobian matrix at the origin. Use Theorem 8.4.3.]

   (a) $x' = -x + y^2$, $y' = -8y + x^2$

(b) $x' = 2x + y + x^4, \quad y' = x - 2y + x^3 y$

(c) $x' = 2x - y + x^2 - y^2, \quad y' = x - y$

(d) $x' = e^{x+y} - \cos(x - y), \quad y' = -\sin x$

2. (*Jacobian Matrices and Stability*). Find the equilibrium points and discuss the stability properties of the system at each of these points. [*Hint*: If p is an equilibrium point of $x' = F(x)$, calculate the signs of the real parts of the eigenvalues of the Jacobian matrix $[\partial F/\partial x]_{x=p}$ and use Theorem 8.4.3 when applicable.]

   (a) $x' = y, \quad y' = -6x - y - 3x^2$

   (b) $x' = y^2 - x, \quad y' = x^2 - y$

   (c) $x' = -x - x^3, \quad y' = y + x^2 + y^2$

   (d) $x' = -y - x(x^2 + y^2), \quad y' = x - y(x^2 + y^2)$ [*Hint*: If $V = x^2 + y^2$, find V'.]

   (e) $x' = x + xy^2, \quad y' = x$

   (f) $x' = -x + y^2, \quad y' = x + y$

3. Plot several orbits of each of the systems in Problems 3(a)–(f) in the neighborhood of each equilibrium point. From the visual appearance of these orbits, what are your conclusions about the nature of the eigenvalues of the Jacobian matrix at each equilibrium point? Assume in parts (a), (b), and (c) that you know in advance that none of the eigenvalues has 0 real part. (Warning! Your conclusions based solely on the graphical evidence could be wrong since the higher-order terms may be decisive in parts (d), (e), (f).)

4. (*Solving the Lyapunov Equation*). In each case construct a Lyapunov asymptotic stability function $V = [x\,y]B[x\,y]^T$ at the origin by solving the Lyapunov equation (6) for the symmetric matrix B, where A is the matrix of coefficients of the linear terms.

   (a) $x' = -x - y, \quad y' = x - y$ **(b)** $x' = x - 3y, \quad y' = 4x - 6y$

   (c) $x' = -7x + x^3 y, \quad y' = -y + x^{10}$ **(d)** $x' = -3y, \quad y' = 2x - 5y + x^3 y$

5. (*Zero Eigenvalue*). Show that the linear approximation at the origin to the system $x' = -x, \; y' = ay^3$ has eigenvalues -1 and 0. Show that the system is asymptotically stable at the origin if a is a negative constant, unstable if a is a positive constant. What happens if $a = 0$? Plot the orbits in all three cases. Why does Theorem 8.4.3 not apply? [*Hint*: Try Lyapunov functions of the form $V = \alpha x^2 + \beta y^2$.]

6. (*Planar Portrait*). Consider the system, $x' = x - y + x^2 - xy, \; y' = -y + x^2$.

   (a) Find the equilibrium points.

   (b) Using Jacobian matrices and their eigenvalues, determine the stability properties of the system at each equilibrium point.

   (c) Plot orbits in a neighborhood of each equilibrium point. Each of these plots should be restricted to a "small" region containing a single equilibrium point. [*Hint*: See Atlas Plate PLANAR PORTRAIT A.]

   (d) Plot a portrait of the orbits of the system in a rectangle large enough to contain all of the equilibrium points. [*Hint*: See Atlas Plates PLANAR PORTRAIT B, C.]

7. (*Domain of Attraction*). Computer graphics may be used to determine the approximate boundary of the domain of attraction of a locally asymptotically stable equilibrium point. The problem below illustrates the process.

   (a) Show that the system $x' = y, \; y' = -2x - y - 3x^2$ is asymptotically stable at the origin, but is not globally asymptotically stable.

(b) Use computer graphics to find the approximate region of attraction of the origin. Use the plotting region, $-3 \le x \le 2$, $-3 \le y \le 5$.

8. (*A Perturbed RLC Circuit*). The differential equation

$$Lx'' + Rx' + x/C + g(x, x') = 0$$

governs the behavior of a simple *RLC* electric circuit, where x represents the charge on the capacitor, x' is the current in the loop, and L, S, and C are the positive circuit parameters inductance, resistance, and capacitance. The twice continuously differentiable function $g(x, x')$ represents nonlinearities in the circuit. Suppose that $g(x, x')$ is of order 2 in (x, x') in a neighborhood of $x = 0$, $x' = 0$. Show that the equivalent system, $x' = y$, $y' = -x/LC - Ry/L - g(x, y)/L$ is asymptotically stable at the origin.

9. For each of the systems of the indicated Atlas Plates, find the equilibrium points and the Jacobian matrix and its eigenvalues at each equilibrium point. Use this information and Theorem 8.4.3 to determine the stability properties of the system at each equilibrium point.

(a) BIFURCATION A **(b)** BIFURCATION G (any value of $\varepsilon > -1$)

10. (*Pendulum with a Torque*). The equation of motion of a damped simple pendulum with constant external torque is $x'' + ax' + b \sin x = L$, where a, b, and L are positive constants. If $0 < L < b$, analyze the character of stability of every equilibrium point of the equivalent system, $x' = y$, $y' = -b \sin x - ay + L$. Plot a portrait of orbits in a rectangle that includes at least five equilibrium points.

11. (*Pure Imaginary Eigenvalues*). Find functions P_1 and P_2 of order of magnitude at least 3 such that the system $x' = -y + P_1$, $y' = x + P_2$ has the stability characteristics at the origin as indicated in **(a)–(c)**. [*Hint*: Let $P_i(x, y) = z_i(x^2 + y^2)$, where z_i is $\pm x$ or $\pm y$. Write the system in polar coordinates.]

(a) Globally asymptotically stable.

(b) Unstable.

(c) Neutrally stable.

(d) The perturbation terms P_i may affect the stability properties of the linearized system $x' = y$, $y' = -x$. Why is this not a contradiction of Theorem 8.4.3?

(e) For each of your systems, plot orbits in a region containing the origin.

12. (*Stabilizing a System by "Tuning" a Parameter*). Find all values of α such that the system $x' = z + x^2y$, $y' = x - 4y + xz^2$, $z' = \alpha x + 2y - z + x^2$ is asymptotically stable at the origin. [*Hint*: Use the Routh Criterion of Section 7.6.]

13. (*Asymptotic Stability over a Parameter Range*). Show that the system

$$x_1' = -x_1 + 2kx_4 - 2x_2^2$$

$$x_2' = -x_2 + 2x_1x_2$$

$$x_3' = -3x_3 + x_4 + kx_1$$

$$x_4' = -2x_4 - x_3 - kx_2$$

is asymptotically stable at the origin when $|k|$ is small. [*Hint*: Choose $V = x_1^2 + x_2^2 + x_3^2 + x_4^2$.]

14. Show that the matrix $B = \int_0^\infty e^{tA^T} e^{tA} \, dt$ is symmetric if the improper integral converges.

15. (*Stable Cooperation Model*). Construct a Lyapunov asymptotic stability function for the system $x' = (4 - 2x + y)x$, $y' = (4 + x - 2y)y$ at the equilibrium point $(4,4)$. Plot orbits in the region $0 \leq x \leq 10$, $0 \leq y \leq 10$; plot the corresponding component graphs. Then find the rate equations of the linearized system and plot orbits of that system using the same initial points as for the orbits of the nonlinear system. Is there much difference between the orbits of the two systems? [*Hint*: Find the Jacobian matrix A of the system at $(4,4)$. The Lyapunov function is $[u \quad v]B[u \quad v]^T$ where $u = x - 4$, $v = y - 4$, and B solves the Lyapunov equation (6). The linearized system at $(4, 4)$ is $u' = a_{11}u + a_{12}v$, $v' = a_{21}u + a_{22}v$, where a_{ij} are the entries in A.]

16. (*The New Zealand Possum Model*). Consider the New Zealand possum population/disease dynamic model: $dx/ds = cx - y$, $dy/ds = y(x - y) - ry$, outlined in Section 6.5. Assume that $0 < c < 1$ and $r > 1$.

    (a) Use Theorem 8.4.3 to determine the stability properties of the system at each equilibrium point. Introduce coordinates based on the equilibrium point inside the population quadrant and write the nonlinear system and its linearization in these new coordinates.

    (b) When $c = 0.05$ and $r = 1.1$, plot and describe the behavior of the orbits of the linearized system near the equilibrium point $(r/(1 - c), cr/(1 - c))$; repeat for orbits of the nonlinear system.

17. (*Interacting Species Models*). Consider the interacting species model $x' = (2 + ax + by)x$, $y' = (2 + cx + dy)y$ in the cases indicated below. Give a complete stability analysis of each system at each equilibrium point. Plot orbits in the population quadrant.

    - Cooperation leading to explosive growth model: $a = d = -1$ and $b = c = 2$. See Example 6.3.1.

    - Stable competition model: $a = d = -2$ and $b = c = -1$. See Example 6.3.3.

    - Competitive exclusion model: $a = d = -1$ and $b = c = -2$. See Example 6.3.4.

    - Fix the parameters $b = c = -1$ and choose any nonzero values for a and d you want to and analyze the resulting model.

Perturbed Planar Systems: Stability

A nonlinear planar autonomous system with an equilibrium point at the origin has the form

$$x' = ax + by + P(x, y)$$
$$y' = cx + dy + Q(x, y)$$

where a, b, c, and d are real constants and P and Q are continuously differentiable and of order at least two in x, y at the origin. The linear approximation at the origin and the corresponding system matrix A are

$$x' = ax + by \qquad A = \begin{bmatrix} a & b \\ c & d \end{bmatrix}$$
$$y' = cx + dy$$

The eigenvalues of A are denoted by λ_1 and λ_2, and it is assumed that neither is zero because the stability properties in that case are very difficult to determine.

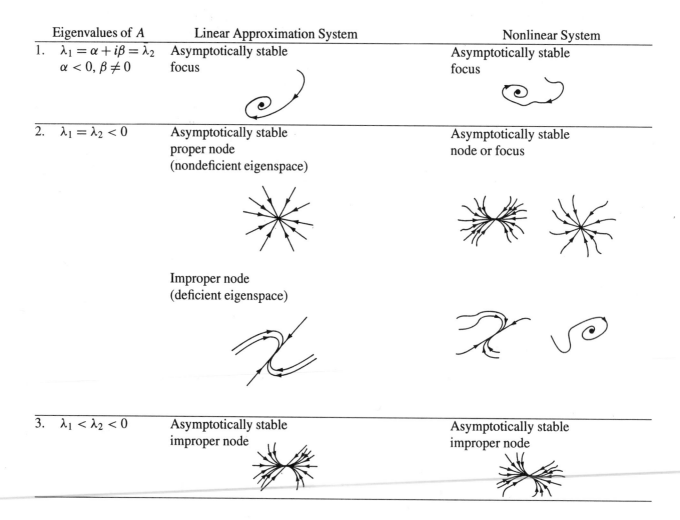

Eigenvalues of A	Linear Approximation System	Nonlinear System
1. $\lambda_1 = \alpha + i\beta = \bar{\lambda}_2$ $\alpha < 0, \beta \neq 0$	Asymptotically stable focus	Asymptotically stable focus
2. $\lambda_1 = \lambda_2 < 0$	Asymptotically stable proper node (nondeficient eigenspace) Improper node (deficient eigenspace)	Asymptotically stable node or focus
3. $\lambda_1 < \lambda_2 < 0$	Asymptotically stable improper node	Asymptotically stable improper node

Eigenvalues of A	Linear Approximation System	Nonlinear Planar System

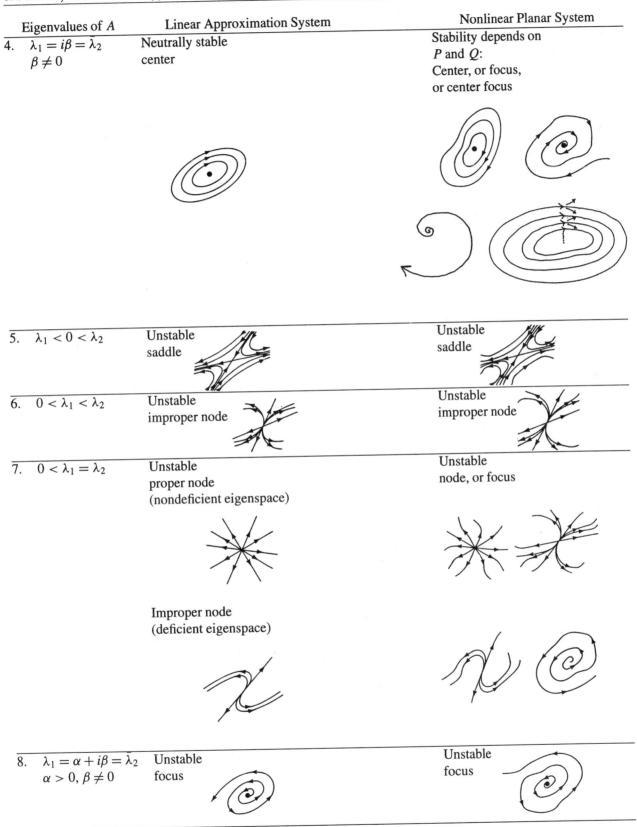

4. $\lambda_1 = i\beta = \lambda_2$
   $\beta \neq 0$

Neutrally stable center

Stability depends on P and Q:
Center, or focus, or center focus

5. $\lambda_1 < 0 < \lambda_2$

Unstable saddle

Unstable saddle

6. $0 < \lambda_1 < \lambda_2$

Unstable improper node

Unstable improper node

7. $0 < \lambda_1 = \lambda_2$

Unstable proper node (nondeficient eigenspace)

Unstable node, or focus

Improper node (deficient eigenspace)

8. $\lambda_1 = \alpha + i\beta = \lambda_2$
   $\alpha > 0, \beta \neq 0$

Unstable focus

Unstable focus

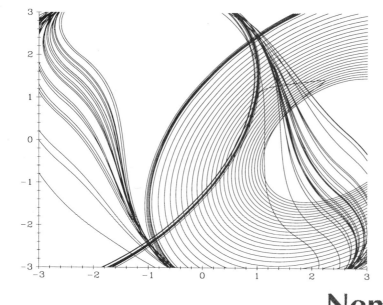

Nonlinear Systems: Cycles, Bifurcations, and Chaos

For autonomous dynamical systems the simplest equilibrium states are those in which the system is at rest, the equilibrium points. In this chapter another type of equilibrium state is studied. This type consists of the cycles in state space, corresponding to periodic solutions of the dynamical equations. Of special importance are the attracting cycles that pull all nearby orbits toward themselves as time advances. The first part of the chapter is devoted to the study of cycles and a variant called cycle-graphs. In the second part we take up the sensitivity question. What happens as the value of a parameter in a rate function is pushed up or down? If the change is small, one expects the new orbits to look like the old, and that is often true. However, for many systems as a parameter moves through a critical value, the behavior suddenly changes. For example, an asymptotically stable equilibrium point destabilizes and ejects an attracting cycle, or the period of a cycle suddenly doubles. Or a cycle loses its periodic character and becomes chaotic. The last two sections take up examples of sensitivity to changes in parameters or initial data. Most of the systems considered are autonomous, nearly all are nonlinear. The chapter cover figure (and the figure on the book cover) shows a projection of a chaotically wandering orbit of the scroll circuit system (Problem 4, Section 9.4).

9.1 Limit Cycles: van der Pol Systems

The aeolian harp, a pneumatic hammer, the scratching noise of a knife on a plate, the waving of a flag in the wind, the humming noise sometimes made by a water-tap, the squeaking of a door, the tetrode multivibrator . . . , the intermittent discharge of a condenser through a neon tube, the periodic recurrence of epidemics and of economic crises, the periodic density of an even number of species of animals living together and the one species serving as food for the other, the sleeping of flowers, the periodic recurrence of showers behind a depression, the shivering from cold, menstruation, and finally, the beating of a heart.[1]

In this quotation, van der Pol has grouped the most varied phenomena into a single category of systems that exhibit periodic oscillations, or cycles, even though there are no external, nonconstant periodic forces causing such oscillations. One wonders, however, about some of van der Pol's systems; for example, the sleeping of flowers is surely a response to the diurnal cycle of darkness and light. According to van der Pol, most of these systems can be modeled more or less accurately by a pair of first-order nonlinear ordinary differential equations (ODEs) of the form

$$x' = f(x, y), \qquad y' = g(x, y) \tag{1}$$

The symbols x and y denote the magnitudes of the principal quantities being modeled. For example, x might denote the voltage across the capacitor of a neon tube and y the rate at which that voltage changes. Or x and y might denote the population densities of a predator and its prey. In this section, however, we will assume that the system (1) is given and look for any cycles that may be present in the orbit set. From now on we use the term *cycle* to mean the orbit of a nonconstant periodic solution of system (1). We assume that the rate functions f and g are continuous and continuously differentiable in some region S of the xy-plane, and so the Existence/Uniqueness/Extension/Continuity Theorem 6.2.1 holds. In particular, this means that the region S is filled up by the orbits of system (1) and that no two orbits can intersect.

Cycles for Planar Linear Systems

The nonconstant orbits of an autonomous planar linear system whose constant system matrix A has a pair of conjugate pure imaginary eigenvalues are cycles. If the eigenvalues of A are $i\beta$ and $-i\beta$, $\beta \neq 0$, with corresponding eigenvectors v and $\bar{v}$, then the general real-valued solution of the system is

$$c_1 \operatorname{Re}\left[v e^{i\beta t} \right] + c_2 \operatorname{Im}\left[v e^{i\beta t} \right]$$

as was shown in Section 7.3, Problem 5, and illustrated in Section 7.4. In more familiar terms, the components of every nonconstant solution are linear combinations of the sinusoids $\sin \beta t$ and $\cos \beta t$ of period $2\pi/\beta$.

EXAMPLE 9.1.1

Simple Harmonic Oscillator

The general real-valued solution of the harmonic oscillator system

$$x' = y, \qquad y' = -k^2 x$$

where k is a nonzero real constant is

$$x = A\cos(kt + \varphi), \quad y = -kA\sin(kt + \varphi)$$

---

[1]Balthazar van der Pol, *Phil. Mag.*, 6 (ser 7)(1928), pp. 763–775.

where A and φ are arbitrary real constants. These solutions have common period $2\pi/k$ and define a family of elliptical cycles

$$x^2 + y^2/k^2 = A^2$$

that enclose the equilibrium point at the origin and fill up the xy-plane. Note that the eigenvalues of the system matrix here are $\pm ik$.

There is nothing special about the harmonic oscillator and its elliptical cycles. As was shown in Section 7.4 in the discussion of the equilibrium point called a center, the properties of the harmonic oscillator are shared by all planar linear systems whose system matrix has a pair of conjugate pure imaginary eigenvalues. In particular, all cycles of such a system have the same period and there are no isolated cycles. We emphasize these points now, because when we turn to nonlinear systems a much more complex and interesting picture of cycle behavior emerges.

Cycles: Limit Cycles

We confine our attention to planar system (1) and to cycles of that system. We make no assumptions about the linearity or nonlinearity of (1), but the focus will be on nonlinear systems. A cycle Γ may be part of a family of cycles, i.e., some neighborhood of Γ is filled with cycles of the system. Or Γ may be *isolated* in the sense that there is a neighborhood of Γ containing no cycles other than Γ. See the margin for sketches of both types of cycles.

Each cycle Γ has the following properties:

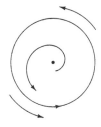

- Γ divides the xy-plane into a bounded interior region and an unbounded exterior.
- The interior region of Γ contains at least one equilibrium point.

The proofs of these two properties are omitted.[2] The properties have important consequences. First of all, an orbit of system (1) that begins in the interior region of a cycle must remain there forever; likewise an orbit that passes through a point of the exterior region of a cycle always remains outside the cycle. Otherwise, an orbit would have to cross the cycle, thereby violating the Uniqueness Theorem. One consequence of the second property is that there is no point in searching for cycles in a simply connected region free of equilibrium points.[3]

Although families of cycles appear in several important models (e.g., the linearized harmonic oscillator, the simple pendulum, the Lotka-Volterra predator-prey model), they are not the topic of this subsection. It is the isolated, attracting cycles that are of interest to us here. Such cycles correspond to "robust" periodic motion, periodic motion that will show up even if initial data are changed.

❖ **Limit Cycle.** Let Γ be a cycle of system (1). Γ is a *limit cycle* if there is a point Q not on Γ such that the orbit of the solution through Q approaches Γ either as $t \to +\infty$ or as $t \to -\infty$.

---

[2]The first property is a special case of the Jordan curve theorem. See M. H. A. Newman, *Elements of the Topology of Plane Sets of Points*, 2nd ed., (Cambridge: Cambridge University Press, 1951) for a proof. A proof of the second property may be found in M. W. Hirsch and S. Smale, *Differential Equations, Dynamical Systems, and Linear Algebra* (New York: Academic Press, 1974).

[3]A *simply connected* planar region is a region without any holes.

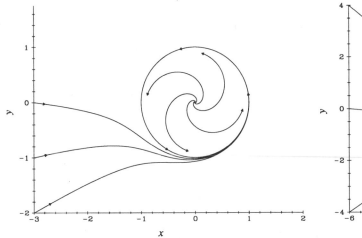

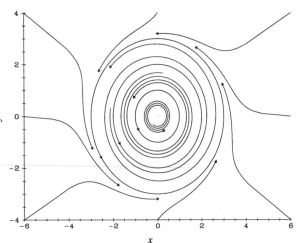

FIGURE 9.1.1 Attracting limit cycle. See Example 9.1.2.

FIGURE 9.1.2 One repelling and two attracting limit cycles. See Example 9.1.3.

❖ **Attractor**. Let Γ be a limit cycle of system (1). Γ is an *attractor* if there is a neighborhood of Γ with the property that any orbit which penetrates that neighborhood approaches Γ as $t \to +\infty$. The *region of attraction* of Γ is the set of all points Q for which the orbit through Q tends to Γ as $t \to +\infty$.

Although not proven here, it can be shown that an orbit that tends to a limit cycle does so by spiraling toward it as time increases; if one orbit spirals toward a limit cycle as time increases, all nearby orbits on the same side do so as well. A *repelling limit cycle* Γ has a similar definition, but nearby orbits tend to Γ as $t \to -\infty$. Examples 9.1.2–9.1.4 show that limit cycles may be present in fairly simple nonlinear systems.

| EXAMPLE 9.1.2 | **An Attracting Limit Cycle** |

An Attracting Limit Cycle

The system

$$x' = x - y - x(x^2 + y^2), \quad y' = x + y - y(x^2 + y^2)$$

has a single equilibrium point at the origin and a unique attracting limit cycle Γ of period 2π defined by $x^2 + y^2 = 1$. The simplest way to see this is to introduce polar coordinates $r^2 = x^2 + y^2$, $\tan \theta = y/x$ (see Section 6.2), obtaining, after differentiating and simplifying, the system

$$r' = r(1 - r^2), \quad \theta' = 1$$

whose general solution is

$$r(t) = r_0[r_0^2 + (1 - r_0^2)e^{-2t}]^{-1/2}, \quad \theta(t) = t + \theta_0$$

The function $r(t)$ is found by solving the Bernoulli equation $r' - r = -r^3$. A solution corresponding to $r_0 = 1$ is $r = 1$, $\theta = t$; in xy-coordinates, we have $x = \cos t$, $y = \sin t$, a solution of period 2π. Since $\lim_{t\to\infty} r(t) = 1$ if $r_0 > 0$, the unit circle Γ is an attracting limit cycle (see Figure 9.1.1) whose region of attraction is the plane with the origin deleted.

It is not often that a solution formula for a nonlinear system can be used to show the existence of a limit cycle, as was done in Example 9.1.2. The next example uses Sign Analysis of Section 2.2 in the search for limit cycles.

EXAMPLE 9.1.3

Three Limit Cycles in One Eye
The system in polar coordinates

$$\theta' = 1, \quad r' = r(1 - r^2)(4 - r^2)(9 - r^2)/1000$$

has two attracting limit cycles ($r = 1, r = 3$) and a repelling limit cycle ($r = 2$). These results are suggested by the changes of sign of r' as r increases through the values 1, 2, and 3 and the techniques of Sign Analysis. See Figure 9.1.2.

EXAMPLE 9.1.4

A Limit Cycle in Each Eye
The quadratic system

$$x' = -y - y^2, \quad y' = 0.5x - 0.2y + xy - 1.2y^2$$

has a repelling limit cycle enclosing the equilibrium point at the origin and an attracting limit cycle around the equilibrium point $(-2, -1)$, but we shall not prove this. See Atlas Plate FIRST ORDER I for a portrait of the orbits.

There is a reason that the systems of Examples 9.1.2– 9.1.4 with limit cycles are nonlinear. We have already seen why this is so in the case of a planar linear system, but the result is true in any dimension.

THEOREM 9.1.1

> No Limit Cycles. Autonomous, first-order linear systems have no isolated cycles.

Proof. Suppose that the autonomous linear system, $x' = Ax$, has a nonconstant periodic solution $x(t)$. Then $cx(t)$ is a nonconstant periodic solution for every nonzero value of the constant c because

$$(cx(t))' = cx'(t) = cAx(t) = A(cx(t))$$

Thus, the cycle defined by $x = x(t)$ is not isolated from other cycles. Note that $x(t)$ and $cx(t)$ have the same period.

Oscillator Circuits: van der Pol Systems and Limit Cycles

Suppose that we are asked to design an electrical circuit with a periodic output current of prescribed amplitude and period. Moreover, whatever active state the circuit is in at time 0, we require that the circuit must approach the periodic state quickly. Van der Pol encountered the problem in his study of the multioscillator circuits of the first commercial radios of the 1920s. His analysis of the corresponding mathematical model is still the basis of many of the studies of nonlinear oscillations. The van der Pol circuit is a simple *RLC* loop (see Section 4.3) in which the passive resistor is replaced by an active element.

Passive *RLC* Circuit

First we shall review the equations of the passive *RLC* circuit, but from a state-variable and system point of view. Suppose that a source of constant voltage is attached to the circuit and then withdrawn. How do current and voltages across the elements attenuate with the passage of time? According to Kirchhoff's Voltage Law, the voltage drops across the three circuit elements satisfy

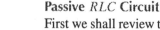

$$V_{13} = V_{12} + V_{23} \tag{2}$$

where V_{ij} denotes the voltage drop from node i to node j. The separate voltages are related to the corresponding circuit elements by the laws

$$V_{12} = LI', \qquad V_{23} = RI \qquad V'_{13} = -\frac{1}{C}I \qquad (3)$$

where the minus sign in the last equation follows from the relation $V_{13} = -V_{31}$. The relations given in (3) are, respectively, Faraday's Law, Ohm's Law, and the differential form of Coulomb's Law. From (2) and (3), we see that the two state variables I and $V = V_{13}$ are enough to characterize the dynamics of the circuit:

$$I' = \frac{1}{L}V_{12} = \frac{1}{L}(-V_{23} + V) = \frac{1}{L}(-RI + V)$$

$$V' = -\frac{1}{C}I \qquad (4)$$

If R, L, and C are positive constants, every solution $(I(t), V(t))$ of the linear system (4) decays to the equilibrium point $(0, 0)$ of the IV-state plane because the eigenvalues of the system matrix have negative real parts.

The physical reason for the orbital decay of the circuit is that the resistor dissipates energy. In fact, the energy of the circuit is the positive definite function

$$E = \frac{1}{2}I^2 + \frac{C}{2L}V^2 \qquad (5)$$

and so the rate of change of E following the motion of (4) is given by

$$E' = I \cdot I' + \frac{C}{L}V \cdot V' = I \cdot \frac{1}{L}(-RI + V) + \frac{C}{L}V \cdot \left(-\frac{1}{C}I\right) = -\frac{R}{L}I^2 \le 0 \quad (6)$$

Thus, $E(t)$ is a nonincreasing function of time along an orbit, and hence is a Lyapunov stability function for the equilibrium point at the origin (see Section 8.2). In fact, because the eigenvalues of the system matrix of the linear system (4) have negative real parts, it follows that $I(t)$ and $V(t)$ tend to 0 as $t \to +\infty$, and so does $E(t)$. Thus, there are no attracting limit cycles, no cycles of any type.

Active *RLC* Circuit

The circuit is active instead of passive if energy is pumped into the circuit whenever the amplitude of the current falls below some level. One way to inject energy is to replace the resistor by an active element which acts as a "negative resistor" at low current levels but dissipates energy at high levels. In van der Pol's time there were vacuum tubes and now there are semiconductor devices that do just this. See the margin sketches for an active circuit and the nonlinear voltage-current relationship of a typical semiconductor device.

The corresponding rate equations for the active circuit are obtained from system (4) for the passive circuit by replacing RI by $F(I)$:

$$I' = \frac{1}{L}(-F(I) + V)$$

$$V' = -\frac{1}{C}I \qquad (7)$$

If time, current, and voltage are rescaled to dimensionless variables, system (7) takes on a simpler form. In particular, let

$$\tau = t/\sqrt{LC}, \qquad x = I/I_0, \qquad y = V\sqrt{C/L}/I_0$$

where I_0 is some positive reference current. Then system (7) becomes the *dimension-*

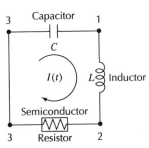

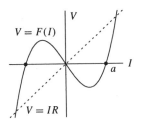

less van der Pol system

$$\frac{dx}{d\tau} = y - \mu f(x), \qquad \frac{dy}{d\tau} = -x \tag{8}$$

where μ is the positive constant $\sqrt{C/L}/I_0$ and $f(x) = F(I_0 x)$. For convenience, we continue to refer to τ as time, x as current, and y as voltage.

In terms of x and y the energy function E defined in (5) has the form

$$E = (\frac{1}{2}x^2 + \frac{1}{2}y^2)I_0^2 \tag{9}$$

Thus $\sqrt{2E}/I_0$ has the magnitude of the Euclidean distance of the point (x, y) from the origin. Let's see what happens to the values of E/I_0^2 as we move along an orbit of system (8) with increasing time. The derivative of E following the motion of system (8) is

$$\frac{dE}{d\tau} = \left[x\frac{dx}{d\tau} + y\frac{dy}{d\tau}\right]I_0^2 = \left[x(y - \mu f(x)) + y(-x)\right]I_0^2 = -\mu x f(x)I_0^2 \tag{10}$$

If the voltage-current relationship has the general shape shown in the sketch in the margin, then $-\mu x f(x)I_0^2$ is negative for $|x| > a$, but positive for $0 < |x| < a$. We see from (10) that energy is lost along arcs of an orbit Γ of system (8) in the regions $x > a$ and $x < -a$, while energy is injected into the system along arcs of Γ lying in the region $-a < x < a$. This suggests that there may be a cycle Γ for which the net energy change over one period is zero. Γ would cut across regions of energy loss and gain in such a way that the loss would exactly balance the gain. Orbits in the zone, $-a < x < a$, would move outward from the point of minimal energy at the origin, while orbits that spend most of the time in the energy drain region, $|x| > a$, presumably suffer a net loss of energy and, over time, tend to move inward.

The above analysis is informal, but suggests that the active *RLC* circuit may indeed have a unique, attracting periodic orbit, i.e., an attracting limit cycle.

The van der Pol Cycle

Van der Pol and other physicists, engineers, and mathematicians have taken the ideas outlined above and come up with the following theorem.

THEOREM 9.1.2

> The van der Pol Cycle. Suppose that the continuous and piecewise smooth function $f(x)$ is defined for all x and has the properties (a) $f(-x) = -f(x)$; (b) for some positive constant a, $f(x) < 0$ if $0 < x < a$ and $f(x) > 0$ if $x > a$; (c) $f(x) \to +\infty$ as $x \to +\infty$. Then for each positive value of μ the van der Pol system, $x' = y - \mu f(x)$, $y' = -x$, has a unique limit cycle, and this cycle attracts all orbits (except for the repelling equilibrium point at the origin).

See Problem 11 for the main steps of the proof of this theorem.

EXAMPLE 9.1.5

A Particular van der Pol System
Van der Pol originally used the function

$$f(x) = x^3/3 - x$$

which meets the conditions of Theorem 9.1.2 if $a = \sqrt{3}$. Figures 9.1.3 and 9.1.4 show the graph of $y = \mu f(x)$ (dashed curve), an orbit of the van der Pol system winding onto

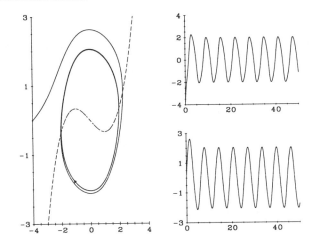

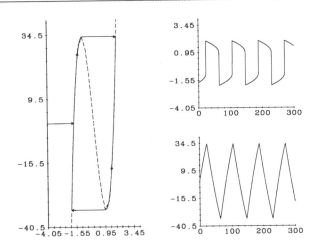

FIGURE 9.1.3 Van der Pol limit cycle, $\mu = 0.5$. See Example 9.1.5.

FIGURE 9.1.4 Van der Pol relaxation oscillation, $\mu = 50$. See Example 9.1.5.

the limit cycle, and x (top right) and y (lower right) component graphs for $\mu = 0.5, 50$, respectively. For values of $\mu \geq 3$, the limit cycle has slanting sides nearly along arcs of the graph of $y = \mu f(x)$. These sides are traced out rather slowly as time increases, while the "top" and "bottom" of the limit cycle are traversed almost instantaneously (see Figure 9.1.4). The limit cycle resembles a slightly tilted parallelogram for values of $\mu \geq 3$. It is said to represent a *relaxation oscillation*; the current slowly declines from its extreme values as it almost coincides with the graphs of $y = \mu f(x)$ on which the rate of change of the scaled current is zero, and then suddenly the current "relaxes" and reverses direction.

Atlas Plates LIMIT CYCLE A–C illustrate the relaxation phenomenon for the van der Pol circuit with a piecewise-linear current/voltage relation (see also Problem 8(**e**)).

Comments

Attracting limit cycles are analogous to asymptotically stable equilibrium points. Each depicts the long-term behavior of nearby orbits. From a practical point of view, a nonlinear system with an attracting limit cycle that possesses a large region of attraction and given period and amplitude may be preferable to a linear harmonic oscillator whose nonconstant orbits are all cycles of fixed period. In the real world initial conditions are rarely precisely known and shocks and disturbances are inevitable. The orbits of the nonlinear systems may quickly return to the periodic state even if forced away from it temporarily. However, the orbits of the harmonic oscillator, once driven off of the desired orbit, stay on a new cycle with a new amplitude. For the same reason, the nonlinear predator-prey system of Section 6.4 may not be a good model because the nonconstant orbits in the population quadrant are nonattracting cycles of varying amplitudes and periods. Disturbances of the system would make it impossible to maintain a cycle of given amplitude and period.

Solution formulas for limit cycles of real systems are rarely available. The presence of a cycle must be detected by computer simulation or by reference to a result like the van der Pol Theorem. Additional theorems that guarantee the existence of a limit cycle are presented in the next section.

PROBLEMS

1. (*Cycles in Polar Coordinates*). Find the equilibrium points and the cycles of the following systems written in polar coordinates. Determine whether the equilibrium point at the origin, $r = 0$, is asymptotically stable, neutrally stable, or unstable; determine whether each cycle is a limit cycle, and, if it is, whether it is attracting or repelling. Sketch the cycles and other orbits in the xy-plane by hand, inserting arrowheads to show the direction of increasing time.

   (a) $r' = 4r(4 - r)(5 - r)$, $\theta' = 1$

   (b) $r' = r(r - 1)(2 - r^2)(3 - r^2)$, $\theta' = -3$

   (c) $r' = r(1 - r^2)(4 - r^2)$, $\theta' = 1 - r^2$ [*Hint*: r' and θ' are 0 if $r = 1$.]

   (d) $r' = r(1 - r^2)(9 - r^2)$, $\theta' = 4 - r^2$ [*Hint*: $\theta' = 0$ but $r' \neq 0$ if $r = 2$. The origin is a center-focus; see also Problem 3.]

   (e) $r' = r \cos \pi r$, $\theta' = 1$

   (f) $r' = r \sin(\pi/r)$, $\theta' = -2$, where r' is defined to be 0 if $r = 0$. [*Hint*: Use the definition of stability in Section 8.1 to show that the system is stable at the origin.]

2. (*More Cycles*). Find all cycles and identify each as attracting or repelling. [*Hint*: Write the system in polar coordinates (see Section 6.2).]

   (a) $x' = y - x(x^2 + y^2)$, $y' = -x - y(x^2 + y^2)$

   (b) $x' = x + y - x(x^2 + y^2)$, $y' = -x + y - y(x^2 + y^2)$

   (c) $x' = 2x - y - x(3 - x^2 - y^2)$, $y' = x + 2y - y(3 - x^2 - y^2)$

3. (*Center-Focus*). Consider the system in polar coordinates, $r' = r^3 \sin(1/r)$, $\theta' = 1$, where r' is defined to be 0 at $r = 0$. Note that there are infinitely many limit cycles and they are alternately attracting and repelling. The origin of the corresponding xy-system is said to be a *center-focus*. Explain the name. [*Hint*: See also Problem 1(f) and the table at the end of Chapter 8.]

4. (*Nonisolated, Nonlinear Cycles*). Show that the nonconstant orbits of the system $x' = y^3$, $y' = -x^3$ are cycles which enclose the equilibrium point at the origin and fill the xy-plane. Plot the orbits and component curves corresponding to initial points $(0, 0)$, $(0.5, 0)$, $(1, 0)$, $(2, 0)$. Are the periods of distinct cycles the same?

5. (*Semistable Cycles*). Some cycles repel on one side and attract on the other; they are one kind of *semistable* cycle. Find all cycles of $r' = r(1 - r^2)^2(4 - r^2)(9 - r^2)$, $\theta' = 1$ and identify each as attracting, repelling, or semistable. Sketch the orbits.

6. (*A Strange Cycle*). Explain why the system $r' = r(r - 1)^2 \sin(\pi/(r - 1))$, $\theta' = 1$, where r' is defined to be 0 if $r = 1$, has a cycle $r = 1$ which is neither isolated from other cycles nor is it a limit cycle. Explain how nearby orbits behave. [*Hint*: As $r \to 1$, $(r - 1)^2 \sin(\pi/(r - 1)) \to 0$; $r' = 0$ at $r = 1$ and $r = 1 \pm 1/n$.]

7. (*Cycles in $\mathbf{R}^3$*). Determine the behavior of orbits near the cycle $r = 1$, $z = 0$ of the system in cylindrical coordinates $r' = r(1 - r^2)$, $\theta' = 25$, $z' = \alpha z$, where α is a constant. Consider separately the three cases $\alpha < 0$, $\alpha = 0$, $\alpha > 0$. Plot orbits in xyz-space for $\alpha = -0.5, 0, 1$.

8. (*Van der Pol Systems*). Verify that each system satisfies the conditions of the van der Pol Cycle Theorem (Theorem 9.1.2). Plot the cycle and some orbits that are attracted to it. Estimate the period, and the x- and the y-amplitude of the cycle.

   (a) $x' = y - \mu(x^3 - 10x)$, $y' = -x$; $\mu = 0.1, 2$

   (b) $x' = y - \mu x(|x| - 1)$, $y' = -x$; $\mu = 0.5, 5, 50$

   (c) $x' = y - \mu x(x^4 + x^2 - 1)/10$, $y' = -x$; $\mu = 0.1, 1$

(d) $x' = y - \mu x(2x^2 - \sin^2 \pi x - 2)$, $y' = -x$; $\mu = 0.5, 5$

(e) $x' = y - \mu(x - |x+1| + |x-1|)$, $y' = -x$; $\mu = 0.5, 5, 50$ [*Hint:* See Atlas Plates LIMIT CYCLE A–C.]

9. (*Van der Pol Cycles*). Plot an orbit and its component graphs for solutions near the cycle of the van der Pol system, $x' = y - \mu f(x)$, $y' = -x$ for a range of values of μ, $1/10 \le \mu \le 10$, for each of the functions given in **(a)** and **(b)** below. Use the tx or ty component graphs to estimate the period of the cycle for each value of μ you used. Any conclusions about how the period changes as μ increases? How does the maximal amplitude of the tx-graph change as μ increases? The maximal amplitude of the ty-graph?

   (a) $f(x) = x^3 - x$ **(b)** $f(x) = |x|x^3 - x$

10. (*Liénard Equation*). The nonlinear, second-order ODE, $x'' + f(x)x' + g(x) = 0$, where $f(x)$ and $g(x)$ are continous and piecewise smooth, is called the *Liénard equation* after the French mathematician and applied physicist, Alfred Liénard (1869–1958), who first studied it.

   (a) Show that if $y = x' + F(x)$, where $F(x) = \int_0^x f(s)\,ds$, then the Liénard equation can be written in system form as $x' = y - F(x)$, $y' = -g(x)$. The xy-plane is called the *Liénard plane*.

   (b) Plot the orbit of the periodic solution of $x'' + (x^2 - 1)x' + x = 0$ both in the xx'-plane and in the Liénard xy-plane.

   (c) (*Rayleigh Equation*). The *Rayleigh equation* is $z'' + \mu[(z')^2 - 1]z' + z = 0$. Differentiate the Rayleigh equation with respect to t and set $x = \sqrt{3}z'$ and show that $x'' + \mu(x^2 - 1)x' + x = 0$. Show that the ODE in x reduces to a form of system (8) if we set $y = x' + \mu(x^3/3 - x)$. Show that the Rayleigh equation has a unique attracting limit cycle for each $\mu > 0$. Plot the Rayleigh cycle and orbits through initial points $(1.5, 1.5)$ and $(0.5, 0.5)$ in the zz'-plane for each of the following values of μ: $\mu = 1/10, 1, 5, 10$.

11. (*Van der Pol Cycle Theorem*). Prove part of Theorem 9.1.2 by answering the following questions.

   - Explain why if $x = x(t)$, $y = y(t)$ is a solution of $x' = y - \mu f(x)$, $y' = -x$, then so is $x = -x(t)$, $y = -y(t)$. Why does this imply that the orbits in the left half-plane are the reflections through the origin of the orbits in the right half-plane?

   - Why must a limit cycle (if there is one) enclose the origin and cut the y-axis at points $(0, y_0)$ and $(0, -y_0)$ equidistant from the origin?

   - Why must each orbit starting on the positive y-axis move to the right and downward as t increases, and then cut the curve K defined by $y = \mu f(x)$ vertically and move downward and to the left, and eventually cut the negative y-axis?

   - If $E = x^2/2 + y^2/2$, and if an orbit starts at $(0, y_0)$ where $y_0 > 0$ and then next cuts the y-axis at $(0, y_1)$ where $y_1 < 0$, why is the function $\delta E(y_0) = E(0, y_1) - E(0, y_0)$ positive and bounded if y_1 is small enough, and a strictly decreasing function of y_0 if y_0 is large enough? Why does $\delta E(y_0) \to -\infty$ as $y_0 \to +\infty$?

   - Why do the above properties of E show that the van der Pol system has a cycle?

   The above steps outline a proof that there is a cycle, but more refined analysis is needed to show that there is a unique cycle and that it attracts all orbits except the equilibrium point at the origin.

9.2 The Poincaré-Bendixson Alternatives

The current and voltage of the van der Pol circuit of Section 9.1 approach a periodic steady state. The angle and angular velocity of a swinging pendulum tend to a constant steady state as friction dissipates energy. The orbits of the planar autonomous systems that model these physical phenomena remain bounded as time advances. Can a bounded orbit of such a system do anything else except tend to a constant steady state (an equilibrium point) or a periodic state (a cycle)? Henri Poincaré[4] and the Swedish mathematician Ivar Bendixson (1861–1935) discovered a third alternative, the cycle-graph. A cycle-graph is a strange hybrid of cycle and equilibrium point that is neither constant nor periodic. Some definitions are needed before we can take up the Poincaré and Bendixson discoveries and their consequences.

Bounded Orbits

Suppose that $f(x, y)$ and $g(x, y)$ are continuously differentiable on $\mathbf{R}^2$. Then by Theorem 6.2.1, every IVP for the system

$$x' = f(x, y),$$
$$y' = g(x, y) \tag{1}$$

has a unique, maximally extended solution. From this point on all solutions of all IVPs and ODEs are assumed to be maximally extended.

❖ **Bounded Orbits.** A solution $x(t)$, $y(t)$ of system (1) and the orbit Γ it generates are *positively bounded* if for some point $(x(t_0), y(t_0))$ on Γ the orbital arc $\Gamma_{t_0}^+ = \{(x(t), y(t)) : t \geq t_0\}$ lies in a rectangle. The solution and orbit are *negatively bounded* if $\Gamma_{t_0}^- = \{(x(t), y(t)) : t \leq t_0\}$ lies in a rectangle. The solution and orbit are *bounded* if they are positively and negatively bounded.

Note that the boundedness of an orbit is independent of the choice of the point on the orbit.

EXAMPLE 9.2.1

Boundedness
The system

$$x' = x - 10y - x(x^2 + y^2)$$
$$y' = 10x + y - y(x^2 + y^2) \tag{2}$$

has some orbits that are bounded, others that are only positively bounded. One way to show this is to apply the techniques of Sign Analysis (Section 2.2) to the polar

---

[4]The French mathematician Jules Henri Poincaré (1854–1912) is regarded as the "last universalist," the last person to understand all of the mathematics (and much of the physics) of his era. His professional life was spent mostly at the University of Paris and the École Polytechnique, where he was professor of mathematical physics, probability, celestial mechanics, and astronomy. Every year he lectured on a different subject, his students taking notes which were later published. His research covered most of the areas of the mathematics of his time: groups, number theory, theory of functions, algebraic geometry, algebraic topology (which he invented), mathematical physics, celestial mechanics, partial differential equations, and (above all) ordinary differential equations. His first research paper (1878) and his last (1912) were on that topic. Poincaré and A. M. Lyapunov created the modern approach to ODEs with its emphasis on the general behavior of orbits and solutions, rather than on solution formulas. Poincaré had a talent for good exposition, wrote many popular books on mathematics and science, and was the first (and, so far, the only) mathematician elected to the literary section of the French Institute.

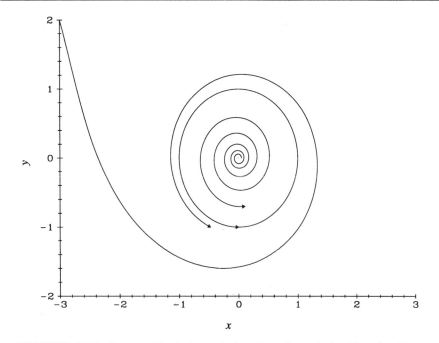

FIGURE 9.2.1 One positively bounded and two bounded orbits. See Examples 9.2.1 and 9.2.2.

coordinate version of system (2):

$$r' = r(1 - r^2)$$
$$\theta' = 10 \tag{3}$$

The unit circle $r = (x^2 + y^2)^{1/2} = 1$ is a bounded orbit (see Figure 9.2.1) and is generated by a periodic solution of period $2\pi/10$. Any orbit that starts inside the circle is bounded because it must remain inside for all time. On the other hand, an orbit outside the unit circle is positively bounded because at every point of the orbit r' is negative and the orbit moves inward as time increases. However, as time decreases along that orbit, $r(t)$ tends to infinity, and so the orbit cannot be negatively bounded (see Example 9.2.2 for a formal proof).

Limit Sets

The circular orbit of Figure 9.2.1 represents the limiting behavior of the other two nonconstant orbits in the figure as time tends to infinity. The next definition makes this idea precise.

❖ **Limit Sets of an Orbit.** Suppose that Γ is a positively bounded orbit defined by a solution $x = x(t)$, $y = y(t)$ of the system $x' = f(x, y)$, $y' = g(x, y)$. The *omega limit set* of Γ, $\omega(\Gamma)$, is the set of all points Q in the xy-plane for which

$$\lim_{n \to \infty} (x(t_n), y(t_n)) = Q$$

for some increasing and divergent sequence $\{t_n\}$ of times [i.e., $t_0 < t_1 < \cdots < t_n < \cdots$, and $t_n \to +\infty$ as $n \to \infty$]. The *alpha limit set*, $\alpha(\Gamma)$, of a *negatively bounded orbit* Γ is defined similarly, except that $\{t_n\}$ is decreasing and divergent [i.e. $t_0 > t_1 > \cdots$, and $t_n \to -\infty$ as $n \to \infty$].

The omega limit set of an orbit represents the ultimate fate of the orbit, while its alpha limit set encodes the orbit's origins.

EXAMPLE 9.2.2

Alpha and Omega Limit Sets

Denote the circular orbit of system (2) by Γ_2, a nonconstant orbit inside the circle by Γ_1, and an orbit outside the circle by Γ_3. The limit sets of the orbits Γ_1, Γ_2, and Γ_3 are not hard to find:

- $\alpha(\Gamma_1)$ consists of the origin, $\omega(\Gamma_1)$ is Γ_2
- $\alpha(\Gamma_2) = \omega(\Gamma_2) = \Gamma_2$
- $\omega(\Gamma_3) = \Gamma_2$ [$\alpha(\Gamma_3)$ doesn't exist because Γ_3 is not negatively bounded]

All of this may be proven by first separating variables in the rate equation for r in system (3) and then solving to obtain

$$r(t) = r_0[r_0^2 + (1 - r_0^2)e^{-2t}]^{-1/2}, \quad \theta(t) = 10t + \theta_0 \tag{4}$$

where $r_0 = r(0)$ and $\theta_0 = \theta(0)$. To show that $\omega(\Gamma_1) = \omega(\Gamma_2) = \omega(\Gamma_3) = \Gamma_2$ we proceed in the following way. Suppose that a point Q on the unit circle has polar coordinates $r = 1, \theta = \theta_1$. Then let

$$t_n = (\theta_1 - \theta_0 + 2n\pi)/10$$

and observe that $\{t_n\}$ is an increasing sequence that tends to ∞ as $n \to \infty$. Moreover, $r(t_n) \to 1$ because from the first equation of (4) we see that $r(t) \to 1$ as $t \to \infty$ as long as $r_0 > 0$. In addition $\theta(t_n) = \theta_1 + 2n\pi$, which is equivalent to θ_1 because the difference is a multiple of 2π. Thus Q belongs to the omega limit sets of Γ_1, Γ_2, and Γ_3, and we have that $\omega(\Gamma_1) = \omega(\Gamma_2) = \omega(\Gamma_3) = \Gamma_2$. The reader may show in a similar fashion that $\alpha(\Gamma_1)$ consists of the equilibrium point at the origin.

That $\alpha(\Gamma_3)$ does not exist follows from the fact that if $r_0 > 1$, then $r(t) \to \infty$ as t decreases from 0 to the finite escape time $t_1 = -0.5 \ln(r_0^2/(r_0^2 - 1))$ where the value of t_1 is calculated by setting the bracketed term in the formula for $r(t)$ in (4) equal to zero and then solving for t. Thus, Γ_3 is not negatively bounded.

The limit sets of system (2) consist of full orbits. This is not a fluke because it has been proven (but not here) that a limit set of an orbit of system (1) consists of one or more orbits of (1). This is the so-called *invariance property* of a limit set. It means that if an orbit has a point in common with a limit set, then that orbit lies entirely in the limit set, a result that follows from the Uniqueness Theorem. Observe in Example 9.2.2 that the orbits Γ_1 and Γ_3 never actually enter their common omega limit set, the cycle Γ_2 on the unit circle; they can only spiral ever closer to it as time advances. Likewise, although Γ_1 spirals toward the equilibrium point at the origin (which is Γ_1's alpha limit set) as time decreases to $-\infty$, Γ_1 cannot reach that point. Finally, observe that orbits such as the origin and the unit circle are their own alpha and omega limit sets for system (2), as, indeed, are the equilibrium points and cycles of any system.

The next example gives visual evidence of the existence of a limit set that is neither empty, nor a cycle, nor an equilibrium point.

EXAMPLE 9.2.3

A Strange Omega Limit Set

Consider the planar autonomous system

$$\begin{aligned} x' &= x(1 - x - 15y/4 + 2xy + y^2) \\ y' &= y(-1 + y + 15x/4 - 2x^2 - xy) \end{aligned} \tag{5}$$

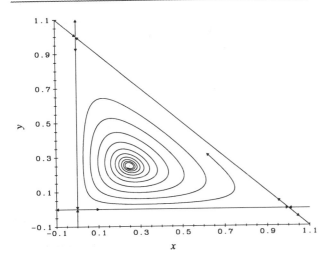

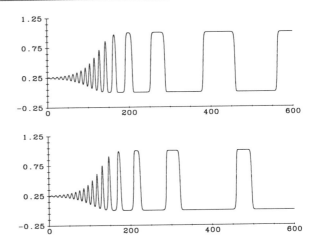

FIGURE 9.2.2 A triangular cycle-graph and omega limit set. See Example 9.2.3.

FIGURE 9.2.3 The x and y component graphs of orbit spiraling out to triangular cycle-graph, fast near the sides and slow near the corners. See Example 9.2.3.

The x and y axes are unions of orbits since $x' = 0$ if $x = 0$ and $y' = 0$ if $y = 0$. The line $x + y = 1$ is also a union of orbits because $x' + y' = 0$ along the line, as a straightforward, but rather long, calculation shows. The axes and this line intersect at the points $(0, 0)$, $(1, 0)$, and $(0, 1)$, which are equilibrium points for system (5). Figure 9.2.2 shows the direction of motion along the sides of the orbital triangle that connects these points. The triangle is a directed graph whose counterclockwise orientation is determined by the motion of the edge orbits as time increases. Each edge orbit tends to an equilibrium point as $t \to \infty$ and as $t \to -\infty$. The triangle cannot be traversed in finite time since it takes infinitely long to trace out each side. Thus, the triangle is not a periodic cycle of system (5).

There is an equilibrium point at $(1/4, 1/4)$ inside the triangle. Every nonconstant orbit inside the triangle has that equilibrium point as its alpha limit set, while its omega limit set is the triangle itself. We shall not give a formal proof of this fact, but the visual evidence shown in Figure 9.2.2 is quite convincing. Figure 9.2.3 shows the x-component graph (upper) and the y-component graph (lower) of the spiraling orbit of Figure 9.2.2. The time-intervals where both graphs are horizontal correspond to the slow motion near vertex equilibrium points. The nearly vertical segments correspond to rapid changes in x or y away from the equilibrium points.

The orbital triangle of Example 9.2.3 is an example of a cycle-graph. The discovery of cycle-graph limit sets led to the Poincaré and Bendixson Theorems below.

Cycle-Graphs and the Poincaré-Bendixson Alternatives

We need a formal definition of a cycle-graph.

❖ **Cycle-Graph.** A *cycle-graph* (or *polycycle*) of a planar autonomous system is a graph in state space that closes on itself and consists of N vertices ($N \geq 1$), which are equilibrium points, and at least N edges, which are nonconstant orbits whose alpha and omega limit sets are the vertices. The graph must be coherently oriented along the edges by time's increase (i.e., the graph can be traversed by following the arrows).

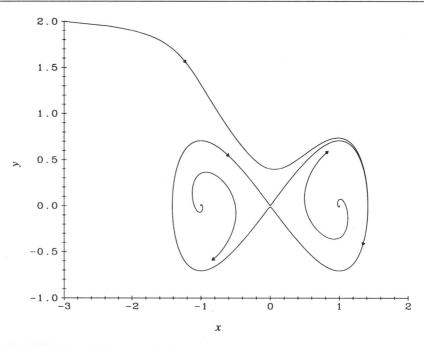

FIGURE 9.2.4 The lazy eight cycle-graph and other orbits. See Example 9.2.4.

The cycle-graph of Example 9.2.3 has three vertices and three edges and is oriented counterclockwise. Cycle-graphs may have more edges than vertices and the margin sketches show some of the many possibilities. It would be hard to come up with systems for which the sketches are indeed cycle-graphs, but the next example shows (at least computationally) that some exotic cycle-graphs can indeed arise in specific systems. A cycle-graph need not be the limit set of any orbit; however, the cycle-graphs of Examples 9.2.3 and 9.2.4 are also limit sets.

EXAMPLE 9.2.4

A Lazy Eight Cycle-Graph and Omega Limit Set
Consider the planar autonomous system

$$x' = y + x(1 - x^2)(y^2 - x^2 + x^4/2)$$
$$y' = x - x^3 - y(y^2 - x^2 + x^4/2) \tag{6}$$

The orbit Γ_1 that enters from the top left corner of Figure 9.2.4 appears to have as its omega limit set the lazy eight cycle-graph composed of the orbits Γ_2 (left lobe) and Γ_3 (right lobe) and the equilibrium point at the origin. The left lobe of the lazy eight is itself a cycle-graph consisting of Γ_2 and the origin; it is the omega limit set of all nonconstant orbits inside the left lobe. The alpha limit set of these nonconstant orbits is the equilibrium point $(-1, 0)$. A similar conclusion applies to the orbit Γ_5 inside the right lobe.

The lobe cycle-graphs of Example 9.2.4 are examples of the simplest kind of cycle-graph with a single edge and vertex. Nonconstant orbits such as these are *homoclinic orbits*, the alpha and omega limit sets of each edge orbit being a common equilibrium point. See also Atlas Plate POINCARÉ-BENDIXSON B.

Finally, we come to the main subject of the section, the Poincaré-Bendixson Alternatives.

THEOREM 9.2.1

> The Poincaré-Bendixson Alternatives. Suppose that $f(x, y)$ and $g(x, y)$ are continuously differentiable functions in a region R of the xy-plane. Suppose that Γ is a positively bounded orbit in R of the system, $x' = f(x, y)$, $y' = g(x, y)$, and that some segment Γ_T^+ of Γ lies entirely inside a rectangle that contains only a finite number of equilibrium points. Then the omega limit set of Γ is precisely one of the following:
>
> - An equilibrium point
> - A cycle
> - A cycle-graph
>
> There are no other alternatives. A similar result holds for the alpha limit set of a negatively bounded orbit.

The proof of the theorem is omitted.[5] See the margin for some of the limit set possibilities for a bounded orbit. It should be noted that an orbit can approach a limit cycle or a cycle-graph only in a spiral fashion, circling toward it with infinitely many revolutions as time goes to infinity.

The Poincaré-Bendixson Alternatives are simple to state, but there are profound implications. The next two theorems give two of them.

THEOREM 9.2.2

> Unbounded Orbits. If the system $x' = f(x, y)$, $y' = g(x, y)$ has no equilibrium points, then all orbits become unbounded both as time increases and as time decreases.

Proof. Suppose that Γ is a positively bounded orbit. Then $\omega(\Gamma)$ must be a cycle by the Poincaré-Bendixson Theorem. It was noted in Section 9.1 that the interior region of a cycle must contain at least one equilibrium point, but there are none. This contradiction implies that Γ cannot, after all, be positively bounded. Similarly, Γ cannot be negatively bounded.

The next result follows immediately from the Poincaré-Bendixson Theorem.

THEOREM 9.2.3

> Bounded Orbits. Every bounded orbit of the system $x' = f(x, y)$, $y' = g(x, y)$ has an equilibrium point or a cycle in its alpha and omega limit sets.

If we want to see what a bounded orbit does as time tends to $+\infty$ or to $-\infty$, we should first locate all equilibrium points, cycles, and cycle-graphs. Finding the equilibrium points is not hard; it is the algebraic problem of solving the equations $f(x, y) = 0$, $g(x, y) = 0$ simultaneously. Many computer solvers use Newton's Method to approximate the coordinates of the equilibrium points. Finding cycles and cycle-graphs is, in general, a harder task, and it is time now to find some techniques for doing just that.

---

[5]See M. W. Hirsch and S. Smale, *Differential Equations, Dynamical Systems, and Linear Algebra* (New York: Academic Press, 1974.)

Tests for Cycles and Cycle-Graphs

The first test is for the existence of a cycle.

THEOREM 9.2.4

> **Poincaré-Bendixson Cycle Test.**   Suppose that Γ is a positively bounded orbit of the system $x' = f(x, y)$, $y' = g(x, y)$ and that its omega limit set contains no equilibrium points. Then the omega limit set is a cycle.

This result is an immediate consequence of Theorem 9.2.1. The test may be applied even if all that is known about Γ is that one of its arcs Γ_T^+ lies in a closed and bounded region S without equilibrium points. Then the only alternative for $\omega(\Gamma)$ is that it be a cycle in S. In many applications S is an annulus (possibly distorted) whose inner and outer edges are intersected by orbits moving into S as time increases. If such an annulus S does exist, then there must be, of course, equilibrium points in the "hole," and at least one cycle in S that also encloses the hole.

EXAMPLE 9.2.5

Locating a Cycle
The system

$$x' = y + x(1 - 2x^2 - y^2/2)$$
$$y' = -x + y(1 - 2x^2 - y^2/2) \tag{7}$$

has a single equilibrium point at the origin. If the system has a cycle at all, it must enclose the origin. We shall construct an annulus around the origin with the property that orbits cross the perimeters moving inward into the annulus with the advance of time. Because there are no equilibrium points in the annulus itself, the Poincaré-Bendixson Cycle Test tells us that every orbit Γ entering the annulus has a cycle as its omega limit set. Here is how the perimeters of an annulus with the desired properties may be constructed. Let $V = x^2 + y^2$. The derivative of V following the motion of system (7) is

$$V' = 2xx' + 2yy' = 2(x^2 + y^2)(1 - 2x^2 - y^2/2)$$

Observe that $V' > 0$ if $x^2 + y^2 \leq 1/3$, and $V' < 0$ if $x^2 + y^2 \geq 3$ because

$$1 - 2x^2 - y^2/2 > 1 - 3(x^2 + y^2) \geq 0 \qquad \text{if } x^2 + y^2 \leq 1/3$$
$$1 - 2x^2 - y^2/2 < 1 - (x^2 + y^2)/3 \leq 0 \qquad \text{if } x^2 + y^2 \geq 3$$

These are rough estimates, but they do give us the circular outer and inner perimeters, $x^2 + y^2 = 3$ and $x^2 + y^2 = 1/3$, of an annulus with the property that orbits touching either perimeter must move into the annulus as time increases. Figure 9.2.5 shows how orbits move into the annulus across its edges. Two orbits are plotted as well. Since both orbits seem to have the same omega limit set, which must be a cycle, there appears to be a unique limit cycle in the annulus.

The next example has an interpretation in terms of competing species.

EXAMPLE 9.2.6

Competing Species
The system

$$x' = x(1 - x/10 - y/10), \quad y' = y(2 - x/20 - y/40) \tag{8}$$

models the populations of two competing species (Section 6.3). The equilibrium points $(0, 0)$, $(0, 80)$, $(10, 0)$, and $(70, -60)$ lie on the boundary of, or outside the population

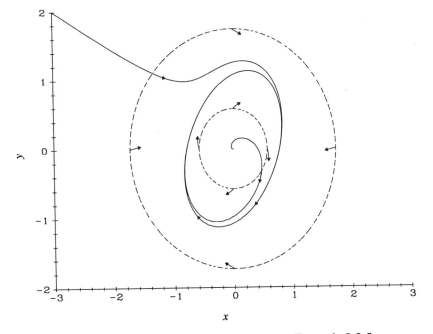

FIGURE 9.2.5 Limit cycle inside an annulus. See Example 9.2.5.

quadrant $x \geq 0$, $y \geq 0$. There can be no cycles inside the population quadrant, no cycles that intersect that quadrant. The first assertion follows from the fact that a cycle inside the quadrant must enclose an equilibrium point. The second is a consequence of the fact that the x and the y axes are unions of orbits; no orbits can enter the first quadrant from any other quadrant.

Bendixson discovered a criterion that is widely used for the absence of cycles and cycle-graphs in a simply connected region of the xy-plane.

THEOREM 9.2.5

> Bendixson's Negative Criterion. Suppose that the divergence $\partial f/\partial x + \partial g/\partial y$ of the field vector $f(x, y)\mathbf{i} + g(x, y)\mathbf{j}$ of the system $x' = f(x, y)$, $y' = g(x, y)$ has a fixed sign in a simply connected region R of the xy-plane. Then R contains no cycle or cycle-graph.

Proof. Suppose, for example, that the divergence $\partial f/\partial x + \partial g/\partial y > 0$ throughout a simply-connected region R. Suppose that a cycle or cycle-graph Γ of (1) is such that Γ and its interior lie in R. We shall show that this leads to a contradiction. Applying Green's Theorem (Appendix C.5) to $\partial f/\partial x + \partial g/\partial y$ over the region S consisting of Γ and its interior, we have that

$$0 < \int_S \int (\partial f/\partial x + \partial g/\partial y)\, dx\, dy \quad \text{[since } \partial f/\partial x + \partial g/\partial y > 0 \text{ in } S\text{]}$$

$$= \oint_\Gamma (g\, dx - f\, dy) \quad \text{[Green's Theorem]}$$

But $dx = f\, dt$ and $dy = g\, dt$ along Γ because Γ is an orbit or a union of orbits of (1). This implies that

$$g\,dx - f\,dy = gf\,dt - fg\,dt = 0$$

everywhere on Γ. We are left with the contradiction

$$0 < \oint_{\Gamma} (g\,dx - f\,dy) = 0$$

Hence, there can be no cycle or cycle-graph with the property that it and its interior lie in the region R where the divergence is positive. A similar proof applies if the divergence is negative.

Here is one practical application of Bendixson's Negative Criterion.

EXAMPLE 9.2.7

Friction Means No Cycles or Cycle-Graphs

The equation $x'' = g(x, x')$ is a generalization of the equations of motion of a pendulum (Section 4.1), a body on the end of a spring (Section 3.1), and the current in a simple circuit (Section 4.3). The corresponding system is

$$x' = y, \qquad y' = g(x, y)$$

The divergence of the vector field is $\partial(y)/\partial x + \partial(g(x, y))\partial y = \partial g/\partial y$. Thus, if $\partial g/\partial y$ has fixed sign in the xy-plane, there can be no cycles or cycle-graphs. For example, the system that models a damped simple pendulum is

$$x' = y, \quad y' = -a\sin x - by, \qquad a \text{ and } b \text{ positive constants}$$

The term $-by$ models the effect of friction on the motion of the pendulum, and the divergence of the vector field is the negative constant $-b$. Energy dissipates, and periodic or cycle-graph behavior cannot occur.

Comments

The tests and alternatives of this section are qualitative, not quantitative. The asymptotic behavior of orbits of planar systems is described rather than solution formulas given or estimates made of growth or decay. Results such as these are appropriate at the early stages of an analysis of a complex system modeling some physical phenomenon. They are also useful in designing a system, if for no other purpose than to tell us what cannot be done. For example, if the aim is to construct a planar autonomous system with a cycle or a cycle-graph, we must allow for an equilibrium point. In analyzing the orbital structure of a specific system it is often easiest to use a numerical solver. Attracting (or repelling) cycles, cycle-graphs, and equilibrium points are usually quite obvious on the screen, and we have visual evidence of the Poincaré-Bendixson results.

PROBLEMS

1. (*Limit Sets*). Using either analytical or graphical techniques, find the alpha and the omega limit sets of the orbits through the points indicated. Which nonempty limit sets are equilibrium points, cycles, or cycle-graphs?

   (a) $x' = y$, $y' = -x$; $(0, 0)$, $(1, 1)$ (b) $x' = y$, $y' = x$; $(1, 1)$, $(1, -1)$, $(1, 0)$

   (c) (*Van der Pol*). $x' = y - (x^3/3 - x)$, $y' = -x$; $(0.1, 0)$, $(3, 0)$ [*Hint*: See Section 9.1.]

(d) (*Simple Pendulum*). $x' = y$, $y' = -\sin x$; $(0, 1)$, $(0, \sqrt{2})$, $(0, 2)$ [*Hint*: See Section 4.1.]

(e) (*Polar Form*). $r' = r(1 - r^2)(4 - r^2)$, $\theta' = 5$; $r = 1/2, 3/2, 5/2$; $\theta = 0$

(f) $x' = [x(1 - x^2 - y^2) - y][(x^2 - 1)^2 + y^2]$, $y' = [y(1 - x^2 - y^2) + x][(x^2 - 1)^2 + y^2]$; $(1/2, 0)$, $(0, 1)$, $(3/2, 0)$ [*Hint*: Use a numerical solver, and note that the unit circle appears to be a cycle-graph consisting of two equilibrium points and two arcs joining them.]

2. (*Bendixson Negative Criterion*). Show that each ODE has no periodic solutions and no cycle graphs. [*Hint*: Convert each ODE to a system.]

   (a) (*Damped Simple Pendulum*). $mLx'' + cLx' + mg \sin x = 0$, where m, L, c, and g are positive constants.

   (b) (*Damped Nonlinear Spring*). $mx'' + ax' + bx + cx^3 = 0$, where m, a, and b are positive constants and c is any constant.

   (c) $x'' + (2 + \sin x)x' + g(x) = 0$, where $g(x)$ is any continuously differentiable function.

   (d) $x'' + f(x)x' + g(x) = 0$, where f and g are continuously differentiable and $f(x)$ has fixed sign.

3. (*Bendixson Negative Criterion*). Show that each system has no cycle or cycle-graph in the region indicated.

   (a) $x' = 2x - y + 36x^3 - 15y^2$, $y' = x + 2y + x^2y + y^5$; xy-plane

   (b) $x' = 12x + 10y + x^2y + y\sin y - x^3$, $y' = x + 14y - xy^2 - y^3$; the disk, $x^2 + y^2 \le 8$

   (c) $x' = x - xy^2 + y\sin y$, $y' = 3y - x^2y + e^x \sin x$; the open disk, $x^2 + y^2 < 4$

4. (*No Limit Sets*). Explain why none of the orbits of the following systems have alpha or omega limit sets. Plot some of the orbits and observe that as t increases or decreases orbits exit the screen. [*Hint*: Use Theorem 9.2.2.]

   (a) $x' = x - y + 10$, $y' = x^2 + y^2 - 1$ (b) $x' = y$, $y' = -\sin x - y + 2$

   (c) $x' = x^2 + 2y^2 - 4$, $y' = 2x^2 + y^2 - 16$

5. (*Any Cycles?*). Determine whether the following systems have cycles. Find all cycles (graphically or analytically), if any exist. If there are no cycles, prove it.

   (a) $x' = e^x + y^2$, $y' = xy$ [*Hint*: Does x ever decrease?]

   (b) $x' = 2x^3y^4 + 5$, $y' = 2ye^x + x^3$ [*Hint*: Use Bendixson's Negative Criterion.]

   (c) (*Polar Coordinates*). $r' = r\sin r^2$, $\theta' = 1$

6. (*The Divergence Averaged*). Prove that the average value of the divergence $\partial f/\partial x + \partial g/\partial y$ on the region inside a cycle Γ of the system $x' = f(x, y)$, $y' = g(x, y)$ must be zero. [*Hint*: Apply Green's Theorem to $\oint_\Gamma (g\,dx - f\,dy)$, as in the proof of Bendixson's Negative Criterion.]

7. (*Competing Species*). Explain the meaning of each nonlinear term in the rate functions of system (8), using the vocabulary of a model for competing species. Plot orbits in the population quadrant, and then make a complete analysis of the ultimate fate of each species.

8. (*Negative Divergence and a Cycle in a Region*). The divergence of the vector field of the system $x' = x - 10y - x(x^2 + y^2)$, $y' = 10x + y - y(x^2 + y^2)$ of Example 9.2.1 is $2 - 4x^2 - 4y^2$, which is negative in the region R defined by $x^2 + y^2 > 0.5$. As shown in the example, the unit circle is a cycle that lies entirely in R. Why is this not a contradiction of Bendixson's Negative Criterion?

9. (*Triangle Cycle-Graph*). System (5) appears to have a triangular cycle-graph that is the omega limit set of at least one orbit inside the triangle.

(a) Plot a comprehensive portrait of the orbits of (5) in the region $-2 < x < 2$, $-2 < y < 2$. Mark the six equilibrium points in that region. On the basis of what the nearby orbital arcs look like, label each point as stable or unstable, and name each as a node (stable, unstable?), a saddle, or a focus (stable, unstable?).

(b) Use the Jacobian technique of Section 8.4 to verify that the equilibrium points $(0, 0)$, $(1/4, 1/4)$, $(4/3, 4/3)$ have the characteristics claimed in part (a).

(c) (*Cycle-Graph Sensitivity*). Replace the term $-15y/4$ in the first equation of (5) by cy, but leave the second equation as is. For $c = -4$, $-3/2$, plot a portrait of the orbits. On the basis of your graphs, what do you think happens to the orbits as c increases through the "critical" value $c = -15/4$?

10. (*Lazy Eight Cycle-Graphs*). Consider the orbital portrait begun in Figure 9.2.4.

(a) Complete the portrait of the orbits of system (6) by plotting one more orbit in each lobe, and three other orbits outside the lazy eight. From your graphs what do you believe the omega and alpha limit sets of each new orbit are?

(b) Duplicate the Atlas Plates POINCARÉ-BENDIXSON A–C. Add an orbit inside the left node in each plate and an orbit outside all other orbits. Explain what you see.

11. (*Scaling Time*). In the system $x' = f(x, y)$, $y' = g(x, y)$, scale the time variable by $s = s(t)$ where $dt/ds = h(x(t), y(t))$, and h has a fixed sign. The new system has the form $dx/ds = fh$, $dy/ds = gh$.

(a) If $h(x, y)$ is positive everywhere, explain why the two systems have identical orbits with the same orientation induced by time's advance. What happens if $h(x, y)$ is everywhere negative?

(b) Suppose that $h(x, y)$ is positive everywhere except that h is zero at a finite number of points $p_1, \ldots, p_N$ which lie on the respective distinct nonconstant orbits $\Gamma_1, \ldots, \Gamma_n$ of the unscaled system. What can you say about the corresponding orbits of the scaled system?

(c) Let $f = x - 10y - x(x^2 + y^2)$, $g = 10x + y - y(x^2 + y^2)$, $h = 1 - \exp[-10(x - 1)^2 - 10y^2]$. Explain why the unscaled system has an attracting limit cycle, and the scaled system an attracting cycle-graph on the unit circle. Plot the orbit through the point $(0.5, 0)$ and the corresponding component curves for the original system. Repeat for the scaled system. [*Hint*: See Examples 9.2.1 and 9.2.2 and Atlas Plates POINCARÉ-BENDIXSON D–F.]

12. (*Dulac's Criterion*). Let R be a simply connected region in the plane. Suppose that f, g, K are continuously differentiable in R. Suppose that $\partial(Kf)/\partial x + \partial(Kg)/\partial y$ has a fixed sign in R. Prove that the system $x' = f$, $y' = g$ cannot have a cycle in R. [*Hint*: Replace f and g by Kf and Kg, respectively, in the proof of the Bendixson Negative Criterion.]

13. (*Ragozin's Negative Criterion for Interactive Species*). Suppose that a two-species interaction is governed by the model system $x' = xF(x, y)$, $y' = yG(x, y)$. The two species are said to be *self-regulating* if $\partial F/\partial x$ and $\partial G/\partial y$ are each negative throughout the population quadrant. Observe that if the x-species, say, is self-regulating, then the negativity of $\partial F/\partial x$ implies that the per unit growth rate F diminishes as x increases. Use Dulac's Criterion [Problem 12] to show that the system of ODEs of a pair of self-regulating species has no cycles in the population quadrant. [*Hint*: Let $K = 1/xy$ for $x > 0$, $y > 0$.]

14. Use ODEs to draw the face of a cat.

9.3 Bifurcations

Cycles and equilibrium points are distinctive features of the portrait of the orbits of planar autonomous systems. If the field vector of the system is changed slightly, one expects the new portrait to resemble the old; a cycle might contract a little, an equilibrium point shift somewhat, a spiral tighten or loosen, but the dominant features of the portrait would remain—or so one might think. This seems reasonable, but it is not necessarily true. Actually, small changes in a coefficient in a rate function may mean the disappearance of one feature and the sudden appearance of another which is altogether different. For example, as a parameter changes, an equilibrium point may suddenly appear and then spawn a second equilibrium point. Or a stable equilibrium point may destabilize and eject an attracting limit cycle. These are examples of bifurcations. "Bifurcate" means "to divide into two branches," but in the context of ODEs the word has come to mean any marked changes in the structure of the orbits of a nonlinear system as a parameter passes through a critical value, the "bifurcation point." Atlas Plates BIFURCATION A–L show some of the varieties of bifurcations.

Saddle-Node, Transcritical, Pitchfork Bifurcations

The simplest bifurcations involve the appearance, disappearance, or splitting of an equilibrium point as a parameter ε changes. Here are three examples. In every case the nature of each equilibrium point when ε is not at the bifurcation value 0 is determined by the eigenvalues of the system's Jacobian matrix evaluated at the point.

EXAMPLE 9.3.1

Saddle-Node Bifurcation

The system

$$x' = \varepsilon + x^2, \qquad y' = -y \tag{1}$$

has no equilibrium points if the parameter ε is positive. See Figure 9.3.1 for orbits where $\varepsilon = 1$. If $\varepsilon = 0$, then a strange hybrid of saddle point and stable node suddenly appears at the origin. It resembles a node on the left, a saddle on the right (Figure 9.3.2); it is an example of a *saddle-node* equilibrium point. It is not an elementary equilibrium point (Section 7.4) because 0 is an eigenvalue of the Jacobian matrix of the rate functions of (1) at the origin. As ε decreases below 0, the origin splits into two equilibrium points $(\pm\sqrt{-\varepsilon}, 0)$, one a saddle (unstable) and the other a (stable) node, that move in opposite directions away from the origin (Figure 9.3.3). Now imagine ε to change from negative values through zero and into the positive range; stable node and unstable saddle move toward each other, collide, and vanish.

The nature of the bifurcation in system (1) is pictured in the diagram of Figure 9.3.4. That diagram shows the x-coordinates of the equilibrium points as functions of ε, the solid curve denoting the x-coordinate of the stable equilibrium point and the dashed curve the unstable equilibrium point. This diagram is an example of a *bifurcation diagram*, which is a way of describing the evolution of the orbits of a system as a parameter traverses a bifurcation value.

EXAMPLE 9.3.2

Transcritical Bifurcation

If ε is positive, the system

$$x' = \varepsilon x + x^2, \qquad y' = -y \tag{2}$$

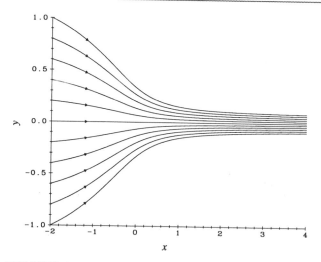

FIGURE 9.3.1 Before the appearance of the saddle node bifurcation: $\varepsilon = 1$. See Example 9.3.1.

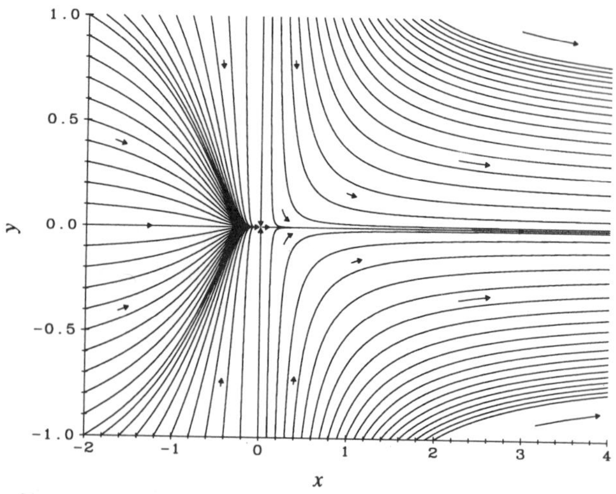

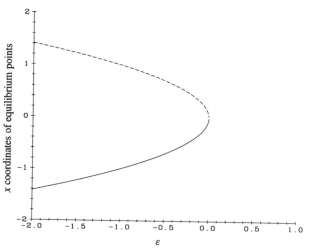

FIGURE 9.3.2 Saddle node at the bifurcation value $\varepsilon = 0$. See Example 9.3.1.

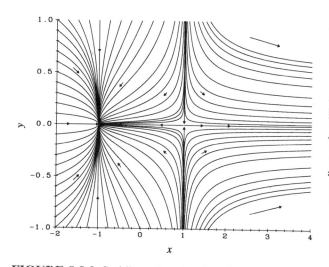

FIGURE 9.3.3 Saddle and node after the saddle node bifurcation: $\varepsilon = -1$. See Example 9.3.1.

FIGURE 9.3.4 A saddle node bifurcation diagram. See Example 9.3.1.

has a stable node at $(-\varepsilon, 0)$ and a saddle at the origin. The two equilibrium points merge into a saddle node at the origin if ε is decreased to 0. As ε moves away from 0 and becomes negative the equilibrium point $(-\varepsilon, 0)$ and the origin exchange orbital characters, $(-\varepsilon, 0)$ becoming the saddle and the origin the stable node. This is an example of a *transcritical bifurcation*. Atlas Plates BIFURCATION D–F show that the node and the saddle exist throughout the process, but merge at the critical value of $\varepsilon = 0$. This contrasts with the system of Example 9.3.1 where a saddle and a node are created at the critical value of ε.

EXAMPLE 9.3.3

Pitchfork Bifurcation
The system

$$x' = \varepsilon x - x^3, \qquad y' = -y \tag{3}$$

has an asymptotically stable equilibrium point at the origin if $\varepsilon \leq 0$ (Atlas Plate BI-FURCATION A). The equilibrium point splits into three equilibrium points $(\pm\sqrt{\varepsilon}, 0)$ and $(0, 0)$ as ε becomes positive. The outer two are stable nodes and the origin is a saddle (Atlas Plate BIFURCATION B). The bifurcation diagram of Atlas Plate BIFURCATION C justifies the name of the phenomenon.

The bifurcation diagrams of the systems of these examples partially explain the use of the term "bifurcation," although the diagram for the pitchfork is really a "trifurcation." There are other kinds of equilibrium point bifurcations which we leave to the problems. The phenomena of Examples 9.3.1–9.3.3 are essentially one-dimensional. The y-rate equations in the three examples are used mostly to heighten the pictorial effect of the bifurcation. Note again that all of the systems are nonlinear.

The Hopf Bifurcation

The following example illustrates a bifurcation that requires two dimensions because a cycle is involved.

EXAMPLE 9.3.4

Bifurcation to a Limit Cycle
For all values of ε the origin is an equilibrium point of the system:

$$\begin{aligned} x' &= \varepsilon x + 5y - x(x^2 + y^2) \\ y' &= -5x + \varepsilon y - y(x^2 + y^2) \end{aligned} \tag{4}$$

The linear approximation at the origin to the nonlinear system (4) is the linear system

$$\begin{aligned} x' &= \varepsilon x + 5y \\ y' &= -5x + \varepsilon y \end{aligned} \tag{5}$$

The eigenvalues of the system matrix of (5) are $\varepsilon \pm 5i$. We see that as the parameter ε increases through 0, the equilibrium point of system (5) at the origin changes from an asymptotically stable focus, to a neutrally stable center, and then to an unstable focus. From section 8.4 we see that for the nonlinear system (4) the origin is also an asymptotically stable focus if $\varepsilon < 0$ (Figure 9.3.5) and unstable if $\varepsilon > 0$ (Figure 9.3.7). Something quite different happens as ε increases through 0: at $\varepsilon = 0$ the origin is still an asymptotically stable focus, but the inward motion of the spiral is very slow (Figure 9.3.6). As ε becomes positive, the origin destabilizes and emits an attracting limit cycle which is a circle of radius $\sqrt{\varepsilon}$ (Figure 9.3.7). These results are easily seen by using the polar rate equations for system (4) (see Section 6.2).

$$r' = r(\varepsilon - r^2), \qquad \theta' = -5$$

Figure 9.3.8 shows a bifurcation diagram for the phenomenon. The horizontal ε-axis is first "solid" for the asymptotically stable equilibrium point at the origin and then "dashed" because the origin is unstable for positive ε. The "solid" curve $\sqrt{\varepsilon}$ depicts the amplitude of the attracting limit cycle.

The phenomenon illustrated in Example 9.3.4 is a special case of a general mechanism for generating limit cycles. Consider the system

$$\begin{aligned} x' &= \alpha(\varepsilon)x + \beta(\varepsilon)y + P(x, y, \varepsilon) \\ y' &= -\beta(\varepsilon)x + \alpha(\varepsilon)y + Q(x, y, \varepsilon) \end{aligned} \tag{6}$$

where P and Q are at least second order in (x, y) and twice continuously differentiable in (x, y, ε) throughout a region in $\mathbf{R}^3$ containing the point $(0, 0, 0)$, and $\alpha(\varepsilon)$ and $\beta(\varepsilon)$ are continuously differentiable functions of ε on an interval enclosing $\varepsilon = 0$.

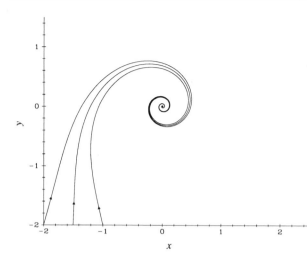

FIGURE 9.3.5 Before the Hopf bifurcaton: $\varepsilon = -1$. See Example 9.3.4.

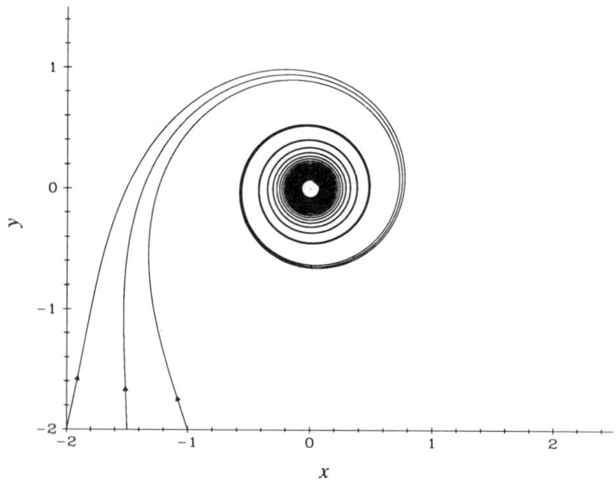

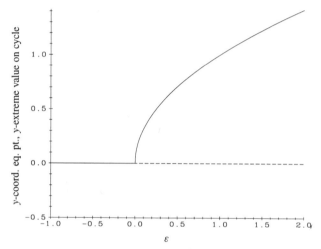

FIGURE 9.3.6 Asymptotic stability at the Hopf bifurcation value: $\varepsilon = 0$. See Example 9.3.4.

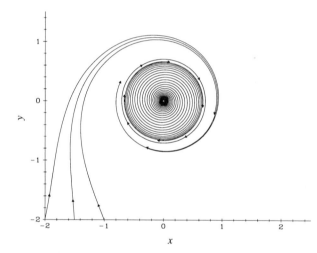

FIGURE 9.3.7 Attracting limit cycle $r = \varepsilon^{1/2}$ after the Hopf bifurcation: $\varepsilon = 0.5$. See Example 9.3.4.

FIGURE 9.3.8 Hopf bifurcation diagram: stable equilibrium point or cycle (solid line), unstable equilibrium point (dashed line). See Example 9.3.4.

Under these conditions, the origin in the xy-plane is an isolated equilibrium point of system (6). Observe that the eigenvalues of the linear system matrix of (6) at the origin (i.e., the Jacobian matrix of system (6) at the origin) are $\alpha(\varepsilon) \pm i\beta(\varepsilon)$.

In 1942 the German mathematician Eberhard Hopf showed that, as ε increases through zero, cycles suddenly appear among the orbits of system (6). The following is one version of Hopf's result (α' and β' denote derivatives with respect to ε).

THEOREM 9.3.1

> Hopf Bifurcation. Suppose that $\alpha(0) = 0$, $\alpha'(0) > 0$, $\beta(0) \neq 0$, and that system (6) is asymptotically stable at the origin for $\varepsilon \leq 0$. Then as ε increases through 0 the origin destabilizes and ejects an attracting limit cycle of amplitude $A(\varepsilon)$, where $A(0) = 0$ and $A(\varepsilon)$ is a continuous and increasing function of ε for $\varepsilon \geq 0$ and small.

Supercritical Hopf bifurcaton is the name given to this transition from an asymptotically stable equilibrium point to an unstable equilibrium point enclosed by an attracting limit cycle. A formal proof of the Hopf theorem is omitted, but here is an informal argument for its validity. System (6) is assumed to be asymptotically stable for $\varepsilon = 0$. By the Asymptotic Stability Test (Theorem 8.2.4), there is a Lyapunov asymptotic stability function $V(x, y)$ defined on a region containing the origin and such that orbits of system (6) when $\varepsilon = 0$ cut across level sets $V = C$ nontangentially and moving inward. The portion of the level set $V = C$ in a region about the origin is a simple closed curve with the origin inside. Fix the value of C at C_0 and suppose that ε is small and positive. Because the field vector of system (6) is a continuous function of ε, orbits of (6) for small positive ε still cross the level set $V = C_0$ transversely and moving inward. But the origin is an unstable focus if ε is small and positive because the eigenvalues $\alpha(\varepsilon) \pm i\beta(\varepsilon)$ have positive real parts. Thus, orbits of (6) that cut across the level set $V = C_0$ as time increases must remain in the interior region bounded by $V_0 = C$. Since these orbits are positively bounded and cannot approach the origin (which is unstable for $\varepsilon > 0$) and since there are no equilibrium points other than the origin inside $V_0 = C$, the Poincaré-Bendixson Alternatives (Theorem 9.2.1) imply that the orbits must approach a limit cycle. This turns out to be the Hopf cycle.

EXAMPLE 9.3.5

Revisiting Example 9.3.4
System (4) is a special case of system (6) with $\alpha(\varepsilon) = \varepsilon$, $\beta(\varepsilon) = 5$, $P = -x(x^2 + y^2)$, $Q = -y(x^2 + y^2)$. P and Q are of order 3 and are certainly twice continuously differentiable, and the other conditions, $\alpha(0) = 0$, $\alpha'(0) > 0$, $\beta(0) \neq 0$ are evidently satisfied. Thus the attracting cycle in Figure 9.3.7 is a Hopf limit cycle.

In a *subcritical Hopf bifurcation* an unstable focus stabilizes and spawns a repelling limit cycle as the parameter ε changes (here $\alpha'(0) < 0$). There are also other types of the Hopf bifurcation in which the equilibrium point moves as ε changes and the form of the system is not as specific as (6). Computer programs are available that will test for the presence of a Hopf bifurcation, but we shall be satisfied with the visual evidence provided by numerical solvers. The next example illustrates this process.

EXAMPLE 9.3.6

Hopf Bifurcation and Satiable Predation
When food is plentiful, a predator's appetite is soon satiated, and an increase in the prey population has little effect on the interaction terms in the model rate equations. Following the discussion in Section 6.3, we see that one model for the interactions of a satiable predator species and a prey species susceptible to overcrowding is

$$x' = (-a + by/(c + ky))x$$
$$y' = (d - ey - fx/(c + ky))y \tag{7}$$

where x is the predator population, y is the prey population, and a, b, c, d, e, f, and k are positive constants. For certain ranges of values of the coefficients Hopf bifurcations occur. In Figures 9.3.9–9.3.13, $a = 0.5$, $b = d = e = f = 1$, $c = 0.3$, and k is taken to be the bifurcation parameter. Observe that the larger the value of k the more rapidly the predator's appetite satiates as y increases. Figures 9.3.9–9.3.12 show how the population orbits behave as time advances for three values of k: $k = 0$ (no satiation effect), $k = 0.9$, and $k = 1.35$. In the first and third cases, all orbits in the population quadrant spiral toward asymptotically stable equilibrium points. In the second case, orbits are pulled toward an attracting limit cycle. See Figure 9.3.12 for graphs of orbits from a common initial point $(0.45, 0.225)$ and tending to equilibrium points ($k = 0$, 1.35) or to a cycle ($k = 0.9$). Although we do not prove it, as k increases through 0.5 (approximately) a Hopf bifurcation takes place as an internal equilibrium point destabilizes and spawns an attracting limit cycle. As k increases the cycle's amplitude first

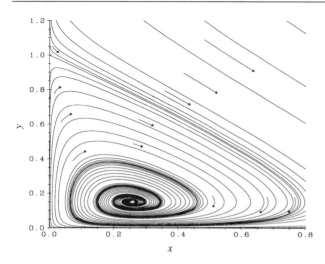

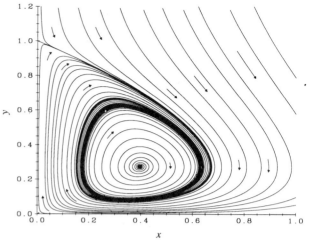

FIGURE 9.3.9 Predator and prey approach an equilibrium point before a Hopf bifurcation: $k = 0$. See Example 9.3.6.

FIGURE 9.3.10 Predator and prey approach a limit cycle after Hopf bifurcation: $k = 0.9$. See Example 9.3.6.

increases, but around $k = 0.85$ the amplitude begins to shrink. By $k = 1.2$ (approximately) the internal equilibrium point has restabilized and "absorbed" the cycle in a "reverse" Hopf bifurcation. As k increases beyond 1.2, only the asymptotically stable equilibrium point remains.

Figure 9.3.13 shows a bifurcation diagram for this particular system; the vertical axis represents the y-coordinate of each of the three equilibrium points, and the y-amplitude of cycles. The equilibrium points $(0, 0)$ and $(0, 1)$ are always unstable (they are nonlinear saddle points), and are represented by dashed lines. The internal equilibrium point has y-coordinate given by $y = 0.15/(1 - 0.5k)$. As k increases from 0, the coordinate increases. For $0 \leq k \leq 0.5$, and for $1.2 \leq k < 2$ the internal equilibrium point is asymptotically stable (solid curve), but for $0.5 < k < 1.2$ the point is unstable (dashed curve) and there is an attracting limit cycle represented in Figure 9.3.13 by the solid "hump" of the extreme y-value on the cycle.

In Example 9.3.4 the amplitude of the Hopf cycle expands to infinity as the parameter increases. In Example 9.3.6 the amplitude first expands but then decreases as the cycle eventually is reabsorbed into an equilibrium point. The next example shows a third way the Hopf cycle can behave after the bifurcation.

EXAMPLE 9.3.7

Homoclinic Bifurcation

The system to be considered is

$$x' = \varepsilon x - y + x^2, \qquad y' = x + \varepsilon y + x^2 \tag{8}$$

where $\varepsilon > -1$. Although not easily shown, there is a Hopf bifurcation at the origin when $\varepsilon = 0$. The origin is asymptotically stable for $\varepsilon \leq 0$ and unstable for $\varepsilon > 0$. The other equilibrium point

$$\left(\frac{-1 - \varepsilon^2}{1 + \varepsilon}, \frac{(1 - \varepsilon)(1 + \varepsilon^2)}{(1 + \varepsilon)^2} \right) \tag{9}$$

is a nonlinear saddle point. As ε increases from 0 the Hopf limit cycle expands. As ε increases toward $\varepsilon_0 \approx 0.10895$ the cycle gets closer and closer to the saddle point (which moves down and to the right as ε increases). At ε_0 the cycle loses its periodicity

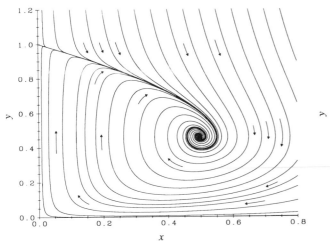

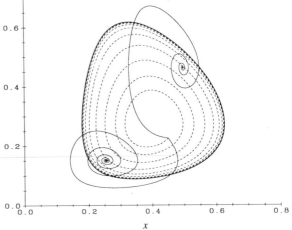

FIGURE 9.3.11 Predator and prey tend to an equilibrium point after reverse Hopf bifurcation: $k = 1.35$. See Example 9.3.6.

FIGURE 9.3.12 Predator/prey orbits from a common initial point for 3 satiation levels: $k = 0, 0.9, 1.35$. See Example 9.3.6.

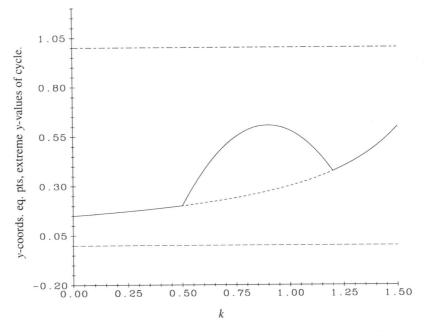

FIGURE 9.3.13 Bifurcation diagram for satiable predation. See Example 9.3.6.

and becomes a homoclinic orbit (see Section 9.2) that exits the saddle point (9) as t increases from $-\infty$, makes a turn around the origin, and returns to the saddle point as t approaches $+\infty$ (see Atlas Plate BIFURCATION H). This homoclinic orbit attracts internal orbits, which spiral outward toward it as t increases. As ε increases from 0.10895, the homoclinic orbit "breaks" and the spiraling orbits "escape" and become unbounded as time increases. Atlas Plates BIFURCATION G–I illustrate the sequence of events, but in reverse order. The sudden creation of a limit cycle out of a homoclinic orbit as a parameter changes (in this case as ε decreases through the value 0.10895) is called a *homoclinic bifurcation*.

Comments

Contemporary research on bifurcations suggests that mysterious oscillations in physical systems and their sudden appearance and disappearance may often be explained in terms of a Hopf bifurcation. Oscillations in the concentrations of chemicals in a chemical reactor, the life cycles of a cell, the van der Pol limit cycle in a electrical circuit, all seem to be triggered as a system parameter is changed and a Hopf bifurcation appears. Other kinds of bifurcations model other physical phenomena.

It is not always a simple matter to verify the conditions for a bifurcation. Once one has identified a possible bifurcation parameter in the system, computer graphics can be used to display visual evidence that a bifurcation event has occurred.

PROBLEMS

1. Describe the bifurcations in each of the following systems. Find the values of ε at which a bifurcation occurs, and identify the bifurcation as of saddle-node, transcritical, or pitchfork type. Plot graphs of orbits for values of ε before, at, and after the bifurcation. [*Hint*: Use values of ε near 0.] Sketch bifurcation diagrams.

   (a) $x' = \varepsilon + 10x^2$, $y' = x - 5y$ (b) $x' = \varepsilon x - x^2$, $y' = -2y$

   (c) $x' = \varepsilon x + 10x^2$, $y' = x - 2y$ (d) $x' = \varepsilon x - 10x^3$, $y' = -5y$

   (e) $x' = \varepsilon x + x^5$, $y' = -y$

2. (*Hopf Bifurcation*). Show that each system experiences a Hopf bifurcation at $\varepsilon = 0$. Plot orbits before, at, and after the bifurcation. Draw a bifurcation diagram for (a) and (b) showing the y-amplitude of the limit cycle and the y-coordinate of each equilibrium point as functions of ε. [*Hint*: Write the ODEs in polar coordinates.]

   (a) $x' = \varepsilon x + 2y - x(x^2 + y^2)$, $y' = -2x + \varepsilon y - y(x^2 + y^2)$

   (b) $x' = \varepsilon x - 3y - x(x^2 + y^2)^3$, $y' = 3x + \varepsilon y - y(x^2 + y^2)^3$

   (c) $x' = y - x^3$, $y' = -x + \varepsilon y - y^3$

   (d) (*Rayleigh's Equation*). $x' = y$, $y' = -x + \varepsilon y - y^3$

3. (*Subcritical Hopf Bifurcation*). Show that the system $x' = -\varepsilon x + y + x(x^2 + y^2)$, $y' = -x - \varepsilon y + y(x^2 + y^2)$ bifurcates as ε increases through 0 from an unstable focus to a stable focus surrounded by a repelling limit cycle. Plot orbits and sketch a bifurcation diagram.

4. (*Satiable Predation*). This problem continues the exploration of system (7) given in Example 9.3.6. Let $a = 0.5$, $d = e = f = 1$, $c = 0.3$, $k = 0.9$, and let b be the bifurcation parameter. Explore what happens as b increases from 0.75 to 3. Any Hopf bifurcations? Explain what you see and sketch bifurcation diagrams.

5. (*Hopf Bifurcation: Moving Equilibrium Point*). The system

   $$x' = \varepsilon(x - 5\varepsilon) + (y - 5\varepsilon) - (x - 5\varepsilon)[(x - 5\varepsilon)^2 + (y - 5\varepsilon)^2]$$

   $$y' = -(x - 5\varepsilon) + \varepsilon(y - 5\varepsilon) - (y - 5\varepsilon)[(x - 5\varepsilon)^2 + (y - 5\varepsilon)^2]$$

   has a single equilibrium point $P(5\varepsilon, 5\varepsilon)$.

   (a) Rewrite the system in terms of coordinates centered at P: $u = x - 5\varepsilon$, $v = y - 5\varepsilon$. Then rewrite the new system in polar coordinates, $r = (u^2 + v^2)^{1/2}$, $\theta = \arctan(v/u)$.

(b) Using **(a)**, show that the system has a supercritical Hopf bifurcation at $\varepsilon = 0$. Show that for $\varepsilon > 0$, the circle of radius $\varepsilon^{1/2}$ centered at $P(5\varepsilon, 5\varepsilon)$ is an orbit. [*Hint*: See Example 9.3.4.]

(c) Plot orbits for $\varepsilon = -0.5, 0, 0.5, 1, 1.5$ and describe what is happening. [*Hint*: see Atlas Plates BIFURCATION J–L.]

6. (*Homoclinic Bifurcation*). Example 9.3.7 gives a system with a homoclinic bifurcation. The problems below explore that system.

(a) Verify that for each $\varepsilon > -1$, $x = -(1+\varepsilon^2)/(1+\varepsilon)$, $y = (1-\varepsilon)(1+\varepsilon^2)/(1+\varepsilon)^2$ are the coordinates of an equilibrium point of the system.

(b) Find the Jacobian matrix of the system at the equilibrium point of part **(a)** and show that for $\varepsilon \geq 0$ its eigenvalues have opposite signs.

(c) Plot orbits in the rectangle $|x| \leq 2$, $|y| \leq 2$ for $\varepsilon = -0.5, 0, 0.05, 0.1, 0.10895$, $0.2, 0.5$. Try to plot the homoclinic orbit when $\varepsilon = 0.10895$. Explain what you see. See also Atlas Plates BIFURCATION G–I.

7. (*The Autocatalator*). The autocatalator models a chemical reaction in which the concentration of a precursor R decays exponentially, generating a new species A in the process. Species A decays to B and at the same time reacts autocatalytically with B, the latter reaction creating more B than is consumed. B decays in turn to C. The net rate equations for the process are based on the scheme:

$$R \overset{k_1}{\rightarrow} A, \quad A \overset{k_2}{\rightarrow} B, \quad A + 2B \overset{k_3}{\rightarrow} 3B, \quad B \overset{k_4}{\rightarrow} C$$

The third step is the autocatalytic step. State variables x_1, x_2, x_3, and x_4 denote the respective dimensionless concentrations of the species R, A, B, and C. The corresponding rate equations are

$$x_1' = -k_1 x_1$$
$$x_2' = k_1 x_1 - k_2 x_2 - k_3 x_2 x_3^2$$
$$x_3' = k_2 x_2 - k_4 x_3 + k_3 x_2 x_3^2$$
$$x_4' = k_4 x_3$$

where the rate coefficients k_1, k_2, k_3, and k_4 are positive parameters, and the independent variable is dimensionless time. The concentrations of the chemical intermediates A and B undergo violent oscillations for certain ranges of the parameters and certain levels of species R. The goal of this project is to understand the behavior of the reaction, given various sets of data and parameters. See S.K. Scott, *Chemical Chaos* (Oxford: Clarendon Press, 1991), for more material on the autocatalator. First read Examples 6.1.5 and 6.1.6 in Section 6.1. Then try to duplicate Atlas Plates AUTOCATALATOR A–C. In your report on the autocatalator, address the following points:

- The system can be treated as a nonautonomous planar system by setting $x_1 = x_1(0)e^{-k_1 t}$ and ignoring the rate equation for x_4. This reduction may help with your solver graphics.

- It is reasonable to set $x_2(0) = x_3(0) = x_4(0) = 0$.

- After some initial oscillations in x_2 and x_3, all four concentrations behave as expected as time increases if $x_1(0) = 500$, $k_1 = 0.002$, $k_2 = 0.14$, $k_3 = k_4 = 1.0$, $0 \leq t \leq 1500$. What levels do the four concentrations approach as t gets large?

- Repeat with $k_2 = 0.08$ instead of 0.14, and $x_1(0) = 5000$. What do you think is happening?

- Repeat with $k_2 = 0.02$ and $k_1 = 0.001, 0.003, 0.008$, $k_3 = k_4 = 1.0$; set $x_1(0) = 5000$. Now what is happening?

- In all of the above, plot component curves. If possible with your solver graphics, plot projected orbits in $x_2x_3x_4$-space or in any other space of three variables selected from t, x_1, x_2, x_3, x_4. Explain each graph.

- Plot orbits in x_2x_3-space.

- Set $k_1 = 0.002$, $k_2 = 0.08$, $k_3 = k_4 = 1.0$, $x_1(0) = 500$. For various values of k_2 above and below 0.08, plot $x_2(t)$ against t and $x_2(t)$ against $x_3(t)$. Try to explain in terms of a Hopf bifurcation for the system $x_2' = C - k_2x_2 - k_3x_2x_3^2$, $x_3' = k_2x_2 - k_4x_3 + k_3x_2x_3^2$, where C is some positive constant.

9.4 The Lorenz System and Chaos

The Poincaré-Bendixson Alternatives (Theorem 9.2.1) characterize the asymptotic behavior of bounded orbits of planar autonomous systems, but the alternatives only apply if there are no more then two state variables. If there are three or more state variables, then the long-term behavior of bounded orbits may evolve in completely unexpected and strange ways. In particular, apparently random (but still determinate) behavior may appear, behavior that is said to be chaotic. In this section we look at the Lorenz system, the best known example of chaos in an autonomous system of ODEs.

In 1963 the American mathematician and meteorologist, E.N. Lorenz, published an account[6] of the peculiarities he had observed among the computed solutions of a system of three autonomous, first-order, nonlinear ODEs. The ODEs model thermal variations in an air cell below a thunderhead. For certain values of the parameters in the rate equations of the system, solutions behave in an almost random way. The corresponding orbits remain bounded as time advances, but do not tend to an equilibrium point, nor to a cycle, nor to a cycle-graph, but instead wander between regions of oscillatory behavior. Small changes in the initial data evolve into large and apparently unpredictable differences in subsequent states. These properties persist when the parameters of the system are changed slightly. Lorenz could not find an explanation for this strange behavior. Although the problem is still not completely understood, it has led to myriad studies, both theoretical and applied, by scientists, engineers, mathematicians, economists, and others trying to formulate a good definition and theory for chaotic, but deterministic dynamics; "chaotic" because of the difficulty of predicting future states of the system, given the present state, and "deterministic" because the Existence and Uniqueness Theorem 6.2.1 guarantees that each set of initial values determines a unique solution. In this section we shall present some of the features of the Lorenz system.

The Lorenz System: Equilibrium Points

The *Lorenz system* in dimensionless state variables is

$$x' = -\sigma x + \sigma y$$
$$y' = rx - y - xz \tag{1}$$
$$z' = -bz + xy$$

---

[6]E.N. Lorenz, "Deterministic Nonperiodic Flow", *J. Atmos. Sci.*, **20**, (19) pp. 130–141.

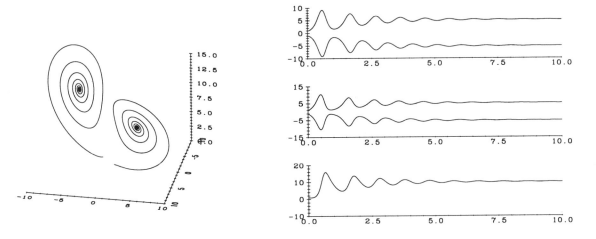

FIGURE 9.4.1 Orbits of the Lorenz system (1) through $(1, 1, 1)$ and $(-1, -1, 1)$: $r = 11, \sigma = 10, b = 8/3$.

FIGURE 9.4.2 Component graphs of the orbits of Figure 9.4.1. Top to bottom: x, y, z vs t.

where σ, r, and b are positive parameters that denote physical characteristics of fluid flow, x is the amplitude of the convective currents in the air cell, y is the temperature difference between rising and falling currents, and z is the deviation from the normal temperatures in the cell. The equations are derived by applying Fourier series techniques to the nonlinear Navier-Stokes PDEs of turbulent flow and keeping only the lowest order terms. Studies of the Lorenz system focus on the effects of changing one of the parameters (usually r) while keeping the other two fixed.

For all positive values of the parameters σ, r, and b the Lorenz system has an isolated equilibrium point at the origin. The eigenvalues of the Jacobian matrix of system (1) at the origin are given by

$$\lambda_1, \lambda_2 = \frac{1}{2}[-\sigma - 1 \pm ((\sigma - 1)^2 + 4\sigma r)^{1/2}], \quad \lambda_3 = -b < 0 \tag{2}$$

If $0 < r < 1$, then λ_1 and λ_2 are negative, and the system may be shown to be globally asymptotically stable at the origin. If $r > 1$, then λ_2 and λ_3 are still negative, but λ_1 is positive; the origin is no longer stable according to Theorem 8.4.3.

If $0 < r < 1$, there are no equilibrium points other than the origin. As the parameter r is increased through the value 1, the origin ejects two new equilibrium points,

$$P_1 = \left(-(br - b)^{1/2}, -(br - b)^{1/2}, r - 1\right), \quad P_2 = \left((br - b)^{1/2}, (br - b)^{1/2}, r - 1\right) \tag{3}$$

The eigenvalues of the Jacobian matrix of the right-hand side of system (1) at P_1 (and P_2) are the roots of the cubic,

$$\lambda^3 + (1 + b + \sigma)\lambda^2 + b(\sigma + r)\lambda + 2\sigma b(r - 1) \tag{4}$$

For $r > 1$ at least one of the roots is negative because a cubic polynomial with positive coefficients must have a negative real root. An application of the Routh Criterion of Section 7.6 to the cubic shows that all the other roots are also negative (or have negative real parts) if r lies between 1 and a *critical value* r_c

$$1 < r < r_c = \frac{\sigma(\sigma + b + 3)}{\sigma - b - 1} \tag{5}$$

For these *subcritical* values of r, P_1 and P_2 are locally asymptotically stable, as the results of Section 8.4 imply. Figures 9.4.1 and 9.4.2 show the orbits, their projections, and the component curves defined by the solutions through $(1, 1, 1)$ and $(-1, -1, 1)$ if

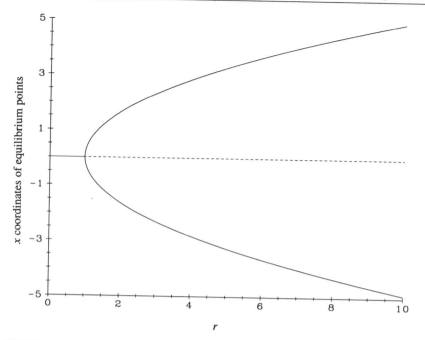

FIGURE 9.4.3 Bifurcation diagram for the equilibrium points of the Lorenz system: $0 \leq r \leq r_c$.

$\sigma = 10, b = 8/3$, and $r = 11$. Note the symmetries in the x- and y-component graphs, and that the z-component graphs are the same for the two graphs. For the given values of σ and b, we have that $r_c = 470/19$, and so the value $r = 11$ is subcritical. Orbital behavior is regular and uncomplicated.

What actually happens as r increases through $r = 1$ and then through $r_c = 470/19$? As r goes through 1, a pitchfork bifurcation occurs (Section 9.3). The stable origin (solid line segment) splits into three equilibrium points (Figure 9.4.3) as it destabilizes (dashed line), and transfers its stability to the new equilibrium points P_1 and P_2 (solid curves). As r increases through r_c, a subcritical Hopf bifurcation takes place as P_1 and P_2 lose their stability by "swallowing" repelling limit cycles, and taking on the unstable behavior of the cycles. These and many other properties of the Lorenz system are treated by Sparrow.[7]

All sorts of interesting things happen when the value of r exceeds r_c, but we put that discussion off until we have proved the "squeeze" property and shown the existence of an "attractor" set of orbits.

The Lorenz Squeeze

Suppose R is a closed region of positive volume in the xyz-state space of the Lorenz system. For example, R might be a solid spherical ball or a solid ellipsoid. Let t be a positive number, and follow each point P of R for t units of time along the orbit which passes through P at time 0 to the point P_t on the orbit. The resulting set R_t

---

[7]Colin Sparrow, *The Lorenz Equations: Bifurcations, Chaos, and Strange Attractors*, vol. 41 of the series Applied Mathematical Sciences, New York: Springer-Verlag, 1982.

of all points P_t for P in R in state space may no longer resemble R very much if t is large, but one would expect R_t to be very close to R if t is close to 0 since $R_0 = R$. The point transformation $T_t : R \rightarrow R_t$ which carries each point P to P_t is continuous since solutions of the Lorenz system are continuous in the initial data. T_t is the *orbit map* of the Lorenz system. The orbit map T_t is defined for all $t \geq 0$ for the Lorenz system since, as we shall see below, every orbit of (1) is defined for all $t \geq 0$. For fixed positive σ, r, b, the transformation T_t, $t > 0$, "squeezes" any region R with positive volume into a region R_t which has smaller volume.

To show the squeeze property we argue informally along the following lines. Suppose R is a bounded region of state space with positive volume $V(R)$ and bounding surface S, and suppose that S is smooth, connected, simple, and closed. Let S_t denote the bounding surface of R_t, and let $V(R_t)$ denote the volume of the region R_t. Interpreting the Lorenz equations as the equations of motion of the particles of a gas or a fluid, we have that

$$\frac{dV(R_t)}{dt} = \left\{ \begin{array}{c} \text{net volume of fluid} \\ \text{crossing } S_t \text{ in the} \\ \text{unit outward normal} \\ \text{direction per unit of} \\ \text{time} \end{array} \right\} = \iint_{S_t} \mathbf{F} \cdot \mathbf{n} \, dA \tag{6}$$

where $\mathbf{F}$ is the vector field defined by the right-hand side of (1), $\mathbf{n}$ is a unit outward normal vector field to the bounding surface S_t, and the integral is a surface integral. By the Divergence Theorem of vector calculus (see Appendix C.5), the surface integral becomes the volume integral of the divergence over R_t:

$$\iint_{S_t} \mathbf{F} \cdot \mathbf{n} \, dA = \iiint_{R_t} \nabla \cdot \mathbf{F} \, dV \tag{7}$$

The divergence of $\mathbf{F}$ is given by

$$\nabla \cdot \mathbf{F} = \frac{\partial F_1}{\partial x} + \frac{\partial F_2}{\partial y} + \frac{\partial F_3}{\partial x}$$

where F_1, F_2, F_3 are given by the right-hand sides of the equations of system (1). We have that $\nabla \cdot \mathbf{F} = -\sigma - 1 - b$. From (6) and (7) we have that

$$\frac{dV(R_t)}{dt} = \iiint_{R_t} -(\sigma + 1 + b) \, dV = -(\sigma + 1 + b)V(R_t) \tag{8}$$

Thus, upon solving the scalar ODE $V' = -(\sigma + 1 + b)V$, with initial condition $V = V(R)$ when $t = 0$, we have that

$$V(R_t) = V(R)e^{-(\sigma + 1 + b)t} \tag{9}$$

We see that as $t \rightarrow \infty$, $V(R_t) \rightarrow 0$.

We call this phenomenon of contracting volumes the *Lorenz squeeze*. This does not necessarily mean that the region R_t squeezes down to a single point as $t \rightarrow \infty$, but it does mean that R_t somehow "thins out" with increasing time and tends toward a subset of $\mathbf{R}^3$ whose volume is zero.

The Lorenz Attractor

Every orbit of the Lorenz system enters an ellipsoidal region with the advance of time and remains there forever after. In particular, all solutions are bounded in positive time. We may determine the ellipsoidal region in the following way. Let the positive definite function U [relative to the point $(0, 0, 2r)$] be defined by

$$U = rx^2 + \sigma y^2 + \sigma(z - 2r)^2$$

Calculating the derivative $U'(t)$ following the motion of the Lorenz system, we have

$$U' = 2rxx' + 2\sigma yy' + 2\sigma(z - 2r)z' = 2\sigma(-rx^2 - y^2 - bz^2 + 2brz)$$

Thus, U' is nonnegative only inside the solid ellipsoidal region A defined by $rx^2 + y^2 + bz^2 \leq 2brz$, i.e., by

$$rx^2 + y^2 + b(z - r)^2 \leq br^2$$

Suppose that M is the maximum value of the scalar function U on A. Then consider the ellipsoidal solid E defined by $U \leq M + 1$. We claim that all orbits of (1) eventually enter and remain within E. If P is a point outside E, then $U(P) > M + 1$ and, hence, $U'(P) < 0$. In fact, for some positive a, we must have $U'(P) \leq -a$ for all P outside of E. As t increases from 0, $U(T_t P)$ must decrease, and within a finite time the orbit through P enters E across its bounding surface. Since $U' < 0$ on the surface, orbits of (1) cut across the surface moving inward, and none can escape.

Thus, every orbit of system (1) is bounded with the advance of time. This, together with the previously derived squeeze property, implies that for every orbit Γ of system (1), the omega limit set of Γ (as defined in Section 9.2) is nonempty, bounded, and has volume 0.

The volume of the set $T_t(E)$ squeezes to 0 as $t \to \infty$. Since every orbit eventually enters E, it makes sense to define the *Lorenz attractor* Z to be the intersection of all regions $T(E_t)$, $t \geq 0$, or more simply:

$$Z = E \cap T(E_1) \cap T(E_2) \cap \cdots$$

Z is a nonempty, closed, bounded, invariant set of zero volume.[8] It may be shown that if $0 < r < 1$, then $Z = \{0\}$ (see the problem set). In fact, in this case, the Lorenz system is globally asymptotically stable at 0. For values of $r > 1$ there is only partial information about the nature of the orbits inside Z, and in any case Z itself changes if r (or σ or b) is changed. All orbits tend to Z with advancing time (i.e., Z is a *global attractor*), but the nature of that approach (and Z itself) may be quite strange.

The Lorenz Portraits

First, it should be noted that any orbit beginning on the z-axis remains there for all t:

$$x = 0, \qquad y = 0, \qquad z = z_0 e^{-bt}$$

is a solution of the Lorenz system. If x and y are replaced by $-x$ and $-y$, respectively, in the Lorenz system, the rate equations do not change. Hence, if $x(t)$, $y(t)$, $z(t)$ is a solution, so is $-x(t)$, $-y(t)$, $z(t)$, and so orbits are symmetric through the z-axis. This explains why there is only one z-component curve in Figure 9.4.1.

It is known that for $r > 1$ there are exactly two orbits of the system that exit from the origin as time advances. They are called the *unstable manifolds*, and one is the reflection through the z-axis of the other. As time advances an unstable manifold from 0 may tend to P_1, to P_2, or may jump back and forth from a neighborhood of one point

---

[8]That is, Z contains all of its limit points, Z lies within a bounded region of $\mathbf{R}^n$, Z has the property that any orbit intersecting Z stays in Z for all of t, and Z contains at least one point.

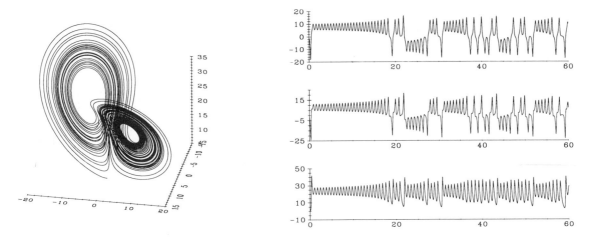

FIGURE 9.4.4 The orbit with initial point $(-1, -1, 1)$ quickly approaches the Lorenz attractor: $r = 25, \sigma = 10$, $b = 8/3, 0 \le t \le 60$.

FIGURE 9.4.5 The x, y, and z component curves of the orbit in Figure 9.4.4.

to a neighborhood of the other point. There are corresponding *stable manifolds* into P_1 and P_2 for $r > r_c$. The Lorenz attractor Z contains all of these manifolds.

The Lorenz attractor set Z may contain stable and unstable periodic orbits. In fact, Z may contain infinitely many of the latter, depending upon the values of r, σ, and b.

The parameters σ and b are fixed at the values

$$\sigma = 10, \qquad b = 8/3$$

respectively, in all of the portraits in this section, but r varies. For all but Figures 9.4.1–9.4.3, 9.4.16, and 9.4.17, r exceeds r_c. One of the unsettling characteristics of the Lorenz model is that orbital structure may change considerably with only a small change in the initial point. Thus, as time advances, the single orbit shown in each portrait may bear little resemblance to any of its neighboring orbits in state space. Each orbit is computed over a finite time interval, and its asymptotic behavior as $t \to \infty$ can only be inferred. All of the portraits in Figures 9.4.4–9.4.6 depict orbits and component graphs near the Lorenz attractor Z (the vertical axis in the three-dimensional graphs is the z-axis unless noted otherwise). Look at the pictures first and then we shall say something about them and what they seem to imply.

What do the Lorenz Portraits Say?

For values of r larger than r_c orbits head rapidly to the Lorenz attractor Z, or to a part of it. What is visible in Figures 9.4.4 and 9.4.6 can be interpreted as pieces of Z for two different values of r. Z seems to consist of two almost planar pieces at an angle to one another and each containing oscillatory arcs near the equilibrium points P_1, P_2. The orbit wanders from one piece to the other, but unpredictably, as the component graphs in Figure 9.4.5 show. In Figures 9.4.4 and 9.4.5, $r = 25$, just above the critical value of $r_c = 470/19$. In Figures 9.4.6 and 9.4.7, $r = 70$, far above the critical value. The portion of Z shown in Figure 9.4.6 for $r = 70$ resembles that shown in Figure 9.4.4 for $r = 25$, but clearly there have been changes. Figure 9.4.7 shows a distinctive feature of chaotic dynamics: extreme sensitivity to small changes in the initial data. This means that it is not possible to make accurate long-term predictions about the state of the

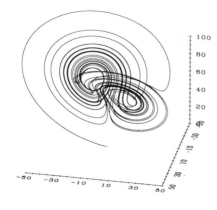

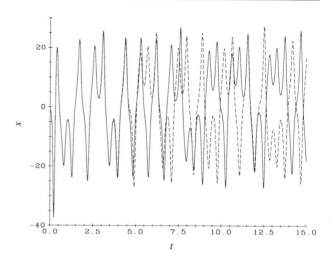

FIGURE 9.4.6 An xyz-view of an orbit with initial point $(-1, -1, 1)$ very near the Lorenz attractor: $r = 70$, $\sigma = 10$, $b = 8/3$, $0 \leq t \leq 20$.

FIGURE 9.4.7 Solutions are sensitive to changes in data: x-component curves for initial data $x_0 = y_0 = -1$ and $z_0 = 1$ (solid), $z_0 = 1.01$ (dashed): $r = 70$, $\sigma = 10$, $b = 8/3$.

system even if the values of the parameters σ, r, b and the initial values x_0, y_0, z_0 are known quite accurately.

The Lorenz system appears to have attracting periodic orbits for certain values of r. Figure 9.4.8 shows part of the x_1-component graph of a periodic orbit for $r = 230$. Figures 9.4.9, 9.4.10, and 9.4.11 show what happens when r diminishes from 230 to $r = 220$, 217, and finally 210. These graphs of parts of the x_1 component curves of the periodic orbits indicate that the corresponding period doubles, then doubles again, and finally the solution becomes an aperiodic, seemingly chaotic, but still organized oscillation. Figures 9.4.8–9.4.11 are obtained by selecting an initial point (in this case, the point $(15, 15, 100)$) near a periodic solution, waiting until the corresponding orbit is nearly "on" the attracting periodic orbit, and then plotting the x-component graph. It is widely believed that there are many sequences of values of r which correspond to period doubling ending in chaos. The period doubling sequence above earlier starts at $r = 230$. The reader may find "shorter" period doubling sequences that start at the same value of r. Such sequences are called *period doubling bifurcations*, and also *routes to chaos*.

Figures 9.4.12–9.4.15 show a better way to view orbital tangles. In a *Poincaré time section* of an orbit the orbital points $(x(t), y(t), z(t))$ are shown only for a discrete set of values of t, in this case at time intervals with $\Delta t = 0.005$, starting at $t = 0$ and ending at $t = 80$. It is as if a strobe light flashes every Δt seconds as the orbit is being traced, and the resulting dots appear on a photographic plate.

Chaos and Strange Attractors

The Lorenz attractor set Z appears to be the key element in understanding the asymptotic behavior of the orbits of a Lorenz system. With σ and b fixed positive constants, as the parameter r is stepped up from the value 0 through the value r_c the set Z becomes increasingly more complex. For $0 < r < 1$, Z is the origin, which is also the omega limit set of every orbit. As r increases from 1 through r_c, the nature of Z becomes more intricate and less understood. It is thought that with the advance of r

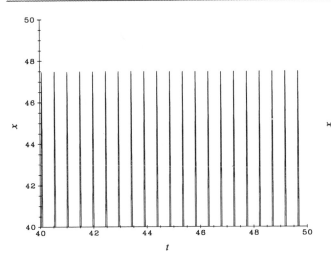

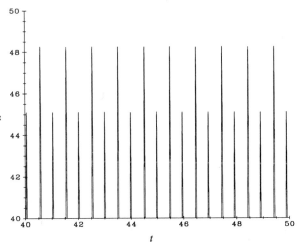

FIGURE 9.4.8 Part of the x-component of a periodic orbit; $r = 230$, $\sigma = 10$, $b = 8/3$, $x_0 = 15 = y_0$, $z_0 = 100$.

FIGURE 9.4.9 The period doubles: $r = 220$, $\sigma = 10$, $b = 8/3$, $x_0 = 15 = y_0$, $z_0 = 100$.

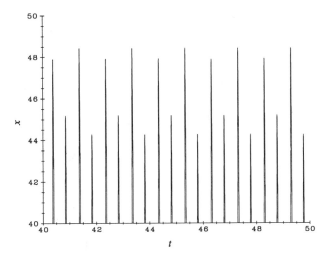

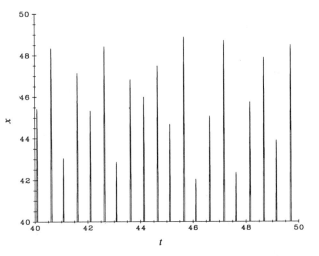

FIGURE 9.4.10 The period doubles again: $r = 217$, $\sigma = 10$, $b = 8/3$, $x_0 = 15 = y_0$, $z_0 = 100$.

FIGURE 9.4.11 Apparently chaotic motion: $r = 210$, $\sigma = 10$, $b = 8/3$, $x_0 = 15 = y_0$, $z_0 = 100$.

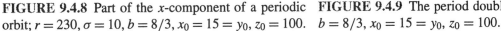

unstable periodic orbits will be created and then disappear inside Z, that there will be ranges of r for which Z contains infinitely many periodic orbits with distinct periods. Such "chaotic" behavior within Z is thought to induce the turbulent behavior of orbits as they wind down toward Z with the advance of time.

There is not yet any standard definition of chaos, but we offer an informal definition here. A nonempty, closed, and bounded invariant set A for an autonomous system of ODEs is *chaotic* if it contains a dense orbit (i.e., an orbit which passes arbitrarily close to every point of A as time advances), contains at least two orbits (otherwise a single cycle or equilibrium point would fit the definition), and if orbits in A are sensitive to changes in initial data. An attractor set A is said to be *strange attractor* if it has no proper subsets that are both attracting and invariant, and (most important) if it is chaotic. Observe that in this definition chaos is a property of a set, not of a single orbit, and not of orbits outside the set. The problem with this definition (and with others known to the authors) is that it is nearly impossible to verify that any particular

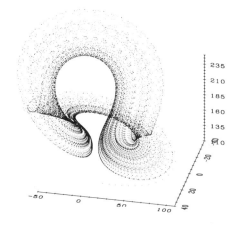

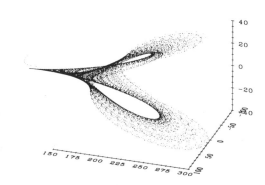

FIGURE 9.4.12 A Poincaré time section of an orbit near a strange attractor, z-axis is vertical: $r = 200$, $\sigma = 10$, $b = 8/3$, $x_0 = y_0 = 15$, $z_0 = 100$.

FIGURE 9.4.13 Another view of the Poincaré time section of Figure 9.4.12, x-axis is vertical.

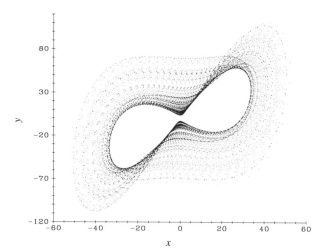

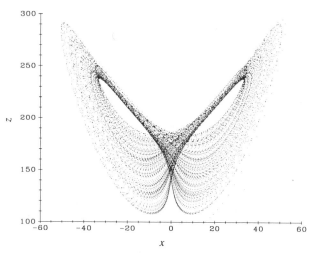

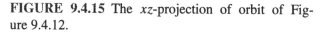

FIGURE 9.4.14 The xy-projection of orbit of Figure 9.4.12.

FIGURE 9.4.15 The xz-projection of orbit of Figure 9.4.12.

system (such as the Lorenz system) possesses a strange attractor.

Uncertainties abound. Because the certainties of the Poincaré-Bendixson Alternatives are no longer valid if there are three or more state variables, there is little theoretical guidance. Computed orbits suggest that for some values of $r > r_c$ the Lorenz attractor Z contains a strange attractor A, but (so far) no one has been able to prove it.

Computational Sensitivity

Computers truncate, chop, round-off, and otherwise reduce infinite strings of digits (such as decimal expansions) to finite strings. The effect on computed solutions of ODEs over short time intervals is (usually) negligible. However, in cases of extreme sensitivity to changes in data, computer truncations may result in a graph that is very different from the "true" graph. Figure 9.4.16 shows the graphs of the x-component

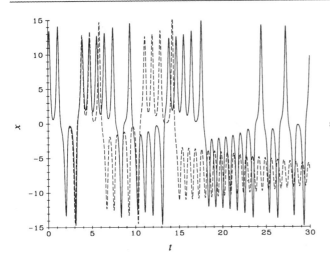

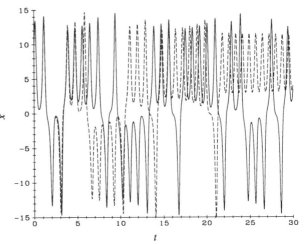

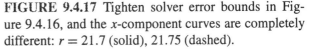

FIGURE 9.4.16 The x-component curves for orbits of Lorenz system through $(10, 10, 10)$: $r = 21.7$ (solid), 21.75 (dashed).

FIGURE 9.4.17 Tighten solver error bounds in Figure 9.4.16, and the x-component curves are completely different: $r = 21.7$ (solid), 21.75 (dashed).

curves of two orbits of system (1). Both orbits start at $(10, 10, 10)$, but $r = 21.7$ for the solid graph and $r = 21.75$ for the dashed graph. Note how the slight change in the r-value from 21.7 to 21.75 leads within a few units of time to a radical change in the component curves (i.e., extreme sensitivity to changes in r). Now look at Figure 9.4.17 where the same initial data are used but the absolute and relative local error bounds used by the authors' solver are tightened by an order of magnitude. For $t \geq 15$, the component curves are quite different from those in Figure 9.4.13. It makes one wonder if chaos is only a computational artifact!

However, there is strong, but indirect evidence that the sensitivity in the Lorenz system is real and not just computational. In addition, it is now known that close to each "numerical" orbit computed over a finite time span there lies a "true" orbit that closely resembles what has been computed.[9]

Comments

The Lorenz system has other curious features. There appear to be many homoclinic bifurcations (see Section 9.3) to unstable limit cycles as r is varied in the neighborhood of the value 13.24 (even before the critical value of $r = 470/19$). The homoclinic orbits in this case that emerge from the origin shoot out toward P_1 or P_2 and then fall back into the origin as time increases. Another odd feature (but believed to be characteristic of chaos) is that the dimension of the Lorenz attractor Z and strange attractor A inside Z for many values of r (e.g., $r = 25, 70, 200$) is not 1 or 2 but probably some real number a little larger than 2. Dimension must be defined and in this setting it is what is called the Hausdorf dimension. The Hausdorf dimension of a smooth surface is 2 and of a smooth curve is 1, but a noninteger dimension can be assigned to many other "crinkly" sets.[10]

---

[9]This is called the *shadowing property*. See, for example, S. M. Hammel, J. A. Yorke, C. Grebogi, "Numerical orbits of chaotic processes represent true orbits," *Bull. Amer. Math. Soc.*, **19** (1988), pp. 465–469.

The Lorenz system was originally proposed as a mathematical model of the thermal turbulence under a thunderhead. Although it is now known that it is not a very good model for turbulence, the Lorenz system is an excellent model for the general phenomenon of chaotic, but deterministic behavior. Engineers, scientists, mathematicians, social scientists, economists, even novelists now use the system as a paradigm for chaos. Clearly, we are only at the beginning of a new theory, and that theory is to take many twists and turns before it is put on a solid foundation.

PROBLEMS

1. (*Lorenz System*). In the following problems you are asked to verify assertions made in the section about the equilibrium points of the Lorenz system (1).

   (a) Show that the origin and the points P_1, P_2 given in (3) are the only equilibrium points (the latter two only for $r > 1$).

   (b) Construct the Jacobian matrix of the Lorenz system at the origin and show that the eigenvalues are as given in (2). Prove that all eigenvalues are negative if $0 < r < 1$, but that one eigenvalue is positive if $r > 1$.

   (c) Construct the Jacobian matrix of the Lorenz system at P_1 and at P_2 for $r > 1$. Show that the eigenvalues of the matrix at P_1 (or P_2) are roots of the cubic given in (4). Explain why one root is negative for all $r > 1$, while the other two roots are negative or have negative real parts if and only if r satisfies (5). [*Hint*: Use the Routh Criterion (Theorem 7.6.7).]

   (d) Show that a cubic $\lambda^3 + A\lambda^2 + B\lambda + C$, $B > 0$, has a pair of pure imaginary roots if and only if $C = AB$. Apply this result to the Lorenz system with $\sigma = 10$, $b = 8/3$ to find the value of r for which there are pure imaginary eigenvalues at the equilibrium points P_i.

2. (*Lorenz System*). The aim of this group project is to generate a gallery of portraits of orbits, projections, and their component curves for several values of r (set $\sigma = 10$, $b = 8/3$). Explain what you see. Look particularly for periodic orbits, period doubling sequences, and (of course) strange attractors similar to those shown in Figures 9.4.4, 9.4.6, or 9.4.12. Be sure to check sensitivity to changes in initial data. Find a copy of Sparrow's book (see footnote 7) and try to duplicate some of his pictures. If your ODE solver is not adequate, you might replace the Lorenz system by its approximation via Euler's Method (see Section 2.5):

$$x_{n+1} = x_n + h(-\sigma x_n + \sigma y_n)$$

$$y_{n+1} = y_n + h(r x_n - y_n - x_n z_n)$$

$$z_{n+1} = z_n + h(-b z_n + x_n y_n)$$

where h is the step size. The Euler approximation is surprisingly good for $h = 0.01$.

---

[10] For the Hausdorf dimension see the book by M. Martelli, *Discrete Dynamical Systems and Chaos* (New York: Longman/Wiley, 1992).

3. (*Rössler System*). The Rössler system[11] is

$$x' = -y - z$$

$$y' = x + ay$$

$$z' = b - cz + xz$$

where a, b, and c are positive parameters. O.E. Rössler wanted to build the "simplest possible" system that would model the phenomena displayed by the somewhat complex Lorenz system. The system, now named after the inventor, is the result. In the parameter region of interest, this "model of a model" has two equilibrium points and one nonlinear term rather than the three equilibrium points and two nonlinear terms of the Lorenz system. As in the Lorenz system, there appears to be a "folded" strange attractor, spiraling, chaotic wandering, and period doubling. There are sequences of values of a which (for $b = 2$ and $c = 4$) correspond to period doubling and apparently terminate with the creation of a strange attractor. The orbits and component curves of the Rössler system are to be studied as one of the three parameters is varied. For most of the study, fix the parameters b and c at respective values 2 and 4, and vary a. Look for period doubling sequences, perhaps interspersed with values of a for which there are orbits of other periods. Check for chaotic spiraling, and especially for the folded strange attractor. If possible, plot orbits in xyz-space from different points of view. Plot orbital projections on various component planes and use component plots to "see" periodic orbits. If possible, plot only over an end segment of the time interval for which the system is actually solved, (e.g., if the system is solved for $0 \leq t \leq 1000$, plot for $900 \leq t \leq 1000$). This way any transient behavior is likely to have disappeared and the orbit will be essentially in the strange attractor. Some other suggestions:

- Duplicate the graphs of Atlas Plates STRANGE ATTRACTOR A–C.

- Plot orbits and component curves for the sequence of values of $a = 0.3, 0.35,$ $0.375, 0.386, 0.3909, 0.398, 0.4, 0.411$. Use $x_0 = 0$, $y_0 = 2$, $z_0 = 0$ and identify periodic orbits and their periods, period doubling, chaotic spiraling, and any other significant phenomena. What happens if different initial data are used?

- For arbitrary positive values of a, b, c, find the equilibrium points of the system. For $b = 2$, $c = 4$, and the values of a given above, study orbital behavior near each equilibrium point.

- Explain why, if $|z|$ is very small, the subsystem $x' = -y$, $y' = x + ay$ has an unstable focus at the origin if $0 < a < 2$. Explain why this leads to a spreading apart of nearby orbits (a central part of chaotic wandering). This spreading is not unbounded, however. Explain from the third equation of the Rössler system how if $x < c$, then the coefficient of z is negative and the z-system stabilizes near $b/(c - x)$. If $x > c$, the z-system "diverges" (assuming $b > 0$). Now look at the x-system and explain why orbits alternately lie in a plane parallel to the xy-plane, then are thrown upward, and in turn are folded back and reinserted into the plane, but closer to the origin.

- Vary b and c as well as a and analyze the effect on the orbits. Any period doubling, folding, chaotic wandering? Explain.

[11]The description of the Rössler System, the Scroll Circuit (Problem 4), and Duffing's Equation (Problem 5) are adapted by permission from R.L. Borrelli, C. Coleman, and W.E. Boyce, *Differential Equations Laboratory Workbook* (New York: Wiley, 1992).

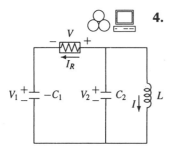

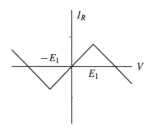

4. (*Scroll Circuit*). The diagram in the margin shows a circuit with two capacitors (one with a negative capacitance), a nonlinear resistor, and an inductor. Because of the shape of the projected orbit shown in Atlas Plate STRANGE ATTRACTOR E, it is now called the *scroll circuit*.

The circuit is modeled by the system of dimensionless ODEs:

$$x' = -cf(y - x)$$
$$y' = -f(y - x) - z$$
$$z' = ky$$

where $x = V_1/E_1$, $y = V_2/E_1$, $z = I(C_2E_1)$, $c = C_2/C_1$, $k = 1/(LC_2)$, $f(v) = -m_0v/C_2 + 0.5(m_0 + m_1)(|v + 1| - |v - 1|)/C_2$, and $-m_0$ and m_1 are the respective slopes of the falling and the rising segments of the I_RV-characteristic in the margin. With the parameters fixed except for the dimensionless parameter c, the circuit may be "tuned" by changing c. The resulting graphs display period doubling, chaotic motion, and a stable periodic orbit that bifurcates first to a torus, and then to a solid torus that appears to contain a strange attractor. In analyzing the scroll system you may want to consult the paper by Matsumoto *et al.* and the book by Parker and Chua.[12]

In carrying out this project you will need to solve the ODEs over a fairly long time span, e.g., $0 \le t \le 1000$. Whenever appropriate, plot the graphs only over an end segment of that span, e.g., $900 \le t \le 1000$. Address the following points:

- Using the basic circuit laws of Section 4.3, derive the equations

$$C_1V_1' = -g(V_2 - V_1)$$
$$C_2V_2' = -g(V_2 - V_1) - I$$
$$LI' = V_2$$

  where $g(v) = -m_0v + 0.5(m_0 + m_1)(|v + E_1| - |v - E_1|)$. Then scale time and the state variables V_1, V_2, and I appropriately to obtain the scroll system.

- Duplicate Atlas Plates STRANGE ATTRACTOR D,E,F. What are the corresponding values of the parameter c? Draw similar graphs for $c = 1, 6, 20, 50$. Use these graphs to estimate the value of c at which a bifurcation to a strange attractor occurs.

- Let $a = m_0/C_2 = 0.07$, $b = m_1/C_2 = 0.1$ in the formula for $f(v)$; let $k = 1$. Solve numerically the scroll system over the time interval $[0, 1000]$, but plot over the time interval $[500, 1000]$ to eliminate transient initial behavior. Use $x_0 = 1$, $y_0 = 1$, $z_0 = 0$, and the following values for the parameter c: $0.1, 0.5, 2, 8.8, 9.6, 10.8, 12, 13, 13.4, 13.45, 13.52, 15, 33$. Graph the projections of the orbits on the xz-plane and in xyz-space (if possible). Then graph the components against t to pick out periodic behavior. Identify periods and any period doubling you detect, as well as the approach to a strange attractor. The background figure on the bookcover and the Chapter 9 cover figure are projections onto the xz-plane of the orbit corresponding to $c = 33$.

- With $a = 0.07$, $b = 0.1$, $c = 0.1$, $k = 1$, $x_0 = 1$, $y_0 = 1$, $z_0 = 0$, decrease time from $t = 0$ and see if you pick up a repelling toroidal limit set.

[12]T. Matsumoto,, L.O. Chua, and R. Tokunaga, "Chaos via Torus Breakdown," *IEEE Trans. Circuits Syst. CAS-34*(3) (March 1987), pp. 240–254. T.S. Parker, and L.O. Chua, *Practical Numerical Algorithms for Chaotic Systems* (New York: Springer-Verlag, 1989).

5. (*Duffing's Equation*).   Chaos in differential systems needs at least three dimensions, and the autonomous Lorenz, Rössler, and scroll systems all have three state variables. However, a somewhat different kind of chaotic wandering may appear with just two state variables as long as the system is nonautonomous; in this setting time itself may be considered to be the third dimension. Duffing's equation and system, and the equation and system of the driven pendulum (Problem 6) provide examples of apparently chaotic behavior in two state dimensions with a sinusoidal driving term that is periodic in the time variable.

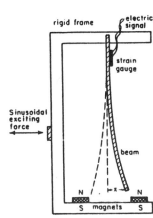

rigid frame

electric signal

strain gauge

Sinusoidal exciting force

beam

N   ⊢x⊣   N

S   magnets   S

The motion of a long, thin, elastic steel beam with its top end embedded in a pulsating frame and the bottom end hanging just above two magnets (see margin sketch) can be modeled by the *driven damped Duffing's equation*:

$$x'' + cx' - ax + bx^3 = A\cos\omega t \qquad (10)$$

where a, b, c, A, and ω are nonnegative constants. The equivalent *driven damped Duffing's system* is

$$x' = y$$
$$y' = ax - bx^3 - cy + A\cos\omega t$$

Orbits of the system have a remarkable variety of behaviors, ranging from oscillatory, to periodic, to chaotic. For some sets of values of the parameters and some sets of initial data, orbits appear to approach a strange attractor. The sources cited below give additional background.[13] Warning: numerical solutions of chaotic systems over long time spans are almost certain to be inaccurate. Nevertheless, the shadowing property mentioned in footnote 9 gives us some confidence that what the numerical computations reveal actually do approximate reality. In carrying out the Duffing project, follow the outline below:

- Analyze and graph the orbits of an undriven, undamped Duffing system: $x' = y$, $y' = 0.5x - x^3$. Find the three equilibrium points. Find an orbital figure eight cycle-graph that consists of the origin and two other orbits. Why are all but five orbits periodic? What happens to the periods of the orbits inside a lobe of the figure eight as the amplitudes diminish to zero? As the periodic orbits approach the boundary of the lobe? Analyze the periods of the orbits outside the lobes.

- Add damping and obtain the system of Atlas Plate DUFFING A in which orbits are graphed for $x_0 = 1$, $0.8 \le y_0 \le 1.0$. Some orbits tend to $(-1, 0)$, others to $(1, 0)$ as $t \to +\infty$. Fix x_0 at 1 and plot orbits for several values of y_0 from 0 to 5.0. For each y_0 solve over the t-interval, $0 \le t \le 200$. Do you see any pattern? Construct a table having the values of y_0 in the first column, and the number of times the orbit starting at $(1, y_0)$ crosses the y-axis as t increases from 0 to 200 in the second. From the table, estimate the length of each y_0 interval that generates orbits with a given number of crossings. Explain why all orbits starting inside any one of these y_0 intervals approach one of the equilibrium points $(-1, 0)$, $(1, 0)$, while the orbits starting in adjacent intervals approach the other equilibrium point. What happens to the lengths of these y_0 intervals as y_0 increases?

[13] J. Guckenheimer, and P. Holmes, *Nonlinear Oscillations, Dynamical Systems, and Bifurcations of Vector Fields* (New York: Springer-Verlag, 1983); P. Holmes, and F.C. Moon, *J. Appl. Mech.* **108**, (1983), pp. 1021–1032; J.M.T. Thompson, and H.B. Stewart, *Nonlinear Dynamics and Chaos* (New York: Wiley, 1986).

- Change the damping a little and add a periodic driving force to obtain the Duffing system of Atlas Plate DUFFING B. What is the period of the periodic orbit shown? Is it a stable attractor (i.e., do nearby orbits approach it as t increases)?

- Change the damping back to that of Atlas Plate DUFFING A but keep the periodic driving force to obtain the system in Atlas Plate DUFFING C. What is the period of the enclosing periodic orbit visible in the picture? What is the other orbit doing? These orbits are solved for $0 \le t \le 1000$ but plotted only for $500 \le t \le 1000$. What happens if the driving and plotting are done over other time spans? Duplicate Atlas Plates DUFFING A–C. Explain the information they contain.

- Atlas Plate DUFFING D shows Poincaré time sections ($\Delta t = 2\pi$) of two orbits. The one with all the dots is the section of the chaotic orbit in Atlas Plate DUFFING C. What is the other orbit doing? Duplicate Plate D.

- Plot Poincaré time sections with $\Delta t = 2\pi$ over $0 \le t \le 10,000$ for several initial points near $(-1, 1)$. Use the Duffing system of Atlas Plate DUFFING D. Explain what you see.

- Explain what you see in Atlas Plates DUFFING E, F. Why is it a good idea to plot the orbits in the toroidal coordinates of Atlas Plate DUFFING E?

- Rescale x and t and reduce the number of parameters in (10) from five (c, a, b, A, ω) to three. Then carry out a study of what happens as one or more of the three parameters is varied. Use Poincaré time sections if appropriate. Explain what you see. Do you see a strange attractor?

6. (*Driven Pendulum*). The system for a sinusoidally driven pendulum with a constant torque has the form

$$x' = y$$
$$y' = -ay - b\sin x + c\cos(\omega t) + d$$

where a, b, c, d, and ω are nonnegative parameters. Explain the model. Search for chaotic wandering, and plot and discuss your results.

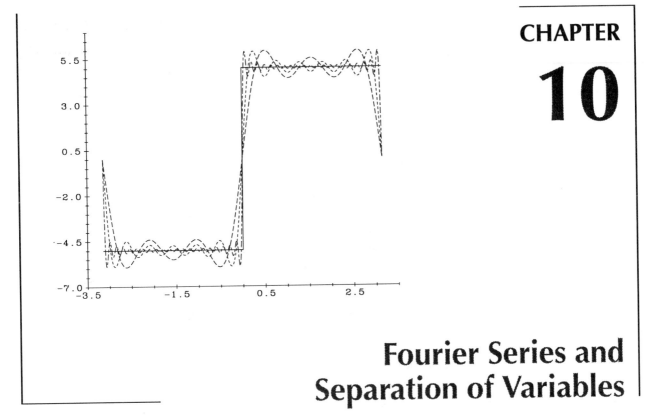

Fourier Series and Separation of Variables

The search for solution formulas for partial differential equations had been going on for a very long time when, in the early 1800s a significant breakthrough occurred. A method of separation of variables was devised which, when coupled with a special kind of trigonometrical series, could provide a solution formula in the form of a series for any reasonable boundary and initial data. The series were first used decisively by Joseph Fourier in a heat flow problem and hence are named in his honor. In this chapter we will examine these two partners, Fourier Series and Separation of Variables, in some detail for the wave equation and leave for Chapter 12 the extension of this approach to the diffusion equation and Laplace's equation. The figure above shows a square wave and several of the partial sums of the approximating Fourier sine series (see Example 10.3.2).

10.1 Wave Motion: Vibrations of a String

Periodic disturbances and oscillations play an important role in many diverse areas in science and engineering. Examples are the motion of a pendulum in a gravitational field, the motion of celestial bodies, and oscillations in an electrical circuit. Oscillations which travel in space are known under the collective title of "wave motion." Some familiar examples of wave motion are the motion of ripples on the surface of a pond, sound waves, and the transverse vibrations of a string.

We shall analyze the basic system that involves wave motion: the transverse vibrations of a taut, flexible string. This special case appears in many other physical phenomena and can be treated by the same mathematical model.

Linear Model for the Vibrating String

Suppose that a string is stretched between the points $x = 0$ and $x = L$ of the x-axis. We shall construct a model for the motion of the string under the action of the tension alone. To simplify the problem, we assume that the string is *flexible*; that is, that the force of tension in the string acts tangentially to the string and hence the string offers no resistance to bending. Since the equilibrium position of the string under the action of tension alone is the interval $0 \le x \le L$, we may describe the motion of the string by specifying three functions, $u_1(x, t)$, $u_2(x, t)$, and $u_3(x, t)$, the spatial coordinates at time t of the point x on the string in equilibrium position. On physical grounds, the motion of the string depends not only on the initial deflection and velocity of the string but also on how the string is restrained at the boundary points $x = 0$ and $x = L$. We discuss boundary conditions later in this section. It will be helpful to review the continuity classes for functions of several variables.

> ❖ **Continuity Classes**. A function $u(x, y, z)$ defined on an open (or closed) region S in $\mathbf{R}^3$ is said to belong to the *continuity class* $C^k(S)$, for some nonnegative integer k, if all partial derivatives of u of order less than or equal to k exist as continuous functions on S.[1] There are similar definitions for functions of n variables, $n \ge 2$.

We make some further simplifying assumptions in order to obtain a simple, linear partial differential equation (PDE) describing the motion of the string:

- The motion of the string takes place in a fixed plane, and points on the string are constrained to move only in a direction transverse to the string. Thus the motion of the string may be described by a single function $u(x, t)$ defined in the region $R = \{(x, t) : 0 < x < L, t > 0\}$.

- The motion u belongs to the classes $C^2(R)$ and $C^1(\text{cl } R)$, and only "small" deflections from equilibrium are allowed in the sense that second-order terms in u_x may be ignored when compared to terms of lower order.[2]

- The string tension $\mathbf{T}(x, t)$ is of class $C^1(\text{cl } R)$. The string's linear density ρ is constant.

- There are no external transverse forces acting on the string. In particular, the weight of the string is neglected.

To derive the equation of motion of the string, let us find the forces acting on the segment $[x, x']$ of the string. The forces acting on this segment are due to tension alone. We shall find the parallel and transverse components of the tension forces acting on $[x, x']$ at each of the endpoints. Referring to Figure 10.1.1, we have that (the labeling in Figure 10.1.1 defines the symbols used):

$$\|\mathbf{T}_u(x', t)\| = \|\mathbf{T}(x', t)\|\,|\sin \alpha|; \quad \|\mathbf{T}_x(x', t)\| = \|\mathbf{T}(x', t)\|\,|\cos \alpha|$$
$$\|\mathbf{T}_u(x, t)\| = \|\mathbf{T}(x, t)\|\,|\sin \beta|; \quad \|\mathbf{T}_x(x, t)\| = \|\mathbf{T}(x, t)\|\,|\cos \beta|$$

---

[1] Recall that a *region* R in $\mathbf{R}^n$ is a connected (i.e., not in disjoint pieces) set consisting only of interior points. The *closed region*, denoted by cl R, is the set consisting of the region R together with its boundary points.

[2] A function $h(z)$ is nth *order* in z, denoted by $h = O(|z|^n)$ as $z \to 0$, if there is a positive constant c such that for all $|z|$ small enough, $|h(z)| \le c|z|^n$. Thus for small $|z|$ the lower-order terms in a sum of nonnegative powers of z are dominant, and for simplicity, higher-order terms may be dropped. For example, for small $|z|$, $1 + 3z + 5z^2$ may be replaced by $1 + 3z$, or even by 1.

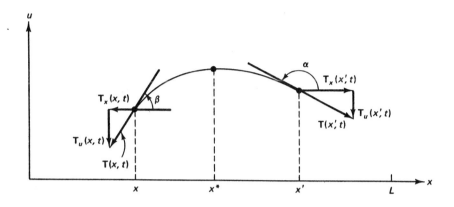

FIGURE 10.1.1 Profile on $[x, x']$ of the string at time t.

But since we have that

$$|\sin \alpha| = \frac{|u_x(x', t)|}{\sqrt{1 + |u_x(x', t)|^2}}, \qquad |\cos \alpha| = \frac{1}{\sqrt{1 + |u_x(x', t)|^2}}$$

and similar formulas for $\sin \beta$, $\cos \beta$, we see from the second assumption above that

$$|\sin \alpha| = |u_x(x', t)|, \qquad |\sin \beta| = |u_x(x, t)|, \qquad |\cos \alpha| = 1, \qquad |\cos \beta| = 1$$

where second- and higher-order terms in u_x have been ignored.[3] Thus, supplying the proper orientations to the components, we see that the transverse and parallel components for the net tension forces acting on the segment $[x, x']$ are, respectively,

$$\|\mathbf{T}(x', t)\| u_x(x', t) - \|\mathbf{T}(x, t)\| u_x(x, t) \qquad \text{(transverse)}$$

and

$$\|\mathbf{T}(x', t)\| - \|\mathbf{T}(x, t)\| \qquad \text{(parallel)}$$

But since we assumed that only transverse motions of the string were possible, we conclude that $\|\mathbf{T}(x', t)\| = \|\mathbf{T}(x, t)\|$, and since x and x' were arbitrary we must have that $T = \|\mathbf{T}\|$ is independent of x in $[0, L]$. Now applying Newton's Second Law to the segment $[x, x']$ of mass $\rho \cdot (x' - x)$, we have the equation

$$\rho \cdot (x' - x) u_{tt}(x^*, t) = T\{u_x(x', t) - u_x(x, t)\} \tag{1}$$

where x^* is the x-coordinate of the center of mass of the string segment between x and x'. Dividing (1) through by $\rho(x' - x)$, denoting T/ρ by c^2, and taking limits as $x' \to x$,

---

[3]We have expanded $(1 + |u_x|^2)^{-1/2}$ into a binomial series, $1 - \frac{1}{2}|u_x|^2 + \frac{3}{8}|u_x|^4 - \cdots$, and then kept only the first term, 1.

we obtain the *wave equation in one space dimension,*

$$u_{tt} = c^2 u_{xx} \tag{2}$$

which must be satisfied for all points (x, t) in R. We shall assume that c^2 is constant. Note that PDE (2) is a second-order homogeneous linear partial differential equation and that any consistent set of units can be used. The variables u and x carry units of length, t of time, and the constant c has units of length/time.

The wave equation (2) has many solutions. For example, if F and G are any C^2-functions, then $u = F(x - ct) + G(x + ct)$ is a solution. This may be shown by repeated application of the Chain Rule to calculate partial derivatives; for example, if r denotes $x - ct$, and we set $u = F(x - ct)$, then

$$u_t = \frac{dF}{dr}\frac{\partial r}{\partial t} = F'(r)(-c) \quad \text{and} \quad u_{tt} = \frac{d}{dr}(F'(r))\frac{\partial r}{\partial t}(-c) = c^2 F''(r)$$

Similarly, we have that

$$u_x = \frac{df}{dr}\frac{\partial r}{\partial x} = F'(r) \quad \text{and} \quad u_{xx} = \frac{d}{dr}(F'(r))\frac{\partial r}{\partial x} = F''(r)$$

and hence $u_{tt} = c^2 u_{xx}$. Physical intuition for the circumstances of the vibrating string suggests that *initial conditions* such as prescribing $u(x, 0)$ and $u_t(x, 0)$, $0 < x < L$, together with *boundary conditions* such as prescibing $u(0, t)$ and $u(L, t)$, $t \geq 0$, are needed to obtain a unique and physically plausible solution.

Boundary/Initial Value Problems for the Wave Equation

The effect of the boundary points of a finite string were not considered at all in deriving the wave equation. We are now ready to consider the effect of various conditions at the boundaries of a finite string. It is assumed that the string has length L (i.e., that $0 \leq x \leq L$) and that time is measured forward from the initial time $t = 0$.

Conditions at the Ends of the String: Boundary Operators

Only transverse motions of the string are considered. Thus if the endpoints of the string (i.e., $x = 0$, $x = L$) are allowed to move, we may imagine the string to be looped around transverse rods at these two ends, but otherwise free to move (other interpretations are, of course, possible). There are three types of boundary conditions of interest. The simplest is when the transverse displacement of an endpoint is a prescribed function of time. For example, at $x = 0$ we may require that

$$u(0, t) = \alpha(t), \qquad t \geq 0 \tag{3}$$

where $\alpha(t)$ is a given function. If $\alpha(t) = 0$, the string is fastened securely to the x-axis at $x = 0$. This special case is called a *fixed boundary condition*. A similar condition might be imposed at $x = L$.

In the second kind of boundary condition the endpoint is acted on by a transverse force, say $F = \gamma(t)$. Taking the endpoint to be $x = 0$ and applying this force and the transverse component of the tension forces described in the opening paragraphs of this section to the segment $0 \leq x \leq h$ of the string, we have by Newton's Second Law that $\rho h u_{tt}(x^*, t) = T u_x(h, t) + \gamma(t)$, where ρ is the density of the string, x^* is the center of mass of the string segment $[0, h]$, and T is the (constant) tension in the string. Let h decrease to 0 in this equation. Since $u_{tt}(x, t)$ is bounded in x for each $t \geq 0$, the left-hand side $\rho h u_{tt}(x^*, t)$ tends to 0. Hence this condition is

$$T u_x(0, t) + \gamma(t) = 0, \qquad t \geq 0 \tag{4}$$

where $\gamma(t)$ is given. If $\gamma(t) = 0$ for all t, it is a *free boundary condition*. A similar condition may be imposed at the endpoint $x = L$.

Finally, suppose that the left endpoint of the string is connected to the point $x = 0$ on the x-axis by a Hooke's Law spring with spring constant k (see Section 3.1). Then the function $\gamma(t)$ in (3) must be $-ku(0, t)$. Hence we have that

$$Tu_x(0, t) - ku(0, t) = 0, \qquad t \geq 0 \tag{5}$$

which is called an *elastic boundary condition*. A similar condition could be imposed at the other endpoint.

The boundary conditions above may be described in terms of *boundary operators*. For example, the condition $u_x(0, t) - 4u(0, t) = 0$ may be written as $B[u(x, t)] = 0$, where the operator B is defined by $B[u(x, t)] = u_x(0, t) - 4u(0, t)$. Note that B may be applied to any function for which $u(0, t)$ and $u_x(0, t)$ are defined. Note also that B is a linear operator: $B[au + bv] = (au + bv)_x(0, t) - 4(au + bv)(0, t) = a[u_x(0, t) - 4u(0, t)] + b[v_x(0, t) - 4v(0, t)] = aB[u] + bB[v]$. The operators associated with the other boundary conditions are also linear. All of this may be formalized in the following way.

❖ **Boundary Conditions Expressed with Operators.** The *fixed, free*, and *elastic boundary conditions* at $x = 0$ described in (3)–(5) above may be expressed as $B_0[u] = 0$, where the boundary operator B_0 is defined respectively by:

(a) *Fixed Boundary:* $B_0[u(x, t)] = u(0, t)$
(b) *Free Boundary:* $B_0[u(x, t)] = u_x(0, t)$
(c) *Elastic Boundary:* $B_0[u(x, t)] = Tu_x(0, t) - ku(0, t)$

Similar definitions hold for fixed, free, or elastic boundary conditions at $x = L$ with boundary operators B_L.

Since the boundary operators B_0 (and B_L) are linear operators, it follows that the null space of any one of them is closed under linear combinations. This fact will be of some importance later. The operators above may also be used to express nonhomogeneous boundary conditions; for example, the boundary condition (4) implies that $B_0[u(x, t)] = -\gamma(t)/T$, where the action of B_0 is given by $B_0[u] = u_x(0, t)$.

The Mixed Initial/Boundary Value Problem for the Vibrating String

As we saw above, the deflection of a vibrating string, $u(x, t)$ must satisfy the *wave equation* (1) in the region $R = \{(x, t) : 0 < x < L, t > 0\}$, provided that $u(x, t)$ in $C^2(R)$ satisfies some simplifying conditions. We saw that if u is in $C^1(\mathrm{cl}\ R)$, it is appropriate to demand that u and its first partial derivative with respect to x satisfy certain conditions at the boundaries $x = 0$ and $x = L$ for all $t \geq 0$ (depending on the character of the restraint of the string at the endpoints). Of course, the initial deflection $u(x, 0)$ and the initial velocity $u_t(x, 0)$ of the string may also be specified. Thus, in mathematical terms, the motion of a vibrating string with given initial and endpoint conditions will be a function u in $C^2(R)$ and $C^1(\mathrm{cl}\ R)$ which satisfies all of the relations below inside the region R or on its boundaries:

$$
\begin{aligned}
\text{(PDE)} \quad & u_{tt} - c^2 u_{xx} = 0 & & \text{for } 0 < x < L, \quad t > 0 \\
\text{(IC)} \quad & u(x, 0) = f(x), & & 0 \leq x \leq L \\
& u_t(x, 0) = g(x), & & 0 \leq x \leq L \\
\text{(BC)} \quad & B_0[u] = \phi(t), & & t \geq 0 \\
& B_L[u] = \mu(t), & & t \geq 0
\end{aligned}
\tag{6}
$$

where $f(x)$, $g(x)$, $\phi(t)$, and $\mu(t)$ are given functions, and the boundary operators B_0 and B_L each may be any of the three types of boundary operators mentioned above. The problem (6) with *initial conditions* (IC) and *boundary conditions* (BC) can be shown to have no more than one solution for any appropriately smooth *initial data* [$f(x)$ and $g(x)$] and *boundary data* [$\phi(t)$ and $\mu(t)$] (see Problem 6).

Standing Waves and Separated Solutions

There is a class of solutions for the wave equation (1) which have considerable significance in the theory of acoustics. These are the *standing waves* $X(x)T(t)$, where $T(t)$ is a periodic function of time. More accurately, $X(x)$ is the standing wave *with amplitude modulated* by $T(t)$. Substituting $X(x)T(t)$ into (2) and separating variables, we see that if $X(x) \neq 0$ and $T(t) \neq 0$, then

$$\frac{X''(x)}{X(x)} = \frac{1}{c^2}\frac{T''(t)}{T(t)}$$

holds for all $0 < x < L$ and $t > 0$. Since x and t are independent variables, the only way this identity could hold is for each side to be equal to the same constant. Calling this constant λ, the identity can be separated to yield the ordinary differential equations (ODEs)

$$X''(x) = \lambda X(x), \qquad T''(t) = c^2\lambda T(t) \tag{7}$$

for the constant λ. Actually, if any value of λ is used in (7) and $X(x)$ and $T(t)$ are any solutions of the ODEs, then $X(x)T(t)$ is a solution of the wave equation (1). Such solutions are generally called *separated solutions* for the PDE (1). In acoustics, standing waves are separated solutions where $T(t)$ is periodic. Thus for standing waves we must have $\lambda < 0$ for $T(t)$ to be periodic.

Let us consider the special case of a guitar string of length L with tension T, density ρ, and $c^2 = T/\rho$. If the motion is not too vigorous, the deflection $u(x, t)$ will be a solution to the problem

$$\begin{aligned} u_{tt} &= c^2 u_{xx} & 0 &< x < L, \quad t > 0 \\ u(0, t) &= 0, \quad u(L, t) = 0, & t &\geq 0 \\ u(x, 0) &= f(x), & 0 &\leq x \leq L \\ u_t(x, 0) &= 0, & 0 &\leq x \leq L \end{aligned} \tag{8}$$

where $f(x)$ is the initial deflection of the string, and the boundary operators B_0 and B_L have the action $B_0[u] = u(0, t)$ and $B_L[u] = u(L, t)$. Since the string is clamped at both ends, any standing wave supported by the string must satisfy the boundary conditions $X(0) = X(L) = 0$. Thus to find all standing waves, we see from (7) that we must first determine all constants λ such that the problem

$$X''(x) = \lambda X(x), \qquad X(0) = X(L) = 0 \tag{9}$$

has a nontrivial C^2-solution. This is an example of a *Sturm-Liouville Problem* (more on such problems in Section 10.5). Notice that if $X(x)$ is any C^2-function on $[0, L]$ such that $X(0) = X(L) = 0$, integration by parts shows that

$$\int_0^L X(x)X''(x)\,dx = -\int_0^L (X'(x))^2\,dx \leq 0 \tag{10}$$

Thus if $X(x)$ is a solution of (9) for some constant λ, (10) implies that

$$\lambda \int_0^L (X(x))^2\,dx = \int_0^L X(x)X''(x)\,dx \leq 0$$

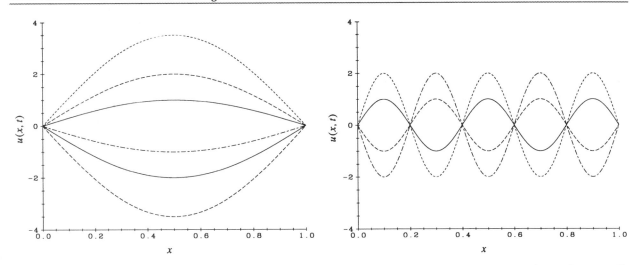

FIGURE 10.1.2 Standing wave on guitar string, $n = 1$.   **FIGURE 10.1.3** Standing wave on guitar string, $n = 5$.

and hence $\lambda \le 0$. Now clearly (9) does not have a nontrivial solution for $\lambda = 0$, so let us try $\lambda = -k^2$ for some $k > 0$. The general solution of $X'' + k^2 X = 0$ is $X(x) = A\cos kx + B\sin kx$ for arbitrary constants A and B. Thus the condition $X(0) = 0$ implies that $A = 0$, and the condition $X(L) = 0$ implies that $B\sin kL = 0$. Wishing to avoid the trivial solution, we assume that $B \ne 0$ and hence we must take $kL = n\pi$, where n is a nonnegative integer. Thus for any value $\lambda_n = -(n\pi/L)^2$, $n = 1, 2, \dots$, problem (9) has the corresponding solution

$$X_n(x) = \sin(n\pi x/L)$$

up to an arbitrary multiplicative constant.

Now, from the second ODE in (7), we see that

$$T_n(t) = A_n \cos \frac{n\pi c}{L} t + B_n \sin \frac{n\pi c}{L} t$$

since $\lambda_n = -(n\pi/L)^2$. Thus, all standing waves for the guitar string have the form

$$u_n(x, t) = \sin \frac{n\pi x}{L}\left(A_n \cos \frac{n\pi c t}{L} + B_n \sin \frac{n\pi c t}{L}\right), \qquad n = 1, 2, 3, \dots \quad (11)$$

where A_n and B_n are arbitrary constants. If time is measured in seconds, the *period* of the standing wave u_n is $2L/nc$ seconds (per cycle), and the standing wave recovers its initial sinusoidal profile $u_n(x, 0) = A_n \sin(n\pi x/L)$ at $t = k(2L/nc)$ seconds for every integer k. The *frequency* of u_n is the reciprocal of the period: $f_n = nc/2L$ cycles/second (or hertz). The numbers f_n are sometimes said to be the *natural frequencies* supported by the string. Observe that the points $x = 0, L/n, \dots, (n-1)L/n, L$, are rest points, or *nodes*, of the standing wave u_n. In Figures 10.1.2 and 10.1.3 we have sketched two standing waves for several values of t. The standing wave u_n is a solution of problem (6) with fixed boundary conditions, $u(0, t) = u(L, t) = 0$, and initial conditions

$$u(x, 0) = A_n \sin \frac{n\pi x}{L}, \qquad u_t(x, 0) = \frac{n\pi c}{L} B_n \sin \frac{n\pi x}{L}$$

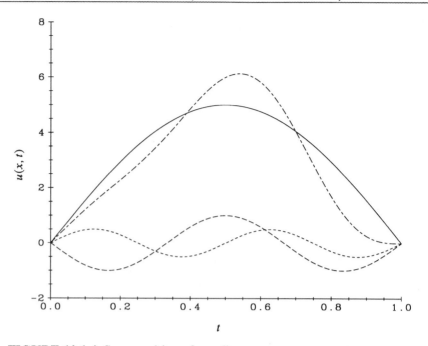

FIGURE 10.1.4 Superposition of standing waves.

A Musical Interlude

The standing waves (11) and their frequencies have acquired musical names because of their association with musical tones and instruments. The standing wave

$$u_1(x, t) = \sin \frac{\pi x}{L} \left(A_1 \cos \frac{\pi c t}{L} + B_1 \sin \frac{\pi c t}{L} \right)$$

is called the *fundamental* or *first harmonic*, while the standing waves of higher frequency are the *overtones* or *higher harmonics*. For example,

$$u_3(x, t) = -2 \sin \frac{3\pi x}{L} \cos \frac{3\pi c t}{L}$$

is a second overtone (or third harmonic). The waveform (long/short dashes) sketched in Figure 10.1.4 is the profile of a fundamental (solid line) plus a third (long dashes) and a fourth harmonic (short dashes). Each stringed instrument produces its own characteristic combinations of fundamentals and overtones and it is the variety of these combinations that distinguishes a guitar from a harp, or even one guitar from another.

The comments above seem to imply that it is the physical vibrations of the guitar string, violin string, or piano wire that we actually hear. Of course, what really happens is that these mechanical vibrations of the wire cause corresponding variations in the air pressure, variations that we hear as various musical sounds. In fact, it may be shown that air pressure itself is a solution of a wave equation, where the constant c^2 is a ratio of specific heats and of standard air pressures and densities. Thus what our ears actually detect are air-pressure waves set in motion by the vibrations of a string, a larynx, or some other source of sonic energy.

The Plucked Guitar String: Superposition of Standing Waves

It is a remarkable fact that problem (8) can be solved as a superposition of the standing waves $u_n(x, t)$ defined in (11), where the constants A_n, B_n are determined appropriately for all $n = 1, 2, \ldots$. This technique can be generalized to an important method for solving boundary/initial value problems for PDEs. Indeed, suppose that a solution $u(x, t)$ of problem (8) has the form $u(x, t) = \sum_{n=1}^{\infty} u_n(x, t)$, where the constants A_n, B_n are such that the series converges at each point in the closed region $S = \{(x, t) : 0 \leq x \leq L, t \geq 0\}$. We now show how to find the values that the coefficients A_n, B_n must have in order that the series $\sum_{n=1}^{\infty} u_n(x, t)$ converge to a "solution" of problem (8). First note that each u_n satisfies the endpoint conditions, and hence so does the sum $\sum u_n$. If the partial derivatives $\partial^2 u/\partial x^2$ and $\partial^2 u/\partial t^2$ can be computed by differentiation of the series $\sum u_n$ term by term, then $u = \sum u_n$ satisfies the wave equation because each u_n does. Thus the constants A_n, B_n must be determined so that the series $\sum u_n$ satisfies the initial conditions. This implies that

$$f(x) = u(x, 0) = \sum_{n=1}^{\infty} A_n \sin \frac{n\pi x}{L}, \qquad 0 \leq x \leq L \qquad (12)$$

$$0 = u_t(x, 0) = \sum_{n=1}^{\infty} \frac{n\pi c}{L} B_n \sin \frac{n\pi x}{L}, \qquad 0 \leq x \leq L \qquad (13)$$

Now comes a step which is of crucial importance in determining the constants A_n, B_n in (12) and (13). For all integers m and n we have that [see Problem 4]:

$$\int_0^L \sin \frac{n\pi x}{L} \sin \frac{m\pi x}{L} \, dx = \begin{cases} 0, & m \neq n, \text{ or } m = n = 0 \\ L/2, & m = n \neq 0 \end{cases}$$

$$\int_0^L \cos \frac{n\pi x}{L} \cos \frac{m\pi x}{L} \, dx = \begin{cases} 0, & m \neq n \\ L, & m = n = 0 \\ L/2, & m = n \neq 0 \end{cases} \qquad (14)$$

where the latter equality is included here for reference purposes. Multiplying both sides of (12) and (13) by $\sin(m\pi x/L)$ and integrating over $0 \leq x \leq L$ (assuming that integration and infinite summation can be interchanged), we obtain

$$\int_0^L f(x) \sin \frac{m\pi x}{L} \, dx = A_m \left(\frac{L}{2} \right), \qquad m = 1, 2, \ldots$$

$$0 = \frac{m\pi c}{L} B_m \left(\frac{L}{2} \right), \qquad m = 1, 2, \ldots$$

which uniquely determines A_m and B_m. Hence if our assumptions are valid, we are led to the characterization of the solution of problem (8) as the series

$$u(x, t) = \sum_{n=1}^{\infty} A_n \sin \frac{n\pi x}{L} \cos \frac{n\pi c t}{L} \qquad (15)$$

where

$$A_n = \frac{2}{L} \int_0^L f(x) \sin \frac{n\pi x}{L} \, dx \qquad (16)$$

We shall show later in Section 10.6 that if the data function $f(x)$ is "nice" enough, then (15) with coefficients as in (16) does indeed solve problem (8), and if we know that this problem has no more than one solution (see Problem 6), then (15), (16), must be *the* solution of problem (8).

Comments

As we hinted above, boundary/initial value problems can be solved by the technique of superposition of standing waves. Although this technique is rather limited in its scope, it is very important since it provides explicit solution formulas for a wide class of commonly occurring boundary/initial value problems. It is called the Method of Separation of Variables, and we will discuss this approach in more detail in Section 10.6 after we have developed the necessary machinery.

PROBLEMS

1. (*Guitar String*). The guitar string modeled by system (8) has length $L = \pi$, tension T, and density ρ. Find the solution in the form of the series (15) with specific values for the coefficients A_n if the initial profile $f(x)$ is as given.

    (**a**) $f(x) = \sin x$ (**b**) $f(x) = 0.1 \sin 2x$ (**c**) $f(x) = \sum_{n=1}^{10} \frac{1}{n} \sin nx$

    (**d**) Explain in mathematical terms why the tauter the string, the higher the pitch. Now replace the string by a heavier string and pluck it again. What do you hear now? Explain in mathematical terms.

2. (*Standing Waves*).

    (**a**) Find all standing waves for the vibrating string of length L with free boundary conditions at the endpoints.

    (**b**) Use a superposition of standing-waves technique to solve problem (6) with the free boundary conditions, $u_x(0, t) = 0$, $u_x(L, t) = 0$.

3. Verify directly that the Sturm-Liouville Problem (9) has no nontrivial solutions when either $\lambda = 0$, or $\lambda > 0$.

4. Verify the integration in (14). [*Hint*: Use the identity $\sin \alpha \sin \beta = \frac{1}{2}[\cos(\alpha - \beta) - \cos(\alpha + \beta)]$.]

5. (*Plucked Guitar String*). A guitar string of length π initially has the shape

    $$f(x) = \begin{cases} x, & 0 \le x \le \pi/2 \\ \pi - x, & \pi/2 \le x \le \pi \end{cases}$$

    and is released from rest. Graph the profile of the string at $t = 1$, $t = 5$, and $t = 10$. [*Hint*: Use a computer algebra solver (CAS) package to calculate enough coefficients A_n in the solution formula (15) to give a reasonably good approximation. Then graph the truncated series for the various t-values using a CAS package again.]

6. (*Uniqueness Theorem*). Let R be the region $0 < x < L$, $t > 0$, and suppose that $F(x, t)$ is continuous on cl R, $\alpha(t)$, $\beta(t)$ are continuous on $t \ge 0$, and that $\gamma(x)$, $\delta(x)$ are continuous on $0 \le x \le L$. Following the outline below, show that the problem below cannot have more than one solution $u(x, t)$ that is in $C^2(R)$ and $C^1(\text{cl } R)$:

(PDE)	$u_{tt} = c^2 u_{xx} + F(x, t)$	in R
(BC)	$u(0, t) = \alpha(t)$,	$u(L, t) = \beta(t)$, $t \ge 0$
(IC)	$u(x, o) = \gamma(x)$,	$u_t(x, 0) = \delta(x)$, $0 \le x \le L$

    - Suppose the given problem has two solutions u and v. Put $U = u - v$ and show that U satisfies the same problem, but with the data F, α, β, γ, and δ all set equal to zero.

- Define $W(t) = \frac{1}{2} \int_0^L (U_t^2 + c^2 U_x^2)\, dx$, for any $t \geq 0$. Take advantage of the identity $(U_x U_t)_x = U_{xx} U_t + U_x U_{tx}$ to show that

$$W'(t) = U_x(L, t) U_t(L, t) - U_x(0, t) U_t(0, t)$$

- Show that $U_t(L, t) = 0$, and $U_t(0, t) = 0$, for $t \geq 0$, and hence that $W'(t) = 0$ for $t \geq 0$. [*Hint*: Recall the definition of partial derivatives.]

- Conclude that $U_t(x, t) = 0$ and $U_x(x, t) = 0$, for all (x, t) in cl R, and hence that $U = $ constant on R.

- Show that actually $U = 0$ on R and conclude that $u = v$ on R.

10.2 Orthogonal Functions and Fourier Series

In Section 4.1 we saw that the dot (or scalar) product for vectors in $\mathbf{R}^3$ brought not only a certain geometric sense to the construction of models but significant computational simplifications as well because of orthogonality. In this section we extend the notion of a scalar product to more general linear spaces than $\mathbf{R}^3$. Then we go on to set up specific scalar products on commonly occurring linear spaces of functions and show how orthogonality provides a valuable new tool for the approximation of functions. Later, in Section 10.6 we shall show in detail how to use these concepts to solve boundary/initial value problems for PDEs.

Linear Spaces

$\mathbf{R}^3$ is an example of a linear space, a collection of objects, called vectors, for which vector addition and scalar multiplication are defined and satisfy the properties:

❖ **Vector Space Properties.** In the nonempty collection V of objects (called *vectors*) there are two operations defined which satisfy the properties listed below. Denote by "+" the operation of *addition* which assigns a vector $u + v$ to every pair of vectors u and v. For a given set of *scalars*, the operation of *scalar multiplication* assigns to every scalar α and every vector u the vector αu. With these operations V has the properties:

- (A1) *Addition is order independent*. A finite collection of vectors can be added together in any order without affecting the sum.

- (A2) *Existence of zero element*.[4] There is an additive *zero* element in V, denoted by 0, with the property that $u + 0 = u$ for all vectors u.

- (A3) *Existence of additive inverses*. Every vector u has an *additive inverse*, denoted by $-u$, with the property that $u + (-u) = 0$.

- (M1) *Scalar multiplication is distributive*. For any vectors u and v and any scalars α and β, we have that $\alpha(u + v) = \alpha u + \alpha v$ and $(\alpha + \beta)u = \alpha u + \beta u$.

- (M2) *Scalar multiplication is associative*. For any vectors u and any scalars α and β, we have that $\alpha(\beta u) = (\alpha \beta)u$.

- (M3) For the scalar 1 we have that $1u = u$, for all vectors u.

---

[4]It is traditional to use the same symbol 0 for the zero scalar and the zero vector—context distinguishes one from the other.

The Vector Space Properties define the abstract notion of a linear (or vector) space.

❖ **Linear Space.** Let V be a set of objects and $\mathbf{F}$ a set of scalars ($\mathbf{F}$ is either $\mathbf{R}$, the real numbers, or $\mathbf{C}$, the complex numbers) with two operations, "addition" and "scalar multiplication," defined in V such that the Vector Space Properties hold. Then V is a *real linear* (or *vector*) *space* if $\mathbf{F} = \mathbf{R}$, and a *complex linear* (or *vector*) *space* if $\mathbf{F} = \mathbf{C}$. The elements of V are sometimes called *vectors*.

A feature of this abstract approach to linear spaces is that we can deduce from the defining properties (A1)–(A3) and (M1)–(M3) some additional properties which must then hold in *any* linear space. Some of them are as follows:

THEOREM 10.2.1

Further Properties of Linear Spaces. Let V be any linear space.

1. There is only one zero vector in V.

2. The vector αu is zero if either the scalar $\alpha = 0$ or the vector $u = 0$. Conversely, if $\alpha u = 0$, then either $\alpha = 0$ or $u = 0$.

3. Every vector v in V has a *unique* additive inverse, $-v$; moreover $-v = (-1)v$ for any vector v.

Proof. Each assertion is proven below:

1. Let 0 and $0'$ be two zero vectors for V, then by (A1) and (A2) we see that $0 = 0 + 0' = 0' + 0 = 0'$, so there is only one zero in V.

2. For any u note that $u = (1 + 0)u = u + 0u$ using (M1) and (M3). Adding $-u$ to both sides of this equation, we obtain by (A3) that $0 = 0u$, as asserted. Next, we assume that $\alpha \neq 0$. Then for any vector u, $\alpha 0 + u = \alpha 0 + (\alpha \cdot \alpha^{-1})u = \alpha(0 + \alpha^{-1}u) = \alpha(\alpha^{-1}u) = 1u = u$, where we have used (M1)–(M3). Thus $\alpha 0$ acts like a zero vector, and since there is only one such vector, we must have $\alpha 0 = 0$. Finally, suppose that $\alpha u = 0$ for some scalar α and some vector u. If $\alpha \neq 0$, then $u = \alpha^{-1}(\alpha u) = \alpha^{-1}0 = 0$.

3. Let v and w be two additive inverses for the vector u. Then $v = v + 0 = v + (u + w) = (v + u) + w = 0 + w = w$ where we have used (A1) and (A2). Hence u has a unique inverse. Now since $0 = 0u = (1 - 1)u = 1u + (-1)u = u + (-1)u$, we see that $-u = (-1)u$, finishing the proof.

It is worth noting that there is a linear space V with only one element. In view of the result above, we see that that single element must be the zero vector, and hence V is written as $\{0\}$ and is referred to as the *trivial* linear space.

EXAMPLE 10.2.1

Real n-tuple Space
Let $\mathbf{R}^n$ for any $n = 1, 2, \ldots$ be the collection of all ordered n-tuples of real numbers, $x = (x_1, x_2, \ldots, x_n)$. Note that x here denotes the n-tuple and the x_i are real numbers. Two elements x and y in $\mathbf{R}^n$ are identified if and only if $x_i = y_i$, $i = 1, 2, \ldots, n$. Using the real numbers $\mathbf{R}$ as scalars and defining the operations of addition and multiplication by scalars component-wise by

$$x + y = (x_1 + y_1, \ldots, x_n + y_n), \qquad \alpha x = (\alpha x_1, \ldots, \alpha x_n) \tag{1}$$

for all x, y in $\mathbf{R}^n$ and all scalars α, it is straightforward to verify that $\mathbf{R}^n$ is a real linear space. Note that $\mathbf{R}^1$ can be identified with $\mathbf{R}$, and hence $\mathbf{R}$ itself can be considered a real linear space.

EXAMPLE 10.2.2

Complex *n*-tuple Space

Let $\mathbf{C}^n$ for any $n = 1, 2, \ldots$ be the collection of all ordered n-tuples of complex numbers $z = (z_1, z_2, \ldots, z_n)$. Using the complex numbers $\mathbf{C}$ as scalars and defining operations of addition and multiplication by scalars as in (1), it is straightforward to show that $\mathbf{C}^n$ is a complex linear space. Note that $\mathbf{C}^1$ can be identified with $\mathbf{C}$.

EXAMPLE 10.2.3

Linear Spaces of Functions

For any interval I and any integer $k \geq 0$ the continuity class $C^k(I)$ of real-valued functions is a real linear space if real scalars and the operations of addition and scalar multiplication are given by

$$(f + g)(t) = f(t) + g(t), \qquad (\alpha f)(t) = \alpha f(t), \quad \text{all } t \text{ in } I \tag{2}$$

If $C^k(I)$ consists of complex-valued functions, then $C^k(I)$ is a complex linear space when complex scalars and the operations (2) are used.

Scalar Products

It turns out that the basic properties of the scalar product on $\mathbf{R}^3$ are symmetry, bilinearity, and positive definiteness (see Problem 2, Section 4.1). This leads us to ask whether or not on *any* given linear space we can find a way to associate a scalar with each (ordered) pair of vectors so that these three properties are satisfied. When this is possible, we may imitate what was done for the dot product to develop notions of "orthogonality," "distance," and "convergence of a sequence" in the linear space.

❖ **Scalar Product, Euclidean Space.** Let V be a linear space, and suppose that $\langle \cdot, \cdot \rangle$ is a function which associates with every pair of vectors u, v the scalar $\langle u, v \rangle$. Then $\langle \cdot, \cdot \rangle$ is a *scalar product* if the following properties are satisfied for all vectors u, v, w, and all scalars α, β:

$$(Symmetry) \qquad \langle v, u \rangle = \overline{\langle u, v \rangle}, \text{ the complex conjugate of } \langle u, v \rangle \tag{3a}$$

$$(Bilinearity) \qquad \langle \alpha u + \beta v, w \rangle = \alpha \langle u, w \rangle + \beta \langle v, w \rangle \tag{3b}$$

$$(Positive\ Definiteness) \qquad \langle u, u \rangle \geq 0 \text{ and } = 0 \text{ if and only if } u = 0 \tag{3c}$$

A linear space V with a scalar product is called a *Euclidean space*.

There are many scalar products for the same linear space; examples follow. It is easy to check that $\langle x, y \rangle = \sum_1^n x_k y_k$ is a scalar product on $\mathbf{R}^n$; this *standard scalar product* makes $\mathbf{R}^n$ into a real Euclidean space and generalizes the dot product on $\mathbf{R}^3$. Similarly, $\langle x, y \rangle = \sum_1^n x_k \overline{y}_k$ makes $\mathbf{C}^n$ into a complex Euclidean space and is the *standard scalar product* on $\mathbf{C}^n$.

Since the scalar product of any vector with itself is always a nonnegative number (even for complex linear spaces), we can define the length of a vector in the same way as for the dot product.

❖ **Norm.** In a linear space V with scalar product $\langle \cdot, \cdot \rangle$ we define the *norm* (or *length* or *magnitude*) of a vector u by the nonnegative quantity

$$\|u\| = (\langle u, u \rangle)^{1/2} \tag{4}$$

This norm is called *the norm induced by the scalar product* (other norms are possible). Vectors of length 1 are called *unit vectors*.

Dividing any nonzero vector by its length produces a unit vector. This process is called *normalization*.

❖ **Distance Between Two Vectors.** In a linear space V with scalar product $\langle \cdot, \cdot \rangle$ we define the *distance* between any two vectors u and v in V to be $\|u - v\|$, where $\| \cdot \|$ is the norm induced by $\langle \cdot, \cdot \rangle$.

Recall that a function $f(x)$ on a closed interval $[a, b]$ is said to be *piecewise continuous* if there are finitely many points $c_1, c_2, \ldots, c_k$ such that $a < c_1 < c_2 < \cdots < c_k < b$, f is continuous on each open interval (a, c_1), (c_1, c_2), ..., (c_{k-1}, c_k), (c_k, b), and the one-sided limits of f exist at all the points $x = a, c_1, \ldots, c_k, b$. Examples are step functions. [*Note*: The actual values of f at the points $x = a, c_1, \ldots, c_k, b$ play no role in the definition. In fact, f need not even be defined at these points.] The set of all piecewise continuous functions on $[a, b]$ is denoted by $PC[a, b]$. Using the usual notion of addition of functions and multiplication of a function by a scalar, we see that $PC[a, b]$ is a linear space (a real linear space if the functions are real-valued and the scalars are real numbers).

EXAMPLE 10.2.4

The Euclidean Space $PC[a, b]$

For the linear space $PC[a, b]$ it is easy to see that

$$\langle f, g \rangle = \int_a^b f(x)\overline{g(x)}\,dx, \qquad f, g \text{ in } PC[a, b] \tag{5}$$

is a scalar product if we identify functions in $PC[a, b]$ which differ at only a finite set of points [otherwise (3c) fails to hold]. We will use the same symbol $PC[a, b]$ for this Euclidean space, and the same functions, but simply remember that the statement $f = 0$ means that f vanishes in $[a, b]$ except at most on a finite set. The scalar product (5) is called the *standard scalar product* on $PC[a, b]$ (there are infinitely many others). Note that (5) is also a scalar product on $C''[a, b]$. The norm induced by (5) has the form

$$\|f\| = \left[\int_a^b |f|^2\,dx\right]^{1/2}, \qquad f \text{ in } PC[a, b] \tag{6}$$

Any scalar product $\langle \cdot, \cdot \rangle$ and its associated norm $\| \cdot \|$ have the following useful properties (proofs omitted):

THEOREM 10.2.2

Properties of Scalar Product. Let V be a linear space with scalar product $\langle \cdot, \cdot \rangle$ and corresponding norm $\| \cdot \|$. For all u, v in V, and scalars α the following hold:

(*Cauchy-Schwarz inequality*) $|\langle u, v \rangle| \leq \|u\|\|v\|$.
(*Positive definiteness*) $\|u\| \geq 0$, and $\|u\| = 0$ if and only if $u = 0$.
(*Homogeneity*) $\|\alpha u\| = |\alpha|\|u\|$.
(*Triangle inequality*) $\|\|u\| - \|v\|\| \leq \|u + v\| \leq \|u\| + \|v\|$.

These properties are very useful in calculations and in producing error estimates in Euclidean spaces, as we shall see presently.

Orthogonality

A scalar product is used to define the orthogonality of vectors.

> ❖ **Orthogonality.** Two vectors u, v in a Euclidean space V are said to be *orthogonal* if and only if $\langle u, v \rangle = 0$. A subset S in V not containing the zero vector is an *orthogonal subset* if $\langle u, v \rangle = 0$ for any two vectors u, v in S.

Of course, the zero vector is orthogonal to all elements of V, but it is the only element with that property. This notion of orthogonality extends the familiar notion of perpendicularity in the standard Euclidean spaces $\mathbf{R}^3$ and $\mathbf{R}^2$. For instance, $(a, b) \neq (0, 0)$ in $\mathbf{R}^2$ is orthogonal to $(-b, a)$, and this coincides with the familiar fact that a pair of lines intersects orthogonally if their slopes are negative reciprocals. The notion of orthogonality in a function space such as $C^0[a, b]$ or $PC[a, b]$ has no such simple geometric interpretation.

EXAMPLE 10.2.5

An Orthogonal Set in PC$[-\pi, \pi]$
The set $\Phi = \{1, \cos x, \sin x, \ldots, \cos nx, \sin nx, \ldots\}$ is an orthogonal subset of $PC[-\pi, \pi]$ under the standard scalar product. Indeed, using a trigonometric identity we can show that for any integers m and n,

$$\langle \sin nx, \cos mx \rangle = \int_{-\pi}^{\pi} \sin nx \cos mx \, dx$$

$$= \int_{-\pi}^{\pi} \frac{1}{2}[\sin(n+m)x + \sin(n-m)x] \, dx = 0$$

The other orthogonality relations follow in a similar fashion.

EXAMPLE 10.2.6

An Orthogonal Set in the Complex Euclidean Space PC$[-T, T]$
Let $PC[-T, T]$ consist of complex-valued functions. The set $\Psi = \{e^{ik\pi/T} : k = 0, \pm 1, \pm 2, \ldots\}$ is an orthogonal set in $PC[-T, T]$. Using the standard scalar product in $PC[-T, T]$, we have for $k \neq m$,

$$\langle e^{ik\pi x/T}, e^{im\pi x/T} \rangle = \int_{-T}^{T} e^{ik\pi x/T} e^{-im\pi x/T} \, dx = \frac{Te^{i(k-m)\pi x/T}}{i(k-m)\pi} \bigg|_{-T}^{T} = 0$$

Mean Approximation

Let f be a given element in a Euclidean space E. If g in E is considered as an approximation to f, then $\|f - g\|$ is known as the *error in the sense of the mean*, or simply *error in the mean*. Suppose that $\Phi = \{\phi_1, \phi_2, \ldots\}$ is a finite or infinite orthogonal set in the Euclidean space E. Then for a given element f in E we may ask the following question: What linear combination of the first n elements of Φ is "closest" to f in the sense of the distance measure on E (i.e., *in the mean*)? In symbols: For what values of the constants $c_1, c_2, \ldots, c_n$ is $\|f - \sum_{k=1}^{n} c_k \phi_k\|$ a minimum? The answer to this question follows:

THEOREM 10.2.3

Mean Approximation Theorem. For any orthogonal set $\Phi = \{\phi_1, \phi_2, \ldots\}$ in a Euclidean space E, any f in E, and any positive integer n, the problem

$$\left\| f - \sum_{k=1}^{n} c_k \phi_k \right\| = \text{minimum}$$

has the unique solution

$$(c_k)_{\min} = \langle f, \phi_k \rangle / \|\phi_k\|^2, \qquad k = 1, 2, \ldots, n \tag{7}$$

Moreover,

$$\left\| f - \sum_{k=1}^{n} (c_k)_{\min} \phi_k \right\|^2 = \|f\|^2 - \sum_{k=1}^{n} |\langle f, \phi_k \rangle|^2 / \|\phi_k\|^2 \tag{8}$$

Proof. For simplicity of notation we assume that E is a real Euclidean space. Now notice that by the bilinearity of the scalar product $\langle \cdot, \cdot \rangle$ in E,

$$\left\| f - \sum_{k=1}^{n} c_k \phi_k \right\|^2 = \left\langle f - \sum_{k=1}^{n} c_k \phi_k, \, f - \sum_{k=1}^{n} c_k \phi_k \right\rangle$$

$$= \|f\|^2 - 2 \sum_{k=1}^{n} c_k \langle f, \phi_k \rangle + \sum_{k=1}^{n} c_k^2 \|\phi_k\|^2 \tag{9}$$

Completing the square on the quadratic polynomial in c_k in (9), we see that

$$\|\phi_k\|^2 c_k^2 - 2\langle f, \phi_k \rangle c_k = \|\phi_k\|^2 \left(c_k - \frac{\langle f, \phi_k \rangle}{\|\phi_k\|^2} \right)^2 - \frac{|\langle f, \phi_k \rangle|^2}{\|\phi_k\|^2}$$

and hence (9) becomes

$$\left\| f - \sum_{k=1}^{n} c_k \phi_k \right\|^2 = \|f\|^2 + \sum_{k=1}^{n} \|\phi_k\|^2 \left(c_k - \frac{\langle f, \phi_k \rangle}{\|\phi_k\|^2} \right)^2 - \sum_{k=1}^{n} \frac{|\langle f, \phi_k \rangle|^2}{\|\phi_k\|^2} \tag{10}$$

To minimize the right-hand side of (10), the c_k must be chosen as in (7), and when this is done, (8) results from (10), finishing the proof.

EXAMPLE 10.2.7

Mean Approximation

Direct calculation verifies that $\Phi = \{\sin \frac{k\pi x}{L} : k = 1, 2, \ldots\}$ is an orthogonal set in PC$[0, L]$, for any positive constant L, and that $\| \sin \frac{k\pi}{L} \|^2 = L/2$, for any $k = 1, 2, \ldots$.
    Let

$$f(x) = \begin{cases} x, & 0 \le x \le L/2 \\ 0, & L/2 \le x \le L \end{cases}$$

We shall find the best mean-square approximation to f in the form $c_1 \sin(\pi x/L) + c_2 \sin(2\pi x/L) + c_3 \sin(3\pi x/L)$. Using (7), we see that for any $k > 0$:

$$(c_k)_{\min} = \frac{2}{L} \int_0^{L/2} x \sin \frac{k\pi x}{L} \, dx = \frac{2L}{k\pi} \left(\frac{1}{k\pi} \sin \frac{k\pi}{2} - \frac{1}{2} \cos \frac{k\pi}{2} \right)$$

Thus $(c_1)_{\min} = 2L/\pi^2$, $(c_2)_{\min} = L/2\pi$, $(c_3)_{\min} = -2L/9\pi^2$, and hence

$$g_{\min} = \frac{2L}{\pi^2} \sin \frac{\pi x}{L} + \frac{L}{2\pi} \sin \frac{2\pi x}{L} - \frac{2L}{9\pi^2} \sin \frac{3\pi x}{L}$$

is the best approximation in the mean to f of the desired form. The square of the mean error (or the *mean-square error*) of this approximation is given by (7) and (8) as

$$\|f - g_{\min}\|^2 = \|f\|^2 - \sum_{k=1}^{3} (c_k)^2_{\min} \| \sin \frac{k\pi x}{L} \|^2 = \frac{L^3}{24} - \frac{L}{2} \left(\frac{4L^2}{\pi^4} + \frac{L^2}{4\pi^2} + \frac{4L^2}{81\pi^4} \right)$$

Convergence in the Mean

As we saw earlier, the induced norm $\| \cdot \|$ on a Euclidean space E measures the "distance" between two elements u, v in E as $\|u - v\|$. We can use this "distance" measure to define convergence of infinite sequences and series in E.

❖ **Convergence in the Mean.** Let $\| \cdot \|$ be the induced norm on a Euclidean space E. We say that the sequence $\{f_n\}$, $n = 1, 2, \ldots$, in E *converges in the mean* if and only if there exists an element f in E such that $\|f_n - f\| \to 0$, as $n \to \infty$. For g_k in E, $k = 1, 2, \ldots$, the series $\sum_{k=1}^{n} g_k$, $n = 1, 2, \ldots$, converges in the mean if and only if the sequence of partial sums $f_n = \sum_{k=1}^{n} g_k$, $n = 1, 2, \ldots$, converges in the mean. If $\sum_{k=1}^{n} g_k \to f$ as $n \to \infty$, then we write $f = \sum_{k=1}^{\infty} g_k$.

As in ordinary convergence, a mean convergent sequence $\{f_n\}$ can have only one limit.

THEOREM 10.2.4

> **Mean Convergence Theorem.** For any sequence $\{f_n\}$ in a Euclidean space E there is not more than one element f in E such that $\|f_n - f\| \to 0$ as $n \to \infty$.

Proof. Suppose that $f_n \to g$ and $f_n \to h$ as $n \to \infty$, where $g \neq h$. Now $\|g - h\| > 0$, and from the Triangle Inequality we have that

$$0 < \|g - h\| = \|(g - f_n) + (f_n - h)\| \leq \|f_n - g\| + \|f_n - h\|, \qquad \text{for all } n$$

Thus it is impossible that both $\|f_n - g\| \to 0$, and $\|f_n - h\| \to 0$, unless $g = h$, establishing our claim.

Mean convergence has another important property (see Problem 8 for the proof):

THEOREM 10.2.5

> **Continuity of the Scalar Product.** Suppose that $f_n \to f$, and $g_m \to g$ as $n, m \to \infty$ in a Euclidean space. Then $\langle f_n, g_m \rangle \to \langle f, g \rangle$, as $n, m \to \infty$.

EXAMPLE 10.2.8

Mean Convergence versus Pointwise Convergence
Mean convergence of a sequence $\{f_n(x)\}$ in PC$[a, b]$ implies the existence of an element f in PC$[a, b]$ such that $\int_a^b (f_n(x) - f(x))^2 \, dx \to 0$ as $n \to \infty$. Consider the sequence g_k in PC$[0, 1]$, $k = 1, 2, \ldots$ defined in Figure 10.2.1. We claim that $\{g_k\}$ does *not* converge in the mean. Indeed, say $g_k \to g$, for some g in PC$[0,1]$. Writing $g_k = (g_k - g) + g$ and applying the Triangle Inequality, we have $\|g_k\| \leq \|g_k - g\| + \|g\|$, and hence the sequence $\{\|g_k\|\}$ is bounded. But since $\|g_k\|^2 = \int_0^1 |g_k|^2 \, dx = 2k/3 \to \infty$, as $k \to \infty$, it follows that $\{\|g_k\|\}$ is unbounded and hence $\{g_k\}$ cannot converge.

On the other hand, consider the sequence h_k in PC$[0,1]$, $k = 1, 2, \ldots$ defined in Figure 10.2.2. Note that h_k, for $k = 2^n + m$, has the value 1 inside the interval $[(m -$

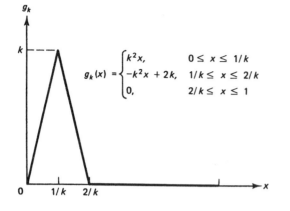

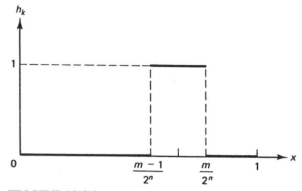

FIGURE 10.2.1 Sequence $\{g_k\}$ in PC[0,1]. See Example 10.2.8.

FIGURE 10.2.2 Sequence $\{h_k\}$ in PC[0,1]. See Example 10.2.8.

$1)/2^n, m/2^n]$ and vanishes otherwise in $[0,1]$. Thus $\int_0^1 |h_{2^n+m}|\,dx = 1/2^n$ for all $n = 1, 2, \ldots$, and all $m = 1, 2, \ldots, 2^n - 1$. It follows that $\|h_k\|^2 = \|h_k - 0\|^2 \to 0$ as $k \to \infty$. Thus $\{h_k\}$ converges in the mean to the zero function of $[0,1]$, even though for each k there is an interval on which h_k has the value one.

Mean, Pointwise, and Uniform Convergence

Mean convergence for sequences in PC$[a, b]$ is not to be confused with the usual notions of pointwise and uniform convergence for function sequences. We now review pointwise and uniform convergence for a sequence $\{f_n(x)\}$, $n = 1, 2, \ldots$, in PC$[a, b]$ (see also the discussion in Appendix A.2). The sequence $\{f_n(x)\}$ is said to converge *pointwise* at x_0 in $[a, b]$ if there is a value y_0 such that $f_n(x_0) \to y_0$ as $n \to \infty$ in the ordinary sense of number sequences. Restated, this means that for any interval of length $2\epsilon > 0$ centered at y_0, the values of $f_n(x_0)$ "eventually" all remain in this interval, that is, for all $n \geq N$ for some positive integer N (see Figure 10.2.3).

Uniform convergence of the function sequence $\{f_n(x)\}$ on a closed interval $[a, b]$ is *more* than just saying that $\{f(x)\}$ converges pointwise at *each* x_0 in $[a, b]$. Uniform convergence of $\{f_n(x)\}$ to a function $g(x)$ on $[a, b]$ means that for any "tube" with center spine $g(x)$ and vertical diameter $2\epsilon > 0$, the graphs of $f_n(x)$ are "eventually" all contained in this tube, for all $n \geq N$ for some positive integer N (see Figure 10.2.4).

EXAMPLE 10.2.9

Pointwise, Uniform, and Mean Convergence

The first sequence $\{g_k\}$ in Example 10.2.8 converges pointwise to the zero function on $[0,1]$. On the other hand, since $g_k(1/k) = k$ it is clear that the graphs of the $g_k(x)$ cannot eventually all lie in any tube about the zero function on $[0,1]$. Thus $\{g_k\}$ does *not* converge uniformly on $[0,1]$. It is true, however, that $\{g_k\}$ converges uniformly to the zero function on any interval $[\delta, 1]$, where $0 < \delta < 1$. The second sequence $\{h_k\}$ considered in Example 10.2.8 does not converge pointwise for *any* point in $[0,1]$. Indeed, for any x_0 in $[0,1]$ and any positive integer K, there are integers $r > K, s > K$ such that $h_r(x_0) = 0$ and $h_s(x_0) = 1$. Thus $\{h_k(x_0)\}$ cannot converge.

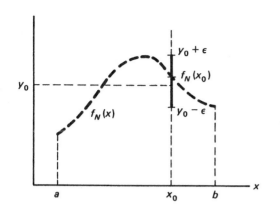

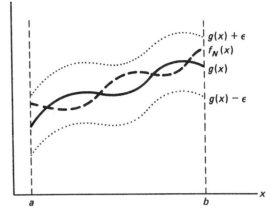

FIGURE 10.2.3 Geometry for pointwise convergence.   **FIGURE 10.2.4** Uniform convergence.

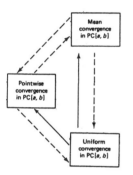

Finally, we see that if the sequence $\{g_k\}$ in PC$[a, b]$ converges uniformly to g on $[a, b]$, then $g_k \to g$ in the mean (over PC$[a, b]$). This fact and Examples 10.2.8 and 10.2.9 summarize the general relationship between the three modes of convergence in PC$[a, b]$ considered here. See the margin sketch. Uniform convergence implies mean convergence only if $-\infty < a < b < \infty$ (see Problem 12).

Orthogonal Series: Bases

Let $\Phi = \{\phi_1, \phi_2, \ldots\}$ be an orthogonal set in a Euclidean space E. The series given by $\sum_{k=1}^{\infty} c_k \phi_k$, for any scalars $c_1, c_2, \ldots$, is called an *orthogonal series* (whether or not it converges). By definition, the series $\sum_{k=1}^{\infty} c_k \phi_k$ converges (in the mean) to the element f in E if and only if $\| \sum_{k=1}^{n} c_k \phi_k - f \| \to 0$ as $n \to \infty$.

In finite-dimensional linear spaces we have seen how useful it is to have a set with the property that every element in the space can be uniquely written as a linear combination over this set. There is an analog of this concept in infinite-dimensional Euclidean spaces like PC$[a, b]$ which makes use of the convergence property to attach a meaning to the sum of an orthogonal series: An orthogonal set $\Phi = \{\phi_k : k = 1, 2, \ldots\}$ in a Euclidean space E is called a *basis* for E if and only if for every f in E there exists a unique set of scalars $\{c_k\}_1^{\infty}$ such that $f = \sum_{k=1}^{\infty} c_k \phi_k$.

The continuity of the scalar product (Theorem 10.2.5) implies that if $f = \sum_{k=1}^{\infty} c_k \phi_k$, then necessarily $c_n = \langle f, \phi_n \rangle / \|\phi_n\|^2$ for each $n = 1, 2, \ldots$.

THEOREM 10.2.6

> *Fourier-Euler Theorem.* If for an orthogonal set $\Phi = \{\phi : k = 1, 2, \ldots\}$ in the Euclidean space E we have that $f = \sum_{k=1}^{\infty} c_k \phi_k$, then the constants c_k are given by the *Fourier-Euler Formulas*
>
> $$c_k = \frac{\langle f, \phi_k \rangle}{\|\phi_k\|^2} \qquad \text{for each } k = 1, 2, \ldots \tag{11}$$

Proof. From the continuity of the scalar product, we see that since $\sum_{k=1}^{N} c_k \phi_k \to f$ as $N \to \infty$, it follows that $\langle \sum_{k=1}^{N} c_k \phi_k, \phi_n \rangle \to \langle f, \phi_n \rangle$, for each (fixed) $n = 1, 2, \ldots$. But by orthogonality, $\langle \sum_{k=1}^{N} c_k \phi_k, \phi_n \rangle = c_n \langle \phi_n, \phi_n \rangle$, if $N > n$, and the assertion is established.

❖ **Fourier Coefficient.** The constant $c_k = \langle f, \phi_k \rangle / \|\phi_k\|^2$ given by (11) is called the kth *Fourier* (or *Fourier-Euler*) *coefficient* of f over the orthogonal set Φ.

This result and the Mean Approximation Theorem 10.2.3 give the following equivalent (and more convenient) definition of a basis:

❖ **Basis.** An orthogonal set $\Phi = \{\phi_k : k = 1, 2, \ldots\}$ is a basis for the Euclidean space E if and only if for each f in E

$$\sum_{k=1}^{\infty} \frac{\langle f, \phi_k \rangle}{\|\phi_k\|^2} \phi_k = f \qquad \text{in the sense of mean convergence} \qquad (12)$$

An important property of bases is given by the following result:

THEOREM 10.2.7

> Totality Theorem. Let $\Phi = \{\phi_k : k = 1, 2, \ldots\}$ be a basis for the Euclidean space E. If f in E is such that $\langle f, \phi_k \rangle = 0$, for all $k = 1, 2, \ldots$, then $f = 0$.

Proof. Follows immediately from (12).

Fourier Series

Orthogonal series constructed using coefficients given by (11) have a special name.

❖ **Fourier Series.** Let $\Phi = \{\phi_k : k = 1, 2, \ldots\}$ be an orthogonal set in the Euclidean space E. Then the orthogonal series

$$\text{FS}[f] = \sum_{k=1}^{\infty} \frac{\langle f, \phi_k \rangle}{\|\phi_k\|^2} \phi_k, \quad \text{for any } f \text{ in } E \qquad (13)$$

is called *the Fourier series of f over* Φ. The symbol $\text{FS}[f]$ stands for the formal series in (13); no implication is made about convergence.

The same function may have any number of Fourier Series, depending on the particular choice of an orthogonal set Φ in E. A basis can now be defined as follows: An orthogonal set Φ in a Euclidean space E is a basis for E if and only if every f in E is the sum of its Fourier Series over Φ. The results below will be used later.

THEOREM 10.2.8

A Trio of Theorems. Let $\Phi = \{\phi_k : k = 1, 2, \ldots\}$ be an orthogonal set in the Euclidean space E, but not necessarily a basis. Then for any f in E, the series $\sum_{k=1}^{\infty} |\langle f, \phi_k \rangle|^2 / \|\phi_k\|^2$ converges, and we have the trio of properties:

Bessel Inequality: We have that

$$\|f\|^2 \geq \sum_{k=1}^{\infty} \frac{|\langle f, \phi_k \rangle|^2}{\|\phi_k\|^2} \qquad \text{for all } f \text{ in } E \tag{14}$$

Decay of Coefficients: If $c_k = \langle f, \phi_k \rangle / \|\phi_k\|^2$, the kth Fourier coefficient of f, and Φ is nonfinite, then

$$\|\phi_k\| \cdot |c_k| \to 0 \qquad \text{as } k \to \infty \tag{15}$$

Parseval Relation: Φ is a basis for E if and only if

$$\|f\|^2 = \sum_{k=1}^{\infty} \frac{|\langle f, \phi_k \rangle|^2}{\|\phi_k\|^2} \qquad \text{for all } f \text{ in } E \tag{16}$$

Proof. The identity (8) in the Mean Approximation Theorem holds for all positive integers n. Since the left-hand side of (8) can never be negative, it follows that

$$\|f\|^2 \geq \sum_{k=1}^{n} \frac{|\langle f, \phi_k \rangle|^2}{\|\phi_k\|^2} \qquad \text{for all } n \tag{17}$$

Thus the sequence $\left\{ \sum_{k=1}^{n} |\langle f, \phi_k \rangle|^2 / \|\phi_k\|^2 \right\}$, $n = 1, 2, \ldots$, is non-decreasing and bounded from above, and hence the infinite series $\sum_{k=1}^{\infty} |\langle f, \phi_k \rangle|^2 / \|\phi_k\|^2$ is convergent. The inequality (14) now follows immediately from (17) by taking the limit as $n \to \infty$. Because the infinite series in (14) is convergent, the kth term converges to zero. The property (15) follows from the fact that $|\langle f, \phi_k \rangle|^2 / \|\phi_k\|^2 = \|\phi_k\|^2 |c_k|^2$, where c_k is the Fourier coefficient given by (11). Finally, the Parseval Relation follows immediately from (7) and (8) of the Mean Approximation Theorem.

EXAMPLE 10.2.10

Fourier Series in PC$[-\pi, \pi]$

Consider the orthogonal set Φ in PC$[-\pi, \pi]$ given in Example 10.2.5. We now use the definition (13) to calculate the Fourier Series of the function $f(x) = x$ on $[-\pi, \pi]$. Notice first that $\int_{-\pi}^{\pi} \sin^2 nx \, dx = \int_{-\pi}^{\pi} \cos^2 nx \, dx = \pi$, for all $n \geq 1$, and that $\int_{-\pi}^{\pi} 1^2 \, dx = 2\pi$. Now FS$[f]$ has the form

$$\text{FS}[f] = A_0 + \sum_{k=1}^{\infty} (A_k \cos kx + B_k \sin kx) \tag{18}$$

where $A_0 = \frac{1}{2\pi} \int_{-\pi}^{\pi} 1 \cdot x \, dx = 0$, $A_k = \frac{1}{\pi} \int_{-\pi}^{\pi} x \cos kx \, dx = 0$ (since $x \cos kx$ is an odd function on $[-\pi, \pi]$), and $B_k = \frac{1}{\pi} \int_{-\pi}^{\pi} x \sin kx \, dx = 2(-1)^{k+1}/k$. So we have that

$$\text{FS}[f] = 2 \sum_{k=1}^{\infty} \frac{(-1)^{k+1}}{k} \sin kx \tag{19}$$

Figure 10.2.5 shows three partial sums (dashed) of the Fourier Series (19) and the graph (solid line) of $f(x) = x$ on the interval $[-\pi, \pi]$.

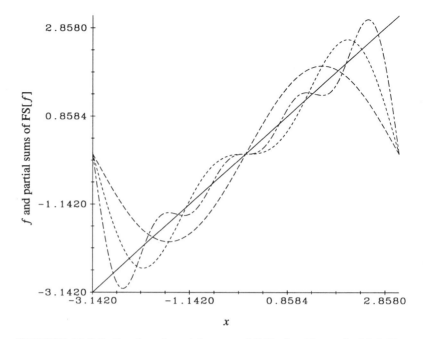

FIGURE 10.2.5 Graphs of partial sums of (19). See Example 10.2.10.

PROBLEMS

1. (*Subspaces*). A subset A of a linear space V is called a *subspace* of V if for any scalar, α, and any pair of vectors u, v in A, we have that αu and $u + v$ is in A. For any scalars $\alpha_1, \alpha_2, \ldots, \alpha_n$ and any vectors $u_1, u_2, \ldots, u_n$ in A, the sum $\alpha_1 u_1 + \cdots + \alpha_n u_n$ is said to be a *finite linear combination* over A. The *span* of A, denoted by $\mathrm{Span}(A)$, is the collection of all finite linear combinations over A.

   (a) For any linear space V, show that the set of all vectors in V and the set consisting of the zero vector alone are both subspaces of V.

   (b) Show that a set of vectors A in a linear space V is a subspace of V if and only if $\mathrm{Span}(A) = A$.

   (c) For a closed interval $I = [a, b]$ in $\mathbf{R}$ and any nonnegative integers m, n, suppose that the continuity sets $C^n(I)$ and $C^m(I)$ are both real or both complex linear spaces of functions. Show that $C^n(I)$ is a subspace of $C^m(I)$ if $n \geq m$.

   (d) Let $C^\infty(I)$ be the set of all real-valued functions on an interval I which have derivatives of all orders on I. Show that $C^\infty(I)$ is a real linear space.

2. (*A Weighted Scalar Product*). For f and g in the real space $C^0[0, 1]$ define $\langle f, g \rangle = \int_0^1 f(x)g(x)e^x \, dx$. The function e^x is said to be a *weight* or *density* factor.

   (a) Show that $\langle \cdot, \cdot \rangle$ defines a scalar product.

   (b) Calculate the scalar product of each pair of functions: $f = 1 - 2x$, $g = e^{-x}$; $f = x^2$, $g = e^x$.

   (c) Calculate the scalar product of each pair of functions: $f = x$, $g = 1 - x$; $f = e^{-x/2}\sin(\pi x/2)$, $g = e^{-x/2}\sin(3\pi x/2)$; $f = \cos(\pi x/2)$, $g = 1$.

3. For any closed interval $I = [a, b]$ and any nonnegative integer n, show that

$$\langle f, g \rangle = \int_a^b f(x)\overline{g(x)}\,dx$$

is a scalar product on $C^n(I)$.

4. Let $\langle \cdot, \cdot \rangle$ be a scalar product on a real linear space V, and let $\| \cdot \|$ be the norm induced by the scalar product.

(a) Show that $\|u + v\|^2 + \|u - v\|^2 = 2\|u\|^2 + 2\|v\|^2$, for all u, v in V.

(b) (*Cauchy-Schwarz Inequality*). Show that $|\langle u, v \rangle| \leq \|u\| \cdot \|v\|$, for all u, v in V. [*Hint:* Use the fact that $\|\alpha u - \beta v\|^2 \geq 0$ for *all* scalars α and β.]

(c) Show that $\langle u, v \rangle = (\|u + v\|^2 - \|u - v\|^2)/4$, for all u, v in V.

5. (*"Little Ell Two": l^2*). Let $\mathbf{R}^\infty$ be the set of all sequences of real numbers, $(x_n)_{n=1}^\infty$, abbreviated simply as (x_n). Define addition ("+") and multiplication by reals ("·") in $\mathbf{R}^\infty$ termwise; that is, $(x_n) + (y_n) = (x_n + y_n)$, $\alpha \cdot (x_n) = (\alpha x_n)$. Let $l^2 = \{(x_n)$ in $\mathbf{R}^\infty$ with $\sum_{n=1}^\infty |x_n|^2 < \infty\}$; l^2 is called "little ell two." Show that

(a) For any x, y in l^2, $\sum_{i=1}^\infty x_i y_i < \infty$.

(b) $\langle x, y \rangle = \sum_{i=1}^\infty x_i y_i$ is a scalar product on l^2.

6. (*Orthogonal Projection*). Let E be a Euclidean space and S a subspace of E spanned by the finite orthogonal set $\{u^1, u^2, \ldots, u^n\}$. For any u in E the vector $\text{proj}_S(u) = \sum_{i=1}^n [\langle u, u^i \rangle / \|u^i\|^2] u^i$ is called the *orthogonal projection* of u onto S. [*Note:* $\text{proj}_S(u)$ does not depend on the orthogonal basis of S chosen.] Show that

(a) For every u in E there is a unique decomposition $u = v + w$ with v in S and w orthogonal to S.

(b) For any u in E, $\|u - \text{proj}_S(u)\| \leq \|u - v\|$, for all v in S, and equality holds if and only if $v = \text{proj}_S(u)$.

(c) In $\mathbf{R}^3$, find the distance from the point $(1,1,1)$ to the plane: $x + 2y - z = 0$.

7. (*Best Approximation*). In the Euclidean space $\text{PC}[-\pi, \pi]$, consider the subspace S_N spanned by the set $\Phi_N = \{1, \cos x, \sin x, \ldots, \cos Nx, \sin Nx\}$. Find the element in S_N that is closest to the element $f(x) = x$, $|x| < \pi$, in $\text{PC}[-\pi, \pi]$.

8. (*Continuity of Scalar Product*). Prove the continuity of the scalar product (Theorem 10.2.5).

9. (*Uniqueness*). Let Φ be a basis for a Euclidean space E. Show that if two elements f and g in E have the same Fourier Series over Φ, then $f = g$.

10. (*Calculating Fourier Series*). $\Phi = \{1, \cos x, \sin x, \ldots, \cos nx, \sin nx, \ldots\}$ is an orthogonal subset of the Euclidean space $\text{PC}[-\pi, \pi]$. Find the Fourier Series of the following elements of $\text{PC}[-\pi, \pi]$ with respect to Φ.

(a) *Sawtooth*

$$f(x) = \begin{cases} -A(1 + x/\pi), & -\pi \leq x < 0 \\ A(1 - x/\pi), & 0 < x < \pi \end{cases}$$

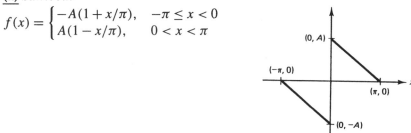

(b) *Triangular*

$$f(x) = \begin{cases} A(1+x/\pi), & -\pi \le x \le 0 \\ A(1-x/\pi), & 0 < x < \pi \end{cases}$$

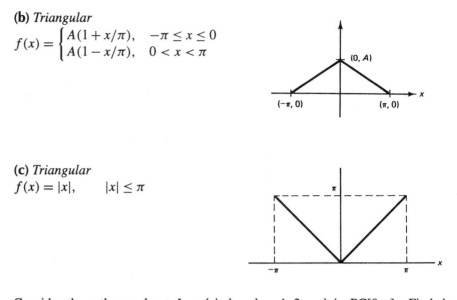

(c) *Triangular*

$$f(x) = |x|, \qquad |x| \le \pi$$

11.   Consider the orthogonal set $\Phi = \{\sin kx : k = 1, 2, \ldots\}$ in $\text{PC}[0,\pi]$. Find the Fourier Series of the following elements of $\text{PC}[0,\pi]$ with respect to Φ.

   (a) $f(x) = x, \quad 0 \le x \le \pi$ **(b)** $f(x) = 1 + x/\pi, \quad 0 \le x \le \pi$

   (c) $f(x) = 1, \quad 0 \le x \le \pi$

12.   **(a)** Show that the collection $E = \{f \text{ in } C^0(\mathbf{R}) : \int_{-\infty}^{\infty} |f|^2 \, dx < \infty\}$ is a linear space if addition and multiplication by scalars are defined by (2).

   (b) Show that $\langle f, g \rangle = \int_{-\infty}^{\infty} fg \, dx$ is a scalar product on E.

13.   Let E be the Euclidean space in Problem 12, and consider the sequence

$$f_n(x) = n^{-1/2} e^{-x^2/n^2}, \quad n = 1, 2, \ldots$$

   (a) Show that $\{f_n\}$ converges uniformly to the zero function of $\mathbf{R}$.

   (b) Show that $\{f_n\}$ does not converge in the mean in E. [*Hint:* If $\{f_n\}$ is mean convergent, the limit must be the zero function (why?), but the continuity of the scalar product (Theorem 10.2.5) contradicts this.]

10.3   Fourier Trigonometric Series

Some techniques of applied analysis crucially depend on the possibility of writing a given real-valued function as the sum of specialized trigonometrical series over some interval. In fact, this situation arises so frequently in the applications that a considerable body of literature has been generated on this topic. In this section we pursue these ideas in a systematic fashion.

Using some trigonometric identities (see Appendix C.4) before calculating the integrals, we can verify directly that

$$\Phi_T = \left\{ 1, \cos \frac{\pi x}{T}, \sin \frac{\pi x}{T}, \ldots, \cos \frac{n\pi x}{T}, \sin \frac{n\pi x}{T}, \ldots \right\} \tag{1}$$

is an orthogonal set in the Euclidean space $\text{PC}[-T, T]$ under the standard scalar product $\langle f, g \rangle = \int_{-T}^{T} fg \, dx$. Notice that

$$\left\| \cos \frac{k\pi x}{T} \right\|^2 = \int_{-T}^{T} \cos^2 \frac{k\pi x}{T} \, dx = \begin{cases} 2T & \text{if } k = 0 \\ T & \text{if } k > 0 \end{cases}$$

$$\| \sin \frac{k\pi x}{T} \|^2 = \int_{-T}^{T} \sin^2 \frac{k\pi x}{T} \, dx = T \qquad k > 0$$

Thus for any f in $PC[-\pi, \pi]$, we see (see Section 10.2) that the Fourier Series of f over Φ_T is

$$FS[f] = A_0 + \sum_{k=1}^{\infty} \left(A_k \cos \frac{k\pi x}{T} + B_k \sin \frac{k\pi x}{T} \right) \qquad (2)$$

where the coefficients are given by the Fourier-Euler formulas,

$$A_0 = \frac{1}{2T} \int_{-T}^{T} f(x) \, dx, \quad A_k = \frac{1}{T} \int_{-T}^{T} f(x) \cos \frac{k\pi x}{T} \, dx \quad k > 0$$

$$B_k = \frac{1}{T} \int_{-T}^{T} f(x) \sin \frac{k\pi x}{T} \, dx, \quad \text{for } k > 0 \tag{3}$$

This particular orthogonal series (2) is called a *Fourier Trigonometrical Series*[5] on $[-T, T]$ and evidently consists of a superposition of a constant term and sinusoids whose frequencies are multiples (harmonics) of a basic (fundamental) frequency of $1/(2T)$ hertz. By popular usage, however, the series (2), (3) is commonly called the *Fourier Series* of f, and not its more accurate name. We shall adopt this popular name whenever no confusion can arise.

Although it is not evident, the orthogonal set Φ_T is a basis for $PC[-T, T]$.

THEOREM 10.3.1

> Fourier Trigonometrical Basis Theorem. The subset Φ_T given by (1) is a basis for $PC[-T, T]$, and so for any f in $PC[-T, T]$, we have that $FS[f]$ converges in the mean to f.

See Problem 11 for a proof in a special case.

Calculation of Fourier Trigonometrical Series

Given a function f in $PC[-T, T]$, the process of setting up the trigonometrical series (2) with coefficients (3) is known as "expanding f into a Fourier Trigonometrical Series," and the values of A_k and B_k in (3) are "the Fourier coefficients of f." Calculating the Fourier coefficients of a function f can be a lengthy exercise in integration techniques. Keep in mind, however, that two functions f and g in $PC[-T, T]$ whose

---

[5]The trigonometric series encountered in this chapter are called Fourier Series in honor of the French physicist Jean Baptiste Joseph Fourier [1768–1830] who used them in the study of heat flow. From about 1740 onward, physicists and mathematicians such as Bernoulli, D'Alembert, Lagrange, and Euler had engaged in heated discussions about the possibility of representing a more or less arbitrary periodic function as a sum of a trigonometric series. In fact, in 1777 Euler had discovered the integral formulas for what are now called the Fourier coefficients. Fourier added his thoughts, perhaps prematurely, at the Paris Academy of Sciences in 1807 and 1811. In the papers on heat conduction he presented there, he suggested that a completely arbitrary function could be represented as the sum of a trigonometric series. The referees criticized him for a lack of rigor. Lagrange, who was at that time one of the referees, denied vehemently Fourier's contention, and these papers were never published. Fourier continued his work, however, and in 1822 he published *La Théorie Analytique de la Chaleur*, throughout which he made use of Fourier series in his analysis of heat flow. Besides its widespread use in the study of the partial differential equations modeling wave motion, heat flow, and electrostatics, the theory of Fourier series has directly affected such diverse areas of mathematics as modern algebra and Cantor's work on set theory.

values differ at only a finite number of points in $[-T, T]$ have the *same* Fourier coefficients. Thus, changing the values of a function at a finite number of points in $[-T, T]$ does not change its Fourier coefficients.

The process of calculating Fourier coefficients can be simplified if the function has certain symmetry properties. A function f in $PC[-T, T]$ is *even* if $f(-x) = f(x)$ for $|x| < T$ (except possibly a finite number of points), while f is *odd* if $f(-x) = -f(x)$, for $|x| < T$. It is easy to show that $f(x)g(x)$ is even if f and g are both even or both odd, and $f(x)g(x)$ is odd if one factor is even and the other odd. Note that $\int_{-a}^{a} f(x)\, dx = 0$ if f is odd, and $\int_{-a}^{a} f(x)\, dx = 2\int_{0}^{a} f(x)\, dx$ if f is even. Thus if f in $PC[-T, T]$ is odd, the Fourier coefficients $A_k = 0$, since $f(x)\cos(k\pi x/T)$ is odd for each $k = 0, 1, 2, \ldots$. Similarly, $B_k = 0$ for $k = 1, 2, \ldots$ if f is even. See the problems for more "tricks" of this kind.

EXAMPLE 10.3.1

Fourier Series of an Even Function

Let us find the Fourier (Trigonometrical) Series of the function $f(x) = |x|$, $-T \leq x \leq T$. Since f is even, all the coefficients B_k vanish. Now we have

$$A_0 = \frac{1}{T} \int_0^T x\, dx = \frac{T}{2}, \quad A_k = \frac{2}{T} \int_0^T x\cos\frac{k\pi x}{T}\, dx = \frac{2T}{\pi^2 k^2}(\cos k\pi - 1), \quad \text{for } k > 0$$

Note that since $\cos k\pi = (-1)^k$, $A_k = 0$ when k is a positive even integer. Setting $k = 2m+1$, $m = 0, 1, \ldots$, we see that $A_{2m+1} = -4T/\pi^2(2m+1)^2$. Thus

$$\text{FS}[f] = \frac{T}{2} + \sum_{m=0}^{\infty} \frac{-4T}{\pi^2(2m+1)^2} \cos\left(\frac{(2m+1)\pi x}{T}\right)$$

Convergence Properties

Since Φ_T is a basis for $PC[-T, T]$, we know that for any f in $PC[-T, T]$, the Fourier Trigonometrical Series (2) and (3) converges in the mean to f; that is,

$$\left\| A_0 + \sum_{k=1}^{n}(A_k \cos kx + B_k \sin kx) - f \right\| \to 0, \qquad \text{as } n \to \infty$$

But what about the pointwise or uniform convergence properties of Fourier Series?

EXAMPLE 10.3.2

Fourier Series of an Odd Function; Loss of Uniform Convergence

The function

$$f(x) = \begin{cases} 5, & 0 < x < \pi \\ -5, & -\pi < x < 0 \end{cases}$$

is in $PC[-\pi, \pi]$ and since it is an odd function, its Fourier Series contains no cosine terms (or constant term). A straightforward calculation shows that

$$\text{FS}[f] = \frac{20}{\pi} \sum_{k=0}^{\infty} \frac{\sin(2k+1)x}{2k+1} \tag{4}$$

In Figure 10.3.1, the graphs of the partial sums of the series (4) using (3), (9), and (15) terms are compared with the graph of f. Apparently, we have pointwise convergence of FS[f] to f except at the endpoints $x = \pm\pi$, and at the "jump" discontinuity at $x = 0$. The convergence also appears to be uniform on any closed interval not containing any of these three exceptional points.

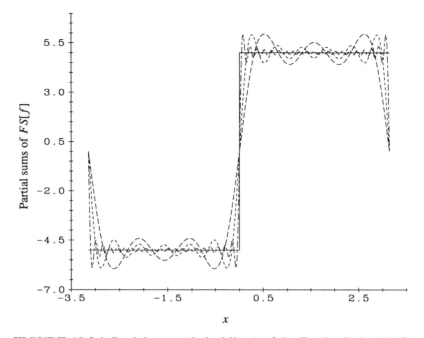

FIGURE 10.3.1 Partial sums (dashed lines) of the Fourier Series (4) for the square wave (solid line): 3 terms; 9 terms; 15 terms. See Example 10.3.2.

The behavior of the partial sums near $x = 0$ and $x = \pm\pi$ is somewhat mysterious. Apparently, the "hump" in the graph of the partial sums near these points does not diminish in size but merely "moves" over closer to the exceptional points, $x = 0$ and $x = \pm\pi$, as more terms are included in the partial sums. This is the *Gibbs phenomenon* and always occurs at jump discontinuities. We now examine the precise pointwise convergence properties of the Fourier Trigonometrical Series, but first we gather some useful information.

❖ **Piecewise Smoothness, PS$[-T, T]$.** A function f on $[-T, T]$ is *piecewise smooth* if it is piecewise continuous and is differentiable at all but at most a finite set of points in $(-T, T)$, and f' is piecewise continuous on $[-T, T]$. The set of all piecewise smooth functions on $[-T, T]$ is denoted by PS$[-T, T]$. Thus f is in PS$[-T, T]$ if and only if both f and f' are in PC$[-T, T]$.

THEOREM 10.3.2

> **Properties of Piecewise Continuous Functions.**
>
> **Decay of Coefficients:** Let f be in PC$[-T, T]$, and let FS$[f]$ be the Fourier Series of f given by (2) and (3). Then $A_k \to 0$, $B_k \to 0$, as $k \to \infty$.
>
> **Integration of Periodic Functions:** Let f be any periodic function on **R** with period $2T$ which is piecewise continuous over the interval $[-T, T]$. Then
>
> $$\int_{-T}^{T} f(x)\,dx = \int_{a}^{a+2T} f(s)\,ds, \qquad \text{for any real } a \qquad (5)$$

Proof. Notice that $\| \cos(k\pi x/t) \|^2 = \| \sin(k\pi x/T) \|^2 = T$ for $k > 0$, and from the Decay of Coefficients Theorem (Theorem 10.2.8) we see that $A_k \to 0$, $B_k \to 0$, as $k \to \infty$. The integration formula (5) is treated in Problem 7.

We will also need the notion of a periodic extension $\tilde{f}$ of a function f in PC$[-T, T]$ to the whole real line. It is just what one would expect it to be: $\tilde{f}$ is a periodic function of period $2T$ whose graph on $-T < x < T$ is exactly that of the given function f in PC$[-T, T]$. More precisely:

❖ **Periodic Extension:** $\tilde{f}$. Let $f(x)$ be a piecewise continuous function on $-T \leq x \leq T$. The periodic extension $\tilde{f}$ of f into **R** with period $2T$ is defined as follows: For x in **R** which is not an odd multiple of T there is a unique integer n such that $|x - 2Tn| < T$. Define $\tilde{f}(x) = f(x - 2Tn)$, for $2Tn - T < x < 2Tn + T$. For x in **R** which are odd multiples of T, we leave $\tilde{f}$ undefined.

Notice that $\tilde{f}$ is piecewise continuous on **R**. No harm is done in leaving $\tilde{f}$ undefined at odd multiples of T since integrals involving $\tilde{f}$ are not affected in any way by the values of $\tilde{f}$ at these points. Also, notice that the limits $\tilde{f}(x_0^+)$ and $\tilde{f}(x_0^-)$ are not affected by the value of $\tilde{f}$ at x_0.

In what follows we set $T = \pi$ for convenience; it will be clear how the results can be modified for general T.

THEOREM 10.3.3

Dirichlet Representation Theorem. Let f be in PC$[-\pi, \pi]$, and let $\tilde{f}$ be the periodic extension of f in **R** with period 2π. Let $S_n(x)$ denote the nth partial sum of FS$[f]$ as in (2) and (3); that is, $S_n(x) = A_0 + \sum_{k=1}^{n}(A_k \cos kx + B_k \sin kx)$, where A_k, B_k are as in (3). Then we have the *Dirichlet Representation* for $S_n(x)$,

$$S_n(x) = \int_{-\pi}^{\pi} \tilde{f}(x+s) \frac{\sin(n+\frac{1}{2})s}{2\pi \sin(s/2)}\, ds, \qquad x \text{ in } \mathbf{R}, \quad n = 0, 1, 2, \ldots \quad (6)$$

The function $D_n(s) = \sin((n+\frac{1}{2})s)/(2\pi \sin(s/2))$ which appears in the integrand of (6) is called the *Dirichlet kernel* and has the property that

$$\int_{-\pi}^{\pi} D_n(s)\, ds = 1, \qquad \text{for all } n = 0, 1, 2, \ldots \quad (7)$$

Proof. Recalling the definition of the coefficients A_k, B_k given in (3) and substituting in $S_n(x)$, we obtain that (after interchanging summation and integration and using an identity from Appendix C.4 for $\cos(\alpha - \beta)$):

$$S_n(x) = \frac{1}{\pi} \int_{-\pi}^{\pi} f(t) \left[\frac{1}{2} + \sum_{k=1}^{n} \cos k(t-x) \right] dt, \qquad \text{for any } n = 0, 1, 2, \ldots$$

Using the identity $\frac{1}{2} + \sum_{k=1}^{n} \cos kx = \sin((n+\frac{1}{2})s)/(2 \sin(s/2))$ (see Problem 8 for a proof), we have that

$$S_n(x) = \frac{1}{\pi} \int_{-\pi}^{\pi} f(t) \frac{\sin(n+\frac{1}{2})(t-x)}{2 \sin\frac{1}{2}(t-x)}\, dt \quad (8)$$

Now after replacing f by $\tilde{f}$ in (8), making the change of variables $t - x = s$, and applying the Integration Theorem for Periodic Functions, the desired formula (6) for $S_n(x)$ is obtained. The identity (7) results immediately from application of (6) to the function $f = 1$ for all x, and we are done.

Now we are ready to state a major result of this section.

THEOREM 10.3.4

> **Pointwise Convergence Theorem for Fourier Series.** Let f be piecewise smooth on $[-\pi, \pi]$, with $\tilde{f}$ the periodic extension of f into **R** with period 2π. Then for each x_0 in **R**, FS$[f]$ evaluated at x_0 converges to $\frac{1}{2}[\tilde{f}(x_0^+) + \tilde{f}(x_0^-)]$.

Proof. Writing $S_n(x)$ in (6) as

$$S_n(x) = \int_{-\pi}^0 \tilde{f}(x+s)D_n(s)\,ds + \int_0^\pi \tilde{f}(x+s)D_n(s)\,ds$$

we will show that

$$\int_0^\pi \tilde{f}(x_0+s)D_n(s)\,ds \to \frac{1}{2}\tilde{f}(x_0^+), \qquad \text{as } n \to \infty$$

$$\int_{-\pi}^0 \tilde{f}(x_0+s)D_n(s)\,ds \to \frac{1}{2}\tilde{f}(x_0^-), \qquad \text{as } n \to \infty \tag{9}$$

which implies the assertion. We prove only the first statement in (9). Note that $D_n(s)$ is an even function and so from (7), $\int_0^\pi D_n(s)\,ds = \int_{-\pi}^0 D_n(s)\,ds = \frac{1}{2}$. Using this expression and an identity from Appendix C.4 for $\sin(\alpha + \beta)$, we have

$$\int_0^\pi \tilde{f}(x_0+s)D_n(s)\,ds - \frac{1}{2}\tilde{f}(x_0^+) = \int_0^\pi [\tilde{f}(x_0+s) - \tilde{f}(x_0^+)]D_n(s)\,ds$$

$$= \frac{1}{2\pi}\int_{-\pi}^\pi \{H(s)[\tilde{f}(x_0+s) - \tilde{f}(x_0^+)]\}\cos ns\,ds \tag{10}$$

$$+ \frac{1}{\pi}\int_{-\pi}^\pi \left\{ H(s) \cdot \frac{\tilde{f}(x_0+s) - \tilde{f}(x_0^+)}{s} \frac{\frac{1}{2}s}{\sin(s/2)} \cos\left(\frac{s}{2}\right) \right\} \sin ns\,ds$$

where $H(s)$ is the piecewise constant function whose value is 1, $0 < s < \pi$, and 0, $-\pi < s < 0$. Now $H(s)[\tilde{f}(x_0+s) - \tilde{f}(x_0^+)]$ belongs to the class PC$[-\pi, \pi]$ as a function of s, and hence the Decay of Coefficients Theorem implies that the first integral in (10) tends to zero as $n \to \infty$. Now if we can show that the function in braces $\{\cdot\}$ in the second integral in (10) is also in PC$[-\pi, \pi]$ as a function of s, the Decay of Coefficients Theorem also implies that this integral tends to zero as $n \to \infty$, finishing the proof. Observe that $H(s)$, $\frac{1}{2}s/\sin(s/2)$, and $\cos(s/2)$ are all piecewise continuous on $-\pi \le s \le \pi$. Now since $H(s) = 0$ for $-\pi \le s < 0$, it remains only to show that $[\tilde{f}(x_0+s) - \tilde{f}(x_0^+)]/s$ is piecewise continuous on $0 \le s \le \pi$. Evidently, the only trouble point is $s = 0$. Recall that f is piecewise smooth on $[-\pi, \pi]$; thus $\tilde{f}(x_0+s)$ is also piecewise smooth for $-\pi \le s \le \pi$. Thus for all sufficiently small positive s, we have from the Mean Value Theorem that

$$\frac{\tilde{f}(x_0+s) - \tilde{f}(x_0^+)}{s} = \tilde{f}'(x_0+s^*) \qquad \text{for some } s^* \text{ with } 0 < s^* < s \tag{11}$$

Since $\tilde{f}'(x_0+s^*) \to \tilde{f}'(x_0^+)$ as $s \to 0^+$, we see from (11) that the function

$[\tilde{f}(x_0 + s) - \tilde{f}(x_0^+)]/s$ has passed the last test to show that it is piecewise continuous on $0 \le s \le \pi$, and hence we are done.

Thus we see that Fourier Series of piecewise smooth functions converge pointwise for all x in **R**, and that the sum at a point x_0 where $\tilde{f}$ is continuous is just $\tilde{f}(x_0)$, whereas at both endpoints $x = \pm\pi$ the sum has the same value, $\frac{1}{2}[f(\pi^-) + f(-\pi^+)]$.

EXAMPLE 10.3.3

Pointwise Convergence of Fourier Series

Let us apply the Pointwise Convergence Theorem to the function in Example 10.3.2. For $-\pi < x < \pi$ we easily calculate that

$$\frac{1}{2}[\tilde{f}(x_0^+) + \tilde{f}(x_0^-)] = \begin{cases} 5, & 0 < x_0 < \pi \\ 0, & x_0 = 0 \\ -5, & -\pi < x_0 < 0 \end{cases}$$

But what about the endpoints $x_0 = \pm\pi$? A little thought shows that $\tilde{f}(\pi^+) = f(-\pi^+) = -5$ and $\tilde{f}(\pi^-) = f(\pi^-) = 5$, and that $\tilde{f}(-\pi^+) = f(-\pi^+) = -5$ and $\tilde{f}(-\pi^-) = f(\pi^-) = 5$. Thus in our case, as it should be,

$$\frac{1}{2}[\tilde{f}(x_0^+) + \tilde{f}(x_0^-)] = 0, \qquad \text{for } x_0 = \pm\pi$$

Uniform Convergence of Fourier Series

If we assume a bit more smoothness for f, FS[f] will converge uniformly and not just pointwise. The function in Example 10.3.2 shows that something more than merely piecewise smoothness is required. First, however, we need an effective test for uniform convergence. The simplest, perhaps, is the following (proof omitted).

THEOREM 10.3.5

> **Weierstrass M-Test.** Let the functions $g_k(x)$, $k = 1, 2, \ldots$, be continuous on the interval I, and suppose that the positive constants M_k are such that $|g_k(x)| \le M_k$ for all x in I, and $\sum_{k=1}^{\infty} M_k < \infty$. Then there exists a continuous function $g(x)$ on I such that the series $\sum_{k=1}^{\infty} g_k(x)$ converges uniformly to g on I.

Now we can prove the following theorem

THEOREM 10.3.6

> **Uniform Convergence Theorem for Fourier Series.** Let f be in PS$[-\pi, \pi]$ such that $\tilde{f}$ is in $C^0(\mathbf{R})$. Then FS[f] converges uniformly to $\tilde{f}$ on **R**, where $\tilde{f}$ is the periodic extension of f into **R** with period 2π.

Proof. $\tilde{f}$ will be in $C^0(\mathbf{R})$ if f is in $C^0[-\pi, \pi]$ and $f(-\pi) = f(\pi)$. The Cauchy-Schwarz inequality applied to $A_k \cos kx + B_k \sin kx$, the kth term in FS[f], yields the inequality

$$|A_k \cos kx + B_k \sin kx| \le (A_k^2 + B_k^2)^{1/2}, \qquad \text{for all } x \text{ in } \mathbf{R}$$

Thus if $M_k = (A_k^2 + B_k^2)^{1/2}$ and if $\sum_{k=1}^{\infty} M_k < \infty$, the Weierstrass M-test provides the desired result. The rest of the proof is aimed toward showing that

$\sum_{k=1}^{\infty} M_k < \infty$. Since both f and f' are in PC$[-\pi, \pi]$, we may write

$$FS[f] = \frac{A_0}{2} + \sum_{k=1}^{\infty}(A_k \cos kx + B_k \sin kx)$$

$$FS[f'] = \frac{A_0'}{2} + \sum_{k=1}^{\infty}(A_k' \cos kx + B_k' \sin kx)$$

where the coefficients A_k, B_k, A_k', B_k' are given by the Euler formulas (3). Integration by parts in the Euler formulas for A_k' and B_k' implies that

$$A_0' = 0, \quad A_k' = kB_k, \quad B_k' = -kA_k, \quad k = 1, 2, \ldots \tag{12}$$

Thus the Bessel Inequality applied to FS$[f']$ implies that $\sum k^2(A_k^2 + B_k^2) < \infty$. Using this and the Cauchy-Schwarz Inequality, we obtain that

$$\sum_{k=1}^{n} \frac{1}{k}[k^2 A_k^2 + k^2 B_k^2]^{1/2} \le \left[\sum_{k=1}^{\infty} \frac{1}{k^2}\right]^{1/2}\left[\sum_{k=1}^{\infty} k^2(A_k^2 + B_k^2)\right]^{1/2}$$

The series $\sum[A_k^2 + B_k^2]^{1/2}$ converges because its partial sums form a non-decreasing sequence bounded from above, and so we are done.

Decay Estimates

There is a precise estimate on how fast the coefficients in FS$[f]$ decay, but first a definition is needed. A function $A(k)$ is said to be $O(1/k)$ as $k \to \infty$ [A is *big O of* $1/k$] if there are positive constants c, K such that $|A(k)| \le c/k$ for all $k \ge K$.

THEOREM 10.3.7

Decay of Fourier Coefficients of Piecewise Smooth Functions. Let g be in PS$[-\pi, \pi]$. Then the coefficients in FS$[g]$ are $O(1/k)$, as $k \to \infty$. If $\tilde{g}$, the periodic extension of g, is not in $C^0(\mathbf{R})$, then this estimate cannot be improved.

Proof. Since g is in PS$[-\pi, \pi]$ we can partition $[-\pi, \pi]$ with points x_n, $-\pi = x_0 < x_1 < x_2 < \cdots < x_N < x_{N+1} = \pi$, such that g is continuously differentiable on each subinterval $[x_n, x_{n+1}]$. The coefficients A_k and B_k of FS$[g]$ are given by the Fourier-Euler formulas (3) [replacing f by g, and T by π, in those formulas]. Now we turn to FS$[g']$. We may integrate the coefficients of the sine terms in FS$[g']$ by parts as follows:

$$\frac{1}{\pi}\int_{-\pi}^{\pi} g'(x)\sin kx\, dx = \frac{1}{\pi}\sum_{n=0}^{N}\int_{x_n}^{x_{n+1}} g'(x)\sin kx\, dx$$

$$= \frac{1}{\pi}\sum_{n=0}^{N} g(x)\sin kx\Big|_{x_n^+}^{x_{n+1}^-} - \frac{k}{\pi}\sum_{n=0}^{N}\int_{x_n}^{x_{n+1}} g(x)\cos kx\, dx \tag{13}$$

$$= G - k\cdot\frac{1}{\pi}\int_{-\pi}^{\pi} g(x)\cos kx\, dx = G - kA_k$$

Since G is bounded and the coefficients of FS$[g']$ decay to zero as $k \to \infty$, we see that the coefficients A_k of the cosine terms in FS$[g]$ are $O(1/k)$, as $k \to \infty$.

A similar computation holds for the coefficients of the sine terms in FS[g]. We omit the proof that this decay rate cannot be improved.

Now let us assume that f in PC$[-\pi, \pi]$ is such that $\tilde{f}$ is in $C^m(\mathbf{R})$, and $f^{(m)}$ is in PS$[-\pi, \pi]$. Set

$$FS[f^{(j)}] = A_0^j + \sum_{k=1}^{\infty}(A_k^j \cos kx + B_k^j \sin kx), \qquad \text{for } j = 0, 1, \ldots, m+1$$

where j is a superscript, not a power. Using the definition of the coefficients A_k^j, B_k^j in (3) and integration by parts we have the recursion relation

$$A_k^{j+1} = kB_k^j, \qquad B_k^{j+1} = -kA_k^j, \qquad k > 0, \text{ and } j = 0, \ldots, m \qquad (14)$$

Now let us assume in addition just a bit more smoothness; namely, that $f^{(m+1)}$ is *piecewise smooth*. Taking $g = f^{(m+1)}$ we conclude from the previous result that the coefficients in FS[$f^{(m+1)}$] are $O(1/k)$, as $k \to \infty$. Putting this fact together with (14) produces the following estimate.

THEOREM 10.3.8

> Decay Theorem for Fourier Trigonometrical Coefficients. Let f belong to PC$[-\pi, \pi]$, $\tilde{f}$ belong to $C^m(\mathbf{R})$, and $f^{(m+1)}$ belong to PS$[-\pi, \pi]$ for some $m = 0, 1, 2, \ldots$ If A_k, B_k are the Fourier coefficients of f given by (3), then
>
> $$A_k = O(1/k^{m+2}), \qquad B_k = O(1/k^{m+2}), \qquad \text{as } k \to \infty \qquad (15)$$
>
> Also, if $\tilde{f}$ is *not* in $C^{m+1}(\mathbf{R})$, the estimates (15) cannot both be improved.

Proof. The decay estimates (15) follow from the formulas given in (14) because

$$A_k^{j+1} = kB_k^j = -k^2 A_k^{j-1} = \cdots = -(k)^{j+2}A_k, \qquad j \text{ odd}$$

Similar formulas can be obtained if j is even and also for B_k. We omit the proof that these estimates are best possible.

The Decay Theorems are a good check when calculating Fourier coefficients.

EXAMPLE 10.3.4

Decay of Fourier Coefficients for Function in PC$[-\pi, \pi]$
The "square wave" of Example 10.3.2 does not come under the Decay Theorem above since its periodic extension is not continuous. The square wave is, however, in PS$[-\pi, \pi]$, and from the first decay theorem (Theorem 10.3.7) we see that the Fourier coefficients decay precisely like $O(1/k)$, which we observe from (4) is the case.

PROBLEMS

1. Use trigonometric identities, common sense, or previous results to find the Fourier Series on $[-\pi, \pi]$ of each function below *without* using the Euler formulas (3).

(a) $5 - 4\cos 6x - 7\sin 3x - \sin x$

(b) $\cos^2 7x + (\sin \frac{1}{2}x)(\cos \frac{1}{2}x) - 2\sin x \cos 2x$

(c) $\sin^3 x$

2. Calculate FS[f] for the functions below, each assumed to have domain $[-\pi, \pi]$.

   (a) $2x - 2$ (b) x^2 (c) $a + bx + cx^2$

   (d) $\sin \pi x$ (e) $|x| + e^x$ (f) $|x(x^2 - 1)|$

   (g) *(Rectangular Pulse)*. For a positive number $B \le \pi$,

   $$f(x) = \begin{cases} A, & |x| < B \\ 0, & \pi > |x| > B \end{cases}$$

   (h) *(Alternating Pulse)*. For a positive number $B \le \pi$,

   $$f(x) = \begin{cases} 0, & -\pi \le x \le -B \quad \text{or} \quad B \le x \le \pi \\ A, & -B < x < 0 \\ -A, & 0 < x < B \end{cases}$$

3. Compute the Fourier Series of each of the following functions on the indicated interval. As you will see, the Fourier Series of a fixed function is highly dependent on the interval of definition of the function.

   (a) $f(x) = \sin x$, $-\pi \le x \le \pi$

   (b) $f(x) = \sin x$, $-\pi/2 \le x \le \pi/2$ [*Hint:* Use (2), (3) with $T = \pi/2$]

   (c) $f(x) = \sin x$, $-3\pi/2 \le x \le 3\pi/2$ [*Hint:* Use (2), (3) with $T = 3\pi/2$]

4. Let $f(x)$ be the function $|x|$ on $[-\pi, \pi]$.

   (a) Using FS[f] and Theorem 10.3.4 with $x_0 = 0$, show that

   $$\frac{\pi^2}{8} = \sum_{k=0}^{\infty} \frac{1}{(2k+1)^2}$$

   (b) Discuss the convergence properties of FS[f] on all of **R**.

5. *(Not All Trigonometric Series Are Fourier Series)*. Show that

   $$\sum_{1}^{\infty} \frac{1}{n^{1/4}} \cos nx, \qquad \sum_{1}^{\infty} (\sin n) \sin nx, \qquad \sum_{2}^{\infty} \frac{1}{\ln(n)} \sin nx$$

   are *not* the Fourier Series of any functions in PC$[-\pi, \pi]$. [*Hint:* Use the Parseval Theorem (Theorem 10.2.8) where Φ is given by (1) with $T = \pi$.]

6. Evaluate $\lim_{n \to \infty} \int_{-\pi}^{\pi} e^{\sin x}(x^5 - 7x + 1)^{52} \cos nx \, dx$. [*Hint:* Relate the integral to a Fourier coefficient of a portion of the integrand.]

7. *(Integrating a Periodic Function)*. Show that $\int_{-T}^{T} f(x) \, dx = \int_{a}^{a+2T} f(s) \, ds$ for all real numbers a if f is periodic of period $2T$ and piecewise continuous on every finite interval. [*Hint:* Break up the integral on the right and change the integration variable.]

8. *(Lagrange's Identity)*. Show that

   $$\frac{1}{2} + \sum_{k=1}^{n} \cos ks = \frac{\sin(n + \frac{1}{2})s}{2 \sin(s/2)}.$$

   [*Hint:* Use L'Hôpital's Rule if $s = 2m\pi$, m an integer. For all other s, use the identity $2 \sin(s/2) \cos ks = \sin[k + (1/2)]s - \sin[k - (1/2)]s$.]

9. (a) Prove that $\{1, \cos x, \dots, \cos nx, \dots\}$ is *not* a basis of PC$[-\pi, \pi]$.

   (b) Prove that $\{1, \cos x, \dots, \cos nx, \dots\}$ *is* a basis of the subspace of all even functions in PC$[-\pi, \pi]$.

   (c) Formulate and prove similar statements for $\{\sin x, \dots, \sin nx, \dots\}$.

10. Prove that $\|f - S_n\| \le \|f - S_m\|$ if $n \ge m$, where S_n and S_m are the corresponding partial sums of FS[f].

11. (*A Basis Theorem*). Let S be the linear space of functions f in PS$[-T, T]$ whose periodic extensions are in $C^0(\mathbf{R})$. Show that Φ_T as defined in (1) is a basis for S. [*Hint*: Use the Uniform Convergence Theorem 10.3.6.]

10.4 Half-Range and Exponential Fourier Series

Half-Range Expansions

There are some standard ways of constructing orthogonal sets (and even bases) for the Euclidean space PC[0, L] with the standard scalar product.

We begin by finding bases for PC[0, L] which consist only of sine functions or only of cosine functions. Let f in PC[0, L] be given and put

$$f_{\text{even}}(x) = \begin{cases} f(x), & 0 \le x \le L \\ \\ f(-x), & -L \le x \le 0 \end{cases}$$

$$f_{\text{odd}}(x) = \begin{cases} f(x), & 0 < x \le L \\ \\ -f(-x), & -L < x < 0 \end{cases}$$

Observe that f_{odd} is odd about $x = 0$ and f_{even} is even about $x = 0$, and that f_{odd} and f_{even} are in PC$[-L, L]$. The functions f_{odd} and f_{even} are called the *odd* and *even extensions*[6], respectively, of f to the interval $[-L, L]$. Observe that FS[f_{odd}] contains only sine functions and FS[f_{even}] contains only cosine functions. We shall define the *Fourier Sine Series* and *Fourier Cosine Series of f*, denoted by FSS[f] and FCS[f], as follows:

$$\text{FSS}[f] = \text{FS}[f_{\text{odd}}], \qquad \text{FCS}[f] = \text{FS}[f_{\text{even}}] \tag{1}$$

Thus, for example, for f in PC[0, L], we have that

$$\text{FSS}[f] = \sum_{k=1}^{\infty} b_k \sin \frac{k\pi x}{L}, \qquad b_k = \frac{2}{L} \int_0^L f(x) \sin \frac{k\pi x}{L} dx, \quad k = 1, 2, \dots \tag{2}$$

Our knowledge of the convergence properties of Fourier Series tells us much about Fourier Sine Series. In particular the functions

$$\Phi_s = \{\sin(k\pi x/L) : k = 1, 2, \dots\}$$

form an orthogonal set in PC[0, L] since for all $k \ne m$,

$$0 = \int_{-L}^{L} \sin \frac{k\pi x}{L} \sin \frac{m\pi x}{L} dx = 2 \int_0^L \sin \frac{k\pi x}{L} \sin \frac{m\pi x}{L} dx$$

It is not difficult to show that Φ_s is a basis of PC[0, L] since the Fourier Sine Series of any f in PC[0, L] converges in the mean to f. The convergence properties of

---

[6]Note that f_{odd} is not defined at $x = 0$. This will have no effect at all on what follows. In general, the actual value of a piecewise continuous function at a discontinuity is immaterial when calculating Fourier coefficients.

FSS[f] can be traced back to those criteria for the Fourier Series FS[f_{odd}]. Similarly, the Fourier Cosine Series of a function f in PC[0, L] is given by

$$FCS[f] = \frac{a_0}{2} + \sum_{k=1}^{\infty} a_k \cos \frac{k\pi x}{L}$$

$$a_k = \frac{2}{L} \int_0^L f(x) \cos \frac{k\pi x}{L} dx, \quad k = 0, 1, \ldots$$

(3)

Again, the convergence properties of FCS[f] are known from the behavior of FS[f_{even}]. In particular we may verify that the set

$$\Phi_c = \{\cos(k\pi x/L) : k = 0, 1, 2, \ldots\}$$

is an orthogonal set in PC[0, L] and is a basis for that space.

The series FSS[f] and FCS[f] in (2) for f in PC[0, L] are frequently referred to as *half-range expansions* in the applications.

EXAMPLE 10.4.1

Half-Range versus Full-Range Expansions

Consider the four functions f, g, f_{odd}, and f_{even} defined below:

$$f(x) = e^x, \quad 0 \le x \le \pi \qquad\qquad g(x) = e^x, \quad -\pi \le x \le \pi$$

$$f_{odd}(x) = \begin{cases} -e^{-x}, & -\pi \le x < 0 \\ e^x, & 0 < x \le \pi \end{cases} \qquad f_{even}(x) = \begin{cases} e^{-x}, & -\pi \le x \le 0 \\ e^x, & 0 \le x \le \pi \end{cases}$$

Observe that f_{odd} is the *odd*, and f_{even} is the *even*, *extension* of f to $(-\pi, \pi)$. See Figure 10.4.1 for a sketch of the graphs.

Notice that the four functions have the same values on $(0, \pi)$. Carrying out the necessary calculations to find the Fourier series of g, f_{odd}, and f_{even}, we obtain,

$$FS[g] = \frac{2\sinh\pi}{\pi} \sum_0^{\infty} (-1)^k \frac{\cos kx - k\sin kx}{1 + k^2}$$

$$FS[f_{odd}] = \frac{2}{\pi} \sum_1^{\infty} \frac{k[1 - (-1)^k e^{\pi}]}{1 + k^2} \sin kx$$

$$FS[f_{even}] = \frac{2}{\pi} \sum_0^{\infty} \frac{e^{\pi}[(-1)^k - 1]}{1 + k^2} \cos kx$$

Now from the definition of half-range expansions (1) above we know what FSS[f] and FCS[f] converge to on the entire real line. From the convergence theorem in the previous section we see that FCS[f] and FSS[f] both converge to $f(x) = e^x$ on $0 < x < \pi$, but on $-\pi < x < 0$, FCS[f] and FSS[f] converge to distinct functions, neither of which is e^x. Indeed FCS[f] converges to e^{-x} on $-\pi \le x \le 0$, while FSS[f] converges to $-e^{-x}$ on $-\pi < x < 0$. Looking at the periodic extensions $\tilde{f}_{odd}$ and $\tilde{f}_{even}$ we discover (again from the Pointwise Convergence Theorem 10.3.4) that

$$FSS[f](0) = FSS[f](\pi) = FSS[f](-\pi) = 0, \quad FCS[f](x) = f_{even}(x) \text{ on } |x| \le \pi$$

Fourier Exponential Series

If the functions in PC[$-\pi, \pi$] are complex-valued, we may use the orthogonal set

$$\Psi = \{e^{ikx} : k = 0, \pm 1, \pm 2, \ldots\}$$

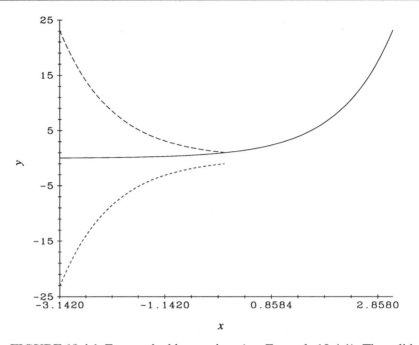

FIGURE 10.4.1 Even and odd extensions (see Example 10.4.1). The solid line is e^x. The long-dashed and the short-dashed lines are, respectively, the even and odd extensions of $f(x) = e^x$, $0 \le x \le \pi$.

Recall that the scalar product in $PC[-\pi, \pi]$ is $\langle f, g \rangle = \int_{-\pi}^{\pi} f\overline{g}\, dx$, where $\overline{g}$ denotes the complex conjugate of g. Note that $\|e^{ikx}\|^2 = \int_{-\pi}^{\pi} e^{ikx} e^{-ikx}\, dx = 2\pi$, for all k. The Fourier Series of f in $PC[-\pi, \pi]$ with respect to Ψ is the *Fourier Exponential Series*

$$\mathrm{FS}[f] = \sum_{n=-\infty}^{\infty} c_n e^{inx},$$

$$c_n = \frac{\langle f, e^{inx} \rangle}{\|e^{inx}\|^2} = \frac{1}{2\pi} \int_{-\pi}^{\pi} f(x) e^{-inx}\, dx, \qquad n = 0, \pm 1, \pm 2, \dots \tag{4}$$

Leaving aside the question of ordering, there are advantages to grouping the nth and the $-n$th terms together to obtain

$$\mathrm{FS}[f] = c_0 + \sum_{n=1}^{\infty} (c_{-n} e^{-inx} + c_n e^{inx}) \tag{5}$$

but it is customary to write $\mathrm{FS}[f]$ in the two-sided form (4).

Observe that if f in $PC[-\pi, \pi]$ happens to be real-valued, it still can be expanded in a Fourier Exponential Series. In that case we have

$$c_{-n} = \overline{c}_n \quad \text{and} \quad c_n = \frac{1}{2}(A_n - iB_n) \qquad \text{for all } n > 0 \tag{6}$$

where A_n and B_n are the respective Fourier cosine and sine coefficients in the real Fourier series of f. Hence the series in (5) reduces to the Fourier Trigonometric Series $\mathrm{FS}[f]$ given in the previous section. We have intentionally used the same symbol, $\mathrm{FS}[f]$, to denote both the real and complex forms of Fourier Series. There is little risk of confusion in this practice because of the special relation between the orthogonal sets Φ and Ψ used in these Fourier Series.

EXAMPLE 10.4.2

A Fourier Exponential Series

Let us find the exponential form of $FS[f]$ when $f(x) = x$, $|x| \le \pi$. Clearly, $c_0 = 0$, and for $n \ne 0$, $c_n = \frac{1}{2\pi} \int_{-\pi}^{\pi} x e^{-inx} dx = (i/n) \cos n\pi = i(-1)^n/n$. Thus

$$FS[f] = i \sum_{n=1}^{\infty} \left[\frac{(-1)^n}{n} e^{inx} - \frac{(-1)^n}{n} e^{-inx} \right]$$

Recalling that $e^{i\theta} = \cos\theta + i\sin\theta$, we easily see that also

$$FS[f] = 2 \sum_{n=1}^{\infty} \frac{(-1)^{n+1}}{n} \sin nx$$

which we could have obtained by expanding f directly into a Fourier Series over the trigonometrical set Φ.

EXAMPLE 10.4.3

Another Fourier Exponential Series

Let us compute $FS[f]$ for $f(x) = e^{i\omega x}$, $|x| \le \pi$, where ω is not an integer.

$$c_n = \frac{1}{2\pi} \int_{-\pi}^{\pi} e^{i\omega x} e^{-inx} dx = \frac{1}{2\pi} \frac{1}{i(\omega - n)} e^{i(\omega-n)x} \Big|_{-\pi}^{+\pi} = \frac{\sin(\omega - n)\pi}{(\omega - n)\pi}$$

Thus

$$FS[f] = \frac{\sin \omega\pi}{\omega\pi} + \sum_{n=1}^{\infty} \left[\frac{\sin(\omega - n)\pi}{(\omega - n)\pi} e^{inx} + \frac{\sin(\omega + n)\pi}{(\omega + n)\pi} e^{-inx} \right]$$

Using the properties of Fourier Series developed in Section 10.2 and 10.3, we can show that Ψ is a basis for the complex linear space $PC[-\pi, \pi]$, and that the Pointwise and Uniform Convergence Theorems hold for $FS[f]$ when f is complex-valued.

Application to an RLC Circuit

Consider the simple RLC circuit shown in the margin where R, L, and C are positive constants. As we saw in Section 4.3, the charge $q(t)$ on the capacitor satisfies the equation

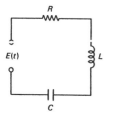

$$Lq''(t) + Rq'(t) + \frac{1}{C}q(t) = E(t) \tag{7}$$

Although we have already solved such ODEs for arbitrary piecewise continuous input voltages $E(t)$, the Fourier Series approach is more natural if $E(t)$ is periodic and we are searching for a periodic solution. Suppose, then, that $E(t)$ has period 2π and that the restriction of $E(t)$ to $[-\pi, \pi]$ is piecewise continuous.

We want to find all functions $q(t)$ of period 2π for which

$$q \text{ is in } C^1(\mathbf{R}), \qquad q'' \text{ is in } PC[-\pi, \pi] \tag{8a}$$

$$Lq'' + Rq' + \frac{1}{C}q - E = 0 \qquad \begin{array}{l} \text{on } [-\pi, \pi] \text{ except possibly} \\ \text{for a finite number of points} \end{array} \tag{8b}$$

Suppose that ODE (7) has a solution q meeting these requirements. First put

$$FS[q] = \sum_{k=-\infty}^{\infty} c_k e^{ikt} \quad \text{and} \quad FS[E] = \sum_{k=-\infty}^{\infty} b_k e^{ikt} \tag{9}$$

where the c_k are to be found, while the b_k are known. Using the smoothness conditions of (8a), (8b) and integration-by-parts, we have that

$$\text{FS}[q'] = \sum ikc_k e^{ikt}, \qquad \text{FS}[q''] = \sum (ik)^2 c_k e^{ikt} \qquad (10)$$

Expanding both sides of ODE (7) in Fourier Exponential Series and using (9), (10), we have

$$\sum [P(ik)c_k - b_k]e^{ikt} = 0, \qquad \text{where } P(x) = Lx^2 + Rx + \frac{1}{C} \qquad (11)$$

Since $\{e^{ikt} : k = 0, \pm1, \pm2, \ldots\}$ is a basis, it follows from (11) that $P(ik)c_k - b_k = 0$, $k = 0, \pm1, \ldots$. But $P(ik) \neq 0$, $k = 0, \pm1, \ldots$, and hence $c_k = b_k/P(ik)$, all k. Thus

$$\text{FS}[q] = \sum_{k=-\infty}^{\infty} \frac{1}{P(ik)} b_k e^{ikt} \qquad (12)$$

We omit the proof that the series in (12) converges uniformly to a function $q(t)$ satisfying (8a). By construction the Fourier Series of $Lq'' + Rq' + (1/c)q - E$ is the same as the Fourier Series of the zero function. Hence by the Totality Property (Theorem 10.2.7), we have that $Lq'' + Rq' + (1/C)q - E = 0$ on $[-\pi, \pi]$ except possibly at a finite number of points. Thus (8b) holds and we are done.

We have shown that the *RLC* circuit has a unique periodic response $q(t)$ to a periodic input voltage $E(t)$. The response $q(t)$ is known as a *forced oscillation*. Thus *every* solution of ODE (7) is the superposition of damped exponentials (the *transients*) and the steady-state periodic solution just constructed (see also Sections 1.2 and 7.6).

PROBLEMS

1. (*Fourier Sine and Cosine Series*). Find the Fourier Sine Series, FSS[f], of each of the following functions. Then find the Fourier Cosine Series, FCS[f].

   (a) $f(x) = 1$, $\quad 0 < x < \pi$　　(b) $f(x) = \sin x$, $\quad 0 \le x \le \pi$

2. (*Another Kind of Fourier Sine Series*). Let f be in PC[0, c], $f(x) = f(c - x)$ for all x, $c/2 < x < c$ (i.e., f is even about $x = c/2$).

   (a) Prove that FSS[f] contains only terms of the form $\sin[(2k+1)\pi x/c]$.

   (b) Reflect the function $\sin x$, $0 \le x \le \pi/4$, about $\pi/4$ to obtain a function f defined on $[0, \pi/2]$ which is even about $x = \pi/4$. Use part (a) to find FSS[f].

3. (*Another Kind of Fourier Cosine Series*). Let f be in PC[0, c], $f(x) = -f(c-x)$, $c/2 < x < c$ (i.e., f is "odd" with respect to $x = c/2$).

   (a) Prove that FCS[f] contains only terms of the form $\cos[(2k+1)\pi x/c]$.

   (b) Use part (a) to find FCS[f] for the function f defined by reflecting x^2, $0 \le x < 1$, "oddly" through $x = 1$.

4. (*Fourier Exponential Series*). Find the Fourier Exponential Series of each of the following functions defined on $[-\pi, \pi]$.

   (a) $\cos x$　　(b) x^2　　(c) $|x| + ix$

5. (*Electrical Circuit*). Find the steady-state charge in the capacitor in an *RLC* loop if $R = 10 \ \Omega$, $L = 0.5$ H, and $C = 10^{-4}$ F; let the impressed electromotive force $E(t)$, measured in volts, be described by the figure in the margin.

6. (*Driven Hooke's Law Spring*). In a spring-mass system, let the spring constant be 1.01 newtons/meter and the weight of the mass be 1 kilogram. Suppose that while in motion, the mass is acted on by viscous damping with coefficient equal to

0.2, and also suppose that the mass is driven by the force (measured in newtons) described in the figure. Find the steady-state motion of the system.

10.5 Sturm-Liouville Problems

The results below are important for the Method of Separation of Variables for solving boundary/initial value problems for PDEs. Orthogonality and orthogonal series play a key role in these results.

Let M be the operator defined on $C^2[a, b]$ by

$$M[y] = \frac{1}{\rho(x)} \left\{ \frac{d}{dx} \left[p(x) \frac{dy}{dx} \right] + q(x)y \right\} \tag{1}$$

where

$$\rho \text{ is in } C^0[a, b], \quad \rho \neq 0 \text{ on } (a, b), \quad q \text{ is in } C^0[a, b], \tag{2}$$
$$p \neq 0 \text{ on } (a, b), \quad p \text{ is in } C^1[a, b]$$

The operator M in (1) is not as special as it seems. Indeed, the general normal second-order linear differential operator $P(D) = a_2(x)D^2 + a_1(x)D + a_0(x)$, where $a_i(x)$ is in $C^0[a, b]$, $i = 0, 1, 2$, and $a_2(x) \neq 0$ on $[a, b]$, can be written in the form (1) if we let

$$p(x) = e^{\alpha(x)}, \quad q(x) = \frac{a_0(x)}{a_2(x)} e^{\alpha(x)}, \quad \rho(x) = \frac{e^{\alpha(x)}}{a_2(x)} \tag{3}$$

where $\alpha(x) = \int^x [a_1(x)/a_2(x)]\,dx$. Using the *weighted scalar product on $C^0[a, b]$ with weight factor ρ*, we have that $\langle f, g \rangle = \int_a^b \rho(x) f(x) \overline{g}(x)\,dx$. Then integration-by-parts yields the relation

$$\langle Mu, v \rangle = p(u'v - uv') \big|_a^b + \langle u, Mv \rangle \quad \text{for all } u, v \text{ in } C^2[a, b] \tag{4}$$

Now we define an operator L whose action is given by M in (1) and whose domain Dom (L) is a restriction of $C^2[a, b]$ defined by conditions at the endpoints $x = a$ and $x = b$. We will be interested in symmetric operators.

❖ **Symmetric Operators.** An operator L defined on a subspace of a Euclidean space E and taking values in E is said to be *symmetric* if

$$\langle Lu, v \rangle = \langle u, Lv \rangle, \quad \text{for all } u, v \text{ in Dom } (L).$$

Now let us look more carefully at boundary conditions appropriate for our differential operator M.

❖ **Boundary Conditions.** For u in $C^2[a, b]$, the conditions

$$\alpha u'(a) + \beta u(a) = 0, \quad \gamma u'(b) + \delta u(b) = 0 \tag{5}$$

are called *separated* (or *unmixed*) *boundary conditions* at $x = a$ and $x = b$, respectively, where the constants α, β, γ, δ are such that $\alpha^2 + \beta^2 \neq 0$ and $\gamma^2 + \delta^2 \neq 0$. The conditions

$$u(a) = u(b), \quad u'(a) = u'(b) \tag{6}$$

are referred to as *periodic boundary conditions*.

We have the following results whose proof follows from (4).

THEOREM 10.5.1

Symmetric Operator Theorem. Let L be an operator with action given by M in (1) where the domain Dom (L) is one of the two types:

- Dom $(L) = \{u$ in $C^2[a, b] : u$ satisfies a separated boundary condition at each endpoint where $p \neq 0\}$.

- Dom $(L) = \{u$ in $C^2[a, b] : u(a) = u(b), u'(a) = u'(b)$ and $p(a) = p(b)\}$.

Then L is a *symmetric operator* in $C^0[a, b]$ under the weighted scalar product $\langle f, g \rangle = \int_a^b \rho(x) f(x) \overline{g(x)} \, dx$.

Symmetric linear operators such as L in the theorem above are defined on a subspace Dom (L) of $C^0[a, b]$ and take values in $C^0[a, b]$. The eigenvalues and eigenspaces of such operators play an important role in applied mathematics. The problem of finding all the eigenvalues and the corresponding eigenspaces of such operators L is called a *Sturm-Liouville Problem*.[7] If conditions (2) on the action operator M are satisfied then the Sturm-Liouville problem for L is said to be *regular*. In general, we have the following definition for the eigen elements of any operator L:

❖ **Eigenvalues, Eigenspaces.** Let S be a subspace of a linear space V and let $L : S \rightarrow V$ be a linear operator. A scalar λ (a real or complex number according as V is a real or complex linear space) is an *eigenvalue* for L if there exists a nonzero v in S such that $Lv = \lambda v$. The vector v is an *eigenvector* of L corresponding to the eigenvalue λ. For any eigenvalue λ of L the set V_λ, of all vectors in V that satisfy the equation $Lv = \lambda v$ is the *eigenspace* of L corresponding to λ.

A similar definition for operators on $\mathbf{R}^n$ was given in Section 7.2. Since our symmetric operators act on functions in $C^2[a, b]$, their eigenvectors are often called *eigenfunctions*. Notice that for any eigenvalue λ, the eigenspace V_λ contains the eigenvectors for L corresponding to λ. (The zero vector can never be an eigenvector of any linear operator.) Thus V_λ is comprised of all the eigenvectors of L corresponding to λ with the zero vector thrown in. Two eigenspaces V_λ and V_μ of a linear operator with $\lambda \neq \mu$ can only have the zero vector in common.

Eigenspaces and eigenvalues for our symmetric differential operators have several important properties, which we list below.

THEOREM 10.5.2

Eigenspace Properties. Let L be any differential operator of the type defined in the Symmetric Operator Theorem. Then the following properties hold:

Orthogonality of Eigenspaces: Any two eigenspaces of L, V_λ, and V_μ, corresponding to two distinct eigenvalues λ and μ, must be orthogonal; that is, any element of V_λ is orthogonal to any element of V_μ under the scalar product $\langle f, g \rangle = \int_a^b \rho(x) f(x) \overline{g(x)} \, dx$.

Simplicity of Eigenspaces: If a separated boundary condition is used in defining Dom (L), then all eigenspaces of L are one-dimensional.

---

[7] Jacques Charles François Sturm (1803–1855) and Joseph Liouville (1809–1882) jointly introduced these ideas to resolve problems concerning the solutions of the partial differential equations of wave motion and thermal diffusion.

Proof. The orthogonality of eigenspaces follows immediately from the symmetry relation $\langle Lu, v \rangle = \langle u, Lv \rangle$ for all u, v in Dom (L). The only choices for the dimension of any eigenspace are one or two. Now if the dimension of the eigenspace corresponding to an eigenvalue λ is two, *all* solutions of the homogeneous equation $(p(x)y')' + q(x)y = \lambda\rho(x)y$ must satisfy the endpoint conditions. This is impossible if Dom (L) is defined with a separated condition at either $x = a$ or $x = b$. Thus the dimension of the eigenspace is one, as asserted.

THEOREM 10.5.3

> Nonpositivity of Eigenvalues. Let L be any differential operator of the type defined in the Symmetric Operator Theorem such that $q(x) \le 0$ on $[a, b]$, and, if either of the separated conditions (5) is used to define Dom (L), then $\alpha\beta \le 0$, and $\gamma\delta \ge 0$. Then the eigenvalues of L are all nonpositive.

Proof. Use integration-by-parts to show that $\langle Lu, u \rangle \le 0$, for all u in Dom (L). The assertion follows immediately from this inequality.

EXAMPLE 10.5.1

Sturm-Liouville Problem: Separated Conditions

Suppose that the operator L has action $Lu = u''$ and domain Dom $(L) = \{u$ in $C^2[0, T] : u(0) = u(T) = 0\}$. Then L is a symmetric differential operator with $a = 0$, $b = T$, $\rho(x) = p(x) = 1$, $q(x) = 0$ on $[0, T]$, and separated boundary conditions (5) with $\alpha = 0$, $\beta = 1$, $\gamma = 0$, $\delta = 1$. Thus all the eigenvalues of L are nonpositive, and all eigenspaces are one-dimensional and mutually orthogonal with the scalar product $\langle f, g \rangle = \int_0^T f(x)g(x)\,dx$. To find the eigenvalues and eigenspaces, first try $\lambda = 0$ in the eigenvalue equation $u'' = \lambda u$. The general solution of $u'' = 0$ is $u = Ax + B$ for arbitrary constants A, B. The condition $u(0) = u(T) = 0$ in this case yields that $A = B = 0$, and hence $\lambda = 0$ cannot be an eigenvalue of L. Now try $\lambda = -k^2$ for some positive constant k. The general solution of the eigenvalue equation $u'' = -k^2 u$ is $u = A\cos kx + B\sin kx$. The conditions $u(0) = u(T) = 0$ imply that $A = 0$ and $k = n\pi/T$. Thus L has the eigenvalues $\lambda_n = -(n\pi/T)^2$, with corresponding eigenspaces spanned by the respective eigenfunctions $u_n = \sin n\pi x/T$, $n = 1, 2, \ldots$.

EXAMPLE 10.5.2

Sturm-Liouville Problem: Periodic Boundary Conditions

As we saw in Section 10.2,

$$\Phi_T = \{1, \cos \pi x/T, \sin \pi x/T, \ldots, \cos n\pi x/T, \sin n\pi x/T, \ldots\}$$

is an orthogonal set in the Euclidean space $C^0[-\pi, \pi]$ under the scalar product $\langle f, g \rangle = \int_{-\pi}^{\pi} fg\,dx$. The elements of this set Φ_T turn up as eigenfunctions for the Sturm-Liouville problem with periodic boundary conditions:

$$y'' = \lambda y, \qquad y(-T) = y(T), \quad y'(-T) = y'(T) \tag{7}$$

Indeed, associated with (7) is the symmetric operator $Ly = y''$ with domain defined by Dom $(L) = \{y$ in $C^2[-T, T] : y(-T) = y(T), y'(-T) = y'(T)\}$. From Theorem 10.5.3 we see that the eigenvalues of L are nonpositive. Note that $\lambda = 0$ is an eigenvalue and the corresponding eigenspace V_0 is spanned by the constant function 1. The other eigenvalues are $\lambda_n = -(n\pi/T)^2$, $n = 1, 2, \ldots$; and each corresponding eigenspace V_n is two-dimensional and is spanned by the orthogonal set $\{\cos(n\pi x/T), \sin(n\pi x/T)\}$.

We end now with a fundamental property of Sturm-Liouville systems.

THEOREM 10.5.4

Sturm-Liouville Theorem (in the Regular Case). Let the operator L have action given by M in (1) and (2) and domain in $C^2[a, b]$ characterized by separated conditions (5) at both $x = a$ and $x = b$. Then

- The eigenvalues of L form a sequence λ_n, $n = 1, 2, \ldots$, with $|\lambda_n| \to \infty$.
- The corresponding eigenspaces V_n are one-dimensional and mutually orthogonal under the scalar product $\langle f, g \rangle = \int_a^b \rho f g \, dx$.
- If Φ is any set consisting of precisely one eigenfunction from each V_n, then Φ is a basis for PC$[a, b]$.
- The Fourier Series over Φ of any function u in Dom (L) converges uniformly to u on $[a, b]$.

EXAMPLE 10.5.3

Basis for Piecewise Continuous Functions
The Sturm-Liouville system in Example 10.5.1 is in the regular case. Thus the Sturm-Liouville Theorem 10.5.4 implies that the set $\Phi = \{\sin n\pi x/T : n = 1, 2, \ldots\}$ is a basis for PC$[0, T]$ under the standard scalar product. For example, the function $v = -(x - T/2)^2 + T^2/4$ is in the domain of the associated operator L. Thus, from Theorem 10.5.4, the Fourier series of v converges uniformly to v on $[0, T]$.

We treat singular (i.e., the operator L is not regular) Sturm-Liouville problems in Chapter 12, where they arise in separating the variables for PDEs in non-Cartesian coordinates.

PROBLEMS

1. In each of the regular Sturm-Liouville systems below identify the operator L associated with it, find the eigenvalues and eigenspaces of L, and state the orthogonality and basis properties of the eigenfunctions.

    (a) $y'' = \lambda y$; $y(0) = 0$, $y(\pi/2) = 0$
    (b) $y'' = \lambda y$; $y(0) = 0$, $y'(T) = 0$
    (c) $y'' = \lambda y$; $y'(0) = 0$, $y'(T) = 0$
    (d) $y'' = \lambda y$; $y(-T) = y(T)$, $y'(-T) = y'(T)$
    (e) $y'' = \lambda y$; $y(0) = 0$, $y(\pi) + y'(\pi) = 0$
    (f) $y'' - 4y' + 4y = \lambda y$; $y(0) = 0$, $y(\pi) = 0$

10.6 Separation of Variables and Eigenfunction Expansions

The Method of Separation of Variables used in Section 10.1 to construct a solution of the problem of the plucked guitar string has general application to a wide class of boundary/initial value problems for linear partial differential equations. We shall build on the theory and examples of earlier sections to outline this general method. We shall also generalize the method to solve boundary/initial value problems involving nonhomogeneous linear partial differential equations.

The Method of Separation of Variables

Although we shall work out the method only in the context of a specific problem, the method itself applies to many other problems (see Sections 12.3–12.5).

Consider the transverse motion of a taut, flexible string of length L introduced in Section 10.1. Assuming that the endpoints are held fixed, we saw that the deflection $u(x, t)$ of the string at the point x and time t is the unique solution of the boundary/initial value problem given below.

Plucked Guitar String Problem

For $G = \{(x, t) : 0 < x < L, t > 0\}$ we seek a function u which is both in $C^2(G)$ and in $C^1(\text{cl } G)$, and such that

$$
\begin{array}{lll}
\text{(PDE)} & u_{tt} - c^2 u_{xx} = 0 & \text{in } G \\
\text{(BC)} & u(0, t) = 0, \quad u(L, t) = 0, \quad t \geq 0 & \\
\text{(IC)} & u(x, 0) = f(x), \quad u_t(x, 0) = g(x), \quad 0 \leq x \leq L &
\end{array}
\tag{1}
$$

where c is a positive constant and the functions $f(x)$ and $g(x)$ represent the initial deflection and initial velocity, respectively, of the string. A solution of (1) that satisfies the stated smoothness conditions is called a *classical solution*.

We solve the boundary/initial value problem (1) by completing a series of steps.

I. Set Up Mathematical Model as Boundary/Initial Value Problem: We already did this in (1). If (PDE) is not homogeneous or if the boundary conditions are not homogeneous, the techniques at the end of this section must be applied.

II. Choose Independent Variables Appropriate to Shape of Domain: In general, solutions $u = u(x, y, z, t)$ of PDEs of physical interest are defined on regions of the form $G = H \times \{t \geq 0\}$, where H is a region in xyz-space. The Method of Separation of Variables will only work when, in the coordinates used to describe H, each part of the boundary of H is a set on which one of these coordinates is constant (i.e., a coordinate level set). In the case of problem (1), H is the interval $0 \leq x \leq L$ whose endpoints $x = 0$ and $x = L$ are coordinate level "sets" for the cartesian coordinate on the real line. If H were, for example, a circular disk, then polar coordinates should be used to express the space derivatives in the PDE (see Section 12.4).

III. Separate the Variables: Look for all separated solutions. Suppose that $u = X(x)T(t)$ satisfies (PDE). Substituting into (PDE), we have

$$
X(x)T''(t) - c^2 X''(x)T(t) = 0
\tag{2}
$$

or, after dividing through by $X(x)T(t)$,

$$
\frac{X''(x)}{X(x)} = \frac{1}{c^2} \frac{T''(t)}{T(t)} \quad \text{in } G
\tag{3}
$$

The name *Method of Separation of Variables* derives from the fact that equation (2) can be put into the variables-separated form (3). For $t_0 > 0$ where $T(t_0) \neq 0$ we have

$$
\frac{X''(x)}{X(x)} = \frac{1}{c^2} \frac{T''(t_0)}{T(t_0)}, \quad \text{for } 0 \leq x \leq L
$$

and so the ratio $X''(x)/X(x)$ is constant for $0 \leq x \leq L$; call the constant λ. Then, repeating the argument for some $0 < x_0 < L$ with $X(x_0) \neq 0$, we see that the ratio

$T''(t)/c^2T(t)$ is also equal to that same constant λ for $t \geq 0$. Thus there exists a *separation constant* λ such that

$$\frac{X''(x)}{X(x)} = \frac{1}{c^2}\frac{T''(t)}{T(t)} = \lambda \qquad (4)$$

for all (x, t) in cl G. Thus, the functions $X(x)$ and $T(t)$ that solve (2) must necessarily satisfy the pair of equations

$$\begin{aligned} X'' - \lambda X &= 0, \quad 0 < x < L \\ T'' - \lambda c^2 T &= 0, \quad t > 0 \end{aligned} \qquad (5)$$

for some constant λ. Conversely, any solution of the ODEs (5) for any real constant λ must be such that $X(x)T(t)$ is a solution of (PDE) in G. This separation argument is the same every time the Method of Separation of Variables is used.

IV. Set Up a Sturm-Liouville Problem: Now we require, in addition, that the solution $X(x)T(t)$ of (PDE) generated by (5) also satisy (BC). This amounts to a restriction on the choice of λ. For $X(x)T(t)$ to satisfy the (BC), we must have that

$$X(0)T(t) = 0, \quad X(L)T(t) = 0, \qquad \text{all } t \geq 0$$

Now if either $X(0) \neq 0$ or $X(L) \neq 0$, we would have that $T(t) = 0$ for all t, and hence $v(x, t) = X(x)T(t)$ would be the trivial solution. Thus, to ensure a nontrivial solution, we must take $X(0) = X(L) = 0$. But recall that $X(x)$ satisfies the ODE $X'' - \lambda X = 0$ on $0 < x < L$; and so we have the Sturm-Liouville Problem

$$X'' = \lambda X, \qquad X(0) = X(L) = 0 \qquad (6)$$

Recall from Section 10.5 that (6) is an eigenvalue problem for an operator with action d^2/dx^2 and domain $\{X \text{ in } C^2[0, L] : X(0) = X(L) = 0\}$. An eigenfunction of this operator [i.e., a solution of (6)] is often called a *normal mode*.

V. Solve the Sturm-Liouville Problem: We studied (6) in Section 10.5 and found that it has a nontrivial solution if and only if $\lambda = \lambda_n = -(n\pi/L)^2, n = 1, 2, \ldots$; the λ_n are the eigenvalues of the operator associated with the Sturm-Liouville Problem (6). The general solution of (6) corresponding to $\lambda = \lambda_n$ is given by

$$X_n(x) = C_n \sin\frac{n\pi x}{L}, \qquad n = 1, 2, \ldots \qquad (7)$$

where C_n is an arbitrary real constant. The X_n are the eigenfunctions of the operator associated with (6). Now inserting $\lambda = \lambda_n$ in the second equation in (5) gives the equation $T'' + (n\pi c/L)^2 T = 0, n = 1, 2, \ldots$, whose general solution is given by

$$T_n(t) = \alpha_n \cos\frac{n\pi ct}{L} + \beta_n \sin\frac{n\pi ct}{L}, \qquad n = 1, 2, \ldots$$

where the constants α_n and β_n are arbitrary real numbers for each n. Thus we have found all the standing waves $v_n(x, t) = X_n(x)T_n(t)$ supported by the string, that is, all separated solutions of (PDE) which satisfy (BC):

$$v_n(x, t) = \sin\frac{n\pi x}{L}\left(A_n \cos\frac{n\pi ct}{L} + B_n \sin\frac{n\pi ct}{L}\right), \qquad n = 1, 2, \ldots$$

where A_n and B_n are arbitrary constants. The numbers $\lambda_n = n\pi c/L$ are the *natural frequencies*.

VI. Construct the Formal Solution: We shall look for a solution of (1) as a linear combination of the standing waves $\{v_n\}$. That is, assume that the solution of (1) has the form

$$u(x, t) = \sum_{n=1}^{\infty} \sin\frac{n\pi x}{L}\left(A_n \cos\frac{n\pi ct}{L} + B_n \sin\frac{n\pi ct}{L}\right) \qquad (8)$$

for some choice of the constants A_n and B_n. Let us now restrict ourselves to only those choices for A_n and B_n for which all the series derived by termwise differentiation of (8) up to two times with respect to t and x converge uniformly in cl G. (It would be very difficult to determine in advance the conditions on A_n and B_n to ensure the uniform convergence of the series mentioned above, but fortunately as we shall see, there is no need to do so.) Thus for *any* such choice of the A_n and B_n, $u(x, t)$ satisfies the (PDE) and both boundary conditions. Hence it remains only to determine values for A_n and B_n in (8) such that $u(x, t)$ also satisfies the initial conditions. Now $u(x, t)$ will satisfy (IC) in problem (1) if for $0 \leq x \leq L$

$$f(x) = \sum_{n=1}^{\infty} A_n \sin \frac{n\pi x}{L} \qquad \text{[set } t = 0 \text{ in (8)]}$$

$$g(x) = \sum_{n=1}^{\infty} \frac{n\pi c}{L} B_n \sin \frac{n\pi x}{L}, \qquad \text{[differentiate (8) termwise, set } t = 0\text{]}$$

Recall from Example 10.5.3 that the set of functions $\Phi_S = \{\sin n\pi x/L : n = 1, 2, \ldots\}$ is a basis for PC$[0, L]$. Thus it follow that for $n = 1, 2, \ldots$,

$$A_n = \frac{\langle f, \sin(n\pi x/L)\rangle}{\|\sin(n\pi x/L)\|^2} = \frac{2}{L} \int_0^L f(x) \sin \frac{n\pi x}{L} \, dx$$

and

$$\frac{n\pi c}{L} B_n = \frac{\langle g, \sin(n\pi x/L)\rangle}{\|\sin(n\pi x/L)\|^2} = \frac{2}{L} \int_0^L g(x) \sin \frac{n\pi x}{L} \, dx$$

Hence the *formal solution* of (1) by the Method of Separation of Variables is

$$u(x, t) = \sum_{1}^{\infty} \sin \frac{n\pi x}{L} \left(A_n \cos \frac{n\pi c t}{L} + B_n \sin \frac{n\pi c t}{L} \right)$$

$$A_n = \frac{2}{L} \int_0^L f(x) \sin \frac{n\pi x}{L} \, dx, \qquad\qquad n = 1, 2, \ldots \qquad (9)$$

$$B_n = \frac{2}{n\pi c} \int_0^L g(x) \sin \frac{n\pi x}{L} \, dx, \qquad\qquad n = 1, 2, \ldots$$

where we have simply *assumed* that the series in (9) and its derived series up to second order converge uniformly. Although $u(x, t)$ obviously satisfies (BC), we need the stated uniform convergence properties to verify (PDE) and (IC).

VII. Determine If the Formal Solution is a Classical Solution: To verify that the formal solution $u(x, t)$ as defined by (9) is a *classical solution* [i.e., that u satisfies (PDE) and (IC) in (1) as well as (BC)], we must first show that u belongs to $C^2(G)$ and to $C^1(\text{cl } G)$. To do this, we must impose smoothness conditions on the initial data f and g. Let $\tilde{f}$ and $\tilde{g}$ denote the odd extensions of f and g into the interval $[-L, L]$, which are then extended periodically to **R**. We shall use these extensions in the proof of the following theorem.

THEOREM 10.6.1

> Classical Solution. Let f and g satisfy the conditions:
> (a) f is in $C^2[0, L]$, f''' is in PC$[0, L]$, $f(0) = f(L) = f''(0) = f''(L) = 0$;
> (b) g is in $C^1[0, L]$, g'' is in PC$[0, L]$, $g(0) = g(L) = 0$.
> Then $u(x, t)$ defined by (9) is a classical solution of problem (1).

Proof. We calculate the Fourier Sine Series of f and g on $[0, L]$:

$$\text{FSS}[f](x) = \sum_{n=1}^{\infty} \alpha_n \sin \frac{n\pi x}{L}, \qquad \text{FSS}[g](x) = \sum_{n=1}^{\infty} \beta_n \sin \frac{n\pi x}{L}$$

The hypotheses imply that $\tilde{f}$ lies in $C^2(\mathbf{R})$, $\tilde{g}$ in $C^1(\mathbf{R})$, while $\tilde{f}'''$ and $\tilde{g}''$ are in $PC(I)$ for every interval I. Apply the Decay of Coefficients Theorem 10.3.8 to $\text{FSS}[f]$ and $\text{FSS}[g]$ on $[0, L]$; we have that

$$\alpha_n = O(1/n^4), \quad \beta_n = O(1/n^3)$$

and so

$$\sum_{1}^{\infty} n^2 |\alpha_n| < \infty \quad \text{and} \quad \sum_{1}^{\infty} n|\beta_n| < \infty \tag{10}$$

Since $A_n = \alpha_n$ and $n\pi c B_n/L = \beta_n$ are Fourier Sine Series coefficients of f and g, respectively, we have from (10) that

$$\sum_{1}^{\infty} n^2 |A_n| < \infty, \quad \sum_{1}^{\infty} n^2 |B_n| < \infty \tag{11}$$

Using (11) and the Weierstrass M-test (Theorem 10.3.5), we see that the series obtained by differentiating the series in (9) term-by-term up to second order in x and t all converge uniformly on cl G. Thus $u(x, t)$ as defined by (9) belongs to $C^2(\text{cl } G)$ and all derivatives of u up to second order can be computed by term-by-term differentiation. Thus (9) defines a classical solution of problem (1).

Often it is enough to find a formal solution such as (9) without bothering to check that the initial data are smooth enough to produce a classical solution. Indeed, some of the most interesting behavior, at least for the wave equation, occurs precisely in problems where the initial data are not smooth but have "corners" which propagate into G. In this case the formal solution cannot possibly be a classical solution since at the interior points where $u(x, t)$ has "corners," u is not even differentiable.

The Method of Eigenfunction Expansions

The Method of Separation of Variables applies if the PDE and boundary conditions involved are homogeneous [as in (1)], but what do we do if this is not the case? We saw how to do this in Section 10.4 for the ODE modeling the charge on the capacitor in a driven electrical circuit. A solution technique was employed that involved the expansion of the driving force in an eigenfunction series of an associated Sturm-Liouville Problem. First we describe the process for an ODE, using our experience with the circuit problem as a guide. Then we adapt the method to PDEs.

Suppose that an ordinary differential operator has action given by

$$M[y] = \frac{1}{\rho(x)} \left\{ \frac{d}{dx} \left[p(x) \frac{dy}{dx} \right] + q(x)y \right\}$$

where the coefficients satisfy the conditions (2) of Section 10.5. The regular Sturm-Liouville problem with separated boundary conditions

$$\begin{cases} M[y] = \lambda y \\ B_a[y] = \alpha y(a) + \beta y'(a) = 0, \quad \alpha^2 + \beta^2 \neq 0 \\ B_b[y] = \gamma y(b) + \delta y'(b) = 0, \quad \gamma^2 + \delta^2 \neq 0 \end{cases} \tag{12}$$

has eigenvalues λ_i, $i = 1, 2, \ldots$, and a corresponding collection of eigenfunctions $\Phi = \{y_n(x)\}$ which is a basis of PC[a, b]. Now consider the boundary value problem

$$M[y] = f, \qquad B_a y = 0, \qquad B_b y = 0 \tag{13}$$

where f is a given function belonging to PC[a, b]. We look for a solution to this problem in the form $y(x) = \sum_{k=1}^{\infty} c_k y_k$. The function $y(x)$ always satisfies the boundary conditions for any choice of the constants c_k, so it remains just to choose the c_k such that $M[y] = f$ is satisfied (if we can). Expanding f in a Fourier series with respect to the basis $\Phi = \{y_k\}$ of eigenfunctions of the Sturm-Liouville System (12), we have $f = \sum a_k y_k$. Substitution into the ODE of (13) yields

$$M[y] = M \sum_{k=1}^{\infty} c_k y_k = \sum_{k=1}^{\infty} c_k M[y_k] = \sum_{k=1}^{\infty} c_k \lambda_k y_k = \sum_{k=1}^{\infty} a_k y_k$$

where we have *assumed* that we can interchange M and $\sum$. Now since $\Phi = \{y_n(x)\}$ is a basis for PC[a, b] it follows by matching coefficients with the same index that $c_j \lambda_j = a_j$, $j = 1, 2, \ldots$. If no $\lambda_j = 0$, we only need take $c_j = a_j/\lambda_j$, $j = 1, 2, \ldots$, and obtain the solution of (13) in the form

$$y(x) = \sum_{k=1}^{\infty} \frac{a_k}{\lambda_k} y_k \tag{14}$$

Nonhomogeneous Partial Differential Equations

Now we shall adapt the Method of Eigenfunction Expansions to boundary/initial value problems for nonhomogeneous PDEs. We begin with the following problem:

$$\begin{cases} u_{tt} - c^2 u_{xx} = F(x, t), & 0 < x < L, \quad t > 0 \\ u(0, t) = u(L, t) = 0, & t \geq 0 \\ u(x, 0) = f(x), & 0 \leq x \leq L \\ u_t(x, 0) = g(x), & 0 \leq x \leq L \end{cases} \tag{15}$$

where $F(x, t)$ is in $C^0(G)$, $G = \{(x, t) : 0 < x < L, t > 0\}$ and f, g belong to PC[0, L]. Now we know that the solution of the homogeneous version of (15) with $F = 0$ [i.e., problem (1)] is a superposition of functions from the basis $\{\sin(n\pi x/L)\}$ with time-varying coefficients. In the *Method of Eigenfunction Expansions* for (15) with F not identically zero, the solution $u(x, t)$ is expressed as a superposition $\sum U_n(t) \sin(n\pi x/L)$, for some coefficients $\{U_n(t)\}$. This coefficient set is determined by a technique of undetermined coefficients. We carry out the detailed calculations below.

All the data functions F, f, g may be expanded into Fourier Sine Series in x (assuming throughout that the functions are sufficiently smooth):

$$\text{FSS}[f](x) = \sum_{n=1}^{\infty} A_n \sin \frac{n\pi x}{L}, \qquad A_n = \frac{2}{L} \int_0^L f(x) \sin \frac{n\pi x}{L} \, dx \tag{16a}$$

$$\text{FSS}[g](x) = \sum_{n=1}^{\infty} B_n \sin \frac{n\pi x}{L}, \qquad B_n = \frac{2}{L} \int_0^L g(x) \sin \frac{n\pi x}{L} \, dx \tag{16b}$$

$$\text{FSS}[F](x, t) = \sum_{n=1}^{\infty} C_n(t) \sin \frac{n\pi x}{L}, \qquad C_n(t) = \frac{2}{L} \int_0^L F(x, t) \sin \frac{n\pi x}{L} \, dx \tag{16c}$$

Now expand the solution $u(x, t)$ of (15) in a Fourier Sine Series in x:

$$\text{FSS}[u](x, t) = \sum_{n=1}^{\infty} U_n(t) \sin \frac{n\pi x}{L}, \qquad (x, t) \text{ in } G \qquad (17)$$

where $U_n(t) = (2/L) \int_0^L u(x, t) \sin(n\pi x/L)\, dx$. Proceeding formally, we assume that $u(x, t)$ is the sum of its Fourier Sine Series $\text{FSS}[u](x, t)$, insert (17) into the partial differential equation of (15), and use (16) to calculate the set of coefficient functions $\{U_n(t)\}$. We have [using Eq. (16c)]

$$\sum_{n=1}^{\infty} \left\{ U_n''(t) + c^2 \left(\frac{n\pi}{L}\right)^2 U_n(t) - C_n(t) \right\} \sin \frac{n\pi x}{L} = 0, \qquad (x, t) \text{ in } G$$

Since $\Phi = \{\sin n\pi x/L : n = 1, 2, \ldots\}$ is a basis for PC$[0, L]$ we have that

$$U_n''(t) + \left(\frac{n\pi c}{L}\right)^2 U_n(t) = C_n(t), \qquad t > 0, \quad n = 1, 2, \ldots \qquad (18)$$

Using (16a), (16b) to determine initial conditions for each $U_n(t)$, we have that

$$u(x, 0) = f(x) = \sum_{n=1}^{\infty} U_n(0) \sin \frac{n\pi x}{L} = \sum_{n=1}^{\infty} A_n \sin \frac{n\pi x}{L}, \quad 0 \le x \le L$$

$$u_t(x, 0) = g(x) = \sum_{n=1}^{\infty} U_n'(0) \sin \frac{n\pi x}{L} = \sum_{n=1}^{\infty} B_n \sin \frac{n\pi x}{L}, \quad 0 \le x \le L$$

Thus the initial conditions on U_n are

$$U_n(0) = A_n, \qquad U_n'(0) = B_n \qquad (19)$$

Using the Variation of Parameters Method of Section 3.6 to solve the IVP (18), (19), we see that

$$U_n(t) = A_n \cos \frac{n\pi c t}{L} + \frac{LB_n}{\pi n c} \sin \frac{n\pi c t}{L} + \frac{L}{\pi n c} \int_0^t \sin\left[\frac{n\pi c}{L}(t - s)\right] C_n(s)\, ds \qquad (20)$$

Thus the *formal solution* of (15) is given by (17) and (20), with A_n, B_n, and C_n given by (16a), (16b), and (16c). Observe that this solution can be written as

$$u(x, t) = u_1(x, t) + u_2(x, t) \qquad (21)$$

where

$$u_1(x, t) = \sum_{n=1}^{\infty} \sin \frac{n\pi x}{L} \left(A_n \cos \frac{n\pi c t}{L} + \frac{LB_n}{\pi n c} \sin \frac{n\pi c t}{L} \right)$$

$$u_2(x, t) = \sum_{n=1}^{\infty} \sin \left[\frac{n\pi x}{L}\right] \left\{ \frac{L}{\pi n c} \int_0^t \sin\left[\frac{n\pi c}{L}(t - s)\right] C_n(s)\, ds \right\} \qquad (22)$$

Thus $u_1(x, t)$ represents the response of the string to the initial conditions, while $u_2(x, t)$ is the response to the external force $F(x, t)$.

    Remark. If the "driving force" $F(x, t)$ is periodic of a frequency near a "natural frequency," one might expect resonance to occur. This is indeed the case. See Problem 7 for a related problem.

Shifting the Data

The Method of Separation of Variables handles initial conditions together with homogeneous boundary conditions and a homogeneous linear PDE. The Method of Eigenfunction Expansions extends the solution process to a nonhomogeneous PDE. Now we show how to reduce a problem where *all* the conditions are nonhomogeneous to one or the other of the problems above. As always, the discussion is in terms of the wave equation on a finite interval.

Consider the vibrating string moving under the influence of a vertically acting external force $F(x, t)$, subject to the usual initial conditions and with time-varying boundary conditions:

$$\begin{align}
&\text{(PDE)} && u_{tt} - c^2 u_{xx} = F(x, t), && 0 < x < L, \quad t > 0 \\
&\text{(BC)} && u(0, t) = \alpha(t), && u(L, t) = \beta(t), \quad t \geq 0 \\
&\text{(IC)} && u(x, 0) = f(x), && u_t(x, 0) = g(x), \quad 0 \leq x \leq L
\end{align} \quad (23)$$

Observe that the function $v(x, t) = \alpha(t) + (x/L)(\beta(t) - \alpha(t))$ satisfies the boundary conditions $v(0, t) = \alpha(t)$, $v(L, t) = \beta(t)$. Let $w(x, t)$ be a solution of the problem

$$\begin{align}
&w_{tt} - c^2 w_{xx} = F(x, t) - (v_{tt} - c^2 v_{xx}) \\
&w(0, t) = w(L, t) = 0 \\
&w(x, 0) = f(x) - v(x, 0) \\
&w_t(x, 0) = g(x) - v_t(x, 0)
\end{align} \quad (24)$$

We have that $u(x, t) = w(x, t) + v(x, t)$ is a solution of (23).

What we have done is to introduce the *boundary function* $v(x, t)$, which satisfies both boundary conditions of (23) but none of the other conditions. The effect of letting $w = u - v$ is to introduce a boundary/initial value problem for w in which the boundary data have been *shifted* away from the endpoints $x = 0$ and $x = L$ and attached in altered form to the right-hand sides of the PDE and the initial conditions. But a problem such as (24) for w can be solved by the Method of Eigenfunction Expansions outlined earlier in this section.

In many special cases of practical interest, nonzero data can be "shifted" in such a way as to make the modified problem have a homogeneous PDE *and* homogeneous boundary conditions (but at the expense of altered initial data), which can then be treated directly by the Method of Separation of Variables. For illustrations of this technique, see the problem set.

Comments

The Method of Separation of Variables and the Method of Eigenfunction Expansions are based on the same idea—that of expanding functions in a series of eigenfunctions of an appropriate Sturm-Liouville Problem. The Sturm-Liouville Problem itself is inherent in the geometry and the partial differential and boundary operators of the boundary/initial value problem being solved.

Different operators and different regions lead to different Sturm-Liouville Problems, to different eigenfunctions and eigenvalues, and to different bases of different spaces. The possibilities seem unlimited. In Sections 12.3–12.5 we explore some other types of Sturm-Liouville Problems.

PROBLEMS

1. Use the Method of Separation of Variables to solve (1) under the following conditions.

(a) $f(x) = 0, \quad g(x) = 3\sin(\pi/L)x, \quad 0 \leq x \leq L$

(b) $f(x) = g(x) = \begin{cases} x, & 0 \leq x \leq L/2 \\ L - x, & L/2 \leq x \leq L \end{cases}$

(c) $f(x) = x(L - x) = -g(x), \quad 0 \leq x \leq L$

2. **(a)** Construct the Sturm-Liouville Problem associated with the boundary/initial value problem $u_{tt} - c^2 u_{xx} = 0$, $0 < x < L$, $t > 0$; $u(0, t) = 0$, $u_x(L, t) = -hu(L, t)$, $t \geq 0$, h a positive constant; $u(x, 0) = f(x)$, $u_t(x, 0) = g(x)$, $0 \leq x \leq L$.

   (b) Show that the eigenvalues of the Sturm-Liouville operator of part **(a)** are $\lambda_n = -s_n^2/L^2$, $n = 1, 2, \ldots$, where s_n is the nth consecutive positive zero of the equation $s + hL \tan s = 0$. Show that there are infinitely many zeros, s_n, but do not evaluate them. Find a corresponding basis of eigenfunctions for PC[0, L].

3. (*Damped Wave Equation*). The equation $u_{tt} + b^2 u_t - a^2 u_{xx} = 0$, $0 < x < L$, $t > 0$, models a vibrating string, taking into account air resistance. Find the formal solution $u(x, t)$ of the boundary problem of the damped wave equation:

$$u_{tt} + b^2 u_t - a^2 u_{xx} = 0, \qquad\qquad 0 < x < L, \quad t > 0$$
$$u(0, t) = u(L, t) = 0, \quad t \geq 0$$
$$u(x, 0) = f(x), \qquad\qquad u_t(x, 0) = 0, \quad 0 \leq x \leq L$$

   where $b^2 < 2\pi a/L$ and a are positive constants and f and g belong to PC[0, L].

4. (*Shifting Data*). The boundary/initial value problem

$$u_{tt} - c^2 u_{xx} = g, \qquad\qquad 0 < x < L, \quad t > 0$$
$$u(0, t) = u(L, t) = 0, \quad t \geq 0$$
$$u(x, 0) = u_t(x, 0) = 0, \quad 0 \leq x \leq L$$

   models the vertical displacement $u(x, t)$ of a taut flexible string tied at both ends, with vanishing initial data, and acted on by gravity (g is the constant gravitational acceleration). The outline below shows how to shift the nonhomogeneity in the partial differential equation onto an initial condition.

   (a) Find a function $v(x)$ such that $-c^2 v_{xx} = g$, $0 < x < L$, $t > 0$; $v(0) = v(L) = 0$, $t > 0$. [*Hint*: Try $v(x) = Ax + Bx^2$, where A, B are constants.] Note that $v(x)$ is the steady-state sag of the string under the force of gravity.

   (b) Let $w = u - v$ and show that w satisfies the same equations as u, but with g replaced by 0 and the condition $u(x, 0) = 0$ replaced by $w(x, 0) = -gx(L - x)/2c^2$.

   (c) Find $u(x, t)$.

5. (*Shifting Boundary Data Onto Initial Data*). A string of unit length with $c^2 = 1$ clamped at one end, driven by $\sin \pi t/2$ at the other end, and given an initial velocity, is modeled by $u_{tt} - u_{xx} = 0$, $0 < x < 1$, $t > 0$; $u(0, t) = \sin \pi t/2$, $u(1, t) = 0$, $t \geq 0$; $u(x, 0) = f(x)$, $u_t(x, 0) = g(x)$, $0 \leq x \leq 1$. The following steps show how to shift the boundary data $\sin \pi t/2$ onto an initial condition.

   (a) Let v be any solution of $v_{tt} - v_{xx} = 0$, $v(0, t) = \sin \pi t/2$, $v(1, t) = 0$, $t \geq 0$. Let $w = u - v$. Show that w is a solution of the same problem as u except the boundary condition that $u(0, t) = \sin \pi t/2$ is replaced by $w(0, t) = 0$, and the data $f(x)$ and $g(x)$ are replaced by $f(x) - v(x, 0)$, and $g(x) - v_t(x, 0)$.

   (b) Find $v(x, t)$ in the form $X(x)T(t)$. [*Hint*: Let $v = X(x)\sin(\pi t/2)$.]

   (c) Solve the problem if $g(x) = 0$ for all x.

6. Solve the following problems.

   (a) Use the Method of Eigenfunction Expansions to solve

$$u_{tt} - u_{xx} = 6x, \qquad\qquad 0 < x < 1, \quad t > 0$$
$$u(0, t) = u(1, t) = 0, \quad t \geq 0$$
$$u(x, 0) = u_t(x, 0) = 0, \quad 0 \leq x \leq 1$$

(b) First solve the problem $v_{tt} - v_{xx} = 0$, $v(0, t) = \sin(3\pi t/2L)$, $v(L, t) = 0$. Then use v to shift the boundary data onto the initial conditions and solve the following problem:

$$u_{tt} - u_{xx} = 0, \qquad\qquad\qquad 0 < x < L, \quad t > 0$$

$$u(0, t) = \sin\frac{3\pi t}{2L}, \quad u(L, t) = 0, \quad t \geq 0$$

$$u(x, 0) = u_t(x, 0) = 0, \qquad\qquad 0 \leq x \leq L$$

7. (*Breaking a String by Shaking It*). Consider the problem of a string of length L with one end fastened and the other driven:

$$u_{tt} - c^2 u_{xx} = 0, \quad 0 < x < L, \quad t > 0$$

$$u(0, t) = 0, \quad u(L, t) = \mu(t), \quad t \geq 0$$

$$u(x, 0) = 0, \quad u_t(x, 0) = 0, \quad 0 \leq x \leq L$$

Show that there is a periodic function $\mu(t)$ such that the string will eventually break. [*Hint*: What if $\mu(t) = A \cos \omega t$ for ω near a natural frequency $n\pi c/L$?]

8. Use the Method of Eigenfunction Expansions to solve the problem

$$u_{tt} - c^2 u_{xx} = F(x, t), \qquad\qquad 0 < x < L, \quad t > 0$$

$$u(0, t) = u_x(L, t) = 0, \qquad\qquad t \geq 0$$

$$u(x, 0) = f(x), u_t(x, 0) = g(x), \quad 0 \leq x \leq L$$

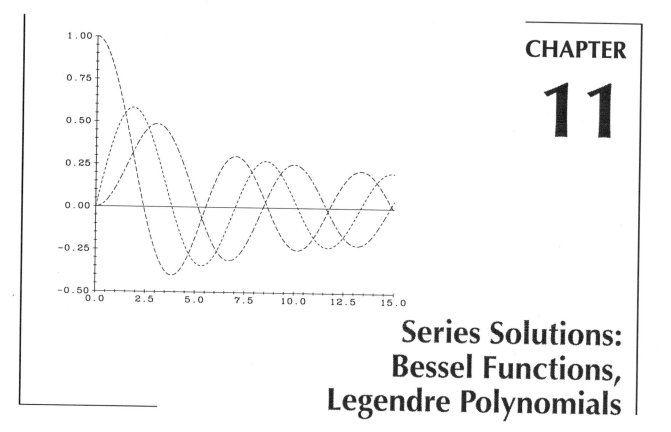

CHAPTER

11

Series Solutions: Bessel Functions, Legendre Polynomials

Linear ordinary differential equations that model physical phenomena or that arise in separating variables in a partial differential equation often have nonconstant coefficients, and the techniques of Chapters 1– 5 for finding solutions in closed form may not apply. If the coefficient functions are polynomials or convergent power series, however, the solutions may be written as infinite series that are closely related to power series. Many of the most important functions of mathematics, science, and engineering have been introduced this way, i.e., as series solutions of linear second-order ordinary differential equations with nonconstant coefficients. In this chapter, we shall show how these series solutions are constructed, and in the process introduce Bessel functions, Legendre polynomials, and other special functions. The cover figure shows the graphs of three of the Bessel functions (see Section 11.6).

11.1 Introduction: Aging Springs and Steady Temperatures

The differential equations considered in this chapter have the form

$$a_2(x)y''(x) + a_1(x)y'(x) + a_0(x)y(x) = 0 \tag{1}$$

We assume throughout that the coefficient functions a_2, a_1, and a_0 are either polynomials or convergent power series. Examples 11.1.1 and 11.1.2 indicate some of the sources that lead to equations like ODE (1).

EXAMPLE 11.1.1

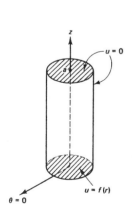

Aging Springs

The elastic coefficient of a spring will gradually diminish with advancing time until the spring stretches like chewing gum and, finally, snaps. Springs are the core elements in openers, closers, dampers, and controllers, and their failure may have disastrous consequences. For this reason good models for the motion of a spring are of considerable importance. We may begin with the dynamical model of the motion of a body of mass m suspended by an "ageless" Hooke's law spring (Section 3.1)

$$my''(t) + ky(t) = 0$$

where the spring constant $k > 0$ does not depend on time, and the displacement $y(t)$ is measured from the spring's unstretched (and uncompressed) state. Gravity is ignored. The solutions of this ODE are sinusoids that oscillate with period $2\pi(m/k)^{1/2}$. Now replace the elastic constant k by a decaying function of time, $ke^{-\varepsilon t}$, $\varepsilon > 0$, that models the gradual loss of elasticity:

$$my''(t) + ke^{-\varepsilon t}y(t) = 0 \qquad (2)$$

What are solutions $y(t)$ of ODE (2) like? We might expect $y(t)$ to oscillate as long as the elastic coefficient $ke^{-\varepsilon t}$ remains significant. Eventually, however, the elastic restoring force will be so weak that the energy of motion will overwhelm it, and $y(t)$ will decrease without bound until the spring breaks or the model is no longer valid.

All of this may be made precise if we knew more about the properties of solutions of ODE (2) or had solution formulas. Although ODE (2) is a linear equation, the solution methods of Chapters 3, 4, and 5 do not directly apply, because the coefficients are not constant. In Section 11.2 we shall expand $ke^{-\varepsilon t}$ in a convergent Taylor series and then use a method of undetermined coefficients to find solutions of ODE (2) as power series

$$y(t) = a_0 + a_1 t + a_2 t^2 + \cdots + a_n t^n + \cdots$$

where $a_0 = y(0)$, $a_1 = y'(0)$. In Section 11.7 we return to the problem of the aging spring and express solutions of ODE (2) in terms of Bessel functions and derive both oscillatory properties and the ultimate destruction of the aging spring.

The following example is quite different.

EXAMPLE 11.1.2

Steady Temperatures in a Cylinder

A solid iron cylinder of radius 1 and height a is enclosed in a blanket of ice that keeps the top and the sidewall at temperature zero. A heat source is fixed to the bottom and keeps the temperatures there at levels that depend only on the radial distance from the cylinder's center line. The problem is to find the steady-state temperature at each point inside the cylinder.

The cylinder is given in cylindrical coordinates by $0 \leq r \leq 1$, $-\pi \leq \theta < \pi$, $0 \leq z \leq a$. If the temperature at the point (r, θ, z) is denoted by $u(r, \theta, z)$, then the corresponding boundary problem (see Section 10.6) is

$$u_{rr} + \frac{1}{r}u_r + \frac{1}{r^2}u_{\theta\theta} + u_{zz} = 0, \qquad 0 < r < 1, \quad 0 < z < a, \quad -\pi < \theta < \pi$$
$$u(1, \theta, z) = 0, \qquad 0 \leq z \leq a, \quad -\pi \leq \theta < \pi$$
$$u(1, \theta, a) = 0, \qquad 0 \leq r \leq 1, \quad -\pi \leq \theta < \pi \qquad (3)$$
$$u(r, \theta, 0) = f(r), \qquad 0 \leq r \leq 1, \quad -\pi \leq \theta < \pi$$

Since the boundary data do not depend on θ we suspect that the solution u of the boundary value problem (3) is a function of r and z only, $u = u(r, z)$.

According to the separation of variables technique described in Chapter 10, we look for solutions of $\nabla^2 u = 0$ having the form $u = R(r)Z(z)$. Inserting this form of u into $\nabla^2 u = 0$ and recognizing that $u_{\theta\theta} = 0$, we must have that

$$R''(r)Z(z) + \frac{1}{r}R'(r)Z(z) + R(r)Z''(z) = 0$$

After dividing by RZ, separating the variables, and introducing the separation constant $-\lambda$, we see that

$$\frac{1}{R}R'' + \frac{1}{rR}R' = -\frac{Z''}{Z} = -\lambda$$

In particular, we have that $R'' + R'/r = -\lambda R$, i.e.,

$$rR'' + R' + \lambda rR = 0 \tag{4}$$

This equation has the form of ODE (1), and its solutions may be written as power series

$$R(r) = b_0 + b_1 r + b_2 r^2 + \cdots + b_n r^n + \cdots$$

See Example 11.7.3 for details and Section 12.5 for a full treatment of boundary value problem (3).

Series Approach to Solving ODE (1)

Examples 11.1.1 and 11.1.2 show how second-order linear ordinary differential equations with nonconstant coefficients may arise in modeling physical phenomena. The following example uses a simpler problem, with constant coefficients to show how solutions may be constructed in the form of series. See Appendix C.2 for a short review of power series.

EXAMPLE 11.1.3

Power Series Approach to Solving ODEs

The initial value problem (IVP)

$$y''(x) + y(x) = 0, \qquad y(0) = 1, \quad y'(0) = 0 \tag{5}$$

has the unique solution $y(x) = \cos x$ (see Section 3.4). Let us see how we can use the differential equation itself to construct a power series solution. Because the solution of (5) is unique, the power series must converge to $\cos x$.

Suppose that IVP (5) has a solution in the form of a power series

$$y = a_0 + a_1 x + a_2 x^2 + \cdots + a_n x^n + \cdots = \sum_{n=0}^{\infty} a_n x^n$$

that converges on an interval containing $x = 0$. The problem is to determine the coefficients a_n. The first two coefficients a_0 and a_1 are determined by the initial data:

$$\begin{aligned} 1 = y(0) = a_0 + a_1 \cdot 0 + a_2 \cdot 0^2 + \cdots = a_0 \\ 0 = y'(0) = a_1 + 2a_2 \cdot 0 + 3a_3 \cdot 0^2 + \cdots = a_1 \end{aligned} \tag{6}$$

where we have used the fact that a power series may be differentiated term-by-term inside its convergence interval. Differentiating $y = \sum_{n=0}^{\infty} a_n x^n$ within its convergence interval, we have that

$$y' = \sum_{n=0}^{\infty} na_n x^{n-1} = \sum_{n=1}^{\infty} na_n x^{n-1}$$

and

$$y'' = \sum_{n=1}^{\infty} n(n-1)a_n x^{n-2} = \sum_{n=2}^{\infty} n(n-1)a_n x^{n-2}$$

where terms with a coefficient of zero have been dropped from the sums.

Inserting these sums into the ODE of IVP (5) we have

$$y'' + y = \sum_{n=2}^{\infty} n(n-1)a_n x^{n-2} + \sum_{n=0}^{\infty} a_n x^n = 0$$

Reindexing the first summation by lowering the index range by 2 and (to compensate) increasing each n by 2 inside the summation:

$$y'' + y = \sum_{n=0}^{\infty} (n+2)(n+1)a_{n+2} x^n + \sum_{n=0}^{\infty} a_n x^n$$

$$= \sum_{n=0}^{\infty} [(n+2)(n+1)a_{n+2} + a_n]x^n = 0$$

The insertion of the final (and trivial) summation $\sum_{n=0}^{\infty} 0x^n = 0$ allows us to use the Identity Theorem of Appendix C.2 and conclude that

$$(n+2)(n+1)a_{n+2} + a_n = 0, \quad n = 0, 1, 2, \ldots$$

Thus we have a *recursion formula* for the coefficients

$$a_{n+2} = -\frac{a_n}{(n+2)(n+1)}, \quad n = 0, 1, 2, \ldots \quad (7)$$

Using (7) and $a_0 = 1$ (from (6)) we have that

$$a_2 = -\frac{a_0}{2 \cdot 1} = -\frac{1}{2}, \quad a_4 = -\frac{a_2}{4 \cdot 3} = \frac{1}{24}, \quad \cdots$$

$$a_{2k} = -\frac{a_{2k-2}}{2k(2k-1)} = \frac{a_{2k-4}}{2k(2k-1)(2k-2)(2k-3)} = \cdots = \frac{(-1)^k}{(2k)!}a_0 = \frac{(-1)^k}{(2k)!}$$

for $k = 1, 2, \ldots$. Thus, all of the coefficients with even subscripts have been determined. For the coefficients with odd subscripts we use (7) and $a_1 = 0$:

$$a_3 = -\frac{a_1}{3 \cdot 2} = 0, \quad a_5 = -\frac{a_3}{5 \cdot 4} = 0, \quad \cdots, \quad a_{2k+1} = 0, \quad \cdots$$

for $k = 1, 2, \ldots$, and all of these coefficients are zero.

Hence, we have the series solution of (5)

$$y(x) = 1 - \frac{1}{2}x^2 + \frac{1}{24}x^4 - \frac{1}{720}x^6 + \cdots + \frac{(-1)^k}{(2k)!}x^{2k} + \cdots = \sum_{k=0}^{\infty} \frac{(-1)^k}{(2k)!}x^{2k} \quad (8)$$

and this is the Taylor series of $\cos x$ about the point $x_0 = 0$.

The Ratio Test may be applied to determine the convergence interval:

$$\lim_{k \to \infty} \left| \frac{(-1)^{k+1}x^{2k+2}/(2k+2)!}{(-1)^k x^{2k}/(2k)!} \right| = \lim_{k \to \infty} \frac{x^2}{(2k+2)(2k+1)} = 0$$

for all x. Hence the series converges for all x, and as noted above, the series converges to $\cos x$. Other solutions of $y'' + y = 0$ can be obtained as series in the same way.

The key steps in this *Method of Undetermined Coefficients* are as follows:

- Insert the "solution" $y = \sum_{n=0}^{\infty} a_n x^n$ with undetermined coefficients a_n into the differential equation.
- Reindex each summation as needed so that all have a common index range, for example, $\sum_{n=0}^{\infty}$.
- Write the differential equation in the series form

$$\sum_{n=0}^{\infty} [\text{terms involving various } a_n] x^n = 0$$

- Note that $a_0 = y(0)$ and $a_1 = y'(0)$.
- Use the Identity Theorem to equate the bracketed term $[\cdots]$ to 0; this equation gives the recursion formula for generating a_n, $n \geq 2$, from a_0 and a_1.

In specific problems, the details may differ somewhat from those described above, but the basic process is always the same.

The series approach outlined in Example 11.1.3 may be used in cases where (unlike that example) there are no known closed-form solution formulas. Solutions in series form have several advantages; for example, they often may be used to estimate the values of a solution, as we now show.

EXAMPLE 11.1.4

How Good is a Series Approximation?

The alternating series (8) represents the solution of Initial Value Problem (5). Suppose that we want to use the sum of the first three terms, $w(x) = 1 - x^2/2 + x^4/24$, of the alternating series to approximate the solution $y(x)$ over the interval $|x| \leq 0.1$. On that interval, each term of the alternating series is smaller in magnitude than the preceding term, and the limiting value as $k \to \infty$ of the magnitude of the kth term, $x^{2k}/(2k)!$, is zero for each fixed value of x. Thus, all of the conditions of the Alternating Series Test (Appendix C.2) are met, and we have that

$$|y(x) - w(x)| \leq \left| \frac{-x^6}{720} \right| \leq 1.4 \cdot 10^{-9}, \qquad \text{for } |x| \leq 0.1$$

where $-x^6/720$ is the next term in the series after the three terms of $w(x)$. The biquadratic $w(x)$ is indeed a good approximation to $y(x)$ if $|x| \leq 0.1$.

Because of the convergence properties of power series, we would expect the sum of the first several terms in the series to be a useful approximation near the base point x_0 of the expansion ($x_0 = 0$ in Examples 11.1.3 and 11.1.4), and less useful at some distance from that point. See Figure 11.1.1 for a graph of the solution $y(x)$ of IVP (5) and overlays of some of the polynomial approximations,

$$y_{(2k)}(x) = 1 - \frac{1}{2}x^2 + \frac{1}{24}x^4 + \cdots + \frac{(-1)^{2k}}{(2k)!} x^{2k}$$

As k increases, the curves "follow" the exact solution over a longer x-interval before diverging from it. Observe that k must be quite large to obtain a good approximation on the full interval $0 \leq x \leq 8$.

Another reason that power series solutions are important is that computers can be used to approximate solutions of differential equations. Most digital computations are based on the four arithmetical operations, and a polynomial approximation like $w(x)$ or $y_{(2k)}(x)$ is well suited for such calculations. Power series techniques are some of the most widely used methods for constructing solutions of linear ODEs.

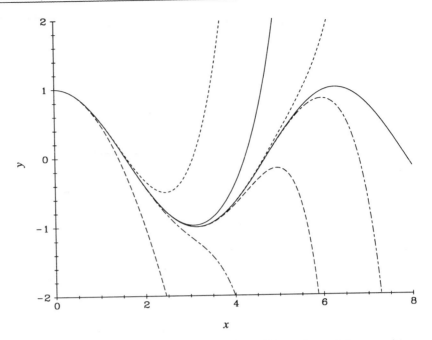

FIGURE 11.1.1 The solution of IVP (5) and its polynomial approximations, $y_{2k}(x)$, $k = 1, \ldots, 7$.

PROBLEMS

1. Determine the interval of convergence for each of the following series by using the Ratio Test. [*Hint*: The Ratio Test is stated in Appendix C.2.]

(a) $\displaystyle\sum_{n=0}^{\infty} \frac{x^n}{n!}$ (b) $\displaystyle\sum_{n=0}^{\infty} \frac{nx^n}{2^n}$ (c) $\displaystyle\sum_{n=1}^{\infty} \frac{x^{2n+1}}{(2n+1)!}$

2. Use the formula for a Taylor series to expand each function about the given point, and determine the interval of convergence. Over the given interval plot the graphs of the function and its polynomial approximations that use, respectively, the first 2, 3, and 4 nonzero terms of the series (use overlays if possible). [*Hint*: See Appendix C.2 for a brief review of Taylor series.]

(a) e^x, $x_0 = 0$, $|x| \le 2$ (b) $\sin x$, $x_0 = 0$, $|x| \le \pi$

(c) $(1 + x)^{-1}$, $x_0 = 0$, $|x| < 1$ (d) $\sqrt{x}$, $x_0 = 1$, $|x - 1| < 1$

3. Reindex each of the following as a series of form $\sum[\cdots]x^n$

(a) $\displaystyle\sum_{n=0}^{\infty} \frac{2(n+1)}{n!} x^{n+1}$ (b) $\displaystyle\sum_{n=2}^{\infty} n(n-1)a_n x^{n-2}$ (c) $\displaystyle\sum_{n=1}^{\infty} (-1)^{n-1} \frac{x^{n+1}}{n(n+1)}$

4. Use the Taylor series of e^x, $\sin x$, $\cos x$, and $1/(1 + x)$ about the base point $x_0 = 0$ to obtain the Taylor series expansions of each of the following functions.

(a) $e^x + e^{-x}$ (b) $\sin x - \cos x$

(c) $\sin x \cos x$ (d) $e^x/(1 + x)$

5. Use the Identity Theorem to determine a recursion formula for a_n in terms of n and a_k, $k < n$. Then "solve" each recursion formula to express each a_n in terms of n, and a_0 or a_1.

    (a) $\sum_{n=1}^{\infty}(na_n - n + 2)x^n = 0$ (b) $\sum_{n=1}^{\infty}[(n+1)a_n - a_{n-1}]x^n = 0$

    (c) $\sum_{n=0}^{\infty}[(n+2)(n+1)a_{n+2} - a_n]x^n = 0$

6. Show that $y = \sum_0^{\infty} x^n/n!$ is the solution of $y'' - y = 0$, $y(0) = y'(0) = 1$ by direct substitution. How is this solution expressed in terms of elementary functions?

7. (a) Use series to solve the problem $y'' - 4y = 0$, $y(0) = 2$, $y'(0) = 2$. Express the solution in terms of exponentials, and describe the connection between this form of the solution and the series.

    (b) Plot the solution and the graphs of the polynomial approximations that use the first 2, 3, and 4 nonzero terms of the series over the interval $|x| \le 1$.

8. Use the Identity Theorem to solve the initial value problem $y'' - y = \sum_{n=0}^{\infty} x^n$, $y(0) = 0$, $y'(0) = 0$, where $|x| < 1$. [*Hint*: Try $y = \sum_2^{\infty} a_n x^n$.]

9. Find the coefficient a_{100} in the series $\sum_{n=0}^{\infty} a_n x^n$ if it is known that $\sum_{n=0}^{\infty}[(n+1)^2 a_{n+2} - n^2 a_{n+1} + (n-1)a_n]x^n = 0$ and that $a_0 = a_1 = 1$.

10. Find a series solution of ODE (4) with $\lambda = 1$ such that $R(0) = 1$, $R'(0) = 0$. Find a recursion formula and the first four nonzero terms of the series expansion.

11. (*Aging Spring*). Create a model of an aging spring for which the elastic coefficient is zero after a finite time. Include the following in the analysis of your model. Ignore gravity.

    - What would you expect to happen to the spring after that time?

    - Compare your expectation with the actual solutions of a model equation with a zero elastic coefficient.

    - How can the damping effect of friction be included in the model of an aging spring?

    - Plot graphs of solutions of each model created, and explain the behavior displayed with the graph.

11.2 Series Solutions Near an Ordinary Point

So far we have only found solution formulas for second-order linear ODEs with constant coefficients. If the coefficients are not constant but functions of the independent variable, then in many cases a method that uses power series can still be applied to construct solutions. As with Example 11.1.3, a solution in the form of a power series is assumed, and the coefficients of the series are determined after substituting the series into the ODE.

Dividing by the leading coefficient $a_2(x)$, the differential equation $a_2(x)y'' + a(x)y' + a_0(x)y = 0$ can be written in the *normalized form*,

$$y'' + P(x)y' + Q(x)y = 0 \qquad (1)$$

which is a more convenient form for expressing the results of this section. In this section we present a method for finding power series solutions of the form $\sum a_n(x - x_0)^n$ for ODE (1) when the coefficients $P(x)$ and $Q(x)$ can be written as convergent power series in powers of $x - x_0$, i.e., when P and Q are real analytic at x_0 (see Appendix C.2). At the end of this section we use the method to find a series solution for the equation of the aging spring.

Ordinary Points, Singular Points

We shall be interested in finding power series solutions of ODE (1) in powers of $x - x_0$, where x_0 is an *ordinary point* of the ODE. An ordinary point is defined as follows:

❖ **Ordinary Point.** The point x_0 is an *ordinary point* of ODE (1) if both $P(x)$ and $Q(x)$ are real analytic at x_0.

The opposite of "ordinary" is "singular":

❖ **Singular Point.** x_0 is a *singular point* of ODE (1) if $P(x)$ or $Q(x)$ is not real analytic at x_0.

If x_0 is an ordinary point, $P(x)$ and $Q(x)$ have Taylor series

$$P(x) = \sum_0^\infty p_n(x - x_0)^n, \qquad Q(x) = \sum_0^\infty q_n(x - x_0)^n \tag{2}$$

which converge inside a common interval centered at x_0. The coefficients p_n and q_n of the Taylor series (2) are

$$p_n = \frac{P^{(n)}(x_0)}{n!}, \qquad q_n = \frac{Q^{(n)}(x_0)}{n!}$$

The identification of the ordinary points of (1) is the first step in constructing series solutions.

EXAMPLE 11.2.1

Ordinary Points
Every point x_0 is an ordinary point for the constant coefficient equation $y'' + p_0 y' + q_0 y = 0$. Every point x_0 is also an ordinary point if $P(x)$ and $Q(x)$ are polynomials in x. For instance, the polynomial $P(x) = 1 - x^2$ can be written as a finite Taylor series about x_0:

$$P(x) = P(x_0) + P'(x_0)(x - x_0) + \frac{1}{2}P''(x_0)(x - x_0)^2$$

$$= (1 - x_0^2) - 2x_0(x - x_0) - (x - x_0)^2$$

EXAMPLE 11.2.2

More Ordinary Points
Every point x_0 is an ordinary point of $y'' + e^{-x}y = 0$ since $Q(x) = e^{-x}$ has a convergent Taylor series in powers of $x - x_0$ given by

$$e^{-x} = \sum_{n=0}^\infty \frac{1}{n!}(-1)^n e^{-x_0}(x - x_0)^n$$

a series that converges to e^{-x} for all x.

EXAMPLE 11.2.3

An Ordinary Point and a Singular Point

The point $x_0 = 0$ is an ordinary point of $y'' + [1/(1 - x)]y = 0$ since $1/(1 - x)$ has a convergent geometric series expansion,

$$\frac{1}{1 - x} = \sum_{n=0}^{\infty} x^n, \quad \text{where} \quad |x| < 1$$

However, $x = 1$ is a singular point since $(1 - x)^{-1}$ is not defined at $x = 1$.

EXAMPLE 11.2.4

Legendre's Equation

The equation $(1 - x^2)y'' - 2xy' + p(p + 1)y = 0$, where p is a nonnegative constant, is called *Legendre's equation of order p*. In normalized form (1) the coefficients are

$$P(x) = -2x(1 - x^2)^{-1}, \qquad Q(x) = p(p + 1)(1 - x^2)^{-1}$$

The equation has two singularities, $x = 1$ and $x = -1$; all other points are ordinary points.

EXAMPLE 11.2.5

A Singular Point

The equation $y'' + |x|y = 0$ has a singularity at 0 because $|x|$ is not differentiable at $x_0 = 0$; hence $|x|$ has no Taylor series expansion about $x_0 = 0$. All other points x_0 are ordinary because $|x|$ is real analytic at x_0 in any interval centered at x_0 and not containing 0.

Sometimes an apparent singularity can be converted to an ordinary point. For example, the ODE $y'' + [(x - 1)/(x^2 - 1)]y = 0$, has an apparent singularity at $x = 1$, but this is easily removed by factoring $x^2 - 1$ and canceling the common factor $x - 1$ to obtain $y'' + y/(x + 1) = 0$. Thus, $x = 1$ is a *removable singularity* (actually, an ordinary point), for this equation. Of course, $x = -1$ is a singularity however the differential equation is written and, thus, is not a removable singularity. Removable singularites are treated as ordinary points.

Series Solutions Near an Ordinary Point

Let x_0 be an ordinary point of the ODE $y'' + P(x)y' + Q(x)y = 0$. Theorem 3.3.7 guarantees that the solution space of the ODE (1) in a neighborhood of x_0 is two-dimensional and is spanned by any pair of independent solutions. We shall actually show that all solutions of the ODE near x_0 have the form of power series, $\sum_0^{\infty} a_n(x - x_0)^n$, and hence are real analytic at x_0. For the moment we assume that solutions of this form do exist and find necessary conditions on the coefficients a_n which must be satisfied, as in the following example.

EXAMPLE 11.2.6

Hermite's Equation

The second-order ODE

$$y'' - 2xy' + 2y = 0 \tag{3}$$

is known as *Hermite's equation of order 2*. In this ODE we see that $P(x) = -2x$ and $Q(x) = 2$. Evidently, all points x_0 are ordinary and we look for series solutions of the form $\sum a_n(x - x_0)^n$. If initial data $y(x_0) = y_0$, $y'(x_0) = y'_0$, are given, then note that $a_0 = y_0$ and $a_1 = y'_0$. For simplicity we set $x_0 = 0$ and find solutions of the form

$y = \sum_{n=0}^{\infty} a_n x^n$. We must find the undetermined coefficients a_n, $n = 2, 3, 4, \dots$ in terms of a_0 and a_1. Differentiating the series term-by-term, we have that inside the interval of convergence,

$$y' = \sum_{n=1}^{\infty} n a_n x^{n-1} \quad \text{and} \quad y'' = \sum_{n=2}^{\infty} n(n-1) a_n x^{n-2}$$

Hence,

$$\sum_{n=2}^{\infty} n(n-1) a_n x^{n-2} - 2x \sum_{n=1}^{\infty} n a_n x^{n-1} + 2 \sum_{n=0}^{\infty} a_n x^n = 0 \tag{4}$$

Reindexing the first two sums, we obtain that

$$\sum_{n=2}^{\infty} n(n-1) a_n x^{n-2} = \sum_{n=0}^{\infty} (n+2)(n+1) a_{n+2} x^n$$

and that

$$-2x \sum_{n=1}^{\infty} n a_n x^{n-1} = \sum_{n=1}^{\infty} 2(-n) a_n x^n = \sum_{n=0}^{\infty} 2(-n) a_n x^n$$

All of the series in (4) can be combined into a single sum as

$$\sum_{n=0}^{\infty} [(n+2)(n+1) a_{n+2} - 2n a_n + 2 a_n] x^n = 0 \tag{5}$$

By the Identity Theorem,

$$(n+2)(n+1) a_{n+2} - 2n a_n + 2 a_n = 0, \qquad n = 0, 1, 2, \dots$$

and so we have the *recursion formula* for the coefficients:

$$a_{n+2} = \frac{2(n-1) a_n}{(n+2)(n+1)}, \qquad n = 0, 1, 2, \dots \tag{6}$$

Given a_0 and a_1, we may use the recursion formula (6) to determine all the other coefficients a_n. For example,

$$a_2 = \frac{-2a_0}{2 \cdot 1} = -a_0, \quad a_4 = \frac{2a_2}{4 \cdot 3} = -\frac{1}{6} a_0, \quad a_6 = \frac{6a_4}{6 \cdot 5} = -\frac{1}{30} a_0,$$

$$a_3 = \frac{0 a_1}{3 \cdot 2} = 0, \quad a_5 = \frac{4a_3}{5 \cdot 4} = 0, \dots, \quad a_{2k+1} = 0, \quad n \geq 1$$

The vanishing of all but the first of the odd-indexed coefficients is a fortunate occurrence here, but such simplicity is not common. Solutions of ODE (3) may be written as

$$y = a_0 \left[1 - x^2 - \frac{1}{6} x^4 - \frac{1}{30} x^6 - \cdots \right] + a_1 x \tag{7}$$

where a_0 and a_1 are arbitrary constants. In particular, if we let $a_0 = 1$, $a_1 = 0$ and then $a_0 = 0$, $a_1 = 1$, we have two solutions of ODE (3):

$$y_1 = 1 - x^2 - \frac{1}{6} x^4 - \frac{1}{30} x^6 - \cdots, \qquad y_2 = x \tag{8}$$

Observe that $y_1(x)$ and $y_2(x)$ are independent solutions of (3) because their Wronskian $W[y_1, y_2](0) \neq 0$ (see Section 3.3). Thus, every solution of ODE (3) is a linear combination of the solutions $y_1(x)$ and $y_2(x)$.

There remain two unresolved questions about the solutions of ODE (3) of the form $\sum_{n=0}^{\infty} a_n x^n$, where the a_n are determined by the recursion formula (6). What is the interval of convergence of the series solution? What is the specific value of the nth coefficient a_n? Both questions may be answered quite simply.

EXAMPLE 11.2.7

Interval of Convergence: Continuation of Example 11.2.6
We find the interval of convergence of the series solution which has the form $\sum_{k=0}^{\infty} a_{2k} x^{2k}$, where

$$a_{2k+2} = \frac{2(2k-1)a_{2k}}{(2k+2)(2k+1)}, \qquad k = 0, 1, \ldots \qquad (9)$$

which we obtain from the recursion formula (6) by setting $n = 2k$. Applying the Ratio Test, we have that

$$\lim_{k \to \infty} \left| \frac{a_{2k+2} x^{2k+2}}{a_{2k} x^{2k}} \right| = x^2 \lim_{k \to \infty} \left| \frac{2(2k-1)}{(2k+2)(2k+1)} \right| = 0, \quad \text{all } x$$

Thus the series solution $y = \sum_{k=0}^{\infty} a_{2k} x^{2k} + a_1 x$, where a_0 and a_1 are arbitrary constants and a_{2k} is determined by the recursion formula (6), converges for all x. The differentiation of this solution term-by-term is now justified since the series converges everywhere, and can be seen to define a solution of ODE (3).

Occasionally, it is feasible to "solve" a recursion formula for the coefficients of the power series expansion of a solution.

EXAMPLE 11.2.8

Finding the Series Solution: Solving the Recursion Identity
We have from (9) that

$$a_{2k+2} = \frac{2(2k-1)a_{2k}}{(2k+2)(2k+1)} = \frac{2(2k-1)}{(2k+2)(2k+1)} \frac{2(2k-3)a_{2k-2}}{2k(2k-1)} = \cdots$$

$$= \frac{2^{k+1}(2k-1)(2k-3)\cdots 5 \cdot 3 \cdot 1 \cdot (-1)a_0}{(2k+2)(2k+1)\cdots 4 \cdot 3 \cdot 2 \cdot 1}$$

$$= -\frac{2^{k+1}(2k-1)(2k-3)\cdots 5 \cdot 3 \cdot 1}{(2k+2)!} a_0$$

$$= -\frac{2(2k)!}{k!(2k+2)!} a_0 = -\frac{2a_0}{k!(2k+2)(2k+1)}, \quad k = 0, 1, 2, \ldots$$

since

$$2^k(2k-1)(2k-3)\cdots 3 \cdot 1 = \frac{(2k)(2k-1)(2k-2)(2k-3)\cdots 3 \cdot 2 \cdot 1}{k(k-1)\cdots 4 \cdot 2}$$

$$= \frac{(2k)!}{k!}$$

Hence the general solution of Hermite's ODE (3) may be written as

$$y = a_0 \left\{ 1 - \sum_{k=0}^{\infty} \frac{2}{k!(2k+2)(2k+1)} x^{2k+2} \right\} + a_1 x \qquad (10)$$

Although we have solved the recursion formula in this case, it is not always useful to do so, even when easy. The "unsolved" recursion formula (6) is already in good form for a computer program to evaluate as many coefficients as desired. The computer must be provided with a_0 and a_1, the formula itself, and a range for n (e.g., $n = 2$ to

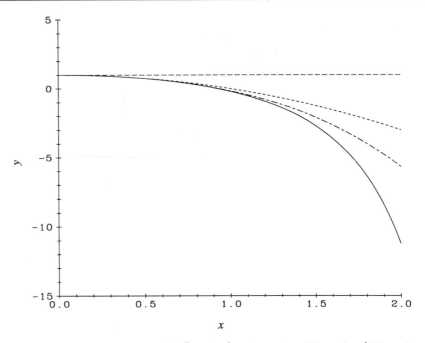

FIGURE 11.2.1 Solution of $y'' - 2xy' + 2y = 0$, $y(0) = 1$, $y'(0) = 0$ (solid line), and three of its approximating polynomials (dashed). See Example 11.2.6.

100). The computer will do all the work simply by iterating (6) for successive even values of n, beginning with $n = 0$.

Each value of a_0 and a_1 in (10) yields a new function, a function defined by the series. Each of these functions is real analytic for all x; moreover, the functions are easy to differentiate or integrate—just differentiate or integrate the series term-by-term. The first few terms of the series can be used as a polynomial approximation to the solution. Figure 11.2.1 shows the computed solution (solid curve) of (3) with $y(0) = 1$, $y'(0) = 0$ and the polynomial approximations, $y_{(0)} = 1$ (long dashes), $y_{(2)} = 1 - x^2$ (short dashes), and $y_{(4)} = 1 - x^2 - x^4/6$ (long/short dashes). The approximations are quite good near $x_0 = 0$, but not so good at a distance.

Convergence

We have carried out the details of the example above to show how the Method of Undetermined Coefficients works in practice. Of course, all the work would be in vain if the constructed series failed to converge to a solution, or perhaps did not converge at all. The following result assures that the method always succeeds. It is stated for the initial value problem

$$y'' + P(x)y' + Q(x)y = 0 \qquad y(0) = a_0, \quad y'(0) = a_1 \qquad (11)$$

THEOREM 11.2.1

Convergence Theorem. Let $P(x) = \sum_{n=0}^{\infty} p_n x^n$, $Q(x) = \sum_{n=0}^{\infty} q_n x^n$, where the series converge on a common interval $J = (-\rho, \rho)$, $\rho > 0$. Then IVP (11) has the unique solution $y = \sum_{n=0}^{\infty} a_n x^n$, where the coefficients a_n, $n \geq 2$, are determined by the *recursion formula*,

$$(n+2)(n+1)a_{n+2} + \sum_{k=0}^{n} [p_{n-k}a_{k+1}(k+1) + q_{n-k}a_k] = 0 \qquad (12)$$

The series $\sum_{n=0}^{\infty} a_n x^n$ is real analytic on J.

The proof of the convergence theorem is omitted. Observe that the recursion formula expresses a_{n+2} in terms of a_k, $k < n + 2$, and thus (12) may be used to find all coefficients once a_0 and a_1 are given. In practice, (12) is rarely used for a specific equation. It is much easier to insert the unknown series $\sum a_n x^n$ and its respective derivative series into the differential equation, group the coefficients of like powers of x as we did above, and derive the recursion formula directly.

Finally, observe that if $y_1(x)$ and $y_2(x)$ are solutions of (11) which, respectively, satisfy the initial data $y_1(0) = 1$, $y_1'(0) = 0$, and $y_2(0) = 0$, $y_2'(0) = 1$, then all solutions of $y'' + Py' + Qy = 0$ are linear combinations of $y_1(x)$ and $y_2(x)$. This follows because the Wronskian $W[y_1, y_2](0) = y_1(0)y_2'(0) - y_1'(0)y_2(0) = 1 \neq 0$; hence $\{y_1, y_2\}$ is a basis of the solution space of $y'' + Py' + Qy = 0$. Observe that the solution of IVP (11) is $y = a_0 y_1(x) + a_1 y_2(x)$.

Aging Springs

We will apply the technique of power series to find solutions of the equation of an aging spring and some approximate solution formulas.

EXAMPLE 11.2.9

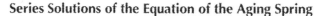

Series Solutions of the Equation of the Aging Spring
An ODE for the vibrations of an undamped but aging spring (Section 11.1) is $my'' + ke^{-\varepsilon t}y = 0$, where m, k, and ε are positive constants. Since the exponential coefficient is real analytic on the entire t-axis, we may replace that coefficient by its Taylor series. The equation of motion meets all conditions of the Convergence Theorem with, say, $t_0 = 0$ as the ordinary point about which solutions will be expanded in power series. Suppose that $k/m = 1$ and that the spring is stretched 1 unit from its equilibrium position and then released:

$$y''(t) + e^{-\varepsilon t}y(t) = y''(t) + \left[\sum_0^{\infty} \frac{(-\varepsilon t)^n}{n!}\right]y(t) = 0 \qquad (13)$$
$$y(0) = 1, \qquad y'(0) = 0$$

The solution of IVP (13) is a power series, $\sum_0^{\infty} a_n t^n$, which converges for all t. Thus

$$\sum_{n=2}^{\infty} n(n-1)a_n t^{n-2} + \left[\sum_{n=0}^{\infty} \frac{(-\varepsilon)^n}{n!}t^n\right]\left[\sum_{n=0}^{\infty} a_n t^n\right] = 0$$

Reindexing the first series and applying the product formula (see Appendix C.2) to the second and third yields

$$\sum_{n=0}^{\infty}(n+2)(n+1)a_{n+2}t^n + \sum_{n=0}^{\infty}\left[\sum_{k=0}^{n}\frac{(-\varepsilon)^k}{k!}a_{n-k}\right]t^n = 0 \tag{14}$$

From this equation and the Identity Theorem, we have that

$$a_{n+2} = -\frac{1}{(n+2)(n+1)}\sum_{k=0}^{n}\frac{(-\varepsilon^k)}{k!}a_{n-k}$$

This recursion formula cannot easily be solved for a_{n+2} in terms of a_0 or a_1, but, as we noted before, this does not restrict its usefulness.

To find the solution of IVP (13), we set $a_0 = 1$ and $a_1 = 0$ since we require that $y(0) = 1$ and $y'(0) = 0$. We may use (14) to find $a_2, a_3, \ldots$.

$$a_0 = 1, \qquad a_1 = 0, \qquad a_2 = -\frac{1}{2}\frac{(-\varepsilon)^0}{0!}a_0 = -\frac{1}{2},$$

$$a_3 = -\frac{1}{3\cdot2}\sum_{k=0}^{1}\frac{(-\varepsilon)^k}{k!}a_{1-k} = -\frac{1}{6}[a_1 - \varepsilon a_0] = \frac{\varepsilon}{6}$$

Similarly,

$$a_4 = \frac{1}{24}(1-\varepsilon^2), \qquad a_5 = \frac{\varepsilon}{120}(4-\varepsilon^2),$$

$$a_6 = -\frac{1}{720}(1-11\epsilon^2+\epsilon^4)$$

Thus, the solution of (13) is

$$y(t) = 1 - \frac{1}{2}t^2 + \frac{\varepsilon}{6}t^3 + \frac{1}{24}(1-\varepsilon^2)t^4 + \cdots \tag{15}$$

If we set $\varepsilon = 0$ in (15), then we obtain the Taylor series for $\cos t$ about $t_0 = 0$, which is the solution of IVP (13) if $\varepsilon = 0$.

Figure 11.2.2 and 11.2.3 overlay the computed solution of IVP (13) if $\varepsilon = 0.002$ with the graph of the polynomial approximation consisting of the terms in the series expansion of $y(t)$ through t^6. The approximation is good for $0 \le t \le 1$, but not good over the longer time interval. Observe that the computed solution shows the spring stretching and then compressing, while the approximation vastly exaggerates the amount of stretching and delays the onset of compression. Clearly, many more terms in series (15) must be used even to begin to correct these exaggerations. As noted previously, a polynomial approximation derived from a power series may be accurate only near the base point of the series expansion.

Comments

Examples 11.2.6–11.2.8 and Example 11.2.9 illustrate both the advantages and the defects of solving IVPs by series expansions. On the one hand, we do have solution formulas which may be used to evaluate the solutions, at least approximately, for specific values of the independent variable. On the other hand, although the series solutions may converge quite rapidly near the ordinary point at which the expansions are based, the convergence may be very slow farther away. Second, solutions may not even be expandable as power series if one of the coefficient functions $P(x)$ or $Q(x)$

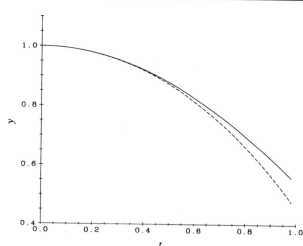

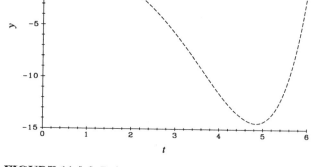

FIGURE 11.2.2 Computed solution of IVP (13) (solid) and a polynomial approximation (dashed).

FIGURE 11.2.3 Polynomial approximation (dashed) is a bad fit to solution curve (solid line) of IVP (13) at a distance from $t = 0$.

is not real analytic. For example, as noted in Example 11.2.5 the method fails to produce a solution of the equation $y''(x) + |x|y(x) = 0$ on any interval containing $x = 0$ since the coefficient function $|x|$ is not real analytic at 0. Nevertheless, power series solutions are useful and appear frequently in both theory and applications.

PROBLEMS

1. Determine whether the given point x_0 is ordinary or singular.

   (a) $y'' + k^2 y = 0$, x_0 arbitrary (k is a constant)

   (b) $y'' + \dfrac{1}{1+x} y = 0$, $x_0 = -1$

   (c) $y'' + \dfrac{1}{x} y' - y = 0$, $x_0 = 0$

2. (*Singular Points*). Find all singular points of the following equations.

   (a) $y'' + (\sin x)y' + (\cos x)y = 0$ **(b)** $y'' - (\ln|x|)y = 0$

   (c) $y'' + |x|y' + e^{-x}y = 0$ **(d)** $(1-x)y'' + y' + (1-x^2)y = 0$

   (e) $(1-x^2)y'' + xy' + y = 0$ **(f)** $x^2 y'' + xy' - y = 0$

3. (*Series Solutions About Ordinary Points*). Verify that $x = 0$ is an ordinary point and find all solutions as a power series in x. Determine the interval of convergence of the series. Plot the solution satisfying $y(0) = y'(0) = 1$. [*Hint*: Use the Taylor series about $x_0 = 0$ for the familiar solutions of the ODEs in parts **(a)** and **(b)**.]

   (a) $y'' + 4y = 0$ **(b)** $y'' - 4y = 0$

   (c) $y'' + xy' + 3y = 0$ **(d)** $(x^2 + 1)y'' - 6y = 0$, $y(0) = y'(0) = 1$

4. (*Airy's Equation*). *Airy's equation* is $y'' - xy = 0$.

   (a) Find all the solutions of the equation $y'' - xy = 0$ as power series in x. (Certain solutions of the equation are called Airy functions and are used to model the diffraction of light.)

   (b) Plot the solution satisfying $y(0) = 1$, $y'(0) = 0$ on the interval $-20 \le x \le 2$.

(c) Plot the sums of the first 2, 4, 6, 10, 14, and 18 nonvanishing terms of the series expansion of the solution of $y'' - xy = 0$, $y(0) = 1$, $y'(0) = 0$.

(d) Why would you expect the solution to oscillate for $x < 0$ but not for $x > 0$? [*Hint*: Consider the solutions of the IVPs, $y'' + \alpha y = 0$ ($x \le 0$), and $y'' - \alpha y = 0$ ($x \ge 0$), $y(0) = 1$, $y'(0) = 0$ and α a positive constant.]

5. (*Mathieu Equation*). The *Mathieu equation* is $y'' + (a + b\cos\omega x)y = 0$. Calculate the first four nonvanishing coefficients in the power series expansion of the solution of the Mathieu equation; use $b = 2$, $\omega = 1$, and $y(0) = 1$, $y'(0) = 0$.

6. (*Series Solutions of Nonlinear ODEs*). Assume a solution of the form $y = \sum_0^\infty a_n x^n$ and find $a_0, a_1, \ldots, a_7$ for the following IVPs that involve nonlinear ODEs.

(a) $y' = 1 + xy^2$, $\quad y(0) = 0$ **(b)** $y' = x^2 + y^2$, $\quad y(0) = 0$

7. (*Calculating the Leading Coefficients*). Calculate the first five nonzero terms in the power series expansion of the general solution of the equation $y'' + (1 - e^{-x^2})y = 0$.

8. (*Estimating the Value of a Solution*). Estimate $y(1)$ to three decimal places if $y(x)$ is the solution of the IVP, $y'' + x^2 y' + 2xy = 0$, $y(0) = 1$, $y'(0) = 0$.

11.3   Legendre Polynomials

The linear equation

$$(1 - x^2)y'' - 2xy' + p(p + 1)y = 0 \tag{1}$$

where p is any nonnegative constant, is called *Legendre's equation of order p*. Equation (1) is actually a family of differential equations, one equation for each value of p. The term "order p" refers to a factor in the coefficient of the y-term in (1), not to the "order of the differential equation," which is, of course, 2. Legendre's equation arises in an amazing variety of physical phenomena ranging from the calculation of gravitational and electrostatic potential to wave functions for the hydrogen atom.

The equation is of particular interest if p is a nonnegative integer, say n, since in that case the equation has a polynomial solution $P_n(x)$ of degree n. Initially, however, we do not restrict p to integer values.

Solutions of Legendre's Equation

The Legendre equation in normalized form is

$$y'' - \frac{2x}{1 - x^2}y' + \frac{p(p + 1)}{1 - x^2}y = 0$$

The point $x_0 = 0$ is an ordinary point because the coefficients $-2x(1 - x^2)^{-1}$ and $p(p + 1)(1 - x^2)^{-1}$ are real analytic in the interval $I = (-1, 1)$. This may be shown by using geometric series (see Appendix C.2) to expand $-2x(1 - x^2)^{-1}$ in a convergent power series

$$-2x(1 - x^2)^{-1} = -2x(1 + x^2 + x^4 + \cdots + x^{2k} + \cdots), \quad |x| < 1$$

Thus, according to the Convergence Theorem of Section 11.2 solutions of Legendre's equation may be written as convergent power series

$$y = \sum_{j=0}^{\infty} a_j x^j$$

The index j is used rather than n, because the latter symbol is reserved in this section to denote an integer value of the constant p. In the steps below the Method of Undetermined Coefficients is used to find the coefficients a_j, each of which depends upon the value of p.

Substituting $\sum_{j=0}^{\infty} a_j x^j$ into (1), we have that $\sum_{j=0}^{\infty} a_j x^j$, is a solution for $|x| < 1$ if and only if

$$(1 - x^2) \sum_{j=0}^{\infty} j(j-1)a_j x^{j-2} - 2x \sum_{j=0}^{\infty} ja_j x^{j-1} + p(p+1) \sum_{j=0}^{\infty} a_j x^j = 0$$

or, bringing the factors inside the summations,

$$\sum_{j=0}^{\infty} j(j-1)a_j x^{j-2} - \sum_{j=0}^{\infty} j(j-1)a_j x^j - \sum_{j=0}^{\infty} 2ja_j x^j + \sum_{0}^{\infty} p(p+1)a_j x^j = 0 .$$

Reindexing the first sum, after noting that the terms corresponding to $j = 0, 1$ may be dropped, we have that

$$\sum_{0}^{\infty} \left\{ (j+2)(j+1)a_{j+2} + [-j(j-1) - 2j + p(p+1)]a_j \right\} x^j = 0$$

Since $-j(j-1) - 2j + p(p+1) = (p+j+1)(p-j)$, we have the *Legendre recursion formula*

$$a_{j+2} = -\frac{(p+j+1)(p-j)}{(j+2)(j+1)} a_j, \quad j = 0, 1, 2, \ldots \tag{2}$$

Observe that the value of a_0 determines $a_2, a_4, \ldots$, while a_1 determines $a_3, a_5, \ldots$. Using (2) and an argument by induction, we have that

$$a_{2k} = (-1)^k \frac{(p+2k-1) \cdots (p+1)p \cdots (p-2k+2)}{(2k)!} a_0 \tag{3a}$$

$$a_{2k+1} = (-1)^k \frac{(p+2k) \cdots (p+2)(p-1) \cdots (p-2k+1)}{(2k+1)!} a_1 \tag{3b}$$

The convergence for $|x| < 1$ of any infinite series solution of (1) follows from the Ratio Test because

$$\left| \frac{a_{j+2} x^{j+2}}{a_j x^j} \right| = \left| \frac{(p+j+1)(p-j)}{(j+2)(j+1)} \right| x^2$$

which approaches x^2 as $j \to \infty$.

For each value of p a basis $\{y_1(x), y_2(x)\}$ for the solution space of the Legendre equation may be constructed by first setting $a_0 = 1$ and $a_1 = 0$ to find a solution y_1, and then reversing the values to $a_0 = 0, a_1 = 1$, to find a second solution y_2:

$$y_1(x) = 1 - \frac{(p+1)p}{2 \cdot 1} x^2 + \cdots + (-1)^k a_{2k} x^{2k} + \cdots \tag{4}$$

$$y_2(x) = x - \frac{(p+2)(p-1)}{3 \cdot 2} x^3 + \cdots + (-1)^k a_{2k+1} x^{2k+1} + \cdots \tag{5}$$

where a_{2k} and a_{2k+1} are given in (3a) and (3b). Observe that $y_1(x)$ is an even function, while $y_2(x)$ is odd. Since the Wronskian $W[y_1(0), y_2(0)] = 1$, the pair $\{y_1, y_2\}$ is an

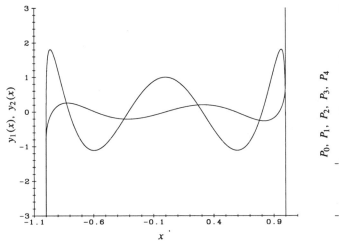
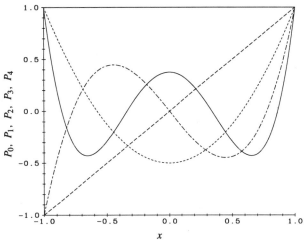

FIGURE 11.3.1 The graphs of y_1 and y_2 for $p = 4.5$ (equations (4) and(5)).

FIGURE 11.3.2 P_0 and P_4 (solid), P_1 (long dashes), P_2 (short dashes), P_3 (long/short dashes).

independent set. The general solution of Legendre's equation for $|x| < 1$ is

$$y = c_1 y_1(x) + c_2 y_2(x)$$

Observe that $c_1 = y(0)$ and $c_2 = y'(0)$. Finally observe that $\{y_1, y_2\}$ is a family of basis sets (just as (1) is a family of differential equations), one set for each value of p. Figure 11.3.1 displays the graphs of $y_1(x)$ (the "bumpier" graph) and $y_2(x)$ in the case $p = 4.5$; the series for $y_1(x)$ and $y_2(x)$ diverge for $|x| = 1$, and this shows up in the graphs.

Legendre Polynomials

Assume now that $p = n$ where n is a nonnegative integer. From (2) we see that $a_{n+2} = 0$ and, hence, $a_{n+4} = a_{n+6} = \cdots = 0$. The Legendre equations for $n = 2k$ and $n = 2k + 1$ have the respective polynomial solutions of degree $2k$ and $2k + 1$:

$$y = q_{2k}(x) = 1 - \frac{(2k+1)2k}{2}x^2 + \cdots + (-1)^k a_{2k}x^{2k} \tag{6}$$

$$y = q_{2k+1}(x) = x - \frac{(2k+3)2k}{6}x^3 + \cdots + (-1)^k a_{2k+1}x^{2k+1} \tag{7}$$

where a_{2k} and a_{2k+1} are given by (3a) and (3b) with $p = 2k$ and $p = 2k + 1$, respectively. Any nonzero constant multiple of a polynomial solution is another polynomial solution of the same degree. To normalize those polynomials, it is customary to divide each by its value at $x = 1$. As we shall see, none of the polynomials has a root at $x = 1$ and, hence, the division is legitimate. The *Legendre polynomials* are given by

$$P_{2k}(x) = \frac{q_{2k}(x)}{q_{2k}(1)}, \quad P_{2k+1}(x) = \frac{q_{2k+1}(x)}{q_{2k+1}(1)}, \qquad k = 0, 1, 2, \ldots \tag{8}$$

There is a single formula for these polynomials

$$P_n(x) = \frac{1}{2^n} \sum_{j=0}^{[n/2]} (-1)^j \frac{(2n-2j)!}{j!(n-j)!(n-2j)!} x^{n-2j} \tag{9}$$

where $[n/2]$ denotes the largest integer not larger than $n/2$, i.e., the integer part of the number $n/2$. We have from (6)–(8) or from (9) that

$$P_0 = 1, \quad P_1 = x, \quad P_2 = \frac{1}{2}(3x^2 - 1), \quad P_3 = \frac{1}{2}(5x^3 - 3x)$$

$$P_4 = \frac{1}{8}(35x^4 - 30x^2 + 3), \quad P_5 = \frac{1}{8}(63x^5 - 70x^3 + 15x)$$

See Figure 11.3.2. Equations (6)–(8) or (9) are not efficient ways to construct the Legendre polynomials. Fortunately, there is a recursive process that will do the trick.

THEOREM 11.3.1

Legendre Recursion Relation. For $n = 2, 3, \ldots,$

$$P_n(x) = (2 - \frac{1}{n})x P_{n-1}(x) - (1 - \frac{1}{n})P_{n-2}(x) \tag{10}$$

Formula (10) may be verified by using (9) to express each Legendre polynomial and then observing that the terms on the left and right sides of (10) (after combining terms of like powers of x) are identical.

Recall from Section 10.3 that the family of sine functions $\{\sin nx : n = 1, 2, \ldots\}$ have the "orthogonality property" over the interval $0 \leq x \leq \pi$:

$$\int_0^\pi \sin nx \sin mx \, dx = \begin{cases} 0, & n \neq m \\ \pi/2, & n = m \end{cases}$$

The family of Legendre polynomials has a similar property for $-1 \leq x \leq 1$.

THEOREM 11.3.2

Orthogonality of Legendre Polynomials. The Legendre polynomials are orthogonal for $|x| \leq 1$:

$$\int_{-1}^1 P_n(x) P_m(x) \, dx = \begin{cases} 0, & n \neq m \\ \dfrac{2}{2n+1}, & n = m \end{cases} \tag{11}$$

Proof. $P_n(x)$ and $P_m(x)$ are solutions of the ODEs:

$$(1 - x^2)P_n'' - 2x P_n' + n(n+1)P_n = 0$$

$$(1 - x^2)P_m'' - 2x P_m' + m(m+1)P_m = 0$$

If the first equation is multiplied by P_m and the second by P_n, subtraction yields

$$\begin{aligned}(1 - x^2)[P_n'' P_m - P_m'' P_n] - 2x[P_n' P_m - P_m' P_n] = \\ +[m(m+1) - n(n+1)]P_n P_m\end{aligned} \tag{12}$$

The left side of (12) may be expressed as the derivative

$$\frac{d}{dx}[(1 - x^2)(P_n' P_m - P_m' P_n)] \tag{13}$$

Integrating both sides of (12) from -1 to 1 and using (13), we have

$$[(1 - x^2)(P_n' P_m - P_m' P_n)]\Big|_{x=-1}^{x=1} = [m(m+1) - n(n+1)]\int_{-1}^1 P_n(x) P_m(x) \, dx$$

Since the left-hand side of this equality vanishes and $n \neq m$, orthogonality follows. The proof that $\int_{-1}^{1} P_n^2(x)\,dx = 2/(2n+1)$ is omitted.

The orthogonality of the Legendre polynomials has the same importance in solving certain boundary problems for Laplace's equation in spherical coordinates (Section 12.5) that the orthogonality of the family of sine functions has in solving boundary problems for the diffusion of heat in a rod (Section 12.3). Legendre polynomials satisfy many other relations, and Table 11.3.1 shows the most important of these.

The proofs of many of the identities in Table 11.3.1 are omitted, but we will derive the "Roots and Bounds" property.

THEOREM 11.3.3

Roots and Bounds. The Legendre polynomials have the following properties

(a). $P_n(x)$, $n > 0$, has n distinct real roots and all lie in the interval $(-1, 1)$.

(b). $|P_n(x)| \leq 1$ if $|x| \leq 1$ and $n = 0, 1, 2, \ldots$.

Proof. (a) Since $\int_{-1}^{1} P_n(x)\,dx = \int_{-1}^{1} P_n(x)P_0(x)\,dx = 0$ if $n > 0$ (by (11)), $P_n(x)$ must take on both positive and negative values on $(-1, 1)$; since $P_n(x)$

TABLE 11.3.1 Legendre Polynomials

Differential equation	$(1 - x^2)y'' - 2xy' + n(n+1)y = 0$
Polynomial expansion	$P_n(x) = \dfrac{1}{2^n} \sum_{j=0}^{[n/2]} \dfrac{(-1)^j (2n - 2j)!}{j!(n-j)!(n-2j)!} x^{n-2j}$
Rodrigues's formula	$P_n(x) = \dfrac{1}{2^n n!} \dfrac{d^n}{dx^n}\left[(x^2 - 1)^n \right]$
First six polynomials	$P_0(x) = 1$ $\qquad\qquad\qquad P_1(x) = x$ $P_2(x) = (3x^2 - 1)/2$ $\qquad P_3(x) = (5x^3 - 3x)/2$ $P_4(x) = (35x^4 - 30x^2 + 3)/8$ $\quad P_5(x) = (63x^5 - 70x^3 + 15x)/8$
Recursion formula	$P_n = (2 - 1/n)xP_{n-1} - (1 - 1/n)P_{n-2}$
Orthogonality	$\int_{-1}^{1} P_n(x)P_m(x)\,dx = \begin{cases} 0, & n \neq m \\ 2/(2n+1), & n = m \end{cases}$
Roots and bounds	P_n has n real roots, all in $(-1,1)$; $|P_n(x)| \leq 1$, x in $[-1, 1]$
Values	$P_n(1) = 1, \qquad P_n(-1) = (-1)^n,$ $P_{2n+1}(0) = 0, \quad P_{2n}(0) = (-1)^n \dfrac{(2n)!}{2^{2n}(n!)^2}$
Identities	$P_n' = nP_{n-1} + xP_{n-1}' \qquad\qquad xP_n' - nP_n = P_{n-1}'$ $nP_n = nxP_{n-1} + (x^2 - 1)P_{n-1}' \quad (1 - x^2)P_n' + nxP_n = nP_{n-1}$

is continuous it has at least one real root in $(-1, 1)$. Let $x_1, \ldots, x_k$ be the real roots of $P_n(x)$ in $(-1, 1)$. These roots are simple. For if x_j were a multiple root, then $P_n(x_j) = P_n'(x_j) = 0$. However, by the Existence and Uniqueness Theorem, these conditions determine a unique solution of Legendre's equation of order n. Since the trivial solution is one solution of this initial value problem and P_n is another, we must have $P_n(x) = 0$ for all x. This contradiction shows that the roots are simple. In order to show that $k = n$, let $Q_k(x)$ be the polynomial of degree k with simple roots at $x_1, \ldots, x_k$, defined by $Q_k(x) = (x - x_1) \cdots (x - x_k)$. Since $P_n(x)$ and $Q_k(x)$ each change sign at each x_j, we have that $P_n(x)Q_k(x)$ has fixed sign on $[-1, 1]$ and, hence, $\int_{-1}^{1} P_n(x)Q_k(x)\,dx \neq 0$. However, if $Q_k(x)$ has degree less than n, then using the result of Problem 3 in the problem set we see that $\int_{-1}^{1} P_n(x)Q_k(x)\,dx = 0$. This contradiction shows that $Q_k(x)$ has degree at least n. Consequently, P has exactly n distinct roots in $(-1, 1)$ since, as a polynomial of degree n, P_n cannot have more than n distinct roots.

(b) Squaring the identity (see Table 11.3.1) $P_n' = nP_{n-1} + xP_{n-1}'$, multiplying through by $(1 - x^2)/n^2$, and adding it to the square of the identity $P_n = xP_{n-1} + [(x^2 - 1)/n]P_{n-1}'$ (see Table 11.3.1 again) we obtain, for $|x| \leq 1$,

$$0 \leq \frac{1-x^2}{n^2}(P_n')^2 + P_n^2 = P_{n-1}^2 + \frac{1-x^2}{n^2}(P_{n-1}')^2 < P_{n-1}^2 + \frac{1-x^2}{(n-1)^2}(P_{n-1}')^2 \tag{14}$$

Thus, if we have (for $n = 1$) an upper bound for the right side of the inequalities in (14), that bound provides an upper bound for the second term in (14) for all $n \geq 1$. But the right side of (14) for $n = 1$ and for any x, $|x| \leq 1$, has value 1, since $P_0 = 1$ and $P_0' = 0$. Thus,

$$0 \leq \frac{1-x^2}{n^2}(P_n'(x))^2 + P_n^2(x) \leq 1, \qquad |x| < 1$$

and, hence, $|P_n(x)| \leq 1$ for $|x| \leq 1$.

Figure 11.3.2 shows the graphs of P_0, P_1, P_2, P_3, P_4 for $|x| \leq 1$.

Comments

The family of Legendre polynomials is not the only set of orthogonal polynomials that may arise when finding the series solutions of a family of differential equations. Hermite polynomials and Chebyshev polynomials arise in much the same way, but from different differential equations and with a different interpretation of orthogonality. See the Problem Set for some of their properties. As is true for the Legendre polynomials, Hermite and Chebyshev polynomials are often used in solving boundary problems for partial differential equations, and they also appear in many other applications.

PROBLEMS

1. (*Legendre's ODE and the Wronskian Method*). Legendre's equation of orders 0 and 1 have respective solutions $P_0 = 1$ and $P_1 = x$. Use the Wronskian Reduction Method (Problem 13, Section 3.3) to find a second independent solution for each equation in terms of elementary functions. Plot the second solution for $|x| < 1$.

2. (*Values of Legendre Polynomials*). Show that

$$P_{2n+1}(0) = 0, \quad P_{2n}(0) = (-1)^n \frac{(2n)!}{2^{2n}(n!)^2}, \quad P_n(-1) = (-1)^n,$$

[*Hint*: Use induction on n and an identity in Table 11.3.1 to prove the third formula.]

3. (*An Orthogonality Property*). Show that $\int_{-1}^{1} P_n(x)Q(x)\,dx = 0$ if $n > 0$ and $Q(x)$ is any polynomial of degree $k < n$. [*Hint*: Use induction on k to show that there are constants c_j such that $Q(x) = c_0 P_0 + c_1 P_1(x) + \cdots + c_k P_k(x)$ (explain). Then use the Orthogonality Property of Theorem 11.3.2.]

4. (*Legendre Polynomials and Zonal Harmonics*). The boundary value problem

$$\sin\phi(\rho^2 u_\rho)_\rho + (\sin\phi u_\phi)_\phi = 0, \quad \rho < 1,\ 0 < \phi < \pi$$

$$u(1,\phi) = f(\phi), \quad 0 \le \phi \le \pi$$

models the steady-state temperatures $u(\rho, \phi)$ in the unit ball $\rho \le 1$ if the temperature at the point with spherical coordinates (ρ, ϕ, θ) depends only on ρ and ϕ, but not on θ. See Sections 12.4, 12.5 for the derivation of these equations.

(a) Show that the Method of Separation of Variables (see Section 10.6) implies that solutions $u = R(\rho)\Phi(\phi)$ of the partial differential equation (PDE) of the boundary value problem satisfy

$$\rho^2 R'' + 2\rho R' - \lambda R = 0, \quad 0 < \rho < 1,$$

$$\Phi'' + \cot\phi\Phi' + \lambda\Phi = 0, \quad 0 < \phi < \pi$$

where λ is a separation constant, which is assumed to be nonnegative.

(b) Show that the change of variables $x = \cos\phi$ changes the ODE for $\Phi(\phi)$ in (a) to Legendre's equation

$$(1 - x^2)y'' - 2xy' + p(p+1)y = 0$$

where $y(x) = \Phi(\phi)$ and $p = (1/2)(-1 + \sqrt{1 + 4\lambda})$. Explain why, for physical reasons, p must be a nonnegative integer. [*Hint*: If p is not an integer, show that $\Phi(\phi)$ becomes unbounded as $\phi \to 0$ or π.]

5. (*Orthogonal Polynomials*). Let $\{R_n(x) : n = 0, 1, 2, \ldots\}$ be a family of polynomials indexed by degree (i.e., deg $R_n = n$). Let $\rho(x)$ be positive and continuous on an open interval $I = (a, b)$, and assume that $\lim_{x \to a^+} \rho(x)$ and $\lim_{x \to b^-} \rho(x)$ exist (although the limiting values may be infinite). The family of polynomials is said to be *orthogonal on I with respect to the density* ρ (or *weight*, or *scale factor*) if the *scalar product*

$$\langle R_n, R_m \rangle = \int_a^b R_n(x) R_m(x) \rho(x)\, dx \quad \text{is zero for all } n, m, n \ne m$$

(a) What are the values of a, b, and ρ for the family of Legendre polynomials?

(b) Show that if $\{R_n : n = 0, 1, 2, \ldots\}$ is a family of orthogonal polynomials on I with respect to ρ, then there are constants a_n, b_n, c_n such that the *three-term recursion relation*

$$R_n = (a_n x + b_n) R_{n-1} + c_n R_{n-2}, \quad n = 2, 3, \ldots$$

is satisfied. [*Hint*: $R_n = \alpha_n x R_{n-1} + \alpha_{n-1} R_{n-1} + \alpha_{n-2} R_{n-2} + \cdots + \alpha_0 R_0$ for some constants α_i. Then calculate the scalar products $\langle R_n, R_j \rangle$, $j \le n - 2$.]

(c) (*Recursion Formula for Legendre Polynomials*). Use (10) to find the values of a_n, b_n, and c_n in the three-term recursion formula for Legendre polynomials.

6. (*Hermite's Equation and Polynomials*). *Hermite's equation* of order n, n a non-negative integer, is $y'' - 2xy' + 2ny = 0$. [See Examples 11.2.6–11.2.8 for Hermite's equation of order 1 and its solutions.] Show that the equation has solutions which are polynomials of degree n, and that these polynomials are even functions if n is even, odd functions if n is odd. The *Hermite polynomials H_n* are determined by selecting (for each n) the polynomial solution with leading coefficient 2^n : $H_n(x) = 2^n x^n + (\cdot) x^{n-2} + \cdots$. Using the notation of Problem 5, show that the family of Hermite polynomials is orthogonal on the real line with respect to the density $\rho = e^{-x^2}$. It can be shown that the three-term recursion formula for the Hermite polynomials is $H_n = 2xH_{n-1} - 2(n-1)H_{n-2}$. Using $H_0 = 1$, $H_1 = 2x$ and this formula, construct $H_2, \ldots, H_5$. Plot $H_2, \ldots, H_5$ for $|x| \le 3$, and make a conjecture about the number and the nature of the roots of $H_n(x)$.

7. (*Chebyshev's Equation and Polynomials*). *Chebyshev's equation* of order n, n a nonnegative integer, is $(1 - x^2)y'' - xy' + n^2 y = 0$. Show that the equation has solutions which are polynomials of degree n, and that these polynomials are even functions if n is even, odd functions if n is odd. The *Chebyshev polynomials T_n*, $n > 0$, are determined by selecting the polynomial solution with 2^{n-1} as coefficient of x^n and $T_0 = 1$. Show that the family of Chebyshev polynomials is orthogonal on $I = [-1, 1]$ with respect to the density $(1 - x^2)^{-1/2}$. It is known that the three-term recursion formula for the Chebyshev polynomials is $T_n = 2xT_{n-1} - T_{n-2}$. Use the formula and $T_0 = 1$, $T_1 = x$ to construct $T_2, \ldots, T_5$. Plot $T_0, \ldots, T_5$. Make a conjecture about the number, location, and nature of the real roots of $T_n(x)$. [In fact, $T_n(x) = \cos(n \arccos x)$, but we do not prove that here.]

11.4 Regular Singular Points: Euler Equations

The Method of Undetermined Coefficients is successful in constructing a convergent power series solution of a second-order linear differential equation on an interval centered at an ordinary point of the equation. If the point is not ordinary, but singular, the method may fail. Since a number of equations that come up in modeling phenomena of wave motion, heat flow, or electric potential are of interest primarily in the neighborhoods of singular points, new methods of finding solutions are needed. In the 150 years or so since the need was recognized, methods have been found that will handle most of the singularities likely to occur in practice, at least if the singularities are "regular" in a sense to be defined in this section. The simplest class of equations with regular singularities may be solved explicitly without series, and we shall do so here. Solutions behave strangely near a regular singularity, as might be expected from the name, but not too strangely (i.e., they are "regular").

Regular Singular Points

The ODE

$$a_2(x)y'' + a_1(x)y' + a_0(x)y = 0 \tag{1}$$

may be normalized by dividing by $a_2(x)$ to obtain the ODE

$$y'' + P(x)y' + Q(x)y = 0 \tag{2}$$

where $P = a_1/a_2$ and $Q = a_0/a_2$. Our goal in the remainder of this chapter is to modify the series and undetermined coefficient techniques of Section 11.2 so that we can construct solutions of (2) in the neighborhood of a "regular" singular point x_0 of (2). Recall from Section 11.2 that x_0 is a singular point of (2) if P or Q fails to be real analytic at x_0. "Regularity" is defined as follows.

❖ **Regular Singular Points.** A singular point x_0 of ODE (2) is *regular* if the functions $p(x) = (x - x_0)P(x)$ and $q(x) = (x - x_0)^2 Q(x)$ are real analytic at x_0.

It is assumed in the definition that the (finite) values of $p(x_0)$ and $q(x_0)$ are given, respectively, by $\lim_{x \to x_0} (x - x_0)P(x)$ and $\lim_{x \to x_0} (x - x_0)^2 Q(x)$. A singular point of (2) that is not regular is said to be *irregular*. If x_0 is a regular singular point of (2), then (2) may be multiplied by $(x - x_0)^2$ and the equation written in *standard form*

$$L[y](x) = (x - x_0)^2 y'' + (x - x_0)p(x)y' + q(x)y = 0 \qquad (3)$$

where p and q are real analytic at x_0 and L is the linear operator $(x - x_0)^2 D^2 + (x - x_0)p(x)D + q(x)$. The following examples show that it is usually not difficult to classify the singularities of an ODE as regular or irregular and to write an equation in standard form with respect to each regular singularity.

EXAMPLE 11.4.1

A Regular Singular Point

The point $x_0 = 0$ is a singular point of the normalized ODE

$$y'' - 2x^{-1}y' + 2x^{-2}y = 0$$

because both $P(x) = -2x^{-1}$ and $Q(x) = 2x^{-2}$ are infinite at $x = 0$. However, the singularity is regular because the (constant) functions $p(x) = x \cdot P(x) = -2$ and $q(x) = x^2 \cdot Q(x) = 2$ are real analytic at $x_0 = 0$. The standard form of the ODE is $x^2 y'' - 2xy' + 2y = 0$, and there are no singular points other than $x_0 = 0$.

EXAMPLE 11.4.2

Legendre Equation and Regular Singular Points

Legendre's equation of order p

$$(1 - x^2)y'' - 2xy' + p(p + 1)y = 0$$

has singularities at $x = \pm 1$. Dividing by $1 - x^2$, we obtain the ODE

$$y'' - \frac{2x}{1 - x^2}y' + \frac{p(p + 1)}{1 - x^2}y = 0$$

Thus, $P(x) = -2x(1 - x^2)^{-1}$ and $Q(x) = p(p + 1)(1 - x^2)^{-1}$. With $x_0 = 1$, we have (don't confuse the function $p(x)$ with the constant p in the ODE):

$$p(x) = (x - 1)P(x) = 2x(1 + x)^{-1},$$

$$q(x) = (x - 1)^2 Q(x) = -p(p + 1)(x - 1)(1 + x)^{-1}$$

The functions $p(x)$ and $q(x)$ are real analytic at $x_0 = 1$ because we can use a geometric series expansion to obtain

$$\frac{1}{1 + x} = \frac{1}{2} \cdot \frac{1}{1 + (x - 1)/2} = \frac{1}{2} \sum_{n=0}^{\infty} (-1)^n \frac{1}{2^n} (x - 1)^n$$

which converges for $|x - 1| < 2$. Thus, Legendre's equation of order p has a regular singularity at $x = 1$. Similarly, $x = -1$ is a regular singularity.

EXAMPLE 11.4.3

An Irregular Singular Point
The ODE

$$y'' + 2x^{-3}y = 0$$

has an irregular singularity at $x_0 = 0$ because $Q(x) = 2x^{-3}$ is infinite at 0 (so 0 is a singularity) and $q(x) = x^2 \cdot 2x^{-3} = 2x^{-1}$ is still infinite at 0 (hence, q is not real analytic at 0).

EXAMPLE 11.4.4

Regular and Irregular Singular Points
The ODE

$$x^3(x-1)y'' + y = 0$$

may be normalized to

$$y'' + x^{-3}(x-1)^{-1}y = 0$$

Because $Q(x) = x^{-3}(x-1)^{-1}$ is infinite at $x = 0$ and $x = 1$, both are singular points. Because $x^2 \cdot x^3(x-1)^{-1} = x^{-1}(x-1)^{-1}$ is infinite at $x = 0$, the origin is an irregular singular point of the ODE. However, $(x-1)^2 \cdot x^{-3}(x-1)^{-1} = x^{-3}(x-1)$ is real analytic at $x = 1$ (it has the value 0 at that point), and, hence, 1 is a regular singularity of the equation. The standard form of the ODE $y'' + x^{-3}(x-1)^{-1}y = 0$ for the regular singular point, $x = 1$, is $(x-1)^2 y'' + x^{-3}(x-1)y = 0$.

EXAMPLE 11.4.5

Another Irregular Singular Point
The point $x_0 = 0$ is an irregular singularity of

$$y'' + |x|y' + y = 0$$

because $P(x) = |x|$ and $p(x) = x^2 \cdot P(x) = x^2|x|$ are not real analytic at 0.

All other points in the examples above are ordinary points, and the methods of Section 11.2 may be used to find real analytic solutions in the form of power series convergent in the neighborhood of such a point. Our goal here and in the rest of the chapter is to solve (2) or (3) near a regular singular point. We shall suppose from now on that 0 is a regular singular point of (2) or (3). If $x_0 \neq 0$ is a regular singular point, it may be translated to 0 by the change of variable $\bar{x} = x - x_0$, and the resulting differential equation in $\bar{x}$ has 0 as a regular singular point.

Euler ODEs

The simplest kind of ODE of form (2) which has a regular singular point at 0 is the *Euler* (or *Cauchy*, or *Cauchy-Euler*, or *equidimensional*) equation first encountered in the problem set of Section 3.6,

$$L[y](x) = x^2 y'' + p_0 xy' + q_0 y = 0 \tag{4}$$

where p_0 and q_0 are real constants. Euler ODEs can be solved completely in terms of elementary functions. For now it is assumed that $x > 0$.

The form of ODE (4) suggests that $y = x^r$ may be a solution for some value of r. We have that

$$L[x^r] = x^2 r(r-1)x^{r-2} + p_0 xrx^{r-1} + q_0 x^r$$

$$= [r(r-1) + p_0 r + q_0] x^r$$

Hence, $y = x^r$ is a solution of ODE (4), i.e., $L[x^r] = 0$, if and only if r is a root of the *indicial polynomial* $f(r) = r^2 + (p_0 - 1)r + q_0$ for (4). These roots, r_1 and r_2, are

$$r_1, r_2 = -\frac{1}{2}(p_0 - 1) \pm \frac{1}{2}\sqrt{(p_0 - 1)^2 - 4q_0} \qquad (5)$$

and are called the *indicial roots* of ODE (4). We consider separately the three cases where r_1 and r_2 are real and distinct, real and equal, or complex conjugates.

Roots are Real and Distinct

If the roots r_1 and r_2 are real and distinct, then a basis for the solution space of ODE (4) consists of $y_1 = x^{r_1}$ and $y_2 = x^{r_2}$. The reason for this is that each is a solution and the pair forms an independent set because the Wronskian $W[x^{r_1}, x^{r_2}] = x^{r_1} r_2 x^{r_2 - 1} - r_1 x^{r_1 - 1} x^{r_2} = (r_2 - r_1) x^{r_1 + r_2 - 1} \neq 0$, for $x > 0$. Thus, the general solution of ODE (4) in this case is

$$y = c_1 x^{r_1} + c_2 x^{r_2}, \quad x > 0 \qquad (6)$$

Observe that if r is irrational (e.g., $r = \sqrt{2}$), then x^r is defined to be $e^{r \ln x}$.

EXAMPLE 11.4.6

Euler ODE: Distinct Real Roots
The equation $x^2 y'' - 2xy' + 2y = 0$ is an Euler ODE with indicial polynomial, $f(r) = r^2 - 3r + 2$, and indicial roots $r_1 = 1$, $r_2 = 2$. Observe that in this case the restriction to $x > 0$ is unnecessary, and the general solution, $y = c_1 x + c_2 x^2$, is defined for all x. The general solution consists of a family of lines ($c_2 = 0$) and parabolas ($c_2 \neq 0$), all of which go through the origin. Thus, the "singular behavior" in this case lies in the fact that the initial value problem

$$x^2 y'' - 2xy' + 2y = 0 \qquad y(0) = a, \quad y'(0) = b \qquad (7)$$

has no solution at all if $a \neq 0$, and infinitely many if $a = 0$. This does not violate the Existence and Uniqueness Theorem for IVPs (Theorem 3.2.1), because the normalized equation $y'' - 2x^{-1} y' + 2x^{-2} y = 0$ does not satisfy the hypotheses of that theorem at $x = 0$. See Fig. 11.4.1 for solutions of (7) in the case $a = 0$, $b = 2$. In this case solutions have the form $y = 2x + c_2 x^2$, where c_2 is an arbitrary constant; solutions are plotted for three negative and three positive values of c_2 as well as for $c_2 = 0$.

Equal Roots

If the roots r_1 and r_2 of the indicial polynomial are equal, then $y_1 = x^{r_1}$ is one solution, but we must do some work to find a second, independent solution. The Wronskian Reduction Method introduced in Problem 15 of Section 3.3 applies in this case. That method asserts that the Wronskian determinant of two solutions y_1 and y_2 of any normalized ODE such as $y'' + P(x)y' + Q(x)y = 0$ satisfies the ODE

$$W[y_1, y_2] = y_1 y_2' - y_1' y_2 = C e^{-\int^x P(s)\,ds}$$

where C is a constant ($C = 0$, if y_1 and y_2 are dependent, $C \neq 0$ otherwise). With $y_1 = x^{r_1}$, $C = 1$, and $P(x) = p_0 x^{-1}$ from the normalized form of ODE (4), we have a

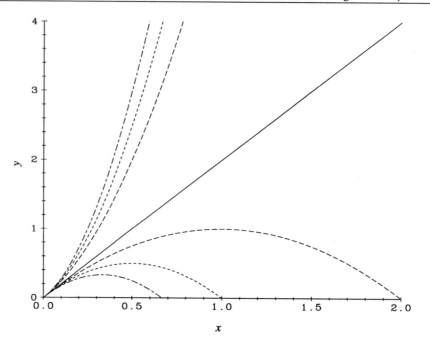

FIGURE 11.4.1 Solution curves of the Euler equation of Example 11.4.6.

first-order linear equation in y_2,

$$x^{r_1} y_2' - r_1 x^{r_1 - 1} y_2 = e^{-p_0 \int_0^x s^{-1} ds} = e^{-p_0 \ln x} = x^{-p_0} \tag{8}$$

Since the indicial polynomial $f(r) = r^2 + (p_0 - 1)r + q_0 = (r - r_1)^2 = r^2 - 2r_1 r + r_1^2$, we must have that $p_0 - 1 = -2r_1$, i.e., $p_0 = 1 - 2r_1$. Hence, using this value for p_0 and dividing each side of (8) by x^{r_1}, we have that

$$y_2' - r_1 x^{-1} y_2 = x^{r_1 - 1} \tag{9}$$

Using the integrating factor x^{-r_1}, we have that

$$(y_2 x^{-r_1})' = x^{-1}$$

from which we conclude that

$$y_2 = x^{r_1}(C + \ln x)$$

We may as well set $C = 0$ and let $y_2 = x^{r_1} \ln x$ and thus obtain the general solution in the case of equal roots:

$$y = C_1 x^{r_1} + C_2 x^{r_1} \ln x \tag{10}$$

We leave it to the reader to show that y_1 and y_2 are independent on $x > 0$ by showing that their Wronskian is not zero.

EXAMPLE 11.4.7

Euler ODE: Equal Roots
The indicial polynomial of the Euler ODE, $x^2 y'' + y/4 = 0$, is $f(r) = r^2 - r + 1/4 = (r - 1/2)^2$. The functions $y_1 = x^{1/2}$ and $y_2 = x^{1/2} \ln x$ form a pair of independent solutions on the interval $x > 0$. See Figure 11.4.2 for graphs of y_1, y_2 (solid curves), and some linear combinations of y_1 and y_2 (dashed curves).

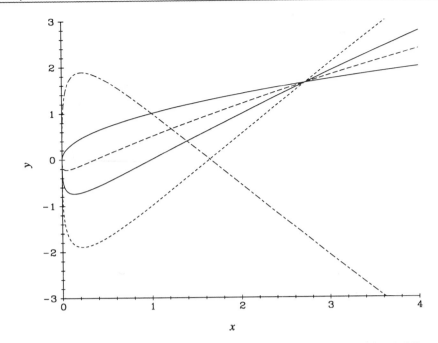

FIGURE 11.4.2 Solution curves of the Euler equation of Example 11.4.7.

Complex Conjugate Roots

Finally, suppose that the roots of the indicial polynomial are $r_1, r_2 = \alpha \pm i\beta, \beta \neq 0$. This will give as solutions the complex-valued functions $x^{\alpha \pm i\beta}$. We have from Appendix C.3 that

$$x^{\alpha \pm i\beta} = e^{(\alpha \pm i\beta)\ln x} = e^{\alpha \ln x}(\cos(\beta \ln x) \pm i \sin(\beta \ln x))$$

$$= x^{\alpha} \cos(\beta \ln x) \pm ix^{\alpha} \sin(\beta \ln x)$$

Much as in the discussion of complex roots of the characteristic polynomial of a constant-coefficient operator in Section 3.4, we see that there is a pair of independent real-valued solutions

$$y_1 = x^{\alpha} \cos(\beta \ln x), \qquad y_2 = x^{\alpha} \sin(\beta \ln x)$$

that form a basis for the solution space in this setting of complex conjugate indicial roots. Again, the reader may show the independence of y_1 and y_2 by showing that their Wronskian is not zero for $x > 0$.

EXAMPLE 11.4.8

Euler ODE: Complex Conjugate Roots
The Euler ODE $x^2 y'' + (1 - 2\alpha)xy' + (\alpha^2 + 400)y = 0$, where α is a real constant, has indicial polynomial $r^2 - 2\alpha r + \alpha^2 + 400$, whose roots are $r_1, r_2 = \alpha \pm 20i$. Thus, $y_1 = x^{\alpha} \cos(20 \ln x)$ and $y_2 = x^{\alpha} \sin(20 \ln x)$ generate the solution space. Figures 11.4.3–11.4.5 show the graphs of $y_1(x)$ for $\alpha = 0.5$, 0, and -0.5, respectively. The rapidly oscillating behavior near the origin corresponds to the fact that $\ln x \to -\infty$ as x decreases toward 0, which implies that the frequency of the cosine term tends to infinity. The factor x^{α} controls the amplitude of the oscillations.

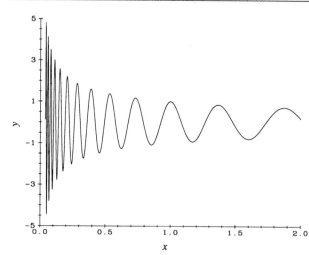

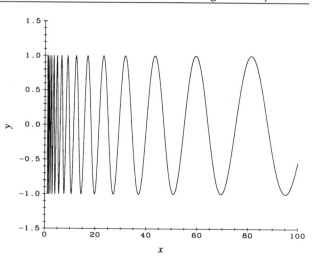

FIGURE 11.4.3 A solution curve of the Euler equation of Example 11.4.8: $\alpha < 0$.

FIGURE 11.4.4 A solution curve of the Euler equation of Example 11.4.8: $\alpha = 0$.

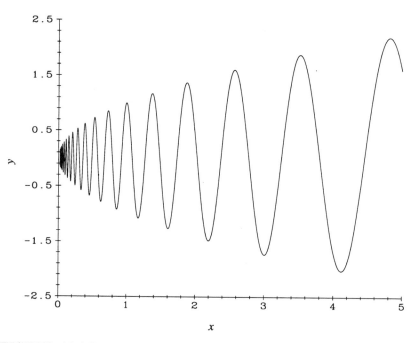

FIGURE 11.4.5 A solution curve of the Euler equation of Example 11.4.8: $\alpha > 0$.

Comments

The examples and figures of this section show that solutions of an Euler equation for $x > 0$ may, or may not, behave in unusual ways near the regular singularity at the origin. The behavior is completely determined by the nature of the roots of the indicial polynomial. The analysis may be extended to negative values of x by replacing x by $|x|$ in all of the solution formulas. Observe that in some cases, solutions are defined at $x = 0$, e.g., for the equations of Examples 11.4.6, 11.4.7, and 11.4.8 (but only if $\alpha > 0$). However, if $\alpha \le 0$ in Example 11.4.8, the solutions are undefined at $x = 0$.

Summarizing, we have the following formulas for the general solution of the Euler ODE (4) with real coefficients in terms of the roots r_1 and r_2 of the indicial polynomial $r^2 + (p_0 - 1)r + q_0$:

$$y = \begin{cases} c_1|x|^{r_1} + c_2|x|^{r_2} & \text{if } r_1 \neq r_2, \quad r_1, r_2 \text{ real} \\ c_1|x|^{r_1} + c_2|x|^{r_1}\ln|x| & \text{if } r_1 = r_2 \\ |x|^{\alpha}[c_1\cos(\beta\ln|x|) + c_2\sin(\beta\ln|x|)] & \text{if } r_1 = \bar{r}_2 = \alpha + i\beta \end{cases} \qquad (11)$$

where $x \neq 0$.

The extension of these ideas to finding solutions near a regular singularity of an equation not of Euler form is taken up in the remaining sections of this chapter.

PROBLEMS

1. (*Regular and Irregular Singularities*). Determine if the given point is a regular or an irregular singular point for the corresponding equation.

   (a) $x^2 y'' + xy' + y = 0$, $\qquad x_0 = 0$

   (b) $xy'' + (1 - x)y' + xy = 0$, $\qquad x_0 = 0$

   (c) $x(1 - x)y'' + (1 - 2x)y' - 4y = 0$, $\qquad x_0 = 1$

   (d) $x^2 y'' + 2y'/x + 4y = 0$, $\qquad x_0 = 0$

2. (*Classifying Singular Points*). Find and classify all singular points of the following equations.

   (a) $(1 - x^2)y'' - xy' + 2y = 0$

   (b) $y'' + \left(\dfrac{x}{1-x}\right)^2 y' + (1 + x)^2 y = 0$

   (c) $x(1 - x^2)^3 y'' + (1 - x^2)^2 y' + 2(1 + x)y = 0$

   (d) $x^3(1 - x^2)y'' - x(x + 1)y' + (1 - x)y = 0$

3. (*Euler ODEs*). Find the general, real-valued solution (for $x > 0$) of each of the following equations. Plot some solutions of each equation.

   (a) $x^2 y'' - 6y = 0$ $\qquad\qquad$ (b) $x^2 y'' + xy' - 4y = 0$

   (c) $x^2 y'' + xy' + 9y = 0$ $\qquad\quad$ (d) $x^2 y'' + xy'/2 - y/2 = 0$

   (e) $xy'' - y' + (5/x)y = 0$ $\qquad$ (f) $x^2 y'' + 7xy' + 9y = 0$

4. (*Nonhomogeneous Euler ODEs*). Find the general solution for $x > 0$ of each of the following ODEs. [*Hint*: Use Undetermined Coefficients to find a particular polynomial solution.]

   (a) $x^2 y'' - 4xy' + 6y = 2x + 5$ $\qquad$ (b) $x^2 y'' - 6y = 2x^2$

5. (*Decaying Solutions of Euler ODEs*). Show that every solution $y(x)$ of the ODE $x^2 y'' + p_0 xy' + q_0 y = 0$, p_0, q_0 real, has the property that $|y(x)| \to 0$ as $x \to \infty$ if and only if $p_0 > 1$ and $q_0 > 0$.

6. (*The nth-order Euler ODE*). The *nth-order Euler ODE* is

$$x^n y^{(n)} + a_{n-1}x^{n-1}y^{(n-1)} + \cdots + a_1 xy' + a_0 y = 0, \quad x > 0 \qquad (12)$$

   where $a_0, a_1, \ldots, a_{n-1}$ are real constants. The *indicial polynomial* is $p(r) = r(r - 1) \cdots (r - n + 1) + r(r - 1) \cdots (r - n + 2)a_{n-1} + \cdots + ra_1 + a_0$.

   (a) Show that $y = x^{r_1}$, $x > 0$, is a solution of (12) if and only if r_1 is a root of $p(r)$.

   (b) Use part (a) to find three independent solutions of

$$x^3 y''' + 4x^2 y'' - 2y = 0, \qquad x > 0$$

11.5   Series Solutions Near Regular Singular Points, I

Equations with regular points are pervasive in the mathematical modeling of physical phenomena. If the equation has Euler form, it may be solved explicitly in terms of powers, logarithms, or trigonometric functions of logarithms of the independent variable. If the equation has a regular singularity but is not of Euler form, we must look elsewhere in our search for solution formulas. As it turns out, a combination of the power series of Section 11.2 and the Euler solutions of the preceding section gives us just what we want, solutions such as

$$x^r \sum_0^\infty a_n x^n, \quad x^r \ln x \sum_0^\infty a_n x^n, \quad \text{or} \quad x^\alpha \cos(\beta \ln x) \sum_0^\infty a_n x^n$$

In this section we focus our attention on solutions of the first type, leaving solutions of the second type to Section 11.7 and not considering the trigonometric solutions at all since they do not come up very often in applications. The approach we shall use is a special case of techniques developed during the nineteenth century by Sturm, Fuchs, Frobenius, and many others. Later the theory was extended by Weyl to irregular singular points.

Indicial Polynomials and Frobenius Series

Suppose that $x = 0$ is a regular singular point of the ODE

$$a_2(x)y'' + a_1(x)y' + a_0(x)y = 0$$

Multiplying the equation by x^2/a_2, we obtain the ODE in standard form

$$x^2 y'' + xp(x)y' + q(x)y = 0 \tag{1}$$

where $p = xa_1/a_2$ and $q = x^2 a_0/a_2$. The functions p and q are assumed to be real analytic at 0 and to have power series expansions of the form

$$p(x) = \sum_0^\infty p_n x^n, \qquad q(x) = \sum_0^\infty q_n x^n \tag{2}$$

which converge on a common interval J centered at 0.

❖ **Indicial Polynomial.** The *indicial polynomial* of (1) is given by $f(r) = r^2 + (p_0 - 1)r + q_0$. Its roots r_1 and r_2 are the *indicial roots* of ODE (1) at $x = 0$.

Observe that if all the coefficients of the series expansions of $p(x)$ and $q(x)$ vanish (except, possibly, p_0 and q_0), then (1) is an Euler equation, and the definition above agrees with that given in the preceding section. Our goal is to find solutions of (1) in the form of a *generalized power series*, or *Frobenius series*,

$$y = x^r \sum_0^\infty a_n x^n \tag{3}$$

As we shall see, r will turn out to be an indicial root r_1 or r_2 and the coefficients $a_0, a_1, \ldots$ may be found by an extension of the Method of Undetermined Coefficients used in Section 11.2. The following example shows how solutions in the form of a Frobenius series may be determined.

EXAMPLE 11.5.1

Finding the Recursion Relation for a Frobenius Series

Consider the equation

$$4xy'' + 2y' + y = 0 \tag{4}$$

Multiplying by $x/4$ to get the equation into the standard form (1), we have

$$x^2 y'' + \frac{x}{2} y' + \frac{x}{4} y = 0 \tag{5}$$

from which we see that $p(x) = 1/2$, $q(x) = x/4$. Thus, 0 is a regular singular point. The indicial polynomial is $r^2 - r/2$ since $p_0 = 1/2$, $q_0 = 0$. The indicial roots are $1/2$ and 0. In what follows it will be more convenient to use (4) rather than the standard form. Let x be positive in the analysis below.

Ignoring convergence questions for the time being, we assume that (4) has a solution of the form $y = \sum_0^\infty a_n x^{n+r}$, insert it into (4), and use the Method of Undetermined Coefficients to calculate the coefficients a_n, $n = 0, 1, 2, \ldots$. Then

$$4x \sum_0^\infty (n+r)(n+r-1)a_n x^{n+r-2} + 2 \sum_0^\infty (n+r)a_n x^{n+r-1} + \sum_0^\infty a_n x^{n+r} = 0$$

that is,

$$\sum_0^\infty 4(n+r)(n+r-1)a_n x^{n+r-1} + \sum_0^\infty 2(n+r)a_n x^{n+r-1} + \sum_0^\infty a_n x^{n+r} = 0 \tag{6}$$

Reindexing the last sum, we have $\sum_0^\infty a_n x^{n+r} = \sum_1^\infty a_{n-1} x^{n+r-1}$. Isolating the first term ($n = 0$) in each of the other two sums so that all three begin at $n = 1$, we may replace (6) by

$$4(0+r)(0+r-1)a_0 x^{r-1} + 2(0+r)a_0 x^{r-1}$$

$$+ \sum_1^\infty [4(n+r)(n+r-1)a_n + 2(n+r)a_n + a_{n-1}]x^{n+r-1} = 0$$

or, after some algebraic simplification,

$$x^{r-1} \left\{ 4r(r-1/2)a_0 + \sum_1^\infty [2(n+r)(2n+2r-1)a_n + a_{n-1}]x^n \right\} = 0$$

By the Identity Theorem for power series, this implies that the coefficient of x^n, $n = 0, 1, 2, \ldots$, must vanish. We have the following recursion relations:

$$n = 0: \quad 4r\left(r - \frac{1}{2}\right)a_0 = 0 \tag{7a}$$

$$n \geq 1: \quad 2(n+r)(2n+2r-1)a_n + a_{n-1} = 0 \tag{7b}$$

From the first relation we conclude that either $r = 1/2$ or 0, or $a_0 = 0$. Observe that

$$f(r) = r(r - 1/2)$$

is the indicial polynomial and $r_1 = 1/2$ and $r_2 = 0$ are the indicial roots. If we set $r = 1/2$ or 0, a_0 is an arbitrary constant. The remaining coefficients are functions of r and a_0: they may be determined from the general recursion formula

$$a_n(r) = \frac{-a_{n-1}(r)}{2(n+r)(2n+2r-1)}, \quad n = 1, 2, \ldots \tag{8}$$

which is a rearrangement of (7b).

Finding the recursion relation for a Frobenius series solution is the critical step, and that has now been accomplished with the derivation of formula (8). Now we can construct a pair of independent solutions of ODE (4).

EXAMPLE 11.5.2

Constructing a Frobenius Series

Returning to ODE (4), its indicial polynomial, and its recursion formula (8), we note that if r is either of the roots $1/2, 0$ of the indicial polynomial, then the denominator in (8) never vanishes. Using (8) to obtain a_n in terms of a_0, we have

$$
\begin{aligned}
a_n(r) &= \frac{-a_{n-1}(r)}{2(n+r)(2n+2r-1)} \\
&= \frac{a_{n-2}(r)}{2(n+r)(2n+2r-1)2(n-1+r)(2n-2+2r-1)} = \cdots \\
&= \frac{(-1)^n a_0}{2^n(n+r)(n+r-1)\cdots(r+1)(2n+2r-1)(2n+2r-3)\cdots(2r+1)}
\end{aligned}
\tag{9}
$$

Observe that $a_0 = 0$ produces either the trivial solution $y = 0$ for all x, or else the recursion process breaks down if a factor of the denominator in (9) vanishes. For $r = 1/2$ or 0, this difficulty cannot occur. We have that

$$
a_n(0) = \frac{(-1)^n a_0}{2^n n!(2n-1)(2n-3)\cdots 1} = \frac{(-1)^n a_0}{(2n)!}
$$

$$
\begin{aligned}
a_n\left(\frac{1}{2}\right) &= \frac{(-1)^n a_0}{2^n(n+1/2)(n-1/2)\cdots(3/2)(2n)(2n-2)\cdots 2} \\
&= \frac{(-1)^n a_0}{(2n+1)(2n-1)\cdots 3(2n)(2n-2)\cdots 2} \\
&= \frac{(-1)^n a_0}{(2n+1)!}
\end{aligned}
$$

We have constructed a pair of independent solutions of (4)

$$
y_1 = x^{1/2}\sum_0^\infty \frac{(-1)^n x^n}{(2n+1)!}, \qquad y_2 = \sum_0^\infty \frac{(-1)^n x^n}{(2n)!}
\tag{10}
$$

where we have taken $a_0 = 1$. More precisely, the solution $y_1(x)$ is defined for $x > 0$ or for $x < 0$ if we replace $x^{1/2}$ by $|x|^{1/2}$. But $y_1(x)$ is not a solution at $x = 0$, since the term $x^{1/2}$ is not differentiable there. The second solution is defined for all x.

The power series in (10) converge everywhere (use the Ratio Test). It is straight-forward now to justify all our earlier steps and check that the two functions defined in (10) actually are solutions of (4). This can be done by direct insertion of y_1 and y_2 into (4). Of course, the recursion relations were chosen specifically so that y_1 and y_2 are solutions. Since the power series parts of y_1 and y_2 converge for all x, it is all right to differentiate term-by-term. The functions y_1 and y_2 are independent since $y_2(0) = 1$, while $|y_1(x)| \to 0$ as $x \to 0$. Thus $\{y_1, y_2\}$ is a basis of the solution space of (4). See Figure 11.5.1 for graphs of $y_1(x)$ and $y_2(x)$.

Observe that even if we had not determined the indicial polynomial and indicial roots before beginning the process of determining the coefficients, the polynomial (or a multiple of it) appears as the coefficient of a_0 in (7a), and we could at that point have chosen r to be a root of the polynomial. We will find this useful fact to hold for the general equation (1) as well.

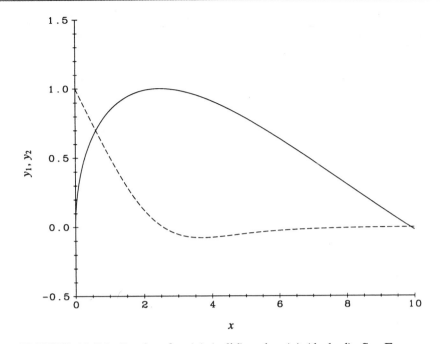

FIGURE 11.5.1 Graphs of $y_1(x)$ (solid) and $y_2(x)$ (dashed). See Example 11.5.1.

Method of Frobenius: Theory

The Method of Undetermined Coefficients used in the example above is called the *Method of Frobenius*, and the series solutions obtained are Frobenius series. A Frobenius series may be a standard power series (see y_2 in (10)) or involve x to a power that is not an integer (see y_1 in (10)). The technique will always generate at least one nontrivial solution defined near (but perhaps not at) the regular singular point. Many times the process will also produce a second independent solution as in the example above—but not always. Thus we have the following basic theorem.

THEOREM 11.5.1

Frobenius's Theorem I. Suppose that $x_0 = 0$ is a regular singular point for the ODE

$$x^2 y'' + x p(x) y' + q(x) y = 0$$

and that the indicial polynomial

$$f(r) = r^2 + (p_0 - 1)r + q_0$$

has real roots, r_1 and r_2, with $r_2 \leq r_1$. Then ODE (1) has a solution of the form

$$y_1 = |x|^{r_1} \sum_0^\infty a_n(r_1) x^n, \qquad x \neq 0 \qquad (11)$$

with $a_0 \neq 0$, where the coefficients are determined by the *recursion relation*,

$$f(r_1 + n) a_n(r_1) = - \sum_{k=0}^{n-1} [(k + r_1) p_{n-k} + q_{n-k}] a_k(r_1), \qquad n \geq 1 \qquad (12)$$

where $a_0 = 1$. The series in (11) converges to a real analytic function on the interval J. Moreover, if $r_1 - r_2$ is not an integer, a second independent solution is given by the series

$$y_2 = |x|^{r_2} \sum_0^\infty a_n(r_2) x^n, \qquad x > 0, \text{ or } x < 0 \qquad (13)$$

where the coefficients are determined by (12), with r_2 replacing r_1, and $a_0 \neq 0$. The series in (13) converges on J to a real analytic function.

The proof of the theorem is omitted. It must be emphasized that usually the most efficient way to solve a specific ODE is to proceed by direct substitution and the Method of Undetermined Coefficients as in the earlier example, rather than to use the recursion relations of (12) directly.

When $r_1 - r_2$ is an integer, some ingenuity must be used to construct a solution that is independent of y_1. One can see from (12) just where the recursion formula breaks down in this case. For example, suppose that $r_1 - r_2 = 3$. Then the indicial polynomial $f(r) = r^2 + (p_0 - 1)r + q_0$ factors into $f(r) = (r - r_2)(r - r_2 - 3)$, and we see that $f(r_2 + 3) = 3 \cdot 0 = 0$. Hence, the recursion formula for $r = r_2$ and $n = 3$ becomes $0 \cdot a_3 = - \sum_{k=0}^{2} [\cdots] a_k$, and we cannot in general determine a_3. We resolve this difficulty in Section 11.7.

Comments

We have seen that the construction of a Frobenius series solution $y = x^r \sum_0^\infty a_n x^n$ of the differential equation consists of the following steps in the case that the indicial roots r_1 and r_2 are real, $r_1 \geq r_2$.

- Check that the differential equation may be written in the form

$$x^2 y'' + x p(x) y' + q(x) y = 0$$

where $p = \sum_0^\infty p_n x^n$ and $q = \sum_0^\infty q_n x^n$, each series converging on an open interval containing 0. The point $x = 0$ is then a regular singularity and the method of Frobenius will apply. This step must not be omitted since you will need the numbers $p_0 = p(0)$ and $q_0 = q(0)$.

- Write out the indicial polynomial $f(r) = r^2 + (p_0 - 1)r + q_0 = 0$ and find its roots, i.e., the indicial roots r_1 and r_2, $r_1 \geq r_2$.

- Clear any denominators from the differential equation and cancel common factors. For example, rewrite

$$x^2 y'' + xy' + [x^3/(1+x)]y = 0$$

as

$$x(1+x)y'' + (1+x)y' + x^2 y = 0$$

The resulting equation need not have the form $x^2 y'' + xp(x)y' + q(x)y = 0$.

- Insert the series $\sum_0^\infty a_n x^{n+r}$ into the ODE found above.

- Reindex and rearrange the resulting series until you have

$$f(r)a_0 x^r + \sum_{n=1}^\infty [\cdots] x^{n+r} = 0$$

In reindexing and rearranging, be particularly careful not to "lose" terms at the head of the series (e.g., terms corresponding to $n = 0$ or 1). Equating the bracketed expression to zero gives the recursion formula for the coefficients.

- Solve the recursion formula if it is simple enough to do so, and find the coefficients a_n in terms of a_0. Otherwise, use the recursion formula to find the first few coefficients $a_1, a_2, \ldots, a_N$ in terms of a_0.

- Let $r = r_1$ in the recursion formula to find a solution $y_1 = x^{r_1} \sum_0^\infty a_n x^n$ with $a_0 \neq 0$.

- If $r_1 - r_2$ is not an integer, repeat the process with $r = r_2$ to find a second, independent solution $y_2 = x^{r_2} \sum_0^\infty b_n x^n$ with $b_0 \neq 0$.

- The numbers a_0 and b_0 are arbitrary, but to obtain specific solutions set a_0 and b_0 equal to specific constants.

- If $r_1 - r_2$ is an integer, refer to Section 11.7 for a method of finding a second independent solution corresponding to the indicial root r_2.

We have assumed that 0 is a regular singular point of the equation. If it is an ordinary point, the series methods of Section 11.2 should be used instead. If 0 is an irregular singular point, series methods may or may not work, but are worth a try. If $x_0 \neq 0$ is a regular singularity of (1), then replace x by $x = \bar{x} + x_0$ in (1) and use the steps described above for the ODE in $\bar{x}$.

PROBLEMS

1. (*Frobenius Series*). Find solutions $y_1 = x^{r_1} \sum_0^\infty a_n x^n$ and $y_2 = x^{r_2} \sum_0^\infty b_n x^n$ where r_1 and r_2 are the indicial roots, $r_1 > r_2$, and $a_0 = 1$. Find the recursion formula for the coefficients.

   (a) $9x^2 y'' + 3x(x+3)y' - (4x+1)y = 0$

   (b) $4x^2 y'' + x(2x+9)y' + y = 0$

   (c) $x^2 y'' + x(1-x)y' - 2y = 0$

   (d) $x^2(1-x^2)y'' + x(x-1)y' + 8y/9 = 0$

2. (*Nonhomogeneous ODEs*). Find the first four nonzero terms of the power series expansion of a particular solution of the following ODE:

$$x^2 y'' + x(1-x)y' - 2y = \ln(1+x)$$

Then find the general solution. [*Hint*: Expand $\ln(1+x)$ in a Taylor series about $x_0 = 0$ (see Appendix C.2). Then assume a particular solution of the form $y_p = \sum_0^\infty a_n x^n$ and use Undetermined Coefficients to find the a_ns. Refer to Problem 1(c) for the homogeneous ODE.]

<u>3.</u> (*Frobenius Series*). Solve the ODE $3x^2 y'' + 5xy' - e^x y = 0$ by expanding e^x in a Taylor series about $x_0 = 0$ and recalling the formula for the product of two series (Appendix C.2). You need only find the first four terms in the Frobenius series.

4. (*Frobenius Series and Nonuniqueness*). Find a solution of $xy'' - y' - 4x^3 y = 0$ for which $y(0) = y'(0) = 0$, but $y(x)$ not identically zero. Why does this not contradict the Uniqueness Theorem?

5. Show that the ODE $x^3 y'' + y = 0$ has no nontrivial solutions of the form $y = x^r \sum_{n=0}^\infty a_n x^n$. Why does this not contradict the results of this section?

6. (*Laguerre's Equation and Polynomials*). The ODE $xy'' + (1-x)y' + py = 0$ is *Laguerre's equation of order p.*

(a) Show that 0 is a regular singular point.

(b) Find the recursion formula for the coefficients of a Frobenius series solution. Solve to find the coefficient a_{n+1} of the series as a multiple of a_0. Set $a_0 = 1$ and write out the corresponding Frobenius series solution of Laguerre's equation of order p.

(c) Show that if p is a nonnegative integer, then there are polynomial solutions of degree p.

(d) The *Laguerre polynomials* are chosen to be those polynomial solutions $y = L_p(x)$ for which $L_p(0) = 1$. Show that the family of Laguerre polynomials $\{L_p : p = 0, 1, 2, \ldots\}$ is orthogonal on the interval $[0, \infty)$ with density $\rho(x) = e^{-x}$. Find the first five Laguerre polynomials.

7. (*Frobenius Series and Legendre's Equation*). The Legendre equation of order p is $(1-x^2)y'' - 2xy' + p(p+1)y = 0$.

<u>(a)</u> Find the indicial polynomial and the indicial roots at the regular singularity $x = 1$.

(b) Find a series solution in powers of $x - 1$. [*Hint*: Let $s = x - 1$ and rewrite the equation in sy-variables.]

11.6 Bessel Functions

The most important differential equation with a regular singularity is *Bessel's equation*[1] *of order p,*

$$x^2 y'' + xy' + (x^2 - p^2)y = 0 \tag{1}$$

where p is a nonnegative real constant. The term "order p" refers to a parameter in Bessel's equation, not to the order of (1) as a differential equation. Actually, ODE (1) is not one differential equation, but a family of equations, one for each value of p. Solutions of ODE (1) are called *Bessel functions of order p.* After the trigonometric, exponential and logarithmic functions, Bessel functions are the most widely used transcendental functions in applied mathematics. Arising in studies of vibrational, thermal, elastic, and gravitational phenomena, Bessel functions have been widely studied and their values tabulated.[2] Many software packages have Bessel functions in their lists of predefined functions. Bessel functions of assorted kinds and orders are studied in this section and the next.

Solving Bessel's Equation

Bessel's equation of order $p \geq 0$ has a regular singularity at $x = 0$; all other points are ordinary. The indicial polynomial at the singularity is $f(r) = r^2 - p^2$, whose roots are p and $-p$. Frobenius's Theorem I in Section 11.5 guarantees the existence of a series solution, $y_1 = x^p \sum_0^\infty a_n x^n$, $x > 0$, with $a_0 \neq 0$. To determine the coefficients, insert $y_1 = \sum_0^\infty a_n x^{n+p}$ and the corresponding derivative series for y_1' and y_1'' into Bessel's equation, obtaining

$$x^2 \sum_0^\infty (n+p)(n+p-1)a_n x^{n+p-2} + x \sum_0^\infty (n+p)a_n x^{n+p-1}$$

$$+ x^2 \sum_0^\infty a_n x^{n+p} - p^2 \sum_0^\infty a_n x^{n+p} = 0$$

Bring the factors to the left of the summation symbols to the inside of the summations, take the factor x^p outside, and combine the first, second, and fourth summations into a single sum to obtain

$$x^p \left\{ \sum_0^\infty [(n+p)(n+p-1) + (n+p) - p^2]a_n x^n + \sum_0^\infty a_n x^{n+2} \right\} = 0$$

---

[1] Friedrich Wilhelm Bessel (1784–1846) was a German mathematician, astronomer, and celestial mechanist who introduced the equation to model perturbations of planetary orbits. Bessel was the first person to calculate accurately the distance to a fixed star and thereby set the scale for all stellar distances. Like John Couch Adams and Urbain Jean Leverrier, Bessel tried to calculate the orbit of a suspected (but at that time undetected) planet by its gravitational effect on the orbit of the planet Uranus. But Bessel did not finish his calculations, and Adams and Leverrier made the mathematical discovery of the planet Neptune, a discovery that was soon verified by telescope.

[2] See *Handbook of Mathematical Functions*, M. Abramowitz and I.A. Stegun, National Bureau of Standards, Washington, D.C., 1964, Dover Publications, New York, 1965; *An Atlas of Functions*, J. Spanier and K.B. Oldham, Hemisphere, New York, 1987; *CRC Standard Mathematical Tables and Formulae*, 29th Ed., W.H. Beyer (ed.) , CRC Press, Boca Raton, Fla., 1991.

Reindex the final summation to the form $\sum_2^\infty a_{n-2}x^n$ and separate out the terms corresponding to $n = 0$ and $n = 1$ in the first summation to obtain

$$x^p \left\{ [p(p-1) + p - p^2]a_0 + [(1+p)(1+p-1) + (1+p) - p^2]a_1 x \right\}$$
$$+ x^p \sum_2^\infty \left\{ [(n+p)(n+p-1) + (n+p) - p^2]a_n + a_{n-2} \right\} x^n = 0 \qquad (2)$$

or

$$(1+2p)a_1 x + \sum_2^\infty [n(n+2p)a_n + a_{n-2}]x^n = 0 \qquad (3)$$

The coefficient of a_0 vanishes in (2) because its value is $f(p) = p^2 - p^2 = 0$. The *recursion formulas* for the coefficients a_n in the $\sum_0^\infty a_n x^{n+p}$ follow from (3) by equating the coefficients of the various powers of x to zero:

$$a_1 = 0$$
$$a_n = \frac{-a_{n-2}}{n(n+2p)}, \qquad n = 2, 3, \ldots \qquad (4)$$

It follows from (4) that $a_3 = a_5 = a_7 = \cdots = 0$, while the even-indexed coefficients are given by

$$a_{2k} = \frac{-a_{2k-2}}{2k(2k+2p)} = \frac{a_{2k-4}}{2k(2k+2p)(2k-2)(2k+2p-2)}$$

$$= \cdots = \frac{(-1)^k a_0}{2^{2k}k!(k+p)(k-1+p)\cdots(1+p)}$$

The corresponding series solutions of (1) are given by

$$y_1 = a_0 x^p \left\{ 1 + \sum_{k=1}^\infty \frac{(-1)^k x^{2k}}{2^{2k}k!(1+p)\cdots(k+p)} \right\}, \qquad x > 0 \qquad (5)$$

where a_0 is any constant. The construction of an independent second solution is carried out in Section 11.7, although it should be noted here that, if p is not an integer and $a_0 \neq 0$, then a second solution independent of y_1 is defined by replacing p by $-p$ in (5).

Frobenius's Theorem I implies that the power series in (5) converges for all x; this may also be shown directly by using the Ratio Test. Formula (5) may be extended to $x \leq 0$ by replacing the factor x^p by $|x|^p$. The "normalizing" constant a_0 plays a curiously important role; a judicious choice of a_0 will result in some remarkable formulas, as we will now see.

Bessel Functions of the First Kind of Integer Order

Suppose that p is a nonnegative integer n. Then define the coefficient a_0 in (5) by

$$a_0 = \frac{1}{2^n n!}$$

After writing $n!(1+n)(2+n)\cdots(k+n)$ as $(k+n)!$ in (5), we obtain a Bessel function with a special name.

❖ **Bessel Functions of the First Kind of Integer Order.** The *Bessel function of the first kind of integer order n* is

$$J_n(x) = \left(\frac{x}{2}\right)^n \sum_{k=0}^{\infty} \frac{(-1)^k}{k!(k+n)!} \left(\frac{x}{2}\right)^{2k} \tag{6}$$

Since n is a nonnegative integer, (6) defines $J_n(x)$ for all x. It follows from (6) that

$$
\begin{array}{ll}
J_0(0) = 1, & J_0'(0) = 0 \\
J_1(0) = 0, & J_1'(0) = 1/2 \\
J_2(0) = 0, & J_2'(0) = 0 \\
\quad\vdots & \quad\vdots \\
J_n(0) = 0, & J_n'(0) = 0
\end{array}
$$

EXAMPLE 11.6.1

Three Bessel Functions
The first three Bessel functions of the first kind are

$$J_0(x) = 1 - \frac{x^2}{4} + \frac{x^4}{64} - \frac{x^6}{2304} + \cdots + (-1)^k \frac{x^{2k}}{(k!)^2 2^{2k}} + \cdots \tag{7}$$

$$J_1(x) = \frac{x}{2}\left\{1 - \frac{x^2}{8} + \frac{x^4}{192} - \frac{x^6}{9216} + \cdots + (-1)^k \frac{x^{2k}}{k!(k+1)!2^{2k}} + \cdots\right\} \tag{8}$$

$$J_2(x) = \frac{x^2}{4}\left\{\frac{1}{2} - \frac{x^2}{24} + \frac{x^4}{768} - \frac{x^6}{46080} + \cdots + (-1)^k \frac{x^{2k}}{k!(k+2)!2^{2k}} + \cdots\right\} \tag{9}$$

See Figure 11.6.1 for the graphs, which resemble decaying sinusoids. This is not surprising because the normalized form of Bessel's equation

$$y'' + \frac{1}{x}y' + \left(1 - \frac{p^2}{x^2}\right)y = 0$$

resembles (for large x) the equation with constant coefficients

$$y'' + \frac{1}{a}y' + \left(1 - \frac{p^2}{a^2}\right)y = 0 \tag{10}$$

whose solutions (for sufficiently large values of a) are linear combinations of $e^{\alpha x}\cos\beta x$ and $e^{\alpha x}\sin\beta x$, where $\alpha = -1/(2a)$ and $\beta = (1 - p^2/a^2 - 1/(4a^2))^{1/2}$. See Figure 11.6.2 for a comparison of $J_0(x)$ with a solution of $y'' + 1/10y' + y = 0$ (take $p = 0$, $a = 10$ in ODE (10)): both functions oscillate and both decay.

Truncations of a convergent, alternating series of terms whose magnitudes decrease as n increases are useful in calculating approximate values for the sum of the series. The following example shows how this may be done for the values of a Bessel function of the first kind of integer order.

EXAMPLE 11.6.2

Approximating $J_0(x)$
The terms in the alternating series (7) for $J_0(x)$ diminish in magnitude as k increases if $|x| < 1$. In this case, the magnitude of the difference between the value of $J_0(x)$ and a partial sum of the series is no more than the magnitude of the first term omitted from the partial sum. For example, if $x = 1/10$, then from (7)

$$\left|J_0\left(\frac{1}{10}\right) - \left[1 - \frac{1}{4}\left(\frac{1}{10}\right)^2\right]\right| \le \frac{1}{64}\left(\frac{1}{10}\right)^4 \approx 1.6 \times 10^{-6}$$

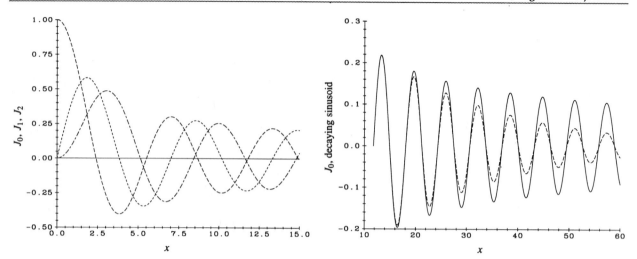

FIGURE 11.6.1 Graphs of J_0 (long dashes), J_1 (short dashes), and J_2 (long/short dashes). See Example 11.6.1.

FIGURE 11.6.2 J_0 (solid) and an exponentially decaying sinusoid (dashes). See Example 11.6.1.

and so $1 - 1/400 = 399/400$ is a good approximation to the value of $J_0(1/10)$. Of course, if x is large, then k may have to be very large before the magnitudes of the terms in (7) begin to diminish. Other methods are generally used to approximate $J_n(x)$ for large x.

The Gamma Function

With a bit of mathematical trickery, we will show that the formula (6) makes sense when n is replaced by any nonnegative number p, not necessarily an integer. This involves introducing the gamma function, which was invented by Euler, to solve the problem of constructing an infinitely differentiable factorial function defined for all real $p \geq 0$ that interpolates the integer factorials.

❖ **Gamma Function.** The *gamma function* is defined for $z > 0$ by

$$\Gamma(z) = \int_0^\infty x^{z-1} e^{-x} \, dx \tag{11}$$

Although the integral in (11) is improper, it may be shown that $\Gamma(z)$ and all of its derivatives with respect to z,

$$\Gamma'(z) = \int_0^\infty (x^{z-1} \ln x) e^{-x} \, dx, \quad \Gamma''(z) = \int_0^\infty [x^{z-1} (\ln x)^2] e^{-x} \, dx, \quad \dots$$

are convergent improper integrals for $z > 0$. This means that $\Gamma(z)$ is continuous and so are all of its derivatives for $z > 0$. It may also be shown that $\Gamma(z) \to \infty$ as $z \to 0^+$. The gamma function has other properties that justify its alternate name as "the interpolated factorial."

THEOREM 11.6.1

Properties of the Gamma Function. The infinitely differentiable function $\Gamma(z)$ satisfies the formulas

$$\Gamma(z+1) = z\Gamma(z), \qquad z > 0 \tag{12}$$

$$\Gamma(n+1) = n!, \qquad n = 0, 1, 2, \ldots \tag{13}$$

$$\Gamma(n+\frac{1}{2}) = \frac{(2n)!\pi^{1/2}}{2^{2n}n!}, \qquad n = 0, 1, 2, \ldots \tag{14}$$

Proof. Replace z by $z+1$ in (11) and integrate by parts:

$$\Gamma(z+1) = \int_0^\infty x^z e^{-x}\,dx = -x^z e^{-x}\Big|_{x=0}^{x=\infty} + \int_0^\infty zx^{z-1}e^{-x}\,dx$$

$$= 0 + z\int_0^\infty x^{z-1}e^{-x}\,dx = z\Gamma(z), \qquad z > 0$$

because $-x^z e^{-x}$ is zero at $x = 0$ and approaches 0 as x approaches ∞. This proves formula (12).

Formula (13) is proved by induction on n. First note that

$$\Gamma(1) = \int_0^\infty e^{-x}\,dx = 1 = 0!$$

Then suppose that $\Gamma(n+1) = n!$ and prove that $\Gamma(n+2) = (n+1)!$:

$$\Gamma(n+2) = \int_0^\infty x^{n+1}e^{-x}\,dx = -x^{n+1}e^{-x}\Big|_{x=0}^{x=\infty} + (n+1)\int_0^\infty x^n e^{-x}\,dx$$

$$= (n+1)\Gamma(n+1) = (n+1)n! = (n+1)!$$

where integration by parts and the supposition that $\Gamma(n+1) = n!$ have been used. By the Principle of Induction, (13) holds for all n. Observe that the value of $\Gamma(n+1)$ is $n!$, not $(n+1)!$ as one might have anticipated. The proof of (14) is left to the reader (Problem 7).

Property (12) written as $\Gamma(z) = \Gamma(z+1)/z$ is used to define the gamma function on $-1 < z < 0$ by means of its values for $0 < z < 1$. For example, $\Gamma(-1/2) = \Gamma(1/2)/(-1/2) = -2\pi^{1/2}$ since $\Gamma(1/2) = \pi^{1/2}$ by (14). Then the values of $\Gamma(z)$ for $-1 < z < 0$ are used to define $\Gamma(z)$ for $-2 < z < -1$, and so on. In this way, the gamma function is defined for all z except the negative integers (the above process breaks down at negative integers since $\Gamma(0)$ is infinite). The extended gamma function is differentiable to all orders wherever it is defined. See Figure 11.6.3 for its graph.

Bessel Functions of the First Kind of Any Order

The gamma function is used to extend the definition of a_0 in (5) to any choice of $p \geq 0$:

$$a_0 = \frac{1}{2^p\Gamma(p+1)}$$

Use this value of a_0 in (5) and the fact that $\Gamma(p+1)(1+p)\cdots(k+p) = \Gamma(k+p+1)$, which follows from repeated application of identity (12) to $\Gamma(k+p+1)$, to define $J_p(x)$ for any real $p \geq 0$:

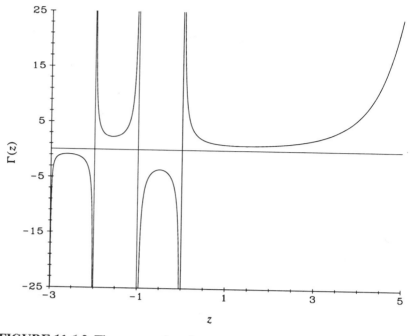

FIGURE 11.6.3 The gamma function.

❖ **Bessel Functions of the First Kind of Order** p. The *Bessel function of the first kind of order* $p \geq 0$ is defined by

$$J_p(x) = \left(\frac{x}{2}\right)^p \sum_{k=0}^{\infty} \frac{(-1)^k}{k!\Gamma(k+p+1)} \left(\frac{x}{2}\right)^{2k}, \quad x > 0 \qquad (15)$$

If $x < 0$, replace $(x/2)^p$ by $(|x|/2)^p$. Formula (15) reduces to (6) if p is an integer n. Formula (15) looks exotic because of the appearance of the gamma function, but the behavior of $J_p(x)$ and its graph resembles that of $J_n(x)$ for n a positive integer "near" p. See Figure 11.6.4. Finally, observe that if p is replaced by $-p$ in (15), then a second solution, J_{-p}, of Bessel's equation (1) is defined as long as the positive number p is not an integer, and this solution is independent of J_p.

Bessel functions of the first kind satisfy several recursion formulas that are somewhat similar to the Legendre recursion identities of Section 11.3. Formula (16) below is particularly important because it can be used to generate J_{p+1} from J_p and J_{p-1}.

THEOREM 11.6.2

Recursion Formulas. Bessel functions of the first kind satisfy the identities

$$J_{p+1} = \frac{2p}{x} J_p - J_{p-1}, \qquad p \geq 1 \qquad (16)$$

$$J_{p+1} = -2J_p' + J_{p-1}, \qquad p \geq 1 \qquad (17)$$

$$(x^p J_p)' = x^p J_{p-1}, \qquad p \geq 1 \qquad (18)$$

$$(x^{-p} J_p)' = -x^{-p} J_{p+1}, \qquad p \geq 0 \qquad (19)$$

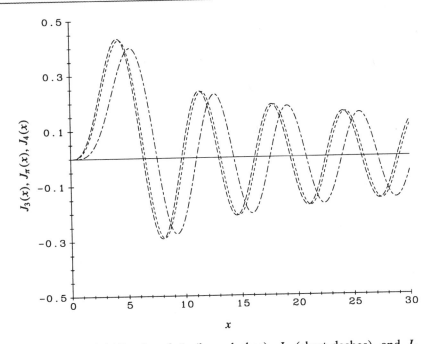

FIGURE 11.6.4 Graphs of J_3 (long dashes), J_π (short dashes), and J_4 (long/short dashes).

Proof. Formula (18) is proved first. Multiply each side of (15) by x^p, differentiate term-by-term, and use the fact that $\Gamma(k+p+1) = \Gamma(k+p)(k+p)$:

$$[x^p J_p]' = \frac{d}{dx}\left\{\sum_{k=0}^{\infty} \frac{(-1)^k 2^p}{k!\Gamma(k+p+1)}\left(\frac{x}{2}\right)^{2k+2p}\right\}$$

$$= \sum_{k=0}^{\infty} \frac{(-1)^k(k+p)2^p}{k!\Gamma(k+p+1)}\left(\frac{x}{2}\right)^{2k+2p-1}$$

$$= x^p\left(\frac{x}{2}\right)^{p-1}\sum_{k=0}^{\infty} \frac{(-1)^k}{k!\Gamma(k+p)}\left(\frac{x}{2}\right)^{2k}$$

$$= x^p J_{p-1}$$

The proof of (19) is similar (see Problem 8). To prove (16), multiply each side of (18) by x^{-p}, each side of (19) by x^p, and subtract the two:

$$x^{-p}[x^p J_p]' - x^p[x^{-p}J_p]' = J_{p-1} + J_{p+1}$$

$$x^{-p}[px^{p-1}J_p + x^p J_p'] - x^p[-px^{-p-1}J_p + x^{-p}J_p'] = J_{p-1} + J_{p+1}$$

from which (16) follows after a little algebraic rearrangement. Formula (17) is derived similarly (see Problem 8). Note that (16)–(19) also hold for the functions J_{-p} if p is positive and not an integer (Problem 8(c)).

EXAMPLE 11.6.3

Half-Integer Bessel Functions of the First Kind
The function $J_{1/2}(x)$ is not nearly as formidable as its definition via (15) makes it

appear. In fact, $J_{1/2}$ is a sine function with a decaying amplitude

$$J_{1/2}(x) = \left(\frac{2}{\pi x}\right)^{1/2} \sin x, \quad x > 0 \tag{20}$$

This is shown by letting $p = 1/2$ in (15):

$$J_{1/2}(x) = \left(\frac{x}{2}\right)^{1/2} \sum_{k=0}^{\infty} \frac{(-1)^k}{k!\Gamma(k+3/2)} \left(\frac{x}{2}\right)^{2k} = \left(\frac{x}{2}\right)^{1/2} \sum_{k=0}^{\infty} \frac{(-1)^k 2^{2k+2}(k+1)!}{k!(2k+2)!\pi^{1/2}} \left(\frac{x}{2}\right)^{2k}$$

$$= \left(\frac{2}{\pi x}\right)^{1/2} \sum_{k=0}^{\infty} \frac{(-1)^k}{(2k+1)!} x^{2k+1} = \left(\frac{2}{\pi x}\right)^{1/2} \sin x \tag{21}$$

where the last series is the Maclaurin series for $\sin x$. In the second step of the derivation, (14) is used with $n = k + 1$:

$$\Gamma(k+3/2) = \frac{(2k+2)!\pi^{1/2}}{2^{2k+2}(k+1)!}$$

The third summation in (21) is obtained from the second by rearranging and canceling some terms.

By the same kind of argument it may be shown that

$$J_{3/2}(x) = \left(\frac{2}{\pi x}\right)^{1/2} \left[-\cos x + \frac{\sin x}{x}\right] \tag{22}$$

Recursion formula (16) may be used to define the remaining *half-integer Bessel functions of the first kind*, $J_{n+1/2}$, $n = 2, 3, \ldots$. For example, by (16), we have

$$J_{5/2} = \frac{3}{x} J_{3/2} - J_{1/2} = \left(\frac{2}{\pi x}\right)^{1/2} \left[\left(\frac{3}{x^2} - 1\right) \sin x - \frac{3}{x} \cos x\right]$$

The graphs of $J_{1/2}$, $J_{3/2}$, and $J_{5/2}$ are displayed in Figure 11.6.5.

Properties of Bessel Functions

The functions $J_p(x)$ seem to behave like decaying sinusoids such as $f(x) = e^{-x} \sin x$. The aim now is to describe the similarities more precisely. The nonnegative zeros of the decaying sinusoid are at π, 2π, 3π, $\ldots$, each zero is simple (i.e., $f'(n\pi) \neq 0$), the zeros are isolated from one another, successive zeros are π units apart, and the increasing sequence of zeros diverges to $+\infty$. The following theorems show that each nontrivial Bessel function (and not just $J_p(x)$) has properties that are similar to those of $f(x)$, but not precisely the same.

THEOREM 11.6.3

> **Zeros of Bessel Functions.** Let $y(x)$ be any nontrivial solution of Bessel's equation (1) for $x > 0$. Then the positive zeros λ_n of $y(x)$ are simple, are isolated from each other, form an increasing sequence, $0 < \lambda_1 < \lambda_2 < \cdots < \lambda_n < \cdots$ that diverges to ∞, and $\lambda_{n+1} - \lambda_n \to \pi$ as $n \to \infty$.

The proof is omitted.

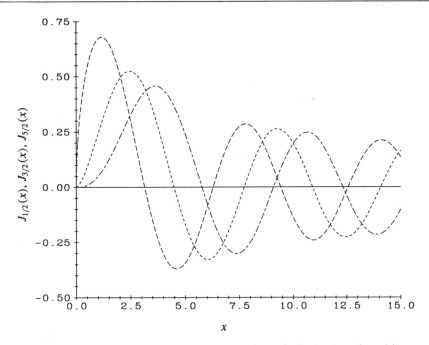

FIGURE 11.6.5 $J_{1/2}$ (long dashes), $J_{3/2}$ (short dashes), $J_{5/2}$ (long/short dashes). See Example 11.6.3.

EXAMPLE 11.6.4

The Zeros of J_0

The list below gives several pairs of zeros λ_n and λ_{n+1} of J_0; note how the difference between successive zeros is close to π if n is large. The zeros of J_0 through J_{20} have been calculated to several decimal places because of their importance in analysis. For one of these applications, see Example 11.7.3 in Section 11.7.

λ_n	λ_1	λ_2	λ_5	λ_6	λ_{19}	λ_{20}
Value	2.4048	5.5201	14.9309	18.0711	58.9070	62.0485
$\lambda_{n+1} - \lambda_n$		3.1153		3.1402		3.1415

The oscillatory character of nontrivial Bessel functions follows from the fact that each function is continuous and has a sequence of zeros as described in Theorem 11.6.3. The zeros are simple and this implies that the graph of a nontrivial Bessel function crosses the x-axis at nonzero angles at the zeros, rising at one zero, falling at the next, just as the graphs indicate. It may be shown that between its successive zeros, a Bessel function takes on exactly one local extreme value, and this is again suggested (but not proved) by the visual evidence of the graphs in this section.

The zeros of $J_p(x)$ play an important role in the orthogonality of Bessel functions on the interval $[0, \ 1]$ with weight function x.

THEOREM 11.6.4

Orthogonality. Let $p \geq 0$, and let λ and μ be positive zeros of $J_p(x)$. Then

$$\int_0^1 x J_p(\lambda x) J_p(\mu x) \, dx = \begin{cases} 0, & \lambda \neq \mu \\ \frac{1}{2}(J_p'(\lambda))^2, & \lambda = \mu \end{cases} \tag{23}$$

Proof. Let λ be any positive zero of J_p. The Chain Rule may be used to show that if L is the *Bessel operator*, $L = x^2 d^2/dx^2 + xd/dx + (x^2 - p^2)$, then

$$L[J_p(\lambda x)] = x^2 \frac{d^2}{dx^2} J_p(\lambda x) + x\frac{d}{dx} J_p(\lambda x) + (\lambda^2 x^2 - p^2) J_p(\lambda x) = 0 \quad (24)$$

A straightforward calculation shows that if λ and μ are zeros of J_0, then

$$\frac{1}{x}[J_p(\mu x)L[J_p(\lambda x)] - J_p(\lambda x)L[J_p(\mu x)]]$$

$$= \frac{d}{dx}\left\{x[J_p(\mu x)J_p'(\lambda x) - J_p'(\mu x)J_p(\lambda x)]\right\}$$

$$+ x(\lambda^2 - \mu^2)J_p(\lambda x)J_p(\mu x) = 0$$

Integrating each side of the equation

$$\frac{d}{dx}\{\cdot\} + x(\lambda^2 - \mu^2)J_p(\lambda x)J_p(\mu x) = 0$$

from 0 to 1 gives

$$\left\{x[J_p(\mu x)J_p'(\lambda x) - J_p'(\mu x)J_p(\lambda x)]\right\}\Big|_{x=0}^{x=1} + (\lambda^2 - \mu^2)\int_0^1 xJ_p(\lambda x)J_p(\mu x)dx = 0$$

The desired result for $\lambda \neq \mu$ follows. The proof of (23) for $\lambda = \mu$ is omitted.

The orthogonality property looks strange, but it is analogous to the orthogonality property of the functions $\sin n\pi x$ and $\sin m\pi x$ on the interval $[0, 1]$:

$$\int_0^1 \sin(n\pi x) \sin(m\pi x)\, dx = \begin{cases} 0, & n \neq m \\ 1/2, & n = m \end{cases}$$

Just as $n\pi$ and $m\pi$ are zeros of $\sin x$, λ and μ are zeros of $J_p(x)$. The difference is that the orthogonality condition for a Bessel function uses a weight factor x. See Figures 11.6.6 and 11.6.7 for graphs of $J_0(\lambda_2 x)$, $J_0(\lambda_5 x)$, and $xJ_0(\lambda_2 x)J_0(\lambda_5 x)$, $0 \leq x \leq 1$, where λ_2 and λ_5 are the second and fifth zeros of $J_0(x)$ (see the table in Example 11.6.4 for their values). In Figure 11.6.7 the area of the region above the x-axis equals the area of the region below since the integral of $xJ_0(\lambda_2 x)J_0(\lambda_5 x)$ must be zero.

The next theorem shows that nontrivial solutions of Bessel's equation do indeed decay, but like $x^{-1/2}$, not like $e^{-\alpha x}$.

THEOREM 11.6.5

Decay. For every nontrivial solution $y(x)$ of Bessel's equation there are constants A and φ and a bounded function $h(x)$ such that

$$y(x) = \left(\frac{2}{\pi x}\right)^{1/2}\left[A\sin(x + \varphi) + \frac{1}{x}h(x)\right], \qquad x \geq 1 \quad (25)$$

The proof is omitted.

EXAMPLE 11.6.5

Decay of $J_{1/2}$ and $J_{3/2}$
Using the formulas of Example 11.6.3 for $J_{1/2}$ and $J_{3/2}$, we see that $J_{1/2}$ already is in the form (25) with $A = 1$, $\varphi = 0$, and $h = 0$. It is not hard to show from (22) that

$$J_{3/2}(x) = \left(\frac{2}{\pi x}\right)^{1/2}\left[\sin(x + \pi/2) + \frac{1}{x}\sin x\right]$$

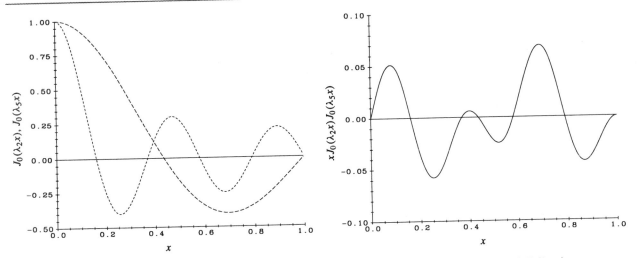

FIGURE 11.6.6 $J_0(\lambda_2 x)$ (long dashes) and $J_0(\lambda_5 x)$ (short dashes), $0 \leq x \leq 1$.

FIGURE 11.6.7 Graph of $x J_0(\lambda_2 x) J_0(\lambda_5 x)$.

with $A = 1$, $\varphi = \pi/2$, and $h = \sin x$.

The form of (25) indicates that all nontrivial solutions of Bessel's equation decay as $x \to \infty$ in much the same way as the half-integer Bessel functions decay.

Comments

Although the formula (15) that defines each Bessel function of the first kind is formidable, the properties of the functions themselves are not very different from those of the familiar decaying sinusoids that appear as solutions of second-order differential equations with constant coefficients. The Recursion Formulas of Theorem 11.6.2 and the formulas given in Problem 9 show that a differential and integral calculus of the functions $J_p(x)$ can be constructed. Bessel functions of the second kind are defined in the next section. They, too, have many (but not all) of the features of the Bessel functions of the first kind.

PROBLEMS

1. Find the first four terms in the series for $J_3(x)$.

2. (*Properties of $J_3(x)$*). The first few terms of the series for $J_3(x)$ can be used to determine some of its properties.

(a) Let $f(x)$ be the ninth-degree polynomial of the first four terms of the series for $J_3(x)$. Plot $f(x)$ for $0 \leq x \leq 2$. Explain why $|J_3(x) - f(x)| < 0.02$ for $0 \leq x \leq 4$.

(b) Explain why the first positive zero of $J_3(x)$ is larger than 4.

(c) Explain why the graph of $J_3(x)$ is concave upward at $x = 0$.

3. (*General Solution of Bessel's Equation*). Use the Wronskian Reduction of Order Method (Problem 13, Section 3.3) to show that the general solution of Bessel's equation of order p can be written in the form

$$y = c_1 J_p(x) + c_2 J_p(x) \int^x \frac{ds}{s[J_p(s)]^2}, \qquad c_1 \text{ and } c_2 \text{ any constants}$$

4. (*Convergence*). Use the Ratio Test to show that the series for $J_p(x)$ converges for all x.

5. Plot J_0, J_1, J_2, and J_3. Then plot $J_{1/3}$, $J_{4/3}$, $J_{7/3}$, $0 \le x \le 20$.

6. (*Zeros of Bessel Functions*).

   (a) Locate the zeros of J_0, J_1, J_2 and $J_{1/2}$, $J_{3/2}$, $J_{5/2}$ in Figures 11.6.1 and 11.6.5, and make a conjecture about the relative positions of the positive zeros of J_p and of J_{p+1}.

   (b) Prove your conjecture.

7. (*Values of* $\Gamma(n+1/2)$). Formula (14) for $\Gamma(n+1/2)$ may be proved by completing parts (a) and (b) below.

   (a) Prove that $\Gamma(1/2) = \pi^{1/2}$ by carrying out the following steps. First make a change of variable in the integral that defines $\Gamma(1/2)$ and show that $\Gamma(1/2) = 2\int_0^\infty e^{-u^2}\, du$. Then show that $(\Gamma(1/2))^2 = 4\int_0^\infty \int_0^\infty e^{-u^2-v^2}\, du\, dv$, and evaluate that double integral by switching to polar coordinates and integrating over the region $r \ge 0$, $0 \le \phi \le \pi/2$.

   (b) Use (a), the property $\Gamma(z+1) = z\Gamma(z)$, and the Principle of Induction to show that

   $$\Gamma\left(n + \frac{1}{2}\right) = \frac{(2n)!\pi^{1/2}}{2^{2n}n!} \text{ for } n = 0, 1, 2, \ldots$$

   (c) Use the result in (a) to find $\Gamma(3/2)$ and $\Gamma(5/2)$. Then find $\Gamma(-3/2)$ and $\Gamma(-5/2)$.

8. (*Recursion Properties of Bessel Functions*).

   (a) Prove Recursion Formula (19) and then (17). [*Hint*: Look at the proofs of (18) and (16).]

   (b) Express J_2, J_3, and J_4 in terms of J_0 and J_1 by using Recursion Formula (16).

   (c) Show that (16)–(19) hold for the function J_{-p}, where $p > 0$, but p not an integer.

9. (*Integrating* $J_p(x)$). The Recursion Formulas (16)–(19) and Integration by Parts may be used to derive integration formulas for Bessel functions of the first kind. The symbol C in the formulas below denotes an arbitrary constant of integration.

   (a) Show that $\int^x J_1(t)\, dt = -J_0(x) + C$ and that $\int^x t J_0(t)\, dt = x J_1(x) + C$.

   (b) Show that $\int^x J_0(t) \sin t\, dt = x J_0(x) \sin x - x J_1(x) \cos x + C$ and

   $$\int^x J_0(t) \cos t\, dt = x J_0(x) \cos x + x J_1(x) + x J_1(x) \sin x + C$$

   [*Hint*: Show that the derivative of the right side is the given integrand.]

   (c) Show that $\int^x t^2 J_0(t)\, dt = x^2 J_1(x) + x J_0(x) - \int^x J_0(t)\, dt$. [*Hint*: Use part (a).]

   (d) Show that $\int^x t^5 J_2(t)\, dt = x^5 J_3(x) - 2x^4 J_4(x) + C$. [*Hint*: $\int^x t^5 J_2\, dt = \int^x t^2(t^3 J_2)\, dt = \int^x t^2(t^3 J_3)'\, dt$.]

10. Show that

$$J_{-1/2} = \left(\frac{2}{\pi}\right)^{1/2} \frac{\cos x}{x^{1/2}}$$

is a solution of Bessel's equation of order $1/2$ and that $\{J_{-1/2}, J_{1/2}\}$ is an independent set of solutions of that equation for $x > 0$.

11. Plot $J_1(x)$ and find (graphically) the first pair of its successive zeros λ_n, λ_{n+1} for which $|(\lambda_{n+1} - \lambda_n) - \pi| < 0.0001$.

12. (*Modified Bessel Equations*). Equations related to Bessel's equation often appear in the applications, but the properties of their solutions may be quite different from those of the Bessel functions.

(a) Replace the independent variable x in Bessel's equation of order p by the complex quantity $-it$, and obtain the *modified Bessel equation of order p*:

$$t^2 y'' + t y' - (t^2 + p^2) y = 0$$

(b) Show that

$$I_0(t) = \sum_{n=0}^{\infty} \frac{1}{(n!)^2} \left(\frac{t}{2}\right)^{2n}$$

is a solution to the modified Bessel equation of order 0, $t \geq 0$.

(c) (*Modified Bessel Function of the First Kind*). Show that

$$I_p(t) = \left(\frac{t}{2}\right)^p \sum_{n=0}^{\infty} \frac{1}{n!\,\Gamma(n+p+1)} \left(\frac{t}{2}\right)^{2n}$$

is a solution of the modified Bessel equation of order $p \geq 0$. I_p is called the *modified Bessel function of the first kind of order p*.

11.7 Series Solutions Near Regular Singular Points, II

Bessel functions were constructed early in the nineteenth century by the method of generalized power series, but, as noted in Section 11.6, the method does not always produce a second, independent solution of Bessel's equation. The puzzle of finding a second solution remained unresolved until 1867, when C.G. Neumann (1832–1925) published his pioneering work on the matter. Frobenius extended Neumann's methods to a complete treatment of all solutions near any regular singular point of any second-order linear ODE. The extended theory of Frobenius always allows us to find a second solution even when the roots of the indicial polynomial differ by an integer. In this section, the extended theory is outlined, Bessel functions of the second kind of any order p are defined, and all solutions of the problem of the aging spring are expressed in terms of Bessel functions.

The Extended Method of Frobenius

To review briefly, the goal is to find a pair of independent solutions near the regular singular point $x = 0$ of the second-order linear ODE

$$x^2 y'' + x p(x) y' + q(x) y = 0 \tag{1}$$

where $p(x)$ and $q(x)$ are real analytic on an interval J centered at the origin, $p(x) = \sum_0^{\infty} p_n x^n$, $q(x) = \sum_0^{\infty} q_n x^n$. Associated with (1) are the quadratic indicial polynomial $f(r) = r^2 + (p_0 - 1)r + q_0$ and its indicial roots r_1 and r_2. We will assume that the indicial roots r_1 and r_2 are real and that $r_2 \leq r_1$. The Method of Frobenius outlined in Section 11.5 always yields a nontrivial solution of ODE (1) of the form

$$y_1 = |x|^{r_1} \sum_0^\infty a_n x^n$$

However, the method may not always yield a second solution when $r_1 = r_2$, or even when $r_1 - r_2$ is a positive integer. In the former case, experience with Euler equations suggests that the second solution should involve a logarithm, but the case where $r_1 - r_2$ is a positive integer is somewhat mysterious. What form do the second solutions have in that situation, and how may they be found? The following complete version of Frobenius's Theorem clears up all of these questions.

THEOREM 11.7.1

Frobenius' Theorem II. Assume that the indicial roots r_1 and r_2 for the ODE $x^2 y'' + xp(x)y' + q(x)y = 0$ are real and that $r_1 \geq r_2$. Then the Method of Undetermined Coefficients produces a nontrivial solution y_1 of the ODE and y_1 has the form:

$$y_1(x) = |x|^{r_1} \sum_0^\infty a_n x^n \tag{2}$$

where a_0 is arbitrary. The a_ns are determined in terms of a_0 by the recursion formula that makes use of the indicial polynomial $f(r) = r^2 + (p_0 - 1)r + q_0$:

$$f(r_1 + n)a_n = -\sum_{k=0}^{n-1}[(k + r_1)p_{n-k} + q_{n-k}]a_k, \qquad n = 1, 2, \dots \tag{3}$$

A second independent solution y_2 of the ODE may also be determined by the Method of Undetermined Coefficients. It has one of the forms described below and is defined for $x > 0$, or $x < 0$:

- **Case I:** If $r_1 - r_2$ is not an integer, then

$$y_2 = |x|^{r_2}\left(1 + \sum_1^\infty b_n x^n\right) \tag{4}$$

where the b_ns are determined by the same recursion formula (3) as the a_ns, with r_2 replacing r_1.

- **Case II:** If $r_1 = r_2$, then

$$y_2 = y_1 \ln|x| + |x|^{r_1} \sum_0^\infty c_n x^n \tag{5}$$

- **Case III:** If $r_1 - r_2$ is a positive integer, then

$$y_2 = \alpha y_1 \ln|x| + |x|^{r_2}\left(1 + \sum_1^\infty d_n x^n\right) \tag{6}$$

where α is a constant (possibly 0).

The power series in formulas (4)–(6) converge in an interval J centered at 0.

The proof of Frobenius's Theorem II is omitted. It is clear from the recursion formula (3) just where the difficulty lies. In Case II, if r_1 in formula (3) is replaced by r_2, no new solution independent of y_1 results. In Case III, if $r_1 - r_2$ is the positive integer m, then (3) may not produce a second solution independent of y_1 when r_1 is replaced by r_2 because $f(r_2 + m) = f(r_1) = 0$, and (3) may not be solvable for a_m.

The coefficients c_n, d_n, and α in (5) and (6) may be determined by substituting the appropriate form for y_2 into ODE (1) and matching coefficients of like powers of x to obtain recursion relations. The first example illustrates Case II.

EXAMPLE 11.7.1

Case II: Equal Indicial Roots
The ODE

$$xy'' + y' + 2y = 0$$

has a regular singular point at 0. Its indicial roots are $r_1 = r_2 = 0$ (Case II of Frobenius's Theorem II). The solution space is generated by a Frobenius series

$$y_1 = \sum_0^\infty a_n x^n$$

and a function with a logarithmic term,

$$y_2 = y_1 \ln |x| + \sum_0^\infty c_n x^n$$

The Method of Undetermined Coefficients may be used to find the values of a_n in y_1:

$$x \sum_0^\infty n(n-1)a_n x^{n-2} + \sum_0^\infty n a_n x^{n-1} + 2 \sum_0^\infty a_n x^n = 0$$

$$\sum_1^\infty [n(n-1)a_n + na_n + 2a_{n-1}]x^{n-1} = 0$$

which leads to the recursion relation $n^2 a_n + 2a_{n-1} = 0$. Hence

$$a_n = \frac{-2a_{n-1}}{n^2} = \frac{4a_{n-2}}{n^2(n-1)^2} = \cdots = \frac{(-1)^n 2^n a_0}{(n!)^2}$$

Setting $a_0 = 1$, we obtain

$$y_1 = 1 + \sum_1^\infty \frac{(-1)^n 2^n x^n}{(n!)^2} \tag{7}$$

For convenience, we put $L[y] = xy'' + y' + 2y$, and turn to the calculation of y_2 for $x > 0$, where y_2 has the form given in formula (5). Using the fact that L is a linear operator, we have

$$0 = L[y_2] = L[y_1 \ln x] + L\left[\sum_0^\infty c_n x^n \right]$$

$$= x \left(y_1'' \ln x + \frac{2y_1'}{x} - \frac{y_1}{x^2} \right) + \left(y_1' \ln x + \frac{y_1}{x} \right)$$

$$+ 2y_1 \ln x + \sum_1^\infty (n^2 c_n + 2c_{n-1})x^{n-1}$$

After collecting the terms including $\ln x$, the above expression may be rearranged:

$$0 = L[y_2] = L[y_1]\ln x + 2y_1' + \sum_1^\infty (n^2 c_n + 2c_{n-1})x^{n-1}$$

$$= 2y_1' + \sum_1^\infty (n^2 c_n + 2c_{n-1}) x^{n-1} \qquad \text{(since } L[y_1] = 0)$$

$$= \sum_1^\infty \frac{(-1)^n 2^{n+1} x^{n-1}}{(n-1)!n!} + \sum_1^\infty (n^2 c_n + 2c_{n-1}) x^{n-1}$$

where it is only in the final step that the actual series (7) for y_1 is used. This leads to the recursion relation

$$n^2 c_n + 2c_{n-1} + \frac{(-1)^n 2^{n+1}}{(n-1)!n!} = 0, \qquad n = 1, 2, \ldots \qquad (8)$$

It is hard to solve this set of recursion relations to find c_n in terms of c_0. Instead, we shall find the coefficients c_1, c_2, and c_3 in terms of c_0, using the recursion (8) with $n = 1, 2, 3$:

$$\begin{array}{ll} c_1 + 2c_0 - 4 = 0, & c_1 = 2(2 - c_0) \\ 4c_2 + 2c_1 + 4 = 0, & c_2 = -3 + c_0 \\ 9c_3 + 2c_2 - 4/3 = 0, & c_3 = 22/27 - 2c_0/9 \end{array}$$

For simplicity, set $c_0 = 0$ and obtain $c_1 = 4$, $c_2 = -3$, $c_3 = 22/27$. Then

$$y_2 = y_1 \ln x + \sum_1^\infty c_n x^n = y_1 \ln x + 4x - 3x^2 + \frac{22}{27} x^3 + \cdots \qquad (9)$$

where the c_ns are defined recursively by (8) with $c_0 = 0$.

Note once more that for all practical purposes an unsolved recursion relation is just as useful as a solved relation. If a computer is being used to calculate the coefficients, the unsolved form is to be preferred since it is almost always simpler. Initializing with c_0 and then iterating with (8) did the trick in Example 11.7.1.

The following example illustrates Case III of Frobenius' Theorem II.

EXAMPLE 11.7.2

Case III: Indicial Roots Differ by a Positive Integer
The ODE

$$xy'' - y = 0$$

has a regular singular point at 0, and its indicial polynomial has roots $r_1 = 1$ and $r_2 = 0$. The equation has a Frobenius series solution of the form

$$y_1 = x \sum_0^\infty a_n x^n = \sum_0^\infty a_n x^{n+1}, \qquad a_0 \neq 0$$

where the a_ns may be found in the usual way. The Method of Undetermined Coefficients with $a_0 = 1$ may be applied to show that

$$y_1 = x \sum_0^\infty \frac{x^n}{(n+1)!n!} = \sum_0^\infty \frac{x^{n+1}}{(n+1)!n!}$$

From Frobenius's Theorem II, a second solution, independent of y_1, has the form

$$y_2 = \alpha y_1 \ln |x| + \sum_0^\infty d_n x^n, \qquad d_0 = 1$$

Let x be positive and apply to y_2 the linear operator L defined by $L[y] = xy'' - y$:

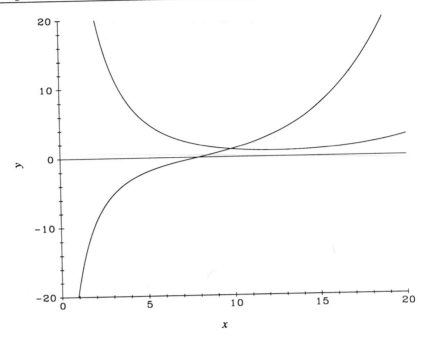

FIGURE 11.7.1 A pair of independent solutions of $xy'' - y = 0$. See Example 11.7.2.

$$0 = L[y_2] = \alpha L[y_1 \ln x] + L\left[\sum_0^\infty d_n x^n\right]$$

$$= \alpha \ln x L[y_1] + \alpha\left(2y_1' - \frac{y_1}{x}\right) + \sum_0^\infty [n(n+1)d_{n+1} - d_n]x^n$$

$$= \alpha \sum_0^\infty \frac{2n+1}{(n+1)!n!}x^n + \sum_0^\infty [n(n+1)d_{n+1} - d_n]x^n$$

where we have used the fact that $L[y_1] = xy_1'' - y_1 = 0$. This gives the recursion relation

$$n(n+1)d_{n+1} - d_n = \frac{-\alpha(2n+1)}{(n+1)!n!}, \quad n = 0, 1, 2, \dots \qquad (10)$$

The first three equations are

$$-d_0 = -\alpha, \quad 2d_2 - d_1 = \frac{-3\alpha}{2}, \quad 6d_3 - d_2 = \frac{-5\alpha}{12}$$

Since $d_0 = 1$ is already built into (6), we see that $\alpha = 1$. Thus,

$$\alpha = d_0 = 1, \quad d_2 = d_1/2 - 3/4, \quad d_3 = d_1/12 - 7\alpha/36$$

Letting $d_1 = 0$ (or any other convenient value), the recursion formula (10) may be used to determine d_n, $n = 2, 3, \dots$. The independent solutions of $xy'' - y = 0$ corresponding to the sets of initial data, $y'(10) = 1$, $y'(10) = 1$, and $y(10) = 1$, $y'(10) = -1$ are graphed in Figure 11.7.1.

Although $\alpha \neq 0$ in Example 11.7.2, it may happen that α vanishes. In such a case, the second solution would be a generalized power series like the first, and there would be no logarithmic term.

Bessel Functions of the Second Kind of Integer Order

We now apply Frobenius's Theorem II to the problem of finding a second solution of Bessel's equation of order n

$$x^2 y'' + xy' + (x^2 - n^2)y = 0 \tag{11}$$

where n is a nonnegative integer. One solution, $J_n(x)$, of the Bessel equation (11) was found in Section 11.6.

Because the difference of the two roots n and $-n$ of the indicial polynomial $f(r) = r^2 - n^2$ is an integer, Frobenius's Theorem I does not apply. There are two cases to consider.

Case 1: If $n = 0$, then Case II of Frobenius's Theorem II applies. A second solution of Bessel's equation of order 0 independent of J_0 has the form

$$y_2 = y_1 \ln |x| + \sum_0^\infty c_n x^n$$

and the c_ns may be determined by the Method of Undetermined Coefficients. This can be done, but it requires ingenuity and patience. In practice, the second solution is taken to be a certain linear combination of $J_0(x)$ and y_2. This is the *Bessel function of the second kind* (or *Weber function*) *of order* 0 and is defined by

$$Y_0(x) = \frac{2}{\pi}\left(\gamma + \ln\left|\frac{x}{2}\right|\right)J_0(x) - \frac{2}{\pi}\sum_{k=0}^\infty \frac{(-1)^k h(k)}{(k!)^2}\left(\frac{x}{2}\right)^{2k}$$

where $h(k) = 1 + 1/2 + \cdots + 1/k$ and γ is *Euler's constant*,[3]

$$\gamma = 1 + \sum_2^\infty\left(\frac{1}{n} + \ln\frac{n-1}{n}\right) = \lim_{k\to\infty}[h(k) - \ln k]$$

Case 2: If $n > 0$, then Case III of Frobenius's Theorem II applies. A second solution of Bessel's equation for $n > 0$, independent of J_n, may also be obtained using Frobenius's Theorem II. In practice, however, the second solution is taken to be

$$\begin{aligned}Y_n(x) = {} & \frac{2}{\pi}\left(\gamma + \ln\left|\frac{x}{2}\right|\right)J_n(x) - \frac{1}{\pi}\left|\frac{x}{2}\right|^{-n}\sum_{k=0}^{n-1}\frac{(n-k-1)!}{k!}\left(\frac{x}{2}\right)^{2k} \\ & - \frac{1}{\pi}\left|\frac{x}{2}\right|^n\sum_{k=0}^\infty\frac{(-1)^k[h(k) + h(k+n)]}{k!(n+k)!}\left(\frac{x}{2}\right)^{2k}\end{aligned} \tag{12}$$

It is called the *Bessel function of the second kind* (or a *Weber function*) *of order n*.

Although the first and third terms in the expression for Y_n remain bounded near 0, the second does not, and $Y_n(x) \to -\infty$ as $x \to 0^+$. See Figure 11.7.2 for the graphs of the first three functions of integer order, Y_0, Y_1, and Y_2. The zeros and the oscillatory character of Y_n, and indeed of any nontrivial solution of Bessel's equation, have been treated in Section 11.6.

---

[3]Euler's constant is 0.5772156649...; it is not known whether the number is rational, algebraic, or transcendental.

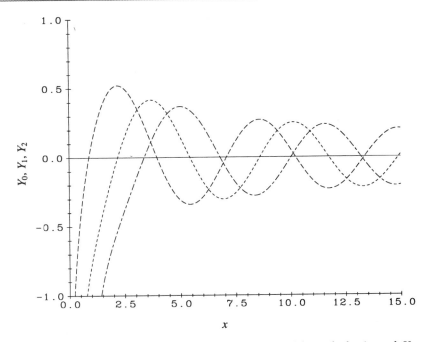

FIGURE 11.7.2 Graphs of Y_0 (long dashes), Y_1 (short dashes), and Y_2 (long/short dashes).

Bessel Functions of the Second Kind of Noninteger Order

Now suppose that the order p of Bessel's equation (11) is nonnegative, but not an integer. Then one solution is given by

$$J_p(x) = \left(\frac{x}{2}\right)^p \sum_{k=0}^{\infty} \frac{(-1)^k}{k!\,\Gamma(k+p+1)} \left(\frac{x}{2}\right)^{2k} \tag{13}$$

As noted in Section 11.6, a second independent solution, $J_{-p}(x)$, is easily constructed by replacing p by $-p$ in (13). However, it is customary not to use J_{-p}, but a certain linear combination of J_p and J_{-p} as the second independent solution of (11).

❖ **The Bessel Function of the Second Kind of Order** p. For p nonnegative and not an integer

$$Y_p(x) = \frac{\cos(p\pi)\,J_p(x) - J_{-p}(x)}{\sin p\pi} \tag{14}$$

is the *Bessel function of the second kind of order p.*

Since it may be shown from formula (14) that

$$\lim_{p \to n} Y_p(x) = Y_n(x)$$

where $Y_n(x)$ is defined by formula (12), formula (14) may be considered to define Y_p for all values of p. See Figure 11.7.3 for the graphs of J_0 and Y_0, $J_{1/2}$ and $Y_{1/2}$.

The functions $Y_p(x)$ satisfy the same Recursion Formulas given in Theorem 11.6.2 for $J_p(x)$ and have the same oscillatory behavior and structure of the set of zeros as J_p. The most notable visual difference between the graphs of $J_p(x)$ and $Y_p(x)$ is the behavior at the singularity $x = 0$: $J_p(0)$ is 1 (if $p = 0$) or 0 (if $p > 0$), but

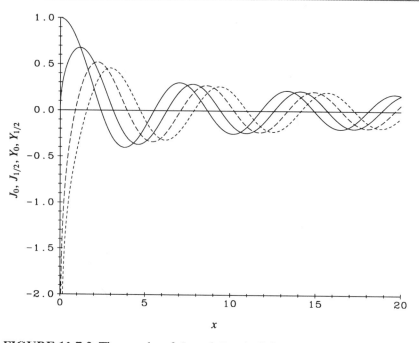

FIGURE 11.7.3 The graphs of J_0 and $J_{1/2}$ (solid), Y_0 and $Y_{1/2}$ (dashed).

$\lim_{x \to 0^+} Y_p(x) = -\infty$ for every nonnegative p. That unbounded behavior of $Y_p(x)$ at $x = 0$ plays a central role in the problem of the aging spring, as we shall now see.

Equations Reducible to Bessel Equations: The Aging Spring

It often happens that an equation superficially quite unlike a Bessel equation may be transformed to a Bessel equation by a judiciously chosen change of variable. Problems 3 and 4 list two families of differential equations that may be transformed to Bessel equations, and Example 11.7.3 below transforms the equation $rR'' + R' + \lambda r R = 0$ of Section 11.1 to Bessel's equation of order 0.

Here we examine the ODE of the aging spring

$$my''(t) + ke^{-\varepsilon t}y(t) = 0 \tag{15}$$

that was first introduced in Section 11.1. Let us introduce a new measure of time, $s = \alpha e^{\beta t}$, where α and β are constants to be chosen later, and see what happens when we change from ty-variables to sy-variables in equation (15). An exponential measure of time is suggested by the form of the elastic coefficient in (15). Since $ds/dt = \beta \alpha e^{\beta t} = \beta s$, and $(s/\alpha)^{-\varepsilon/\beta} = e^{-\varepsilon t}$, we have by the Chain Rule that

$$\frac{dy}{dt} = \frac{dy}{ds}\frac{ds}{dt} = \frac{dy}{ds}\beta s$$

$$\frac{d^2y}{dt^2} = \frac{d}{dt}\left(\frac{dy}{dt}\right) = \frac{d}{ds}\left(\frac{dy}{dt}\right)\frac{ds}{dt} = \frac{d}{ds}\left(\frac{dy}{ds}\beta s\right)\frac{ds}{dt} = \frac{d^2y}{ds^2}(\beta s)^2 + \frac{dy}{ds}\beta^2 s$$

Substituting into ODE (15), we have

$$\frac{d^2y}{dt^2} + \frac{k}{m}e^{-\varepsilon t}y = \beta^2 s^2\frac{d^2y}{ds^2} + \beta^2 s\frac{dy}{ds} + \frac{k}{m}\left(\frac{s}{\alpha}\right)^{-\varepsilon/\beta}y = 0$$

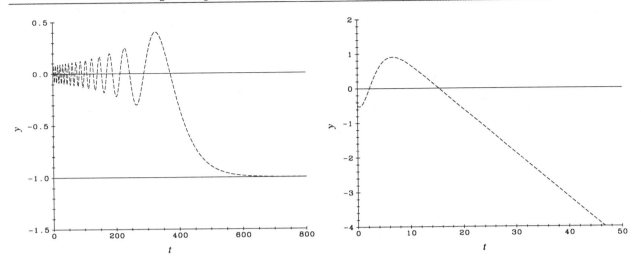

FIGURE 11.7.4 The aging spring: stretching toward a finite extension.   **FIGURE 11.7.5** A rapidly aging spring.

Divide the above ODE in y and s by β^2, choose α and β as follows

$$\beta = -\frac{\varepsilon}{2}, \qquad \alpha = \left(\frac{2}{\varepsilon}\right)\left(\frac{k}{m}\right)^{1/2}$$

and obtain the Bessel equation of order 0

$$s^2\frac{d^2y}{ds^2} + s\frac{dy}{ds} + s^2y = 0 \tag{16}$$

The general solution of (16) is

$$y = c_1 J_0(s) + c_2 Y_0(s) \tag{17}$$

where c_1 and c_2 are arbitrary constants. In terms of the time variable t

$$y = c_1 J_0\left(\frac{2a}{\varepsilon}e^{-\varepsilon t/2}\right) + c_2 Y_0\left(\frac{2a}{\varepsilon}e^{-\varepsilon t/2}\right), \quad \text{where } a = \left(\frac{k}{m}\right)^{1/2} \tag{18}$$

As time t increases, the argument of J_0 and of Y_0 in (18) tends to 0, and the asymptotic behavior of the aging spring is modeled by the behavior of solutions of Bessel's equation of order 0 near the regular singularity at the origin. There are two distinct possibilities for this behavior. First, if $c_2 = 0$, then as $t \to \infty$ the displacement from the rest position (as modeled by (18)) asymptotically approaches the value c_1 since $J_0(0) = 1$.

It is more likely that c_2 is nonzero and that the displacement tends to $c_1 J_0(0) + c_2 Y_0(0)$, i.e., to $\mp\infty$ depending on whether c_2 is positive or negative. If c_2 is negative, then the spring oscillates with increasing amplitude and eventually begins to stretch without bound. In reality, the spring will stretch beyond its elastic limit and snap, or else will behave in a completely inelastic matter which is not modeled by ODE (15) at all. Figures 11.7.4, 11.7.5, and 11.7.6 illustrate some of the possibilities. The case $c_2 < 0$ is not illustrated since it corresponds to the situation of approach to infinite compression. In fact, for any real spring there is a limit for the amount the spring can be compressed, $y(t) \le M$, and the model ODE (15) is not valid when $y(t)$ approaches M.

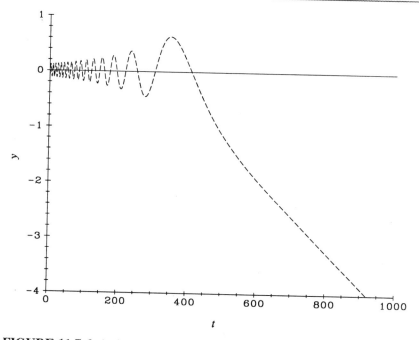

FIGURE 11.7.6 A slowly aging spring.

Equations Reducible to Bessel Equations: Steady Temperatures

Finally, the chapter closes with a reconsideration of the equation connected with the model for the steady temperatures in a solid cylinder (see Example 11.1.2).

EXAMPLE 11.7.3

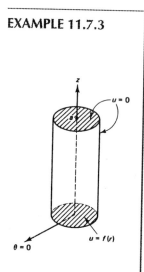

Steady Temperatures in a Cylinder

A solution $R(r)$ of the equation

$$rR''(r) + R'(r) + \lambda r R(r) = 0, \qquad 0 \le r \le 1 \tag{19}$$

is one factor of the temperature function $u(r, z) = R(r)Z(z)$ constructed in Example 11.1.2 as part of a model for steady temperatures. The full set of model equations is given by Boundary Value Problem (3) in Section 11.1. The parameter λ in ODE (19) is a separation constant and is assumed to be positive. The conditions given in Example 11.1.2 imply that $R(r)$ must be bounded for $0 \le r \le 1$, while the second and third boundary conditions of Boundary Value Problem (3) in Section 11.1 imply that the separation constant λ in (19) must be chosen so that $R(1) = 0$. The problem, then, is to find all positive values of λ such that

$$rR''(r) + R'(r) + \lambda r R(r) = 0, \qquad 0 \le r \le 1$$

$$R(1) = 0, \qquad |R(r)| \text{ bounded for } 0 \le r \le 1 \tag{20}$$

First, the ODE in (19) may be transformed to the Bessel equation of order 0 by multiplying the equation by r and introducing the variable changes $r = x\lambda^{-1/2}$ and $y(x) = R(x\lambda^{-1/2})$. Using the Chain Rule, we have that:

$$\frac{dy}{dx} = \frac{dR}{dr}\frac{dr}{dx} = R'(r)\lambda^{-1/2}, \qquad \frac{d^2y}{dx^2} = \frac{d}{dr}\left(R'(r)\lambda^{-1/2}\right)\frac{dr}{dx} = R''(r)\lambda^{-1}$$

Thus, ODE (19) becomes

$$x^2 y'' + xy' + x^2 y = 0 \tag{21}$$

which is Bessel's equation of order 0. The general solution of ODE (21) is

$$y = c_1 Y_0(x) + c_2 J_0(x)$$

where c_1 and c_2 are constants.

In xy-variables, problem (20) becomes

$$x^2 y'' + xy' + x^2 y = 0, \qquad 0 \le x \le \lambda^{1/2}$$

$$y(\lambda^{1/2}) = 0, \qquad |y(x)| \text{ bounded for } 0 \le x \le \lambda^{1/2}$$

(22)

Because the general solution of the ODE (21) involves the function $Y_0(x)$ which is infinite at $x = 0$, that function must be excluded on physical grounds, and the desired solution of (22) has the form

$$y = c_2 J_0(x)$$

The boundary condition $y(\lambda^{1/2}) = 0$ implies that $\lambda^{1/2}$ must be any one of the positive zeros λ_n of J_0 studied in Section 11.6. The acceptable solutions of (20), then, are

$$R_n(r) = c_2 J_0(r\lambda_n), \qquad \lambda_n \text{ a positive zero of } J_0 \tag{23}$$

where c_2 is any constant. Solutions (23) of (20) will be used in Section 12.5 in the solution of several boundary value problems.

PROBLEMS

1. (*Frobenius's Theorem II: Cases II, III*). Check that 0 is a regular singular point of each equation and find a basis for the solution space on the interval $(0, \infty)$. Plot a pair of independent solutions on the interval $(0, 5)$. [*Hint*: In parts (a), (b), (c), the solution y_1 is easily found in closed form. Then use the Wronskian Reduction Method of Problem 13 of Section 3.3 to find a second independent solution.]

   (a) $xy'' + (1 + x)y' + y = 0$ (b) $x^2 y'' + x(x-1)y' + (1-x)y = 0$

   (c) $xy'' - xy' + y = 0$ (d) $xy'' - x^2 y' + y = 0$

2. (*Bessel Functions of the Second Kind*). If p is not an integer, use the formula

$$Y_p = \frac{\cos(p\pi) J_p - J_{-p}}{\sin p\pi}$$

   to show that Bessel functions of the second kind of noninteger order satisfy the recursion formulas (16)–(19) in Section 11.6.

3. (*Equations Reducible to Bessel's Equation*).

   (a) If $w(s)$ is the general solution of Bessel's equation of order p, $s^2 w'' + sw' + (s^2 - p^2)w = 0$, show that $y(x) = e^{ax}w(bx)$ is the general solution of the equation

$$x^2 y'' + x(1 - 2ax)y' + [(a^2 + b^2)x^2 - ax - p^2]y = 0$$

   where a and b are real constants, $b \ne 0$.

   (b) Find the general solution of $x^2 y'' + x(1 - 2x)y' + (2x^2 - x - 1)y = 0$.

4. (*Equations Reducible to Bessel's Equation*).

   (a) Show that if $y(x)$ is a solution of Bessel's equation of order p, then $w(z) = z^{-c}y(az^b)$ is a solution of

$$z^2 w'' + (2c + 1)zw' + [a^2 b^2 z^{2b} + (c^2 - p^2 b^2)]w = 0$$

(b) (*Airy's Equation*). Use part **(a)** to show that the general solution of *Airy's equation* (see Section 11.2, Problem 4), $y'' - xy = 0$, is

$$y = |x|^{1/2} \left[c_1 J_{1/3} \left(\frac{2|x|^{3/2}}{3} \right) + c_2 J_{-1/3} \left(\frac{2|x|^{3/2}}{3} \right) \right]$$

The *Airy functions* $Ai(x)$ and $Bi(x)$ are respectively defined for $x < 0$ by setting $c_1 = c_2 = 1/3$, and $c_1 = -c_2 = -1/\sqrt{3}$. Show that $\{Ai(x), Bi(x)\}$ is an independent set. [*Hint*: First introduce a new variable $x = -z$.]

(c) Find the general solution of $x^2 y'' + \left(\frac{1}{8} + x^4 \right) y = 0$.

(d) Find the general solution of $y'' + x^4 y = 0$.

5. (*Aging Spring*). Suppose that $y = J_0(6e^{-t})$ is the solution of a model of an aging spring. Plot y as a function of t, $0 \leq t \leq 6$. What happens as $t \to \infty$?

6. (*Aging Spring*). Summarize what has been shown in this chapter about the problem of the aging spring. Critique the model's validity. Propose and analyze other models. In particular, consider the model

$$y'' + (b^2 + a^2 e^{-\varepsilon t}) y = 0$$

where a, b, and ε are constants. Show that solutions are given by

$$y = c_1 J_p(s) + c_2 Y_p(s)$$

where $p = 2ib/\varepsilon$ and $s = 2a\varepsilon^{-1} e^{-\varepsilon t/2}$. Critique this model. Plot solutions for various values of a, b, ε, and interpret these solutions in terms of motions of the aging spring.

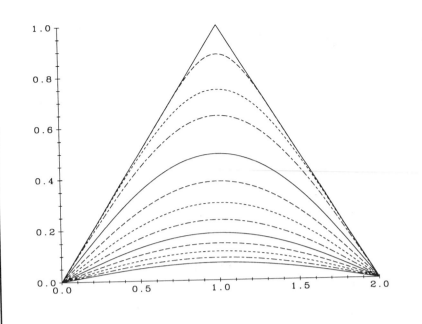

CHAPTER

12

Partial
Differential Equations

The internal temperatures of a solid body whose boundary temperatures are controlled, voltage along a transmission line, the displacement of a vibrating string—these physical systems and many more can be modeled by partial differential equations with appropriate boundary and initial conditions. Our aim in this chapter is to study in some depth a few simple physical systems which have partial differential equations as models. The Method of Separation of Variables introduced in Section 10.6 will be the principal tool for producing solution formulas for partial differential equations. That method rests on the theory and techniques of Fourier Series and of Sturm-Liouville boundary value problems for ordinary differential equations. Consequently, a brief review of these topics in Chapter 10 is highly recommended. The cover figure shows the decaying temperature in a rod (see Section 12.3).

12.1 Introduction to Partial Differential Equations

Partial differential equations (abbreviated as PDEs) are equations that involve the partial derivatives of functions of several variables. Using the "variable subscript" notation for the partial derivatives of the real-valued function $u(x, y, t)$ we shall write $u_x, u_y, u_t, u_{xx}, u_{yx}, u_{xxx}$, and so on, for the respective partial derivatives $\partial u/\partial x$, $\partial u/\partial y$, $\partial u/\partial t$, $\partial^2 u/\partial x^2$, $\partial^2 u/\partial x \partial y = \partial(\partial u/\partial y)/\partial x$, $\partial^3 u/\partial x^3$, and so on. The fact that $u_{xy} = u_{yx}$, if both are continuous, is usually proved in a multivariable calculus course. We shall always assume enough smoothness of derivatives so that the order of differentiation does not matter. The partial differential equations that occur most frequently in the applications are listed below.

$$\text{(Laplace or Potential Equation)} \quad u_{xx} + u_{yy} + u_{zz} = 0$$

$$\text{(Wave Equation)} \quad u_{tt} = c^2\{u_{xx} + u_{yy} + u_{zz}\}$$

$$\text{(Heat or Diffusion Equation)} \quad u_t = K\{u_{xx} + u_{yy} + u_{zz}\}$$

where c and K are positive constants whose values are determined by the physical properties of the materials being modeled. The variables x, y, and z are space variables and t is time. The wave equation arises in models involving wave propagation (hence the name), while Laplace's equation appears in models of steady-state phenomena, and the heat equation is connected with diffusion processes such as heat flow.

A convenient operator notation ∂_t, ∂_x, ∂_{xx}, and so on, is frequently used to denote the operations of differentiation $\partial/\partial t$, $\partial/\partial x$, $\partial^2/\partial x^2$, and so on. Thus, defining the *Laplacian* operator $\nabla^2 = \partial_{zz} + \partial_{yy} + \partial_{zz}$, we observe that the Laplace, wave, and heat equations can be written, respectively, as

$$\nabla^2 u = 0, \qquad (\partial_{tt} - c^2\nabla^2)u = 0, \qquad (\partial_t - K\nabla^2)u = 0$$

The *order* of a partial differential equation is the order of the highest derivative appearing in the equation. All the equations above are second order. Notice that the Laplacian ∇^2 is a linear operator since $\nabla^2[\alpha u + \beta v] = \alpha\nabla^2 u + \beta\nabla^2 v$, for any constants α, β, and any functions u, v which are twice continuously differentiable. The operators ∂_t and ∂_{tt} are linear as well. Thus the basic equations above are linear because they have the form $L[y] = f$, where L is a linear operator. Since our basic equations are homogeneous (i.e., the driving term f is identically zero, and the equations have the form $L[u] = 0$), it follows that a linear combination of solutions to any one of them is again a solution of that same equation. For example, let v and w be solutions of the homogeneous wave equation $Lu = 0$, where L is the operator $\partial_{tt} - c^2\nabla^2$. Then $L[v] = 0$, $L[w] = 0$. Now if a and b are any real numbers, then

$$L[av + bw] = \partial_{tt}(av + bw) - c^2\nabla^2(av + bw)$$

$$= a(\partial_{tt}v - c^2\nabla^2 v) + b(\partial_{tt}w - c^2\nabla^2 w)$$

$$= aL[v] + bL[w] = 0$$

and so $av + bw$ is also a solution of the PDE $(\partial_{tt} - c^2\nabla^2)u = 0$. This simple but important fact will be used over and over again in this chapter (it is sometimes called the *Principle of Superposition of Solutions*).

We now try to learn something about the solvability of partial differential equations by looking at some examples. For simplicity, however, we shall consider PDEs with only one or two space variables.

Examples with Solutions

In contrast with the situation in the theory of ordinary differential equations (ODEs), there is not yet anything like a comprehensive theory concerning basic existence or uniqueness questions for partial differential equations—not even for linear equations. The following examples indicate some situations that can occur when dealing with partial differential equations.

EXAMPLE 12.1.1

A Simple PDE

The simple second-order equation

$$u_{yx} = 0$$

for the unknown function $u(x, y)$ can be solved if we rewrite the equation in the form $(u_y)_x = 0$. Now if u is any solution of $u_{yx} = 0$ belonging to $C^2(\mathbf{R}^2)$, evidently u_y must be a C^1-function independent of x, say $c(y)$. Now the equation $u_y = c(y)$ can be considered as an ordinary differential equation in y for each fixed value of x. Thus antidifferentiation yields that $u = f(x) + g(y)$, where f and g are C^2-functions in x and y, respectively. Conversely, any function of this form for arbitrary C^2-functions f and g is a C^2-solution of $u_{yx} = 0$. Thus we have found all C^2-solutions of $u_{yx} = 0$, namely

$$u = f(x) + g(y)$$

for arbitrary C^2-functions f and g. It is interesting to recall that the general solution of a second-order ordinary differential equation might involve two arbitrary constants, whereas two arbitrary functions seem to be involved in the general solution of second-order partial differential equations. Observe that prescribing the value of u and any number of its derivatives at a point would not be enough to single out a specific solution of $u_{yx} = 0$.

EXAMPLE 12.1.2

Solving a PDE Using ODE Techniques

Sometimes a partial differential equation can be solved by an iteration of the integrating factor technique of Section 1.2. Let $u(x, y)$ be a C^2-solution of the second-order PDE

$$u_{xy} + y u_x = 0$$

Putting $v(x, y) = u_x(x, y)$ we see that $v_y + yv = 0$. But this equation can be treated as a first-order linear ODE in the independent variable y for each fixed value of x. With $e^{y^2/2}$ as an integrating factor, this equation becomes $(e^{y^2/2}v)_y = 0$. Thus $e^{y^2/2}v = c(x)$, where c is a C^1-function, so $v = u_x = c(x)e^{-y^2/2}$, which can be considered an ODE in x for each fixed value of y. It follows that any C^2-solution of $u_{xy} + yu_x = 0$ has the form $u = f(x)e^{-y^2/2} + g(y)$, where f and g are C^2-functions. Conversely, for any choice of C^2-functions f and g,

$$u = f(x)e^{-y^2/2} + g(y)$$

is a C^2-solution of $u_{xy} + yu_x = 0$. So, we have found all C^2-solutions of $u_{xy} + yu_x = 0$.

EXAMPLE 12.1.3

Solving a First-Order Linear PDE

A change of variables technique can be used to find the general solution of the linear first-order PDE

$$u_t + c u_x = 0$$

where c is a nonzero constant. Introducing the change of independent variables

$$y = x + ct, \qquad s = x - ct$$

and using the Chain Rule to compute $\partial_t[u(y, s)]$ and $\partial_x[u(y, s)]$, we have that

$$u_t = c u_y - c u_s, \qquad u_x = u_y + u_s$$

After substitution, the original PDE becomes $u_y = 0$, from which we conclude that $u(y, s) = f(s)$, where f is an arbitrary function. Substituting back the original variables, we have the general solution

$$u(x, t) = f(x - ct)$$

In the next example we go from an overabundance of solutions to a scarcity.

EXAMPLE 12.1.4

A PDE With Just One Solution

Consider the first-order equation

$$xu_x + yu_y + u = 0, \qquad x, y \text{ in } \mathbf{R}^2 \tag{1}$$

We shall show that the only solution of (1) that belongs to $C^1(\mathbf{R}^2)$ is $u = 0$ on $\mathbf{R}^2$. Let $u(x, y)$ be a C^1-solution of (1) and suppose that m is the minimum value of u on the square $S = [-1, 1] \times [-1, 1]$. If $u = m$ at an interior point of S, then $u_x = u_y = 0$ at that point, and hence $u = -xu_x - yu_y = 0$ there. Thus $m = 0$ in this case. Now suppose that $u = m$ at a boundary point of S, say at $(x_0, -1)$. Then certainly, $u_y(x_0, -1) \geq 0$, and the minimality of u at $(x_0, -1)$ implies that

$$u_x(x_0, -1) \begin{cases} = 0 & \text{if } |x_0| < 1 \\ \geq 0 & \text{if } x_0 = -1 \\ \leq 0 & \text{if } x_0 = 1 \end{cases} \tag{2}$$

From (1) and (2) we have that

$$u(x_0, -1) = m = -x_0 u_x(x_0, -1) + u_y(x_0, -1) \geq 0$$

A similar argument applies if u takes on the value m anywhere on the boundary of S. Thus $u \geq m \geq 0$ on S.

Now the same argument can be applied to $-u$, which is also a solution of (1). Thus $M = \min(-u(x, y)) \geq 0$, over (x, y) in S, which implies that $u(x, y) \leq -M \leq 0$ for all (x, y) in S. Hence $m = M = 0$ and $u = 0$ on S. Clearly, this extends to any square $|x| \leq a, |y| \leq a$, and hence $u = 0$ on $\mathbf{R}^2$.

As if this were not bad enough, it was discovered in 1955 that there are very simple *linear* nonhomogeneous PDEs with polynomial coefficients that have no solution at all in any neighborhood of a given point!

What have we learned from these examples? We have seen that certain PDEs can be solved by techniques developed for ordinary differential equations, and that the solutions involve arbitrary functions (in contrast to solutions of ODEs which involve only arbitrary constants). But the most telling difference between ordinary and partial differential equations is that apparently no simple but comprehensive description of the solution sets of partial differential equations has yet been given. It will come as no surprise at all, however, that all the partial differential equations which arise in models of physical phenomena behave "nicely," and hence we need not be unduly worried. On the other hand, it is clear that we cannot attempt anything like a theory of partial differential equations which parallels that developed in earlier chapters for ordinary differential equations.

We next take up some useful "tricks" for finding exponential solutions and separated solutions for linear homogeneous equations, and we end this section with a discussion of initial value problems for partial differential equations of second order.

Exponential Solutions

The second-order linear homogeneous PDE with constant real coefficients

$$Au_{xx} + Bu_{xy} + Cu_{yy} + Du_x + Eu_y + Fu = 0 \tag{3}$$

always has complex-valued exponential solutions of the form $e^{\alpha x}e^{\beta y}$ for complex numbers α, β. Indeed, substituting $e^{\alpha x}e^{\beta y}$ for u into (3) we see that for any solution pair (α, β) of the quadratic equation

$$A\alpha^2 + B\alpha\beta + C\beta^2 + D\alpha + E\beta + F = 0$$

the exponential $e^{\alpha x + \beta y}$ is a (possibly complex-valued) solution of (3). Since (3) has real coefficients, the real and imaginary parts of the exponential solution $e^{\alpha x + \beta y}$ are *real*-valued solutions of (3).

EXAMPLE 12.1.5

Periodic Exponential Solutions: Wave Equation
What are the exponential solutions $e^{\alpha t}e^{\beta x}$ of the wave equation

$$u_{tt} = u_{xx}$$

By substitution we see that the condition $\alpha^2 = \beta^2$ must be satisfied. Thus $\beta = \pm\alpha$, so $e^{\alpha t}e^{\pm\alpha x}$ is a solution for any constant α, real or complex. If we seek a periodic solution of the wave equation, we must take $\alpha = i\omega$ for some real ω. Then $\beta = \pm i\omega$, so

$$e^{i\omega t}e^{\pm i\omega x} = e^{i\omega(t \pm x)} = \cos\omega(t \pm x) + i\sin\omega(t \pm x)$$

whose real and imaginary parts are real-valued periodic solutions of $u_{tt} = u_{xx}$.

EXAMPLE 12.1.6

Periodic Exponential Solutions: Heat Equation
Direct substitution shows that $e^{\alpha t}e^{\beta x}$ is a solution of the heat equation

$$u_t = u_{xx}$$

if $\alpha = \beta^2$. Thus $e^{\beta^2 t}e^{\beta x}$ is a solution of $u_t = u_{xx}$ for any real or complex constant β. If we want a periodic solution of the diffusion equation, we need only choose β such that $\beta^2 = i\omega$ for any positive ω. Thus β must be one or the other of the complex numbers $\pm(\omega/2)^{1/2}(1 + i)$ (see Appendix C.3 for a technique to compute square roots of complex numbers), so if we put $\gamma = (\omega/2)^{1/2}$, then

$$e^{i\omega t}e^{\pm\gamma(1+i)x} = e^{\pm\gamma x}e^{i(\omega t \pm \gamma x)} = e^{\pm\gamma x}\cos(\omega t \pm \gamma x) + ie^{\pm\gamma x}\sin(\omega t \pm \gamma x)$$

whose real and imaginary parts are real-valued solutions of $u_t = u_{xx}$.

Separated Solutions

There is another approach to finding solutions of (3) which generalizes the exponential solution approach and sometimes is successful even when the coefficients in (3) are nonconstant. In this approach we use a *Method of Separation of Variables* to find so-called *separated solutions* of (3) [i.e., solutions having the form $X(x)Y(y)$]. An example will explain the simple procedure.

EXAMPLE 12.1.7

Separated Solutions
Let us find all solutions of the equation

$$x^2 u_{xx} + xu_x = u_{yy}$$

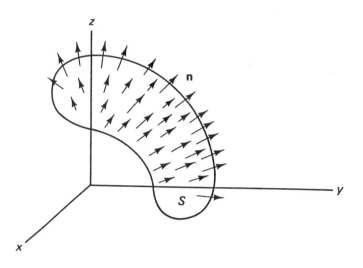

FIGURE 12.1.1 Geometry for an initial manifold.

which have the "separated" form $X(x)Y(y)$ (see also Section 10.1). Substituting $X(x)Y(y)$ for u in the equation and then dividing by $X(x)Y(y)$, we can write the resulting equation in the "variables separated" form

$$\frac{x^2 X''(x) + x X'(x)}{X(x)} = \frac{Y''(y)}{Y(y)} \tag{4}$$

The left-hand side of (4) depends only on the variable x, whereas the right-hand side depends only on y; thus the equation can hold only if each side is equal to the same constant, say λ, called the *separation constant*. Hence $X(x)$, $Y(y)$ must satisfy the pair of ordinary differential equations

$$x^2 X'' + x X' - \lambda X = 0 \tag{5}$$

$$Y'' - \lambda Y = 0 \tag{6}$$

for the same value of λ. Observe that (5) is an Euler equation whose indicial roots satisfy the equation $r^2 - \lambda = 0$ (see Problem 4, Section 3.6 and Section 11.5). For example, if $\lambda = 1$, the general solution of (5) is $X(x) = Ax + Bx^{-1}$. For $\lambda = 1$, (6) has the general solution $Y(y) = Ce^y + De^{-y}$. Hence

$$X(x)Y(y) = (Ax + Bx^{-1})(Ce^y + De^{-y})$$

is a separated solution of $x^2 u_{xx} + x u_x = u_{yy}$ for all values of the constants A, B, C, D.

Initial Value Problems

When a partial differential equation has many solutions, what sort of conditions can be imposed which will "select" a unique solution from among the class of all solutions? Our experience with ordinary differential equations suggests that initial conditions can be used for this purpose, but what form do such conditions take for partial differential equations? Since there is no natural "normal" form for a partial differential equation, initial conditions must be expressed in the following general fashion. If L is a second-order partial differential operator in the three independent variables (x, y, z), an *initial manifold* is a smooth surface S in $\mathbf{R}^3$ (see Figure 12.1.1). For a smooth unit normal vector field $\mathbf{n}$ to S, the *initial value problem* for L on S is given by

$$
\begin{aligned}
L[u] &= h, & &\text{for all } (x, y, z) \text{ "near" } S \\
u(x, y, z) &= F(x, y, z), & &\text{all } (x, y, z) \text{ on } S \\
\left.\frac{du}{d\mathbf{n}}\right|_{(x,y,z)} &= G(x, y, z), & &\text{all } (x, y, z) \text{ on } S
\end{aligned}
\tag{7}
$$

where the "driving term" h and the *initial data* F and G are given, and $du/d\mathbf{n}$ (also written as $\partial u / \partial n$) denotes the directional derivative of u in the direction of $\mathbf{n}$. Thus there is a great deal of choice for the initial manifold S in problem (7). As we shall see in the example below, the properties of the initial value problem (7) for a given partial differential operator L depend crucially on the choice of S. There is no analog of this fact in ordinary differential equations.

EXAMPLE 12.1.8

Effect of Initial Manifold on Solvability

As we saw in Example 12.1.2, the general C^2-solution of $u_{xy} + yu_x = 0$ is

$$
u = f(x)e^{-y^2/2} + g(y)
$$

where f and g are arbitrary C^2-functions of the single variables indicated. First we take the x-axis as the initial manifold S. Then IVP (7) takes the form

$$
\begin{aligned}
u_{xy} + yu_x &= 0 \\
u(x, 0) = F(x), \qquad u_y(x, 0) &= G(x)
\end{aligned}
\tag{8}
$$

where we have set the driving term h equal to zero. Using the general solution, we see that (8) is solvable if and only if f and g can be found such that $f(x) + g(0) = F(x)$ and $g'(0) = G(x)$. Thus (8) has no solution at all, unless G is a constant function, say c, and in that case $g(y) = cy + d$, for an arbitrary constant d, and $f(x) = F(x) - d$. Hence if $G = c$, (8) has infinitely many solutions

$$
u = (F(x) - d)e^{-y^2/2} + (cy + d), \qquad d \text{ an arbitrary constant}
$$

Thus, with the x-axis as the initial manifold, either the initial value problem (8) has *no solution* at all, or else it has *infinitely many*.

Next we consider the initial value problem for $u_{xy} + yu_x = 0$, where the initial manifold S is the line $x = y$. Thus $\mathbf{n} = (1/\sqrt{2}, -1/\sqrt{2})$ is a unit normal field to S, and the problem becomes

$$
\begin{aligned}
u_{xy} + yu_x &= 0 \\
u(x, x) = F(x), \qquad u_x(x, x) - u_y(x, x) &= \sqrt{2}G(x)
\end{aligned}
\tag{9}
$$

Recalling that $u = f(x)e^{-y^2/2} + g(y)$ is the general solution to $u_{xy} + yu_x = 0$, we must choose f and g such that

$$
f(x)e^{-x^2/2} + g(x) = F(x)
$$

$$
f'(x)e^{-x^2/2} + xf(x)e^{-x^2/2} - g'(x) = \sqrt{2}G(x)
$$

Differentiating the first relation and adding to the second, we obtain $2f'(x)e^{-x^2/2} = F'(x) + \sqrt{2}G(x)$. Thus we see that the initial value problem (9) has the *unique* solution $u = [f(x) - f(y)]e^{-y^2/2} + F(y)$, where $f(x)$ is chosen so that $f'(x) = \frac{1}{2}e^{x^2/2}(F'(x) + \sqrt{2}G(x))$ and $g(y) = F(y) - f(y)e^{-y^2/2}$.

Comparing problems (8) and (9), we see that the choice of an initial manifold can seriously affect the solvability of an IVP for a given partial differential equation.

PROBLEMS

1. (*Exponential Solutions*). Find all exponential solutions of the following equations.
   (a) $u_{tt} = u_{xx} - 4u_x + 2u$ (b) $u_{xx} + u_{yy} = 0$

2. (*Separated Solutions*). Find all separated solutions of the following equations.
   (a) $u_{tt} = c^2 u_{xx}$ (b) $u_t = K u_{xx}$
   (c) $u_{xx} + u_{yy} = 0$ (d) $xu_x + y^2 u_{yy} = 0$
   (e) $u_{xx} - y u_y + u = 0$ (f) $u_{yy} = a^2 u_{xx} - 2h u_y$; a, h positive constants

3. Find all C^2-solutions $u(x, y)$ of the equation $u_{xy} + (yu)_y + xu_x + xyu = 0$. [*Hint:* Show that for $v = u_x + yu$ the equation assumes the form $v_y + xv = 0$.]

4. (*Solvability*). Find all solutions (or discuss the solvability) of the initial value problems that arise from the equation $u_{xy} + yu_x = 0$ with the given initial conditions. [*Hint:* See Examples 12.1.2 and 12.1.8.]
   (a) $u(0, y) = F(y)$, $u_x(0, y) = G(y)$
   (b) $u(x, x) = x^2$, $u_x(x, x) - u_y(x, x) = 0$

5. (*PDE With Only One Solution*). Let a, b, c be functions in $C^0(S)$, where S is the square $|x| \le 1$, $|y| \le 1$, and assume that for $|x| \le 1$ and $|y| \le 1$, $a(-1, y) \ge 0$, $a(1, y) \le 0$, $b(x, -1) \ge 0$, $b(x, 1) \le 0$, $c(x, y) < 0$. Using the techniques of Example 12.1.4, show that $u = 0$ is the only solution of $au_x + bu_y + cu = 0$ that is defined for all (x, y) in S.

12.2 Properties of Solutions of the Wave Equation

In Section 10.1 the linear wave equation was derived under appropriate conditions for the transverse vibrations of a taut, flexible string. We saw that the transverse displacement $u(x, t)$ at the point x on the vibrating string and at time t, is given by

$$u_{tt} - c^2 u_{xx} = 0 \tag{1}$$

where c is a positive constant. Since we now wish to study wave motion on a string without concern for the boundaries, we shall consider an idealized taut, flexible string of infinite length and derive the fundamental properties of all solutions of PDE (1).

Characteristics

The following observations will be useful for finding solutions of the wave equation:

THEOREM 12.2.1

> Constancy Theorem. Let Ω be a region in the xt-plane and let u in the class $C^2(\Omega)$ be any solution of the wave equation (1) in Ω. Then
>
> 1. $u_t - cu_x$ is constant on any line segment in Ω of the form $x - ct = $ constant.
> 2. $u_t + cu_x$ is constant on any line segment in Ω of the form $x + ct = $ constant.

Proof. We prove only assertion 1.; the proof of 2. is similar. Observe that PDE (1) can be written as $v_t + cv_x = 0$, where $v(x, t) = u_t(x, t) - cu_x(x, t)$. But notice that $v_t + cv_x$ is proportional to the directional derivative of v in

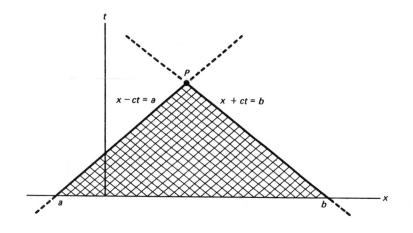

FIGURE 12.2.1 Characteristic triangle $\Delta(a, b)$ for the wave equation (1).

the xt-plane in the direction defined by the vector with x-coordinate c and t-coordinate 1. Thus the line $x - ct =$ constant in Ω is parallel to this direction and the function $v = u_t - cu_x$ has a vanishing directional derivative in the direction of the line. The desired result follows.

The two families of lines $x - ct =$ constant and $x + ct =$ constant have special significance for the wave equation (1).

❖ **Characteristics.** Each line of the form $x + ct =$ constant or $x - ct =$ constant is called a *characteristic* for the wave equation (1).

❖ **Characteristic Triangle.** Let $a < b$ be any two points in **R** and denote by $\Delta(a, b)$ the interior of the triangle whose base is the interval (a, b) and whose sides are segments of the characteristic lines $x - ct = a$, $x + ct = b$. The region $\Delta(a, b)$ is called a *characteristic triangle* for (1) (see Figure 12.2.1).

We now exploit the special property of characteristics for deriving solutions of the wave equation. Let H be the half-space $-\infty < x < +\infty$, $t > 0$. Suppose that $u(x, t)$ is a C^2-solution of the wave equation (1) in H such that u and u_t are continuous on cl $H : -\infty < x < +\infty$, $t \geq 0$. If we put $u(x, 0) = f(x)$ and $u_t(x, 0) = g(x)$, we shall show that $u(x, t)$ is uniquely determined by the *initial data* $f(x)$ and $g(x)$. More generally, we shall show that given any choice of initial data f in $C^2(\mathbf{R})$ and g in $C^1(\mathbf{R})$, the *initial value* (or *Cauchy*) *problem*

$$u_{tt} = c^2 u_{xx}, \quad x \text{ in } \mathbf{R}, \quad t > 0$$
$$u(x, 0) = f(x), \quad x \text{ in } \mathbf{R} \tag{2}$$
$$u_t(x, 0) = g(x), \quad x \text{ in } \mathbf{R}$$

has a unique C^2-solution u in H for which u and u_t are C^0-functions on cl H. First we show uniqueness.

THEOREM 12.2.2

Uniqueness for Problem (2). Let u and v be two solutions of the initial value problem (2). Then $u = v$ in the half-space H.

Proof. Let $w = u - v$. Then w solves problem (2), with initial data $f(x) = 0$, $g(x) = 0$. Thus w_t and w_x both vanish for every point on the x-axis. Let P be a given point (x, t) in the half-space H; we shall show that $w_t(P) = 0$ and $w_x(P) = 0$. Let $\Delta(a, b)$ be the characteristic triangle formed by following the two characteristic lines through P back to where they cross the x-axis, say at a and b, with $a < b$ (see Figure 12.2.1). Now the Constancy Theorem implies that

$$w_t(P) - cw_x(P) = w_t(a, 0) - cw_x(a, 0) = 0$$

$$w_t(P) + cw_x(P) = w_t(b, 0) + cw_x(b, 0) = 0$$

Therefore, $w_t(P) = w_x(P) = 0$. But P in H was arbitrary; thus $w_t = 0$, $w_x = 0$ in H. It follows that w is constant on H.[1] But since w vanishes on the x-axis, the constant must be zero, and hence $w = 0$ on H. This implies that $u = v$ on H.

[*Note*: Actually, the proof above shows that if two solutions u and v of the wave equation (1) have initial data that coincide on an interval (a, b) of the x-axis, then $u = v$ on the characteristic triangle $\Delta(a, b)$.]

Now we turn to the question of the existence of a solution for (2).

THEOREM 12.2.3

D'Alembert's Solution of the Initial Value Problem (2). Let $f(x)$ be in the class $C^2(\mathbf{R})$, and $g(x)$ in the class $C^1(\mathbf{R})$. Then the function

$$u(x, t) = \frac{f(x + ct) + f(x - ct)}{2} + \frac{1}{2c} \int_{x-ct}^{x+ct} g(s) \, ds, \qquad (x, t) \text{ in } H \quad (3)$$

is the unique solution of the problem (2) with the required smoothness.

Proof. A direct calculation using the Leibniz Rule (Appendix C.5) shows that $u(x, t)$ in (3) is indeed a solution of problem (2). By the Uniqueness Theorem, (3) is the only solution and we are done.

Actually, we can show much more. We can derive formula (3) from the Constancy Theorem by carrying the initial data forward along characteristics. Here is how this is done. Let (x, t) be a given point in H, and denote by X_1 and X_2, $X_1 < X_2$, the intersections of the x-axis and the two characteristic lines through (x, t). Clearly, $X_1 = x - ct$, $X_2 = x + ct$. From the Constancy Theorem we conclude that

$$u_t(x, t) + cu_x(x, t) = u_t(X_2, 0) + cu_x(X_2, 0)$$
$$u_t(x, t) - cu_x(x, t) = u_t(X_1, 0) - cu_x(X_1, 0) \tag{4}$$

But since $u_t(X_2, 0) = g(X_2)$, $u_t(X_1, 0) = g(X_1)$, $u_x(X_2, 0) = f'(X_2)$, $u_x(X_1, 0) = f'(X_1)$, we see that (4) becomes

$$u_t(x, t) + cu_x(x, t) = g(x + ct) + cf'(x + ct)$$
$$u_t(x, t) - cu_x(x, t) = g(x - ct) - cf'(x - ct) \tag{5}$$

---

[1] The following theorem is a natural extension of the Vanishing Derivative Theorem of Section 1.1: Let R be a region in $\mathbf{R}^n$ and let f in $C^1(R)$ be such that $\partial f/\partial x_i = 0$ in R, for $i = 1, 2, \ldots, n$. Then f is constant in R.

Solving the identities (5) for u_t and u_x we obtain that

$$2u_t(x, t) = [g(x + ct) + g(x - ct)] + c[f'(x + ct) - f'(x - ct)]$$

$$2u_x(x, t) = \frac{1}{c}[g(x + ct) - g(x - ct)] + [f'(x + ct) + f'(x - ct)] \qquad (6)$$

Now observe that if we put

$$U(x, t) = u(x, t) - \frac{f(x + ct) + f(x - ct)}{2} - \frac{1}{2c}\int_{x-ct}^{x+ct} g(s)\, ds$$

we can use (6) to show that $U_t = 0$, $U_x = 0$ in H. Hence, $U(x, t)$ is a constant in H, but that constant is 0 since $U(x, 0) = \mu(x, 0) - f(x) = f(x) - f(x) = 0$. So (3) holds and we are done.

Note that the wave equation has no other solutions than those given by the form (3) for suitable "arbitrary" functions $f(\cdot)$ and $g(\cdot)$. Thus (3) is the *general solution* of the wave equation. The wave equation is one of the very few second-order partial differential equations for which a general solution can be conveniently found.

Properties of Solutions of the Wave Equation

Having the explicit solution of problem (2), we can now examine not only its mathematical properties but also its significance for the problem of the vibrating string. Examples of solutions appear in Section 10.1.

(A) Superposition of Traveling Waves
First observe that the function $u(x, t)$ in (3) can be written as

$$u(x, t) = P(x + ct) + Q(x - ct) \qquad (7)$$

where

$$P(s) = \frac{1}{2}f(s) + \frac{1}{2c}\int_0^s g(s')\, ds', \qquad Q(s) = \frac{1}{2}f(s) - \frac{1}{2c}\int_0^s g(s')\, ds' \qquad (8)$$

Thus (7) implies that to find the profile $\mu(x, T)$ of our string at the given time T, we need only take the graph of P on the x-axis, translate it to the left by cT units, and then add it to the translation of Q on the x-axis to the right by cT units. Thinking of the graphs of P and Q on the x-axis as wave forms, we see that $P(x + ct)$ is the wave form P moving uniformly in time to the left with velocity c. Similarly, $Q(x - ct)$ is just the wave form Q moving uniformly in time to the right with velocity c. Thus the solution of the initial value problem (2) is the sum of two *traveling waves*, one moving to the right, the other to the left [as is evident from (7)].

(B) Huygens' Principle
Say that the string is undeflected and at rest initially except on the bounded interval J of the x-axis, where the string is deflected and given some initial velocity [i.e., f and g in (2) vanish identically everywhere but on J]. It is of interest to determine how much time must pass for a point x_0 not in J to detect the disturbance on the set J. From (3) it is clear that $u(x, t)$ will vanish in any characteristic triangle $\Delta(x_0 - ct, x_0 + ct)$ whose base does not meet J. If $J = [a, b]$ and a is the nearest endpoint of J to x_0, the earliest time that x_0 can feel the disturbance on J is $t = |a - x_0|/c$. This observation agrees with (A) in that the wave forms P and Q both travel with speed c. Thus x_0 perceives a sharp beginning for the waves generated by the disturbance on J. But observe from (3) that if the initial velocity g is nonzero on J, then x_0 will not in general perceive a sharp end to the waves generated by the disturbance on J, even when J is chosen arbitrarily

small in length. Physicists express this peculiar property of the problem (2) by saying that the Cauchy Problem (2) does not satisfy *Huygens' Principle*, since that principle requires both a sharp beginning and a sharp end to any disturbance passing through an observer's location at x_0.

(C) Continuity in the Data

The concept of a well-set (or well-posed) problem for a PDE is the same as that for an ODE (see Section 2.2). Thus the Cauchy Problem (2) is *well-set* if it has a unique solution for all appropriately smooth data, and if the solution changes "continuously" with respect to changes in the initial data f and g. This last condition needs some interpretation: Let $u(x, t)$ be the solution of problem (2) for data $f(x)$ and $g(x)$. Choose a fixed $T > 0$, and define the strip $S : -\infty < x < \infty, 0 \leq t \leq T$. Now suppose that the data are bounded and suppose that M_f and M_g are the smallest constants such that $|f(x)| \leq M_f$ and $|g(x)| \leq M_g$ for all x in $\mathbf{R}$. then according to (3), we have that

$$|u(x, t)| \leq M_f + T M_g, \qquad \text{all } (x, t) \text{ in } S \tag{9}$$

Now suppose that f^*, g^* is another data pair which is bounded, and that $u^*(x, t)$ is the corresponding solution of problem (3). Then $u - u^*$ is the solution of problem (3) with the data pair $f - f^*$ and $g - g^*$, as is easy to verify. Thus the estimate (9) applies and we have that

$$|u(x, t) - u^*(x, t)| \leq M_{f-f^*} + T M_{g-g^*} \qquad \text{for all } (x, t) \text{ in } S \tag{10}$$

Hence it follows from (10) that the difference $|u(x, t) - u^*(x, t)|$ may be made uniformly small over S if the differences $|f(x) - f^*(x)|$ and $|g(x) - g^*(x)|$ are made sufficiently small over $\mathbf{R}$. Thus the Cauchy Problem (2) is well-set in the sense outlined above.

(D) Propagation of Singularities

What can we say about the solvability of the Cauchy Problem (2) when the data are only piecewise smooth on $\mathbf{R}$? The D'Alembert formula (3) still makes sense for such data. Notice that from (7) if the data are such that $P(s)$ in (8) has a discontinuity at some s_0, this discontinuity "propagates" in the half-space H along the characteristic line $x + ct = s_0$. A similar statement holds for $Q(s)$ and the characteristic line $x - ct = s_0$. Thus in particular we observe that if either $g(s) + cf'(s)$ or $g(s) - cf'(s)$ is discontinuous at s_0, then u_t and u_x cannot be continuous along the characteristic lines $x \pm ct = s_0$. This situation is summarized by saying that the wave equation propagates singularities in the data into the interior along characteristic lines. Observe that the function u defined by (3) cannot satisfy the wave equation at interior singularities since the second derivatives u_{tt} and u_{xx} may not even exist at such points.

(E) The Light Cone

Imagine a signal which is propagated from the point $P(x_0, t_0)$ in space-time and whose intensity $u(x, t)$ satisfies the wave equation. The two lines, $x + ct = x_0 + ct_0$ and $x - ct = x_0 - ct_0$, through P bound sectors forward and backward in time from P. As time increases from t_0 the signal can only be detected at points in the forward sector. Conversely, an observer at P can only detect signals generated in the backward sector. These two sectors form the *light cone* at P (or the *cone of dependence and of influence*).

The Method of Images

We shall show how the D'Alembert solution of the Cauchy Problem (2) can be used to solve the mixed initial/boundary value problem (6) in Section 10.1 for the case where the boundary data vanish [i.e., $\theta(t) = \mu(t) = 0$]. An example will clarify the procedure, called the *method of images*. Let us consider the case of a taut flexible string of length L which is fastened at both endpoints, $x = 0$, L. Then the motion of the string is characterized by the fixed boundary conditions

$$u(0, t) = 0, \quad u(L, t) = 0 \quad \text{for all } t \geq 0 \tag{11}$$

It is remarkable that we can construct a suitable Cauchy Problem (2) for an infinite string whose solution will turn out to be a solution of this finite string problem as well. Let $\overline{f}(x)$ and $\overline{g}(x)$ be the initial data for the finite string problem. Observe that if initial data $f(x)$, $g(x)$ used in the Cauchy Problem (2) are periodic with period $2L$ and odd about $x = 0$ and $x = L$, then the D'Alembert formula (3) implies that the fixed conditions (11) are satisfied.[2] The question then is whether or not we can find such data $f(x)$, $g(x)$ on $\mathbf{R}$ which coincide with $\overline{f}(x)$, $\overline{g}(x)$ on $0 \leq x \leq L$. This may easily be done as follows. Extend $\overline{f}$, $\overline{g}$ into $[-L, 0]$ as odd functions about $x = 0$, and then define f, g to be the periodic extension of these functions on $[-L, L]$ into $\mathbf{R}$ with period $2L$ (see Figures 12.2.2 and 12.2.3). That the data f and g constructed in this manner satisfy the stated properties is left to the reader. Now in order to guarantee the required smoothness properties for f and g, it is necessary and sufficient that

$$\overline{f} \text{ is in } C^2[0, L], \qquad \overline{g} \text{ is in } C^1[0, L]$$

$$\overline{g}(0) = \overline{g}(L) = 0 \quad \text{and} \quad \overline{f}(0) = \overline{f}(L) = \overline{f}''(0^+) = \overline{f}''(L^-) = 0$$

The proof of this fact is omitted. Using the D'Alembert formula, we easily verify that the solution $u(x, t)$ of the Cauchy Problem with the data f, g constructed above does in fact solve the finite string problem with fixed endpoints.

Finite string problems but with free endpoint conditions (or one endpoint fixed and the other free) can be treated similarly. We may also seek solutions of the finite string problem when the data $\overline{f}$ and $\overline{g}$ only satisfy the smoothness requirements piecewise. The procedure above works perfectly well in this case and it is not hard to see that discontinuities in the data propagate into the region $\{(x, t) : 0 < x < L, t > 0\}$ along characteristic line segments $x \pm ct = s_0$ (where s_0 is a point of discontinuity for the data) and their "reflections" in the lines $x = 0$, $x = L$.

EXAMPLE 12.2.1

A Plucked Guitar String

Suppose that a string of length L clamped at both ends is "plucked" and released from rest. What will the subsequent motion be? If the string is not plucked too vigorously then the deflection $u(x, t)$ will be a solution to the problem

$$
\begin{aligned}
u_{tt} &= c^2 u_{xx}, & 0 < x < L, \quad t > 0 \\
u(0, t) &= u(L, t) = 0, & t \geq 0 \\
u(x, 0) &= f(x), & 0 \leq x \leq L \\
u_t(x, 0) &= 0, & 0 \leq x \leq L
\end{aligned}
\tag{12}
$$

---

[2] A function $h(x)$ on $\mathbf{R}$ is *odd* about a point x_0 if $h(x_0 - r) = -h(x_0 + r)$ for all $r \geq 0$ (i.e., h takes opposite values at points which are symmetric about x_0. Note that $h(x_0) = 0$ if h is continuous.

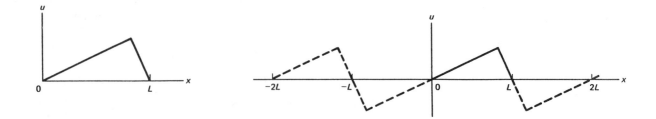

FIGURE 12.2.2 Initial profile of plucked string.   FIGURE 12.2.3 Extension of plucked initial string profile required by the method of images.

where $f(x)$ is the function defined graphically in Figure 12.2.2. Let $\overline{f}(x)$ be the periodic extension into $\mathbf{R}$ of the odd extension of f into $[-L, L]$ (see Figure 12.2.3).

Using the D'Alembert formula, we see that

$$u(x, t) = \frac{1}{2}\overline{f}(x + ct) + \frac{1}{2}\overline{f}(x - ct) \tag{13}$$

which displays the solution of (12) as a sum of two traveling waves, one moving to the left and the other to the right with velocity c. In Figure 12.2.4 we have used the solution formula (13) to plot a profile of the string at a time T (solid line) where $cT = L/2$. See also Section 10.1.

Comments

We have studied the Cauchy Problem (2) for the wave equation in some detail. The key tool was a simple trick (the Constancy Theorem) that enabled us to find the general solution of the wave equation. We saw that the Cauchy Problem (2) was a model for the motion of infinitely long flexible strings under tension. In spite of the fact that one does not often encounter infinitely long (or even *very* long) strings, we found this physically idealistic case useful in understanding solutions of the more realistic problem of finite strings.

PROBLEMS

1. Find the D'Alembert solution of (2) with $c = 1$ for each of the following sets of initial data f, g. Sketch the profile of the string at $t = 1$.

    (a) $f(x) = \begin{cases} \cos x, & |x| < \pi/2 \\ 0, & \text{otherwise} \end{cases}$ $g(x) = 0$,   all x

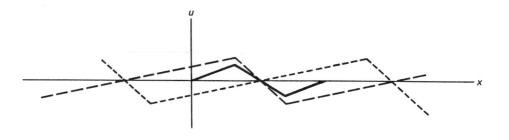

FIGURE 12.2.4 Profile of a plucked string at time $t = L/(2c)$.

(b) $f(x) = \begin{cases} 1 - |x|, & |x| < 1 \\ 0, & \text{otherwise} \end{cases}$ $g(x) = 0, \quad \text{all } x$

(c) $f(x) = 0, \quad \text{all } x;$ $g(x) = \begin{cases} 1, & |x| \le 1 \\ 0, & \text{otherwise} \end{cases}$

2. Find some particular solutions of $u_{xx} + u_{yy} = 0$ other than linear polynomials in x or y. [*Hint*: Write the equation as $u_{xx} - i^2 u_{yy} = 0$, which is the wave equation with $c = i$, and recall that real and imaginary parts of complex-valued solutions must be real-valued solutions of $u_{xx} + u_{yy} = 0$.]

3. **(a)** Prove part (2) of the Constancy Theorem, Theorem 12.2.1.

   (b) Prove part (1) of the Constancy Theorem directly by using the change of variables $x = k + cs, t = s$, for $s_1 \le s \le s_2$ and the Chain Rule.

4. (*Bessel Function Solutions*). Show that the boundary problem

   $$u_{xy} = -au, \quad u(x, 0) = 1, \quad u(0, y) = 1, \quad x \ge 0, \quad y \ge 0$$

   where a is a positive constant, has a solution $u = J_0(2(axy)^{1/2})$, where J_0 is the Bessel function of the first kind of order zero.

5. (*Ill-Posed*). Show that the following initial value problem is not well-posed:
   $$u_{tt} + u_{xx} = 0, \quad x \text{ in } \mathbf{R}, \quad t > 0$$
   $$u(x, 0) = f(x), \quad x \text{ in } \mathbf{R}$$
   $$u_t(x, 0) = g(x), \quad x \text{ in } \mathbf{R}$$

   [*Hint*: Observe that $u = 0$ is a solution if $f = g = 0$, and observe that $u_m = m^{-2}(\sin mx)(\sinh mt)$ is a solution of the initial value problem with data $f_m = 0$ and $g_m(x) = m^{-1} \sin mx$ for each positive integer m, but that $u_m \not\to 0$ as $m \to \infty$ even though f_m and $g_m \to 0$ as $m \to \infty$.]

6. (*Uniqueness Theorem for the Boundary/Initial Value Problem (6) in Section 10.1*).
    Consider the region $R = \{(x, t) : 0 < x < L, t > 0\}$, and let u, v in the classes
    $C^2(R)$ and $C^1(\text{cl } R)$, respectively, be solutions of the wave equation $w_{tt} = c^2 w_{xx}$
    in R such that $u(x, 0) = v(x, 0)$ and $u_t(x, 0) = v_t(x, 0)$ for all $0 < x < L$, and
    $B_0[u] = B_0[v]$, $B_L[u] = B_L[v]$ for all $t > 0$, where B_0, B_L are any of the three
    boundary operators defined in this section. Show that $u = v$ in R. [*Hint*: Use the
    Constancy Theorem to show that $w = u - v$ vanishes in R.]

7. (*The Wave Equation*). Show that the general solution of the wave equation $u_{tt} -$
    $c^2 u_{xx} = 0$ is given by $u(x, t) = F(x + ct) + G(x - ct)$, where F and G are arbitrary
    C^2-functions of a single variable. [*Hint*: See Example 12.2.1.]

12.3 The Heat Equation: Optimal Depth for a Wine Cellar

The heat equation is a linear PDE which models both the flow of heat and a host of
other physical, chemical, and biological phenomena involving diffusion processes. We
derive the heat equation as a mathematical model of thermal conduction and then solve
some simple heat flow problems. Finally, we discuss the mathematical properties of
solutions of the heat equation and the physical meaning of these properties.

Heat Conduction

Molecular vibrations of a material body generate energy which we feel as heat. Heat
flows from warm to cool parts of the body by *conduction*, a process in which the collision of neighboring molecules transfers thermal energy from one molecule to another.
It is conduction that is modeled below. Flow of heat also takes place through *convection*, molecules moving from region to region carrying their thermal energies along
with them, but we shall not model convection here.

The thermal state of a material body at each of its points P at time t is measured by
the *temperature* $u(P, t)$. The units of temperature are degrees centigrade (or Celsius).
Heat itself is measured in *calories* (or *joules*), 1 calorie being the energy needed to
raise the temperature of 1 gram of water by 1 degree (1 Joule $= 0.239$ calorie). Each
substance has a characteristic *specific heat* c, which is the energy required to raise the
temperature of 1 gram of the substance 1 degree (thus the specific heat of water is 1).
The *heat* in a portion D of a material body at time t is given by

$$H(t) = \int_D c\rho u(P, t) \, dP \tag{1}$$

where ρ is the density of the material and the integration is over D. The integral
in (1) is single or multiple according to the spatial dimension of D. The quantity H
as defined by (1) has the units of energy. For simplicity in the derivations below we
take D to be three-dimensional. The one- and two-dimensional cases are treated in the
same way.

The conductive flow of heat through D is governed by the balance equation:

| Rate of change of heat in D | = | Heat generated or consumed inside D per unit time | + | Heat moving across the boundary of D per unit time | (2) |

We shall show how (2) may be represented mathematically by the partial differential
equation of heat flow. From (1) we see that the left side of the balance equation has

the form

$$\frac{dH(t)}{dt} = \frac{d}{dt}\int_D c\rho u(P,t)\,dt = \int_D \frac{\partial}{\partial t}[c\rho u(P,t)]\,dP \qquad (3)$$

where c, ρ, and u are assumed to be continuously differentiable functions of t and continuous in P.

The first term on the right of the balance equation (2) may be expressed as

$$\int_D F(P,t)\,dP \qquad (4)$$

where $F(P,t)$ is the heat generated or consumed per unit volume per unit time at the point P at time t. The point P is a *source* if $F(P,t)$ is positive, a *sink* if $F(P,t)$ is negative.

The last term on the right-hand side of (2) may be written as an integral over the boundary ∂D of D. The integrand is the amount of heat moving across the boundary per unit of surface area per unit of time at the point P on the boundary at time t,

$$\int_{\partial D} k\nabla u(P,t)\cdot\mathbf{n}\,dS \qquad (5)$$

where ∇u is the spatial gradient of u, $\mathbf{n}$ is the unit outward normal to the boundary ∂D at P, and dS denotes integration over the boundary surface. The coefficient k in (5) is called the *thermal conductivity* and measures the ability of the body to conduct heat. Specifically, k is the time rate of change of heat through unit thickness per unit of surface area per unit of temperature. The integral in (5) is the mathematical representation of the *Euler-Fourier Law of Heat Conduction* (*Fick's Law* in the case of gas or liquid diffusion): Heat flows in the direction of maximal temperature drop per unit of distance at a rate proportional to the magnitude of that drop. Since ∇u points in the direction of maximal rise in temperature, heat flow is actually in the direction of $-\nabla u$. Note that if $\nabla u \cdot \mathbf{n}$ is positive at a point P on the boundary of D, the temperature outside D and near P is higher than that at P, and hence thermal energy flows into D through P. Using the Divergence Theorem (Appendix C.5) the surface integral in (5) may be replaced by a volume integral over D:

$$\int_{\partial D} k\nabla u \cdot \mathbf{n}\,dS = \int_D \operatorname{div}(k\nabla u)\,dP \qquad (6)$$

In Cartesian x, y, z coordinates,

$$\nabla u = \frac{\partial u}{\partial x}\mathbf{i} + \frac{\partial u}{\partial y}\mathbf{j} + \frac{\partial u}{\partial z}\mathbf{k}$$

$$\operatorname{div}(k\nabla u) = \frac{\partial}{\partial x}\left(k\frac{\partial u}{\partial x}\right) + \frac{\partial}{\partial y}\left(k\frac{\partial u}{\partial y}\right) + \frac{\partial}{\partial z}\left(k\frac{\partial u}{\partial z}\right)$$

Inserting the expressions (3)–(6) into the balance equation (2), we have, after some rearranging, that

$$\int_D \left[\frac{\partial}{\partial t}(c\rho u) - F - \operatorname{div}(k\nabla u)\right]dP = 0 \qquad (7)$$

If the integrand is continuous in P and in t, (7) is valid for all regions D if and only if the integrand vanishes. This gives the *heat* (or *diffusion*) *equation*,

$$\frac{\partial}{\partial t}(c\rho u) - \operatorname{div}(k\nabla u) = F \qquad (8)$$

at each point P inside the body for all time t for which the model is valid. The *temperature equation* would be a more precise name for (8). Note that (8) is linear in the unknown u and its derivatives.

If there are no internal sources or sinks, F vanishes and we have the *homogeneous* (or *source-free*) *heat equation,*

$$\frac{\partial}{\partial t}(c\rho u) - \text{div}(k\nabla u) = 0 \tag{9}$$

If the material coefficients c, ρ, and k are constants, the *diffusivity* $K = k/c\rho$ may be defined. The units of diffusivity typically are cm^2/s and its values range from 0.0014 for water (a good insulator, but a poor conductor) up to 1.71 for silver (a good conductor). In this case (9) can be written in terms of the Laplacian operator ∇^2 as

$$u_t - K\nabla^2 u = u_t - K(u_{xx} + u_{yy} + u_{zz}) = 0 \tag{10}$$

Boundary and Initial Conditions

On physical grounds one would not expect the heat equation to be enough to determine uniquely the temperatures $u(P, t)$ within a material body M. Initial and boundary conditions are also needed. The *initial condition* may be written as

$$u(P, 0) = f(P), \quad \text{all } P \text{ in } M \tag{11}$$

where f is a given function. Unlike the situation with the wave equation, it is *not* necessary to prescribe $u_t(P, 0)$ since the heat equation itself does that.

There are three common types of thermal conditions imposed on the boundary of M. The first has to do with the *prescribed boundary temperatures*:

$$u(P, t) = g(P, t), \quad P \text{ in } \partial M, \quad t \geq 0 \tag{12}$$

where g is a given function. For example, the body may be submerged in an ice bath, and hence $u(P, t) = 0$ on the boundary, for all $t \geq 0$.

Alternatively, the body may be wrapped with *thermal insulation*, which affects the flow of heat across the boundary:

$$\frac{\partial u(P, t)}{\partial n} = \nabla u \cdot \mathbf{n} = h(P, t), \quad P \text{ in } \partial M, \quad t \geq 0 \tag{13}$$

where h is a prescribed function. If $h = 0$, the insulation is said to be *perfect*, and there is no heat flow through the boundary.

A third type of boundary condition is given by *Newton's Law of Cooling,*

$$\frac{\partial u(P, t)}{\partial n} = r[U(t) - u(P, t)], \quad P \text{ in } \partial M, \quad t \geq 0 \tag{14}$$

where $U(t)$ is the given *ambient temperature* outside M and r is a given *heat transfer coefficient*. If $r > 0$ and if the ambient temperature is higher than the boundary temperature, $\partial u/\partial n$ is positive, indicating a flow of heat into M across the boundary.

There are other types of boundary conditions, but these three cover most cases. Each boundary condition may be written in terms of a linear boundary operator acting on u and $\partial u/\partial n$. For example, (14) may be written as $B[u] = rU$, where B is the linear operator $(\partial/\partial n) + r$ acting on the linear space of sufficiently smooth functions $u(P, t)$ defined for P in ∂M and $t \geq 0$. In a given problem, part of the boundary may be subject to one condition, another part to a different condition. For example, one end of an iron bar may be immersed in an ice bath, the other end in boiling water, while the middle is covered with perfect insulation.

With these remarks in mind, we may formulate a typical *boundary/initial value problem* for the heat equation in a material body M of constant diffusivity K and with no internal sources and sinks: Find $u(P, t)$ such that

(PDE) $u_t - K\nabla^2 u = 0$, P inside M, $t > 0$
(BC) $\alpha \partial u/\partial n + \beta u = f$, P in ∂M, $t \geq 0$ (15)
(IC) $u = g$, P in M, $t = 0$

where α, β, f are prescribed functions of P and t, and g is a function of P. Continuity and smoothness conditions may also be imposed on α, β, f, g and on ∂M so that (15) is well-set (see the end of this section). Rather than consider general problems, we take up and solve two simple and illustrative special cases of problem (15).

Temperature in a Rod

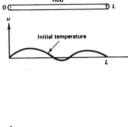

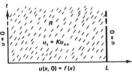

Suppose that a straight rod of constant diffusivity has uniform cross sections. Suppose, moreover, that the lateral surface of the rod is wrapped in perfect insulation, while ice packs are held against the two ends, which then are maintained at a temperature of 0°. Suppose that at some initial time the temperature distribution is known. The problem is to determine the temperature distribution within the rod at later times.

We may take the central axis of the rod to be the x-axis. Because of the cross-sectional symmetry and the insulation on the lateral walls, we shall assume there is no temperature variation in the y and z directions. Denoting the diffusivity by K and the length of the rod by L, we see that we must solve the following boundary/initial value problem ((BC)=boundary condition; (IC)=inital condition) for the temperature $u(x, t)$:

(PDE) $u_t - Ku_{xx} = 0$, $0 < x < L$, $t > 0$
(BC) $u(0, t) = 0$, $u(L, t) = 0$, $t \geq 0$ (16)
(IC) $u(x, 0) = f(x)$, $0 \leq x \leq L$

where the initial temperature distribution f is given. See the margin for a sketch of $R: \ 0 < x < L, \ t > 0$, the region in which (PDE) is to be solved. The shape of R and the nature of the problem itself suggest that the Method of Separation of Variables might be used to construct the solution.

First we look for separated solutions $u = X(x)T(t)$ of (PDE) and of (BC) in (16), leaving the satisfaction of (IC) to a subsequent superposition of separated solutions. Inserting $X(x)T(t)$ into (PDE), we now have that $X(x)T'(t) = KX''(x)T(t)$. Separating variables in the usual way, we have a function of x equaling a function of time t, which can happen only if both equal a *separation constant* λ:

$$\frac{X''(x)}{X(x)} = \frac{1}{K}\frac{T'(t)}{T(t)} = \lambda \qquad (17)$$

Conditions (BC) impose additional restrictions on $X(x)$, but not on $T(t)$. Combining these restrictions with the equations of (17), we see that $X(x)$ and $T(t)$ must be solutions of

$$X''(x) - \lambda X(x) = 0, \qquad X(0) = 0, \quad X(L) = 0 \qquad (18a)$$

$$T'(t) - \lambda KT(t) = 0 \qquad (18b)$$

The values of λ for which the Sturm-Liouville problem (18a) is solvable and the solutions $X(x)$ were determined in Section 10.1:

$$\lambda_n = -(n\pi/L)^2, \qquad X_n(x) = A_n \sin(n\pi x/L), \quad n = 1, 2, 3, \ldots \qquad (19)$$

where A_n is any constant. Corresponding solutions of (18b) are

$$T_n(t) = B_n \exp[-K(n\pi/L)^2 t]$$

where B_n is any constant. Hence solutions of (PDE) and (BC) in (16) are given by

$$u_n = X_n(x)T_n(t) = C_n \sin\left(\frac{n\pi x}{L}\right)\exp\left[-K\left(\frac{n\pi}{L}\right)^2 t\right], \quad n = 1, 2, \ldots \quad (20)$$

where $C_n = A_n B_n$ is an arbitrary constant.

Since (PDE) and (BC) are homogeneous in this problem, any superposition of functions of the form given in (20) is again a solution of (PDE) and (BC). We shall determine constants C_n so that the superposition

$$u = \sum_{n=1}^{\infty} C_n \sin\left(\frac{n\pi x}{L}\right)\exp\left[-K\left(\frac{n\pi}{L}\right)^2 t\right] \quad (21)$$

is also a solution of the initial condition (IC). That is, we must choose C_n so that

$$u(x, 0) = f(x) = \sum_{n=1}^{\infty} C_n \sin\left(\frac{n\pi x}{L}\right), \quad 0 \le x \le L$$

This suggests a Fourier Sine Series. The coefficients C_n may then be found:

$$C_n = \frac{\langle f, \sin(n\pi x/L)\rangle}{\|\sin(n\pi x/L)\|^2} = \frac{2}{L}\int_0^L f(x)\sin\frac{n\pi}{L}x\,dx, \quad n = 1, 2, \ldots \quad (22)$$

Ignoring questions of convergence, the formal solution of (16) is given by (21) and (22). The series defines a classical solution, $u(x, t)$, which belongs to $\mathbf{C}^2(R)$ and to $\mathbf{C}^0(\text{cl } R)$ and satisfies (PDE) in R and (BC) and (IC) on ∂R if the data f are continuous, piecewise smooth, and $f(0) = f(L) = 0$. See the end of this section for more on the classical solution.

The calculation of (22) may be carried out explicitly in the particular case of a rod of length 2, diffusivity 1, and initial temperature

$$f(x) = \begin{cases} x, & 0 \le x \le 1 \\ 2 - x, & 1 \le x \le 2 \end{cases} \quad (23)$$

In fact, $C_n = (-1)^{(n-1)/2}8/(n\pi)^2$ for n odd, $C_n = 0$ for n even. Thus the temperature function in this case is

$$u(x, t) = \frac{8}{\pi^2}\sum_{\text{odd } n}(-1)^{(n-1)/2}\frac{1}{n^2}\sin\left(\frac{n\pi x}{2}\right)\exp\left[\frac{-n^2\pi^2 t^2}{4}\right] \quad (24)$$

Several time profiles of this temperature function are sketched in Figure 12.3.1. Note the "smoothing property" of the heat operator, and note also the decay of the temperature down from the initial angular profile as the heat in the rod "leaks" out through the ends as time increases.

Optimal Depth for a Wine Cellar

One of the first applications of Fourier Series was to model heat flow through soil and rock, Fourier himself taking up this question. We shall consider a simple case here. First, however, we solve a heat problem different from (16):

(PDE)	$u_t - Ku_{xx} = 0,$	$0 < x < \infty,$	$-\infty < t < \infty$	
(BC)	$u(0, t) = A_0 e^{i\omega t},$		$-\infty < t < \infty$	(25)
(Boundedness)	$\|u(x, t)\| < C,$	$0 \le x < \infty,$	$-\infty < t < \infty$	

where K, A_0, ω, and C are assumed to be positive constants. The xt-region, R, is defined by $0 < x < \infty$, $-\infty < t < \infty$, and the data of the problem are sketched in the margin. The use of the complex exponential in the boundary condition of (25) is for

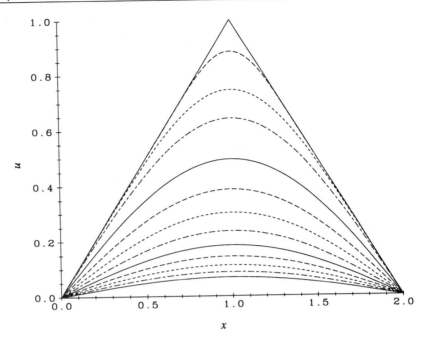

FIGURE 12.3.1 Decaying temperatures in a rod.

convenience in calculation. It may be shown that (25) has no more than one solution that belongs to $C^0(\text{cl } R)$ and satisfies (PDE) throughout R. We shall find a solution, which then must be the only one.

The exponential "input" $A_0 e^{i\omega t}$ along the boundary suggests that the solution of (25) might have the separated form $u = A(x)e^{i\omega t}$. The amplitude $A(x)$ may be found by inserting u into the equations of (25):

$$i\omega A(x)e^{i\omega t} = K A''(x)e^{i\omega t}, \qquad A(0) = A_0, \quad |A(x)| < C \qquad (26)$$

Canceling $e^{i\omega t}$ from the first equation of (26) and solving for $A(x)$ [using the fact $(i\omega/k)^{1/2} = (1+i)(\omega/2K)^{1/2}$], we have that

$$A(x) = C_1 e^{\alpha(1+i)x} + C_2 e^{-\alpha(1+i)x}, \qquad \alpha = \left(\frac{\omega}{2K}\right)^{1/2} > 0$$

where C_1 and C_2 are arbitrary constants. The condition $|A(x)| < C$ implies that $C_1 = 0$ since $e^{\alpha x}$ becomes unbounded as $x \to \infty$. The condition $A(0) = A_0$ implies that $C_2 = A_0$. Thus the solution of (25) is

$$u = A_0 e^{-\alpha(1+i)x} e^{i\omega t} = A_0 e^{-\alpha x} e^{i(\omega t - \alpha x)}, \qquad \alpha = \left(\frac{\omega}{2K}\right)^{1/2}, \quad (x, t) \text{ in cl } R \qquad (27)$$

Problem (25) and its solution (27) may be interpreted in terms of finding the optimal depth for locating a storage cellar. In this setting x is the depth below the surface of the earth, while $A_0 \cos \omega t$ (the real part of $A_0 e^{i\omega t}$) is a crude model of surface temperature normalized about a mean of $0°$. Suppose that the period of the surface wave is 1 year (i.e., $2\pi/\omega = 1 \text{ yr} = 3.25 \times 10^7 \text{ s}$). Then, taking only the real part of $u(x, t)$ from (27), the temperature at depth x at time t is

$$u = A_0 e^{-\alpha x} \cos(\omega t - \alpha x), \qquad \omega = \frac{2\pi}{3.15 \times 10^7}, \qquad \alpha = \left(\frac{\omega}{2K}\right)^{1/2}$$

The *optimal depth* for the storage cellar is defined to be the smallest positive x at which the cellar's "seasons" are 6 months out of phase with the surface seasons. At this depth

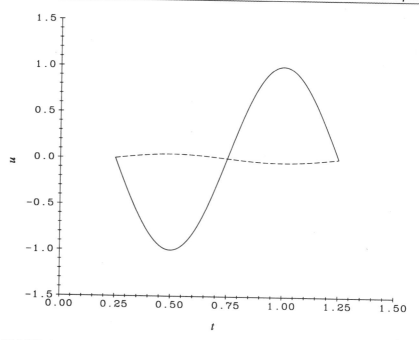

FIGURE 12.3.2 Surface temperature (solid line) and optimal cellar temperature (dashed line) from October ($t = 0.25$) through the following September ($t = 1.25$).

thermal convection currents from the surface tend to move the cellar temperature even closer to the mean. Thus the optimal depth satisfies $\alpha x = \pi$. Now the diffusivity K of average dry soil is 0.002 cm²/s. Hence the optimal depth is

$$x = \frac{\pi}{\alpha} = \pi \left(\frac{2K}{\omega} \right)^{1/2} = \pi(0.004)^{1/2} \left(\frac{3.15 \times 10^7}{2\pi} \right)^{1/2} \approx 445 \text{ cm or } 4.45 \text{ m}$$

The amplitude of the surface wave has dropped from A_0 to $A_0 e^{-x\alpha} = A_0 e^{-\pi} \approx A_0/25$ at the optimal depth. This 25-fold reduction in the amplitude coupled with the reversal of the seasons implies a nearly constant temperature in the cellar; see Figure 12.3.2.

Note that there is no need for an initial condition in this problem (indeed, there is no initial time). Note also that the model is valid near the surface of the earth in a region where there are no subsurface thermal sources or sinks.

Properties of Solutions of the Heat Equation

Boundary/initial value problems for the heat equation have physical significance only if they are *well-set* in the sense that each problem has a solution, exactly one solution, and the solution changes continuously with the data. Separation of variables may often be used to construct a formal series solution as we saw above in the problem of the rod with zero boundary data. However, to show that this series solution has enough convergence properties to be a classical solution requires a detailed study of the series (see below). The other two aspects of a well-set problem (uniqueness and continuity) are somewhat easier to resolve. We shall consider the following problem in this regard:

$$
\begin{array}{llll}
\text{(PDE)} & u_t - Ku_{xx} = 0, & 0 < x < L, & t > 0 \\
\text{(BC)} & u(0, t) = g_1(t), & u(L, t) = g_2(t), & t \geq 0 \\
\text{(IC)} & u(x, 0) = f(x), & 0 \leq x \leq L
\end{array}
\qquad (28)
$$

where g_1 and g_2 belong to $C^0[0, \infty)$ and are bounded for $0 \leq t < \infty$, f belongs to $C^0[0, L]$, $g_1(0) = f(0)$, and $g_2(0) = f(L)$. The equations of (28) model the temperature in a uniform rod with perfect insulation on the lateral boundary, a rod whose endpoints have prescribed (but varying) temperatures. The conditions on f and g ensure the joint continuity of the boundary and initial data.

The first result refers to certain space-time regions shown in the margin. These are defined as follows: $R : 0 < x < L$, $t > 0$; $R_t : 0 < x < L$, $0 < \bar{t} \leq t$; Γ_t, the boundary of R_t with the upper edge deleted. The following basic principle holds for each region R_t. The proof is omitted.

THEOREM 12.3.1

> Maximal Principle for the Heat Equation. Let $u(x, t)$ be any solution of (PDE) in (28) for which u is in $C^2(R)$ and also in $C^0(\text{cl } R)$. Then for every point (x, t) in cl R_t, $|u(x, t)| \leq \max |u(\bar{x}, \bar{t})|$, where the maximum is taken over Γ_t.

The principle has a clear physical interpretation: In the absence of internal sources and sinks, the magnitudes of the temperatures in the rod at time t do not exceed the extreme magnitudes of the end temperatures up to time t and the extreme magnitudes of the initial temperatures.

The Maximal Principle implies uniqueness and continuity in the data:

THEOREM 12.3.2

> Uniqueness. The initial boundary problem (28) has no more than one solution $u(x, t)$.

Proof. Suppose that u_1 and u_2 are both solutions of (28). Then $w = u_1 - u_2$ satisfies (PDE) and the homogeneous conditions, $w(0, t) = w(L, t) = 0$, $w(x, 0) = 0$, $t \geq 0$, $0 \leq x \leq L$. Since $w = 0$ on the three segments of Γ_t (see Figure 12.3.2), the Maximal Principle implies that $|w(x, t)| \leq \max |w(\bar{x}, \bar{t})| = 0$, where $(\bar{x}, \bar{t})$ is taken over Γ_t. Hence $w(x, t) = 0$ and $u_1(x, t) = u_2(x, t)$. Thus if (28) has a solution at all with the required smoothness, the solution is unique.

THEOREM 12.3.3

> Continuity in the Data. Every solution of (28) changes continuously with respect to changes in the data f, g_1, and g_2.

Proof. Let u be a solution of (28), $\bar{u}$ a solution of (28) with $\bar{f}$, $\bar{g}_1$, $\bar{g}_2$ replacing f, g_1, g_2 throughout. Then $u - \bar{u}$ is a solution of (PDE) with respective initial and boundary data $f - \bar{f}$, $g_1 - \bar{g}_1$, $g_2 - \bar{g}_2$. Hence, by the Maximal Principle, for all $t \geq 0$ and all $0 \leq x \leq L$,

$$|u(x, t) - \bar{u}(x, t)| \leq \max_{\substack{0 \leq \bar{x} \leq L \\ 0 \leq \bar{t} \leq t}} \left\{ |f(x) - \bar{f}(x)|, \; |g_1(\bar{t}) - \bar{g}_1(\bar{t})|, \; |g_2(\bar{t}) - \bar{g}_2(\bar{t})| \right\}$$

(29)

Thus, small changes in boundary and initial data mean at most small changes in the temperature; (29) is the mathematical version of this assertion.

for simplicity we shall now set the boundary data g_1 and g_2 equal to 0, reducing boundary/initial value problem (28) to (16). The series given in (21) with coefficients defined by (22) provides a formal solution to the boundary/initial value problem (16). The following results give additional properties of that solution.

THEOREM 12.3.4

> Smoothing Properties. Suppose that the initial data belong to PC[0, L]. Then the formal solution $u(x, t)$ of (16) defined by (21), (22) belongs to $\mathbf{C}^\infty(R)$ and satisfies (PDE) in R.

Proof. The proof of the Smoothing Properties rests on the Weierstrass M-test. First observe from (22) that there is a positive constant A such that $|C_n| \leq A$, $n = 1, 2, \ldots$. Hence, for $N = 1, 2, \ldots$ and all (x, t) in the set S_{t_0} described by $0 \leq x \leq L, 0 < t_0 \leq t$,

$$C_n \sin\left(\frac{n\pi}{L}x\right) \exp\left[-K\left(\frac{n\pi}{L}\right)^2 t\right] \leq A \exp\left[-K\left(\frac{n\pi}{L}\right)^2 t_0\right]$$

Now the series $\sum \exp[-K(n\pi/L)^2 t_0]$ converges by the Ratio Test. Hence by the Weierstrass M-test, the series in (21) converges uniformly in S_{t_0} to a function which, of course, we call $u(x, t)$. If each term of the series in (21) is differentiated k times (r times in x and $k - r$ times in t, say), the nth term of the derived series is no larger in magnitude than a term of the form $Bn^{k-r} \exp[-K(n\pi/L)^2 t_0]$ for all (x, t) in S_{t_0}, for some constant B (which may depend on k, but not on n). But the series $\sum_n n^{k-r} \exp[-K(n\pi/L)^2 t_0]$ also converges by the Ratio Test. Hence by the Weierstrass M-test and a basic theorem on uniform convergence, the derived series converges uniformly to a kth derivative of $u(x, t)$ (i.e., to $\partial^k u/\partial x^r \partial t^{k-r}$) on the closed region S_{t_0}. Since all of the arguments above hold for every $t_0 > 0$ and for every $k = 1, 2, \ldots$, it follows that $u(x, t)$ belongs to $\mathbf{C}^\infty(R)$. A direct calculation involving the term-by-term differentiation of the series for u shows that (21) satisfies (PDE) in R.

That $u(x, t)$ possesses all derivatives of all orders is quite remarkable since the initial data are only required to be piecewise continuous. Visual evidence of the smoothing properties may be seen in Figure 12.3.1 where the sharp corner in the initial data is immediately rounded off for $t > 0$. Under additional assumptions on the initial data, the series in (21) defines a function $u(x, t)$ which is a classical solution of (16), that is, satisfies the initial conditions of (16) [as well as (PDE)].

THEOREM 12.3.5

> Classical Solution. Let $f(0) = f(L) = 0$ and suppose that $f(x)$ is continuous and piecewise smooth, $0 \le x \le L$. Then the function $u(x, t)$ defined by (21) and (22) belongs to $\mathbf{C}^\infty(R)$ and to $\mathbf{C}^0(\mathrm{cl}\ R)$. Moreover, $u(x, t)$ is a solution of (16).

The proof is an extension of the series techniques used above; details omitted.

Comments

The "smoothing property" of the *heat operator* $(\partial/\partial t) - K(\partial^2/\partial x^2)$ is quite unlike any property of the wave operator, $(\partial^2/\partial t^2) - c^2(\partial^2/\partial x^2)$. In fact, smoothing gives a direction to time that the wave operator cannot. We may argue as follows in the context of (23) and (24) corresponding to piecewise linear initial data with a "corner." By the smoothing character of the heat operator, the temperature function lies in $\mathbf{C}^\infty(R)$ for all $t > 0$. Thus the "corner singularity" in the initial data does not propagate from ∂R into R, and the process $f(x) \to u(x, t)$ is irreversible. The sequence of temperature profiles in Figure 12.3.1 cannot be read backward, and *time has an arrow for heat conduction*.

Corner singularities in initial data do propagate with the wave operator, and, in fact, are repeated periodically, as we saw in Section 12.2. The sequence of displacement profiles for a plucked guitar string can be read forward or backward in time without distinction. *Time has no arrow for the wave equation in one spatial dimension*.

There is more. As we saw from D'Alembert's formula, initial disturbances propagate with speed c under the influence of the wave operator. It may be shown, but we shall not do so here, that the *speed of propagation of an initial temperature disturbance is infinity*. In fact, suppose that the initial temperature of a uniform rod is $0°$ except for a temperature $T_0 > 0$ in some segment of small length at the middle of the rod. Then (with end temperatures maintained at $0°$) for any positive t, no matter how small, the solution $u(x, t)$ of the corresponding boundary/initial value problem is positive for every point x inside the rod. The thermal disturbance at the center of the rod has traveled infinitely fast and raised the temperature everywhere inside the rod.

The infinite speed of propagation for the heat operator and the periodic repetition of "corners" on initial displacements for the plucked guitar string and the wave operator show the defects of the respective mathematical models. Thermal diffusion and wave motion cannot be perfectly modeled by the boundary/initial value problems constructed in this chapter. However, the mathematical models presented here are sufficiently accurate in most regards that they continue to be the models of first choice in the treatment of simple heat and wave phenomena.

PROBLEMS

1. (*Temperature in a Rod*). Use the Method of Separation of Variables technique to construct a series solution of the one-dimensional heat equation $u_t - Ku_{xx} = 0$, where $0 < x < L$, $t > 0$ and the following initial and boundary conditions are imposed. Describe how the temperature function $u(x, t)$ behaves as $t \to +\infty$. [*Hint*: See the subsection on temperature in a rod.]

   (a) $u(0, t) = u(L, t) = 0$, $\quad u(x, 0) = \sin(2\pi/L)x$

(b) $u(0, t) = u(L, t) = 0, \quad u(x, 0) = x$

(c) $u(0, t) = u(L, t) = 0, \quad u(x, 0) = u_0 > 0, u_0$ a constant

(d) $u(0, t) = u(L, t) = 0, \quad u(x, 0) = \begin{cases} u_0, & 0 \leq x \leq L/2 \\ 0, & L/2 < x \leq L \end{cases}$

(e) $u(0, t) = 0, u_x(L, t) = 0, \quad u(x, 0) = \sin(\pi x/(2L))$

(f) $u(0, t) = 0, u_x(L, t) = 0, \quad u(x, 0) = x$

(g) $u_x(0, t) = u_x(L, t) = 0, \quad u(x, 0) = x$

2. (*Constant End Temperatures*).

    (a) Find the series solution of the problem $u_t - KU_{xx} = 0, 0 < x < 1, t > 0$; $u(0, t) = 10, u(1, t) = 20, t \geq 0; u(x, 0) = 0, 0 < x < 1$. [*Hint*: Write $u(x, t) = A(x) + v(x, t)$, where $A''(x) = 0$, $A(0) = 10$, $A(1) = 20$, and $v_t - Kv_{xx} = 0$, $0 < x < 1, t > 0; v(0, t) = 0, v(1, t) = 0, t \geq 0; v(x, 0) = -A(x), 0 \leq x \leq 1$. Then find v in the usual way once $A(x)$ has been found.]

    (b) Show that the *steady-state solution* in part **(a)** is $u = A(x)$ in the sense that $v(x, t) \to 0$ as $t \to \infty$.

3. (*Variable End Temperatures*). Let u be a solution to the problem $u_t - Ku_{xx} = 0$, $0 < x < 1, t > 0; u(0, t) = g_1(t), u(1, t) = g_2(t), t \geq 0; u(x, 0) = f(x), 0 \leq x \leq 1$.

    (a) If $V = g_1(t) + [g_2(t) - g_1(t)]x$ and if $U(x, t)$ satisfies the equations $U_t - KU_{xx} = -KV_t, 0 < x < 1, t > 0; U(0, t) = U(1, t) = 0, t \geq 0; U(x, 0) = f(x) - V(x, 0), 0 \leq x \leq 1$, show that $u = V(x, t) + U(x, t)$.

    (b) Use the Eigenfunction Expansion Method of Section 10.6 to find $U(x, t)$ if $g_1(t) = \sin t$ and $g_2(t) = 0$.

    (c) Find $u(x, t)$. [*Hint*: Use your answers to parts **(a)** and **(b)**.]

4. (*Internal Sources/Sinks*). Use the Eigenfunction Expansion Method of Section 10.6 to solve the problem $u_t - Ku_{xx} = 3e^{-2t} + x, 0 < x < 1, t > 0; u(0, t) = u(1, t) = 0$, $t \geq 0; u(x, 0) = 0, 0 \leq x \leq 1$.

5. (*Wine Cellars*). Read the material in this section concerning the optimal depth of a wine cellar.

    (a) Let the surface temperature wave be $T_0 + A_0 \cos \omega t$, where A_0, T_0, and ω are positive constants. Find the optimal depth of the wine cellar. What is that depth if ω corresponds to 1 day instead of 1 year as in the example given in the text?

    (b) Find the temperature function $u(x_0, t)$ at the optimal depth x_0.

    (c) Formulate and solve the wine cellar problem where the surface wave has the form $T_0 + A_1 \cos \omega_1 t + A_2 \cos \omega_2 t$, where ω_1 corresponds to 1 year and $\omega_2 = 365\omega_1$ corresponds to 1 day.

6. (*Time's Arrow*).

    (a) Prove that the series (24) diverges for every $x, 0 < x < 2$, if $t = t_0 < 0$.

    (b) Explain this in terms of time's arrow.

    (c) Find an initial data function $f(x) \neq 0$ so that (21) and (22) do give solutions defined for all time, even for $t < 0$. [*Hint*: Consider $f(x) = \sin(\pi x/L)$.]

7. (*Variable Boundary Temperatures*). Suppose that the temperature $u(x, t)$ in a rod of length 1 and diffusivity 1 satisfies the equations $u_t - u_{xx} = 0, 0 < x < 1, t > 0$; $u(0, t) = te^{-t}, t \geq 0; u(1, t) = 0, t \geq 0; u(x, 0) = 0.01x(1-x), 0 \leq x \leq 1$. Show that $|u(x, t)| \leq 1/e$ for $0 \leq x \leq 1, t \geq 0$.

12.4 Laplace's Equation and Harmonic Functions

Laplace's Equation is the homogeneous, linear, second-order PDE

$$\nabla^2 u = 0 \tag{1}$$

where ∇^2 is the Laplacian operator. Laplace's equation models steady-state temperatures in a body of constant material diffusivity. By "steady state" we mean that the temperature function u does not change with time, although it may vary from point to point within the body. Laplace's equation also models the gravitation and magnetic potentials in empty space, electric potential, and the velocity potential of ideal fluids. For these reasons, (1) is also called the *potential equation*.

The Laplacian operator in the Cartesian rectangular coordinates of 3-space is $\nabla^2 = \partial^2/\partial x^2 + \partial^2/\partial y^2 + \partial^2/\partial z^2$. The operator has the following forms in other coordinate systems:

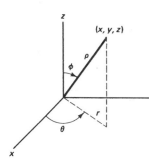

Polar coordinates:

$$u_{xx} + u_{yy} = u_{rr} + \frac{1}{r}u_r + \frac{1}{r^2}u_{\theta\theta} \tag{2}$$

Cylindrical coordinates:

$$u_{xx} + u_{yy} + u_{zz} = u_{rr} + \frac{1}{r}u_r + \frac{1}{r^2}u_{\theta\theta} + u_{zz} \tag{3}$$

Spherical coordinates:

$$u_{xx} + u_{yy} + u_{zz} = \frac{1}{\rho^2}(\rho^2 u_\rho)_\rho + \frac{1}{\rho^2 \sin\phi}(\sin\phi\, u_\phi)_\phi + \frac{1}{\rho^2 \sin^2\phi}u_{\theta\theta} \tag{4}$$

Recall that Cartesian coordinates (x, y, z) in $\mathbf{R}^3$ are related to spherical coordinates (ρ, θ, ϕ) as follows:

$$x = \rho\sin\phi\cos\theta$$

$$y = \rho\sin\phi\sin\theta$$

$$x = \rho\cos\phi$$

Warning: Some authors interchange θ and ϕ.

Laplace's equation has many solutions; for example,

$$u = c_1 e^{-x}\cos y + c_2 z + c_3 e^{-4z}\cos 4x$$

gives solutions in Cartesian coordinates for all constants c_1, c_2, c_3, while

$$u = c_1 r\cos\theta + c_2 r^2 \sin 2\theta$$

gives solutions of the two-dimensional Laplace's equation in polar coordinates for all c_1 and c_2. Boundary or boundedness conditions are needed to select a unique solution. We shall solve representative problems in various coordinate systems and derive the fundamental properties of these solutions.

The Dirichlet Problem: Harmonic Functions

Let G be a region in $\mathbf{R}^2$ or $\mathbf{R}^3$, and let h be a piecewise continuous function defined on ∂G, the boundary of G. Then the *Dirichlet Problem* for G with *boundary data h* is defined as the following boundary value problem. Find a function u in $C^2(G)$ such that $\nabla^2 u = 0$ in G with the additional property that if P is a point of ∂G where h is continuous, then $\lim_{x \to P} u(x) = h(P)$, where x is in G. With this understanding we shall briefly write the Dirichlet Problem for G with boundary data h as

$$\begin{array}{llll} \text{(PDE)} & \nabla^2 u = 0 & \text{in } G \\ \text{(BC)} & u = h & \text{on } \partial G \end{array} \tag{5}$$

A solution of Laplace's equation is called a *harmonic function* or a *potential function*. Observe that when h lies in $\mathbf{C}^0(\partial G)$, a solution u of (5) belongs to $\mathbf{C}^0(\text{cl } G)$ and $u(P) = h(P)$ for all P in ∂G. In this case (5) is called the *classical* Dirichlet Problem and u is called a *classical solution* for the Dirichlet Problem.

Steady-State Temperatures in the Unit Disk: Circular Harmonics

Let $G = \{(x, y) : x^2 + y^2 < 1\}$ and suppose that h is a given piecewise continuous function of ∂G. We shall construct a solution of the Dirichlet Problem (5) for this region, G, and the given function, h, by using the Method of Separation of Variables. If the region G and the conditions (PDE) and (BC) are expressed in Cartesian coordinates, separation of variables will *not* provide us with a formal solution to this problem because ∂G is not composed of level curves in the Cartesian coordinate system. From this point of view, if we find a problem that is equivalent to (5) but expressed in terms of polar coordinates, separation of variables would have some chance for success. Using (2) we see that (5) is equivalent to the problem

$$\begin{array}{llll} \text{(PDE)} & w_{rr} + (1/r)w_r + (1/r^2)w_{\theta\theta} = 0 & \text{for } \theta \text{ in } \mathbf{R}, \quad 0 < r < 1 \\ \text{(BC)} & w(1, \theta) = f(\theta) & \text{for } \theta \text{ in } \mathbf{R} \end{array} \tag{6}$$

where f is the function h expressed in terms of the polar angle θ, and $w(r, \theta) = u(r\cos\theta, r\sin\theta)$. Observe that f belongs to $PC[-\pi, \pi]$ and that w is required to be periodic in θ with period 2π and twice continuously differentiable in the "strip," $0 < r < 1$, $-\infty < \theta < \infty$. Moreover, to ensure that u is twice continuously differentiable in a region containing the origin, we must impose the condition that for any θ_0 in $\mathbf{R}$, the limits of all derivatives of w up to second order exist as $(r, \theta) \to (0, \theta_0)$ and are independent of θ_0. We may interpret (6) as the model for steady-state temperatures in a thin homogeneous disk whose top and bottom faces are perfectly insulated and for which there are prescribed edge temperatures.

Following the method of Separation of Variables to construct a formal solution of (6), we first look for all twice continuously differentiable solutions of (PDE) which have the form $R(r)\Theta(\theta)$. After substituting $R(r)\Theta(\theta)$ into (PDE), the variables can be separated and we are led to consider the separated ordinary differential equations

$$\Theta'' - \lambda\Theta = 0, \quad r^2 R'' + rR' + \lambda R = 0$$

where λ is the separation constant. We must demand that the solution $R(r)\Theta(\theta)$ be periodic in θ and smooth across $\theta = \pi$; thus we must necessarily have the conditions $\Theta(-\pi) = \Theta(\pi)$, $\Theta'(-\pi) = \Theta'(\pi)$. Hence we are led to consider the Sturm-Liouville problem with periodic boundary conditions

$$\Theta'' - \lambda\Theta = 0, \quad \Theta(-\pi) = \Theta(\pi), \quad \Theta'(-\pi) = \Theta(\pi)$$

But we have considered this problem in Example 10.5.2 and found that it has a non-trivial solution if and only if $\lambda = \lambda_n = -n^2$, $n = 0, 1, 2, \ldots$. The solutions of this Sturm-Liouville problem are given by

$$\Theta_0(\theta) = 1, \quad \Theta_n(\theta) = A_n \cos n\theta + B_n \sin n\theta, \quad n = 1, 2, \ldots$$

where A_n and B_n are arbitrary real numbers. Replacing λ by λ_n in the other separated equation, we obtain the differential equations

$$r^2 R'' + r R' - n^2 R = 0, \quad n = 0, 1, 2, \ldots$$

For each n this is an Euler equation (see Section 11.4 or Problem 4 in Section 3.6) and has the general solution

$$R(r) = A r^n + B r^{-n} \quad \text{when } n = 1, 2, \ldots \tag{7a}$$

and

$$R(r) = A + B \ln r \quad \text{when } n = 0 \tag{7b}$$

But since R must be well-behaved as $r \to 0^+$, we must take $B = 0$ for any $n = 0, 1, 2, \ldots$, and hence we have that

$$R_n = r^n, \quad n = 0, 1, 2, \ldots$$

Thus we look for a formal solution to (6) in the form

$$w(r, \theta) = \frac{A_0}{2} + \sum_{n=1}^{\infty} r^n (A_n \cos n\theta + B_n \sin n\theta) \tag{8}$$

We compute the A_n and B_n now by imposing the boundary condition

$$f(\theta) = \frac{A_0}{2} + \sum_{n=1}^{\infty} (A_n \cos n\theta + B_n \sin n\theta) \quad \text{for } -\pi \leq \theta \leq \pi$$

Recalling that $1, \cos x, \sin x, \ldots$ is a basis for PC$[-\pi, \pi]$, we have that

$$\begin{aligned} A_n &= \frac{1}{\pi} \int_{\pi}^{\pi} f(\theta) \cos n\theta \, d\theta, \quad n = 0, 1, 2, \ldots \\ B_n &= \frac{1}{\pi} \int_{\pi}^{\pi} f(\theta) \sin n\theta \, d\theta, \quad n = 1, 2, \ldots \end{aligned} \tag{9}$$

The representation (8) of the Dirichlet Problem (6) is an expansion in *circular harmonics*.

The solution (8) with coefficients defined by (9) is only a formal solution to the Dirichlet Problem, since nothing has been said as yet about convergence properties. Although we will not prove it, if the boundary function h is piecewise continuous on ∂G, then the function w defined by (8) and (9) belongs to $C^\infty(G)$ and satisfies (PDE). Thus steady-state temperature functions have the same strong smoothness properties as the time-dependent temperature functions of Section 12.3. If, in addition, h is continuous on ∂G and also piecewise smooth, then it can be shown that (8) and (9) define a classical solution of (6).

If the disk has radius $r_0 > 0$ rather than radius 1, then (8) and (9) define the solution of the corresponding Dirichlet Problem if r in (8) is replaced by r/r_0.

Properties of Harmonic Functions

Harmonic functions have a number of distinctive properties. For simplicity all of the results below are stated and proved only for planar regions.

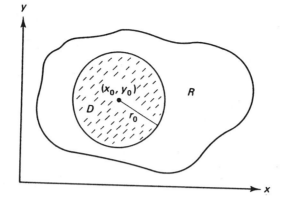

FIGURE 12.4.1 Geometry for the Mean Value Property

THEOREM 12.4.1

> Mean Value Property. Suppose that $u(x, y)$ is a solution of Laplace's equation in a planar region R. Let (x_0, y_0) be a point of R, and D a disk of radius r_0 centered at (x_0, y_0) and lying entirely in R. Then
>
> $$u(x_0, y_0) = \frac{1}{2\pi} \int_0^{2\pi} u(x_0 + r_0 \cos\theta, y_0 + r_0 \sin\theta)\, d\theta$$
>
> That is, the value of a harmonic function at a point is the average of its values around any circle centered at the point.

Proof. Figure 12.4.1 illustrates the geometry. Let the polar coordinates r, θ be referred to the point P. Then using (8) with r replaced by r/r_0, we have that at $r = 0$

$$w(0, \theta) = \frac{A_0}{2}$$

where by (9) A_0 is given by

$$\frac{1}{\pi} \int_{-\pi}^{\pi} u(x_0 + r_0 \cos\theta, y_0 + r_0 \sin\theta)\, d\theta$$

and we are done.

Harmonic functions also satisfy a Maximum Principle.

THEOREM 12.4.2

> Maximum Principle for Harmonic Functions. Unless the harmonic function u is a constant function, u cannot attain either its maximum or its minimum value inside a region R of $\mathbf{R}^2$. That is, if R is open and u is continuous on cl R, then the extreme values of a nonconstant harmonic function u are attained only on the boundary of R.

Proof. Suppose that u attains its maximum in R at some interior point P of R. Then this maximum value is the average of the values of u around the edge of any disk centered at P and lying in R (Theorem 12.4.1). But u is continuous, and so $u(P)$ cannot be both an average and a maximum unless the values of u are constant. A similar argument applies to the minimum.

One of the central questions is whether a problem has exactly one solution and whether the solution changes continuously with the data. If so, the problem is said to be *well-set*. The existence of a solution for an arbitrary Dirichlet problem is not easily shown, although for regions of simple shape solutions may be constructed by the method of Separation of Variables (e.g., the Dirichlet Problem solved above). However, the Maximum Principle may be applied to derive the other two aspects of a well-set Dirichlet Problem (i.e., no more than one solution and continuity in the data).

THEOREM 12.4.3

Uniqueness and Continuity. The Dirichlet Problem (5) for a region G in $\mathbf{R}^2$ has no more than one continuous solution u in cl G if the boundary data h are continuous on ∂G. Moreover, the solution (if it exists) varies continuously with respect to the data.

Proof. Suppose that $\nabla^2 u = \nabla^2 v = 0$ in G while $u = h$ and $v = h + \epsilon$ on ∂G, where h and ϵ are continuous on ∂G. We shall show that

$$\min_{P \text{ on } \partial G} \epsilon(P) \leq v - u \leq \max_{P \text{ on } \partial G} \epsilon(P) \tag{10}$$

for all values of $v(Q) - u(Q)$, Q in cl G. Thus if $|\epsilon(P)|$ is small, v is close to u and we have continuity in the data. Now to show the inequality (10), let $w = v - u$. Then $\nabla^2 w = \nabla^2 v - \nabla^2 u = 0$ in G. By the Maximum Principle (and the corresponding Minimum Principle),

$$\min_{P \text{ on } \partial G} w(P) \leq w(Q) \leq \max_{P \text{ on } \partial G} w(P)$$

for all values $w(Q)$, Q in cl G. But $w = v - u = h + \epsilon - h = \epsilon$ on ∂G, and we are done. Next, we use the inequality (10) to show that the problem (5) has no more than one solution. Suppose that u_1 and u_2 are both solutions of (5). Then $U = u_1 - u_2$ satisfies (PDE) in G but with $U = 0$ on ∂G. Hence by the same argument as above, $0 \leq U \leq 0$, for all values of $U(Q)$, Q in cl G. Hence $U = 0$ and $u_1 = u_2$. Thus (5) has no more than one solution.

Comments

From the results above we see that harmonic functions have strong and distinctive properties. These properties may be interpreted in terms of steady-state temperatures within a thin plate R whose top and bottom faces are perfectly insulated and whose edge temperatures are prescribed.

PROBLEMS

1. (*Plate Temperatures*).

   (a) Show that the solution of the steady-state temperature problem in the rectangular plate G: $0 < x < L$, $0 < y < M$, modeled by

(PDE)	$u_{xx} + u_{yy} = 0$	in G
(BC)$_1$	$u(0, y) = \alpha(y)$,	$0 \leq y \leq M$
(BC)$_2$	$u(L, y) = 0$,	$0 \leq y \leq M$
(BC)$_3$	$u(x, 0) = 0$,	$0 \leq x \leq L$
(BC)$_4$	$u(x, M) = 0$,	$0 \leq x \leq L$

is given by

$$u_\alpha(x, y) = \sum_{n=1}^{\infty} A_n \sin \frac{n\pi y}{M} \sinh \frac{n\pi}{M}(L - x)$$

where

$$A_n \sinh \frac{n\pi L}{M} = \frac{\langle \alpha, \sin(n\pi y/M) \rangle}{\| \sin(n\pi y/M) \|^2} = \frac{2}{M} \int_0^M \alpha(y) \sin \frac{n\pi y}{M} \, dy$$

(b) Show that the formal solution of the temperature problem in G with general prescribed boundary temperatures

(PDE)	$u_{xx} + u_{yy} = 0$	in G
(BC)$_1$	$u(0, y) = \alpha(y),$	$0 \le y \le M$
(BC)$_2$	$u(L, y) = \beta(y),$	$0 \le y \le M$
(BC)$_3$	$u(x, 0) = \gamma(x),$	$0 \le x \le L$
(BC)$_4$	$u(x, M) = \delta(x),$	$0 \le x \le L$

is $u = u_\alpha + u_\beta + u_\gamma + u_\delta$, where u_α is given in part **(a)** and u_β, u_γ, u_δ are defined analogously.

(c) Repeat part **(a)** with (BC)$_1$ replaced by the "insulation" condition $u_x(0, y) = \tilde{\alpha}(y)$, $0 \le y \le M$.

(d) Repeat part **(a)** with (BC)$_3$ and (BC)$_4$ replaced by the "perfect insulation" conditions $u_y(x, 0) = 0 = u_y(x, M)$, $0 \le x \le L$.

2. (*Temperatures in an Annulus*). Find the steady temperature in a thin annulus $G = \{(r, \theta) : \rho < r < R, \quad -\pi \le \theta \le \pi\}$ whose faces are insulated and whose boundaries are maintained at the temperatures $f(\theta)$ and $g(\theta)$ at $r = \rho$ and $r = R$, respectively. [*Hint*: Proceed as in the text for a disk, but keep both terms in formulas (7a), (7b). Thus, in determining the superposition constants one will have

$$f(\theta) = A_0 + B_0 \ln \rho + \sum \rho^n[A_n \cos n\theta + B_n \sin n\theta]$$
$$+ \sum \rho^{-n}[C_n \cos n\theta + D_n \sin n\theta]$$

and a similar expression with f replaced by g, and ρ by R.]

3. (*Temperatures in a Disk*).

   (a) Solve the Dirichlet Problem for steady-state temperatures in the unit disk if $f(\theta) = 3 \sin \theta$ on the edge. Find the maximum and minimum temperatures in the disk.

   (b) Repeat part **(a)** if $f(\theta) = \theta + \pi$ for $-\pi \le \theta < 0$; $f(\theta) = -\theta + \pi$ for $0 \le \theta \le \pi$.

   (c) Repeat part **(a)** if $f(\theta) = 0$ for $-\pi < \theta < 0$; $f(\theta) = 1$ for $0 < \theta \le \pi$.

12.5 Steady Temperatures in Spheres and Cylinders

The main topic of this section will again be Laplace's equation, $\nabla^2 u = 0$, but over spheres and cylinders. It was pointed out earlier that the Method of Separation of Variables fails when the physical boundaries of the region where the solution u is defined does not consist of level surfaces of the coordinate system in which $\nabla^2 u$ is expressed. We experienced this in Section 12.4 when the region was a thin circular plate and we needed to express $\nabla^2 u$ in terms of polar coordinates. As we shall see in this section, when the region in which u is defined is a sphere or a cylinder, then the PDE $\nabla^2 u = 0$ will have to be expressed in spherical coordinates or cylindrical coordinates, respectively, before the Method of Separation of Variables can be applied. In these new coordinate systems the Sturm-Liouville Problem encountered will no longer be a regular one, so we will take care of this difficulty first.

Singular Sturm-Liouville Problems

As defined in Section 10.5, if the functions $\rho(x)$, $p(x)$, and $q(x)$ are such that

$$\rho \text{ is in } C^0[a,b], \quad \rho \neq 0 \text{ on } (a,b), \quad q \text{ is in } C^0[a,b]$$

$$p \neq 0 \text{ on } (a,b), \quad p \text{ is in } C^1[a,b]$$

then a regular Sturm-Liouville Problem asks us to find all values of λ (called eigenvalues) for which the ODE

$$(py')' + qy = \lambda\rho y \tag{1}$$

has nontrivial solutions (called eigenfunctions) in $C^2[a,b]$ which satisfy separated boundary conditions at the endpoints $x = a$, $x = b$ where $p \neq 0$. For example, a separated condition at $x = a$ has the form $\alpha u'(a) + \beta u(a) = 0$, where the constants α and β are not both zero. It was shown that eigenfunctions corresponding to distinct eigenvalues are orthogonal with respect to the scalar product

$$\langle f, g \rangle = \int_a^b \rho f g \, dx, \quad f, g \text{ in } C^2[a,b]$$

If $p(a) = 0$ and/or $p(b) = 0$, then the ODE (1) has a singularity at one or both endpoints of the interval $a \leq x \leq b$, and so the Sturm-Liouville Problem is not a regular one. Several examples below show what happens in this case. It would be helpful to review Sections 11.3 (Legendre Polynomials), 11.6 (Bessel Functions), and 11.7 before plunging into the material that follows.

EXAMPLE 12.5.1

Bessel Functions of Order 0

Consider the eigenvalue equation

$$(1/x)(xu')' = \lambda u$$

where u is in $C^2[0, R_0]$, $R_0 > 0$, and $u(R_0) = 0$. Now the associated operator L has $a = 0$, $b = R_0$, $\rho = x$, $p = x$, $q = 0$. Notice that $p(0) = 0$ and that we have a separated condition at $x = R_0$. Thus, L is a symmetric differential operator, and since a separated condition was used, the eigenspaces are all one-dimensional and mutually orthogonal, with the weighted scalar product

$$\langle f, g \rangle = \int_0^{R_0} x f(x) g(x) \, dx \tag{2}$$

and the eigenvalues are nonpositive. First try $\lambda = 0$; the equation $(xu')' = 0$ has the general solution $u = A \ln x + B$. But the conditions u in $C^2[0, R_0]$ and $u(R_0) = 0$ imply that $A = B = 0$. Thus $\lambda = 0$ is not an eigenvalue. Now try $\lambda = -k^2$, for $k > 0$. The eigenvalue equation $(xu')' + k^2 xu = 0$ has only one solution (up to constant multiples) which belongs to the class $C^2[0, R_0]$; it is $u = J_0(kx)$, where J_0 is the Bessel function of order zero (see Example 11.7.3). The condition $u(R_0) = 0$ implies that $J_0(kR_0) = 0$. But we know from Section 11.6 that J_0 has infinitely many positive zeros: $x_1, x_2, \ldots$. Thus $k_n = x_n/R_0$, $n = 1, 2, \ldots$, so the eigenvalues of L are

$$\lambda_n = -(x_n/R_0)^2, \quad n = 1, 2, \ldots$$

The corresponding eigenspaces are spanned by the eigenfunctions

$$u_n = J_0(x_n x / R_0), \quad n = 1, 2, \ldots$$

$\Phi = \{u_n\}$ is a basis of $PC[0, R_0]$ (PC=piecewise continuous) with the weighted scalar product given by (2), but we omit the proof.

EXAMPLE 12.5.2

Legendre Polynomials

Consider the eigenvalue equation

$$((1 - x^2)u')' = \lambda u$$

for u in $C^2[-1, 1]$. The associated operator L has $a = -1$, $b = 1$, $\rho = 1$, $p(x) = (1 - x^2)$, $q(x) = 0$, and since $p(-1) = p(1) = 0$, $\text{Dom }(L) = C^2[-1, 1]$. Thus L is a symmetric operator under the scalar product

$$\langle f, g \rangle = \int_{-1}^{+1} f(x)g(x)\, dx \tag{3}$$

and moreover the eigenvalues are nonpositive. Although our earlier theorem does not guarantee it, we will see that the eigenspaces are all one dimensional. Of course, the eigenspaces are mutually orthogonal. Observe that for $\lambda = -n(n + 1)$, $n = 0, 1, 2, \ldots$, the eigenvalue equation

$$((1 - x^2)u')' = -n(n + 1)u$$

is Legendre's equation (see Secti on 11.3) whose only solution (up to constant multiples) in $C^2[-1, 1]$ is $P_n(x)$, $n = 0, 1, \ldots$, the Legendre polynomial of order n. It is a fact that $\{P_n(x) : \ n = 0, 1, 2, \ldots\}$ is a basis for $PC[-1,1]$ (although we will not prove it) with the scalar product given by (3). We use this fact to show that L has no other eigenvalues. Assume that $\mu \neq -n(n + 1)$, $n = 0, 1, 2, \ldots$, is an eigenvalue of L with corresponding eigenfunction $v(x)$. Now L is a symmetric operator and so for any $n = 0, 1, 2, \ldots$,

$$\mu \langle v, P_n \rangle = \langle \mu v, P_n \rangle = \langle Lv, P_n \rangle = \langle v, LP_n \rangle = -n(n + 1)\langle v, P_n \rangle$$

Thus $\langle v, P_n \rangle = 0$, for all $n = 0, 1, 2, \ldots$. By the Totality Theorem, Theorem 10.2.7, $v = 0$ and so cannot have been an eigenfunction. Thus the eigenvalues and corresponding eigenfunctions of L are

$$\lambda_n = -n(n + 1), \quad u_n = P_n(x), \quad n = 0, 1, 2, \ldots$$

Hence, if f is in $PC[-1, 1]$, then f has the orthogonal expansion

$$f(x) = \sum_0^\infty A_n P_n(x), \quad A_n = \frac{\langle f, P_n \rangle}{\|P_n\|^2} = \frac{2n + 1}{2} \int_{-1}^1 f(x) P_n(x)\, dx$$

The Sturm-Liouville systems in Examples 12.5.1 and 12.5.2 are not in the regular case. Such systems are said to be *singular Sturm-Liouville systems*. Nevertheless, it is true that conclusions similar to those in the regular Sturm-Liouville Theorem hold. For example, $\Phi = \{J_0(x_n x / R_0) : \ n = 1, 2, \ldots\}$, where x_n is the nth positive zero of J_0, is a basis for $PC[0, R_0]$ under the scalar product $\langle f, g \rangle = \int_0^{R_0} xfg\, dx$. Also, as mentioned earlier, $\Phi = \{P_n(x) : \ n = 0, 1, 2, \ldots\}$, is a basis for $PC[-1, 1]$ under the standard scalar product, where P_n is the Legendre polynomial of order n.

Steady Temperatures in a Ball: Zonal Harmonics and Legendre Polynomials

Now let G be the interior of the unit ball in $\mathbf{R}^3$: $x^2 + y^2 + z^2 < 1$, and consider the Dirichlet Problem

$$
\begin{array}{lll}
\text{(PDE)} & \nabla^2 u = 0 & \text{in } G \\
\text{(BC)} & u = f(\theta, \phi) & \text{on } \partial G
\end{array}
\tag{4}
$$

where the boundary function $f(\theta, \phi)$ is expressed in spherical coordinates and is assumed to be a continuous function. Since the physical boundary of G is a sphere, we would first need to express the PDE in (4) in spherical coordinates if we are to have any hope of being able to use separation of variables. Using (4) from Section 12.4 to transform the problem (4) into spherical coordinates we obtain

$$
\text{(PDE)} \quad \frac{1}{\rho^2}(\rho^2 u_\rho)_\rho + \frac{1}{\rho^2 \sin\phi}(\sin\phi u_\phi)_\phi + \frac{1}{\rho^2 \sin^2\phi}u_{\theta\theta} = 0
\tag{5}
$$

$$
\text{(BC)} \quad u(1, \theta, \phi) = f(\theta, \phi)
$$

We shall assume from here on that f is independent of θ. We shall find a solution u that is also independent of θ. As usual, it is assumed that u belongs to $C^2(G)$ and to $C^0(\text{cl } G)$.

In using the Method of Separation of Variables we need only look for solutions of (PDE) of the form $R(\rho)\Phi(\phi)$ since the boundary data f depend only on ϕ. The variables are easily separated to obtain the ODEs,

$$
\rho^2 R'' + 2\rho R' + \lambda R = 0, \quad 0 < \rho < 1
\tag{6}
$$

$$
\sin\phi \Phi'' + \cos\phi \Phi' - \lambda \sin\phi \Phi = 0, \quad 0 < \phi < \pi
\tag{7}
$$

where λ is the separation constant. Now since we assumed that $R(\rho)\Phi(\phi)$ belongs to $C^2(G)$, we shall look for those constants λ such that the ODEs (6) and (7) have solutions R and Φ with

$$
R \text{ in } C^2[0, 1]
\tag{8a}
$$

$$
\Phi \text{ in } C^2[0, \pi]
\tag{8b}
$$

Observe that the ODEs (6) and (7) are *nonnormal* because the leading coefficients vanish when $\rho = 0$ and when $\phi = 0$ or π; thus the conditions (8a) and (8b) are not redundant. So the Method of Separation of Variables has provided us with a singular Sturm-Liouville Problem to solve.

If we change the independent variable with the mapping $s = \cos\phi$, we see that the interval $0 \le \phi \le \pi$ is mapped one-to-one onto the interval $-1 \le s \le 1$ and (using the Chain Rule) that (7) and (8b) become the singular Sturm-Liouville system

$$
(1 - s^2)\frac{d^2\Phi}{ds^2} - 2s\frac{d\Phi}{ds} - \lambda\Phi = 0, \quad -1 < s < 1, \quad \Phi \text{ in } C^2[-1, 1]
\tag{9}
$$

Now, we have seen the ODE in (9) before; indeed, if λ is replaced by $-n(n+1)$, the ODE (9) is precisely Legendre's equation and has as one of its solutions the Legendre polynomial P_n (n can be any nonnegative integer).

According to Example 12.5.2, we have that $\lambda_n = -n(n+1)$, $n = 0, 1, \ldots$, $\Phi_n(\phi) = P_n(\cos\phi)$, $n = 0, 1, \ldots$.

If we substitute λ_n for λ in (6), we arrive at the problem

$$
\rho^2 R'' + 2\rho R' - n(n+1)R = 0, \quad R \text{ in } C^2[0, 1], \quad n = 0, 1, 2, \ldots
\tag{10}
$$

Now ODE (10) is an Euler equation and is easily seen to have the general solution

$$R(\rho) = A\rho^n + B\rho^{-n-1}, \qquad n = 0, 1, 2, \dots$$

Since R must be in $C^2[0, 1]$ (see (8b)), we must take $B = 0$, and so we have that

$$R_n(\rho) = A\rho^n, \qquad n = 0, 1, 2, \dots$$

Now we look for a solution for (5) in the form

$$u(\rho, \phi) = \sum_{n=0}^{\infty} A_n \rho^n P_n(\cos\phi) \tag{11}$$

Inserting $\rho = 1$ into (11), we have the condition

$$f(\phi) = \sum_{n=0}^{\infty} A_n P_n(\cos\phi), \qquad 0 < \phi < \pi \tag{12}$$

If we make the change of variables $x = \cos\phi$ in (12) and let g be the function on $|x| \le 1$ such that $g(\cos\phi) = f(\phi)$ for all $0 \le \phi \le \pi$, then (12) becomes

$$g(x) = \sum_{n=0}^{\infty} A_n P_n(x), \qquad |x| \le 1 \tag{13}$$

Referring to Example 12.5.2, we see that (13) holds if we choose A_n by using the Fourier-Euler formula appropriate to a Legendre basis of PC$[-1,1]$:

$$A_n = \frac{\langle g, P_n \rangle}{\| P_n \|^2} = \frac{2n+1}{2} \int_{-1}^{1} g(x) P_n(x)\, dx, \qquad n = 0, 1, 2, \dots \tag{14}$$

where $g(x) = f(\arccos x)$. Thus the series (11) with coefficients given by (14) gives a *formal* solution to (5).

Those values of ϕ for which $\cos\phi$ is a root of $P_n(\cos\phi)$ [recall that $P_n(x)$ has n distinct roots, all of which lie in $(-1, 1)$] determine *nodal latitudes* on the unit sphere. The regions between consecutive nodal latitudes are called *zones* and the functions $\rho^n P_n(\cos\phi)$ are called *zonal harmonics*. The solution (11) is a *superposition of zonal harmonics* (see Figure 12.5.1).

The results above may be interpreted in terms of steady-state temperatures in a ball, given boundary temperatures independent of longitude θ.

Steady Temperatures in a Cylinder: Cylindrical Harmonics and Bessel Functions

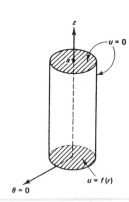

Let C be the cylinder described in cylindrical coordinates by $0 \le 4 \le 1$, $0 \le z \le a$, and let $u(r, \theta, z)$ be the steady temperature in C at the point whose cylindrical coordinates are given by (r, θ, z). We consider the problem of finding the steady temperature in C when the base of the cylinder ($z = 0$) is maintained at the given temperature $f(r)$ degrees, independent of θ, and the rest of ∂C is maintained at $0°$. Thus u is the solution of the boundary problem (see margin figure).

$$
\begin{aligned}
\nabla^2 u &= 0, & 0 \le r < 1, &\quad -\pi \le \theta < \pi, &\quad 0 < z < a \\
u(1, \theta, z) &= 0, & &\quad -\pi \le \theta < \pi, &\quad 0 \le z \le a \\
u(r, \theta, a) &= 0, & 0 \le r \le 1, &\quad -\pi \le \theta < \pi \\
u(r, \theta, 0) &= f(r), & 0 \le r \le 1, &\quad -\pi \le \theta < \pi
\end{aligned}
\tag{15}
$$

where $\nabla^2 u$ is expressed in cylindrical coordinates [see (3) in Section 12.4]. Since the data are a function of the variable r only, we are led to suspect that the solution of (15)

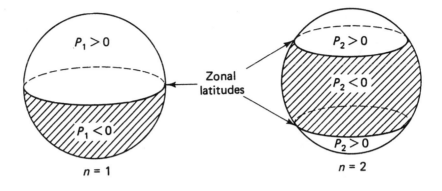

FIGURE 12.5.1 Zones on the surface of the sphere.

is a function of r and z only. Assuming this and separating the variables in the usual way, we obtain the equations

$$R'' + \frac{1}{r}R' - \lambda R = 0, \qquad |R(0+)| < \infty, \qquad R(1) = 0 \tag{16}$$

$$Z'' + \lambda Z = 0, \qquad Z(a) = 0 \tag{17}$$

The singular Sturm-Liouville Problem (16) was treated in Example 12.5.1. The differential equation of (16) is an eigenvalue equation for Bessel's operator of order $p = 0$. Hence the separation constant λ may assume only the values $\lambda = \lambda_n = -x_n^2$, where the x_n are the consecutive positive zeros of $J_0(x)$, $n = 1, 2, 3, \ldots$. The corresponding eigenfunctions for (16) are given by

$$R_n(r) = J_0(x_n r), \qquad n = 1, 2, \ldots$$

Thus (17), with λ replaced by $-x_n^2$, yields

$$Z_n(z) = \sinh x_n(a - z), \qquad n = 1, 2, \ldots$$

Hence we are led to expect a solution of (15) in the form

$$u(r, z) = \sum_{n=1}^{\infty} A_n \sinh x_n(a - z) J_0(x_n r) \tag{18}$$

The remaining condition in (15) which must be satisfied now implies that

$$f(r) = u(r, 0) = \sum_{n=1}^{\infty} A_n \sinh(x_n a) J_0(x_n r)$$

This suggests an orthogonal expansion of f in terms of the basis $\Phi = \{J_0(x_n r), \ n = 1, 2, \ldots\}$ (see Example 12.5.1). Using the Fourier-Euler formula for the coefficients with the inner product given by (2), we have that

$$A_n = \frac{2}{[J_1(x_n)]^2 \sinh(x_n a)} \int_0^1 r f(r) J_0(x_n r) dr \qquad (19)$$

where we have used the formula $\int_0^1 r J_0^2(x_n r) dr = \frac{1}{2} J_1^2(x_n)$ (see Problem 4). The series (18) with coefficients given by (19) defines the formal solution to (15). Although we shall not do it here, it can be shown that this formal solution is indeed a classical solution if $f(r)$ is smooth enough and $f(1) = 0$. The functions $\sinh x_n(a - z) J_0(x_n r)$ are called *cylindrical harmonics*, and (18) is an *expansion in cylindrical harmonics*.

Comments

We have analyzed a sample of Dirichlet Problems which may be solved by the Method of Separation of Variables. As always, the boundary of the region involved must be composed of level sets of the coordinate variables. In practice, this restricts the method to rectangular or boxlike regions, balls, cylinders, disks, or to simple combinations of these regions. We have also showed by two examples that non-Cartesian geometry may lead to singular Sturm-Liouville Problems.

Given the complexities of actually constructing solutions of the Dirichlet Problem, it is remarkable that properties of solutions (e.g., the Maximal Principle) can be proved independently of the particular shape of the region involved. Although we proved the properties only for planar regions, they remain true, after appropriate reformulations, in any number of dimensions.

PROBLEMS

1. (*Temperatures in a Solid Ball*). Find the steady-state temperatures in a ball of unit radius if the surface temperatures f in spherical coordinates are as given. [*Hint*: Use (11), (14), and properties of Legendre given in Table 11.3.1.]

   (a) $f = \cos 2\phi - \sin^2 \phi$

   (b) $f(\phi) = \begin{cases} 1, & 0 < \phi < \pi/2 \\ 0, & \pi/2 < \phi < \pi \end{cases}$ [*Hint*: Use the identity $xP_n' - nP_n = P_{n-1}'$ in Table 11.3.1 and integration by parts to evaluate $\int_{-1}^1 P_n(x)\, dx$.]

   (c) $f(\phi) = |\cos \phi|$

2. (*Temperatures in a Hollow Ball*). Find the steady-state temperatures in a spherical shell, $0 < \rho_1 < \rho < \rho_2$, if the temperatures on the inner sphere are given by $f(\phi)$, on the outer sphere by $g(\phi)$. [*Hint*: Proceed as in the text, but keep both terms in $R(\rho) = A\rho^b + B\rho^{-n-1}$.]

3. (a) Solve (15) if $f(r) = 1$, $0 \le r \le 1$.

   (b) Solve (15) if the side-wall temperature condition $u(1, \theta, z) = 0$ is replaced by the perfect-insulation condition $u_r(1, \theta, z) = 0$, $0 \le z \le a$, $-\pi \le \theta \le \pi$. [*Hint*: Use (19) of Section 11.6 with $p = 0$, and use the identity in Problem 4 with $a = x_n^*$, where x_n^* is a positive zero of J_0' (and so of J_1).]

4. Show that $\int_0^1 r J_0^2(ar) \, dr = \frac{1}{2}(J_0^2(a) + J_1^2(a))$, where a is any real number. [*Hint*: Multiply Bessel's equation of order 0 by $2J_0'(r)$ and rewrite to obtain $[r^2(J_0')^2]' + r^2(J_0^2)' = 0$. Integrate from 0 to a using an integration by parts, and then use the recursion formula (19) of Section 11.6 with $p = 0$: $J_0' = -J_1$.]

5. (*Neumann Problems*). Let G be a bounded planar region whose boundary is a simple smooth closed curve. The problem $\nabla^2 u = 0$ in G, $\partial u / \partial n = f$ on ∂G, where u belongs to $C^2(G)$ and to $C^0(\text{cl } G)$ and f is continuous on ∂G, is called a *Neumann Problem* for G.

   (a) Show that any two solutions of the Neumann Problem differ by a constant. [*Hint*: Let $w = u_1 - u_2$, where u_1 and u_2 solve the Neumann Problem. Apply the Divergence Theorem (Appendix C.5) to the vector $\nabla(w^2)$, and use the facts that $\nabla^2 w = 0$ in G and $\partial w / \partial n = 0$ on ∂G to show that ∇w is the zero vector on cl G.]

   (b) Show that if the Neumann Problem has a solution at all, then $\int_{\partial G} f(s)\, ds = 0$.

   (c) Solve the Neumann Problem above if G is the unit disk in the plane and if $f(\theta) = \sin\theta$, while $u = 0$ at $\theta = 0$, $r = 1$. [*Hint*: Use formula (8), Section 12.4.]

6. (*Singular Sturm-Liouville Problems*). In each of the Sturm-Liouville Problems below identify the operator L, find the eigenvalues and eigenspaces of L, and state the orthogonality and basis properties of the eigenfunctions.

   (a) $x^4 y'' + 2x^3 y' = \lambda y$; $y(1) = 0$, $y(2) = 0$. [*Hint*: To solve the ODE, try the independent variable substitution $s = 1/x$.]

   (b) $xy'' - y' = \lambda x^3 y$; $y(0) = 0$, $y(a) = 0$. [*Hint*: To solve the ODE, try the independent variable substitution $s = x^2$.]

   (c) $\dfrac{1}{x}\left[(xy')' - \dfrac{p^2}{x} y\right] = \lambda y$, y in $C^2[0, R_0]$, $y(R_0) = 0$, where p is a positive number. [*Hint*: This is a Bessel operator of order p.]

APPENDIX A

Basic Theory
of Initial Value Problems

The aim of this Appendix is to present the basic theory for the initial value problem for a single state varable, filling in the proofs that were omitted in Chapter 2. The basic questions of uniqueness, existence, and sensitivity are each taken up in turn. The possibility of extension of solutions is also touched upon in a separate section.

A.1 Uniqueness

The uniqueness question for the initial value problem (IVP)

$$y' = f(t, y), \qquad y(t_0) = y_0 \tag{1}$$

is in some ways the easiest to handle, so we will treat it first. For convenience we first define a *region* in the plane.

❖ **Interior Point, Region.** A point P is an *interior point* of a set R in the ty-plane if for some $\varepsilon > 0$ all points less than ε away from P are also in R. A *region* is a set whose points are all interior points.

THEOREM A.1.1

Uniqueness Principle. In IVP (1) let f and $\partial f/\partial y$ be continuous on some region R containing the interior point (t_0, y_0). Then on any t-interval I containing t_0 there is at most one solution of IVP (1).

Proof. We shall prove the claim only in the case where I is the interval $t_0 \leq t \leq t_0 + a$ and R contains the closed rectangle S: $t_0 \leq t \leq t_0 + a$, $y_0 - b \leq y \leq y_0 + b$, for some positive constants a, b. This special case can be used to establish the more general claim, but we omit the details. Now suppose that $y_1(t)$ and $y_2(t)$ are two solutions of IVP (1) which remain in S over the interval I: $t_0 \leq t \leq t_0 + a$ (see Figure A.1.1). Then

$$[y_1(t) - y_2(t)]' = f(t, y_1(t)) - f(t, y_2(t)), \qquad t_0 < t < t_0 + a \tag{2}$$

Observe that $y_1(t) - y_2(t)$ is continuous on I and vanishes at $t = t_0$. Integrating (2) from t_0 to some t in I, we have that

$$y_1(t) - y_2(t) = \int_{t_0}^{t} [f(s, y_1(s)) - f(s, y_2(s))]\, ds, \qquad t \text{ in } I \tag{3}$$

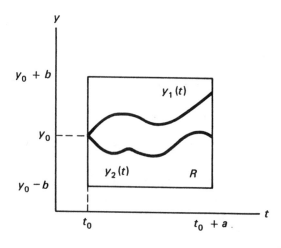

FIGURE A.1.1 Specialized geometry for the IVP (1).

Now since $\partial f/\partial y$ is continuous on S, it follows via the Mean Value Theorem and the Max-Min Value Theorem (see Appendix C.5) that there is a positive constant L such that[1]

$$|f(t, y_1) - f(t, y_2)| \le L|y_1 - y_2|, \qquad \text{for any points } (t, y_1), (t, y_2) \text{ in } S \quad (4)$$

Functions $f(t, y)$ which satisfy the inequality (4) for some constant $L > 0$ are said to satisfy a *Lipschitz condition* in y on S; the constant L is called a *Lipschitz constant*. Thus taking absolute values of each side of (3), denoting $|y_1(t) - y_2(t)|$ by $w(t)$, and estimating the integral in the usual way (see Theorem C.5.9), we obtain the integral inequality

$$0 \le w(t) \le \int_{t_0}^{t} |f(s, y_1(s)) - f(s, y_2(s))|\, ds \le L \int_{t_0}^{t} w(s)\, ds, \qquad \text{all } t \text{ in } I$$

$$(5)$$

We shall show that the only nonnegative solution of the inequality (5) is the function $w(t) = 0$ for all t. Indeed, let $v(t) = \int_{t_0}^{t} w(s)\, ds$; then since $v'(t) = w(t)$ on I we can rewrite (5) as the differential inequality

$$v'(t) - Lv(t) \le 0, \qquad \text{all } t \text{ in } I \qquad (6)$$

Multiplying (6) through by the "integrating" factor $\exp(-Lt)$, we have the inequality $[v(t)\exp(-Lt)]' \le 0$ for all t in I. Thus $v(t)e^{-Lt} \le v(t_0)e^{-Lt_0}$ for all t in I. But since $v(t_0) = 0$, it follows that $v(t) \le 0$ for all t in I. On the other hand, it follows from the definition that $v(t) \ge 0$ for all t in I, and hence $v(t) = 0$. This shows that $w(t) = 0$ as well, so we have shown that $y_1(t) = y_2(t)$ for all t in I. Thus IVP (1) cannot have more than one solution on I, finishing our proof.

---

[1]L may be taken to be max $|\partial f/\partial y|$ for (t, y) in S.

PROBLEMS

1. Consider the function $f(t, y) = |y|$.

   (a) Show that the function f does *not* satisfy the hypotheses of the Uniqueness Principle in any region containing all or part of the t-axis in the ty-plane.

   (b) Show that $y' = |y|$, $y(t_0) = 0$ has a unique solution even so. Is there a contradiction?

   (c) Show that the hypothesis "$\partial f / \partial y$ is continuous" in the Uniqueness Principle can be replaced by "$f(t, y)$ satisfies a *Lipschitz condition* in y". Use this conculsion to prove again that $y' = |y|$, $y(t_0) = 0$ has a unique solution.

2. Let m and n be positive integers without common factors (i.e., relatively prime). Consider the IVP

$$y' = |y|^{m/n}, \qquad y(0) = 0$$

   (a) Show that the IVP has the unique solution $y(t) = 0$ if $m \geq n$.

   (b) Show that the IVP has infinitely many solutions if $m < n$. [*Hint*: See Example 2.2.2.]

A.2 The Picard Process for Solving an Initial Value Problem

We shall now establish the existence of a unique solution of the IVP:

$$y' = f(t, y), \qquad y(t_0) = y_0 \tag{1}$$

where f and $\partial f / \partial y$ are both continuous functions on a region R of the ty-plane containing the interior point (t_0, y_0). The method we shall use is to construct a sequence of iterate functions—called *Picard*[2] *iterates*—which converges to the unique solution of IVP (1).

An Equivalent Integral Equation

To approach the task of finding a solution for IVP (1), we shall first convert it into an "equivalent" integral equation. Let $y(t)$ be a solution of IVP (1) on the interval I containing t_0 such that the graph of $y(t)$ lies in R. Then integrating the relation $y'(t) = f(t, y(t))$ from t_0 to some t in the interval I and using the initial condition, we have

$$y(t) - y_0 = \int_{t_0}^{t} y'(s)\,ds = \int_{t_0}^{t} f(s, y(s))\,ds$$

Hence, if $y(t)$ is a solution of IVP (1) on the interval I containing t_0, then $y(t)$ is a solution of the integral equation

$$y(t) = y_0 + \int_{t_0}^{t} f(s, y(s))\,ds, \qquad t \text{ in } I \tag{2}$$

---

[2]Emile Picard (1856–1941) was an eminent French mathematician and the permanent secretary of the Paris Academy of Science. His mathematical work includes deep results in complex analysis and partial differential equations, as well as in ordinary differential equations.

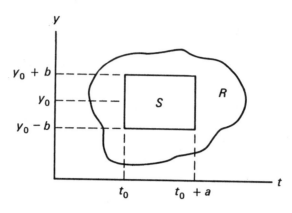

FIGURE A.2.1 Geometry for a forward initial value problem.

Conversely, let $y(t)$ be continuous on I, with its graph lying in R, and satisfy the integral equation (2). Then $y(t_0) = y_0$, and, from the Fundmental Theorem of Calculus (see Appendix C.5), $y(t)$ is differentiable at each interior point of I, and

$$y'(t) = \frac{d(y(t))}{dt} = \frac{d}{dt}\left(y_0 + \int_{t_0}^{t} f(s, y(s))\, ds\right) = f(t, y(t))$$

It follows that $y(t)$ is a solution of IVP (1) on the interval I. This is the sense in which IVP (1) and the integral equation (2) are equivalent. Thus, the solvability of IVP (1) is equivalent to the solvability of the integral equation (2). We shall exploit this equivalence.

Existence of Solution

After this detour, we are ready to prove the existence of a solution to IVP (1). Under the condition that f and $\partial f/\partial y$ are continuous on R, we have already shown that IVP (1) cannot have more than one solution on any interval containing t_0.

THEOREM A.2.1

> **Existence Theorem.** If the functions f and $\partial f/\partial y$ are continuous on a region R of the ty-plane and if (t_0, y_0) is a point of R, then IVP (1) has a (unique) solution $y(t)$ on an interval I containing t_0 in its interior.

Proof. The first step in our proof is to construct in R a closed rectangle S which has the form: $t_0 \leq t \leq t_0 + a$, $y_0 - b \leq y \leq y_0 + b$, where a and b are positive numbers (Figure A.2.1). Such a construction is possible since (t_0, y_0) is an interior point of R. Now the assumptions of the theorem imply that f and $\partial f/\partial y$ are continuous on the closed rectangle S. By Theorem C.5.1 and (4) in Section A.1 there are positive constants M and L such that

$$|f(t, y)| \leq M, \qquad \text{all } (t, y) \text{ in } S$$
$$|f(t, y_1) - f(t, y_2)| \leq L|y_1 - y_2|, \qquad \text{all } (t, y_1), (t, y_2) \text{ in } S \tag{3}$$

Our proof will produce the unique solution, $y(t)$, of the differential equation $y' = f(t, y)$ on the interval $t_0 \le t \le t_0 + c$, where $c = \min\{a, b/M\}$, with $y(t_0) = y_0$. The corresponding solution curve remains in the rectangle S. It may be regarded as a solution of the forward initial value problem associated with IVP (1). The same technique can be used to show the existence of a solution for IVP (1) on an interval: $-d \le t \le t_0$ for some $d > 0$. Putting this "backward" solution together with the "forward" solution, we would then have a "two-sided" solution to IVP (1) on the interval: $t_0 - d \le t \le t_0 + c$, as asserted in the statement of the Existence Theorem. We turn now to the construction of the forward solution.

Construction of the Picard Iterates

We begin with the equivalent formulation of the forward problem for IVP (1) as the integral equation (2). We see that $y(t)$ is a solution of IVP (1) over an interval $t_0 \le t \le t_0 + c$ if and only if $y(t)$ satisfies the integral equation

$$y(t) = y_0 + \int_{t_0}^{t} f(s, y(s)) \, ds, \qquad t_0 \le t \le t_0 + c \qquad (4)$$

We now describe a procedure for generating a sequence of functions $y_0(t)$, $y_1(t)$, $y_2(t), \dots$, all defined and continuous on $t_0 \le t \le t_0 + c$, where $c = \min\{a, b/M\}$, which will be used to generate a solution $y(t)$ of (4) in a sense to be made clear later. The functions $y_n(t)$, $n = 0, 1, 2, \dots$, constructed in this proof arise in a distinctive way and are called *Picard iterates*. They are generated recursively (i.e., $y_{n+1}(t)$ is computed directly from $y_n(t)$, for each $n = 0, 1, 2, \dots$) in the following way. Taking $y_0(t) = y_0, t_0 \le t \le t_0 + c$, we define $y_1(t), y_2(t), \dots$ via the recursion relation

$$y_{n+1}(t) = y_0 + \int_{t_0}^{t} f(s, y_n(s)) \, ds, \qquad n = 0, 1, 2, \dots \qquad (5)$$

with $t_0 \le t \le t_0 + c$. In constructing the Picard iterates $y_n(t)$ via (5), we must take care that at each stage the points $(t, y_n(t))$ remain in S for all $t_0 \le t \le t_0 + c$ before we can go on to compute $y_{n+1}(t)$ via (5). Indeed, the graph of $y_0(t)$ belongs to S. Recalling that $c = \min\{a, b/M\}$, we shall now show by induction that the graph of $y_n(t)$, $n = 0, 1, 2, \dots$ also lies in (5). Assume that for some integer $n \ge 0$, the graph of $y_n(t)$ lies in S for all $t_0 \le t \le t_0 + c$. Using (5), we have the estimate

$$|y_{n+1}(t) - y_0| \le |t - t_0|M \le cM \le b, \qquad t_0 \le t \le t_0 + c \qquad (6)$$

since $c = \min\{a, b/M\}$, and thus the graph of $y_{n+1}(t)$ remains in S for all $t_0 \le t \le t_0 + c$. The induction is complete; hence, there is no difficulty in constructing the Picard iterates via (5) if we restrict t to the interval $[t_0, t_0 + c]$.

Uniform Convergence of a Sequence of Functions

We must interrupt our proof of the Existence Theorem to describe the sense in which the Picard iterates $y_0(t)$, $y_1(t), \dots$ are used to generate the solution of the integral equation (4). Our remarks actually apply to any sequence of functions defined over a common interval, so we shall not restrict ourselves to Picard iterates at the moment.

❖ **Uniform Convergence of a Sequence of Functions.** A sequence of functions $\{y_n(t)\}$, $n = 0, 1, 2, \dots$, all defined on a common interval I, is said to converge

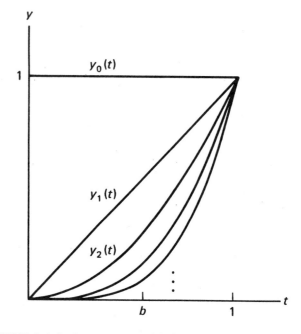

FIGURE A.2.2 Convergence of $\{t^n\}$.

uniformly to the function $y(t)$ on I if given any error tolerance $E > 0$, there exists a positive integer N such that

$$|y_n(t) - y(t)| < E, \quad \text{for all } t \text{ in } I \text{ and all } n \geq N \tag{7}$$

As we shall see, uniform convergence is the way we will generate the solution of the integral equation (4) from the Picard iterates.

As an example, note that the sequence $y_n(t) = t^n$, $n = 0, 1, 2, \ldots$ converges uniformly to the zero function on any closed interval $[0, b]$, where $0 < b < 1$, but does *not* converge uniformly on the closed interval $[0, 1]$ (see Figure A.2.2). It is sometimes useful to view uniform convergence graphically as follows. The sequence $\{y_n(t)\}$ converges to $y(t)$ uniformly over I if for every $E > 0$ the graphs of $y_n(t)$ eventually all remain inside a "tube" of radius E about the graph of $y(t)$.

The inequality (7) is not a very useful test for uniform convergence of a sequence $y_n(t)$ because one must "guess" the limit function $y(t)$ in advance before the test can be applied. As with sequences of numbers, there is a *Cauchy Test for uniform convergence*. It has the advantage that one need not know the limit function in advance before the test can be applied; on the other hand, the Cauchy Test has the disadvantage that when it succeeds one does not necessarily know the limit function.

THEOREM A.2.2

> Cauchy Test for Uniform Convergence.   The sequence $\{y_n(t)\}$ defined over a common interval I is uniformly convergent to some function $y(t)$ if, given any $E > 0$, there is a positive integer N such that
>
> $$|y_n(t) - y_m(t)| < E \quad \text{for all } t \text{ in } I, \quad \text{all } n, m \geq N \tag{8}$$

Note that although the limit is not known from the Cauchy Test, the limit function is unique and can be uniformly approximated by the sequence element $y_n(t)$ as closely as desired by choosing n large enough. Another fact we accept without proof is that the uniform limit of a sequence of continuous functions on a common interval I is also continuous on that same interval. Figure A.2.2 provides an example of this phenomenon on the interval $[0, 1/2]$, as well as an example of what goes wrong when the convergence is not uniform (look at the closed interval $[0,1]$).

Convergence of the Picard Iterates

Now we show that the sequence of Picard iterates $\{y_n(t)\}$ converges uniformly to a solution of the integral equation (4) on the interval $[t_0, t_0 + c]$. First recall from (3) that

$$|f(t, y_1) - f(t, y_2)| \le L|y_1 - y_2|, \quad \text{all } (t, y_1), (t, y_2) \text{ in } I \qquad (9)$$

Thus, using (5) and (9), we have the estimate

$$|y_{n+1}(t) - y_n(t)| \le L \int_{t_0}^{t} |y_n(s) - y_{n-1}(s)| \, ds, \quad t_0 \le t \le t_0 + c, \quad n = 1, 2, \dots \qquad (10)$$

Now observe from (6) that

$$|y_1(t) - y_0| \le b, \quad t_0 \le t \le t_0 + c \qquad (11)$$

and hence, using (10) with $n = 1$, we have

$$|y_2(t) - y_1(t)| \le Lb(t - t_0), \quad \text{for } t_0 \le t \le t_0 + c \qquad (12)$$

Using (12) and (10) with $n = 2$, we see that

$$|y_3(t) - y_2(t)| \le L^2 b \frac{(t - t_0)^2}{2}, \quad \text{for } t_0 \le t \le t_0 + c$$

Proceeding along in this manner, we can use (10) to establish the estimate

$$|y_{n+1}(t) - y_n(t)| \le L^n b \frac{(t - t_0)^n}{n!}, \quad t_0 \le t \le t_0 + c, \quad n = 0, 1, 2 \dots \qquad (13)$$

So for any $n > m$, and using (13) and the triangle inequality, we have the estimate

$$|y_n(t) - y_m(t)| \le |y_n(t) - y_{n-1}(t)| + \cdots + |y_{m+1}(t) - y_m(t)|$$
$$\le b \sum_{k=m}^{n-1} \frac{L^k (t - t_0)^k}{k!} \qquad (14)$$

If we denote by $S_n(t)$ the nth partial sum of $\exp[L(t - t_0)]$, that is,

$$S_n(t) = \sum_{k=0}^{n-1} \frac{L^k (t - t_0)^k}{k!}$$

then (14) can be written as

$$|y_n(t) - y_m(t)| \le b[S_n(t) - S_m(t)], \quad t_o \le t \le t_0 + c \qquad (15)$$

Now since the partial sums $\{S_n(t)\}$ converge uniformly to $\exp[L(t - t_0)]$ on any interval of the t-axis, from what was said earlier it follows that $\{S_n(t)\}$ satisfies the Cauchy Test (8) and hence, from (15), so does the sequence of Picard iterates $\{y_n(t)\}$ over the interval $[t_0, t_0 + c]$. Thus, it follows that $y_n(t)$ converges uniformly to a continuous function $y(t)$ over $[t_0, t_0 + c]$.

To show that $y(t)$ satisfies the integral equation (4) on $[t_0, t_0 + c]$, we need the following result.

THEOREM A.2.3

Integral Convergence Theorem. Let the sequence $\{y_n(t)\}$ of continuous functions converge uniformly to $y(t)$ on $t_0 \le t \le t_0 + c$ and be such that $|y_n(t) - y_0| \le b$ for all $t_0 \le t \le t_0 + c$, where b, c are defined as above. Let $f(t, y)$ satisfy the conditions of IVP (1). Then, as $n \to \infty$, we have

$$\int_{t_0}^{t} f(s, y_n(s)) \, ds \to \int_{t_0}^{t} f(s, y(s)) \, ds, \quad t_0 \le t \le t_0 + c \qquad (16)$$

Proof. Here we prove that the Picard iterates converge to a solution of IVP (1). Observe that the sequence of Picard iterates $\{y_n(t)\}$ satisfies the conditions leading to (16), and also the relation (5) over $[t_0, t_0 + c]$. Thus, taking limits of each side of (5) for each fixed t in $[t_0, t_0 + c]$, we see via (16) that $y(t)$ satisfies the integral equation (4), and hence, the forward IVP (1), finishing the proof.

EXAMPLE A.2.1

Consider the initial value problem

$$y' = ty + 1, \qquad y(0) = 1$$

Comparing this IVP with IVP (1), we see that $f(t, y) = ty + 1$, $t_0 = 0$ and $y_0 = 1$. We may as well take the region R where f is defined to be the whole ty-plane. Let us take the rectangle S to be $0 \le t \le a$, $1 - b \le y \le 1 + b$, where a, b are arbitrary positive numbers. Now $|ty + 1| \le |ty| + 1 \le a(b+1) + 1$ on S, so we may take $M = a(b + 1) + 1$, thus

$$c = \min\left\{ a, \frac{b}{a(b+1)+1} \right\}$$

Because $f(t, y)$ is defined and well-behaved over the entire ty-plane, (5) defines the Picard iterates for all t, but our proof only guarantees convergence of iterates over the interval $0 \le t \le c$. We have

$$y_0(t) = 1$$

$$y_1(t) = 1 + \int_0^t f(s, y(s)) \, ds = 1 + \int_0^t [sy_0(s) + 1] \, ds$$

$$= 1 + \int_0^t (s + 1) \, ds = 1 + t + t^2/2$$

$$y_2(t) = 1 + \int_0^t [sy_1(s) + 1] \, ds$$

$$= 1 + \int_0^t \left[s\left(1 + s + \frac{s^2}{2}\right) + 1 \right] ds$$

$$= 1 + t + \frac{t^2}{2} + \frac{t^3}{3} + \frac{t^4}{8}$$

and so on. Taking $a = 1$, $b = 8$, we see that $c = \min\{1, 0.8\} = 0.8$, and hence the Picard iterates converge in the prescribed sense at least over the interval $0 \le t \le 0.8$ (see Figure A.2.3 where the first five Picard iterates are shown).

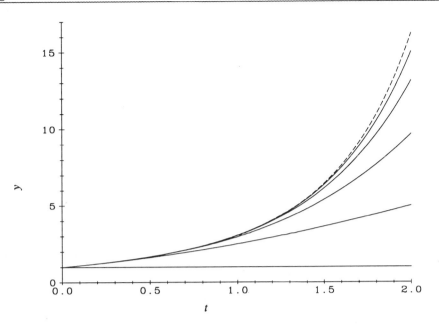

FIGURE A.2.3 First five Picard iterates (solid) and the "true" solution (dashed) of $y' = ty + 1$, $y(0) = 1$. See Example A.2.1.

Comments

Suppose that all the conditions of the Existence and Uniqueness Theorem (EUT) for the IVP (1) are met. The problem has exactly one solution, $y = y(t)$, say. But how can you find it? Only for very special types of functions f is there any hope of finding an explicit formula for the solution y. Alternatively, a computer can be used to come up with an approximate graph of the solution, or a table of its approximate values for a given set of values of t. The computer will use some numerical approximation method to do this. Contemporary computer methods are remarkably robust and accurate, and are now the first choice in solving such problems. But for most users the computer is a "black box" that generates graphs or tables by magic. The Picard iteration scheme, on the other hand, both illuminates what is going on when the initial value problem is solved and lies at the heart of the proof of the EUT. These two assets of the scheme may outweigh its wild impracticality. The problems below illustrate these points. Summarizing, we have the

❖ **Picard Iteration Scheme.** The *Picard Iteration Scheme* for the initial value problem, $y' = f(t, y)$, $y(t_0) = y_0$, consists of constructing the sequence of iterates:

$$y_0(t) = y_0, \ldots, y_{n+1}(t) = y_0 + \int_{t_0}^{t} f(s, y_n(s)) \, ds, \quad n = 0, 1, 2, \ldots$$

According to the Existence and Uniqueness Theorem, $y_n(t) \to y(t)$ (the actual, but unknown solution of the IVP), as $n \to \infty$ for each t in an appropriately defined interval containing t_0. See Atlas Plates PICARD A–C for examples.

PROBLEMS

1. The parts below have to do with Picard iterates for IVP: $y' = -y$, $y(0) = 1$.

   (a) Find the Picard iterates y_1, y_2, y_3.

   (b) Plot the graphs of these iterates for $|t| < 2$.

   (c) Find a formula for the nth iterate y_n by induction.

   (d) Find the exact solution $y(t)$ and sketch its graph along with those of y_0, y_1, y_2, and y_3.

   (e) Show that $y_n(t)$ is a partial sum of the Maclaurin series for the solutiion $y(t)$ of the IVP.

2. Repeat Problem 1 for $y' = 2ty$, $y(0) = 2$.

3. The parts below have to do with Picard iterates for the IVP: $y' = y^2$, $y(0) = 1$.

   (a) Find the Picard iterates y_1, y_2, and y_3.

   (b) Find the solution $y(t)$ directly and plot y_0, y_1, y_2, y_3, and y on their maximal domains.

   (c) Prove by induction that the nth Picard iterate $y_n(t)$ is a polynomial of degree $2^n - 1$.

   (d) Show that every Picard iterate is defined for all t, but that the exact solution is defined only for $t < 1$. Then show that the Maclaurin series of the exact solution converges only for $|t| < 1$. Thus Picard iterates, the exact solution, and the Maclaurin series of the exact solution may all be defined on different intervals.

4. The Picard iteration scheme is sturdy enough to survive a bad choice for the first approximation $y_0(t)$. Show that the Picard iterates for the problem $y' = -y$, $y = 1$ when $t = 0$, converge to the exact solution e^{-t} even if the wrong starting "point" $y_0(t) = 0$ is used to start the iteration scheme. [*Hint*: The Picard iteration scheme becomes in this case $y_{n+1}(t) = 1 + \int_0^t (-y_n(s))\, ds$, for $n = 0, 1, 2, \dots$.]

5. Repeat Problem 4 but with the absurd first approximation $y_0(t) = \sin t$. Show that the sequence of iterates still converges to e^{-t}.

6. The parts below have to do with Picard iterates for IVP: $y' = 1 + y^2$, $y(0) = 0$.

   (a) Find the Picard iterates y_1, y_2, and y_3.

   (b) Prove by induction that $y_n(t)$ is a polynomial of degree $2^{n+1} - 1$.

   (c) Find the exact solution $y(t)$ and identify the maximal interval on which it is defined.

   (d) Does $y_n(t) \to y(t)$ for all t? Explain.

7. How many Picard iterates can you calculate for $y' = \sin(t^3 + ty^5)$, $y(0) = 1$? What is the practical difficulty in finding the iterates?

8. Let $y_0(t) = 3$, $y_1(t) = 3 - 27t$, $y_n(t) = 3 - \int_0^t y_{n-1}^3(s)\, ds$, $n = 0, 1, 2, \dots$ Find $\lim_{n \to \infty} y_n(t)$. [*Hint*: Consider a certain IVP related to the Picard iteration scheme.]

9. (*Uniform Convergence of a Sequence of Functions*).

   (a) Show that the sequence $\{(1/n) \sin(nt)\}_{n=1}^{\infty}$ converges uniformly to the zero function on any interval I. [*Hint*: $|(1/n) \sin(nt) - 0| \leq 1/n$.]

   (b) Show that the sequence $\{t^n\}_{n=1}^{\infty}$ converges uniformly to the zero function on the interval $0 \leq t \leq 0.5$, while it converges, but not uniformly, on the interval $0 \leq t \leq 1$ to the function $f(t) = 0, 0 \leq t < 1$, $f(1) = 1$.

A.3 Extension of Solutions

A solution of the IVP

$$y' = f(t, y), \quad y(t_0) = y_0 \tag{1}$$

defined on an interval can be extended outside that interval when the conditions on the data allow it. This brings up the question of the relationship between the maximally extended solutions of $y' = f(t, y)$ and the region R, where f is defined. In the proof of the Existence Theorem, we showed that for any *closed* rectangle S in the ty-plane where f and $\partial f/\partial y$ are continuous, the solution curve that "enters" S at the midpoint of a vertical side will "exit" through one of the other sides. The result below shows that this behavior of solution curves is typical.

THEOREM A.3.1

> *Extension Principle.* Let R be a region in the ty-plane and suppose that f and $\partial f/\partial y$ are continuous on R. Let D be a closed bounded region contained in R and consider IVP (1) with (t_0, y_0) a point in the interior of D. Then the solution of IVP (1) can be extended forward and backward in t until its solution curve exists through the boundary of D.

Proof. We shall consider only the forward extension assertion since the backward assertion is proved in the same way. Now since D is bounded there is a *smallest* value τ_1 such that no point of D lies in the half-plane $t > \tau_1$. Thus, if the forward solution curve exits D through its boundary, then there is nothing to prove. To have something to prove, let us assume that the forward maximally extended solution of IVP (1) remains in the interior of D and is defined for all t in the interval $t_0 \le t \le \tau_2 \le \tau_1$. We shall show that this assumption leads to a contradiction, and hence the forward solution curve cannot both lie inside D and to the left of some value $\tau_2 < \tau_1$ (see Figure A.3.1). Thus, the solution curve must indeed exit D. To show this, recall that IVP (1) is equivalent to the integral equation:

$$y(t) = y_0 + \int_{t_0}^{t} f(s, y(s)) \, ds$$

and hence,

$$y(t_2) - y(t_1) = \int_{t_1}^{t_2} f(s, y(s)) \, ds \qquad t_0 \le t_1, t_2 \le \tau_2 \tag{2}$$

Since f is continuous on a closed bounded region there is a constant $M > 0$ such that $M \ge |f(t, y)|$ for all (t, y) in D. Thus (2) implies via the standard estimate for integrals (see Theorem C.5.9) that

$$|y(t_1) - y(t_2)| \le M|t_1 - t_2|, \quad t_0 \le t_1, t_2 < \tau_2 \tag{3}$$

From (3) and the Cauchy Test for convergence, it follows that the limit $y(\tau_2^-)$ exists. Now the point $(\tau_2, y(\tau_2^-))$ cannot lie in the interior of D, for then the method described in the proof of the Existence Theorem would allow us to extend $y(t)$ beyond $t = \tau_2$. Thus, $(\tau_2, y(\tau_2^-))$ lies on the boundary of D and we are done.

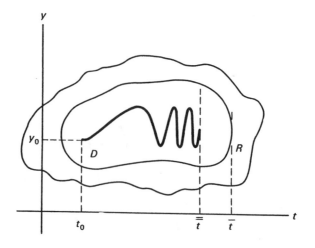

FIGURE A.3.1 Geometry for the Extension Principle.

Comments

The nonlinear problem $y' = y^2$, $y(0) = y_0$, where $y_0 \neq 0$, has the unique solution $y(t) = y_0(1 - ty_0)^{-1}$ defined on the interval $(-\infty, y_0^{-1})$ if $y_0 > 0$, and on the interval (y_0^{-1}, ∞) if $y_0 < 0$. This problem illustrates the fact that it is not easy to predict from the rate function itself just how a solution will approach the boundary of the region in which the existence and uniqueness hypotheses hold; nevertheless, for *linear* problems the situation is much simpler (Problem 3).

PROBLEMS

1. Find a solution formula for each IVP below, and determine the largest t-interval on which the solution is defined. What happens to the solution as t approaches each endpoint of the interval?

   (a) $y' = -y^3$, $y(0) = 1$ **(b)** $y' = -te^{-y}$, $y(0) = 2$

2. Find a first-order ODE $y' = f(t, y)$, each solution of which is defined only on a finite t-interval. [*Hint*: See Section 2.2.]

3. Show that the solution of the IVP $y' + p(t)y = q(t)$, $y(t_0) = y_0$, where $p(t)$ and $q(t)$ are continuous for all t in an open interval I containing t_0, is defined for all t in I. [*Hint*: Use the solution formula (11) of Section 1.2.]

A.4   Sensitivity of Solutions to the Data

There remains a last question concerning the solution of IVP

$$y' = f(t, y), \quad y(t_0) = y_0 \tag{1}$$

Loosely phrased, the question amounts to this: Is it always possible to find bounds on the determination of the data $f(t, y)$ and y_0 in IVP (1) which will guarantee that the corresponding solution will be within prescribed error bounds over a given t-interval? If this question can be answered in the affirmative, one consequence is that any "small" enough change in the data of an initial value problem produces only a "small" change in the solution. Or, stated another way: In spite of the fact that the initial value problem arising from a model has empirically determined components (and therefore can never be known precisely), the model is still relevant and useful.

In our study of IVP (1), we have been guided by the four questions posed in Section 2.2: Does IVP (1) have any solution? How many? What are they? How do solutions respond to changes in the data? If the functions f and $\partial f/\partial y$ are continuous in some region R in the ty-plane and (t_0, y_0) is a point of R, we gave satisfactory answers to the first, second, and third questions. In this section we show that these same simple conditions on the data also lead to a satisfactory answer to the last question.

In addressing the last question, it would be extremely helpful to have a formula for the solution of IVP (1) in which the data appear explicity. But for general nonlinear differential equations, there rarely is a solution formula for IVP (1) in which the data appear explicitly. Thus we shall have to find some other way to answer the fourth question in the general case.

To estimate the change in the solution to IVP (1) as the data $f(t, y)$ and y_0 are modified, it will be very helpful to have an estimate of the type given below.

THEOREM A.4.1

Basic Estimate. For given constants A, B, and L with $L > 0$, suppose that $z(t)$ is a continuous function on the interval $[a, b]$ which satisfies the integral inequality

$$z(t) \leq A + B(t - a) + L \int_a^t z(s)\,ds \quad \text{for all } a \leq t \leq b \qquad (2)$$

Then $z(t)$ satisfies the estimate

$$z(t) \leq Ae^{L(t-a)} + \frac{B}{L}(e^{L(t-a)} - 1) \quad \text{for all } a \leq t \leq b \qquad (3)$$

Proof. Let us put

$$Q(t) = \int_a^t z(s)\,ds \quad \text{for } a \leq t \leq b$$

Then Q is continuous on $[a, b]$, and the Fundamental Theorem of Calculus implies that $Q'(t) = z(t)$ for $a < t < b$; so (2) can be written as

$$Q'(t) - LQ(t) \leq A + B(t - a), \quad a < t < b \qquad (4)$$

Multiply each side of (4) by the positive factor e^{-Lt} and then use the fact that $\left[e^{-Lt}Q\right]' = e^{-Lt}Q' - e^{-Lt}LQ$ to obtain

$$\left[Q(t)e^{-Lt}\right]' \leq [A + B(t - a)]e^{-Lt}, \quad a < t < b \qquad (5)$$

Since $F(t) = -L^{-2}[B + LA + LB(t - a)]e^{-Lt}$ is an antiderivative to the right-hand side of (5), we see that (5) can be written as

$$\left[Q(t)e^{-Lt} - F(t)\right]' \leq 0, \quad a < t < b \qquad (6)$$

The function $G(t) = Q(t)e^{-Lt} - F(t)$ is continuous on the interval $[a, b]$, and hence because of (6), we see that $G(t) \leq G(a)$ for $a \leq t \leq b$, or after simplification [note that $Q(a) = 0$],

$$A + B(t - a) + LQ(t) \leq Ae^{L(t-a)} + \frac{B}{L}(e^{L(t-a)} - 1), \quad a \leq t \leq b \qquad (7)$$

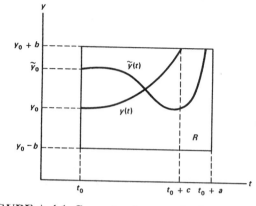

FIGURE A.4.1 Geometry for perturbation estimate.

Now from (2), we see that

$$z(t) \leq A + B(t - a) + LQ(t), \quad a \leq t \leq b$$

which together with (7) yields the inequality (3), and we are done.

There is also a "backward" form of the Basic Estimate, which under the same hypotheses states that the inequality

$$z(t) \leq A + B(b - t) + L \int_t^b z(s)\, ds, \quad a \leq t \leq b \tag{8}$$

implies that

$$z(t) \leq A e^{L(b-t)} + \frac{B}{L}(e^{L(b-t)} - 1), \quad a \leq t \leq b \tag{9}$$

The proof of (9) from (8) is very similar to the foregoing proof for (3) from (2), and hence is omitted (see Problem 1).

We need some estimate of the change in the solution of IVP (1) which occurs when the data $f(t, y)$ and y_0 are modified (or, "perturbed").

THEOREM A.4.2

> Perturbation Estimate. Let the function f in IVP (1) be continuous along with $\partial f / \partial y$ in a rectangle R described by the inequalities: $t_0 \leq t \leq t_0 + a$, $|y - y_0| \leq b$. Suppose that $g(t, y)$ and $\partial g(t, y)/\partial y$ are also continuous functions on R and that on some common interval: $t_0 \leq t \leq t_0 + c$, with $c \leq a$, the solution $y(t)$ of IVP (1) and the solution $\tilde{y}(t)$ of the "perturbed" IVP
>
> $$y' = f(t, y) + g(t, y), \quad y(t_0) = \tilde{y}_0 \tag{10}$$
>
> both have solution curves which lie in R (see Figure A.4.1). Then we have the estimate
>
> $$|y(t) - \tilde{y}(t)| \leq |y_0 - \tilde{y}_0| e^{L(t - t_0)} + \frac{M}{L}(e^{L(t - t_0)} - 1), \quad t_0 \leq t \leq t_0 + c \tag{11}$$
>
> where L and M are any numbers such that
>
> $$M \leq |g(t, y)|, \quad L \leq \left| \frac{\partial f}{\partial y} \right|, \quad \text{all } (t, y) \text{ in } R$$

TABLE A.4.1 Data for IVP (14)

$\tilde{y}(t)$	$\bar{a}$	$\bar{b}$	$\tilde{y}_0$
1	0	0	−3.11
2	0.7	24	−3.11
3	0.5	60	−3.13
4	0.7	36	−3.14
5	0.5	48	−3.15
6	0	0	−3.16
7	0.7	36	−3.16

Proof. The proof is an immediate consequence of the Basic Estimate above. Using the fact proven in Section A.2 that the solution of an IVP also solves an equivalent integral equation, we see that $y(t)$ and $\tilde{y}(t)$ must satisfy the integral equations (for $t_0 \le t \le t_0 + c$)

$$y(t) = y_0 + \int_{t_0}^{t} f(s, y(s))\, ds$$

$$\tilde{y}(t) = \tilde{y}_0 + \int_{t_0}^{t} [f(s, \tilde{y}(s)) + g(s, \tilde{y}(s))]\, ds$$

Subtracting the second equation from the first, we have for any $t_0 \le t \le t_0 + c$ that

$$y(t) - \tilde{y}(t) = y_0 - \tilde{y}_0 + \int_{t_0}^{t} [f(s, y(s)) - f(s, \tilde{y}(s))]\, ds - \int_{t_0}^{t} g(s, \tilde{y}(s))\, ds \tag{12}$$

Putting $z(t) = |y(t) - \tilde{y}(t)|$ and using the inequality (4) in Section A.1 to estimate the integrals in (12), we have the inequality

$$z(t) \le z(t_0) + M(t - t_0) + L \int_{t_0}^{t} z(s)\, ds, \quad t_0 \le t \le t_0 + c \tag{13}$$

Comparing (13) with (12) and applying the Basic Estimate with A and B replaced by $|y_0 - \tilde{y}_0|$ and M, respectively, we obtain the desired estimate (11), concluding our proof.

EXAMPLE A.4.1

As an illustration of the Perturbation Estimate, we graph in Figure A.4.2 solutions of the initial value problems

$$\begin{cases} y' = y \sin y \\ y(0) = y_0 \end{cases} \qquad \begin{cases} y' = y \sin y + \bar{a} \sin(\bar{b}ty) \\ y(0) = \tilde{y}_0 \end{cases} \tag{14}$$

for various values of the constants $\tilde{y}_0$, $\bar{a}$, and $\bar{b}$. The solid lines are the solution curves of $y(t)$ that correspond to the data in Table A.4.1.

The dashed lines come from the error bounds given by the Perturbation Estimate (11). Notice that the error bound is not very sharp as we move away from $t_0 = 0$, and hence is not really all that practical in estimating errors.

Now we are in a position to answer the last of the basic questions.

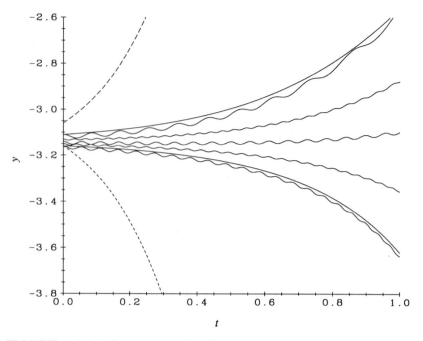

FIGURE A.4.2 Solution curves for (14).

THEOREM A.4.3

Continuity in the Data. Let f, $\partial f/\partial y$, g, and $\partial g/\partial y$ be continuous functions of t and y on the rectangle R defined by $t_0 \leq t \leq t_0 + a$, $|y - y_0| \leq b$. Let $E > 0$ be a given error tolerance. Then there exist positive constants $H < b$ and $c \leq a$ such that the respective solutions $y(t)$ and $\tilde{y}(t)$ of the IVPs

$$\text{(a)} \begin{cases} y' = f(t, y) \\ y(t_0) = y_0 \end{cases} \qquad \text{(b)} \begin{cases} y' = f(t, y) + g(t, y) \\ y(t_0) = \tilde{y}_0 \end{cases} \tag{15}$$

satisfy the inequality

$$|y(t) - \tilde{y}(t)| \leq E, \quad t_0 \leq t \leq t_0 + c \tag{16}$$

for any choice of $\tilde{y}_0$ for which $|y_0 - \tilde{y}_0| \leq H$.

Proof. Let H be any positive constant with $H < b$, and let $\tilde{M}$ be such that $\max |g(t, y)| \leq \tilde{M}$ for all (t, y) in R. Then according to the Existence and Uniqueness Theorem, the solution $\tilde{y}(t)$ of the second IVP in (15) must be defined at least over the interval: $t_0 \leq t \leq t_0 + c$, with

$$c = \min \left\{ a, \frac{b - H}{M + \tilde{M}} \right\} \tag{17}$$

where M is a positive constant such that $|f(t, y)| \leq M$ for all (t, y) in R. Now by the Perturbation Estimate, we have that

$$|y(t) - \tilde{y}(t)| \leq H e^{Lc} + \frac{\tilde{M}}{L}(e^{Lc} - 1), \quad t_0 \leq t \leq t_0 + c \tag{18}$$

where $L > 0$ is any number such that $|\partial f/\partial y| \leq L$, for all (t, y) in R. The positive constants H and c can be determined to make the right-hand side of (18) less than, or equal to, E. Hence, (16) holds and we are done.

Properties of Solutions

Putting together the results of this section with those of previous sections, we now have a rather comprehensive view of how solutions of IVPs such as IVP (1) behave. If $f(t, y)$ and $\partial f/\partial y$ are continuous on a closed region R of the ty-plane, then:

1. Through each point in the interior of R, there is one and only one solution curve of the equation $y' = f(t, y)$, and this solution curve can be extended up to the boundary of R (from the Existence Theorem, the Uniqueness Principle, and the Extension Principle).

2. The change in the solution of IVP (1) is small if the changes in the data $f(t, y)$ and y_0 are sufficiently small (Continuity in the Data Theorem).

The first property above shows that an IVP is an effective tool in solving problems that involve dynamical systems. The second property indicates that models involving IVPs are not invalidated even though empirically determined elements of the model do not have their "true" values.

PROBLEMS

1. Show that the "backward" estimate (9) follows from the inequality (8).

2. Use the Perturbation Estimate to approximate $|\tilde{y}(t) - y(t)|$ if $\tilde{y}(t)$ solves the IVP $y'(t) = (e^{-y^2} + 0.1)y$, $y(0) = \tilde{a}$, and $y(t)$ solves the IVP $y'(t) = 0.1y$, $y(0) = a$.

3. Let $y(t)$ be a continuously differentiable function on the real line such that $|y'(t)| \leq |y(t)|$ for all t. Show that either $y(t)$ never vanishes or else that $y(t) = 0$ for all t.

4. (*Perturbation Estimate When Only Initial Data Changes*). Suppose that the rate function $f(t, y)$ is defined in some rectangle $R : a \leq t \leq b, c \leq y \leq d$ and that f, f_y are continuous on R. If K and L are constants such that $K \leq f_y(t, y) \leq L$, and if $y(t)$ and $\bar{y}(t)$ are any solutions on $a \leq t \leq b$ of the ODE $y' = f(t, y)$ whose solution curves lie in R, then follow the outline below to show that

$$|y(a) - \bar{y}(a)|e^{K(t-a)} \leq |y(t) - \bar{y}(t)| \leq |y(a) - \bar{y}(a)|e^{L(t-a)}, \quad \text{for } a \leq t \leq b$$

- If $K = 0$, show that $|y(a) - \bar{y}(a)| \leq |y(t) - \bar{y}(t)|$, for all $a \leq t \leq b$. If $L = 0$ show that $|y(t) - \bar{y}(t)| \leq |y(a) - \bar{y}(a)|$, for all $a \leq t \leq b$. [*Hint*: First convert the ODE $y' = f(t, y)$ to an equivalent integral equation.]

- Let $0 < K < L$ and follow the proof of the Basic Estimate (Theorem A.4.1) to show the asserted estimate. What goes wrong with this approach when either K or L is negative? [*Hint*: Note that if $y(a) > \bar{y}(a)$, then $y(t) > \bar{y}(t)$ for all $t \geq a$.]

- If either $K < 0$ or $L < 0$ show that the appropriate part of the asserted estimate is valid. [*Hint*: Use simple iteration on (2) in the Basic Estimate (Theorem A.4.1).]

- Consider the function $f(t, y) = (1/10)(-y + \ln(1 + y))$ defined on the rectangle $R: 0 < t < \infty$, $y \geq 0$. Find the sharpest possible values for the constants K and L such that $K \leq f_y(t, y) \leq L$ for all (t, y) in R. Let $y(t)$ be the solution of the IVP: $y' = f(t, y)$, $y(0) = 4$, and let $\bar{y}(t)$ be the solution of the IVP: $y' = f(t, y)$, $y(0) = 1$. Show by direct measurement at $t = 10, 20, 30$, and 40 that the asserted estimate holds.

- Repeat the analysis above for the solution $y(t)$ of the IVP: $y' = \sin t - (1/10)(.5\cos y + y)$, $y(0) = 5$, and the solution $\bar{y}(t)$ of the IVP: $y' = \sin t - (1/10)(.5\cos y + y)$, $y(0) = -5$.

APPENDIX B

Numerical Methods

In the paragraphs below we shall elaborate on the methods introduced in Section 2.5 for finding approximate numerical solutions for IVPs.

B.1 One-Step Methods

Convergence Properties of Euler's Method

We shall derive bounds for the discretization and the total errors incurred when Euler's Method is applied to the initial value problem (IVP):

$$y' = f(t, y), \quad y(t_0) = y_0 \tag{1}$$

where f and $\partial f/\partial y$ are continuous in a rectangle R centered at (t_0, y_0). As noted in Appendix A.1, there is a positive constant L such that the following Lipschitz condition holds:

$$|f(t, y_1) - f(t, y_2)| \leq L|y_1 - y_2| \tag{2}$$

for all points (t, y_1) and (t, y_2) in R. Euler's Method with step size h rests on the formula

$$y_{j+1} = y_j + f(t_j, y_j)h, \quad j = 0, \ldots, N - 1 \tag{3}$$

We have the following result concerning the global discretization error E_N when Euler's Method is used to approximate the solution of IVP (1) at $T = t_0 + Nh$.

THEOREM B.1.1

Discretization Error of Euler's Method. If $f(t, y)$ is continuously differentiable in the rectangle R, the discretization error of the Euler algorithm (3) for solving IVP (1) is first order in h; that is, for all N, $E_N \leq Mh$, for some constant M.

Proof. Suppose that the function f is continuously differentiable in t and y in the rectangle R. Then the solution $y(t)$ has a continuous second derivative, as well as a continuous first derivative. Indeed, using the Chain Rule and the differential equation, we see that

$$\frac{d^2 y(t)}{dt^2} = \frac{\partial f(t, y(t))}{\partial t} + \frac{\partial f(t, y(t))}{\partial y} \frac{dy(t)}{dt}$$

$$= \frac{\partial f}{\partial t} + \frac{\partial f}{\partial y} f(t, y)$$

and hence $y(t)$ is twice continuously differentiable.

According to Taylor's Theorem with the Lagrange form of the remainder (see Appendix C.2), there are numbers c_j in $[t_j, t_{j+1}]$ such that

$$y(t_{j+1}) = y(t_j) + y'(t_j)(t_{j+1} - t_j) + y''(c_j)(t_{j+1} - t_j)^2/2 \qquad (4)$$

or

$$y(t_{j+1}) = y(t_j) + f(t_j, y(t_j))h + y''(c_j)h^2/2$$

Subtracting (3) from (4), we have that

$$y(t_{j+1}) - y_{j+1} = y(t_j) - y_j + [f(t_j, y(t_j)) - f(t_j, y_j)]h + y''(c_j)h^2/2 \quad (5)$$

Taking the absolute value of each side of (5) and using the Triangle Inequality and the Lipschitz condition (2), we have that the global discretization errors $E_j = |y(t_j) - y_j|$ satisfy the relations

$$E_{j+1} \le E_j + |f(t_j, y(t_j)) - f(t_j, y_j)|h + |y''(c_j)|h^2/2$$
$$\le E_j + |y(t_j) - y_j|hL + |y''(c_j)|h^2/2$$
$$\le (1 + hL)E_j + Kh^2$$

where L is a Lipschitz constant for f on R, and K is an upper bound for $|y''(t)|/2$ (recall that every continuous function such as $|y''(t)|/2$ has a maximum value on a closed interval).

Thus we have shown that for $j = 0, 1, 2, \ldots, N - 1$,

$$E_{j+1} \le \alpha E_j + Kh^2 \qquad (6)$$

where $\alpha = 1 + hL$. Since E_0 vanishes, we have from (6) that

$$E_1 \le Kh^2, \quad E_2 \le \alpha E_1 + Kh^2 \le (\alpha + 1)Kh^2, \ldots,$$

$$E_N \le \left(\alpha^{N-1} + \cdots + \alpha + 1\right) Kh^2 = \frac{\alpha^N - 1}{\alpha - 1} Kh^2 = \frac{\alpha^N - 1}{hL} Kh^2 \qquad (7)$$

Using the fact that $\alpha = 1 + hL = 1 + (T - t_0)L/N$ and the fact that for any positive u, $\{(1 + u/N)^N\}_1^\infty$ is an increasing sequence which converges to e^u, we have from (7) that

$$E_N \le \frac{K}{L}\left[e^{(T-t_0)} - 1\right]h \qquad (8)$$

Thus, Euler's algorithm is first order in the step size, where the constant M is the factor multiplying h in (8).

Since (8) implies that $E_N \to 0$, as $h \to 0$, we see that if exact arithmetic were possible, Euler's Method provides as good an approximation of $y(T)$ as desired, provided that h is chosen small enough. But as we know, exact arithmetic is not a practical possibility, and our next result indicates that round-off may have serious consequences for the convergence of Euler's Method.

We may carry out a similar analysis of the cumulative round-off errors when Euler's Method is implemented by computer or calculator, or by hand. Suppose that at each step j there are various round-off and function evaluation errors whose totality is denoted by r_j. Thus, we do not actually see the numbers y_j of the theoretical Euler Method, but instead we see the numbers $\tilde{y}_j$, where

$$\tilde{y}_0 = y_0 + r_0$$
$$\tilde{y}_{j+1} = \tilde{y}_j + f(t_j, \tilde{y}_j)h + r_j, \quad j = 0, 1, 2, \ldots, N - 1 \qquad (9)$$

For simplicity, suppose that the round-off errors are all bounded in magnitude from above by a positive constant r. Define $\varepsilon_j = |y(t_j) - \tilde{y}_j|$, $j = 0, 1, 2, \ldots, N$.

Now we are ready to prove the following result.

THEOREM B.1.2

> Bound for Total Errors in Euler's Method. With the definition of terms above, an upper bound for the total error $\varepsilon_N = |y(t_N) - \tilde{y}_N|$ in Euler's Method is given by
>
> $$\varepsilon_N \le e^{(T-t_0)L}\varepsilon_0 + \frac{1}{L}\left(e^{(T-t_0)L} - 1\right)\left(Kh + \frac{r}{h}\right) \qquad (10)$$

Proof. The estimate (10) is proven by an argument similar to that used above for the discretization errors E_j. First, let us put $\varepsilon'_j = |y_j - \tilde{y}_j|$, where the y_j and $\tilde{y}_j$ are the Euler and "rounded" Euler iterates given by (3) and (9), respectively. Observe that $\varepsilon'_0 = \varepsilon_0$. Now subtracting (9) from (3) and using the Triangle Inequality, we obtain the estimate

$$\varepsilon'_{j+1} \le \alpha\varepsilon'_j + r, \quad j = 0, 1, 2, \ldots, N-1 \qquad (11)$$

where $\alpha = 1 + hL$ and r is an upper bound for the magnitudes of the round-off errors r_j. Using (11) iteratively, we find that

$$\varepsilon_j \le \alpha^j\varepsilon_0 + (\alpha^{j-1} + \alpha^{j-2} + \cdots + 1)r \qquad (12)$$

Since $1 + \alpha + \cdots + \alpha^{j-1} = (\alpha^j - 1)/(\alpha - 1)$ and $\{(1 + u/m)^m\}_{m=1}^{\infty}$ is an increasing sequence converging to e^u, we have from (12) that

$$\varepsilon'_N \le e^{(T-t_0)L}\varepsilon_0 + \frac{r}{hL}\left(e^{(T-t_0)L} - 1\right) \qquad (13)$$

Using the Triangle Inequality, we have that

$$\varepsilon_N \le E_N + \varepsilon'_N$$

and hence using (8) and (13), the desired estimate follows.

The term r/h in (10) shows the difficulty encountered when the step size h is reduced in an attempt to decrease the upper bound on the discretization error. The upper bound on the error due to round-off may actually increase as a result. In fact, the factor $Kh + r/h$ in (10) has its minimum value for positive h when $h = h_0 = (r/K)^{1/2}$. Thus, the upper bound on the error ε_N decreases as the step size is reduced until the step size falls below h_0. Futher reductions in h may lower the upper bound on the discretization error E_N, but would have the opposite effect on the upper bound for ε_N. Hence, it is not easy to say just what step size would minimize the total of all the errors.

One-Step Methods for Systems: Euler's Method, RK4

The numerical methods for computing an approximate solution of an IVP for a single first-order ordinary differential equation (ODE) can be extended to an IVP for a system of first-order ODEs. For simplicity, we extend only Euler's Method and the fourth-order Runge-Kutta Method (RK4), and then only to the IVP

$$\begin{aligned} x' &= f(t, x, y), \quad x(t_0) = x_0 \\ y' &= g(t, x, y), \quad y(t_0) = y_0 \end{aligned} \qquad (14)$$

where f and g satisfy the conditions of the Existence and Uniqueness Theorem (see, e.g., Theorem 6.2.1 in Section 6.2). Euler's Method for IVP (14) has the same interpretation as for the case of the single ODE (see Section 2.5). Calculate $f(t_0, x_0, y_0)$ and $g(t_0, x_0, y_0)$ and move away from the point (x_0, y_0) along the field vector $(f(t_0, x_0, y_0), g(t_0, x_0, y_0))$ of the system in (14) to obtain an Euler approximation (x_1, y_1) to $(x(t_0 + h), y(t_0 + h))$:

$$x_1 = x_0 + hf(t_0, x_0, y_0), \qquad y_1 = y_0 + hg(t_0, x_0, y_0)$$

where $h > 0$ is the step size. To determine approximations to the solutions of IVP (14) at $t_n = t_0 + nh$, we have the Euler algorithm

$$x_{n+1} = x_n + hf(t_n, x_n, y_n)$$
$$y_{n+1} = y_n + hg(t_n, x_n, y_n)$$

(15)

where $h > 0$ and $n = 0, 1, 2, \ldots$

RK4 applied to IVP (14) is the following algorithm:

$$x_{n+1} = x_n + \frac{h}{6}(k_1 + 2k_2 + 2k_3 + k_4)$$

$$y_{n+1} = y_n + \frac{h}{6}(p_1 + 2p_2 + 2p_3 + p_4)$$

(16)

where

$$k_1 = f(t_n, x_n, y_n), \qquad\qquad p_1 = g(t_n, x_n, y_n)$$
$$k_2 = f(t_n + \tfrac{h}{2}, x_n + \tfrac{h}{2}k_1, y_n + \tfrac{h}{2}p_1), \quad p_2 = g(t_n + \tfrac{h}{2}, x_n + \tfrac{h}{2}k_1, y_n + \tfrac{h}{2}p_1)$$
$$k_3 = f(t_n + \tfrac{h}{2}, x_n + \tfrac{h}{2}k_2, y_n + \tfrac{h}{2}p_2), \quad p_3 = g(t_n + \tfrac{h}{2}, x_n + \tfrac{h}{2}k_2, y_n + \tfrac{h}{2}p_2)$$
$$k_4 = f(t_n + h, x_n + hk_3, y_n + hp_3), \qquad p_4 = g(t_n + h, x_n + hk_3, y_n + hp_3)$$

(17)

RK4 is probably the most widely used method for the numerical solution of IVPs.

B.2   Multistep Numerical Methods

An important class of methods for producing approximate solutions to the IVP:

$$y' = f(t, y), \qquad y(t_0) = y_0$$

(1)

is known collectively as *linear multistep methods*. One-step methods ignore the history of the approximation and use only values calculated in the immediately preceding step. In contrast, multistep methods recall values calculated in several earlier steps and use these values to predict the next approximation. Thus the approximation y_{n+k} (k fixed) to the solution of IVP (1) at $t = t_{n+k}$ is given in terms of the k foregoing approximations $y_n, y_{n+1}, \ldots, y_{n+k-1}$ to the solution at $t_n, t_{n+1}, \ldots, t_{n+k-1}$. In general, a *linear k-step method* for the step size h has the general form (k a fixed positive integer)

$$y_{n+k} = \sum_{j=0}^{k-1} \alpha_j y_{n+j} + h \sum_{j=0}^{k} \beta_j f(t_{n+j}, y_{n+j}), \quad n = 0, 1, 2, \ldots$$

(2)

where the α_j and β_j are given constants. Techniques for computing reasonable values for the constants α_j and β_j are given in Appendix B.3. Note that if $\beta_k \neq 0$, then (2) defines y_{n+k} only implicitly and hence such multistep methods are known as *implicit methods*. If $\beta_k = 0$, the method is called an *explicit method*. A disadvantage of multistep methods is that the data points $y_0, y_1, \ldots, y_{k-1}$ at $t_0, t_1, \ldots, t_{k-1}$ are needed to start the methods, whereas IVP (1) gives only the one data point y_0 at t_0. The other needed data points are usually generated by a one-step method (which, for this reason, is sometimes called a "starting" method).

We shall describe one of the most popular explicit multistep methods due to Adams and Bashforth and an improvement developed by Moulton.[1]

❖ **Three-Step and Four-Step Adams-Bashforth Methods.**

$$y_{n+1} = y_n + \frac{h}{12}(23f_n - 16f_{n-1} + 5f_{n-2}) \tag{3}$$

$$y_{n+1} = y_n + \frac{h}{24}(55f_n - 59f_{n-1} + 37f_{n-2} - 9f_{n-3}) \tag{4}$$

where $f_j = f(t_j, y_j)$ and h is the common step size.

There is a k-step Adams-Bashforth method for each $k = 2, 3, \ldots$. The four-step method (4) is widely used, giving as it does reasonably good approximations without an excessive number of calculations. The *order of a multistep method* is defined analogously to that of one-step methods. As with all the Adams-Bashforth algorithms, the order of the method is equal to the number of steps k if the slope function $f(t, y)$ is sufficiently differentiable. Thus the algorithm defined by (4) should be comparable in accuracy to RK4. However, (4) has the added advantage of fewer calculations and function evaluations than the corresponding Runge-Kutta method since the numbers f_n, f_{n-1}, f_{n-2}, and f_{n-3} have already been calculated at previous steps. Thus multistep methods are generally more efficient than one-step methods of the same order. A drawback of every multistep Adams-Bashforth method is that it is not self-starting from the initial values t_0 and y_0. Adams-Bashforth methods are usually initialized by a Runge-Kutta algorithm (of the same order), which runs just long enough to produce the starting values needed.

Adams-Moulton Methods: Prediction and Correction

The Adams-Bashforth method for calculating y_{n+1} makes no use of the values of the slope function $f(t, y(t))$ for $t > t_n$, although one would expect improved accuracy if some of these values were used. Adams-Moulton methods do just that because they have the form (2) with the coefficient $\beta_k \neq 0$.

❖ **Three-Step Adams-Moulton Method.**

$$y_{n+1} = y_n + \frac{h}{24}(9f_{n+1} + 19f_n - 5f_{n-1} + f_{n-2}) \tag{5}$$

where $f_j = f(t_j, y_j)$ and h is the common step size.

See Appendix B.3 for the derivation of this and other Adams-Moulton methods. The three-step method (5) determines y_{n+1} only implicitly because y_{n+1} appears on both sides of the equation.

---

[1]J.C. Adams (1819–1892) was a noted English applied mathematician and astronomer who predicted in 1845 the existence and the orbit of the planet Neptune, his predictions being based on a mathematical analysis of perturbations in the orbit of Uranus. In later years Adams devoted much time to approximating the values of mathematical constants and functions. He and F. Bashforth made use of the algorithm now known by their names in a joint study of capillary action. During World War I, F.R. Moulton (1872–1952) adapted the algorithm to estimate the solutions of ballistics problems.

Note that (5) has the form $y_{n+1} = F(y_{n+1})$, and hence its solutions are called *fixed points* of the function F. The function F in this case might be quite a complicated function of y_{n+1} since it contains the term $f(t_{n+1}, y_{n+1})$, so the task of solving the relation $y_{n+1} = F(y_{n+1})$ for y_{n+1} explicitly is, in general, hopeless. A technique often used to calculate a fixed point y of the function F is to make an initial guess for y, call it $y^{(0)}$, and then calculate the sequence $y^{(m)}$ by the recursion relation $y^{(m)} = F(y^{(m-1)})$. Called the method of *simple iteration*, it can be shown that the *iterates* $y^{(m)}$ converge to the fixed point y if $|F'(y)| < 1$ and the initial guess $y^{(0)}$ is close enough to y. The derivative condition is nearly always satisfied for an Adams-Moulton method if the step size h is small enough, but how does one obtain a good initial guess for y_{n+1} from knowledge of the previous y_j? Moulton's idea was that an Adams-Bashforth method could be used for this purpose, so we have the basic apparatus for a *predictor-corrector method*.

The most widely used multistep predictor-corrector method uses the four-step Adams-Bashforth algorithm (4) to calculate the "predicted" value $y_{n+1}^{(p)}$, which is then used as an initial guess for y_{n+1} in solving the three-step Adams-Moulton method (5) by the method of successive iteration. This "correction" process is usually halted after one iteration (but may continue if error-control procedures so indicate) and this iterate is taken as the value of y_{n+1}. From here on the process is repeated. Thus, we have the

❖ **Adams-Bashforth-Moulton Predictor-Corrector Method**. Using a common step size h, first compute a "predicted" value for y_{n+1} denoted by $y_{n+1}^{(p)}$, with

$$y_{n+1}^{(p)} = y_n + \frac{h}{24}(55 f_n - 59 f_{n-1} + 37 f_{n-2} - 9 f_{n-3}) \qquad (6)$$

Then, setting $f_{n+1}^{(p)} = f(t_{n+1}, y_{n+1}^{(p)})$, the "corrected" value y_{n+1} is calculated with

$$y_{n+1} = y_n + \frac{h}{24}(9 f_{n+1}^{(p)} + 19 f_n - 5 f_{n-1} + f_{n-2}) \qquad (7)$$

The correction procedure can be repeated as often as requried.

The Adams algorithms are used extensively in software packages for solving initial value problems. Both the step size h and the number of steps may be changed automatically to improve efficiency.

EXAMPLE B.2.1

The initial value problem

$$y' = y^2, \qquad y(0) = -1$$

has the exact solution $y = -(1 + t)^{-1}, t > -1$. Table B.2.1 compares the exact solution with approximate solutions computed by using the Adams-Bashforth-Moulton method above and by using RK4 (which was also used to start the Adams algorithm).

Numerical Instability

Our next example shows that the wrong discretization process could lead to a numerically unstable algorithm. Some linear multistep methods are capable of rather peculiar behavior. Indeed, it can happen that the errors in the approximations $y_1, y_2, \ldots$ remain under control at first, but then suddenly begin to increase exponentially. When this happens the method is said to be *numerically unstable*.

TABLE B.2.1 Comparison of Methods for Solving $y' = y^2$, $y(0) = -1$

t	Runge-Kutta	Adams-Bashforth-Moulton	Exact
0.0	−1.000000	−1.000000	−1.000000
0.5	−0.666677	−0.666667	−0.666667
1.0	−0.500029	−0.500000	−0.500000
1.5	−0.400025	−0.400000	−0.400000
2.0	−0.333353	−0.333333	−0.333333
2.5	−0.235730	−0.285714	−0.285714
3.0	−0.250012	−0.250000	−0.250000
3.5	−0.222222	−0.222222	−0.222222
4.0	−0.200003	−0.200000	−0.200000
4.5	−0.181825	−0.181818	−0.181818

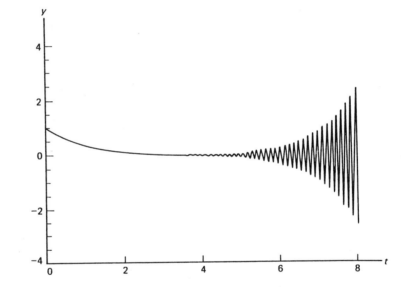

FIGURE B.2.1 Numerical instability of the multistep method (9).

EXAMPLE B.2.2

Numerical Instability: Central Differences

Consider the initial value problem

$$y' = -y, \qquad y(0) = 1 \tag{8}$$

whose true solution is easily seen to be $y(t) = e^{-t}$. In an attempt to discretize the differential equation in (8), let us evaluate both sides at $t = t_n$ and approximate $y'(t_n)$ by the *central difference* $(y(t_{n+1}) - y(t_{n-1}))/2h$, where we have assumed a uniform step size h. This central difference can, in turn, be approximated by $(y_{n+1} - y_{n-1})/2h$ and hence we arrive at the linear multistep method

$$y_{n+1} = y_{n-1} - 2hy_n, \qquad y_0 = 1 \tag{9}$$

Choosing $h = 0.06$ and using Euler's Method to calculate y_1 to start the scheme (9), we obtain the surprising result illustrated in Figure B.2.1.

PROBLEMS

1. Approximate $y(1)$ if $y' = -y$, $y(0) = 1$, where $h = 0.2$, by using
    (a) The three-step Adams-Bashforth method.
    (b) The four-step Adams-Bashforth method.
    (c) The four-step Adams-Bashforth-Moulton method.

2. Repeat Problem 1 for the IVP $y' = -y^3 + t^2$, $y(0) = 1$, $h = 0.2$, using the indicated methods.

3. Find $y(1)$ for the IVP $y' = y^3 + t^2$, $y(0) = 0$ by using an Adams-Bashforth-Moulton method with $h = 0.001$.

4. (*Numerical Instability*). The following steps show that Figure B.2.1 is indeed the polygonal graph of the approximate solution of $y' = -y$, $y(0) = 1$, obtained by using the *linear difference equation* (9), $y_{n+1} = y_{n-1} - 2hy_n$, with initial data $y_0 = 1$, $y_1 = 1 - h$, where y_1 is obtained by Euler's Method.

    (a) Why are $y_2, y_3, \ldots, y_n, \ldots$ uniquely determined by (9) and the initial data?

    (b) (*Solving a Difference Equation*). Show that $y_n = C_1 r_1^n + C_2 r_2^n$, where C_1 and C_2 are arbitrary constants, $r_1 = -h + (h^2 + 1)^{1/2}$ and $r_2 = -h - (h^2 + 1)^{1/2}$, satisfies (9) for $n = 1, 2, \ldots$. [*Hint:* First set $y_n = r^n$ in (9) and then determine the unique pair of values r_1 and r_2 of r for which (9) holds. Then verify that $y_n = C_1 r_1^n + C_2 r_2^n$ is also a solution of (9) if r_1^n and r_2^n are solutions.]

    (c) Show that $y_n = C_1 r_1^n + C_2 r_2^n$, with r_1 and r_2 as in part (b), $C_1 = (1 - h - r_2)/(r_1 - r_2)$ and $C_2 = (1 - h - r_1)/(r_2 - r_1)$ satisfies (9) and the initial data.

    (d) According to the Binomial Theorem (see Appendix C.2), $(h^2 + 1)^{1/2} \approx 1 + h^2/2$, for small h. Thus, show that $r_1 \approx 1 - h + h^2/2$, and $r_2 \approx -1 - h - h^2/2$, or, to first order in h, $r_1 \approx 1 - h$, and $r_2 \approx -1 - h$. Let t be a fixed positive number, $t = nh$. Show that $r_1^n \to e^{-t}$, and $(-1)^n r_2^n \to e^t$, as $n \to \infty$. [*Hint:* $e^z = \lim_{n \to \infty}(1 + z/n)^n$, for any z.]

    (e) On the basis of part (d), we have that $y_n(t) \approx C_1 e^{-t} + C_2 e^t$ for large n and small h such that $nh = t$. Show that $C_1 \to 1$ and $C_2 \to 0$ as $n \to \infty$. Thus, show that the approximation to $y_n(t)$ tends to the exact solution $y = e^{-t}$ of $y' = -y$, $y(0) = 1$, as $n \to \infty$.

    (f) Explain why the approximation y_n may be considered to be the superposition of the exact solution of the IVP and an exponentially growing numerical error with alternating sign.

    (g) Reproduce Figure B.2.1 using (9) with $h = 0.06$, $0 \le t \le 6$, $y_0 = 1$, $y_1 = 1 - h$.

B.3 Adams-Bashforth Methods: Interpolating Polynomials

In this section we will discuss some techniques for deriving multistep method formulas for solving the IVP:

$$y' = f(t, y), \quad y(t_0) = y_0 \tag{1}$$

Suppose that $y(t)$, $t_0 \le t \le T$, is the solution of the IVP (1). Our goal, as always, is to approximate $y(T)$. First subdivide the interval $[t_0, T]$ into N subintervals $[t_j, t_{j+1}]$ of equal length $h = (T - t_0)/N$ and then move from t_0 to t_1, from t_1 to $t_2, \ldots$, from t_{N-1} to $t_N = T$, approximating $y(t_j)$ by a number y_j, for each $j = 0, 1, 2, \ldots, N$. In particular, we shall determine a polynomial $p(t)$ which approximates $f(t, y(t))$ on the interval $[t_j, t_{j+1}]$ and then use that polynomial to replace the exact algorithm

$$y(t_{j+1}) = y(t_j) + \int_{t_j}^{t_{j+1}} y'(t)\,dt = y(t_j) + \int_{t_j}^{t_{j+1}} f(t, y(t))\,dt \tag{2}$$

by the approximating algorithm

$$y_{j+1} = y_j + \int_{t_j}^{t_{j+1}} p(t)\,dt \tag{3}$$

Let K be a nonnegative integer. We shall take for $p(t)$ the unique *interpolating polynomial* of degree no more than K which passes through the $K + 1$ points,

$$(t_{j-K}, f_{j-K}), \ldots, (t_j, f_j), \quad \text{where } f_j = f(t_j, y_j)$$

Thus $p(t)$ is determined by the numbers $y_{j-K}, \ldots, y_j$ calculated at the $K + 1$ steps preceding the $(j + 1)$st step.[2] For example, if K is 0, then $p(t)$ is the constant polynomial passing through the point (t_j, f_j). Hence, $p(t) = f(t_j, y_j)$, and (3) becomes

$$y_{j+1} = y_j + \int_{t_j}^{t_{j+1}} f(t_j, y_j)\,dt = y_j + f(t_j, y_j)h$$

which is Euler's Method. If $K = 1$, then $p(t)$ is the linear function through the two points, (t_{j-1}, f_{j-1}) and (t_j, f_j), and we have that

$$p(t) = -\frac{1}{h}(t - t_j)f_{j-1} + \frac{1}{h}(t - t_{j-1})f_j \tag{4}$$

Using this $p(t)$ in (3), we obtain the *two-step Adams-Bashforth* algorithm

$$y_{j+1} = y_j + \int_{t_j}^{t_{j+1}} p(t)\,dt = y_j + \frac{h}{2}(3f_j - f_{j-1}) \tag{5}$$

For $K \geq 2$, the coefficients of the polynomial

$$p(t) = a_0 + a_1 t + \cdots + a_K t^K$$

are found by solving the $K + 1$ equations

$$f_i = a_0 + a_1 t_i + \cdots + a_K t_i^K, \quad i = j - K, \ldots, j \tag{6}$$

for $a_0, a_1, \ldots, a_K$. Then $p(t)$ is integrated from t_j to t_{j+1} and (3) yields y_{j+1}. When $K = 2$ and 3, we obtain the respective *three-* and *four-step Adams-Bashforth algorithms*,

$$y_{j+1} = y_j + \frac{h}{12}(23f_j - 16f_{j-1} + 5f_{j-2}) \tag{7}$$

$$y_{j+1} = y_j + \frac{h}{24}(55f_j - 59f_{j-1} + 37f_{j-2} - 9f_{j-3}) \tag{8}$$

[2]The proof of the existence and uniqueness of the interpolating polynomial may be found, for example, in the book by J. M. Ortega and W. G. Poole, *Numerical Methods for Differential Equations* (London: Pitman, 1981).

Adams-Bashforth-Moulton Methods: Prediction and Correction

The idea is to use the Adams-Bashforth method based on $f_{j-K}, \ldots, f_j$ to *predict* a value $y_{j+1}^{(p)}$ and then use $f_{j-K+1}, \ldots, f_j, f_{j+1}$, where $f_{j+1} = f(t_{j+1}, y_{j+1}^{(p)})$, to determine a new interpolating polynomial $\tilde{p}(t)$, which via (3) will produce a *corrected* value y_{j+1}. We shall illustrate how these methods work for some low values of K.

If $K = 1$, then as we saw in (5), the Adams-Bashforth method predicts the value

$$y_{j+1}^{(p)} = y_j + \frac{h}{2}(3f_j - f_{j-1})$$

Then set $f_{j+1}^{(p)} = f(t_{j+1}, y_{j+1}^{(p)})$ and use (4) to obtain the linear function $\tilde{p}(t)$ through the points (t_j, f_j) and $(t_{j+1}, f_{j+1}^{(p)})$,

$$\tilde{p}(t) = -\frac{1}{h}(t - t_{j+1})f_j + \frac{1}{h}(t - t_j)f_{j+1}^{(p)} \tag{9}$$

Thus, inserting $\tilde{p}(t)$ from (9) into (3) and integrating, we have the Adams-Bashforth-Moulton corrected value,

$$y_{j+1} = y_j + \int_{t_j}^{t_{j+1}} \tilde{p}(t)\, dt = y_j + \frac{h}{2}(f_j + f_{j+1}^{(p)})$$

as a straightforward integration of $\tilde{p}(t)$ shows.

The most widely used multistep predictor-corrector method of this type uses the four-step Adams-Bashforth algorithm (8) to calculate the predicted value $y_{j+1}^{(p)}$, then finds the interpolating cubic polynomial $\tilde{p}(t)$ through the four points

$$(t_{j-2}, f_{j-2}), \quad (t_{j-1}, f_{j-1}), \quad (t_j, f_j), \quad (t_{j+1}, f_{j+1})$$

and calculates the corrected value of y_{j+1} by the formula

$$y_{j+1} = y_j + \int_{t_j}^{t_{j+1}} \tilde{p}(t)\, dt$$

We have the "predictor"

$$y_{j+1}^{(p)} = y_j + \frac{h}{24}(55f_j - 59f_{j-1} + 37f_{j-2} - 9f_{j-3}) \tag{10}$$

which is then "corrected" by

$$y_{j+1} = y_j + \frac{h}{24}(9f_{j+1}^{(p)} + 19f_j - 5f_{j-1} + f_{j-2}) \tag{11}$$

where $f_{j+1}^{(p)} = f(t_{j+1}, y_{j+1}^{(p)})$. We omit the lengthy but straightforward derivation of the coefficients in the formulas.

APPENDIX C

Background Information

This Appendix has been organized into seven sections for convenience in locating the referenced item. The sections are as follows:

C.1 Engineering Functions
Description of piecewise-continuous functions (square waves, step functions, etc.) which are commonly used in science and engineering.

C.2 Power Series
Summary of the properties of power series. Includes convergence properties, tests for convergence, basic Maclaurin series, Taylor's formula with remainder, real analytic functions, and the calculus of power series.

C.3 Complex Numbers and Complex-Valued Functions
Review of the arithmetic of complex numbers and the calculus of complex-valued functions of a real variable.

C.4 Useful Formulas
Review of trigonometric identities, hyperbolic functions, polynomials, partial fractions.

C.5 Theorems from Calculus
Tabulation of useful results from single- and multi-variable calculus.

C.6 Review of Matrix Theory
Review of matrix operations, determinants, linear equations, inequalities.

C.7 Scaling and Units
Description of commonly used unit systems and conversion factors, scaling and non-dimensionalizing ODEs.

C.1 Engineering Functions

The eight functions defined and graphed in Figures C.1.1–C.1.8 are frequently used in science and engineering applications. These functions are divided into three groups: *step* or *stair functions*, *pulses*, and *wavetrains*. They are all defined on the entire t-axis and their "on-time" (where the function is nonzero) is described as follows: The step and stair functions and the square pulses are first turned on at $t = 0$. The step and stair functions remain on for $t \geq 0$. The square pulse switches off at $t = a$ and remains off. The "on-time" for sawtooth and triangular pulses is $0 < t < a$. The on-time of a square wave over the period $0 \leq t \leq p$ is $0 \leq t < (d/100)p$, where the *duty cycle d* is the percentage of on-time in a cycle. The on-time for triangular and sawtooth waves over a period $0 \leq t \leq p$ is $0 < t < (d/100)p$, again where d is the duty cycle.

$\mathrm{sqw}(t, d, p)$:	periodic square wave, duty cycle $d\%$, period p
$\mathrm{trw}(t, d, p)$:	periodic triangular wave, duty cycle $d\%$, period p
$\mathrm{sww}(t, d, p)$:	periodic sawtooth wave, duty cycle $d\%$, period p
$\mathrm{step}(t)$:	unit step at $t = 0$
$\mathrm{sqp}(t, a)$:	unit square pulse over interval $0 \leq t \leq a$
$\mathrm{trp}(t, a)$:	unit triangular pulse over interval $0 \leq t \leq a$
$\mathrm{trp}(t, a)$:	unit sawtooth pulse over interval $0 \leq t \leq a$
$\mathrm{stair}(t, a)$:	stair, rise of 1 (starting at $t = 0$), run of a

- If the value of an engineering function is constant over a time interval of length T, it may be a good idea to change the maximum internal step size of your computer ODE solver to no more than $T/10$. If this isn't done, an adaptive solver may overlook the points where the function values suddenly jump.

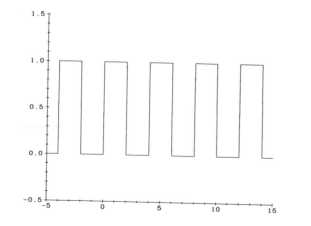

FIGURE C.1.1 Periodic square wave $\mathrm{sqw}(t, 50, 4)$.

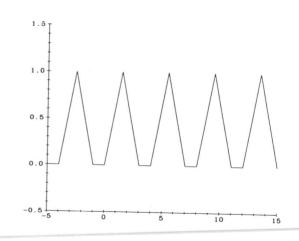

FIGURE C.1.2 Periodic triangular wave $\mathrm{trw}(t, 75, 4)$.

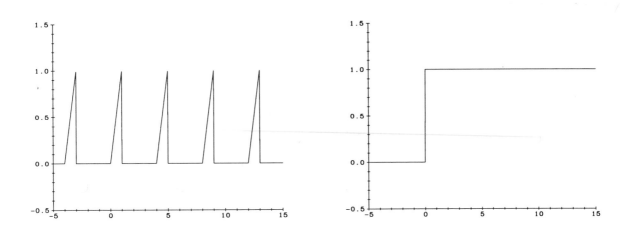

FIGURE C.1.3 Periodic sawtooth wave sww(t, 25, 4).   FIGURE C.1.4 Unit step function step(t).

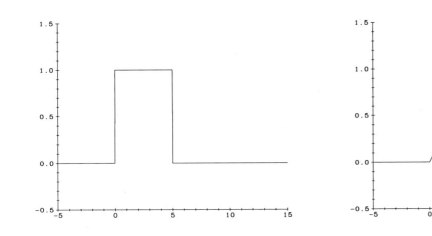

FIGURE C.1.5 Unit square pulse sqp(t, 5).   FIGURE C.1.6 Unit triangular pulse trp(t, 8).

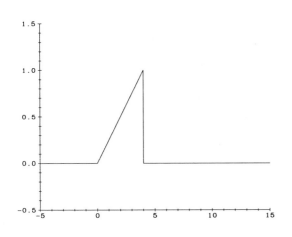
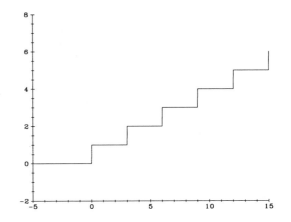
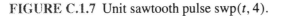

FIGURE C.1.7 Unit sawtooth pulse swp(t, 4).   FIGURE C.1.8 Stair function stair(t, 3).

C.2 Power Series

Power series have some remarkable properties that make them especially well suited for certain types of applications. We shall list these properties below, without proof, in a form that allows easy reference.

1. *Power Series.* The series

$$a_0 + a_1(x - x_0) + a_2(x - x_0)^2 + \cdots + a_n(x - x_0)^n + \cdots = \sum_{n=0}^{\infty} a_n(x - x_0)^n$$

where x_0 and the coefficients a_n are real, is a *power series* based at x_0. We treat only series of the form $\sum_{n=0}^{\infty} a_n x^n$ since $\sum_{n=0}^{\infty} a_n(x - x_0)^n$ may be converted to that form by replacing $x - x_0$ by a new variable. We write $\sum_0^{\infty} a_n x^n$ or $\sum a_n x^n$ for $\sum_{n=0}^{\infty} a_n x^n$ if there is no danger of confusion.

2. *Intervals of Convergence.* $\sum a_n x^n$ converges at a given point x if $\lim_{N \to \infty} \sum_{n=0}^{N} a_n x^n$ exists. Unless $\sum a_n x^n$ converges only at $x = 0$, there is a largest positive number R (possibly infinity) such that $\sum a_n x^n$ converges for all $|x| < R$. R is the *radius of convergence* and $J = (-R, R)$ is the *open interval of convergence*. The series diverges (i.e., does not converge) if $|x| > R$ and may or may not converge if $x = \pm R$. We shall only work with a series inside its open interval of convergence.

3. *Ratio Test.* The *Ratio Test* is a useful way to find the radius of convergence of a power series. For example, suppose that $\lim_{n \to \infty} |a_{n+1}/a_n| = a$. Then

$$\lim_{n \to \infty} \left| \frac{a_{n+1} x^{n+1}}{a_n x^n} \right| = a|x|$$

According to the Ratio Test, we have convergence if the limit $a|x| < 1$ and divergence if $a|x| > 1$. Thus the radius of convergence R is $1/a$ and $\sum a_n x^n$ converges for all $|x| < 1/a$. Note that $R = \infty$ if $a = 0$.

4. *Calculus of Power Series.* The function $f(x) = \sum_{n=0}^{\infty} a_n x^n$, x in J, is continuous and possesses derivatives of all orders. Moreover, $f'(x) = \sum_{n=1}^{\infty} n a_n x^{n-1}$ and this series also has J as its open interval of convergence. In general, $f^{(k)}(x) = \sum_{n=k}^{\infty} n(n-1) \cdots (n-k+1) a_n x^{n-k}$, where the series converges for x in J. In addition, the series

$$\sum_0^{\infty} \frac{a_n}{n+1} (x^{n+1} - b^{n+1})$$

converges to $\int_b^x \sum_0^{\infty} (a_n s^n)\, ds$ for all b and x in J.

5. *Identity Theorem.* If $\sum_{n=0}^{\infty} b_n x^n = \sum_{n=0}^{\infty} a_n x^n$ for all x in an open interval about 0, then $b_n = a_n$ for $n = 0, 1, 2, \ldots$. In particular, if $\sum_{n=0}^{\infty} a_n x^n = 0$ for all x near 0, then $a_n = 0$, $n = 0, 1, 2, \ldots$. This is called the *Identity Theorem* for power series.

6. *Algebra of Power Series.* Convergent power series may be added or multiplied for all x in a common interval of convergence. The resulting *sum* or *product* series converges on that interval. Specifically,

(i) $$\sum_{n=0}^{\infty} a_n x^n + \sum_{n=0}^{\infty} b_n x^n = \sum_{n=0}^{\infty} (a_n + b_n) x^n$$

(ii) $$\left(\sum_{n=0}^{\infty} a_n x^n \right) \left(\sum_{n=0}^{\infty} b_n x^n \right) = \sum_{n=0}^{\infty} c_n x^n \quad \text{where } c_n = \sum_{k=0}^{n} a_k b_{n-k}$$

(iii)   If $\sum_{n=0}^{\infty} b_n x^n \neq 0$ on the interval J, the quotient series,

$$\left(\sum_{n=0}^{\infty} a_n x^n \right) \Bigg/ \left(\sum_{n=0}^{\infty} b_n x^n \right) = \sum_{n=0}^{\infty} d_n x^n$$

converges on that interval. Using (ii), coefficients d_n can be found by successively solving for the coefficients d_n from the formulas $a_n = d_0 b_n + d_1 b_{n-1} + \cdots + d_{n-1} b_1 + d_n b_0$.

7. *Reindexing.* A series may be reindexed in any convenient way. For example, the following three summations are equivalent:

$$\sum_{n=0}^{\infty} a_n x^n, \qquad \sum_{n=2}^{\infty} a_{n-2} x^{n-2}, \qquad \sum_{n=-3}^{\infty} a_{n+3} x^{n+3}$$

Writing out the first few terms of a series explicitly sometimes helps in determining the equivalence of indexing schemes. Sometimes it is convenient to introduce negative indexed coefficients, but it is understood that they have value zero. For example, we could rewrite $\sum_{n=0}^{\infty} a_n x^n + \sum_{n=2}^{\infty} b_{n-2} x^n$ either as $a_0 + a_1 x + \sum_{n=2}^{\infty} (a_n + b_{n-2}) x^n$ or as $\sum_{n=0}^{\infty} (a_n + b_{n-2}) x^n$, where it is understood that $b_{-2} = 0$, and $b_{-1} = 0$.

8. *Real Analytic Functions.* A function $f(x)$ which has a *Taylor series expansion,*

$$\sum_{n=0}^{\infty} \frac{f^{(n)}(x_0)}{n!} (x - x_0)^n$$

converging to $f(x)$ on an interval $(x_0 - R, x_0 + R)$, is said to be *real analytic* on that interval, or, alternatively, at x_0. If $x_0 = 0$, the Taylor series is said to be the *Maclaurin series* of f. By the Identity Theorem, a real analytic function f at x_0 has a unique power series in powers of $x - x_0$; hence, this must be the Taylor (or Maclaurin) series of f. The definitions above extend to functions of several variables. For example, the Maclaurin series of $f(x, y)$ is

$$\sum_{n=0}^{\infty} \sum_{i=0}^{n} \binom{n}{i} \frac{1}{n!} \frac{\partial^n f(0, 0)}{\partial x^{n-i} \partial y^i} x^{n-i} y^i =$$

$$f(0, 0) + \frac{\partial f(0, 0)}{\partial x} x + \frac{\partial f(0, 0)}{\partial y} y$$

$$+ \frac{1}{2} \frac{\partial^2 f(0, 0)}{\partial x^2} x^2 + \frac{\partial^2 f(0, 0)}{\partial x \partial y} xy + \frac{1}{2} \frac{\partial^2 f(0, 0)}{\partial y^2} y^2 + \cdots$$

where $\binom{n}{i}$ is the binomial coefficient, $n! / i!(n - i)!$, and $\binom{n}{0} = 1$.

9. Some Maclaurin series are used so often that it is convenient to list them here.

(*Sine*) $\sin x = x - \dfrac{x^3}{3!} + \dfrac{x^5}{5!} - \cdots + (-1)^n \dfrac{x^{2n+1}}{(2n+1)!} + \cdots$,   all x

(*Cosine*) $\cos x = 1 - \dfrac{x^2}{2!} + \dfrac{x^4}{4!} - \cdots + (-1)^n \dfrac{x^{2n}}{(2n)!} + \cdots$,   all x

(*Exponential*) $e^x = 1 + x + \dfrac{x^2}{2!} + \cdots + \dfrac{x^n}{n!} + \cdots$,   all x

(*Geometric*) $\dfrac{1}{1-x} = 1 - x + x^2 - \cdots + (-1)^n x^n + \cdots$, $|x| < 1$

(*Binomial*) $(1 + x)^{\alpha} = 1 + \alpha + \dfrac{\alpha(\alpha - 1)}{2} x^2 + \cdots$
$$+ \frac{\alpha(\alpha - 1) \cdots (\alpha - n + 1)}{n!} x^n + \cdots, \qquad |x| < 1$$

10. *Error Estimate for Alternating Series.* Let $\alpha_n > 0$, $n = 0, 1, 2, \ldots$. The *alternating series*, $\sum_{n=0}^{\infty}(-1)^n\alpha_n$ (not necessarily a power series) converges if $\lim_{n\to\infty}\alpha_n = 0$ and $\alpha_n \geq \alpha_{n+1}$ for every n. The *error* in using the partial sum $\sum_{n=0}^{N}(-1)^n\alpha_n$ to approximate the true sum $\sum_{n=0}^{\infty}(-1)^n\alpha_n$ has magnitude no greater than α_{n+1}:

$$\left| \sum_{0}^{\infty}(-1)^n\alpha_n - \sum_{0}^{N}(-1)^n\alpha_n \right| \leq \alpha_{N+1}$$

11. *Taylor's Formula with Remainder.* Let $f(x)$ be in $C^{N+1}(x_0 - R, x_0 + R)$. Then for all x with $x_0 - R < x < x_0 + R$,

$$f(x) = \sum_{n=0}^{N} \frac{f^{(n)}(x_0)}{n!}(x - x_0)^n + \frac{f^{(N+1)}(c)}{(N+1)!}(x - x_0)^{N+1} \tag{1}$$

where c is some point between x and x_0; the last term on the right is the *Lagrange form of the remainder.*

The following examples illustrate the properties given above.

EXAMPLE C.2.1

Ratio Test

$\sum_{n=0}^{\infty} n^5(x^n/2^n)$ converges for $|x| < 2$ by the Ratio Test since

$$\lim_{n\to\infty}\left| \frac{(n+1)^5 x^{n+1}/2^{n+1}}{n^5 x^n/2^n} \right| = |x| \lim_{n\to\infty}\frac{(n+1)^5}{2n^5} = \frac{|x|}{2}\lim_{n\to\infty}(1+\frac{1}{n})^5 = \frac{|x|}{2}$$

and $|x|/2 < 1$ for all $|x| < 2$.

EXAMPLE C.2.2

Differentiating of Power Series

$$-\frac{1}{(1+x)^2} = \frac{d}{dx}\left(\frac{1}{1+x}\right) = \frac{d}{dx}(1 - x + x^2 + \cdots + (-1)^n x^n + \cdots)$$

$$= -1 + 2x + \cdots + (-1)^n n x^{n-1} + \cdots$$

if $|x| < 1$ (differentiating a geometric series within the interval of convergence).

EXAMPLE C.2.3

Product of Power Series

$$\left(\sum_{n=0}^{\infty} nx^n\right)\left(\sum_{n=0}^{\infty}\frac{x^n}{n+1}\right) = \sum_{n=0}^{\infty}\left(\sum_{k=0}^{n}\frac{k}{n-k+1}\right)x^n$$

$$= 0 + \left(\sum_{k=0}^{1}\frac{k}{1-k+1}\right)x + \left(\sum_{k=0}^{2}\frac{k}{2-k+1}\right)x^2 + \cdots$$

$$= x + \frac{5}{2}x^2 + \cdots$$

The product series converges to the product of the two factor series for $|x| < 1$ since each factor series converges for $|x| < 1$.

EXAMPLE C.2.4 | **Reindexing**

$$\sum_{n=0}^{\infty} \frac{x^n}{n+3} + \sum_{n=2}^{\infty} x^n = \frac{1}{3} + \frac{1}{4}x + \sum_{n=2}^{\infty} \left(\frac{1}{n+3} + 1 \right) x^n \quad \text{(reindexing and adding)}$$

C.3 Complex Numbers and Complex-Valued Functions

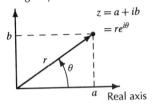

Imaginary axis

It is customary to write *complex numbers* in the form $a + ib$, where a and b are real numbers. Complex numbers are introduced to facilitate calculation of roots of polynomials of degree 2 or higher. We review below the basic features of the complex number system. Our interest, however, goes beyond the calculation of roots of polynomials. To develop an elegant and simple solution technique for linear differential equations with constant coefficients, we shall need to develop the notion of differentiability of complex-valued functions of a single real variable. At first glance this approach may seem like a far cry from our goal of studying real-world dynamical systems, but in fact complex-valued functions are so useful that they are now widely infused into the scientific literature.

If $z = a + ib$, the real number a is called the *real part* of z and is denoted by $\text{Re}[z]$; the real number b is called the *imaginary part* of z and is denoted by $\text{Im}[z]$. Real numbers are, of course, identified with those complex numbers whose imaginary parts vanish. The complex number $a - ib$ is called the *complex conjugate* of $z = a + ib$ and is denoted by $\bar{z}$. Complex numbers are added and multiplied as linear polynomials in the variable i with real coefficients, remembering to replace i^2 by -1. This definition extends to complex numbers the corresponding operations for real numbers. Sometimes the representation $z = a + ib$ is called the *Cartesian* form of the complex number z since we can associate with z the point (a, b) in the Cartesian plane. For this reason we also sometimes refer to the collection of all complex numbers as the *complex plane*. Observe that $\overline{z + w} = \bar{z} + \bar{w}$ and $\overline{vw} = \bar{z}\bar{w}$.

Using polar coordinates for $z = a + ib$, we see that $a = r\cos\theta$ and $b = r\sin\theta$, where $r = (a^2 + b^2)^{1/2}$, $0 \le \theta < 2\pi$; hence, we have the alternative representation $z = r(\cos\theta + i\sin\theta)$, called the *polar* form for z. For any real number θ we shall write

THEOREM C.3.1

> Euler's Formula. For any real number θ,
>
> $$e^{i\theta} = \cos\theta + i\sin\theta \tag{1}$$

Hence, the polar form of a complex number z can be written as $z = re^{i\theta}$. The nonnegative real number r is called the *absolute value* or the *modulus* of z, denoted also by $|z|$, and θ is called the *argument* or *angle* of z. Notice that $|z|$ agrees with the usual notion of absolute value if z is a real number. The complex number zero (denoted simply by 0) is given by $0 + i0$; also note that $z = 0$ if and only if $|z| = 0$. Observe that $z\bar{z} = |z|^2 = r^2 = a^2 + b^2$.

It is a consequence of the addition laws for the sine and cosine functions that for any complex numbers in polar form $z = re^{i\theta}$, $w = \rho e^{i\phi}$,

$$z^{-1} = \frac{1}{z} = \frac{e^{-i\theta}}{r}, \quad \text{if } z \ne 0$$

$$zw = r\rho e^{i(\theta + \phi)}$$

We can use these identities to prove the following useful result.

THEOREM C.3.2

> De Moivre's Formula. For any integer n, positive, negative, or zero,
> $$z^n = r^n e^{in\theta} = r^n[\cos n\theta + i \sin n\theta]$$

The advantage of the polar form for complex numbers is the simple geometric characterization of the product and quotient of complex numbers.

Just as for real numbers, absolute value can be used to measure the distance between complex numbers; indeed, the *distance* between the complex numbers z and w is defined to be $|z - w|$. The sequence $\{z_n\}$ is said to *converge* if and only if there exists a z_0 in the complex plane such that $|z_n - z_0| \to 0$. Now suppose that f is a complex-valued function defined over an interval I on the real t-axis; that is, $w = f(t)$ is a complex number for each t in the interval I. We can define the notion of a derivative for such a function in a manner consistent with the usual notion of derivative for real-valued functions.

❖ **Derivative of a Complex-Valued Function.** We say that the complex-valued function $f(t)$ defined on the real t-interval I is differentiable at some t_0 in I if and only if there exists a complex number z_0 such that

$$\left| \frac{f(t_0 + h) - f(t_0)}{h} - z_0 \right| \to 0 \quad \text{as } h \to 0$$

If this is so, z_0 is said to be the derivative of f and is denoted by $f'(t_0)$

Note that if f is real-valued, the definition above reduces to the usual notion of differentiation for real-valued functions. Hence no confusion can arise if we do not know in advance whether a function is real- or complex-valued. Since differentiation of complex-valued functions of a real variable is defined in precisely the same formal manner as for real-valued functions of a real variable, it is not surprising that all the usual differentiation formulas (such as the Chain Rule, differentiation of sums, products, etc.) hold for these more general functions also.

We now present a useful computational device for computing derivatives of complex-valued functions of a real variable. For each t in I, let $\alpha(t)$ and $\beta(t)$ be the real and imaginary parts of $f(t)$, respectively. Then f can be uniquely written as $f(t) = \alpha(t) + i\beta(t)$, where $\alpha(t)$ and $\beta(t)$ are real-valued functions defined over interval I.

THEOREM C.3.3

> Differentiation Principle for Complex-Valued Functions. The complex-valued function of a real variable $f(t) = \alpha(t) + i\beta(t)$ is differentiable at t_0 if and only if $\alpha(t)$ and $\beta(t)$ are both differentiable at t_0. If $f'(t_0)$ exists, then
> $$f'(t_0) = \alpha'(t_0) + i\beta'(t_0) \tag{2}$$

The proof of this result is a straightforward result of applying the definitions and will be omitted.

Antidifferentiation or integration of $f(t) = \alpha(t) + i\beta(t)$ is completely equivalent to performing the same operation on both $\alpha(t)$ and $\beta(t)$, and conversely.

EXAMPLE C.3.1

Differentiation of Exponentials

The function $f(t) = e^{ikt} = \cos kt + i \sin kt$, where k is a real constant, is differentiable for all t. Using (2) and the definition of e^{ikt},

$$(e^{ikt})' = -k \sin kt + ik \cos kt = ik(\cos kt + i \sin kt) = ike^{ikt} \qquad (3)$$

Formula (3) is easy to remember because it says that the complex-valued function e^{ikt} follows the usual differentiation rule for exponential functions. Let $z = a + ib$ be a (fixed) complex number; then e^{zt} is defined to be $e^{(a+ib)t} = e^{at}e^{ibt}$. Using (2) and (3) together with the product rule, we see that

$$(e^{zt})' = ae^{at}e^{ibt} + e^{at}ibe^{ibt} = (a+ib)e^{at}e^{ibt} = ze^{zt} \qquad (4)$$

a result that is equally easy to remember.

C.4 Useful Formulas

Throughout the text, many formulas are assumed in order to solve the problems. Below are some of these frequently encountered formulas.

- *Logarithms and Exponentials.* The following identities hold for all $A \neq 0$, $B \neq 0$, and for all a, b.

  1. $\ln|A| + \ln|B| = \ln|AB|$
  2. $\ln|A| - \ln|B| = \ln|A/B|$
  3. $a \ln|B| = \ln(|B|^a)$

  4. $e^{a+b} = e^a e^b$
  5. $e^{a-b} = e^a/e^b$
  6. $e^{a\ln|B|} = |B|^a$

- *Trigonometric Identities.*

  1. **Sum and Difference**
     (a) $\sin(\alpha \pm \beta) = \sin\alpha\cos\beta \pm \cos\alpha\sin\beta$
     (b) $\cos(\alpha \pm \beta) = \cos\alpha\cos\beta \mp \sin\alpha\sin\beta$

  2. **Product**
     (a) $\sin\alpha\sin\beta = \frac{1}{2}\{\cos(\alpha-\beta) - \cos(\alpha+\beta)\}$
     (b) $\sin\alpha\cos\beta = \frac{1}{2}\{\sin(\alpha-\beta) + \sin(\alpha+\beta)\}$
     (c) $\cos\alpha\cos\beta = \frac{1}{2}\{\cos(\alpha-\beta) + \cos(\alpha+\beta)\}$

  3. **Double Angle**
     (a) $\sin(2\alpha) = 2\sin\alpha\cos\alpha$
     (b) $\cos(2\alpha) = \cos^2\alpha - \sin^2\alpha$

  4. **Symmetry, Phase-shift**
     (a) $\sin(\theta) = \cos(\pi/2 - \theta)$, $\quad \cos(\theta) = \sin(\pi/2 - \theta)$
     (b) $\sin(-\theta) = -\sin(\theta)$, $\quad \cos(-\theta) = \cos(\theta)$
     (c) $A\sin\theta + B\cos\theta = \sqrt{A^2 + B^2}\cos(\theta - \delta)$, where $\delta = \tan^{-1}(A/B)$ and $|\delta| < \pi/2$

  5. **Law of Cosines**
     If $a, b,$ and c are the three sides of a triangle and θ is the angle opposite c, then

     $$c^2 = a^2 + b^2 - 2ab\cos(\theta)$$

6. Lagrange Identity

$$\frac{1}{2} + \sum_{k=1}^{n} \cos(ks) = \frac{\sin((n+1/2)s)}{2\sin(s/2)}$$

7. Euler's Formulas
 (a) $e^{i\theta} = \cos\theta + i\sin\theta,$ $e^{-i\theta} = \cos\theta - \sin\theta$
 (b) $\cos\theta = (e^{i\theta} + e^{-i\theta})/2,$ $\sin\theta = i(e^{i\theta} - e^{-i\theta})/2$

- *Hyperbolic Trigonometric Functions.*

 1. $e^x = \cosh x + \sinh x,$ $e^{-x} = \cosh x - \sinh x$
 2. $\cosh x = (e^x + e^{-x})/2,$ $\sinh x = (e^x - e^{-x})/2,$ $\tanh x = (e^x + e^{-x})/(e^x - e^{-x})$

- *Polynomials.*

 1. Fundamental Theorem of Algebra (Factorization Formula)
 Let $P(x) = x^n + a_{n-1}x^{n-1} + \cdots + a_0$, where n is a positive integer and $a_0, \cdots, a_{n-1}$ are real or complex constants. Then there is a unique set of numbers $\{r_1, \ldots, r_k\}$, some of which may be complex, and a unique set of positive integers $\{m_1, \ldots, m_k\}$ such that

$$P(x) = (x - r_1)^{m_1} \cdots (x - r_k)^{m_k}$$

 where $m_1 + \cdots + m_k = n$. The numbers $r_1, \ldots, r_k$ are the *zeros* (or *roots*) of $P(x)$ and the positive integers $m_1, \ldots, m_k$ are the corresponding *multiplicities*.

 2. Suppose $s_1, \ldots, s_n$ are the n roots (not necessarily distinct) of the polynomial $x^n + a_{n-1}x^{n-1} + \cdots + a_0$, with each root listed as often as its multiplicity. Then

$$s_1 + s_2 + \cdots + s_n = -a_{n-1}, \qquad s_1 \cdot s_2 \cdots s_n = (-1)^n a_0$$

 3. Partial Fractions
 Let $P(s)$ and $Q(s)$ be polynomials, where $\deg P < \deg Q$ and $Q(s) = \prod_{k=1}^{p}(s - r_k)^{m_k}$. Then there exist unique constants a_{jk} such that

$$\frac{P(s)}{Q(s)} = \sum_{k=1}^{p}\left\{ \sum_{j=1}^{m_k} \frac{a_{jk}}{(s - r_k)^j} \right\}$$

 The terms $a_{jk}/(s - r_k)^j$ are the *partial fractions*. The summation form of this expression indicates its utility in the Laplace transform calculus. For, by the linearity of $\mathcal{L}^{-1}$ (assuming that $\mathcal{L}^{-1}$ exists) and using II.3 in the transform tables at the end of the book, we have

$$\mathcal{L}^{-1}\left[\frac{P(s)}{Q(s)}\right] = \sum_{k=1}^{p}\left\{ \sum_{j=1}^{m_k} a_{jk}\mathcal{L}^{-1}\left[\frac{1}{(s - r_k)^j}\right] \right\}$$

$$= \sum_{k=1}^{p}\left\{ \sum_{j=1}^{m_k} a_{jk}\left[\frac{t^{j-1}}{(j-1)!}e^{r_k t}\right] \right\}$$

and we have the exact form of the inverse transform.

C.5 Theorems from Calculus

In this section, we state without proof the theorems of single- and multi-variable calculus most frequently encountered in the study of differential equations.

Results from Single Variable Calculus

In the theorems that follow f is a function of a single variable. We represent the closed interval from a to b by $[a, b]$, and the open interval by (a, b).

THEOREM C.5.1

> Maximum/Minimum Value Theorem. Let f be continuous on $[a, b]$. Then f is bounded on $[a, b]$ and attains its maximum and its minimum values at points in $[a, b]$. If one of these points is in (a, b) and if f is differentiable on (a, b), then $f' = 0$ at the point.

THEOREM C.5.2

> L'Hôpital's Rule. Let $f(x)$ and $g(x)$ be differentiable and such that $\lim_{x \to a} f(x) = \lim_{x \to a} g(x) = 0$, or $\lim_{x \to a} f(x) = \lim_{x \to a} g(x) = \pm\infty$, where a may be $\pm\infty$. If $\lim_{x \to a} f'(x)/g'(x)$ exists, then so does $\lim_{x \to a} f(x)/g(x)$, and
> $$\lim_{x \to a} \frac{f(x)}{g(x)} = \lim_{x \to a} \frac{f'(x)}{g'(x)}$$

THEOREM C.5.3

> Intermediate Value Theorem. Let f be continuous on $[a, b]$ and let c be any real number between $f(a)$ and $f(b)$ inclusive. Then there exists a point t^* in $[a, b]$ such that $f(t^*) = c$.

THEOREM C.5.4

> Mean Value Theorem.
>
> **Integral Form:** Let f be continuous on $[a, b]$. Then there exists a point t^* in $[a, b]$ such that $\int_a^b f(t)\, dt = f(t^*)(b - a)$.
>
> **Differential Form:** Let f be differentiable on (a, b) and continuous on $[a, b]$. Then there exists a point t^* in (a, b) such that $f(b) - f(a) = f'(t^*)(b - a)$.

THEOREM C.5.5

Fundamental Theorem of Calculus.

1. Let f be continuous on $[a, b]$ and let F be an antiderivative of f. Then $\int_a^b f(t)\, dt = F(b) - F(a)$.

2. Let f be continuous on $[a, b]$ and let c be any point in $[a, b]$. If $F(t) = \int_c^t f(s)\, ds$, then $F'(t) = f(t)$ at each point t in (a, b).

THEOREM C.5.6

Leibniz's Rule. Let R be the closed rectangle in the st-plane given by $a \le t \le b$, $c \le s \le d$. Suppose that $\alpha(t)$, $\beta(t)$ are continuously differentiable functions on $[a, b]$, whose graphs lie in R, and $f(s, t)$, $f_t(s, t)$ are continuously differentiable on R. Then $\int_{\alpha(t)}^{\beta(t)} f(s, t)\, ds$ is a continuously differentiable function of t and

$$\frac{d}{dt} \int_{\alpha(t)}^{\beta(t)} f(s, t)\, ds = \beta'(t) f(\beta(t), t) - \alpha'(t) f(\alpha(t), t) + \int_{\alpha(t)}^{\beta(t)} \frac{\partial}{\partial t} f(s, t)\, ds$$

THEOREM C.5.7

The Chain Rule. Let $g(t)$ be differentiable at the point t and let $f(s)$ be differentiable at the point $s = g(t)$. Then the composition $f(g(t))$ is differentiable at the point t, and $(f(g(t))' = f'(g(t)) \cdot g'(t)$. Stated another way: Put $s = g(t)$, and $y = f(s)$, then the composition $y = f(g(t))$ is differentiable and

$$\frac{dy}{dt} = \left.\frac{dy}{ds}\right|_{g(t)} \cdot \frac{ds}{dt}$$

THEOREM C.5.8

Integration by Parts. Let $f(x)$ and $g(x)$ be differentiable functions on $[a, b]$. Then

$$\int_a^b f(x)g'(x)\, dx = f(x)g(x)|_a^b - \int_a^b f'(x)g(x)\, dx$$

THEOREM C.5.9

Integral Estimate. Let f be continuous on $[a, b]$. Let M be any real number such that $|f(t)| \le M$ for all $a \le t \le b$. Then,

$$\left| \int_a^b f(t)\, dt \right| \le \int_a^b |f(t)|\, dt \le (b - a)M$$

Results from Multi-Variable Calculus

In the theorems that follow, the function f has two or more real variables.

THEOREM C.5.10

Maximum/Minimum Value Theorem. Let $f(x, y)$ be continuous on a closed, bounded region R. Then f is bounded on R and attains its maximum and its minimum values at some points in R. If one of these points is in the interior of R and if f is differentiable, then $\partial f/\partial x$ and $\partial f/\partial y$ are both 0 at the point.

THEOREM C.5.11

Fubini's Theorem: Interchanging the Order of Integration. Let $f(x, y)$ be continuous on the region R described by $a \leq x \leq b, c \leq y \leq d$. Then

$$\int_R f \, dA = \int_a^b \int_c^d f(x, y) \, dy \, dx = \int_c^d \int_a^b f(x, y) \, dx \, dy$$

THEOREM C.5.12

Green's Theorem. Let R be a simply connected planar region whose boundary is a closed, piecewise-smooth curve C traversed counterclockwise. If $f(x, y)$ and $g(x, y)$ have continuous first partial derivatives on some open set containing R, then

$$\int_C f(x, y) \, dx + g(x, y) \, dy = \iint_R \left(\frac{\partial g}{\partial x} - \frac{\partial f}{\partial y} \right) dA$$

If $f(x, y, z)$ is a continuous function, its *gradient* is the vector

$$\nabla f = \frac{\partial f}{\partial x}\hat{\mathbf{i}} + \frac{\partial f}{\partial y}\hat{\mathbf{j}} + \frac{\partial f}{\partial z}\hat{\mathbf{k}}$$

At each point of a level set $f(x, y, z) = C$, the gradient vector is orthogonal to the tangent to the level set and points in the direction of the maximal increase in the values of f. If $\mathbf{F}(x, y, z) = f(x, y, z)\hat{\mathbf{i}} + g(x, y, z)\hat{\mathbf{j}} + h(x, y, z)\hat{\mathbf{k}}$, then the *divergence* of $\mathbf{F}$ is

$$\nabla \cdot \mathbf{F} = \frac{\partial f}{\partial x} + \frac{\partial g}{\partial y} + \frac{\partial h}{\partial z}$$

THEOREM C.5.13

Divergence Theorem. Let G be a solid with boundary surface ∂G oriented by outward unit normals $\mathbf{n} = n_1\hat{\mathbf{i}} + n_2\hat{\mathbf{j}} + n_3\hat{\mathbf{k}}$. If $\mathbf{F}(x, y, z) = f(x, y, z)\hat{\mathbf{i}} + g(x, y, z)\hat{\mathbf{j}} + h(x, y, z)\hat{\mathbf{k}}$, where f, g, and h have continuous first partial derivatives on some region containing G, then

$$\iint_{\partial G} \mathbf{F} \cdot \mathbf{n} \, dS = \iiint_G \nabla \cdot \mathbf{F} \, dV$$

That is,

$$\iint_{\partial G} [f(x, y, z)n_1 + g(x, y, z)n_2 + h(x, y, z)n_3] \, dS = \iiint_G \left(\frac{\partial f}{\partial x} + \frac{\partial g}{\partial y} + \frac{\partial h}{\partial z} \right) dV$$

THEOREM C.5.14

> Chain Rule. Let $x = x(t)$ and $y = y(t)$ be differentiable at t and let $z = f(x, y)$ be differentiable at the point $(x(t), y(t))$. Then,
> $$\frac{dz}{dt} = \frac{\partial z}{\partial x}\frac{dx}{dt} + \frac{\partial z}{\partial y}\frac{dy}{dt}$$

THEOREM C.5.15

> Taylor's Theorem for a Function of Two Variables. Let the function $f(x, y)$ be twice continuously differentiable in a region S containing a point (x_0, y_0). Then there is a rectangle R, $|x - x_0| < a$, $|y - y_0| < b$, such that for all (x, y) in R
> $$f(x, y) = f(x_0, y_0) + f_x(x_0, y_0)(x - x_0) + f_y(x_0, y_0)(y - y_0) + P$$
> where P is at least second-order in $[(x - x_0)^2 + (y - y_0)^2]^{1/2}$.

Rectangular, Cylindrical, Spherical Coordinates

Let G be a three-dimensional region in $\mathbf{R}^3$. If G and its boundary surfaces are described in rectangular coordinates x, y, z, then we shall denote G by G_1; in cylindrical coordinates by G_2; in spherical coordinates by G_3.

- *Rectangular Coordinates*: Let G be a simple solid and let $f(x, y, z)$ be continuous on G. Then
$$\int_G f\,dV = \iiint_{G_1} f(x, y, z)\,dz\,dy\,dx$$

- *Cylindrical Coordinates*: Let G be a simple solid and let $f(x, y, z)$ be continuous on G. Then,
$$\int_G f\,dV = \iiint_{G_2} f(r, \theta, z)r\,dz\,dr\,d\theta$$

- *Spherical Coordinates*: Let G be a simple solid and let $f(x, y, z)$ be continuous on G. Then,
$$\int_G f\,dV = \iiint_{G_3} f(\rho, \theta, \phi)\rho^2 \sin\phi\,d\rho\,d\phi\,d\theta$$

The Laplacian Operator

Laplace's equation is the homogeneous linear second-order PDE
$$\nabla^2 u = 0$$
where ∇^2 is called the *Laplacian operator.*

The Laplacian operator in the Cartesian rectangular coordinates of 3-space is
$$\nabla^2 = \frac{\partial^2}{\partial x^2} + \frac{\partial^2}{\partial y^2} + \frac{\partial^2}{\partial z^2}$$

The operator has the following form in other coordinate systems:

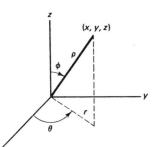

- Polar Coordinates

$$\nabla^2 u = u_{xx} + u_{yy} = u_{rr} + (1/r)u_r + (1/r^2)u_{\theta\theta}$$

- Cylindrical Coordinates

$$\nabla^2 u = u_{xx} + u_{yy} + u_{zz} = u_{rr} + (1/r)u_r + (1/r^2)u_{\theta\theta} + u_{zz}$$

- Spherical Coordinates

$$\nabla^2 u = u_{xx} + u_{yy} + u_{zz} = \frac{1}{\rho^2}(\rho^2 u_\rho)_\rho + \frac{1}{\rho^2 \sin\phi}(\sin\phi u_\phi)_\phi + \frac{1}{\rho^2 \sin^2\phi}u_{\theta\theta}$$

C.6 Review of Matrix Theory

In this section, we review some basic concepts from matrix theory, focusing on those ideas which are useful in the study of systems of linear ODEs.

❖ **Matrix.** An $m \times n$ matrix A is a rectangular array of numbers a_{ij} with m rows and n columns

$$\begin{bmatrix} a_{11} & a_{12} & \cdots & a_{1n} \\ a_{21} & a_{22} & \cdots & a_{2n} \\ \vdots & \vdots & & \vdots \\ a_{m1} & a_{m2} & \cdots & a_{mn} \end{bmatrix}$$

Note the first index of a_{ij} indicates the row, and the second the column in which the number a_{ij} appears. We often abbreviate the notation for A by $[a_{ij}]$. Two $m \times n$ matrices A and B are *equal* if $a_{ij} = b_{ij}$ for all i, j. If $m = n$, then A is called a *square matrix* of order n. If $n = 1$ ($m = 1$), we often think of the column (row) of numbers as the coordinates of a vector, and refer to the matrix as a *column (row) vector*. In systems of linear ODEs, we will be concerned primarily with square matrices and column vectors.

Matrix Operations

Several operations defined for matrices are given below.

❖ **Addition, Multiplication by a Scalar.** Let $A = [a_{ij}]$, $B = [b_{ij}]$ be two $m \times n$ matrices, and let α be a scalar. Then

$$A + B = [a_{ij} + b_{ij}] \quad \text{and} \quad \alpha A = [\alpha a_{ij}]$$

EXAMPLE C.6.1

Let $A = \begin{bmatrix} 2 & -1 \\ 0 & 4 \end{bmatrix}$, $B = \begin{bmatrix} 4 & 7 \\ 9 & 3 \end{bmatrix}$, and $\alpha = 3$. Then $A + B = \begin{bmatrix} 6 & 6 \\ 9 & 7 \end{bmatrix}$ and $\alpha A = \begin{bmatrix} 6 & -3 \\ 0 & 12 \end{bmatrix}$

EXAMPLE C.6.2

Commutativity, Associativity of Addition
Notice that $A + B = B + A$ and $A + (B + C) = (A + B) + C$.

❖ **Matrix Products.** Let $A = [a_{ij}]$ be an $m \times n$ matrix and $B = [b_{ij}]$ be an $n \times k$ matrix. Then AB is defined to be the $m \times k$ matrix $AB = [c_{ij}]$ where $c_{ij} = \sum_{s=1}^{n} a_{is} b_{sj}$.

EXAMPLE C.6.3

Matrix Product

Let $A = \begin{bmatrix} 2 & 1 & -1 \\ 0 & 3 & 1 \end{bmatrix}$ and $B = \begin{bmatrix} -3 \\ 4 \\ 2 \end{bmatrix}$. Then

$$AB = \begin{bmatrix} 2 & 1 & -1 \\ 0 & 3 & 1 \end{bmatrix} \begin{bmatrix} -3 \\ 4 \\ 2 \end{bmatrix} = \begin{bmatrix} (2)(-3) + (1)(4) + (-1)(2) \\ (0)(-3) + (3)(4) + (1)(2) \end{bmatrix} = \begin{bmatrix} -4 \\ 14 \end{bmatrix}$$

In the definition of a matrix product the number of columns in the matrix A must equal the number of rows in matrix B. Thus, in Example C.6.3 the product BA is not defined. In general, matrix multiplication is not commutative, even if the matrices are both $n \times n$ so that both products are defined and of the same size. However, for any matrices A, B, and C with compatible dimensions we do have

$$\begin{aligned} A(BC) &= (AB)C & &\text{Associativity} \\ A(B + C) &= AB + AC & &\text{Distributivity} \end{aligned}$$

If we write an expression without mentioning the dimensions of the matrices involved, then it is assumed that the matrices are of compatible dimensions.

We now define two special matrices.

❖ **Zero Matrix.** The $m \times n$ matrix all of whose entries are zero is denoted by 0 (regardless of its dimension) and is called the *zero matrix*.

❖ **Identity Matrix.** The $n \times n$ matrix with $a_{ii} = 1$ for all i and $a_{ij} = 0$ whenever $i \neq j$, is called the *identity matrix* and is denoted by I_n. We usually write it simply as I.

EXAMPLE C.6.4

Let $A = [a_{ij}]$ be an $m \times n$ matrix. Then it can be directly verified that $0A = 0$ (where the dimensions of the zero vector are compatible with the dimensions of A). A similar statement holds for the product $A0$. Also, $A + 0 = 0 + A = A$. Finally, $I_m A = A I_n = A$.

❖ **Transpose.** Given any $m \times n$ matrix $A = [a_{ij}]$, the $n \times m$ matrix obtained from A by interchanging the rows and columns is called the *transpose* of A and is denoted by A^T. Note that the ij-th entry of A^T is a_{ij}, and for that reason A^T is sometimes denoted by $[a_{ji}]$.

EXAMPLE C.6.5

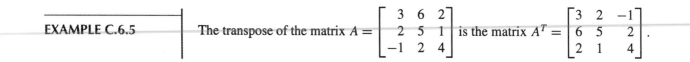

The transpose of the matrix $A = \begin{bmatrix} 3 & 6 & 2 \\ 2 & 5 & 1 \\ -1 & 2 & 4 \end{bmatrix}$ is the matrix $A^T = \begin{bmatrix} 3 & 2 & -1 \\ 6 & 5 & 2 \\ 2 & 1 & 4 \end{bmatrix}$.

It may be shown that the following properties of matrix transposition hold:

$$(A^T)^T = A$$
$$(A + B)^T = A^T + B^T$$
$$(AB)^T = B^T A^T$$

Note that the transpose of a row vector is a column vector.

Systems of Linear Equations

The system of n linear equations in n variables $x_1, \ldots, x_n$

$$a_{11}x_1 + \ldots + a_{1n}x_n = b_1$$
$$\vdots \qquad \qquad \vdots$$
$$a_{n1}x_1 + \ldots + a_{nn}x_n = b_n$$

can be written in matrix form as

$$
\begin{bmatrix} a_{11} & \cdots & a_{1n} \\ \vdots & & \vdots \\ a_{n1} & \cdots & a_{nn} \end{bmatrix}
\begin{bmatrix} x_1 \\ \vdots \\ x_n \end{bmatrix}
=
\begin{bmatrix} b_1 \\ \vdots \\ b_n \end{bmatrix}
$$

or more compactly as $A\mathbf{x} = \mathbf{b}$, where $\mathbf{x}$ and $\mathbf{b}$ are in boldface to emphasize that they are column vectors.

We solve this system by *row reducing* it to a simpler but equivalent system using a sequence of operations chosen from the three *elementary row operations*:

1. Interchange two rows.

2. Multiply one row by a nonzero number.

3. Add a multiple of one row to a different row.

See Example C.6.7. We will come back to elementary row operations after discussing inverses and determinants.

Invertible Matrices

In what follows, we will deal exclusively with square matrices, since it is these matrices that occur most often in the study of systems of differential equations.

❖ **Invertibility.** Let A be an $n \times n$ matrix. If there exists an $n \times n$ matrix B such that $AB = BA = I$, then $B = A^{-1}$ is said to be the *multiplicative inverse of A* and A is said to be *invertible*.

EXAMPLE C.6.6

Let $A = \begin{bmatrix} 3 & 1 \\ 5 & 2 \end{bmatrix}$ and $B = \begin{bmatrix} 2 & -1 \\ -5 & 3 \end{bmatrix}$. The product $AB = BA = \begin{bmatrix} 1 & 0 \\ 0 & 1 \end{bmatrix} = I$, so A is invertible with $A^{-1} = B$.

If a matrix A does not have an inverse, it is said to be *singular*. We now briefly describe a method for calculating the inverse of a square matrix, assuming that the inverse exists. It is possible using the three elementary row operations to reduce any $n \times n$ matrix A to the identity matrix I, as long as the matrix A is invertible.

Now, it can be shown that the sequence of elementary row operations that carries the invertible matrix A to I carries I to A^{-1}. This method is illustrated below.

EXAMPLE C.6.7

We wish to solve the system

$$3x_1 + x_2 = 3$$

$$5x_1 + 2x_2 = 4$$

equivalent to the matrix equation $A\mathbf{x} = \mathbf{b}$ where

$$A = \begin{bmatrix} 3 & 1 \\ 5 & 2 \end{bmatrix}, \quad \mathbf{x} = \begin{bmatrix} x_1 \\ x_2 \end{bmatrix}, \quad \mathbf{b} = \begin{bmatrix} 3 \\ 4 \end{bmatrix}$$

If A is invertible, we can simultaneously find A^{-1} and solve for $\mathbf{x}$. Let

$$[A|b||I] = \begin{bmatrix} 3 & 1 & 3 & 1 & 0 \\ 5 & 2 & 4 & 0 & 1 \end{bmatrix}$$

The steps are as follows:

$$\begin{bmatrix} 6 & 2 & 6 & 2 & 0 \\ 5 & 2 & 4 & 0 & 1 \end{bmatrix} \qquad \begin{array}{l} \text{(2 Row 1)} \\ \text{(Row 2)} \end{array}$$

$$\begin{bmatrix} 1 & 0 & 2 & 2 & -1 \\ 5 & 2 & 4 & 0 & 1 \end{bmatrix} \qquad \begin{array}{l} \text{(Row 1 } - \text{ Row 2)} \\ \text{(Row 2)} \end{array}$$

$$\begin{bmatrix} 1 & 0 & 2 & 2 & -1 \\ 0 & 2 & -6 & -10 & 6 \end{bmatrix} \qquad \begin{array}{l} \text{(Row 1)} \\ \text{(Row 2 } - \text{ 5 Row 1)} \end{array}$$

$$\begin{bmatrix} 1 & 0 & 2 & 2 & -1 \\ 0 & 1 & -3 & -5 & 3 \end{bmatrix} \qquad \begin{array}{l} \text{(Row 1)} \\ \text{(1/2 Row 2)} \end{array}$$

So $A^{-1} = \begin{bmatrix} 2 & -1 \\ -5 & 3 \end{bmatrix}$, which is exactly the matrix B of Example C.6.6. The first two sections of the reduced matrix represent the system $x_1 = 2$, $x_2 = -3$, which is equivalent to our original system. Therefore, $\mathbf{x} = \begin{bmatrix} 2 \\ -3 \end{bmatrix}$ is the unique solution to $A\mathbf{x} = \mathbf{b}$.

For a 2×2 matrix $A = \begin{bmatrix} a & b \\ c & d \end{bmatrix}$, it can be shown using the method given above, that if $ad - bc \neq 0$, then

$$A^{-1} = \frac{1}{ad - bc} \begin{bmatrix} d & -b \\ -c & a \end{bmatrix}$$

Determinants

Every square matrix A has a number associated with it called its *determinant* and denoted by det A. For $n = 2$, we have

$$\det \begin{bmatrix} a & b \\ c & d \end{bmatrix} = ad - bc$$

Assume now that det A has been defined for all square matrices of size $(n-1) \times (n-1)$ or smaller. Let A be an $n \times n$ matrix. The ijth *minor matrix*, denoted by A_{ij} is the

$(n-1) \times (n-1)$ matrix obtained by striking out the ith row and the jth column of A (indicated in bold type) to obtain

$$A_{ij} = \begin{bmatrix} a_{11} & \cdots & \boldsymbol{a_{1j}} & \cdots & a_{1n} \\ \vdots & & \vdots & & \vdots \\ \boldsymbol{a_{i1}} & \cdots & \boldsymbol{a_{ij}} & \cdots & \boldsymbol{a_{in}} \\ \vdots & & \vdots & & \vdots \\ a_{n1} & \cdots & \boldsymbol{a_{nj}} & \cdots & a_{nn} \end{bmatrix}$$

The number $(-1)^{i+j} \det A_{ij}$ is called the *ijth cofactor*, denoted by $\operatorname{cof} A_{ij}$.

❖ **Cofactor Expansion.** Let A be an $n \times n$ matrix. Then the *cofactor expansions* on the ith row and the jth column, respectively, are given by

$$\sum_{k=1}^{n} a_{ik} \operatorname{cof} A_{ik}, \quad \sum_{k=1}^{n} a_{kj} \operatorname{cof} A_{kj}$$

It may by shown that, given the matrix A, all of its cofactor expansions are equal. This leads us to the definition of the determinant.

❖ **Determinant.** Let A be an $n \times n$ matrix. Then the *determinant* of A, denoted by $\det A$, is the scalar defined by any one of the cofactor expansions.

EXAMPLE C.6.8

Let $A = \begin{bmatrix} 3 & 6 & 2 \\ 2 & 5 & 1 \\ -1 & 2 & 4 \end{bmatrix}$. We will expand on the 1st row. Then

$$\det A = 3 \det \begin{bmatrix} 5 & 1 \\ 2 & 4 \end{bmatrix} - 6 \det \begin{bmatrix} 2 & 1 \\ -1 & 4 \end{bmatrix} + 2 \det \begin{bmatrix} 2 & 5 \\ -1 & 2 \end{bmatrix}$$
$$= 3(20-2) - 6(8+1) + 2(4+5) = 18$$

Although not applicable in Example C.6.8, a good strategy to follow in calculating the determinant of a matrix is to do a cofactor expansion on the row or column with the largest number of zeros.

The following are useful properties of determinants. For any $n \times n$ matrices A, B

- $\det A = \det A^T$
- If a row (column) of A is multiplied by a scalar λ, the determinant of that matrix is $\lambda \det A$. Therefore, if A has a row (column) of zeros, then $\det A = 0$.
- If B is the matrix that arises from A by adding a multiple of a row (column) to a different row (column), then $\det A = \det B$.
- If B is the matrix which arises from A by interchanging two rows (or columns), then $\det A = -\det B$.
- $\det AB = (\det A)(\det B) = (\det B)(\det A) = \det BA$

- If A is the *upper triangular* matrix

$$A = \begin{bmatrix} a_{11} & a_{12} & \cdots & a_{1n} \\ 0 & a_{22} & & a_{2n} \\ \vdots & 0 & \ddots & \vdots \\ 0 & \cdots & 0 & a_{nn} \end{bmatrix}$$

  then det A is simply the product $a_{11}a_{22}\cdots a_{nn}$ of the diagonal elements.

- A is invertible if and only if det $A \neq 0$; $\det(A^{-1}) = 1/\det A$.

More on Systems of Linear Equations

Recall the system of linear equations represented by the matrix equation $A\mathbf{x} = \mathbf{b}$. The system has a unique solution $\mathbf{x} = A^{-1}\mathbf{b}$ if and only if det $A \neq 0$, i.e., if and only if the homogeneous system $A\mathbf{x} = \mathbf{0}$ has only the trivial solution $\mathbf{x} = \mathbf{0}$. If det $A = 0$, then A is not invertible, and the system $A\mathbf{x} = \mathbf{b}$ has either no solutions or an infinite number of solutions. The homogeneous system $A\mathbf{x} = \mathbf{0}$ in this case has an infinite number of solutions.

Matrices of Functions

We often wish to take scalar functions as the elements of a matrix. The matrix itself then becomes a function. To evaluate the matrix function at a certain point, we evaluate each element at that point, yielding a matrix of constants.

We differentiate matrix functions elementwise. For example, if $A(t) = [a_{ij}(t)]$, then $A'(t) = [a'_{ij}(t)]$. It is straightforward to prove $(A + B)' = A' + B'$, and $(AC)' = A'C + AC'$, where A, B, and C are any matrices of compatible dimensions.

Inequalities

We end our discussion of matrices and vectors with some frequently used inequalities:

- *Triangle inequality*: If a and b are scalars,

$$||a| - |b|| \leq |a + b| \leq |a| + |b|$$

  If $\mathbf{u}$ and $\mathbf{v}$ are vectors,

$$|\|\mathbf{u}\| - \|\mathbf{v}\|| \leq \|\mathbf{u} + \mathbf{v}\| \leq \|\mathbf{u}\| + \|\mathbf{v}\|$$

  where $\|\mathbf{u}\|$ is the Euclidean length of the vector $\mathbf{u}$.

- *Cauchy-Schwarz inequality*: If $\mathbf{u}$ and $\mathbf{v}$ are vectors,

$$|\mathbf{u} \cdot \mathbf{v}| \leq \|\mathbf{u}\| \, \|\mathbf{v}\|$$

  where $\mathbf{u} \cdot \mathbf{v}$ is the dot product of the vectors $\mathbf{u}$ and $\mathbf{v}$. See Section 4.1.

- *Matrix Norm*: The *norm* $\|A\|$ of a matrix A is defined by $\|A\| = \sum_{i,j} |a_{ij}|$.

- *Matrix Estimate*: If A and B are matrices for which the product AB is defined, then $\|AB\| \leq \|A\| \cdot \|B\|$.

TABLE C.7.1 Comparison of Basic Units in Various Systems.

System	Length (L)	Force (ML/T^2)	Mass (M)	Acceleration (L/T^2)	Energy (ML^2/T^2)
SI	meter (m)	newton (N)	kilogram (kg)	m/s^2	joule (J)
CGS	centimeter (cm)	dyne	gram (gm)	cm/s^2	erg
BE	foot (ft)	pound (lb)	slug	ft/s^2	BTU

C.7 Scaling and Units

Units

Three systems of units are used throughout the text—SI units (the International System of units, also known as MKS (meter-kilogram-second) units), CGS (centimeter-gram-second) units, and British Engineering units. While the SI system is most frequently used in the scientific community, the text also uses the British Engineering System (BE), due to its continued widespread use in the United States. Table C.7.1 summarizes the units of length, force, mass, acceleration, and energy in these three systems. In CGS and SI units, the three fundamental physical units are length (L), mass (M), and time (T). In BE units, the three fundamental physical dimensions are length, force (ML/T^2), and time.

When working with physical systems throughout the text involving these units of measurement, the following points are useful to keep in mind.

- Although any system of units may be used, it is crucial to be consistent in the use of a single system for any given problem.
- Time is measured in seconds (s), hours (h), days (d), or years (y).
- The dimensions of force are the same as those of mass times acceleration. Thus, for example, 1 N=1 kg·m/s^2 and 1 lb=1 slug·ft/s^2. Also, 1 N=0.2248 lb.
- The acceleration due to gravity is denoted by g and has a magnitude near the earth's surface of 9.8 m/s^2 in SI units, 980 cm/s^2 in CGS units, or 32 ft/s^2 in BE units. The *weight* of an object is the gravitational force exerted on it by the earth, and is equal to the product mg of the mass m of the object and the acceleration g due to gravity.
- Although the SI unit of energy is the joule (1 N·m), the calorie is still occasionally used as well (1 cal=4.184 J= 3.968×10^{-3} BTU (British thermal units)).
- A useful method of ensuring that you are using the units correctly is to check that the dimensions are the same on both sides of any equation. Consider, for example, the undamped spring equation $my'' = -ky$. Both sides of the equation should have the dimensions of force. Using the BE system, m is measured in slugs and y'' is in ft/s^2. Thus, my'' is measured in slug·ft/s^2 (lb). Similarly, k is measured in slug/s^2 and y is measured in ft, so ky is also measured in slug·ft/s^2. In the CGS system, m is measured in grams and y'' has units of cm/s^2. Thus, my'' has units of gm·cm/s^2. Since y has units of cm, k must have units of gm/s^2.

The problem of units can be avoided by rescaling equations and using dimensionless variables.

Scaling; Dimensionless Variables and Parameters

An IVP may involve many variables and parameters. For example, the IVPs that model radioactive decay, the motion of an oscillating spring, and the changing concentrations in a chemical reactor include a host of variables, physical parameters, and initial data. Even an "abstract" IVP that does not model a physical system may be written in terms of several variables and parameters. As seen above, variables and parameters can be measured in any one of several systems of units. Is there a way to scale the state variables and time so that the variables of the rescaled IVP are dimensionless, and, hence, any study of the solutions of the dimensionless system would give information about the behavior of solutions regardless of the units used? Is there a way to study the sensitivity of an IVP to changes in some of the parameters or initial data so that the results apply independently of the system of units used? These two questions are clearly related, and the three examples given below show how a suitable rescaling of the state variables and time resolves the questions. In the process, we see that rescaling reduces the number of parameters that we have to work with—always an advantage. Finally, scaling a problem before computing is essential to avoid awkward scales, e.g., an interval 10^5 units long on the x-axis and 2 units long on the y-axis. One important consequence of rescaling to dimensionless quantities is that the results of the sensitivity study of the dimensionless IVP are inherent in the IVP and are not just a consequence of using a particular system of units, scales, and dimensions.

Radiocarbon Dating (Example 1.3.3)

The model IVP for a radioactive carbon-14 (^{14}C) dating process is

$$q'(t) = -kq(t), \quad q(0) = q_0, \quad T \leq t \leq 0 \tag{1}$$

where $q(t)$ is the (dimensionless) fraction of ^{14}C in the mass of all forms of carbon in a sample of once-living material, k is a positive decay constant (measured in reciprocal years), $q(0)$ is the current fraction of ^{14}C (at $t = 0$), T is the time in years in the past when the material was last alive, and q_T is the fraction of ^{14}C in the sample at time T. Observe that $q(t)$ is already dimensionless since it is the ratio of the mass of ^{14}C in the sample to the total mass of carbon. The four parameters here are k, q_0, T, and q_T. The fraction $q(t)$ is very small; to avoid working with small numbers it is convenient to rescale $q(t)$ by dividing it by its current value q_0. In dating problems the time span may be very large (thousands of years). Time can be rescaled (and nondimensionalized) by dividing by the half-life of ^{14}C, $\tau = 5568$ years. Denote the rescaled time by the dimensionless time variable $s = t/\tau$. In terms of the rescaled state variable, $y = q/q_0$, and of the rescaled time s, we have

$$y(s) = \frac{q(t)}{q_0} = \frac{q(\tau s)}{q_0} \tag{2}$$

From (1), (2), and the Chain Rule we have (since $dt/ds = \tau$)

$$\frac{dy(s)}{ds} = \frac{1}{q_0} \frac{dq(t)}{dt} \frac{dt}{ds} = \frac{\tau}{q_0} \frac{dq(t)}{dt}$$

$$= \frac{\tau}{q_0}(-kq(t)) = \frac{\tau}{q_0}(-kq_0 y(s)) = -k\tau y(s) = -(\ln 2)y(s)$$

where we have made use of the half-life relationship $k\tau = \ln 2$. After setting $S = T/\tau$, IVP (1) becomes

$$\frac{dy(s)}{ds} = -(\ln 2)\, y(s), \quad y(0) = 1, \quad S \le s \le 0 \tag{3}$$

Observe that by rescaling the state variable and time we have reduced the number of parameters from four to two, S and $y_S \, (= q_T/q_0)$. The radioactive dating problem now reduces to finding the value $S < 0$ for which $y(S) = q_T/q_0$. Observe also that by rescaling time by dividing by the half-life, the time scale has been reduced from thousands of years (possibly) to a dimensionless time range (usually) of 4 or 5.

Soft Spring (Section 3.1)

The IVP

$$m\frac{d^2 y(t)}{dt^2} + c\frac{dy(t)}{dt} + ky(t) - jy^3(t) = A_0 \cos \omega t, \quad y(0) = 0, \quad y'(0) = 0 \tag{4}$$

models the changing displacement $y(t)$ from equilibrium of a body of mass m attached to a "soft" spring whose action on the mass is modeled by the spring force terms $ky(t) - jy^3(t)$ and which is subject to the external driving force $A_0 \cos \omega t$. The term $c \cdot dy(t)/dt$ represents a frictional damping force. The variables y and t have respective units of length and time. The parameters m, c, k, j, A_0, and ω also have physical units. By rescaling y and t we can simultaneously introduce dimensionless variables and parameters and reduce the number of parameters from six to three. Let $u = y/A$ and $s = t/B$, where the scale factors A and B are to be determined. By the Chain Rule we have

$$\frac{dy(t)}{dt} = A\frac{du}{ds}\frac{ds}{dt} = \frac{A}{B}\frac{du}{ds}$$

$$\frac{d^2 y(t)}{dt^2} = \frac{d}{dt}\left(\frac{dy(t)}{dt}\right) = \frac{d}{ds}\left(\frac{A}{B}\frac{du}{ds}\right)\frac{ds}{dt} = \frac{A}{B^2}\frac{d^2 u}{ds^2}$$

Hence, the differential equation in IVP (4) transforms to

$$m\frac{A}{B^2}\frac{d^2 u}{ds^2} + c\frac{A}{B}\frac{du}{ds} + kAu - jA^3u^3 = A_0 \cos(\omega Bs)$$

which is normalized by dividing by mA/B^2 to yield the IVP

$$\frac{d^2 u}{ds^2} + \frac{c}{m}B\frac{du}{ds} + \frac{k}{m}B^2 u - \frac{j}{m}(AB)^2 u^3 = \frac{A_0}{m}\frac{B^2}{A}\cos(\omega Bs), \quad u(0) = u'(0) = 0 \tag{5}$$

Recalling that the scale factors A and B are yet to be determined, we decide which effects we want to focus on in a computer study of the solution of IVP (4). For example, we may want to see how the frictional force, the soft spring force, and the frequency of the driving force affect the solution. In that case, the coefficients of the terms involving du/ds (friction), $-u^3$ (soft spring), and ω (the frequency) will be the new parameters, while A and B will be chosen to set the remaining coefficients at some convenient value, say 1. Thus, choose A and B so that

$$\frac{k}{m}B^2 = 1, \quad \frac{A_0}{m}\frac{B^2}{A} = 1$$

TABLE C.7.2 Dimensions of Soft Spring Parameters

Parameter	Dimension		Parameter	Dimension
y	L		t	T
dy/dt	L/T		d^2y/dt^2	L/T^2
m	M		c	M/T
k	M/T^2		j	$M/(LT)^2$
A_0	ML/T^2		ω	$1/T$
$A = A_0/k$	L		$B = \sqrt{m/k}$	T

that is,

$$B = \sqrt{\frac{m}{k}}, \quad A = \frac{A_0}{k}$$

Renaming the other three coefficients as $c_1 = cB/m$, $j_1 = j(AB)^2/m$, $\omega_1 = \omega B$, we have the IVP

$$\frac{d^2u}{ds^2} + c_1\frac{du}{ds} + u - j_1u^3 = \cos(\omega_1 s), \quad u(0) = 0, \quad u'(0) = 0 \qquad (6)$$

with three parameters c_1, j_1, and ω_1 (instead of the original six).

In fact, u, s, c_1, j_1, and ω_1 are all dimensionless quantities. To show this, we begin by identifying the physical dimensions of the parameters and variables in (4), using M for mass, L for length, and T for time dimensions. All the summands in the ODE of IVP (4) must have the force dimensions ML/T^2 of the first term on the left. This establishes the dimensions of the parameters and thus of A and B. The appropriate dimensions for the parameters and variables are tabulated in Table C.7.2.

From Table C.7.2, it is straightforward to show the nondimensionality of the variables and parameters of IVP (6). For example, $u = y/A$ has the dimension of y divided by that of A, in other words L/L, which is dimensionless, and $c_1 = cB/m$ has the dimensions of c times the dimension of B divided by that of m (i.e., $(M/T)(T/M)$), which is again dimensionless.

Thus a computer study of the effect of the damping force, the soft spring force, and the driving frequency on solution behavior should use IVP (6), varying the dimensionless parameters c_1, j_1, and ω_1 from some standard values and computing the dimensionless displacement u and velocity du/ds over some dimensionless time interval. The results of such a study then apply to the behavior of the solutions of any damped, driven soft spring equation in any units.

The Autocatalator (Section 6.1)

The IVP for the changing concentrations $x_1, \ldots, x_4$ of four chemical species in an autocatalytic reaction are (Example 6.1.5)

$$\begin{aligned}
x_1' &= -k_1x_1, & x_1(0) &> 0 \\
x_2' &= k_1x_1 - k_2x_2 - k_3x_2x_3^2, & x_2(0) &= 0 \\
x_3' &= k_2x_2 - k_4x_3 + k_3x_2x_3^2, & x_3(0) &= 0 \\
x_4' &= k_4x_3, & x_4(0) &= 0
\end{aligned} \qquad (7)$$

where $k_1, \ldots, k_4$ are positive rate constants. Observe that the total concentration remains fixed at the value $x_1(0)$ since $(x_1 + x_2 + x_3 + x_4)' = 0$. The units of $x_1, \ldots, x_4$

are concentrations C (e.g., moles/liter), and the unit T of time is (usually) seconds. Thus, the five parameters of the problem have units:

$$k_1, \ k_2, \ k_4 \text{ in } T^{-1} \tag{8}$$

$$k_3 \text{ in } T^{-1}C^{-2} \tag{9}$$

$$x_1(0) \text{ in } C \tag{10}$$

Rescaling the four concentrations and time will simultaneously nondimensionalize the concentrations and time and reduce the number of parameters from five to any three of our choice.

Let

$$y_i = x_i/a_i, \ i = 1, \ 2, \ 3, \ 4; \quad s = t/a_5 \tag{11}$$

where the scale constants $a_1, \ldots, a_4$ have the dimension C of concentration and a_5 has the dimension T of time. Changing the variables of IVP (7) accordingly, we have

$$\frac{dy_1}{ds} = -a_5 k_1 y_1, \qquad\qquad\qquad y_1(0) = x_1(0)/a_1$$

$$\frac{dy_2}{ds} = \frac{a_1 a_5}{a_2} k_1 y_1 - a_5 k_2 y_2 - a_3^2 a_5 k_3 y_2 y_3^2, \quad y_2(0) = 0$$

$$\frac{dy_3}{ds} = \frac{a_2 a_5}{a_3} k_2 y_2 - a_5 k_4 y_3 + a_2 a_3 a_5 k_3 y_2 y_3^2, \quad y_3(0) = 0$$

$$\frac{dy_4}{ds} = a_5 k_4 y_3, \qquad\qquad\qquad y_4(0) = 0$$

Setting

$$a_1 = a_2 = a_3 = a_4$$

we have that $d(y_1 + y_2 + y_3 + y_4)/ds = 0$, and the total scaled concentration, $y_1(s) + y_2(s) + y_3(s) + y_4(s)$ remains fixed at $y_1(0) = x_1(0)/a_1$. Suppose that we want to study the sensitivity of the system to changes in the initial concentration of the first chemical species, as well as to changes in the first and second rate constants. Then we should take as the parameters

$$\alpha = y_1(0) = x_1(0)/a_5, \quad \beta = a_5 k_1, \quad \gamma = a_5 k_2 \tag{12}$$

For convenience set $a_5 k_4$ and $a_3^2 a_5 k_3$ equal to 1; then we have that

$$a_5 = 1/k_4, \quad a_1 = a_2 = a_3 = a_4 = (k_4/k_3)^{1/2}$$

Observe that a_5 has the units T of time, while a_1, a_2, a_3, and a_4 have the units of $(k_4/k_3)^{1/2}$, i.e., of $(T^{-1}/T^{-1}C^{-2})^{1/2} = C$. Thus, from (11), the scaled variables s, y_1, y_2, y_3, y_4 are dimensionless.

In the dimensionless variables y_i and s and the new parameters α, β, γ, the IVP becomes

$$\frac{dy_1}{ds} = -\beta y_1, \qquad\qquad y_1(0) = \alpha$$

$$\frac{dy_2}{ds} = \beta y_1 - \gamma y_2 - y_2 y_3^2, \quad y_2(0) = 0$$

$$\frac{dy_3}{ds} = \gamma y_2 - y_3 + y_2 y_3^2, \quad y_3(0) = 0$$

$$\frac{dy_4}{ds} = y_3, \qquad\qquad y_4(0) = 0$$

There is nothing unique about this rescaling or choice of parameters. The particular factors a_i used here were chosen to facilitate sensitivity studies in $x_1(0)$, k_1, and k_2

(rescaled as $y_1(0)$, β, and γ). It would be equally straightforward to choose, say, k_1, k_3, and k_4 for a sensitivity study.

APPENDIX D

Atlas

The figures in this collection were assembled to provide visual evidence for the properties of solutions of differential equations. They are used in a variety of ways. The plates are often referenced to back up claims made in the text and to give hints on how to attack some homework problems.

Contents

Atlas Plates

Autocatalator A–C	Chemical Oscillations
Bifurcation A–C	Pitchfork Bifurcation
Bifurcation D–F	Transcritical Bifurcation
Bifurcation G–I	Homoclinic Bifurcation
Bifurcation J–L	Hopf Bifurcation
Conservative System A–C	Integrals and Orbits
Conservative System D–F	Rotational Stability
Duffing A–C	Nonlinear Oscillations and Chaos
Duffing D–F	Poincaré Time Sections
First Order A–C	Solution Curves and Integral Curves
First Order D–F	Sensitivity to Changes in Data
First Order G–I	Integral Curves and Orbits
First Order J–L	Unusual Behavior
Limit Cycle A-C	Van der Pol Limit Cycle
Pendulum A–C	Effect of Damping
Pendulum D–F	Double Pendulum, Sensitivity
Picard A–C	Convergence of Iterates
Planar Portrait A–C	Orbits and Equilibria
Poincaré-Bendixson A–C	Approach to Cycle and Cycle-Graph
Poincaré-Bendixson D–F	Creating a Cycle-Graph
Predator-Prey A–C	Harvested Species
Strange Attractor A–C	Rössler Attractor
Strange Attractor D–F	Scroll Attractor

Autocatalator: On/Off Oscillations

In an autocatalytic reaction chemicals X_1, X_2, X_3, and X_4 react as follows:

$$X_1 \overset{k_1}{\to} X_2, \quad X_2 \overset{k_2}{\to} X_3, \quad X_2 + 2X_3 \overset{k_3}{\to} 3X_3, \quad X_3 \overset{k_4}{\to} X_4$$

where the k_i are rate constants. The rate equations for the concentrations x_i are

$$x_1' = -k_1 x_1, \quad x_1(0) = \alpha; \quad x_2' = k_1 x_1 - k_2 x_2 - k_3 x_2 x_3^2, \quad x_2(0) = 0$$
$$x_3' = k_2 x_2 - k_4 x_3 + k_3 x_2 x_3^2, \quad x_3(0) = 0; \quad x_4' = k_4 x_3, \quad x_4(0) = 0$$

Reference: Examples 6.1.5 and 6.1.6; Problem 7 of Section 9.3

AUTOCATALATOR A   Set $k_1 = 0.002$, $k_2 = 0.08$, $k_3 = k_4 = 1$, $\alpha = 500$. This plate shows the projection of the orbit into the three-dimensional space of x_2, x_3, and x_4 for $0 \leq t \leq 800$.

AUTOCATALATOR B   The oscillations in the concentrations of the intermediates appear to be due to a Hopf bifurcation. The oscillation in Atlas PLATE A can be turned off by increasing k_2. This plate shows what happens when k_2 is increased from 0.08 (solid curve) to 0.10 (long dashes) and finally to 0.14 (the short-dashed curve with no oscillation).

AUTOCATALATOR C   Now let $\alpha = 5000$, $k_2 = 0.01$ and let $k_1 = 0.001$, 0.005, and 0.010. There is a long sequence of oscillations in x_2 if $k_1 = 0.001$ (solid curve), a shorter sequence if $k_1 = 0.005$ (long dashes), and a short-lived burst of oscillations if $k_1 = 0.010$.

AUTOCATALATOR A   ## Projected Orbit in $X_2 X_4 X_3$-Space

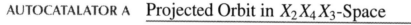

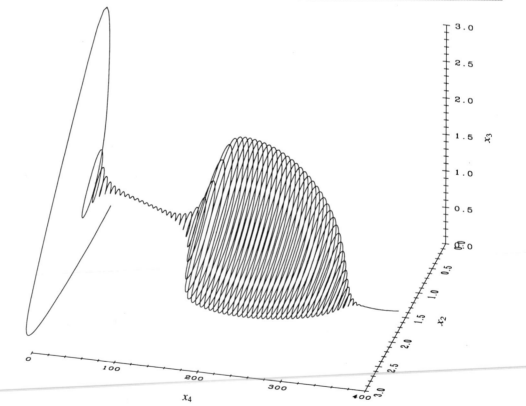

AUTOCATALATOR B Turning Off Oscillations

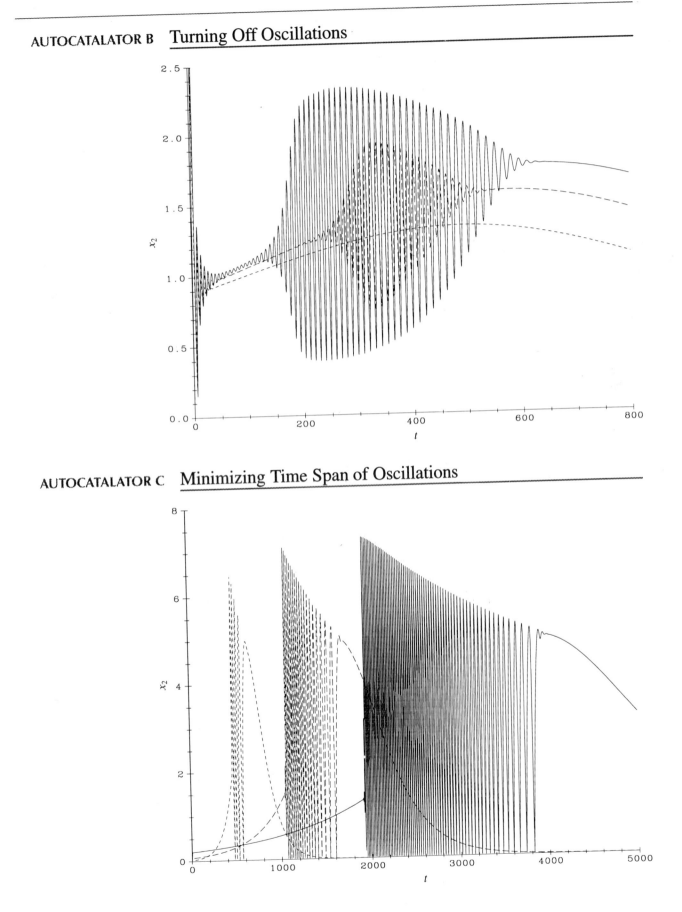

AUTOCATALATOR C Minimizing Time Span of Oscillations

Bifurcation of a Planar System: Pitchfork Bifurcation

In a bifurcation a distinctive feature of the portrait of the orbits of a system suddenly changes into something quite different as a system parameter passes through a critical value. In a pitchfork bifurcation an asymptotically stable equilibrium point splits into three equilibrium points, two stable and one unstable.

Reference: Example 9.3.3; Problem 1, Section 8.1; Problem 3, Section 8.2; Problem 9, Section 8.4

BIFURCATION A   The nonlinear system

$$x' = \varepsilon x - x^3$$
$$y' = -y$$

has an asymptotically stable node at the origin if the parameter ε is negative; $\varepsilon = -1$ in the figure.

BIFURCATION B   As the parameter ε passes through the value 0, the equilibrium point at the origin suddenly destabilizes into a saddle as it ejects two asymptotically stable nodes at $(\pm\sqrt{\varepsilon}, 0)$, $\varepsilon > 0$; $\varepsilon = +1$ in the figure.

BIFURCATION C   In the pitchfork bifurcation diagram, the x-coordinates of the equilibrium points are plotted as functions of ε. Solid lines denote stable points, dashed lines unstable points. This could also be called the pitchfork "trifurcation" since the path splits into three branches, not two.

BIFURCATION A   Before Pitchfork Bifurcation

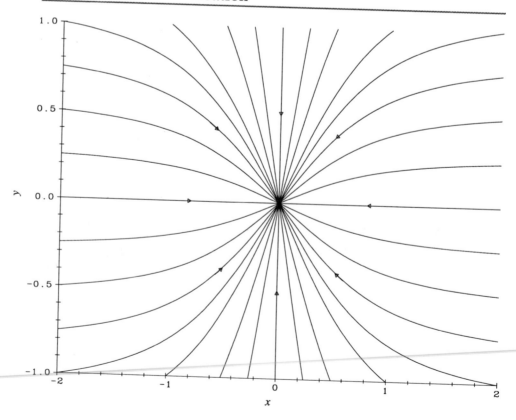

BIFURCATION B   After Pitchfork Bifurcation

BIFURCATION C   Pitchfork Bifurcation Diagram

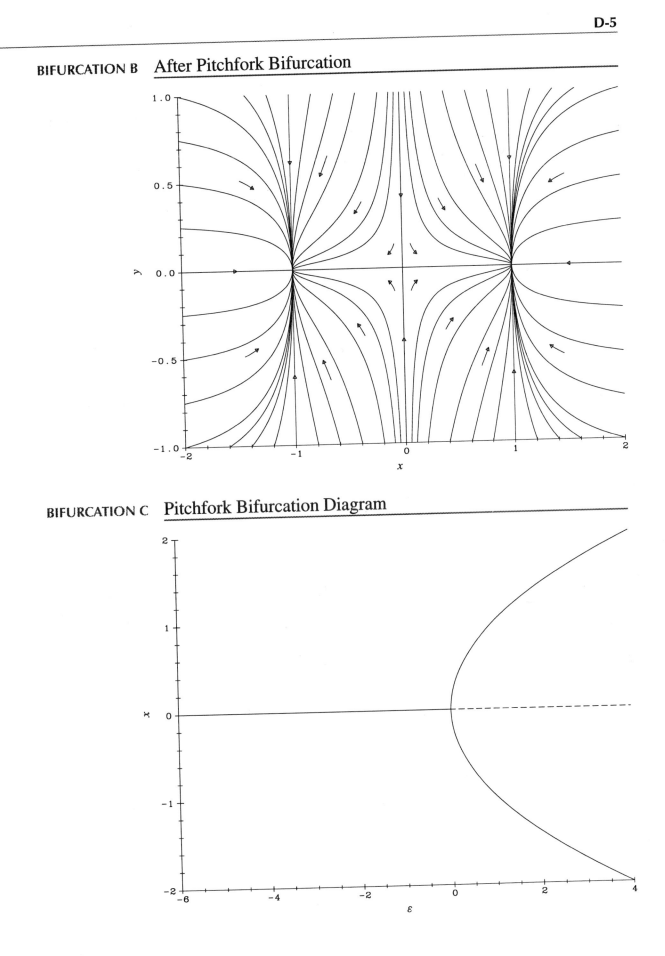

Bifurcation of a Planar System: Transcritical Bifurcation

As a system parameter falls to a critical value in a transcritical bifurcation, an asymptotically stable node and a saddle point merge into a saddle-node that attracts on one side and repels on the other. As the parameter moves below the critical value, the saddle-node splits apart into a stable node and a new saddle point.
Reference: Example 9.3.2; Problem 1, Section 8.1; Problem 3, Section 8.2

BIFURCATION D The nonlinear system

$$x' = \varepsilon x + x^2$$
$$y' = -y$$

for $\varepsilon > 0$ has an asymptotically stable node at the point $(-\varepsilon, 0)$ and a saddle at the origin; $\varepsilon = +1$ in the figure.

BIFURCATION E At the critical value of $\varepsilon = 0$, the node and the saddle form a hybrid saddle-node that attracts orbits from the left half-plane, but sends orbits on the right out to infinity as time increases.

BIFURCATION F As ε decreases through 0, the stable node moves to the right until it collides with the saddle at the origin. The node sticks to the origin for all negative ε, but the saddle moves off to the right as ε decreases; $\varepsilon = -1$ in the figure.

BIFURCATION D ## Before Transcritical Bifurcation

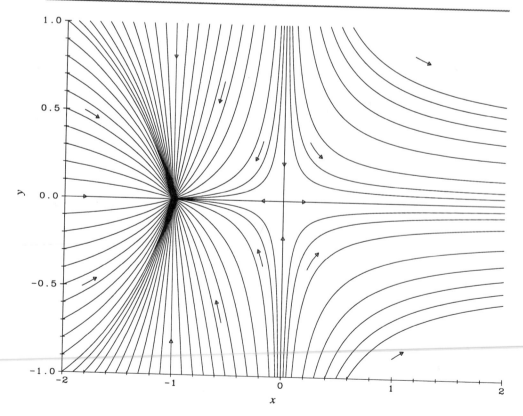

BIFURCATION E Saddle-Node

BIFURCATION F After Transcritical Bifurcation

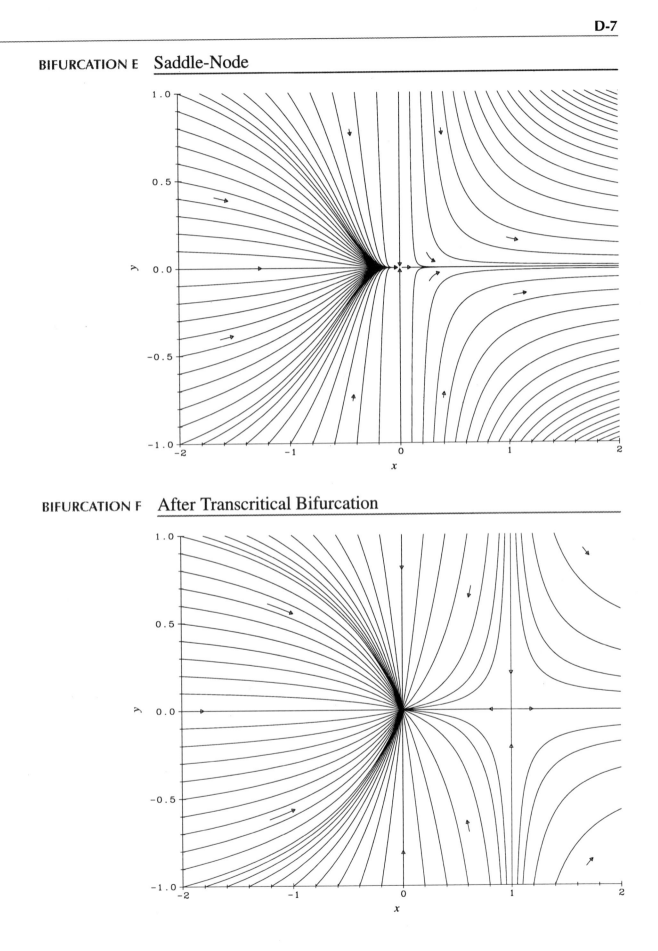

Bifurcation of a Planar System: Homoclinic Bifurcation

A nonlinear system with a saddle point and a nearby unstable focus may undergo a homoclinic bifurcation as a system parameter passes through a critical value. At that value an orbit exits the saddle point as time increases from $-\infty$, turns once around the focus, and reenters the saddle as time increases to $+\infty$; this is the homoclinic orbit. As the parameter goes beyond the critical value, the homoclinic connection breaks into two arms that "enclose" a newly-born limit cycle.

Reference: Example 9.3.7 and Problem 6 of Section 9.3; Problem 1, Section 8.1; Problem 3, Section 8.2; Problem 9, Section 8.4

BIFURCATION G   The nonlinear system (with $\varepsilon > 0$)

$$x' = \varepsilon x - y + x^2, \qquad y' = x + \varepsilon y + x^2$$

has a saddle point for $x = (-1 - \varepsilon^2)/(1 + \varepsilon)$, $y = (1 - \varepsilon)(1 + \varepsilon^2)/(1 + \varepsilon)^2$, while the origin is an unstable focus. One orbit spirals out from the focus and enters the saddle as time tends to $+\infty$; $\varepsilon = 0.2$ in the figure.

BIFURCATION H   At the critical value $\varepsilon_0 \approx 0.10895$ there is a homoclinic loop (cycle-graph) that encloses the unstable focus; the saddle point is at the vertex of the loop.

BIFURCATION I   As ε decreases through ε_0, an attracting limit cycle suddenly appears and the homoclinic loop breaks apart into two orbits, one of which is attracted to the new limit cycle; $\varepsilon = 0.07$ in the figure.

BIFURCATION G   Before Homoclinic Bifurcation

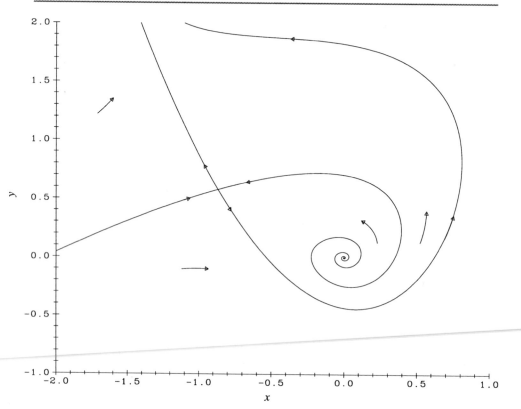

BIFURCATION H Homoclinic Orbit at Bifurcation

BIFURCATION I Limit Cycle after Homoclinic Bifurcation

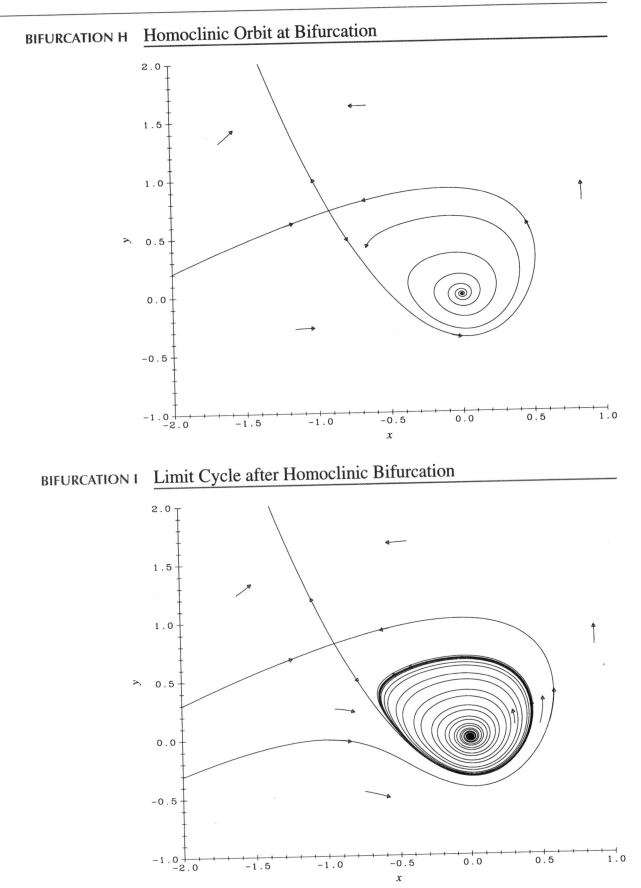

Bifurcation of a Planar System: Hopf Bifurcation

In a Hopf bifurcation an asymptotically stable focus destabilizes and spawns an attracting limit cycle as a system parameter passes through a critical value.
Reference: Problem 5 of Section 9.3

BIFURCATION J   The planar system

$$x' = \varepsilon(x - 5\varepsilon) + (y - 5\varepsilon) - (x - 5\varepsilon)[(x - 5\varepsilon)^2 + (y - 5\varepsilon)^2]$$

$$y' = -(x - 5\varepsilon) + \varepsilon(y - 5\varepsilon) - (y - 5\varepsilon)[(x - 5\varepsilon)^2 + (y - 5\varepsilon)^2]$$

has an equilibrium point at $(5\varepsilon, 5\varepsilon)$. The point is an asymptotically stable focus if $\varepsilon < 0$; $\varepsilon = -0.1$ in the figure.

BIFURCATION K   At the critical value $\varepsilon = 0$ the equilibrium point is still an asymptotically stable focus, but the spiraling approach of the orbit is very slow near the origin; the coils are so close together that it is hard to distinguish one from another.

BIFURCATION L   As ε increases through the critical value 0, the Hopf bifurcation occurs and an attracting limit cycle $r = \sqrt{\varepsilon}$ appears. The figure shows limit cycles for $\varepsilon = 0.1, 0.4, 0.9$, and outward spiraling orbits inside each cycle.

BIFURCATION J   Attracting Focus before Hopf Bifurcation

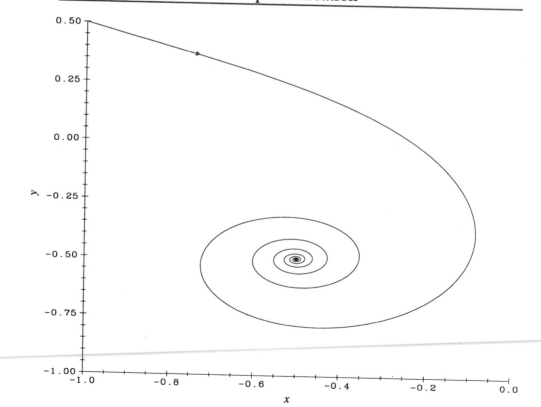

BIFURCATION K   Attracting Focus at Hopf Bifurcation

BIFURCATION L   Three Hopf Limit Cycles

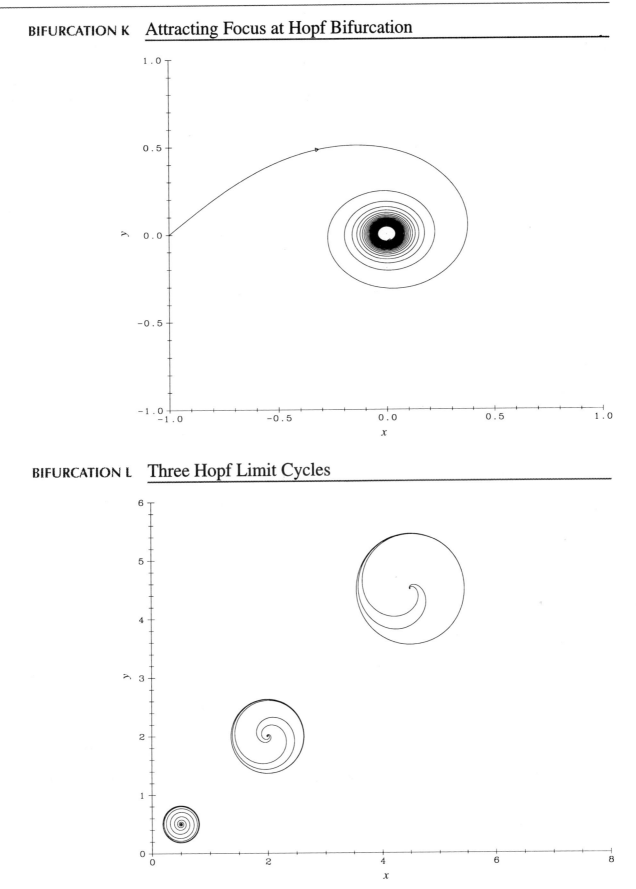

Conservative versus Nonconservative Systems

Each orbit of a conservative system in $\mathbf{R}^3$ with integral $K(x, y, z)$ "lives" for all time on an integral surface $K(x, y, z) = \text{constant}$. The system, $x' = xz - 10yz$, $y' = 10xz + yz$, $z' = -x^2 - y^2$, is conservative with the integral $K = x^2 + y^2 + z^2$, and each orbit stays on a sphere $x^2 + y^2 + z^2 = \text{constant}$. Each sphere in PLATES A, B has an unstable equilibrium point at its north pole $(0, 0, z_0)$, $z_0 > 0$, and a neutrally stable equilibrium point at its south pole $(0, 0, z_0)$, $z_0 < 0$. Nonconstant orbits on a sphere spiral downward away from the north pole toward the south pole.

Reference: Section 8.3, Problem 9 and Example 8.3.3

CONSERVATIVE SYSTEM A   The orbit in the lower right graph moves down the sphere $x^2 + y^2 + z^2 = 4$ through the point $(2/\sqrt{3}, 2/\sqrt{3}, 2/\sqrt{3})$. The three component graphs track this downward spiraling motion. Observe the reversal in the y motion when $z = 0$.

CONSERVATIVE SYSTEM B   The orbit of PLATE A is projected onto the xy, xz, yz state planes and displayed in PLATE B along with the original orbit.

CONSERVATIVE SYSTEM C   If the rate equation for z is changed slightly to $z' = -x^2 - y^2 + 0.1\,\mathrm{abs}(z)$, the new system is no longer conservative. The orbit shown leaves the origin as t increases from $-\infty$, rises up to the "arctic" region near the north pole, spirals outward until it reaches $(2/\sqrt{3}, 2/\sqrt{3}, 2/\sqrt{3})$, and then begins to spiral inward and downward. Near the south pole it turns back and rises toward the origin from below.

CONSERVATIVE SYSTEM A   ## Orbit on Integral Surface, Component Graphs

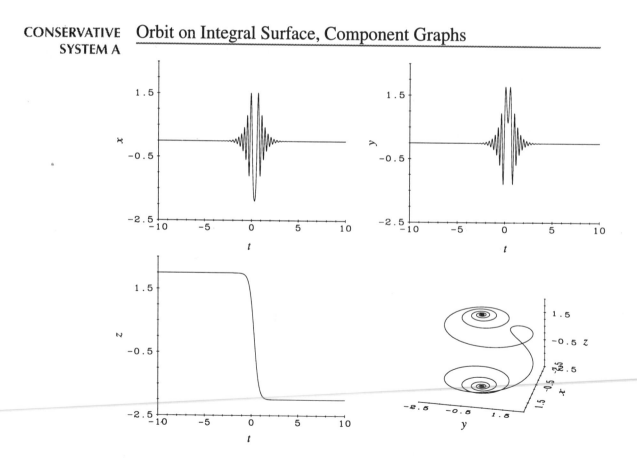

CONSERVATIVE
SYSTEM B

Orbit on Integral Surface, Projections

CONSERVATIVE
SYSTEM C

Nonconservative System: Orbit and Component Graphs

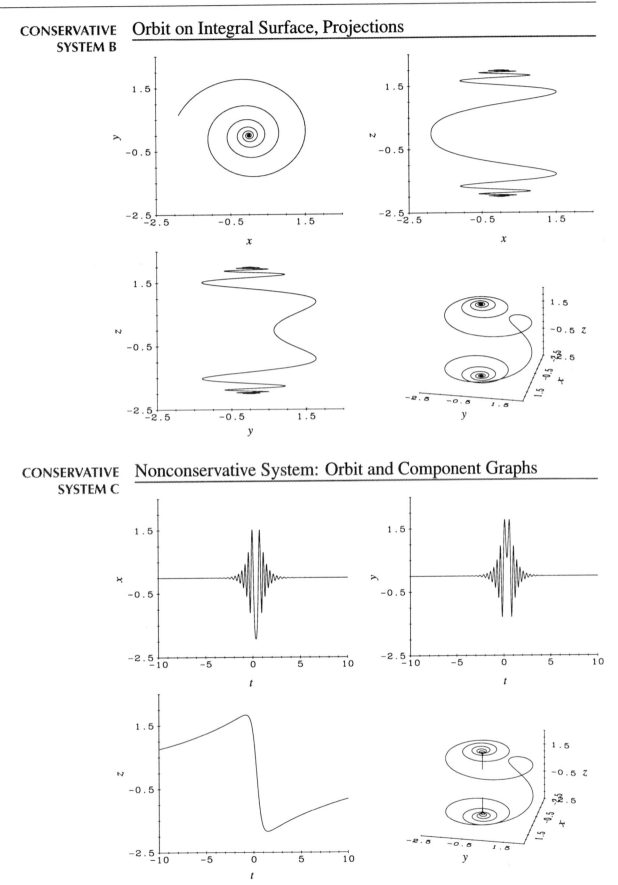

Stability of Steady Rotations of a Rigid Body

These three plates give an insightful view of the behavior of orbits of a conservative first-order system. The system models the rotational motion of a rigid body with distinct principal inertias about a principal axis when no external torques act on the body. The state variables are the components of the angular velocity vector with respect to the principal axes of the body. The model system of ODEs is taken to be $x' = -yz$, $y' = xz/3$, $z' = xy$.

Reference: Section 8.3, Problem 8

CONSERVATIVE SYSTEM D

The ellipsoid $4x^2 + 3y^2 + 3z^2 = 10$ is an integral surface for the system; the ellipsoid's axes are the principal axes of the body and are lined up with the coordinate axes. The vertices of the ellipsoid are equilibrium points of the system and correspond to steady pure rotations of the body about that axis. From the sample orbits on the ellipsoid notice that steady rotations about the axes of largest and smallest inertias are stable whereas steady rotation about the other principal axis is unstable.

CONSERVATIVE SYSTEM E

A view of the unstable orbits of PLATE D. Notice how even small perturbations away from the equilibrium point carry the state of the system all the way over to a neighborhood of the "oppositely directed" equilibrium point.

CONSERVATIVE SYSTEM F

Component graphs of two of the four orbits shown in PLATE E are plotted. Notice that initially the body is undergoing an almost pure rotation about the x-axis, but after a while destabilizes and rotates almost in the opposite direction about the x-axis. A little later the system returns to its initial state.

CONSERVATIVE SYSTEM D

Orbits on an Integral Surface

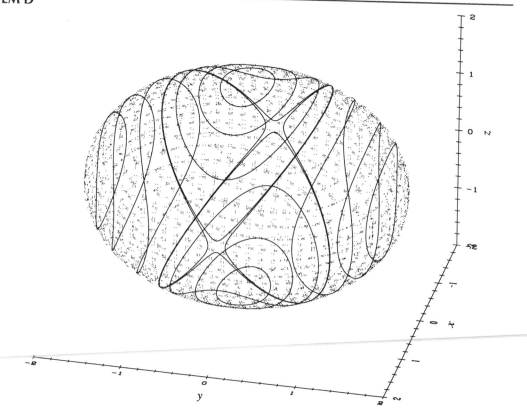

CONSERVATIVE SYSTEM E Orbits of Unstable Rotations

CONSERVATIVE SYSTEM F Two Orbits and Their Component Graphs

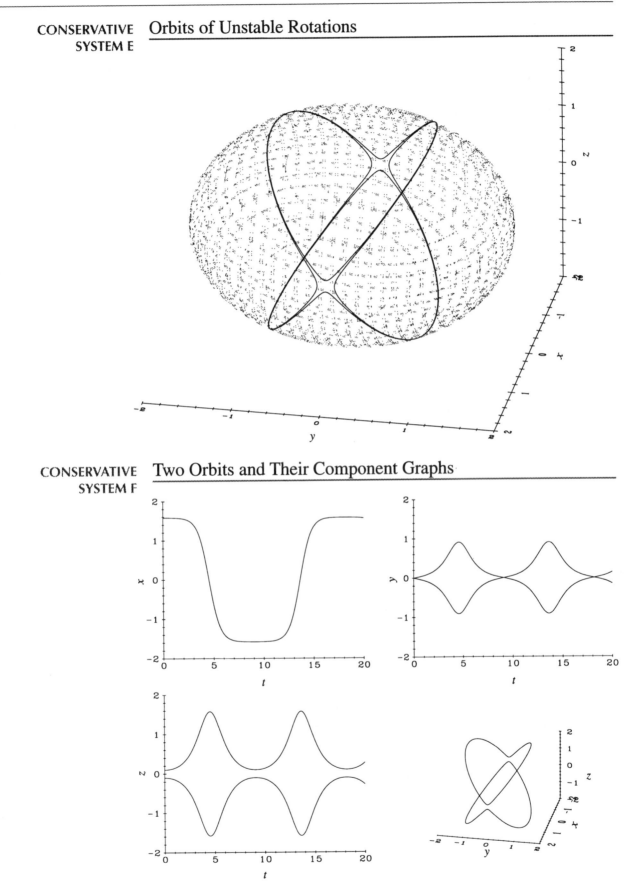

Chaos and the Duffing Equation

If a, b, c, A, ω are nonnegative constants, then Duffing's system

$$x' = y, \qquad y' = ax - bx^3 - cy + A\cos\omega t$$

models the behavior of a damped ($c \neq 0$), nonlinear ($b \neq 0$), periodically driven ($A \neq 0$, $\omega \neq 0$) oscillator. Under certain circumstances, orbits $x = x(t)$, $y = y(t)$ appear to wander chaotically. We have set $a = b = 1$.

Reference: Section 9.4, Problem 5; Section 8.1, Problem 1

DUFFING A In the absence of a driving force (i.e., set $A = 0$), orbits of Duffing's system with damping constant $c = 0.15$ tend to the asymptotically stable equilibrium points $x = \pm 1$, $y = 0$ or (exceptionally) to the unstable, saddle point at the origin. The orbits shown start at $x_0 = 1$, $y_0 = 0.82, 0.84, \ldots, 1$.

DUFFING B Increase the damping constant c to 0.22 and drive the system with the term $0.3\cos t$, and we observe a periodic solution. The initial data are $x_0 = 0.5$, $y_0 = 0.4$, the solve-time is 200 time units, but the graphs are only plotted for $150 \leq t \leq 200$; in this way initial transient wandering is suppressed from the graphs.

DUFFING C Set the damping constant $c = 0.15$ and drive the system with the function $0.3\cos t$. If $x_0 = -0.2$, $y_0 = 1.4$, the solution appears to settle down to a periodic oscillation (outer closed curve), but if we take $x_0 = -1$, $y_0 = 1$, the orbit appears to be chaotic. Both orbits are solved for $0 \leq t \leq 1000$, but plotted for $500 \leq t \leq 1000$.

DUFFING A ## Sensitivity to Initial Data

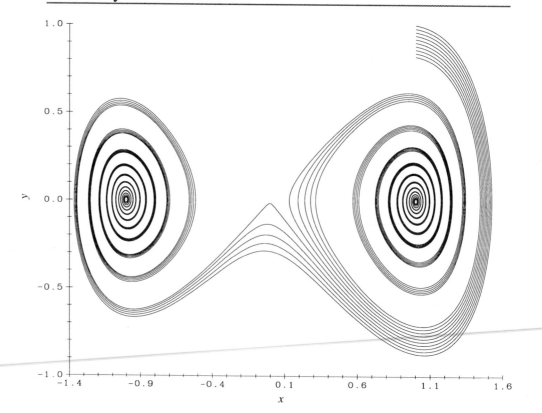

DUFFING B   Periodic Orbit

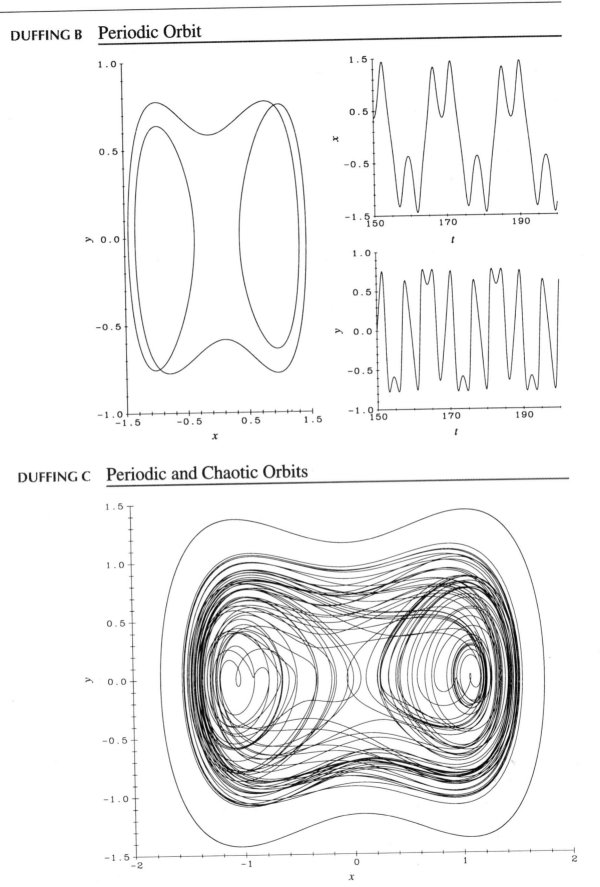

DUFFING C   Periodic and Chaotic Orbits

Patterns Within Chaos: Driven Duffing Equation

Some solutions of the Duffing equation

$$x' = y, \qquad y' = x - x^3 - 0.15y + 0.3\cos t$$

wander chaotically, some are periodic. These plates show patterns within the chaos.
Reference: Section 9.4, Problem 5

DUFFING D   The apparent tangle of a chaotic orbit can be resolved by "photographing" the moving orbital point in Duffing C only at multiples of the period 2π of the driving force (a Poincaré time section). The striking figure of this plate is the result ($0 \le t \le 20000$). The "fly specks" near the point $(0.6, 1.4)$ are the Poincaré time section of the periodic orbit.

DUFFING E   We introduce toroidal coordinates in which time "repeats" every 2π units

$$X = y(t), \qquad Y = (R + rx(t))\cos t, \qquad Z = (R + rx(t))\sin t$$

where R and r are the radii of the equatorial circle and the cross-sectional circle of a torus. We set $R = 4$, $r = 1$, so that the arcs of the solution curve do not interfere with one another. Still, there is a tangle; the curve in this "Duffing donut" is plotted for $0 \le t \le 500$.

DUFFING F   Extend the time interval to $0 \le t \le 10,000$ and slice Duffing's donut by taking Poincaré time sections with $\Delta t = 2\pi/27$ (one slice is magnified and displayed in PLATE D).

DUFFING D   Poincaré Time Section of Chaotic Orbit

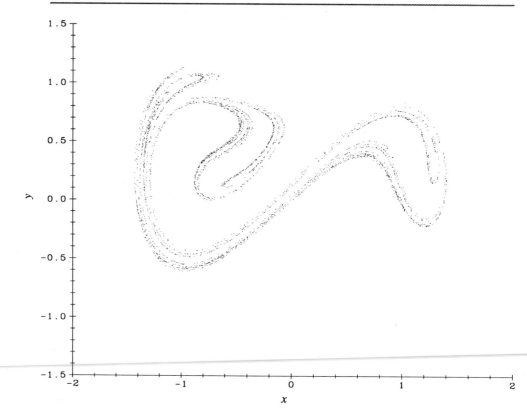

DUFFING E   Duffing's Donut

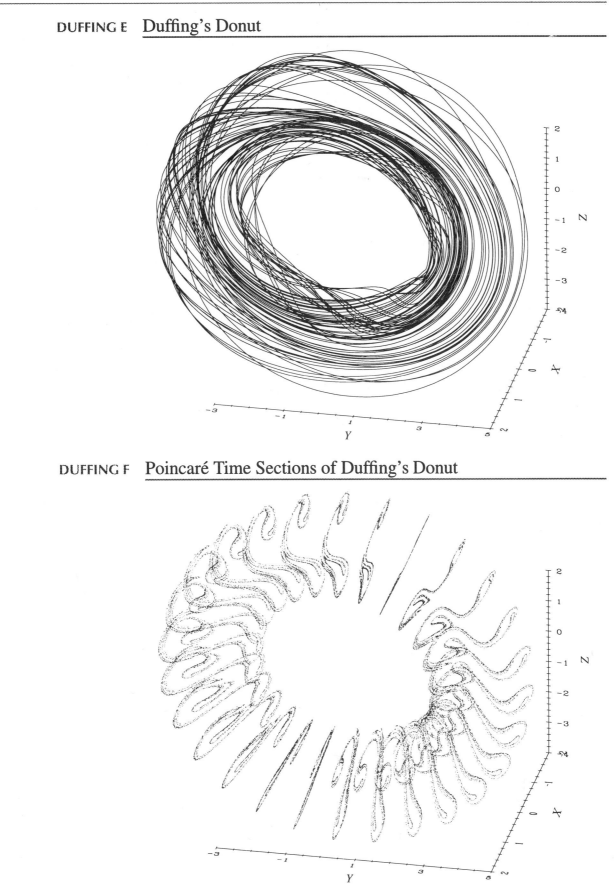

DUFFING F   Poincaré Time Sections of Duffing's Donut

Solution and Integral Curves for First-Order ODEs

Solution curves of first-order ODEs exhibit some remarkable and unusual behavior not easily detected from solution formulas. As illustrated in PLATE B, solution curves always "follow" direction field lines, but as PLATES A and C show, the solution curves can still behave in odd ways.

FIRST ORDER A   The solution curves of the linear ODE of PLATE A seem to merge into a single rapidly oscillating curve as t increases. Remarkably, every solution approaches the same limiting oscillation as t goes to infinity.
Reference: Section 1.2, Problem 6, Problem 17 (for contrast); Theorem 2.2.3

FIRST ORDER B   The solution curves of the linear ODE of PLATE B "fit" the line element field (called a direction field). Notice how the direction field seems to direct the "flow" of the solution curves.
Reference: Section 1.1, after Example 1.1.4; Section 2.1, Problem 5(**a**), Theorem 2.2.3

FIRST ORDER C   The integral curves of the separable ODE displayed in PLATE C have interesting properties. Some are closed curves that surround a point where the coefficients of dx and dy simultaneously vanish. Other integral curves appear to extend indefinitely in both directions, approaching the solution curve $y = x$. Observe that some solution curves, $y = y(x)$, are only defined on a bounded x-interval.
Reference: Section 1.4, Problem 2; Theorem 2.2.3; Section 6.2, Problem 4; Problem 1, Section 8.1

FIRST ORDER A   ## Forced Oscillation

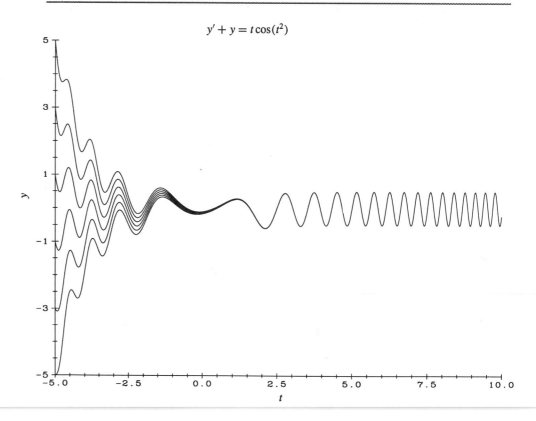

$$y' + y = t\cos(t^2)$$

FIRST ORDER B Direction Field

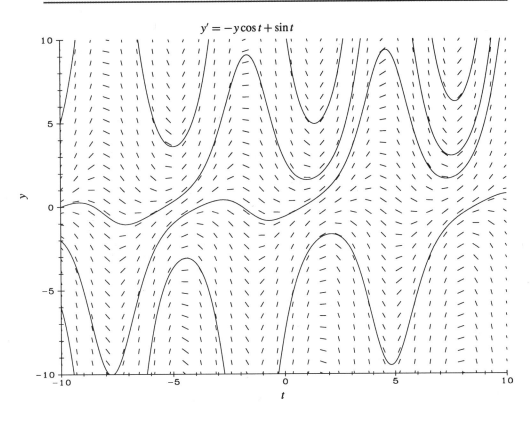

$$y' = -y\cos t + \sin t$$

FIRST ORDER C Integral Curves

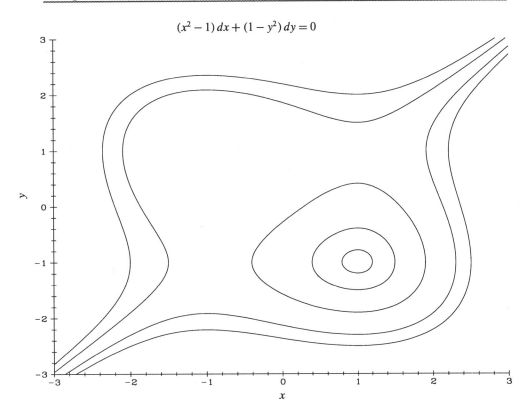

$$(x^2 - 1)\,dx + (1 - y^2)\,dy = 0$$

Sensitivity to the Data

These three plates show how solution curves of an Initial Value Problem change as the data of the problem change. That is, "sensitivity to the data" is examined.

FIRST ORDER D   The salt concentration of the input stream into a tank of brine varies sinusoidally about a mean of 0.2 pound of salt per gallon. If the amplitude of that sinusoid is varied, then the pattern of responses shown in PLATE D appears.
Reference: Section 1.2, Example 1.2.5

FIRST ORDER E   PLATE E shows the effects of seasonal harvesting on a logistic population. A continual high harvest rate soon leads to the extinction of the population (bottom curve), as does a high harvest rate over an eight-month season. However, the population survives if the high harvest rate is limited to a four-month season (upper toothed curve).
Reference: Section 1.5, Problem 10

FIRST ORDER F   PLATE F shows the sensitivity of a scaled model of the "escape velocity" IVP (5) from Section 1.8 to changes in the initial velocity. IVP (5) is scaled in Problem 7 of that section.
Reference: Section 1.8, Problem 7

FIRST ORDER D ## Sensitivity to Input

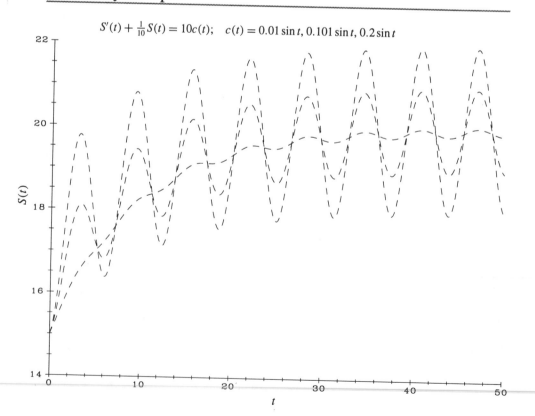

$$S'(t) + \tfrac{1}{10}S(t) = 10c(t); \quad c(t) = 0.01\sin t,\, 0.101\sin t,\, 0.2\sin t$$

FIRST ORDER E <u>Engineering Function Inputs</u>

$$P' = (1 - P/12)P - 6\,\text{sqw}(t, d, 1); \quad d = 0,\ 100/3,\ 200/3,\ 100$$

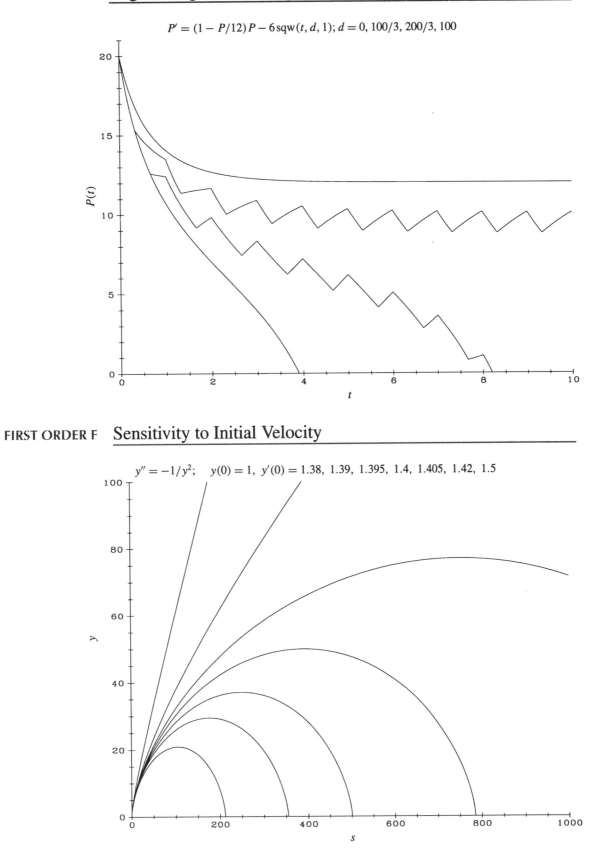

FIRST ORDER F <u>Sensitivity to Initial Velocity</u>

$$y'' = -1/y^2; \quad y(0) = 1,\ y'(0) = 1.38,\ 1.39,\ 1.395,\ 1.4,\ 1.405,\ 1.42,\ 1.5$$

Integral Curves and Orbital Portraits

The orbital portraits of Atlas Plates FIRST ORDER G, H, and I are so striking that one might expect to find some clue to their character in the ODEs themselves. Each ODE can be written in the form $M(x, y)\,dx + N(x, y)\,dy = 0$, or as the equivalent planar system $x' = N(x, y)$, $y' = -M(x, y)$.

FIRST ORDER G   The ODE for PLATE G is $dy/dx = (2xy + y^2)/(2x^2 + 2xy)$; the right side is a homogeneous function of order zero, so we can find a solution formula. The curves are plotted as orbits of the equivalent system $dx/dt = 2x^2 + 2xy$, $dy/dt = 2xy + y^2$.
Reference: Section 1.9, Problem 10(**b**)

FIRST ORDER H   The ODE $(1 + 2xy^2)\,dx + (2x^2y - 2y)\,dy = 0$ of PLATE H is exact. The integral curves are implicitly defined by the equation $x + x^2y^2 - y^2 = -9, -1, 0, 5$, but are plotted as unions of orbits of the equivalent system, $dx/dt = 2x^2y - 2y$, $dy/dt = -1 - 2xy^2$.
Reference: Section 1.9, Problem 9 and Examples 1.9.1, 1.9.2, 1.9.4

FIRST ORDER I   The "staring face" of PLATE I is formed from orbits of the system equivalent to the inexact ODE $(-0.5x + 0.2y - xy + 1.2y^2)\,dx + (-y - y^2)\,dy = 0$. Each "eye" is surrounded by a cycle that is approached asymptotically by nearby orbits.
Reference: Example 6.2.5; Example 9.1.4

FIRST ORDER G   Integral Curves

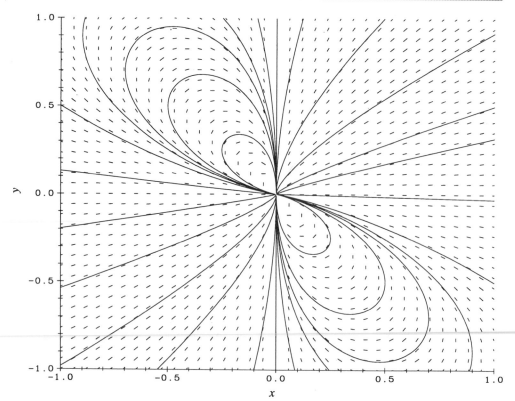

FIRST ORDER H Level Sets

FIRST ORDER I Limit Cycles

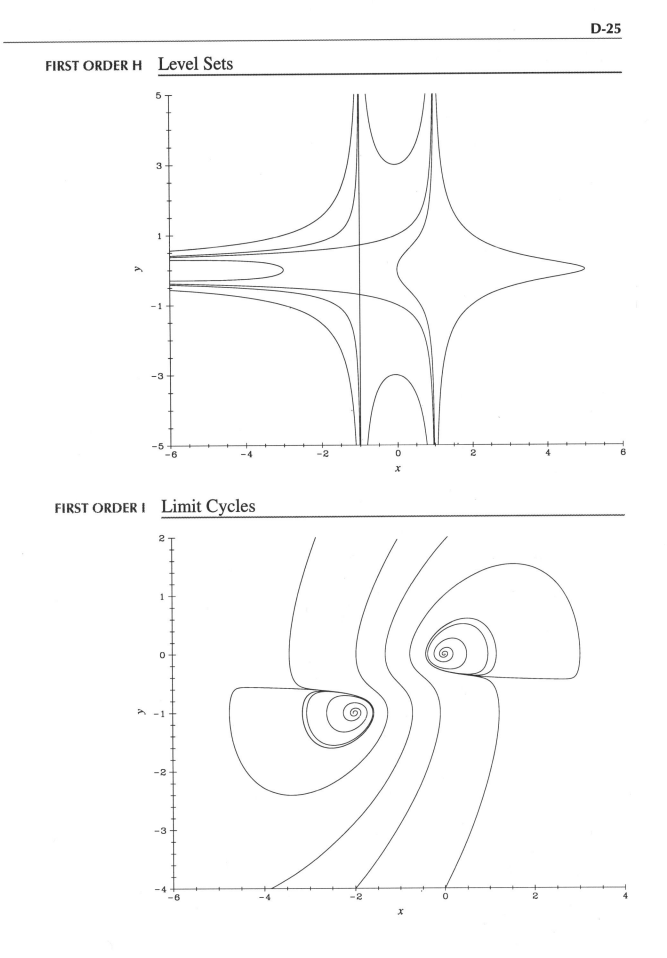

Tracking Solution Curves Numerically

Some first-order differential equations produce solutions with unusual properties that make it difficult to track solution curves numerically.

FIRST ORDER J The solution curves of the ODE $y' = 3y \sin y - t$ merge and diverge in interesting patterns. A numerical solver may not be able to render the curves accurately when solving backwards in time.
Reference: Section 2.3, Problem 5; Section 2.6, Problem 2(**e**)

FIRST ORDER K The first-order linear ODE $y' + cy = q(t)$, for constant $p \neq 0$ and a piecewise continuous driving term $q(t)$ of period T has a unique periodic solution (a forced oscillation) of the same period. This periodic response is easy to detect if $c > 0$ (as is the case in this plate) because it "attracts" all solution curves as $t \to +\infty$. We have set $c = 1$ and $q = 2 \operatorname{sqw}(t, 50, 3)$.
Reference: Section 1.2, Problem 17; Section 2.3, Problem 7; Table, inside front cover sheets

FIRST ORDER L The first-order ODE of this plate is written in the form $M(x, y)\, dx + N(x, y)\, dy = 0$, and the curves depicted are integral curves rather than solution curves of this ODE. Solution curves are arcs of integral curves with no vertical tangent. The exponentials in M and N make the rates so high that it is very hard to plot integral curves in the open areas on the right of the graph.
Reference: Section 2.6, Problem 7

FIRST ORDER J ## Streaming of Solution Curves

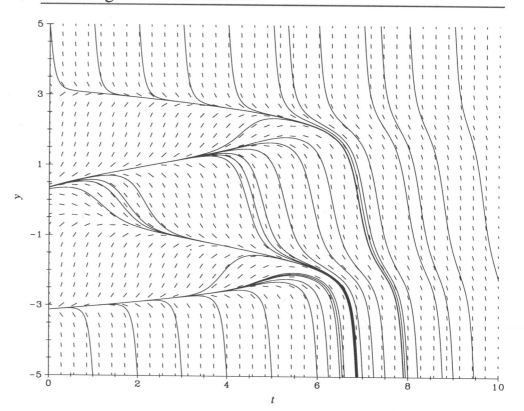

FIRST ORDER K Forced Oscillation

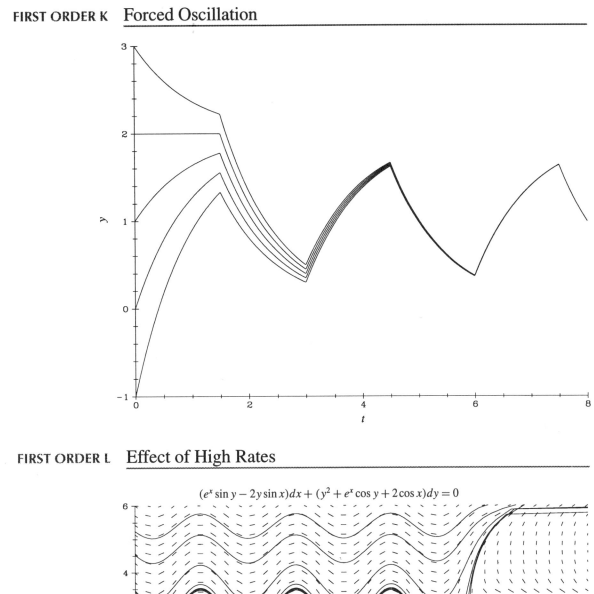

FIRST ORDER L Effect of High Rates

$$(e^x \sin y - 2y \sin x)dx + (y^2 + e^x \cos y + 2\cos x)dy = 0$$

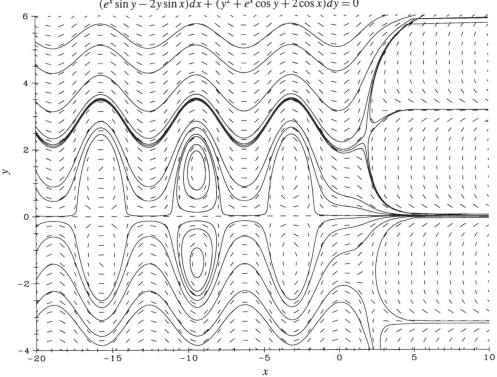

The Van der Pol System: A Sensitivity Study

The van der Pol system, $x' = y - \mu f(x)$, $y' = -x$ models the behavior of the dimensionless current $x(t)$ and the voltage $y(t)$ across a nonlinear resistor in a simple circuit shown in Section 9.1. The parameter $\mu = \sqrt{C/L}/I_0$, where C is capacitance, L is inductance, and I_0 is a positive reference current. The piecewise-linear current-voltage characteristic across the resistor is given by $y = \mu f(x) = \mu(x - |x + 1| + |x - 1|)$.
Reference: Section 9.1, Problem 8**(e)**

LIMIT CYCLE A   If the parameter μ is small (e.g., $\mu = 0.5$), the limit cycle is nearly the circle of radius 2 about the origin.

LIMIT CYCLE B   If $\mu = 5$, the cycle resembles a parallelogram whose slanting slides nearly coincide with the rising arcs of the graph of $y = \mu f(x)$.

LIMIT CYCLE C   For $\mu = 50$ the periodic orbit rises and falls very slowly along the arcs of $y = \mu f(x)$ of positive slope, but traces the horizontal parts of the cycle almost instantaneously. This kind of slow-fast parallelogram cycle is called a *relaxation oscillation*.

LIMIT CYCLE A ## Small Amplitude Limit Cycle

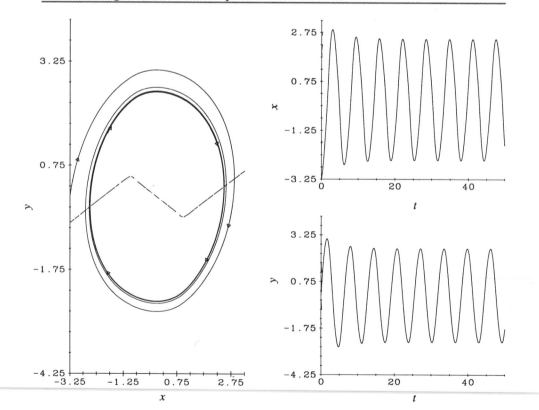

LIMIT CYCLE B   Van der Pol Relaxation Cycle

LIMIT CYCLE C   Another van der Pol Relaxation Cycle

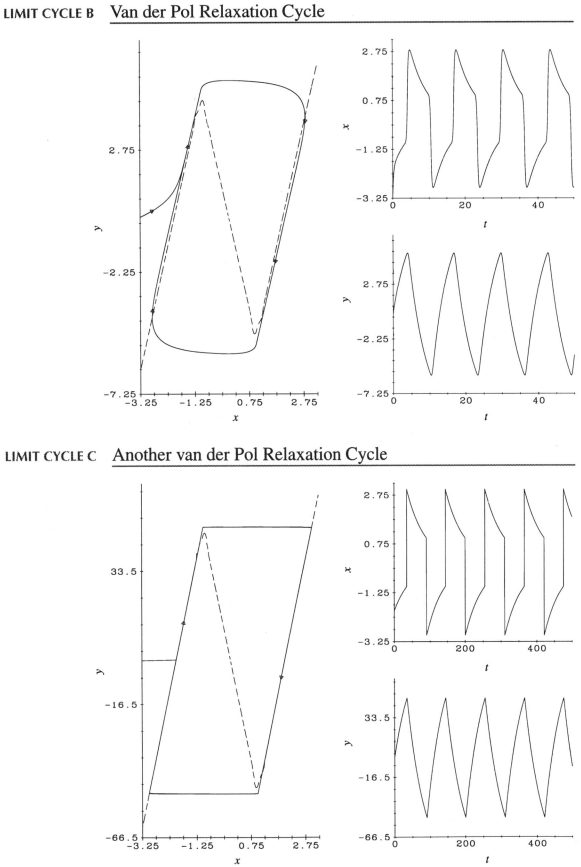

The Damped Pendulum: Basin of Attraction

If a damped, simple pendulum at rest is given enough of a kick, it will swing over the top several times. Eventually, damping will slow it down and it will oscillate with diminishing amplitude about a downward position.
Reference: Sections 4.1; 8.1, Problem 1; 8.2, Examples 8.2.3, 8.2.4, 8.2.6, Problem 6; 8.3, Example 8.3.4; 8.4, Example 8.4.5 and Problem 10

PENDULUM A Each of the asymptotically stable equilibrium points $(-2\pi, 0)$, $(0, 0)$, and $(2\pi, 0)$ of the damped simple nonlinear pendulum system, $x' = y$, $y' = -10\sin x - y$, attracts orbits that go over the top clockwise, or counterclockwise. Which orbit is attracted by which equilibrium point depends on the initial angle and the direction and magnitude of the initial velocity. Orbits and component graphs are shown.

PENDULUM B The damped simple pendulum system has unstable, saddlelike, equilibrium points at $((2k + 1)\pi, 0)$ that correspond to the pendulum balanced vertically upward. The slightest change in kinetic energy will push the pendulum to one side or the other, after which it will oscillate while decaying into an asymptotically stable equilibrium point. All nonconstant orbits shown start at angle $x = 0$, but with initial velocities (hence, kinetic energies) slightly different. After only 3 units of time, the orbits have widely diverged, as the "dancer" orbital portrait shows.

PENDULUM C The "tiger claw" shows parts of the regions of attraction of the three asymptotically stable equilibrium points $(6\pi, 0)$, $(8\pi, 0)$, $(10\pi, 0)$.

PENDULUM A 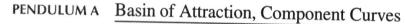 Basin of Attraction, Component Curves

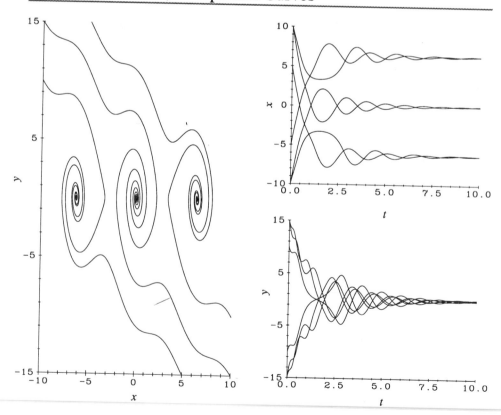

PENDULUM B Orbits Near a Separatrix

PENDULUM C Three Attracting Equilibria

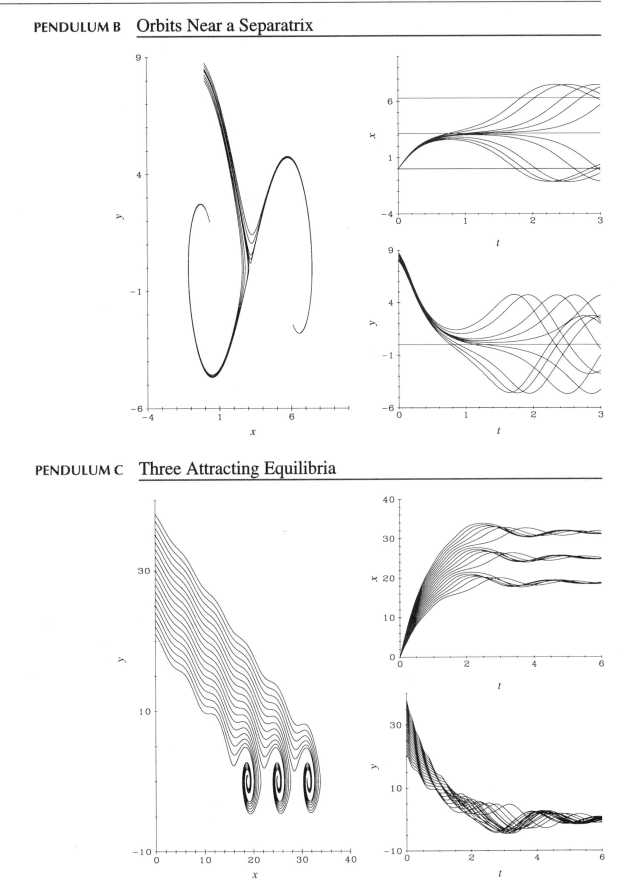

The Double Pendulum: A Sensitivity Study

The motions of a double pendulum can be quite intricate. If time is scaled and if $\alpha = (m_2/m_1)(1 + m_2/m_1)^{-1}$ denotes the reduced mass of the two pendulum bobs, then the linearized and undamped system of approximating ODEs is $x_1' = x_2$, $x_2' = -x_1 + \alpha x_3$, $x_3' = x_4$, $x_4' = x_1 - x_3$, where x_1 and x_3 are the respective angles of the upper and lower pendulums measured clockwise from the downward vertical. The lengths of the support rods are taken to be equal to g, and $x_1(0) = \pi/10$, $x_2(0) = 10$, $x_3(0) = 0$, $x_4(0) = -5$. How sensitive is the system to variations in α?

Reference: Section 7.4, Example 7.4.8, Problem 8; Section 8.3, Example 8.3.5 and Problem 7

PENDULUM D   If $m_2 = 3m_1/7$, then $\alpha = 0.3$. In this case, the upper mass (solid) and lower mass (dashed) seem to move somewhat independently.

PENDULUM E   If the lower pendulum bob is 99 times heavier than the upper bob, then $\alpha = 0.99$. Now the two act much like a simple pendulum, tracing out a high-amplitude, low-frequency path. On top of this motion, however, are small-amplitude, high-frequency oscillations with the two pendulums nearly out of phase.

PENDULUM F   If $m_2 = 3m_1/997$, then $\alpha = 0.003$. The upper pendulum (solid) oscillates almost normally, but when the gyrations of the slower pendulum (dashed) are extreme, much of the kinetic energy of the system is clearly in the lower bob and, consequently, the motions of the upper bob almost come to a halt.

PENDULUM D   ## Double Pendulum: Intermediate Reduced Mass

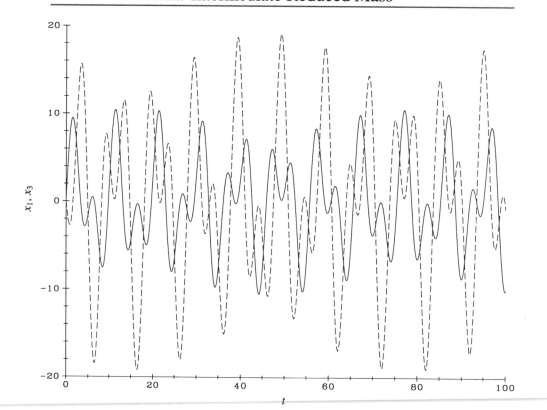

PENDULUM E Double Pendulum: Large Reduced Mass

PENDULUM F Double Pendulum: Small Reduced Mass

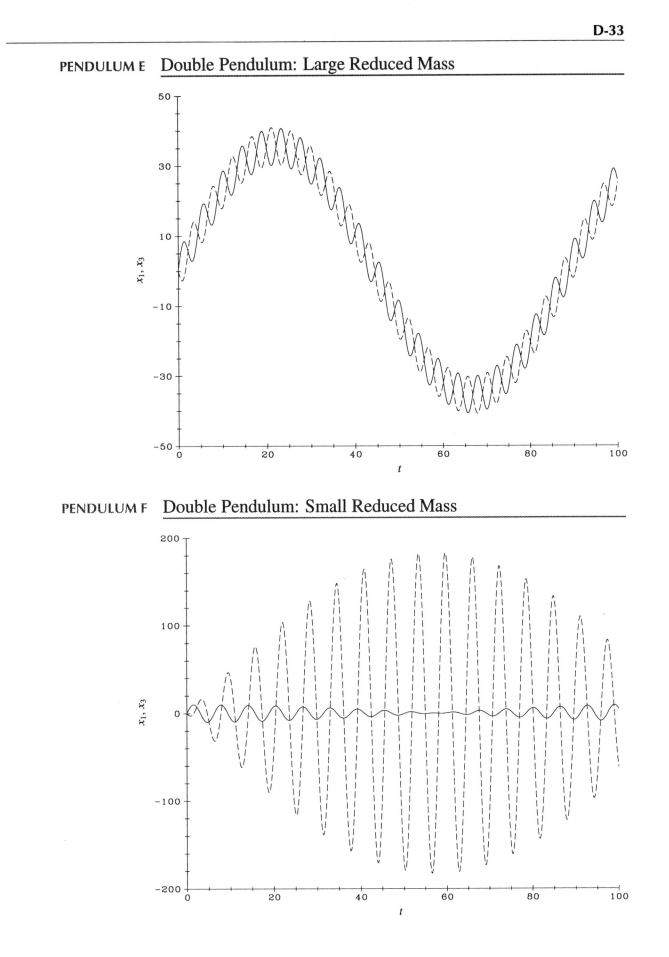

Convergence of Picard Iterates

The Picard Iteration Scheme for constructing a solution of an IVP is quite straightforward, although not particularly efficient in a computational sense. If the IVP is $y' = f(t, y)$, $y(t_0) = y_0$, then the iteration scheme is

$$y_0(t) = y_0, \ldots, y_{n+1}(t) = y_0 + \int_{t_0}^{t} f(s, y_n(s)) \, ds$$

Under appropriate conditions, $y_n(t) \to y(t)$, the desired solution of IVP, over some time interval containing t_0.
Reference: Appendix A.2

PICARD A   The "true" solution of the IVP, $y' = -y$, $y(0) = 3$ is $y = 3e^{-t}$ (dashed). The Picard iterates $y_0, \ldots, y_{10}$ (solid) bracket the true solution. The iterate $y_{10}(t)$ appears to be quite accurate on the interval $0 \le t \le 3$, but not so good on $0 \le t \le 5$.

PICARD B   Although a formula for the "true" solution of the IVP, $y' = (t^2 + y^2 + 1)^{1/2} \sin(ty)$, $y(0) = 1$, is not known, the dashed line is a good approximation computed by a powerful ODE solver. The alternate Picard iterates $y_0, y_2, y_4, \ldots, y_{20}$ seem to fold down onto the solution from above.

PICARD C   The exact solution of the IVP, $y = y^2$, $y(0) = 5$ is $y = 5(1 - 5t)^{-1}$, $t < 0.20$, which escapes to infinity as $t \to 0.20$. The Picard iterates $y_0, \ldots, y_{12}$ are polynomials and do not even "see" the problem at $t = 0.20$, although as n increases y_n turns upward more and more sharply as soon as t exceeds 0.20.

PICARD A   ## Alternating Convergence of Picard Iterates

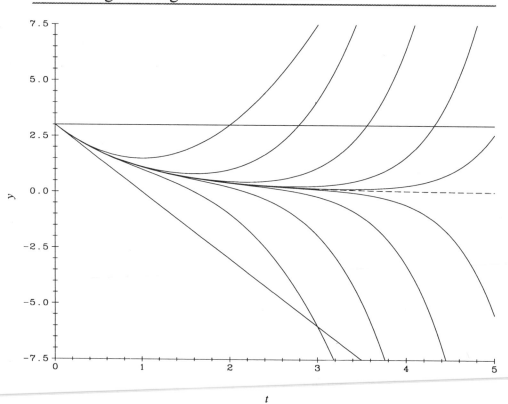

PICARD B Local Convergence of Picard Iterates

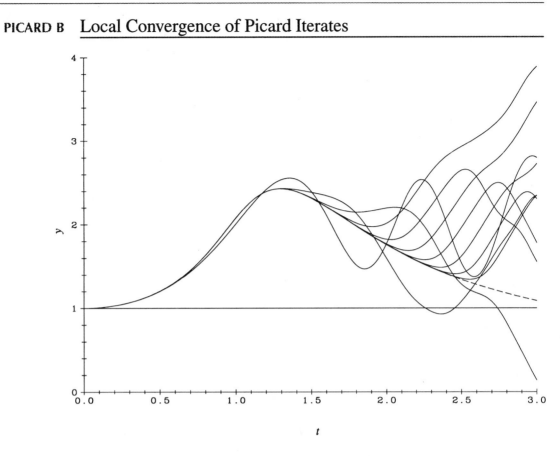

t

PICARD C Convergence of Iterates for Finite Escape Time

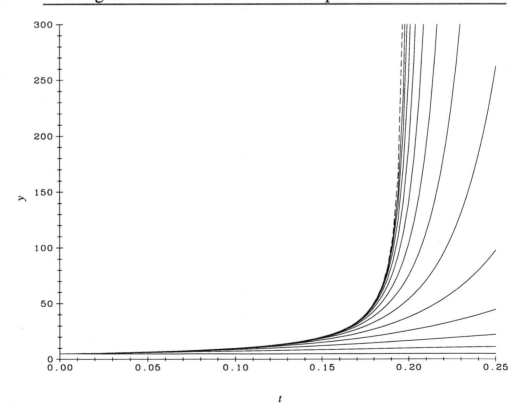

t

Local Orbital Behavior Near Equilibria

Near an equilibrium point of a nonlinear system orbital behavior is usually determined by the local linear approximation to the system. Near the origin the linear approximation to the nonlinear system $x' = x - y + x^2 - xy$, $y' = -y + x^2$, is $x' = x - y$, $y' = -y$. The linear approximation has a saddle equilibrium point at the origin. We expect the nonlinear system to have orbits with the same "hyperbolic" look to them as the orbits of the linearized system. However, nonlinear behavior dominates away from the equilibrium points, and strange things may happen.
Reference: Section 8.4, Problem 6

PLANAR PORTRAIT A   Near the origin the orbits of the nonlinear system (solid) and the linear system (dashed) are much alike.

PLANAR PORTRAIT B   The nonlinear system has equilibrium points at $(0, 0)$, $(1, 1)$, $(-1, 1)$. Away from the origin the orbits that look hyperbolic near the origin bend away from or toward the equilibrium points $(1, 1)$, $(-1, 1)$ in distinctly nonhyperbolic ways. Time is run both forward and backward from the four initial points $(\pm 0.1, \pm 0.1)$.

PLANAR PORTRAIT C   Add a few more orbits, and the portrait of the system in state space emerges.
Reference: Problem 1, Section 8.1

PLANAR PORTRAIT A   ## Linear/Nonlinear Saddle

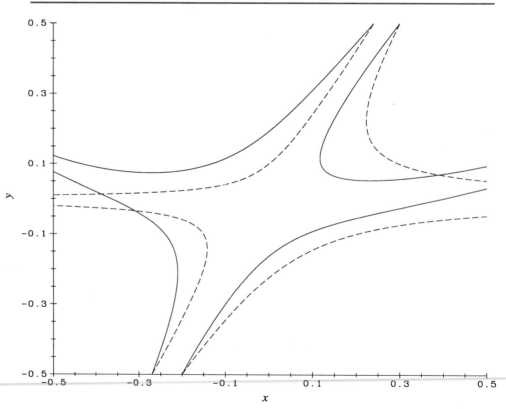

PLANAR PORTRAIT B Extended Nonlinear Orbits of A

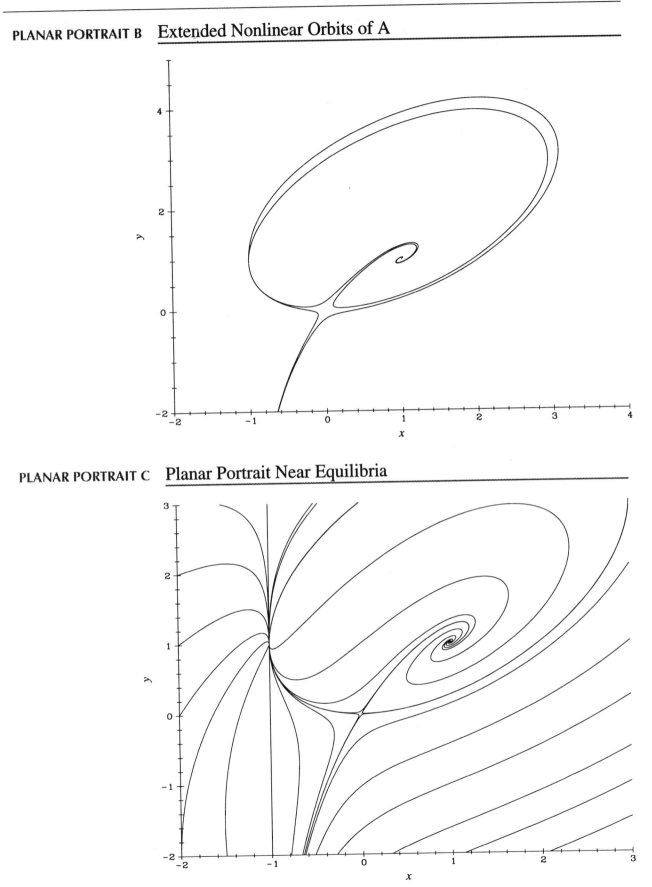

PLANAR PORTRAIT C Planar Portrait Near Equilibria

Poincaré-Bendixson Alternatives

The Poincaré-Bendixson alternatives assert that a positively bounded orbit of a planar autonomous system with finitely many equilibrium points can do only one of three things as $t \to \infty$:

- Approach an equilibrium point
- Approach a periodic orbit (a cycle)
- Approach a cycle-graph (a closed, oriented graph of equilibrium points and orbits)

The Atlas Plates illustrate some of these alternatives.
Reference: Problem 1, Section 8.1

POINCARÉ-
BENDIXSON A

If $\varepsilon = -0.09$ in the system $x' = y$, $y' = x + \varepsilon y - x^3 + 0.1x^2 y$, then the interior orbit shown spirals away from the "outer" limit cycle and toward a "small" limit cycle around the equilibrium point $(-1, 0)$.

POINCARÉ-
BENDIXSON B

As ε increases to -0.08 the limit cycle around $(-1, 0)$ merges with the limit cycle around $(1, 0)$ and forms a clockwise-oriented cycle-graph with vertex at the origin. An orbit spirals inward from the outer limit cycle toward the lazy eight cycle-graph.

POINCARÉ-
BENDIXSON C

As ε increases to -0.079 the cycle-graph opens up into a limit cycle which, like the outer cycle, surrounds all three equilibrium points $(-1, 0)$, $(0, 0)$, $(1, 0)$. The orbit spirals inward from the outer cycle toward the inner cycle.

POINCARÉ-
BENDIXSON A

Orbit Spirals from One Limit Cycle to Another

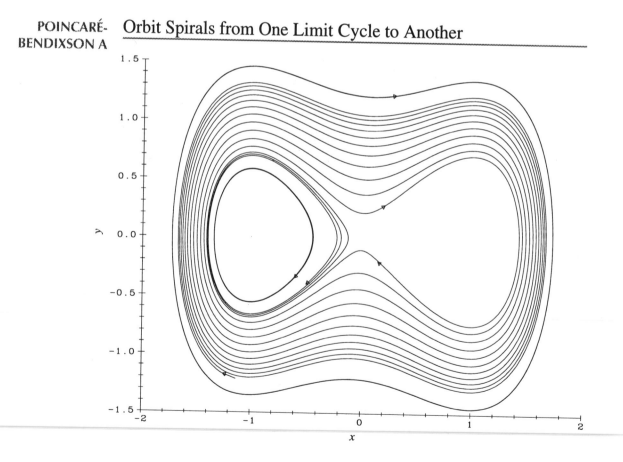

POINCARÉ-
BENDIXSON B

Orbit Spirals to Lazy Eight Cycle-Graph

POINCARÉ-
BENDIXSON C

Lazy Eight Opens up to a Limit Cycle

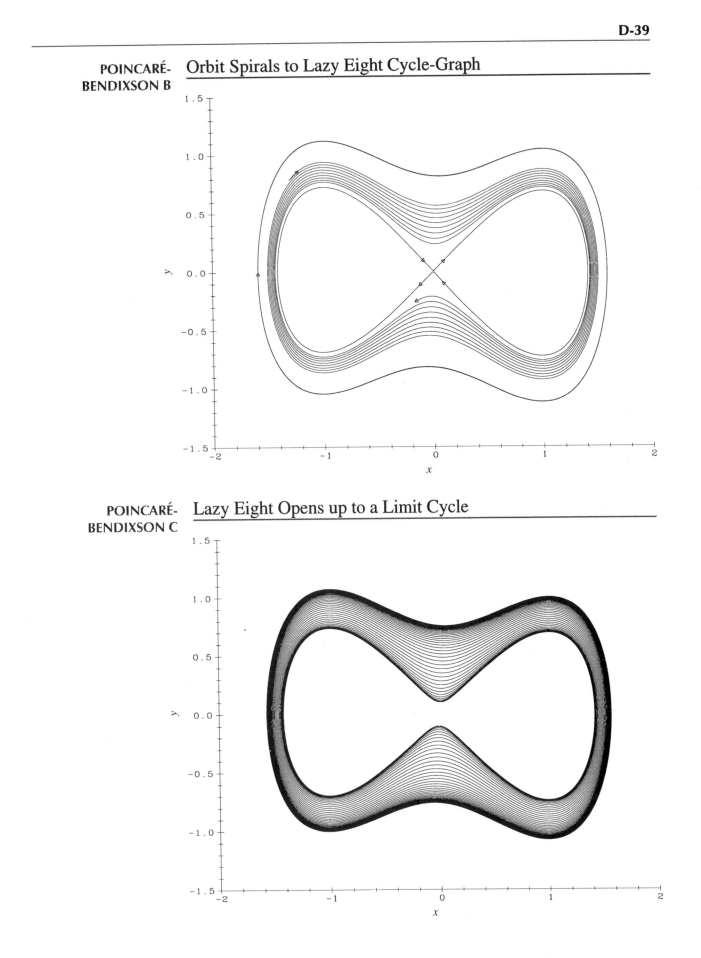

Poincaré-Bendixson Alternatives (continued)

A cycle-graph can be created from a limit cycle of a planar autonomous system if the rate functions are scaled by multiplying by a continuous factor which is positive except at one point of the cycle where the factor has value 0. The orbits of the original and the scaled system appear to be identical, but component graphs reveal that the scaling factor makes drastic changes in the solution functions.
Reference: Problem 11(**c**) of Section 9.2

POINCARÉ-BENDIXSON D   The system $x' = x - 10y - x(x^2 + y^2)$, $y' = 10x + y - y(x^2 + y^2)$ has a limit cycle on the unit circle, $x^2 + y^2 = 1$. Inner orbits spiral outward toward the cycle.

POINCARÉ-BENDIXSON E   Multiply each rate function of the system in plate D by $h = 1 - \exp[-10(x-1)^2 - 10y^2]$, and a new equilibrium point $(1, 0)$ appears. This point lies on the unit circle, which is a cycle-graph (not a cycle) for the system scaled by h.

POINCARÉ-BENDIXSON F   An orbit inside the cycle-graph spirals outward as before, but its behavior in time is very different. The orbit slows down as it nears the new equilibrium point and speeds up as it moves away.

POINCARÉ-BENDIXSON D   ## Approach to Circular Limit Cycle

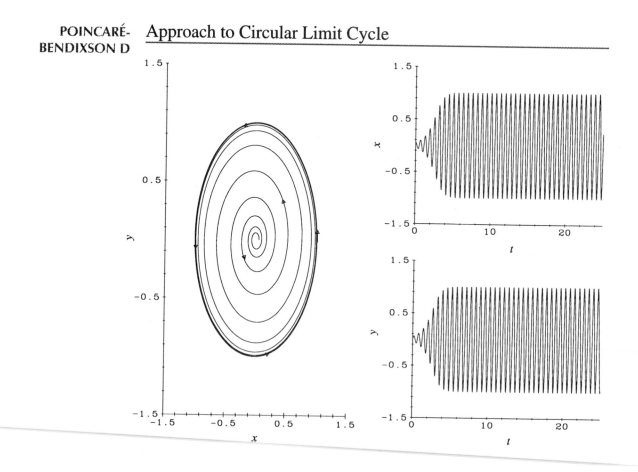

POINCARÉ-BENDIXSON E   Cycle-Graph on Unit Circle

POINCARÉ-BENDIXSON F   Approach to Circular Cycle-Graph

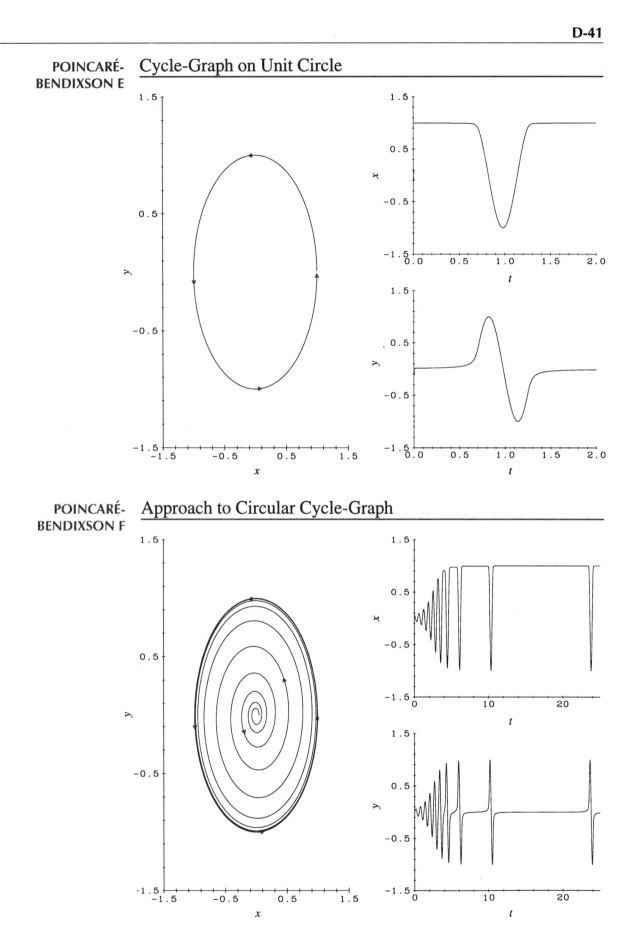

Volterra's Models for Predator-Prey Dynamics

Volterra's ODEs for predator-prey dynamics are based on real data. The rate equations for the populations $x(t)$ and $y(t)$ of the predator and prey are:

$$x' = -ax + bxy - H_1(t)x, \qquad y' = cy - dxy - H_2(t)y$$

where a, b, c, and d are positive rate constants and H_1 and H_2 are the coefficients of constant-effort harvesting. Suppose that the harvesting is restricted to one month per year. Let $a = b = c = d = 1$, $x(0) = 1$, $y(0) = 2$, $H_1 = 4 \, \mathrm{sqw}(t, 100/12, 1) = 0.5 H_2$. See Appendix C.1 for the function sqw. The no-fishing case and the case of continuous, high-rate, constant-effort harvesting ($H_1 = 4$, $H_2 = 8$) are also considered. **Reference:** Section 6.4, Problem 7

PREDATOR-PREY A The oval orbit corresponds to no fishing at all, and the jagged curve to on-off seasonal harvesting. The dashed curve shows that continuous high-rate, constant-effort harvesting kills off both the predator and the prey species.

PREDATOR-PREY B With no fishing, the predator population changes periodically (solid). High-rate, but seasonal, harvesting reduces the average predator population (jagged curve). Continuous high-rate, constant-effort harvesting leads to extinction (dashed curve).

PREDATOR-PREY C With no fishing, the prey population oscillates periodically (solid). Seasonal, high-rate harvesting is beneficial for the prey (jagged), because the predator is also harvested. Continuous high-rate harvesting brings prey extinction (dashed curve).

PREDATOR-PREY A ## Orbits of Harvested Species

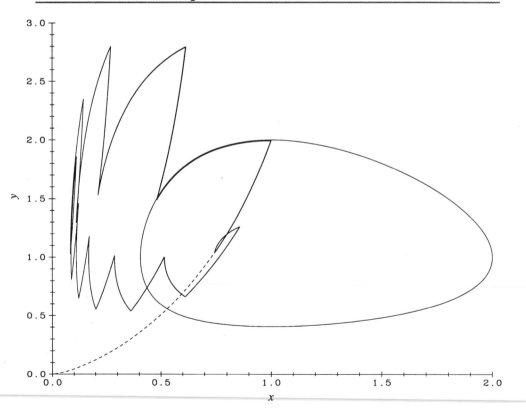

PREDATOR-PREY B Harvested Species: Predator

PREDATOR-PREY C Harvested Species: Prey

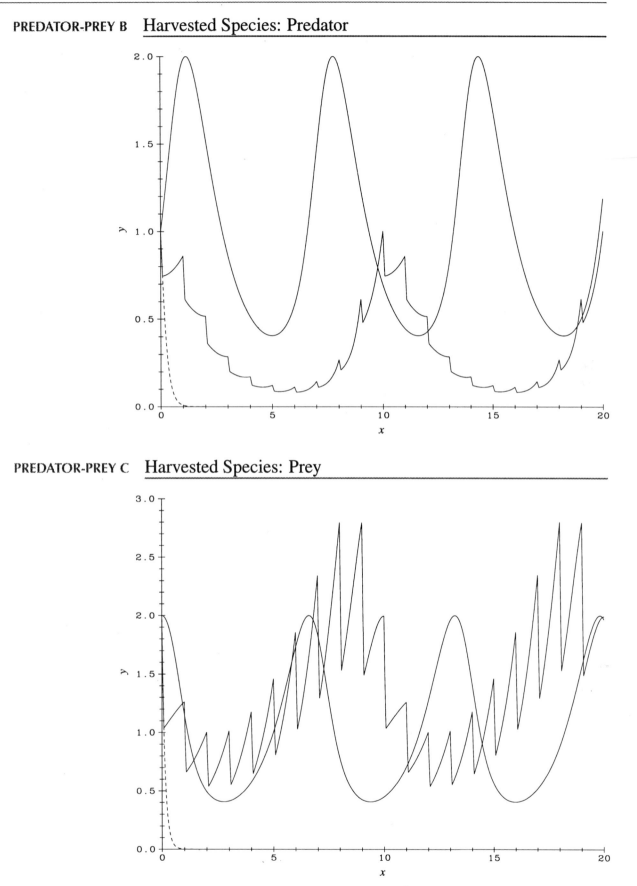

The Rössler System: A Strange Attractor

We set $b = 2, c = 4$ in the Rössler system

$$x' = -y - z$$

$$y' = x + ay$$

$$z' = b - cz + xz$$

where a is a positive constant. The solution is found for $0 \le t \le 1000$, but orbits are plotted for $800 \le t \le 1000$ after initial transient behavior has died out. Each orbit displays a folding behavior like that of the Lorenz orbits of Section 9.4. Initial point: $x(0) = z(0) = 0$, $y(0) = 2$; $a = 0.410, 0.400, 0.395$.
Reference: Problem 3, Section 9.4

STRANGE
ATTRACTOR A
The orbit is very close to an attracting periodic orbit: $a = 0.410$.

STRANGE
ATTRACTOR B
Rössler orbits are sensitive to small changes in a: a is now 0.400, and the strands of the periodic orbit have multiplied.

STRANGE
ATTRACTOR C
The strange attractor seems to appear as a decreases from 0.400 to 0.395. The strands multiply again, the folding is evident, and a strange attractor of nonperiodic, chaotic (but regular) wandering seems to have been created.

STRANGE
ATTRACTOR A
Stable Periodic Orbit

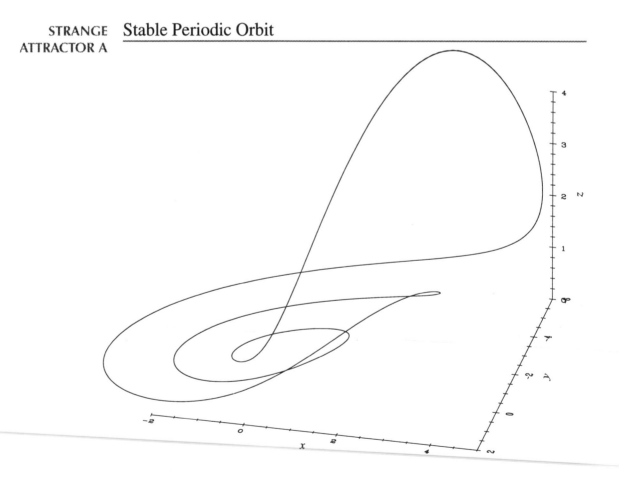

STRANGE
ATTRACTOR B

Strands Multiply

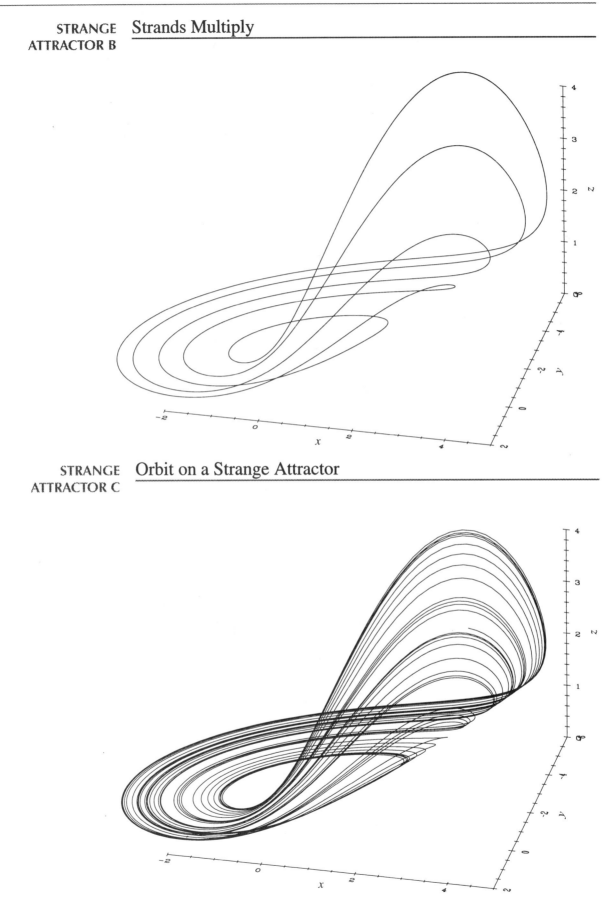

STRANGE
ATTRACTOR C

Orbit on a Strange Attractor

The Scroll Circuit: A Strange Attractor

The scroll ODEs model the voltages and current in an electric circuit, which has an inductor and two capacitors in parallel. A nonlinear resistor with a current-voltage characteristic $f(v)$ is in series with the capacitors. The dimensionless IVP for voltages x and y across the capacitors and the current z through the inductor are:

$$
\begin{aligned}
x' &= -cf(y - x) & x(0) &= 1 \\
y' &= -f(y - x) - z & y(0) &= 1 \\
z' &= y & z(0) &= 1
\end{aligned}
$$

where c is a positive circuit parameter, and

$$f(v) = -0.07v + 0.085(|v + 1| - |v - 1|)$$

Reference: Problem 4 of Section 9.4; cover of book

STRANGE ATTRACTOR D   If $c = 0.1$, the orbit of voltages and current in xyz-state space spirals inside the lip of the "mushroom" as time increases.

STRANGE ATTRACTOR E   The projection of the orbit onto the xz-plane resembles a scroll: $c = 0.1$. The shape of the orbit and its projections changes dramatically as c changes.

STRANGE ATTRACTOR F   By the time the parameter c has increased to 33 a strange attractor containing a chaotic orbit seems to have appeared. The scroll in the xz-plane has evolved into the figure shown.

STRANGE ATTRACTOR D

Nonlinear Circuit: Voltages, Current

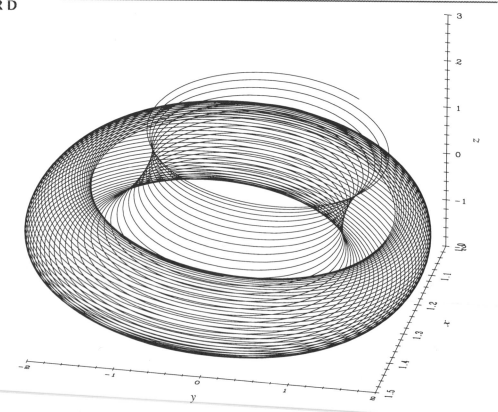

STRANGE
ATTRACTOR E

Projection in *xy*-plane

STRANGE
ATTRACTOR F

Projected Strange Attractor

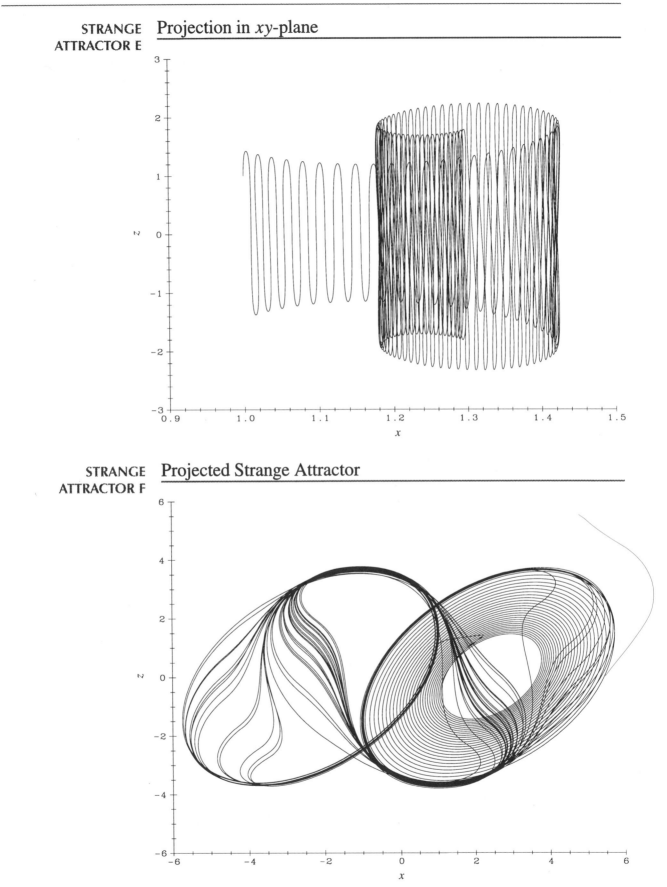

Answers to Selected Problems

Section 1.1   Page 9.

1(a). First order due to y'.

1(c). Second order due to y''.

1(e). The normal linear form is $y'' + e^{-t}\sin(t)y' + 3e^{-t}y = 5$. Second order due to y''.

2(a). $r = -3$. **2(c).** $r = 0, \sqrt{2}, -\sqrt{2}, 1$ or -1.

3(b). $c = 2/29$. **4(a).** $r = (-3 \pm \sqrt{5})/2$.

6(a). $y = 1 + Ce^{-t}$, C any constant.

6(c). $y = t + Ce^{-t}$, C any constant.

6(e). $y = \pm\sqrt{t^2/2 + C}$, C any constant and $t^2/2 > -C$.

10(a). $N(t) = N_0 e^{-kt}$. **10(c).** 81% remains.

11(a). $y' = 2xy$.

13(a). $y = C_1 t + C_2$. **13(c).** $y = t^3/3 + C_1 t^2 + C_2 t + C_3$.

Section 1.2   Page 20.

1(a). $y = Ce^{t^2} - 1/2$, where C is any constant and $-\infty < t < \infty$.

1(c). $y = C\exp(\cos t) + 1$. **1(e).** $y = Ce^{-t^2/2} + 1/2$.

2(a). $y = Ct^{-2} + t^2/4$, where either $t > 0$ or else $t < 0$.

2(e). $t = |y - 2|^{-3}[c + 2y^2 - 4y^3/3 + y^4/4]$, where $y < 2$ or $y > 2$.

4(a). $y = (t + 1)e^{-t}$, $-\infty < t < \infty$.

4(c). $y = 1$, $-\infty < t < \infty$.

4(e). $y = (1/t^2)\sin t - (1/t)\cos t$, $t > 0$.

4(g). $y = \sin t + 2\csc t$, $0 < t < \pi$.

10(a). 12 lb salt. **10(b).** 85.3 lb salt.

Section 1.3   Page 34.

3. $t = 278.5$ days. **5(a).** 2041 cm at $t = 2.04$ sec.

7(a). $H = 62.8$ meters. **10(a).** 9.42%.

13. 2028 B.C., approximately. **15.** $u^* = 98°$ F.

Section 1.4   Page 47.

3(a). $y = x$, $-1 < x < \infty$. **3(c).** $y = \exp(2 - e^{-x})$, for all x.

3(e). $y = (5 - x^2)^{1/2}$, $|x| \le 5^{1/2}$. **7(a).** 3.29 sec.

Section 1.5   Page 56.

1. 2026 A.D. **3(a).** 7.5 years.

5(b). $P(3) \approx 1030$ individuals. **8(b).** $H = abc$, $K = b + c$.

Section 1.6   Page 66.

1(a). $x'(t) = -3x$, $y'(t) = x$ and $z'(t) = 2x$. The solution is $x(t) = e^{-3t}$, $y(t) = (1 - e^{-3t})/3$, and $z(t) = 2(1 - e^{-3t})/3$.

1(c). $x' = 1 + \sin t - 3x$, $y' = 3x - y$. The solution is
$x = 1/3 + (3\sin t - \cos t)/10 - 7e^{-3t}/30$,
$y = 1 + 3(\sin t - 2\cos t)/10 + 7e^{-3t}/20 - 3e^{-t}/4$.

3(a). $x = Ae^{-k_1 t}$, $y = Ak_1(e^{-k_1 t} - e^{-k_2 t})/(k_2 - k_1) + Be^{-k_2 t}$.

5(a). $x = 0.72$ (GI-tract), $y = 7.2$ (bloodstream) for the decongestant. For the antihistamine, $x = 1.44$ and $y = 43$.

8(a). $K(t) = K_0 e^{-kt}$, $A(t) = (K_0 k_1/k)(1 - e^{-kt})$,
$C(t) = (K_0 k_2/k)(1 - e^{-kt})$.

Section 1.7   Page 76.

1(a). $y = x\ln|x| + Cx$, where $x > 0$ or $x < 0$.

1(c). $x^3(x - 2y)(x + y)^2 = C$, where C is a constant.

3(a). $y = -x + 2\arctan(x + C) + n\pi$, where C is any constant and n is any integer.

3(c). $3(x + 2y - 2\ln|x + 2y + 2|) = x + C$, where C is any constant and $|x + 2y + 2| \neq 0$.

8. $y = \pm x\sqrt{C - 2\ln|x|}$, where C is any constant and x lies in an interval for which $C > 2\ln|x|$.

9(a). $dr/d\theta = 2r$. $r = Ce^{2\theta}$ (where $C = e^{C_1} > 0$).

12(c). $y = e^t + 1/z = e^t + [Ce^{-3t} - e^{-t}/2]^{-1}$, where C is any constant and $2Ce^{-3t} \neq e^{-t}$.

13(a). $(\sqrt{2}(x^2 + y^2)^{1/2} + x + y)^2 = a^{\sqrt{2}}(3 + 2\sqrt{2})|x + y|^{2-\sqrt{2}}$.

Section 1.8   Page 84.

2(a). $y = (-3t/2 + 1)^{2/3}$, $t < 2/3$.

4. $y = -t^{-2}\sin t + C_1 t^{-1} + C_2 t^{-2}$, $t > 0$.

Section 1.9   Page 94.

1(c). $xy^2 e^y + x^2 = C$. **3(a).** $x^2/2 + 2xy + y^2/2 = C$.

3(c). $x\cos 2y - x^3 y^2 + \sin 2y/2 = C$.

3(e). $x^2 - 3xy + y^2 = C$.

6(a). $x^2 + y^2 - xy = 3$. **6(c).** $x^2 y^3 = 1$.

8(b). $F(x, y) = e^x \sin y + 2y\cos y + y^3/3$.

11(a). $3\sin x + y\cos^2 x = C$.

Section 2.1   Page 102.

2(a). $y = Ce^{-t^2} + e^{-t^2}\int^t e^{s^2}s\cos s\, ds$.

6(a). Rising: the region, $-8 < y < 0$, and either $-6 < t < -3\pi/2$, or $-\pi/2 < t < \pi/2$, or $3\pi/2 < t < 6$; and the region $0 < y < 8$, and either $-3\pi/2 < t < -\pi/2$, or $\pi/2 < t < 3\pi/2$. Falling elsewhere.

Section 2.2   Page 111.

6. Infinitely many solutions; e.g., $y(t) = -1$ for $t < (2n - 1)\pi$, $y(t) = 1$ for $t > 2n\pi$, $y(t) = \cos t$, for $(2n - 1)\pi \le t \le 2n\pi$, n a nonpositive integer.

8. $t_0 = (-1/2)\ln(-c) = (-1/2)\ln\frac{1+y_0}{y_0-1}$.

Section 2.3   Page 119.

1(b). $|y_c(t) - y_5(t)| = |c - 5|(1 - e^{-t})$, $t \ge 0$.

Section 2.4   Page 127.

1. $P(t) \le e^{-r_0 t}P_0 + \frac{R_0}{r_0}(1 - e^{-r_0 t}) \le \max\{P_0, R_0/r_0\}$, for $t \ge 0$.

Section 2.5   Page 134.

1(a). $y(1) \approx y_{10} = 0.348678$.

Section 2.6   Page 139.

1(a). $y = (1 - 2t)^{-1/2}$, $-\infty < t < 1/2$.

3(a). $y = Ce^{-2t} + 0.4\cos t + 0.2\sin t$, where C is an arbitrary constant. As $t \to \infty$, $y(t) \to y_p(t) = 0.4\cos t + 0.2\sin t$.

5(a). $y = \frac{y_0}{y_0 + (1 - y_0)e^{-t}}$ if $y_0 > 1$.

Section 2.7   Page 151.

3. $a = 2.3$: 2-cycle about the equilibrium. $a = 2.5$: 4-cycle about the equilibrium.

Section 3.1   Page 162.

4. $mz'' = -kz - \alpha/(b + z)^2$, where α is a positive constant and $z > -b$.

Section 3.2   Page 170.

1(a). Single equilibrium solution, $y(t) = 0$, all t.

1(c). $y(t) = 0$, all t.

1(e). Three equilibrium solutions, $y = 0, -1, 1$, all t.

2(a). All solution curves are concave down everywhere.

5. $y(t) = 1/(1 - t)$, $t < 1$.

Section 3.3   Page 182.

1(a). $y = C_1 t + C_2$, C_1 and C_2 arbitrary constants.

2(a). $y = C_1 \ln t + C_2$, $0 < t < \infty$, C_1 and C_2 arbitrary constants.

4(a). $y(t) = e^{2t} - e^t$.   **8(a).** $y(t) = t^2 - 3t/2$.

10(a). $y = C_1 e^{-t} + C_2 e^{3t}$ gives all real-valued solutions.

10(c). $y = C_1 e^{-2t}\cos 3t + C_2 e^{-2t}\sin 3t$ gives all real-valued solutions.

Section 3.4   Page 192.

1(a). $y(t) = C_1 e^{-t} + C_2 e^{2t}$, where C_1 and C_2 are real constants.

2(a). $y'' + 7y' + 10y = 0$.

2(c). $y'' + 8y' + 17 = 0$.

7(a). $y(t) = k_1 e^{-2it} + k_2 e^{it}$, where k_1 and k_2 are any complex constants.

8(a). $y(t) = Ce^{-t}$, where C is any real constant.

Section 3.5   Page 204.

1(a). $[i(e^{it} - e^{-it}) + (e^{2it} - e^{-2it})]/2$.

1(c). $-(1 + t)/2 - (1 + t + 2it^2)e^{2it}/4 + (2it^2 - t - 1)e^{-2it}/4$.

1(e). $(1 - t)(e^{4it} + e^{-2it})/2$.

3(a). $(D^2 + 1)^3$.   **3(c).** $D^2(D^2 + 1)$.

5(a). $y(t) = e^{-t}/3 - e^{t/2}\cos(\sqrt{3}t/2)/3 + e^{t/2}\sin(\sqrt{3}t/2)/\sqrt{3}$.

6(a). $y(t) = C_1 + C_2\cos t + C_3\sin t + C_4 t\cos t + C_5 t\sin t$, where C_1–C_5 are arbitrary real numbers.

7(a). $y(t) = e^{at}(A\cos at + B\sin at) + e^{-at}(C\cos at + E\sin at)$, for arbitrary reals A, B, C, and E.

7(c). $y(t) = Ae^{-t} + Be^{t} + Ce^{-3t}$, for arbitrary reals A, B, and C.

7(e). $y(t) =$
$-t + C_1 e^t + C_2 t e^t + C_3 t^2 e^t + C_4 e^{-t} + C_5 t e^{-t} + C_6 t^2 e^{-t}$,
for arbitrary reals C_1–C_6.

8(a). $y(t) = -e^{-2t}/2 + e^{2t} + 2t - 1/2$.

8(c). $y(t) = -2\cos t - 4\sin t + 2e^{2t}$.

Section 3.6   Page 211.

1(a). $y(t) = (C_1\cos t + C_2\sin t) + \left[(s\sin t + \cos t\ln|\cos s|)|_{s=t_0}^{s=t} \right]$.

1(c). $y(t) = (C_1 e^{-4t} + C_2 e^{3t}) -$
$\frac{1}{7}\left[\left(\frac{1}{5}e^{3t-5s} + \frac{2}{3}e^{3t-3s} + e^{3t-s} + \frac{1}{2}e^{2s-4t} + \frac{1}{2}e^{4s-4t} + \frac{1}{6}e^{6s-4t} \right)\Big|_{s=t_0}^{s=t} \right]$.

3. $y(t) = \int_0^t \sin(t - s)f(s)\,ds$.   **4(c).** $y(t) = t^2 + t$, for all t.

7(a). All t, s on an interval where $a(t)$ and $b(t)$ are continuous.

Section 4.1   Page 221.

3(a). $w_1 - 7w_2 - 5w_3 = 0$.

5. Vertical component of position: 8.8 miles; horizontal component: 79.5 miles; northward: 39.5 miles; eastward: 69.0 miles.

6(a). $-0.33G\hat{\mathbf{i}} - 0.80G\hat{\mathbf{k}}$.   **11(a).** $T = 2\pi\sqrt{L/g}$.

Section 4.2   Page 236.

4. $y_p(\omega) = F_0 M(\omega)\cos(\omega t + \varphi(\omega))$, where $M(\omega)$ and $\varphi(\omega)$ are as given in the text.   **9(a).** $2c = 0.613/\text{sec}$.

Section 4.3   Page 249.

1(a). $I(t) = \exp(-t)[\cos t - \sin t]$.

2(a). $I(t) = k_1 e^{r_1 t} - k_1 e^{r_2 t}$,
$q(t) = 10^{-3} + k_1[(e^{r_1 t} - 1)/r_1 - (e^{r_2 t} - 1)/r_2]$, where $r_1 \approx -2.33$, $r_2 \approx -47.7$, $k_1 \approx -5.04$.

4. $q(t) = \frac{1}{26}[-e^{-t}(9\cos 3t + 7\sin 3t) + 6\sin 2t + 9\cos 2t]$.

Section 5.1   Page 255.

1(a). $\frac{3}{s^2} - \frac{5}{s}$.   **1(c).** $n!/s^{n+1}$, $s > 0$.

1(e). $1/(s - a)^2$, $s > a$.   **2(a).** $\frac{1 + e^{-s\pi}}{1 + s^2}$.

2(c). $\frac{1}{s^2}[1 - (1 + s)e^{-s}]$, $s > 0$.   **3(a).** $y = e^{-2t}$.

Section 5.2   Page 265.

1(a). $a/(s^2 - a^2)$.   **1(c).** $2/(s - a)^3$.

1(e). $-\varphi'(s - 2)$, where $L[f](s) = \varphi(s)$.

1(g). $e^{-s}(1/s^2 + 2/s)$.

1(i). $e^{-(s-a)}/(s - a) - e^{-2(s-a)}/(s - a)$.

2(a). III.7: $L[f] = 3(e^{-2s} - e^{-5s})/s$.

2(c). II.3: $L[f] = 12!/(s - 5)^{13}$.

2(e). II.14: $L[f] = 1/\sqrt{s^2 + 1}$.

2(g). II.17: $f = (e^{2t} - e^{-5t})/t$.

3(a). $y(t) = (e^{3t} + 4e^{-2t})/5$.   **3(c).** $y(t) = e^t\sin t$.

3(e). $y(t) = e^t(1 - t) + H(t - 1)[1 + (t - 2)e^{t-1}]$.

4(a). $\frac{1}{s}[\frac{1}{(s-1)^2} - \frac{1}{(s-1)}]$.   **4(c).** $\frac{1}{\sqrt{2}s}[\frac{1}{(s-1)^2+1} - \frac{s-1}{(s-1)^2+1}]$.

7(c). $q_a(t) = H(t - a)CE_0\left[1 + \frac{r_2}{r_1 - r_2}e^{r_1(t-a)} + \frac{r_1}{r_2 - r_1}e^{r_2(t-a)}\right]$, where $q_a(t)$ is the charge at time $t \geq 0$ if the switch is turned on at time $t = a$.

Section 5.3   Page 271.

1(a). $\frac{1}{1+s^2}$.   **1(c).** $\frac{s^2+2}{s(s^2+4)}$.   **1(e).** $-\frac{2a(3s^2-a^2)}{(a^2+s^2)^3}$.

1(g). $\frac{1}{\sqrt{2}}\frac{s+1}{(s+3)^2+4}$.   **2(a).** $1 + H(t - 1)$.   **2(c).** $\frac{t^{n-1}e^t}{(n-1)!}$.

2(e). $t + (2 - t)H(t - 1)$.

3(a). $1 - e^{-t}$.   **3(c).** $\frac{1}{a-b}[e^{at} - e^{bt}]$.

3(e). $-\frac{4}{5}e^{-t}\cos t + \frac{3}{5}e^t\sin t + \frac{4}{5}e^t$.

Section 5.4   Page 278.

1(a). Collision occurs in every case.

Section 5.5   Page 282.

2(a). $\int_0^t \cos(t - u)\sin u\,du$.   **2(c).** $\int_0^t (t - u)e^{-u}du$.

2(e). $\int_0^t [2e^{-2(t-u)} - e^{-(t-u)}]ue^{-2u}\,du$.

2(g). $\int_0^t f(t - u)\sin u\,du$.

3(a). $\int_0^t \sin(t-u)[uH(u-1)]\,du.$

3(c). $\frac{1}{3}\int_0^t f(t-u)[e^{u/2}-e^{-u}]\,du.$

4(a). $g(t)=(1/2)e^{-3t}\sin 2t.$ **5(a).** $y=(e^{-3t}\sin 2t)/2 * f(t).$

Section 5.6 Page 289.

1(a). $y(t)=-H(t-\pi)e^{-(t-\pi)}\sin t.$

1(c). $y(t)=[-3\cos t+\sin t+5e^{-t}-2e^{-2t}]/5+H(t-\pi)[e^{-(t-\pi)}-e^{-2(t-\pi)}].$

1(e). $y=[e^t-\cos t-\sin t]/2+H(t-1)\sin(t-1).$

Section 6.1 Page 300.

1(a). $x_1=C_1e^{-2t}+C_2e^{2t},\ x_2=-2C_1e^{-2t}+2C_2e^{2t}.$

1(c). $x_1=C_1e^{-4t}+C_2e^{-t},\ x_2=-4C_1e^{-4t}-C_2e^{-t}.$

1(e). $x_1=C_1\cos 4t+C_2\sin 4t,\ x_2=-4C_1\sin 4t+4C_2\cos 4t.$

2(a). $x(t)=e^{-t/2}\left[\cos(\sqrt{3}t/2)+\frac{1}{\sqrt{3}}\sin(\sqrt{3}t/2)\right],$
$y(t)=\frac{1}{\sqrt{3}}e^{-t/2}\sin(\sqrt{3}t/2).$

2(c). $x(t)=\frac{\sqrt{5}+3}{2\sqrt{5}}e^{(-3+\sqrt{5})t/2}+\frac{\sqrt{5}-3}{2\sqrt{5}}e^{(-3-\sqrt{5})t/2},$
$y(t)=-\frac{\sqrt{5}}{5}e^{(-3+\sqrt{5})t/2}+\frac{\sqrt{5}}{5}e^{(-3-\sqrt{5})t/2}.$

3(a). $x_1(t)=10e^{-3t},\ x_2(t)=10(1-e^{-3t})/3,\ x_3=20(1-e^{-3t})/3;$
the limiting values as $t\to\infty$ are $x_1=0,\ x_2=10/3,$
$x_3=20/3.$

3(c). $x_1(t)=10e^{-t},\ x_2(t)=5e^{-t}+15e^{-3t};$ the limiting values as
$t\to\infty$ are $x_1=x_2=0.$

5(a). $x_2=Cx_1^{-3}.$

6(a). $x_1=C_1e^{-2t}+3C_2e^{2t},\ x_2=-C_1e^{-2t}+C_2e^{2t}.$

7(a). $x_1(t)=-9+3t+(9\cos t-3\sin t),$
$x_2(t)=-15+5t+15\cos t.$

Section 6.2 Page 312.

2(a). $(0,0)$ and $(1,1).$ **2(c).** No equilibrium points.

3(a). $x(t)=C_1e^{3t},\ y(t)=C_2e^{-t},\ -\infty<t<\infty.$

3(c). $x(t)=C_1(1+2C_1^2t)^{-1/2},\ y(t)=t+C_2,\ t>-1/2C_1^2.$

4(a). $(\pm 1,\pm 1).$

Section 6.3 Page 320.

2(a). x-nullclines: $x=0,\ y=x-5;$ y-nullclines: $y=0,$
$y=x/5+2;$ equilibrium points: $(0,0),\ (0,2),\ (5,0),$
$(35/4,15/4).$ The species cooperate, and approach the
equilibrium point $(35/4,15/4)$ as $t\to\infty.$

4(a). $2a$ measures the effectiveness of the y-species in "using" the
x-species to promote growth in $y.$

Section 6.4 Page 329.

1(a). $x(t)$ is the predator populatiom, $y(t)$ is the prey population.
The average populations are both 1. The equilibrium points are
$(0,0)$ and $(1,1).$ The orbits turn clockwise as time increases.

3(a). $X'=(bc/d)Y+bXY,\ Y'=-(ad/b)X+bXY.$

Section 6.5 Page 334.

3. $dz/ds=z(c-y)-(1-c-r)y,\ dy/ds=(z-r)y.$ The
equilibrium points are $(0,0),\ (r,cr/(1-c)).$

Section 7.1 Page 341.

1(a). $x_1=1800,\ x_2=699,\ x_3=200583.$

4(a). $x_1'(t)=-20x_1+4x_2+x_3,\ x_2'(t)=x_1-4x_2+2x_3,$
$x_3'(t)=3x_1-8x_3+1.$

5(b). $x_1=I_1/k_{01},\ x_2=(k_{21}/k_{12})(I_1/k_{01}).$

Section 7.2 Page 351.

1(a). 1 is a double eigenvalue; $[0\ \ 1]^T$ is a basis.

1(c). The eigenvalues are 3 and -3 and are simple. V_3 is spanned by
$[1\ \ 1]^T;$ V_{-3} is spanned by $[1\ \ -1]^T.$

1(e). The eigenvalues are $3+5i$ and $3-5i$ and are simple. V_{3+5i} is
spanned by $[1\ \ -i]^T;$ V_{3-5i} is spanned by $[1\ \ i]^T.$

1(g). 1 is a simple eigenvalue and 2 is a double eigenvalue. V_1 is
spanned by $[3\ \ -1\ \ 3]^T$ and V_2 has basis
$\{[2\ \ 1\ \ 0]^T,\ [2\ \ 0\ \ 1]^T\}.$

1(i). -1 is a double eigenvalue and 5 is a simple eigenvalue. V_{-1} is
spanned by $[1\ \ 0\ \ 0]^T$ and V_5 is spanned by $[262\ \ 27\ \ 6]^T.$

Section 7.3 Page 360.

1(a). $x(t)=c_1\begin{bmatrix}1\\-1\end{bmatrix}e^{2t}+c_2\begin{bmatrix}3\\-1\end{bmatrix}e^{4t},\ c_1,c_2$ arbitrary reals.

1(c). $x(t)=c_1\begin{bmatrix}2\\1\end{bmatrix}e^{4t}+c_2\begin{bmatrix}1\\-2\end{bmatrix}e^{9t},\ c_1,c_2$ arbitrary reals.

3(a). $x=c_1e^{-4t}[1\ \ -4]^T+c_2e^{2t}[1\ \ 2]^T,$ where c_1 and c_2 are
arbitrary reals.

3(c). $x=c_1e^{2t}[2\ \ 1]^T+c_2e^{-t}[1\ \ 2]^T,$ where c_1 and c_2 are
arbitrary reals.

3(e). $x=c_1e^{5t}[1\ \ 3]^T+c_2e^{3t}[1\ \ 1]^T,$ where c_1 and c_2 are
arbitrary reals.

3(g). $x=c_1e^t[1\ \ 1\ \ 1]^T+c_2e^{(-1/2+\sqrt{3}/2)t}[1\ \ -1/2+$
$\sqrt{3}/2\ \ -1/2-\sqrt{3}/2]^T+c_3e^{(-1/2-\sqrt{3}/2)t}[1\ \ -1/2-$
$\sqrt{3}/2\ \ -1/2+\sqrt{3}/2]^T,$ where $c_1,\ c_2,$ and c_3 are arbitrary
reals.

7(a). $x=c_1e^t[1\ \ -1\ \ 1]^T+c_2e^{-t}[0\ \ 1\ \ -1]^T+$
$c_3e^{-t}([1\ \ 1\ \ 0]^T+t[0\ \ 1\ \ -1]^T),$ where $c_1,\ c_2,$ and c_3 are
arbitrary reals.

Section 7.4 Page 369.

2(a). $p(\lambda)=\lambda^2+\lambda-6;\ \lambda_1=2,\ \lambda_2=-3;\ v^1=[1\ \ 1]^T,$
$v^2=[1\ \ -4]^T.\ [x\ \ y]^T=c_1v^1e^{2t}+c_2v^2e^{-3t},$ for arbitrary
reals $c_1,c_2.$ Saddle.

2(c). $p(\lambda)=\lambda^2-13\lambda+30;\ \lambda_1=10,\ \lambda_2=3;\ v^1=[2\ \ 1]^T,$
$v^2=[-3\ \ 2]^T.\ [x\ \ y]^T=c_1v^1e^{10t}+c_2v^2e^{3t},$ for arbitrary
reals $c_1,c_2.$ Improper node.

2(e). $p(\lambda)=\lambda^2+2\lambda+5;\ \lambda_1=-1+2i,\ \lambda_2=-1-2i;$
$v^1=[2\ \ -i]^T,\ v^2=[2\ \ i]^T.\ [x\ \ y]^T=$
$c_1e^{-t}[2\cos 2t\ \ \sin 2t]^T+c_2e^{-t}[2\sin 2t\ \ -\cos 2t]^T,$ for
arbitrary reals $c_1,c_2.$ Focus.

7(a). $[x\ \ y]^T=c_1[2\cos 2t\ \ -\sin 2t]^T+c_2[2\sin 2t\ \ \cos 2t]^T,$ for
arbitrary reals $c_1,c_2.$

Section 7.5 Page 382.

1(a). $e^{tA}=\frac{1}{4}\begin{bmatrix}e^{-2t}+3e^{2t} & -3e^{-2t}+3e^{2t}\\ -e^{-2t}+e^{2t} & 3e^{-2t}+e^{2t}\end{bmatrix};$
$x(t)=\frac{1}{4}[-5e^{-2t}+9e^{2t}\ \ 5e^{-2t}+3e^{2t}]^T.$

1(c). $e^{tA}=\begin{bmatrix}e^{-t} & te^{-t}\\ 0 & e^{-t}\end{bmatrix};\ x(t)=e^{-t}[1+2t\ \ 2]^T.$

2(a). $e^{tA}=\begin{bmatrix}1 & t & t+t^2/2\\ 0 & 1 & t\\ 0 & 0 & 1\end{bmatrix};$
$x(t)=[1+5t+3t^2/2\ \ 2+3t\ \ 3]^T.$

4(a). $e^{tA}x^0=\begin{bmatrix}a\cos 2t+b\sin 2t\\ b\cos 2t-a\sin 2t\end{bmatrix};\ e^{-sA}F(s)=\begin{bmatrix}\cos 2s\\ \sin 2s\end{bmatrix}.$

4(c). $e^{tA}x^0=\begin{bmatrix}a\cos t+(2a-5b)\sin t\\ (a-2b)\sin t+b\cos t\end{bmatrix};$
$e^{-sA}F(s)=\begin{bmatrix}\cos^2 s+2\sin s\cos s\\ \sin s\cos s\end{bmatrix}.$

5(a). $x = \begin{bmatrix} e^{it} & e^{-it} \\ (2-i)e^{it} & (2+i)e^{-it} \end{bmatrix} c + \begin{bmatrix} 3e^t/2 - 1 \\ 5e^t/2 - 2 \end{bmatrix}$, where c is a constant column vector.

Section 7.6 Page 393.

1(a). No. **2(a).** $\alpha < -3/2$.
8(a). All roots have negative real parts. **8(c).** No.

Section 7.7 Page 400.

5(c). $a \approx 0.0102$.

Section 8.1 Page 413.

1(a). Bifurcation A: asymptotically stable at $(0, 0)$. Bifurcation B: asymptotically stable at $(-1, 0)$, $(1, 0)$; unstable at $(0, 0)$.
1(c). Bifurcation G, H, I: unstable at the "saddle point" and at the "focal point."
1(e). First Order C: neutrally stable at $(1, -1)$ and $(-1, 1)$; unstable at $(1, 1)$ and $(-1, -1)$.
1(g). Planar Portrait C: unstable at $(0, 0)$, $(1, 1)$; asymptotically stable at $(-1, 1)$.

Section 8.2 Page 428.

1(a). Asymptotically stable. **1(c).** Asymptotically stable.
1(e). Asymptotically stable. **1(g).** Unstable.
3(a). Asymptotically stable. **3(c).** Unstable.

Section 8.3 Page 435.

1(a). $K(x, y) = y^3 x$. The origin is an unstable saddle point.
1(c). $K(x, y) = \frac{1}{4}(x^4 + y^4)$. The origin is neutrally stable.

Section 8.4 Page 446.

1(a). Asymptotically stable. **1(c).** Unstable.
2(a). Asymptotically stable at $(0, 0)$, unstable at $(-2, 0)$.
2(c). Unstable at $(0, 0)$, asymptotically stable at $(0, -1)$.
2(e). Unstable at each point on the y-axis.
4(a). $V = (x^2 + y^2)/2$. **4(c).** $x^2/14 + y^2/2$.
6(a). $(0, 0)$, $(1, 1)$, $(-1, 1)$.
9(a). Asymptotically stable at the origin.

Section 9.1 Page 460.

1(a). $(0, 0)$, unstable; attracting limit cycle $r = 4$ and repelling limit cycle $r = 5$.
1(c). The equilibrium points are $(0, 0)$ (unstable) and all points on the circle $r = 1$ (stable); repelling limit cycle $r = 2$.
   1(e). $(0, 0)$, unstable; attracting limit cycles $r = 2n + 1/2$ and repelling limit cycles $r = 2n + 3/2$.
2(a). Asymptotically stable equilibrium point $(0, 0)$.

Section 9.2 Page 470.

1(a). For the orbit through the equilibrium point $(0, 0)$, the alpha and omega limit sets are the point $(0, 0)$ itself. The alpha and omega limit sets of the orbit through $(1, 1)$ are the orbit itself, which is a cycle.
1(e). The alpha and omega limit sets of the orbit through $r = 1/2$, $\theta = 0$ are, respectively, the origin and the cycle $r = 1$, where the origin is an equilibrium point. The alpha and omega limit sets of the orbit through $r = 3/2$, $\theta = 0$ are, respectively, the cycles $r = 2$ and $r = 1$. The alpha and omega limit sets of the

orbit through $r = 5/2$, $\theta = 0$ are, respectively, the empty set and the cycle $r = 2$. **5(a).** No cycles.

Section 9.3 Page 480.

1(a). Saddle-node bifurcation at $\varepsilon = 0$.
1(c). Transcritical bifurcation at $\varepsilon = 0$.
1(e). Pitchfork bifurcation at $\varepsilon = 0$.
5(a). $u' = \varepsilon u + v - u(u^2 + v^2)$ and $v' = -u + \varepsilon v - v(u^2 + v^2)$, where $u = x - 5\varepsilon$ and $v = y - 5\varepsilon$; $r' = \varepsilon r - r^3$, $\theta' = -1$.

Section 9.4 Page 492.

1(b). $\begin{bmatrix} -\sigma & \sigma & 0 \\ r-z & -1 & -x \\ y & x & -b \end{bmatrix}$.

Section 10.1 Page 506.

1(a). $u(t, x) = \sin x \cos ct$.
2(a). $u_n(x, t) = \cos \frac{n\pi x}{L} \left(A_n \cos \frac{n\pi ct}{L} + B_n \sin \frac{n\pi ct}{L} \right)$.

Section 10.2 Page 518.

2(b). 0 for the first pair of functions; $\frac{1}{4}(e^2 - 1)$ for the second.
7. $\sum_{k=1}^{N} \frac{(-1)^{k+1}}{k} \sin kx$.
10(a). $(2A/\pi) \sum_{k=1}^{\infty} (\sin kx)/k$.
11(a). $2 \sum_{k=1}^{\infty} (-1)^{k+1} (\sin kx)/k$.

Section 10.3 Page 528.

1(a). $A_0 = 5$, $B_1 = -1$, $B_3 = -7$, $A_6 = -4$, all other coefficients are zero.
2(a). $-2 + 4\sum_{k=1}^{\infty} \frac{(-1)^{k+1}}{k} \sin kx$.
2(c). $\left(a + \frac{\pi^2}{3}c\right) + \sum_{k=1}^{\infty} \left(\frac{4c(-1)^k}{k^2} \cos kx + \frac{2b(-1)^{k+1}}{k} \sin kx \right)$.
2(e). $\frac{\pi}{2} - \frac{4}{\pi} \sum_{k=1}^{\infty} \frac{1}{(2k+1)^2} \cos(2k+1)x$.
2(g). $\frac{AB}{\pi} + \frac{2A}{\pi} \sum_{k=1}^{\infty} \frac{\sin kB}{k} \cos kx$.
3(a). $\sin x$.

Section 10.4 Page 534.

1(a). $\frac{4}{\pi} \sum_{odd\ k} \frac{1}{k} \sin kx$; 1.
2(b). $B_{2m} = 0$; $B_{2m-1} = \frac{4}{\pi} \left[\frac{\sin(m-3/4)\pi}{4m-3} - \frac{\sin(m-1/4)\pi}{4m-1} \right]$, for $m = 1, 2, \ldots$.
4(a). $(e^{ix} + e^{-ix})/2$.

Section 10.5 Page 538.

1(a). $L[y] = y''$; $\text{Dom}(L) = \{y \text{ in } C^2[0, \pi/2] : y(0) = y(\pi/2) = 0\}$. $\lambda_n = -4n^2$ for $n = 1, 2, \ldots$; basis $\Phi = \{\sin 2nx : n = 1, 2, \ldots\}$.
1(c). $L[y] = y''$; $\text{Dom}(L) = \{y \text{ in } C^2[0, T] : y'(0) = y'(T) = 0\}$. $\lambda_n = -(n\pi/T)^2$, $n = 1, 2, \ldots$; basis $\Phi = \{\cos(n\pi x/T) : n = 0, 1, 2, \ldots\}$.
1(e). $L[y] = y''$,
$\text{Dom}(L) = \{y \text{ in } C^2[0, \pi] : y(0) = y(\pi) + y'(\pi) = 0\}$. $\lambda_n = -(r_n/\pi)^2$, where the r_n satisfy $\tan r_n = -\pi r_n$; basis $\Phi = \{\sin(r_n x/\pi) : n = 1, 2, \ldots\}$.

Section 10.6 Page 545.

1(a). $u(x, t) = \frac{3L}{\pi c} \sin \frac{\pi ct}{L} \sin \frac{\pi x}{L}$.
2(a). The SL problem is as follows: find all λ such that $X'' = \lambda X$, $X(0) = 0$, $X'(L) + hX(L) = 0$ has a nontrivial solution in $C^2[0, L]$.

4(a). $gx(L-x)/(2c^2)$.

6(a). $u(x,t) = \sum_{n=1}^{\infty} \frac{12(-1)^n}{n^3\pi^3 c^2}(1-\cos n\pi ct)\sin n\pi x$.

Section 11.1   Page 553.

1(a). $-\infty < x < \infty$.   **3(a).** $\sum_{n=1}^{\infty} 2nx^n/(n-1)!$.

4(a). $\sum_{k=0}^{\infty} 2x^{2k}/(2k)!$.

4(c). $\frac{1}{2}\sum_{k=0}^{\infty}(-1)^k(2x)^{2k+1}/(2k+1)!$.

5(a). $na_n - n + 2 = 0$, for $n = 1, 2, \ldots$; $a_n = 1 - 2/n$.

8. $y(x) = x^2/2 + x^3/6 + \sum_4^{\infty} a_n x^n$, where
$a_n = \sum_{k=1}^{[n/2]}(n-2k)!/n!$.

10. $R(r) = 1 - r^2/4 + r^4/(2\cdot 4)^2 - r^6/(2\cdot 4\cdot 6)^2 + \cdots$

Section 11.2   Page 562.

1(a). Ordinary.   **2(a).** No singular points.

2(c). $x = 0$.   **2(e).** $x = \pm 1$.

4(a). $y = a_0\left[1 + \sum_{k=1}^{\infty}\frac{x^{3k}}{(3k)(3k-1)(3k-3)(3k-4)\cdots 3\cdot 2}\right] + a_1\left[x + \sum_{k=1}^{\infty}\frac{x^{3k+1}}{(3k+1)(3k)(3k-2)(3k-3)\cdots 4\cdot 3}\right]$, where a_0 and a_1 are arbitrary.

5. $y = 1 - \frac{3}{2}x^2 + \frac{13}{48}x^4 - \frac{23}{288}x^6 + \cdots$.

6(a). $y = x + \frac{1}{4}x^4 + \frac{1}{14}x^7 + \cdots$.

8. $y(1) \approx 1 - \frac{1}{3} + \frac{1}{18} - \frac{1}{162} + \frac{1}{1944} \approx 0.7166$.

Section 11.3   Page 568.

1. $v_0 = \frac{C_1}{2}\ln\frac{1+x}{1-x} + C_2$; $v_1 = x[C_3 + C_1(-\frac{1}{x} + \frac{1}{2}\ln\frac{1+x}{1-x})]$.

5(a). $a = -1$, $b = 1$.

Section 11.4   Page 577.

1(a). Regular.   **1(c).** Regular.

2(a). $x = \pm 1$ are regular singular points.

2(c). $x = 0, -1$ are regular singular points; $x = 1$ is an irregular singular point.

4(a). $y(x) = c_1 x^3 + c_2 x^2 + x + 5/6$, where c_1 and c_2 are any constants.

6(b). $y = x^{-1}$, $y = x^{-\sqrt{2}}$, $y = x^{\sqrt{2}}$.

Section 11.5   Page 583.

1(a). $y_1 = x^{1/3} + x^{4/3}/5$, $y_2 = x^{-1/3}\sum_{n=0}^{\infty} a_n x^n$ where $a_n = (-1)^n\frac{10}{3^n n!(3n-2)(3n-5)}a_0$, $n \geq 2$.

1(c). $y_1 = x^{\sqrt{2}}\sum_{n=0}^{\infty} a_n(\sqrt{2})x^n$, $y_2 = x^{-\sqrt{2}}\sum_{n=0}^{\infty} a_n(-\sqrt{2})x^n$
where $a_0 = 1$ and $a_n(\pm\sqrt{2}) = \frac{(n\pm\sqrt{2}-1)(n\pm\sqrt{2}-2)\cdots(\pm\sqrt{2})a_0}{n!(n\pm 2\sqrt{2})(n-1\pm 2\sqrt{2})\cdots(1\pm 2\sqrt{2})}$.

3. $y_1 = x^{1/3} + x^{4/3}/7 + 9x^{7/3}/280 + 227x^{10/3}/33600 + \cdots$,
$y_2 = x^{-1} - 1 - x/8 - 11x^2/360 + \cdots$.

7(a). The indicial polynomial is r^2 with roots $r_1 = r_2 = 0$.

Section 11.6   Page 595.

1. $J_3(x) = x^3/48 - x^5/768 + x^7/30720 - x^9/2211840 + \cdots$.

Section 11.7   Page 607.

1(a). $y_1 = e^{-x}$, $y_2 = e^{-x}[\ln x + x + x^2/4 + \cdots + x^n/(n!n) + \cdots]$.

1(c). $y_1 = x$, $y_2 = x[-1/x + \ln x + \cdots + x^{n-1}/(n!(n-1)) + \cdots]$.

3(b). $y(x) = e^x(c_1 J_1(x) + c_2 Y_1(x))$.

4(c). $y = x^{1/2}[c_1 J_p(x^2/2) + c_2 J_{-p}(x^2/2)]$, $p = 1/(4\sqrt{2})$, $x > 0$.

Section 12.1   Page 616.

1(a). $u = e^{\alpha t}e^{(2\pm\sqrt{\alpha^2+2})x}$, for any complex constant α.

2(a). $X(x)T(t) = (C_1 e^{kx} + C_2 e^{-kx})(C_3 e^{ckt} + C_4 e^{-ckt})$.

2(c). $X(x)Y(y) = (C_1 e^{kx} + C_2 e^{-kx})(C_3 e^{iky} + C_4 e^{-iky})$.

2(e). $X(x)Y(y) = (C_1 e^{kx} + C_2 e^{-kx})y^{1+k^2}$.

4(a). $u(x,y) = h(x)e^{-y^2/2} + F(y)$, where h is any $\mathbf{C}^2$-function with $h(0) = 0$ and $h'(0) = k$.

Section 12.2   Page 622.

1(a). $u(x,y) = (\cos^*(x+t) + \cos^*(x-t))/2$, where
$\cos^* s = \begin{cases} \cos s, & if\, |s| \leq \pi/2 \\ 0, & \text{otherwise.} \end{cases}$

Section 12.3   Page 633.

1(a). $u(x,t) = \sin\frac{2\pi x}{L}\exp\left[-K\left(\frac{2\pi}{L}\right)^2 t\right]$.

1(c). $u(x,t) = \frac{4u_0}{\pi}\sum_{odd\, n}\frac{1}{n}\sin\frac{n\pi x}{L}\exp\left[-K\left(\frac{k\pi}{L}\right)^2 t\right]$.

1(e). $u(x,t) = \sin\frac{\pi x}{2L}\exp\left[-K\left(\frac{\pi}{2L}\right)^2 t\right]$.

2(a). $u(x,t) = 10(1+x) - \sum_{n=1}^{\infty}\frac{20[1-2(-1)^n]}{n\pi}\sin n\pi x\exp[-K(n\pi)^2 t]$.

3(b). $U(x,t) = \sum_{n=1}^{\infty} e^{-K(n\pi)^2 t}\left[2\int_0^1 f(x)\sin n\pi x\, dx - \frac{2k}{n\pi}\int_0^t \cos\tau e^{K(n\pi)^2\tau}\, d\tau\right]\sin n\pi x$.

4. $U(x,t) = 2\sum_{n=1}^{\infty}\left\{\frac{(-1)^n}{K(n\pi)^3}\left[e^{-K(n\pi)^2 t} - 1\right] + \frac{3(-1)^n - 3}{n\pi(Kn^2\pi^2-2)}[e^{-K(n\pi)^2 t} - e^{-2t}]\right\}\sin n\pi$.

5(b). $u = T_0 + A_0 e^{-\pi}\cos(\pi - \omega t)$.

Section 12.4   Page 639.

3(a). $u(r,\theta) = 3r\sin\theta$. The maximum and minimum temperatures in the disk are ± 3, respectively.

Section 12.5   Page 646.

1(a). $u = -1 + 2P_2(\cos\phi)\rho^2$.

3(a). $u(r,z) = \sum_{n=1}^{\infty}\frac{2}{x_n^2}\cdot\frac{\sinh x_n(a-z)}{\sinh(x_n a)}\cdot\frac{J_0(x_n r)}{J_1(x_n}$.

6(a). $L[y] = (x^2 y')'/x^{-2}$;
Dom$(L) = \{y$ in $\mathbf{C}^2[1, 2]:\ y(1) = y(2) = 0\}$. $\lambda_n = -4n^2\pi^2$, $n = 1, 2, \ldots$. $\Phi = \{\sin(2n\pi/x):\ n = 1, 2, \ldots\}$ is a basis.

Section A.1   Page A-3.

2(b). One infinite set of solutions is $y(t) = \begin{cases} (at-ac)^{1/a}, & t \geq c \\ 0, & t < c. \end{cases}$

Section A.2   Page A-10.

1(a). $y_0(t) = 1$, $y_1(t) = 1 - t$, $y_2(t) = 1 - t + t^2/2$,
$y_3(t) = 1 - t + t^2/2 - t^3/6$.

6(a). $y_0(t) = 0$, $y_1(t) = t$, $y_2(t) = t + t^3/3$,
$y_3(t) = t + t^3/3 + 2t^5/15 + t^7/63$.

Section A.3   Page A-12.

2. One such equation is $y' = 2ty^2$.

Section A.4   Page A-17.

2. $|y(t) - \tilde{y}(t)| \leq |a - \tilde{a}|e^{t/10} + 5(e^{t/10} - 1)$, $0 \leq t$, where we have chosen M to be $1/2$.

Section B.2   Page B-8.

3. $y(1) \approx y_{1000} = 1.43100$.

Index

Abel's Formula, 196
Abel's Theorem, 174
Adams, J.C., B-5
Adams-Bashforth method, B-5
 interpolating polynomials, B-8
Adams-Bashforth-Moulton
  predictor-corrector
  method, B-6, B-10
Adams-Moulton method, B-5
Aging spring, 154, 549, 554, 560,
 604, 608
Airy's Equation, 562, 608
Amplitude, 189
Annihilator, 193, 194
Approximate slope function, 130, 131
Archimedes' Buoyancy Principle,
 236
Attractor, 455
 strange, 489
Autocatalator, 297, 481
Autonomous ODEs, 101, 109
 definition of, 40
 equilibrium solutions, 101, 110
 sign analysis, 110
 translation property of
  solutions, 110
Autonomous systems of ODEs
 asymptotic stability, 409
 cycles, 304
 definition of, 293
 equilibrium points, 304, 405
  attractors, 410
 Maximal Extension Theorem,
  304
 planar, 305
  x-nullclines, 306
  y-nullclines, 306
 Separation of Orbits Theorem,
  304
 stability of, 405
  neutral, 412
 Translation Property of Orbits,
  294

Backward IVP, 31
Balance Law, 16
Basic Questions
 for IVPs, 104
Basic Sensitivity Estimate, 116
Basis, 344, 515

Beats, 230–231
Bendixson's Negative Criterion, 469
Bernoulli's Equation, 69, 78
Bessel function of the first kind, J_p
 decay of, 594
 integration of, 596
 modified, I_p, 597
 of any order, 590
 of integer order, 587
 orthogonality of, 593
 oscillation of, 593
 recursion formulas for, 590
 solving a PDE, 645
 zeros of, 592
Bessel function of the second kind,
  Y_p
 of any order, 603
 of integer order, 602
Bessel's Equation, 585
 modified, 597
 solution of, 585
Bessel, Friedrich Wilhelm, 585
bifurcation diagram, 473
Bifurcations
 homoclinic, 478, D-8
 Hopf, 475, D-10
 period doubling, 488
 pitchfork, 474, D-4
 saddle-node, 473
 subcritical Hopf, 477
 supercritical Hopf, 477
 transcritical, 474, D-6
Bodé plots, 234
Boundary condition
 elastic, 501
 fixed, 500
 free, 501
 operator formulation of, 501
 periodic, 535
 separated, 535
 unmixed, 535
Boundary function, 545
Bounded Input-Bounded Output
  Principle, 121, 123
Boundedness
 of Legendre polynomials, 567
Brahe, Tycho, 81

Calorie, 624
Capacitance, 240

Carrying capacity, 51
Cascades
 linear cascade, 64
Cauchy Uniform Convergence Test,
 A-6
Cauchy-Schwarz Inequality, 222,
 C-20
Chain Rule, C-12, C-14
Chaos, 489
Chaotic wandering, 148
Characteristic polynomial
 of a square matrix, 348
 of P(D), 186
Characteristic Polynomial Theorem,
 348
Chebyshev polynomials, 570
Chebyshev's Equation, 570
Circuits, electrical, 238
 LC, 266
 LC oscillator, 244
  driven, 268
 active RLC, 457
 charge in, 238
 Coulomb's Law, 240
 current in, 238
 electromotive force, 239
 elements of, 239
  capacitor, 240
  inductor, 239
  resistor, 239
 Faraday's Law, 240
 Kirchhoff's Laws
  current, 248
  voltage, 241
 low-pass filter, 395
 multiloop, 248
 Ohm's Law, 239
 scroll circuit, 494
 simple RLC, 241, 456
  driven, 244
  free solutions of, 243
 tuning of, 246
 voltage drop (or potential
  difference), 239
Circular harmonics, 637
Clearance coefficient, 61
Cobweb diagrams, 143
Coefficient Test, 389
Cofactor Expansion, C-19
Combat model, 44

Compartment model, 64
  closed, 402
  linear, 401
  open, 401
Complex numbers, C-7
  complex-valued functions
    derivative of, C-8
    differentiation principle for,
      C-8
    De Moivre's Formula, C-8
    Euler's formula, C-7
Component graph, 40, D-15
Computer implementation, 135
Conduction, 624
Constant steady state, 384
Continuity class, 498
Continuity sets
  C^0, 164
  C^1, 164
Contour, 88
Convergence
  in the mean, 513
  Integral convergence theorem,
    A-8
  interval of, C-4
  radius of, C-4
  ratio test for, C-4
  uniform of a sequence of
    functions, A-5
    Cauchy test for, A-6
Convergence Theorem, 560
Convolution
  definition of, 279
  properties of, 280
Convolution Theorem, 280
Coordinate systems
  cylindrical, C-14
  rectangular, C-14
  spherical, C-14
Coulomb's Law, 240
Critically damped, 226
Cycle-graph, 465
Cycles, 305, 453
  attractor, 455
  isolated, 454
  limit cycle, 454
  tests for
    Bendixson's Negative
      Criterion, 469
    Poincaré-Bendixson Cycle
      Test, 468
Cylindrical harmonics, 646

D'Alembert's Formula, 618
Damping constant, 154
Decay estimate, 387
Definiteness, 418
  planar quadratic forms theorem,
    419

quadratic forms theorem, 428
  testing for, 419
Delta function
  as a limit, 286
  definition of, 284
  properties of, 284
Derivative following the motion, 419
Determinant, C-18, C-19
De Moivre's Formula, C-8
Difference equations, 266
Differential equation, see ODE
Differential form of an ODE, 92
Differential operator D, see Operators
Differential-delay equation, 274
Diffusion equation, 625
Dimension
  Hausdorf, 491
  of a linear space, 345
Dirac delta function, see Delta
    function
Direction field, 6, 99, 305, D-21
  aspect ratio, 100
  method of isoclines, 103
Dirichlet kernel, 524
Dirichlet Problem, 635
  boundary data for, 635
  classical, 636
    solution for, 636
  expansion in circular
    harmonics, 637
  well-set, 639
Dirichlet representation, 524
Distribution, 286
Divergence, C-13
Divergence Theorem, C-13
Doubling time, 51
Driving term, 11, 29
Duffing's Equation, 495
Dulac's Criterion, 472
Duty cycle, C-2

Eigenbasis, 347
Eigenfunction, 536
Eigenspace, 346, 536
  deficient, 350
  nondeficient, 350
Eigenspace Dimension Theorem, 350
Eigenvalue, 346, 536
Eigenvector, 346, 536
  generalized, 351
Einstein's Field Equations, 85
Elastic coefficient, 549
Elementary row operations, C-17
Engineering functions, C-2
Equidimensional equation, see Euler
    Equation
Equilibrium point, 219, 405
  attractor, 410
  classification of, 363

center (or vortex), 363
  deficient improper, 363
  focus (or spiral), 363
  improper, 363
  proper (or star), 363
  saddle, 363
Equilibrium solutions, 51, 110, 158,
    305
Errors, 131
  global (or accumulated)
    discretization, 131
  of Euler's Method, B-1
  in the mean, 511
  local discretization (or formula
    or truncation), 131
  mean-square, 513
Escape velocities, 82
Euclidean space, 509
Euler Equation, 211, 572, 637
Euler's formulas, C-7, C-10
Euler's Method, 128, 141–150
  bound for total errors, B-3
  chaotic wandering, 148
  discretization error, B-1
Euler-Fourier Law of Heat
    Conduction, 625
EUT, see Existence and Uniqueness
    Theorems
Exact equation
  definition of, 87
  solutions of, 87
Existence and Uniqueness Theorems
  for a system of ODEs, 353
  for ODEs
    nth-order equation, 195
    first-order equations, 105
    second-order equations, 165
Existence Theorem, A-4
Exponential growth, 50
Exponential identities, C-9
Exponential order, functions of, 255
Exponential, matrix, see Matrix
    exponential
Extension Principle, 107, 353, A-11

Faraday's Law, 240
Fick's Law, 625
First-order linear ODE
  Bounded Input-Bounded Output
    Principle, 121
  coefficient $p(t)$, 11
  continuity with respect to the
    data, 118
  decomposition of solutions, 116
  definition of, 11
  driving term $q(t)$, 11
  General Solution Theorem, 12
  integrating factor for, 12
  IVP, 11

solution of, 16
sensitivity estimate, 116
Flight paths, 75
Forced oscillation, 22, 385, 534,
        D-20, D-27
Formal solution, 541
Forward IVP, 31
Fourier coefficient, 516
Fourier series, 516, 521
        exponential, 532
        odd/even extensions, 530
        trigonometrical, 521
Fourier-Euler formulas, 515
Free equation, 13
Frequency, 189
        circular, 189
Frobenius series, 578
Frobenius' Theorem, 582, 598
Fubini's Theorem, C-13
Functions
        even, 522
        homogeneous of order zero, 73
        level curves of, 88
        odd, 522
        of exponential order, 255
        order of, 498
        piecewise continuous, 19
        polynomial-exponential, 194
Fundamental matrix, 374
Fundamental set, 176, 196, 357
Fundamental Theorem of Algebra,
        C-10
Fundamental Theorem of Calculus,
        C-12

Galilei, Galileo, 23
Gamma function
        definition of, 588
        properties of, 589
General Solution Theorems
        nth-order, constant-coefficient
                complex-valued, 197
                real-valued, 199
        for $P(D)[y] = f(t)$, 178
        for $y' + p(t)y = q(t)$, 12
        for the system $x' = A(t)x$, 376
        second-order
                complex-valued, 186
                real-valued, 188
Generalized function, 286
Generalized power series, *see*
        Frobenius series
Geometry
        of orbits
                boundedness, 467
                direction fields, 99
                Dulac's Criterion for cycle
                        nonexistence, 472

Poincaré-Bendixson Cycle
        Test, 468
        Ragozin's Negative
                Criterion, 472
        separation of orbits in
                autonomous systems, 304
        unboundedness, 467
of solution curves, 99
        converging flows in, 103
        direction fields, 99
        diverging flows in, 103
        flow lines in, 103
        method of isoclines, 103
Gerschgorin disks, 391–393
Gibbs phenomenon, 523
Gradient, C-13
Green's function, 282
        as a limit, 287
Green's Kernel, 207
Green's Theorem, C-13

Half-life, 25
Harmonic function, 636
        Maximum Principle, 638
        Mean Value Property, 638
        uniqueness of solutions, 639
Harmonic motion
        natural circular frequency of,
                225
        resonance, 227
        simple, 190, 225
Harvesting, Law of, 327
Hausdorf dimension, 491
Heat equation, 625
        classical solution, 633
        continuity in the data, 631
        homogeneous, 626
        Maximal Principle, 631
        smoothing properties, 632
        uniqueness of solutions, 631
Heaviside function, 260
Heaviside, Oliver, 260
Hermite polynomials, 570
Hermite's Equation, 556, 570
Hertz, 189
Heun's Method, 132
Homogeneous equation, 13, 173
Homogeneous of order zero, 73
Hooke's Law, 154
Hopf bifurcation, 475
        subcritical, 477
        supercritical, 477
Hopf Bifurcation Theorem, 476
Hopf, Eberhard, 476
Huygen's Principle, 620
Hyperbolic trigonometric functions,
        C-10

Images, Method of, 621

Implicit Solution Theorem, 87
Impulse, 210, 288
Indicial polynomial, 578
        roots of, 578
Inductance, 240
Initial condition, 11, 31
Initial manifold, 615
Initial value problem, *see* IVP
Input data, 29
Integral convergence theorem, A-8
Integral curve, 37, 88, D-21, D-24
Integral Estimate, C-12
Integral functions, 430
Integral of an ODE, 37, 88
Integrating factor, 12, 92
Integration by parts, C-12
Interior point, 302, A-1
Intermediate Value Theorem, C-11
Interpolating polynomial, B-9
Intrinsic rate coefficients, 314
Invariance
        of a limit set, 464
Isocline, 101
        c-isocline, 103
        nullcline, 6, 101
Isolation of the population quadrant
        theorem, 314
IVP, 11, 31
        approximation of solutions, 128
                Euler's Method, 141–150
                Heun's Method, 132
                problems in implementing
                        methods, 135
                Runge-Kutta Methods, 132
        backward, 31, 293
        basic questions for, 104
        existence and uniqueness of
                solution, 98
        Existence and Uniqueness
                Theorem, 105
        for first order linear ODEs
                continuity with respect to
                        data, A-16
        for first-order linear ODEs
                Basic Sensitivity Estimate,
                        116, A-13
                continuity with respect to
                        data, 118
        for first-order ODEs
                general, 104
                sensitivity, 112
        for first-order system of ODEs,
                293
        forward, 31, 293
        Picard process for solving, A-3
        properties of solutions, A-17
        two-sided, 31, 293
        well-posed, 104

Jordan Curve Theorem, 454
Joule, 624

Kepler, Johannes, 81
Kernel function, 207
Kinematics of motion, 216
Kirchhoff's Laws
    current, 248
    voltage, 241

L'Hôpital's Rule, C-11
Lagrange Identity, C-10
Lagrange's Identity, 529
Laguerre polynomials, 584
Laguerre's Equation, 584
Lanchester, Frederick William, 44
Laplace Transform
    action of, 252
    Convolution Theorem, 280
    inverse of, 253
    inverse transform of the nth
        derivative, 258
    linearity of, 253
    of a periodic function, 267
    of an integral, 259
    of the nth derivative, 257
    Shifting Theorem for, 261
    smoothness and decay property
        for, 256
Laplace's Equation, 635, C-14
Laplacian, 610
Laplacian operator, C-14
Law of Cosines, C-9
Leakey, Louis and Mary, 68
Legendre polynomials
    boundedness of, 567
    definition of, 565
    orthogonality of, 566
    recursion relation for, 566
    roots of, 567
    table of properties for, 567
Legendre's Equation, 563, 571, 584
Level curves, 88
    nullcline, 101
Level set, D-25
level set, 88
Libby, William, 32
Liebniz's Rule, C-12
Lienard Equation, 461
Light cone, 620
Limit cycle, 454, D-25
    attracting, 455
    repelling, 455
Limit sets, 463
    alpha, 463
    invariance property of, 464
    omega, 463
Limiting velocity, 28
Linear approximation, 160

to systems of ODEs, 437
    asymptotic stability theorem,
        439
    instability theorem, 443
Linear combination, 173
Linear dependence, 185
Linear differential equation, 4
    first-order, 11
        general solution, 16
Linear independence, 185, 344
Linear space, 185, 343–344, 508
    basis for, 344
    definition of, 343
    dimension of, 345
    linear independence in, 344
    spanning set for, 343
    subspace of, 343, 518
Linearity Theorem for $P(D)$, 173
Linearization, 161
Lipschitz condition, A-2
Loading dose, 68
Logarithmic identities, C-9
Logistic equation, 51, 111, 141
Logistic growth, 51, 125
Lorenz system, 482
    Lorenz squeeze, 485
    orbit map of, 485
    shadowing property, 491
Lorenz, E.N., 482
Low-pass channel, 248
Low-pass filter, 292, 295, 395
    passbands, 395
    stopbands, 395
Lyapunov functions, 426
    asymptotic stability, 423
    instability, 425
    stability, 420
Lyapunov method, 426
Lyapunov, Alexander Mikhailovich,
    414

Maclaurin series, C-5
Malthus, Thomas, 50
Manifold
    stable, 487
    unstable, 486
Mass Action, Chemical Law of, 296
Mathieu Equation, 563
Matrices, 342, C-15
    characteristic polynomial, 348
    determinant, C-18, C-19
    eigenvalues of, 346
    elementary row operations,
        C-17
    identity, C-16
    invertibility of, C-17
    invertible, 350
    operations, C-15
    singular, 350, C-17

transpose of a, C-16
upper triangular, C-20
zero, C-16
Matrix Estimate, C-20
Matrix exponential
    definition of, 379
    properties of, 380
Matrix Norm, C-20
Maximally extended solutions, 107,
    108, 166
Maximum/Minimum Value Theorem,
    C-11, C-13
Mean Value Theorem, C-11
Method of Annihilators, 200–202
Method of Frobenius, 581
Method of Undetermined
    Coefficients, 551
Method of Variation of Parameters,
    see Variation of
    Parameters
Method of Varied Parameters, 84
Modeling principles
    Balance Law, 16
    Coulomb's Law, 240
    definition of, 23
    Euler-Fourier Law of Heat
        Conduction, 625
    Faraday's Law, 240
    Fick's Law, 625
    Kirchhoff's Laws
        current, 248
        voltage, 241
    Law of Vertical Motion, 2
    Newton's First Law of Motion,
        216
    Newton's Law of Cooling, 36,
        626
    Newton's Law of Universal
        Gravitation, 81
    Newton's Second Law of
        Motion, 27, 217
    Newton's Third Law of Motion,
        217
    Occam's Razor, 50
    Ohm's Law, 239
    Radioactive Decay Law, 24
    spring force, 154
    viscous damping force, 154
Models, 23
    Aleutian carbon ecosystem, 403
    autocatalator, 481
    car-following, 273–277
        reciprocal separation model,
            278
        separation control model,
            278
    chemical reactor, 124
        dynamics in, 295
    cold pills, 58

antihistamine, 58
    decongestant, 58
combat, 44
compartment model, 64, 337
competition model, 318
construction of, 30
cooperation model, 317
definition of, 23
Einstein's field equations, 85
electrical circuits, *see* Circuits,
    electrical
elements of, 28
escape velocity, 82
exponential growth, 50
falling body
    moving ball, 2, 8
    Newtonian damping, 42
    parachutist, 48
frequency response, 231
    curves of, 234
harmonic oscillator, 306
harvesting a population, 53, 266
    critical harvesting, 57
    seasonal harvesting, 58
    subcritical harvesting, 57
    supercritical harvesting, 57
heat conduction, 624
interest rates, 35
lead in the human body, 341,
    397
linear cascade, 64
logistic growth, 51
    with migration, 125
low-pass filter, 292, 295
New Zealand possum plague,
    331–333
optimal depth for a wine cellar,
    629
pendulum motion, *see*
    Pendulum
population, 315
    computer simulations,
        317–319
    harvesting, 315, 327
    Principle of Competitive
        Exclusion, 319
potassium-argon dating, 68
predator-prey, 323
pursuit, 75
radioactivity, 24, 266
    radiocarbon dating, 33
    Uranium series, 65
range of validity of, 30
salt accumulation, 16, 17
SIR disease model, 334
springs, 154
    aging, 157
    damped Hooke's Law, 294
    hard, 157

Hooke's Law spring, 155
    soft, 157
temperature in a cylinder, 549,
    606
temperature in a rod, 627
tetracycline in the body, 68
validation of, 30
vibrating string, 498
Multiplicity
    of a root, 197
    of an eigenvalue, 349
Multistep numerical methods, B-4
    Adams-Bashforth, B-5
        interpolating polynomials,
            B-8
    Adams-Bashforth-Moulton
        predictor-corrector, B-6,
        B-10
    Adams-Moulton, B-5
    numerical instability of, B-6

Natural frequency, 503
Natural law, 29
Newton's Law of Cooling, 36, 626
Newton's Laws of Motion, 27, 154,
    216, 217
Newton, Isaac, 214
Newtonian mechanics
    inertial frames, 217
    momentum, 217
    Newton's Laws of Motion, 216,
        217
Nilpotent, 381
Nodal latitude, 644
Node, 503
Norm, 509
Normal form, 4
Normal mode, 540
Normal mode vectors, 369
Null space of $P(D)$, 173
    closure property for, 173
Nullcline, *see* Isocline
Numerical methods
    Euler's method, 128
    Heun's method, 132
    one-step methods, 130
    Runge-Kutta method, 132
Numerical solvers
    generating solution curves, 98

Occam's Razor, 50
Occam, William, 50
ODE, 2
    Airy's Equation, 562
    Bernoulli's Equation, 69, 78
    Bessel's Equation, 585
    Chebyshev's Equation, 570
    differential form, 92
    Duffing's Equation, 495

Euler Equation, 211, 572, 637
exact equation, 86
first-order linear
    Bounded Input-Bounded
        Output Principle, 121, 123
first-order nonlinear
    bounded input, 124
general nth-order differential
    equation, 3
Hermite's Equation, 556, 570
Laguerre's Equation, 584
Legendre's Equation, 563, 571
Lienard Equation, 461
linear, 4, 11
logistic equation, 51
Mathieu Equation, 563
maximally extended solutions,
    108
normal form, 4
    first-order, 97
Rayleigh Equation, 461, 480
Riccati's Equation, 78
separable, 36
solution
    Extension Principle, 107
    maximally extended, 107
Ohm's Law, 239
Olduvai, bones of, 68
One-step method, 130
    order of, 131
Operators, 155
    $P(D)$
        annihilator for, 194
        characteristic polynomial for,
            186
        definition of, 172
        linearity of, 173
        monic, 195
        Null Space Closure Property
            of, 173
    action of, 172
    codomain of, 172
    domain of, 172
    formulation of an IVP, 22
    Laplace Transform operator, *see*
        Laplace Transform
    Laplacian, 610
    linear, 155, 354
    range of, 172
    symmetric, 535
Orbits, 40, 159, 166
    of an undamped simple
        pendulum, 219
Order, 3
    of a function, 438
Ordinary differential equation, *see*
    ODE
Ordinary point, 555
Orthogonal projection, 519

Orthogonality, 511
   with respect to a density, 569
Output, 30
Overdamped, 226

Painlevé, Paul, 170
Parallelogram Law, 215
Parallelogram property of norms, 221
Partial differential equation, *see* PDE
Partial fractions, 262, C-10
Particular solution, 13
PDE, 2
   diffusion equation, 625
   heat equation, 625
      homogeneous, 626
   Laplace's Equation, 635
   separated solution, 502
Pendulum, 218–221
   damped simple, 219, 442, 444,
      471
   double, 368, 434
   driven, 496
   driven damped linear, 221
   effects of changing mass, 309
   equation of the double, D-32
   orbits of, 219, 224
   simple undamped, 434
   simple with viscous damping,
      223
Period, 189
   fundamental, 189
Periodic Cycle, Law of, 325
Periodic function, 189
Periodic steady state, 384
Phase plane, 40
Picard iterates, A-5
Picard Iteration Method, A-9, D-34
Piecewise continuity, 19, 510
Piecewise smooth, 523
Poincaré time section, 488
Poincaré, Jules Henri, 462
Poincaré-Bendixson Alternatives,
      467, D-38
Poincaré-Bendixson Cycle Test, 468
Polar coordinates, 71, 310, 471
Polycycle, *see* Cycle-graph
Polynomial-exponential function, 194
Population quadrant, 314
Portraits
   gallery of planar orbital
      portraits, 362
Potential function, 636
Power series, C-4
Predator-prey interaction, 315
   growth coefficients, 315
   Law of Averages, 325
   natural decay, 315
   satiable predation, 316, 477
   Volterra's Principles, 322

Pulse function, 260

Rössler system, 493, D-44
Radioactive Decay Law, 24
Ragozin's Negative Criterion, 472
Rate equations
   intrinsic rate coefficients, 314
   isolation of the population
      quadrant theorem, 314
   population quadrant, 314
Rate function, 5
Ratio test, C-4
Rayleigh Equation, 461, 480
Reaction rate constant, 24
Real analytic functions, C-5
Recursion formula, 551
Reduced mass, D-32
Reduction of order, 79, 183
Region, 302, A-1
Reindexing a series, C-5
Relaxation oscillation, 459, D-28
Removable singularity, 556
Resistance, 239
Resonance, 227
   frequency of, 235
   in damped systems, 235
   pure, 228
Response, 30
Riccati's Equation, 78
Rodrigues's Formula, 567
Routh array, 389
Routh Criterion, 389
Runge-Kutta Methods, 132

Saturation population, 51
Sawtooth wave function, C-2
Scalar product, 509
   standard, 509
   weighted, 518
Scaling, 58, 70, 335, 472, C-22
Scroll circuit, 494, D-46
Second-order linear ODEs, 172
   Abel's Theorem, 174
   constant-coefficient, 185
   Existence and Uniqueness
      Theorem, 174
   forced oscillation, 212
   free solutions, 225
   frequency response modeling,
      231
   homogeneous, 185
   operational notation for, 172
   Variation of Parameters, 206
Sensitivity, D-23
   computational, 490
   to changes in data, 112
      basic estimate theorem, A-13
      perturbation estimate, A-14
   to parameter changes, 395–399

Separable equation, 36
Separation constant, 540
Separation of Variables, 613
Series
   Fourier, 516
   Fourier Cosine, 530
   Fourier sine, 530
   Fourier trigonometrical, 521
   half-range expansions, 531
   orthogonal, 515
Series solution technique, 550
   Convergence Theorem, 560
   Frobenius series, 578
   Frobenius' method, 581
   Frobenius's method
      extended, 597
   indicial polynomial, 578
   Method of Undetermined
      Coefficients, 551
   near a singular point, 578, 597
   near an ordinary point, 556
   recursion formula, 551
Shifting data, 544
Shifting Theorem, 261
Sign analysis, 110
Simple harmonic oscillator, *see*
      Harmonic motion, 453
Singular point, 555
   regular, 571
Solution curves, 4, 166
   non-intersection of, 105
Solution of an IVP, 165
Solutions of ODEs
   nth-order, 197, 199, 282, 285
   backward, 293
   Bessel's Equation, 595
   definition of, 4, 40
   driven systems, 378
   exact equation, 87
   first-order
      table of techniques, ii
   first-order linear equation, 16
   forward, 293
   free, 225
      transients, 226, 232
   linear system, 356
   second-order, 186, 188
   steady-state, 234
   two-sided, 293
Span, 343, 518
Sparrow, Colin, 484
Spitznagel, Edward, 58
Springs, 170
   aging, 154, 549, 554, 560, 604,
      608
   damped, 471
   damped Hooke's Law, 291, 294
   hard, 154
   Hooke's Law spring, 154

soft, 154
Square Law of Combat, 44
Square wave function, C-2
Stability, 405
  asymptotic, 409
    of linear approximations, 439
  instability of linear
    approximations, 443
  neutral, 412
  testing for
    asymptotic theorem, 423
    instability, 425
    Lyapunov method, 426
    Lyapunov stability functions,
      420
Stair function, C-2
Standing wave
  first harmonic, 504
  frequency of, 503
  fundamental, 504
  higher harmonics, 504
  node of, 503
  overtones, 504
  period of, 503
State plane, 40
State portrait, 159
State space, 40, 159
State variable, 1, 29, 290
State vector, 339
static deflection, 155
Stationary solution, 51
Steady state, 384
Steady-state, 234
Step function, 260
Strange attractor, 489
Sturm-Liouville Problem, 536
Sturm-Liouville problem
  regular, 536
Sturm-Liouville system
  singular, 642
Superposition of Solutions, Principle
  of, 610
System matrix, 339
System of ODEs, 9
  autonomous, *see* Autonomous
    systems of ODEs
  coefficients, 353
  conservative, 430
  driving term, 353
  first-order, 9, 40, 92
    component graph, 294
    converting scalar equation to
      system, 291
    Existence, Uniqueness,
      Extension, and Continuity

Theorem, 302
  IVPs for, 293
  Laplace transform technique,
    301
  linear, 294, 338
  orbit (or trajectory), 294
  planar, 40
  portrait, 294
  rate functions, 293
  solution curve, 294
  state (or phase) space, 294
  state variables, 290
  input functions, 353
  input vector, 353
  integrals and orbits theorem,
    431
  linear
    stability theorem, 412
  Lorenz system, 482
  no attractors theorem, 433
  Rössler system, 493
  stability properties, 444
  van der Pol system, 458
System parameter, 29
System variables, 28

Taylor series expansion, C-5
  with remainder, C-6
Taylor's Theorem, C-14
Thermal conductivity, 625
Time section, *see* Poincaré time
    section
Time-state curve, 165
Total differential, 91
Total time derivative, 420
Trace test, 390
Trajectory, 166
Transfer function, 233
Transition matrix
  definition of, 374
  properties of, 376
Triangle Inequality, C-20
Triangular wave function, C-2
Trigonometric identities, C-9
Two-sided IVP, 31

Underdamped, 226
Undetermined Coefficients Theorem,
  205
Uniqueness Principle, A-1
Unit sawtooth pulse, C-2
Unit square pulse function, C-2
Unit step function, C-2
Unit triangular pulse function, C-2
Units

CGS system, 217
MKS system, 217
SI system, 217
Universal Gravitation, Newton's Law
  of, 81

Van der Pol system, 458, D-28
Van der Pol, Balthazar, 453
Vanishing Data Theorem, 175
Vanishing Derivative Theorem, 7
Variation of Parameters, 206–209
  for driven systems, 377
Vector space, 507
Vectors, 215, 343, 507
  geometric
    angle between, 215
    basic operations for, 215
    coordinate frame, 215
    derivative of, 216
    dot product of, 215, 221
    magnitude, 215
    norm (or length) of, 215
    orthogonality of, 216
    Parallelogram Law, 215
    position, 216
  linear combination of, 343
  normal mode, 369
  normalization of, 510
  operations
    addition, 507
    scalar multiplication, 507
  perpendicular, 216
Verhulst, Pierre-Francois, 51
Vertical Motion, Law of, 2
Viscous damping force, 154
Volterra's Laws, 325

Wave equation, 500
  characteristic for, 617
  characteristic triangle, 617
  standing wave, 502
Weber function, *see* Bessel function
  of the second kind
Weierstrass *M*-Test, 526
Well-posed IVP, 104
Well-set, 639
Well-set problem, 620
Wronski, Höene, 176
Wronskian, 356
  Abel's Formula, 196
  definition of, 176, 196
  reduction of order technique,
    183

Zonal harmonics, 569, 644

Tables of Laplace Transforms
I. General Properties

$f(t)$	$g(s) = \mathcal{L}[f] = \int_0^\infty e^{-st} f(t)\, dt$
1. $f(at)$	$\frac{1}{a} g(s/a)$
2. $e^{at} f(t)$	$g(s-a)$
3. $H(t-a) f(t-a)$	$e^{-as} g(s)$
4. $f^{(n)}(t)$	$s^n g(s) - s^{(n-1)} f(0) - s^{(n-2)} f'(0) - \cdots - f^{(n-1)}(0)$
5. $t^n f(t)$	$(-1)^n g^{(n)}(s)$
6. $\int_0^t f(u)\, du$	$\dfrac{g(s)}{s}$
7. $\int_0^t (t-u)^{n-1} f(u)\, du$	$(n-1)! \dfrac{g(s)}{s^n}$
8. $\int_0^t f_1(t-u) f_2(u)\, du$	$g_1(s) g_2(s)$
9. $\dfrac{f(t)}{t}$	$\int_s^\infty g(u)\, du$
10. $f(t)$ is periodic with period p	$\dfrac{1}{1-e^{-ps}} \int_0^p e^{-su} f(u)\, du$
11. $\dfrac{1}{\sqrt{\pi t}} \int_0^\infty e^{-u^2/4t} f(u)\, du$	$\dfrac{g(\sqrt{s})}{\sqrt{s}}$
12.† $t^{n/2} \int_0^\infty u^{-n/2} J_n(2\sqrt{ut}) f(u)\, du$	$\dfrac{1}{s^{n+1}} g(1/s), \quad n \geq 0$
13. $\displaystyle\sum_{k=1}^n \dfrac{P(\alpha_\kappa)}{Q_\kappa(\alpha_\kappa)} e^{\alpha_\kappa t}$	$\dfrac{P(s)}{Q(s)}$

P is a polynomial, degree $< n$;
$Q(s) = (s - \alpha_1) \cdots (s - \alpha_n)$, $\alpha_1, \ldots, \alpha_n$, distinct; $Q_k(\alpha_k) = \prod_{i=1, i \neq k}^n (\alpha_k - \alpha_i)$

† J_n is the Bessel function of the first kind of order n. See Section 11.6.

II. Special Laplace Transforms

	$f(t)$	$g(s) = \mathcal{L}[f] = \int_0^\infty e^{-st} f(t)\, dt$
1.	t^{n-1}	$(n-1)!/s^n, \quad n = 1,\, 2,\, 3,\, \ldots$
2.	t^{p-1}	$\Gamma(p)/s^p, \quad p > 0$
3.	$t^{n-1}e^{at}$	$\dfrac{(n-1)!}{(s-a)^n}, \quad n = 1,\, 2,\, 3 \ldots$
4.	$t^{p-1}e^{at}$	$\dfrac{\Gamma(p)}{(s-a)^p}, \quad p > 0$
5.	$\sin at$	$a(s^2 + a^2)^{-1}$
6.	$\cos at$	$s(s^2 + a^2)^{-1}$
7.	$\sinh at$	$a(s^2 - a^2)^{-1}$
8.	$\cosh at$	$s(s^2 - a^2)^{-1}$
9.	$e^{bt} - e^{at}$	$\dfrac{b-a}{(s-a)(s-b)}, \quad a \neq b$
10.	$be^{bt} - ae^{at}$	$\dfrac{(b-a)s}{(s-a)(s-b)}, \quad a \neq b$
11.	$\sin at - at\cos at$	$2a^3(s^2 + a^2)^{-2}$
12.	$t\sin at$	$2as(s^2 + a^2)^{-2}$
13.	$\dfrac{e^{-at} - e^{-bt}}{2(\pi t^3)^{1/2}}$	$\sqrt{s+b} - \sqrt{s+a}$
14.†	$J_n(t), \quad n > -1$	$(s^2 + 1)^{-1/2}[s + (s^2 + 1)^{1/2}]^{-n}$
15.	$f(t) = \displaystyle\sum_{k=1}^{[t]} r^{k-1}$	$\dfrac{1}{s(e^s - r)} = \dfrac{e^{-s}}{s(1 - re^{-s})}$
	where $[t]$ = greatest integer $\leq t$	
16.	$\dfrac{a}{2\sqrt{\pi t^3}}e^{-a^2/4t}$	$e^{-a\sqrt{s}}, \quad a > 0$
17.	$\dfrac{e^{-bt} - e^{-at}}{t}$	$\ln\left(\dfrac{s+a}{s+b}\right)$
18.	$\delta(t-a)$	e^{-as}
19.	$H(t-a)$	e^{-as}/s

III. Transforms of Graphically Defined Functions

$$f(t) \qquad\qquad g(s) = \mathcal{L}[f] = \int_0^\infty e^{-st} f(t)\, dt$$

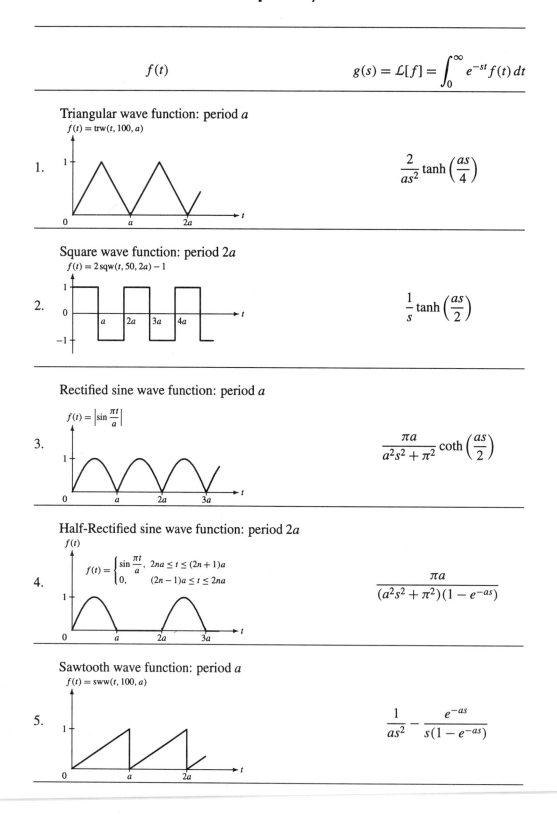

1. **Triangular wave function: period a**
   $f(t) = \mathrm{trw}(t, 100, a)$

   $$\frac{2}{as^2} \tanh\left(\frac{as}{4}\right)$$

2. **Square wave function: period $2a$**
   $f(t) = 2\,\mathrm{sqw}(t, 50, 2a) - 1$

   $$\frac{1}{s} \tanh\left(\frac{as}{2}\right)$$

3. **Rectified sine wave function: period a**
   $f(t) = \left|\sin \frac{\pi t}{a}\right|$

   $$\frac{\pi a}{a^2 s^2 + \pi^2} \coth\left(\frac{as}{2}\right)$$

4. **Half-Rectified sine wave function: period $2a$**
   $f(t)$

   $$f(t) = \begin{cases} \sin \dfrac{\pi t}{a}, & 2na \le t \le (2n+1)a \\ 0, & (2n-1)a \le t \le 2na \end{cases}$$

   $$\frac{\pi a}{(a^2 s^2 + \pi^2)(1 - e^{-as})}$$

5. **Sawtooth wave function: period a**
   $f(t) = \mathrm{sww}(t, 100, a)$

   $$\frac{1}{as^2} - \frac{e^{-as}}{s(1 - e^{-as})}$$

| $f(t)$ | $g(s) = \mathcal{L}[f] = \displaystyle\int_0^\infty e^{-st} f(t)\,dt$ |

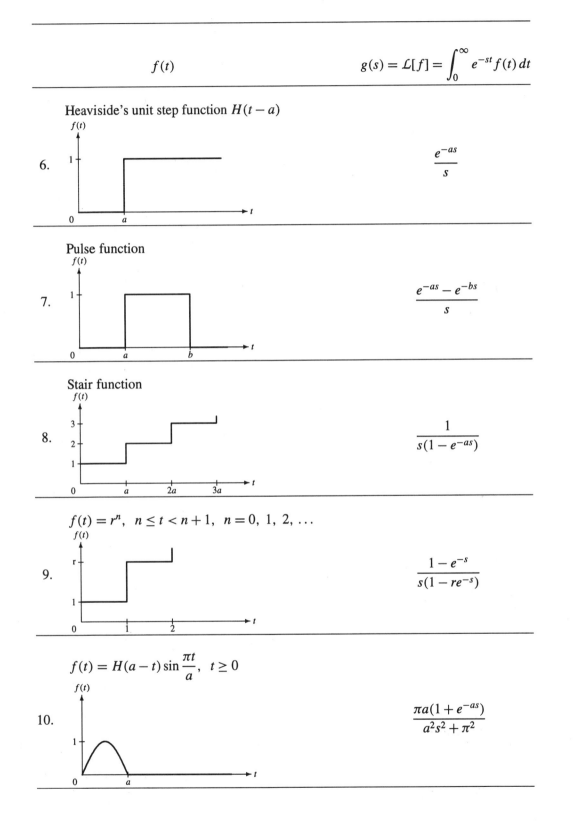

Heaviside's unit step function $H(t-a)$

6. $\dfrac{e^{-as}}{s}$

Pulse function

7. $\dfrac{e^{-as} - e^{-bs}}{s}$

Stair function

8. $\dfrac{1}{s(1 - e^{-as})}$

$f(t) = r^n, \quad n \le t < n+1, \quad n = 0,\ 1,\ 2,\ \ldots$

9. $\dfrac{1 - e^{-s}}{s(1 - re^{-s})}$

$f(t) = H(a - t)\sin\dfrac{\pi t}{a}, \quad t \ge 0$

10. $\dfrac{\pi a(1 + e^{-as})}{a^2 s^2 + \pi^2}$